April 21–24, 2013,
Prague, Czech Republic

Association for
Computing Machinery

Advancing Computing as a Science & Profession

ICPE'13

Proceedings of the 2013 ACM/SPEC International Conference on Performance Engineering

Sponsored by:

ACM SIGMETRICS, ACM SIGSOFT, and SPEC Research

Supported by:

SAP, CQK Holding, and Hewlett Packard

Association for Computing Machinery

Advancing Computing as a Science & Profession

The Association for Computing Machinery
2 Penn Plaza, Suite 701
New York, New York 10121-0701

ISBN: 978-1-4503-2121-1 (Digital)

ISBN: 978-1-4503-2438-0 (Print)

Additional copies may be ordered prepaid from:

ACM Order Department
PO Box 30777
New York, NY 10087-0777, USA

Phone: 1-800-342-6626 (USA and Canada)
+1-212-626-0500 (Global)
Fax: +1-212-944-1318
E-mail: acmhelp@acm.org
Hours of Operation: 8:30 am – 4:30 pm ET

Printed in the USA

General Chairs' Welcome

Welcome to the ICPE 2013 Conference! ICPE is an annual meeting that provides a forum for the integration of theory and practice in the field of performance engineering. It brings together researchers and industry practitioners to share ideas, discuss challenges, and present results of both work-in-progress and state-of-the-art research on performance engineering of software and systems. Now at its 4th installation, it has grown out of the ACM Workshop on Software Performance (WOSP) and the SPEC International Performance Engineering Workshop (SIPEW).

We would like to thank the Program Co-Chairs, José Nelson Amaral and Tony Field, for the tremendous effort they have put into making sure this year's conference has a high-quality, thought-provoking research track program. The Industrial Chair, Seetharami R. Seelam, has done a wonderful job in selecting a strong set of industry research papers. The Organizing Committee has been invaluable in running a smooth process. We want to thank Lubomír Bulej, for his outstanding job as Finance Chair; the Tutorial Chair, Mirco Tribastone, for his thoughtful efforts in assembling the tutorial program; Lucy Cherkasova and Vittorio Cortellessa, the conference Award Chairs, for proposing and running the award process; Kaustubh Joshi, the Poster and Demo Chair, for his tireless efforts in selecting and putting together the poster and demo exhibit; Evgenia Smirni, for her wonderful activity in putting together the proceedings as the Publication Chair; our Publicity Chairs, John Murphy and Kai Sachs, for a broad and timely advertisement of the conference across many publicity channels; Juan F. Pérez, who did a great job as the Registration Chair; Weikun Wang, the Webmaster, for his continuous help and quick responses to our demands; and Milena Zeithamlová, for handling the multitude of local arrangements.

We would like to thank Andre Bondi and Samuel Kounev, for their continuous support throughout the preparation of this conference. Their guidance has been tremendously helpful in ensuring that this year's conference could adopt the best practices of the past editions.

New this year are the workshops co-located with the main conference, HotTopiCS and Multi-Cloud, both focusing on emerging topics in cloud computing. Given the importance of performance engineering in this area, we believe that these events will help broaden the ICPE community and inspire new research directions.

Our sincere thanks go to the corporate supporters, who at the time we write include SPEC, SAP, and HP Labs. Their generous financial contributions have made this conference possible. We are also thankful to SPEC and ACM, through its SIGSOFT and SIGMETRICS special interest groups, for a continuous support of the ICPE conference.

A warm thanks goes also to the conference keynote speakers, Len Bass, Yuqing Gao, and Prashant Shenoy. We are sure that conference attendees will enjoy their presentations.

On behalf of the organizing committee, we welcome you to the beautiful city of Prague and hope you will enjoy the conference and your stay.

Petr Tůma
Charles University, Czech Republic
ICPE 2013 General Co-Chair

Giuliano Casale
Imperial College London, U.K.
ICPE 2013 General Co-Chair

Program Chairs' Welcome

We are delighted to bring you an outstanding technical program to *2013 International Conference on Performance Engineering - ICPE'13* in Prague. The main Research track for the conference attracted 42 submissions. Thanks to the diligent efforts of the members of the Program Committee each paper received a minimum of four reviews. After extensive deliberation the Program Committee decided to accept 20 submissions as regular papers and two as short papers.

The Industry and Experience track focuses on the application of research results to industrial performance engineering problems and addresses in particular innovative implementations, the novel application of performance-related technologies and the reporting of insightful performance results. This track received 22 submissions of which 8 were selected for presentation at the conference.

The papers accepted to the Research track and to the Industry and Experience track cover several topics such as software development and various flavors of modeling, including performance, survivability and scalability modeling. The development of representative workloads and benchmarks is also well represented. There are then a number of papers that focus on performance aspects of cloud-related systems and more general aspects of scheduling and load balancing.

The Vision/Work-in-Progress track is a feature of ICPE that allows researchers to present and discuss ideas that they are still working on or that they are planning to work on in the near future. It is a great forum for learning about the direction of research in the area. This year we received 18 submissions to this track and were able to accommodate 10 short presentations in the conference program. The topics covered by this track are similar to those in the main Research track, which suggests that they are likely to feature again at ICPE in the near future.

We would like to thank the authors for their diligence in preparing their manuscripts and camera-ready papers, the members of the Program Committee for the many hours that they spent reviewing papers and participating actively in the discussions after all the reviews were submitted. A special thanks goes to the authors of those papers that were not selected for presentation at the conference. We would like to encourage them to use the feedback received through the reviewing process and continue their work on performance engineering. Finally, we would like to thank the two general chairs for their valuable assistance on organizing the conference program.

In summary, there were 81 submissions in total across the three tracks, of which 38 were selected for presentation. We are now looking forward to several days of great presentations and stimulating discussions at ICPE 2013 in beautiful Prague. It has been a privilege and a pleasure for us to be involved.

J. Nelson Amaral
ICPE'13 Program Co-Chair
University of Alberta, Canada

Tony Field
ICPE'13 Program co-Chair
Imperial College London, England

Seetharami Seelam
ICPE'13 Industrial Chair
IBM Corporation, USA

Table of Contents

Session 6: Performance Analysis and Benchmarking I

Session 7: Keynote Address II

Session 8: Performance Analysis and Benchmarking II

Session 9: Performance Modeling and Prediction

Session 10: Performance in Cloud, Virtualized and Multi-Core Systems

Session 11: Vision and Work-in-Progress Papers

Session 12: Keynote Address III

ICPE 2013 Conference Organization

General Chairs: Petr Tůma *(Charles University, Czech Republic)*
Giuliano Casale *(Imperial College London, UK)*

Program Chairs: Tony Field *(Imperial College London, UK)*
José Nelson Amaral *(University of Alberta, Canada)*

Industrial Chair: Seetharami R. Seelam *(IBM T.J. Watson Research Center, USA)*

Finance Chair: Lubomír Bulej *(Charles University, Czech Republic)*

Tutorial Chair: Mirco Tribastone *(LMU Munich, Germany)*

Publication Chair: Evgenia Smirni *(College of William and Mary, USA)*

Local Arrangements: Milena Zeithamlová *(Action M, Czech Republic)*

Publicity Chairs: John Murphy *(University College Dublin, Ireland)*
Kai Sachs *(SAP AG, Germany)*

Poster and Demo Chair: Kaustubh Joshi *(AT&T Labs Research, USA)*

Award Chairs: Lucy Cherkasova *(HP Labs, USA)*
Vittorio Cortellessa *(Universita dell'Aquila, Italy)*

Registration Chair: Juan F. Pérez *(Imperial College London, UK)*

Webmaster: Weikun Wang *(Imperial College London, UK)*

ICPE'13 Program Committee: ***Research Track:***

Alberto Avritzer *(Siemens Corporate Research, USA)*
Martin Arlitt *(HP Labs, USA, and University of Calgary, Canada)*
Steffen Becker *(University of Paderborn, Germany)*
Anne Benoit *(ENS Lyon, France)*
Andre Bondi *(Siemens Corporate Research, USA)*
Edson Borin *(Universidade de Campinas, Brazil)*
Jeremy Bradley *(Imperial College London, UK)*
Lydia Chen *(IBM Zurich, Switzerland)*
Lucy Cherkasova *(Hewlett-Packard Laboratories, USA)*
Lawrence Chung *(University of Texas at Dallas, USA)*
Susanna Donatelli *(Universita' di Torino, Italy)*
Sandhya Dwarkadas *(University of Rochester, USA)*
Dick Epema *(Delft University of Technology, Netherlands)*
Dror Feitelson *(Hebrew University, Israel)*
Sebastian Fischmeister *(University of Waterloo, Canada)*
Stephen Gilmore *(University of Edinburgh, UK)*
Lars Grunske *(TU Kaiserslautern, Germany)*
Wilhelm Hasselbring *(University of Kiel, Germany)*
Michael Hind *(IBM T.J. Watson Research Center, USA)*
Robert Hundt *(Google Inc., USA)*
Alexandru Iosup *(TU Delft, The Netherlands)*
Stephen Jarvis *(University of Warwick, UK)*
Carlos Juiz *(University of the Balearic Islands, Spain)*
David Kaeli *(Northeastern University, USA)*
William Knottenbelt *(Imperial College London, UK)*
Samuel Kounev *(Karlsruhe Institute of Technology, Germany)*
Heiko Koziolek *(ABB Corporate Research, Germany)*
Anirban Mahanti *(NICTA, Australia)*
Pat Martin *(Queen's University, Canada)*
Daniel Menasce *(George Mason University, USA)*
Raffaela Mirandola *(Politecnico di Milano, Italy)*
David Pearce *(Victoria University of Wellington, New Zealand)*
Dorina Petriu *(Carleton University, Canada)*
Meikel Poess *(Oracle Corporation, USA)*
Lawrence Rauchwerger *(Texas A&M University, USA)*
Ralf Reussner *(Karlsruhe Institute of Technology, Germany)*
George Riley *(Georgia Institute of Technology, USA)*
Alma Riska (*EMC, USA)*
Jerry Rolia *(HP Labs, USA)*
Peter Sweeney *(IBM T.J. Watson Research Center, USA)*
Mirco Tribastone *(LMU Munich, Germany)*
Catia Trubiani *(Universita' dell'Aquila, Italy)*
Carey Williamson *(University of Calgary, Canada)*
Murray Woodside *(Carleton University, Canada)*
Peng Wu *(IBM T. J. Watson Research Center, USA)*
Xiaoyun Zhu *(VMware, USA)*

ICPE'13 Program Committee (continued):

Industrial Track:
Walter Bays *(Oracle Corporation, USA)*
Andrew Bond *(Red Hat, USA)*
Winnie Cheng *(American Express, USA)*
Trilce Estrada *(University of Delaware, USA)*
Pankaj K. Garg *(ZeeSource, USA)*
Klaus-Dieter Lange *(HP, USA)*
Che-Rung Lee *(National Tsing Hua University, Taiwan)*
Elmoustapha Ould-Ahmed-Vall *(Intel, USA)*
Meikel Poess *(Oracle Corporation, USA)*
Kai Sachs *(SAP AG, Germany)*
Connie Smith *(L&S Technology LLC, USA)*
Ian Whalley *(Google, USA)*
Rui Zhang *(PARC, USA)*

Additional reviewers:

Matthias Becker
Fabian Brosig
Erik Burger
Zoya Durdik
Justin Frye
Bogdan Ghit
Katja Gilly
Carlos Guerrero
Sebastian Lehrig
Aleksandar Milenkoski
Vahid Mohammadi
Samaneh Navabpour
Qais Noorshams
Alexandru Olteanu
Michal Papez
Claudia Priesterjahn
Chris Rathfelder
Thomas Reidemeister
Siqi Shen
Timmie Smith
Nathan Thomas
André van Hoorn
Jan Waller
Wei Tan

ICPE 2013 Sponsors & Supporters

Sponsors:

ACM SIGMETRICS

SIGSOFT

SPEC Research

Gold Corporate Supporter:

SAP

Corporate Supporter:

CQK HOLDING

HP

Holistic Optimization of Distribution Automation Network Designs Using Survivability Modeling and Power Flow Equations

Anne Koziolek
University of Zurich
Zurich, Switzerland

Alberto Avritzer
Siemens Corporation
Princeton, NJ 08540

Daniel Sadoc Menasché
Federal University of Rio de Janeiro, Brazil

Categories and Subject Descriptors

C.4 [**Performance of Systems**]: Modelling Techniques

Keywords

Survivability; Transient Analysis; Smart Grid; Fault Tolerance

Abstract

Smart grids are fostering a paradigm shift in the realm of power distribution systems. Whereas traditionally different components of the power distribution system have been provided and analyzed by different teams, smart grids require a unified and holistic approach taking into consideration the interplay of distributed generation, distribution automation topology, intelligent features, and others.

In this tutorial, we present how transient survivability metrics can be used to create better distribution automation network designs. Our approach combines survivability analysis and power flow analysis to assess the survivability of the distribution power grid network.

We first describe the reliability challenges power utilities face and discuss the additional challenges and opportunities introduced by a smarter grid with features such as fault detection, isolation, and restoration. We then lay the foundations by introducing relevant reliability metrics and by putting survivability metrics into the broader context of dependability metrics.

We then describe how two complex systems such as a telecommunications server and a smart grid distribution automation grid could be modelled using survivability metrics. Specifically, we show how a smart grid distribution automation grid can be modelled in an abstract way using a Markov reward model and how this model is parameterized with results from power flow analysis and measured data from the utility.

Finally, we present an initial approach to automatically optimize available investment decisions with respect to survivability and investment costs and highlight its benefits by applying it to the design of a real distribution automation circuit.

ICPE'13, April 21–24, 2013, Prague, Czech Republic.
ACM 978-1-4503-1636-1/13/04.

Use Case-Driven Performance Engineering without "Concurrent Users"

Morten Heine Sørensen
Formalit
Byenden 32, 4660 Store Heddinge
Denmark
+45 3031 2923
mhs@formalit.dk

ABSTRACT

The concept of *concurrent users* often causes confusion when used to define performance requirements in industrial software projects. The term is frequently used to state performance requirements without clarification of *what* the users will be doing, or *how often*. This paper offers a thorough analysis of the concept and related notions.

Despite the confusion surrounding it, the concept of concurrent users – in a precise form – is advocated in the community for stating performance requirements. However, we argue in this paper that, even when stated in precise terms, this approach has drawbacks. Indeed, a system may perform *better* than expected, even if the number of concurrent users it can handle is *worse* than expected. A better suited notion is that of *through-put.*

But even when basing performance requirements on clear, well-suited concepts, there appears to be no uniform format in the literature for such requirements. In particular, the requirements are sometimes stated in general, rather than for the specific areas of functionality of the system. As a consequence, the point may be missed that the through-put may be unevenly distributed over the functionality of the system. In this paper we therefore advocate the format of *performance-annotated use cases*, adding requirements on through-put and response-time to the traditional use case.

It is well-known how *functional* test cases are developed from use cases. In contrast, less has been said about the generation of performance test cases. Therefore, we show how the enriched use cases not only provide precise and meaningful requirements, but also yield detailed specification of the performance test set-up which can be directly input as configuration of load test clients. As a bonus, initial configuration of the system's capacity for handling concurrent users and requests is also provided.

Finally we outline an overall approach to performance test based on the above ideas. The approach has been followed in several industrial projects.

Categories and Subject Descriptors

D.2.8 **[Software Engineering]** Metrics – Performance measures.

Keywords

Use cases, agile development, performance requirements, concurrent users, through-put, response time, think time, load test.

1. CONCURRENT USERS: CONFUSION

In a recent project in which the author participated, the four most central performance requirements were stated roughly as follows:

R1. The system must support 1500 concurrent user sessions.

R2. The response time for any user action must be at most 1 sec.

R3. The response time for a service call must be at most 0.1 sec.

R4. The system must handle 1000 events per day.

The requirements pertain to a system for which users can log in, perform user actions, and log out again. The system also receives service calls from external systems. Finally, events arrive at the system through either user actions or service calls, see Figure 1.

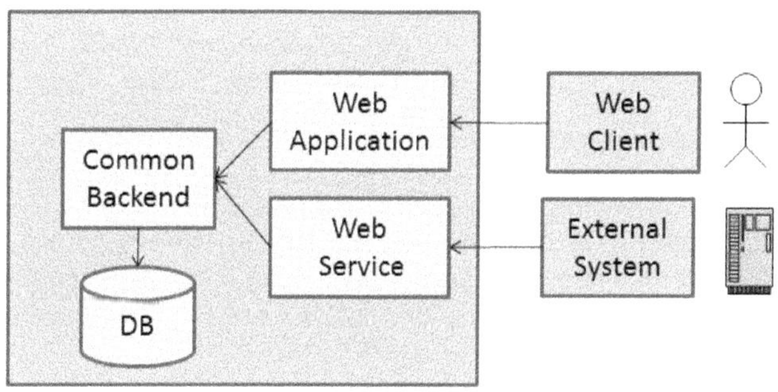

Figure 1. An interactive system also called by external systems.

Requirement **R1**, about *concurrent users,* is unsatisfactory because it does not state what the users do or how often they do it.

Handling 1500 users who log in and do nothing more is very different from accomodating the same number of users repeatedly requesting a heavy operation every second. A system may easily be capable of the former, while unable to live up to the latter. Whether a system satisfies the requirement is therefore heavily dependent upon interpretation.

Requirement **R2**, about *response time*, may be misleading because it fails to state that the response times must hold while the system is running at the expected *load.*

Responding promptly to a single user is an entirely different task than maintaining reasonable response-times for a large number of users simultaneously burdening the system.

Requirement **R3**, also about response time, is vague for the same reason as the second requirement: it fails to refer to the expected load.

In fact, both service call response times and response times experienced by users should respect the stated upper bounds while the system is burdened with the combined expected load from users *and* external systems, because both types of requests may share the same resources.

Finally, requirement **R4**, about *through-put,* is vague, because it does not state what functions are executed in the system at what frequency.

The event handling may have different flows, involving a different number of steps, depending on the type of event. Handling 1000 single-step flows may be a lot easier than 1000 flows each with 10 steps, and if the latter is closer to what actually happens when the system is in production, measuring the requirement in the former manner says little about what will happen during production.

Thus, the point is: *Requirements for response times are relative to some load; and if the load is defined in terms of concurrent users, it should explain what the users do and how often they do it.*

The above review of the requirements **R1-R4** concerns one specific project, of course, but the point is valid more generally. It frequently occurs in industrial projects that the client has difficulty expressing performance requirements in a precise manner, if they are stated at all.

This may be due to the fact that the performance requirements are on the border between the business domain and the technical domain. For instance, as will be argued in Section 2, the concept of concurrent users may be a purely technical term, or part business, part technical, depending on its precise understanding.

People writing requirements are often business people with limited training in IT, at least when there is a clear separation between the vendor and the client, and especially if the project is fully described by the client prior to one or more vendors making bids for the contract. These people often lack the necessary training or experience for stating precise performance requirements.

Another reason may be that there is little tradition for writing precise performance requirements in many organizations. As a consequence, the habit is not picked up by the participants of the organization's different projects.

As a further illustration, we refer to the standard contract ("K02") for IT projects delivering systems to public organizations in Denmark [18], see Figure 2.

K-1 The response times for any part or parts of the Delivery must comply with the following requirements:

Transaction	Description/If applicable assumptions	Service level goals for compliance rate, in %	Service level goals for maximum response time, in seconds
Simple	[...]	[...]	[...]
General	[...]	[...]	[...]
Complex	[...]	[...]	[...]
[If applicable, more specific transaction]	[...]	[...]	[...]
[...]	[...]	[...]	[...]

Figure 2. Performance requirements in K02.

Note that the requirement only states the maximum response time (last column) and the percentage of requests/transactions that must be below the maximum response time (third column).

The guide to the standard contract mentions the following example:

"Within any half-hour period, the response-time must be kept within the following limits:

- 99.5% of the response times must be less than 20 seconds.
- 98% of the response times must be less than 8 seconds.
- 95% of the response times must be less than 5 seconds.
- The average response times must be less than 3 seconds."

While very explicit on response times and fractiles, nothing is said about load. It must probably be understood that such details fall under "assumptions" in the second column.

A new version ("K03") has recently been developed, accommodating agile thinking [19]. However, the handling of performance requirements is not significantly changed.

It could be argued that load is only relevant for a particular kind of application – one used by multiple users simultaneously. For an application running entirely on a stand-alone PC making no request to other systems, the concept is not relevant and should therefore not be adopted in a standard for contracts covering all kinds of applications. However, the multi-user scenario is a very common (and the only one considered in this paper), and for this reason the argument is not convincing.

In order to revise the requirements **R1-R4** according to the point made earlier, we must define precisely the concepts of concurrent users, response time, and load – the latter will be rephrased as through-put. We begin with the service calls, because they constitute the simplest case.

Consider a software system responding to individual, unrelated requests; we denote this kind of system *session-less*. For instance, it could be an application server exposing web services.

Define the following notions:

- *Average response time* (ART): average time in minutes from the system receives a request, until the response is sent. This is a measure of the system's *speed.*[1]
- *Through-put* (TP): The number of requests finished in a minute. This is a measure of the system's *capacity*.
- *Concurrent requests* (CR): The number of requests being served by the system at any given moment.

The three notions are related to each other via the following rule, known as *Little's Law:*

$$TP = CR/ART.$$

Note in particular that if we know two of the values, we can compute the third.

Example 1. A system providing a public web service with currency exchange rates is observed over a minute to have a through-put of 60 requests and an average response time of 6 seconds (0.1 minute). It follows that, on average, there were 6 concurrent requests along the way, see Figure 3.

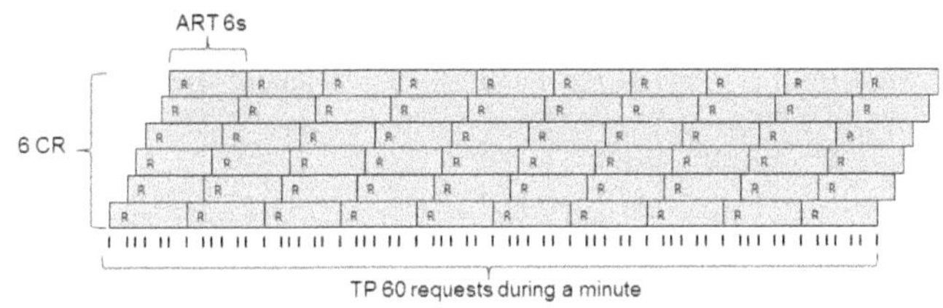

Figure 3. Little's law.

The small vertical bars at the bottom "|" indicate the arrival of requests. The solid yellow blocks indicate the time during which the system serves a request. Each level in the figure indicates a request waiting to complete or, equivalently, a server thread handling a request. At each level, whenever one request finishes, a new one is served.

Note that CR, like ART, is an *average* number. If, during the minute, the response times go up and down, so will the number of concurrent requests. In fact, it is unlikely that the number of concurrent requests will be precisely 6 all the time.

As mentioned, Little's law lets us compute one value from the two others. Another relationship between the three notions, may illustrate the concepts further. It concerns the situation where we test a system with a specified number of test clients that repeatedly fire requests—of course, this number is what we have called CU. If we begin with a small number, the system will probably have no difficulty serving the requests with some ART and TP. As we increase CU, initially ART will not be affected, but TP will grow, since we are getting more work done. At some point, the server's capacity for handling requests in parallel will be exhausted, and from this point on TP will, ideally remain the same, whereas ART will grow linearly, because the additional requests must now wait in line for their turn. The situation is illustrated in Figure 4.

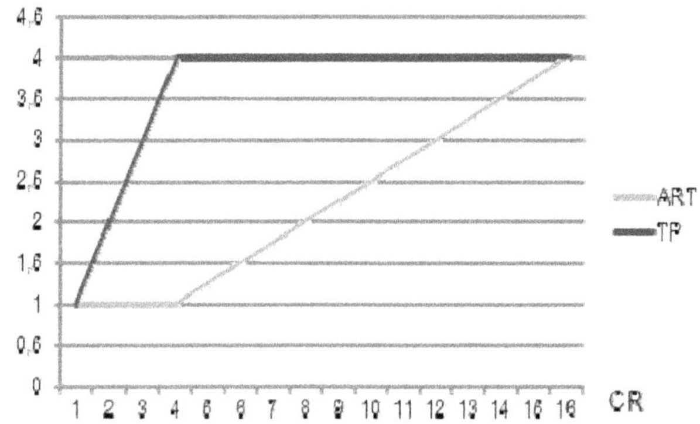

Figure 4. Effect on ART and TP of increasing CU.

In reality the graph may look different, because the queue mechanism often does not work in the ideal manner, see Figure 5.

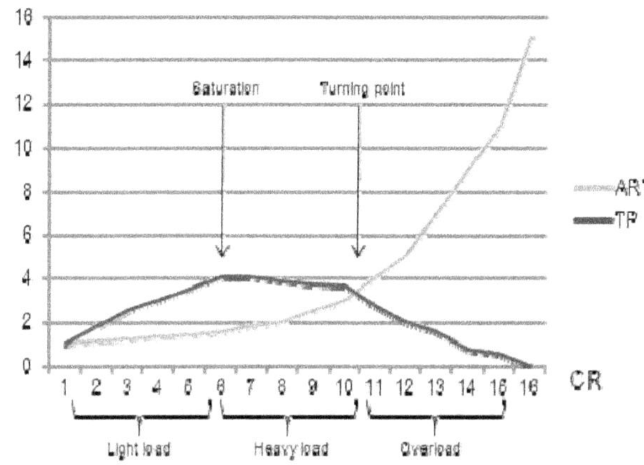

Figure 5. More realistic scenario.

For a heavily loaded server, it is an interesting exercise to configure the server's number of concurrent threads high enough to exploit the server's resources and low enough to avoid overloading the server (the additional requests will then wait in line in the server's execute queue, instead of burdening the server's resources).

Next consider a system against which a user logs on, performs a number of actions, and logs off again; we call this type of system *session-full*. The notion of logging in may not actually consist in providing identity and credentials; it could amount to an anonymous user having a session identified by a cookie returned by the system at the first request, and resubmitted in each subsequent request by the browser. Similarly, the notion of logging out may simply amount to the session timing out.

The concepts ART and TP carry over to the new setting, but CR is split into two separate concepts CU and CAU (the latter being the closest relative to CR); in addition a new notion concerning the user *think time* becomes relevant:

[1] A similar definition is to measure the time from a client makes a request until the client receives the response. This includes network time. During a performance test it may be relevant to ensure that the experienced network time is actually realistic. A further variation is to include also the client's time to render the response.

- *Concurrent logged in users* (CU): The number of users that have logged in, but not yet out. In other words, the number of users that have a *session* with the application.
- *Concurrent active users* (CAU): The number of users waiting for a response from the system at any given moment. In other words, the number of *threads* currently in use by the system to serve user requests.
- *Average think time* (ATT): The time taken from a user receives a response from the system until he or she sends the next request within the session. There is no think time after the last response is received; but if there is no explicit log out, the think time after the last step is considered to be the same as the time until the session times out.

Note that think time is meaningless in the setting of a session-less system. There is no concept of user, only individual requests, and each request is unrelated to all others. It is irrelevant whether subsequent requests are made from the same user/system.

What makes sense, though, is through-put – a certain number of requests are expected per minute. Moreover, we can discuss the *distribution* of the requests within the through-put. For instance, the through-put may be 120 requests per hour, but it could be that 90 requests are received in the first half hour and the rest in the second half hour.

When discussing a fixed through-put we always assume the distribution is more or less even (perhaps with some amount of randomization) within the period. Thus, if testing a system at peak load 120 requests per hour, where in reality the distribution is uneven as described above, we should rather test it at a peak load of 90 requests per half-hour, i.e. 180 requests per hour.

Coming back to the setting of session-full systems, the two new notions CU and ATT are related to TP and ART by the following generalization of Little's law, also called the *Response Time Law*:

$$TP = CU / (ART + ATT).$$

Example 2. A world-wide insurance company has a system for reporting damages online. A through-put of 8 reports per minute is observed at peak.

The customer logs in and reports the damage, and is automatically logged out at the end after the last step. The process has 5 user steps (log in, click "Report damage", enter details part I, enter details part II, confirm). The system has an average response time of 5 seconds for each request. Suppose that any user "thinks" 5 seconds before each next action; this includes the time to actually enter the information to the system.

Since the total ART for the whole report is 25 seconds and the total ATT is 20 seconds, we have a total ART + ATT of 45 seconds, i.e. 0.75 minutes, so CU is 6. The situation is illustrated in Figure 6.

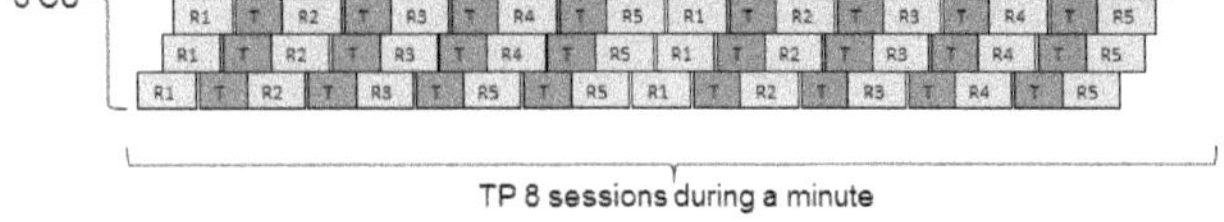

Figure 6. Response Time Law.

Each yellow block again represents a request, and now the number (R1-R5) indicates which request inside the flow is meant. The blue blocks indicate the user think time.

Note that TP on the one hand and ART plus ATT on the other hand must be understood relative to the same unit of work, which can be individual steps or the whole session. In the latter case ART and ATT is the sum of all ARTs and ATTs for the session.

Example 3. Counting requests instead of reports/sessions in the preceding example, we have 5 requests per report, i.e. a TP of 40 requests per minute. ART is 5 seconds. Since the user is immediately logged out after the last step, there is no think time after that step, but 5 seconds after the first four steps. This yields an ATT of 4 seconds per step. Thus ART + ATT is 9 seconds, i.e. 0.15 minutes. In conclusion, CU is again 6.

The density of active users within logged in users is given by:

$$CAU = CU \cdot ART / (ART+ATT).$$

In other words, the factor between CU and CAU is the time spent waiting for the system's response over the total time, i.e. user waiting time plus user thinking time (the time the user is idle before making the next request).

Example 4. Continuing the preceding example, recall that the total ART for the whole report was 25 seconds and the total ATT was 20 seconds. Finally CU was 6, so CAU is 3.3.

Note that ATT=0 if, and only if, CAU=CU. This happens if the session has just one step after which it immediately ends. This is what happens in the session-less case; so the latter can be considered a special case of the session-full case.

At any point, the active users are a subset of the logged in users which, in turn, are a subset of the *registered users,* if the latter concept makes sense for the system under consideration. It is not considered important whether users in subsequent sessions are actually the same or different registered users.

A few remarks about sessions may be in order. Above we assumed that users log in, do their work, and log out again. This type of session might be termed *dense,* since the user's activities happen in short time within the session, each step is separated in time from its predecessor by some small amount of think time.

However, not all systems with sessions work in this way. Another typical scenario is that users log in during the morning, do their work on and off during the day, and log out again in the evening. This type of session might be called *sparse*. In this scenario the number of concurrent users will typically be a fixed, known fraction of the registered users. There are even *ultra-sparse* sessions where the user can log-on, abandon the application, and hit it again weeks later with the session (basket) intact; for instance, Amazon has this nature. Probably, the session is passivated after a time-out and reactivated when the user returns.

There are two different ways to fit sparse sessions into the framework of dense sessions: we can either count sparse sessions as a variation of the dense sessions with very long thinking times, or we can separate the sparse sessions into sequences of dense sessions. We prefer the latter, though the choice should be made with some care.

The risk in splitting sessions is that the performance test may miss memory leaks, because the system terminates sessions quicker during the test than in reality, thereby yielding falsely low CU. Another risk is that response times may be higher than in reality, because the caching behavior during the test may be unrealistic, because information cached in the session is lost during the test.

In fact, even in the case of session-less systems, it is important to reflect actual caching behavior, if present. Both cached information of relevance for all requests (e.g. reference data and other list of values) as well as information relevant only for individual users, but still stored in a general cache. (In the case of session-full systems, the latter kind of information can be stored in the session.)

We close this section with a reflection on the role of use cases in session-full systems. Recall that every user logs on, performs a number of steps, and logs out again. It is natural to envisage these steps as belonging to one or more use cases. The think time then amounts to the time the user waits within a use case, from one step to the next, after receiving a reply from the system, as well as the time from one use case ends, until the next is initiated. This is most natural when we work with dense sessions.

2. CONCURRENT USERS: CRITIQUE

We now have the apparatus to make the requirements of the preceding section more exact. But before doing so, let us turn to the literature for inspiration.

One precise form in the literature [17] is the following: "for use case ABC the system will respond to a valid user entry within 5 seconds for a median load of 250 [concurrent] active users and 2000 [concurrent] logged in users 95% of the time."

In the author's opinion, this formulation is still not optimal. First of all, a use case usually consists of several steps, and the response time must hold for each step in which an answer is expected from the system.

But, more seriously, it is in fact hard to justify an estimate on the number of concurrent *active* users. Of course, some kind of estimate is involved in any formulation of expected load, but the point is that the number of logged in and active users actually depend on internal attributes of the system itself.

Example 5. Consider two different systems implementing use case ABC mentioned above, and assume the use case has, say, 3 steps:

- System A has, in each step, a response time of 1 second as long as the number of concurrent active users is less than 200, but crashes when the number exceeds 200. It can handle 2000 concurrent logged in users.
- System B has, in each step, a response time of 5 seconds as long as the number of concurrent active users is less than 300, but crashes when the number of active users exceeds 300. It can handle 2000 concurrent logged in users.

Which system satisfies the requirement? Which system is the most desirable one?

Well, imagine the scenario that use case ABC is executed 1.000 times per minute by different users (it is not important if they are actually different from execution to execution of the use case).

If the users adopt system B, each execution involves 3 steps, each with 5 secs. spent by the system, and, say, 35 secs. spent by the user thinking before the next step. Thus each execution of the use case takes 2 minutes. With 1000 executions per minute, there will indeed be 2000 concurrent logged in users. The ratio between user waiting time and total time, i.e. user waiting time plus user think time is 5 to 5+35, i.e. 1 to 8, so among the 2000 logged in users, 250 will be active at any moment. Thus, our scenario corresponds exactly to the requirements and System B meets the criteria.

Now consider System A. It cannot handle the required 250 active users, but consider what happens in our scenario. Each execution of the scenario takes 3 times 1 + 35 secs, ie. 108 secs., i.e. 1.8 minutes. Thus there will only be 1800 logged in users, and the ratio between waiting time and total time is now 1/36, so there will only be around 50 active users at any point.

If we insist on verifying that System A can handle 250 concurrent active users, we have to push a through-put of 5.000 ABC executions per minute through the system, just because it handles requests more efficiently than System B (fewer if the average response time decreases with the increase in tested through-put).

The point is: *The faster the system, the fewer concurrent logged in and active users will be required to obtain a given through-put.*

Since we do not know the response time in advance, we do not know what to require of the system regarding concurrent logged in and active users.

In some situations, the number of logged in users may be fixed. For instance, all employees in a department may log in during the morning and log out in the afternoon. But for a fixed number of executions of use case ABC, the number of concurrent *active* users still depends on the response times of the system. Thus, again we cannot know the number of concurrent active users.

In some cases, the users of a site may abandon it, if it is slow, and this may provide some self-adjusting behavior, where CAU does not increase, despite higher ART than expected, because CU decreases. But of course, this is not a satisfactory way to satisfy the requirements (partially).

We conclude that it is difficult to predict CAU because it depends on ART, which we cannot know before the system is built. Will requests take half a second or a whole second? We cannot know.

In contrast, the through-put for each use case will often be known. For instance, a use case may concern the creation of some kind of business entity in the system, contracts, customers, orders, etc., and it might be known by the business how many of these arrive over a period. It is more difficult to predict the through-put of use cases that do not leave a trace that makes sense in business terms. For instance, if a user can browse entities before selecting one, it may be difficult to know the amount of browsing. In this case, some kind of estimate must be made.

Having fixed TP and ATT to some expected values, and fixed an upper bound for ART, we can compute an upper bound for CAU, and use this value for the relevant purposes. But one might wonder whether we have replaced one problem by a more difficult one: Is it not as difficult to predict ATT as ART? The answer to the last question may be "yes," but note that CAU is independent of ATT. Indeed, it follows from the response time law and the law regarding user density that

$$CAU = ART \cdot TP.$$

Intuitively, ATT does not impact CAU, because the user won't be counted as active during the think time anyway. (Note, incidentally, that the latter rule says that Little's law also holds for the session-full case, when we read CR as CAU.)

However, there are other reasons that it may be important to estimate ATT as precisely as possible. If we run a test without think time, despite having estimated ATT to some value, CU will be lower during the test than in reality, so we may miss memory problems stemming from many or large sessions. Also, varying think times in reality could cause chaotic phenomena causing difficulties for the system that are not revealed during the test. At least this is claimed in the literature.

A similar point pertains to the definition of which use cases a user executes in a scenario. Consider a system for paying bills online. It has use cases Log in, Pay Bill, Log out. In every execution, the user may repeat "Pay Bill" a number of times. The question now arises whether, for a given through-put, it makes any difference how it is distributed over users; that is, whether it makes any difference whether we have, say, 30 users each paying two bills (in a single session) or 60 users each paying one bill.

Example 6. Assume TP is indeed 60 bills paid per hour, i.e. 1 per minute. Users log on, pay one (or two) bills, and log out again. For simplicity we assume each of the three use cases has a single step only. Suppose ART is 5 seconds for of the three use cases and that ATT after Log on and Pay bill is 35 seconds.

In the one-bill per user variant we have ATT+ART = 85 seconds, i.e. 1.4 minutes, so CU = $1 \cdot 1.4 = 1.4$. CAU is 0.25.

In the two-bill per user variant, the through-put of sessions is 30 per minute, i.e. 0.5 per minute and ATT+ART=145 seconds, i.e. 2.4 minutes, and so CU = $0.5 \cdot 2.4 = 1.2$. CAU is 0.17.

The one-bill per user variant has slightly higher CU and CAU, because the overhead of use case Log On and Log Out is higher. In practice, such overhead may be negligible.

However, with each user paying two bills, there could be some benefit of caching, not experienced with separate users. If the application implements a relevant type of caching, this will be missed during the performance test when we run with one bill/per user, which may thus yield falsely high ART.

Above we have assumed that the system under consideration is some kind of multi-user application (where the users may be other systems). The notions of concurrent requests, logged in users, and active users usually will not be interesting for a desk-top application used by a single user.

Now consider a thick client, either the old-fashioned type (i.e. an application written in, say, C++, Visual Basic, or Java, that make server calls) or a modern Ajax-application running in a browser, with, say, HTTP calls mixed with significant portions of Javascript running entirely in the browser without service calls. In this case, only the functionality that actually makes server calls fits into the framework. In other words, the pure client functionality is not relevant for a load test of the server system.

3. ANNOTATED USE CASES

Recall that in Section 1 we complained that requirements R1 and R2 were vague because, although they specify restrictions on response time and the number of concurrent users, they do not explain *what* those users are expected to be doing or *how often*. We subsequently clarified the "how often" part, by introducing precise notions of concurrent user and related concepts. Also, we went on by arguing that through-put was a better suited notion for quantifying "how often."

It remains to address the "what" part. Indeed, even when basing performance requirements on clear, well-suited concepts, such as through-put, there appears to be no widely adopted format in the literature for relating these notions to functionality of the system. Perhaps this is the reason that requirements are sometimes stated in general for "the system", rather than for specific areas of functionality of the system. As a consequence, the point may be missed that the through-put may be unevenly distributed over the functionality of the system. Requirements R1 and R2 also illustrate this.

Use cases, user stories, and other similar techniques for specifying functional requirements are in wide-spread use today. They are ideally suited to support iterative development of a system, because the iteration plan can be arranged according to use cases. Each of the traditional disciplines – requirements, architecture, development, testing – can be carried out in each iteration for some subset of the use cases, and the iteration yields a complete version of the system, with functionality implemented end-to-end.

We therefore suggest the format of *performance-annotated use cases*, adding requirements on through-put and response-times to the traditional description of use cases. For instance, consider the use case description in Figure 7.

The use case allows a bank customer to pay bills from one of his accounts using an online banking system. The last three rows in the template describe *required through-put, required maximum response time,* and *assumed think time.* We discuss each in turn.

Use Case: Pay Bill		
Prerequisites	The customer is logged in.	
Standard flow	The customer	The system
	1. Selects Pay Bill.	2. Requests code line for bill, amount, account and date.
	3. Enters details and clicks OK.	4. Validates details, and requests password.
	5. Enters password and clicks Confirm.	6. Validates password, transfers amount, shows main menu again.
Alternate flows		
Result	Amount deducted from account and transferred to receiver.	
Through-put	3200 per hour (2 per user session).	
Response time	1 sec. for step 2 and 4; 10 secs. for step 6.	
Think time	10 secs before step 3 and 5.	

Figure 7. Performance-annotated use case.

The required through-put describes the through-put *for this particular use case* that the system must be able to support. In this example, the entry also mentions how many executions are contained in the same user session. The reason for this is that, due to caching behavior, it may make a difference to the system whether a single customer is paying several bills or each bill is paid by a separate customer in a separate session, as previously mentioned. The performance test should mimic the actual usage patterns of the system as closely as possible, taking time, money, risk and other relevant factors into account.

The required maximum response time specifies an upper bound for each system step that the system must observe. In a real project, the system will probably only be required to meet the requirement in a specified percentage of the cases, say 95%. It may then be a requirement that in the remaining 5% of the cases, the response time must be at most, say, twice the specified maximum, i.e. 2 seconds after step 1 and 3, and 20 seconds after step 5. For instance, something like this is the case with the standard contract K02 as mentioned earlier.

Also, the actual maxima (1 and 10 seconds) may not be stated explicitly for each use case step; rather, the steps could be categorized into small, medium, and large, and a general maximum response time could then be formulated for each of the three types of step, say 1, 10, and 30 seconds, respectively.

The last entry describes the user think time. As previously explained, the think time may be significant because it drags sessions longer than what follows just from the response-times, thereby increasing the number of concurrent logged in users, and this could have an impact e.g. on memory consumption. In general it is desirable to reflect as closely as possible how the system is actually used in real life, though compromises may be made when balancing resources against risk. As with response times, the think times may be categorized into small, medium, and large (say 1, 10, and 60 seconds, respectively).

Say that we write every medium step with a single underline, a large step receives two underlines, and a small step receives no underline. Then a more quite notation for the above use case could be as shown in Figure 8.

Use Case: Pay Bill		
Prerequisites	The customer is logged in.	
Standard flow	The customer	The system
	1. Selects Pay Bill.	2. Requests code line for bill, amount, account and date.
	3. Enters details and clicks OK.	4. Validates details, and requests password.
	5. Enters password and clicks Confirm.	6. Validates password, transfers amount, shows main menu again.
Alternate flows		
Result	Amount deducted from account and transferred to receiver.	
Through-put	3200 per hour (2 per user session).	

Figure 8. More "quiet" notation.

The question arises what should be done about the alternate flows (or scenarios) of the use case. In practice, some prioritization must be made of use cases and their scenarios. It may easily be the case that not even all use cases are included in the performance test, let alone individual scenarios of a single use case. The decision must necessarily be decided on a case-by-case basis. For instance, if a use case concerns the possibility of doing searches in the system, and the alternate scenarios of the use case correspond to different ways of searching, then more (or all) of them could easily be relevant, so response times for all each of them should be specified. Fortunately, this is easy with the quiet notation.

Note that the use case refers to a *maximum* response time, rather than an *average* response time. Nevertheless, the response time law still provides valuable information. Indeed, if the response time for every system step is *at most* a specified amount, then the *average* response time will be less than this amount, and therefore the through-put will be *at least* the amount specified by the response time law, for a given number of logged in users.

Similarly to the performance requirements for use cases, we may also have performance requirements regarding maximum response time and minimum through-put for web services called externally. We do not have a specific format for this in mind.

4. PERFORMANCE TEST CASES

As indicated earlier, there is a widely adopted practice of generating functional test cases from use cases. Roughly, the test cases can be seen as instantiations of use case scenarios where concrete values are substituted for named concepts.

For instance, whereas a use cases speaks of *a customer* who logs on and pays *a bill* of *some amount*, the test case will involve a *specific customer* known to the system who will pay a *specific bill*, known to the system, of a specified, concrete amount.

These test cases could in principle be extended with information about expected response-time and assumed think time, as well as expected through-put. However, in reality such test cases will not be run repeatedly, manually by humans. Instead they are typically run *once* and recorded with some proxy and then revised so as to use input data from files or a database. After this the test cases can be run automatically by a load test client. Figure 9 below has (in the left pane) a test scenario that amounts to the main success scenario of a the Report Damage use case of the insurance system mentioned earlier.

The test cases are then configured with respect to the actual performance parameters. For instance, in the popular tool *JMeter*, one configures a so-called thread group for each test case, and part of the configuration is to fix a number of threads (users), see Figure 9.

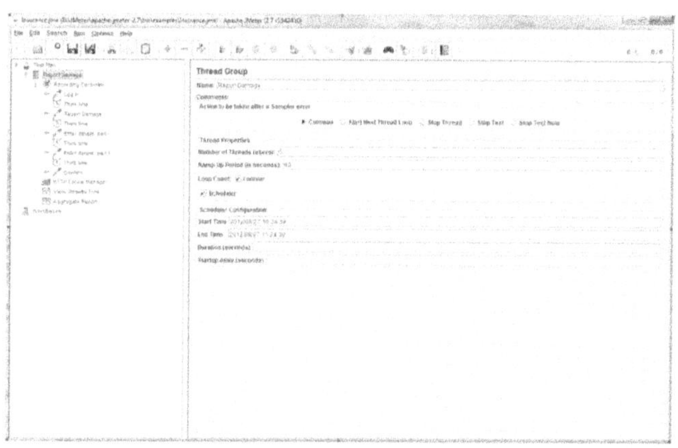

Figure 9. Concurrent logged in users.

This number corresponds to CU and can be computed from the required through-put and maximum response time, plus the assumed average think time. For instance, in the example with the insurance company, we need 6 threads, as mentioned in the last example, if the system precisely meets the requirement for ART.

If more threads turn out to be needed, it must be because the system cannot meet the requirement for ART. If the system's response time is significantly lower than the required ART, we may actually need fewer threads than 6.

In fact, one can configure JMeter to achieve the required TP by using the necessary number of threads (up to 6), whatever that number turns out to be depending on the system's response time behavior, see Figure 10.

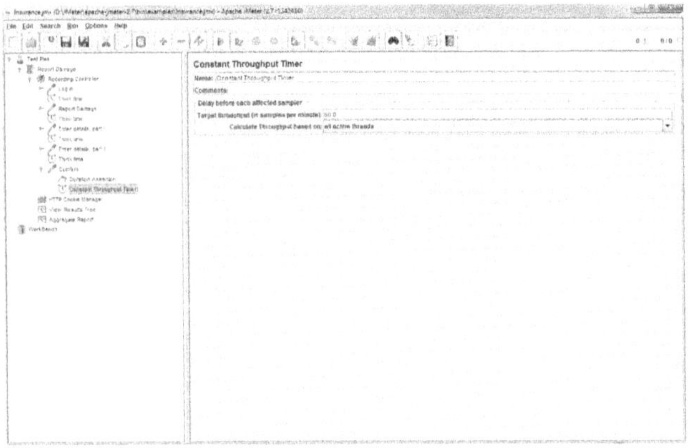

Figure 10. Achieving the required through-put.

In addition to CU we can configure think time, see Figure 11.

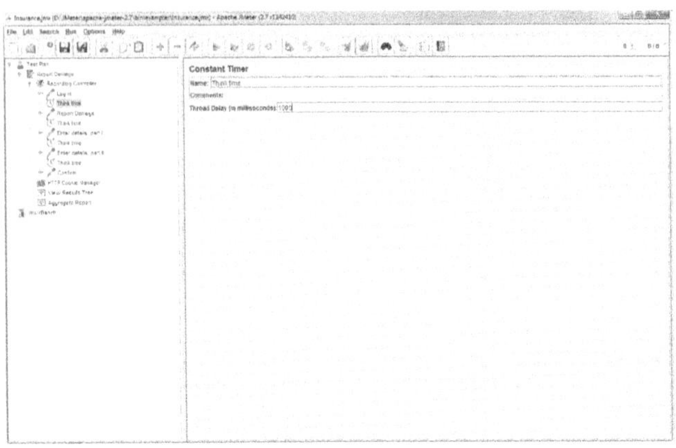

Figure 11. Think time.

Finally we can mark excessive response-times as failures, see Figure 12.

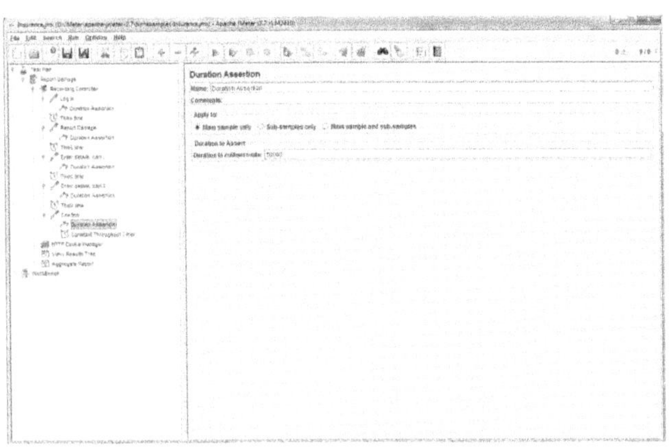

Figure 12. Marking high response time as failure.

Finally, we can ask the tool to produce a report for the test run that displays the response times and through-put, so that we can see whether the requirements are met. See Figure 13.

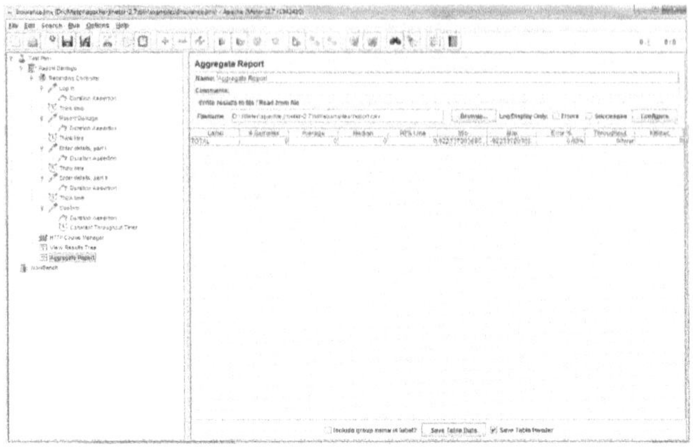

Figure 13. The test run results.

Thus we see that a typical load test tool, exemplified by JMeter, provides facilities for configuring test cases and performance requirements expressed in terms of through-put, response time and think time.

5. OVERALL APPROACH

We finally summarize our approach to performance-testing which comprises a number of activities; these happen partly before we go into the iterations of a project, and partly inside the iterations themselves.

1. Identify the functional requirements.
2. Identify the system and environment.
3. Identify the performance requirements.
4. Prioritize the performance requirements.
5. Choose performance test type.
6. Plan the performance test.
7. Build the performance test.
8. Run the performance test.
9. Analyze the output from performance test.
10. Profile the system.
11. Optimize the system.

In the first activity we identify the functional requirements, i.e. the use cases, or something similar. One may wonder why it is relevant to identify the *functional* requirements, when we are in fact interested in *performance,* i.e. *non-functional* requirements. The answer is that the latter pertain to the former; that is, each use case has its own through-put (and possibly response time) requirements.

In the second activity we get an overview of the architecture of the system and the production environment as well as the performance test environment. It is important to understand the architecture, because this has a bearing on how we formulate and test the performance requirements. For instance, we need to know if the system can be accessed both interactively and through web service calls.

But the architecture is also important for deciding what we measure during the test. For instance, a database server calls for a different kind of attention as compared to an application server. Finally, the environments are important because it should be verified that the performance test environment resembles the production environment as much possible.

The configuration of *threads* through-out the production environment can be based (at least in the first version) on the performance requirements.

Example 7. The insurance company mentioned earlier is building a new system for reporting damages online. The system must support a through-put of 60 reports per hour, and these are submitted in the Report Damage use case. The total maximum permitted response time for the whole use case is 25 seconds and the total think time is 20 seconds, so we have a total, so CU will be at most 45, and CAU will be at most 25. It follows that, in the worst case, the system must support 45 user sessions and 25 simultaneous requests.

If each request on the web server may open at most two connections to the back-end server, then a total of 50 threads will be needed at that level. If each thread on the back-end server, in turn, requires at most two connections to the database server, a total of at most 100 will be needed at that level. Here we have only considered the traffic from the online users. To all this, the threads handling web service requests must be added.

Moving on, we identify, in the third activity, the performance requirements, if they exist, or else contribute to their formulation.

In the fourth activity we decide which performance requirements to include in the performance test, as all can probably not be tested. Or, more concretely, we decide which use cases (and which scenarios of these) must have their performance requirements tested.

In the fifth activity we identify which type of performance test to run, e.g. load test, stress test, soak test, etc. (We omit detailed discussion of the different possibilities, as it is not the main concern in this paper.)

In the sixth activity we plan the performance test. The overall idea is that the performance testing of use cases follows the ordinary testing of use cases, i.e. it occurs in the iterations where the coding for those use cases is accomplished.

In the seventh step we build the test, as indicated in the preceding section. Of course, much scripting is usually needed.

In the eighth activity we run the test and collect statistics along the way. The performance test tool provides details about through-put and response time, as also shown in the preceding section.

In the ninth activity we analyze the output of the test and conclude whether the requirements are actually met.

If not, the tenth activity may include further profiling of the system using various tools to understand e.g. where excessive CPU time or memory consumption occurs in the system.

Once the reasons for inefficiencies are located, the system is improved in the eleventh activity.

Of course, the last four steps are often iterated several times.

6. CONCLUSION

We have provided an analysis of the fundamental concepts relevant for performance testing of multi-user applications: concurrent users, response-time, through-put and think time.

Also, we have criticized the user of "concurrent users" for stating performance test requirements as this notion is a technical attribute of the resulting system, which cannot be known in advance. Instead we have advocated the user of through-put.

We have also demonstrated how performance test requirements can be stated in a compact fashion adding just a few details to the existing description of use cases.

Moreover, we have shown where the identified performance test parameters fit into the construction of an automated performance test using a popular tool.

Finally we have fitted all the pieces into an overall performance testing methodology.

The paper draws on a number of references.

The exposition of fundamental laws connecting the different concepts related to performance requirements is inspired by [9].

As mentioned earlier, the critique of "concurrent users" was inspired by an example in [17].

The positioning of the paper inside a broader tradition of performance *engineering*, rather than merely performance *testing*, is inspired by [14]. Many other text books on performance testing emphasize not only the actual testing activities, but also the precise formulation of performance requirements as well as an overall approach for managing performance, see, e.g. [4], [7], [11] [12] [16].

Nevertheless, the literature distinguishes between performance *testing* and performance *tuning*, see [8], [13] [15] for some popular references in the latter category. The distinction is the same as that between testing for functional errors on the one hand, and fixing code bugs on the other hand. For functional testing, the distinction is often quite sharp in projects between those who look for errors and those who fix them. In the case of performance testing, the same persons are often involved in both activities

Use cases were invented by Ivar Jacobsen and popularized by Cockburn [1] and many others. They are in wide-spread use today.

A less formal approach is the user of *user stories,* see e.g. [2]. These contain fewer details and correspond more (but not exactly) to scenarios of a use case than to the use case itself. Performance requirements may also be attributed to user stories, though the individual steps of the scenarios of use cases may be useful for stating response and think times.

Developing test cases from use cases is a standard practice, see, for instance, [3][5]. It is natural to use the same approach to drive the testing of performance requirements and the idea occurs more or less explicitly in [10] and [17], and probably many other places. The guide for "K03" [19] also mentions the idea informally. Instrumenting use cases with response time requirements is hinted at in [10].

Our examples used the popular, open source tool JMeter [6]. There are numerous commercial tools available as well, as a quick Google search will reveal, but JMeter has also been used for many large-scale industrial projects.

7. REFERENCES

[1] Cockburn, A. 2001. *Writing Effective use Cases.* The Agile Software Development Series. Addison-Wesley.

[2] Cohn, M. 2004. *User Stories Applied for Agile Software Development.* The Addison-Wesley Signature Series. Addison-Wesley.

[3] Collard, R.1999. Test Design: Developing Test Cases from Use Cases. Software Testing & Quality Engineering Magazine.

[4] Haines, S. 2006. Java EE 5 Performance Management and Optimization. Apress.

[5] Heumann, J. 2002. *Generating Test Cases from Use Cases.* Journal of Software Testing Professionals.

[6] JMeter documentation. http://jmeter.apache.org/

[7] Joines, S., Willenborg. R., and Hygh, K. 2002. *Performance Analysis for Java Web Sites.* Addison Wesley.

[8] Killelea, P. 1998. *Web Performance Tuning.* O'Reilly.

[9] Lazowska, E.D., Zahorjan, J., Scott Graham, G. and Sevcik, K.C. 1984. *Quantitative System Performance. Computer System Analysis Using Queueing Network Models.* Prentice-Hall.

[10] Leffingwell, D. and Widrig, D. 2003. *Managing Software Requirements – A Use Case Approach.* Second Edition. Object Technology Series. Addison-Wesley.

[11] Microsoft Corporation. 2007. *Performance Testing Guidance for Web Applications.* Microsoft Press.

[12] Molyneaux, I. 2009. *The Art of Application Performance Testing.* O'Reilly.

[13] Shirazi, J. 2003. *Java Performance Tuning.* O'Reilly.

[14] Smith, C. U. 1990. Performance Engineering of Software Systems. Addison-Wesley.

[15] Tow, D. 2003. *SQL Tuning.*O'Reilly.

[16] Zadrozny, P. 2003. *J2EE Performance Testing.* Apress

[17] Wikipedia entry on performance engineering, http://en.wikipedia.org/wiki/Performance_engineering.

[18] *K02 – Standard contract for long-term IT project.* http://www.digst.dk/Styring/Standardkontrakter/K02-Standardkontrakt-for-laengerevarende-it-projekter.

[19] *K03 - Standard contract for agile IT project.* http://www.digst.dk/Styring/Standardkontrakter/K03-Standardkontrakt-for-agile-projekter

Automated Modeling of Complex Data Center Applications

[Keynote Talk]

Prashant Shenoy
School of Computer Science
University of Massachusetts
Amherst MA 01003
shenoy@cs.umass.edu

ABSTRACT

As online services become increasingly common, the complexity of backend distributed server applications in data centers has also continued to grow. At the same time, there is an increasing need to enhance the manageability of these large applications by automating common management tasks, which requires a good understanding of the run-time behavior of the application under different scenarios. However, the rising complexity of these applications makes the tasks of manually modeling and analyzing their run-time behavior increasingly difficult. In this talk, I will argue for the need to automate the modeling of the run-time performance of distributed data center applications. I will present techniques for automatically deriving application models using statistical methods from machine learning. I will describe how we have put these ideas into practice into two systems that we have built, Modellus and Predico, and will present case studies of using these systems for management tasks such as capacity planning and what-if analysis of data center applications.

Categories and Subject Descriptors

D.4.8 [**Operating Systems**]: Performance—*Modeling and prediction, measurements, queuing theory*

General Terms

Performance

Keywords

Performance modeling, data centers, learning

ICPE'13, April 21–24, 2013, Prague, Czech Republic.
ACM 978-1-4503-1636-1/13/04.

Towards Energy-Proportional Computing for Enterprise-Class Server Workloads

Balaji Subramaniam and Wu-chun Feng
Dept. of Computer Science
Virginia Tech
{balaji, feng}@cs.vt.edu

ABSTRACT

Massive data centers housing thousands of computing nodes have become commonplace in enterprise computing, and the power consumption of such data centers is growing at an unprecedented rate. Adding to the problem is the inability of the servers to exhibit *energy proportionality*, i.e., provide energy-efficient execution under all levels of utilization, which diminishes the overall energy efficiency of the data center. It is imperative that we realize effective strategies to control the power consumption of the server and improve the energy efficiency of data centers. With the advent of Intel Sandy Bridge processors, we have the ability to specify a limit on power consumption during runtime, which creates opportunities to design new power-management techniques for enterprise workloads and make the systems that they run on more *energy proportional*.

In this paper, we investigate whether it is possible to achieve *energy proportionality* for an enterprise-class server workload, namely SPECpower_ssj2008 benchmark, by using Intel's Running Average Power Limit (RAPL) interfaces. First, we analyze the power consumption and characterize the instantaneous power profile of the SPECpower benchmark within different subsystems using the *on-chip energy meters* exposed via the RAPL interfaces. We then analyze the impact of RAPL *power limiting* on the performance, per-transaction response time, power consumption, and energy efficiency of the benchmark under different load levels. Our observations and results shed light on the efficacy of the RAPL interfaces and provide guidance for designing power-management techniques for enterprise-class workloads.

Categories and Subject Descriptors

C.0 [**Computer Systems Organization**]: General—*System architectures*; C.4 [**Computer Systems Organization**]: Performance of systems—*Design studies, Measurement techniques*

General Terms

Measurement, Experimentation

Keywords

Power Limiting, Energy Proportionality, Greenness

ICPE'13, April 21–24, 2013, Prague, Czech Republic.

1. INTRODUCTION

Massive data centers, which house thousands of computing nodes, have become increasingly more common. A large fraction of such data centers' total cost of ownership (TCO) comes from the cost of building and maintaining infrastructure that is capable of powering such large-scale data centers and from the recurring energy costs [13]. Consequently, power and energy have emerged as first-order design constraints in data centers. These issues are further magnified by the inability of servers to provide energy-efficient execution at all levels of utilization. The recent recommendation of *energy proportionality* in servers, i.e., to design servers that consume power proportional to the utilization, is a move in the right direction as it has the potential to double the energy efficiency of servers [10]. However, achieving energy-proportional operation is a challenging task, particularly given that typical servers consume 35-45% of peak power even when idling.

Typically, dynamic voltage and frequency scaling (DVFS) has been used to achieve better energy efficiency as it can potentially give up to cubic energy savings [18, 25, 27]. However, as we will show in this paper, the subsystem affected by DVFS (i.e., the *core*[1]) is already the most energy-proportional part of the system. There are other subsystems, such as the *uncore*,[2] that consume constant power, irrespective of the system utilization. In order to achieve energy proportionality, we need to understand the power consumption of each subsystem at different levels of utilization and to leverage mechanisms that enable us to control the power consumption of these subsystems.

With the advent of Intel Sandy Bridge processors, we have better control over the power consumption of the system via the running average power limit (RAPL) interfaces [12, 1]. RAPL exposes *on-chip energy meters* for the *core* subsystem, processor package, and DRAM and enables the tracking of power consumption at a time resolution (~1 ms) and

[1]The *core* subsystem includes components such as the ALUs, FPUs, L1, and L2 caches [2].

[2]The *uncore* subsystem includes components such as the memory controller, integrated I/O, and coherence engine [2].

system-level granularity that was not possible before. Moreover, it facilitates deterministic control over the power consumption of subsystems through *power-limiting* interfaces. These interfaces allow a user to specify a power bound and a time window over which the bound should be maintained. While this hardware-enforced power limiting is an appealing option, the impact of power limiting on the performance, power, and energy efficiency of enterprise-class server workloads is still not well understood and remains an active area of research.

In this paper, we investigate whether it is possible to achieve *energy-proportional* operation for an enterprise-class server workload, namely the SPECpower_ssj2008 benchmark (henceforth referred to as SPECpower) by using the RAPL interfaces. To this end, this paper makes the following contributions:

- Insights into the mechanisms of power management for enterprise-class server workloads using the RAPL interfaces via an analysis of the SPECpower benchmark by calibrating its input parameters.
- A rigorous quantification of the energy proportionality of each subsystem within a server node via an analysis of power-consumption profiles of the different subsystems when running SPECpower at different load levels.
- The identification of system activity based on performance counter traces that strongly correlates with subsystem-level power consumption captured via the RAPL interfaces.
- An analysis and characterization of the instantaneous-power profiles at different load levels of SPECpower to understand the time resolution at which power limiting should be applied.
- Empirical results on the impact of RAPL power limiting on power, performance, per-transaction response time, and energy efficiency.

Through our contributions, we make the following observations and conclusions on the power management of the SPECpower benchmark using RAPL interfaces. First, the *core* is the most energy-proportional subsystem and the *uncore* is the least. Second, there is the opportunity for limiting the power consumption of the *core* and processor package subsystems at a resolution less than 50 ms for different configurations of SPECpower. Third, as the load level increases, the ability to limit the power consumption at the DRAM-level decreases. Fourth, power limiting at the level of the *core* subsystem is the best option for improving energy efficiency and achieving energy proportionality. Fifth, even when power limiting at the processor package-level, most of the power savings comes from the *core* subsystem. Sixth, better power-management mechanisms are required to achieve energy proportionality at the *uncore* subsystem-level. Seventh, though we were not able to achieve energy proportionality at the full system level, i.e., entire compute node, we show that energy-proportional operation is possible at the granularity of subsystems over which we have control via RAPL power limiting (i.e., *core* subsystem, processor package, and DRAM).

The rest of the paper is organized as follows. In Section 2, we present the details of the SPECpower benchmark and Intel RAPL interfaces. Section 3 describes our analysis and characterization of average and instantaneous power consumption of all subsystems at different load levels in SPECpower and the observations from these experiments. Next, in Section 4, we limit the power consumption of SPECpower within different subsystems and present the impact of power limiting on power, performance, and energy efficiency. In Section 5, we describe the related work, and we conclude in Section 6.

2. BACKGROUND

In this section, we provide an overview of the SPECpower benchmark and its design as well as details into its configurable parameters. We then present the control and capabilities exposed by Intel's RAPL interfaces.

2.1 Overview of SPECpower Benchmark

SPECpower [4] is an industry-standard benchmark that measures both the power and performance of a server node. The benchmark mimics a server-side Java transaction processing application and is based on the SPECjbb2005 benchmark [3]. It stresses the CPU, caches, and memory hierarchy and tests the implementations of the Java virtual machine (JVM), just-in-time (JIT) compiler, garbage collection, and threads. The benchmark requires two systems: (1) the system under test or SUT and (2) the control and collection system (CCS) [6] with communication between the systems established via Ethernet [7].[3] The SUT runs the workload and is connected to a power meter. The power meter, in turn, is connected to the CCS. The CCS collects the performance and power data passed to it by the SUT and power meter, respectively.

The SPECpower benchmark is designed to produce consistent and repeatable performance and power measurements. It executes different type of transactions and the transactions are grouped together in *batches* for scheduling purposes. Each *load-level* is achieved by controlling delay between the arrival of batches.

More specifically, the SPECpower benchmark is a graduated workload, i.e., it runs the workload at different *load-levels* and reports the power and performance at each load-level [9]. The benchmark starts with a calibration phase, which determines the maximum throughput. The calibrated throughput is set as the throughput target for 100% load-level. The throughput target for the rest of the load-levels is calculated as a percentage of the throughput target for 100% load-level. For example, if the throughput target for 100% load-level is 100,000, then the target for 70% load-level is 70,000, 40% is 40,000 and so on. The throughput is measured in server-side Java operations per second (ssj_ops).

The benchmark supports a set of configurable parameters.[4] For example, the maximum target throughput and the batch size can be manually configured. We refer the reader to [8] for further information on configurable parameters. The flexibility, coupled with the consistency and repeatability of SPECpower, allows us to evaluate the applicability of newer power-management interfaces, such as RAPL, to enterprise-class server workloads.

[3] SUT and CCS can be the same system. Communication is established via Ethernet only if the systems are different.

[4] Only a subset of these parameters can be changed for compliant runs.

2.2 Intel's Running Average Power Limit (RAPL) Interfaces

RAPL was introduced in Intel Sandy Bridge processors. The RAPL interfaces provide mechanisms to enforce power consumption limits on a specific subsystem. The only official documentation available for these interfaces is section 14.7 of the Intel software developer's manual [1]. Our experiments deal only with the Sandy Bridge server platforms.

The RAPL interfaces can be programmed using the model-specific registers (MSRs). MSRs are used for performance monitoring and controlling hardware functions. These registers can be accessed using two instructions: (1) *rdmsr*, short for "read model-specific registers" and (2) *wrmsr*, short for "write model-specific registers." The *msr* kernel module can be used for accessing MSRs from *user space* in Linux environments. When loaded, the *msr* module exposes a file interface at */dev/cpu/x/msr*. This file interface can be used to read from or write to any MSR on that CPU.

According to the Intel documentation, RAPL interfaces operate at the granularity of a processor socket. The server platforms provide control over three domains (i.e., subsystems):[5] (1) package (PKG), (2) power plane 0 (PP0), and (3) DRAM. We expect PKG, PP0 and DRAM to represent the processor package (or socket), the *core* subsystem, and memory DIMMs associated with that socket, respectively. The *MSR_RAPL_POWER_UNIT* register contains the units for specifying time, power, and energy, and the values are architecture-specific. For example, our testbed requires and reports time, power, and energy at increments of 976 microseconds, 0.125 watts, and 15.3 microjoules, respectively. Each domain consists of its own set of RAPL MSR interfaces. On a server platform, RAPL exposes four capabilities:

1. Power limiting – Interface to enforce limits on power consumption.
2. Energy metering – Interface reporting actual energy usage information.
3. Performance status – Interface reporting performance impact due to power limit.
4. Power information – Interface which provides value range for control attributes associated with power limiting.

2.2.1 Power Limiting

RAPL maintains an average power limit over a sliding window instead of enforcing strict limits on the instantaneous power. The advantage of having an average power limit is that if the average performance requirement is within the specified power limits the workload will not incur any performance degradation even if the performance requirement well exceeds the power limit over short bursts of time. The user has to provide a power bound and a time window in which the limit has to be maintained. Each RAPL domain exposes a MSR which is used for programming these values. The PKG domain provides two power limits and associated time window for finer control over the workload performance whereas other domains provide only one power limit. The interface provides a *clamping* ability, which when enabled, allows the processor to go below an OS-requested P-state.

[5]Note: We use RAPL domain and subsystem interchangeably in rest of the paper.

Table 1: Per-Socket Parameter Range (MTW = Maximum Time Window, MaxP = Maximum Power, MinP = Minimum Power). Multiply by 2 for Full Two-Socket System.

Domain/Range	MTW	MaxP	MinP
Package	45.89 ms	180 watts	51 watts
DRAM	39.06 ms	75 watts	15 watts

2.2.2 Energy Metering

Each domain exposes a MSR interface that reports the energy consumed by that domain. On a server platform, we expect the that (1) *energy(PKG) = energy consumed by the processor package*, (2) *energy(PP0) = energy consumed by the core subsystem*, and (3) *energy(uncore subsystem) = energy(PKG) − energy(PP0)*.

2.2.3 Performance Status

This MSR interface reports the total time for which each domain was throttled (i.e., functioning below the OS-requested P-state) due to the enforced power limit. This information will be useful in understanding the effects of power limiting on a particular workload.

2.2.4 Power Information

The PKG and DRAM domains expose a MSR interface that provides information on the ranges of values that can be specified for a particular RAPL domain for limiting its power consumption. This includes maximum time window, maximum power, and minimum power. The range of per-socket values on our experimental platform is given in Table 1.

3. AN ANALYSIS OF POWER CONSUMPTION FOR SPECPOWER

In this section, we characterize the power consumption of the SPECpower benchmark and analyze energy proportionality from the perspective of the entire system as well as each RAPL domain. We then analyze performance counter traces to understand the trends seen in power consumption. Finally, we present our characterization of the instantaneous power profile of SPECpower and provide insights into the time window used for limiting the power consumption of the benchmark under different configurations.

We will also show the following: (1) the most and least energy-proportional subsystems are the *core* (PP0) and the *uncore* (Package-PP0), respectively, (2) there is ample opportunity to limit the power consumption of SPECpower at different load-levels below the 50-ms resolution for the package and PP0 domains, and (3) there is little opportunity to limit the power consumption at the DRAM-level as the load-level increases.

3.1 Experimental Setup

The SUT for our experiments is an Intel Xeon E5-2665 processor (Intel Romley-EP). The node has two such processors for a total of 16 cores and 32 cores when hyper-threading is ON, as in our experiments. It has 256 GB of

Table 2: Performance Counters Used

Last-level cache references/second (LLCR)	Last-level cache misses/second (LLCM)
Instructions retire/second (IR)	Unhalted core clock cycles (UC)
Number of offcore outstanding reads/second (NOUT)	Cycles in which there was one or offcore outstanding reads (COUT)
Misses in DTLB loads that cause a page walk/second (DTLBLM)	Misses in DTLB stores that cause a page walk/second (DTLBSM)
Cycles spent in page walks due to DTLB load misses (DTLBLC)	Cycles spent in page walks due to DTLB store misses (DTLBSC)
Misses in ITLB that cause a page walk/second (ITLBM)	Cycles spent in page walks due to ITLB misses (ITLBC)

memory and runs a Linux kernel version 3.2.0. The CCS has an Intel Xeon E5405 processor with 8 cores and 8 GB of RAM. The CCS runs a Linux kernel version 2.6.32. The CCS and SUT were connected through a gigabit Ethernet network. To measure power, we used a Yokogawa WT210 power meter.

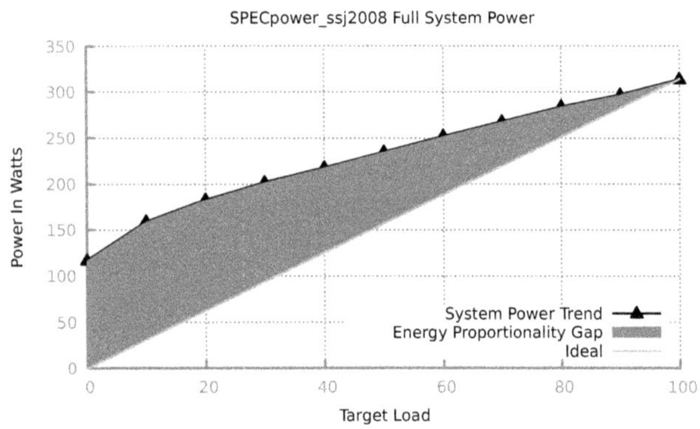

Figure 1: Illustration of Energy Proportionality Gap

We used all the cores in SUT for our experiments. Eight warehouses with four threads for each warehouse were used as the configuration for SPECpower. The four threads in each warehouse were pinned to two adjacent physical cores on the SUT using *numactl*. In order to provide consistent performance results throughout our experiments, we configured the *input.load_level.target_max_throughput* parameter to achieve the same performance for each run. It was set to 115,375 ssj_ops for each warehouse for a total of 923,000 ssj_ops for the entire run. In all our experiments, 100% load-level corresponds to 923,000 ssj_ops. This value was determined by averaging 10 calibration runs.

We changed the runtime for each load-level to 120 seconds using the *input.load_level.length_seconds* parameter and the pre- and post-measurement interval to 15 seconds in order to reduce the total runtime of the benchmark. On average, our testbed consumes 117 watts at idle and 314 watts at 100% load-level of SPECpower. We would like to stress that the system consumes 37.2% of peak power when idling.

3.2 Energy Proportionality Metric

As discussed earlier, we are interested in analyzing the energy proportionality of the system. The deviation of the power curve of the system from the ideal power curve is of particular interest to us. To illustrate with an example, Figure 1 shows the average power consumed by the testbed at each workload and the hypothetical/ideal energy proportional power curve for the system. The red shaded region is the deviation of interest. Henceforth, this area will be referred to as *energy proportionality gap (EPG)*. We quantify the EPG using the EP metric. The EP metric is calculated as one minus the ratio of EPG (the red region in the figure) and the area under the ideal curve [26]. A value of 1 for the metric represents an ideal energy-proportional system. A value of 0 represents a system that consumes a constant amount of power irrespective of the percentage of load-level. In our discussions, we determine the metric by calculating the area under curves. In particular, we calculate the area under the curve as the sum of areas of rectangles.

3.3 Analysis of Subsystem-Level Power Consumption

In this section, we present the details on the power consumption of SPECpower at a subsystem-level. We were able to profile the benchmark at a granularity that has not been possible until the advent of Intel Sandy Bridge by using the on-chip energy meters exposed by the RAPL interfaces. Our results provide insights into the energy proportionality of a system as a whole as well as at the RAPL domain-level. We also present the details of the power consumed using different batching sizes. Batching queries to exploit and create opportunities for power management is a well-researched area [21]. The number of transactions in each batch scheduled of the SPECpower benchmark is calibrated using the *input.scheduler.batch_size* input parameter. We use four different batching sizes (i.e, 1,000, 5,000, 7,500 and 10,000) in each of our experiments.

3.3.1 Methodology

We used the energy meters exposed in each RAPL domain to determine the power dissipated in each domain. In all our results, we report the average power[6] over five runs for the domain-level power consumption. For full-system power measurement, we have followed the power measurement methodology specified and developed by the SPEC organization for the SPECpower benchmark [5]. We also collected specific performance counter data that correspond to RAPL domain power and full system power in order to deliver more insights into the power profile of SPECpower. Table 2 shows the list of performance counters we profiled. We determined these performance counters based on previous work [23, 15], and we use newer ones that might have effects on the power curve of the system.

Because we have access to energy meters at different RAPL domains, we are also interested in finding the performance

[6]Average power is calculated as (initial energy reading - final energy reading)/time.

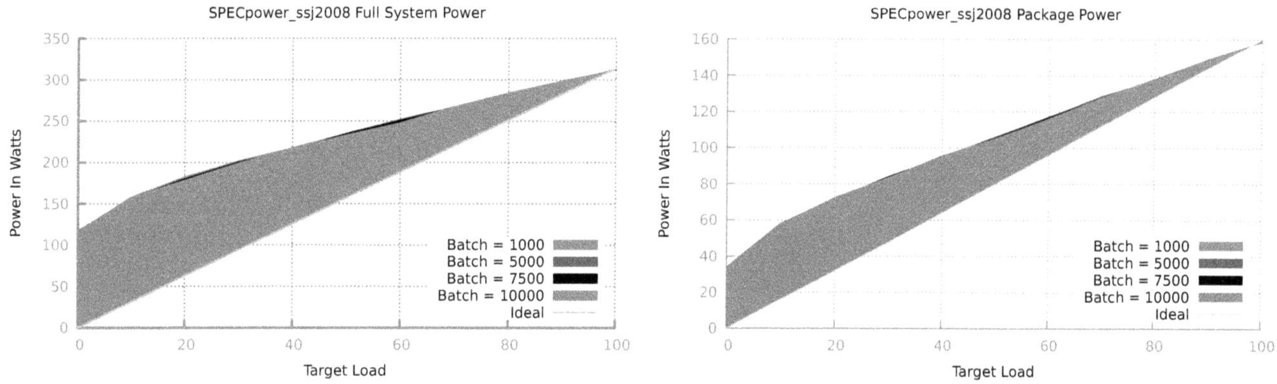

Figure 2: Power Profiles – Left:Full System, Right:Package. Both power trends are similar indicating a strong correlation between full system and package power consumption.

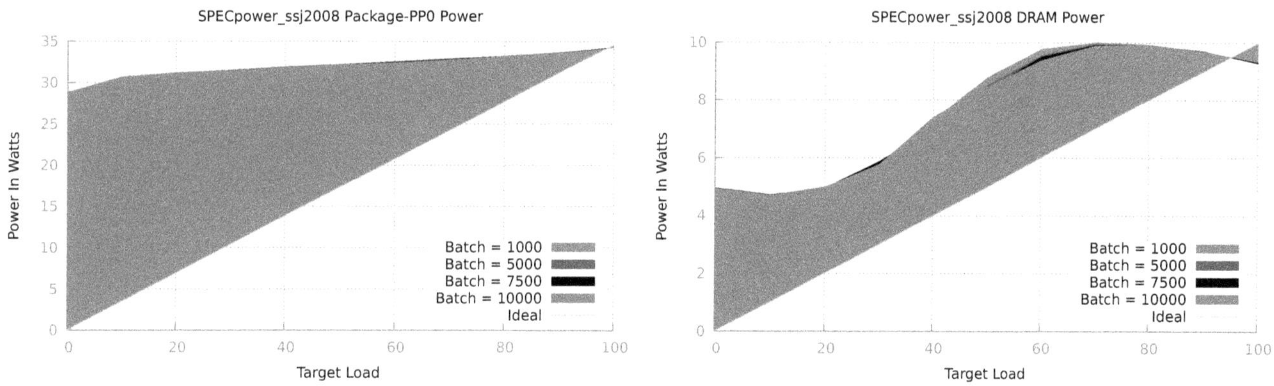

Figure 3: Power Profiles – Left:*Uncore* (Package-PP0), Right:DRAM. *Uncore* power is constant irrespective of the load-level making it the least energy-proportional subsystem.

counters that affect the power curve of that domain. Specifically, we want to find the performance counters that correlate to the power consumption of each domain. This study will help guide the modeling of the power consumption of different load-levels at a subsystem-level. Principal component analysis (PCA) is used to determine the most significant performance counter for a particular RAPL domain as well as the full system. We perform PCA on system activity to provide a concise and statistically meaningful overview of the causes for the trends in the subsystem as well as the full system power trend. Note that the correlation analysis is performed before performing PCA in order to obtain performance counters that show good correlation to power.

PCA is widely used for dimension reduction in many areas. The main concept behind PCA is to reduce the number of dimensions in a data set while preserving the variance of the data set as much as possible. This is achieved by transforming the data into principal components (PCs). The principal components are linear combinations of the original set of variables and are uncorrelated to each other. We first normalize the performance counter data to a unit normal distribution to mitigate the effect of the varying range of values for different performance counters and then apply PCA. We use R to perform these statistical analyses in our experiments.

3.3.2 Profiling of Full System and Package Power

Figure 2 shows the EPGs for the full system and package RAPL domain while running SPECpower using different batch sizes. In both cases, the batch size does not affect the average power profile of the system. It is also worth noting that both the power profile follow a similar trend, indicating a strong correlation between full system and package power. (The Pearson correlation is greater than 0.99.) Consequently, the power consumed by the domains have a strong correlation (> 0.95) with the same performance counters. Both these power profiles have a strong correlation with all counters except the counters related to ITLB in Table 2. Our PCA reveals that five PCs (LLCM, IR, NOUT, DTLBLM and DTLBSM) account for 95% of the variance.

The EP metric value for the full system is 0.54 for all batch sizes whereas the metric value for the package domain is 0.71 for all batch sizes. Comparison of the EP metric shows that EPG for the package domain is smaller than the full system (recall that larger is better for EP and smaller is better for EPG) suggesting that the package domain is more energy-proportional than the full system.

3.3.3 Profiling of Uncore (Package-PP0) and DRAM Power

Figure 3 shows the *uncore* subsystem and DRAM EPGs.

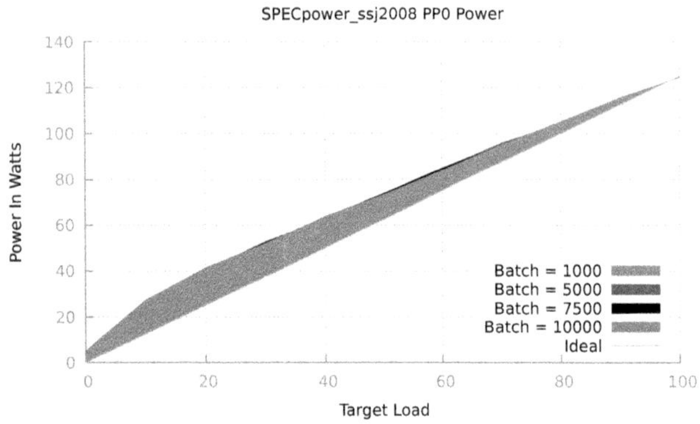

Figure 4: Power Plane 0 (PP0) Power Profile. PP0 subsystem has the smallest EPG making it the most energy-proportional subsystem.

The counters, except the ones related to ITLB and DTLB stores, have a correlation greater than 0.80 with the power consumption of the *uncore* subsystem. Our PCA revealed that four PCs (IR, NOUT, DTLBLM and LLCM) account for 83% of the variance. Our analysis reveals that the counters in Table 2 are insufficient to model the power consumption of *uncore* subsystem. We intend to look at *uncore* performance counters [2] exposed in the Sandy Bridge server platforms to uncover performance counters with better correlation in the future.

The *uncore* subsystem power remains almost constant, irrespective of the percentage of the load-level with an EP metric value of 0.17. The *uncore* subsystem has the greatest EPG, and as a result, exhibits the worst power consumption trend among the full system and RAPL domains from the perspective of energy-proportional power scaling.

Interestingly, the DRAM domain consumed a maximum average power of only 10 watts (i.e., 5 watts per socket). The performance counters related to LLC, DTLB, instructions retired and outstanding bus requests have a correlation greater than 0.91 with the power consumed by DRAM. Four PCs (LLCM, IR, DTLBLM and NOUT) retain 91% of the variance. The DRAM domain's value for the EP metric is 0.47. The SPECpower benchmark only consumes a maximum of 10 watts for the DRAM domain. Moreover, the benchmark consistently consumes its maximum power for the DRAM domain at 70% workload for all batches.

3.3.4 Profiling of the Power Plane 0 (PP0) Power

Figure 4 represents the EPG of PP0 domain. Similar to other domains, the batch size does not affect the average power consumption of the domain. All counters, except the ones related to ITLB and LLC, have strong correlation with the power profile of this domain. Four PCs (IR, NOUT, DTLBLM and DTLBSM) account for 97% of all variance. PP0 is the most energy proportional subsystem with an EP metric value of 0.86 across all batches. Interestingly, the idle power consumption of PP0 domain is 5.27 watts. The PP0 domain has the least EPG and exhibits near energy-proportional power scaling.

3.4 An Analysis of Subsystem-Level Instantaneous Power Consumption

Here we present our results for the instantaneous power profile analysis of the SPECpower benchmark. Our main goal is to visualize the opportunities for power limiting under different time windows while consuming power proportional to the load-level. We collected instantaneous power profile for three load-levels (50%, 60%, and 70%) and four different batch sizes of SPECpower using a loadable kernel module. The resolution of our power profile is limited to 5 milliseconds (ms). Our instrumentation had negligible impact on the power and performance of the SPECpower benchmark.

3.4.1 Methodology

Figures 5, 6, and 7 present the results on the instantaneous power profile analysis of SPECpower. These plots describe the opportunity for power limiting under different time windows while consuming power proportional to the load-level. The x-axes in the graphs is the resolution of time for which we profile the instantaneous power. The y-axes represents the ratio of number of instances for which the instantaneous power was greater than ideal "energy-proportional" power consumption[7] and the total number of instances for that time resolution. Henceforth, we refer to the ideal power consumption for a load-level as LL% of peak power. As an example, in Figure 5 at 50 ms resolution using 1000 as batch size, the instantaneous power was over ideal energy-proportional power consumption for 99.58, 97.25 and 93.25 percent of the total instances for 50%, 60%, and 70% load-levels, respectively.

When considering energy-proportional operation, performance is a implicit metric. Consequently, we are interested in time resolutions that will allow us to achieve energy-proportional operation with minimal or no impact on performance. These graphs can be used to determine the time window(s) useful to limit the power consumption for a certain batch size and load-level to achieve energy proportional operation. Note that we restrict our analysis to 50, 60 and 70% load-level for simplicity. However, our methodology is applicable to all batch sizes and load-levels.

3.4.2 Results

Package: Figure 5 presents the results for the instantaneous power profile analysis of the package RAPL domain. The workload consumes more than LL% of peak power over 95% of the execution time with batch size = 1000 at a resolution greater than 100 ms for all load-levels. This limits the opportunity to limit the power to achieve energy-proportional operation at time windows greater than 100 ms. However, increasing the batch size of SPECpower improves the opportunity to apply power limiting at larger time resolution. We also observe that this improvement is greater for higher load-levels. For example at 100-ms resolution, the workload consumes more than LL% of peak power for 99.83%, 99.66% and 96.16% of the total instances for 50%, 60%, and 70% load-levels with 1,000 as batch size, whereas only 96.16% to 97.16%, 92.00% of the total instances consume more than LL% of peak power when using 10,000 as batch size.

PP0: Figure 6 presents the results for the instantaneous power profile analysis of the PP0 RAPL domain. In general, PP0 provides better opportunities for power limiting

[7]For example, consuming 50% and 70% of the power consumed at 100% load-level is considered ideal energy-proportional power consumption for 50% and 70% load-levels respectively.

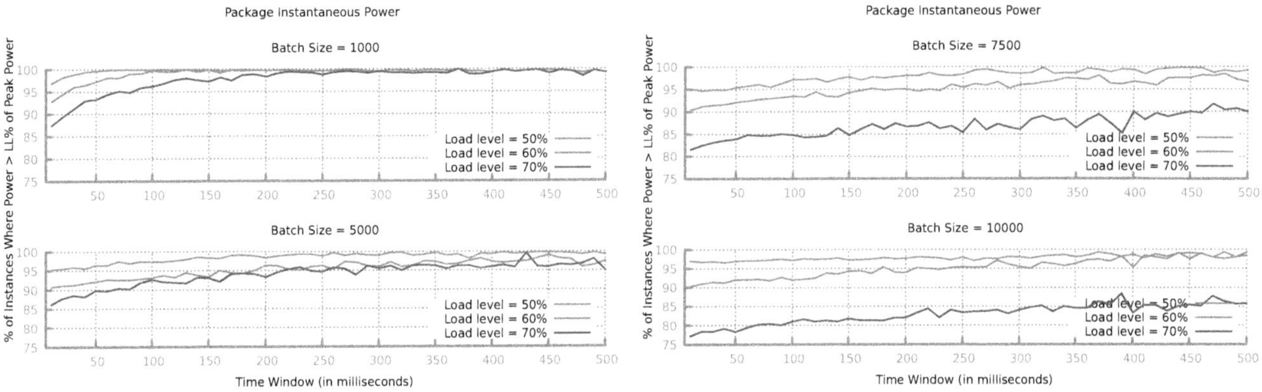

Figure 5: Instantaneous Power Analysis For Package Domain

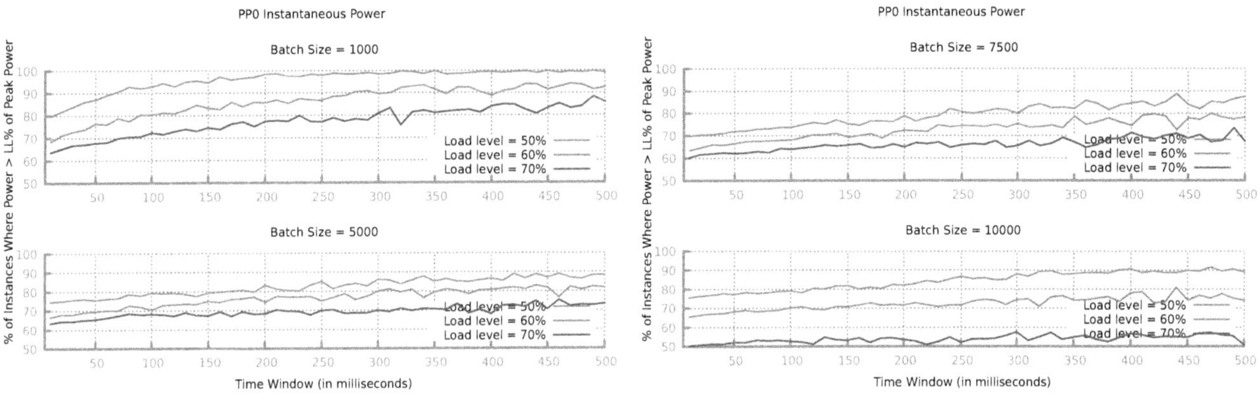

Figure 6: Instantaneous Power Analysis For PP0 Domain

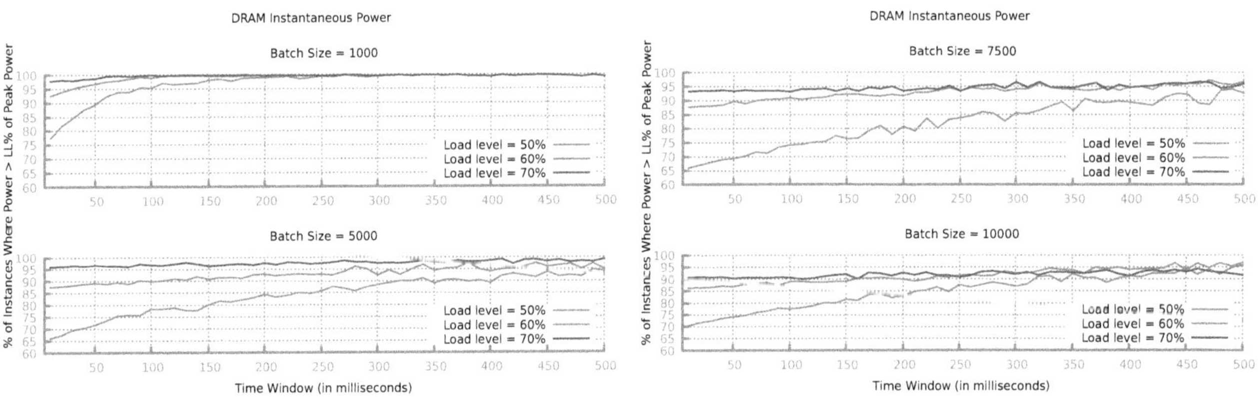

Figure 7: Instantaneous Power Analysis For DRAM Domain

at different time resolutions when compared to the package domain. This is expected as the *uncore* subsystem consumed almost constant power, irrespective of the load-level, as discussed in Section 3.3.3. Even using a batch size of 1,000, there are significant opportunities to limit power to LL% of peak power at granularities less 250 ms for 60% and 70% load-levels. Similar to the package domain, increasing the batch size improves the opportunity to limit power at larger resolution. However, the improvement is greater when compared to the package domain. For example at 100-ms resolution, the workload consumes more than LL% of peak power for 92.83%, 80.33% and 72.33% of the total instances for 50%, 60%, and 70% load-levels with 1,000 as the batch size, whereas only 79.00%, 70.33%, and 52.66% of the total instances consume more than LL% of peak power when using 10,000 as the batch size.

DRAM: Figure 7 shows the results for the instantaneous power profile analysis of the DRAM RAPL domain. In contrast to the package and PP0 RAPL domains, the opportunity for power limiting at larger time resolutions occurs

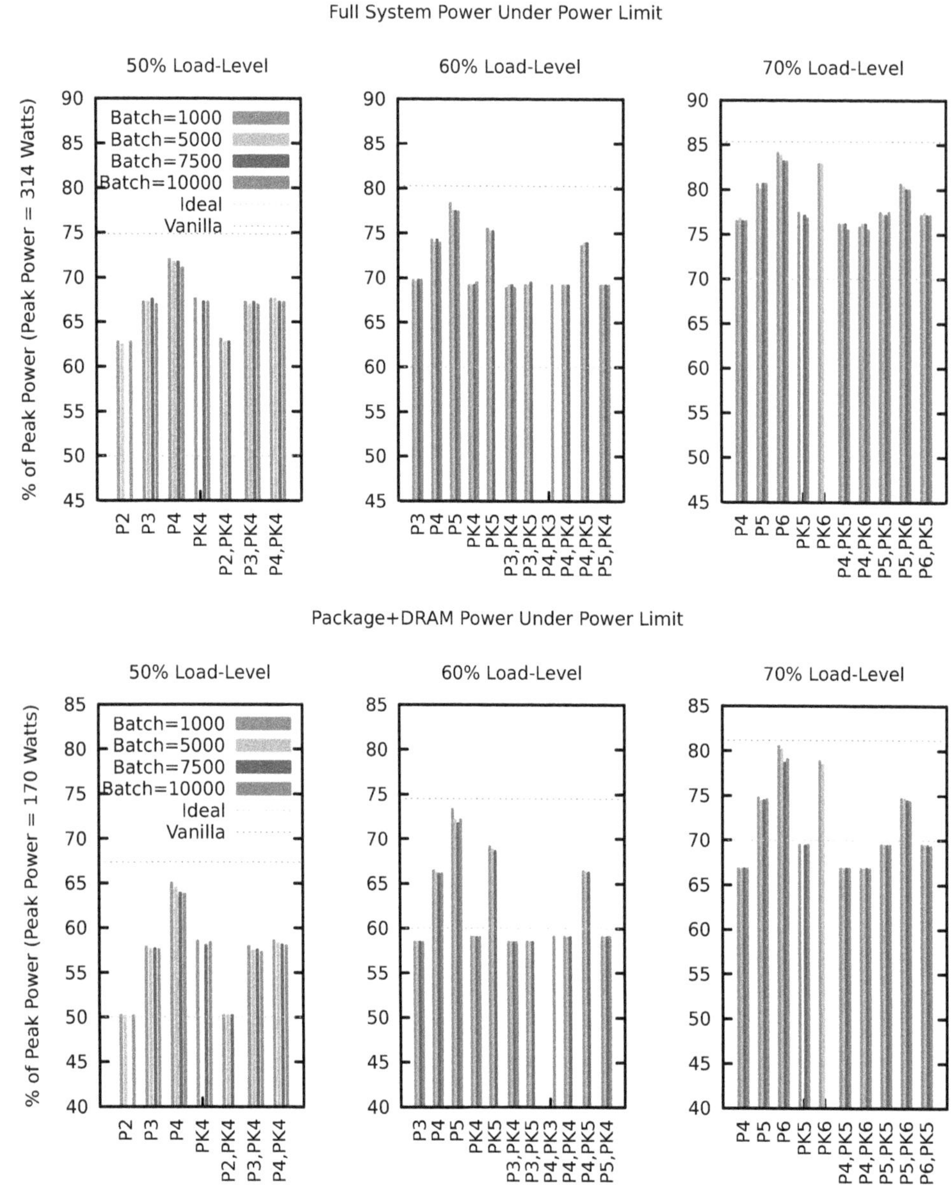

Figure 8: Power Consumption Under Power Limit

more at lower load-levels. We observe that for 60% and 70% load-levels, approximately 90% of the total instances consume more than LL% of peak power for all different batch sizes. This shows that the opportunity to achieve energy-proportional DRAM operation for these load-levels through power limiting is limited. However, 50% load-level is amenable to power limiting. Similar to the Package and PP0 domains, increasing the batch size allows power limiting even at larger time resolutions. However, the improvement is better for lower load-levels, i.e., 50% load-level has more opportunity for power limiting than 60% or 70% load-levels.

4. IMPACT OF POWER LIMITING ON SPECPOWER

In this section, we discuss the effects of power limiting on the performance and power of SPECpower. Specifically, we leverage the RAPL interfaces and analyze whether energy-proportional operation can be achieved. While we are not able to achieve energy proportionality at full system-level, RAPL power limiting can be used to improve energy efficiency without performance degradation. In addition, we are able to achieve energy-proportional operation or better at package- and memory-level (the subsystems over which we have power limiting control) for the load-levels discussed.

4.1 Methodology

We run 50%, 60%, and 70% load-levels of SPECpower under power limits by manually configuring the power limiting interfaces. We were not able to use the power limiting control for the DRAM domain for SPECpower because the maximum average DRAM power consumption (see Section 3.3.3)

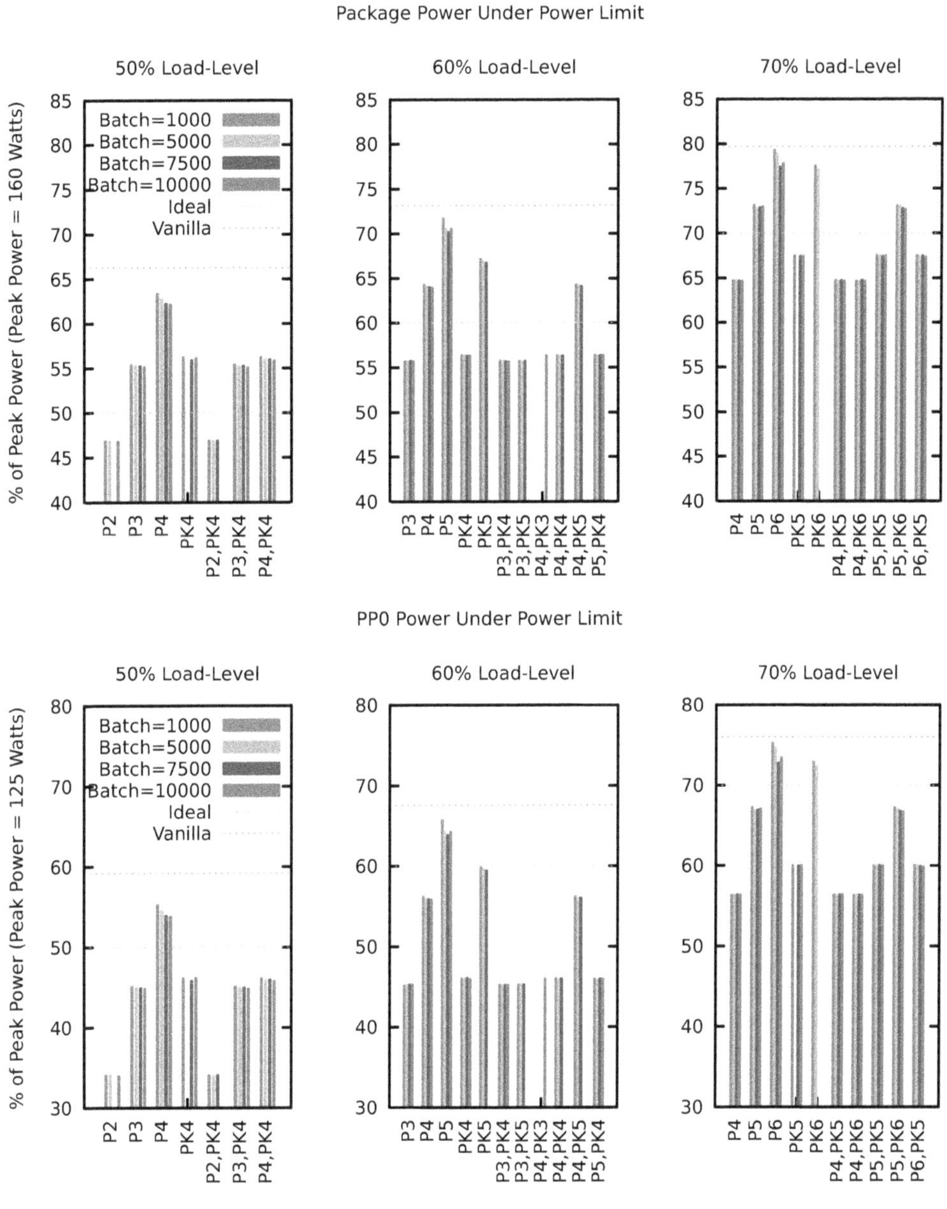

Figure 9: Power Consumption Under Power Limit

is less than the allowed minimum power consumption for the DRAM domain (see Section 2.2.4). Experiments specifying relevant power consumption limits for the DRAM domain for SPECpower resulted in no effect on power. As a result, our experiments will focus only on limiting the power consumption of the Package and PP0 RAPL domains. We manually limit the power consumption of these domains to different values of power for different load-levels, as shown in Table 3.

4.2 Impact on Power Consumption

Figures 8 and 9 show the power consumption of SPECpower under power limits for three different load-levels. We present power consumption results for the full system and three subsystems: (1) full system, (2) Package+DRAM (the domains over which we have power limiting control), (3) Package, and (4) PP0. Each plot presents only the configurations from Table 3 that achieved the particular load-level. For comparison, we also show the average power consumption of the

Table 3: Power Limit Configuration For Each Load-Level (In % of Peak Power For That Domain)

Load-Level/Domain	PP0	Package
50	20, 30, 40, 50(P1, P2, P3, P4)	25, 30, 40, 50(PK1, PK2, PK3, PK4)
60	30, 40, 50, 60(P2, P3, P4, P5)	30, 40, 50, 60(PK2, PK3, PK4, PK5)
70	40, 50, 60, 70(P3, P4, P5, P6)	40, 50, 60, 70(PK3, PK4, PK5, PK6)

subsystem without power bounds (vanilla – red dotted line) and energy-proportional power (ideal – green dotted line). All the results presented in Figures 8 and 9 use the smallest time window for power limiting (i.e., 976 microseconds).

We observe that power limiting does not allow us to enable energy-proportional operation at the full system level. We would like to emphasize that the system consumed 37.2% of peak power (i.e., 117 watts) even when idling. However, we achieved moderate power savings at the full system level *without* any performance degradation by using power limiting. We also observe that the power saving increases with decreased load-level. For example, the best full system power saving achieved at 50% and 70% load-level are 13% and 9% over the vanilla runs of SPECpower, respectively.

We achieved energy-proportional operation at the PKG+DRAM, PKG, and PP0 RAPL domain levels for all load-levels. Power limiting in the PP0 RAPL domain is the best possible mechanism to achieve energy-proportional operation. Interestingly, the combination of a PP0 power limit and package power limit always consumed approximately the same power as a corresponding run with only the PP0 power limit. Our power-limiting experiments with only the package domain revealed that approximately 98.5% of the power savings came from the PP0 domain for all load-levels. The reason for such power profiles was due to the fact that the *uncore* consumes almost constant power, irrespective of the load-level (see Section 3.3.3). As such, we conclude that better mechanisms are needed to control the power consumption of the *uncore* subsystem [2].

4.3 Impact on Response Time

The response time in workloads such as SPECpower can greatly impact user experience. In this section, we measure the impact of power limiting on per-transaction response time. The response time can vary widely even if the throughput (i.e., load-level) is maintained. The arrival time, wait time, and total response time of batches in a run can be logged by setting the *input.scheduler.log_arrival_rates* to true. We are interested in the total response time (wait + execution time) as it is directly related to the performance seen by a user.

Figure 10 shows the average slowdown on per-transaction response time of three different load-levels and four different batch sizes under a power limit. The slowdown is defined as the ratio of response time under the power limit and the response time of a vanilla run. As expected, the response time varies widely within a given load-level. PP0 power limiting gives the best power-performance trade-off; the response time is directly correlated to batch size for all load-levels. For larger batch sizes, the response time for SPECpower decreases. The response time with batch size of 1,000 worsens as we use package power limiting. We also observe that power limiting can have drastic effects on the response time for different batch sizes. Determining the *right* power limit to achieve a certain load-level with acceptable per-transaction response time is an avenue for future work.

4.4 Impact on Energy Efficiency

As discussed earlier, SPECpower is a power and performance benchmark with ssj_ops/watt (i.e., performance-to-power ratio) as the main metric of interest. Figure 11 depicts the energy efficiency improvement under power limits. The improvement is computed as the ratio of the difference

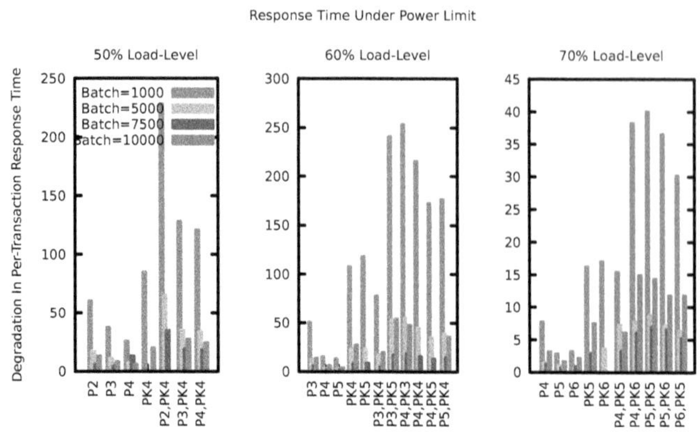

Figure 10: Response Time Under Power Limit

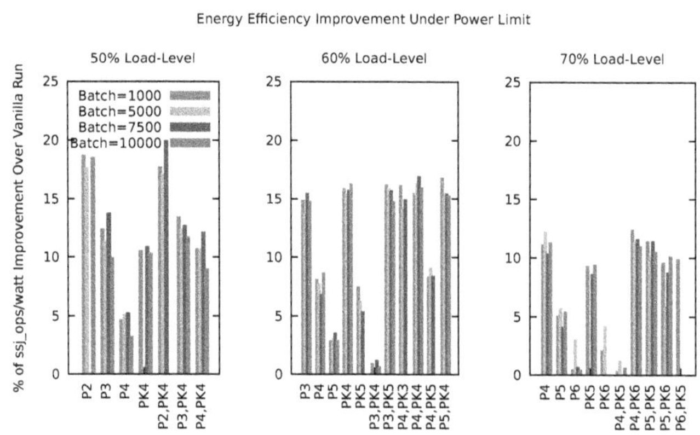

Figure 11: Energy Efficiency Improvement Under Power Limit

between the energy efficiency under a power limit and vanilla run (i.e., no power limit) over the efficiency in a vanilla run. We observe that the percentage gain in energy efficiency decreases with increase in load-level. Power limiting enables us to achieve a maximum energy efficiency improvement of 20%, 16%, and 12% at 50%, 60%, and 70% load-levels, respectively. These results show that PP0 power limiting (in isolation) is the best mechanism to achieve better energy efficiency.

4.5 Impact of a Power-Limiting Time Window on Response Time

Figure 12 shows the impact of a power-limiting time window on the per-transaction response time for running SPECpower with a batch size of 5,000 and setting a power limit for PP0 RAPL domain. We show only one batch size and load-level for simplicity. The rest of the batch sizes and load-levels showed similar trends. We emphasize that the average power consumption of SPECpower remains approximately the same for a specified power limit with different power-limiting time windows. There is a huge improvement moving from a ~2ms to a ~4ms time window. In general, increasing the time window improves the response time of SPECpower, though this improvement asymptotes as the size of the window continues to increase.

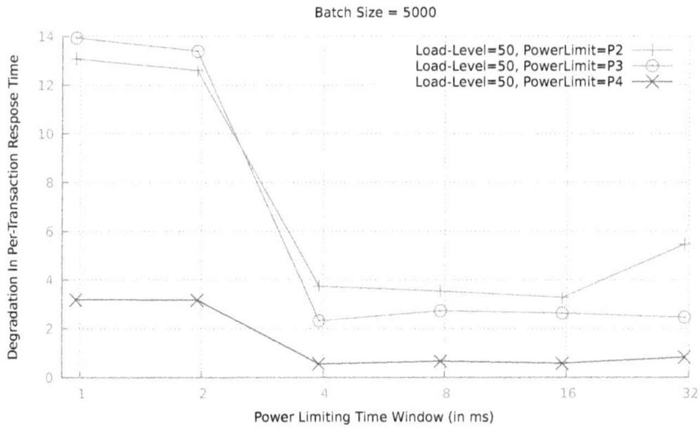

Figure 12: Impact of a Power-Limiting Time Window on Response Time

5. RELATED WORK

We classify the related work in three categories: (1) energy-proportional operation for enterprise-class workloads, (2) characterization and power analysis of the SPECpower benchmark, and (3) power limiting.

5.1 Energy-Proportional Operation For Enterprise Class Workloads

Our work is most related to the work of Meisner et al. [22]. They characterize online data-intensive services (OLDI) to identify opportunities for power management, design a framework that predicts the performance of OLDI workloads and investigate the power and performance trade-offs using the framework. Our work leverages Intel's RAPL interfaces to characterize the power consumption of SPECpower and analyze whether it is possible to achieve energy-proportional operation on a system.

Fan et al. [13] investigate the benefits of energy-proportional systems in improving the efficiency of power provisioning using their power models. They provide evidence that energy-proportional systems will enable improved power-capping at the data-center level. In contrast, we look at leveraging the *power-capping* mechanism to achieve energy-proportional operation for SPECpower.

Tolia et al. [28] proposed that by migrating workloads from under-utilized systems to other systems and turning the under-utilized systems off, energy proportionality can be approximated at an ensemble-level (i.e., for a group of nodes or rack-level). They used virtual machine (VM) migration as a software mechanism to move workloads off of under-utilized systems. In this paper, we use user-defined and hardware-enforced power limiting to achieve energy-proportional systems at the node-level.

5.2 Charecterization and Power Analysis of SPECpower

Fanara et al. [14] and Lange [20] comment on the design and development of the SPECpower benchmark. An overview of the workload, its behavior, and detailed characterization of the SPECpower benchmark under different load-levels is described in [17]. Our goal in this paper is to characterize SPECpower to provide insights into the opportunities for power management for an enterprise-class server workload.

Hsu et al. [19] defined accurate and portable power models that capture the linearity of the power curve under different load-levels of SPECpower. Varsamopoulos et al. [29] proposed the idle-to-power ratio (IDR) and linear deviation ratio (LDR) metrics, analyzed a large number of SPECpower results, and provided insights into server provisioning. Similarly, Ryckbosch et al. [26] proposed the EP metric to quantify the energy proportionality of the system. In this paper, for the first time, we quantify the energy proportionality at a subsystem-level using the on-chip energy meters exposed through RAPL. We also characterize the instantaneous power profile at different load levels of SPECpower.

5.3 Power Limiting

Several mechanisms to *cap* the power consumption of the system have been studied [11, 16]. However, we study the use of RAPL power limiting which is hardware-enforced in this paper. David et al. [12] proposed RAPL and evaluated RAPL for the memory sub-system. They present a model that accurately predicts the power consumed by the DIMMs and use RAPL to *cap* the power consumption. Rountree et al. [24] use RAPL power limiting to study the behavior of performance for benchmarks in the NAS parallel benchmark suite. Specifically, they are interested in the performance of various compute nodes under a power bound. Weaver et al. [30] have have exposed RAPL energy meters through PAPI. We use RAPL interfaces to achieve energy-proportional operation for SPECpower benchmark and to the best of our knowledge, there is no previous study on using RAPL interfaces for enterprise class server workloads.

6. CONCLUSION

The management of power and energy is a key issue for data centers. Efficient power management of enterprise-class server workloads have the potential to greatly reduce energy-related costs and facilitate efficient power provisioning.

Energy proportionality holds the potential to significantly improve the energy efficiency of data centers. Consequently, in this paper, we investigate the potential of achieving energy proportionality for SPECpower benchmarks using RAPL interfaces. Our study sheds light on the mechanisms for power management of enterprise-class server workloads and the efficacy of RAPL interfaces. We identify the least and most energy-proportional subsystem using the on-chip energy meters. Performance counter traces are collected to identify which events have strong correlation with power consumption at a subsystem-level. We also characterize the instantaneous power profile of SPECpower and identify the time resolutions at which there is opportunity to limit the power consumption of the benchmark. We use RAPL power limiting and show that we are able to achieve energy-proportional operation (or better) at the subsystem-level.

7. ACKNOWLEDGMENTS

The authors would like to thank Ashwin Aji, Vedvyas Duggirala, and Hari Pyla for their consultation and the anonymous reviewers for their feedback. This work was supported in part by a gift and equipment donation from Supermicro.

8. REFERENCES

[1] Intel 64 and IA-32 Software Developer Manuals - Volume 3. Available at `www.intel.com/content/www/us/en/processors/architectures-software-developer-manuals.html`.

[2] Intel Xeon Processor E5-2600 Product Family Uncore Performance Monitoring Guide. Available at `http://www.intel.com/content/dam/www/public/us/en/documents/design-guides/xeon-e5-2600-uncore-guide.pdf`.

[3] SPECjbb Benchmark. Available at `http://www.spec.org/jbb2005/`.

[4] SPECpower Benchmark. Available at `http://www.spec.org/power_ssj2008`.

[5] SPECpower Benchmark – Benchmarking Methodology. Available at `http://www.spec.org/power/docs/SPEC-Power_and_Performance_Methodology.pdf`.

[6] SPECpower Benchmark – CCS Design Document. Available at `http://www.spec.org/power/docs/SPECpower_ssj2008-Design_ccs.pdf`.

[7] SPECpower Benchmark – Design Overview. Available at `http://www.spec.org/power/docs/SPECpower_ssj2008-Design_Overview.pdf`.

[8] SPECpower Benchmark – Run Rules. Available at `http://www.spec.org/power/docs/SPECpower_ssj2008-Run_Reporting_Rules.html`.

[9] SPECpower Benchmark – SSJ Workload Design Document. Available at `http://www.spec.org/power/docs/SPECpower_ssj2008-Design_ssj.pdf`.

[10] L. A. Barroso and U. Hölzle. The Case for Energy-Proportional Computing. *IEEE Computer*, 40(12):33–37, 2007.

[11] R. Cochran, C. Hankendi, A. K. Coskun, and S. Reda. Pack & Cap: Adaptive DVFS and Thread Packing Under Power Caps. In *Proc. of the Int'l Symp. on Microarchitecture*, MICRO-44, pages 175–185, 2011.

[12] H. David, E. Gorbatov, U. R. Hanebutte, R. Khanna, and C. Le. RAPL: Memory Power Estimation And Capping. In *Proceedings of the International Symposium on Low Power Electronics and Design*, ISLPED, pages 189–194, 2010.

[13] X. Fan, W.-D. Weber, and L. A. Barroso. Power Provisioning For a Warehouse-Sized Computer. In *Proceedings of the International Symposium on Computer Architecture*, ISCA, 2007.

[14] A. Fanara, E. Haines, and A. Howard. The State of Energy and Performance Benchmarking for Enterprise Servers. In R. Nambiar and M. Poess, editors, *In Performance Evaluation and Benchmarking*, pages 52–66. Springer-Verlag, 2009.

[15] M. Ferdman, A. Adileh, O. Kocberber, S. Volos, M. Alisafaee, D. Jevdjic, C. Kaynak, A. D. Popescu, A. Ailamaki, and B. Falsafi. Clearing the Clouds: A Study of Emerging Scale-out Workloads on Modern Hardware. In *Proceedings of International Conference on Architectural Support for Programming Languages and Operating Systems*, ASPLOS, pages 37–48, 2012.

[16] A. Gandhi, M. Harchol-Balter, R. Das, J. Kephart, and C. Lefurgy. Power Capping Via Forced Idleness. In *Proceedings of Workshop on Energy-Efficient Design*, WEED, 2009.

[17] L. Gray, A. Kumar, and H. Li. Workload Characterization of the SPECpower_ssj2008 Benchmark. In *Performance Evaluation: Metrics, Models and Benchmarks*, Lecture Notes in Computer Science, pages 262–282. 2008.

[18] C. Hsu and W. Feng. A Power-Aware Run-Time System for High-Performance Computing. In *Proceedings of the SC Conference*, 2005.

[19] C. Hsu and S. W. Poole. Power signature Analysis of the SPECpower_ssj2008 Benchmark. In *Proc. of the Int'l Symp. on Performance Analysis of Systems and Software*, ISPASS, pages 227–236, 2011.

[20] K. Lange. Identifying Shades of Green: The SPECpower Benchmarks. *Computer*, 42(3):95–97, Mar. 2009.

[21] D. Meisner, B. T. Gold, and T. F. Wenisch. PowerNap: Eliminating Server Idle Power. In *Proceedings of International Conference on Architectural Support for Programming Languages and Operating Systems*, ASPLOS, 2009.

[22] D. Meisner, C. M. Sadler, L. A. Barroso, W.-D. Weber, and T. F. Wenisch. Power Management of Online Data-Intensive Services. In *Proceedings of the International Symposium on Computer Architecture*, ISCA, pages 319–330, 2011.

[23] S. Rivoire. Models and Metrics for Energy-Efficient Computer Systems. *Ph.D. Dissertation, Department of Electrical Engineering, Stanford*, 2008.

[24] B. Rountree, D. Ahn, B. de Supinski, D. Lowenthal, and M. Schulz. Beyond DVFS: A First Look at Performance Under a Hardware-Enforced Power Bound. In *Proc. of the Int'l Parallel and Distributed Processing Symp. Workshops and PhD Forum*, IPDPSW, pages 947–953, 2012.

[25] B. Rountree, D. K. Lownenthal, B. R. de Supinski, M. Schulz, V. W. Freeh, and T. Bletsch. Adagio: Making DVS Practical For Complex HPC Applications. In *Proc. of the Int'l Conf. on Supercomputing*, ICS, pages 460–469, 2009.

[26] F. Ryckbosch, S. Polfliet, and L. Eeckhout. Trends in Server Energy Proportionality. *IEEE Computer*, (9):69–72, 2011.

[27] D. C. Snowdon, S. M. Petters, and G. Heiser. Accurate On-line Prediction of Processor and Memory Energy Usage Under Voltage Scaling. In *Proceedings of the International Conference on Embedded Software*, EMSOFT, pages 84–93, 2007.

[28] N. Tolia, Z. Wang, M. Marwah, C. Bash, P. Ranganathan, and X. Zhu. Delivering Energy Proportionality with Non Energy-Proportional Systems: Optimizing the Ensemble. In *Proceedings of the Conference On Power Aware Computing and Systems*, HotPower. USENIX Association, 2008.

[29] G. Varsamopoulos, Z. Abbasi, and S. Gupta. Trends and Effects of Energy Proportionality on Server Provisioning in Data Centers. In *Proceedings of the International Conference on High Performance Computing*, HiPC, pages 1–11, 2010.

[30] V. Weaver, M. Johnson, K. Kasichayanula, J. Ralph, P. Luszczek, D. Terpstra, and S. Moore. Measuring Energy and Power with PAPI. In *Int'l Workshop on Power-Aware Systems and Architectures*, 2012.

Automated Root Cause Isolation of Performance Regressions during Software Development

Christoph Heger
Karlsruhe Institute of Technology
Am Fasanengarten 5
76131 Karlsruhe, Germany
christoph.heger@kit.edu

Jens Happe
SAP Research
Vincenz-Priessnitz-Strasse 1
76131 Karlsruhe, Germany
jens.happe@sap.com

Roozbeh Farahbod
SAP Research
Vincenz-Priessnitz-Strasse 1
76131 Karlsruhe, Germany
roozbeh.farahbod@sap.com

ABSTRACT

Performance is crucial for the success of an application. To build responsive and cost efficient applications, software engineers must be able to detect and fix performance problems early in the development process. Existing approaches are either relying on a high level of abstraction such that critical problems cannot be detected or require high manual effort. In this paper, we present a novel approach that integrates performance regression root cause analysis into the existing development infrastructure using performance-aware unit tests and the revision history. Our approach is easy to use and provides software engineers immediate insights with automated root cause analysis. In a realistic case study based on the change history of Apache Commons Math, we demonstrate that our approach can automatically detect and identify the root cause of a major performance regression.

Categories and Subject Descriptors

C.4 [**Performance of Systems**]; D.2.5 [**Software Engineering**]: Testing and Debugging; D.2.8 [**Software Engineering**]: Metrics—*performance measures*

General Terms

Measurement, Performance

Keywords

Performance Regression, Root Cause Isolation

1. INTRODUCTION

Despite a growing awareness for the importance of software quality, performance evaluation is still postponed to late development stages (the "fix-it-later" approach [32]). Performance evaluation is still a manual and time-consuming task. Software engineers need broad knowledge and expertise about the various tools and techniques to analyse the performance of their application. The conducted case studies of [35, 34] show that comparing to functional bugs, performance bugs are more challenging and require experienced software engineers and more code changes to fix.

Due to the high efforts required and the missing experience, software engineers defer performance tests as far as possible. But if a performance problem is found after deployment (usually by customers) it would be too late to fix it efficiently. Zimran and Butchart already showed that the integration of performance engineering into software development improves performance when applied early [36]. In contrast, the costs for fixing problems grow heavily the later they are discovered [10, 11].

Today in software development, many teams contribute to the code base of an application. Unit tests are written to validate the functionality of development artifacts. Revision control systems are employed to manage code artifacts and merge changes in order to enable the development teams to simultaneously contribute to the code base. Automated build infrastructure immediately notifies software engineers when a build is broken after changes have been applied. Performance regression testing is also an established discipline with many facets. Standard performance regression tests are specific for an early defined scenario. Software engineers create tests for the scenario which are then executed continuously. On the unit test level, software engineers are provided with tools like ContiPerf [2] or JUnitPerf [8]. Both tools build on existing JUnit [7] tests to evaluate the performance whenever a build is performed. However, both tools lack support for root cause analysis. In consequence, software engineers still have to identify the responsible revision which means, depending on the revision history, one or more revisions have to be tested for the occurrence of the performance regression. Thereafter, software engineers often create and run problem specific performance tests and manually investigate the functionality in the identified revision in order to find the root cause.

Software Performance Engineering (SPE) provides guidelines and approaches for evaluating software performance throughout the development process. Approaches for model-driven performance analysis allow software architects to create architectures that can satisfy their performance requirements. Since many performance problems are caused by low-level implementation details, they cannot be detected in high-level architecture models which miss the necessary details. Continuous performance tests during development are needed to identify problems that are caused by low level

ICPE'13, April 21–24, 2013, Prague, Czech Republic.

details. Load tests validate the performance of the system as a whole. However, they are usually executed late (e.g. just before a product is shipped) and provide no support for tracking the performance throughout the development process. Specific regression benchmarks [13, 16, 15] built for a specific problem, are proven to be useful to identify performance problems. While this approach provides important insights, it cannot be easily applied to a broader range of products. Automatic detection of performance problems and root cause analysis in load tests using statistical process control charts for performance problem detection and root cause analysis have been presented in [25] but require further research. Nevertheless, the authors show that in most cases automation reduces the root cause analysis time by about 90 percent [25]. Identification of performance bugs (not only performance regression problems) by monitoring deployed applications and providing the information to software engineers helps in finding the root causes [23]. However, load testing cannot be applied during development when no running application is available. None of these approaches provide continuous development-time root cause analysis by making use of unit tests and the revision history of the system under development.

In this paper, we propose a novel approach called PRCA (Performance regression Root Cause Analysis) for automated root cause analysis of performance regressions. PRCA uses unit tests to continuously monitor the application during the development phase, detects performance regressions, and identifies the revision in the history in which the performance regression was introduced. Thereafter, it isolates the root cause with systematic performance measurements based on dynamically extracted call tree information of unit tests. Our approach aims at ease of use and immediate insight for software engineers. We build on existing and commonly used technologies, such as JUnit [7] for unit testing and Git [4] or Subversion [9] as revision control systems known to software engineers. The performance tests can be easily integrated in the development infrastructure and their execution requires no further effort from software engineers. This approach allows us to automatically isolate the root cause of performance regressions in many cases.

We apply our approach to the revision history of Apache Commons Math [1], an open source math library, with the goal of demonstrating that PRCA can isolate the root cause of a known performance regression that took more than 14 month to be resolved. As a result, PRCA correctly identified the root cause of the performance regression.

Overall, we make the following contributions:

- We introduce an algorithm called PRCA to identify the root cause of a performance regression. The algorithm combines performance-aware unit tests with the revision history of the revision control system to isolate the problem's root cause as far as possible.

- We demonstrate that PRCA can isolate the root cause of a performance regression successfully.

The paper is structured as follows: In Section 2, we give an overview of our approach. In Section 3 we discuss our employed performance regression detection approach. In Section 4 we present our performance regression root cause analysis approach including bisection, call tree analysis, change correlation and software engineer feedback. We demonstrate that our approach identifies the root cause of a major performance regression in Section 5. We discuss our approach in Section 6 with potential future work, and present the related work in Section 7. Section 8 concludes the paper.

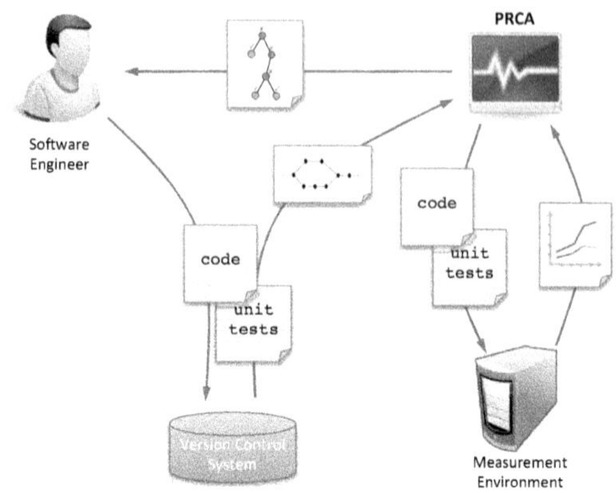

Figure 1: Overview of the PRCA approach

2. OVERVIEW

The main goal of our approach is twofold: to automatically identify performance problems as early as possible and to support software engineers with root cause analysis of the problems in order to facilitate a swift and efficient problem resolution.

To achieve this goal, PRCA combines the concepts of regression testing, bisection and call tree analysis to detect and provide performance regression root cause analysis as early as possible. The idea is to have the approach tightly integrated into the existing development infrastructure for continuous integration and regression testing. The continuity of performance tests allows identifying problems early. The close integration with the development infrastructure simplifies the usage of performance detection method.

The approach assumes the existence of two essential elements in the development environment: (a) unit tests as important artifacts of the target system that indirectly identify the main functionalities of the system, and (b) a code repository that offers change graphs of the artifacts of the system. Hence, the approach is applicable to any Java-based software system that has a proper set of unit tests and its source code is maintained on a revision control system.

Figures 1 and 2 provide an overview of PRCA. At regular intervals, PRCA fetches the latest revision of the software system from the repository, instruments the tests in unit test classes, runs all tests (with repetition) and gathers measurements on performance metrics of interest (step 1). For every test in unit test classes, it then compares the measurement results with the previous measurements done at the last performance test (step 2). If a performance regression is detected for any test, PRCA retrieves the change history graph from the revision control system and, performing a bisect algorithm, identifies one of the (potentially many) changes that introduced the performance regression (step 3). In the next step, the methods called by the affected test are instrumented, the test is executed and a call tree annotated with performance data is produced. The annotated call tree is then analyzed to identify the methods that con-

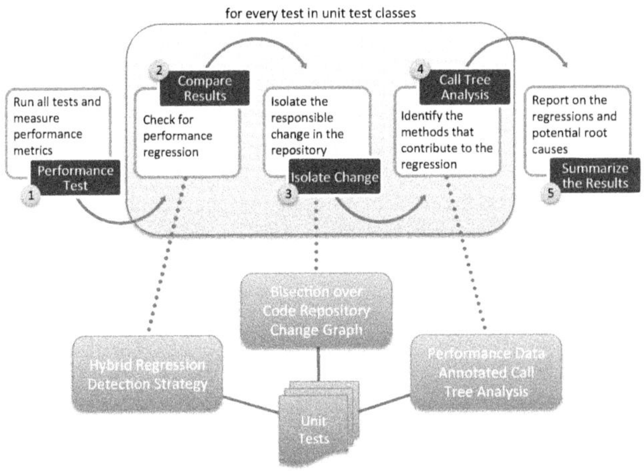

Figure 2: The PRCA process

tribute to the performance regression (step 4). The result of the analysis is then reported to the software engineer (step 5).

In order to achieve the best results, we assume that for the components that have performance requirements, the unit tests accompanying the source code are *performance-aware unit tests*. We define performance-aware unit tests as unit tests that are developed with performance requirements in mind. For example, a unit test for a component that computes the critical path of a graph, may test the component with graph sizes of 1, 5, and 10, focusing only on the graph structure. If the component is expected to be used, in the target application, with graphs of size 50 to 100 nodes, a performance-aware unit test will either test the component for larger graph sizes (such as 30, 60, and 120), or will have an explicit test for at least one large graph size (e.g., size 50) in order to test the component under a "typical" load.

For regression detection (step 2 of Figure 2), we need to properly define what is considered as a performance regression. In Section 3, we briefly discuss various methods of regression detection considered for this approach.

In order to isolate the change in the repository (step 3), we need an efficient strategy to identify the change that introduced the regression. Running performance tests is a time and resource consuming task. Each performance test may require a warm up time and many repetitions in order to achieve stable and reproducible results. As a consequence, performance tests cannot be executed with every build. For the scope of this paper, we assume that performance tests are run in longer but regular intervals (e.g., weekly). If a regression is detected, there may be a long history of changes made to the application. In this step, we apply a variation of the Git bisection algorithm [3] to identify the change(s) that introduced the regression. We discuss this topic in Section 4.

For the call tree analysis (step 4), our approach automatically instruments the methods that are called by the test (and that are part of the application or component interface) and systematically searches for increases in response times (Section 4). It then performs a breath first search that investigates the methods with the largest increase in response time. The measurements are done on both revisions the new and the changed one. If the implementation of a method has been changed compared to the original revision, we consider this method as the root cause of the regression problem.

3. REGRESSION DETECTION

In this work, we consider a performance regression as a significant increase of the response time of a tested method. In detecting regressions based on measurements, we are facing the challenge to report regressions that are relevant in the application context and valuable for the software engineers in improving the performance of the software. Therefore, we have to avoid reporting irrelevant or insignificant regressions (false positives) or miss a critical regression (false negatives). The balance here is to apply a meaningful regression strategy that deals with noise and also distinguishes between minor and major regressions. Existing approaches employ either some kind of a threshold [18, 22, 25] (derived or specified) or historic data [13, 28] each can be considered as a baseline acting as comparison counterpart. Lee et al. [25] state that the main issues in regression detection include the determination of the baseline of performance of the system under test and the deviation from the baseline which is considered as performance regression. PRCA includes no contribution in the field of performance regression testing. We consider the regression detection part as an extension point for future research. Our approach can be enhanced with automated generation of parameterized unit tests [19], test input selection which maximizes resource consumption in order to find performance problems [21] and Stochastic Performance Logic formulas [12] to enable software engineers expressing performance requirements. This could reduce the burden of providing performance-aware unit tests for software engineers.

In PRCA, we currently employ statistically rigorous performance evaluation after the work of Georges et al. [20] for performance regression detection. Since errors in experimental measurements affect the accuracy of the results, we employ a stable measurement environment to avoid that systematic errors introduce a bias into the measurements. Our stable measurement environment is a hardware platform and Java Virtual Machine (JVM) implementation that remains unchanged and is dedicated to performance testing. We run a unit test multiple times obtaining stable measurements in the measurement environment for a certain revision. We do not exclude JIT (re)compilation overhead from our performance measurements because we observed that the first run does the compilation and the performance of subsequent runs suffer less from variability due to JIT (re)compilation. We use the Analysis of Variance (ANOVA) to compare the response time distributions of two revisions. Our decision to use ANOVA is based on the suggestion of Georges et al., who propose ANOVA as a statically rigorous method to compare alternatives wrt. performance: "Also, ANOVA assumes that the errors in the measurements for the different alternatives are independent and Gaussian distributed. ANOVA is fairly robust towards non-normality, especially in case there is a balanced number of measurements for each of the alternatives." [20] After applying an ANOVA test PRCA can conclude if there is a statistically significant difference between the two samples that is a performance regression. Based on a series of experiments, we choose a confidence level of 99% as suitable to deal with false positives and false negatives in most cases. The termination criteria of the unit test run repetition can either be a fixed number of repetitions or a

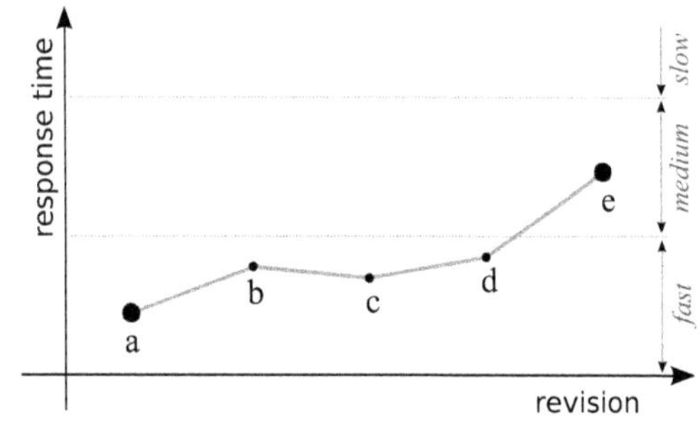

Figure 3: Regression detection strategy example. Category separation is indicated by the horizontal dashed line.

certain confidence level. While confidence levels are more appropriate, it is unclear how many repetitions are actually necessary. In our case studies, we repeated each test run 50 times in order to achieve stable results based on our observation that 50 repetitions with a confidence level of 0.95 can result in a relative measurement error of less than 10% for larger response times (>100 ms). However, this setting is very case specific and needs to be considered carefully.

Furthermore, not every performance degradation is a major performance regression that must be corrected. To support software engineers in understanding the importance of a regression and minimize false positives, we employ a regression detection strategy that classifies performance degradation into minor and major regressions. The classification uses categories (thresholds) as baseline for major performance regression detection and previous measurements as baseline for minor performance regression detection. Categories span a range defining an interval for response time; for example, performance categories "fast", "medium" and "slow" can be defined (cf. Figure 3). Methods are automatically classified based on the measured response time during the first test execution. PRCA reports a warning if it is not the fastest category from a performance perspective. Software engineers can change the assignments as needed. The defined categories act as baselines indicated by the dashed horizontal line in Figure 3. With evolving revisions our approach then reports a major performance regression if the requirements of the assigned category are not satisfied. For example, the changes applied in revision e in Figure 3 change the category of the tested method from "fast" to "medium" leading to a major regression. In contrast, if the category's requirements are satisfied but historic data comparison shows a significant performance regression (cf. revisions a and b in Figure 3), PRCA reports a minor performance regression. The strategy enables software engineers to easily distinguish between minor and major performance regressions and to select only crucial performance regressions for further problem diagnosis. The advantage is that a manually specified threshold is not needed and the risk of detecting a performance regression late is reduced. Furthermore, it overcomes the lack of knowledge software engineers may have particularly with regard to the acceptable response time.

4. ROOT CAUSE ANALYSIS

Our problem diagnosis approach is partitioned into two steps: bisection and call tree analysis. Within the bisection step, we systematically evaluate the performance of revisions in the repository in order to identify the change that leads to the performance regression. The call tree analysis step is then applied to find the root cause of the problem. Therefore, we extract a performance annotated call tree (serving as a performance behaviour model) out of the tested method and apply our analysis to systematically investigate the tested methods call tree.

4.1 Bisection

Detecting the presence of a performance regression alone within a performance test period is not sufficient, especially when there are many revisions in the given period. When a performance regression is detected, it would be desirable to know which change introduced the regression. Hence, identifying the responsible change of a performance regression is the first step of our proposed problem diagnosis approach. It supports and limits the problem diagnosis effort by enabling the software engineer(s) to focus on the responsible change rather than the whole history of the changes within the test period.

In this section, we describe our approach of identifying the responsible change in the revision control system that lead to a performance regression. The approach is based on the Git bisect functionality [3]. This feature of Git offers a binary search over the change history and enables software engineers to identify the change that introduced a problem. The bisection process can also be automated by configuring Git bisect to launch a script or command at each bisection step checking if the current change is already affected by the regression.

We adapt the bisection algorithm [3] of Torvalds and Hamano to find the best intersection point in each step. Change histories in general can be described as directed acyclic graphs (DAG). When a performance regression is detected between a pair of changes (c_s, c_e), where $time(c_s) < time(c_e)$, the actions performed by the algorithm can be summarized as follows:[1]

1. Filter the DAG for relevant revisions and keep only the nodes that are ancestors of the change c_e, excluding c_s and ancestors of c_s.

2. Compute and associate the number of ancestors a_i to each node c_i starting with the leaves of the DAG. This is the first step to compute a weight for each node in the DAG.

3. Compute the corresponding weight

$$w_i = min(a_i + 1, N - (a_i + 1)) \qquad (1)$$

   and associate it with node c_i. Here, N is the number of nodes in the filtered DAG.

4. The next intersection point is the change c_i with the highest weight w_i. We call this change c_c.

We then apply our regression detection strategy of Section 3 to examine the triple of changes (c_s, c_c, c_e). We first check the pair (c_s, c_c) to see if c_c includes a performance

[1]We assume that there are no skipped changes.

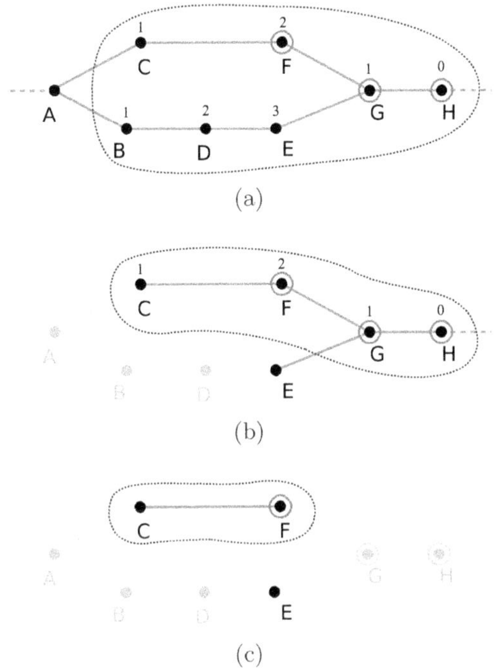

Figure 4: Bisection steps on a change graph

regression. To do so, we test the hypothesis that the mean of the response times μ_{c_s} is equal to that of μ_{c_c}. Therefore, we check if the null hypothesis

$$H_0 : \mu_{c_s} = \mu_{c_c} \tag{2}$$

cannot be rejected. If the response times do not differ significantly or if they differ but $\mu_{c_s} > \mu_{c_c}$, we assume that c_c is not affected by a performance regression and continue the bisection method with the pair (c_c, c_e). Otherwise, c_c includes a performance regression and we continue the bisection method with the pair (c_s, c_c). The method continues until for a given input pair of (c_s, c_e), we have only the nodes c_s and c_e in the graph. We then report c_e as the change responsible for the performance regression.

Figure 4 shows an example of applying bisection to a change graph for the pair of changes (A, H). At each step, the dotted grey line indicates the filtered graph. Assuming that a detectable performance regression is introduced in change F, the regression is propagated into changes G and H (marked with red circles in Figure 4(a)). Comparing changes H and A, our method detects a performance regression at H. At the first step of bisection, node E has the highest weight of $3 = min(3, 4)$ in the filtered graph, so E is selected as the next intersection point. Since E does not introduce a performance regression compared to A, bisection is repeated for the pair of (E, H). The graph is filtered (Figure 4(b)) and F (with a weight of 2) is selected as the next intersection point. Node F shows a performance regression compared to E, hence bisection continues for the pair (E, F) with only the nodes C and F remaining in the filtered graph (Figure 4(c)). Node C does not show a regression compared to E, so the bisection continues with the pair (C, F) and since there would be no more node in the graph except C and F, node F is reported as the node responsible for the introduced regression.

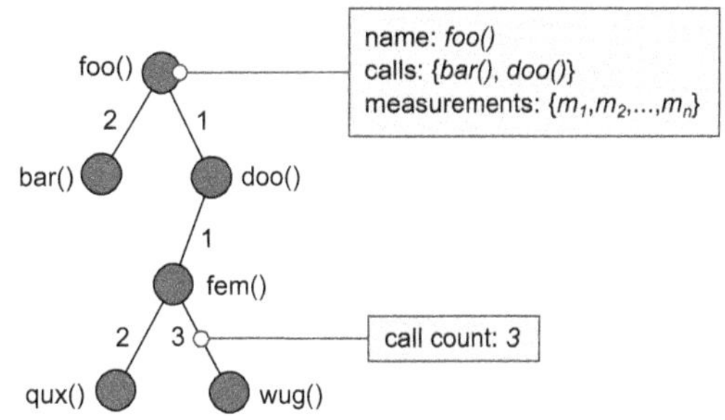

Figure 5: Call tree with performance annotations

Finding the next intersection point is the first step of the bisection process within our approach. The second step is to examine the change by running performance regression tests using our regression detection strategy described in Section 3. In each bisection step, the source code of the change is checked out and an automatic build is executed. In the resulting executable any tested method is instrumented for runtime monitoring. The corresponding unit tests are then executed to gather response time measurements.

4.2 Call Tree Analysis

In this section, we present PRCA's strategy for root cause analysis using call trees. PRCA identifies methods in the call tree and reports paths that are suspected to have caused the performance regressions to software engineers. For this purpose, PRCA examines the call trees of tested methods for which performance regressions have been detected. Furthermore, we explain PRCA's method for extracting call trees annotated with performance data gathered through systematic measurements.

For root cause analysis, we employ a call tree with performance annotations (cf. Figure 5). Methods are represented as nodes (e.g., *foo()*) in the call tree. Each node contains the signature of the method, a list of method calls and a series of response time measurements. Edges connecting two nodes represent method calls (e.g. *foo()* calls *bar()*). The edges are annotated with the number of calls (*call count*) performed during the execution of the corresponding tests. For example in Figure 5, *fem()* calls *wug()* 3 times.

PRCA extracts annotated call trees in two steps. In the first step, we use a full instrumentation of the test case to retrieve the structure of a call tree and the number of method calls. In the second step, we extract performance data of individual methods. This step requires many test executions and a very selective instrumentation in order to produce reliable performance data. In the following, we describe both steps in more detail.

To extract the structure of the call tree (i.e., its nodes and edges), we instrument all methods called during test execution. The instrumentation is fully automated and performed by PRCA requiring no manual intervention. For this purpose, PRCA employs byte code weaving based on Javassist [6, 14] and the monitoring possibilities of Kieker [31]. Java classes are loaded into a class pool, a copy of the byte code is stored in a byte code repository and the monitoring instructions are weaved into the byte code of the method(s) before they are used. The original class, if already loaded into the JVM, is replaced with the instrumented class using the HotSwap mechanism for class file replacement of the

Java SDK. The byte code stored in the repository is later used to undo the instrumentation. Based on the generated trace data, PRCA can extract the structure of the call tree including the number of calls from one method to another. Because the call tree and the call count are not affected by monitoring overhead, all data can be collected in a single test execution.

However, the same approach is not suitable for gathering the performance data. Response times can be strongly affected by measurement overhead introduced by a full instrumentation. The instrumentation of one method can affect the response time of another introducing systematic errors which causes a bias in measurements affecting the accuracy of the results [20]. For example, when *doo()* and *fem()* (cf. Figure 5) are instrumented and monitored in the same test run, the time required to execute the monitoring instructions in *fem()* are part of the response time of *doo()*. If this is not considered, the overall regression detection can easily be misleading. In addition, performance measurements have to be reproducible and minimize potential disturbances.

To address these issues, PRCA breaks down the extraction of response times in multiple steps and repeats measurements several times. In each step, PRCA instruments only the methods on one level of the call tree. With this strategy, PRCA minimizes disturbances as none of the methods called by the instrumented method are monitored. For the example in Figure 5, PRCA first instruments *foo()*, second *bar()* and *doo()*, third *fem()*, and finally *qux()* and *wug()*. For each step, PRCA executes the test multiple times which results in measurements $m_1, \ldots, m_n$ for each method (cf. Figure 5).

The goal of the call tree analysis is to identify paths in the call tree with methods that are suspected to cause the performance regression. For this purpose, the call trees with performance annotations are extracted for the revisions where a regression has been introduced. Figure 6 continues the example introduced in Section 4.1. In this example, the changes made from revision C to F introduced a regression. To identify the methods that potentially cause the performance regression (cf. *fem()* in Figure 6), PRCA compares the nodes in the call tree of revision C with its counterpart in revision F. For each pair of nodes, PRCA conducts an ANOVA test for the response time measurements. Using a breadth-first search, our approach looks for nodes with statistically significant differences between the measurements of both revisions (e.g. *foo()*, *doo()* and *fem()*). Such nodes are flagged and all methods called by them are evaluated in the next step of the breadth-first search. If the ANOVA test cannot conclude that there are statistically significant differences in the response times of a method between the two revisions, our algorithm does not further investigate it. For example, this is the case for *bar()*, *qux()* and *wug()* in Figure 6. Furthermore, if the call tree of a method changed from one revision to the next and its response time increased significantly, our call tree comparison will flag the method as a candidate for the root cause of the regression. Since the call trees of both revisions are different, our algorithm does not further investigate it. After the breath-first search is done, the flagged nodes point to the potential root cause in the call tree (*foo()*→*doo()*→*fem()* in Figure 6). PRCA uses the path and correlates it with the changes in the revision control system.

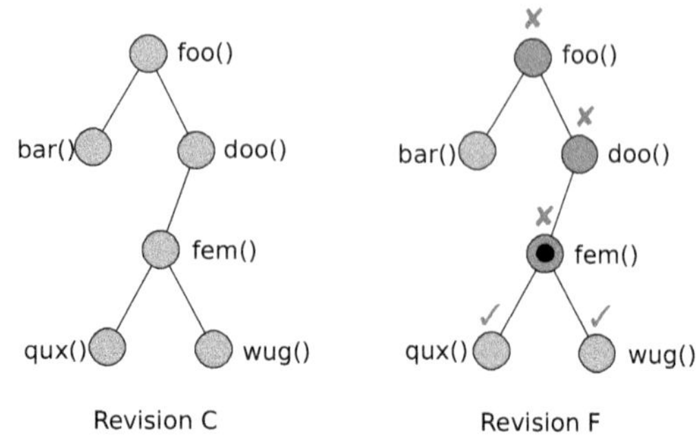

Figure 6: Example of call tree analysis. Methods that are suspected to cause a performance regression are marked with a cross.

4.3 Change Correlation

The goal of change correlation is to identify the methods in a path that are affected by changes that are suspected to cause the performance regression. The challenge here is to track down the changes included in a revision to the methods establishing a path. The changes are provided by and can be queried from the revision control system (e.g. Git). Without employing static code analysis, we are only able to track down changes to class level. Therefore, we extract the class name out of any node in a path (e.g. in Java from the full qualified method name *<packagename>.<classname>.foo()*). We then look up the changes in the revision control system for the responsible revision to check if there are any changes made to that class. Methods that might be affected by changes are marked (cf. *fem()* in Figure 6). This enables us to exclude methods of classes that are unchanged from the feedback we give to the software engineer.

4.4 Software Engineer Feedback

The goal of our approach is to support software engineers in root cause analysis of performance regressions. The challenge of providing feedback on root cause analysis results is the amount and complexity of the presented information. Our goal is to provide self-explanatory feedback (cf. Figure 7) to software engineers that shows the starting point for problem resolution without requiring deep knowledge (e.g. in statistics). Following a lean approach, we hide the complexity as much as possible. For a detected performance regression, we show the name of the unit test and the test method as well as the name of the tested method. We employ two figures to communicate the results of the root cause analysis. We highlight the performance regression between the two revisions (e.g. C and F) as well as the categories. The results of the call tree analysis are shown in a figure that contains the call tree of the identified revision. The call tree is similar to the one for revision F in Figure 6. Methods that are suspected to cause the regression are highlighted. Nodes that are correlated with changes are flagged with a black dot (cf. *fem()* in Figure 7). The feedback includes all necessary information to provide software engineers with a well prepared starting point for problem resolution.

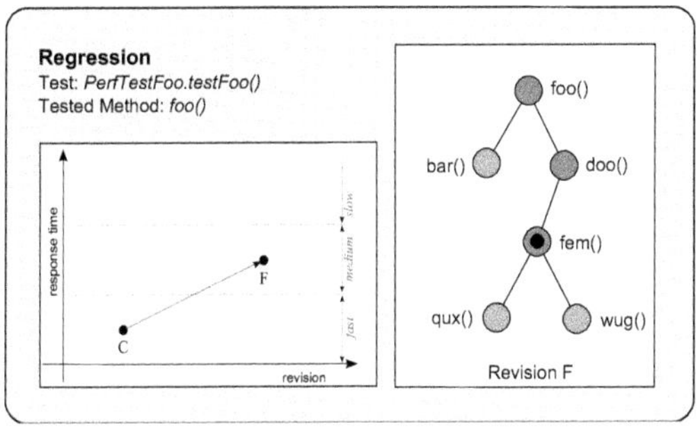

Figure 7: Feedback given to software engineers.

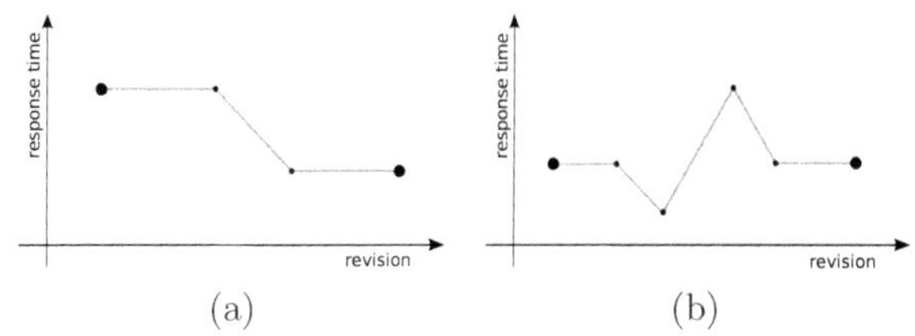

Figure 8: (a) Performance improved; (b) No performance change detected

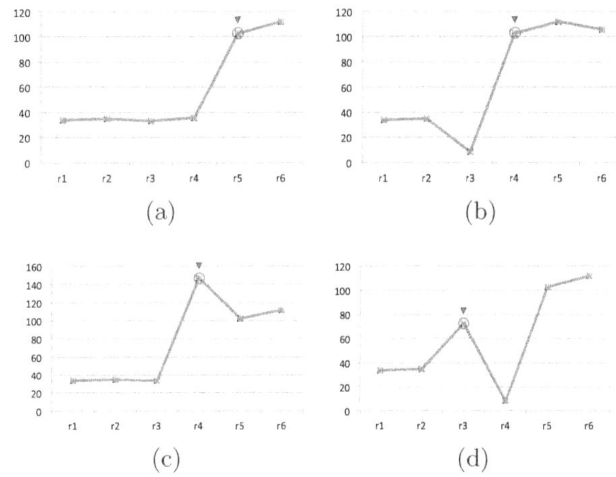

Figure 9: Evaluation of the Bisection Method

5. CASE STUDY

In this section we look into a number of case studies that validate the approach presented in this paper. We begin with four synthetic experiments that cover four cases of a performance regression (see Figure 9) and validate the regression detection approach for each case. We then provide a summary of applying our method to a real-world scenario, the Apache Commons Mathematics Library (Commons Math) [1]. There, we utilized our call tree analysis method to find the root cause of a performance regression that was introduced in a certain revision of Apache Commons Math.

5.1 Controlled Experiments

In order to demonstrate the validity of the regression detection method, we applied our method to four well-defined controlled experiments with artificial performance regressions. With these experiments, we address the overarching question:

Can PRCA identify the injected regressions as expected?

We defined each experiment around six consecutive revisions of a test system, namely revisions $[r_1 \dots r_6]$, where revision r_1 represents the last point in the revision history that performance regression testing was applied and revision r_6 is the latest revision in the repository when the current performance regression test is being performed. Since the goal is not to validate the already established bisection algorithm, we assume a linear order of the changes in each experiment. The system under test is an implementation of the Fibonacci function, which allows us to simply focus on and introduce changes in the response time. We assume that the configuration of the system under test provides the input value to the Fibonacci function, which can vary between revisions. The system comes with a unit test that reads the input value from the configuration file and runs the Fibonacci function for that particular value.

At every performance test, we are facing one of the following three cases: (i) performance is improved since the last tested revision (Figure 8(a)), (ii) performance is almost the same as the last tested revision with no significant change (Figure 8(b)), even though performance oscillation may have occurred between the two revisions which will not be detected, (iii) performance is worsened between the current revision and the last tested revision (examples in Figure 9). For these experiments, we focus only on the latter case where we have a significant performance regression between the two tested revisions.

In order to simulate the performance change in a revision, we change the system configuration and increase or decrease the input value to the Fibonacci function. The input value is changed such that it results to a significant performance change of the system. In the following sections, we describe each experiment in detail.

Simple Regression

The purpose of this experiment is to validate detection of a simple performance regression introduced at some point between revisions 1 and 6 (see Figure 9(a)). We simulate the performance regression by increasing the input value in the configuration at the fifth revision (r_5). The null hypothesis $H_0 : \mu_{r_1} = \mu_{r_6}$ is rejected during the performance regression test and the bisection starts. In the first bisection step $H_0 : \mu_{r_1} = \mu_{r_3}$ is not rejected. The bisection continues until the test $H_0 : \mu_{r_4} = \mu_{r_5}$ is rejected as a result of performance regression. Hence, the bisection identifies r_5 as the responsible change (indicated by the arrow in Figure 9(a)).

Improvement and Regression

Next, we validate detection of a performance regression introduced after a performance improvement (see Figure 9(b)). We simulate a leading performance improvement in front of a performance regression. Therefore, we decrease the value of the input parameter at r_3 before we use fault injection to increase it in r_4. The performance regression detection rejects $H_0 : \mu_{r_1} = \mu_{r_6}$ and starts bisection. In the following bisection steps $H_0 : \mu_{r_1} = \mu_{r_3}$ is rejected due to a performance improvement, and $H_0 : \mu_{r_3} = \mu_{r_6}$ is rejected as a result of a performance regression, so bisection continues with r_3 and r_6. In the last bisection step, $H_0 : \mu_{r_3} = \mu_{r_4}$ is rejected due to performance regression and as a result, the bisection identifies r_4 as the source of performance regression.

The reverse of the story is also presented in Figure 9(c), where a performance improvement (r_5) follows a performance regression (r_4). The bisection method correctly detects r_4 as the source of regression.

Performance Oscillation

In the last experiment, we alternate performance regressions and improvements to simulate an oscillating performance (see Figure 9(d)). We use fault injection in revisions r_3 and r_5 to decrease the performance. We reverse the fault injection in r_4 to temporarily improve the performance. As a result, both revisions r_3 and r_5 are introducing performance regression. Through the bisection process $H_0 : \mu_{r_1} = \mu_{r_3}$ is rejected and the bisection identifies r_3 as responsible change and the bisection ends. We will further discuss, in Section 6, possible improvements to our method with respect to performance oscillation.

5.2 Apache Commons Math

Apache Commons Math (henceforth called *commons math*) is a lightweight mathematics and statistics library for Java that addresses the common mathematical problems for which a suitable solution is not available in the Java programming language [1]. It is a small set of utilities that address programming problems such as fitting a line to a set of data points, curve fitting, solving system of linear equations, random number generation, and other miscellaneous mathematical functions. With a focus on real-world applications, it is apparent that performance of the commons math library in computing mathematical functions is of high importance. Various sources maintain and publish performance comparison of commons math against other available libraries (see [5] for example).

On October 10th, 2010, a code modification submitted to the commons math code repository (revision 1006301) introduced a substantial performance regression in one of the functions. The problem stayed hidden for 7 months until it was first reported[2] on May 16th, 2011. The resolution of the problem took more than 14 months until it was finally resolved on July 22nd, 2012.

There are two interesting aspects in this story: (a) the performance regression introduced by the submitted code stayed unreported for more than half a year; (b) it took more than one additional year for the bug to be resolved. Our proposed approach in this paper targets both of these aspects. Continuous performance monitoring based on performance-aware unit tests reveals performance issues early in the development process, before allowing their causes to get lost in the history of changes or their effects be propagated further into other components. When a performance regression is detected, our proposed call tree analysis method investigates and reports the root cause of the problem, hence considerably facilitating the problem solving process.

Assuming that performance-aware unit tests are available,[3] we applied our approach to the commons math code repository. We supposed a weekly performance regression

[2] `https://issues.apache.org/jira/browse/MATH-578`

[3] We manually extended the unit tests with one that tests the affected functionality of commons math.

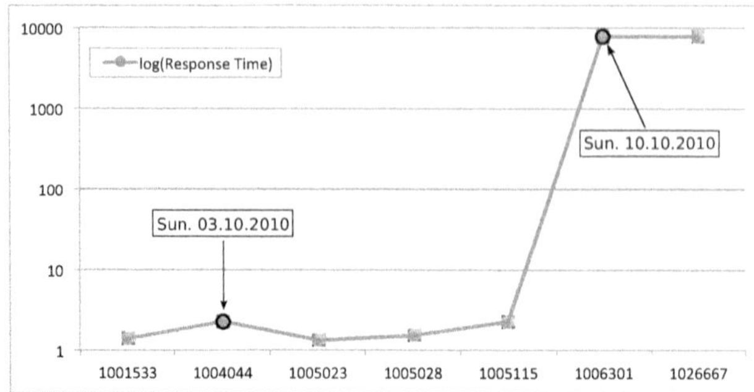

Figure 10: Performance Regression in Apache Commons Math (2010)

routine performed every Monday morning at 00:00 that runs all unit tests and monitors their performance. Applying this method retrospectively to the commons math code repository revisions in October 2010 we observe the performance graph of Figure 10 for our performance-aware variation of the `DescriptiveStatisticsTest` unit test that uses the troubling function. The execution of all unit tests[4] (including repetition) took about 48 minutes for revision 1006301 and 30 minutes for revision 1004044. Our bisection method correctly detects a performance regression and identifies revision 1006301 as the cause of this regression. In this case, this is the right-most revision in testing interval. The residual distribution plots of the two revisions marked in the diagram of Figure 10 are provided in Figure 11. Despite a few outliers, the residuals in both plots are close to the straight line indicating that random errors have a close to normal distribution.

In the next step, PRCA applies the call tree analysis method to find the root cause of the problem in revision 1006301. It instruments the methods that are called in the test that introduced regression and identifies a regression in

`DescriptiveStatistics.getPercentile(double)`

in revision 1006301 compared to 100515. A breadth-first analysis of the call tree (comparing the performance of the methods between the two revisions) is continued until the root cause of the regression is isolated to the call tree of Figure 12. The problem appears to be originated from a change in the following method:

`Percentile.evaluate(double[], int, int, double)`

The call tree analysis of PRCA detects that (a) the call tree under the above method is changed in the problematic revision, and (b) the common parts of that call tree do not show a performance regression. As a result, the call tree leading to that method is reported as a potential cause of the introduced performance regression. This analysis also matches what happened in reality. The revision 1006301 introduced a change in this method that caused the performance of the method to decrease seriously for certain boundary cases.

6. DISCUSSION

There is no need to argue that the proposed combination of automated detection and root cause analysis of perfor-

[4] Unit tests in folder `src/test/java` (excluding 3 tests that we were not able to run properly with our tooling).

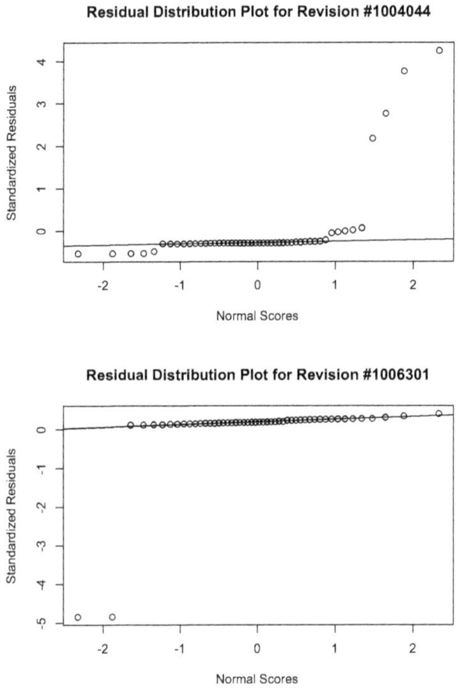

Figure 11: Residual Distribution Plots

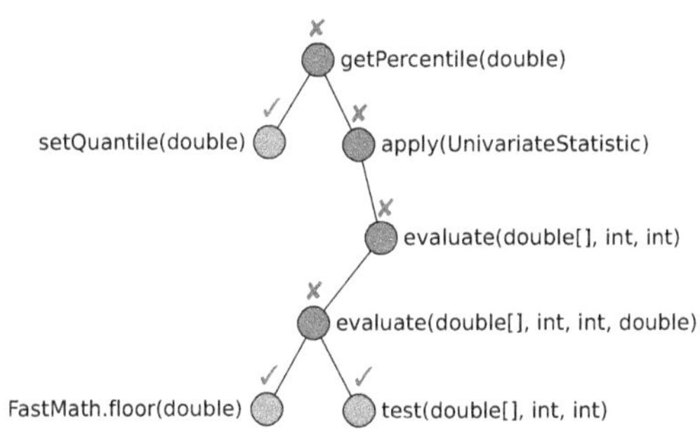

Figure 12: The Problematic Call Tree in Apache Commons Math Revision 1006301

mance regressions early in the software development process can significantly reduce the costs of fixing performance issues and maintaining the quality of the software. Automated regression detection identifies performance issues as soon as possible before they are hidden in the future changes to the software. When a regression is detected, the proposed root cause analysis then helps software engineers to focus on the potential sources of the problem.

We have already applied the idea of continuous performance regression testing and root cause analysis internally at SAP. Over the course of one month, we detected a crucial performance regression which would otherwise have stayed hidden. Despite the encouraging results of applying the method to internal projects and case studies, there are still limitations and issues that require improvement. In the following, we briefly point out these limitations and discuss potential solutions and future improvements.

Although one can apply the proposed method to any software under development that has a proper set of unit tests,

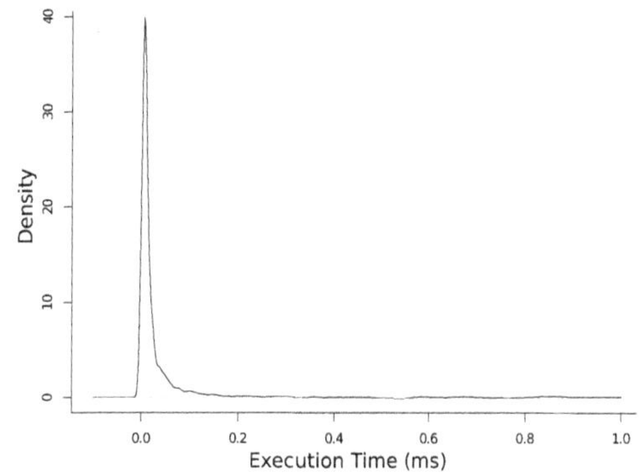

Figure 13: An example of unit test time distribution

regular unit tests are usually not enough to detect performance regressions. For example, Figure 13 shows the distribution of unit test execution times of an internal project at SAP. In this example, the majority of the tests are executed in less than 0.1 milliseconds. The execution times for these tests are too short to allow for a meaningful detection of regressions. In most cases such as this example, regular unit tests focus only on functional testing with usually the minimum set of test cases to test the functionality of the components. In order to detect performance regressions properly, our proposed method requires *performance-aware unit tests* that test the system units not only for functionality, but also for cases that can potentially be critical for the performance of the code in the target application context. This also implies that the performance unit tests have the right granularity and coverage to detect relevant performance regressions. Of course, considering that the goal of this approach is to mostly rely on the existing unit tests and *not* to build specifically designed unit tests that target all the performance-influencing parts of the code, there is no guarantee that all performance regressions will be detected.

The proper definition of *performance regression* and the development and evaluation of good heuristics for its detection are crucial for a broader applicability of performance regression testing. The heuristics for regression detection must minimize false positives (wrong alerts) and false negatives (regressions going undetected) in order to be useful. Software engineers quickly lose interest in performance regression tests if the results are not reliable or they have to provide too much additional information for the tests. In this work, we used a performance regression detection strategy based on historical data and categories. Although this approach worked well in our scenarios, it still requires further evaluation and refinements, such as conducting more experiments with performance regression issues and a thorough analysis of the results of fault positive and fault negative rates.

The current version of our framework has limitations in detecting and properly reporting three cases of performance regression:

1. Performance oscillations within a test period which lead to similar performance at both ends of the period are not detectable with our approach. If this is the result of a performance regression introduced and then resolved within the test period, it can be ignored.

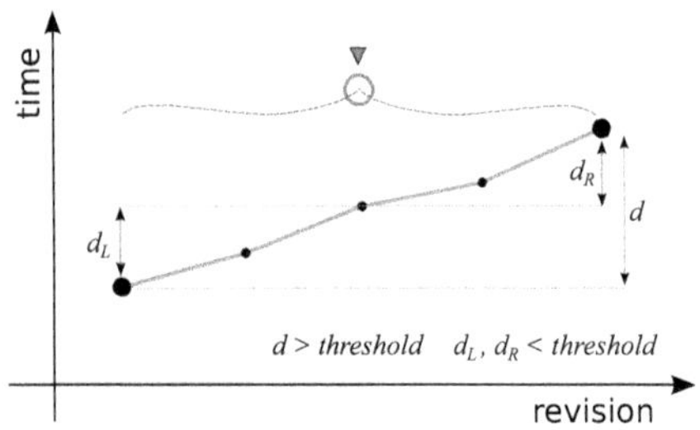

Figure 14: The case of slow performance decay

However, such an oscillation can be caused by a pair of unrelated performance improvement and regression. In such a case, it is desirable to be able to detect the regression introduced in the code. We can reduce the chance of such cases by having more frequent performance regression tests (hence shortening the test periods). However, unless there are enough resources to perform regression tests on every change submitted, we cannot detect the performance regressions introduced in such cases.

2. The second is the case of slow decay of performance within a test period. Using our regression detection strategy, we would ultimately detect such a decay, but the root cause analysis would be difficult. Take the example of Figure 14, in which the performance is decreasing slowly such that a performance regression is detected between the two points of the test interval, but the problem diagnosis method cannot identify a single revision as the cause of the regression since the mid point of the period does not introduce a significant change in performance. In such cases, we would like to report all the changes within the test period as potential sources of performance regression.

3. The third limitation occurs when there is more than one regression introduced within the test period. In this case, we only report one change as the cause of the overall performance regression. We would like to improve our bisection algorithm such that once a performance regression is detected for a testing period, the algorithm would recursively identify and report all the changes that introduce a performance regression within the test period.

We consider the issues and ideas for improvements discussed in this section as topics of future work.

7. RELATED WORK

The approach presented in this paper combines performance regression testing with automatic diagnosis of performance problems. In the following, we discuss existing work that addresses these aspects.

Only few approaches combine performance regression testing with problem diagnosis and root cause analysis. Lee et al. [25] present an approach for automatically detecting performance anomalies in database management systems and isolating their root cause. They developed a framework that significantly reduces the effort of detecting and isolating performance problems for database developers. The approach uses established database benchmarks and profiling methods to compare new commits to the most recent stable state. Lee et al. share our goal of detecting performance problems early and supporting software engineers in isolating the problem's root cause. However, their solution is highly specific to the domain of database management systems and cannot easily be generalized. Mostafa and Krintz [28] developed an approach to automatically identify differences in call trees between two subsequent code revisions. The call tree can also contain performance metrics and include them in the tree comparison. While performance is considered, the authors focus on tree comparison and try to identify changes in the call tree (like added, removed or modified methods). The most important challenges of performance regression testing are not addressed.

Performance regression testing itself has been subject to research for several years [13, 18, 29]. Nguyen et al. [29] use statistical process control charts to analyze performance counters across test runs to check for performance regressions. Bulej et al. [13] use specialized regression benchmarks to detect performance regressions in application middleware. Specific regression benchmarks provide important insights, but cannot be easily applied to a broader range of products. Foo et al. [18] present an automated approach for performance regression testing to uncover performance regressions in situations where software performance degrades compared to previous releases. Performance metrics of a software system are monitored and compared to a set of estimated correlated metrics extracted from previous performance regression testing results. None of the approaches supports the root cause isolation of performance regressions.

In addition, the performance of a software system depends on its input data [24]. Thus, it is important to choose the input data for test cases well. Existing approaches [19, 21] support software engineers in choosing good input data for functional and performance tests. Grechanik *et al.* [21] apply machine learning techniques to find the input set which maximizes resource consumption in order to increase the probability of finding performance problems. Fraser and Zeller [19] present an approach for automatic generation of parameterized functional unit tests. Such approaches complement our work. The combination of input data generation with our approach can increase the efficiency of performance problem detection and root cause analysis.

Existing approaches that focus on performance problem detection during runtime [17, 26, 27, 30, 33] are using monitoring techniques to track a software system's performance and to gather measurement data for performance analysis. Even though these approaches can detect performance problems in a productive system and sometimes diagnose their root causes, they can only be a last resort. Detecting performance problems during operation is way too late to solve them efficiently.

8. CONCLUSIONS

In this paper we presented PRCA, a novel approach that utilizes unit tests and revision history graphs for automatic detection and root cause analysis of performance problems throughout the development process. The approach presented here builds on (i) a hybrid regression detection strategy, (ii) bisection over revision change graphs, and (iii) analysis of performance annotated call trees, to provide a process of continuous performance regression root cause analysis. We evaluated different heuristics for regression detection in order to minimize false positives. We extended Git bisection algorithm to identify the changes that introduced a performance problem. Finally, we designed an approach to systematically analyze call trees to identify the methods and call graphs that most likely caused the performance regression.

Once our methods are integrated in the development infrastructure used at SAP, software engineers can use familiar tools and techniques to write performance-aware unit tests. The tests are automatically executed on a regular basis (e.g. during integration builds). If a regression is detected, PRCA automatically provides information on which change caused the regression and which methods are involved. Using performance regression testing, software engineers can detect performance problems early and fix them swiftly. We have validated and demonstrated the application of PRCA using control experiments and Apache Commons Math as a real-world example.

Based on the promising results of our case study, we plan to apply our approach inside SAP. To achieve this, we need to improve the implementation of PRCA toward offering it as a stand-alone tool and extending its supported performance metrics. In many cases, throughput and resource consumption are also needed to be monitored in addition to the response time.

9. ACKNOWLEDGMENTS

We like to thank Dennis Westermann and Alexander Wert for their constructive comments and feedback. This work is supported by the German Research Foundation (DFG), grant RE 1674/6-1.

10. REFERENCES

[1] Commons math: The apache commons mathematics library, 2012. `http://commons.apache.org/math`.

[2] Contiperf 2, 2012. `http://databene.org/contiperf.html`.

[3] Fighting regressions with git bisect manual page, 2012. `www.git-scm.com/docs/git-bisect-lk2009.html`.

[4] Git, 2012. `http://git-scm.com`.

[5] Java matrix benchmark, 2012. `http://code.google.com/p/java-matrix-benchmark`.

[6] Javassist, 2012. `http://www.csg.is.titech.ac.jp/~chiba/javassist`.

[7] Junit: A programmer-oriented testing framework for java, 2012. `http://kentbeck.github.com/junit`.

[8] Junitperf, 2012. `www.clarkware.com/software/JUnitPerf.html`.

[9] Subversion, 2012. `http://subversion.apache.org`.

[10] B. W. Boehm. *Software Engineering Economics*. Prentice-Hall Advances in Computing Science & Technology Series. Prentice-Hall, Englewood Cliffs, NJ, 1981.

[11] B. W. Boehm. Software engineering economics. *Software Engineering, IEEE Transactions on*, SE-10(1):4 –21, jan. 1984.

[12] L. Bulej, T. Bureš, J. Keznikl, A. Koubková, A. Podzimek, and P. Tůma. Capturing performance assumptions using stochastic performance logic. In *Proceedings of the third joint WOSP/SIPEW international conference on Performance Engineering*, ICPE '12, pages 311–322, New York, NY, USA, 2012. ACM.

[13] L. Bulej, T. Kalibera, and P. Tma. Repeated results analysis for middleware regression benchmarking. *Perform. Eval.*, 60(1-4):345–358, May 2005.

[14] S. Chiba and M. Nishizawa. An easy-to-use toolkit for efficient java bytecode translators. In *Proceedings of the 2nd international conference on Generative programming and component engineering*, GPCE '03, pages 364–376, New York, NY, USA, 2003. Springer-Verlag New York, Inc.

[15] J. Davison de St. Germain, A. Morris, S. G. Parker, A. D. Malony, and S. Shende. Performance analysis integration in the uintah software development cycle. *International Journal of Parallel Programming*, 31:35–53, 2003.

[16] J. de St. Germain, A. Morris, S. Parker, A. Malony, and S. Shende. Integrating performance analysis in the uintah software development cycle. In H. Zima, K. Joe, M. Sato, Y. Seo, and M. Shimasaki, editors, *High Performance Computing*, volume 2327 of *Lecture Notes in Computer Science*, pages 305–308. Springer Berlin / Heidelberg, 2006.

[17] J. Ehlers, A. van Hoorn, J. Waller, and W. Hasselbring. Self-adaptive software system monitoring for performance anomaly localization. In *Proceedings of the 8th ACM international conference on Autonomic computing*, ICAC '11, pages 197–200, New York, NY, USA, 2011. ACM.

[18] K. C. Foo, Z. M. Jiang, B. Adams, A. E. Hassan, Y. Zou, and P. Flora. Mining performance regression testing repositories for automated performance analysis. In *Proceedings of the 2010 10th International Conference on Quality Software*, QSIC '10, pages 32–41, Washington, DC, USA, 2010. IEEE Computer Society.

[19] G. Fraser and A. Zeller. Generating parameterized unit tests. In *Proceedings of the 2011 International Symposium on Software Testing and Analysis*, ISSTA '11, pages 364–374, New York, NY, USA, 2011. ACM.

[20] A. Georges, D. Buytaert, and L. Eeckhout. Statistically rigorous java performance evaluation. *SIGPLAN Not.*, 42:57–76, October 2007.

[21] M. Grechanik, C. Fu, and Q. Xie. Automatically finding performance problems with feedback-directed learning software testing. In *Software Engineering (ICSE), 2012 34th International Conference on*, pages 156 –166, june 2012.

[22] Z. M. Jiang, A. E. Hassan, G. Hamann, and P. Flora. Automated performance analysis of load tests. *Software Maintenance, IEEE International Conference on*, 0:125–134, 2009.

[23] M. Jovic, A. Adamoli, and M. Hauswirth. Catch me if you can: performance bug detection in the wild. In *Proceedings of the 2011 ACM international conference on Object oriented programming systems languages and applications*, OOPSLA '11, pages 155–170, New York, NY, USA, 2011. ACM.

[24] H. Koziolek, S. Becker, and J. Happe. Predicting the performance of component-based software architectures with different usage profiles. In

S. Overhage, C. Szyperski, R. Reussner, and J. Stafford, editors, *Software Architectures, Components, and Applications*, volume 4880 of *Lecture Notes in Computer Science*, pages 145–163. Springer Berlin Heidelberg, 2007.

[25] D. Lee, S. Cha, and A. Lee. A performance anomaly detection and analysis framework for dbms development. *Knowledge and Data Engineering, IEEE Transactions on*, 24(8):1345 –1360, aug. 2012.

[26] B. P. Miller, M. D. Callaghan, J. M. Cargille, J. K. Hollingsworth, R. B. Irvin, K. L. Karavanic, K. Kunchithapadam, and T. Newhall. The paradyn parallel performance measurement tool. *Computer*, 28(11):37–46, Nov. 1995.

[27] A. V. Mirgorodskiy and B. P. Miller. Diagnosing distributed systems with self-propelled instrumentation. In *Proceedings of the 9th ACM/IFIP/USENIX International Conference on Middleware*, Middleware '08, pages 82–103, New York, NY, USA, 2008. Springer-Verlag New York, Inc.

[28] N. Mostafa and C. Krintz. Tracking performance across software revisions. In *Proceedings of the 7th International Conference on Principles and Practice of Programming in Java*, PPPJ '09, pages 162–171, New York, NY, USA, 2009. ACM.

[29] T. H. Nguyen, B. Adams, Z. M. Jiang, A. E. Hassan, M. Nasser, and P. Flora. Automated detection of performance regressions using statistical process control techniques. In *Proceedings of the third joint WOSP/SIPEW international conference on Performance Engineering*, ICPE '12, pages 299–310, New York, NY, USA, 2012. ACM.

[30] T. Parsons and J. Murphy. Detecting performance antipatterns in component based enterprise systems. *Journal of Object Technology*, 7(3):55–90, 4 2008.

[31] M. Rohr, A. van Hoorn, J. Matevska, N. Sommer, L. Stoever, S. Giesecke, and W. Hasselbring. Kieker: continuous monitoring and on demand visualization of java software behavior. In *Proceedings of the IASTED International Conference on Software Engineering*, SE '08, pages 80–85, Anaheim, CA, USA, 2008. ACTA Press.

[32] C. Smith. Software performance engineering. *Performance Evaluation of Computer and Communication Systems*, pages 509–536, 1993.

[33] D. Yan, G. Xu, and A. Rountev. Uncovering performance problems in java applications with reference propagation profiling. In *Proceedings of the 2012 International Conference on Software Engineering*, ICSE 2012, pages 134–144, Piscataway, NJ, USA, 2012. IEEE Press.

[34] S. Zaman, B. Adams, and A. Hassan. A qualitative study on performance bugs. In *Mining Software Repositories (MSR), 2012 9th IEEE Working Conference on*, pages 199 –208, june 2012.

[35] S. Zaman, B. Adams, and A. E. Hassan. Security versus performance bugs: a case study on firefox. In *Proceedings of the 8th Working Conference on Mining Software Repositories*, MSR '11, pages 93–102, New York, NY, USA, 2011. ACM.

[36] E. Zimran and D. Butchart. Performance engineering throughout the product life cycle. In *CompEuro '93. 'Computers in Design, Manufacturing, and Production', Proceedings.*, pages 344 –349, may 1993.

Multiple Class G-Networks with Restart

Jean-Michel Fourneau
PRiSM, CNRS / Université de Versailles St Quentin
45, avenue des Etats-Unis
F-78035 Versailles
jean-michel.fourneau@prism.uvsq.fr

Katinka Wolter
Institute of Computer Science
Freie Universität Berlin
Takustr. 9, 14195 Berlin, Germany
katinka.wolter@fu-berlin.de

Philipp Reinecke
HP Labs Bristol
Long Down Avenue, Bristol
BS34 8QZ
philipp.reinecke@hp.com

Tilman Krauß
Institute of Computer Science
Freie Universität Berlin
Takustr. 9, 14195 Berlin
tilman.krauss@fu-berlin.de

Alexandra Danilkina
Institute of Computer Science
Freie Universität Berlin
Takustr. 9, 14195 Berlin
alexandra.danilkina@fu-berlin.de

ABSTRACT

Restart is a common technique for improving response-times in complex systems where the causes of delays can either not be discerned, or not be addressed by the user. With restart, the user aborts a running job that exceeds a deadline, and resubmits it to the system immediately. In many common scenarios, this approach can reduce the response-times that the user experiences. Restart has been well-studied for scenarios where only one user applies restart, and typically in cases where queueing effects can be neglected. In this paper we approach the question of restart in a scenario where restart is applied by many users in a system that can be modelled as an open queueing network. We apply the G-Networks formalism to this problem. We use negative customers to model the abortion and retry of a request. The open G-network uses multiple classes with phase-type distributed service times. This allows the approximation of a preemptive repeat different behaviour as it is natural for multiple restarts of a request. We compute the response time of a request and show that an optimal restart interval can be found. The results are compared with simulation.

Categories and Subject Descriptors

Networks [**Network performance evaluation**]: Network performance modeling; D.2.8 [**Software Engineering**]: Metrics—*complexity measures, performance measures*

General Terms

Keywords

Restart, G-Networks, Phase-type distributions

1. INTRODUCTION

The restart mechanism is known to most people, since there exist many situations in daily life where the impatient customer after some waiting time refuses to wait any longer for completion of his job, cancels the job and restarts it. Internet downloads are the most widely known situation that can benefit from restarts, but many other instances do exist. While restart is often an elegant and simple solution, it may also have a negative effect on the system to which it is applied, as restart can effectively mean an increase of the load on the system, thereby exacerbating the problem it should address. Therefore, a careful choice of the restart interval is often necessary.

Several aspects of the restart problem have been explored in recent years. In [18, 19] a stochastic model for restart to minimise job completion times has been proposed. The probability of job completion under restart has been maximised in [20]. In these works the authors considered an individual user issuing independent jobs that are completed according to some completion-time distribution. It could be shown that with this restriction restart can be successful in reducing the completion-times experienced by the user.

In this paper we address the question whether restart is still beneficial if applied by several users on one or more shared resources. This naturally leads to the formulation of the problem as an open queueing-network model. We utilise the formalism of G-networks, that is, queueing networks with signals. Signals in our G-network model restarts as they remove a random job in the queue. The restarted job returns to the queueing network in a different class which allows us to model a different processing speed upon restart. Our model uses phase-type (PH) distributions for the service-time distributions [14], in order to be able to reflect characteristics of real systems. In particular, the distributions are more general than the exponential distribution.

The remainder of this paper is structured as follows: In Section 2 we provide the necessary background on G-networks. In Section 3 we formulate our model. Section 4 provides a product-form solution for this model. In Section 5 we discuss computation of measures of interest using our solution. We then illustrate the approach in Section 6 and discuss its application in the evaluation of practical questions related to restart in Section 7.

2. G-NETWORKS

In this paper we define a class of G-networks that can model restart. The theory of queues with signals has received considerable attention since the seminal paper on positive and negative customers [6] published by Gelenbe 20 years ago. Traditional queueing networks model systems that are used to represent contention among customers for some resources. Customers move from server to server, they wait for service, but they do not interact among each other. Signals have been used to change these rules. In a network of queues with signals (also denoted as a G-network of queues) customers are allowed to change to signals at the completion of their service and signals interact at their arrival into a queue with customers already present in the queue. Moreover signals are never queued. They try to interact with customers and disappear immediately. Note that they may fail to interact with some probability or due to some conditions which are not satisfied. Despite this deep modification of the model, G-networks still preserve the product-form property for the steady-state distribution of some Markovian queueing networks.

The first type of signal [6] was introduced as a negative customer. A negative customer deletes a positive customer at its arrival at a backlogged queue. A negative customer is never queued. Positive customers are usual customers which are queued and receive service or are deleted by negative customers. Under typical assumptions for Markovian queueing networks (Poisson arrival for both types of customers, exponential service time for positive customers, Markovian routing of customers, open topology, independence) Gelenbe proved that such a network has a product-form solution for its steady-state distribution. It must be clear that the results are more complex than those for Jackson networks. The G-networks flow equations exhibit some uncommon properties: they are neither linear as in closed queueing networks nor contracting as in Jackson queueing networks. Therefore the existence of a solution had to be proved [10] by new techniques from the theory of fixed point equation. A numerical algorithm was developed in [5].

G-networks have also motivated many new important results in the theory of queues. The original proofs were based on the analysis of the global balance equations. Indeed, as negative customers lead to customer deletions, the original description of quasi-reversibility by arrivals and departures did not hold anymore and new versions have been proposed. At the time being, the description proposed by Chao and his co-authors in [2] looks sufficient to study queues with customers and signals. A completely different approach, based on Stochastic Process Algebra, was proposed by Harrison [12, 11]. The main results (CAT and RCAT theorems and their extensions [1, 12, 11]) give some sufficient conditions for product-form stationary distributions. This technique clearly has a different range of applications as it allows to represent component-based models which are much more general and more detailed than networks of queues.

These techniques have been used to study many new signals which all lead to product-form solutions: triggers which redirect other customers among the queues, catastrophes which flush all the customers out of a queue [8, 7], resets [9], synchronised arrivals in a set of queues [4], signals which change the class of the customer in service [16]. Here we present a new result for open G-networks where the effect of the signal is to restart a customer service. The service-times follow PH distributions, which are class dependent.

3. THE MODEL

We investigate generalized networks with an arbitrary number N of queues. We consider K classes of positive customers and only one class of negative customers (also denoted as signals). The external arrivals to the queues follow independent Poisson processes. The external arrival rate to queue i is denoted by $\lambda_i^{(k)}$ for positive customers of class k and Λ_i^- for signals.

The customers are served according to the processor sharing (PS) policy. The service times are assumed to be phase-type distributed. At phase p, the intensity of service for customers of class k in queue i is denoted as $\mu_i^{(k,p)}$. The transition probability matrix $H_i^{(k)}$ describes how, at queue i, the phase of a customer of class k evolves. Without loss of generality we assume that the PH distributions which describe the service times follow these rules:

- The initial state has index 1.
- The exit state has index 0.

Thus the service in queue i is an excursion from state 1 to state 0 following matrix $H_i^{(k)}$ for a customer of class k and we have:

$$\forall i, k, p \quad \sum_{q=0}^{P} H_i^{(k)}[p,q] = 1. \tag{1}$$

At its service completion time in queue i (i.e. transition from phase p to phase 0 in $H_i^{(k)}$), and according to a Markovian transition matrix, a customer of class k may join queue j as a positive customer of class l with probability $P_{i,j}^{+(k,l)}$. It may also leave the network with probability $d_i^{(k)}$. We assume that a customer cannot return to the queue it has just left: $P_{i,i}^{+(k,l)} = 0$ for all i, k and l. As usual, we have:

$$\forall i, k \quad \sum_{j=1}^{N}\sum_{l=1}^{K} P_{i,j}^{+(k,l)} + d_i^{(k)} = 1. \tag{2}$$

Signals do not stay in the network. At its arrival into a queue, a signal interacts with a selected customer and then vanishes instantaneously. If, at its arrival, the queue is already empty, it also vanishes instantaneously. The selected customer is chosen at random following a state-dependent distribution which mimics the PS scheduling. At state $\vec{x}_i$, the probability for a customer to be selected is $\frac{x_i^{(k,p)}}{|\vec{x}_i|}\mathbb{1}_{\{|\vec{x}_i|>0\}}$ and the signal has an effect with probability $\alpha_i^{(k,p)}$. The effect is the restarting of the customer: this customer (remember it has class k and phase p) is routed as a customer of class l at phase 1 with probability $R_i^{(k,l)}$. We assume for all k, $R_i^{(k,k)} = 0$. Of course we have:

$$\forall k \quad \sum_{l=1}^{K} R_i^{(k,l)} = 1. \tag{3}$$

The state of the queueing network is represented by the vector $\vec{x} = (\vec{x}_1, \vec{x}_2, \ldots, \vec{x}_N)$, where the component $\vec{x}_i$ denotes the state of the queue i. As usual with multiple class PS queues with Markovian distribution of service, the state of queue i is given by the vector $(x_i^{(k,p)})$, for all class index k and phase index p. Furthermore, we note $|\vec{x}_i|$ the total number of customers in queue i.

Clearly $\vec{x}$ is a Markov chain.

Note that at the end of a service, we do not allow that a customer becomes a signal. This is not needed in our model because we do not want to represent the joint restart of a group of customers.

4. PRODUCT-FORM THEOREM

Let $p(\vec{x})$ be the stationary probability distribution of the network state if it exists. The following result establishes the existence of a product-form solution if a fixed point system has a solution which satisfies the stationarity constraints.

THEOREM 1. *Consider an arbitrary open G-network with p classes of positive customers and a single class of negative customers the effect of which is to restart one customer in the queue. If the system of linear equations:*

$$\rho_i^{(k,1)} = \frac{\lambda_i^{(k)} + \sum_{o=1}^{P} \mu_i^{(k,o)} \rho_i^{(k,o)} H_i^{(k)}[o,1] + \nabla_i^{k,1} + \Delta_i^{k,1}}{\mu_i^{(k,1)} + \Lambda_i^- \alpha_i^{(k,1)}}, \tag{4}$$

where

$$\Delta_i^{k,1} = \sum_{p=1}^{P} \sum_{l=1}^{K} \Lambda_i^- \alpha_i^{(l,p)} \rho_i^{(l,p)} R_i^{(l,k)}, \tag{5}$$

$$\nabla_i^{k,1} = \sum_{j=1}^{N} \sum_{l=1}^{K} \sum_{q=1}^{P} \mu_j^{(l,q)} \rho_j^{(l,q)} H_j^{(l)}[q,0] P_{j,i}^{+(l,k)}, \tag{6}$$

and,

$$\forall p > 1, \quad \rho_i^{(k,p)} = \frac{\sum_{o=1}^{P} \mu_i^{(k,o)} \rho_i^{(k,o)} H_i^{(k)}[o,p]}{\mu_i^{(k,p)} + \Lambda_i^- \alpha_i^{(k,p)}} \tag{7}$$

has a positive solution such that for all station i:

$$\sum_{k=1}^{K} \sum_{p=1}^{P} \rho_i^{(k,p)} < 1, \tag{8}$$

then the system stationary distribution exists and has product form:

$$p(\vec{x}) = \prod_{i=1}^{N} p_i(\vec{x}_i), \tag{9}$$

where

$$p_i(\vec{x}_i) = (1 - \sum_{k=1}^{K} \sum_{p=1}^{P} \rho_i^{(k,p)}) |\vec{x}_i|! \prod_{k=1}^{K} \prod_{p=1}^{P} \frac{(\rho_i^{(k,p)})^{x_i^{(k,p)}}}{x_i^{(k,p)}!}. \tag{10}$$

Before we can give the proof, we introduce some usual notations. We denote by $(\vec{x}_i + e_i^{(k,p)})$ (resp. $(\vec{x}_i - e_i^{(k,p)})$) the state of queue i obtained by adding (resp. suppressing) one customer of class k at phase p of service. We note $M_i^{(k,p)}(\vec{x}_i)$ the service rate of customers of class k in phase p at queue i. Since the service discipline considered is processor sharing, $M_i^{(k,p)}(\vec{x}_i)$ can be written as a function of $\mu_i^{(k,p)}$:

$$M_i^{(k,p)}(\vec{x}_i) = \mu_i^{(k,p)} \frac{x_i^{(k,p)}}{|\vec{x}_i|} 1\!\!1_{\{|\vec{x}_i|>0\}}. \tag{11}$$

As the selection of customers mimics the PS discipline, the probability of restarting a customer of class k at step p when the queue is at state $\vec{x}_i$ is:

$$N_i^{(k,p)}(\vec{x}_i) = \alpha_i^{(k,p)} \frac{x_i^{(k,p)}}{|\vec{x}_i|} 1\!\!1_{\{|\vec{x}_i|>0\}}. \tag{12}$$

The proof consists mainly of algebraic manipulations of the Chapmann-Kolmogorov equation for steady-state distribution:

$$\begin{aligned}
p(\vec{x}) \quad & \textstyle\sum_{i=1}^{N} \sum_{k=1}^{K} \Big[\lambda_i^{(k)} + \\
& \textstyle\sum_{p=1}^{P} M_i^{(k,p)}(\vec{x}_i) 1\!\!1_{\{x_i^{(k,p)}>0\}} + \\
& \textstyle\sum_{p=1}^{P} \Lambda_i^- N_i^{(k,p)}(\vec{x}_i) 1\!\!1_{\{x_i^{(k,p)}>0\}}\Big] \\
= \quad & \sum_{i=1}^{N} \sum_{k=1}^{K} \lambda_i^{(k)} p(\vec{x} - e_i^{(k,1)}) 1\!\!1_{\{x_i^{(k,1)}>0\}} \qquad [1] \\
+ \quad & \sum_{i=1}^{N} \sum_{k=1}^{K} \sum_{p=1}^{P} M_i^{(k,p)}(\vec{x}_i + e_i^{(k,p)}) d_i^{(k)} * \\
& \quad H_i^{(k)}[p,0] p(\vec{x} + e_i^{(k,p)}) \qquad [2] \\
+ \quad & \sum_{i=1}^{N} \sum_{k=1}^{K} \sum_{p=1}^{P} \sum_{q=1}^{P} M_i^{(k,p)}(\vec{x}_i + e_i^{(k,p)} - e_i^{(k,q)}) * \\
& \quad p(\vec{x} + e_i^{(k,p)} - e_i^{(k,q)}) H_i^{(k)}[p,q] 1\!\!1_{\{x_i^{(k,q)}>0\}} \qquad [3] \\
+ \quad & \sum_{i=1}^{N} \sum_{k=1}^{K} \sum_{j=1}^{N} \sum_{l=1}^{K} \sum_{p=1}^{P} M_i^{(k,p)}(\vec{x}_i + e_i^{(k,p)}) P_{i,j}^{+(k,l)} * \\
& \quad p(\vec{x} + e_i^{(k,p)} - e_j^{(l,1)}) H_i^{(k)}[p,0] 1\!\!1_{\{x_j^{(l,1)}>0\}} \qquad [4] \\
+ \quad & \sum_{i=1}^{N} \sum_{k=1}^{K} \sum_{p=1}^{P} \Lambda_i^- N_i^{(k,p)}(\vec{x}_i + e_i^{(k,p)}) \sum_{l=1}^{K} R_i^{(k,l)} * \\
& \quad p(\vec{x} + e_i^{(k,p)} - e_i^{(l,1)}) 1\!\!1_{\{x_i^{(l,1)}>0\}}. \qquad [5]
\end{aligned} \tag{13}$$

Let us give some explanations about the right-hand side of this equation. The first term corresponds to external arrivals of customers. The second term is used to describe an end of service and departure to the outside while the third one is associated to the completion of phase p of service. With the fourth term we consider the case where a customer of class k leaves queue i to queue j as a customer of class l. The last term is associated with the restart: a signal arriving at queue i restart a customer of class k and phase p which joins queue i as a customer of class l at step 1 (term [5]). For the sake of readibility the proof of the theorem is transferred into an appendix. But it is also important to prove that the necessary conditions of the theorem do not exclude the existence of a feasible solution for Equations 4 to 7. Such a question is easy for Jackson's networks because the necessary conditions are equivalent to a liear system which is proved to be non singular or contracting depending on the network topology. For G-networks, the problem is much more complex: see for instance [10].

4.1 Existence of a solution to the f ow equation

We now have to prove that there exists a non-empty region of the set of parameters where the flow equation has a solution. We reorganize the equations to simplify the presentation.

Let $\vec{q}$ be the vector of values $\rho_i^{(k,p)}\mu_i^{(k,p)}$ for all indices i, k, p. Thus $\vec{q}$ takes value in $\mathbb{R}^{NKP}$. The system of equation can easily be written as a linear system:

$$\vec{q} = \vec{q}G + \vec{a},$$

where $\vec{a}$ is the renormalized arrival rate (i.e. $\lambda_i^{(k)} \frac{\mu_i^{(k,p)}}{\mu_i^{(k,p)}+\Lambda_i^- \alpha_i^{(k,p)}}$) to take into account the signal. Proving that $(Id - G)$ is not singular is sufficient to guarantee the existence of $\vec{q}$. Thus we now consider matrix G and check if the system has a unique solution. Note that we do not check if the solution satifies Equation 8. We propose two solutions to prove that $(Id - G)$ is not singular.

The first idea consists in numerically compute matrix G from the parameters. Then we can check that the matrix is strictly sub-stochastic. Indeed, remember that if G is sub-stochastic and does not have any recurrent class, then $(Id - G)$ is not singular. The strict sub-stochastic property (every lines has a sum strictly smaller than one) implies that the matrix does not have any recurrent class and it is easier to check than the existence of recurrent classes. This is stated in the following two properties.

PROPERTY 1. *If G is strictly sub-stochastic, then Equations 4 to 7 have a solution in R^{NKP}.*

PROPERTY 2. *If G is sub-stochastic and does not contain any recurent class, then Equations 4 to 7 have a solution in R^{NKP}.*

Let us now consider a slightly more complex technique based on Brouwer's theorem to establish the existence of a fixed point system. We can easily decomposed matrix G into two matrices:

$$G = G1 + G2,$$

where Matrix $G1$ models the PS queues where all the entries have been multiplied by a damping factor equal to $\frac{\mu_i^{(k,p)}}{\mu_i^{(k,p)}+\Lambda_i^- \alpha_i^{(k,p)}}$ and $G2$ represents the terms associated to the arrival of restart signal. More precisely, the entries of $G1$ and $G2$ are described by:

- Matrix $G1$ (row (i,k,1), column (j,l,q))

$$\frac{\mu_i^{(k,1)}}{\mu_i^{(k,1)} + \Lambda_i^- \alpha_i^{(k,1)}} (H_i^{(k)}[q,1] + H_i^{(k)}[q,0] P_{j,i}^{+(l,k)}).$$

- Matrix $G1$ (row (i,k,p), column (i,k,o))

$$H_i^{(k)}[o,p] \frac{\mu_i^{(k,p)}}{\mu_i^{(k,p)} + \Lambda_i^- \alpha_i^{(k,p)}}.$$

- Matrix $G2$ (row (i,k,1), column (i,l,p))

$$\frac{\alpha_i^{(l,p)}}{\mu_i^{(l,p)}} R_i^{(l,k)} \frac{\mu_i^{(k,1)}}{\mu_i^{(k,1)} + \Lambda_i^- \alpha_i^{(k,1)}} \Lambda_i^- .$$

The system can be written as:

$$\vec{q}(Id - G1) = \vec{q}G2 + \vec{a},$$

where $\vec{a}$ has been previously defined. Let us now define $\mathcal{S}$ as the stability set (i.e. the set of non negative values of values of $\rho_i^{(k,p)}$ which satisfy the constraints in Equation 8). We have the following properties:

PROPERTY 3. *Assume that Matrix $(Id - G1)$ is not singular. One can find $\vec{b}$ such that $\vec{q}\, G2 \leq \vec{b}$ for all vectors $\vec{q}$ in $\mathcal{S}$.*
Furthermore let $\vec{q_0}$ be the vector such that:

$$\vec{q_0} = (\vec{a} + \vec{b})\ (Id - G1)^{-1}.$$

$\vec{q_0}$ exists because matrix $(Id - G1)$ is not singular.

Proof: Indeed Matrix $G2$ is non negative and $\mathcal{S}$ is a closed and convex subset of R^{NKP}.

DEFINITION 1. *The network is superstable if $\vec{q_0}$ is in $\mathcal{S}$.*

Intuitively, superstability means that the network where signals are considered as arrival of fresh customers at the maximum rate allowed by the model is stable when considered as an ordinary network of PS queues with ordinary customers. Of course we expect that superstability implies stability. This is proved in Property 4. We begin by Brouwer's theorem.

THEOREM 2. *(Brouwer) Let F be a fixed point system, (i.e $x = F(x)$) in R^n. If*

1. *function F is continuous on a subset H of R^n*
2. *H is non empty*
3. *H is a compact and convex subset,*
4. *$F(H) \subseteq H$*

then $x = F(x)$ has at least one solution.

PROPERTY 4. *If the network is superstable, then there exist a solution to the equations 4 to 7 which satisfies the stability constraint of Equation 8.*

Proof: We will build a set $\mathcal{S}2$ such that the assumptions of Brouwer's theorem will be satisfied by the fixed point function on $\mathcal{S}2$. First note that $F(\vec{q}) = \vec{q}G + \vec{a}$. Therefore F is continuous on any subset of R^{NKP}. Now let $\mathcal{S}2$ be the set of non negative vectors $\vec{r}$ such that $\vec{r} \leq \vec{q_0}$ component-wise. $\mathcal{S}$ is compact, convex and non empty because $\vec{q_0}$ is non zero. Furthermore G is non negative, therefore $\vec{q} \leq \vec{q_0}$ implies that $\vec{q}G \leq \vec{q_0}G$ and $F(\vec{q}) \leq F(\vec{q_0}) = \vec{q_0}$. And $F(\vec{q})$ is clearly non negative for a non negative vector $\vec{q}$. Therefore $F(\mathcal{S}2) \subseteq \mathcal{S}2$. All the assumptions of the theorem are satisfied. There exists a fixed point solution for F in $\mathcal{S}2$.

5. COMPUTATION OF MEASURES

Remember that the proof based on global balance has shown that

$$p(\vec{x}) = \prod_{i=1}^{N} p_i(\vec{x}_i), \tag{14}$$

where

$$p_i(\vec{x}_i) = C(1 - \sum_{k=1}^{K}\sum_{p=1}^{P} \omega_i^{(k,p)})|\vec{x}_i|! \prod_{k=1}^{K}\prod_{p=1}^{P} \frac{(\omega_i^{(k,p)})^{x_i^{(k,p)}}}{x_i^{(k,p)}!}.$$

where C is a normalisation constant.

PROPERTY 5. *The normalization constant C is equal to 1.*

Proof: As the network is separable and all the states are reachable, we just have to verify that for each queue the sum of the probability is equal to one.

$$\sum_{\vec{x}_i} p_i(\vec{x}_i) = C\sum_{\vec{x}_i}(1-\sum_{k=1}^{K}\sum_{p=1}^{P} \rho_i^{(k,p)})|\vec{x}_i|! \prod_{k=1}^{K}\prod_{p=1}^{P} \frac{(\rho_i^{(k,p)})^{x_i^{(k,p)}}}{x_i^{(k,p)}!}.$$

We partition the summation on $\vec{x}_i$ according to the norm of $\vec{x}_i$.

$$\sum_{\vec{x}_i} p_i(\vec{x}_i) = C\sum_{m=0}^{\infty}(1 - \sum_{k=1}^{K}\sum_{p=1}^{P} \rho_i^{(k,p)}) * \sum_{\vec{x}_i/|\vec{x}_i|=m} |\vec{x}_i|! \prod_{k=1}^{K}\prod_{p=1}^{P} \frac{(\rho_i^{(k,p)})^{x_i^{(k,p)}}}{x_i^{(k,p)}!}.$$

Substitute $|\vec{x}_i|$ by m in the previous equation. Remember the definition of the multinomial

$$\sum_{\vec{x}_i/|\vec{x}_i|=m} m! \prod_{k=1}^{K}\prod_{p=1}^{P} \frac{(\rho_i^{(k,p)})^{x_i^{(k,p)}}}{x_i^{(k,p)}!} = \left[\sum_{k=1}^{K}\sum_{p=1}^{P} \rho_i^{(k,p)}\right]^m$$

After substitution we obtain:

$$\sum_{\vec{x}_i} p_i(\vec{x}_i) = C\sum_{m=0}^{\infty}(1 - \sum_{k=1}^{K}\sum_{p=1}^{P} \rho_i^{(k,p)}) \left[\sum_{k=1}^{K}\sum_{p=1}^{P} \rho_i^{(k,p)}\right]^m$$

As by assumption we have $\left[\sum_{k=1}^{K}\sum_{p=1}^{P} \rho_i^{(k,p)}\right]^m < 1$, the sum converges and we get $\sum_{\vec{x}_i} p_i(\vec{x}_i) = C$. Thus the normalization constant in the product form is equal to one.

PROPERTY 6. *The former proof also establishes that the probability to have exactly m customers in the queue is equal to:*

$$Pr(m\ customers) = (1 - \sum_{k=1}^{K}\sum_{p=1}^{P} \rho_i^{(k,p)}) \left[\sum_{k=1}^{K}\sum_{p=1}^{P} \rho_i^{(k,p)}\right]^m.$$

This allows to get the expectation of the number of customers.

PROPERTY 7. *The expected number of customers in the queue is equal to:*

$$E[N] = \frac{\sum_{k=1}^{K}\sum_{p=1}^{P} \rho_i^{(k,p)}}{1 - \sum_{k=1}^{K}\sum_{p=1}^{P} \rho_i^{(k,p)}}. \tag{15}$$

To get the average queuing delay, one usually applies Little's law. However, one must be very careful when describing a G-network with signals in this context. Indeed, signals may increase the queue size (see for instance the joint arrival described in [4] or the resets). In general one must take into account all these increases of the queue size when computing an "artificial" arrival rate into the queue. Here the situation is relatively simple. Signals do not increase the number of customers. Thus the arrival rate in Little's law is exactly the arrival of real customers.

6. EXAMPLES

We will now illustrate the approach using two examples. We start exploring the modelling power of the G-networks with restart using a relatively simple example. Throughout this section we use our Mathematica implementation of Eqn. 15.

6.1 Cycle of Erlangs

The first queueing-network we study consists of one queue with $K = 5$ classes. Jobs arrive only to the first class at rate $\lambda = 0.1$. In each class $k = 1, \ldots, 5$ service times follow an Erlang distribution of length 2. Subsequent classes have increasingly higher rates. For a job in $k = 1, \ldots, 4$, restart causes the job to move from class k to class $k + 1$, which implies that service gets faster upon restart. For jobs in class 5, restart means that the job is moved back to class 1. This type of model is appropriate to study the effect of restart in scenarios where the service becomes faster (or, equivalently, service-demand is reduced) on subsequent trials because the service was already partially completed on previous attempts, but where the user aborts and repeats the complete request, including all partial results, after several restarts. For instance, this is often the case with downloads of files or websites from the Internet. A similar model could also be used to study restart in a scenario with load-balancing, where incoming and restarted jobs are assigned to servers with different processing speeds.

The parameters are defined as follows: Jobs only arrive to class 1 at rate $\lambda_1 = 0.1$. The arrival rate to all other classes $k > 1$ equals zero, i.e. $\lambda_k = 0$, for $k > 1$. The rates of the Erlang service-time distributions are defined as $\mu_1^{(1,p)} = 1, \mu_1^{(2,p)} = 2, \mu_1^{(3,p)} = 3, \mu_1^{(4,p)} = 4, \mu_1^{(5,p)} = 5$. Signals arrive at rate Λ to the queue, which we vary across the experiments. Signals always lead to abort and restart, hence $\alpha_i^{(k,p)} = 1, \forall i, k, p$. The matrix H is for the Erlang distribution in all k classes $H_1^k = \begin{bmatrix} 1 & 0 & 0 \\ 0 & 0 & 1 \\ 1 & 0 & 0 \end{bmatrix}$. The routing matrix R sends each job after restart to the next class, where it completes the remainder of its service. If a job does not meet the restart deadline in the final class it is reissued and starts all over from scratch. The matrix R describes the change of class from k_1 to k_2 as a job is restarted.

$$R_i[k_1, k_2] = \begin{bmatrix} 0 & 1 & 0 & 0 & 0 \\ 0 & 0 & 1 & 0 & 0 \\ 0 & 0 & 0 & 1 & 0 \\ 0 & 0 & 0 & 0 & 1 \\ 1 & 0 & 0 & 0 & 0 \end{bmatrix}.$$

Completed jobs leave the network with probability $d_i^k = 1$. In consequence, the probability Pr that a job remains in the network after completion is zero. The parameters are not motivated by experimental data, they purely illustrate the dynamics of the model.

The scenario is illustrated in Figure 1.

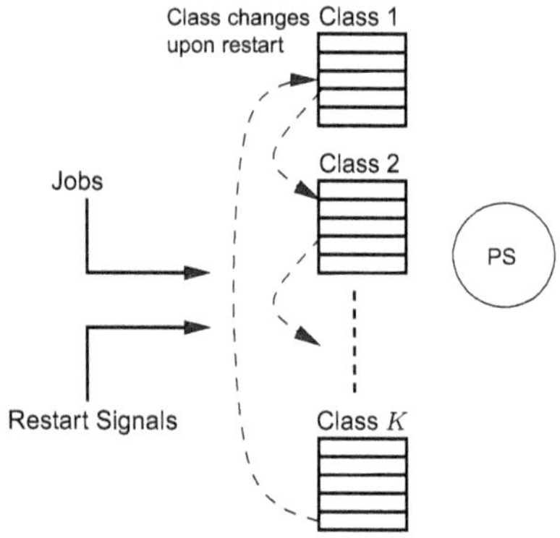

Figure 1: Multiple class G-network with restart

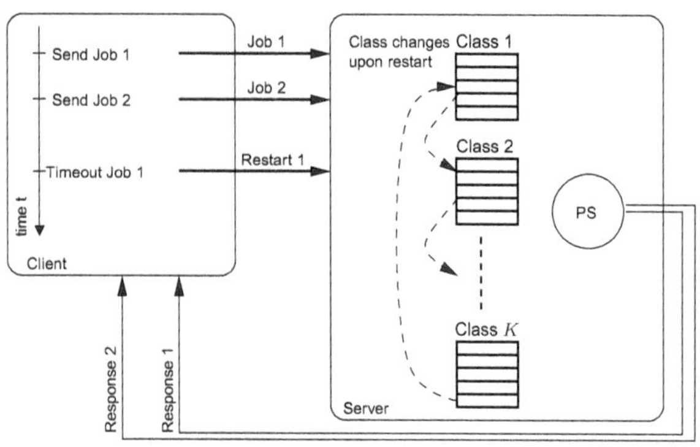

Figure 2: Simulation scenario

Figure 3 shows the results from this model. Several observations can be made. Even though restarts increase the load, the system remains stable for the range of parameters we investigate. Even stronger, for zero restart rate the system behaves as a system without restart, which is the well-known M/PH/1 queue with processor sharing discipline [?]. We can observe that for a suitably chosen restart rate the utilisation of the system can be reduced with restart. Further, we can see that the queue length and the waiting time are a function of the utilisation ρ and hence both have the same minimum as the utilisation with respect to the restart rate. The waiting time is computed using Little's law where the restart rate is not included in the arrival rate of jobs. Obviously, restart does not increase the number of jobs that enter the system, but it does manipulate the time spent in the system by each job.

In order to validate the results, we simulated the scenario using the SFERA framework [3]. SFERA is a framework for evaluation of restart algorithms using discrete-event simulation. We implemented the model shown in Figure 2: The client generates jobs as a Poisson process and sends them to the server, where they are processed according to the processor-sharing strategy. Upon completion of a job, the server sends a response back to the client. Before sending each job to the server, the client draws a restart interval for that job from an exponential distribution. When the interval elapses, the client sends a restart signal for this job to the server. The server aborts the processing of the job and restarts it in the next class, according to the configuration of the restart acceptance and routing probabilities. We configured the classes to form a circle, as for the G-network. Note that this model is much more realistic than the G-network model, in that it allows restart intervals specific to each job. In contrast, in the G-network model the restart signals arrive independently of jobs and affect random jobs.

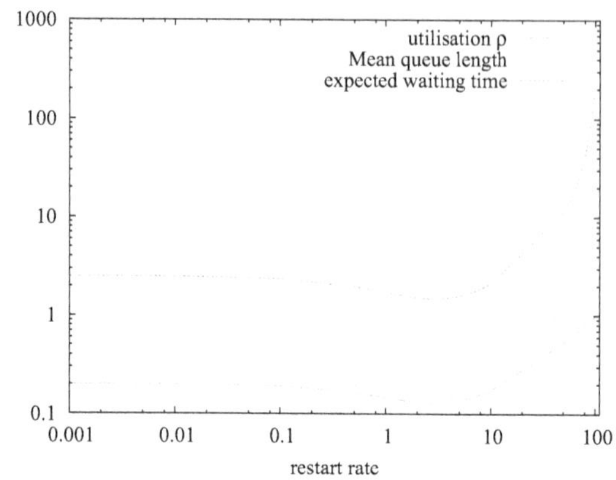

Figure 3: Expected queue length and expected waiting time for different values of the restart rate

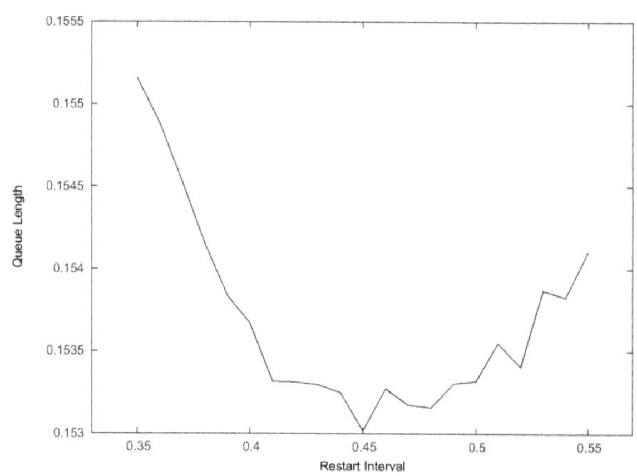

Figure 4: Average queue length and average response time for the simulation

With SFERA we can observe the average queue length in the server and the completion-times, as experienced by the client. We show the results of the simulation in Figure 4. It is obvious that the simulation results also indicate the existence of an optimal restart rate at which the queue length and completion-times are minimised. Please note that the optimal restart interval in Fig. 4 of approximately 0.45 corresponds to an optimal restart rate of 2.2, which is close to the optimal value shown in Figure 3. For restart intervals larger than this value, the measures approach the case without restart, while for intervals below it both measures increase because the server cannot complete jobs within the allotted restart intervals. We also note that the optimal value of the restart rate is slightly different from the simulation results. We attribute this to the differences in the models, since the simulation model includes dependencies between the job and the signal arriving processes, which are not present in the G-network model. In the simulation model each arriving job is assigned its individual restart timeout that is sampled from an exponential distribution at that time. The timeout could also be drawn from an arbitrary distribution (including the deterministic). Each event in the simulation, i.e. the arrival or departure of a job, leads to a reschedule of all present deadlines but the timeout is not resampled. In contrast, the semantics of the analytical model imply that timeouts are not stored. Instead, the time between signals (restarts) is a randomly sampled sequence of non-overlapping time intervals.

While in the simulation a job at the head of the queue experiences expiry of its timeout with higher probability than a job at the end of the queue, in the G-network each job in the queue is equally likely to be hit by a signal and restarted. Through the decreasing timeout value the simulation includes some notion of age of a job, which is not present in the G-network.

Taking into account that in real-world scenarios restart is typically triggered by a job-specific timeout, rather than an independent stream of signals, our simulation results thus also illustrate that the predictions of the theoretical model hold reasonably well in realistic scenarios.

6.2 A more complex model

In our second example we formulate a more complex, and more realistic, model. The model again has only one queue, which is a limitation in the implementation, not in the formalism. The model has 5 classes of customers. Jobs arrive at rate $\lambda = 0.004$ to class 1 and at rate $\lambda = 0.01$ to class 2. We make the following assumptions about the restart:

- Classes 1 and 2 represent "fresh" customers while class 3 (resp. class 4 and 5) represent customers of class 1 (resp. class 2 and 3) which have been restarted once.
- It is not possible to restart a customer of class 4.
- When we restart a customer of class 5, it becomes a customer of class 1 again.

We assume the following distributions for the service (the first state of the PH has state number zero and it represents the end of the PH (see page 1).

- Class 1 $H_1^1 = \begin{bmatrix} 1 & 0 & 0 & 0 \\ 0 & 0 & 0.9 & 0.1 \\ 1 & 0 & 0 & 0 \\ 1 & 0 & 0 & 0 \end{bmatrix}$ with $\mu_1^{1,1} = 5$, $\mu_1^{1,2} = 2$ and $\mu_1^{1,3} = 0.1$. Thus the PH describes an exponential of rate 5 followed by either an exponential of rate 2 with probability 0.9 or an exponential of rate 0.1 (which means a long average time compared to the other transitions).

  We also assume the following about the effect of the restart: if the customer selected at the queue by the restart signal has class k and is in phase p, the restart succeeds with probability $\alpha_i^{(k,p)}$. We model the following typical assumptions: restarts do only succeed with probability 1/2 for customer of class 1 when it is in phase 1 or 2. And it succeeds when it is in phase 3.

  $$\alpha_1^{(1,1)} = 0.5,\ \alpha_1^{(1,2)} = 0.5,\ \alpha_1^{(1,3)} = 1.$$

- Class 2 $H_1^2 = \begin{bmatrix} 1 & 0 & 0 & 0 \\ 0 & 0 & 2/3 & 1/3 \\ 3/4 & 1/4 & 0 & 0 \\ 1 & 0 & 0 & 0 \end{bmatrix}$ with $\mu_1^{1,1} = 1$, $\mu_1^{1,2} = 1$ and $\mu_1^{1,3} = 0.1$. Thus the PH has a directed loop (this is possible; we are not restricted to acyclic PH). Again phase 3 has a long average time compared to the other transitions). Restarts do not succeed for customer of class 2 when it is in phase 1 or 2, $\alpha_1^{(2,1)} = \alpha_1^{(2,2)} = 0$. A restart succeeds with probability 1 when it is in phase 3.

  $$\alpha_1^{(1,3)} = 1.$$

- Class 3 represents restarted class 1 customers. We assume an Erlang 3 with rate equal to 0.5. This is modelled by matrix $H_1^3 = \begin{bmatrix} 1 & 0 & 0 & 0 \\ 0 & 0 & 1 & 0 \\ 0 & 0 & 0 & 1 \\ 1 & 0 & 0 & 0 \end{bmatrix}$ with $\mu_1^{3,1} = \mu_1^{3,2} = \mu_1^{3,3} = 0.5$. We assume that it is possible to restart class 3 customers.

  $$\alpha_1^{(3,1)} = 1,\ \alpha_1^{(3,2)} = \alpha_1^{(3,3)} = 2/3$$

- Class 4 represents restarted class 2 customers. We assume an Erlang-3 with rate 0.5. This is modelled by matrix $H_1^4 = \begin{bmatrix} 1 & 0 & 0 & 0 \\ 0 & 0 & 1 & 0 \\ 0 & 0 & 0 & 1 \\ 1 & 0 & 0 & 0 \end{bmatrix}$ with $\mu_1^{4,1} = 0.5$, $\mu_1^{4,2} = 0.5$, $\mu_1^{4,3} = 0.5$. We assume that it is unlikely to restart class 4 customers in phase 1 but quite likely to restart them successfully in phase 2 and 3. Thus

  $$\alpha_1^{(4,1)} = 0.1,\ \alpha_1^{(4,2)} = 2/3,\ \alpha_1^{(4,3)} = 2/3$$

- Class 5 represent customers which have been restarted twice. The service is identical to that for class 4.

  The customer can be restarted: $\alpha_1^{(5,1)} = 0.5$, $\alpha_1^{(5,2)} = 0.5$ $\alpha_1^{(5,3)} = 1$.

The routing matrix is as follows.

$$R_i[k_1, k_2] = \begin{bmatrix} 0 & 0 & 1 & 0 & 0 \\ 0 & 0 & 0 & 1 & 0 \\ 0 & 0 & 0 & 0 & 1 \\ 0 & 0 & 0 & 0 & 0 \\ 1 & 0 & 0 & 0 & 0 \end{bmatrix}.$$

We show the analytical results for this model in Figure 5. Again, we observe the existence of an optimal restart rate, which for this model is around 10, and the typical increase in the measures above and beyond this value.

7. APPLICATIONS

The G-Networks-based approach to the analysis of restart we presented here can be used to evaluate the effects of restart in systems that can be modelled as queueing networks, such as service compositions in service-oriented systems. The approach can be applied both by the designer or operator of such systems and, to a limited degree, by the service user.

In the following we sketch application of the approach in the analysis of a practical system. We use a service-oriented

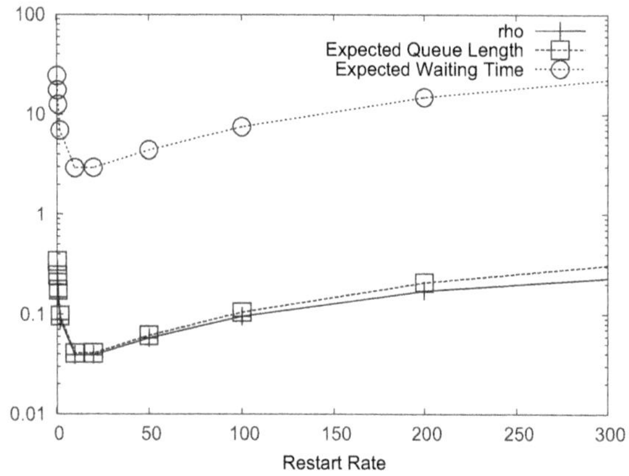

Figure 5: Load ρ, average queue length and average waiting time for the second model.

system as our example, where we assume that several services are combined to form a service-composition. Each service is implemented with a queue for incoming requests and processing threads to handle the requests. The first step in the evaluation then consists in obtaining a structural model of the system. This structural model represents the different services as queues with associated servers, and captures the request flows between services in the routing matrix. In the second step, the processing characteristics of the services must be captured and modelled as phase-type distributions, which will then be included as the service-time distributions in the model. Adequate service-time distributions would typically be obtained by measuring processing times and fitting phase-type distributions to the samples, using one of several well-known tools [13, 17, 15]. The impact of restart can then be evaluated in the third step using Equation 15. For instance, the designer or operator of the system may be interested in whether the system will be stable under restart, and for which range of restart intervals stability holds. This is of particular importance with systems whose users are likely to restart requests due to impatience, such as online shopping. If the evaluation indicates that restart can endanger the stability of the system, countermeasures may be applied, such as increasing the system's capacity or preventing restart. On the other hand, application of Equation 15 allows a user of the system to compute an optimal restart timeout. However, this application of the method is currently limited to cases where the service-time distributions and the system architecture are published by the services.

The approach may also be employed in the development and evaluation of algorithms to compute restart timeouts. Here, the fact that subsequent restarts may encounter different service-time distributions can be modelled explicitly, as in our first example. Furthermore, the implications of many users applying restart can be incorporated by increasing job and signal arrival rates.

It should be mentioned that modelling restarts with G-networks may come with some hardships. The possibility of infinitely many restarts cannot directly be incorporated into the model. We solve it with creating a cycle from the first to the last queue. This only leads to an exact representation if all service time distributions are equal. Then the model can be reduced to two classes and jobs alternate between them.

We did not present models for a finite number of restarts as we did not obtain optimal restart intervals for those and found them otherwise difficult to validate. It is unclear whether or not such an optimal restart interval exists with finite trials. The formulation and solution of the model is straight forward.

8. CONCLUSIONS

In this paper we considered multiple-class G-networks with restart as an approach for studying the effectiveness of the restart method in systems that can be modelled as queueing-networks with multiple classes of users with class-dependent service-times. The G-network formalism allowed us to show that such systems can remain stable, and indeed superstable, under restart. Furthermore, we showed that in this case there exists an optimal restart rate at which the system load, and consequently other measures of interest (such as the waiting time), is minimised. Our comparison of the analytical results to simulation results indicates that this is also the case in more realistic scenarios where the arrival process of restart signals is not independent of the job arrival process.

Acknowledgements

Katinka Wolter and Philipp Reinecke are partly supported by the German Science Foundation (DFG) under grant No. Wo 898/3-1 and by DAAD PROCOPE programme for co-operation with French research institutions.

9. REFERENCES

[1] Simonetta Balsamo, Peter G. Harrison, and Andrea Marin. A unifying approach to product-forms in networks with finite capacity constraints. In Vishal Misra, Paul Barford, and Mark S. Squillante, editors, *SIGMETRICS 2010, Proceedings of the 2010 ACM SIGMETRICS International Conference on Measurement and Modeling of Computer Systems, New York*, pages 25–36. ACM, 2010.

[2] X. Chao, M. Miyazawa, and M. Pinedo. *Queueing Networks Customers, Signals and Product Form solutions*. John Wiley & Sons, 1999.

[3] Alexandra Danilkina, Philipp Reinecke, and Katinka Wolter. Sfera: A simulation framework for the performance evaluation of restart algorithms in service-oriented systems. *Electronic Notes in Theoretical Computer Science*, 291:3–14, 2013. <ce:title>Second Workshop on Quantitative Models for Performance and Dependability (QMPD 2012)</ce:title>.

[4] Thu-Ha Dao-Thi, J.-M. Fourneau, and Minh-Anh Tran. G-networks with synchronised arrivals. *Perform. Eval.*, 68(4):309–319, 2011.

[5] J.-M. Fourneau. Computing the steady-state distribution of networks with positive and negative customers. In *$13^{t}h$ IMACS World Congress on Computation and Applied Mathematics, Dublin*, 1991.

[6] E. Gelenbe. Product-form queuing networks with negative and positive customers. *Journal of Applied Probability*, 28:656–663, 1991.

[7] E. Gelenbe. G-networks with instantaneous customer movement. *Journal of Applied Probability*, 30(3):742–748, 1993.

[8] E. Gelenbe. G-networks with signals and batch removal. *Probability in the Engineering and Informatonal Sciences*, 7:335–342, 1993.

[9] E. Gelenbe and J.-M. Fourneau. G-networks with resets. *Perform. Eval.*, 49(1-4):179–191, 2002.

[10] E. Gelenbe and R. Schassberger. Stability of g-networks. *Probability in the Engineering and Informational Sciences*, 6:271–276, 1992.

[11] Peter G. Harrison. Compositional reversed Markov processes, with applications to G-networks. *Perform. Eval.*, 57(3):379–408, 2004.

[12] P.G. Harrison. Turning back time in Markovian process algebra. *Theoretical Computer Science*, 290(3):1947–1986, 2003.

[13] András Horváth and Miklós Telek. PhFit: A General Phase-Type Fitting Tool. In *TOOLS '02: Proceedings of the 12th International Conference on Computer Performance Evaluation, Modelling Techniques and Tools*, pages 82–91, London, UK, 2002. Springer-Verlag.

[14] Marcel F. Neuts. *Matrix-Geometric Solutions in Stochastic Models: An Algorithmic Approach.* The Johns Hopkins University Press, 1981.

[15] Philipp Reinecke, Tilman Krauss, and Katinka Wolter. Hyperstar: Phase-type fitting made easy. In *Quantitative Evaluation of Systems (QEST), 2012 Ninth International Conference on*, pages 201 –202, sept. 2012.

[16] Thu-Ha Dao Thi, J.-M. Fourneau, and Minh-Anh Tran. Networks of symmetric multi-class queues with signals changing classes. In Khalid Al-Begain, Dieter Fiems, and W. J. Knottenbelt, editors, *Analytical and Stochastic Modeling Techniques and Applications, 17th International Conference, ASMTA 2010, Cardiff, UK*, volume 6148 of *Lecture Notes in Computer Science*, pages 72–86. Springer, 2010.

[17] Axel Thümmler, Peter Buchholz, and Miklos Telek. A novel approach for phase-type fitting with the em-algorithm. *IEEE Transactions on Dependable and Secure Computing*, 3:245–258, 2006.

[18] Aad van Moorsel and Katinka Wolter. Analysis and Algorithms for Restart. In *Proc. 1st International Conference on the Quantitative Evaluation of Systems (QEST)*, pages 195–204, Twente, The Netherlands, September 2004. Best paper award.

[19] Aad P. A. van Moorsel and Katinka Wolter. Analysis of restart mechanisms in software systems. *IEEE Transactions on Software Engineering*, 32(8):547–558, August 2006.

[20] Aad P.A. van Moorsel and Katinka Wolter. Optimal restart times for moments of completion time. *IEE Proceedings Software*, 151(5):219– 223, October 2004.

Appendix: Proof of Theorem 1:

First remember the Chapman-Kolmogorov equation for the steady-state distribution:

$$
\begin{aligned}
p(\vec{x}) \quad & \sum_{i=1}^{N} \sum_{k=1}^{K} \Big[\lambda_i^{(k)} + \\
& \sum_{p=1}^{P} M_i^{(k,p)}(\vec{x}_i) \mathbb{1}_{\{x_i^{(k,p)}>0\}} + \\
& \sum_{p=1}^{P} \Lambda_i^{-} N_i^{(k,p)}(\vec{x}_i) \mathbb{1}_{\{x_i^{(k,p)}>0\}} \Big] \\
= \quad & \\
& \sum_{i=1}^{N} \sum_{k=1}^{K} \lambda_i^{(k)} p(\vec{x} - e_i^{(k,1)}) \mathbb{1}_{\{x_i^{(k,1)}>0\}} \\
+ \quad & \sum_{i=1}^{N} \sum_{k=1}^{K} \sum_{p=1}^{P} M_i^{(k,p)}(\vec{x}_i + e_i^{(k,p)}) d_i^{(k)} * \\
& \qquad H_i^{(k)}[p,0] p(\vec{x} + e_i^{(k,p)}) \\
+ \quad & \sum_{i=1}^{N} \sum_{k=1}^{K} \sum_{p=1}^{P} \sum_{q=1}^{P} M_i^{(k,p)}(\vec{x}_i + e_i^{(k,p)} - e_i^{(k,q)}) * \\
& \qquad p(\vec{x} + e_i^{(k,p)} - e_i^{(k,q)}) H_i^{(k)}[p,q] \mathbb{1}_{\{x_i^{(k,q)}>0\}} \\
+ \quad & \sum_{i=1}^{N} \sum_{k=1}^{K} \sum_{j=1}^{N} \sum_{l=1}^{K} \sum_{p=1}^{P} M_i^{(k,p)}(\vec{x}_i + e_i^{(k,p)}) P_{i,j}^{+(k,l)} * \\
& \qquad p(\vec{x} + e_i^{(k,p)} - e_j^{(l,1)}) H_i^{(k)}[p,0] \mathbb{1}_{\{x_j^{(l,1)}>0\}} \\
+ \quad & \sum_{i=1}^{N} \sum_{k=1}^{K} \sum_{p=1}^{P} \Lambda_i^{-} N_i^{(k,p)}(\vec{x}_i + e_i^{(k,p)}) \sum_{l=1}^{K} R_i^{(k,l)} * \\
& \qquad p(\vec{x} + e_i^{(k,p)} - e_i^{(l,1)}) \mathbb{1}_{\{x_i^{(l,1)}>0\}}.
\end{aligned}
$$

We divide both sides by $p(\vec{x})$ and we take into account the following simplications:

$$\frac{p(\vec{x} - e_i^{(k,1)})}{p(\vec{x})} = \frac{x_i^{(k,1)}}{\rho_i^{(k,1)} |\vec{x}_i|} \tag{16}$$

$$\frac{p(\vec{x} + e_i^{(k,p)})}{p(\vec{x})} = \frac{\rho_i^{(k,p)}(|\vec{x}_i| + 1)}{x_i^{(k,p)} + 1} \tag{17}$$

$$\frac{p(\vec{x} + e_i^{(k,p)} - e_i^{(k,q)})}{p(\vec{x})} = \frac{\rho_i^{(k,p)} x_i^{(k,q)}}{\rho_i^{(k,q)}(x_i^{(k,p)} + 1)} \tag{18}$$

$$M_i^{(k,p)}(\vec{x}_i + e_i^{(k,p)}) = \mu_i^{(k,p)} \frac{x_i^{(k,p)} + 1}{|\vec{x}_i| + 1} \tag{19}$$

$$M_i^{(k,p)}(\vec{x}_i + e_i^{(k,p)} - e_i^{(k,q)}) = \mu_i^{(k,p)} \frac{x_i^{(k,p)} + 1}{|\vec{x}_i|} \mathbb{1}_{\{|\vec{x}_i|>0\}} \tag{20}$$

$$N_i^{(k,p)}(\vec{x}_i + e_i^{(k,p)}) = \alpha_i^{(k,p)} \frac{x_i^{(k,p)} + 1}{|\vec{x}_i| + 1}. \tag{21}$$

Combining these relations we get: $\frac{p(\vec{x} + e_i^{(k,p)} - e_j^{(l,q)} + e_j^{(l,o)})}{p(\vec{x})} = \frac{\rho_i^{(k,p)}(|\vec{x}_i|+1)}{x_i^{(k,p)}+1} \frac{\rho_j^{(l,o)} x_j^{(l,q)}}{\rho_j^{(l,q)}(x_j^{(l,o)}+1)}$

After substitution and simplification we get:

$$\sum_{i=1}^{N}\sum_{k=1}^{K}\Big[\lambda_i^{(k)}+ \sum_{p=1}^{P}(\mu_i^{(k,p)}+\Lambda_i^{-}\alpha_i^{(k,p)})* \frac{x_i^{(k,p)}}{|\vec{x}_i|}\mathbb{1}_{\{|\vec{x}_i|>0\}}\mathbb{1}_{\{x_i^{(k,p)}>0\}}\Big]$$
$$= \sum_{i=1}^{N}\sum_{k=1}^{K}\lambda_i^{(k)}\frac{x_i^{(k,1)}}{|\vec{x}_i|\rho_i^{(k,1)}}\mathbb{1}_{\{x_i^{(k,1)}>0\}}$$
$$+ \sum_{i=1}^{N}\sum_{k=1}^{K}\sum_{p=1}^{P}\mu_i^{(k,p)}\rho_i^{(k,p)}H_i^{(k)}[p,0]d_i^{(k)}$$
$$+ \sum_{i=1}^{N}\sum_{k=1}^{K}\sum_{p=1}^{P}\sum_{q=1}^{P}\mu_i^{(k,p)}\frac{\rho_i^{(k,p)}}{\rho_i^{(k,q)}}\frac{x_i^{(k,q)}}{|\vec{x}_i|}* H_i^{(k)}[p,q]\mathbb{1}_{\{x_i^{(k,q)}>0\}}$$
$$+ \sum_{i=1}^{N}\sum_{k=1}^{K}\sum_{j=1}^{N}\sum_{l=1}^{K}\sum_{p=1}^{P}\mu_i^{(k,p)}\frac{\rho_i^{(k,p)}}{\rho_j^{(l,1)}}* H_i^{(k)}[p,0]P_{i,j}^{+(k,l)}\frac{x_j^{(l,1)}}{|\vec{x}_j|}\mathbb{1}_{\{x_j^{(l,1)}>0\}}$$
$$+ \sum_{i=1}^{N}\sum_{k=1}^{K}\sum_{p=1}^{P}\sum_{l=1}^{K}\Lambda_i^{-}\alpha_i^{(k,p)}\rho_i^{(k,p)}* R_i^{(k,l)}\frac{x_i^{(l,1)}}{\rho_i^{(l,1)}|\vec{x}_i|}\mathbb{1}_{\{x_i^{(l,1)}>0\}} \quad (22)$$

First remark that $\frac{x_i^{(k,p)}}{|\vec{x}_i|}\mathbb{1}_{\{x_i^{(k,p)}>0\}} = \frac{x_i^{(k,p)}}{|\vec{x}_i|}\mathbb{1}_{\{|\vec{x}_i|>0\}}$. Thus we can simplify the left hand side of the equation.

Now consider the right hand side. We exchange some indices to simplify the expressions:

- queues indices i and j in the fourth term
- class indices k and l in the fourth and the fifth terms
- and phases indices p and q in the third and fourth terms.

It is now easy to factorize the first term with the term and the fifth term. We obtain:

$$\sum_{i=1}^{N}\sum_{k=1}^{K}\frac{x_i^{(k,1)}}{|\vec{x}_i|\rho_i^{(k,1)}}\mathbb{1}_{\{|\vec{x}_i|>0\}}\Big[\lambda_i^{(k)}+ \sum_{j=1}^{N}\sum_{l=1}^{K}\sum_{q=1}^{P}\mu_j^{(l,q)}\rho_j^{(l,q)}H_j^{(l)}[q,0]P_{j,i}^{+(l,k)}+ \sum_{p=1}^{P}\sum_{l=1}^{K}\Lambda_i^{-}\alpha_i^{(l,p)}\rho_i^{(l,p)}R_i^{(l,k)}\Big]$$

Now we remember that:

$$\Delta_i^{k,1} = \sum_{p=1}^{P}\sum_{l=1}^{K}\Lambda_i^{-}\alpha_i^{(l,p)}\rho_i^{(l,p)}R_i^{(l,k)}$$

and,

$$\nabla_i^{k,1} = \sum_{j=1}^{N}\sum_{l=1}^{K}\sum_{q=1}^{P}\mu_j^{(l,q)}\rho_j^{(l,q)}H_j^{(l)}[q,0]P_{j,i}^{+(l,k)}$$

We can now substitute all these relations in the balance equation.

$$\sum_{i=1}^{N}\sum_{k=1}^{K}\lambda_i^{(k)} + \sum_{i=1}^{N}\sum_{k=1}^{K}\sum_{p=1}^{P}\Big[\mu_i^{(k,p)}+\Lambda_i^{-}\alpha_i^{(k,p)}\Big]* \frac{x_i^{(k,p)}}{|\vec{x}_i|}\mathbb{1}_{\{|\vec{x}_i|>0\}}$$
$$= \sum_{i=1}^{N}\sum_{k=1}^{K}\frac{x_i^{(k,1)}}{|\vec{x}_i|\rho_i^{(k,1)}}\mathbb{1}_{\{|\vec{x}_i|>0\}}* \Big[\lambda_i^{(k)}+\nabla_i^{k,1}+\Delta_i^{k,1}\Big]$$
$$+ \sum_{i=1}^{N}\sum_{k=1}^{K}\sum_{p=1}^{P}\mu_i^{(k,p)}\rho_i^{(k,p)}H_i^{(k)}[p,0]d_i^{(k)}$$
$$+ \sum_{i=1}^{N}\sum_{k=1}^{K}\sum_{p=1}^{P}\frac{x_i^{(k,p)}}{|\vec{x}_i|\rho_i^{(k,p)}}\mathbb{1}_{\{x_i^{(k,p)}>0\}}* \sum_{q=1}^{P}\mu_i^{(k,q)}\rho_i^{(k,q)}H_i^{(k)}[q,p] \quad (23)$$

Taking into account equations 4 to 7, all the state-dependent terms cancel and we obtain:

$$\sum_{i=1}^{N}\sum_{k=1}^{K}\lambda_i^{(k)} = \sum_{i=1}^{N}\sum_{k=1}^{K}\sum_{p=1}^{P}\mu_i^{(k,p)}\rho_i^{(k,p)}H_i^{(k)}[p,0]d_i^{(k)} \quad (24)$$

This equation is a flow equation. Indeed the l.h.s. is the flow of positive customers entering the system while the r.h.s represents the customers leaving the network. This is formally proved in the next section.

9.1 Proof of the f ow equation

Let us now consider Equations 4 to 7 to prove formally that equation 24 is a flow equation. Multiply $\rho_i^{(k,p)}$ and $\rho_i^{(k,1)}$ by the denominator in equation 7 or 4 and make the summation for all p.

$$\sum_{p=1}^{P}\rho_i^{(k,p)}(\mu_i^{(k,p)}+\Lambda_i^{-}\alpha_i^{(k,p)})$$
$$= \lambda_i^{(k)} + \sum_{p=1}^{P}\sum_{q=1}^{P}\mu_i^{(k,q)}\rho_i^{(k,q)}H_i^{(k)}[q,p] + \sum_{p=1}^{P}\sum_{l=1}^{K}\Lambda_i^{-}\alpha_i^{(l,p)}\rho_i^{(l,p)}R_i^{(l,k)} + \sum_{j=1}^{N}\sum_{l=1}^{K}\sum_{q=1}^{P}\mu_j^{(l,q)}\rho_j^{(l,q)}* H_j^{(l)}[q,0]P_{j,i}^{+(l,k)}. \quad (25)$$

Note that for all i, k, q we have:

$$\sum_{p=1}^{P}H_i^{(k)}[q,p] = 1 - H_i^{(k)}[q,0].$$

Thus equation 25 becomes after substitution and cancellation :

$$\sum_{p=1}^{P} \rho_i^{(k,p)} \Lambda_i^- \alpha_i^{(k,p)} \quad + \\ \sum_{q=1}^{P} \rho_i^{(k,q)} \mu_i^{(k,q)} H_i^{(k)}[q,0] \quad = \\ \lambda_i^{(k)} \quad + \quad \sum_{p=1}^{P}\sum_{l=1}^{K} \Lambda_i^- \alpha_i^{(l,p)} \rho_i^{(l,p)} R_i^{(l,k)} \quad + \\ \sum_{j=1}^{N}\sum_{l=1}^{K}\sum_{q=1}^{P} \mu_j^{(l,q)} \rho_j^{(l,q)} H_j^{(l)}[q,0] P_{j,i}^{+(l,k)}. \tag{26}$$

We now sum over all values of k and we take into account that for all l we have $\sum_{k=1}^{K} R_i^{(l,k)} = 1$. We cancel some terms.

$$\sum_{k=1}^{K}\sum_{q=1}^{P} \rho_i^{(k,q)} \mu_i^{(k,q)} H_i^{(k)}[q,0] \quad = \\ \sum_{k=1}^{K} \lambda_i^{(k)} \quad + \\ \sum_{k=1}^{K}\sum_{j=1}^{N}\sum_{l=1}^{K}\sum_{q=1}^{P} \mu_j^{(l,q)} \rho_j^{(l,q)} H_j^{(l)}[q,0] P_{j,i}^{+(l,k)}. \tag{27}$$

We now sum over all values of queue index i and we remember that for all j and l, we have:

$$\sum_{i=1}^{N}\sum_{k=1}^{K} P_{j,i}^{+(l,k)} = 1 - d_j^{(l)}.$$

After substitution in Equation 27 and cancellation, we get:

$$\sum_{i=1}^{N}\sum_{k=1}^{K} \lambda_i^{(k)} \quad = \\ \sum_{j=1}^{N}\sum_{l=1}^{K}\sum_{q=1}^{P} \rho_j^{(l,q)} \mu_j^{(l,q)} H_j^{(l)}[q,0] d_j^{(l)},$$

which is the same as Equation 24. Thus it is a flow equation which is consistent with Equations 4 to 7

Mean-field Analysis of Data Flows in Wireless Sensor Networks

Marcel C. Guenther
Imperial College London
180 Queen's Gate
London SW7 2RH, United Kingdom
mcg05@doc.ic.ac.uk

Jeremy T. Bradley
Imperial College London
180 Queen's Gate
London SW7 2RH, United Kingdom
jb@doc.ic.ac.uk

ABSTRACT

Wireless Sensor Networks (WSNs) are often used for environment monitoring, an application which requires reliable routing of messages from source to sink nodes via multi-hop networks. Prior to installing such WSNs, engineers commonly analyse the network using discrete event simulation (DES). Whilst sophisticated simulators such as Castalia and TOSSIM take into account many low-level features of WSNs, their biggest drawback is the lack of scalability. This inhibits design-time system optimisation for large or complex networks. In this paper, we discuss how Population CTMC (PCTMC) models, used in conjunction with mean-field analysis, can be used to mitigate this problem. To illustrate the potential of PCTMC models in the WSN domain, we present a PCTMC model for a failsafe, dynamic routing protocol, which we implemented in Castalia. We show that the mean-field solution for the model yields good qualitative agreement with corresponding low-level simulations, but at a fraction of the computational cost. In particular we see good agreement for average metrics describing buffer occupancy and data flow behaviour. Moreover, our PCTMC model produces good results when packets are lost due to channel interference, an important consideration for WSNs.

Categories and Subject Descriptors

C.4 [**Performance of systems**]: Modeling techniques

General Terms

Application of modelling formalisms

Keywords

PCTMC, Mean-field analysis, Spatial modelling

1. INTRODUCTION

With the increasing availability of inexpensive wireless sensor hardware, the number of application areas for wireless sensor networks (WSNs) has continually grown over the last couple of years. One such application is environment monitoring, an application in which WSN nodes periodically capture local information such as temperature, humidity or luminosity and forward the information to a sink node via a multi-hop *ad hoc* network. In addition to periodical samples, some of these applications further allow event notification. This is beneficial, for instance, in the case of forest fire being detected by an environment monitoring WSN or where an intruder has entered an area surveilled by a security monitoring WSN. In the literature there are numerous examples of WSNs for different applications such as fire detection [3], landslide detection [25] and agricultural area monitoring [30] as well as structural monitoring and controlling [32, 2].

While all of the above examples use periodic and event based communication patterns, the requirements in terms of energy efficiency, reliability or throughput vary considerably between different deployments. An outdoor application for instance usually requires better energy efficiency as battery replacement is more expensive in locations that are difficult to access. An indoor intrusion detection system on the other hand may require low latency and high throughput to transmit pictures of an intruder. As a consequence, many MAC and network protocols have been developed for WSNs over the last decade in order to cater for a large variety of Quality of Service (QoS) demands [4]. However, it is difficult to decide which protocol to use for a given application, as the performance depends on the WSN topology and the configuration of the protocol stack. A recent study [10] came to the conclusion that protocol comparison literature generally provides insufficient information regarding experimental setups. Moreover, the study claims that too many protocol comparisons do not adequately tune protocols for their test environment. Being left with a range of candidate protocols for a specific application, engineers need to test a large number of different protocol stacks with different parameter configurations in order to optimise their WSN before deployment.

The most widely used low-level WSN simulators for this task are TOSSIM [21], Castalia [6] (an extension to the Omnet++ [20] simulation environment) and ns-2/3 [22]. These tools provide state-of-the-art models for simulating wireless networks, enabling engineers to test complete protocol stacks under fairly realistic conditions [5]. The major downside of these discrete event simulation (DES) frameworks, however, is the lack of scalability, so that the analysis of large wireless networks with hundreds of nodes becomes computationally expensive [8]. This makes it virtually impossible to test a large range of protocol stacks and con-

ICPE'13, March 21–24, 2013, Prague, Czech Republic.

figurations in an effort to optimise the software for an application. Population CTMC (PCTMC) analysis methods potentially provide an alternative to DES when investigating particular aspects of WSNs. PCTMCs are continuous time Markov chains that keep track of the evolution of populations over time. Their main advantage is that the population moments of the underlying stochastic process, such as mean and variance of the populations, can be approximated by ordinary differential equations (ODEs) [18, 19]. This efficient form of analysis, which is also referred to as mean-field analysis [23], has made PCTMCs a popular paradigm for modelling large client-server systems. Naturally one of the major constraints of PCTMCs is that the delay between any two events must be negatively exponentially distributed or alternatively belong to a distribution which can be approximated by a phase-type distribution. In the past this may have deterred some engineers from using PCTMC models to represent WSN protocols, which often exhibit deterministically timed behaviour. As a consequence PCTMC-style models for large WSNs are hardly covered in the literature[1]. Nevertheless, a promising example of a PCTMC-style model was given in [7], where the authors showed how mean-field analysis can be used to investigate the swarm intelligence based network reorganisation of a failsafe network originally described in [24]. Another example was the use of PCTMC-style models to investigate different sleep policies for WSN nodes [13]. In [16] we reviewed existing PCTMC WSN modelling techniques, presented the simple interference-free PCTMC WSN model that is reviewed in Section 3.2 and discussed open challenges. Moreover, in [15] we showed that for the resulting models, recently developed mean-field analysis tools are capable of efficiently approximating higher-order statistics for various WSN specific metrics. This is particularly useful when analysing transient behaviour of WSNs during network topology changes.

Though self-contained, this paper builds upon the work presented in [15, 16]. The contributions of this paper are as follows. In Section 3.3 this paper addresses the missing representation of channel interference in the simple PCTMC model described in [16] (cf. Section 3.2). This may appear to be a mere incremental improvement, however, it is generally a non-trivial problem to incorporate and analyse interference in mathematical models with complex topologies. Moreover, interference representation is an essential feature that is required if we aim to promote PCTMC modelling as a tool in the engineering community. In addition to this, we show that mean-field analysis is a tractable, fast solution technique for PCTMC WSN models even when simultaneously considering dynamic routing, message exchange and interference effects. Finally, in contrast to [15, 16] this paper compares the steady-state mean-field results of our PCTMC models with the simulation results obtained from a low-level Castalia protocol stack implementation and shows that the PCTMC model can be seen as a limiting case when message sampling and forwarding rates in the network become large.

Our paper is organised as follows; in Section 2 we provide background on PCTMC and spatial PCTMC modelling and evaluation techniques. In particular we introduce the so-called mean-field ODE analysis technique [23]. Section 3 subsequently describes a Castalia and a PCTMC model implementation of a data gathering WSN with dynamic routing. In Section 4 we then compare the data flow results from the Castalia simulation and the PCTMC ODE analysis for different topologies and communication constraints. In particular we compare our model to the Castalia simulation under both ideal and more realistic wireless channels. Finally, Section 5 concludes our work and suggests further research opportunities.

[1]By PCTMC-style we mean models that have been expressed in any formalism for which a lossless mapping to PCTMCs exists.

2. BACKGROUND

In this section we first introduce the notion of population modelling, which has inspired the development of population CTMCs (PCTMCs). Subsequently, we formally review PCTMCs and the mean-field moment approximation method for PCTMCs. We then introduce some extra notation for PCTMCs to simplify the representation of spatially distributed populations, which will later be used for describing our WSN model. Thereafter, we look at basic WSN concepts and terminology.

2.1 PCTMCs

In population models it is assumed that a large number of identical individuals belonging to a particular population interact with individuals from other populations. While we generally assert that individuals within a population have the same behaviour, this does not prevent populations of different individuals from interacting with one another. Interactions either result in a proportion of individuals changing their state locally or in births and deaths of individuals. The abstraction from individuals to populations can help to reduce the complexity and the state-space of the underlying model. Moreover, in the case of CTMCs, PCTMCs arise from a lumpable state-space, which can be efficiently analysed using mean-field techniques discussed below. Common examples of population models are chemical reaction models [12], where populations represent molecule concentrations, ecology models [31] describing the behaviour of groups of animals or plants and software performance models [29] capturing the interactions between components in massively parallel systems.

Population Continuous-Time Markov Chains (PCTMCs) consist of a finite set of populations S, $n = |S|$ and a set E of transition classes. States are represented as an integer vector $\vec{P}(t) = (P_1(t), \dots, P_n(t)) \in \mathbb{Z}^n$, with the ith component being the current population level of species $S_i \in S$ at time t. A transition class $(r_e, \vec{c}_e) \in E$ for an event e describes a transition with negatively exponentially distributed delay D at rate $r_e : \mathbb{Z}^n \to \mathbb{R}$ which changes the population vector $\vec{P}(t+D)$ to $\vec{P}(t) + \vec{c}_e$. The analogue to PCTMCs in the systems biology literature are Chemical Reaction Systems, were $\vec{P}(t)$ describes a molecule count vector and transition classes represent chemical reactions between the molecules with r_e being the reaction rate function and $\vec{c}_e$ the stoichiometric vector for a specific reaction. For notational convenience we write an event/reaction e as

$$\underbrace{S_* + \cdots + S_*}_{\text{in}} \to \underbrace{S_* + \cdots + S_*}_{\text{out}} \qquad \text{at } r_e(\vec{P}(t)) \quad (1)$$

where $S_* \in S$ represent different species that are involved in the event. The corresponding change vector is $\vec{c}_e = (s_1^{\text{out}} - s_1^{\text{in}}, \dots, s_n^{\text{out}} - s_n^{\text{in}}) \in \mathbb{Z}^n$ where s_i^{in} represents the number of occurrences of a particular species $S_i \in S$ on the left hand side of the event and s_i^{out} its number of occurrences on the right hand side. The event rate is

$$\begin{cases} r_e(\vec{P}(t)) & \text{if } P_i(t) \geq s_i^{\text{in}} \text{ for all } i = 1, \ldots, n \\ 0 & \text{otherwise} \end{cases} \quad (2)$$

An important feature of PCTMC models is that approximations to the evolution of population moments of the underlying stochastic process can be represented by the following system of ODEs [18]

$$\frac{\mathrm{d}}{\mathrm{d}t}\mathbb{E}[T(\vec{P}(t))] = \sum_{e \in E} \mathbb{E}[(T(\vec{P}(t)+\vec{c}_e)-T(\vec{P}(t)))r_e(\vec{P}(t))] \quad (3)$$

To obtain the ODE describing the evolution of the mean of a population, all we need to do is to substitute $T(\vec{P}(t)) = P_i(t)$ in the above equation, where $P_i(t)$ is the random variable representing the population count of species S_i at time t. In the literature the resulting ODEs are often referred to as mean-field approximations [23]. Similarly ODEs for higher joint moments can be obtained by choosing adequate $T(\vec{P}(t))$, e.g. $T(\vec{P}(t)) = (P_i(t) - \mu_i)^2$ for the variance of $P_i(t)$. Alternatively stochastic simulation [12] can be used to evaluate PCTMCs. Like discrete event simulation for low-level protocol models, this latter simulation technique captures the stochastic behaviour of PCTMCs exactly, but does not scale for models with larger populations.

2.2 Spatial PCTMCs

When modelling spatially distributed networks, it is often easier to use Spatial PCTMCs (SPCTMCs). SPCTMCs have a discrete, finite number of locations each with a finite number of population for different agent states. When evaluating an SPCTMC we keep track of the evolutions of all agent state populations in all locations. Aside from helping modellers to describe spatial processes more easily, the extra spatial information can occasionally be used to speed-up higher-order moment ODE analysis [27]. When describing SPCTMCs we usually first describe local agent behaviour. Later a composite model which describes topology, initial agent state populations for all locations and interactions between neighbouring populations is created. For the remainder of this paper we refer to a species S at location l at time t as $S@l(t)$ [11]. Moreover, $S@l_*(t)$ is used as a placeholder for defining events that occur in all locations.

2.3 Wireless sensor networks

Nodes in WSNs are small, embedded, battery powered devices with significant processing, bandwidth, radio and energy constraints [1]. The radio range heavily depends on the environment in which the network is deployed [26]. As for bandwidth, nodes such as the MicaZ can transmit up to 250 Kbps [1], although in many applications the actual throughput is much lower because of channel contention and other communication overheads. Since many types of nodes are battery powered, energy has to be used efficiently. To overcome these QoS related challenges a vast number of protocols have been suggested over the last decade [34], each of which aims to optimally balance different QoS constraints.

Figure 1 gives a high-level overview over the basic software architecture of wireless sensor applications. A more detailed representation can be found in [33]. The **Application layer** contains the logic required for data acquisition and data processing. A simple application might measure quantities such

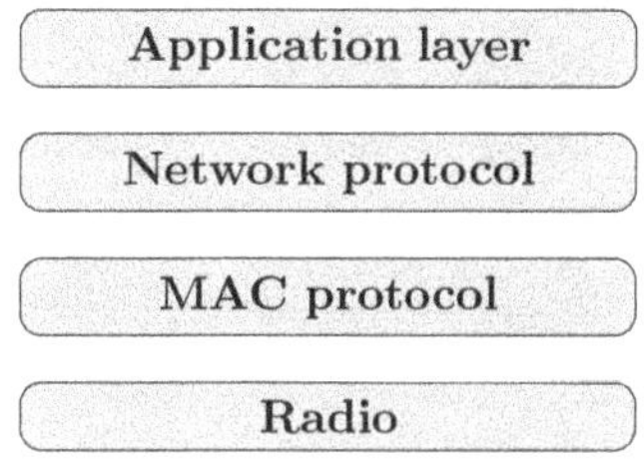

Figure 1: A simple wireless sensor network protocol stack [16].

as temperature, humidity or luminosity in regular intervals and forward the data to a sink node. Other applications might also process measured data, serve data requests or send messages in response to external events. Moreover, applications need to decide which nodes to forward their data to. This can either be specific nodes or a high-level destinations such as data sinks. The **Network layer** [34] is responsible for ensuring that data from the application layer is routed towards its destination. A common communication pattern is *convergecast*, where all nodes in the network sample information and forward the data to dedicated sink nodes via multi-hop routes. In multi-hop networks, routing protocols therefore need to relay incoming packets from other nodes in addition to handling packets coming from their own application layer. Network protocols are either centralised or decentralised. A centralised routing protocol elects one or several nodes which control the routing behaviour of the network, whereas decentralised protocols let nodes decide autonomously where to forward messages to. Protocols in the latter category are sometimes referred to as swarm intelligence based protocols [24]. **MAC layer** protocols on the other hand, control the communication with actual neighbouring nodes. Additionally, MAC-protocols are in charge of managing the node's duty-cycle behaviour. When duty-cycled, nodes turn off their radio units according to a sleep policy. As the name suggests this happens in a cyclic manner such that over a period T a node has its radio turned on x% of the time. The lower x, the longer the network lifetime will be. There are two classes of MAC protocols, contention based protocols and schedule based protocols. In contention based protocols such as CSMA (*Carrier Sense Multiple Access*), nodes can send messages at any time provided the channel is clear. On the other hand schedule based protocols like TDMA (*Time Division Multiple Access*) allocate a specific time window to each node, during which it can transmit messages [4]. Moreover, many hybrid protocols, which use elements of both contention and schedule based MAC protocols, have been suggested. Finally the **Radio layer** controls nodes' radio hardware and can be used to configure signal modulation, frequency or transmission power.

3. A DYNAMIC WSN DATA FLOW MODEL

In the following we outline a failsafe WSN routing protocol, i.e. a protocol that can automatically detect failing nodes and dynamically close gaps in multi-hop routes, similar to the ones described in [15, 16]. Our analysis focuses on the data flow from source to sink nodes in different topologies. Figure 2 shows a simple example of a WSN. Each node has up to four neighbours and communication is done in a

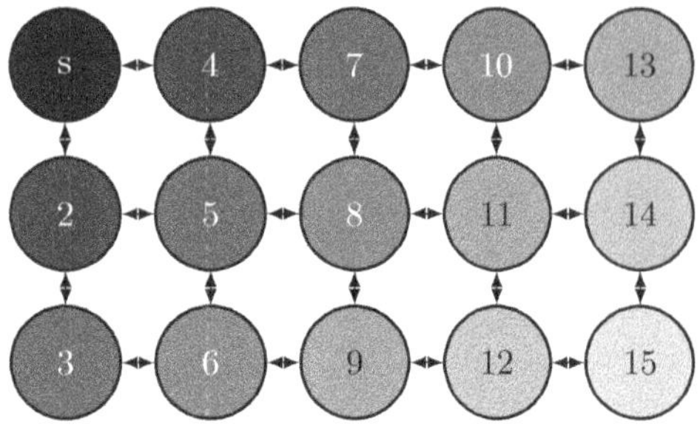

Figure 2: Node 's' is the sink and has the highest pheromone and buffer level. Both levels decrease with increasing hop distance from the sink [16].

Manhattan style pattern, i.e. we assume a unit disk radio model. Note that this is only for illustration purposes, in practice we could incorporate less ideal radio assumptions. We further assume that the grid setup is *ad hoc*, so that no node knows its position in the network in advance. However, some nodes are sink nodes to which all other nodes forward messages, possibly via intermediate relay nodes. To allow nodes to route messages towards the sink, each sink in the network sends a beacon message in regular intervals containing a pheromone value. From zoology, pheromone is a hormone laid down by colony-based insects, to indicate popular routes to food sources or new nest sites. In a similar manner pheromone gradients have been adapted in the WSN literature as an abstract means of studying the evolution of routes from source to sink nodes [7]. Upon reception of a pheromone beacon, a node updates its own pheromone level and broadcasts it. So long as the pheromone level update function generates levels that are strictly lower than the largest level received through incoming beacons, a pheromone gradient will emerge that is highest at sink nodes and lowest at nodes that are farthest away from any sink. Each node can subsequently relay messages to the sink via neighbours that have higher pheromone levels than itself, knowing that the message thereby travels closer to a sink. If nodes fail, they will stop emitting pheromone beacons and neighbours will hence learn to avoid them until they are back online.

While such self-organising routing protocols are preferred over centralised protocols in failsafe WSNs, it can happen that pheromone based routing decisions produce unwanted congestion in the network. During our model analysis we want to study the impact of the pheromone based routing approach on the buffer level of different nodes in the network. To do this we will measure the average buffer occupancy at each node and find out whether dynamic pheromone based routing distributes load in a fair manner. Naturally nodes closer to sink nodes usually have a higher buffer occupancy than more distant nodes, since they need to relay more messages to the adjacent sink. However, in large WSNs with multiple sink nodes we would like to ensure that all nodes that are at a certain distance from the sink will experience similar message loads. If this is the case then we can consider our network fair and well-structured. In the next section we describe our Castalia implementation of the failsafe WSN, followed by our corresponding PCTMC model.

3.1 Castalia protocol model

Our Castalia simulation consists of the following protocol stack. The application layer is a simple sensor application which periodically generates sensor readings, say at *sampleRate* samples per second and forwards them to a sink node. Our routing layer is based on concepts similar to those presented in [24], however, in our protocol the routing decisions are probabilistic. For the MAC layer our network uses CSMA. However, we assume that after having sent a message, nodes wait for a fixed period of time before sending the next message and listen for incoming messages in the meantime. Naturally, in a live application we would use a more energy efficient MAC protocol. While beyond the scope of this paper, one of our future research goals is the integration of realistic duty-cycle behaviour (cf. Section 2.3) into our mathematical WSN models, possibly using deterministic delays such as the ones suggested in [17].

Our pheromone network protocol implementation is as follows. At a regular time interval, every node broadcasts a pheromone beacon to all its neighbours. A neighbour that receives this message will store it in a neighbourhood table, along with the *id* of the sending node. If the pheromone level contained in the beacon is larger than the receiving node's current pheromone level, the receiver will update its pheromone level. If node i receives a beacon from node j and if the pheromone level of j is higher than that of node i, then

$$ph@l_i = (ph@l_i + ph@l_j)/2 \tag{4}$$

where $ph@l_*$ are floating point values stored by each individual node[2]. To ensure that pheromone evaporates over time, a non-sink node i always decreases its pheromone level before broadcasting its own pheromone beacon

$$ph@l_i = ph@l_i - \sqrt{ph@l_i} \tag{5}$$

Sink nodes, however, being a source of pheromone, have constant pheromone levels. Given its own pheromone level and the levels of its neighbours, a node can make informed routing decisions for application layer packets. We assume that in regular intervals of x milliseconds a node attempts to forward a message from its buffer to the next link in the multi-hop route. x is chosen so that $1000msec/x$ yields the desired number of *fwRate* messages a node can send every second. We assume that this number is much lower than the physical bandwidth of the communication channel such that no message has to be discarded. Buffers contain both messages from the node's own application layer as well as messages from other nodes that it needs to forward to the sink. To decide which node to relay a message to, nodes use a probabilistic routing approach. Let $T_i = \{ph@l_1, \ldots, ph@l_m\}$ be the set of all pheromone levels $ph@l_j$ of neighbours of node i such that $ph@l_j > ph@l_i$, i.e. T_i contains the pheromone levels of all neighbouring nodes of i that are closer to a sink. The probability that node i routes a message via node k then becomes

$$\mathbb{P}(send\ msg\ to\ k) = ph@l_k \ / \sum_{ph@l_j \in T_i} (ph@l_j - ph@l_i) \tag{6}$$

which is a discrete distribution we can easily sample from. When simulating the protocol stack in Castalia we regularly

[2]These should not be confused with species labels $Ph@l_i$ used in Section 3.2, which are part of the PCTMC model definition.

measure the number of messages in every node's buffer. For sink locations we assume that messages are processed immediately and thus we are not concerned about their buffer status. The source code of the resulting simulation as well as detailed simulation parameter configurations can be found on our website [14]. Later, when we discuss our Castalia simulation results we will distinguish between collision free communication and the more realistic collision model that simulates channel interference [6]. Moreover, we assume a sharp noise threshold for the radio reception as well as constant radio signal strength to recreate the unit disk effect.

3.2 PCTMC protocol model

This section reviews the interference-free version of our PCTMC model, which was taken from [16]. Our PCTMC model has been implemented in a stochastic process algebra that is part of the GPA-analyser tool [28]. For illustration purposes, we will represent the core elements of the model using reactions in the style of Eq. (1). In the following we implicitly assume that random variables are time-dependent and thus drop the (t) parameter used in Eq. (1). To avoid confusion between a species label and the random variable that describes the population of a given species we use $State@l_i$ and $\#State@l_i$ respectively.

We will deal with 3 types of nodes: functioning sink and non-sink nodes as well as broken nodes. Non-sink nodes forward their data samples to sink nodes. When a node stops communicating with other nodes it is considered broken. Each node is further assumed to have a finite buffer that can hold up to m messages. The buffer holds a node's own samples as well as messages that it needs to relay. We start by describing the PCTMC events for working sink nodes and non-sink nodes. Clearly, broken nodes do not interact with other nodes and thus we do not need to represent them in the model. The reactions for the pheromone spread for non-sink nodes are as follows

$$\begin{array}{lll} \emptyset & \rightarrow Ph@l_* & \text{at } pheroIncRate@l_* \\ Ph@l_* & \rightarrow \emptyset & \text{at } pheroDecRate@l_* \end{array} \quad (7)$$

where $pheroIncRate@l_*$ is the sum of the difference between a node's pheromone level and that of its neighbours. For sink nodes the pheromone population level remains constant, so no events need to be defined. Assuming we have a neighbourhood structure as shown in Figure 2, we would obtain the following reaction rates for the pheromone changing events of the node at location 3

$$\begin{aligned} pheroIncRate@l_3 = c_1{}^* \, [(&\max(0, \#Ph@l_2 - \#Ph@l_3)+ \\ &\max(0, \#Ph@l_6 - \#Ph@l_3)] \end{aligned} \quad (8)$$

where c_1 is a constant that regulates the rate at which pheromone levels change. In general $pheroIncRate@l_*$ is defined as

$$\sum_{l_j \in neighbours\ of\ l_*} c_1{}^* \max(0, \#Ph@l_j - \#Ph@l_*) \quad (9)$$

Moreover,

$$pheroDecRate@l_* = c_2{}^* \min(0.1, \#Ph@l_* - 2) \quad (10)$$

describes the rate at which pheromone evaporates, i.e. the PCTMC analogue of Eq. (5), with c_2 being another constant that is chosen so that the pheromone spread is quite linear. The min term ensures that pheromone levels will remain positive. Although this pheromone gradient only encodes a node's distance from the sink, it is generally possible to incorporate other neighbourhood information, such as buffer levels or battery status, in the pheromone concentration in case further QoS constraints have to be met by the protocol. Figure 3 illustrates the pheromone mean-field results when applying the above reactions to the topology shown in Figure 2 for a given initial pheromone population at sink location l_1.

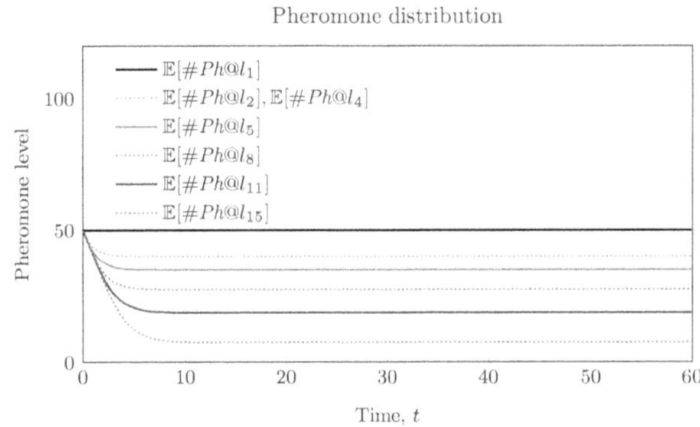

Figure 3: Pheromone distribution in the network depicted in Figure 2.

Having described PCTMC events governing the dynamics of pheromone agents $Ph@l_*$, we now need to represent the routing probability distribution used by the Castalia protocol stack (cf. Eq. (6)). To avoid reactions that have fractions of populations in their reaction rates, we use a routing species which can be used to infer routing probabilities from pheromone levels. Take for instance the node at location l_6, which can route up messages to location l_5, route left to l_3 and route right to l_9. As in Eq. (6) we want nodes with higher pheromone levels to be used more frequently. Hence, we need to increase the proportion of messages sent to the left, as the pheromone level of node 3 becomes larger than that of nodes 5 and 9

$$\begin{aligned} &RouteU@l_6 \rightarrow RouteL@l_6 \\ &\quad \text{at } \#RouteU@l_6{}^* \max(0, \#Ph@l_3 - \#Ph@l_6) \\ &RouteR@l_6 \rightarrow RouteL@l_6 \\ &\quad \text{at } \#RouteR@l_6{}^* \max(0, \#Ph@l_3 - \#Ph@l_6) \end{aligned} \quad (11)$$

where $\max(0, \#Ph@l_3 - \#Ph@l_6)$ is the pheromone excess of node 3 over 6. Clearly, if location 3 has a lower pheromone level than location 6, node 6 will not route messages via node 3 since all transitions that increase $\#RouteL@l_6$ have a 0 reaction rate. If node 3 has a higher pheromone level, the population $\#RouteL@l_6$ will be non-zero and thus a proportion of messages from node 6 is relayed to the sink via location 3. All reactions for $RouteU@l_*$, $RouteD@l_*$, $RouteL@l_*$ and $RouteR@l_*$ follow the same pattern. We define the constant c_3 to be a normalising constant that allows us to compute the proportion of messages that are forwarded to a particular neighbour

$$\begin{aligned} c_3 = \ & \#RouteU@l_* + \#RouteD@l_* + \\ & \#RouteL@l_* + \#RouteR@l_* \end{aligned} \quad (12)$$

We then use the resulting normalised distribution to model a similar routing behaviour as described in Eq. (6). The routing population reactions are the same for sink and non-sink nodes and broken nodes are automatically avoided due to their constant pheromone level of 0.

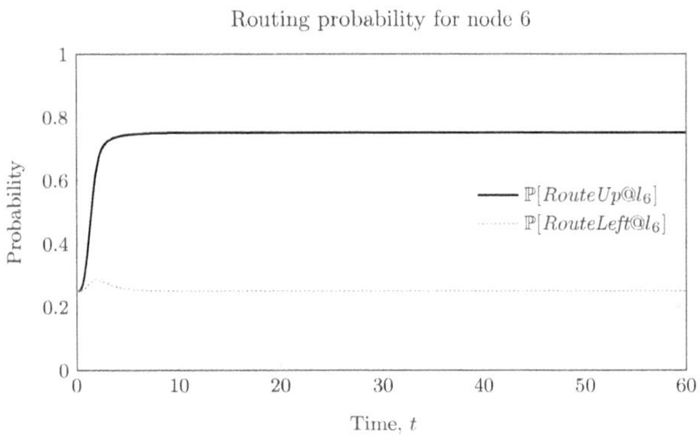

Figure 4: Routing distribution for node 6 in Figure 2.

To express the effects of dynamic routing in the message exchange, we need to feed back the routing information into the PCTMC events that describe message exchanges between nodes. For the message buffer we decided to represent different buffer levels as individual states in our PCTMC. A single node has buffer states $\{Buf_0, \ldots, Buf_m\}$, so that if a node is in state Buf_i its buffer contains i messages. In all locations we chose m to be sufficiently high, so state Buf_m would hardly ever be reached, i.e. we do not need to handle buffer overflow scenarios. Buffers are affected by two different types of reactions. The simpler one is the increase that occurs when the sensor of a non-sink node generates a new sample

$$Buf_i@l_* \rightarrow Buf_{i+1}@l_* \quad \text{at } \#Buf_i@l_* {}^* sampleRate \tag{13}$$

Change also occurs due to message exchange with neighbouring nodes. Take for example the message exchange between nodes 2 and 3

$$\begin{aligned} &Buf_i@l_2 + Buf_j@l_3 \rightarrow Buf_{i+1}@l_2 + Buf_{j-1}@l_3 \\ &\quad \text{at } \#Buf_i@l_2{}^*\#Buf_j@l_3{}^* \\ &\quad \#RouteU@l_3/c_3{}^*fwRate \\ &Buf_j@l_2 + Buf_i@l_3 \rightarrow Buf_{j-1}@l_2 + Buf_{i+1}@l_3 \\ &\quad \text{at } \#Buf_j@l_2{}^*\#Buf_i@l_3{}^* \\ &\quad \#RouteD@l_2/c_3{}^*fwRate \end{aligned} \tag{14}$$

where $0 \leq i < m$, $0 < j \leq m$. $fwRate$ represents the number of packets that each node can forward per time unit. Similar reactions can describe all other unicast message exchanges in the network. Moreover, we have

$$\sum_{i=0}^{m} \mathbb{E}[\#Buf_i@l_*] = 1 \tag{15}$$

In other words each node only has a single buffer instance. While sink nodes receive messages like non-sink nodes, we assume that they are connected to an uplink to which they immediately forward incoming messages

$$Buf_j@l_* \rightarrow Buf_{j-1}@l_* \text{ at } \#Buf_j@l_* {}^* 4 {}^* fwRate \tag{16}$$

For any node, the average number of elements in the buffer is given by

$$\sum_{j=1}^{m} j {}^* \mathbb{E}[\#Buf_j@l_*] \tag{17}$$

A complete implementation of the model can be found in the model files provided on [14].

3.3 Modelling channel interference

It is not hard to see that the simple WSN model shown in Section 3.2 is only accurate in the absence of channel interference. By channel interference, we mean interference that is caused by the hidden terminal problem. Imagine two nodes A and C that have a common neighbour B with which they can both communicate, while at the same time A and C cannot directly communicate. Before sending a message using CSMA, both nodes check whether the channel is currently occupied. If it is free they send a message, otherwise they try again later. The hidden terminal problem occurs when C assesses the channel while A is transmitting a message to B. Since C cannot sense A's signal it assumes that the channel is clear and sends data to B. However, B now receives two signals simultaneously and can at best demodulate one of them. This effect is generally unavoidable in WSNs. Unlike wired networks, WSNs share a limited frequency band and hence no physically separated channel can be guaranteed for any pair of senders and receivers. Although newer protocols utilise multiple frequencies to reduce the number of nodes sharing a particular frequency, messages sent over a particular frequency can still be received by other nodes in hearing distance. As a consequence, unicast and broadcast of small packets essentially work the same in WSNs, the only difference being that in unicast mode nodes discard messages that are not intended for them.

Modelling channel interference in PCTMCs is generally nontrivial as it requires us to introduce further state variables in order to keep track of every node's radio state. To add these extra states to the model without increasing model complexity unnecessarily, we found that further simplifying independence assumptions were required. While it was possible to accurately reflect correlations between senders and receivers in the PCTMC events for the idealised network presented in Section 3.2, this becomes a lot harder as we introduce interference effects. To faithfully represent interference effects in a PCTMC model we need to incorporate the state of all 1- and 2-hop neighbours of the receiver when expressing message exchange events like in Eq. (3.2). However, this would result in an extremely large number of complex reactions involving 10 or more species each. Fortunately, as we only intend to use fast first-order mean-field approximation techniques to analyse the model, simplifying assumptions can be made, which allow us to incorporate channel interference behaviour into our model without resorting to complex and verbose model descriptions.

While mean-field analysis often outperforms simulation even for moderate population sizes, it is important to bear in mind that although the number of first-order ODEs grows linearly in the number of PCTMC states, the number of moments required to approximate higher-order moments increases exponentially in the order of the highest moment that we require [18]. In practice this limits mean-field analysis of large WSNs to first-order moment approximations. However, depending on the nature of the reactions, we require both first- and higher-order moments just to obtain accurate first-order mean-field results [9, 14]. Technically this is the case whenever the PCTMC has reactions such as Eq. (14) where rates are non-linear in population terms. In such cases the only way to obtain a closed system of first-order approximating mean-field ODEs is to assume statistical independence between populations. Although accurate correlations are generally vital for models, empirical evidence suggests that for many PCTMC models the independence assumption required for first-order mean-field analysis often has a limited effect on the accuracy of the mean approximation [7, 13]. In the following we will therefore describe interference dynamics in our PCTMC model assuming statistical independence between populations. Naturally, the resulting simplified PCTMC model can only be used for a first order mean-field approximation of the semantically more accurate PCTMC model whose reactions capture all correlations.

For our interference model we assume that a node's radio is either in *RX*, *TX* or *Idle* state. When a node is *Idle* it can either receive a new message and switch to *RX* until the exchange has completed or decide to send a message and switch to *TX* for the duration of the transfer. Hence, a node can only send a new message when it is in *Idle* state. As we assumed the total bandwidth to be much larger than the number of messages a node attempts to forward every second (cf. Section 3.2), the introduction of the radio state will have no visible impact on the rate of the event defined in Eq. (3.2). The following reactions describe the local transition of a node's radio state

$$\begin{array}{lll} RX@l_* & \rightarrow Idle@l_* & \text{at } \#RX@l_*{}^*900 \\ TX@l_* & \rightarrow Idle@l_* & \text{at } \#TX@l_*{}^*900 \end{array} \tag{18}$$

such that the transmission of a single messages takes about 1/900 seconds. Moreover, we need to have

$$\#RX@l_* + \#TX@l_* + \#Idle@l_* = 1 \tag{19}$$

The following events describe state changes due to pheromone beacon broadcasts, assuming a single beacon is sent every second

$$\begin{array}{lll} Idle@l_* & \rightarrow RX@l_* & \text{at } \sum_{n \in N@l_*} \#Idle@l_* \\ Idle@l_* & \rightarrow TX@l_* & \text{at } \#Idle@l_* \end{array} \tag{20}$$

where $N@l_*$ is the set of 1-hop neighbours of the node. Similarly a node changes from *Idle* to *RX* when one of its neighbours is transmitting a message and to *TX* when sending a message itself

$$\begin{array}{l} Idle@l_* \rightarrow RX@l_* \\ \quad \text{at } \#Idle@l_*{}^* \sum_{n \in N@l_*} (1 - \#Buf_0@l_n)^* fwRate \\ Idle@l_* \rightarrow TX@l_* \\ \quad \text{at } \#Idle@l_*{}^*(1 - \#Buf_0@l_*)^* fwRate \end{array} \tag{21}$$

Having defined these transitions we can now easily describe interference as using the following reactions. The first one describes interference of data message transfers caused by pheromone beacons

$$\begin{array}{l} Buf_j@l_* \rightarrow Buf_{j-1}@l_* \\ \quad \text{at } \sum_{n \in N2@l_*} \#RX@l_*{}^*\#Buf_j@l_* \end{array} \tag{22}$$

where $0 < j \leq m$ and $N2@l_*$ is the set of all 1-hop and 2-hop neighbours of the receiving node. Although we assume that we generally cannot receive messages from 2-hop neighbours, transmission of messages by 2-hop neighbours can still cause interference. Similarly to the pheromone messages above, data messages sent by the receiving node's 1-hop and 2-hop neighbours may also result in message loss

$$\begin{array}{l} Buf_j@l_* \rightarrow Buf_{j-1}@l_* \\ \quad \text{at } \#RX@l_*{}^* \sum_{n \in N2@l_*} (1 - \#Buf_0@l_n)^* fwRate \end{array} \tag{23}$$

Finally, we need to add one more interference event which ensures that when 1-hop neighbours interfere with the node that they want to send a message to, two messages are lost

$$\begin{array}{l} Buf_j@l_* \rightarrow Buf_{j-1}@l_* \\ \quad \text{at } \#RX@l_*{}^* fwRate^* \\ \qquad \sum_{n \in N@l_*} (1 - \#Buf_0@l_n)^* \#Routel_*@l_n/c_3 \end{array} \tag{24}$$

where $\#Routel_*@l_n/c_3$ is the routing probability from node n to the node located at l_*.

4. ANALYSIS

In this section we compare the normalised buffer levels of nodes in a WSN running the failsafe application described in Section 3. The aim of our analysis and the way it is presented is to show that there is good qualitative agreement between the Castalia simulations results and their PCTMC mean-field approximations, even when the number of messages sent per second is low. If we can show such agreement with low-level WSN simulation results in a range of scenarios, we will have gained a measure of belief in the mean-field approach for spatial WSN analysis. To obtain the buffer levels at steady-state from Castalia, we first allowed the network a 60 second period to establish an initial pheromone gradient and subsequently measured the average buffer level at each node over a period of 240 seconds. For each configuration we then took the average over 50 such simulations to obtain the mean buffer level at each location. The resulting 95% confidence intervals for the mean buffer levels lay within < 3% for all locations. Similarly, we evaluated the mean buffer level using mean-field ODEs and subsequently took the average buffer occupancy at each location in the network after 60 seconds. Throughout our experiments we found that the absolute buffer levels predicted by the PCTMC and Castalia model were quite different. Most likely this is due to the simplifying assumptions that we made in order to create a PCTMC model that is suitable for fast ODE solution techniques. While this error is of interest for future work, in the following we only study the normalised mean buffer levels at steady-state. By normalised we mean that all locations are normalised by the non-sink node with the highest buffer level, which is then assumed to have a buffer level of 100%. We use the busiest non-sink node rather than the busiest sink node as we do not want to make additional assumptions about the uplink capacity of sink nodes. To visualise hot spots and idle locations we use spatial heat maps. In these heat maps a node with a relative buffer level of $x\%$ is coloured black at $x\%$ opacity. Sink nodes are coloured at 100% opacity and are marked 's', whereas broken nodes are marked 'x' and coloured white. In addition to these heat maps, we also deploy difference maps to visualise relative errors between the PCTMC and the Castalia models. To obtain these we simply subtract normalised PCTMC buffer levels from normalised Castalia ones for each location and divide by the normalised Castalia buffer levels. In contrast to the mean buffer level heat maps the difference maps show the resulting error in % for each location. While the heat maps support our claim that the

PCTMC and the Castalia results match qualitatively, i.e. relative differences between busy and idle locations are preserved, the difference plots capture the quantitative accuracy better. Unless stated otherwise we conduct the below experiments with a forward rate of up to 20 messages per second and message production rate of 1 message per second.

4.1 Ideal communication

First, we will investigate the Castalia and the PCTMC predictions for the data flow assuming ideal communication, i.e. without interference or radio signal strength variation. Figure 5 shows the results for a network with 49 nodes and 2 sinks. In both models the nodes under high load are the same, despite the fact that the average absolute buffer level in the Castalia level is roughly 50% higher than in the PCTMC model. Note, however, that the Castalia simulation heat map has a much higher contrast. In other words the actual magnitude of the difference between busy and idle locations approximated by the mean-field analysis is lower than the Castalia simulation results suggest.

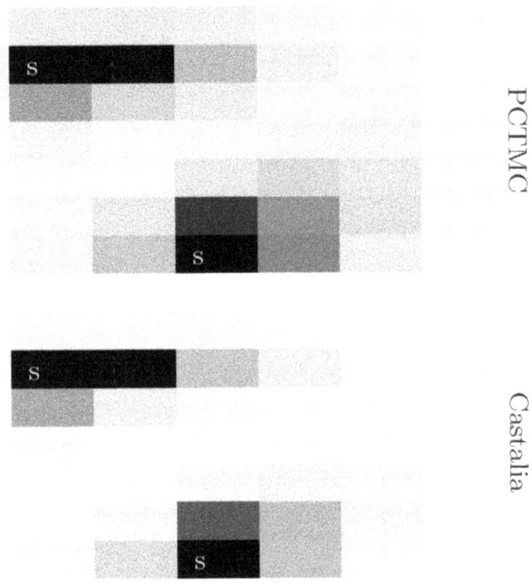

Figure 5: Data flow in a network with 49 nodes, 2 sinks, a forward rate of 20 messages per second and a message production rate of 1 message per second. The top diagram is based on the steady-state PCTMC mean-field approximation, the bottom one on the averaged Castalia simulation results under ideal radio conditions.

Our second example in Figure 6 illustrates the effect on normalised buffer levels as we increase the message production rate to 1.5 messages per second without changing the network topology. The increase in contrast in the heat map of the PCTMC model indicates that the relative difference between the buffer level of the busiest node compared to the other nodes has become larger. This is what we expected to happen, as the busiest nodes will naturally experience much higher buffer occupancy under load. In Section 4.2 we will show that interference slows down this effect since busier nodes start to lose more packets. The difference plot in Figure 6 shows that the results of the PCTMC mean-field analysis and the Castalia simulation model are much better in locations with high traffic than in areas of low traffic. This is not surprising since mean-field analysis is generally

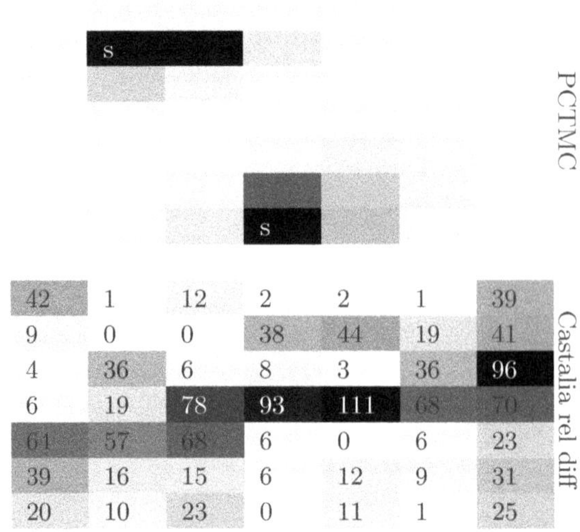

Figure 6: Data flow in a network with 49 nodes and 2 sinks with ideal communication and 1.5x message production rate.

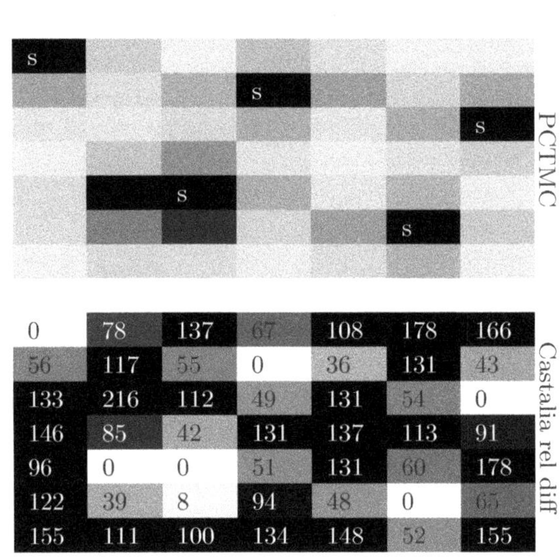

Figure 7: Data flow in a network with 49 nodes and 5 sinks with ideal communication.

most accurate when events rates, i.e. the message production and forwarding rate in this example, are high. If our PCTMC model truly represents the process described by the Castalia model, the error should thus decrease as we scale both message production and forwarding rate. To illustrate this effect, our next example looks at the effect on the error in a model with 49 nodes and 5 sinks as we scale the message production rate of 1 message per second and the forwarding rate of 20 messages per second (cf. Figure 7), by a factor of 10 (cf. Figure 8). As can be seen in the difference plots, the error between the PCTMC mean-field and the Castalia simulation results decreases drastically as the rates are increased. Therefore, in the ideal communication scenario the normalised PCTMC mean-field buffer levels can be seen as a limiting case for the network and the protocol when both forward and sampling rate are extremely high.

In practice, individual WSN nodes fail eventually, requir-

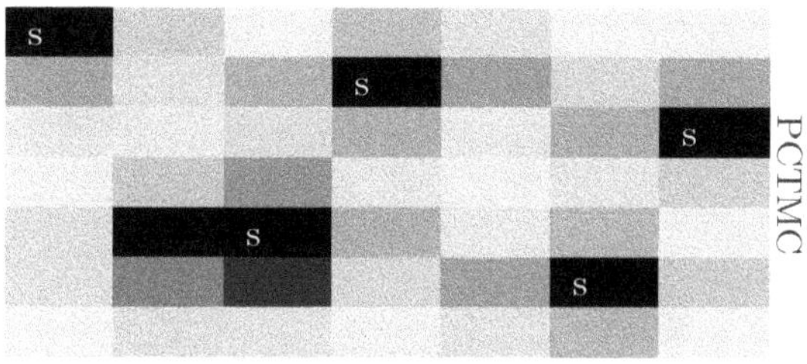

0	3	2	1	3	15	11	Castalia rel diff
3	4	4	0	3	16	1	
9	31	1	0	5	1	0	
2	7	3	8	2	5	3	
7	0	0	0	8	6	15	
16	7	4	1	4	0	6	
6	18	1	3	12	3	6	

Figure 8: Data flow in a network with 49 nodes and 5 sinks with ideal communication, 10x forward and message production rate.

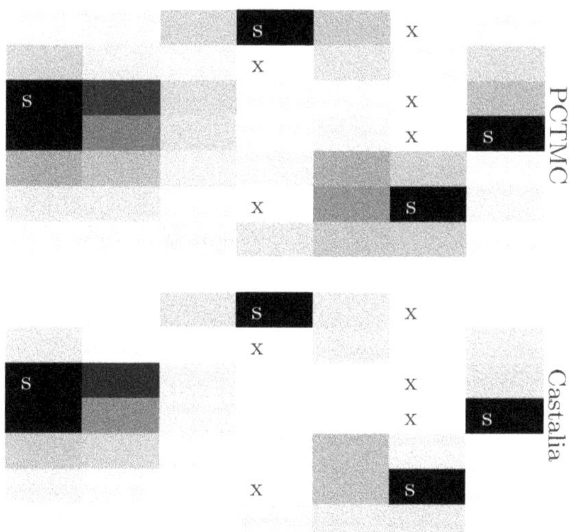

Figure 9: Data flow in a network with 49 nodes and 4 sinks and 5 broken nodes with ideal communication.

ing the network to reconfigure itself. In Figure 9, we can see that even in presence of broken nodes, both models produce qualitatively similar results. This confirms that the failsafe aspects of the pheromone model are captured well in both instances. Finally, in order to illustrate that our PCTMC model is indeed scalable, we created a large model with 100 nodes, several sinks and broken nodes. The data flow created during the subsequent analysis is shown in Figure 10. As can be seen even in this large network the ODE approximation produces useful data flow predictions.

4.2 Interference

Having compared at the Castalia simulation results with PCTMC mean-field predictions for the data flow in an idealised communication scenario, we now repeat the analysis taking signal interference into account. Apart from this, no changes were made with respect to the analysis. In particular we still assert fixed radio signal strength throughout our experiments. The unit disk model has been configured such that nodes can only detect a busy nodes if they are within their transmission distance, i.e. 1-hop neighbours. However, we assume that the signal strength of 2-hop neighbours is strong enough to cause interference.

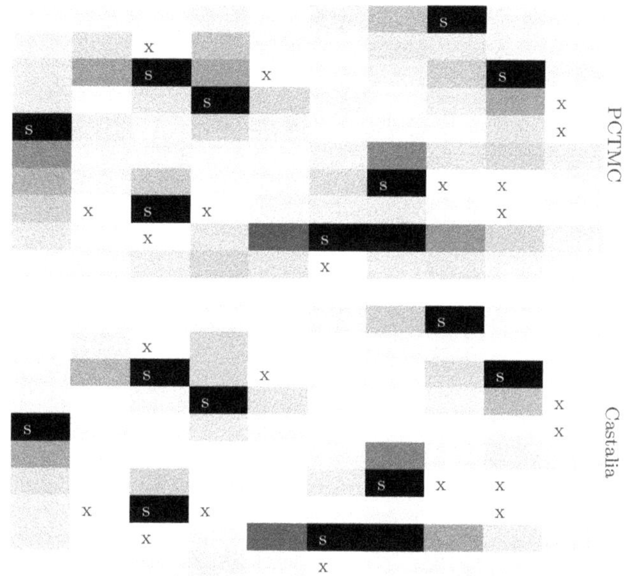

Figure 10: Data flow in a network with 100 nodes several sinks and broken nodes with ideal communication.

Interference is most likely to cause message loss in areas of high traffic, i.e. among nodes close to a sink. As a consequence we expect that under equal network load the relative differences between buffer levels become less extreme than in the examples shown in Section 4.1. Indeed, comparing the interference-free results shown in Figures 5, 6 and 10 with those in Figures 11, 12 and 13 we can see that the contrast of the heat maps is visibly lower when we model interference. In Figures 11, 12 and 13 it is also noticeable that the contrast of the cells in the PCTMC heat maps are lower than in the Castalia case, which implies that the channel interference effect is less severe in the realistic low-level simulation. All four examples with interference produce decent qualitative results, although the relatively small message load causes qualitative differences in some places. We further compared the PCTMC mean-field without interference to the Castalia results with interference for Figures 11, 12 and 13. When comparing the resulting relative errors of the 5 busiest locations in both cases, it turns out that the maximum error of our extended PCTMC model with interference (cf. Section 3.3) produces between 3 to 10 times more accurate results than the simple PCTMC model without interference (cf. Section 3.2).

Like the comparison between Figure 7 and Figure 8 in the previous section, the difference between Figure 13 and Figure 14 shows that also in the presence of interference our extended PCTMC model approaches the Castalia model as we scale the message forwarding and generation rate.

Most interestingly the quality of the prediction does not visibly deteriorate as we add interference effects. This is es-

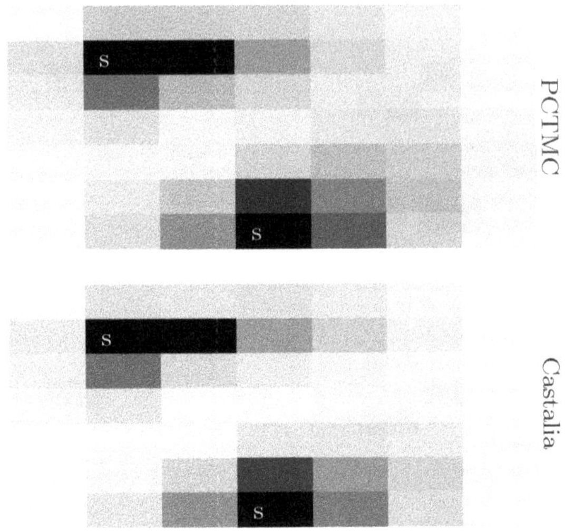

Figure 11: Data flow in a network with 49 nodes, 2 sinks, a forward rate of 20 messages per second and a message production rate of 1 message per second. The heat map at the top is based on the PCTMC mean-field approximation, the bottom one on the Castalia simulation results. We assume communication with interference without variation in radio signal strength.

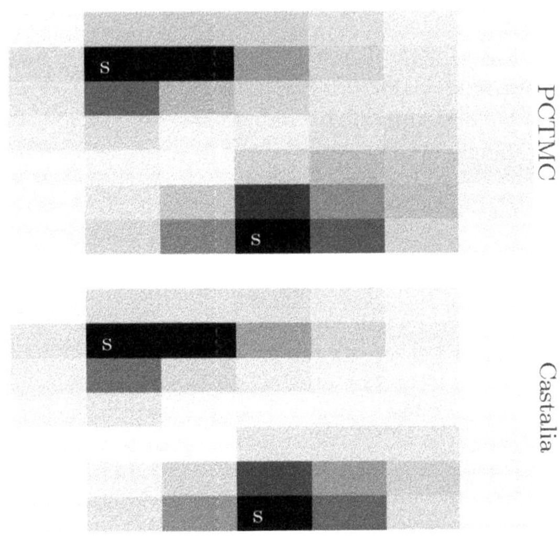

Figure 12: Data flow in a network with 49 nodes and 2 sinks with signal interference and 1.5x message production rate.

pecially surprising since we compare a simple mathematical PCTMC model to simulation results from a sophisticated low-level packet collision model. However, we do need to stress that we are just looking at first-order moments and that in its current form our model cannot be used to approximate higher-order moments of the underlying stochastic process.

4.3 Low-level Simulation vs high-level ODEs

One of our main arguments for the use of mean-field techniques over discrete event simulation is that the computa-

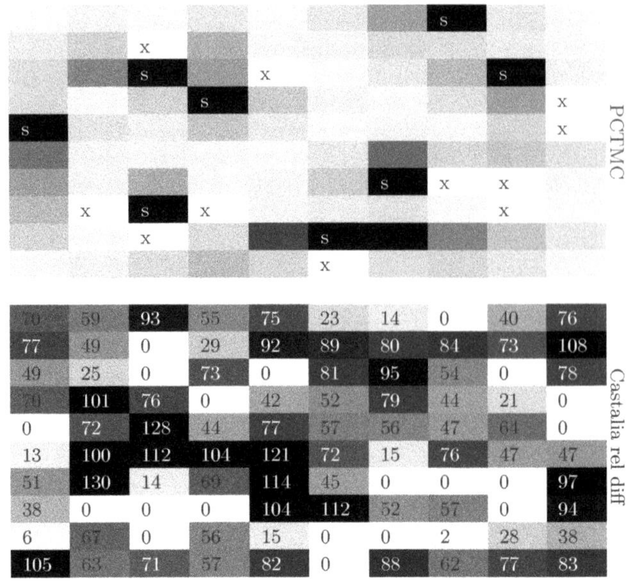

Figure 13: Data flow in a network with 100 nodes several sinks and broken nodes in presence of interference.

tional burden of simulations increases drastically as we make our model larger. The generation and the analysis of the ODEs representing the first order moments of the PCTMC models generally took $1 - 2$ minutes for the 49/100 node models respectively, while 50 Castalia simulation runs took between $5 - 15$ minutes on a modern quad-core desktop computer, depending on the message load and forwarding rate used in a given scenario.

While it is not unlikely that the simulation time grows linearly with the model size, it has to be taken into account that different setups, for instance a higher message production rate, slows down the simulation, whereas the evaluation time of the ODEs is not affected by such parameter changes unless the system becomes stiff[3]. However, for a fair comparison between simulation and ODE analysis speed, we need to take into account confidence intervals for the simulation results of the buffer size in order to determine the optimal length for each simulation run and the optimal number of replications required in order to obtain good sample mean estimates. Here, our decision compute the average buffer levels using 50 simulation produced tight 95% confidence intervals were ($< 3\%$ of the actual mean buffer size for all locations). The reason we only need a small number of repetitions to obtain tight intervals is that the buffer level, computed for each location in each simulation run, represents a steady-state average over a period of 240 seconds rather than a single measurement at the end of each simulation run. Needless to say that for a WSN with a single busy location we would thus require even less simulation runs to identify the node with the highest load in the network. However, the more marginal the differences between busy locations are, the more expensive the Castalia simulation will

[3] A stiff system of ODEs can be solved using slower solvers, which dynamically adapt the integration step size according to the relative differences between gradients in order to avoid large numerical errors.

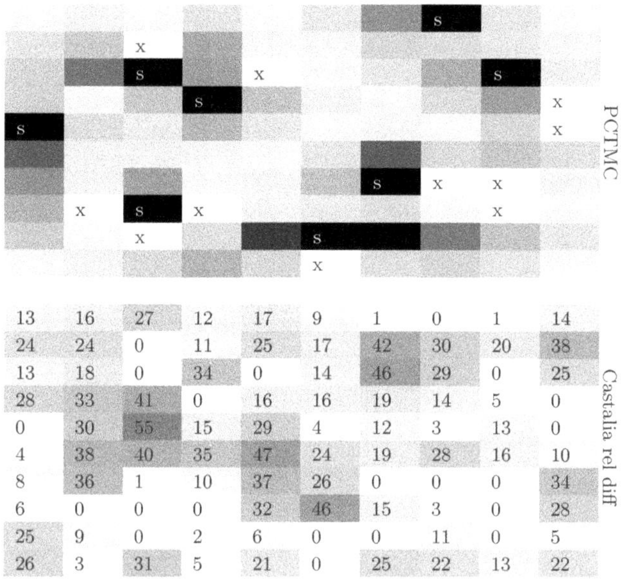

Figure 14: Same setup as in Figure 13 but with 2x forward and message production rate.

become, since we need more repetitions in order to further reduce the confidence interval width. This is especially the case when optimising parameters in a protocol, as we have to ensure that a change in parameter actually manifests in a measurable statistical improvement.

Finally, in order to obtain a fair comparison between the a simulation and a mean-field based evaluation approach we would also need to take into account the framework implementation. Our ODE analysis was conducted using the GPA-analyser tool [28], which is implemented in Java, whereas Castalia is implemented in C++. As this discussion illustrates, an accurate benchmark is itself a difficult challenge and beyond the scope of this paper. However, it is true to say that for the above data flow experiments, the mean-field technique is a fast, scalable evaluation method. Given that low-level WSN simulations are computationally expensive, the mean-field method presents a very attractive approach, not as a replacement analysis technique but rather in unison with simulation. This is particularly useful when the mean-field technique is considered as a heuristic methodology for reducing the design parameter search space that can then be explored in more detail by simulation.

5. CONCLUSIONS

In this paper we have presented a PCTMC model for the data flow analysis of large, distributed WSNs running a fail-safe pheromone based routing protocol. The data flow results presented in Section 4, are the result of a large number of interacting processes including pheromone spread, dynamic routing decisions and buffer level estimation. Moreover, we illustrated that our model produces good qualitative results even in abundance of channel interference or when message volumes are low. Furthermore, we showed that as we scale up the number of messages sampled by each node and the rate at which messages are forwarded, the mean-field approximation becomes even more accurate.

Despite encouraging results, further research is needed to improve the representation of WSN features, such as interference and duty-cycle behaviour, in PCTMC models. The availability of such key concepts is essential if PCTMC modelling aims to become an established analysis tool for the WSN community. While some aspects of WSN protocols might be hard to represent using only exponential delays, we further intend to study mean-field techniques for Generalised Semi-Markov Processes (GSMPs) [17] which are likely to work better under certain circumstances. However, given the variety of tools for mean-field evaluation of PCTMC models, it is worth trying to develop further PCTMC modelling techniques for WSNs first.

6. ACKNOWLEDGEMENTS

Jeremy Bradley is supported in part by EPSRC on the AMPS project, the Analysis of Massively Parallel Stochastic Systems, ref. EP/G011737/1.

7. REFERENCES

[1] Crossbow: Crossbow datasheet on MicaZ, 2006.

[2] M. R. Akhondi, A. Talevski, S. Carlsen, and S. Petersen. Applications of Wireless Sensor Networks in the Oil, Gas and Resources Industries. *2010 24th IEEE International Conference on Advanced Information Networking and Applications*, pages 941–948, 2010.

[3] M. S. Al-Fares and Z. Sun. Self-Organizing Routing Protocol to achieve QoS in Wireless Sensor Network for Forest Fire Monitoring. *Systems Research*, pages 211–216, 2009.

[4] A. Bachir, M. Dohler, T. Watteyne, and K. K. Leung. MAC Essentials for Wireless Sensor Networks. *IEEE Communications Surveys Tutorials*, 12(2):222–248, 2010.

[5] L. Bergamini, C. Crociani, A. Vitaletti, and M. Nati. Validation of WSN simulators through a comparison with a real testbed. In *Proceedings of the 7th ACM workshop on Performance evaluation of wireless ad hoc sensor and ubiquitous networks*, pages 103–104. ACM, 2010.

[6] A. Boulis. Castalia: revealing pitfalls in designing distributed algorithms in WSN. In S. Jha, editor, *Proceedings of the 5th international conference on Embedded networked sensor systems*, pages 407–408. ACM, 2007.

[7] D. Bruneo, M. Scarpa, A. Bobbio, D. Cerotti, and M. Gribaudo. Markovian agent modeling swarm intelligence algorithms in wireless sensor networks. *Performance Evaluation*, 69(3-4):135–149, Mar. 2012.

[8] E. Egea-Lopez, J. Vales-Alonso, A. Martinez-Sala, P. Pavon-Mario, and J. Garcia-Haro. Simulation Scalability Issues in Wireless Sensor Networks. *IEEE Communications Magazine*, 44(7):64–73, 2006.

[9] S. Engblom. Computing the moments of high dimensional solutions of the master equation. *Applied Mathematics and Computation*, 180(2):498–515, 2006.

[10] A. Förster and A. L. Murphy. A Critical Survey and Guide to Evaluating WSN Routing Protocols. In *The First International Workshop on Networks of Cooperating Objects (CONET)*, Stockholm, 2010.

[11] V. Galpin. Towards a spatial stochastic process algebra. In *Proceedings of the 7th Workshop on Process Algebra and Stochastically Timed Activities (PASTA)*, Edinburgh, 2008.

[12] D. T. Gillespie. Exact stochastic simulation of coupled chemical reactions. *Journal of Physical Chemistry*, 81(25):2340–2361, 1977.

[13] M. Gribaudo, D. Cerotti, and A. Bobbio. Analysis of On-off policies in Sensor Networks Using Interacting Markovian Agents. *6th IEEE International Conference on Pervasive Computing and Communications PerCom (2008)*, pages 300–305, 2008.

[14] M. C. Guenther. PCTMC model and Castalia simulation code (http://www.doc.ic.ac.uk/~mcg05/wsnrouting), 2012.

[15] M. C. Guenther and J. T. Bradley. Mean-field performance analysis of a hazard detection Wireless Sensor Network. In *VALUETOOLS'12*, 2012.

[16] M. C. Guenther and J. T. Bradley. PCTMC models of Wireless Sensor Network protocols. In *UKPEW'12, The 28th UK Performance Engineering Workshop*, 2012.

[17] R. A. Hayden. Mean-field approximations for performance models with generally-timed transitions. *ACM SIGMETRICS Performance Evaluation Review*, 2011.

[18] R. A. Hayden and J. T. Bradley. A fluid analysis framework for a Markovian process algebra. *Theoretical Computer Science*, 411(22-24):2260–2297, 2010.

[19] J. Hillston. Fluid flow approximation of PEPA models. *Second International Conference on the Quantitative Evaluation of Systems QEST05*, pages 33–42, 2005.

[20] R. Hornig and A. Varga. AN OVERVIEW OF THE OMNeT++ SIMULATION ENVIRONMENT. *Proceedings of the First International ICST Conference on Simulation Tools and Techniques for Communications Networks and Systems*, pages 1–10, 2008.

[21] P. Levis and N. Lee. TOSSIM : A Simulator for TinyOS Networks. *UC Berkeley September*, pages 1–17, 2003.

[22] Ns. The Network Simulator - ns-2, 2002.

[23] M. Opper and D. Saad. *Advanced Mean Field Methods: Theory and Practice*. The MIT Press, 2001.

[24] M. Paone, L. Paladina, D. Bruneo, and A. Puliafito. A Swarm-based Routing Protocol for Wireless Sensor Networks. *Sixth IEEE International Symposium on Network Computing and Applications NCA 2007*, 31(Nca):265–268, 2007.

[25] M. Ramesh, N. Vasudevan, and J. Freeman. Real Time Landslide Monitoring via Wireless Sensor Network. *Geophysical Research Abstracts*, 11(EGU2009-14061):14061–14061, 2009.

[26] K. Sohrabi, B. Manriquez, and G. J. Pottie. Near ground wideband channel measurement in 800-1000 MHz. *1999 IEEE 49th Vehicular Technology Conference Cat No99CH36363*, 1(3):571–574, 1999.

[27] A. Stefanek, M. C. Guenther, and J. T. Bradley. Normal and inhomogeneous moment closures for stochastic process algebras. In *10th Workshop on Process Algebra and Stochastically Timed Activities (PASTA'11)*, Ragusa, 2011.

[28] A. Stefanek, R. A. Hayden, and J. T. Bradley. A new tool for the performance analysis of massively parallel computer systems. *Eighth Workshop on Quantitative Aspects of Programming Languages QAPL 2010 March 2728 2010 Paphos Cyprus*, 2010.

[29] A. Stefanek, R. A. Hayden, and J. T. Bradley. Fluid computation of the performance-energy trade-off in large scale Markov models. *SIGMETRICS Performance Evaluation Review*, 2011.

[30] S. Verma, N. Chug, and D. V. Gadre. Wireless Sensor Network for Crop Field Monitoring. In *2010 International Conference on Recent Trends in Information, Telecommunication and Computing*, pages 207–211. IEEE, Mar. 2010.

[31] P. J. Wangersky. Lotka-Volterra population models. *Annual Review of Ecology and Systematics*, 9(1):189–218, 1978.

[32] N. Xu, S. Rangwala, K. K. Chintalapudi, D. Ganesan, A. Broad, R. Govindan, and D. Estrin. A wireless sensor network For structural monitoring. *Proceedings of the 2nd international conference on Embedded networked sensor systems SenSys 04*, 20(7):13–24, 2004.

[33] P. Yadav. *Cross-Layer Protocols to Support Periodic Data Collection and Event Driven Wireless Sensor Network Applications*. Phd thesis, Imperial College, 2011.

[34] J. Yick, B. Mukherjee, and D. Ghosal. Wireless sensor network survey. *Computer Networks*, 52(12):2292–2330, 2008.

When Spatial and Temporal Locality Collide: The Case of the Missing Cache Hits

[Experience Paper]

Mattias De Wael
Software Languages Lab
Vrije Universiteit Brussel
madewael@vub.ac.be

David Ungar
Watson Research Center
IBM Research
dungar@ibm.com

Tom Van Cutsem
Software Languages Lab
Vrije Universiteit Brussel
tvcutsem@vub.ac.be

ABSTRACT

Even the simplest hardware, running the simplest programs, can behave in the strangest of ways. Tracking down the cause of a performance anomaly without the complete hardware reference of a processor is a prime example of black-box architectural exploration. When doubling the work of a simple benchmark program, that was run on a single core of Tilera's TILEPro64 processor, did not double the number of consumed cycles, a mystery was unveiled. After ruling out different levels of optimization for the two programs, a cycle-accurate simulation attributed the sub-optimal performance to an abnormally high number of L1 data cache misses. Further investigation showed that the processor stalled on every Read-After-Write instruction sequence when the following two conditions were met: 1) there are 0 or 1 instructions between the write and the read instruction and 2) the read and the write instructions target distinct memory locations that share an L1 cache line. We call this performance pitfall a *RAW hiccup*. We describe two countermeasures, memory padding and the explicit introduction of pipeline bubbles, that sidestep the RAW hiccup.

This experience paper serves as a useful troubleshooting guide for uncovering anomalous performance issues when the hardware design under study is unavailable.

Categories and Subject Descriptors

D.2.8 [**Software Engineering**]: Metrics—*Performance measures*

Keywords

L1 data cache, TilePro64, padding, pipeline bubble

ICPE'13, April 21–24, 2013, Prague, Czech Republic.

1. INTRODUCTION

In the context of our research in parallel algorithms for many-core architectures, we chose to explore the TilePro64 processor by writing a simple program and measuring its absolute and relative performance on a single core. To verify the measured results, we doubled the work of our program expecting the running time to double as well.

The experiment, however, revealed that *duplicating the work did not double the execution time.* What could possibly be causing this anomalous performance? And how does one track down such changes in performance efficiency? We conducted a series of experiments, timing and simulating different code sequences. Each step answered one question only to ask another. Finally, we were able to pinpoint the instruction and memory reference sequence that was responsible. Our hunt for the performance anomaly is a nice example of black-box architectural exploration, as we did not have a access to a complete hardware design reference.

The rest of this paper is organized as follows: First the used hardware is discussed, followed by the presentation of the two benchmark programs that are used throughout the text. Section 4 elaborates on the *expected* and *measured* performance of both programs, revealing a significant discrepancy between the two. In Section 5 the compiler generated instruction sequences of the benchmark programs are discussed. Section 6 focusses on processors stalls and how they map to the source code. Section 7 and 8 each describe a countermeasure that sidesteps the RAW hiccup in the benchmark programs. Finally, in Section 9 we compare the RAW hiccup to similar performance pitfalls.

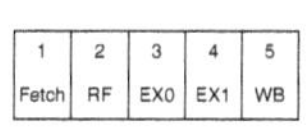

(a) 5 Stage Pipeline of the TILEPro64

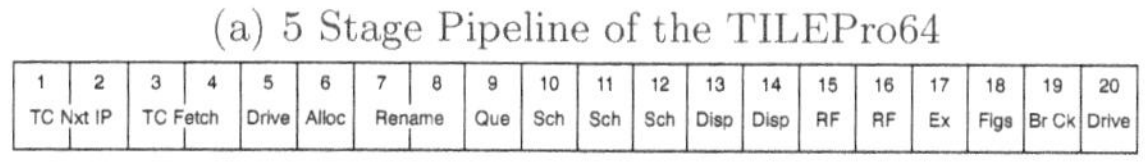

(b) 20 Stage pipeline of the Intel Pentium IV.

Figure 1: The Tilera TilePro64 processor has a much more shallow pipeline than for instance Intel's Pentium IV. Therefore, understanding programs and predicting instruction timings on the TilePro64 processor should be relatively easy

```
1  for (  long i = 0; i < N; ++i ) {
2    long sum0 = 0;
3
4    for (  long j0 = 0; // init
5           j0 < i;      // test
6           ++j0 ) {     // incr
7
8      sum0 += j0;       // body
9
10   }
11   total += sum0;
12
13
14 }
```

```
1  for (  long i = 0; i < N; ++i ) {
2    long sum0 = 0;
3
4    for (  long j0 = 0; // init
5           j0 < i;      // test
6           ++j0 ) {     // incr
7
8      sum0 += j0;       // body
9
10   }
11   total += sum0;
12
13   long sum1 = 0;
14
15   for (  long j1 = 0; // init
16          j1 < i;      // test
17          ++j1 ) {     // incr
18
19     sum1 += j1;       // body
20
21   }
22   total += sum1;
23
24
25 }
```

Figure 2: Two simple benchmark programs, Program 1 and Program 2 respectively, where the second program has exactly twice the work to do compared to the first program. Program 2 is expected to consume twice as many cycles as Program 1.

2. THE PLATFORM: TILEPRO64 PROCESSOR

At first sight, Tilera's TILEPro64 [9] might look like a complex many-core processor chip. Its 8x8 mesh network connects 64 processing cores, keeps two levels of distributed cache coherent, and supports inter-core communication. Each core utilizes a three-way Very Long Instruction Word (VLIW) architecture to support explicit Instruction Level Parallelism (ILP) by executing up to three *bundled instructions* simultaneously by one of the three pipelines of a single core.

But, when looked at in isolation, a single pipeline of a single core of the TILEPro64 processor has a relatively simple architecture. The in-order pipelines execute an instruction in 5 stages (Figure 1a): a rather shallow pipeline compared to contemporary hardware that have pipelines as deep as 20 stages (Figure 1b) [5]. With this short, in-order pipeline the TILEPro64 aims at low branch and low load-to-use latencies. Thus when considering a single pipeline on a single core, the TILEPro64 can arguably be described as simple hardware and performance prediction of any sequential programs it runs should be fairly easy. But even the simplest hardware running the simplest programs, can behave in the strangest of ways. How simple could those programs be?

3. THE PROGRAM: TETRAHEDRAL NUMBERS

We chose a very simple algorithm for our benchmark: the computation of the nth tetrahedral number. To compute the nth tetrahedral number it suffices to accumulate the n first triangular numbers [7]. And to compute the nth triangular number it suffices to accumulate the n first nonzero integers [8]. Mathematically this could be written as $\sum_{i=0}^{N}\sum_{j=0}^{i} j$ and Program 1 (Figure 2) is the straightforward translation of this formula into C-code. Thus, Program 1 consists of one outer loop (lines 1-14 in Figure 2) with one inner loop (lines 4-10 in Figure 2), of which the body consists of a single statement accumulating a counter (line 8 in Figure 2). Program 2 (Figure 2) just about doubles the work of Program 1 by repeating the inner loop (lines 4-10 and lines 15-21 in Figure 2). When measurements revealed that *doubling the work did not double the execution time*, we were mystified.

4. MEASURED PERFORMANCE

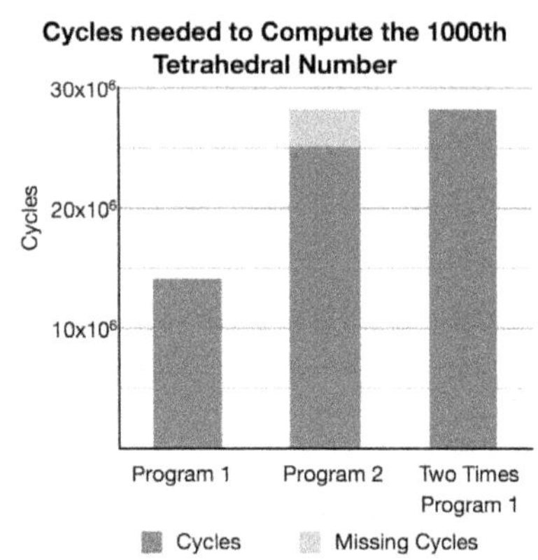

Figure 3: Cycles consumed by Program 1 and Program 2, compared to the number of expected cycles for Program 2.

"To measure is to know", but in this case measuring the performance of both benchmark programs raised more questions than it resolved. We measured the absolute performance of both programs in *cycles needed to complete the outer loop*. Computing the 1000th tetrahedral number by running Program 1 requires 14×10^6 cycles. Doubling the work by running Program 1 twice, requires 28×10^6 cycles. Surprisingly, when running Program 2 which also does twice the work as Program 1, only 25×10^6 cycles were consumed. The difference of 3×10^6 cycles between expected and measured performance, depicted as the lighter colored box in 3,

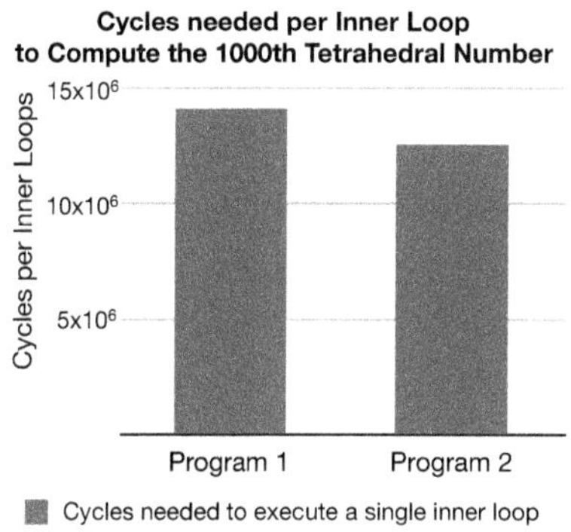

Figure 4: Average number of cycles needed by Program 1 and Program 2 to compute a single inner loop. Surprisingly, Program 2 needs fewer cycles than Program 1.

is too large to be attributed to the overhead of running two outer loops when running Program 1 twice.

Put differently, as depicted in Figure 4, when looking at the average number of cycles needed to complete a single inner loop, we see that Program 1 needs 14×10^6 cycles, as opposed to the 12.5×10^6 cycles needed by Program 2. These data suggest that the loops in Program 2 run more efficiently. But why would the same source code run faster?

5. COMPARING INSTRUCTION SEQUENCES

The execution of the inner loops require quadratic time, actually the number of additions needed to compute the nth tetrahedral number is equal to the nth triangular number, and thus indeed *quadratic* in function of N. Since the outer loops induce only linear overhead, *the inner loops dominate the overall performance* of the benchmark programs, and we focused on these.

Could it be that the compiler was generating different instruction sequences for syntactically equal inner loops? If this were true, the assumption that Program 2 did exactly twice the work of Program 1 would be false, and the unexpected execution times could be explained. So the first step in unraveling the mystery was to guarantee that the generated machine instruction sequences for the two programs were similar for all inner loops.

We used Tilera's gcc compiler 2.0.2 with optimization flag -O0 which compiled the inner loop of Program 1 into the 21 instructions shown in Figure 5. The first two instructions initialize the inner loop and are only executed a linear number of times. More interesting were the remaining 19 instructions, which formed the heart of the computation and were executed $\frac{N\times(N-1)}{2}$ times each. Ignoring the small constant and linear overhead induced by the inner and outer loops, $19 \times \frac{N\times(N-1)}{2}$ was a fair approximation of the total number of executed instructions for Program 1 when N was sufficiently large.

This instruction sequence can be further optimized. The most invasive and effective optimization would be to reduce the program to compute the formula $\frac{N\times(N+1)\times(N+2)}{N}$ which would be semantically equivalent to Program 1 but would require only constant time for any input. But even if we wanted to keep the structure of the computation, many optimizations were possible. For example, the computation

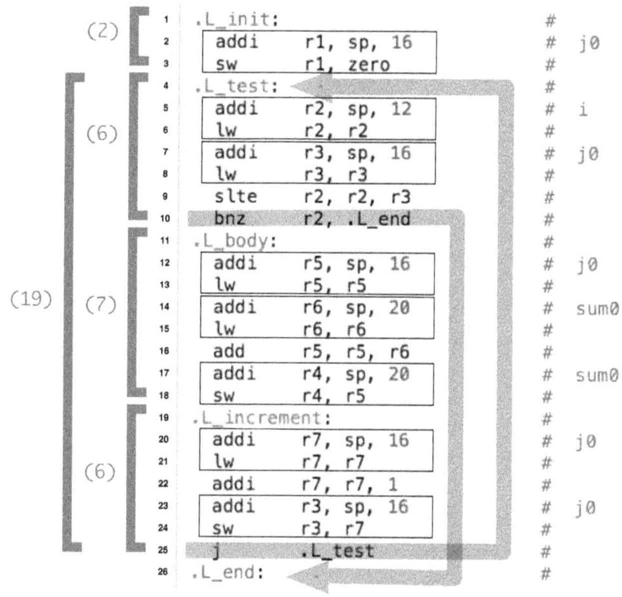

Figure 5: The inner loop of Program 1 compiled into assembler code (-O0). The last 19 instructions dominate the performance of Program 1, and are executed $\frac{N\times(N-1)}{2}$ times.

in Program 1 used only 4 variables and the value of the stack pointer (register) is not changed during the execution. Therefore it would be possible to compute the absolute addresses of the four variables once and store them in a dedicated register. This optimization could save 7 addi instructions in each iteration. The difference in relative performance would be accounted for if the compiler had applied this or any other optimization to the inner loops of Program 2, but not to the inner loop of Program 1. But a comparison of the generated assembler instruction sequences for both Program 1 and Program 2 revealed that *all three inner loops were compiled into the same 21 instructions.* The instruction sequences for each loop differed only in the relative addresses of the variables. At this point, it was clear that twice as many assembler instructions were executed for Program 2, than for Program 1. Yet, Program 2 was executing those instructions more efficiently. What could explain this behavior?

If an instruction sequence gets more densely packed into instruction bundles such that the instruction level parallelism supported by the VLIW architecture gets exploited, it will run faster. Could it be possible that the instructions of Program 2 got bundled more efficiently than those for Program 1? The assembly-level instruction sequence, as shown in Figure 5, did not make these bundles explicit. And since the processor cores operate in-order they are not responsible for any implicit instruction level parallelism themselves. When we decompiled the machine instructions back into assembler code we saw that all instructions were wrapped in a single bundle and thus issued to be executed sequentially without any instruction level parallelism. *Bundles were not the answer.*

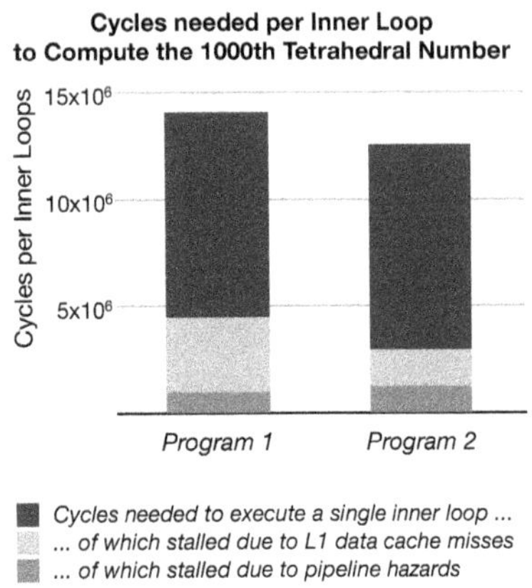

Figure 6: Relative to the amount of work, Program 1 suffered from many more L1 data cache misses than Program 2. This discrepancy explained the difference in performance, but left us wondering about the cause of the discrepancy.

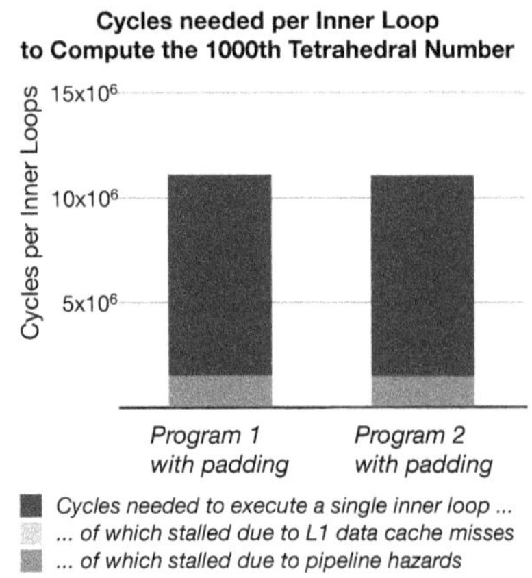

Figure 8: Adaptations of Program 1 and Program 2 that add padding around the inner loop counters make all L1 data cache misses disappear. Consequently, also the change in efficiency is gone and both programs behave as expected.

Because all executed bundles were single-instructions, the processor executed the exact same instruction sequence once for Program 1 and twice for Program 2. Why did the processor work at different speeds in each program?

6. CACHE MISSES CAUSED THE PROCESSOR TO STALL

If a processor consumes a different amount of cycles for equal instruction sequences, it must be stalling somewhere. A cycle-accurate simulated execution of the applications reported that the processor was indeed stalling, and attributed these stalls to L1 and L2 instruction- and data-cache misses, to pipeline hazards, and to mispredicted branches. Of these categories, only the *L1 data cache misses* and the *pipeline hazards* were numerous enough to observably impact the performance. If these stalls were the true culprits, then the anomalous increase in efficiency of Program 2 over Program 1 should also have been reflected in the number of corresponding stalls. For stalls attributed to pipeline hazards, doubling the work also doubled the number of stalls. A regular evolution, so that category was exonerated. Remarkably, *the number of stalls caused by L1 data cache misses was not affected by doubling the work.*

Even considering each program in isolation, the number of L1 data cache misses was unexpectedly high. Because both benchmark programs used only 4 and 6 variables respectively, we *expected no misses* at all. But the clue to the mystery was in the observation that the number of L1 data cache misses did not increase when the amount of work was doubled. Figure 6 shows these data relative to the number of inner loops: on average a single inner loop of Program 2 suffered from half as many L1 data cache stalls as the loop of Program 1. Why was Program 2 more efficient?

More rigorous simulation revealed that all the L1 data cache misses were read misses. This commonality limited the instructions where the misses could occur to the lw (load word) instructions of the program (lines 6, 8, 13, 15, and 21 in Figure 5). On the C-code level, the cycle-accurate simulator indicated that the loop-increment operations on $j0$ (line 6 in Figure 7a) in Program 1, and on $j1$ (line 17 in Figure 7b) in Program 2 were *responsible for almost all stalls*, as is shown in Figure 7 where the source lines of Program 1 and Program 2 are annotated with the number of observed cache misses. The synthesis of these two pieces of evidence allowed us to identify the slow instruction in each program. In Program 1 this would be the load instruction lw r7 r7 shown on line 21 of Figure 5. When N is 1000, the number of inner loop iterations is 4950 which was exactly the number of observed L1 data cache misses on line 6 and 17 of Program 1 and Program 2 respectively.

What we know now is that the inner loop of Program 1 ran slowly because *every iteration induced a cache miss.* One of the inner loops of Program 2 ran equally slowly for the same reason, but on average Program 2 was more efficient because the other inner loop never suffered from a cache miss. But why did the processor have to wait for data that should have been in its L1 cache in the first place?

7. PADDING RESOLVES THE CACHE MISSES

Recall that the difference in performance of our two benchmark programs was caused by an abnormally large number of L1 data cache misses that were not expected and moreover did not increase with the amount of work. A cache miss occurs when a processing unit fails to access a piece of data in the cache which results in a much more expensive operation that reroutes the memory instruction to the next level of memory. In this case reading from the L1 data cache fails, causing a load from the L2 cache which is 7 cycles away on the TILEPro64.

Consider Program 1, which used only 4 variables for its entire computation. By the time the instruction on line 15 in Figure 5 had been executed for the first time, we expected all the variables to reside in the L1 cache and to stay there for the remainder of the computation. This expectation arose because 8KB of L1 data cache is plenty of room for storing 4 values of 4 bytes each: they even fit on a single cache line as the TILEPro64's L1 cache lines are 16 bytes wide [9]. Even for the 6 variables of Program 2 the 8KB should have more than sufficed. Thus any L1 data cache line that was

```
for ( long i = 0; i < N; ++i ) {                      1
  long sum0 = 0;

  for ( long j0 = 0; // init                         199
        j0 < i;      // test
        ++j0 ) {     // incr                        4950

    sum0 += j0;      // body

  }
  total += sum0;


}
```

(a) Program 1: The bulk of L1 data cache misses occur on the increment of the only inner loop.

```
for ( long i = 0; i < N; ++i ) {                    101
  long sum0 = 0;

  for ( long j0 = 0; // init
        j0 < i;      // test
        ++j0 ) {     // incr

    sum0 += j0;      // body

  }
  total += sum0;

  long sum1 = 0;

  for ( long j1 = 0; // init
        j1 < i;      // test
        ++j1 ) {     // incr                        4950

    sum1 += j1;      // body

  }
  total += sum1;


}
```

(b) Program 2: The bulk of L1 data cache misses occur on the increment of the second inner loop.

Figure 7: To understand the reason for all these cache misses, a better view is needed on where they actually occur. Program 2 has one normal and one affected loop. Since these are syntactically equivalent, memory layout of variables is a probable culprit of the discrepancy.

invalidated in the execution of either program qualified as unanticipated behavior.

In parallel computing, a case of unanticipated cache invalidation, called *false sharing* occurs when a computation writes to a memory location that resides on the same cache line of a distinct memory location used by a concurrent computation [6]. False sharing can be reduced by padding the memory layout of variables so that the memory locations used by concurrent computations do not share cache lines [1]. Padding is a low-level programming technique in which, usually unused, memory is allocated around variables to obtain a more suitable layout of variables in the caches and/or memory. Although this line of inquiry seems far-fetched in the case of a sequential program using only 4 variables, we experimented with memory layout anyway because the distributions of variables over cache lines can affect the number of misses.

Figure 9 shows four possible layouts of the variables over different cache lines when padding is introduced. Each line shows three cache lines (alternating colors) of four times 4 bytes each. The first line in Figure 9 shows the memory layout as observed for Program 1 without any padding: the first cache line (white) contains the variables *total* and i, the consecutive cache line (gray) contains the variables $sum0$ and $j0$, finally the third cache line (white) contains no variables relevant to our case. The others three layouts in Figure 9 show how introducing padding before $j0$ moved $j0$ into a different cache line than the other 3 variables.

Program 1, when adapted such that $j0$ resided in its own cache line, consumed only 11×10^6 cycles, as opposed to the 14×10^6 cycles it had consumed before. The cycle-accurate simulation showed that the *padding eliminated all the L1 data cache misses.* Further, when we moved the variable $j1$ of Program 2 to a separate cache line, the number of cycles consumed by Program 2 dropped from 25×10^6 to 22×10^6. The import of this performance improvement is that neither benchmark program suffered from unanticipated L1 data cache misses when a change in the memory layout places the inner loop counter and the inner loop accumulator on different L1 data cache lines.

Summarized, the lw instruction (line 21 in Figure 5) causes the processor to stall if the preceding sw instruction (line 18 in Figure 5) targets the same L1 data cache line. For this reason we call this behavior a *read-after-write hiccup*, or RAW hiccup for short. In our example, the sw and lw instructions are separated by an addi instruction (line 20 in Figure 5) which raised the question if, besides the memory layout, also the addi instruction plays a role in the RAW hiccup?

8. INJECTING PIPELINE BUBBLES

Isolating the missed variable in a different cache line sidestepped the RAW hiccup, but could instruction reordering accomplish the same?

Data hazards occur when subsequent instructions have data dependencies and are executed at the same time in a pipeline. Many contemporary processors use out-of-order execution to avoid these dependencies [4]. The TILEPro64 processor, however, supports only in-order execution. In that case the only way to avoid the data hazard is by introducing a pipeline bubble. Pipeline bubbling is an instruction scheduling technique that prevents data hazards from occurring by delaying the execution of dependent instructions in the pipeline. Typically this is done by the processor's logic by stalling the execution of the depending instruction. A compiler could simulate this behavior by inserting a no-operation instruction (NOP), but what if we hardcoded such a NOP in the slow running inner loops of Program 1 and Program 2?

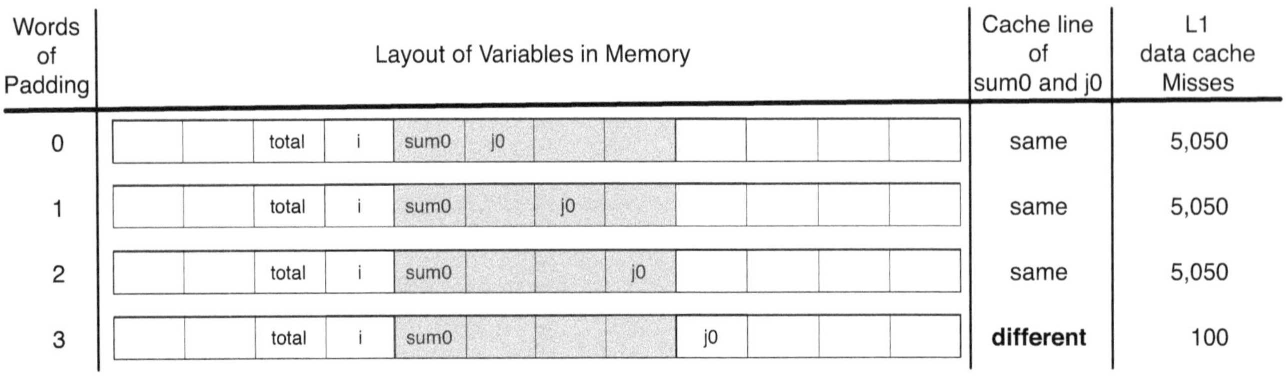

Figure 9: With as few as 4 variables there is no reason to suffer from cache eviction. In parallel computing padding is a tried and true approach to tackle false sharing, a phenomenon where cache eviction is also unanticipated. Adding padding between $sum0$ and $j0$, such that they reside on different cache lines, causes a significant drop in L1 data cache misses.

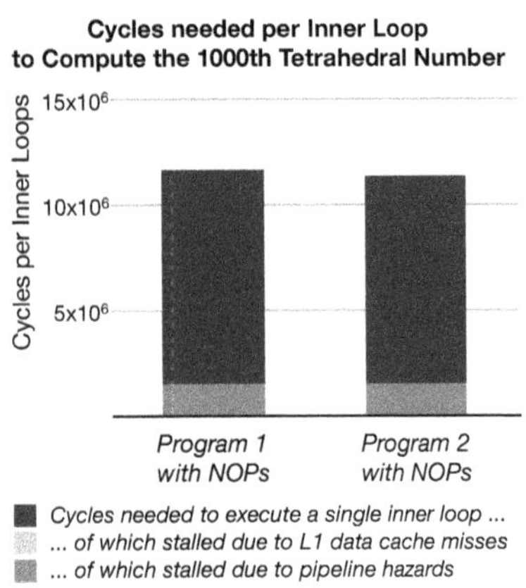

Figure 10: Adding NOPs to the bodies of the slow inner loops of Program 1 and Program 2 make all L1 data cache misses disappear. Consequently, also the change in efficiency is gone and both programs behave as expected.

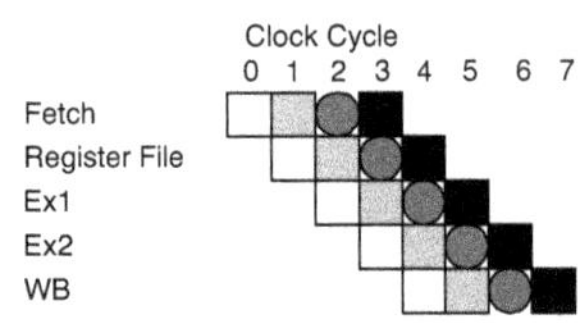

(a) Pipelined execution of 3 instructions where the processor stalls on the Ex1 stage of the second instruction.

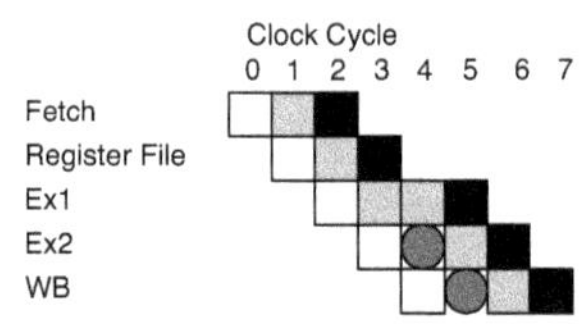

(b) Pipelined execution of 2 instructions, a NOP, and another instruction where the processor does not stall.

Figure 11: The overall performance of a processor stalling for one cycle in a sequence of three instructions, is equivalent to the performance of executing that same sequence with an additional NOP instruction if the processor does not stall.

In our benchmark programs there was no actual data dependency between the memory instructions on lines 18 and 21 (see Figure 5), the targeted memory locations only shared a cache line. Thus, adding a NOP in the body of the slow running inner loop should have made the program run even slower because the inner loop now consisted out of 20 instead of 19 instructions. Perversely, the measured performance was closer to that of the fast programs with padding, than to the execution time of the original benchmark programs with all the cache misses. The cycle-accurate simulator indicated that *adding a NOP to the body of the slow inner loops removed all L1 data cache misses*.

Synthesizing the effects observed when either padding or pipeline bubbles were introduced, allowed us to conclude that *on the TILEPro64 a RAW hiccup occurs in a read-after-write instruction sequence if two conditions are met. First, if both the store and the load instruction target distinct memory locations that share an L1 data cache line, and second if at most one instruction separates the store and the load.* If either condition is eliminated, the RAW hiccup is gone. To overcome this pitfall on the hardware level, a much more complicated power and area consuming micro-architecture would be needed in order to work around the store word inefficiencies.

9. SIMILAR PERFORMANCE PITFALLS

From the software perspective, pinpointing the origin of anomalous performance to a specific instruction and memory reference sequence suffices to render a program more performant simply by avoiding that sequence. From a hardware perspective, the question remains what architectural design choices caused the anomalous performance. To the best of our knowledge the details of hardware implementations we are hitting in this concrete example are not documented by the chip producer. Without these details we can only make an educated guess about the concrete origin of the stalls.

However, three well know performance pitfalls exist that look similar to the RAW hiccup:*false sharing*, *load-hit-store*, and *write misses*.

False sharing *False sharing* only occurs in the case of concurrent processes, and thus does not apply in the case of a RAW hiccup which occurs in a single thread of control [1]. But, besides the number of processes the commonalities are omnipresent: both performance pitfalls occur when a sequence of memory instructions, with at least one write, target unrelated variables that share a cache line.

Load-hit-Store A single process can suffer from the performance pitfall *load-hit-store* (LHS) when a read is issued too soon after a write to read the new value [3]. The RAW hiccup also issues a read too soon after a write, but as opposed to a LHS, the memory instructions target *distinct memory locations*.

Write Stall Finally, when a processor must wait for writes to complete during *write through*, the processor is said to *write stall* [4]. The L1 data cache of the TILEPro64 uses the *write-through policy*, thus what we called a RAW hiccup could actually be a write stall. Other literature, however, refines the definition of write stalls in the context of write-through caches as the delay caused when *a write encounters another write in progress* [2]. Thus excluding read instructions as the origin of a write stall. This makes the instruction sequence causing a RAW hiccup different from the instruction sequence causing a write stall.

Identifying the actual implementation details that caused the RAW hiccup is notoriously difficult without a complete hardware reference. We recognized *false sharing*, *load-hit-stores*, and *write misses* as performance pitfalls similar to the RAW hiccup. They differ, however, from the RAW hiccup in either number of processes, targeted memory locations, or instruction sequence.

10. CONCLUSIONS

A simple C program exhibited a significant discrepancy between its expected and measured performance. The origin of the anomaly was found to be a counter-intuitive multi-cycle processor stall that occurred whenever a store instruction was followed within two instructions by a load instruction targeting the same L1 data cache line. We called this performance pitfall a RAW hiccup. Eliminating either of the two conditions caused the RAW hiccup to disappear: The introduction of sufficient padding on the one hand and the introduction of a manual pipeline bubble on the other hand removed all anomalous L1 data cache misses.

This experience paper reports on the hunt for a performance anomaly observed when studying the behavior of a trivially simple sequential program. The exact instruction and memory reference sequence that was responsible was found only after performing various experiments and simulations each one a step down the ladder of abstraction. This experience raises the question that if we can not predict the performance of trivially simple sequential code on a simple processing unit, how can we hope to understand the performance of parallel applications running on an 8x8 mesh network-on-chip?

11. ACKNOWLEDGMENTS

Mattias De Wael is supported by a doctoral scholarship granted by the Institute for the Promotion of Innovation through Science and Technology in Flanders (IWT Vlaanderen), Belgium.

Tom Van Cutsem is a Postdoctoral Fellow of the Research Foundation, Flanders (FWO).

This research has been partially supported by IBM Research.

12. REFERENCES

[1] S. J. Eggers and T. E. Jeremiassen. Eliminating false sharing. In *ICPP (1)*, pages 377–381, 1991.

[2] J. S. Emer and D. W. Clark. A characterization of processor performance in the vax-11/780. *SIGARCH Comput. Archit. News*, 12(3):301–310, Jan. 1984.

[3] B. Heineman. Common performance issues in game programming, June 2008.

[4] J. L. Hennessy and D. A. Patterson. *Computer Architecture, Fourth Edition: A Quantitative Approach.* Morgan Kaufmann Publishers Inc., San Francisco, CA, USA, 2006.

[5] G. Hinton, D. Sager, M. Upton, D. Boggs, D. P. Group, and I. Corp. The microarchitecture of the pentium 4 processor. *Intel Technology Journal*, 1:2001, 2001.

[6] D. A. Patterson and J. L. Hennessy. *Computer Organization and Design, Fourth Edition, Fourth Edition: The Hardware/Software Interface (The Morgan Kaufmann Series in Computer Architecture and Design).* Morgan Kaufmann Publishers Inc., San Francisco, CA, USA, 4th edition, 2008.

[7] N. J. A. Sloane. Tetrahedral (or triangular pyramidal) numbers, Oct. 2012.

[8] N. J. A. Sloane. Triangular numbers, Oct. 2012.

[9] Tilera. *Tile Processor User Architecture Manual*, June 2010.

APPENDIX

A. TRIPLING THE WORK

Besides the experiments with Program 1 and Program 2, those presented in this paper, six similar programs also contributed to the discovery of the RAW hiccup. Program 2 was our attempt to double the work of Program 1 by duplicating its single inner loop, the remaining four programs were our attempt to multiply the work of Program 1 by 3 trough 8, also by replicating the inner loop. Figure 12 shows what Program 3 looks like.

Recall that the first clue towards the discovery of the RAW hiccup was that number of L1 data cache misses did not increase when the amount of work was doubled. However, when we *tripled the work* of Program 1 *the number of L1 data cache misses did double.* Adding a fourth inner loop again left the number of L1 data cache misses unchanged. This step-wise increase continued for all eight programs, as is shown in Figure 13. The number of cycles stalled because of pipeline hazards on the other hand increased linearly with the number of inner loops. When considering these data relative to the number of inner loops, as is shown in Figure 14, we see that consecutive programs alternate between being less efficient and more efficient.

At the level of C-source code, this translates to one inner loop suffering from the RAW hiccup for every two inner loops in the program ($\lceil \frac{\#\text{inner loops}}{2} \rceil$). Since all inner loops are compiled into the same instruction sequences (Figure 5) the memory layout must be causing the RAW hiccup. Consider a layout of variables where *sum*0 and *j*0 share a cache line as in line 2 of Figure 9. Introducing a new inner loop consequently introduces also two new variables (e.g. *sum*1 and *j*1) that do not share a cache line. Again adding two new variables (e.g. *sum*2 and *j*2) gives rise to a new RAW hiccup because *sum*2 and *j*2 do share a L1 data cache line. This explains why the number of cache misses only increases when adding a new inner loop to a program with an even number of inner loops.

```
for ( long i = 0; i < N; ++i ) {
  long sum0 = 0;

  for ( long j0 = 0; // init
        j0 < i;      // test
        ++j0 ) {     // incr

    sum0 += j0;      // body

  }
  total += sum0;

}
```

```
for ( long i = 0; i < N; ++i ) {
  long sum0 = 0;

  for ( long j0 = 0; // init
        j0 < i;      // test
        ++j0 ) {     // incr

    sum0 += j0;      // body

  }
  total += sum0;

  long sum1 = 0;

  for ( long j1 = 0; // init
        j1 < i;      // test
        ++j1 ) {     // incr

    sum1 += j1;      // body

  }
  total += sum1;

  long sum2 = 0;

  for ( long j2 = 0; // init
        j2 < i;      // test
        ++j2 ) {     // incr

    sum2 += j2;      // body

  }
  total += sum2;


}
```

Figure 12: Program 3 (right) triples the work of Program 1 (left).

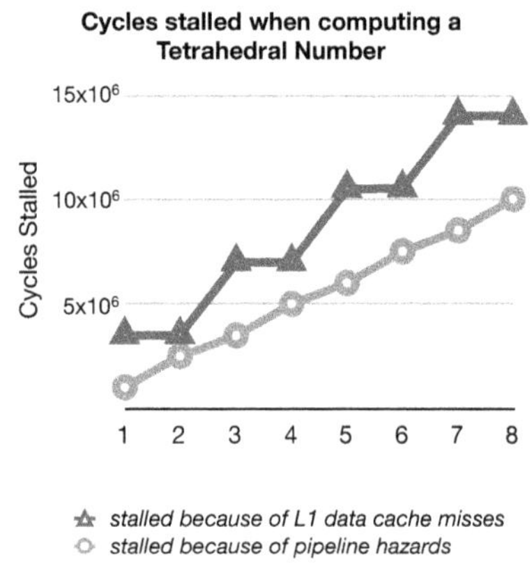

Figure 13: The number of cycles stalled caused by pipeline hazards increases linearly with the amount of work. Remarkably, the number of stalls caused by L1 data cache misses increases in steps.

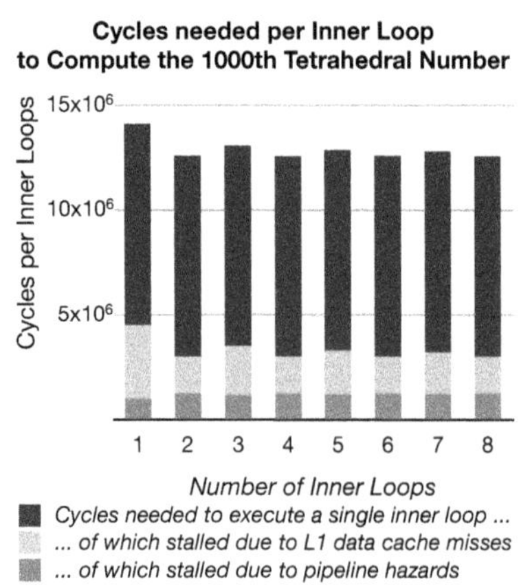

Figure 14: Comparing number of cycles needed to compute a single inner loop shows that programs with an even number of inner loops are more efficient because they suffer from less L1 data cache misses..

On Load Balancing: A Mix-Aware Algorithm for Heterogeneous Systems

Sebastiano Spicuglia
University of Lugano
Lugano, Switzerland
sebastiano.spicuglia@usi.ch

Mathias Björkqvist
IBM Research Zurich Lab
Switzerland
mbj@zurich.ibm.com

Lydia Y. Chen
IBM Research Zurich Lab
Switzerland
yic@zurich.ibm.com

Giuseppe Serazzi
Politecnico di Milano
Milano, Italy
serazzi@elet.polimi.it

Walter Binder
University of Lugano
Lugano, Switzerland
walter.binder@usi.ch

Evgenia Smirni
College of William and Mary
Virginia, US
esmirni@cs.wm.edu

ABSTRACT

Today's web services are commonly hosted on clusters of servers that are often located within computing clouds, whose computational and storage resources can be highly heterogeneous. The workload served typically exhibits disparate computation patterns (e.g., CPU-intensive or IO-intensive), that fluctuate both in terms of volume and mix. The system heterogeneity together with workload diversity further exacerbates the challenge of effective distribution of load within a computing cloud. This paper presents a novel, mix-aware load-balancing algorithm, which aims to distribute requests sent by multiple applications in heterogeneous servers such that the application response times are minimized and system resources (e.g., CPU and IO) are equally utilized. To this end, the presented algorithm tries to not only balance the total number of requests seen by each server, but also to shape the requests received by each server into a certain "mix", that is analytically shown to be optimal for response time minimization. Our experimental results—based both on simulation and on a prototype implementation—show that the mix-aware algorithm achieves robust performance in most workload mixes as well as a consistent performance improvement in comparison with one of the most robust load-balancing schemes of the Apache server.

Categories and Subject Descriptors

H.3.4 [**Systems and Software**]: Performance evaluation—*Algorithms, Experimentation, Measurement*

Keywords

Load balancing algorithm; Heterogeneous system; Cloud computing; Simulation; Prototype

1. INTRODUCTION

To ensure scalability, today's web services are replicated and hosted on distributed systems that experience regular resource upgrades and are thus comprised of heterogeneous components. Web service applications are characterized not only by disparate resource requirements (e.g., a CPU-intensive browsing mix versus an I/O-intensive transaction mix), but also by time-varying request workloads [4, 17]. Consequently, the overall system workloads fluctuate in terms of mixes of dissimilar applications and their volume of requests [16]. The heterogeneity of servers, together with the workload heterogeneity, further exacerbate the challenges of load balancing. Load balancing within a compute cloud is critical for performance and effectiveness of the compute cloud paradigm, see for example the elastic load balancer in Amazon EC2 [1].

There is a large body of load balancing studies [3, 6, 8, 17] that focus on mainly homogeneous systems and consider a single bottleneck resource where queues build up. Dispatching requests to the servers with the least number of outstanding requests, the so-called Join the Shortest Queue (JSQ) policy, has been shown to be theoretically robust [10] and is also used in practice [12] for distributing the entire load across distributed servers. In a heterogeneous system experiencing a time-varying workload mix, such a policy can potentially lead to the situation where servers receive similar amounts of requests but servers with powerful CPU (resp. IO) process a lot of IO- (resp. CPU-) intensive requests. Such unanticipated behavior can be detrimental for the overall system performance as well as for the individual application response times. It is therefore imperative for the load-balancing scheduler to evolve such that it becomes aware of both of the heterogeneity of the workload (e.g., is it IO- or CPU-intensive) but also on the heterogeneity of the architecture of the back-end servers.

In this paper, we aim to provide a robust load balancing solution in a heterogeneous compute cloud. We first focus on the following questions: Is there an optimal application mix for each server which can lead to a low response time such that all resource types equally utilized? Moreover, is there a load-dispatching policy to control the received application mix for each server such that an optimal value is reached? Last, as JSQ has been shown to be very effective in balancing the overall loads on homogeneous servers, can one leverage JSQ when designing a mix-aware load dispatching policy to further balance resource loads on heterogeneous servers?

To this end, we develop a novel mix-aware load-balancing algorithm, which aims to balance the server loads as well as balance bottleneck across the various server resources such that the global response time averaged over all applications is minimized. We explore the JSQ policy and the analytical results in [15], which illustrate via a closed queueing network model that there is an optimal application mix for each server. Our algorithm distributes requests based on two criteria: the number of outstanding requests per server and the desire to balance the load across each resource of each server, by eliminating the existence of a single bottleneck, e.g., CPU or IO. A *balance* metric, that measures the variability of queue lengths for outstanding requests across servers, is used to decide whether it is more beneficial to balance the number of outstanding requests per server using JSQ or to slightly violate the JSQ conditions by aiming to eliminate single resource bottlenecks in each server by directing requests such that each server serves an "optimal" load. The threshold value of the *balance* metric that triggers the above activity is based on a bound analysis of the potential queue lengths at the various servers. We use an event-driven simulator to evaluate the scalability of the proposed algorithm and a prototype implementation to support the applicability of the proposed algorithm on a real system.

The contributions of this paper are both analytical and practical. Our proposed load-balancing algorithm is designed for distributing time-varying requests from multiple applications, hosted on a set of heterogeneous systems with multiple resources. Being aware of the queue length of the servers and of the application mix, our algorithm is able to achieve consistent performance improvements when compared to the standard load-balancing policy available on Apache web server, i.e., the `bybusyness` policy.

This paper is organized as follows: The system architecture is explained in Section 2. The proposed mix-aware load-balancing algorithm is explained in Section 3. Section 4 presents experimental results. Related studies are summarized in Section 5. Section 6 concludes this paper.

2. SYSTEM MODEL

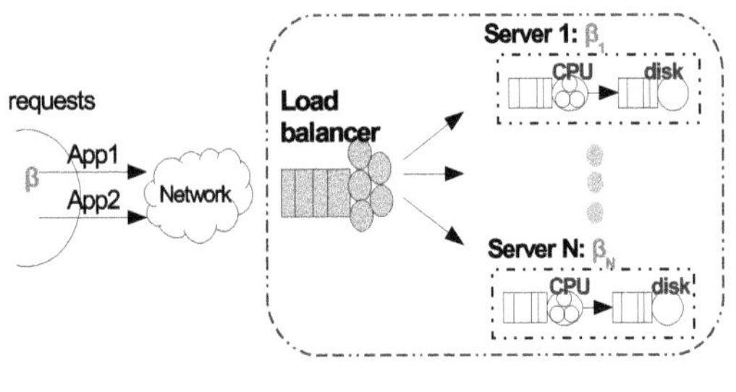

Figure 1: System model

We consider a service hosting system consisting of K heterogeneous servers and one load balancer, as well as *two* types of applications, that we classify as *CPU-intensive* and *IO-intensive*. These two classes of applications are referred with the index $j = \{1,2\}$. We model each server as a two-station queueing model shown in Fig. 1, see also [2, 7] for the effectiveness of such queuing network models in capturing the performance of non-trivial application co-location. The first station marked "CPU" represents the aggregate computational capacity of the available cores in the server, and the second station marked "IO" corresponds to the disk of the server. These two stations are referred with the index $i = \{CPU, IO\}$. The available computational and IO capacity is different across servers. As a result, the execution times of the same application on each server may vary. The threads at the load balancer are concurrently dispatched to the servers, based on a load balancing scheme. Performance statistics (e.g., response times and the characteristics of the workload mix) are collected at the load balancer. We consider that the network delay of dispatching requests from the load balancer to servers to be negligible.

Clients generate requests, which are sent to the load balancer. We assume that the interarrival times of requests follow an exponential distribution with parameter λ. The percentage of the two application classes in the arriving flow of requests is $\{\beta_1, \beta_2\}$ with $\beta_1 + \beta_2 = 1$. Both types of requests can be executed on any server; the execution of a request needs certain CPU time and IO time. Once requests complete execution in both stations, they leave the system.

The cumulative time an application spends on the CPU station includes any memory and cache accessing times. Yet, as shown in [2, 7], although the memory and caches are not explicitly modeled, if the CPU execution times without any queueing are correctly measured, then the performance effects of memory and cache are captured via the queueing delays at the CPU station. The cumulative time spent on the CPU or on the IO resources by an application in the absence of any queueing is referred to as *resource demand*. CPU and IO demands are assumed to be exponentially distributed with mean R_{ij}, with the subscript i referring to the resource and the subscript j referring to the application. Note that the resource demands do not reflect any waiting nor queueing times due to resource contention with other applications.

Resource demands vary on different servers, because of the server heterogeneity. Consequently, to capture the variability of CPU and IO demands on different servers, we introduce the scaling factors $\gamma_{i,k}$, corresponding to each resource i on server $k = \{1 \dots K\}$. In a homogeneous system, the scaling factors equal to one for all servers, implying that the average resource demands of an application are the same across all servers. Based on this convention, the average resource demands of application j on server k are $R_{ij} \cdot \gamma_{i,k}$. Because services are replicated, i.e., each server may be able to execute *any* application, we assume that it serves a percentage of requests β_j of application j. The application mix received by each server depends on the load-balancing policy, and this mix may be very different from the one that is observed by the load balancer (or alternatively, the one generated by the clients).

3. LOAD BALANCING SCHEME

In this section, we first introduce some theoretical background regarding optimization of application mixes for single server systems and load-balancing schemes for multiple servers. Motivated by the advantages of balancing server loads and resource loads on a single server, we propose a new, mix-aware, load-balancing algorithm.

3.1 Background

3.1.1 Optimal Application Mix in a Single Server

Rosti et. al [15] showed that there is an optimal application mix, which can minimize response time while maximizing throughput. The authors of [15] focus on a single server closed system with multiple resources and provide a closed form formula on the application mix, which is derived from the equal utilization point for all system resources.

Intuitively, a system operates at the best performance when the global utilization of the resources is maximized. According to their methodology, in a system with two stations and two classes of applications, the optimal mix of application class j on server k, is:

$$\beta^*_{1,k} = \frac{\log \frac{R_{IO,2}\gamma_{IO,k}}{R_{CPU,2}\gamma_{CPU,k}}}{\log \frac{R_{CPU,1}\gamma_{CPU,k} R_{IO,2}\gamma_{IO,k}}{R_{CPU,2}\gamma_{CPU,k} R_{IO,1}\gamma_{IO,k}}}, \quad (1)$$
$$\beta^*_{2,k} = 1-\beta^*_{1,k}.$$

This optimal mix refers to a single server k. Here the problem we are addressing is significantly more complex because to minimize response times in a system consisting of multiple heterogeneous servers, one needs to reach not only the optimal mix of Eq. 1 for each server, but also minimal queue lengths of outstanding requests across all servers. Clearly, joining the server with the shortest queue of outstanding requests is not a sufficient condition to optimize performance.

3.1.2 *Load Balancing on Multiple Servers*

There is a large body of literature on how to balance loads on multiple servers, especially for web systems. The Apache web server provides three default policies, namely `bybusyness`, `byrequest`, and `bytraffic`. The `byrequest` policy is very similar to the round-robin policy. The `bybusyness` policy is almost identical to JSQ, except when handling the situation where multiple servers have the same number of outstanding requests. Indeed, JSQ is a simple yet powerful policy for balancing homogeneous server loads, its optimality has been shown theoretically [10]. The `bytraffic` policy tries to balance number of bytes transmitted by each server.

3.2 Mix-Aware Algorithm

We now develop a mix-aware policy, leveraging the advantages of the optimal application mix on a single server and the JSQ policy on multiple servers. We first illustrate some bound analysis of the servers which provides us guidance to tune the algorithm parameters and we then present the mix-aware load balancing algorithm.

3.2.1 *Overview*

The proposed mix-aware algorithm tries to balance the queue length of outstanding requests on servers using JSQ and the resource loads on each server by achieving the optimal application mix that equalizes the utilization IO and CPU resources. When servers have very different queue lengths, it is imperative to balance the outstanding requests across servers. When queues of servers are balanced, then the algorithm slightly "unbalances" the queues by aiming to reach the optimal mix as shown by Eq. (1) on each server and achieve equiutilization across its resources. Straddling between the above two competing targets, the mix-aware algorithm aims to achieve overall better performance.

We definite a metric called *balance* that quantifies the degree of (un)balancing across the K servers. We start by defining this metric and discuss its usefulness.

3.2.2 *Balance and Threshold*

The queue length of outstanding requests is an indicator of the server load. The higher the variability of queue length, the higher the performance improvement that can be achieved with JSQ. To identify the best opportunity for applying JSQ, we propose a metric, called *balance*, defined as follows:

$$balance = 1 - \frac{Q^{min}}{\sum_{k=1}^{K} Q_k} \times K. \quad (2)$$

Q_k denotes the queue length of server k and Q^{min} is the minimum queue length across all servers, i.e., $Q^{min} = \min\{Q_k, \forall k\}$. The reasoning behind this is that when comparing Q^{min} to the average queue length, one can obtain not only a rough estimate of the variability but also of the potential performance improvement using JSQ. When all servers have the same queue lengths, Q^{min} is equal to the average value and thus the *balance* value is 0. In contrast, when Q^{min} is significantly lower than the average queue length, *balance* is approaching one and thus using JSQ can improve the overall system performance. A higher *balance* value indicates greater variability across queues as well as the advantage of applying JSQ. The mix-aware algorithm dispatches requests primarily using JSQ but when the *balance* value is low, as a second step it attempts to unbalance the queues at the servers by aiming to equally utilized the CPU and IO resources at the servers. This unbalancing action is bringing higher performance gains on a per-server basis, which is also positively reflected in the overall system performance.

Since *balance* is a metric between zero and one, it is not clear what is its value that should trigger using JSQ or instead the unbalancing the queue lengths while aiming equiutilization of individual server resources. To calculate this threshold $\bar{B}$, we use the upper bound of the queue length variability across servers. Let $Var[Q]$ present the variance of queue lengths.

COROLLARY 1. *The upper bound of $Var[Q]$ is a function of balance:*

$$Var[Q] \le \frac{(K-(1-balance)(K-1))^2-1}{K^2}\left(\sum_{k=1}^{K} Q_k\right)^2. \quad (3)$$

PROOF. Based on the definition of variance, one can write

$$Var[Q] = \frac{\sum_{k=1}^{K} Q_k^2}{K} - \left(\frac{\sum_{k=1}^{K} Q_k}{K}\right)^2. \quad (4)$$

Following simple algebraic manipulations, one can write the upper bound of the queue length of server k as

$$Q_k \le \sum_{h=1}^{K} Q_h - (K-1)Q^{min}, k \in \{1 \ldots K\} \quad (5)$$

As Q^{min} is bounded by the average queue length, then $Q^{min} \in \{0 \ldots \frac{\sum_{k=1}^{K} Q_k}{K}\}$. We further relax the discrete property of Q^{min} and express Q^{min} as a continuous variable using $\alpha \in [0,1]$,

$$Q^{min} = \alpha \times \frac{\sum_{k=1}^{K} Q_k}{K}, \alpha \in [0,1]. \quad (6)$$

Substituting Eq. 6 in Eq. 5, we obtain Eq. 7

$$Q_k \le \left(\frac{K-\alpha(K-1)}{K}\right)\sum_{k=1}^{K} Q_k. \quad (7)$$

Substituting Eq. 7 in Eq. 4 we obtain Eq. 8

$$Var[Q] \le \frac{(K-\alpha(K-1))^2-1}{K^2}\left(\sum_{k=1}^{K} Q_k\right)^2. \quad (8)$$

Combining Eq. 6 and Eq. 2, we straightforwardly obtain

$$balance = 1-\alpha, \alpha \in [0,1]. \quad (9)$$

Substituting Eq. 9 in Eq. 8, we thus can express the upper bound of $Var[Q]$ as a function of *balance* □

To quantitatively gauge the relationship between the queue length variability and *balance*, we plot the upper bound of $Var[Q]$ in Figure 2, assuming a scenario of three servers, i.e., $K = 3$. Clearly, the upper bound of the queue length variability increases as *balance* increases. We can observe that when *balance* values are small, e.g., less than 0.1, such upper bounds are constantly low, independently of the values of $(\sum_{k=1}^{K} Q_k)^2$, as shown by the flat area on the left corner of Figure 2. On the other hand, when *balance* is greater than 0.2, the upper bound varies a lot due to the multiplication of $(\sum_{k=1}^{K} Q_k)^2$ in Eq. 3. We conjecture that a small *balance* value ensures the similarity of queue lengths across servers. Our extensive evaluation of the numerical results of Eq. 3 leads us to set the *balance threshold*, $\bar{B}$, very low, i.e., between 0.05 to 0.1, so as to capture the cases where the queue lengths of all servers are very similar.

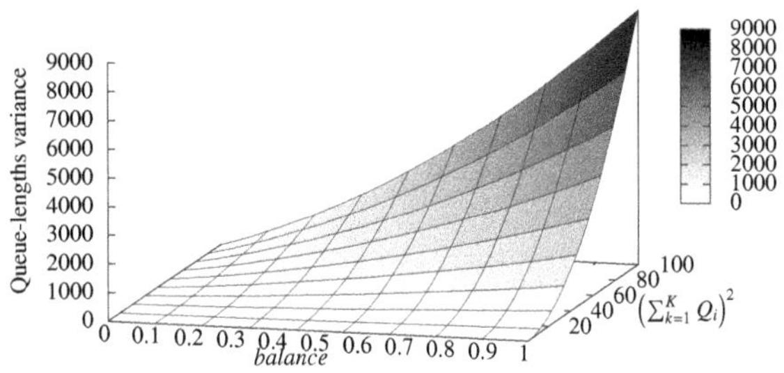

Figure 2: The upper bound of queue length variability of a three server case

3.2.3 Unbalance by Equiutilizing

When dispatching a particular request, the aim is to search for a server, whose optimal application mix (i.e., server equiutilization, as expressed by Eq. 1) can be reached as quickly as possible. When a request from application j arrives, we check the difference between the current application mix, $\beta_{j,k}$, including this request, and the optimal mix on server k, $\beta^*_{j,k}$, for all servers:

$$S_{j,k} = \beta_{j,k} - \beta^*_{j,k}.$$

We chose the server with the minimum $S_{j,k}$ value. When this value is negative, it implies that there is not a sufficient number of requests from application j on server k. The resource utilizations (loads) on server k can be better balanced by processing an extra request from application j. When all servers have positive S values, it implies that all servers have more application j requests than their optimal values, at the time instant the new request arrives. In such a case, choosing a server with the minimum S value can minimize the deviation from the optimal one.

3.2.4 Combining JSQ and Unbalance by Equiutilization

Combining together JSQ and Unbalance by Equiutilizing, we obtain the Mix-Aware Load Balancing Policy. As a first step, the algorithm calculates the *balance* metric and compares it to the $\bar{B}$ threshold. If the value is larger than $\bar{B}$, then the incoming request from application class j is scheduled using JSQ. Otherwise, the $S_{j,k}$ values are computed for all k servers and the server than can best improve its optimal value in Eq. 1 is selected. The required inputs of the algorithm are the class of the current request, the current queue lengths on all servers, and the current application mixes on all servers. The load balancer computes these inputs by keeping counters of outstanding requests for both application classes on each server. When a request arrives or leaves the cluster, the load balancer retrieves its application type from the request header and increments or decrements the corresponding counter. The load balancer is assumed to know the optimal application mix listed in Eq. 1 for all servers. The output of the algorithm is the server to which the arriving request should be dispatched. We assume that the resource demands and scaling factors required in Eq. 1 can be obtained via statistical profiling methods, although the development of such profilers is out of the scope of this paper. We summarize all this in Alg. 1.

Algorithm 1 Mix-Aware load-balancing algorithm.

//j is the class of the arriving request
function MIX-AWARE(j)
  $\mathbf{Q} \leftarrow \{Q_i, \ldots, Q_k\}$
  $\mathbf{S}_j \leftarrow \{(\beta_{j,1} - \beta^*_{j,1}), \ldots, (\beta_{j,K} - \beta^*_{j,K})\}$
  $server \leftarrow k \; such \; that \; Q_k = \min(\mathbf{Q}), k \in \{1 \ldots K\}$ //JSQ
  $balance \leftarrow 1 - \frac{Q_{res}}{\sum_{k=1}^{K} Q_k} \times K$
  if $balance < \bar{B}$ **then** //Unbalance by Equiutilizing
    $server \leftarrow k \; such \; that \; (\beta_{j,k} - \beta^*_{j,k}) = \min\{\mathbf{S}_{j,k}\}, \forall k$
  end if
  return *server* //send request to the selected server
end function

4. EVALUATION

In this section, we evaluate the proposed mix-aware algorithm for service systems, using event driven simulation and a prototype. The performance metrics evaluated are the response times, and difference between CPU and disk utilization of servers. We benchmark the mix-aware load balancing algorithm against purely JSQ and `bybusyness`, under different system sizes and workload mixes.

4.1 Simulation Results

To evaluate the scalability of the mix-aware algorithm on different system sizes, we built an event driven simulator of the model shown in Figure 1 and also a discrete event simulator for JSQ. We simulate the systems with 10, 20 and 30 replicated servers. The workload is generated using the following traffic intensities: interarrival rates are exponentially distributed with $\lambda = 9$, 18 and 27 per second, for the scenarios with 10, 20 and 30 replicas, respectively. We assumed a two-class workload, with 55% of the requests are from class one, and the remaining requestst are from class two. The CPU and disk demands for both applications are also exponentially distributed with averages being $R_{CPU,1} = 0.75s$, $R_{CPU,2} = 0.64s$, $R_{IO,1} = 0.48s$, and $R_{IO,2} = 1.25s$. To enforce server heterogeneity, the scaling factors $\gamma_{i,k}$, are uniformly distributed in $[0.8, 1.2]$.

Our mix-aware algorithm computes the optimal $\beta^*_{j,k}$ for application class j in server k, see Eq. 1. With the above inputs, this value is in $[0.37, 0.89]$. Note that we assume that the resource demands are known and are obtained via a profiling methodology, see [?, ?]. The threshold value of *balance*, $\bar{B}$, is set to 0.1.

We summarize the simulation results in Figure 3. Compared to JSQ, the mix-aware algorithm can achieve a significantly lower (roughly 35%) average response time. As expected, the average response times of mix-aware decreases with an increasing number of servers, whereas JSQ seems to not scale well with the cluster size, as shown by the increase in the response times from the scenario of 10 servers to 20 servers. Focusing on equiutilizing the loads on the server resources is expected to be more effective when there is a larger number of servers, because of a higher chance to attain the optimal mix on individual servers. To validate this reasoning, we compute the difference of average CPU and disk utilization per server and plot the maximum values among all servers, see Figure 3(b). The plotted values show the worst case of unbalance resource utilizations, which results in resource bottlenecks, overall higher queue lengths, and thus higher response times. Clearly, the mix-aware algorithm can achieve more "balanced "resource utilizations than JSQ.

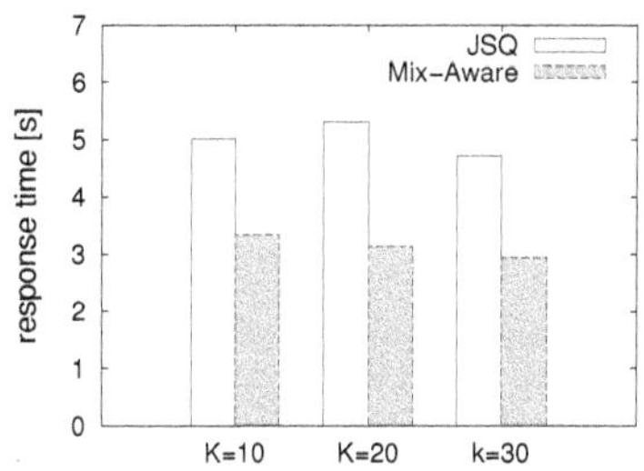

(a) Response times

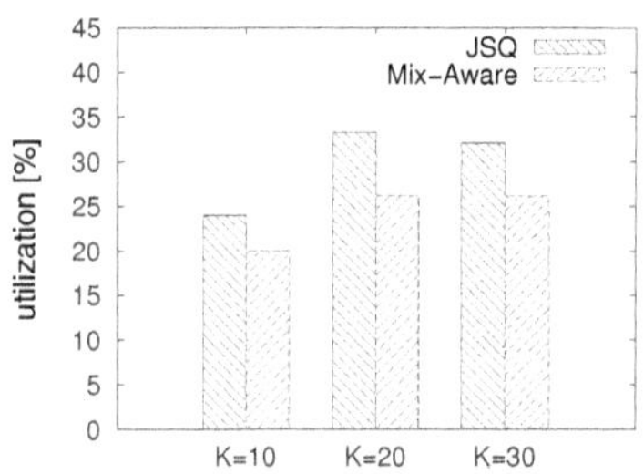

(b) Maximum difference of cpu and disk utilization

Figure 3: Simulation results: comparison of JSQ and mix-aware algorithm on systems with different numbers of servers.

4.2 Prototype

We built a prototype system, consisting of three servers and one load balancer, in the IBM Research Cloud. Each component is based on a virtual instance, equipped with 4GB of RAM, 4 x86_64 CPUs, 55GB of disk, and Fedora 14 the operating system distribution. Requests are generated from the client servers, located at the IBM Zurich Research Laboratory, using the httperf load generation tool [13], and forwarded to the load-balancer, which is based on the Apache web server version 2.4.2. Naturally, we present here experiments with `bybusyness`, the standard Apache load balancing policy that resembles JSQ and our mix-aware policy.

We consider a two-class workload. Class one consists of the `fop` benchmark in the Dacapo suite [11], a CPU intensive benchmark producing PDF documents from XSL-FO files. Class two consists of `luindex`, a disk intensive benchmark that indexes a set of documents. The average inter-arrival rate of arriving requests (`fop` or `luindex`) is $\lambda = 2$ applications per second. We consider two workload scenarios: (1) the two-class mix is *stationary* and kept at 0.5, and the total number of requests (i.e., applications sent for execution) is 2000; (2) the workload mix is *non-stationary*, i.e., changes across time as shown in Figure 4(a).

The challenge of applying the proposed mix-aware algorithm on a real system lies in deciding the optimal application mix for each server. Within a server, the optimal mix is 0.4, i.e., 40% of `fop` and 60% of `luindex`, see [2] for the methodology to compute ths mix. As in the simulation experiments, the threshold value of *balance*, $\bar{B}$, is set to 0.1.

4.2.1 Stationary Workload

In Table 1, we summarize the average, 90th, 95th, 99th percentile and maximum response times for both bybusyness and mix-aware algorithms. To accommodate high performance variability in the computing cloud [5], we present the results of two separate experimental runs. One can see that the statistics of the higher percentiles for a given policy have higher differences from one experimental run to the other. Moreover, one can observe that the average response times of mix-aware are lower than `bybusyness` by roughly 10%, for all presented statistics. The performance improvement of mix-aware is particularly visible for the higher percentiles of response times. For example, in run one, the maximum response time of mix-aware is lower than `bybusyness` by roughly 24%, whereas the performance gain of the 90th−percentile of the mix-aware algorithm is only 10%. This experiment illustrates that the mix-aware algorithm is able to achieve lower response times, especially for the higher percentiles, meaning that the worse performance is better mitigated.

4.2.2 Non-stationary Workload

In this subsection, we evaluate our proposed algorithm on a non-stationary workload mix, as shown in Figure 4(a). Here we focus on presenting the higher percentile of response times computed from 2 minute windows, see Figure 4 (b), (c), and (d). One can observe that mix-aware is able to achieve lower response times, especially for the 99th percentile. Another observation worth mentioning is that the performance gain of the mix-aware algorithm compared to `bybusyness`, is lower in our cloud prototype than in simulation. This can be attributed to applying inaccurate values of optimal application mix on virtual servers, which are known to be highly variable [5], i.e., resource demands may change across time. In our future work, we will address the profiling methodology that is able to compute the optimal mix on the fly. Nonetheless, given the inherent inaccuracy of optimal values applied in the mix-aware algorithm in the cloud environment, we are still able to mitigate the worse performance, compared to the `bybusyness` algorithm.

5. RELATED WORK

There is a large body of related studies of load balancing for various conventional service systems [6, 8, 14, 17] and modern cloud systems [9]. As a detailed survey of the extensive related work is not possible here, we only outline some particularly relevant work. Cardellini et. al [6] qualitatively classified existing load balancing schemes for web server systems into four approaches, namely server-based, client-based, DNS-based, and dispatcher-based. They quantitatively compared the maximum utilization of clusters, via simulation. Cherkasova and Ponnekanti [8] developed FLEX, a locality-aware load balancing solution, especially for efficient memory usage. Zhang et. al [14, 17] proposed ADAPTLOAD for rapidly fluctuating workloads that are characterized in terms of arrival rates and document popularity. Via simulation, they showed that ADAPTLOAD can achieve a low slowdown and resource utilization on a cluster of homogeneous web servers. Björkqvist et. al [3] used a lottery balancing algorithm for distributed stateful service systems. Dejun et. al [9] focused on balancing requests among a set of heterogeneous machine instances in the cloud, based on the profile of response times and request rates for each server.

Singh et. al [16] leveraged the idea of application mixes and proposed a mix-aware resource allocation for data centers. They used a k-means clustering algorithm to automatically determine the workload mix and allocate servers. However, they did not explore the workload mix in their load balancing scheme.

Most of the aforementioned studies focus on balancing loads on homogeneous servers with a single resource type, i.e., CPU. In contrast, our mix-aware algorithm considers multiple resources on a set of heterogeneous servers, and aims at balancing the overall server loads as well as their resource load.

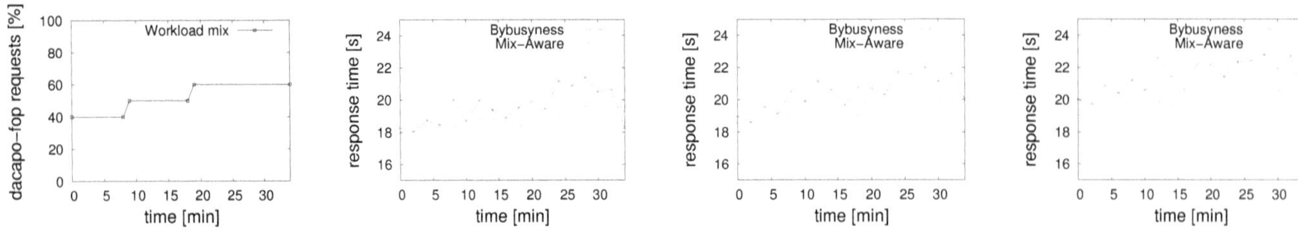

(a) workload mix, β_{ge} (b) 90th percentile (c) 95th percentile (d) 99th percentile

Figure 4: Non-stationary workload mix: comparison of `bybusyness` and mix-aware algorithm on 90th, 95th and 99th response times over the time.

Statistics	mean[s]		90th[s]		95th[s]		99th[s]		Max[s]	
Algorithm	bybusy.	m-w	bybusy.	m-w	bybusy.	m-w	bybusy.	m-w	bybusy.	m-w
run 1	16.40	15.01	19.70	17.30	20.60	17.90	21.80	18.60	24.10	19.54
run 2	16.46	15.08	19.10	17.30	19.90	17.90	21.10	18.80	22.93	19.96

Table 1: Response times statistics of stationary workload mix: comparison of `bybusyness` and mix-aware load balancing.

6. CONCLUSION

In this paper, we propose a mix-aware load-balancing algorithm for web services that are hosted on heterogeneous servers and cater for requests from a dynamically varying application mix. Our algorithm aims at minimizing the response times by dispatching requests in a way that optimizes the application mix for each server. We evaluate the how well the variability of outstanding requests across servers and based on this metric we balance the outstanding requests using JSQ or unbalance the server queues by aiming to equally utilize CPU and disk resources on the servers. Our results obtained with simulation and with a cloud prototype show that our proposed algorithm can scale with an increasing system size, and improve on the worst response times compared to the `bybusyness` policy, one of the most robust default load-balancing algorithms in the Apache web server. Regarding ongoing research, we are exploring efficient on-line profiling methodologies to obtain the optimal application mix, as well as extending our algorithm to a larger number of applications.

7. ACKNOWLEDGMENTS

The research presented in this paper has been supported by the Swiss National Science Foundation (project 200021_141002) and by the European Commission (Seventh Framework Programme grant 287746). The research presented in this paper was conducted while the first author was with IBM Research Zurich Lab. Evgenia Smirni is partially supported by NSF grants CCF-0937925 and CCF-1218758.

8. REFERENCES

[1] Amazon EC2. http://www.amazon.com/.

[2] D. Ansaloni, L. Y. Chen, E. Smirni, and W. Binder. Model-driven Consolidation of Java Workloads on Multicores. In *Proceedings of IEEE/IFIP DSN*, pages 1–12, 2012.

[3] M. Björkqvist, L. Y. Chen, and W. Binder. Load-balancing dynamic service binding in composition execution engines. In *Proceedings of APSCC*, pages 67–74, 2010.

[4] M. Björkqvist, L. Y. Chen, and W. Binder. Dynamic replication in service-oriented systems. In *IEEE/ACM International Symposium on Cluster, Cloud and Grid Computing (CCgrid)*, pages 531–538, 2012.

[5] M. Björkqvist, L. Y. Chen, and W. Binder. Opportunistic service provisioning in the cloud. In *Proceedings of IEEE Cloud*, pages 237–244, 2012.

[6] V. Cardellini, M. Colajanni, and P. S. Yu. Dynamic load balancing on web-server systems. *IEEE Internet Computing*, 3(3):28–39, 1999.

[7] L. Y. Chen, D. Ansaloni, E. Smirni, A. Yokokawa, and W. Binder. Achieving Application-Centric Performance Targets via Consolidation on Multicores: Myth or Reality? In *Proceedings of HPDC*, pages 37–48, 2012.

[8] L. Cherkasova and S. Ponnekanti. Optimizing a 'content-aware' load balancing strategy for shared web hosting service. In *Proceedings of MASCOTS*, pages 492–, 2000.

[9] J. Dejun, G. Pierre, and C.-H. Chi. Resource provisioning of web applications in heterogeneous clouds. In *Proceedings of the 2nd USENIX conference on Web application development*, pages 5–5, 2011.

[10] V. Gupta, K. Sigman, M. Harchol-Balter, and W. Whitt. Insensitivity for ps server farms with jsq routing. *SIGMETRICS Performance Evaluation Review*, 35(2):24–26, 2007.

[11] http://dacapobench.org/. `Dacapo suite`.

[12] http://httpd.apache.org/. `Apache`.

[13] http://www.hpl.hp.com/research/linux/httperf/. `httperf`.

[14] A. Riska, W. Sun, E. Smirni, and G. Ciardo. Adaptload: Effective balancing in custered web servers under transient load conditions. In *Proceedings of ICDCS*, pages 104–111, 2002.

[15] E. Rosti, F. Schiavoni, and G. Serazzi. Queueing Network Models with Two Classes of Customers. In *Proceedings MASCOTS*, pages 229–234, 1997.

[16] R. Singh, U. Sharma, E. Cecchet, and P. Shenoy. Autonomic mix-aware provisioning for non-stationary data center workloads. In *Proceedings of ICAC*, pages 21–30, 2010.

[17] Q. Zhang, A. Riska, W. Sun, E. Smirni, and G. Ciardo. Workload-aware load balancing for clustered web servers. *IEEE Trans. Parallel Distrib. Syst.*, 16(3):219–233, 2005.

Decision Support via Automated Metric Comparison for the Palladio-based Performance Blame Analysis

Frank Brüseke
s-lab – Software Quality Lab
University of Paderborn
Paderborn, Germany
fbrueseke@s-lab.upb.de

Gregor Engels
s-lab – Software Quality Lab
University of Paderborn
Paderborn, Germany
engels@s-lab.upb.de

Steffen Becker
Heinz Nixdorf Institute
University of Paderborn
Paderborn, Germany
steffen.becker@upb.de

ABSTRACT

When developing component-based systems, we incorporate third-party black-box components. For each component, performance contracts have been specified by their developers. If errors occur when testing the system built from these components, it is very important to find out whether components violate their performance contracts or whether the composition itself is faulty. This task is called performance blame analysis. In our previous work we presented a performance blame analysis approach that blames components based on a comparison of response time values from the failed test case to expected values derived from the performance contract. In that approach, the system architect needs to manually assess if the test data series shows faster or slower response times than the data derived from the contract. This is laborious as the system architect has to do this for each component operation. In this paper we present an automated comparison of each pair of data series as decision support. In contrast to our work, other approaches do not achieve fully automated decision support, because they do not incorporate sophisticated contracts. We exemplify our performance blame analysis including the automated decision support using the "Common Component Modeling Example" (CoCoME) benchmark.

Categories and Subject Descriptors

D.2.5 Testing and Debugging, D.2.8 Metrics, G.3 Probability and Statistics

Keywords

Performance blame analysis, CBSE, data series comparison, performance prediction, performance test

1. INTRODUCTION

In component-based software engineering, software architects develop systems by composing third-party black-box components. For each component, functional and non-functional component contracts have been specified by their developers. After composing all components, software architects test the composition in test cases before shipping the system to its end-users. We focus here on a subset of these test cases, which test the fulfillment of the system's performance. If the architect discovers errors while executing such a performance test case, she has to investigate in a so-called performance blame analysis activity whether components violate their performance contracts or whether the composition itself is faulty.

In order to blame components, architects face the problem to identify components violating their performance contract. To tackle this problem, we have proposed a performance blame analysis process in previous work [2] (c.f. Figure 1). It is based on the collection of measured component performance metrics from the failed performance test case and expected performance metrics derived from the component's performance contract (Step 1). Performance contracts are formalized using the Palladio Component Model (PCM) [1]. If the measured performance metrics from the performance test case violate their performance contract, architects must blame the respective component (Step 3). However, in order to compare the measurements to the specification, the architect currently has to compare both performance metrics of each component. The performance metrics are either represented as complex raw data sets or statistical characterizations. This comparison is tedious and error prone. In this paper, we add a novel automated decision support step (Step 2) to our process to speed up the performance blame analysis.

Existing semi-automated decision support approaches for component-based performance blame analysis are limited. If they use components without performance contracts, decision support is restricted to data aggregation only leaving the tedious comparison task to the system architect. In case they do support component performance contracts, these contracts are not parameterized by the component's context (e.g. [15, 17]). The context of a component is defined as the component's usage, its connected external services, and its allocation on execution environments. As the context impacts a component's performance, approaches for non-

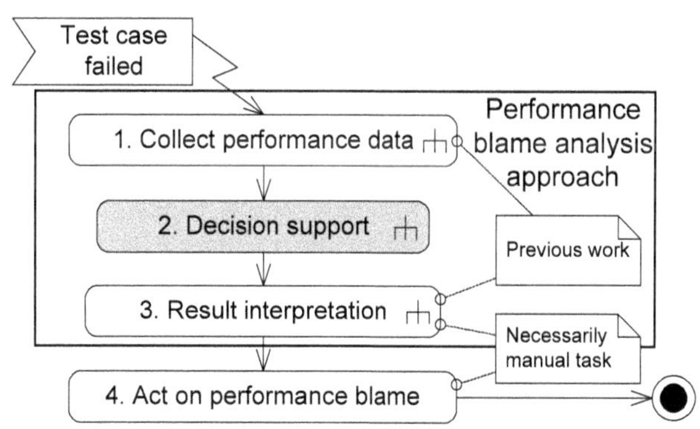

Figure 1: UML 2 activity diagram showing the steps in our previous performance blame analysis approach [2] plus the novel decision support step (inside the box)

parameterized component performance contracts have to replace missing context information by manual or heuristically generated specifications. Our approach is the first to explicitly consider context influences. By exploiting the parameterized component performance contracts, we have achieved a sophisticated automated comparison of performance contracts and performance measurements from failed test cases.

In this paper, we enhance our existing blame analysis process by a novel automated decision support step to drastically reduce the manual analysis effort. In order to automate the decision support step, we have derived a set of statistical indicators that can be used in an automated process to identify components with faulty performance. Our tool computes these indicators for the data collected during the execution of the performance test case and the performance metrics derived from a component's performance contract. Computed indicators are mapped to severity levels which direct the system architect to components violating their contractually specified performance. We include these indicators into a novel visualization based on flame graphs [6]. Our extended flame graphs use a coloring scheme based on the indicated severity to highlight contract violations.

We have validated the novel decision support step on two variants of the Common Component Modeling Example (CoCoME) [9]. This system has been designed as a benchmark for component-based analysis methods. It implements the supply chain management of a supermarket chain. Our evaluation shows that the automated decision support step helps the software architect to easily blame the appropriate components. The blamed components' response time during the test case execution violates the expected response time specification. We claim that our visualization is a concise and easy-to-grasp tool for software architects.

The contribution of this paper is an extension to our performance blame analysis process, which semi-automatically identifies components to blame. It uses a set of derived statistical indicators to feed a novel visualization for blame analysis support. We provide a case study based on CoCoME giving evidence for the effectiveness of our approach.

The remainder of the paper is structured as follows. Section 2 introduces a running example that is used throughout the paper. Section 3 revisits our previous work, the Palladio-based blame analysis for component-based systems. Section 4 derives the automated decision support and introduces indicators and visualizations. Moreover, it also discusses necessary changes to the result interpretation. Section 5 explains the automated decision support and result interpretation in detail. It exemplifies these new elements in detail using the running example. Section 6 describes the limitations of our approach. Section 7 discusses related work from the areas of performance visualization and performance blame analysis. Lastly, Section 8 concludes the paper and gives an outlook on our future work.

2. RUNNING EXAMPLE

Throughout this paper we use a running example. The running example is used to exemplify certain points in the reasoning of this paper. An in-depth evaluation of this example follows in Section 5.

The scenario of the example is the most sophisticated use case of the Common Component Modeling Example (CoCoME) [9]. The CoCoME was specified as a benchmark system to compare different approaches for component-based software architecture analysis, such as performance predictions. It specifies a trading system for a supermarket chain that was inspired by a real example. We refer to the trading system as the CoCoME-

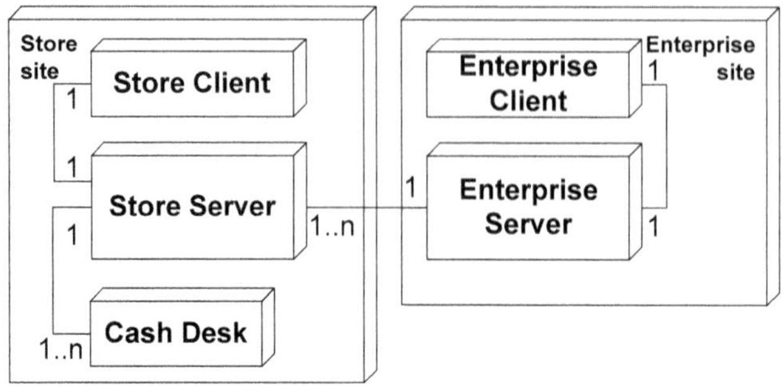

Figure 2: System structure of the CoCoME trading system

system. The CoCoME-system includes the whole infrastructure from the cash desks, where items are sold to customers, to the store servers registering those sales and to an enterprise server aggregating all the information. This system structure of the CoCoME is shown in Figure 2.

The CoCoME architecture is an archetype of a whole class of systems. While the cash desks resemble the embedded system part of the overall system the store and enterprise servers make up the information system part. In this class of systems the data is produced by distributed use of embedded system devices. All data are then stored and processed in the overlaying information system.

This paper focuses on the store server and enterprise server of the CoCoME-system. These make up the information system part of the trading system. In this setting, we investigate the scenario that solely consists of the use case "inter-store exchange" for shipping items among the stores of an enterprise. This use case is the most sophisticated use case of the CoCoME. It is triggered by each sale. A store server checks if the items' stock falls below a minimal threshold. If so, it triggers the enterprise server to cause near-by stores to send some of the missing items to the store in need. It then optimizes the transports such that no store runs out of the transported items and the overall distribution distance is minimal.

Performance blame analysis, as outlined in our previous work [2], deals with the response times of component operations. Figure 3 outlines the calls of the component operation in our running example. Figure 3 shows the participating component objects (meaning an instance of a deployed component; cf. Cheesman and Daniels [3]) being deployed on several nodes. The enterprise server and several store servers participate in the scenario. The solid arrows along the sockets and interfaces represent component operation calls and the dashed arrows stand for operation returns. These messages are numbered in the order of their execution. Since most return messages are given directly after the corresponding operation call these return messages are not numbered.

The scenario in question starts with a store server (node on top in Figure 3) registering a sale (message 1). The origin of message 1 is a test driver component (cf. Figure 3). The registered sale lowers the store's stock, such that it crosses the minimal threshold for the item stock. The store then finds all missing products (message 2) and communicates to the enterprise server (message 3) to initiate a transport that replenishes the store's stock for these products. The `ProductDispatcher` component first determines the relevant enterprise information (message 4). The enterprise information also contains references that enable the `ProductDispatcher` to contact all stores. The `ProductDispatcher` queries the stock information for each missing product in all the other stores (messages 5). Messages crossing the border to the multi-node "Other store servers" stand for several messages. One message is sent to each store node in consideration. The enterprise server then creates an optimized transport plan and lowers the stock for the transported products in those stores that need to send products (messages 6). The enterprise server then returns the amount of transported products back to the initial store server (message 7). Finally, the store then saves the products as incoming.

3. PREVIOUS WORK

The approach presented in this paper improves our previously introduced performance blame analysis approach [2]. Our previous work is based on comparing performance metrics from testing with performance metrics derived from performance contracts. In our previous work and also in this paper, we exclusively deal with the performance metric response time. The response times are derived from the performance contracts that the component developers supply in the form of the Palladio Component Model (PCM) [1]. This partial PCM model can then be incorporated into a system model by the system architect. The first step of our approach (cf. Figure 1) is then the collection of response time measurements from testing and from the PCM system model for a failing test case. To derive expected values from the PCM system model, the system architect uses performance prediction, i.e. the PCM simulation. Next, these two data series are used to produce a histogram. The system architect shall use the histogram to interpret (cf. step 3 in Figure 1) if a particular component operation violates its contract and therefore needs to be blamed. The system architect decides this by judging if the test data series has overall higher response time values than the data series of expected values.

Our blame analysis approach relies on several assumptions. First, the test case and the performance analysis scenario must be equivalent. Our approach does not guarantee this equivalence, but we have suggested including PCM models in the test case specification [2]. Then, the system architect can draw test cases and PCM models for performance prediction from the same source,

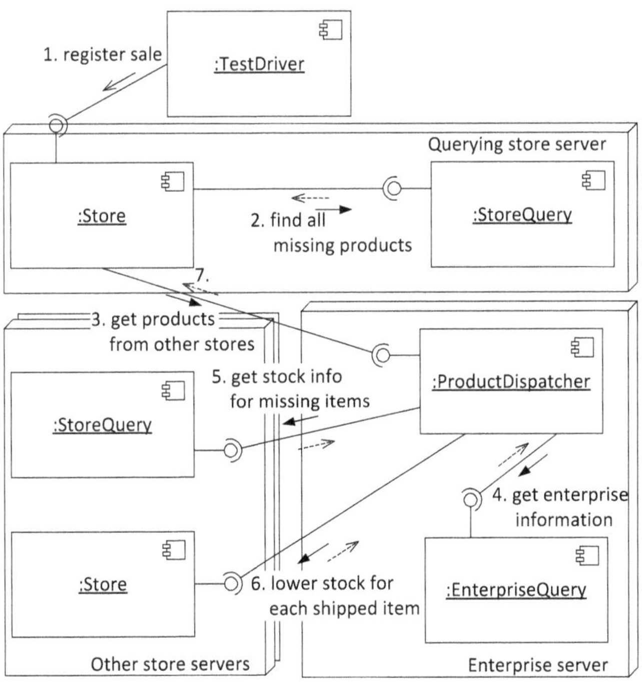

Figure 3: Scenario with all participating components and their exchanged messages

which implies equivalence. Second, the system architects need PCM contracts that adequately reflect the component developer's intentions for the performance of each component. The system architect must then correctly assemble a system model containing those PCM contracts. Moreover, the hardware usages included in the PCM contracts must be modeled in a standardized way. Combined with the hardware definitions used in the project they must result into response time values that can be easily compared to the measurements acquired in testing.

The following subsections elaborate more on the data collection and result interpretation of our previous work. Figure 4 gives a more detailed overview of the performance blame analysis approach. The data collection parts for testing and performance prediction as well as the result interpretation are introduced in turn. Subsection 3.1 introduces the test part of data collection while Subsection 3.2 deals with the performance prediction part. The result interpretation is covered in Subsection 3.3. The result interpretation is more explicitly explained, because this paper introduces an automated decision support speeding up result interpretation.

3.1 Data Collection – Test

As stated before, blame analysis always starts with a failing test case[1]. The system architect collects the performance measurements for the scenario that are included in the test case results. She checks if the data taken while executing the test case are sufficient. For our blame analysis approach the system architect

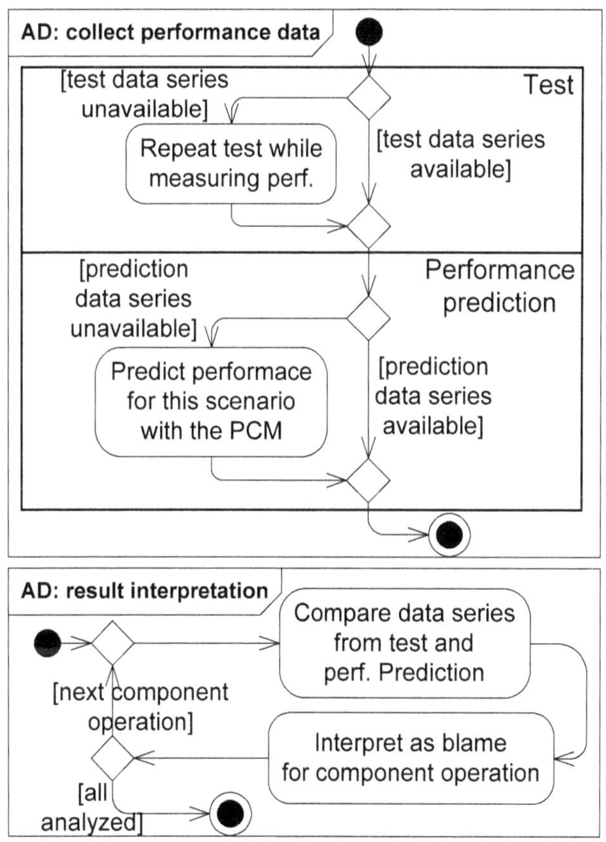

Figure 4: Previous performance blame analysis process [2]

[1] Please note that we assume that failing test cases can be identified with the help of detailed requirements, which specify the expected performance at the system boundary.

must collect the response times for all component operations that are involved in the test case. In our running example she must measure the response time of each operation call indicated in Figure 3. For message 4 "get enterprise information" she has to measure the response time of the call to the operation `queryEnterpriseByID()` in the component `EnterpriseQuery`. If all the needed response time measurements were collected, the system architect proceeds, otherwise she must repeat the test case and take the necessary measurements while doing so.

3.2 Data Collection – Performance Prediction

In our approach, the system architect utilizes the Palladio Component Model (PCM) [1] for performance prediction. We have chosen the PCM for this because of its role model. The role model has an explicit role for the component developer. It specifies which diagrams and model contents have to be delivered by the component developer. Moreover, it states how these parameterized PCM models have to be completed by the system architect, such that it reflects a specific system with specific usage scenarios. So, the component developers are able to specify their components' performance contracts, such that the specification is usable in different systems and for different use cases. This makes PCM performance specifications suited for exchanging between component developers and system architect.

In a full PCM model all parameters have been subsequently filled in by the various roles creating the model [1]. This PCM model explicitly covers the components' context. I.e. it models the system composition including external services, the allocation and the system usage. The usage model specifies the workload produced by virtual users. It includes frequency and order of calls as well as optional parameters like think time and input characterizations (e. g. value or number of elements).

The system architect must supply predicted response times for the failed test case by analyzing such a full PCM model. The system architect checks if they already have matching data series from performance analysis during the design time (cf. Figure 4). If they do not have predicted the performance for this environment setting and scenario yet, they must do so now. The system architect needs the predicted response times for any component operation participating in this scenario. In our running example the system architect must supply a response time data series for each of the operation calls in Figure 3, just like in testing.

For proper comparison, the performance data series from testing on one hand and from performance prediction on the other hand need to stem from the same scenario. In particular, the input workload must be the same performance-wise, and also the deployment must be the same.

3.3 Result Interpretation

Finally, the system architect can proceed to result interpretation (cf. Figure 4). In our previous work, the interpretation of the measured performance metrics from test and performance prediction is manual. To decide if a component operation needs to be blamed, the system architect has to compare the two response time data series. Then, the system architect has to decide whether the component operation exhibited higher or lower response time values than predicted. In our previous work [2], we have suggested that the system architect creates histograms with bins of the same size (e.g. 125 ms each). Each histogram visualizes the relative frequencies of the response time data series from testing and performance prediction side by side.

Figure 5 depicts an example histogram showing the response time distributions for the component operation `queryLowStockItems` of the component `StoreQuery`. This operation call corresponds to message 2 in Figure 3. The histogram in Figure 5 shows the relative frequency of the response time values from testing and the expected response time values from performance prediction. The X-axis exhibits the response time (in ms) and the Y-axis depicts the relative frequency how often this response time occurred. Figure 5 shows the histogram bins not as bars. This is also correct, because all the bins are of the same width, i.e. 125 ms wide. The curves for testing (blue) and performance prediction (red) each have only one peak. So, in this case it is clear that the test response time values are overall lower than predicted. However, the decision is not always trivial. In our experience the decision is hard to make, when the curves intersect more often. Also, if the blue curve had a different shape and had far more frequent occurrences of the response time around 1000 ms, the decision would be quite hard to make. Recognizing that an operation has overall slightly lower or higher response time values only by looking at a histogram such as this is quite hard.

The system architect has to construct and interpret one response time histogram, as shown in Figure 5, for each component operation. In our running example, the system architect needs to analyze 6 different component operation calls (cf. Figure 3). Thus, the comparison step is quite time consuming and tedious. This paper shows how an automated decision support step can look like.

4. DERIVING AN AUTOMATED DECISION SUPPORT SUB-PROCESS

This section discusses what indicators and visualizations are suitable for the automated decision support introduced in this paper. Subsection 4.1 discusses statistical indicators that can decide whether the response time data series from the test has higher values than the response time data series from performance prediction. Subsection 4.2 presents visualizations for these statistical indicators. Subsection 4.3 introduces a decision support subprocess that combines the indicators and visualizations. It also deals with how the obtained results of the decision support need to be interpreted by the system architect to identify the blamed component operations.

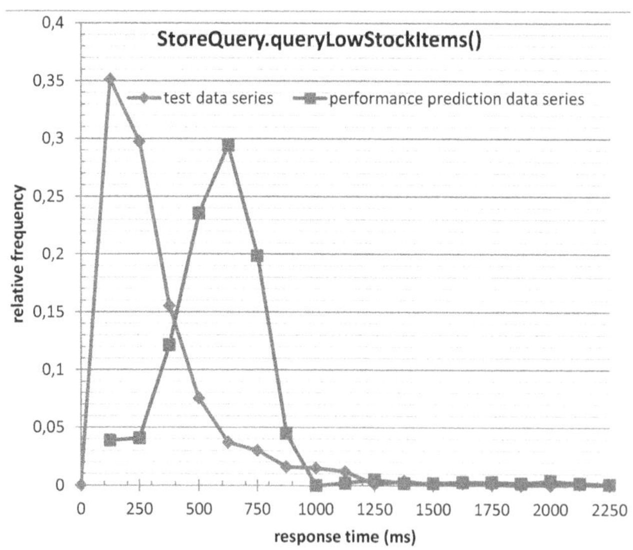

Figure 5: response time chart for manual comparison

4.1 Decision Criteria for an Automated Decision

This subsection develops automatable decision criteria in terms of indicators which help to decide whether the values of one response time data series exceeds the values of another response time data series. The following questions are relevant with respect to the decision criteria:

- Which indicators are suited to automatically decide whether a component operation is to be blamed or not?
- Which of these indicators are most robust (i.e. give the correct result in most/all cases)?

The questions also imply requirements on a suitable indicator or decision criterion: 1.) it must work with response time data series; 2.) it must be able to decide whether one data series has lower or higher values than another one; 3.) the decision must be correct in as many cases as possible. The first two requirements can be checked by consulting the definition of the indicator. This subsection discusses only indicators that match these two requirements. The third and most important requirement can practically only be decided using example data series. We have used data series from our experiments with CoCoME. Our test set included 20 pairs of data series. For each pair of data series, we manually assessed whether the test data series exhibits higher, lower, or about equal values than the performance prediction data series. To decide whether an indicator is suited, we tested whether it decides the same way that we have decided manually. The more often the indicator decides as we would have done, the better it is suited for our performance blame analysis approach. We used this benchmark for all the indicators (point estimators and statistical tests) described in this subsection. We use this example-based benchmark to derive a heuristic that works well for CoCoME and likely also applications similar to CoCoME.

The most obvious comparison indicators for data series are point estimators. Point estimators summarize a data series with a single numerical value. Two data series can then be compared by contrasting the point estimator value for each data series. The most prominent point estimators are minimum, maximum, mean, and the quartiles (i.e. 25%-quartile, the median and the 75%-quartile) [4, 10]. We have tried all these indicators to compare the data series in our test sample. The minimum and maximum are the two point estimators whose decision differed from the manual decisions most often. The minimum had seven and the maximum three false indications. The mean differed less from the manual decisions with only two wrong decisions in close cases. In one example where the mean fails, the test data series has a long tail, i.e. it has a lot of outliers near the maximum. The outliers caused the mean to fail, because in that case the mean and median differed significantly. The 25%-quartile, the median and the 75%-quartile performed best amongst all point estimators. They only failed in one case.

While some point estimators already qualify as comparison indicators, we looked for better indicators in statistical testing. The Kolmogorov-Smirnov-Test (KS-Test) [16] is a test that quantifies whether one data series generally has high or low values. There are two versions of the KS-Test that test if one data series has higher or lower values than another one, respectively. To the best of our knowledge the KS-Test is the only test that can test on lower and higher values. Other statistical test can only test equality (e.g. the Chi^2-test [4, 16]).

We use the "less"- and the "greater"-variant of the KS-Test (as implemented in R [14]). We investigate the hypotheses whether the test data series exhibits *lower* response time measurements than the predicted data series and whether test data series has *higher* measurements. As a consequence, we gather more information than when using a single test. In a clear case we can reject one of the two hypotheses with confidence. When the two data series are very close to one another, both tests will show the same result indicating that the KS-Test cannot tell them apart. This additional information enabled the KS-Test to outperform the point estimators. The KS-Test decided according to our manual decisions for the test data series. However, in two cases it made no decision, i.e. both KS-Tests indicated that their hypothesis could be rejected. Manually, we were able to decide one of the two cases, but could not do so for the other. As a result, the KS-Test was undecided in one case where a decision was possible. Thus, it did a little better than the quartiles, which made one actually wrong decision.

Altogether, our experiments suggest that the three point-estimators 25%-quartile, median, and 75%-quartile are reliable comparison indicators. The KS-Test seems to the best comparison indicator. It does not only decide, but it also indicates cases in which a decision is hard and which therefore need human intervention.

4.2 Visualizations for the Decision Criteria

Subsection 4.1 identified the 25%-quartiles, median, 75%-quartile, and the KS-Test as good indicators for performance blame analysis. Now, we look for a helpful visualization of these indicators. Most importantly, the visualization needs to be easily interpretable in terms of blaming. If possible, the visualizations shall not only present the indicator in question, but also other relevant data. Other relevant data confers to data that quantifies performance or that reflects influences on performance. For response time data, an important supplement is the call tree, the deployment and architecture, and the input data.

A box-and-whiskers-plot [10] can visualize the quartiles as well as the minimum and maximum. Showing two box-and-whiskers-plots (one for each response time data series) in one chart helps comparing whether the values of one of the data series exceed the values of the other one. This comparison can be done by comparing the position of the boxes of the data series. Each box stands for 50% of the measurements. So, if one box is drawn above the other, the data series likely has the higher values. As the quartiles only serve as a secondary indicator, we did not try to couple the quartiles with other relevant data.

An execution of the two variants of the KS-Test gives one of three possible results: 1.) the test data series shows lower response times; 2.) the predicted data series shows lower response times; or 3.) the data series are so close the test cannot decide clearly. These three result types stand for three different severity levels. Having three different severity levels suggests using a "traffic light indicator". Green indicates that the test data series has the lower values, yellow indicates that the KS-Test is unable to decide, and red indicates that the predicted data series has the lower response times. Such a traffic light can visualize the KS-Test outcome for each component operation that contributed in the test case. This traffic light coloring cannot only be used on its own, but it can also be incorporated into other visualizations that cover all the component operations participating in the test case.

Profilers are well matured performance analysis tools. Profilers often visualize caller/callee relationships of operations along with operation response time (cf. Section 7). This data combination is the main data that profilers present for analyzing the response times. We take this as evidence that this data combination can be seen as "other relevant data".

Brendan Gregg [6] presented his "Flame Graphs" which are a variation of the usual profiler charts. This particular variant shows

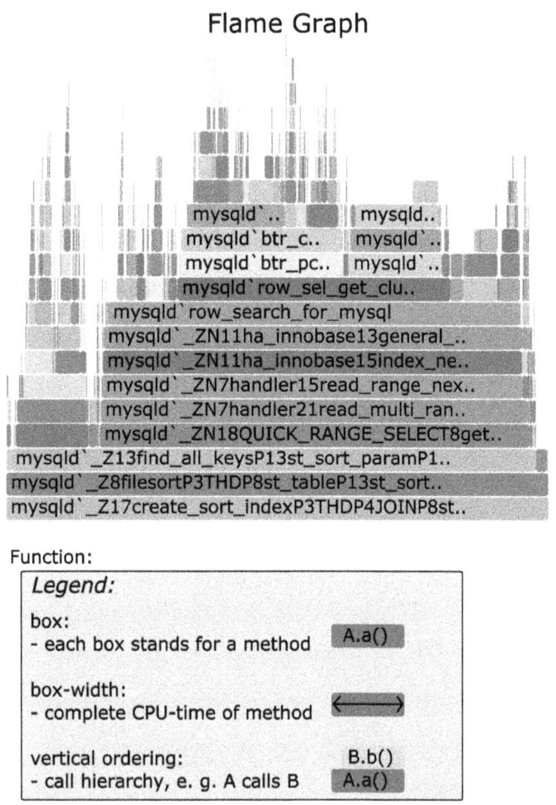

Figure 6: sample "Flame Graph" [6] with additional legend

the mentioned information in a compact form. Figure 6 shows an excerpt from a sample Flame Graph [6] presenting data measured in a MySQL run. A legend has been added in Figure 6 to give additional explanations. The chart shows the CPU-time of functions in MySQL. Each function is represented by a box. The width of the box shows the complete CPU time of the function during the test. The vertical ordering in the chart stands for the call hierarchy. A calls B if the box representing B is drawn on top of the box representing A. The width of a box that is not covered by a box on top of it represents the CPU time consumed by computations inside this function. In the original Flame Graphs the box colors have no meaning. Also the positioning of the boxes from left to right conveys no meaning.

The compact representation of the Flame Graphs requires to omit some information. The names of the functions (or operations) and the exact number for the CPU time cannot be shown in each box (cf. Figure 6). The Flame Graphs circumvent this limitation using interactivity. Flame Graphs are interactive SVG charts. The missing information for a particular box is shown next to the "Function" text at the bottom of Figure 6, when moving the mouse cursor over that box.

We see the call hierarchy in combination with a response time metric as performance relevant information. The Flame Graphs show these metrics. They have the advantage of being compact, when compared with other usual profiler visualizations (cf. Section 7). Moreover, we have found that the KS-Test result can be sensibly visualized using traffic light colors. Until now, the colors in the Flame Graphs have not conveyed meaning. As a consequence, we can adapt Flame Graphs with the KS-Test traffic light coloring as box colors. Moreover, the box width can represent the median response time of the component operation in the test case. It showed that the median is indicative for our performance blame analysis (cf. Subsection 4.1). This way, we get a compact diagram that includes the KS-Test results as well as the median response time and call hierarchy. We call this adapted version of the graph "Blame Graph".

In this subsection, we have introduced the Blame Graph that couples the KS-Test result in the form of traffic light colors with the average response time for each component operation and the position in the call tree. The KS-Test results are our prime decision criterion and are also coupled with other relevant decision criteria. Next, we visualize the quartiles using box-and-whiskers-plots. We have not coupled this secondary indicator with other relevant data.

4.3 Incorporating Decision Support

This subsection presents our decision support sub-process that incorporates the indicators and visualizations that the subsections 4.1 and 4.2 have introduced. Moreover, this subsection discusses how the result interpretation is affected by introducing automated decision support into the performance blame analysis.

The starting point of the decision support sub-process is that the system architect has response time data series from testing and performance prediction available for the test case in question. Figure 7 shows the additional decision support process. It is split in two parts. On the left hand side, Figure 7 shows the action that the system architect needs to perform and on the right hand side Figure 7 shows steps that are automatically computed. We have implemented a Python script that uses the R language [14] for statistical computations.

First, the system architect needs to prepare the computation by mapping component operations to PCM entities. Then the system architect gives conversion factors for the test and the predicted data series. These factors will be used to convert the response time data series to a common time unit. After that, computation begins. The computation takes the configuration mentioned before as well as the various response time data series as input and produces two different outcomes. First, it produces the Blame Graph which summarizes all the component operations

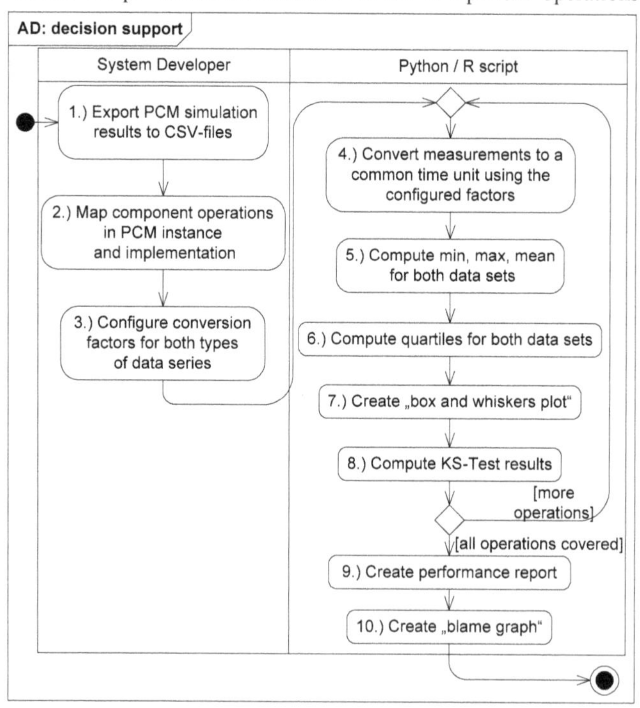

Figure 7: the additional decision support sub-process

involved in the test case in one chart. This chart shows the median response time for each component operation in context of the call hierarchy. We chose the median, as it turned out to be indicative for performance blame analysis (cf. Subsection 4.1).

Moreover, the computation produces a performance report that consists of two sheets for each of the component operations participating in the test case. One sheet shows a box-and-whiskers-plot for the two response time series associated with it, i.e. the data series from test and performance prediction. The next sheet exhibits all the point estimators for the two data series as well as the KS-Test results. The point estimators covered in the performance report are the quartiles, the minimum, the maximum, and the mean value.

The system architect must then look at the blame graph and the performance report to find blamed component operations. Figure 8 shows the updated result interpretation sub-process. In contrast to our previous work (cf. lower part of Figure 4) it relies on the Blame Graph and the performance report. The Blame Graph shows all the component operations in one sheet. All blamed operations are drawn in red. In addition, the system architect can also prioritize the blamed operations by call hierarchy (vertical box order) and median response time (box width). So, the blame graph supplies the system architect with three different relevant decision criteria. The system architect can then look into the performance report for more detailed figures about each component operation. If the system architect is still in doubt whether or not to blame the component operation in question, the system architect should of course make her own analyses.

5. EXEMPLIFIED AUTOMATED DECISION SUPPORT

This section exemplifies the automated decision support as depicted in Figure 7. It covers the configuration by the system architect as well as the automated analysis. This analysis results in the Blame Graph and the performance report. Next, this section discusses how to interpret these two artifacts in terms of blaming (cf. Figure 8). Also, it illustrates how each step of the decision support and the result interpretation has been performed in our running example (cf. Section 2) and its results. Please note that you can find all example data and analysis results along with our analysis scripts on this paper's companion website[2].

Subsection 5.1 first introduces the implementations and the test bed that have been used in the case studies with both CoCoME implementations. This includes the instrumentation used to produce the response time measurements. The Subsections 5.2 to 5.5 introduce the automated decision support and the result interpretation showing the results produced with the CoCoME reference implementation. Subsection 5.2 introduces the mapping from the component operations in the implementation to their counterparts in the model (cf. steps 1 and 2 in Figure 7). It also covers the subsequent conversion of the values of the data rows into a common time unit (cf. steps 3 and 4 in Figure 7). Subsection 5.3 presents the performance report and the statistical computations related to it. That subsection covers the steps 5, 6, 7, 8 and 10 in Figure 7. Subsection 5.4 comprises the creation of the Blame Graph and the related statistical computation. It covers the steps 8, 9 and 10 in Figure 7. Subsection 5.5 discusses the result interpretation by the system architect (cf. Figure 8). Finally, Subsection 5.6 covers the lessons learned from the case study with the CoCoME 2 implementation.

[2] http://homepages.uni-paderborn.de/bruesie/icpe2013/

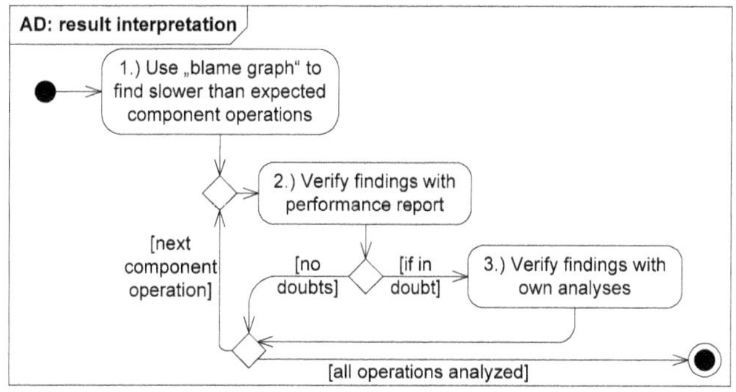

Figure 8: the updated result interpretation sub-process

5.1 Tested Implementation and Test Bed

We have chosen CoCoME as our running example, because it is a good representative of an information system that is fed by data generated by embedded system devices. It consists of several subsystems. The information system that we focus on consists of nine different components. These components are deployed on four different types of nodes, i.e. store server, store client, enterprise client, and enterprise server. Each store servers hosts five component objects and the enterprise server hosts six component objects. The clients are not participating in the test case used in the running example.

The CoCoME was originally released with a reference implementation[3] written in Java. We have fixed some bugs in the reference implementation, such that it can execute the scenario of our running example (cf. Section 2). This fixed CoCoME implementation comprises 6320 lines of Java code. The following refers to this implementation as CoCoME RI. Subsections 5.2 to 5.5 deal with the CoCoME RI case study in detail. Additionally, the original reference implementation has been fixed, improved and released as CoCoME 2[4]. The Karlsruhe Institute of Technology (KIT) created this version for the same reasons which made us develop a fixed version of the reference implementation. The KIT still maintains this version for further usage. This improved implementation comprises 9788 lines of Java code. Subsection 5.6 briefly summarizes the lesson learned in the CoCoME 2 case study.

To test the use case "exchange products among stores", introduced in Section 2, we have adapted the test bed to our needs. We have introduced a load test driver and response time measuring using bytecode instrumentation in both implementations. The instrumented bytecode measures the time between method entrance and return on the server side. This metric also includes possible queuing delays. This is consistent with the performance prediction of the PCM. The measurement bytecode writes this response time to a file and it also includes other important information, namely, the method name, the current time, the current stack trace, and the thread identifier. BTrace[5] supplies the actual bytecode instrumentation, data query, and file writing facilities. We created a short program using BTrace to enable response time measuring. While this custom measurement solution could be more efficient and also could trace cross-process control flow, it suits our purpose. Our measurement solution adds an overhead to

[3] see http://cocome.org/

[4] see http://sdqweb.ipd.kit.edu/wiki/CoCoME2 and http://sourceforge.net/projects/cocome/

[5] see http://kenai.com/projects/btrace

Table 1: mapping between implementation and PCM

Java stack trace	CSV-file from PCM
(9) overall	bookSale0.csv
(8) otherStoreInterchange	orderProductsAvailableAtOtherStores0.csv
(7) markStock	markProductsUnavailableInStock1.csv
(6) solveOptimization	solveOptimization0.csv

the response time of the tested methods. The method directly called by the test driver shows about 34.6 ms (i.e. 6%) slower response times when instrumented. This overhead is tolerable.

The test setup was very simplistic. The test setup was used for either implementation, i.e., CoCoME RI and CoCoME 2. A single dedicated PC (Pentium D 3 GHz single core CPU with two threads, 2 GB RAM, Windows 7 with 64 bit) was used to execute the CoCoME system and also the load test driver. The executed CoCoME system consisted of three store servers and an enterprise server as well as the necessary infrastructure, i.e. the RMI registry, the Active MQ JMS messaging server, and the Apache Derby database included in each CoCoME implementation. The load test driver simulated eight parallel users. Each user queries the same store server 125 times (1000 requests total). This setup should produce a certain amount of contention in the system. However, we did not quantify contention (e.g. by measuring queue length).

5.2 Map Operations and Unit Conversion

First the system architect needs to prepare some configuration data for the Python/R script, which computes the performance report and the Blame Graph. The system architect needs to ensure that there exists a mapping from the component operations in the implementation to their representation in the performance model. The mapping needs to consider each operation and its complete call hierarchy at operation level. The contents of the mapping need to reflect this. The testing side of the mapping consists of all the Java stack traces from the test case that indicate the call of a component operation. On the performance prediction side, the system architect enters the file name of a set of data for this operation invocation. In the PCM, calls to operations are modeled as part of a component's control flow. So, the PCM can distinguish different invocations of the same operation. The PCM's performance prediction yields a separate set of values for each different invocation of an operation.

Because the mapping takes the call hierarchy of a component into account, the mapping needs to be individually constructed for any system. The system architect may use information from the component developers that states what method in the implementation correspond to which model element, but she needs to add information from the component context.

Both the response time measurements from the CoCoME test and also from the PCM need to be represented in a comma-separated values (CSV) file. One line in the CSV file from the test must consist of the response time and the full stack trace. Having both data series in CSV format, we can easily process the data with our analysis.

For better readability, we have compressed the different Java stack traces by numbering them and added a short text for descriptiveness. In the following, we will only exhibit this short form. Table 1 displays an example of the mapping from the compressed Java stack trace to the CSV file with predicted response time values.

As a next step, the system architect needs to convert the response time values from both sources (test and performance prediction) to a common time unit. The system architect gives two conversion factors to the Python/R script, one for the test data series and one for the performance prediction data series. In our running example, the measurements from testing and performance prediction are converted from nanoseconds and seconds to the common unit milliseconds respectively.

The manual steps (cf. steps 1 to 3 in Figure 7) were finished in about one hour for each implementation of CoCoME in the example. These manual steps have not been automated due to the prototype status of our tooling, but are partly automatable. The measurement export and the computation of conversion factors (if the time units are known) can be automated. Creating the mapping can be semi-automated.

5.3 Performance Report

While the Blame Graph is the primary analysis tool, the performance report is the secondary analysis tool that the automated decision support delivers. The performance report gives a detailed overview of each pair of response time data series for all the component operations. It contains the point estimators, the KS-Test results, and a box-and-whiskers-plot for each operation. Specifically, it comprises all computed point estimators: maximum, minimum, quartiles, and mean. In addition, it visualizes the quartiles in a box-and-whiskers-plot that shows the test and performance analysis data series side by side.

After finishing the configuration (cf. Subsections 5.2), the Python/R script computes the point estimators for each component operation. Table 2 lists the point estimators for the method "(8) otherStoreInterchange". The system architect can compare the point estimator values to get an impression, which data series has higher values. In Table 2, all the point estimators in the test data series have higher values than in the performance prediction data series. The Python/R script also draws a box-and-whiskers plot that shows the test and performance analysis response time data series side by side. Figure 9 depicts this box plot. For better readability the box plot does not include the highest 1% of the measurements of each data series, because the test data series has a few

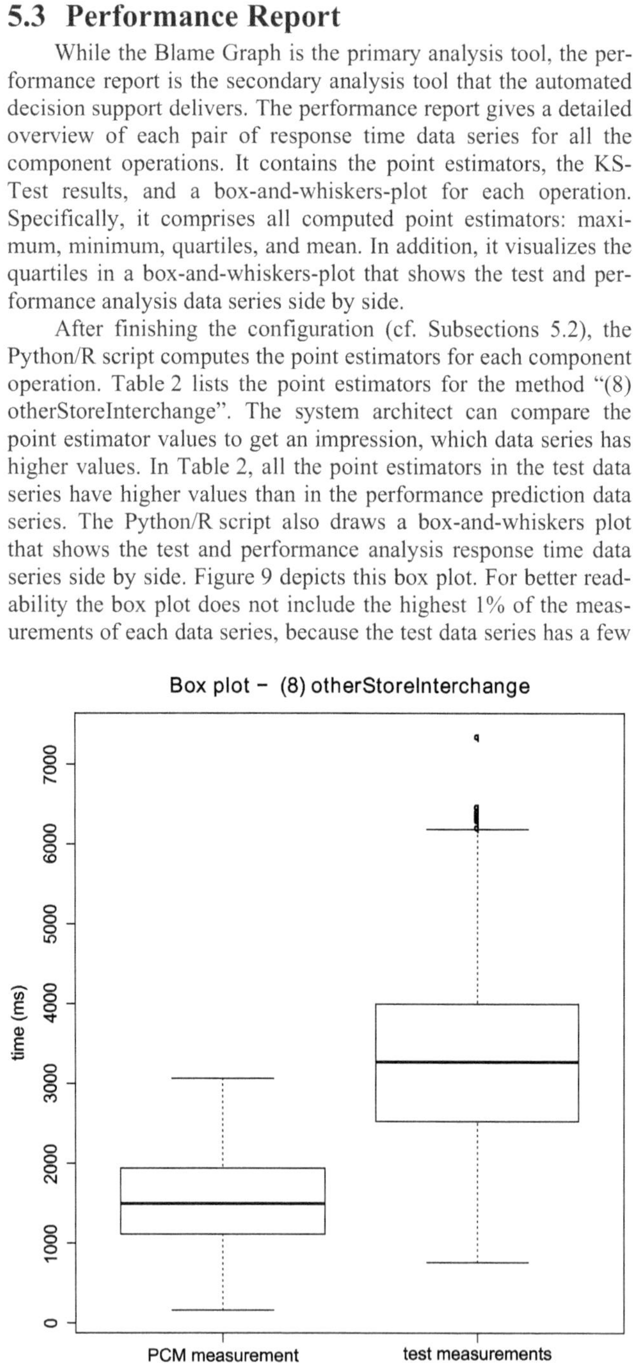

Figure 9:example box-and-whiskers plot

Table 2: point estimators for the operation "(8) otherStoreInterchange"

	PCM – resp. time	Test – resp. time
minimum	163.1 ms	760.7 ms
25%-quartile	1120.8 ms	2528.6 ms
median	1506.2 ms	3279.6 ms
75%-quartile	1952.4 ms	4012.1 ms
maximum	3186.0 ms	24595.4 ms
mean	1500.3 ms	3248.7 ms

outliers that are way higher than most of the measurements. Figure 9 shows that the box for the test data series (which represents 50% of the measurements) lies above the box of the performance prediction data series. This indicates that the values in the test data series are higher than those of the performance prediction data series.

The result of the KS-Test is also shown in the performance report. The report comprises the probability value (p-value [16]) that the "less" and the "greater" variant of the KS-Test result in. Section 5.4 deals with the KS-Test in more detail.

5.4 Blame Graph

The performance report is the secondary indicator for analyzing performance blame, while the Blame Graph is the primary indicator. The Blame Graph depicts operations that are slower than expected as red boxes. The box colors stand for the result of the KS-Test (cf. Subsection 4.3). The KS-Test implementation in R takes both data series as input and results in the p-value [16]. The Python/R script executes the KS-Test and accepts the hypothesis when the p-value is at most 5%. This 5% threshold is a starting value which we will evaluate in future experiments.

The Blame Graph visualizes the results of the KS-tests. It also visualizes the call relationships in the tested implementation and the median response times of each component operation in the test. For the example test case, our script produces the result presented in Figure 10. We added a legend to Figure 10 for better readability. The method names in the Blame Graph are given in terms of the Java implementation of CoCoME. The boxes in Figure 10 are not labeled with fully qualified class names due to the restricted width of the boxes (even the short names may be cut off in small boxes). The fully qualified class name is shown next to the "Function"-text in the original SVG Blame Graph, when moving the mouse over the box representing the operation. In the given Blame Graph the system architect can identify that three methods are blamed, because they are drawn as red boxes. Details on how to interpret the Blame Graph are discussed in Subsection 5.5.

5.5 Result Interpretation

The system architect first looks at the Blame Graph. The system architect identifies the red boxes in the Blame Graph representing the operations that are slower than expected. In our example Blame Graph (cf. Figure 10), there are three blame operations:

- StoreImpl.bookSale(..)
  (also known as "(8) otherStoreInterchange")
- ProductDispatcher.orderProductsAvailableAtOtherStores(..)
- StoreImpl.markProductsUnavailableInStock(..)

Then, the system architect can prioritize the order in which the blamed operations are analyzed by either call hierarchy (vertical box order) or by median response time (box width). For example, the system architect could decide to prioritize the operation "StoreImpl.markProductsUnavailableInStock", as this operation does not call other operations.

The system architect can check the performance report to investigate if the KS-Test has blamed the operations correctly. Using the performance report, the system architect can also prioritize the order to investigate the blamed operations differently. For instance, the system architect could decide to prioritize the component operations according to the difference between the test values and the expected values. If the system architect is still not satisfied, she needs to perform her own analyses.

Compared to the manual process in our previous work, the system architect only needs to configure the Python/R script and can then already interpret the generated results. This saves the system architect the trouble to create the chart by herself. Moreover, the results of the automated decision support can be interpreted easier. The indicators produced by the automated decision support are tailored to judge whether the test data series exhibits the higher values than the performance prediction data series. In our previous work, the system architect needed to judge whether the test data series has the higher values by looking at a specific histogram.

5.6 Case Study: CoCoME 2

While the Subsections 5.2 to 5.5 deal with our running example in the CoCoME RI, this subsection summarizes what we learned in the case study with the CoCoME 2. In this second case study, we could largely confirm what we learned in the CoCoME RI case study. The system architect could employ the automated decision support to get the Blame Graph and the performance report. The Python/R-script quickly produced this result. The

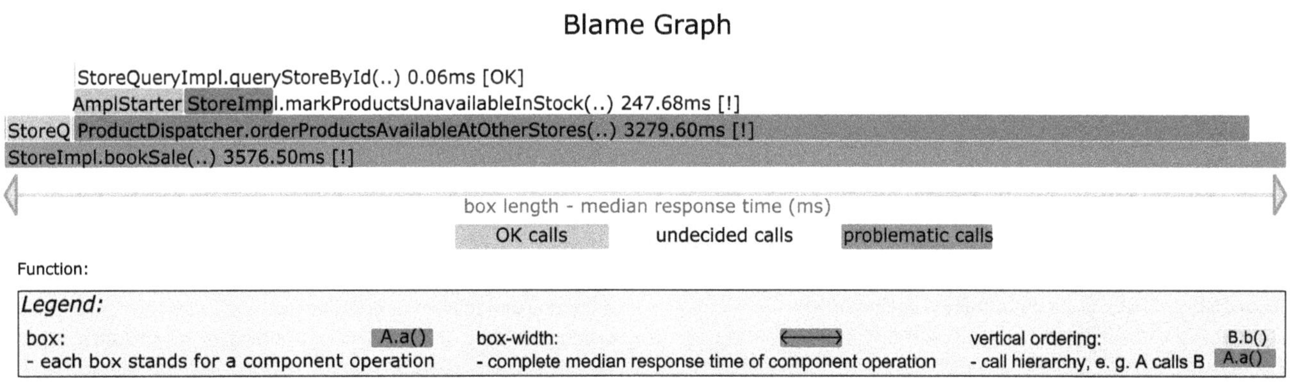

Figure 10: blame graph for the example test case

effort for the analysis was considerably lower than the effort for implementing the test driver and measurement solution.

The Blame Graph for CoCoME 2 gives a quick overview which methods exceed the response time values derived from the PCM contract. But this Blame Graph also shows undecided operation calls that the KS-Test cannot clearly decide. The performance report confirms the results of the Blame Graph, but also shows that one of the undecided results can actually be decided with the performance report (cf. Subsection 4.1). This shows that it is important to also include the performance report as a decision support result. Moreover, the Blame Graph also shows that our current monitoring solution is unable to track requests crossing process boundaries. Operation calls in other processes are drawn just as operations called directly from the test driver.

6. LIMITATIONS

Our approach is currently limited to analyzing the response times of component operations. It can only detect errors in specific component operations. If there are no deviations found for any component operations, our approach assumes that the error is in the composition. However, our approach does not help to detect defects in the architecture, the deployment, etc.

In addition, this approach is based on the differences in the performance contract compared to the actual implementation. If the contracts were correct, comparison between the testing values and expected values from performance prediction will always result in equal values. This is why this approach assumes that the contracts reflect the component developer's intentions of the component's performance.

The approach depends on the quality of the performance contracts that are delivered by the component developers. The component developers could deliver overly simple contracts. Such contracts are only viable in very specific cases and will evaluate to unrealistic expected values in the majority of cases. Moreover, the component developer can betray us with wrong contracts. They could either deliver "safe contracts" or "flattering contracts". "Safe contracts" always result in higher response time values than the values measured in test. This would render this approach useless, but would also lower the chance system architects acquire that component from the market. "Flattering contracts" indicate that the component is performing better than it actually is. Such components are more likely to be acquired from the market, but our approach will also blame such components in case of performance issues. If our approach blames all components, the result is not useful as well. Even when the component developers deliver contracts as intended, the contracts still may not be completely accurate. It is very hard for the component developers to foresee every possible context a component may be used in and to design an according and accurate contract.

The test measurements considered in our analysis are gained with an instrumentation that implies an overhead (cf. Subsection 5.1). This overhead may cause false positives, because it adds to the values in the test data series.

Only measuring response times leaves out certain aspects of the middleware that is used to execute the components. For example the measurements do not differentiate queuing delays and the response time also includes the response time of called components. That means that the blame may travel down the blame graph along the call chain towards the callers. In this case our approach may blame more components than necessary.

The visualization introduced in this paper can also be improved. The Blame Graph does not show process or host boundaries. This is partly due to the monitored data that it visualizes. Our current monitoring solution cannot track requests that cross process boundaries. Thus, the Blame Graph shows internally called methods from another host or process on the same level as calls that are called directly from the test driver. Moreover, the Blame Graph shows the median response time as box-width. It does not cover the amount of times an operation was called and may therefore be deceiving. The number of calls is omitted due to the prototype status of our tooling, but can be easily added to the Blame Graph.

7. RELATED WORK

This paper introduces an automated decision support for performance blame analysis. This decision support delivers the Blame Graph that take the context of the component operations into account. So, Subsection 7.1 discusses other performance blame analysis approaches in terms of their decision support. It evaluates for each approach what their decision support looks like and to what degree it is automated. Moreover, it presents what results the decision support for each approach yields. The set of results shall be easily interpretable and it shall take the context of the results into account.

One contribution of this paper is to present performance metrics visually embedded in the component operation context. Subsection 7.2 presents several visualization techniques from other areas, because visual result presentation is seldom used in performance blame analysis. It covers the areas of performance profiling and software cartography.

7.1 Performance Blame Analysis

One approach that does come from the field of performance blame analysis in component-based systems is presented by Srinivas and Srinivasan [17]. The approach analyzes the call tree of component-based software to perform blame analysis. The call tree is annotated with cumulated percentage costs. The costs in the call tree can be any performance metric such as response times or CPU usages, which are expressed using percentages. The analysis then converts the cumulated costs to absolute costs. The cost conversion takes only a user-defined set of components into account. Calls to other components are defined as in-method computation, and are hence not differentiated in the analysis. Then the analysis computes if there are calls in the altered call tree that exceeds a chosen cost threshold. The result of this approach is a list of component methods with their absolute percentage costs. The threshold can be used to scale the number of results given.

Compared to our approach the approach by Srinivas and Srinivasan [17] has a slightly lower degree of automation. On one hand, the user has to supply several complex configuration settings. The user has to specify a threshold and the set of classes and methods, which are considered in the analysis, as substitute for the contract and the component context, respectively. The configuration settings have massive influence on the quality of results. On the other hand, the decision support as such is automated. The decision support is the computation of the final cost percentages that are higher than the user-defined threshold. The decision support results in an ordered list of methods and cost percentages. In the list the system architect can easily identify the most costly methods. This kind of analysis relies on the observation that often times high cost of a method means that it needs to be improved. However, if the methods that need to be improved are not at the top of the list, they are hard to find. The list can be very extensive if the threshold is set too low. In such a case an additional visualization would be helpful. The cost-metric is not so much dependent on the call tree as pure response time measurements. The algorithm internalizes calls into calling methods, as specified by the user. Then it computes absolute cost values. These absolute values can be interpreted independent of the call tree.

Another approach that deals with performance blame analysis is exhibited by Rutar and Hollingsworth [15]. The approach computes the blame for variables and data structures on the basis of data flow. Every operation on a variable like a simple assignment or an addition of another variable increases the blame of this variable. This is also the case if the variable is used in loops or branches. The blame is also aggregated in data structures. These blame data are gathered by static analysis and by runtime monitoring. The result of this approach is a list of variables and their blame factor.

The approach by Rutar and Hollingsworth works on a different abstraction level than our approach. They are doing a white-box analysis covering the variables and are therefore working on very detailed level. This implies that the source code of all components needs to be available, since static analysis works on the source code. The static analysis is not fully automated and needs to be assisted in complicated cases. Moreover, the result list does neither cover blamed component operations nor components. Therefore, the system architect has to manually map the variables to components and component operation in order to find the blamed components or component operations. This means that the approach has no decision support.

The last approach tests the conformance of a component implementation to its performance specification. Groenda's approach [7, 8] is also based on the PCM. On one hand, Groenda measures the performance of the implementation during test cases derived from PCM instances. On the other hand, he derives performance measurements from PCM performance prediction. These two data series are then compared in terms of equality. The comparison is done on component operation level.

In contrast to our approach Groenda [7, 8] investigates whether the data series for a component operation in a specific test case from testing matches an equivalent data series from performance prediction. He statistically tests the hypotheses whether the two data series deviate or match. He interprets the test result as "yes" or "no" with the help of a user-defined error threshold. While this approach exemplifies the automated use of statistical testing as part of the decision support, it does not elaborate further on this topic. In particular, Groeda does not mention how the single results of the repeated tests are aggregated and how the result is presented.

7.2 Performance Visualization

This subsection deals with the areas of performance profiling and software cartography. It describes which visualizations are used in these areas and how they compare to the Blame Graph. The Blame Graph is the prime visualization introduced in this paper.

The popular open source profiler JFluid [5] (also known as Netbeans Profiler) exhibits a tabular representation based upon the Calling Context Tree. This presentation consists of a table of methods that incorporates a tree view. Each row stands for a called method. The methods it calls are drawn with more indentation in the table rows underneath and the called method's name is connected by a line to its caller's row. Each row is annotated with the information how many times the method was invoked as well as its net and total response time.

The commercial profiler JProfiler [11] exhibits a similar view than JFluid. JProfiler does not have the tree included in a table, but it exhibits a call tree view drawn with vertices and arrows between them. Each vertex carries the same information as the table rows in the JFluid view. It is worth noting that JProfiler features a similar view for memory usage.

The JFluid and JProfiler visualizations are both similar to the blame graph. They display a similar set of metrics. However, the Blame Graph additionally includes the KS-Test result that is visualized using a color code. The Blame Graph also is more compact. Especially when compared to the JProfiler view. Because of its compactness the Blame Graph does not include each method name and operations with very short response time are barely visible.

The "j2eeprof" approach [12] presented by Kłaczewski and Wytrębowicz introduces another view of the call tree. In this view every method called is represented by a box. A box for the method B() is drawn on top of the box for method A() if A calls B. The width of each box then resembles the response time of each respective method. This view is essentially identical to the Blame Graph. The Blame Graph defines the box width as median response time of the test data series and adds the box color as visualization of the KS-Test result.

Software cartography describes how the techniques of cartography, e.g. used for city maps, can be carried over to software. Krogmann et al. [13] have shown that this approach is well suited for the presentation of performance metrics. They have focused on the presentation of resource usages in the context of a component-based system. They have created several views that use the data from several scenarios as overlays. The overlays can be switched on and off at will to allow rapid comparison.

While Krogmann et al. show that the concepts of software cartography are very advanced and allow for good result representation, we cannot directly use their results as they focus on resource usage. However, concepts such as overlays are also desirable for our result representation. For example, the Blame Graph can use overlays in terms of different color codes that stand for different metrics.

8. CONCLUSIONS AND FUTURE WORK

In this paper we have extended our previously introduced performance blame analysis process [2] with an automated decision support step. The decision support step statistically and automatically decides whether the test data series has the overall higher response time values in comparison with the expected values derived from the component's performance contract. The decision support evaluates this by the KS-Test and by comparing point estimators. The automated decision support results in the Blame Graph and the performance report. The Blame Graph concisely presents the blame (from the KS-Test results) for each component operations along with the call hierarchy and the median response time. The Blame Graph is the main tool for the system architect to determine blame. She uses the performance report to back her decision up with detailed results for the KS-Test and the point estimators.

The automated decision support helps the system architect to interpret performance measurements in terms of performance blame analysis. Instead of looking directly at the raw data series or the complex statistical characterizations, the system architect can work with the Blame Graph and the performance report. Both artifacts make it easy to judge whether the test data series has higher or lower response time values compared to the expected values. The Blame Graph exhibits this directly using a color code that stands for the KS-Test results and the performance report visualizes point estimators for both data series side by side.

In the future we will continue to work on the approach presented in this paper. As a next step, we want to evaluate the reasoning of our automated blame analysis with an even more realistic example system. We want to assess if the Blame Graph and the performance report are sufficient decision criteria or if we have to add more criteria (like the difference between the values of the two data series) to form a decision tree. Moreover, we want to set up an empirical case study testing if the proposed visualization is user-friendly and adequate in all circumstances. We might need to add different views to our visualization using the overlay concept from software cartography [13]. In addition, we are developing a test case notation that incorporates PCM instances. Such enriched test cases constructively ensure that the test case scenario is always compatible to the PCM simulation scenario.

9. REFERENCES

[1] Becker, S. et al. 2009. The Palladio component model for model-driven performance prediction. *Journal of Systems and Software*. 82, 1 (Jan. 2009), 3–22.

[2] Brüseke, F. et al. 2011. Palladio-based performance blame analysis. *Proceedings of the 16th International Workshop on Component-oriented programming - WCOP '11* (Boulder, Colorado (USA), 2011), 25–32.

[3] Cheesman, J. and Daniels, J. 2000. *UML Components: A Simple Process for Specifying Component-Based Software*. Addison-Wesley Longman, Amsterdam.

[4] Devore, J.L. and Berk, K.N. 2012. *Modern Mathematical Statistics with Applications*. Springer.

[5] Dmitriev, M. 2004. Selective profiling of Java applications using dynamic bytecode instrumentation. *Performance Analysis of Systems and Software, 2004 IEEE International Symposium on - ISPASS*. (2004), 141–150.

[6] Gregg, B. 2011. Flame Graphs: *http://dtrace.org/blogs/brendan/2011/12/16/flame-graphs/*. Accessed: 2012-10-04.

[7] Groenda, H. 2011. An Accuracy Information Annotation Model for Validated Service Behavior Specifications. *Models in Software Engineering*. J. Dingel and A. Solberg, eds. Springer Berlin / Heidelberg. 369–383.

[8] Groenda, H. 2010. Usage profile and platform independent automated validation of service behavior specifications. *Proceedings of the 2nd International Workshop on the Quality of Service-Oriented Software Systems* (New York, NY, USA, 2010), 6:1–6:6.

[9] Herold, S. et al. 2008. CoCoME - The Common Component Modeling Example. *The Common Component Modeling Example*. A. Rausch et al., eds. Springer Berlin / Heidelberg. 16–53.

[10] Jain, R. 1991. *The Art of Computer Systems Performance Analysis: Techniques for Experimental Design, Measurement, Simulation, and Modeling*. John Wiley & Sons.

[11] Java Profiler - JProfiler: *http://www.ej-technologies.com/products/jprofiler/overview.html*. Accessed: 2012-03-23.

[12] Kłaczewski, P. and Wytrębowicz, J. 2007. j2eeprof — a tool for testing multitier applications. *Software Engineering Techniques: Design for Quality*. Springer Boston. 199–210.

[13] Krogmann, K. et al. 2009. Improved Feedback for Architectural Performance Prediction Using Software Cartography Visualizations. *Architectures for Adaptive Software Systems*. Springer Berlin / Heidelberg. 52–69.

[14] R Development Core Team 2011. *R: A Language and Environment for Statistical Computing*. http://www.R-project.org.

[15] Rutar, N. and Hollingsworth, J.K. 2009. Assigning Blame: Mapping Performance to High Level Parallel Programming Abstractions. *Euro-Par 2009 Parallel Processing*. H. Sips et al., eds. Springer Berlin Heidelberg. 21–32.

[16] Shao, J. 2003. *Mathematical Statistics*. Springer.

[17] Srinivas, K. and Srinivasan, H. 2005. Summarizing application performance from a components perspective. *ACM SIGSOFT Software Engineering Notes* (New York, NY, USA, 2005), 136–145.

Propagation of Incremental Changes to Performance Model due to SOA Design Pattern Application

Nariman Mani, Dorina C. Petriu, Murray Woodside

Carleton University
Department of Systems and Computer Engineering
1125 Colonel By Drive
Ottawa, Ontario, Canada

{nmani | petriu | cmw}@sce.carleton.ca

ABSTRACT

Design patterns for Service Oriented Architecture (SOA) provide solutions to architectural, design and implementation problems, involving software models in different layers of a SOA design. For performance analysis, a performance model can be generated from the SOA design and used to predict its performance. The impact of the design patterns is also reflected in the performance model. It is helpful to be able to trace the causality from the design pattern to its predicted performance impact. This paper describes a technique for automatically refactoring a SOA design model by applying a design pattern and for propagating the incremental changes to its LQN performance model. A SOA design model is expressed in UML extended with two standard profiles: SoaML for expressing SOA solutions and MARTE for performance annotations. The SOA design pattern is specified using a Role Based Modeling Language (RBML) and their application is automated using QVT-O. Automated incremental transformations are explored and evaluated for effectiveness on a case study example.

Categories and Subject Descriptors

H.3.4 **[Systems and Software]**: Performance evaluation (efficiency and effectiveness)

General Terms

Performance, Design, Experimentation, and Verification.

Keywords

Software performance, service-based systems, SOA pattern, model change, change propagation, LQN.

1. INTRODUCTION

Service Oriented Architecture (SOA) is an innovative software architectural paradigm for developing and deploying applications as a set of reusable composable units called services, that can be collectively and repeatedly used for the realization of specific goals [1]. In Model Driven Engineering (MDE), the performance of a SOA design can be evaluated using model transformations to generate a performance model (hereafter called *PModel*) of the SOA system from its software design model (hereafter called *SModel*) extended with performance annotations. In our work, the initial SModel to PModel transformation is performed with the PUMA transformation chain [2], previously developed in our research group, which requires the generation of an intermediate model Core Scenario Model (CSM), as shown in Figure 1.(A). CSM captures the essence of performance specification and estimation as expressed by SModel annotations, and strips away the design detail which is irrelevant to performance analysis. During the initial transformation, trace links expressing the mapping between the SModel and the corresponding PModel elements are created [3].

In this paper we propose to use performance evaluation to screen candidate SOA pattern applications, after an initial SOA design has been modeled and its PModel derived for the first time. We assume that the candidate patterns are selected by the designer in order to improve a certain quality of the design (such as maintainability, robustness, performance, security, etc.). However, the logic behind the pattern selection is not addressed in this paper, being left for future work. The focus of this paper is on the automatic application of a SOA design pattern, followed by the incremental propagation of SModel changes directly to the PModel, as illustrated in Figure 1.(B) (Model modifications are denoted using the "Δ" symbol). SOA design patterns describe generic solutions for different architectural, design and implementation problems[1, 4]. Changes due to the application of such patterns may impact performance and other non-functional properties either positively or negatively.

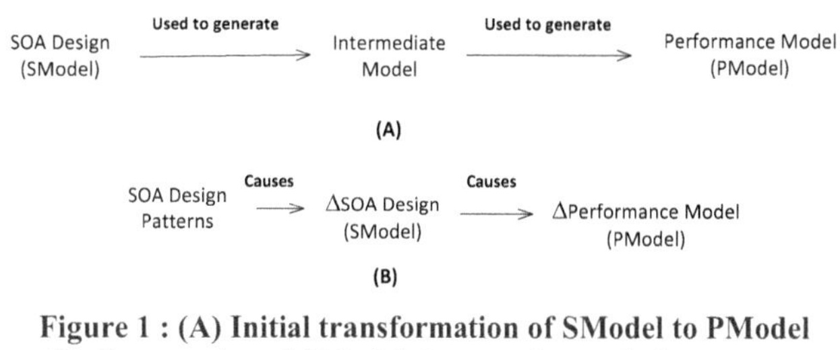

Figure 1 : (A) Initial transformation of SModel to PModel (B) Propagation of SModel changes due to design pattern application directly to PModel

Reconstructing the entire intermediate and performance models from scratch after a pattern application requires a lot of effort. To address this issue, we propose here an incremental approach that allows investigating the impact of changes due to design patterns on the system performance in a time and cost effective way, by identifying the affected performance model elements and

ICPE'13, April 21–24, 2013, Prague, Czech Republic.

parameters without re-constructing the entire performance model. This enables incremental studies of numerous design alternatives when applying a large number of SOA design patterns.

As already mentioned, we use the standard UML profile Service Oriented Architecture Modeling Language (SoaML) [5] to represent the design of a SOA system. SoaML defines extensions to UML 2 to support a range of modeling requirements for service-oriented architectures [6].

Another standard UML profile used in our approach is MARTE (Modeling and Analysis of Real-Time and Embedded Systems) [7], which bridges the gap between the software and the performance domains. The MARTE annotations help us not only in transforming the SModel into a PModel, but also in propagating the modifications due to the application of design patterns.

For studying the impact of design patterns on the system and its performance, the SOA design patterns need to be formally specified. In this paper, we use the Role Based Modeling Language (RBML) [8] to formally define two parts of a SOA design pattern: the set of SModel elements and their relationships that represents the *problem* addressed by the pattern, as well as the subset that constitutes the *solution*. The pattern application will actually replace the elements from the "problem" subset with the "solution" subset, leaving the remaining SModel elements unchanged. The actual implementation of the SModel transformation that corresponds to the pattern application rules is implemented using the standard language QVT (Query, View, and Transformation) Operational (QVT-O) [9]. Trace links between SModel and Pmodel (created during the initial PModel derivation) are used to propagate the changes to the PModel. We illustrate the proposed approach with the Service Façade design pattern applied to a Shopping and Browsing system.

The paper is organized as follows: Section 2 presents related work; Section 3 shows an overview of the approach proposed in this paper; Section 4 presents briefly the derivation of a performance model from the SOA design (i.e., SModel to PModel transformation); Section 5 discusses SOA design pattern specification using RBML; Section 6 presents the pattern application to a SModel using RBML and QVT-O; Section 8 shows how the SModel changes are propagated to the performance model and Section 9 presents the performance analysis results for a case-study system before and after a design patterns is applied. Finally Section 10 concludes the paper.

2. RELATED WORK

There are a significant number of papers on the derivation of a PModel of a system from the design model (SModel). The OMG standards for UML performance annotations, SPT and MARTE, have enabled research to transform UML design specifications into many kinds of performance models, based for example on Queuing Networks [10], Layered Queuing Networks [2, 11, 12] Stochastic Petri nets , etc.

PUMA is a set of transformations which take as input different SModels and transform them into different performance models [2, 11]. PUMA uses a pivot language, the Core Scenario Model (CSM), to extract and audit performance information from different kinds of design models (e.g., different UML versions and types of diagrams) and to support the generation of different kinds of performance models (e.g., QN, LQN, Petri nets, simulation). In [12], a graph-grammar based algorithm was proposed to divide the activity diagram into activity sub-graphs, which are further mapped to LQN phases or activities. In this work we propose to create a mapping between SModel and PModel elements during the PUMA model transformation, to help us propagating incremental modifications of the design model directly to PModel.

Another category of related works studies the impact of design patterns on software performance. Beside the patterns which are recognized as best practices for software development, Smith and Williams [13] introduced general performance anti-patterns, which exclusively focus on performance concerns. Anti-patterns are defined as common design mistakes that consistently occur, causing undesirable results. Cortellessa et al. [14] present an approach based on anti-patterns for identifying performance problems and removing them. The identification of an anti-pattern suggests the architectural alternatives that can remove that specific problem. Arcelli et al. [15] used the anti-pattern detection approach in [14] and also RBML to detect and remove performance anti-patterns from the software architectural model. In their approach [15], RBML is used to present the anti-pattern problems as source and anti-pattern solution as target role models. We use a similar approach for specifying the SOA design pattern, but the difference is that we implement the pattern application in QVT-O and propagate the changes directly to the PModel.

Menascé et al [16] presents a framework called SASSY whose goal is to allow designers to specify the system requirements using a visual activity-based language and to automatically generate a base architecture that corresponds to the requirements. The architecture is optimized with respect to quality of service requirements (i.e. as measured by several performance metrics such as execution time and throughput) through the selection of the most suitable service providers and application of quality of service architectural patterns.

Parsons and Murphy [17] introduce an approach for the automatic detection of performance anti-patterns by extracting the run-time system design from data collected during monitoring by applying a number of advanced analysis techniques.

Xu [18] applied rules to performance model results to diagnose performance problems and to propose solutions for fixing them, which resulted in changes at the PModel level, which could be interpreted to suggest corresponding design changes. However, the changes were not propagated automatically to the SModel.

In this paper, we propose a technique for propagating the design pattern changes to the design of a service based system, which allows us to study the pattern impact on the system performance. We use RBML for the specification of the SOA design patterns and the OMG standard for model transformation, QVT-O for refactoring the SOA design according to the design pattern. To the best of our knowledge, none of the works in literature addresses the problem as proposed in this paper.

3. PROPOSED APPROACH OVERVIEW

Figure 2 shows an overview of the proposed approach. The key areas of the overview are shown in grey boxes (numbered from 1 to 3). Grey area 1 (solid grey) shows a more detailed view of what is shown in Figure 1.(A). As briefly discussed earlier, the performance model of a system (PModel) is generated from its software design model (SModel) extended with performance annotations (i.e. MARTE) using the transformation chain PUMA [2]. PUMA uses an intermediate model called "Core Scenario

Model" (CSM) to bridge the gap between different kinds of software models accepted as input and different kinds of performance models generated as output. CSM bridges the semantic gap between SModel and PModel, and captures only those software aspects that are relevant to performance models. During the initial transformation, trace links expressing the mapping between the SModel and the corresponding PModel elements are created which is used in grey area (3).

We assume that, in order to improve the quality of a SModel, the designer will apply different design patterns, which may impact performance and other non-functional properties, either positively or negatively. A designer selects a design pattern to solve a performance or non-performance problem and decides where to apply it by binding the roles defined in the pattern to specific SModel elements. In this paper we do not discuss the factors involved in selecting a candidate design pattern for SModel design issues, leaving it for future work. The grey areas (2) and (3) (dotted grey) in Figure 2 show the processes involved in applying the pattern and propagating the changes.

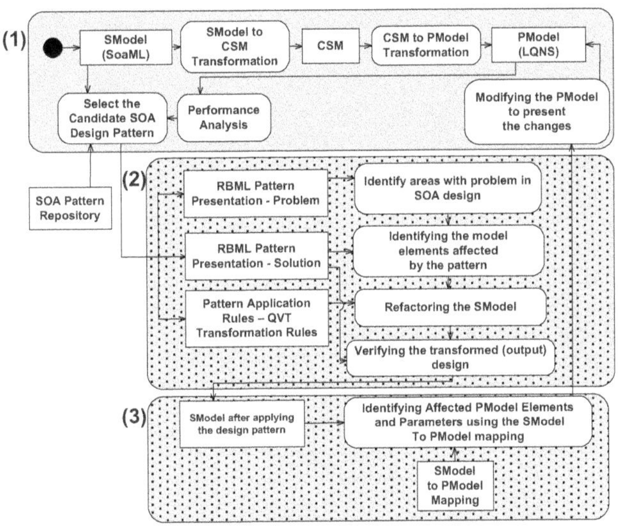

Figure 2 : Proposed Approach Overview

Grey area (2) represents the process for refactoring the SModel by applying a pattern. The process starts when the designer selects a pattern. Next, the pattern specification is retrieved from the SOA pattern repository. As further discussed in Section 5, each design pattern consists of three main parts: problem, solution and application rules. In this paper, we use RBML to formally specify the first two, which are named "RBML Pattern Problem Specification"/"RBML Problem" and "RBML Pattern Solution Specification"/"RBML Solution". Details about RBML and the design pattern specifications are provided in Section 6. We use the QVT-O language for implementing the model transformation that corresponds to the pattern application rules. In Figure 2, area (2) "RBML Pattern Problem Specification" is used for identifying a subset of SModel elements that will be modified by the pattern (i.e., the respective model elements and their relationships must conforms to the RBML problem specification). If a match is found, the respective subset will be replaced by a "solution subset", which must conform to the "RBML Pattern Solution Specification". The actual transformation is described in the form of generic QVT rules, which consist of one or more basic actions, such as Adding, Removing, and Moving a model element or parameter. After performing the model transformation, the outcome is verified by checking whether the newly modified SModel elements conform with the RBML Pattern Solution Specification (see Section 7 for more details).

The SModel changes are propagated to PModel. As shown in Figure 2, area (3), the mapping between the elements of SModel and PModel is used to identify the required performance model changes. Finally, the flow of the processes in Figure 1 goes back to area (1) where the modified PModel after applying the design pattern is being re-analyzed to investigate the impact on the system performance. In Section 9, we discuss the performance results changes due to application of Service Façade design pattern to the SOA system case study.

4. SMODEL TO PMODEL TRANSFORMATION

In this section we present briefly the SModel expressed in UML extended with SoaML and MARTE, and the corresponding PModel using the Layered Queuing Network formalism [2].

4.1 SoaML and MARTE annotations

The SoaML specification is an OMG standard for the design of services within a SOA system [5, 6]. It is defined as a UML profile that provides a standard way to architect and model SOA solutions. In this section we provide a brief description of the SModel of a SOA system used in our approach.

In SOA design using SoaML, the first step is to identify services by analyzing the business goals and objectives of the system, as well as the businesses processes to be implemented for meeting these objectives. Using these processes it is possible to identify other business-relevant services. The model created in this step is called Business Processes Model (BPM) diagram [8]. In this paper we specify the BPM diagram using the UML activity diagram notation [19]. An UML activity diagram representing an example of such a model for a shopping and browsing service is shown in Figure 3. Once the business processes have been identified, each will be specified in a business process diagram and refined in regard to participants, their tasks, and information flow between the participants.

The next step is to define the Service Architecture Model (SEAM) based on the existing BPMs. SEAM is a high level description of how participants work together for a purpose by providing and using services expressed as service contracts. Participants are recognized from pools, participants and lanes specified in the BPM processes. Once the participants are known, the possible interactions between the different participants must be identified and represented as service contracts. Figure 4 represents a SEAM in the form of a UML collaboration diagram with service participants and contracts. Participants are modeled as UML classes stereotyped *«Participant»* and a service contract is a UML collaboration with the stereotype *«ServiceContract»*. In SoaML, the service contracts can be refined further in a Service Contracts diagram [6]. Each participant plays either the role of Provider or of Consumer with respect to a service. For instance, in Figure 4 the Shopping participant provides PlaceRequest and consumes PlaceOrder. A participant may provide or consume any number of services. A service contract represented by a collaboration may contain nested participants and services.

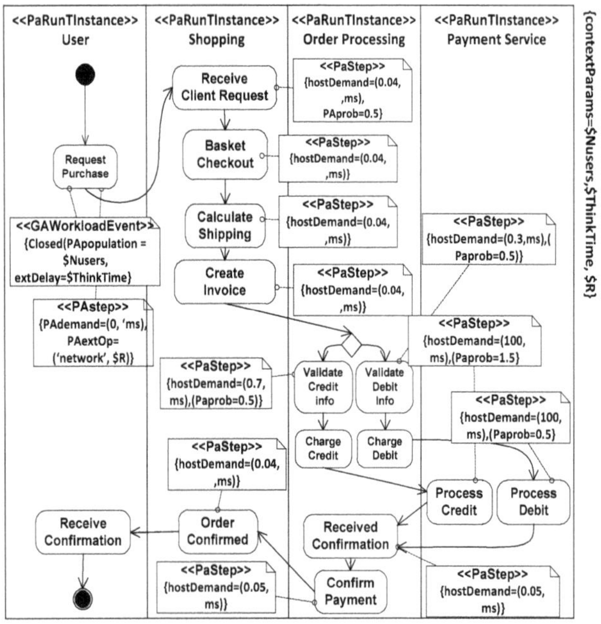

Figure 3: Checkout Business Process Model

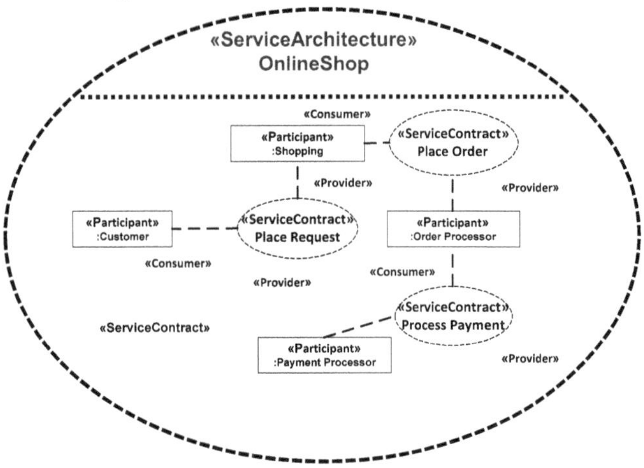

Figure 4: Service Architecture model for the Online Shop case study

Next, the specification of each service is further refined by defining the service interfaces between consumers and providers of the service. The diagram created in this step is called Service Interface diagram. It represents the provided and required interfaces, the roles that the interfaces play in the service specification, and protocols for how the roles interact. The third step, Service Realization, starts with identifying services that each participant provides and/or uses. This process has an important effect on service availability, distribution, security, transaction scopes, and coupling. The third step actually models how each service functional capability is implemented and how the required services are actually used. The forth step, called Service Composition or Components diagram, is about assembling and connecting the service participants models and then choreographing their interactions to provide a complete solution to the business requirements. Finally the last step is Service Implementation.

Among all the SoaML models mentioned in this section, our approach uses BPM (Figure 3) and SEAM (Figure 4) to propagate to PModel the changes made into the SModel by the application of design patterns. The reason is that all the information required to create the PModel can be obtained from these diagrams [3]. (See Section 8 for more details).

We add performance information to a UML+SoaML specification by adding annotations defined in the OMG standard profile MARTE [7]. Figure 3 shows an example of a SoaML behavioral diagram for the checkout operation of a shopping application. MARTE performance annotations describe the behavior as a Scenario, with steps and a workload *«GaWorloadEvent»* attached to the first step. Concurrent runtime instances *«PaRunTInstance»* are identified with swimlane roles. *«PaStep»* represents the execution of an activity or an operation invoked by a message, and has attributes *hostDemand* giving the required execution time and *PAprob* giving the probability if it is an optional step. The workload *«GaWorloadEvent»* defines a closed workload (a set of users) with a population given by the variable *$Nusers* and a think time for each user given by the variable *$ThinkTime*.

For performance analysis, the SOA specification must include the deployment of concurrent runtime component instances (see Figure 5). Where deployment is not specified, some kind of deployment is assumed, such as one host node with one concurrent process for each service. UML deployment diagrams will be used, and performance annotations are also added to this diagram. In Figure 5, the processing nodes are stereotyped as *«GaExecHost»* and the communication network nodes as *«GaCommHost»*, and the stereotypes have attributes for processing capacity, etc. Such annotations help us not only in transforming SModels into PModels, but also in propagating future modifications due to the application of design patterns.

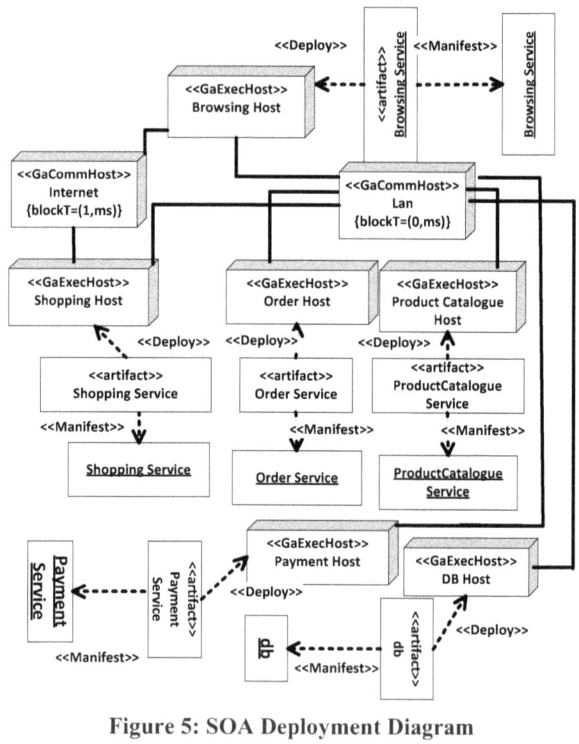

Figure 5: SOA Deployment Diagram

4.2 Performance Model (PModel)

This work constructs performance models (PModels) in an extended queuing format called Layered Queuing Networks (LQNs) [2]. An LQN represents congestion in waiting for service provided by host processors and also by software servers.

Figure 6 shows the LQN model corresponding to the SoaML business process in Figure 3. For each service there is a task, shown as a bold rectangle, and for each of its operations there is an entry, shown as an attached (thin line) rectangle. The task has a parameter for its multiplicity (e.g. {1} for OrderProcessing task in Figure 6) and the entry has a parameter for its host CPU demand, equal to the *hostDemand* of the operation in SoaML (e.g. [0.3 ms]). Calls from one operation to another are indicated by arrows between entries (a solid arrowhead indicates a synchronous call for which the reply is implicit, while an open arrowhead indicates an asynchronous call). The arrow is annotated by the number of calls per invocation of the sender (e.g. (3)). For deployment, the host node is indicated by a round symbol attached to each task.

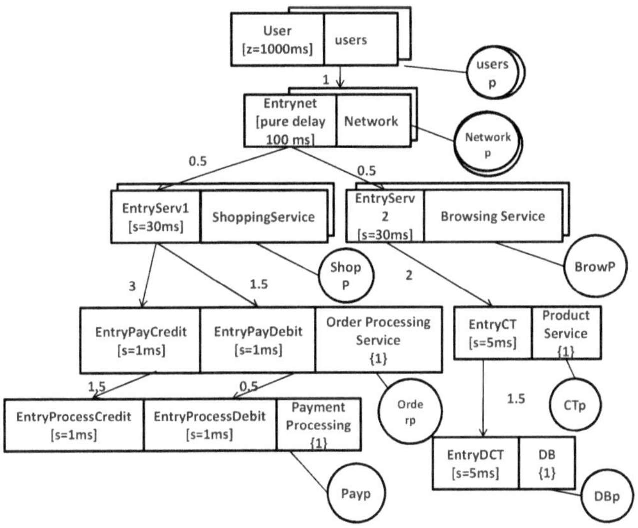

Figure 6 : LQN Performance Model (PModel) corresponding to Figures 3 and 4

We apply existing techniques [2, 11, 12] and also the technique in our previously published paper [3] to convert the UML+SoaML+MARTE SModel to a LQN model and to build the mapping between SModel and PModel elements. The mapping is further discussed in Section 8.

5. PATTERN APPLICATION RULES

A design pattern is defined as a proven design solution for a common design problem that is formally documented in a consistent manner. In the context of SOA, there are many categories of design patterns which address different aspect of a SOA-based systems including but not limited to: service messaging patterns, service implementation patterns, service security patterns, composition implementation patterns, etc. [1]. A SOA design pattern contains the description of a situation or problem where it applies, the solution for it and application rules. The RBML formal specification of these patterns and implementation of their application rules in form of QVT rules proposed in this paper helps the system designers to systematically identify the place in a SModel where a pattern can be applied and to automatically apply the pattern.

This section describes a pattern from the category of service implementation patterns called "Service Façade" [1]. Later this example will be used to show traceability issues and techniques.

Design Pattern Service Façade [1]. In general, the Service Façade addresses the way in which a service can accommodate changes to its contract or implementation while allowing the core service logic to evolve independently.

Problem: Usually a service contains a core logic that is responsible for operating its main capabilities. When a service is subject to change either due to changes in the contract or in its underlying implementation, this core service logic is also prone to modifications to accommodate that change.

Solution: Façade logic is added into the service architecture to create one or more layers of abstraction that can accommodate future changes to the service contract, the service logic, and the underlying service implementation.

Applications Rules: Service façade components can be positioned within the service architecture in different ways, depending on the nature and extent of abstraction required, such as: 1) between the core service logic and the contract to intentionally tightly couple to their respective contracts, allowing the core service logic to remain loosely coupled or even decoupled 2) between the core service logic and the underlying implementation resources to help shield core service logic from changes to the underlying implementation by abstracting backend parts.

As can be understood from the pattern description, it requires a new component to be added to the service architecture wherever there is service core logic that is tightly coupled with its service contract or underlying implementation. The application rules for this pattern are defined as:

- **Conditions:** If there is a core service logic in the SOA design which is identified as tightly coupled (considered as a design-choice by the system designer) with its associated contracts or underlying implementations.
- **Actions:** Add a new façade participant between the core service logic participant and the coupled contracts/underlying resources. Change all the communications in the service core logic and the coupled contract/underlying implementation to consume/provide services from/to the façade participant. Add processes (activities) to the newly added participant as façade to handle the responses/requests from the service core logic and also service contract/underlying implementation. Since the façade participant is created to be coupled with the service core, this newly added participant should be created as a private participant and in the same component (or even the same physical service).

As it can be understood from this pattern, new processes and participants need to be created in the SOA design. Therefore this pattern is targeting the business process in the SOA design. In future sections, we use this design pattern and apply it to the case study shown in Figure 3 and trace its impacts on the PModel of the case study.

There are several ways that the facade pattern could be applied to the SOA specification shown in Figure 3 and Figure 5. In general, if there is a service core which is prone to future changes, a service façade can help to protect it. In the SOA example in this paper, the shopping and browsing services, payment service, and product service are candidates for service façade, but we assume that the shopping service is selected as core service. The Service Façade pattern will be applied to accommodate requests on multiple channels (different types of devices, e.g. desktops, tablets, and smartphones). The capability to handle requests from multiple channels could be created either by adding new activities to each service/task (which may corrupt the core logic), or by applying the Service Façade pattern, so we consider both alternatives. In Section 7 and 8 , the change due to application of the Service Façade pattern is propagated to the PModel and then in Section 9, the system performance is evaluated for both approaches (multi-channel implementation with and without Service Façade).

6. ROLE-BASED SOA DESIGN PATTERN SPECIFICATION

Informal descriptions of the design patterns (similar to the one discussed in section 5) are useful for communicating proven solutions to the development team, but they lack the formality needed to support precise specification of design patterns and the development of automated techniques and supporting tools. Although formal pattern specification languages using a mathematical notation (e.g., see [20, 21]) provide the concepts needed to precisely describe design patterns, applying them requires sophisticated mathematical skills. Therefore in this research, we decided to deploy pattern specification languages that are based on familiar software modeling concepts (e.g. UML) to specify SOA design patterns in our proposed technique in this paper. The notation used is called Role-Based Modeling Language (RBML) [8] , which it was used for a similar purpose (i.e., specifying performance anti-patterns) in [22].

A Role Model is a structure of meta-roles (hereafter called roles), where a role defines properties that determine a family of UML model elements [8]. A UML model element conforms to a role if it is an instance of the role's base and has the properties specified in the role. A Role Model is thus a characterization of UML diagrams and its profiles such as SoaML. Therefore a Role Model realization is a model (e.g., a static structural diagram, sequence diagram) that consists of realizations of the roles in the Role Model. The concept of role in RBML is very helpful to capture the various aspects of the knowledge in the design patterns.

In model specification using RBML, the specification consists of a Static Role Specification (SRMs) that specifies the static structural models, and a set of Interaction Role Models (IRMs) that specifies interactions (i.e. behaviors) [8]. Respectively, we use SRM to represent the static and IRM for interactive (behavior) aspects of SOA design patterns.

In [8], authors provided the SRM notation and IRM samples for UML sequence and collaboration diagrams. Since in this research we use UML+SoaML to present the SOA system under study, we extended the concepts proposed in [8] with IRM for activity diagrams using UML extended with profiles.

6.1 Static Role Models (SRMs) in SOA

Figure 7.(A) shows the structure of a single SRM role. A SRM role characterizes a set of UML static modeling constructs (e.g., class, and association constructs). The top section has three parts: 1) a role base declaration in form of "«Base Role»", where Base is the name of the role's base (i.e., the name of a metamodel class) 2) a role name declaration in form of "/RoleName", where RoleName is the name of the role; and 3) a realization multiplicity that specifies the allowable number of realizations that can exist for the role in a realization of the SRM that includes the role. The second section contains metamodel-level constraints and the third, optional, section contains feature roles that determine a family of application-specific properties (e.g., properties represented by attributes and operations defined in application-specific classes).

In RBML, there are two types of feature roles:

(1) Structural roles that specify state of the properties that are realized by attributes or operations in a SRM role realization.

(2) Behavioral roles specify behaviors that are realized by a single operation or method.

In this paper, we use SRM to represent the static aspects of a SOA design pattern. In pattern specification using SRM, each SOA participant (e.g., providers and consumers) can be considered as a role. An example of service façade SRM is shown in Figure 7.(A) SRM presents a role for coupled core service logic. In Figure 7.(B), the "Shopping" service is playing that role (assuming that this was the decision of the designer).

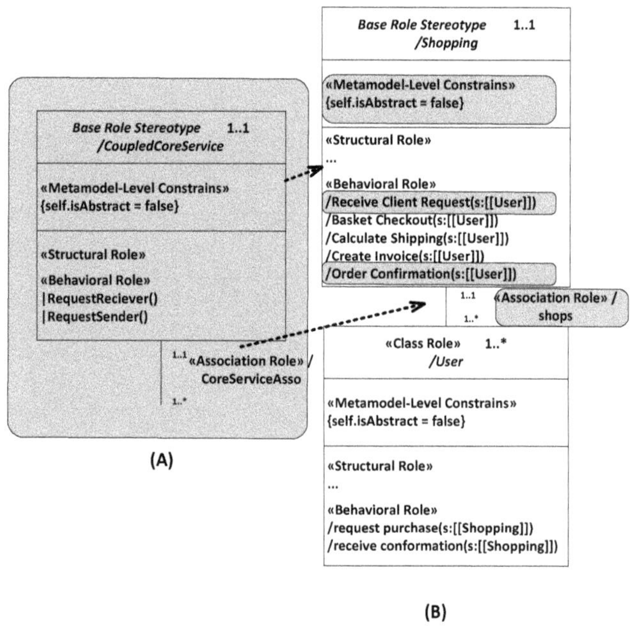

Figure 7 : (A) SRM of a coupled core service in Service Façade (B) Conformance to coupled core service in Shopping example

In SOA systems, the architecture is represented not only with class diagrams, but also with collaboration diagrams, as for example the service architecture. Therefore, we also use SRM specifications in the form of collaborations (see Figure 8).

6.2 Role Relationships

In RBML, a role can be associated with another role in a class diagram, indicating that the realizations of the roles are associated in a manner that is consistent with how the bases of the roles are related in the UML model. RBML uses the UML form of association to represent relationships between roles. Role associations can be named and can have multiplicities associated with their ends. An example of relationship between roles in SoaML is shown in Figure 7.(B). In this figure, the "Shopping" service, which plays the role of Coupled Core Service in the

Service Façade design pattern is in "shop" relationship with "User". The multiplicity on the role relationship shows that more than one user can be in "shops" relationship with the "Shopping" service.

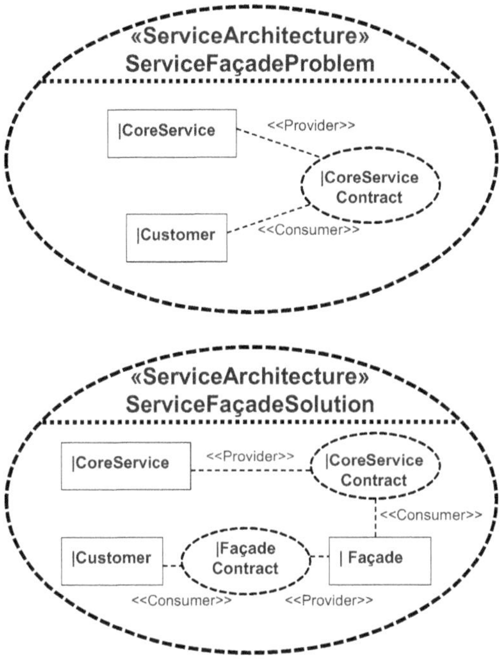

Figure 8: SRM specification for the Service Façade pattern: problem (top) and solution specification (bottom)

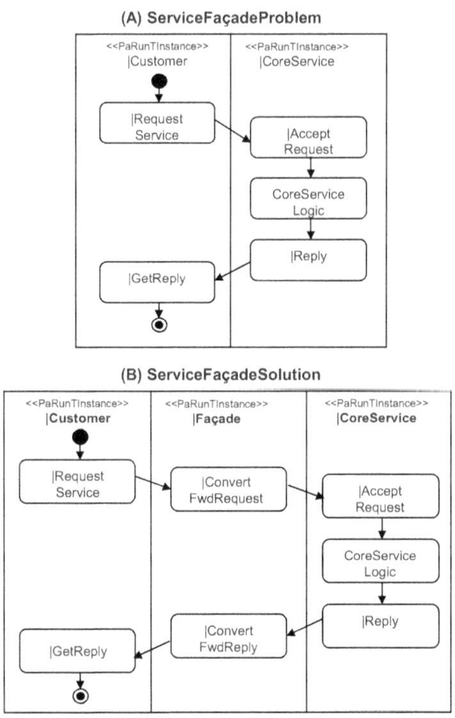

Figure 9: IRM specification for the Service Façade behavior view: (A) pattern problem specification; and (B) pattern solution specification

The RBML specification of the service architecture view of the Façade Pattern is illustrated in Figure 8: the problem specification on the top and the solution on the bottom. The roles played by the participants and the service contracts are represented as formal parameters (by convention the names start with '|'). When the pattern will be applied, the formal parameters |Customer, |CoreService, |CoreServiceContract will be matched and bound to actual participants and service contracts from the SModel service architecture, as described in Section 7.

6.3 Interaction Role Models (IRM)

We use Interaction Role Model (IRM) to present the interactions between participants in SOA behavior diagrams (e.g., a BPM diagram). IRM is defined using the roles specified as formal parameters in the SRM for specifying the participants and may add other roles to specify some of their actions. Figure 9 illustrates the IRM behavior view of the Façade Pattern: (A) the pattern problem and (B) the pattern solution. When the pattern is applied, a new participant |Façade is added. Its activities are located in a new swimlane in the activity diagram in Figure 9.(B). The activities from the Service Façade Problem with names starting with '|' (i.e., formal parameters) will be matched and bound to activities from a concrete BPM diagram, which will be then modified according to the pattern solution. For instance, a participant playing the role of |Customer will need to perform an activity that matches |RequestService and another that matches |GetReply. The activity sending the request will be followed by an edge crossing the swimlane to the |CoreService, while the activity receiveing the reply has an incoming edge from the |CoreService swimlane. Similarly, the activities execute by a participant playing the role of |CoreService will have to be matched with |AcceptRequest and |Reply, respectively. Figure 10 illustrates the refactored BPM after applying the Service Façade; its activities shaded in grey are bound to formal parameters from Figure 9.(B).

6.4 Conformance with Pattern SRM and IRM

An SRM is a generic presentation of the SOA model structure. Each role in a SRM specifies a specialization of its base class in the UML model. In this paper, we consider that a SModel diagram or sub-diagram structure conforms to a pattern SRM if each role defined in the pattern SRM can be played by at least one model element from the SModel. This involves the following:

(i) Each role defined in the pattern SRM has at least one matching SModel element and for each role there is at least one match for metamodel-level constraints specified in SRM.

(ii) Each attribute/behavior defined for the role in pattern SRM has a matching attribute/behavior among the SModel elements.

Also we consider that a SModel behavior diagram conforms to a pattern IRM if: (a) it conforms structurally to the associated pattern SRM and (b) the behavior of roles specified in IRM match the behavior of elements playing SRM roles. The definition of behavioral conformance is dependent to the UML behavioral diagram used to illustrate the scenario. In SoaML, a BPM which is an UML activity diagram is one of the main behavioral diagrams. An activity diagram used in a SOA SModel conforms to IRM if the model elements matching the SRM roles are following the same execution flow and are located in matching swimlanes as the roles in IRM.

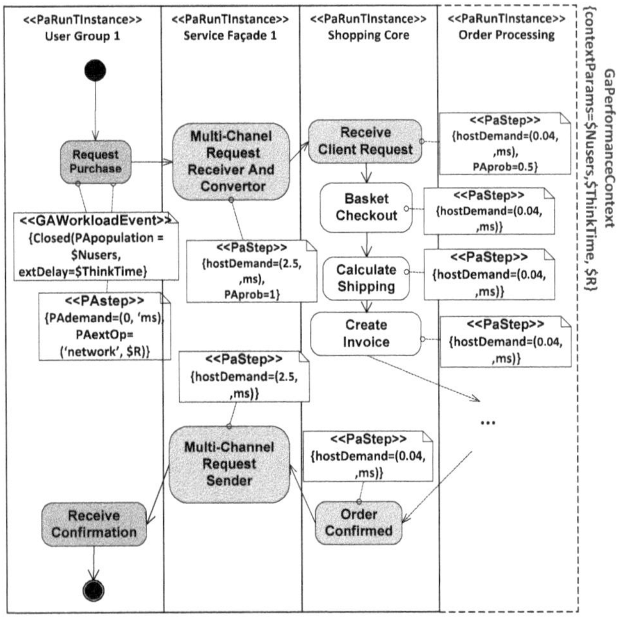

Figure 10 : Refactored BPM after applying the service façade design pattern

7. PATTERN APPLICATION USING RBML SPECIFICATIONS AND QVT-O

In Section 6, we discussed the importance of using a formal UML-based pattern definition language for the specification of the design patterns. Furthermore we discussed how we used RBML to specify the design pattern problem and solution description. In this section we discuss the way that the RBML specification for both solution and problem is being used to apply the design patterns to a SModel. First, the "RBML problem" is used for identifying the subset of SModel elements that will be affected by the pattern, then the "RBML solution" is used to identify the affected model elements due to application of the pattern . Once the impacted elements in the SOA are identified, the QVT-O transformation rules are executed to apply the changes (Section 7.2). Finally, the refactored SModel is being verified by checking its conformance with "RBML solution". Due to lack of space, in this section we only show the application of design patterns to SModel BPM and SEAM. Using the same techniques, the design pattern changes can be reflected to SModel in Deployment Diagram.

7.1 Conformance with RBML Specification

The first step toward applying a design pattern to a SOA design is to identify the places in the SModel where the design pattern can be applied. This can be done by finding a diagram or subset of it which conforms to the RBML Pattern Problem Specification.

In our proposed approach, the place in the SModel where a design pattern should be applied is identified by the software designer. Figure 11.(B) shows the SoaML service architecture diagram for the case study system, and the subset of model elements for applying the Service façade pattern shown in grey does conform to the RBML Pattern Problem Specification shown in Figure 11.(A).

As shown in Figure 11, there might be more than one place in the SModel which conforms to the RBML Pattern Problem Specification, but since Service Façade is supposed to be applied to the core service logic only, the software designer must indicate which service is considered as core service. This is another reason for the direct intervention of software designer in the selection of design patterns and the place where they should be applied.

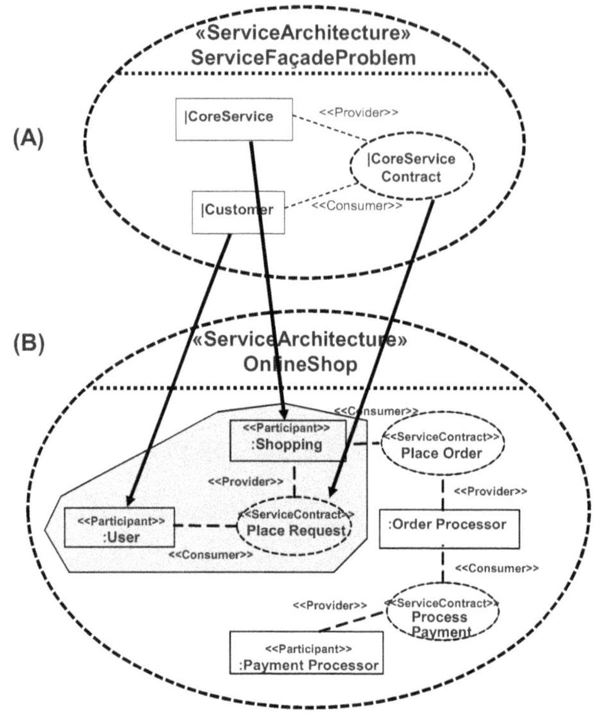

Figure 11: (A) RBML Facade Pattern Problem Specification" (B) SModel Service Architecture diagram

Once the subset of SModel elements corresponding to the pattern problem was identified, the following changes need to take place:

- *Adding a new contract to service architecture associated to the new communication between the service façade participant and service (in this case, "Format Multi-Channel Request" is added as the Façade contract)*
- *Moving all the communication to core service logic contract to the newly added service contract (in this case the communication from participant " User1" is moved to "Format Multi-Channel Request")*
- *Adding a new contract participant as Service Façade ("Multi Channel Façade 1" is added)*
- *Moving all the communications from the core service logic contract to the newly added contract participant (all the communications from "Place Request" are moved to "Multi Channel Façade 1").*

All the above steps are defined as generic steps for adding a service Façade to an existing SOA design. The above steps become specific when they are bound to model elements of a specific pattern application. Therefore similar to RBML Pattern Specifications of the problem and solution, the above can be translated into a form of model transformation rules and be re-used for future application of the Service Façade design pattern to any SModel. In the Section 7.2 we discuss this in more details.

7.2 Applying the Design Pattern using QVT-Operational

The RBML graphical presentation of the design patterns helps the software engineer to identify the elements that need to be added or removed from a SModel, but it does not help in applying the changes to the model. Therefore in this research, we define the model transformation rules corresponding to the application of a pattern in OMG's standard language QVT-Operational [9]. The rules are able to transform (more specifically, refactor) a SModel by adding or removing model elements. QVT defines three domain-specific languages: Relations, Core and Operational. In this research, we chose QVT Operational (QVT-O) because of its flexibility in comparison with the other two languages, due to the fact that it includes constructs commonly found in imperative languages, such as loops and conditions (i.e. Figure 12).

```
1 -- source is SA (Service Architecture)
2 -- target is SA (Service Architecture)
3 modeltype SA uses SoaML('http://www.omg.org/spec/SoaML/1.0/');
4 transformation ServiceFaçade (in source:SA, out target:SA);
5 main() {
6       srcSaDiagram.objectsOfType(SA:SAElement)->  map getCoreServiceToFaçade();
7       srcSaDiagram.objectsOfType(SA:SAElement)->  map addFaçadeService();
8       srcSaDiagram.objectsOfType(SA:SAElement)->  map addNewFaçadeContract();
9 }
10 query SA::SAElement::getCoreServiceToFaçade( ): Void
11 {
12   if (self.isSelectedObject(ToFaçade))  then {
13   objectToFaçade  := self;
14   } endif ;
15 }
16 mapping SA::SAElement::addFaçadeService():  Void
17 {
18  var newService:SAElement:=  serviceToFaçade.deepclone().
   setElementID(serviceToFaçade).oclAsType(SAService);
19  if (serviceToFaçade.sourceEdges->notEmpty  ( ) ) then {
20     newEdge:=  serviceToFaçade.sourceEdges->first().deepclone().oclAsType(Edge) ;
21     newEdge.source:=serviceToFaçade  ;
22     newEdge.target:=  newService ;
23     }else {
24       if(serviceToFaçade.targetEdges->notEmpty())  then {
25       newEdge:=  objecToFaçade.targetEdges->first().deepclone().oclAsType  (Edge);
26       newEdge.source  := objectToFaçade ;
27       newEdge.target:=  newService ;
28    } endif ;
29   }endif ;
30   objectToDivide.sourceEdges->forEach(outgoingEdge  ) {
31   if(outgoingEdge  <> newEdge ) then {
32   outgoingEdge.source:=  newService;
33   umlEdgesToTransform  += outgoingEdge.element.oclAsType(Edge);
34   }endif ;
35  };
36 }
37 mapping SA::SAElement::addNewFaçadeContract():  Void
38 {
39 ...
40 }
41
```

Figure 12: QVT-O rules for Service Façade Pattern

By examining the types of SModel changes needed to apply a design pattern, we introduced the following three refactoring primitives: 1- Adding a model element, 2- Removing a model element, and 3- Moving a model element. Using these refactoring primitives, more complicated refactoring actions (e.g. Merge and Divide) can also be implemented. In the model transformation we use in this paper, it's assumed that the model elements which are not impacted by the patterns are being copied into result model.

Figure 12 shows the QVT-O rules that have been created for the solution statement of the Service Façade pattern. The main function starts in line 5 with *getCoreServiceToFaçade()*. This function retrieves the selected core service logic by the system designer that the Service Façade needs to be applied for. Then a new Service Façade and also an associated service contract are being created by *addFaçadeService()* and *addNewFaçadeContract()* and all the connecting edges (i.e. incoming and outgoing) are being updated to accommodate the new model nodes.

addFaçadeService() adds the new Service Façade node to the target diagram by making a copy of the provided core service logic. Line 18 shows that the new node is being created using the *deepclone()* that QVT-O provides. Lines 19-29 contain conditional statements checking if there exist any incoming or outgoing edges that connect to the core service logic. When found, the first edge in either set is copied with the QVT-O *deepclone()* operation and then connected to the core service logic and the Service Façade (cloned node) by setting the source and target properties appropriately.

In lines 29-36 all outgoing edges from the core service logic are moved to the Service Façade by setting the source property of each edge to point to the new element. *addNewFaçadeContract()* function will also works very similarly to *addFaçadeService()* to create a new service contract for the newly created Service Façade.

The result of executing the QVT-O transformation rules for Service Façade design pattern (i.e. which are partially presented in Figure 12) is shown in Figure 13.(B). The Façade service (i.e. called contract participant in SoaML service architecture diagram) and the Façade contract are both added and all the communications are moved to accommodate the design pattern changes.

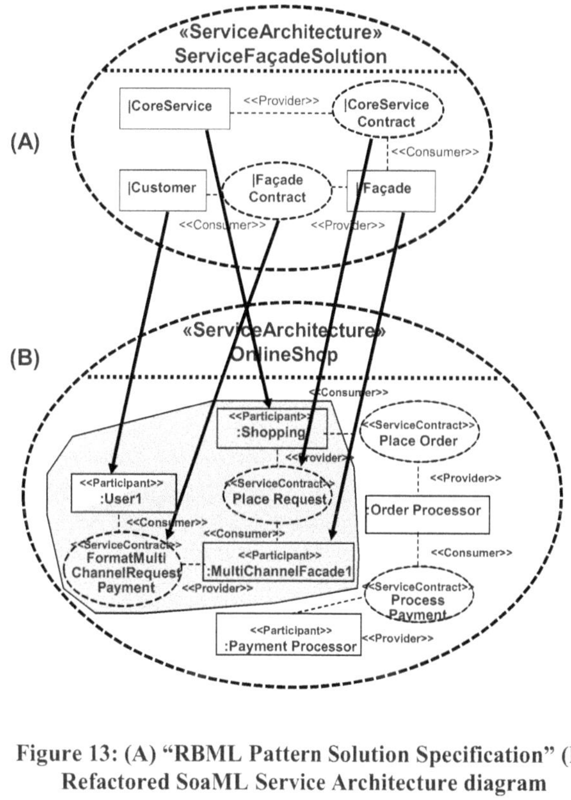

Figure 13: (A) "RBML Pattern Solution Specification" (B) Refactored SoaML Service Architecture diagram

7.3 Verifying the Transformation Output

After executing the model transformation implemented in QVT-O the transformed model can be verified to make sure that it represents the changes correctly. This not only verifies the content of the generated SOA design, but it also verifies the transformation rules which are used to generate the transformed diagrams. The application of the design pattern is implemented in the transformation and thus if the verification detects an error, this can be considered as a bug in the transformation and the transformation rules can be corrected. The step is part of the research (as an end-to-end verification of the transformation) rather than part of the proposed methodology. In this work, we verify the generated SOA design by checking the conformance of the transformed areas with the "RBML Pattern Solution Specification". Figure 13 shows the generated model (Figure 13.(B)) using the QVT-O transformation rules and also the conformance of the modified area with the "RBML Solution" (Figure 13.(A)).

8. IDENTIFY PMODEL CHANGES

In Section 7, we discussed how a design pattern can be applied to an SModel and how the modifications are identified. As a PModel is being generated from the SModel, modifications to SOA design lead to changes into the associated PModel of SOA and the performance analysis results. The LQN performance model is derived from the SOA Service Architecture (Figure 13.(B)), SOA BPM (Figure 10) and the SOA Deployment Diagram (Figure 5). Due to lack of space, in this paper we do not discuss the changes made to deployment diagram due to application of design pattern but for the sake of case study in this paper, the process is performed based on discussions in Sections 6 and 7 and results are used in performance evolution provided in Section 9. The changes to the LQN performance model due to the impact of design pattern are passed through these three models. The key elements of the LQN model are LQN Tasks, Processors, Entries and RequestArcs. The LQN performance model of the SOA under study before applying the Service Façade design pattern is shown in Figure 6 and its structure is discussed in Section 4.2. Therefore, a very straightforward mapping between SModel and PModel (which is determined automatically) is established as follows [3]:

- Mapping deployment diagram Executing Host Nodes to LQN hosts (processors)
- Mapping Service Architecture diagram Contract Participant (i.e. services) to LQN tasks
- Mapping activities in the BPM swimlane associated to the service presented in Service Architecture diagram to LQN entries
- Mapping Swimlane-crossing Edges to LQN Requests Arcs.

Some swimlane-crossing Edges associated to a service in BPM represent outgoing requests for the service and some represent the incoming responses. Since in LQN diagram, the LQN Requests Arcs represent both requests and responses from the LQN task, only the BPM Swimlane-crossing Edges that make the request are mapped into LQN Requests Arcs. A more detailed discussion about mapping Swimlane-crossing Edges to LQN requests can be found in [12]. Based on the SOA design changes discussed in Section 7 and also the mapping showed in this section (see the grey area 3 in Figure 4 approach overview), the newly added Service Façade in service architecture diagram (Figure 13.(B)) in and the activities (annotated by «PaStep») in its associated swimlane (annotated by «PaRunTInstance») in BPM (Figure 10) cause the creation of the "Service Façade" LQN task and its LQN entries, "EntryFaçade" shown in grey in the LQN model in Figure 14. Also, since a new processing node is added to the deployment diagram for the newly added "Service Façade" component, a LQN processor is created in the LQN model and attached to the corresponding LQN task.

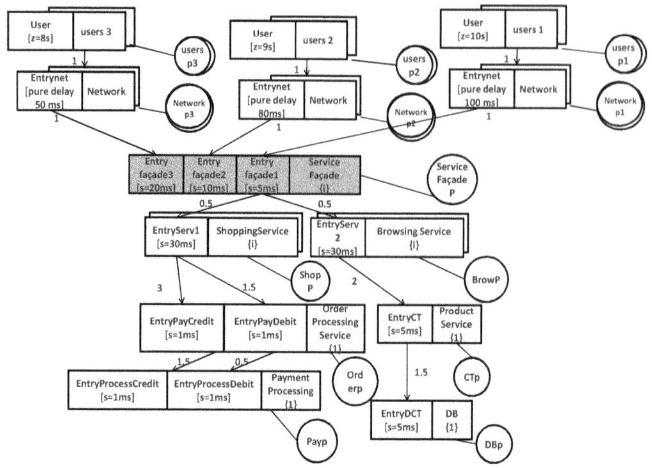

Figure 14: SOA Performance Model after applying the façade design pattern

9. PERFORMANCE ANALYSIS

This section discusses the impact of the pattern application on the performance results of the case study system. We evaluate the performance results of Shopping and Browsing SOA before and after applying the service façade design pattern. In this case study, it is assumed the core service logic is the Shopping and Browsing service and the service façade design pattern is applied to help the core service handle requests from more than one user type by adding a multi-channel service façade to SOA. We evaluate four scenarios named A, B, C and D, to find the system throughput and response time of all three user groups for a range of numbers of users in each group. For N users in "User Group 1", we defined 2N in "User Group 2", and N/2 in "User Group 3". N was varied from 2 to 220, so the total number of the system users (i.e. N+2N+N/2=3.5N) ranged from 7 to 770. In scenario A, the service façade pattern is not applied and instead the multi-channel capability is implemented by adding to each of the shopping and browsing services in the system. The performance model for scenario A is shown in Figure 15. In scenario A, a separate task entry is created for each service as used by each user group (1, 2 and 3). In scenario B the service façade pattern is applied, giving the performance model shown in Figure 14. The new service façade task added to the system has three entries, each responsible for the interface with one group of users. Also the service facade task has its own dedicated processor. Figure 16 compares the system throughputs and Figure 17 compare the system response times of scenarios A and B for each user group of the system. Figure 16 and Figure 17 show that scenario B has better throughput and response time than A for all three user groups. The underlying reason is that in scenario B a new processor is dedicated to the Service Façade task, while in scenario A the new functionality related to the multi-channel interface is running on the existing processors for the Shopping and Browsing services. To make a more fair comparison between the "before" and "after" scenarios, we use our knowledge of how the resources are used in the scenarios A and B to experiment with scenarios C and D.

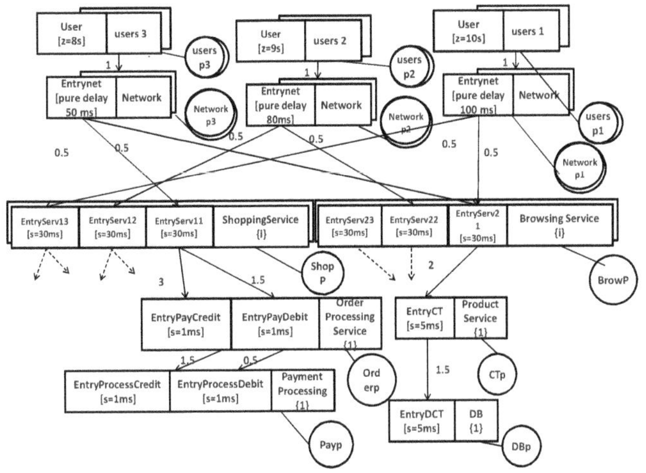

Figure 15: Performance Model after applying the multi-channel capability to each service without applying any Service Façade Design Pattern

Scenario C is obtained from A by changing the processors for the browsing and shopping services from single to dual-core. The reason is that these processors become the bottleneck in case A at high loads. Scenario D is obtained from B by changing the processors for the shopping and browsing services to dual processors and by allocating the service façade task to one of the existing processors that are under-utilized (in this case the processor of the Order Service task). So, the hardware resources for C and D are identical. The results for C and D are shown in Figure 18 and Figure 19.

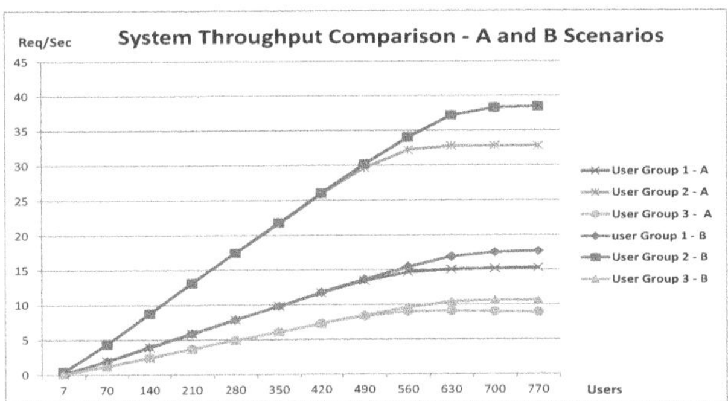

Figure 16: System Throughput for Scenarios A and B

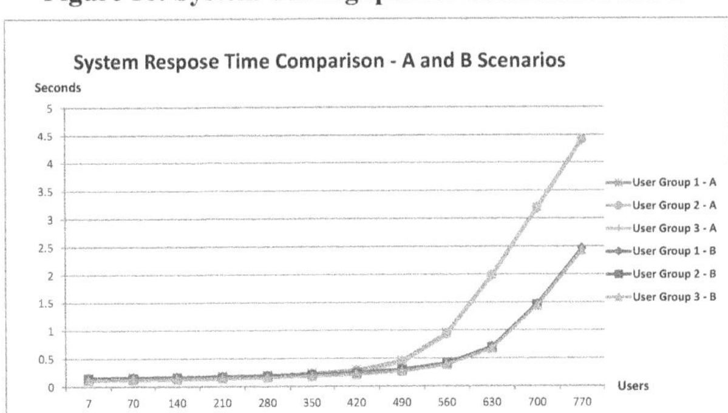

Figure 17: System Response Time for Scenarios A and B

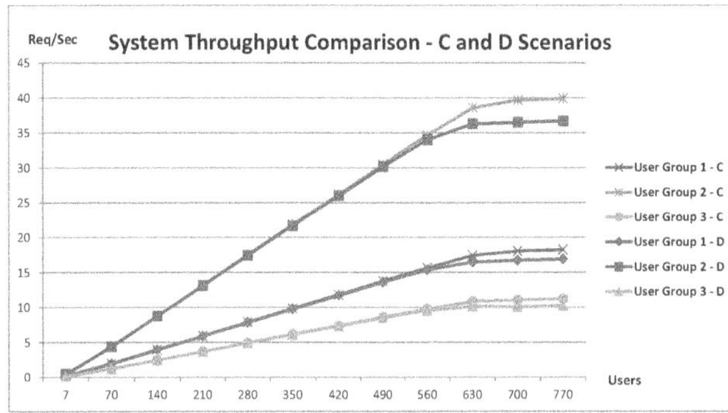

Figure 18: System Throughput for Scenarios C and D

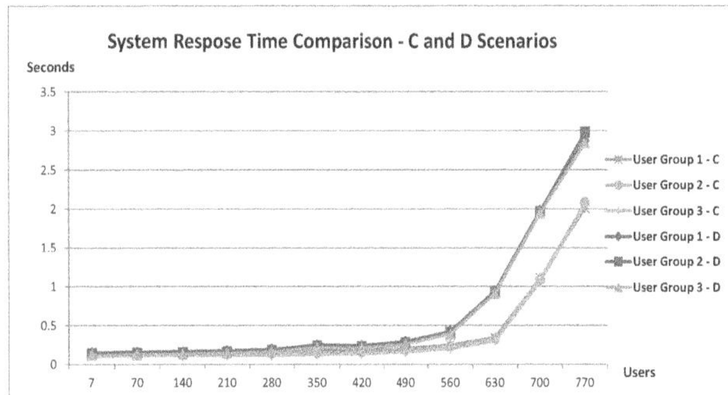

Figure 19: System Response Time for Scenarios C and D

Firstly, the throughput of C is better by about 20% than A due to the extra processing power added to the system. Secondly, the throughput and response time for D (the modified version of B) are only slightly worse than scenario C (the modified version of A) even though now C and D are running on identical hardware resources. The underlying reason is that we made use of our understanding of the resource utilization in the system and deployed the service façade in D on one of the existing but under-utilized processors. In conclusion, we could minimize the negative performance effect of the façade service pattern with the help of the LQN model, while taking advantage of its benefits to the system architecture (i.e. creating one or more layers of abstraction that can accommodate future changes to the service).

10. CONCLUSIONS

This paper traces the change propagation due to applying a SOA design pattern, from the SOA design model (created using SoaML) to its performance model. In illustration and using our approach, an example of SOA design patterns called Service Façade was applied to a Shopping and Browsing SOA case study, to equip it with a multi-channel request handler. In Sections 6 and 7 we discussed how the SOA design patterns can be applied to the SOA design using the RBML pattern presentation and QVT-O application rules. In Section 8 we showed that how the changes to SModel due to the application of a design pattern can be traced into performance model of the SOA using the mapping between SModel and PModel elements.

The overall possibilities of the technique are demonstrated in Section 9 by showing the results of the performance analysis. In Section 9, we evaluate the performance of the Shopping and Browsing service using four different scenarios that represent the Shopping and Browsing SOA with multi-channel capability before and after applying the service façade. The experiment shows the system designer the impact of the design pattern on the performance results of system in a time and cost effective way (without constructing a new performance model for the modified SOA design) and helps to make decisions about applying the design pattern. The results also indicate that the impact of the design pattern is partly determined by how the system is deployed. We are planning on developing QVT rules to also modify the LQN model to reflect the SOA design pattern changes. Furthermore, we are planning to integrate the proposed approach in a methodology for improving the quality of SOA systems, which will address the issue of selecting appropriate design pattern candidates. This will give a complete technique for SOA design improvement.

11. ACKNOWLEDGEMENTS

This work was supported by the Ontario Centers of Excellence through its Centre of Excellence for Research in Adaptive Systems (CERAS), by the Natural Sciences and Engineering Research Council (NSERC) through its Discovery Grant program, and by the Healthcare Support through Information Technology Enhancements (hSITE) Strategic Research Network.

REFERENCES

[1] Erl, T. *SOA Design Patterns* Prentice Hall PTR, Boston, MA, 2009.

[2] Woodside, M., Petriu, D. C., Petriu, D. B., Shen, H., Israr, T. and Merseguer, J. Performance by Unified Model Analysis (PUMA). In *Proceedings of the WOSP '05 Proceedings of the 5th international workshop on Software and performance*. ACM New York, NY, USA, 2005.

[3] Mani, N., Petriu, D. C. and Woodside, M. Studying the Impact of Design Patterns on the Performance Analysis of Service Oriented Architecture. In *Proceedings of the 37th EUROMICRO Conference on Software Engineering and Advanced Applications (SEAA)* (Oulu, Finland). IEEE Computer Society Washington, DC, USA, 2011.

[4] Rotem-Gal-Oz, A., Bruno, E. and Dahan, U. *SOA Patterns (Early Access Edition)*. Manning Publications, 2007.

[5] Object Management Group, *Service oriented architecture Modeling Language (SoaML) URL: http://www.omg.org/spec/SoaML/1.0.1/ [Last time accessed Oct 5th, 2012]*. Version 1.0, formal/2009-11-02.

[6] Elvesæter, B., Carrez, C., Mohagheghi, P., Berre, A., Johnsen, S. G. and Solberg, A. *Model-driven Service Engineering with SoaML*. Springer Vienna, City, 2011.

[7] Object Management Group, *A UML Profile for MARTE (Modeling and Analysis of Real-Time and Embedded systems)*. Version 1.0, formal/2009-11-02 URL: http://www.omg.org/spec/MARTE/1.0/PDF/ [Last time accessed Oct. 5th ,2012].

[8] France, R. B., Kim, D.-K., Ghosh, S. and Song, E. A UML-Based Pattern Specication Technique. *IEEE Trans. Software Eng.*,Vol 30 (3), 2004, 193-206.

[9] Object Management Group, *Meta Object Facility (MOF) 2.0 Query/View/Transformation (QVT) URL : http://www.omg.org/spec/QVT/1.1/ [Last time accessed Oct 5th, 2012]*. Version 1.1, formal/January 2011.

[10] Woodside, M., Petriu, D. C., Petriu, D. B., Xu, J., Israr, T., Georg, G., France, R., Bieman, J. M., Houmb, S. H. and Jürjens, J. Performance analysis of security aspects by weaving scenarios extracted from UML models. *Journal of Systems and Software*,Vol 82 (1), 2009, 56-74.

[11] Petriu, D. C. *Software Model based Performance Analysis*. ISTE Ltd and John Wiley & Sons Inc., City, 2010.

[12] Petriu, D. C. and Shen, H. Applying the UML Performance Profile: Graph Grammar-based Derivation of LQN Models from UML Specifications. In *Proceedings of the TOOLS '02 Proceedings of the 12th International Conference on Computer Performance Evaluation, Modelling Techniques and Tools*. Springer-Verlag, London, UK, 2002.

[13] Smith, C. U. and G.Williams, L. *Performance Solutions : A Practical Guide to Creating. Responsive, Scalable Software*. Addison Wesley, Boston, MA,, 2002.

[14] Cortellessa, V. and Mirandola, R. Deriving a Queueing Network based Performance Model from UML Diagrams. In *Proceedings of the WOSP '00 Proceedings of the 2nd international workshop on Software and performance*. ACM New York, NY, USA, 2000.

[15] Arcelli, D., Cortellessa, V. and Trubiani, C. Antipattern-based model refactoring for software performance improvement. In *Proceedings of the Proceedings of the 8th international ACM SIGSOFT conference on Quality of Software Architectures* (Bertinoro, Italy). ACM New York, NY, 2012.

[16] Menascé, D. A., Ewing, J. M., Gomaa, H., Malex, S. and Sousa, J. P. A framework for utility-based service oriented design in SASSY. In *Proceedings of the WOSP/SIPEW '10 Proceedings of the first joint WOSP/SIPEW international conference on Performance engineering*. ACM New York, NY, USA, 2010.

[17] Parsons, T. and Murphy, J. Detecting Performance Antipatterns in Component Based Enterprise Systems. *Journal of Object Technology*,Vol 7 (3), 2008.

[18] Xu, J. Rule-based automatic software performance diagnosis and improvement. In *Proceedings of the Proceeding 7th Intl Workshop on Software and Performance*, Princeton, NJ, 2008.

[19] IBM, *Modeling with SoaML, the Service-Oriented Architecture Modeling Language*. January 7th , 2010 URL: http://www.ibm.com/developerworks/rational/library/09/modelingwithsoaml-1/ [Last time accessed Oct 5th, 2012].

[20] Eden, A. H., Yehudai, A. and Gil, J. Y. Precise specification and automatic application of design patterns. In *Proceedings of the ASE '97 Proceedings of the 12th international conference on Automated software engineering* (Incline Village, NV). IEEE Computer Society Washington, DC, USA, 1997.

[21] Lano, K., Bicarregui, J. C. and Goldsack, S. Formalising Design Patterns. In *Proceedings of the Proceeding of BCS-FACS Northern Formal Methods Workshop, Electronic Workshops in Computer Science*. British Computer Society Swinton, UK, 1996.

[22] Cortellessa, V., Marco, A. D., Eramo, R., Pierantonio, A. and Trubiani, C. Digging into UML models to remove performance antipatterns. In *Proceedings of the Proceeding of 2010 ICSE Workshop on Quantitative Stochastic Models in the Verification and Design of Software Systems* (Cape Town). ACM New York, NY, 2012.

Rapid Performance Modeling by Transforming Use Case Maps to Palladio Component Models

Christian Vogel*, Heiko Koziolek†, Thomas Goldschmidt†, Erik Burger*
*Software Design & Quality, Karlsruhe Institute of Technology, Germany
E-Mail: sdq@ipd.uka.de,burger@kit.edu
†ABB Corporate Research, Industrial Software Systems Program, Ladenburg, Germany
E-Mail: {heiko.koziolek,thomas.goldschmidt}@de.abb.com

ABSTRACT

Complex information flows in the domain of industrial software systems complicate the creation of performance models to validate the challenging performance requirements. Performance models using annotated UML diagrams or mathematical notations are difficult to discuss with stakeholders from the industrial automation domain, who often have a limited software engineering background. We introduce a novel model transformation from Use Case Maps (UCM) to the Palladio Component Model (PCM), which enables performance modeling based on an intuitive notation for complex information flows. The resulting models can be solved using existing simulators or analytical solvers. We validated the correctness of the transformation with three case study models, and performed a user study. The results showed a performance prediction deviation of less than 10 percent compared to a reference model in most cases.

Categories and Subject Descriptors

D.2.8 [**Software Engineering**]: Metrics—*complexity measures, performance measures*; D.2.11 [**Software Engineering**]: Software Architecture

Keywords

Use case maps; Performance modeling; Palladio Component Model; Model Transformation; User study

1. INTRODUCTION

Industrial software systems, such as distributed control systems, follow complex information flows. For example, they transport sensor data from industrial field devices along multiple hubs to central servers, from which the data can be processed by different kinds of operator stations. Designing software architectures for such systems is hard because of the complex control flows. In addition, these systems have challenging performance requirements for high throughput and short response times. Thus, early performance modeling is desired to identify performance bottlenecks and design efficient information flows.

To enable early performance modeling, numerous mathematical notations (e.g., queuing networks, stochastic Petri nets, stochastic process algebra) and software-related notations (e.g., annotated UML) have been proposed [1, 29, 12]. These notations usually require expertise from the performance domain and their graphical representations are often difficult to discuss with stakeholders from the domain of industrial systems. This complicates reasoning on different alternatives for scalable architectures and leads to a general reluctance to apply performance modeling in this domain.

The notation of Use Case Maps (UCM) [11] has been created by analyzing how software systems are typically sketched in early stages on white boards. UCMs capture high level software components as well as the control flow for specific usage scenarios in an intuitive notation. In former work, Petriu and Woodside [20] proposed a model transformation from UCMs to layered queuing networks (LQN) to allow performance predictions from UCMs. Their transformation did however not result in a component-based model, and the implementation is no longer maintained. Other transformations mapped UML state machines, sequence diagrams, or activity diagrams to stochastic Petri nets [3] or stochastic process algebras [26], but require UML/MARTE [18] and/or performance expertise to be useful.

The contribution of this paper is UCM2PCM, a model transformation from UCMs into the Palladio Component Model (PCM) [2], which enables component-oriented performance modeling as well as multiple solution methods (e.g., discrete event-based simulation [2], layered queuing networks [13], queuing Petri nets [15]). The transformation bridges semantic gaps between the requirements-oriented UCM notation and the component-oriented PCM notation. Users can create UCMs using existing graphical editors and transparently run performance solvers using the PCM tool chain. Our approach supports a two-staged software performance engineering process, where first domain stakeholders can perform initial modeling based on UCMs and second performance experts and architects can refine the resulting PCM models.

In this paper, we evaluate UCM2PCM by transforming UCMs from three existing software systems into respective PCM models. We validate the correct bridging of semantical gaps between the notations by comparing UCM-based simulation results with former purely PCM-based simulation results. In addition, we validate the usability of the models and tooling with an empirical user survey, demonstrating the benefits of the new modeling approach. UCM2PCM [27] is available for free, allows faster modeling, and requires a lower effort for model modifications, thus increasing the motivation to try different design alternatives. UCMs provide a better overview of systems with complex control flows and can be easier discussed with domain stakeholders.

The remainder of this paper is organized as follows: Section 2 briefly introduces UCM and PCM. The mapping from UCMs to PCM is described in Section 3. Section 4 presents the evaluation of the approach, including the results of the survey. Section 5 differentiates our approach from related work. The last Section 6 provides conclusions and future work.

2. FOUNDATIONS

This section provides the backgrounds of this work and introduces the Use Case Maps language and the Palladio Component Model.

2.1 Use Case Maps (UCM)

As part of the User Requirements Notation (URN) specification [11], UCMs are used to visualize how a system works and what the requirements and causal responsibilities are. UCMs are behavioral diagrams and use scenarios. This is similar to UML sequence diagrams, but UCMs remain on a higher abstraction level. In difference to sequence diagrams, UCMs don't show all messages or signals that are exchanged between components or actors, but only control flows with importance for the behavior of a system. Skipping the details enhances the overview and allows the usage of UCM early in the design process, where not much detail is specified yet [5].

The design of UCMs is clear and simple and the notation is intuitive because it is based on sketching a model by hand on a piece of paper. This shall make UCM easy to learn and understandable, also for non-experts.

A UCM diagram consists of paths, showing a possible control flow through a system. In Fig. 1 an example of a UCM diagram for a software implementing a media store can be found. Every path has at least one `Start-` and one `EndPoint`. Along each path so called `Responsibilities` describe the actions that take place on a high abstraction level. `Responsibilities` are symbolized with a cross. Alternatively, `Stubs` can be used. A `Stub` is displayed as a rhombus and represents an own UCM diagram that describes in more details what happens inside.

To model the control flow, forks can be added to a path. An `AndFork` splits a path into two branches, which are executed in parallel. An `OrFork` offers alternative branches. To decide, which branch is taken, conditions or probabilities can be specified. The `OrJoin` joins branches. Additionally, an `AndJoin` contains a barrier, which continues the control flow only when all incoming branches finished executing.

UCMs also support modeling of components. With components, the entities involved in a scenario can be specified and an architectural structure of a system can be defined. Components in UCMs can also be nested. In a diagram, components are represented by boxes. All elements inside the box are bound to that component. Unlike in the PCM, there are no interface definitions for components. Their external behavior can only be derived by the paths entering or exiting the component. To model different design alternatives with UCM, it is often sufficient to keep the paths and only replace or rearrange the components of a model.

2.2 Palladio Component Model (PCM)

The PCM [2] is a domain-specific tool for component based software engineering (CBSE) that is based on the Eclipse Modeling Framework (EMF). Besides other analyses it supports performance prediction for software models [24]. It offers the comparison and selection of several design alternatives of a software system already in the modeling state. The PCM has recently been applied in a number of industrial case studies [14, 8, 23].

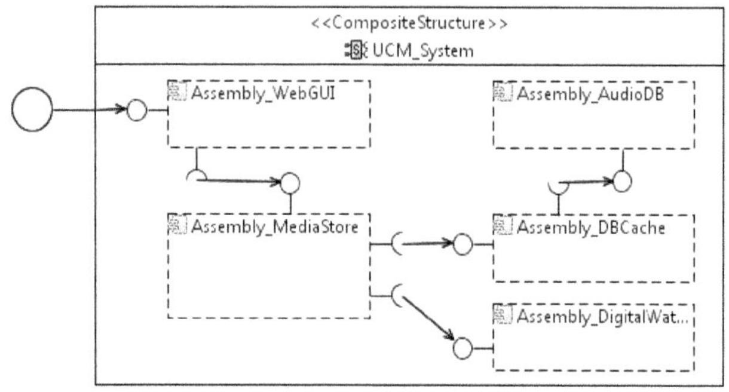

(a) PCM System of the Media Store

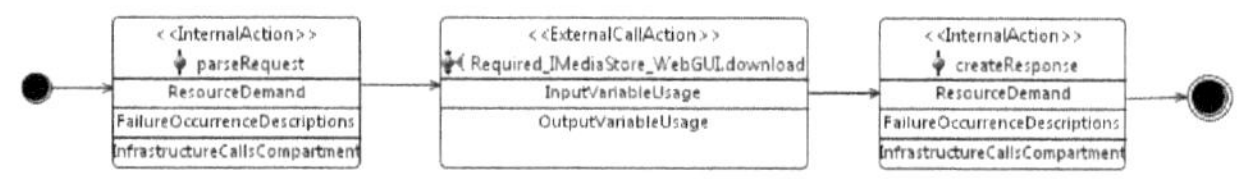

(b) Example PCM SEFF of the Media Store's WebGUI Download Service

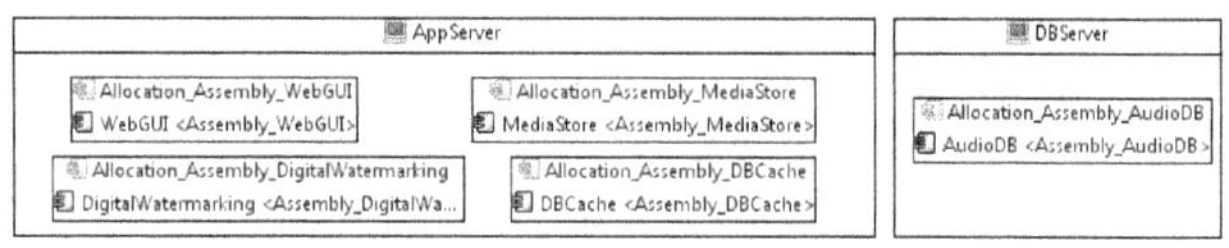

(c) PCM Allocation of the Media Store

Figure 2: Media Store PCM Example

A PCM model consists of four parts that cover different aspects of the performance model:

- **Repository:** The repository contains the component types & interfaces of the system. There are `Basic-` and `CompositeComponents`. `BasicComponents` contain so called `ServiceEffectSpecification (SEFF)` that abstractly describe performance-relevant behavior of a single component service. `CompositeComponents` are composed from inner components. A `SEFF` contains actions that determine the control flow inside the component. For example, Fig. 2b depicts a `SEFF` for the MediaStore's WebGUI service 'download'. Every `SEFF` starts with a `StartAction` and ends with an `EndAction`. `InternalActions` represent computations or other resource consumption inside a component and thus contain execution time annotations that are later transformed into service demands and used

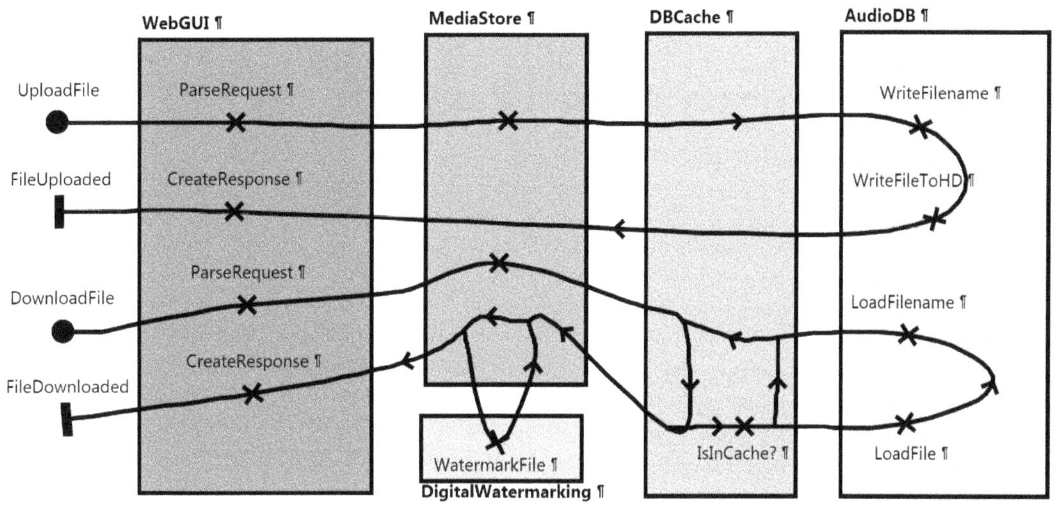

Figure 1: UCM of the Media Store components and inner control flow. The usage model is another UCM belonging to the model, which is omitted here for space reasons.

by the performance simulator. `ExternalCallActions` can be used to call external services that are required. `Fork-`, `Branch-` and `LoopActions` are used to manage the control flow. Components can provide or require interfaces and are bound to them via `Provided-` or `RequiredRoles`.

- **System:** A system is a composite structure and shows how the components in a software system are connected (cf. example in Fig. 2a). It consists of `AssemblyContexts` that represent instances of the components in the repository and thus having the same `Provided-` or `RequiredRoles`. If the `Provided-` and `RequiredRole` of two `AssemblyContexts` match, they can be connected by an `AssemblyConnector`. Like every composite structure, a system itself can also have `Provided-` or `RequiredRoles`. A `SystemProvidedRole` represents services that the software system provides to its environment. With a `ProvidedDelegationConnector` the `SystemProvidedRole` is connected to the `ProvidedRole` of the inner `AssemblyContext` that offers the provided service. If an `AssemblyContext` needs to call a service outside the system, its `RequiredRole` is connected to the `SystemRequiredRole`, using a `RequiredDelegationConnector`. By that, the request for a service is forwarded to the system boundaries.

- **Allocation:** The allocation defines, how component instances (i.e., `AssemblyContexts`) are deployed on hardware resources (cf. example in Fig. 2c). The mapping relation for a component instance is called `AllocationContext`. The hardware resources are stored in a `ResourceEnvironment`. Each hardware resource can have different processing entities, such as CPU or HDD.

- **Usage Model:** In a usage model, several `UsageScenarios` can be created, each describing an expected usage of the system, by a user. For each `UsageScenario` a workload (i.e., open or closed) and a behavior can be defined. This behavior consists of a set of `EntryLevelSystemCalls` that refer to services in `SystemProvidedRoles`. In this set, also sequences, branches and loops can be used.

The PCM Workbench provides multiple tools to conduct the performance predictions. The tool PCM Solver can be used to calculate the execution times and the related probabilities of a workload. The second tool, SimuBench allows simulating different number of runs for a workload. For every call in the model, the execution times can be simulated and visualized in several diagrams. Additionally, diagrams that show the utilization of hardware resources, such as CPUs or hard disks, can be created.

3. UCM TO PCM MAPPING

This section describes how UCM elements or constructs are translated into corresponding PCM elements or constructs, which performance relevant annotations to UCM elements are specified and what limitations exist for the transformation. The main idea of the transformation is to map UCM components to PCM components and then to segment each UCM path per component to create PCM SEFFs.

3.1 Paths

Every path in a UCM corresponds with a `SystemProvidedRole`, as both represent a function that the software system provides to its environment (Fig. 3). In contrast to standard UCMs, UCM paths to be transformed into PCM need exactly one `StartPoint` and one `EndPoint`. Different paths may not join. Otherwise, the mapping could be ambiguous in the UCM2PCM tool. To translate a path into PCM, the UCM2PCM transformation follows a path node by node. In the following, the mapping for the most important UCM constructs is presented.

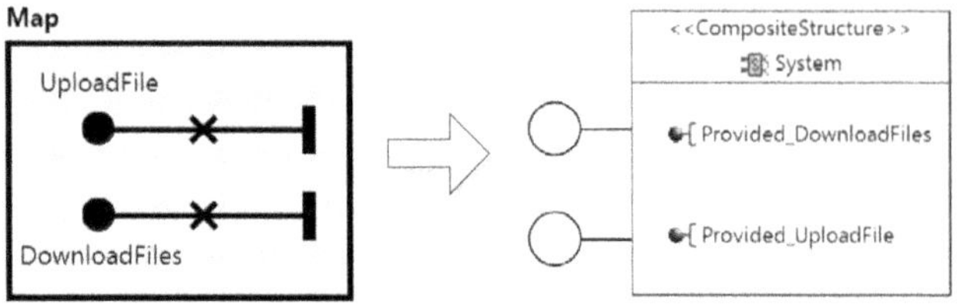

Figure 3: UCM2PCM Mappings - UCM Map

3.2 Hardware resources

To be able to predict the performance of a component, it has to be mapped to an active resource. This is done in PCM with an `AllocationContext`. In UCMs, there is no corresponding construct. Therefore, a performance annotation in the meta data of the UCM component is needed. Here, the allocation to a hardware resource can be specified as a key-value pair as shown in Fig. 4. If the allocation meta data is missing, then the component is allocated to a default hardware resource.

If a PCM hardware resource is not yet present from former transformation steps, the transformation automatically creates it first. All hardware resources are created with a default CPU and a default HDD as processing entities, no parameters can be specified here. This is a deliberate limitation to keep modeling with the UCM2PCM tool simple and allow for rapid performance modeling. However, if servers with custom CPU or HDD are needed, then this can be defined later in PCM after the transformation as manual refinement step.

3.3 Components

In a UCM, components can be reused in different diagrams. That is why the component *definitions* do not appear in a UCM diagram. The boxes in a diagram only are references to UCM components defined elsewhere. This is the same concept as in PCM, where a component of the PCM repository can be reused in several `AssemblyContexts` in a software system. In the transformation therefore, for every reference an own `AssemblyContext` will be created. If the corresponding component already exists in the PCM repository, it will be used; otherwise a new component for that `AssemblyContext` will be created.

In UCM, all component references that are based on the same component share the same name and meta data. These attributes are stored in the component and not in its references. This means that the corresponding `AssemblyContexts`, created by the transformation, also share the same name and are allocated on the same hardware resource. If two `AssemblyContexts`, based on the same component, should be named differently or they should be allocated on different hardware resources, this has be to done manually in the PCM model resulting from the UCM2PCM transformation. We decided that this limitation is acceptable, because to solve this problem the UCM meta model would have to be changed.

Fig. 4 shows, how a normal UCM component reference is translated into a `BasicComponent`. Additionally, a new `SEFF` with a `StartAction` will be created for the component, to follow the path inside the component. If the UCM component reference contains a `Stub`, then a `CompositeComponent`

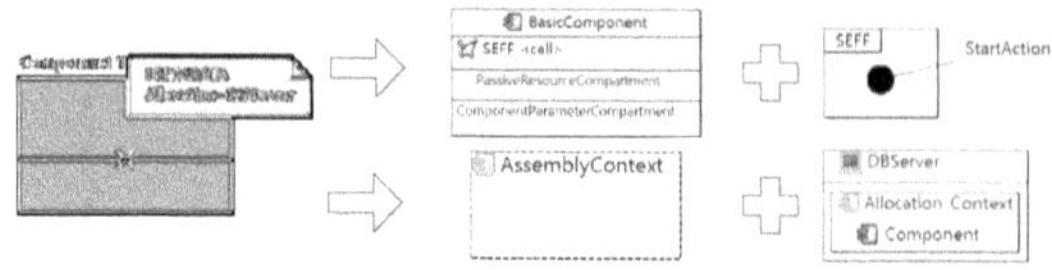

Figure 4: UCM2PCM Mappings - Standard Component

will be created instead in the PCM repository. This can be seen in Fig. 5.

3.4 Composite Components

`Stubs` can be used to create layered UCMs. They contain a complete UCM map. Every path leading inside or outside a `Stub` can be connected to a `Start-` or `EndPoint` inside the Stub. `CompositeComponents` have a similar function in PCM. They are composed from inner components and can be seen as a kind of subsystem. Therefore, a direct mapping between a UCM component containing a `Stub` and a `CompositeComponent` can be defined. A `CompositeComponent` itself may not contain any actions. This is only possible for the inner components. Therefore, in a UCM component containing a `Stub`, also no additional nodes are allowed; otherwise, no mapping is possible and the component can't be translated into PCM.

For all components inside a `Stub`, the allocation meta data will be ignored, because all components inside a `CompositeComponent` in PCM are defined to be allocated on the same hardware resource as the `CompositeComponent`.

In PCM, actions can only occur inside components, while in UCMs there are no dependencies between nodes and components. To be able to map a UCM path into PCM, the UCM2PCM transformation ignores all nodes or constructs, which are located outside a component, except for `Responsibilities` that are mapped as `SystemRequiredCalls`.

By moving a node inside or outside a component, it can quickly be included or excluded from the transformation. This can even accelerate the process of testing and converting different design alternatives into PCM.

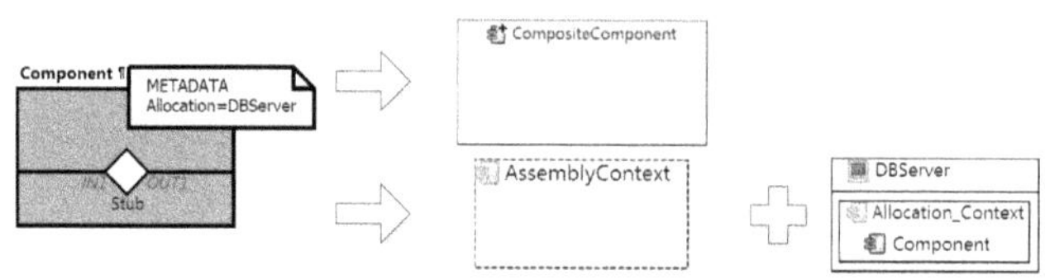

Figure 5: UCM2PCM Mappings - Stub Component

3.5 Responsibilities and Resource Demands

For UCM `Responsibilities`, there is a 1:1 mapping with `InternalActions`, where resources are consumed. Therefore, resource demands performance annotations have to be specified for `Responsibilities`. The UCM2PCM tool supports the CPU & HDD resources of PCM. They can be entered as key-value pairs in the meta data of the `Responsibilities` as shown in Fig. 6. If no resource demand is specified, a default CPU demand will be used.

In PCM models, it can be necessary to model a control flow that enters a component without any `InternalAction` executed. For example if a component forwards a call to another component, instead of processing it internally. In UCMs, this behavior is not possible. If a new component is

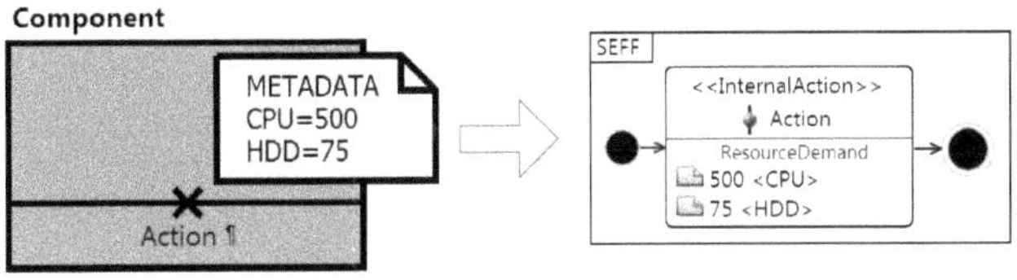

Figure 6: UCM2PCM Mappings - Responsibilities

entered, there always has to be a node inside the component, otherwise it could not be determined if the component has been entered. As a workaround, a **Responsibility** with an explicit CPU resource demand of zero can be used for this purpose. The transformation skips all **Responsibilities** with CPU resource demand of zero.

3.6 Calls

Component calls to other components have no dedicated representation in a UCM but they implicitly occur when the path enters or leaves a component. In UCMs, a call can be determined, by checking if the surrounding component of a node has changed compared to the predecessor node.

PCM implements the concept of interfaces as a contract between different components. Therefore, an interface has to be created for every call in a UCM, by the transformation. In general, a **RequiredRole** targeting this interface has to be created for the caller, and a **ProvidedRole** has to be created for the callee. An example of this is shown in Fig. 7.

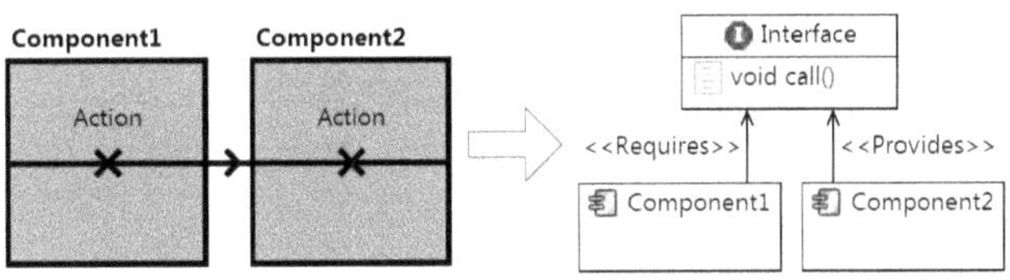

Figure 7: UCM2PCM Mappings - Call Interfaces

The transformation can distinguish different types of calls, depending on whether a component has been entered, been changed, been left or if a Stub has been entered or left. The following enumeration describes the different calls in more detail:

1. **SystemProvidedCall**: Here, the previous node is a **StartPoint** and the path has entered a new UCM component as depicted in Fig. 8. In this case, for the new component & **AssemblyContext** in PCM, a **ProvidedRole** corresponding to the call is created. In addition the **ProvidedRole** will be cloned and added as a **SystemProvidedRole** to the **AssemblyContexts** parent container. This container is the PCM system, or a **CompositeComponent**, if the path resides inside a **Stub**. Then the two **ProvidedRoles** will be connected by a **ProvidedDelegationConnector**.

2. **ComponentCall**: A **ComponentCall** occurs when a UCM path crosses the borders between two UCM components. In this case the component of the current node is different than the component of its predecessor node, as shown in Fig. 9. For the component, which is left a **RequiredRole**, for the component, which is entered a **ProvidedRole**, corresponding to the

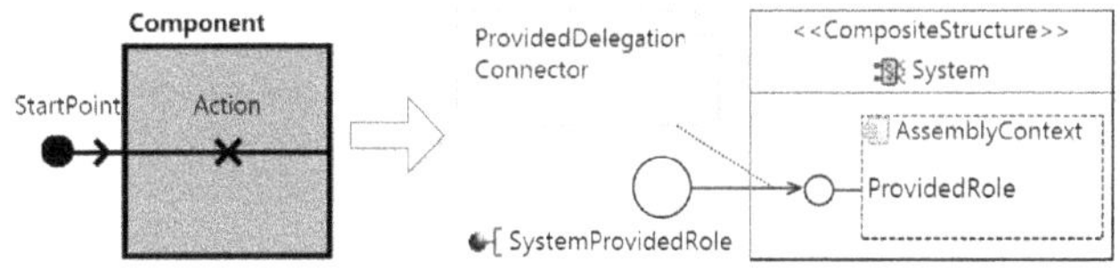

Figure 8: UCM2PCM Mappings - SystemProvidedCall

call, are created. Additionally, the transformation creates an **AssemblyConnector** that connects the created **Provided-** and **RequiredRole**. Finally, in the **SEFF** of the calling component, an **ExternalCallAction** targeting the new **RequiredRole**, is added.

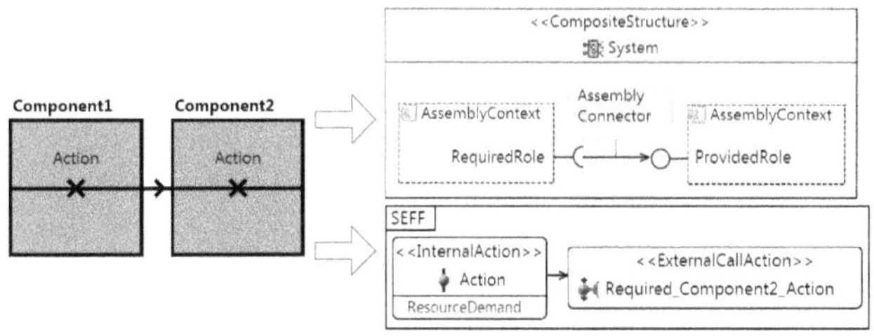

Figure 9: UCM2PCM Mappings - ComponentCall

3. **SystemRequiredCall**: A **SystemRequiredCall** is used, if a component needs to be called, which is outside the modeled software system. This is depicted in Fig. 10. In a UCM, such a call can be modeled by a path leaving a component, followed by a **Responsibility** outside of a component. In the transformation, a new **RequiredRole** for the calling PCM component is created. In addition, the **RequiredRole** is cloned and added to the PCM System as **SystemRequiredRole**. Finally, both **RequiredRoles** will be connected, using a **RequiredDelegationConnector**. In the **SEFF** of the calling component, an **ExternalCallAction** targeting the new **RequiredRole** is added. Inside a **Stub**, a **SystemRequiredCall** is not possible. Here, a **ParentCall** has to be used instead. A **SystemRequiredCall** is the only case, where a node outside a component is used to create PCM elements. In all other cases, nodes outside a component will be ignored by the UCM2PCM transformation.

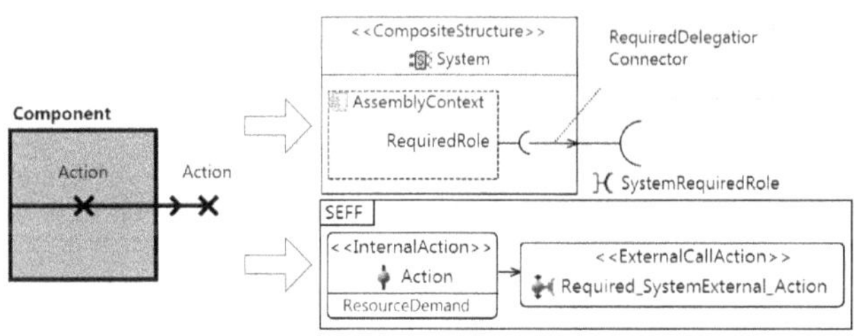

Figure 10: UCM2PCM Mappings - SystemRequiredCall

4. **ParentCall**: A **ParentCall** is used, inside a **Stub**, to call a target, which is outside the **Stub**. This is equivalent to a **SystemRequiredCall**, as in both cases the target of the call is outside the parent container. The syntax of a **ParentCall** is shown in Fig. 11. Inside

a Stub, there has to be an EndPoint inside the calling component. In the same component there also needs to be a StartPoint, where the control flow returns back into the Stub. Both the End- and the StartPoint need to be connected by a path leaving and entering the Stub. In the path outside the Stub, any types of calls can be made. The ParentCall will be translated similar as the SystemRequiredCall. A RequiredRole and a SystemRequiredRole are created and connected via a RequiredDelegationConnector.

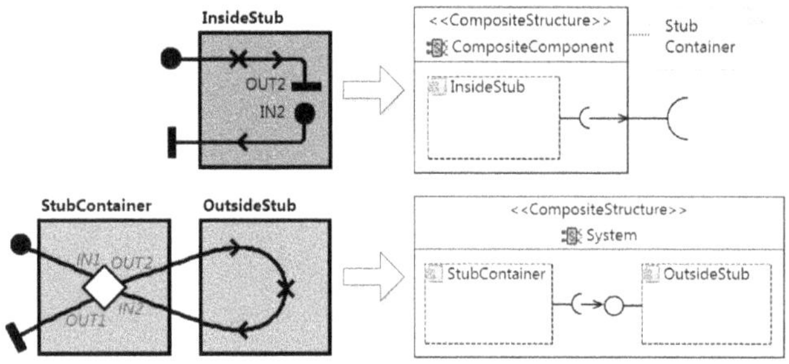

Figure 11: UCM2PCM Mappings - ParentCall

In a PCM model, ExternalCallActions model synchronous calls by default. Therefore, all calls in UCMs are also assumed to be synchronous. If there is an open call, meaning that the control flow has not yet returned to the caller, and the calling component is reentered, then the transformation interprets this as the return of the open call, rather than a new call. Before the call returns, all calls opened in the meantime will automatically return and get closed in the opposite order than they were opened. When a path reaches an EndPoint and is finished, all calls, which are still open, are automatically returned and closed.

However, if it is necessary to model components in UCMs that are calling each other with circular dependency (e.g. in the callback software pattern), then this could be realized using a workaround. To avoid a call being interpreted as return call, an OrFork can be entered before the call. The call needs to be moved to one branch of the OrFork and its probability has to be set to '1'. The probability of the other branch will be set to '0'. The left diagram in Fig. 12 shows the default behavior, while the workaround is shown on the right side.

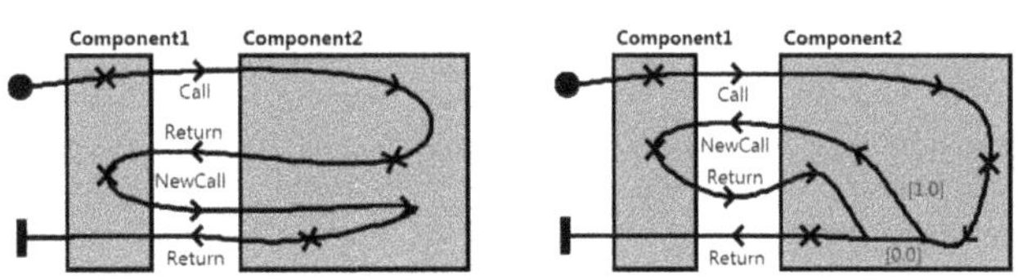

Figure 12: UCM2PCM Mappings - Workaround for Components Calling Each Other

3.7 Branches

Beside the sequential control flow, UCM2PCM also supports modeling all branching constructs of the PCM. In PCM there are Fork-, Branch- and LoopActions. In UCMs, these actions can be modeled using AndForks, AndJoins, OrForks and OrJoins.

1. A ForkAction in PCM consists of several ForkedBehaviors that contain subpaths, which are executed in parallel. In UCM, an AndFork with corresponding AndJoin has the same semantic meaning, therefore there is a 1:1 mapping of both constructs, depicted in Fig. 13.

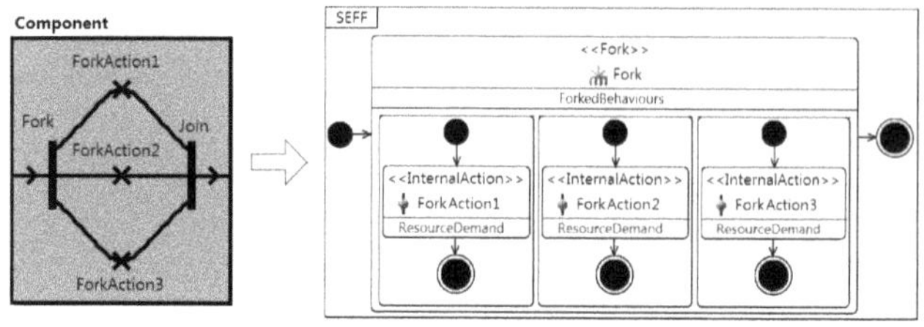

Figure 13: UCM2PCM Mappings - Forks

2. A BranchAction in PCM also splits the control flow, but, in contrast to a ForkAction, only one of the possible paths will be followed. In PCM, there are guarded and probabilistic BranchActions. In guarded BranchActions, the choice of which branch is taken is depends on guard conditions. They are not supported by the UCM2PCM transformation. For probabilistic BranchActions, probabilities have to be specified for each branch. In UCMs, they can be modeled by an OrFork with corresponding OrJoin, as seen in Fig. 14. The probabilities of each branch in UCM have to be specified in the attribute "probability" of the connection, which connects the OrFork with the first node in the subpath in the UCM diagram.

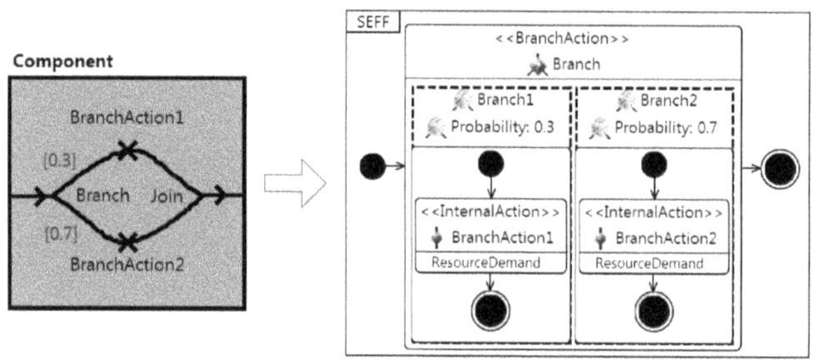

Figure 14: UCM2PCM Mappings - Branches

3. LoopActions can only be emulated in UCMs. There is no special loop construct to model them 1:1. For the UCM2PCM transformation, a loop is defined as an OrJoin, directly followed by an OrFork with two branches. One branch, containing the loop body, leads back to the OrJoin, the other branch continues the path after the loop. From the performance point of view the number of iterations for the loop is relevant. This information can be specified as performance annotation, in the meta data of the OrFork, as shown in Fig. 15.

In the context of the UCM2PCM transformation, the usage of join nodes is affiliated with two problems. Firstly, the fork and the corresponding join node could be located in different components, which is not supported by PCM. To avoid this problem, we decided that the fork node in UCM defines, in which component the resulting action in PCM

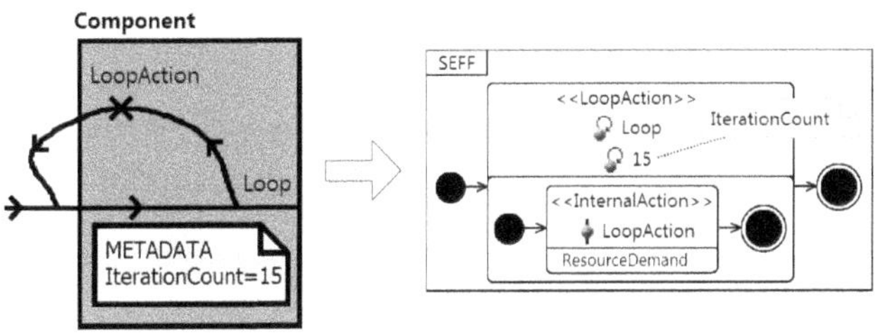

Figure 15: UCM2PCM Mappings - Loops

will be located. The join node is automatically assumed to be located in the same component.

Secondly, there is a problem with nesting. Like in PCM, all branching constructs can be nested. The use of joins in UCM can lead to a "dangling join" problem. Therefore, the transformation is restricted to only hierarchical nesting. This implies that each join, which is not part of a loop, belongs to the latest opened fitting fork. The UCM2PCM transformation contains a special algorithm [27] to check, if the branching in a UCM is valid.

3.8 UsageScenarios

The built-in rudimentary scenario support in UCMs does not match the capabilities of PCM. Therefore, we decided that `UsageScenarios` have to be modeled explicitly with UCMs. This is possible with hardly any additional effort and allows us to also model sequential `EntryLevelSystemCalls` with branches and loops.

A UCM component of type "Actor" represents a `UsageScenario`, as depicted in Fig. 16. In the meta data, the workload can be specified with the same properties as in PCM. If the workload is open, the interarrival time of new users has to be specified; if it is closed, then the population and the think time are necessary parameters.

The whole path of the scenario including the `StartPoint` has to be located inside the "Actor" component. This allows the transformation to distinguish between a normal path through the software system and a `UsageScenario`. Branches and Loops can be built like in a normal UCM path. The `Responsibilities` inside a scenario are translated into `EntryLevelSystemCalls`, instead of `InternalActions`. Therefore their name has to be the same as the name of the `StartPoint` representing the `SystemProvidedRole` that should be called by the `EntryLevelSystemCall`.

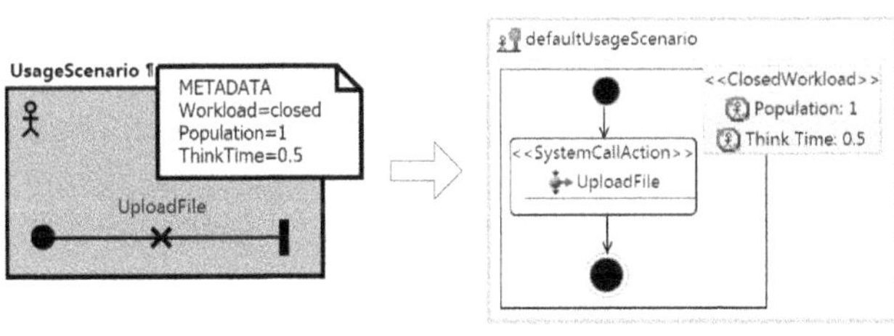

Figure 16: UCM2PCM Mappings - UsageScenarios

3.9 Assumptions and Limitations

There are a few limitations for the use of the UCM2PCM tool that are described in the following:

- **Limited PCM feature support:** Our approach is not intended to fully support all PCM features; instead, we propose a trade-off between ease-of use and complexity of a tool. Thus, PCM variables and guarded branches are not supported. Right now, only fixed values or the PCM "Stochastic Expressions" [24] can be used for performance annotations in UCMs. It is also not possible to create a custom `ResourceEnvironment`. Allocating two `AssemblyContexts`, implementing the same component, onto different hardware resources is infeasible, due to UCM meta model limitations.
- **Limited input assistance:** The jUCMNav editor for creating UCMs [22] has limited input assistance. Performance annotations need to be added in the meta data of each UCM node, as key-value pairs. This is not convenient, as the user needs to know the names of all performance annotations. At design time, in the UCM editor, there is no possibility to validate a UCM model and to indicate errors to a user. This is only possible after modeling, by running the UCM2PCM transformation.
- **No reverse transformation:** After transforming a UCM into a PCM model, this model can be further refined in PCM. Thus, there is no reverse transformation yet, to transfer the changes back into the UCM. However, it would be desirable to regain an easy presentable UCM representation of the refined performance model.

3.10 Implementation

Fig. 17 shows a high-level sketch of the UCM2PCM tool chain. The UCM2PCM prototype is implemented as an Eclipse plug-in. The transformation uses the Eclipse Model to Model (M2M) transformation framework and is implemented in QVT Operational [17] (approx. 3000 lines of code). UCMs can be created with the jUCMNav editor [22] based on the Graphical Modeling Framework (GMF). UCM2PCM can then be invoked from inside Eclipse accepting a UCM model as input. The resulting PCM model can be further refined by editing it with the PCM Workbench [6], which is also based on GMF.

The transformations from the resulting PCM model into the different performance models are out of the scope of this paper. Becker et al. [2] describe the transformation of PCM instances into extended queuing networks, which are then solved using the discrete-event simulator SimuCom. Koziolek et al. [13] map PCM instances to layered queuing networks, which are then solved using the fast, heuristic layered queueing network solver. Meier et al. [15] explain how to map PCM instances to queueing Petri nets, which are solved by approximate analytical solvers or simulation.

The whole tool chain is open source and available free of charge.

4. EVALUATION

To validate UCM2PCM, we need to show that its mapping of model elements is valid and the tool is useful. Thus,

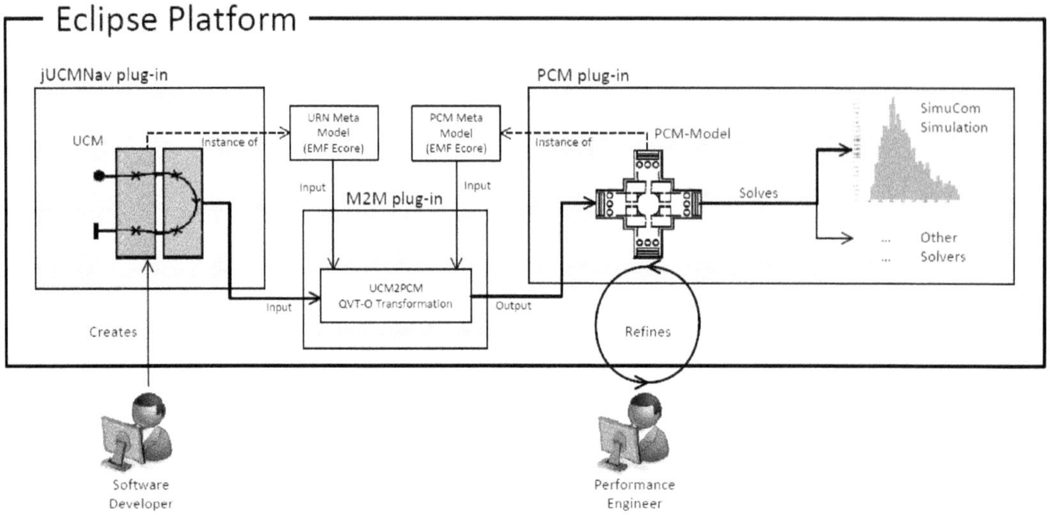

Figure 17: UCM2PCM Toolchain

this section evaluates (a) the accuracy of the transformation (i.e., the correct mapping) and (b) the usability of the UCM model. To address (a), we first ran a number of internal tests with a number of artificial models which are not detailed here. Afterwards, we transformed performance-annotated UCMs of three different case study systems into the respective PCM models (Section 4.1). We then ran a series of experiments (Section 4.2) and compared the simulation results of these models with simulation results from former PCM models of these systems (Section 4.3). A low difference between both simulation results demonstrates that UCM2PCM successfully bridges semantic gaps between UCM and PCM. To address (b), we let six users model a UCM and apply the UCM2PCM transformation (Section 4.4). We then asked them for feedback in a user survey. While the low number of users is not statistically relevant to draw general conclusion, it provides a first hint on the usability of the approach.

4.1 Systems under Study

We applied UCM2PCM on three heterogeneous, mid-sized systems, to increase the external validity of our approach. The models also demonstrate that UCM2PCM can be used to model and process complex control flows and is therefore well-suited for industrial control systems. The analyzed models are:

- **Media Store:** This system (Fig. 1) is a plain Java web application for storing and retrieving audio or video files using a MySQL database. The model reflects a use case where a digital watermark is added to downloaded files for copy protection. The model contains a hard disk resource, which is accessed when retrieving files. Resource demands for the Media Store have been measured using manual instrumentation of the Java implementation [2]. The model comprises 19 resource demands and a closed workload.

- **SPECjAppServer:** SPECjAppServer is an industry-standard benchmark, designed to measure the performance of application servers conforming to the Java EE 5.0 or later specifications. It is modeled after an automobile manufacturer, where dealers place customer orders or interact with suppliers. The system is implemented as a Java Enterprise application, deployed on two servers using an Oracle database. The 15 resource demands in the model have been determined using an estimation technique based on measured response times and resource utilization [4].

- **Business Reporting System:** The BRS (Fig. 18) is loosely modeled after a management information system, formerly analyzed at Carlton University [30]. Users can retrieve live business data from the system and run statistical analyses. The analyzed configuration comprises two open workload usage scenarios, nine components, and four servers. The 37 resource demands of the BRS are based on estimations. Fig. 18 depicts a high-level overview of five of the most important UCM paths in the model. The model includes UCM `Stubs` with nested UCMs that are mapped to PCM `CompositeComponents`. Fig. 19 shows the PCM repository resulting from the UCM2PCM transformation.

4.2 Experiments

Based on the available PCM instances of the systems, we manually created corresponding UCMs, using jUCMNav 4.3.0. We had to resolve parameter dependencies from the models (e.g., converting guards on branches into probabilities or calculating resource demands by substituting the required parameters), since they are not supported by UCM2PCM. The resulting UCMs contained branch probabilities, loop counts, and resource demands.

Because of the missing parameterization concepts in UCM, we used the mean resource demands instead of parameterized resource demands in the model. As discussed in Section 3.9, a mapping of all PCM concepts is neither possible nor intended. Thus, we expect a certain deviation of prediction results.

For the experiments, we transformed the UCMs into PCM models and then ran the discrete-event simulator SimuCom [2]. We also ran SimuCom on the former PCM models and then compared both results. We used Eclipse 3.7, PCM

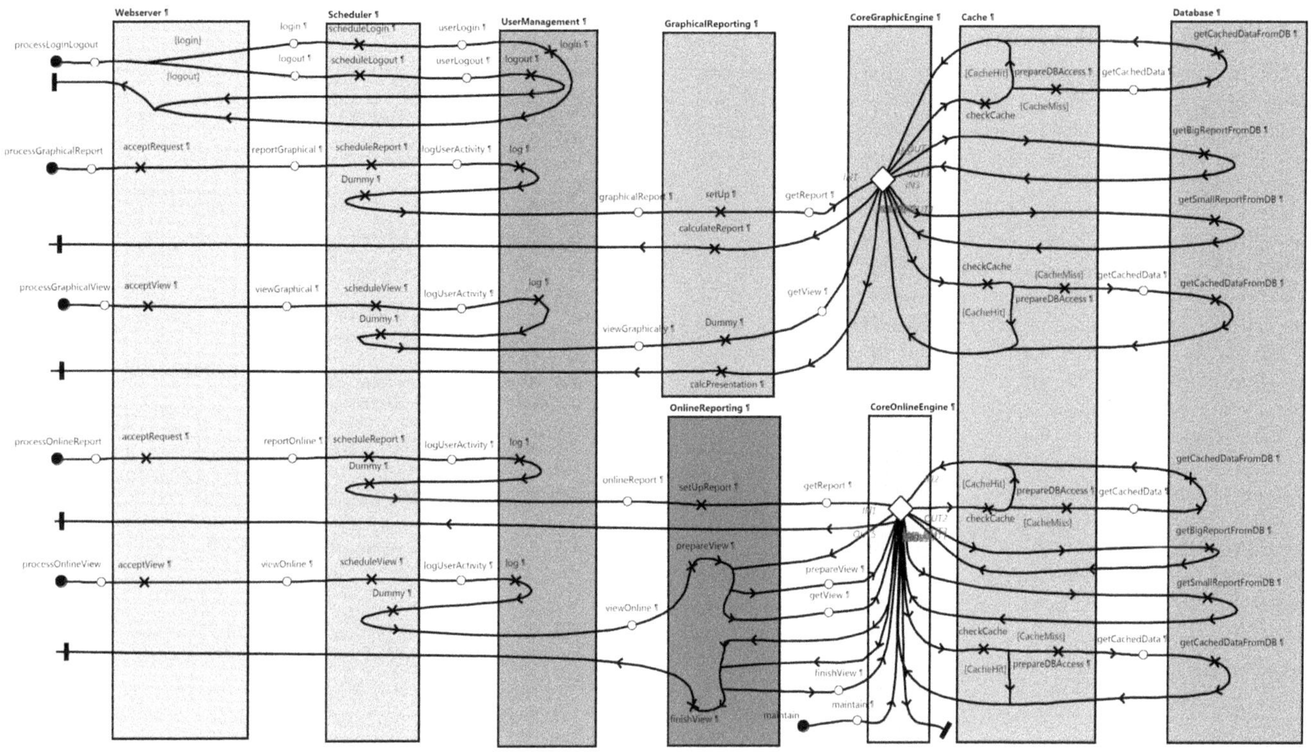

Figure 18: UCM of the high-level Business Reporting System. Nested UCM Stubs and the corresponding UCM usage model are not shown here.

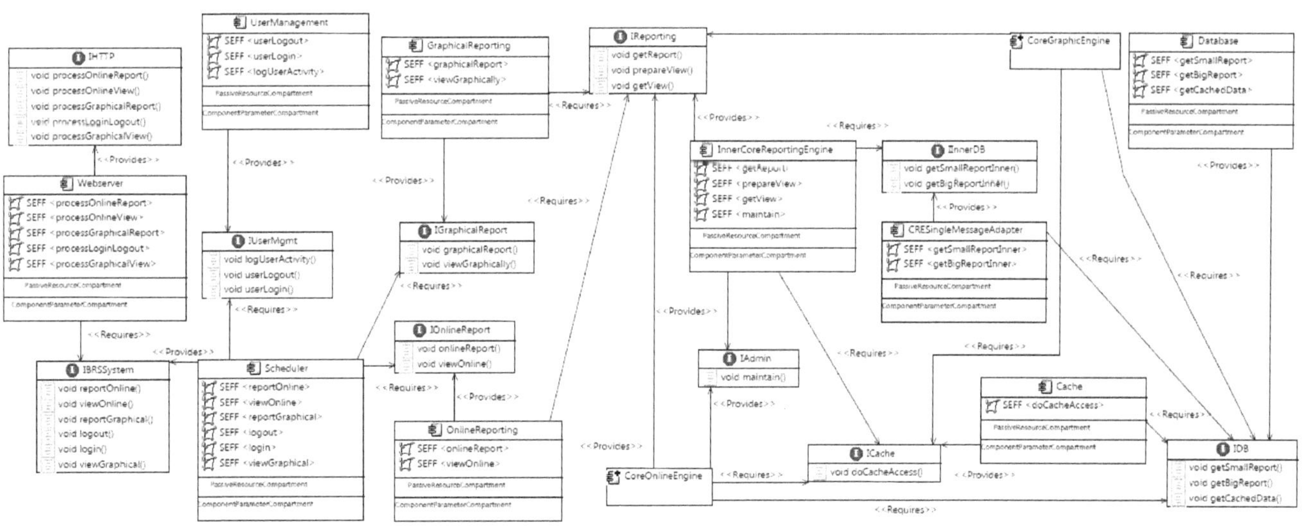

Figure 19: PCM repository resulting from the transformation of the BRS system with UCM2PCM

3.3 running on a Windows 7 PC with 16 GB RAM and an Intel Core i7-2720QM CPU at 2.20 GHz. Each simulation run took less than five minutes to execute.

4.3 Validation of Accuracy

Tab. 1 shows different performance metrics for the systems, both from simulating the original model and from simulating the UCM-based models. The table reports mean and standard deviations of response times and throughputs in seconds, the mean resource utilizations as percentages.

The table shows that the difference between the original model and the UCM-based models is below 10 percent in most cases. The standard deviation of the MediaStore response time shows a difference of 14 percent compared to the original model, which probably could have prevented by using the distribution functions instead of mean values as resource demands. The mean response time for the BRS show an 11 percent higher difference. In this case, the difference resulted from several simplications that had to be introduced into the BRS UCM.

Despite some differences in the simulation results, we deem the accuracy of UCM2PCM sufficient to support early design-time performance decisions. The low differences for different kinds of models demonstrate that UCM2PCM was successful in bridging most semantic differences in these models.

	Original model	UCM model	% Difference
MediaStore			
Mean ResponseTime in sec (Scenario 1)	2,368	2,366	-0,1%
StdDev Response Time in sec (Scenario 1)	1,705	1,496	-14,0%
Throughput per sec (Scenario 1)	1,267	1,268	0,1%
Utilization (App Server CPU)	58,7%	58,8%	0,2%
Utilization (DB Server CPU)	2,0%	2,0%	0,0%
Utilization (DB Server HD)	87,6%	87,6%	0,0%
SPECjAppServer			
Mean ResponseTime in sec (Scenario 1)	0,026	0,026	-1,0%
StdDev Response Time in sec (Scenario 1)	0,012	0,012	2,2%
Throughput per sec (Scenario 1)	10,407	10,092	-3,1%
Mean Response Time in sec (Scenario 2)	1,044	1,050	0,5%
StdDev Response Time in sec (Scenario 2)	0,014	0,013	-1,9%
Throughput per sec (Scenario 2)	9,626	9,817	1,9%
Utilization (WL Server)	34,7%	34,8%	0,3%
Utilization (Oracle Server)	17,5%	17,6%	0,6%
Business Reporting System			
Mean ResponseTime in sec (Scenario 1)	8,853	7,974	-11,0%
StdDev Response Time in sec (Scenario 1)	3,295	3,163	-4,2%
Throughput per sec (Scenario 1)	0,531	0,540	1,7%
Utilization (Server1)	59,4%	60,2%	1,3%
Utilization (Server2)	65,3%	59,5%	-9,7%
Utilization (Server3)	59,4%	60,4%	1,7%

Table 1: Evaluation results showing simulation results from the original PCM models, and the UCM-based PCM models created by UCM2PCM.

4.4 Validation of Usability

Tab. 2 shows the results from a user survey to evaluate the usability of UCM2PCM. We let each participant (i.e., computer science graduate students and PhD students) work through a tutorial and then model the Media Store system as an exercise. Afterwards, the participants used a web tool to anonymously submit their ratings for the approach and the tool on the following scale: (--, -, -, +, ++, +++).

Questions	Participant 1	Participant 2	Participant 3	Participant 4	Participant 5	Participant 6	Participant 7
How long (in minutes) did you evaluate UCM2PCM ?	30	45	40	40	150	30	45
How long (in minutes) did it take you to model the "MediaStore with Cache" as UCM?	45	60	15	15	40	45	120
Speed: How fast basic performance models can be created with the UCM2PCM tool?	++	++	++	++	+	+++	-
Complexity: Is the UCM2PCM tool also suited for creating more complex models?	--	--	++	+	--	-	---
Comprehensibility:Is a UCM model, created with UCM2PCM clear and comprehensible?	++	+	+++	++	+	++	++
Simplification: Is it easier to create performance models using the UCM2PCM tool, compared to modeling with PCM?	-	-	++	++	+	++	+
Speed up: Does the usage of the UCM2PCM tool speeds up performance modeling, compared to modeling with PCM?	+	-	++	++	+	+++	--
Comprehensibility for experts: Are UCMs created with the UCM2PCM tool more comprehensive than PCM models?	++	--	++	-	+	+	+
Comprehensibility for non-experts: Are UCMs created with the UCM2PCM tool more comprehensive than PCM models?	+++	++	+++	++	++	+++	++

Table 2: User Ratings for UCM2PCM

While all of the participants were able to quickly produce a UCM, a majority was afraid that UCMs would not be suitable for complex models. This might be a pointer for future work, to further simplify and support the creation of large models with the tool. The comprehensibility of the models was deemed good or very good by all participants, especially for non-experts. Thus is a first hint for the good applicability of the tool for discussing designs with stakeholders from the industrial automation domain.

The results of this user survey are not statistically significant due to the small sample size. Furthermore, the participants might have been biased towards the novel approach. A future empirical study should investigate a larger sample size, analyze different design alternatives, and compare UCM modeling with other methods.

5. RELATED WORK

Surveys on model-based performance prediction [1], performance evaluation of component-based systems [12] and the future of software performance engineering [29] provide a broad overview of recent approaches for performance modeling. In addition, Smith and Williams [25] described fundamentals of software performance engineering and Menasce et al. [16] detailed on capacity planning with queuing models.

Several other approaches focused on model transformations from performance-annotated UML models to various performance models, such as queuing networks [9], stochastic Petri nets [3, 10], and stochastic process algebra [26]. Petriu and Woodside use the UML Schedulability, Performance, and Time (SPT) profile to enrich UML diagrams with performance annotations [19]. In [28] they show the usage of the UML MARTE (Modeling and Analysis of Real-Time and Embedded Systems) profile, which is the successor of the SPT profile.

As UML is the de-facto standard modeling language in the software industry, these approaches have a higher probability of being broadly adopted in practice. However, the UML specification and the UML MARTE specification for performance annotations are complex and therefore difficult to learn for stakeholders from industrial automation. In contrast, UCMs provide a reduced set of model attributes and intuitive visualizations and are thus more applicable in our context. UML mainly focuses on object-oriented software development, while we target component-based development.

KLAPER [7] is a meta model language for performance prediction of component based systems. With the KLAPER Suite, there is a set of tools to actually create performance prediction from the model. This is comparable to the PCM tool chain, but KLAPER is no language to design component-based models. It does not offer a front-end for software design.

UCMs are more abstract than for example UML sequence diagrams. They do not include all message and signals exchanged between components, but can focus on the important, performance-relevant behaviors. Abstracting from unnecessary details is an essential part of modeling and allows to use the notation during early development stages. In contrast to UML activity diagrams, the UCM mapping of responsibilities to software components is arguably more intuitive and visually more appealing than using the UML swimlane notation.

Petriu and Woodside proposed a specific approach for performance modeling with UCMs [21, 20]. They introduced the scenario to performance (S2P) algorithm to transform UCMs into LQNs. S2P allows to transform modeled asynchronous, synchronous, and forwarding requests between software processes to represent common communication patterns in distributed systems. Compared to our approach S2P resulted in a monolithic performance model (LQN), which does not support a component-based development process. It complicates exchanging individual components in the system because of the lacking interface concept. Our target model PCM supports reusable component models as well as a decoupled resource model. The latter allows quickly attaching different resource environment models to the component model so that sizing and capacity planning questions can be answered quickly. Furthermore, tool support for S2P is no longer available.

6. CONCLUSION & FUTURE WORK

Performance engineering is an important part of improving the quality of service of software systems, but usually requires performance specialists to be involved in the development process. The approach presented in this paper lowers the entrance barrier for performance engineering by combining the intuitively understandable modeling language (UCM) with a sophisticated performance engineering approach (PCM). Software engineers can create UCMs and use the defaults in the presented approach to transform them into PCM instances. The simulation and analysis tools of the PCM workbench deliver performance values that can be used for the evaluation of design alternatives.

If more detailed performance properties are required, the generated PCM instances can be further adapted. Thus, the approach enables software engineers to model performance properties on different levels of abstraction. Data-flow oriented systems can be described more easily and clearly than with PCM alone, without losing the benefits of the PCM simulation and analysis techniques.

The integration of UCMs into the PCM development process and modeling workflow has to be investigated further. A reverse transformation that allows developers to (re)transform PCM models into the more comprehensible UCM is desirable. Furthermore, a validation of UCM models is currently lacking at editing time. Once model transformations are fast enough to check the consistency of input models on-the-fly in a similar way to text-based code assistance in state-of-the-art CASE tools, the approach could also be extended to component models other than the PCM, with UCMs serving as an easily understandable modeling language in early phases of the software design process.

7. ACKNOWLEDGMENTS

The authors would like to thank all participants of the user study for their support.

8. REFERENCES

[1] S. Balsamo, A. Di Marco, P. Inverardi, and M. Simeoni. Model-based performance prediction in software development: A survey. *Software Engineering, IEEE Transactions on*, 30(5):295–310, 2004.

[2] S. Becker, H. Koziolek, and R. Reussner. The Palladio component model for model-driven performance prediction. *JSS*, 82:3–22, 2009.

[3] S. Bernardi and J. Merseguer. Performance evaluation of uml design with stochastic well-formed nets. *Journal of Systems and Software*, 80(11):1843–1865, 2007.

[4] F. Brosig, N. Huber, and S. Kounev. Automated Extraction of Architecture-Level Performance Models of Distributed Component-Based Systems. In *26th IEEE/ACM Intl. Conference On Automated Software Engineering (ASE)*, November 2011.

[5] R. Buhr. Use case maps: A new model to bridge the gap between requirements and design. In *OOPSLA workshop – Requirements Engineering: Use Cases and More, Sunday October 15*, 1995.

[6] Chair Software Design and Quality, Karlsruhe Institute of Technology. Palladio: The software architecture simulator. http://www.palladio-simulator.com/, August 2012.

[7] A. Ciancone, A. Filieri, M. Drago, R. Mirandola, and V. Grassi. Klapersuite: An integrated model-driven environment for reliability and performance analysis of component-based systems. *Objects, Models, Components, Patterns*, pages 99–114, 2011.

[8] T. de Gooijer, A. Jansen, H. Koziolek, and A. Koziolek. An industrial case study of performance and cost design space exploration. In *Proc. 3rd Int. Conf. on Performance Engineering (ICPE'12)*, pages 205–216. ACM, April 2012.

[9] A. Di Marco and P. Inverardi. Compositional generation of software architecture performance qn models. In *Proc. 4th Working IEEE/IFIP Conference on Software Architecture (WICSA'04)*, pages 37–46. IEEE, 2004.

[10] S. Distefano, M. Scarpa, and A. Puliafito. From uml to petri nets: The pcm-based methodology. *IEEE Transactions on Softw*, 37(1):65 –79, jan.-feb. 2011.

[11] International Telecommunication Union (ITU). *User requirements notation (URN) - Language definition*, z.151 edition, 11 2008.

[12] H. Koziolek. Performance evaluation of component-based software systems: A survey. *Performance Evaluation*, 67(8):634–658, 2010.

[13] H. Koziolek and R. Reussner. A model transformation from the palladio component model to layered queueing networks. *Performance Evaluation: Metrics, Models and Benchmarks*, pages 58–78, 2008.

[14] H. Koziolek, B. Schlich, C. Bilich, R. Weiss, S. Becker, K. Krogmann, M. Trifu, R. Mirandola, and A. Martens. An industrial case study on quality impact prediction for evolving service-oriented software. In *Proc. 33rd ACM/IEEE Int. Conf. on Software Engineering (ICSE'11) Software Engineering in Practice Track*, pages 776–785. ACM, May 2011.

[15] P. Meier, S. Kounev, and H. Koziolek. Automated transformation of component-based software architecture models to queueing petri nets. In *Proc. 19th IEEE/ACM Int. Symposium on Modeling, Analysis and Simulation of Computer and Telecommunication Systems (MASCOTS'11)*, July 2011.

[16] D. Menasce, V. Almeida, L. Dowdy, and L. Dowdy. *Performance by design: computer capacity planning by example*. Prentice Hall, 2004.

[17] Object Management Group (OMG). Meta Object Facility (MOF) 2.0 Query/View/Transformation (QVT). http://www.omg.org/spec/QVT/Current, January 2011.

[18] Object Management Group (OMG). The UML Profile for MARTE: Modeling and Analysis of Real-Time and Embedded Systems. http://www.omg.org/spec/MARTE/, September 2011.

[19] D. Petriu and M. Woodside. A metamodel for generating performance models fron uml designs. *Lecture notes in computer science*, pages 41–53, 2004.

[20] D. Petriu and M. Woodside. Software performance models from system scenarios. *Performance Evaluation*, 61(1):65–89, 2005.

[21] D. B. Petriu. Layered software performance models constructed from use case map specifications. Master's thesis, Carleton University Ottawa, 2001.

[22] Project SEG. jUCMNav Project Website. http://goo.gl/4JL3K, May 2012.

[23] C. Rathfelder, S. Becker, K. Krogmann, and R. Reussner. Workload-aware system monitoring using performance predictions applied to a large-scale e-mail system. In *Proceedings of the Joint 10th Working IEEE/IFIP Conference on Software Architecture (WICSA) & 6th European Conference on Software Architecture (ECSA)*, Helsinki, Finland, 2012.

[24] R. Reussner, S. Becker, E. Burger, J. Happe, M. Hauck, A. Koziolek, H. Koziolek, K. Krogmann, and M. Kuperberg. The Palladio Component Model. Technical report, KIT, Fakultät für Informatik, Karlsruhe, 2011.

[25] C. Smith and L. Williams. *Performance Solutions: a practical guide to creating responsive, scalable software*, volume 34. Addison-Wesley Boston, MA;, 2002.

[26] M. Tribastone and S. Gilmore. Automatic extraction of pepa performance models from uml activity diagrams annotated with the marte profile. In *Proceedings of the 7th International Workshop on Software and Performance*, pages 67–78. ACM, 2008.

[27] C. Vogel. A use case map editor for rapid performance modeling and reasoning. Master's thesis, Karlsruhe Institute of Technology (KIT), 2012.

[28] M. Woodside. From annotated software designs (uml spt/marte) to model formalisms. In *Proceedings of the 7th international conference on Formal methods for performance evaluation*, pages 429–467. Springer-Verlag, 2007.

[29] M. Woodside, G. Franks, and D. Petriu. The future of software performance engineering. In *2007 Future of Software Engineering*, pages 171–187. IEEE Computer Society, 2007.

[30] X. Wu and M. Woodside. Performance Modeling from Software Components. In *Proc. 4th International Workshop on Software and Performance (WOSP'04)*, volume 29, pages 290–301, New York, NY, USA, 2004. ACM Press.

Non-Markovian Analysis for Model Driven Engineering of Real-Time Software

Laura Carnevali, Marco Paolieri, Alessandro Santoni, Enrico Vicario
Dipartimento di Ingegneria dell'Informazione, Università di Firenze
3, via di Santa Marta, 50139 Firenze, Italy
{laura.carnevali, marco.paolieri, alessandro.santoni, enrico.vicario}@unifi.it

ABSTRACT

Quantitative evaluation of models with stochastic timings can decisively support schedulability analysis and performance engineering of real-time concurrent systems. These tasks require modeling formalisms and solution techniques that can encompass stochastic temporal parameters firmly constrained within a bounded support, thus breaking the limits of Markovian approaches. The problem is further exacerbated by the need to represent suspension of timers, which results from common patterns of real-time programming. This poses relevant challenges both in the theoretical development of non-Markovian solution techniques and in their practical integration within a viable tailoring of industrial processes.

We address both issues by extending a method for transient analysis of non-Markovian models to encompass suspension of timers. The solution technique addresses models that include timers with bounded and deterministic support, which are essential to represent synchronous task releases, timeouts, offsets, jitters, and computations constrained by a Best Case Execution Time (BCET) and a Worst Case Execution Time (WCET). As a notable trait, the theory of analysis is amenable to the integration within a Model Driven Development (MDD) approach, providing specific evaluation capabilities in support of performance engineering without disrupting the flow of design and documentation of the consolidated practice.

Categories and Subject Descriptors

C.4 [**Performance of Systems**]: Modeling Techniques

General Terms

Performance, Algorithms

Keywords

Software performance engineering, non-Markovian stochastic analysis, model driven development, real-time systems.

1. INTRODUCTION

Model Driven Development (MDD) provides a way to incorporate formal methods in the development process of safety-critical real-time systems, so as to support formal verification of design correctness, automatic compilation of specifications into code skeletons and other artifacts, unit and integration testing [40, 32]. This practice is explicitly encouraged within various regulatory standards whenever the system exhibits complex behaviors mixing concurrency and timing [36, 26].

Various approaches and tools have been developed to support the aim for concurrent systems with non-deterministic temporal parameters. Notable examples include Uppaal [4] with the extensions TIGA [20] and TRON [29], based on the theory of timed automata [1]; Giotto [28], a tool-supported programming methodology for the validation and synthesis of control software; Charon [2], a language for modular specification of interacting hybrid systems, relying on the notions of agent and mode; Simulink [35], a block diagram environment for simulation and model-based design of complex control systems.

The Oris tool [10] supports MDD by leveraging the theory of preemptive Time Petri Nets (pTPNs) [11], an extension of Time Petri Nets (TPNs) [43] that encompasses priority-driven preemptive scheduling with possible suspension of computations, attaining an expressive power that compares with stopwatch automata [21] and Petri nets with hyper-arcs [37]. The MDD approach of [18] integrates the theory of pTPNs and the Oris tool within a tailoring of the V-Model life cycle [14], supporting formal verification of SW design, generation of real-time code, measurement-based profiling of temporal parameters, and testing. As a relevant trait, the designer is allowed to create a model through a semi-formal specification, which captures design choices of the periods and the concurrency structure of tasks, as well as design assumptions on the duration of computations, initially guessed by analogy with previous realizations. During development iterations, the results of formal analysis and execution time profiling are used to refine the specification and implementation until SW requirements are satisfied. The formal methodology of [18] is extended in [25] to enable a smooth integration within an industrial process of SW development, supporting the documentation process prescribed by Military Standard 498 (MIL-STD-498) [42] while avoiding to disrupt consolidated industrial practice.

With the evolution of certification standards for safety-critical software towards RAMS (Reliability, Availability, Maintainability, Safety) practices, safety requirements excluding the *possibility* of anomalous behaviors are often recast into quantitative values of guaranteed *probability* [26, 27]. This motivates the extension of MDD approaches so as to incorporate quantitative evaluation of RAM indices into the formal verification process. For this purpose, modeling and solution techniques should encompass inherent characteristics of real-time software, such as multiple concurrent temporal parameters characterized by general (GEN) distributions over

ICPE'13, April 21–24, 2013, Prague, Czech Republic.

bounded or deterministic supports. Computations usually have execution times firmly constrained within finite intervals; task periods, offsets, and timeouts are necessarily deterministic temporal parameters; jitters and inter-release times of asynchronous tasks are often lower-bounded. At the same time, the analysis techniques must be able to deal with suspension in the advancement of timers, as an effect of preemptive scheduling, which corresponds to the *Preemptive Resume* (PRs) policy, as opposed to cooperative scheduling represented by the *Preemptive Repeat Different* (PRD) policy [7].

In this paper, we present an approximated solution technique for the transient analysis of non-Markovian preemptive models that include deterministic and bounded timers. Transient analysis based on stochastic state classes [30] is extended from a PRD policy (where all enabled timers advance with the same speed and disabled timers are re-sampled at newly-enabling) to a PRs policy (where some enabled timers may be suspended and later resumed). The addition of suspension has a major impact on the analysis technique with respect to the PRD policy: in this case, the geometric shape of the joint support of enabled timers is no more in the form of a Difference Bounds Matrix zone (DBM zone), but becomes a linear convex polyhedron, leading to exponential complexity in the derivation and encoding of the state space. The issue is addresses in [11] for non-deterministic models by resorting to an overapproximation of polyhedral supports with DBM zones. For stochastic models, the problem is exacerbated by the fact that the joint probability density function (PDF) of enabled timers has a piecewise form over a partition of the polyhedral support into polyhedral subdomains. The computational complexity of such partitioning excludes the generalization of approaches based on DBM zone partitioning [44], and requires a solution technique based on the approximation of both supports and distributions. The proposed approach allows to represent deterministic timers that are never suspended, which allow to represent synchronous task releases and timeouts, as resulting from common patterns of real-time programming. In contrast, suspension may occur on stochastic timers, which represent computation times associated with non-pointlike support, both to allow a margin of laxity in the implementation and to be robust with respect to possible changes in the embedding environment.

The rest of the paper is organized as follows: in Sect. 2, we discuss related work; in Sect. 3, we present a novel solution technique for the transient analysis of non-Markovian preemptive models including temporal parameters with bounded and deterministic supports; in Sect. 4, we present experimental results; in Sect. 5, we draw conclusions.

2. RELATED WORK

Quantitative evaluation of real-time systems faces the analysis of models that encompass both stochastic temporal parameters with bounded support and priority-driven preemptive scheduling with suspension and resumption of timers. This mix of requirements is partially fulfilled by a few discrete time approaches and basically unsolved in the continuous time perspective.

The numerical approach presented in [31] for the analysis of Non-Markovian Stochastic Petri Nets (NMSPNs) relies on a time-discretization algorithm that approximates the firing times of transitions through Discrete Phase-Type (DPH) random variables, mapping the evolution of the marking process into an expanded Discrete Time Markov Chain (DTMC). The technique allows any kind of general distribution, combining together the PRs policy and the PRD policy [7]. The inherent state-space explosion problem is alleviated in [39] by using a Kronecker algebra, which avoids an explicit enumeration of timed states while requiring a finite number of reachable markings. A similar approach based on DPHs is proposed in [45] for the model of Discrete Deterministic and Stochastic Petri Nets (DDSPNs), which encompass race policies equivalent to PRs and PRD policies. In [12], a discrete-time variant of time Petri nets [43] supports the representation of preemptive behavior and associates a stochastic characterization with all non-deterministic logical and temporal choices. The analysis technique relies on a maximal step semantics which conveniently separates the effects of logical concurrency from those of non-Markovian timed behavior, circumventing the problems of confusion and non-well-definedness arising in discrete-time models with an interleaved semantics of concurrency.

As a common trait, discretization approaches suffer a trade-off between the accuracy attained by the analysis and the computational complexity, which largely depends on the time scale used to cover the ranges of variation of temporal parameters. The problem is further exacerbated in those methods that approximate the distribution of timers with finite support through DPHs, which may produce unexpected results when the finiteness of behavior actually depends on the sequencing constraints induced by timing.

Managing non-Markovian and preemptive behavior is a much harder challenge in continuous time, where the limits of the analysis reduce the expressive capabilities of the model. In [23, 8], GEN timers are approximated by means of Continuous Phase-Type (CPH) distributions, following an approach that can be regarded as a continuous-time version of the phase-type expansion technique of [31]. The symbolic technique of [33] supports efficient computation and encoding of the reachability graph of the untimed model, although the dimension of the state space still comprises a limitation in the step of solution and the obtained approximants are not able to firmly preserve bounded supports. An analytical approach based on a continuous abstraction of time was proposed in [9] for Deterministic and Stochastic Petri Nets (DSPNs), which include exponential and deterministic timers, and extended in [7] to manage a combined use of different preemption polices. The approach leverages Markov Renewal Theory under the so-called enabling restriction, which rules out concurrent enabling of multiple GEN transitions. As a major limit, this structurally clashes with the need to represent sequencing restrictions due to mutual timing.

The case of multiple concurrent GEN timers with possibly overlapping activity cycles is managed by the method of stochastic state classes. Steady state analysis can be performed exactly in the case that every cyclic behavior that changes the enabling status of GEN transitions visits at least one regeneration point [44]. Transient analysis [30] can be performed exactly provided that the model does not include cycles that must be completed in zero time (no Zeno behaviors), and it can be always guaranteed to be finite in probability under the assumption of an approximation threshold $\epsilon > 0$ on the total error. However, the solution techniques of [44, 30] address models that do not encompass suspension and resumption of timers. In [15], the method of [44] is extended to support steady-state analysis of systems that run under static-priority preemptive scheduling. Nevertheless, this still maintains two major limits in the perspective of application to the development of real-time systems. In fact, the technique is developed under the restriction that all timers have non-pointlike support, which precludes deterministic timers and thus prevents the representation of structural patterns of real-time programming such as timeouts and periodic task releases. In addition, the approach does not support transient analysis, which instead turns out to take major relevance in the verification of most real-time requirements.

3. MODELING AND EVALUATION OF STOCHASTIC REAL-TIME SYSTEMS

We first introduce the model of stochastic preemptive Time Petri Nets, which aims at providing sufficient expressive capabilities for the representation of realistic patterns of real-time programming (Sect. 3.1). We then present a novel solution technique that extends the method of [30] to encompass transient analysis of non-Markovian preemptive models (Sect. 3.2).

3.1 Modeling task-sets through spTPN

A stochastic preemptive Time Petri Net represents a discrete-event system whose state is made of a logic condition called *marking* and a set of continuous timers called *times-to-fire*, each associated with a transition enabled by the marking. Each enabled transition may request one or more resources according to a static priority preemptive scheduling: it is called *progressing* or *suspended* depending on whether it is assigned each of its requested resources or not, respectively. The progressing transition with minimum time-to-fire is selected as the next event and its execution leads to a new state with different marking and times-to-fire. After the marking update, transitions can persist or be newly-enabled, disabled, reset. Times-to-fire of newly-enabled and reset transitions are sampled according to static distributions; times-to-fire of persistent transitions are decreased by that of the fired transition if they are progressing, and maintained if they are suspended; times-to-fire of disabled transitions are discarded.

3.1.1 Syntax

A stochastic preemptive Time Petri Net (spTPN) is a tuple $\langle P; T; A^-; A^+; A^{\cdot}; m_0; EFT^s; LFT^s; Res; Req; Prio; \mathcal{F}; \mathcal{C}\rangle$.

The first eight elements comprise the model of *time Petri nets* [43]. P is a set of places. T is a set of transitions disjoint from P. $A^- \subseteq P \times T$, $A^+ \subseteq T \times P$, and $A^{\cdot} \subseteq P \times T$ are sets of precondition, postcondition, and inhibitor arcs. $m_0 : P \rightarrow \mathbb{N}$ is the initial marking associating each place with an initial non-negative number of tokens. $EFT^s : T \rightarrow \mathbb{R}_0^+$ and $LFT^s : T \rightarrow \mathbb{R}_0^+ \times (\mathbb{R}_0^+ \cup \{\infty\})$ associate each transition with a *static Earliest Firing Time* and a (possibly infinite) static *Latest Firing Time* ($EFT^s(t) \leq LFT^s(t)\ \forall\, t \in T$). As usual in Petri Nets, a place p is said to be an *input*, an *output*, or an *inhibitor* place for a transition t if $\langle p, t\rangle \in A^-$, $\langle t, p\rangle \in A^+$, or $\langle p, t\rangle \in A^{\cdot}$, respectively.

Res, Req, and $Prio$ comprise a mechanism of resource assignment in the style of *preemptive Time Petri Nets* (pTPNs) [11], which makes the progress of timed transitions dependent on the availability of a set of preemptable resources. Res is a set of preemptable resources disjoint from P and T. $Req : T \rightarrow 2^{Res}$ and $Prio : T \rightarrow \mathbb{N}$ associate each transition with a subset of Res representing its resource request and with a static priority level, respectively. We assume that low priority numbers correspond to high priority levels.

As in *stochastic Time Petri Nets* (sTPNs) [44, 16], $\mathcal{F}$ and $\mathcal{C}$ define a measure of probability for non-deterministic choices. Specifically, $\mathcal{C} : T \rightarrow \mathbb{R}^+$ associates each transition with a positive weight and $\mathcal{F} : T \rightarrow F_t^s$ associates each transition with a static Cumulative Distribution Function (CDF) supported over its static firing interval $[EFT^s(t), LFT^s(t)]$. As usual in Stochastic Petri Nets, a transition t is called *immediate* (IMM) if $EFT^s(t) = LFT^s(t) = 0$ and *timed* otherwise. A timed transition t is called *exponential* (EXP) if $F_t^s(x) = 1 - e^{\lambda x}$ over $[0, \infty]$ for some rate $\lambda \in \mathbb{R}_0^+$ and *general* (GEN) otherwise. In particular, a GEN transition t with $EFT^s(t) = LFT^s(t) > 0$ is called *deterministic* (DET). Moreover, for each GEN transition t with $EFT^s(t) \neq LFT^s(t)$, which we call *distributed* transition, we assume that F_t^s is absolutely continuous and, thus, that there exists a Probability Density Function (PDF) f_t^s such that $F_t^s(x) = \int_0^x f_t^s(y)dy$.

3.1.2 Semantics

The *state* of an spTPN is a pair $\langle m, \tau\rangle$, where $m : P \rightarrow \mathbb{N}$ is a marking that associates each place with a non-negative number of tokens and $\tau : T^e(m) \rightarrow \mathbb{R}_0^+$ associates each transition enabled by m with a (dynamic) real-valued time-to-fire, where a transition t is said to be *enabled* by m if each of its input places contains at least one token and none of its inhibitor places contains any token (i.e., $m(p) \geq 1\ \forall\, \langle p, t\rangle \in A^-$ and $m(p) = 0\ \forall\, \langle p, t\rangle \in A^{\cdot}$).

An enabled transition t is *progressing* if every resource it requires is not required by any other enabled transition with a higher level of priority (i.e., $Prio(t) \leq Prio(t')\ \forall\, t' \in T^e(M)$ such that $Req(t) \cap Req(t') \neq \emptyset$); otherwise, it is *suspended*.

A progressing transition t is *firable* in state s if its time-to-fire is not higher than that of any other transition that is progressing in s (i.e., $\tau(t) \leq \tau(t')\ \forall\, t' \in T_p(s)$, where $T_p(s)$ is the set of transitions that are progressing in s). When multiple transitions are firable, the choice is resolved by the random switch determined by $\mathcal{C}$: $Prob\{t \text{ is selected}\} = \mathcal{C}(t)/\sum_{t_i \in T_f} \mathcal{C}(t_i)$.

When a transition t fires, the state $\langle m, \tau\rangle$ is replaced by a new state $s' = \langle m', \tau'\rangle$, which we write as $s \xrightarrow{t} s'$. Marking m' is derived from m by removing a token from each input place of t (i.e., if $\langle p, t\rangle \in A^-$ then $m_{tmp}(p) = m(p) - 1$, else $m_{tmp}(p) = m(p)$) and by adding a token to each output place of t (i.e., if $\langle t, p\rangle \in A^+$ then $m'(p) = m_{tmp}(p) + 1$, else $m'(p) = m_{tmp}(p)$). Transitions that are enabled both by the intermediate marking m_{tmp} and by m' are said *persistent*, while those that are enabled by m' but not by m_{tmp} or m are said *newly-enabled*. If t is still enabled after its own firing, it is always regarded as newly enabled [5, 43].

For any transition t_n newly-enabled after the firing of t, the time-to-fire takes a random value sampled in the static firing interval according to the static probability distribution $F_{t_n}^s$, i.e., $EFT^s(t_n) \leq \tau'(t_n) \leq LFT^s(t_n)$ and $Prob\{\tau'(t_n) \leq x\} = F_{t_n}^s(x)$. For any transition t_p that was progressing in the previous state and is persistent after the firing of t, the time-to-fire is reduced by the time elapsed in the previous state (which is equal to the time-to-fire of t measured at the entrance in the previous state), i.e., $\tau'(t_p) = \tau(t_p) - \tau(t)$. For any transition t_s that was suspended in the previous state and is persistent after the firing of t, the time-to-fire remains unchanged, i.e., $\tau'(t_s) = \tau(t_s)$.

3.1.3 An illustrative example

Fig. 1 shows a periodic real-time task with a semaphore synchronization within a wider (not shown) task-set. A transition with no input places nor resource requests models a task release, and fires repeatedly with inter-firing times falling within its firing interval, e.g., t_{10} models periodic task releases with period of 50 time units. A time-consuming operation performed on one or more processors is represented by a transition associated with a resource request, a static priority, and a firing interval corresponding to its range of execution time, e.g., t_{12} and t_{13} model two computations performed on resource *cpu* with priority level 1 in a time ranging within $[5, 10]$ and $[2, 4]$ time units, respectively. A binary semaphore is represented as a place initially marked with one token, e.g., $mutex$. While a *wait* operation is modeled as a transition having the semaphore place as an input place, a *release* operation is usually represented by a transition that has the semaphore place as an output place and that also accounts for the completion of a synchronized computation, e.g., t_{11} represents a wait operation on $mutex$, while t_{13} models a release operation on $mutex$ as well as the completion of the second computation of the task.

Some temporal parameters of a real-time task-set attain a non-deterministic range of variation (e.g., the execution time of computations and the inter-release time of asynchronous tasks), and they are modeled as transitions that, in principle, can be associated with any kind of distribution (e.g., t_{11}, t_{12}, and t_{13}). Other temporal parameters are intrinsically deterministic (e.g., timeouts and the inter-release time of synchronous tasks), and thus their are modeled as IMM or DET transitions, which are associated with a Dirac delta function (e.g., t_{10} is associated with a Dirac delta function $f^s_{t_{10}}(y) = \delta(y - 50)$, centered at $y = 50$ time units).

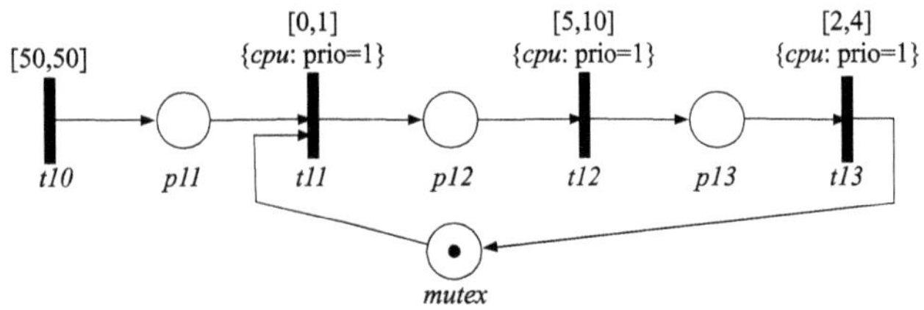

Figure 1: A fragment of the spTPN model of a task-set, representing a periodic real-time task.

In general, the construction of the model of a real-time task-set can be conveniently cast into a minor extension of the development process devised in [18, 25], which was proven viable in the industrial practice [25] but did not include any probabilistic quantification. To this end, during the early stages of SW Design, the distributions associated with the execution times of computations and with the inter-release times of asynchronous tasks are guessed by analogy with previous or similar implementations. Next, during the iterations of the development process, they are progressively refined according to the histograms of temporal parameters observed during the execution of the real-time code, which are obtained as a by-product of the approach implemented in [18, 25]. In doing so, the additional effort demanded to the designer to manage probabilities consists in finding GEN distributions in the class of expolynomial functions that fit the shape and support of the measured histograms.

Note that the spTPN semantics combines the PRD and PRs policies of [7, 9], also known as *enabling memory* policy and *age* policy [22], respectively. In the formulation of [7, 9], the clock of a transition disabled by the lack of a token is reset or maintained depending on whether the transition is associated with PRD or PRs policy, respectively. Conversely, in our treatment, the progress of times-to-fire depends both on the presence of tokens into input places and on the availability of preemptable resources. Specifically, when a transition is disabled by the lack of a token in some of its input places, its time-to-fire is reset, following the PRD policy; otherwise, when a transition is suspended by the lack of some of its required resources, its time-to-fire is maintained and resumed when the transition is assigned the resource again, following the PRs policy. This enables separate modeling of inter-task communication mechanisms from real-time concurrency on resources, facilitating the representation of usual patterns of real-time programming [13]. As shown by the example of Fig. 1, while the presence of tokens into places models logic conditions such as the availability of a semaphore, preemptable resources of the net are used to account for processors that are necessary to perform a computation.

3.2 Transient analysis of spTPN

We recall here the salient aspects of the method of transient stochastic state classes (Sect. 3.2.1), for which we also refer the reader to [30]. We then develop a novel and substantial extension of the approach of [30] that encompasses suspension and resumption of timers by relying on a continuous-time approximation in the state-space (Sects. 3.2.3, 3.2.2, and 3.2.4). We finally discuss how the complexities arising from the interaction of suspension with DET transitions can be managed (Sect. 3.2.5).

3.2.1 Transient classes

The state of an spTPN evolves depending on the times-to-fire sampled by transitions and the resolution of random switches according to transitions weights. Within a given time bound, the method of transient stochastic state classes [30] characterizes the state of an spTPN after each transition firing and the absolute time of the firing. Specifically, a transient stochastic state class provides the joint PDF for the vector of times-to-fire of enabled transitions augmented with a timer, called τ_{age}, which accounts for the time elapsed since when the class was entered. For convenience, τ_{age} encodes the opposite of the elapsed time, so that it evolves with the same slope as that of times-to-fire of progressing transitions and it can be treated as an always-persistent-progressing time-to-fire.

DEFINITION 3.1. *A transient stochastic state class (transient class for short) is a tuple $\langle m, New, D, f_{\langle \tau_{age}, \underline{\tau} \rangle} \rangle$, where: $m : P \to \mathbb{N}$ is a marking; $\langle \tau_{age}, \underline{\tau} \rangle$ is a random variable made of the scalar variable τ_{age} and the vector $\underline{\tau} = \langle \tau_0, \tau_1, ..., \tau_{N-1} \rangle$ of times-to-fire of transitions enabled by m; New is the set of transitions that are newly enabled when the class is entered; and, $f_{\langle \tau_{age}, \underline{\tau} \rangle}$ is the joint PDF of the variable in $\langle \tau_{age}, \underline{\tau} \rangle$ supported over D.*

A reachability relation is defined among transient classes:

DEFINITION 3.2. *We say that $\Sigma' = \langle m', D', New', f_{\langle \tau'_{age}, \underline{\tau}' \rangle} \rangle$ is a successor of $\Sigma = \langle m, New, D, f_{\langle \tau_{age}, \underline{\tau} \rangle} \rangle$ through the firing of t_0 with probability μ_0, which we write $\Sigma \stackrel{t_0, \mu_0}{\Longrightarrow} \Sigma'$, iff, given that the marking of the net is m and the time vector $\langle \tau_{age}, \underline{\tau} \rangle$ is a random variable with support D and PDF $f_{\langle \tau_{age}, \underline{\tau} \rangle}$, then the firing of t_0 occurs with probability μ_0 and leads to a new marking m' and a new time vector $\langle \tau'_{age}, \underline{\tau}' \rangle$ with support D', PDF $f_{\langle \tau'_{age}, \underline{\tau}' \rangle}$, and the set New' of newly-enabled transitions.*

In the initial transient class $\Sigma_0 = \langle m_0, D, f_{\langle \tau_{age}, \underline{\tau} \rangle} \rangle$, τ_{age} is equal to 0 and all enabled transitions are newly-enabled with times-to-fire independently distributed according to their PDFs supported over their firing intervals. According to this:

$$D = [0, 0] \times \prod_{t_i \in T_e(m_0)} [EFT^s(t_i), LFT^s(t_i)],$$

$$f_{\langle \tau_{age}, \underline{\tau} \rangle}(x_{age}, \underline{x}) = \delta(x_{age}) \cdot \prod_{t_i \in T_e(m)} f_{t_i}(x_i),$$

where $\delta(x_{age})$ is the Dirac delta function centered at $x_{age} = 0$. Given an initial transient class Σ_0, the transitive closure of the reachability relation among transient classes enumerates a *transient stochastic tree*, where vertexes are transient classes and edges are labeled with a transition t and a probability μ. This enables the derivation of continuous-time transient probabilities of reachable markings within any given time bound [30].

3.2.2 Transient classes in partitioned form

When the spTPN model includes IMM and DET transitions, a class $\Sigma = \langle m, New, D, f_{\langle \tau_{age}, \underline{\tau} \rangle} \rangle$ may have a domain D with null measure in $\mathbb{R}^{N+1}$, where N is the cardinality of the set $T^e(m)$ of transitions enabled by marking m [44]. To avoid a stochastic characterization of deterministic and deterministically dependent timers, we resort to the $\langle Q, U, V \rangle$ partition of [44], where: Q is the set (of maximal cardinality) of *distributed* variables, i.e., variables that have a non-deterministic value and mutual delay (i.e., $\tau_i \in Q$ iff $b_{i*} + b_{*i} \neq 0 \wedge b_{ij} + b_{ji} \neq 0 \ \forall \tau_j \in Q$ with $i \neq j$); U is the set of *deterministic* variables, i.e., variables that have a deterministic value (i.e., $\tau_h \in U$ iff $b_{h*} + b_{*h} = 0$); and, V is the set of *synchronized* variables, i.e., variables that have a deterministic delay with respect to some distributed variable (i.e., $\tau_k \in V$ iff $b_{k*} + b_{*k} \neq 0 \wedge \exists \tau_{\iota_k} \in Q$ such that $b_{k\iota_k} + b_{\iota_k k} = 0$).

According to this, the normal form of D can be rewritten in a *partitioned form* that combines the set D_Q of constraints limiting the variables of Q with a set of equalities expressing all the remaining variables of U and V:

$$D_Q = \begin{cases} \tau_i - \tau_j \leq b_{ij} \\ \tau_i - \tau_* \leq b_{i*} \\ \tau_* - \tau_i \leq b_{*i} \\ \forall \tau_i, \tau_j \in Q \end{cases} \quad D = \begin{cases} D_Q \\ \tau_h = b_{h*} + \tau_* \\ \tau_k = b_{k\iota_k} + \tau_{\iota_k} \\ \forall \tau_h \in U, \forall \tau_k \in V \end{cases} \tag{1}$$

When domain D is represented in partitioned form, the joint PDF of variables in $Q \cup U \cup V$ is completely defined by the joint PDF of variables in Q. According to this, $f_{\langle \tau_{age}, \underline{\tau} \rangle}$ is completely defined by the joint PDF $f_{\langle \tau_{age}, \underline{\tau}_Q \rangle}$ of τ_{age} and $\underline{\tau}_Q$. According to this, a transient class $\langle m, New, D, f_{\langle \tau_{age}, \underline{\tau} \rangle} \rangle$ will be represented as $\langle m, New, D, f_{\langle \tau_{age}, \underline{\tau}_Q \rangle} \rangle$.

3.2.3 *Enumeration of transient classes*

In the derivation of successor classes, we use the following notational conventions: *i)* given a vector $\underline{x} = \langle x_0, \ldots, x_{N-1} \rangle \in \mathbb{R}^N$ and a scalar $\delta \in \mathbb{R}$, $\underline{x}+\delta$ denotes the vector $\langle x_0+\delta, \ldots, x_{N-1}+\delta \rangle$; *ii)* if $D \subseteq \mathbb{R}^N$ and $n \in \{0, \ldots, N-1\}$, then $D \downarrow_{\tau_n}$ denotes the projection of D that eliminates τ_n: $D \downarrow_{\tau_n} = \{\langle x_0, \ldots, x_{n-1}, x_{n+1}, \ldots, x_{N-1} \rangle \in \mathbb{R}^{N-1} \mid \exists x_n \in \mathbb{R} . \langle x_0, \ldots, x_{n-1}, x_n, x_{n+1}, \ldots, x_{N-1} \rangle \in D\}$; *iii)* $D \downarrow_{A_1, \ldots, A_N}$ denotes the projection of D that eliminates all variables in $A_1 \cup \ldots \cup A_N$; *iv)* given a transient class $\Sigma = \langle m, New, D, f_{\langle \tau_{age}, \underline{\tau} \rangle} \rangle$, the number of enabled transitions (i.e., the number of times-to-fire) is denoted by M; the number of distributed times-to-fire is denoted by N and their vector is denoted by $\underline{\tau}_Q = \langle \tau_0, \tau_1, ..., \tau_{N-1} \rangle$; the time vector is represented as $\langle \tau_{age}, \underline{\tau} \rangle = \langle \tau_{age}, \tau_0, \tau_1, ..., \tau_{N-1}, \tau_N, ..., \tau_{M-1} \rangle$ so as to encode τ_{age} and distributed times-to-fire in the first $N+1$ positions.

Successors detection and calculus of their probability. According to the spTPN semantics, the firing of a transition t_0 is a possible outgoing event from a transient class $\Sigma = \langle m, New, D, f_{\langle \tau_{age}, \underline{\tau} \rangle} \rangle$ iff t_0 is progressing in Σ and has a time-to-fire not higher than that of any other progressing transition. This occurs iff the restricted domain $D^0 = D \cap \{\tau_0 \leq \tau_p \ \forall t_p \in T^{progr}(\Sigma)\}$ is not empty, where $T^{progr}(\Sigma)$ denotes the set of transitions that are progressing in Σ. The probability that t_0 fires is derived in a different manner by distinguishing the following cases:

- If all progressing transitions have deterministic time-to-fire, then t_0 is selected in the random switch with progressing transitions t_p having the same time-to-fire as that of t_0 (i.e., $t_p \in T^{progr}(\Sigma)$ such that $b_{p0} = b_{0p} = 0$):

$$\mu_0 = \frac{\mathcal{C}(t_0)}{\mathcal{C}(t_0) + \displaystyle\sum_{\substack{t_n \neq t_0 \in T^{progr}(\Sigma) \\ \text{with } b_{p0} = b_{0p} = 0}} \mathcal{C}(t_p)}. \tag{2}$$

- If at least a progressing transition has distributed time-to-fire and the projection of D^0 that eliminates all variables in U and V has null measure (i.e., $||D^0 \downarrow_{U,V}|| = 0$), then $\mu_0 = 0$. This occurs iff a time-to-fire exists that was distributed before the conditioning that yields D^0 and becomes synchronized or deterministic after the conditioning.

- If at least a progressing transition has distributed time-to-fire and the projection of D^0 that eliminates all variables in U and V has non-null measure (i.e., $||D^0 \downarrow_{U,V}|| \neq 0$), then μ_0 is the joint probability that the vector of times-to-fire $\underline{\tau}$ takes a value such that τ_0 is not lower than that of any other progressing transition and t_0 is selected in the random switch among all of the progressing transition that have the same time-to-fire as that of t_0:

$$\mu_0 = \frac{\mathcal{C}(t_0)}{\mathcal{C}(t_0) + \displaystyle\sum_{\substack{t_n \neq t_0 \in T^{progr}(\Sigma) \\ \text{with } b_{p0} = b_{0p} = 0}} \mathcal{C}(t_p)} \cdot \int_{D^0 \downarrow_{U,V}} f_{\langle \tau_{age}, \underline{\tau}_Q \rangle}(x_{age}, \underline{x}_Q) dx_{age} d\underline{x}_Q. \tag{3}$$

If $\mu_0 > 0$, the marking m' and the set of newly-enabled transitions are derived according to the spTPN semantics. The support D' of the time vector $\langle \tau'_{age}, \underline{\tau}' \rangle$ after the firing of t_0 and the PDF $f_{\langle \tau'_{age}, \underline{\tau}'_Q \rangle}$ are derived according to the following steps, regarding τ_{age} as a persistent time-to-fire.

Conditioning. The assumption that t_0 is the next transition to fire conditions $\langle \tau_{age}, \underline{\tau} \rangle$ and yields a new random variable $\langle \tau^a_{age}, \underline{\tau}^a \rangle = \langle \tau_{age}, \underline{\tau} \mid \tau_0 \leq \tau_p \ \forall t_p \in T^{progr}(\Sigma) \rangle$ ranging within D^0. Thus, $\langle \tau^a_{age}, \underline{\tau}^a_Q \rangle$ is distributed over $D^0 \downarrow_{U,V}$ according to:

$$f_{\langle \tau^a_{age}, \underline{\tau}^a_Q \rangle}(x_{age}, \underline{x}) = \frac{f_{\langle \tau_{age}, \underline{\tau}_Q \rangle}(x_{age}, \underline{x})}{\mu_0}, \tag{4}$$

where $\underline{x} = \langle x_0, x_1, ..., x_{N-1} \rangle$. The assumption that $\mu_0 > 0$ guarantees that the conditioning leaves the $\langle Q, U, V \rangle$ partition unchanged so that $\langle Q^a, U^a, V^a \rangle = \langle Q, U, V \rangle$.

Time advancement. At the firing of t_0, the age and times-to-fire of progressing transitions are reduced by the time-to-fire value of t_0 given by τ^a_0, while those of suspended transitions remain unchanged. This yields a new random variable $\langle \tau^b_{age}, \underline{\tau}^b \rangle = \langle \tau^a_{age} - \tau^a_0, \tau^a_1 - \tau^a_0, ..., \tau^a_{P-1} - \tau^a_0, \tau^a_P, ..., \tau^a_{N-1}, ... \tau^a_{M-1} \rangle$, where: $\underline{\tau}_P = \langle \tau^a_1, ..., \tau^a_{P-1} \rangle$ is the vector of distributed times-to-fire that are progressing; $\underline{\tau}_S = \langle \tau^a_P, ..., \tau^a_{N-1} \rangle$ is the vector of distributed times-to-fire that are suspended; and, τ^a_N, ..., τ^a_{M-1} are times-to-fire that are either deterministic or synchronized. As synchronized times-to-fire result from multiple progressing deterministic timers that persist at the firing of a distributed timer, each of them is always progressing and synchronized with a distributed time-to-fire.

This step is then developed in a different manner depending on whether τ^a_0 is deterministic, distributed, or synchronized with a distributed variable.

- If τ^a_0 is *deterministic*, the $\langle Q, U, V \rangle$ partition does not change. The support D^b is derived from D^a through a constant shift of progressing timers and $f_{\langle \tau^b_{age}, \underline{\tau}^b_Q \rangle}$ is expressed as:

$$\begin{aligned} & f_{\langle \tau^b_{age}, \underline{\tau}^b_Q \rangle}(x_{age}, \underline{x}) \\ & = f_{\langle \tau^a_{age}, \underline{\tau}^a_Q \rangle}(x_{age} + x_0, x_0, \underline{x}_P + x_0, \underline{x}_S) dx_0, \end{aligned} \tag{5}$$

where $\underline{x} = \langle x_1, x_2, \ldots, x_{N-1} \rangle$, $\underline{x}_P = \langle x_1, x_2, \ldots, x_{P-1} \rangle$, and $\underline{x}_S = \langle x_P, x_{P+1}, \ldots, x_{N-1} \rangle$.

- If τ^a_0 is *distributed*, the $\langle Q, U, V \rangle$ partition may change. Specifically, distributed times-to-fire remain distributed; progressing deterministic times-to-fire become distributed but remain synchronized among themselves, thus one is added to Q^b while the other ones are added to V^b; suspended deterministic times-to-fire remain deterministic; progressing times-to-fire that are synchronized with τ_0^a become deterministic; progressing times-to-fire that are synchronized with a distributed time-to-fire $\tau_h^a \neq \tau_0^a$ remain synchronized with it.

The treatment proceeds in a different manner depending on whether *i)* all distributed timers are progressing, *ii)* all distributed timers are suspended, or *iii)* some distributed timer is progressing and some is suspended. In all these cases, the derivation also distinguishes whether some time-to-fire is deterministic or not.

i) If all distributed timers are progressing and no time-to-fire is deterministic, then $f_{\langle\tau_{age}^b,\underline{\tau}_Q^b\rangle}$ is obtained by shifting all the components by the value of τ_0^a and by eliminating it through a projection:

$$f_{\langle\tau_{age}^b,\underline{\tau}^b\rangle}(x_{age},\underline{x}) = \int_{Min^0(x_{age},\underline{x})}^{Max^0(x_{age},\underline{x})} f_{\langle\tau_{age}^a,\underline{\tau}^a\rangle}(x_{age}+x_0,x_0,\underline{x}_P+x_0)dx_0 \quad (6)$$

with $\underline{x}=\langle x_1,...,x_{P-1}\rangle$ and $[Min^0(x_{age},\underline{x}), Max^0(x_{age},\underline{x})]$ being the support of all possible values of τ_0^a. As in [16, 15], this may partition $D^b\downarrow_{U,V}$ into a finite set of DBM subdomains over which $f_{\langle\tau_{age}^b,\underline{\tau}^b\rangle}$ accepts a piece-wise representation.

Otherwise, if all distributed timers are progressing and some time-to-fire is deterministic, let τ_P^a be the deterministic time-to-fire that becomes distributed. Then $f_{\langle\tau_{age}^b\underline{\tau}_Q^b\rangle}$ has global analytic representation over DBM domain D^b:

$$f_{\langle\tau_{age}^b,\underline{\tau}^b\rangle}(x_{age},\underline{x}) = f_{\langle\tau_{age}^a,\underline{\tau}^a\rangle}(x_{age}-x_P-B_{p*}, -x_P-B_{p*},\underline{x}-x_P-B_{p*}), \quad (7)$$

where $\underline{x}=\langle x_1,\dots,x_{P-1}\rangle$.

ii) If some distributed timer is progressing and some is suspended, D^b takes the form of a linear convex polyhedron.

If no time-to-fire is deterministic, then $f_{\langle\tau_{age}^b,\underline{\tau}^b\rangle}$ accepts a piece-wise representation over a partition of D^b into a finite set of polyhedral subdomains. The problem is overcome by approximating D^b through its tightest embedding DBM and by approximating $f_{\langle\tau_{age}^b,\underline{\tau}^b\rangle}$ through Bernstein polynomials.

Otherwise, if some time-to-fire is deterministic, then a deterministic timer becomes distributed and the other ones become synchronized with it. According to this, $f_{\langle\tau_{age}^b,\underline{\tau}^b\rangle}$ accepts a global analytic representation over D^b:

$$f_{\langle\tau_{age}^b,\underline{\tau}^b\rangle}(x_{age},\underline{x}) = f_{\langle\tau_{age}^a,\underline{\tau}^a\rangle}(x_{age}-x_P, -x_P+B_{p*},\underline{x}_P-x_P+B_{p*},\underline{x}_S), \quad (8)$$

where $\underline{x}=\langle x_1,\dots,x_{P-1}\rangle$. To resort to a global analytic representation over a DBM domain, also in this case D^b and $f_{\langle\tau_{age}^b,\underline{\tau}^b\rangle}$ are approximated as in the previous case.

iii) If all distributed timers are suspended and no time-to-fire is deterministic, then $f_{\langle\tau_{age}^b,\underline{\tau}^b\rangle}$ is derived through a projection that eliminates τ_0^a:

$$f_{\langle\tau_{age}^b,\underline{\tau}^b\rangle}(x_{age},\underline{x}) = \int_{Min^0(x_{age},\underline{x})}^{Max^0(x_{age},\underline{x})} f_{\langle\tau_{age}^a,\underline{\tau}^a\rangle}(x_{age},x_0,\underline{x})dx_0, \quad (9)$$

where $\underline{x}=\langle x_P,\dots,x_{N-1}\rangle$. Also in this case, $f_{\langle\tau_{age}^b,\underline{\tau}^b\rangle}$ accepts a piece-wise representation over a partition of $D^b\downarrow_{U,V}$ into a finite set of DBM subdomains.

Otherwise, if all distributed timers are suspended and some time-to-fire is deterministic, then a deterministic timer becomes distributed and the other ones become synchronized with it. Hence, this case can be reduced to the case where some distributed timer is progressing and some is suspended.

- If τ_0^a is *synchronized* with some $\tau_{\iota_0}^b \in Q^b$, this case can be reduced to the previous one by swapping τ_0^b and $\tau_{\iota_0}^b$ in the partition of the sets V^b and Q^b.

Disabling. At the firing of t_0, multiple transitions may be disabled and a new time vector is derived by repeatedly eliminating the time-to-fire of each of them according to the following step. Let t_d be a disabled transition and $\langle\tau_{age}^c,\underline{\tau}^c\rangle$ be the random variable yielded by the elimination of its time-to-fire τ_d^b. Three cases are distinguished depending on whether τ_d^b is deterministic, or distributed and synchronized with some element of V^b, or distributed and not synchronized with any element of V^b.

- If τ_d^b is deterministic or synchronized, it can be removed without changing $\underline{\tau}_Q^b$ and $f_{\langle\tau_{age}^b,\underline{\tau}_Q^b\rangle}$, i.e., $\underline{\tau}_Q^c = \underline{\tau}_Q^b$ and $f_{\langle\tau_{age}^c,\underline{\tau}_Q^c\rangle} = f_{\langle\tau_{age}^b,\underline{\tau}_Q^b\rangle}$.
- If τ_d^b is distributed and synchronized with some $\tau_{\iota_1}^b \in V^b$, this case can be reduced to the previous one by swapping τ_d^b and $\tau_{\iota_1}^b$ in the partition of the sets Q^b and V^b.
- If τ_d^b is distributed but not synchronized with any element of V^b, then its elimination reduces by one element $\underline{\tau}_Q^b$. Let τ_d^b occupy the first position in $\underline{\tau}_Q^b$, i.e., $\tau_d^b=\tau_1^b$. Its elimination yields a new random variable $\langle\tau_{age}^c,\underline{\tau}^c\rangle = \langle\tau_{age}^b,\tau_2^b,\dots,\tau_{M-1}^b\rangle$ ranging within $D^c = D^b\downarrow_{\tau_1}$ and $f_{\langle\tau_{age}^c,\underline{\tau}_Q^c\rangle}$ is derived by marginalizing $f_{\langle\tau_{age}^b,\underline{\tau}_Q^b\rangle}$ with respect to τ_1^b:

$$f_{\langle\tau_{age}^c,\underline{\tau}^c\rangle}(x_{age},\underline{x}) = \int_{Min^1(x_{age},\underline{x})}^{Max^1(x_{age},\underline{x})} f_{\langle\tau_{age}^b,\underline{\tau}^b\rangle}(x_{age},x_1,\underline{x})dx_1 \quad (10)$$

where $\underline{x}=\langle x_2,\dots,x_{N-1}\rangle$.

As in [16, 15], this step may partition $D^c\downarrow_{U,V}$ into a finite set of DBM subdomains over which $f_{\langle\tau_{age}^c,\underline{\tau}^c\rangle}$ accepts a piece-wise representation.

Newly enabling. At the firing of t_0, multiple transitions may be newly-enabled and a new time vector is derived by adding the time-to-fire of each of them according to the following step. Let t_N be a newly-enabled transition with static PDF $f_N(x_N)$ supported over $[EFT_N, LFT_N]$. The time vector is extended with its time-to-fire τ_N^d, yielding a new random variable $\langle\tau_{age}^d,\underline{\tau}^d\rangle = \langle\tau_{age}^d,\tau_2^d,\dots,\tau_{M-1}^d,\tau_N^d\rangle$ ranging over $D' = D^c\times[EFT_N,LFT_N]$. In particular, if t_N has point-like firing interval (i.e., $EFT_N = LFT_N$), then τ_N^d is added to U' without changing $\underline{\tau}_Q^c$ and $f_{\langle\tau_{age}^c,\underline{\tau}_Q^c\rangle}$, i.e., $\underline{\tau}_Q^d=\underline{\tau}_Q^c$ and $f_{\langle\tau_{age}^d,\underline{\tau}_Q^d\rangle} = f_{\langle\tau_{age}^c,\underline{\tau}_Q^c\rangle}$. Otherwise, if t_N has distributed firing interval (i.e., $EFT_N \neq LFT_N$), then τ_N^d is added to Q' and the joint PDF of τ_{age} and $\underline{\tau}_Q^d$ is obtained as:

$$f_{\langle\tau_{age}^d,\underline{\tau}_Q^d\rangle}(x_{age},\underline{x},x_N) = f_{\langle\tau_{age}^c,\underline{\tau}_Q^c\rangle}(x_{age},\underline{x})\cdot f_N(x_N), \quad (11)$$

where $\underline{x}=\langle x_2,\dots,x_{N-1}\rangle$.

3.2.4 Approximation of transient classes

If the model does not include transitions that can be suspended, the support of the time vector $\langle \tau_{age}, \underline{\tau} \rangle$ can be encoded as a Different Bounds Matrix (DBM) zone [43, 5, 24], i.e., the set of solutions of a system of linear inequalities constraining the difference between the times-to-fire of any two enabled transitions (including τ_{age}, which is regarded as an always-persistent time-to-fire, and a fictitious time-to-fire τ_*, which denotes the ground time of the class): $\tau_i - \tau_j \leq b_{ij}\ \forall\, i \neq j \in \{*, age, 0, 1, ..., N-1\}$ and $b_{ij} \in \mathbb{R} \cup \{\infty\}$. For instance, we consider the example of Fig. 2, where all transitions are assumed to be uniformly distributed over their supports. In the initial class, transitions t_{10}, t_{20}, t_{30},

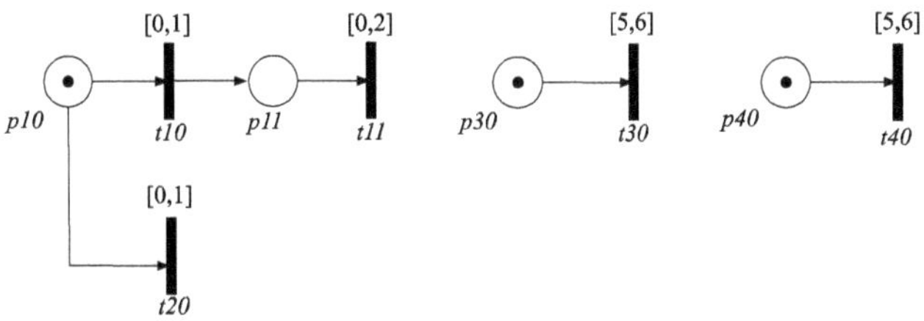

Figure 2: An example with uniformly distributed transitions.

and t_{40} are newly enabled and thus independently distributed. According to this, in the initial stochastic class Σ_0, the marking is $m_0 = p_{10}\ p_{30}\ p_{40}$ and the time vector $\langle \tau_{10}, \tau_{20}, \tau_{30}, \tau_{40} \rangle$ is distributed according to $f^1(\tau_{10}, \tau_{20}, \tau_{30}, \tau_{40}) = 1$ over D^0:

$$D^0 = \begin{cases} 0 \leq \tau_{10} - \tau_* \leq 1 \\ 0 \leq \tau_{20} - \tau_* \leq 1 \\ 5 \leq \tau_{30} - \tau_* \leq 6 \\ 5 \leq \tau_{40} - \tau_* \leq 6 \end{cases}$$

In the initial class, τ_{age} is a deterministic timer (equal to 0) and thus it is not included in the support D^0 which refers to distributed timers, as it will better emerge in Sect. 3.2.5.

The DBM form is closed with respect to the steps of derivation of successor classes, enabling their efficient computation and encoding in polynomial time $O((N+1)^2)$ with respect to the cardinality $(N+1)$ of the time vector $\langle \tau_{age}, \underline{\tau} \rangle$. Moreover, the joint PDF of τ_{age} and $\underline{\tau}$ accepts a piece-wise representation over a partition of the support into DBM sub-domains and its symbolic form can be derived through an efficient closed-form calculus [16] if all transitions of the model have expolynomial distribution. For instance, in the example of Fig. 2, after the subsequent firings of t_{10} and t_{11}, a stochastic class Σ_2 is reached with marking $m_2 = p_{30}p_{40}$. In Σ_2, t_{30} and t_{40} are both persistent while t_{10}, t_{11}, and t_{20} are all disabled. The time vector $\underline{\tau}' = \langle \tau'_{age}, \tau'_{30}, \tau'_{40} \rangle$ turns out to be distributed according to a piece-wise density function f_2 over D^2:

$$D^2 = \begin{cases} -3 \leq \tau'_{age} - \tau'_* \leq 0 \\ 2 \leq \tau'_{30} - \tau'_* \leq 6 \\ 2 \leq \tau'_{40} - \tau'_* \leq 6 \\ -1 \leq \tau'_{30} - \tau'_{40} \leq 1 \\ 5 \leq \tau'_{30} - \tau'_{age} \leq 6 \\ 5 \leq \tau'_{40} - \tau'_{age} \leq 6 \end{cases}$$

$$f^2(\tau'_{30}, \tau'_{40}, \tau'_{age}) = \begin{cases} 0.5 & \text{if } \underline{\tau}' \in D_1^2 \\ -\tau'_{age} - 0.5\,(\tau'_{age})^2 & \text{if } \underline{\tau}' \in D_2^2 \\ 4.5 + 3\,\tau'_{age} + 0.5\,(\tau'_{age})^2 & \text{if } \underline{\tau}' \in D_3^2 \end{cases}$$

with $D_1^2 = D^2 \cap \{-2 \leq \tau'_{age} \leq -1, 3 \leq \tau'_{30} \leq 5, 3 \leq \tau'_{40} \leq 5\}$, $D_2^2 = D^2 \cap \{-1 \leq \tau'_{age} \leq -0, 4 \leq \tau'_{30} \leq 6, 4 \leq \tau'_{40} \leq 6\}$, and, $D_3^2 = D^2 \cap \{-3 \leq \tau'_{age} \leq -2, 2 \leq \tau'_{30} \leq 4, 2 \leq \tau'_{40} \leq 4\}$. As an example, Fig. 3 shows the (DBM) projection of D^2 on the space of timers τ'_{30} and τ'_{40}:

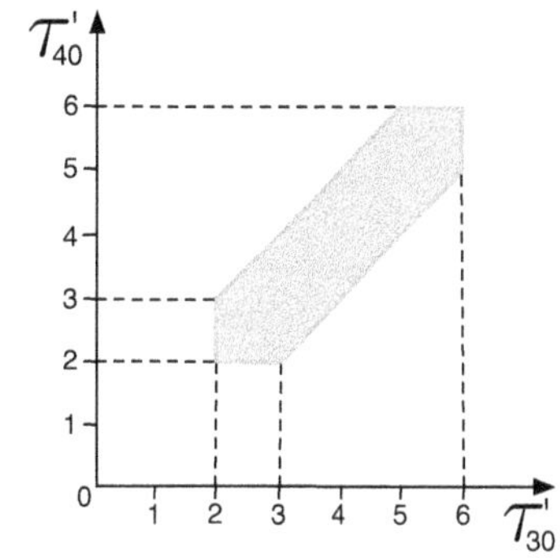

Figure 3: The joint support of $\langle \tau'_{30}, \tau'_{40} \rangle$ of the model of Fig. 2 after the subsequent firings of t_{10} and t_{11}.

Conversely, if the model includes transitions that can be suspended, the support of the time vector $\langle \tau_{age}, \underline{\tau} \rangle$ breaks the DBM structure and takes the form of a linear convex polyhedron [11, 6]. With reference to the example of Fig. 2, we consider the case when t_{30} and t_{40} require a resource res with priority level 1 and 2, respectively. At the firing of t_{10}, transitions t_{30} is progressing while transition t_{40} is suspended, reaching a stochastic class Σ_1 with marking $m_1 = p_{11}\ p_{30}\ p_{40}$ and time vector $\langle \tau'_{age}, \tau'_{11}, \tau'_{30}, \tau'_{40} \rangle$ distributed over the polyhedral support D^1:

$$D^1 = \left\{ \begin{array}{ll} -1 \leq \tau'_{age} \leq 0 & 0 \leq \tau'_{20} \leq 1 \\ 4 \leq \tau'_{30} \leq 6 & 5 \leq \tau'_{40} \leq 6 \\ -6 \leq \tau'_{20} - \tau'_{30} \leq -4 & -7 \leq \tau'_{20} - \tau'_{40} \leq -4 \\ 0 \leq \tau'_{20} - \tau'_{age} \leq 1 & -2 \leq \tau'_{30} - \tau'_{40} \leq 1 \\ 5 \leq \tau'_{30} - \tau'_{age} \leq 6 & -7 \leq \tau'_{age} - \tau'_{40} \leq -5 \\ & \\ -7 \leq -\tau'_{40} - \tau'_{20} \leq -4 & (*) \\ -12 \leq -\tau'_{40} - \tau'_{30} \leq -9 & (*) \\ -6 \leq -\tau'_{40} - \tau'_{age} \leq -4 & (*) \\ 9 \leq -\tau'_{20} + \tau'_{30} + \tau'_{40} \leq 12 & (*) \\ -1 \leq \tau'_{20} - \tau'_{30} + \tau'_{40} \leq 2 & (*) \\ 5 \leq -\tau'_{age} + \tau'_{20} + \tau'_{40} \leq 7 & (*) \\ 4 \leq \tau'_{age} - \tau'_{20} + \tau'_{40} \leq 6 & (*) \\ 10 \leq -\tau'_{age} + \tau'_{30} + \tau'_{40} \leq 12 & (*) \\ -1 \leq \tau'_{age} - \tau'_{30} + \tau'_{40} \leq 1 & (*) \end{array} \right.$$

Specifically, the constraints denoted by $(*)$ are not linear inequalities constraining the difference between two timers, but they rather involve more than two timers or the sum of two timers. This actually makes D^1 break the form of a DBM zone.

In principle, polyhedral domains could be managed through the Parma Polyhedra Library [3], which is currently used by several applications in the field of analysis and verification of hardware and software systems. In the practice, manipulation of polyhedral domains would lead to exponential complexity for derivation and encoding of successor classes, impairing efficient enumeration of the state-space. In [11], in a non-deterministic perspective, the problem of polyhedral domains is circumvented by resorting to an over-approximation that replaces supports with their tightest embedding DBM zones. The analysis technique of [11] also supports the exact identification of feasible timings of selected paths through an algorithm that cleans up false behaviors produced by the approximation, enabling efficient verification of reachability properties under timing constraints and real-time deadlines. Porting the approximation approach of [11] from a non-deterministic to a stochastic perspective comprises a more difficult challenge. In fact, the support takes the form of a linear convex polyhedron and the joint PDF of the time vector $\langle \tau_{age}, \underline{\tau} \rangle$ assumes a global analytic form over a partition of the support into sub-domains that are no longer DBM zones but rather polyhedra themselves. The presence of polyhedral constraints also increases the complexity of the analytic form of PDFs. In fact, in each subdomain, the lower and the upper bound of a timer can be not only a constant or a monovariate linear function of a single component of the time vector, but also a bivariate linear function of two components of the time vector. This takes relevance in the subsequent integrations performed to enumerate successor classes.

We address both aspects of the problem by relying on a twofold approximation of timing domains and density functions. On the one hand, we replace polyhedral supports with their tightest embedding DBM as in the non-deterministic approach of [11]. This adds false behaviors but still preserves all timings that would be feasible for the exact representation of domains. On the other hand, we approximate piece-wise density functions through multivariate Bernstein polynomials [34, 38], extending the approximation technique that was devised and experimented in [16] for steady-state analysis of non-preemptive models. Bernstein approximants can be derived in a straightforward manner by weighting a kernel of multivariate monomials according to the samples of the approximated density functions taken over a regular grid. As a matter of fact, replacing piece-wise density functions with global approximants actually perturbs the distribution of feasible behaviors and also associates false behaviors with a non-null probability. In positive, the probability of false behaviors can be reduced by assigning null value to the samples that belong to the tightest embedding DBM but not to the polyhedron.

In principle, both the limits deriving from the approximation of domains and densities could be overcome by working on the complexity of the Bernstein approximant. In fact, as the number of samples goes to infinity and those samples that belong to the extended DBM support but not to the exact polyhedron are assigned zero value, the Bernstein approximant uniformly converges to the approximated function with an error bounded by a Lipschitz inequality [34], provided that the approximated function is continuous (which is the case of our density functions). Of course, the number of samples used in the approximation is finite. Actually, Bernstein polynomials turn out to be very good at providing a rough approximation of density functions with the benefit of simplicity of derivation, much less in converging to the approximated function with tight accuracy. Yet, other approximants could be considered.

3.2.5 *On the effect of suspension on DET transitions*

In stochastic analysis of non-preemptive models [44, 30], multiple timers $\{\tau_i\}_{i \in I}$ may have a deterministic delay Δ with respect to a distributed timer τ_k and we call them *synchronized* with τ_k: $\tau_i = \tau_k + \Delta \quad \forall\, i \in I$, i.e., $\tau_i - \tau_k = \Delta,\ \forall\, i \in I$. For instance, consider the case when a timer τ_1 takes values over the interval $[0, 3]$ and a timer τ_2 is bound to have a deterministic delay $\Delta = 5$ with respect to τ_1, thus varying within $[5, 8]$ (for illustration purposes, the variable τ_{age} is not considered in the example). The domain D_{12} of the time vector $\langle \tau_1, \tau_2 \rangle$ has the following form:

$$D_{12} = \begin{cases} 0 \leq \tau_1 - \tau_* \leq 3 \\ 5 \leq \tau_2 - \tau_* \leq 8 \\ 5 \leq \tau_2 - \tau_1 \leq 5 \end{cases}$$

where all constraints are in DBM form. In particular, the constraint $\tau_2 - \tau_1 = 5$ is expressed by the couple of linear inequalities $\tau_1 - \tau_2 \leq -5$ and $\tau_2 - \tau_1 \leq 5$. Fig. 4 illustrates domain D_{12}, making evident that it has null measure due to the presence of the synchronized timer τ_2.

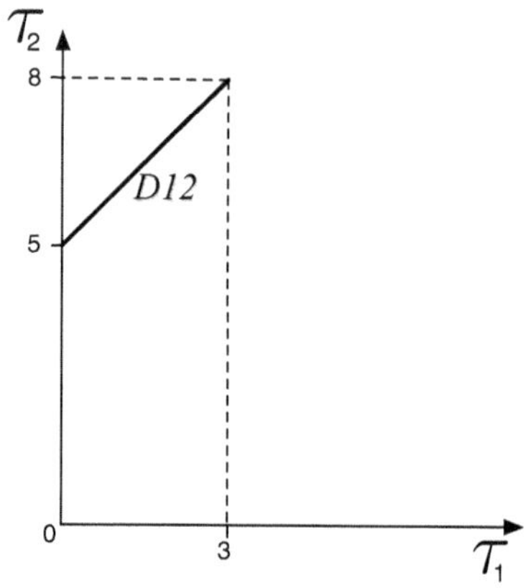

Figure 4: A bivariate DBM domain D_{12} with dimension $||D_{21}|| = 0$ as $\tau_2 - \tau_1$ is bound to assume a deterministic value.

Synchronized timers arise when a set of DET transitions, enabled in the same class or in classes with a deterministic distance between their entering times, persist at the firing of a distributed transition. In the specific application context addressed in this paper, this occurs for transition accounting for periodic task releases, which are simultaneously enabled at each hyper-period and enabled within a deterministic time gap at any other task activation. This condition is illustrated by the model of Fig. 5, which includes two deterministic transitions t_1 and t_2 accounting for the periodic releases of two tasks with period of 5 time units and 10 time units, respectively. Both t_1 and t_2 are enabled at each multiple of 10 time units, and t_1 is also enabled 5 time units after each time it is enabled with t_2. The model also includes a distributed t_3 with support $[2, 5]$, accounting for the release time of a sporadic task with minimum and maximum inter-arrival time equal to 2 and 5, respectively.

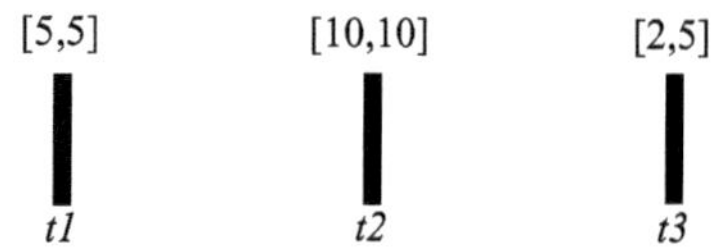

Figure 5: A trivial model with two deterministic transitions t_1 and t_2 and a uniformly distributed transition t_3.

When multiple DET transitions persist at the firing of a distributed transition, one of these DET transitions becomes distributed while the other ones are *synchronized* with it. This is the case that occurs in the example of Fig. 5 when the deterministic transitions t_1 and t_2 persist at the firing of the distributed transition t_3. In this example, in the initial class, each transition takes values within its static firing interval and the domain D_{123} of the time vector $\langle \tau_1, \tau_2, \tau_3 \rangle$ has the following form:

$$D_{123} = \begin{cases} 5 \leq \tau_1 - \tau_* \leq 5 \\ 10 \leq \tau_2 - \tau_* \leq 10 \\ 2 \leq \tau_3 - \tau_* \leq 5 \end{cases}$$

At the firing of t_3, the times-to-fire of t_1 and t_2 are reduced by the time-to-fire of t_3 while t_3 is regarded as newly enabled, leading to a new class where τ_1 is distributed over the support $[5, 8]$, τ_2 is bound to have a deterministic delay equal to 5 with respect to τ_1, and τ_3 is distributed over the interval $[2, 4]$. According to this, the time domain of the class reached through the firing of t_3 is D'_{123}:

$$D'_{123} = \begin{cases} 0 \le \tau_1 - \tau_* \le 3 \\ 5 \le \tau_2 - \tau_* \le 8 \\ 5 \le \tau_2 - \tau_1 \le 5 \\ 2 \le \tau_3 - \tau_* \le 5 \end{cases}$$

which is actually equal to domain D_{12} shown in Fig. 4 plus the inequalities that constrain τ_3 within its static firing interval $[0, 2]$, i.e., $D'_{123} = D_{12} \cup \{2 \le \tau_3 - \tau_* \le 5\}$. It is worth noting that, when a synchronized timer persists at the firing of a distributed timer, then it either becomes deterministic (if it is synchronized with the firing transition), or it remains synchronized with the same distributed transition with the same deterministic delay (otherwise). This maintains synchronization constraints in DBM form across subsequent derivations of successor classes.

As shown by the example of Fig. 4, when a class includes DET timers and/or synchronized timers, its time domain has null measure. The problem is circumvented by restraining the stochastic characterization to a minimal set of timers that are sufficient to determine all feasible values for variables of the time domain. This basically consists in maintaining the joint PDF of distributed timers and τ_{age}, while avoiding a stochastic characterization of deterministic and synchronized timers. The solution also reduces the number of managed variables, limiting the complexity of encoding and computation of classes.

Unfortunately, when the solution technique encompasses suspension and resumption of timers, the way how deterministically dependent timers synchronize with each other becomes much more complicated to manage. In fact, it may be the case that multiple timers become synchronized according to a constraint that breaks the limit of a DBM equality and takes the form of a linear convex polyhedral equality. Specifically, the unfavorable case occurs when multiple DET or synchronized timers persist at the firing of a distributed timer and at least on of them is *i) progressing* and synchronized with a *suspended* (distributed) timer or *ii) suspended* and synchronized with a *progressing* distributed timer that is not coincident with the time-to-fire of the firing transition. For instance, consider the model of Fig. 6, which extends the example of Fig. 5 by chaining transition t_3 with a deterministic transition t_4 requiring a resource res with priority level 1 and with a transition t_5 distributed over $[0, 3]$. Resource res is also required by transition t_2 with (lower) priority level 2. In the initial class, the

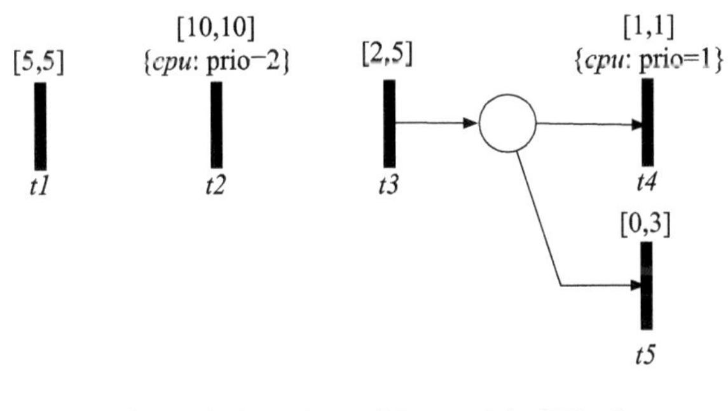

Figure 6: A variant of the model of Fig. 5.

enabled transitions are all progressing (as res is requested only by t_2) and the vector $\langle \tau_1, \tau_2, \tau_3 \rangle$ is supported over D_{123}. When t_3 fires: t_1 and t_2 are persistent and progressing and their times-to-fire are reduced by that of t_3, which is regarded as newly enabled; t_4 and t_5 are also enabled. According to this, in the arrival class, the time vector $\langle \tau_1, \tau_2, \tau_3, \tau_4, \tau_5 \rangle$ is distributed over $D_{1234} = D'_{123} \cup \{1 \le \tau_4 \le 1, 0 \le \tau_5 \le 3\}$. Specifically, in this class: τ_1, τ_3, and τ_5 are progressing distributed timers, τ_2 is a suspended timer synchronized with τ_1, and τ_4 is a deterministic progressing timer. At the subsequent firing of t_5 (which is disabled in the class reached through its firing), the values of τ_1, τ_3, and τ_4 are reduced by the value of τ_1, while τ_2 is not changed, yielding a new time vector $\langle \tau'_1, \tau'_2, \tau'_3, \tau_4 \rangle = \langle \tau_1 - \tau_5, \tau_2, \tau_3 - \tau_5, \tau'_4 - \tau_5 \rangle$. According to this, τ'_2 remains equal to $\tau_1 + 5$. By subsequently applying the substitutions $\tau_1 = \tau'_1 + 5$ and $\tau_5 = \tau_4 - \tau'_4 = 1 - \tau'_4$, we obtain that $\tau'_2 = \tau'_1 - \tau'_4 + 6$, which is not an equality in DBM form but rather a linear convex polyhedral equality. The derivation of the exact form of these synchronizations requires exponential complexity in the number of synchronized timers and would impair efficient enumeration of the state space. Note that these polyhedral constraints are additional with respect to those polyhedral constraints that may involve distributed timers, which are well managed through the approximation of both domains and density functions as sketched in Sect. 3.2.4.

In the context of a model-driven approach to the development of real-time SW components, a timer may arise from a timeout, from the inter-release time of a periodic or a sporadic task, from a deterministic offset or a non-deterministic jitter between subsequent task releases, or from the execution time of a computation. In the practice, DET timers mainly represent temporal parameters that are never suspended (i.e., timeouts, task periods, and release offsets), while distributed timers account both for temporal parameters that are never suspended (i.e., inter-release times of asynchronous tasks and release jitters) and for temporal parameters that can be suspended (i.e., the execution time of a computation). Whereas DET temporal parameters representing the execution time of computations that can be suspended are often considered in the literature, they actually correspond to an approximation abstraction that replaces the actual execution time with its Worst Case Execution Time (WCET). In fact, in the practice, the code of a chunk computation is usually characterized by multiple execution paths depending on the input values, which results in a non-pointlike spectrum of possible execution times. According to this, without loss of generality, we can safely assume that DET transitions are never suspended, i.e., DET transitions do not require resources (or require resources with higher priority level than that of any other transition requiring at least one of those resources). Thus, DET transitions are always progressing in every class where they are enabled. As a direct consequence, any synchronized timer is always progressing and the distributed timer that it is synchronized with is also progressing. In the theoretical perspective, this rules out the case of polyhedral constraints among synchronized timers, maintaining synchronization constraints in the form of DBM equalities which can be efficiently managed as in [44, 30]. In the applicative perspective, this permits to use DET transitions to model timeouts, delays, and synchronous release times, but not computation chunks, which nevertheless does not reduce the expressive capabilities of the approach with respect to the models at hand.

Note that the applicability of the proposed solution technique could be lifted to encompass various other special cases, e.g., a single possibly suspended DET transition (which mainly accounts for a preemptable computation with DET execution time), multiple DET transitions that are bound to be all progressing or all suspended in each class (which primarily model concurrent computations with DET execution time that can undergo preemption only simultaneously), and multiple DET transitions requiring some resources with higher priority level than that of any other transition requiring at least one of those resources (such DET transitions cannot be suspended by construction and represent concurrent computations with a DET execution time running at maximum priority).

4. COMPUTATIONAL EXPERIENCE

We illustrate the potentials of the proposed solution technique in the evaluation of real-time task-sets. The approach supports the derivation of transient probabilities of reachable states of concurrency, which comprise all the possible combinations of task computations that are concurrently ready, running, or blocked. As an example of probabilistic measures that can be evaluated, we derive the transient probability that a task holds a semaphore and the transient probability that a task misses a deadline. To this end, the task-set purposely includes a trivial case of priority inversion, which causes a job of the high-priority task to be released before the previous job is completed. Priority inversion could be avoided through many techniques of program structuring, for instance by applying a priority ceiling emulation protocol [41]. Accuracy of the approximated analysis is assessed by comparing analytic transient probabilities against simulation results. Both analysis and simulation were performed through the Sirio Framework [17, 19] on an Intel Core i7 laptop processor.

4.1 Three concurrent synchronized tasks

Fig. 7 shows the representation of a non-deterministic task-set introduced in [11], which is here extended with a stochastic characterization that associates each transition with a distribution (times are expressed in ms). The task-set includes a high-priority periodic task Tsk_1 with period of 50 ms, a mid-priority sporadic task Tsk_2 with an inter-arrival time falling within $[100, 300]$ ms, and a low-priority periodic task Tsk_3 with period of 150 ms. Moreover, Tsk_1 and Tsk_3 are synchronized on a binary semaphore named $mutex$. To fulfill the assumption that deterministic timers are always progressing, transitions t_{11} and t_{31}, which were IMM in the original model of [11], are here associated with a non-pointlike firing interval. Note that this change increases not only the complexity of the model, but also its validity, as semaphore operations are actually time-consuming. With respect to [11], the task-set complexity is also increased by reducing the maximum inter-release time of Tsk_2 from $+\infty$ to 300 ms. Moreover, the support of the computations (i.e., t_{12}, t_{21}, and t_{32}) is enlarged to increase the probability of a deadline miss of Tsk_1 and observe this behavior also in simulation, obtaining a ground truth for analytical results.

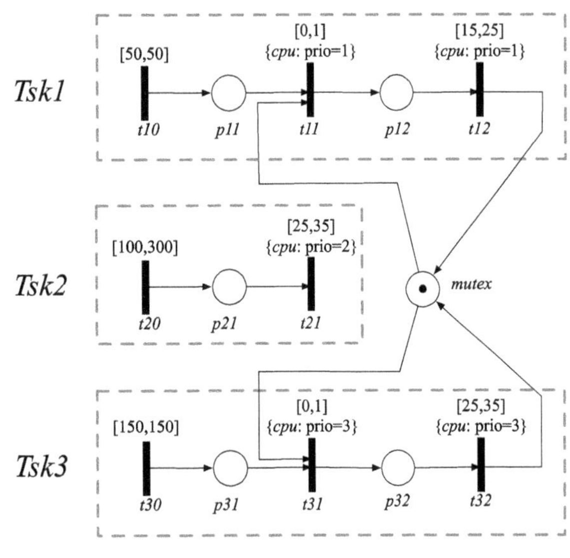

Figure 7: The model of a task-set with three concurrent tasks and a binary semaphore.

In the model of Fig 7, t_{10} and t_{30}, representing periodic task releases of Tsk_1 and Tsk_2, are associated with Dirac delta distributions centered at 50 and 150 ms, respectively. We assume that the remaining transitions have uniform distribution over their firing intervals, i.e., t_{20}, representing sporadic releases of Tsk_2; t_{12}, t_{21} and t_{32}, accounting for the execution times of computations; t_{11} and t_{31}, modeling semaphore acquisition operations.

4.2 Experimental results

Transient analysis of the task-set of Fig. 7 is repeated for values 3, 5, and 7 of the maximum polynomial degree of each variable in the multivariate Bernstein approximation (degree for short), adopting an error threshold of $\epsilon = 0.01$ on the total unallocated probability within the time limit of 300 ms and evaluating transient probabilities at multiples of 0.5 ms. The enumeration yields 938, 935 and 851 transient classes for degree 3, 5, and 7 in 3, 8, and 97 minutes, respectively, while the simulation of 2, 000, 000 runs took about an hour. For increasing values of the degree, the decreasing number of transient classes can be ascribed to the increasing number of samples of the approximated PDFs that belong to the tightest embedding DBM but not to the polyhedron. As these samples are assigned null probability, false behaviors introduced by the approximation of supports have lower probability and, thus, the number of transient classes decreases.

The complexity of the analysis is impacted by the number of transient classes. This depends on the number feasible events within the time bound, which is determined by: the number of concurrent tasks (especially sporadic tasks), the ratio between the minimum temporal parameter and the hyper-period of the task-set, and the mean time-to-fire duration [11]. The complexity of the analysis is also impacted by the complexity of density functions, which is affected by different factors for approximated and exact transient classes. In the former case, it depends on the polynomial complexity of Bernstein approximants, determined by the number of samples. In the latter case, it depends on the complexity of domain partitioning and expolynomial functions [16]. For the task-set of Fig. 7, a significant number of transient classes include preemptive behaviors, and thus the computation time increases with the degree.

As an example, Fig. 8 plots the transient probability that the semaphore $mutex$ is held by Tsk_1, which corresponds to the probability of any marking where place p_{12} contains a non-null number of tokens (and, by construction, place $mutex$ contains no tokens). The analysis follows the trend of the simulation with sufficient accuracy, but the error becomes more relevant for behaviors undergoing repeated approximations of domains and densities.

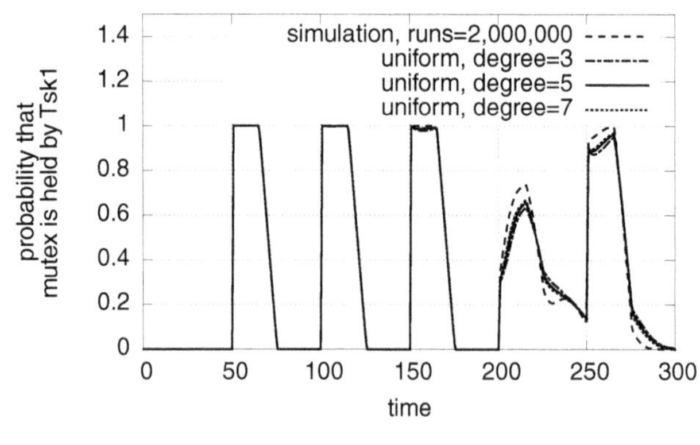

Figure 8: Probability that $mutex$ is held by Tsk_1 for the task-set of Fig. 7 with uniformly distributed execution times.

Transient analysis also evidences traces where the high priority task Tsk_1 undergoes a priority inversion with respect to Tsk_3. This occurs when Tsk_1 is blocked on semaphore $mutex$ (acquired and not yet released by Tsk_3) and a job of Tsk_2 preempts Tsk_3, leading Tsk_1 to miss its deadline. Fig. 9 shows the transient probability that two jobs of Tsk_1 are active, i.e., Tsk_1 is in a marking m such that $m(p_{11}) + m(p_{12}) > 1$. This condition is caused by a deadline miss of Tsk_1, which is deterministically bound to occur at time 250 ms. The probability associated with the peak at 250 ms (0.11 in the simulation and nearly 0.14 in the analysis) corresponds to the probability of having a deadline miss for Tsk_1 before 300 ms. After the deadline miss, the task-set recovers from the condition of having two active jobs of Tsk_1, reaching a marking m' such that $m'(p_{11}) + m'(p_{12}) \leq 1$ before the subsequent release of Tsk_1 at 300 ms.

In this case, the approximation affects the transient reward in a more substantial manner, by prolonging the maximum time required for the recovery, which is 270 ms in the simulation and 290 ms according to the analysis. This is a combined effect of the approximation of supports and density functions, which results in false behaviors with non-null probability that lower the probability of feasible behaviors. As a future direction, the error introduced by the extension of domains could be evaluated on a per-class basis and employed in the computation of an error bound on the analysis results, thus extending the cleanup procedure of [11] in a probabilistic perspective.

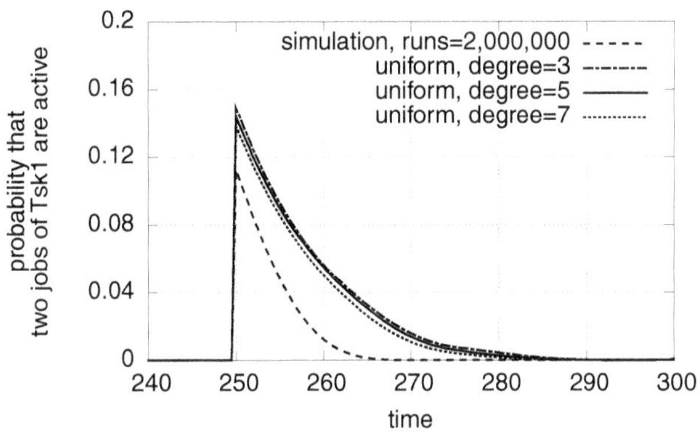

Figure 9: Probability that a deadline of Tsk_1 is missed for the task-set of Fig. 7 with uniformly distributed execution times.

5. CONCLUSIONS

We proposed a novel solution technique for transient analysis of real-time systems having non-Markovian temporal parameters and running under fixed-priority preemptive scheduling. The approach extends [30] so as to encompass suspension and resumption of timers taking values over a non-deterministic support, which enables the representation of concurrent computations with an execution time constrained between a minimum and a maximum value. As a notable trait, the approach addresses models that may also include deterministic timers that are bound to be never suspended, which is crucial to represent key features of real-time programming such as timeouts, offsets, and periodic task releases.

In the theoretical perspective, the approach faces major challenges due to the presence of progressing and suspended non-Markovian timers. This exacerbates the complexity of the analysis, yielding time vectors distributed over polyhedral supports according to probability density functions that assume a piece-wise analytical form over a partition of the support in polyhedral subdomains. The issue is circumvented through the approximation of both domains and density functions, by replacing the exact distributions with global Bernstein approximants supported over the tightest embedding DBM zones of polyhedral domains. It is worth stressing that the proposed solution technique relies on the approximation of joint multivariate PDFs in the state-space rather than on the approximation of monovariate PDFs in the model structure. This opens the way to the analysis of other classes of models that break the structure of DBM time domains, e.g., hybrid systems and, notably, multi-rate systems.

In the applicative perspective, the approach is amenable to smooth integration within a MDD methodology that was proven to be viable in a non-deterministic perspective [18, 25], providing decisive support for schedulability analysis and performance engineering of real-time systems while avoiding to disrupt consolidated practices of design and documentation. Specifically, since the very initial steps of software design, the approach permits to derive a quantitative formal specification of the task-set where unknown distributions of temporal parameters are tentatively guessed by analogy with previous realizations. This supports rapid exploration of the design space and early performance validation of design assumptions, long before the model is actually implemented on a real-time system. During the iterations of the development process, the temporal distributions are refined according to the results of the execution time profiling technique implemented in [18, 25], by deriving expolynomial distributions that fit the measured histograms. As a significant aspect, this comprises the main effort on the part of the developer to manage a stochastic characterization of timers.

6. REFERENCES

[1] R. Alur and D. L. Dill. A theory of timed automata. *Theor. Comput. Sci.*, 126(2):183–235, Apr. 1994.

[2] R. Alur, I. Lee, and O. Sokolsky. Compositional refinement for hierarchical hybrid systems. In *Hybrid Systems: Computation and Control, LNCS 2034*, pages 33–48. Springer–Verlag, 2001.

[3] R. Bagnara, P. M. Hill, and E. Zaffanella. The Parma Polyhedra Library: Toward a complete set of numerical abstractions for the analysis and verification of hardware and software systems. *Science of Computer Programming*, 72(1-2):3–21, 2008.

[4] G. Behrmann, A. David, and K. Larsen. A tutorial on UPPAAL. *Formal methods for the design of real-time systems*, pages 33–35, 2004.

[5] B. Berthomieu and M. Diaz. Modeling and Verification of Time Dependent Systems Using Time Petri Nets. *IEEE Trans. on SW Eng.*, 17(3):259–273, March 1991.

[6] B. Berthomieu, D. Lime, O. H. Roux, and F. Vernadat. Reachability Problems and Abstract State Spaces for Time Petri Nets with Stopwatches. *LAAS Report 04483*, 2004.

[7] A. Bobbio, A. Puliafito, and M. Telek. A modelling framework to implement preemption policies in non-Markovian SPNs. *IEEE Trans. on SW Eng.*, 26(1), January 2000.

[8] A. Bobbio, A. Puliafito, M. Telek, and K. S. Trivedi. Recent Developments in Non-Markovian Stochastic Petri Nets. *Journal of Circuits, Systems, and Computers*, pages 119–158, 1998.

[9] A. Bobbio and M. Telek. Transient Analysis of a Preemptive Resume M/D/1/2/2 through Petri Nets. *Periodica Politechnica*, 41:123–146, September 1997.

[10] G. Bucci, L. Carnevali, L. Ridi, and E. Vicario. Oris: a tool for modeling, verification and evaluation of real-time systems. *Int. Journal of Software Tools for Technology Transfer*, 12(5):391 – 403, 2010.

[11] G. Bucci, A. Fedeli, L. Sassoli, and E. Vicario. Timed state space analysis of real time preemptive systems. *IEEE Trans. SW Eng.*, 30(2):97–111, Feb. 2004.

[12] G. Bucci, L. Sassoli, and E. Vicario. Correctness verification and performance analysis of real time systems using stochastic preemptive time petri nets. *IEEE Trans. on SW Eng.*, 31(11):913–927, November 2005.

[13] G. Buttazzo. *Hard Real-Time Computing Systems*. Springer, 2005.

[14] BWB - Federal Office for Military Technology and Procurement of Germany. *V-Model 97, Lifecycle Process Model-Developing Standard for IT Systems of the Federal Republic of Germany. General Directive No. 250*, June 1997.

[15] L. Carnevali, J. Giuntini, and E. Vicario. A symbolic approach to quantitative analysis of preemptive real-time systems with non-markovian temporal parameters. In *VALUETOOLS*, May 2011.

[16] L. Carnevali, L. Grassi, and E. Vicario. State-Density Functions over DBM Domains in the Analysis of Non-Markovian Models. *IEEE Trans. on SW Eng.*, 35(2):178–194, 2009.

[17] L. Carnevali, L. Ridi, and E. Vicario. A framework for simulation and symbolic state space analysis of non-markovian models. In *SAFECOMP*, pages 409–422, 2011.

[18] L. Carnevali, L. Ridi, and E. Vicario. Putting preemptive Time Petri Nets to work in a V-Model SW life cycle. *IEEE Trans. on SW Engineering*, 37(6), Nov./Dec. 2011.

[19] L. Carnevali, L. Ridi, and E. Vicario. Sirio: A framework for simulation and symbolic state space analysis of non-Markovian models. In *QEST'11*, pages 153–154, 2011.

[20] F. Cassez, A. David, E. Fleury, K. G. Larsen, and D. Lime. Efficient on-the-fly algorithms for the analysis of timed games. In *Proc. Int. Conf. on Concurrency Theory*, pages 66–80. Springer-Verlag, 2005.

[21] F. Cassez and K. G. Larsen. The Impressive Power of Stopwatches. In *Proceedings of the 11th International Conference on Concurrency Theory*, CONCUR '00, pages 138–152, London, UK, UK, 2000. Springer-Verlag.

[22] G. Ciardo, A. Blakemore, P. Chimento, J. Muppala, and K. Trivedi. Automated generation and analysis of Markov reward models using Stochastic Reward Nets. *IMA Volumes in Mathematics and its Applications*, 48:145–191, 1993.

[23] A. Cumani. ESP: A Package for the Evaluation of Stochastic Petri Nets with Phase-Type Distributed Transition Times. In *PNPM*, pages 144–151, 1985.

[24] D. Dill. Timing assumptions and verification of finite-state concurrent systems. *Proc. Workshop on Computer Aided Verification Methods for Finite State Systems*, 1989.

[25] J. B. Dugan. Galileo: A tool for dynamic fault tree analysis. In *Computer Performance Evaluation / TOOLS*, pages 328–331, 2000.

[26] European Committee for Electrotechnical Standardization. *CENELEC EN 50128 Railway applications - Communications, signalling and processing systems - Software for railway control and protection systems*, July 2009.

[27] European Cooperation for Space Standardization. *ECSS-Q-ST-80C Space product assurance - Software product assurance*, March 2009.

[28] T. A. Henzinger, B. Horowitz, and C. M. Kirsch. Giotto: A time-triggered language for embedded programming. In *Proc. of the IEEE*, pages 84–99. IEEE, 2003.

[29] A. Hessel, K. Larsen, M. Mikucionis, B. Nielsen, P. Pettersson, and A. Skou. Testing Real-Time systems using UPPAAL. In *Formal Methods and Testing*, pages 77–117. 2008.

[30] A. Horváth, M. Paolieri, L. Ridi, and E. Vicario. Transient analysis of non-Markovian models using stochastic state classes. *Performance Evaluation*, 2012.

[31] A. Horvath, A. Puliafito, M. Scarpa, and M. Telek. Analysis and Evaluation of non-Markovian Stochastic Petri Nets. In *Proc. of the* 11th *Int. Conf. Computer Performance Evaluation (TOOLS)*, pages 171–187, 2000.

[32] G. Karsai, J. Sztipanovits, A. Ledeczi, and T. Bapty. Model-integrated development of embedded software. *Proc. of the IEEE*, 91:145–164, Jan. 2003.

[33] F. Longo and M. Scarpa. Applying Symbolic Techniques to the Representation of Non-Markovian Models with Continuous PH Distributions. In *Proc. European Perf. Eng. Workshop on Computer Perf. Eng.*, EPEW '09, pages 44–58, Berlin, Heidelberg, 2009. Springer-Verlag.

[34] G. Lorentz. *Bernstein Polynomials*. University of Toronto Press, 1953.

[35] The Mathworks. *Simulink*. www.mathworks.com/products/simulink.

[36] Radio Technical Commission for Aeronautics. *DO-178B, Software Considerations in Airborne Systems and Equipment Certification*, 1992.

[37] O. H. Roux and D. Lime. Time Petri Nets with inhibitor hyperarcs: formal semantics and state-space computation. 25th *Int. Conf. on Theory and Application of Petri Nets*, 3099:371–390, 2004.

[38] T. Sauer. Multivariate Bernstein polynomials and convexity. *Computer Aided Geometric Design*, 8(6):465 – 478, 1991.

[39] M. Scarpa and A.Bobbio. Kronecker Representation of Stochastic Petri Nets With Discrete PH Distribution. In *IPDS Proceedings*, pages 52–62, 1998.

[40] D. C. Schmidt. Model–driven engineering. *IEEE Computer*, pages 1–2, February 2006.

[41] L. Sha, R. Rajkumar, and J. P. Lehoczky. Priority inheritance protocols: An approach to real-time synchronization. *IEEE Trans. Comput.*, 39(9):1175–1185, 1990.

[42] US Department of Defense. MIL-STD-498: Military standard for software development and documentation. Technical report, 1994.

[43] E. Vicario. Static analysis and dynamic steering of time dependent systems using Time Petri Nets. *IEEE Trans. on SW Eng.*, 27(1):728–748, August 2001.

[44] E. Vicario, L. Sassoli, and L. Carnevali. Using stochastic state classes in quantitative evaluation of dense-time reactive systems. *IEEE Trans. SW Eng.*, 35(5):703–719, 2009.

[45] R. Zijal, G. Ciardo, and G. Hommel. Discrete Deterministic and Stochastic Petri Nets. In *In Proc. of Measurement, Modeling, and Valuation of Computer and Communication Systems*, pages 103–117, 1997.

Scalability Testing of MS Lync Services: Towards Optimal Provisioning of Virtualised Hardware

Knut Helge Rygg
IDI, NTNU
Trondheim, Norway
khsrygg@gmail.com

Gunnar Brataas
SINTEF ICT and IDI, NTNU
Trondheim, Norway
Gunnar.Brataas@sintef.no

Geir Millstein
Telenor GID
Fornebu, Oslo, Norway
Geir.Millstein@telenor.com

Terje Molle
Telenor GID
Fornebu, Oslo, Norway
Terje.Molle@telenor.com

ABSTRACT

A method for scalability testing of the Microsoft Lync 2010 communication system is presented, exploring the relation between system size and system load. The method can be used for optimal provisioning, balancing user Quality of Experience (QoE) with equipment volume and energy consumption. Observing a standard edition of Lync, on a virtualised platform using the VMware hypervisor, the method indicated linear scalability. QoE was mainly limited by the Mean Opinion Score (MOS). This MOS limit corresponded to a Lync front end server utilisation of about 60%.

Categories and Subject Descriptors

K.6.2 [**Computing Milieux**]: Performance and Usage Management; H.1.0 [**Information Systems**]: Models and Principles, General

Keywords

Scalability, testing, virtualisation, provisioning, energy efficiency, MS Lync, workload, performance, data centres, telecom industry

1. INTRODUCTION

The objective of this paper is to describe a method for scalability testing of the Microsoft Lync 2010 communication system. In this context, scalability describes the relation between system load and required system resources. When we know the required system load, our method can find the required amount of system resources. Although the word "scalability" is often used, it still has no formal definition [8]. Therefore, methods for scalability testing have not received much attention.

ICPE'13, April 21–24, 2013, Prague, Czech Republic.

Scalability testing may be used as part of optimising provisioning processes. With optimal provisioning, we strike a balance between under- and over-provisioning of computer resources:

- With resource *under-provisioning*, Quality of Experience (QoE) will degrade, or we may even have outages.
- *Over-provisioning* gives low utilisation as well as high equipment, operational and energy costs.

In the ideal case, energy consumption should be proportional to the computer system utilisation [6], but until we reach this ideal, optimal provisioning is important. Even in a virtualised environment, it is beneficial to have provisioning guidance, which helps buying the right type and amount of hardware, by the right time.

This work was performed for Telenor Norway, the largest Norwegian telecom operator and part of the Telenor international operations. Microsoft Lync can be configured in various setups with different services and user scenarios. This paper focuses on *hosted Lync* as it is provided by Telenor. This means that Telenor hosts Lync servers on behalf of customers who do not have the resources to manage their own deployment. Telenor's requirements should also be representative for industry requirements in general.

We have studied the operation of Lync in a virtualised environment, using the VMware hypervisor. As a workload generator, we used the Lync Server 2010 Stress and Performance Tool [3].

This work was performed as a master's thesis [16], which imposed limitations in terms of time and available equipment. These limitations are also typical in industry.

This paper describes related work in Section 2. The scalability baseline is outlined in Section 3. Our method is described in Section 4 and applied to Lync in Section 5. In Section 6, we offer conclusions and outline further work.

2. RELATED WORK

In this description of related work we focus on the issue of Lync provisioning. The Lync Planning Guide by Microsoft [13] is based on physical servers and is crude for virtualisation. The scenario-based capacity planning spreadsheet [15] (also from Microsoft) is a more fine-grained analysis of Lync capacity requirements. The system designer can specify both the number of users and a custom user model by varying a number of different parameters. The output is a recommendation on the number of servers to use and their respective CPU utilisation, memory usage, etc. The calculations are still based on the same standard configuration hardware described in the capacity planning guide and the percentage of external users is hard coded to 30%. This does not fit with Telenor's hosted scenario where all users are external. The resource demand on the edge server will be higher than the calculator falsely indicates. Even though the capacity calculator allows more flexibility, it is not sufficient to provide good estimates for Telenor's deployments.

Although provisioning of Microsoft Lync is described in books explaining Microsoft Lync deployment [20, 11], they only reference Microsoft's own capacity planning guidelines. Therefore, there is little help for those planning to use custom hardware or specialised scenarios.

Bandwidth usage in Microsoft Office Communication Server (OCS) is described in [17]. OCS is a predecessor of Microsoft Lync with much of the same functionality, except for telephone integration. The overall server load in terms of network traffic can be estimated from the connected user endpoints and their respective traffic patterns. Peak network utilisation can be quite different from the average, so using mean values may be inaccurate for bandwidth provisioning. Network usage is proportional to the number of users. However, the report gives little information on provisioning other hardware resources like CPU, memory, or storage.

A sizing study of Microsoft Lync [18] describes a test environment built by Dell PowerEdge R720 servers using LyncPerfTool. The report finds that the number of users scales linearly to the system size in the range of 3,000 to 12,000 users. Compared with the Telenor deployment studied in this report, the differences are:

1. Dell's sizing study uses the enterprise edition of Lync, which allows several servers to be assigned to the same pool. The number of front end server virtual machines are 1, 2, and 4. For each configuration, the maximum number of users are found. Our paper uses the standard edition of Lync which does not support pools.

2. Dell's sizing study uses virtual machines for the front end servers, while the remaining servers are installed on physical machines. In this paper, all Lync servers and databases are virtualised.

3. Dell's sizing study has only internal users. Internal users do not require edge or reverse proxy servers. However, this paper is based on a hosted Lync scenario which has only external users. The edge server and reverse proxy server roles are therefore included in this paper.

3. SCALABILITY BASELINE

This section describes basic scalability concepts which are prerequisites for the method described in Section 4. We build on the scalability framework outlined by Brataas & Hughes et al. [7]. We first describe *workload*. Then we describe *performance* and finally *scalability*.

3.1 Workload = Work + Load

In our framework, we separate workload into work and load. *Work* is about what is done and is determined by the type and nature of the invoked services. A *system* consists of connected components. Each *component* in the system has one or more *services*, which are different types of work it offers to other components. *Load* describes how often the services are performed, and is measured in terms of throughput, i.e., the number of finished tasks/jobs during a given time interval, but can also be the number of users in a closed system. Work and load together constitute workload.

3.2 Performance

The performance of a system may describe the system quality as a function of system load as depicted in Figure 1. With higher load the quality degrade. Implicitly in this figure, work is kept constant. A common measure of system quality is response time. With increasing load, response times typically increase too. The capacity of the system is the load when a defined operating point is reached. This operating point may also be termed service level agreement (SLA) or quality limit.

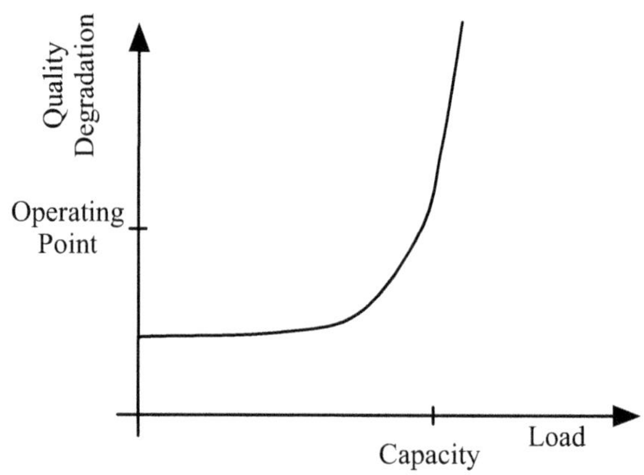

Figure 1: Performance of a system.

3.3 Scalability

System size is normally constant when measuring the performance of a system, but not when we describe the scalability of a system. Scalability is a relation between system size and system capacity, given fixed system quality and work, as described in Figure 2. If system capacity is proportional to system size and the quality of the system is constant, then we have linear scaling. This is not the case in Figure 2, where we clearly see that system capacity reach an upper limit. Further increase in system size actually leads to decreasing capacity.

The concept of range is also important. A system may be linearly scalable within a certain system size range. For example, there may be approaches with large scalability problems which do not matter if these problems only occur in ranges not used.

In our framework, system size has three dimensions, processing speed, storage amount, and connectivity, but in this work, we focus only on processing speed. System size is also hierarchic, and we can typically define the system level, the

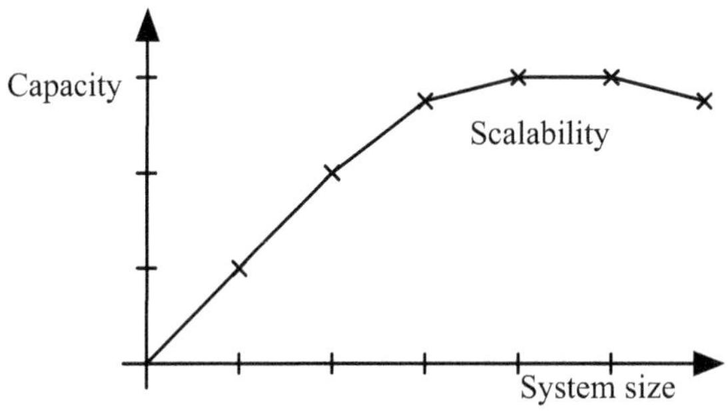

Figure 2: Scalability of a system.

subsystem level, and the device level. On all of these levels, size can be increased by replication. Replication at the subsystem level (adding more servers for each server role like the front end server) is typically called *scale out*, while *scale up* refers to replication at the device level (adding more CPUs, disks, etc.)

4. METHOD

The empirical scalability testing method for finding optimal provisioning builds on the scalability concepts in Section 3. The objective of the empirical tests is to determine the maximum number of users for each system size and user workload scenario. A number of quality metrics are used to assure that the quality degradation of the Lync services are within acceptable limits. The maximum number of users can be found by methods of systematic trial and error and adjusting the number of users for each run. The number of users is increased if quality metrics are within limits, and decreased if the limits are exceeded.

One way of reducing the search space is to use binary search. For each iteration of the binary search, the search space is halved. Interpolation on quality metrics could also be considered, but since there are no guarantees on linearity in the metrics, a binary search is a more robust alternative. The binary search can theoretically continue until the search space has only one remaining user. However, searching with such fine granularity would probably not produce stable results because of the random errors in all measures. Such errors are more dominant in fine granularity tests. Running many iterations also requires more time for testing. In these tests 4 - 6 iterations were used for all system sizes until an acceptable value was found.

The following numbering refers to Section 5, where the method is applied to the Lync test case.

5.1: Understand environment The method relies on a test bed, enabling the use and validation of the method, as well as introducing some limitations and constraints. The test bed has three parts: We start on the bottom with hardware and continue with software and load generation:

- **5.1.1: (Test bed) hardware** where we also outline the virtualisation platform.
- **5.1.2: (Test bed) software** which in our case describes the Lync server roles used.
- **5.1.3: (Test bed) load generation** used to generate test bed workload.

5.2: Identify requirements for scalability testing. There are several types of requirements concerning system size scaling path, work, and quality metrics:

- **5.2.1: System size scaling path** Explore the system size scaling options available and select an enumerated set of these options. Questions to be answered include: Where can we add resources in our configuration so that we can increase the size of the system? What are the practical resource ranges/limits? Which discrete set of resource quantities should be selected for further investigations?
- **5.2.2: Determine work** Which types of services are relevant? Work is further described in Section 3.1.
- **5.2.3: Explore the quality metrics** most applicable for our purposes. In our case, there are many, such as MOS (Mean Opinion Score), CPU utilisation, RTT (Round Trip Time), jitter, and packet loss. The task here is simply to get an overview of all of them.

5.3: Run tests to find the capacity for each system size as outlined in Section 3.2 and Section 3.3. In our case this step involves:

- **5.3.1: Tune configuration** e.g., allocate appropriate number of vCPUs to each of the Lync server roles.
- **5.3.2: Determine run length** For how long do we need to run each experiment?
- **5.3.3: Run experiments** using binary search for each system size.
- **5.3.4: Decide on quality metric** determining if a given load (capacity) is within our quality limits.
- **5.3.5: Find capacity for each system size**

5.4: Analyse results consisting of the following steps:

- **5.4.1: Find bottleneck resource** This gives insight into the system by finding the bottleneck resource, which is useful when we are upgrading the hardware. Monitoring the bottleneck resource is a good way to keep an eye on QoE.
- **5.4.2: Analyse the scalability** i.e., the relation between capacity and system size. Is it linear? In detail, this step involves finding the capacity for each system size using the same operating point. We use binary search to find the capacity for each system size.
- **5.4.3: Extrapolation** Is it possible to extrapolate beyond measured system size range?

Although this method appears to be linear (where we follow the steps in natural order), some degree of iteration (backtracking) will be required. For example, to determine the run length in step 5.3.2, we needed to know the number of IM (Instant Messaging) conference users, which was determined by the next step 5.3.3, running the experiments. In step 5.3.1, we first tried using the default value of 40 GB disk for the SQL server, but then the disk ran full in step 5.3.3, and we had to adjust the disk size to 100 GB.

5. APPLYING THE METHOD

In this section, we apply the method from Section 4 and describe the challenges and results for each step. In Section 5.5, we comment on the amount of human time required to perform these steps.

5.1 Understand Environment

This step focuses on understanding the test bed, which consists of virtualised hardware, Lync software, and a load generator.

5.1.1 Hardware & Virtualisation

In this section, the platform is described on a detailed level. Many of the details will not be essential for the upper layers of the test bed and the further steps of the method.

The hosted Lync deployment is built using components from Cisco Unified Computing System [5] (UCS) architecture. The architecture consists of blade servers that are connected to a Cisco UCS 5108 blade server chassis. A blade server has 96 GB RAM, two Intel Xeon E5650 CPUs, each with 6 cores and hyper-threading, and two 15k rpm SAS disk drives. Each blade server chassis is connected to a 10 Gbps fabric interconnect using a Cisco UCS 2104xp fabric extender. A Cisco UCS 6140 fabric interconnect can be connected to a SAN to provide increased storage for the blade servers. The architecture supports up to 40 blade server chassis on the same fabric interconnect. Each blade server chassis can take up to 8 blade servers. This allows for 320 blade servers in the same infrastructure. A central RAID60 SAN is connected to the fabric interconnect. The SAN is divided into disk groups consisting of 20 disks.

Intel Xeon has 9 discrete power states that can be selected in order to lower processor speed and power usage. The VMware ESXi hypervisor 5.0 has four different host power management policies to choose from. We used a high performance policy where the processor always runs in the highest processor state.

The VMware hypervisor is used as a virtualisation platform. Each blade server (also called host) is running separate instances of ESXi 5.0. VMware vCenter is used to manage all ESXi hosts. Several hosts can be organised into resource pools containing one or more virtual machines for each host. VMs can move from host to host within the resource pool and automatic monitoring can be enabled to maintain resource balancing among hosts. VMs are then automatically migrated from a saturated host to a new host with enough available resources. Virtualisation can therefore provide flexibility along with better hardware utilisation.

Virtualisation has some advantages compared to bare machine performance. VMware ESXi provides a proprietary transparent page-sharing technique that detects identical memory pages used by several VMs. This proves useful when several VMs use the same OS, shared libraries, etc. Pages containing shared libraries and static OS components are normally locked to read-only mode and several VMs can therefore safely access them and avoid keeping their local copy. This allows for memory overcommitment where memory entitlement is larger than available memory.

However, virtualisation can also lead to performance degradation:

1. Even if hardware-assisted CPU virtualisation offers a guest mode, the processor must still exit guest mode and enter root mode, for example when updating page tables.

2. The VMkernel as well as the VMconsole requires CPU processing time.

3. When VMs use physical memory, a double lookup is needed. The OS has its own virtual memory which is mapped to (what the VM thinks is) physical memory. This must again be mapped to the actual machine memory. A shadow lookup table can be used to decrease lookup wait time, but at the cost of maintaining more lookup tables, which consumes RAM. Our Intel Xeon E5650 supports hardware-assisted memory virtualisation where two layers of page tables are supported. Still, the TLB miss latency is higher.

4. Both the VMkernel and the VMconsole, as well as each VM, need to allocate and use memory.

5. Most hypervisors allow virtual network components (hubs, switches, routers, etc.). These are emulated in software and require resources to run. The Cisco UCS [5] technology provides some virtual networking functionality that reduces the need for software emulation.

For more information on the performance impact of virtualisation, consult [9].

Even though this paper focuses on scalability in Microsoft Lync, studying the scalability of virtualised servers can prove useful in a broader perspective because the interest and usage of virtualisation is increasing.

5.1.2 Lync Server Roles

This section describes which server roles we focused on and why. In our experiments, a total of seven Lync servers were used, and together constitute what is called the central site.

- The *reverse proxy* allows external users to download content.
- The *edge server* allows external users to connect to internal servers.
- The (standard edition) *front end server* contains the back end server and the A/V (audio / video) server role.
- The *domain controller* is an active directory (AD) domain controller and Kerberos key distribution centre and authentication server (KDC/AS).
- The *monitoring server* collects data to monitor QoE.
- The *SQL server* is a database for the monitoring server.
- The *mediation server* enables telephone integration via a PSTN gateway.

In a *hosted Lync* deployment a service provider (like Telenor in this case) provides Lync functionality to customers who connect remotely. All servers are administered at a central site and the customers only need to install client software. This is a huge benefit for the customer and many small and mid-sized companies choose this solution when they start using Lync. Lync can be configured in a variety of ways, including other server roles and both internal and external

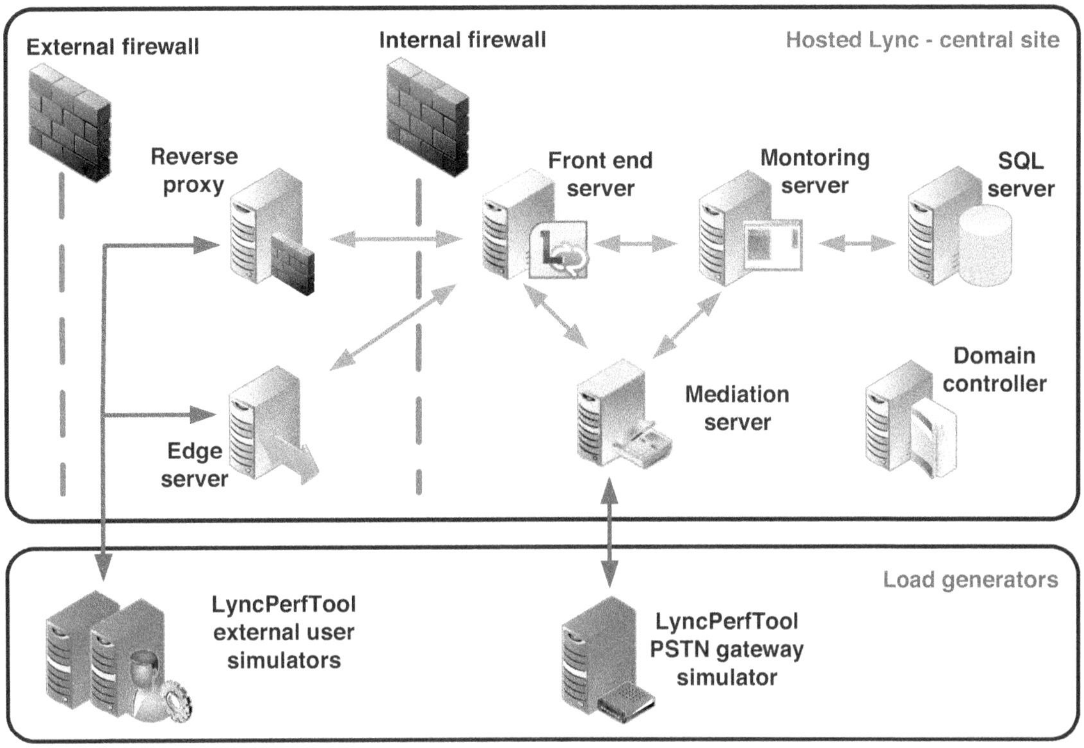

Figure 3: Reference architecture for hosted Lync deployment.

users, but the setup and workload in this test environment is typical for a hosted Lync deployment.

All of these seven servers are shown in Figure 3. The green arrows between the servers indicate which servers communicate with each other in order to process the workload imposed by users. The domain controller is not directly related to Lync, but more to Microsoft active directory. It communicates with all servers except the reverse proxy. The Kerberos part of the domain controller is essential for encrypted communication between servers, but the utilisation of the domain controller is hardly affected by the Lync workload (the total utilisation never exceeds 2%). Details on server-to-server communication can be found in [16].

All Lync servers are setup as VMs running on a blade server (Cisco UCS M200). The user simulators and the PSTN gateway simulator are running on two identical servers, but separate from the Lync server roles to avoid interfering with performance measures.

All VMs are provisioned with (more than) enough memory so it should be possible to keep a complete working copy of the VM image in-memory and occasionally write the changes back to the SAN. The local hard drives on each ESXi host contain swap memory which is used whenever the ESXi host runs out of memory. Esxtop [1] is a performance measurement tool that runs on each ESXi host. Results from this tool showed that swapping never occurred. Communication between the Lync servers is handled using a dedicated VLAN on the link to the fabric interconnect. The fabric interconnect uses 10.0 Gbit/s links and the latency is negligible. All Lync servers shared the same Microsoft Windows Server 2008 R2 operating system.

Other Lync roles like director and archiving were not used in our configuration. The director is an optional server role created to offload the internal servers by authenticating external user requests before sending them to other servers. This would give better performance if the director had dedicated hardware, but as long as the director and front end server share the same resources, there is no performance gain. The archiving server archives IM (Instant Messaging) content, but should only be used by companies that have legal compliance concerns.

In this test setup, user simulators, the PSTN gateway simulator, and the Lync server roles are all connected to the same LAN. In a real configuration, external users would be connected to the edge server and reverse proxy by an IPT connection (WAN). The telephone users would be connected by another IPT connection and a PSTN gateway. The PSTN gateway simulator used fewer resources than both of the two LyncPerfTool Clients, and was therefore put on the same physical server as one of them. As a result, we used three physical servers: one for all of the seven Lync servers, and the remaining two for the LyncPerfTool Client (with the PSTN gateway on one of the latter two servers). Firewalls are used to separate the internal Lync servers from the remaining network. External users can only access the edge server or reverse proxy server.

Our Cisco UCS platform used during measurement is similar to a production Cisco UCS platform with one small exception: The blade servers have a single network link to one fabric interconnect. In production, each blade server would have two links, each going to a different fabric interconnect for better redundancy. A single link in the test setup allowed a single point of monitoring, which was used for controlling the traffic among servers.

5.1.3 *Load Generation*

This section describes the load generation. Ideally, we would like to test each Lync service on its own and we would also like to characterise each Lync component individually to have a clear view of scalability. However, this would require developing a new load generator for Lync. It quickly became clear that this was not possible within the limited project resources. We therefore had to rely on the Lync Server 2010 Stress and Performance Tools [3] (LyncPerfTool).

The tests described in this paper used two Cisco UCS M200 blade servers for running the LyncPerfTool clients. The configuration is shown in Figure 3. There are three drawbacks with the test setup that may create bias in the performance measurements:

1. In a real deployment both clients and the PSTN gateway would be connected by an IPT connection, not by a LAN.
2. In a real deployment, users at a customer site may use wireless networks. Both IPT and a wireless network will introduce significantly more network delay and packet losses compared to a LAN.
3. In a real deployment, all of the network traffic can be (and normally is) encrypted. LyncPerfTool does not support encrypted protocols so in the test environment the traffic to and from the PSTN gateway simulator is unencrypted.

LyncPerfTool can simulate all of the main user workload types, except for these three: (1) group chat (2) video workload for peer-to-peer calls or conferencing, and (3) web conferencing. The group chat functionality enables multiple users to participate in conversations in which they post and access content about specific topics. Each session can be persistent. Enabling group chat requires installation of special group chat communications software and a back end database on separate servers. The group chat functionality is not a part of Telenor's standard hosted configuration. Therefore, the lack of support for this workload type in LyncPerfTool will not affect the measurements.

For handling video conferencing, there are three alternatives:

1. Direct address between the two Lync clients, which is the shortest and fastest option, if allowed by firewalls.
2. Connecting using NAT routers.
3. Relaying the video session through the edge server external network interface.

With only external users, the lack of video simulation in LyncPerfTool will only create a performance bias as long as we either use alternative 1 or 2, and not involve the edge server in alternative 3. In a hosted Lync environment, alternative 1 or 2 will be used. LyncPerfTool only supports web conferencing using the web access client, which does not allow voice or video in the conferences. However, both audio and video is simulated in dedicated audio/video conferences.

5.2 Identify Requirements

This section describes the requirements for the scalability testing. The requirements involve the system size scaling path, the work, and also the quality metrics.

Table 1: Scaling hardware resources

Size	CPU	Network	Memory	Storage
4/4	4/4	4/4	4/4	4/4
2/4	2/4	4/4	4/4	4/4
1/4	1/4*	4/4	4/4	4/4

* The level 3 shared cache is scaled 2/4

5.2.1 *System Size Scaling Path*

We now have to consider where we can increase the system size. Since we studied a Lync deployment installed with a standard edition license, it is not meaningful to add more than one physical server for the edge server role (normally termed *scale out*). However, it is possible to add more resources to the physical server (normally termed *scale up*). The maximum system size was limited by the resources available. In this test setup, only three Cisco UCS 200 M2 blade servers were available for running the virtual Lync servers and load generators.

The load generators required twice the hardware resources of the Lync servers, so only one blade server was used for virtual Lync servers. The system size was therefore scaled by reducing the resources available on the blade server. The CPU resources can be scaled by disabling processors or processor cores. As mentioned in Section 5.1.1 the blade servers use two Intel Xeon E5650 with 6 cores and hyper-threading support. This gives a total of 24 virtual cores that are all enumerated by the hypervisor.

By using CPU affinity setting in VMware vSphere client, it is possible to decide which virtual cores can be used to schedule each vCPU of a VM. For the 2/4 system size, the affinity setting for all VMs was set to 0-11. This effectively disabled one of the two processors. For the 1/4 system size, the affinity was set to 0-5, so that only half of the cores of one processor could be used.

However, while the number of cores is constrained, the processor caches are still widely available. The level 1 and 2 caches are separate for each core, so these scale according to the number of cores, but the level 3 cache is shared among all 6 cores of one processor. This means that for the 1/4 system size, the number of cores is correctly scaled, but the level 3 cache is scaled 2/4, so it is twice as big as it should be. This gives a bias towards better performance estimates because more instructions and data can be stored in the level 3 cache instead of being fetched from main memory, which has a larger access time.

Network resources were not scaled in this test setup. Each virtual machine has access to the 10.0 Gbit/s network interface of the blade server. Still, most of the communication between VMs is internal to the server. Only the edge server, reverse proxy, and mediation server need to use the network interface.

Memory resources or storage resources were not scaled. Table 1 summarises the hardware scaling.

5.2.2 *Determine Work*

As described in Section 5.1.3, Lync has several types of services or work, but as also described in this section, we had to use the LyncPerfTool load generator. However, LyncPerfTool allows both light and heavy work, reflecting different user types. We therefore used both types of work to make our results more robust.

5.2.3 Explore Quality Metrics

Quality metrics were collected from two distinct sources, the monitoring server and the LyncPerfTool clients. The limits in Table 2 are found in Microsoft's recommendations [4] and the Lync sizing study [18]. The SIP limits are LyncPerfTool counters and the rest is monitoring server statistics.

Table 2: Quality metric limits [4, 18]

Metric	Limit
Round trip time (RTT)	100 ms
Jitter	20 ms
Packet loss	0.1 per sec.
MOS degradation	0.5
SIP 503 messages/sec	≈ 0
SIP 504 messages/sec	≈ 0

The front end server and mediation server automatically sample all ongoing sessions and send CDRs to the monitoring server. The monitoring server produces statistics on audio quality using several metrics. Round trip time describes the time it takes from when a request is sent to when the answer is received. A high round trip time can be noticed by the user as a speech delay. Buffering enables acceptable speech quality even with a high round trip time. Jitter measures the variance in round trip time. A high jitter value makes buffering and processing hard because the packets arrive at highly variable times. Packet loss denotes the number of packets that are lost in transmission per second. The real time network protocols used in Lync have several built-in mechanisms for reconstructing the speech with acceptable quality despite some lost packets. However, when the packet loss exceeds 0.1, the packet loss affects call quality.

Mean opinion score (MOS) is a compound metric used to measure end user service quality of experience. MOS is a technically synthesised counterpart to the mean subjective rating of call quality on a scale from one (bad) to five (excellent) [10]. MOS is calculated using both transport layer parameters such as packet loss, jitter, and delay, as well as payload parameters such as audio codec, noise-level, echo, gain, and talk-over effects [12]. Each codec that is used in Lync has a maximum MOS value [2]. The monitoring server records the degradation of MOS value, i.e., the difference between the maximum MOS and measured MOS. The MOS degradation should not exceed the 0.5 limit. There are actually three MOS values, one for conference calls, another for peer-to-peer calls, and then a final one for PSTN calls. We are interested in the maximum degradation, and will therefore use the highest MOS degradation value.

SIP 503 is a reply message indicating that the Lync server is too busy to handle the request. A SIP 504 message is a server timeout message. LyncPerfTool counts the average number of such messages per second. Having some messages is acceptable, but over a longer period of time, the number should be close to zero.

A number of quality metrics could also have been collected from the front end -, edge - or mediation servers. This instrumentation would, however, affect the performance measures and are not included in the tests. The monitoring server reports were collected after each test run and will not affect performance. The metrics collected on the LyncPerfTool clients also did not create any measurement bias.

Table 3: VM resource allocation

			vCPU/syst. size		
VM	Disk	RAM	1/4	2/4	4/4
Front end	40 GB	32 GB	4	7	14
Edge	40 GB	16 GB	2	4	7
SQL	100 GB	32 GB	1	2	3
Mediation	40 GB	16 GB	1	1	2
Monitoring	40 GB	16 GB	1	1	1
Reverse proxy	40 GB	16 GB	1	1	1
Domain contr.	40 GB	16 GB	1	1	1
SUM	340 GB	144 GB	11	17	29

5.3 Run Tests

The overall aim of this step was to find the capacity for each system size in the scaling path. This step consists of several sub-steps described in detail below.

5.3.1 Tuning Virtual Resources

When deploying virtual machines (VMs), it is important to make sure each virtual machine has the resources it needs to operate well. An overview of resource allocations for each virtual machine (VM) is shown in Table 3. The disk drive size of most virtual machines is 40 GB, which is the default allocation size for Microsoft Server 2008 R2 installations. For the SQL server, 100 GB was required to handle all of the call detail records generated for each ongoing audio session. Memory is provisioned to each virtual machine based on the hardware requirements for similar physical machine configuration recommendations from Microsoft [14].

There are several allocation options for provisioning CPU resources. The VMs in the test setup were provisioned using *allocation shares*, in contrast to *allocation reservation* or *allocation limit*, because shares are more flexible, allowing the VMs to share unused resources. Moreover, using allocation shares was also the preferred allocation option in the production environment in Telenor. Using allocation shares, each VM had a normal allocation setting with 1000 shares per vCPU (in contrast to the high value of 2000 shares or the low value of 500 shares), so the actual CPU allocation was defined by the number of vCPUs allocated to each VM. It is worth noting that all of the VMs had the same allocation priority. Therefore, the priority of each VM was determined by the number of vCPUs only. The number of vCPUs that were allocated for each VM was calculated based on the utilisation of each VM, U_{VM}, and the number of available virtual processors U_{total}:

$$\text{vCPU}_{\text{VM}} = \lceil \frac{U_{\text{VM}}}{U_{\text{total}}} \times N_{\text{logproc}} \rceil$$

As shown in the previous section, the number of available logical processors, N_{logproc}, is 6 for 1/4 system size, 12 for 2/4 and 24 for 4/4. The $\lceil$ and $\rceil$ delimiters in the formula simply mean rounding up to the next whole integer. Each server is provisioned with its relative share of available vCPUs. The rounding up makes sure that no server gets less than its relative share, but this means that the total number of allocated vCPUs will be a bit larger than the available logical processors.

Theoretically, each VM could have allocated the maximum number of vCPUs. However, over-provisioning CPU resources may degrade performance. The VMware CPU scheduler uses an algorithm called relaxed co-scheduling [19]. Co-scheduling means that all vCPUs belonging to the same VM should be treated as a group. Therefore, the hypervisor tries to schedule all vCPUs in a group at the same time. Non-continuous scheduling of vCPUs of the VMs would create a different processor experience than on a physical processor. SMP programs that use several threads to complete a compute-intensive task may rely on synchronisation among threads. If some threads are halted because a vCPU is not scheduled, all other threads (and their respective vCPUs) will suffer a delay. The VMware hypervisor therefore implements a co-scheduling algorithm that tries to assign equal time intervals to each vCPU. If a certain vCPU has received more processing time than the other vCPUs in the same group, it is put into a waiting state called CO-WAIT. *Relaxed* co-scheduling means that a vCPU must succeed before a certain time limit in order to be put in the CO-WAIT state. More vCPUs per group increases the chances of one or more vCPUs being put into the CO-WAIT state. Spending time in the CO-WAIT state decreases performance. The vCPU calculation method described above will reduce the CO-WAIT overhead. This method is not limited to this deployment, but can prove useful in other multi-VM environments.

5.3.2 Determine Run Length

LyncPerfTool first signs in all users at a rate of 1 user per second. The different workload types are run as separate processes, so several users are signed in in parallel. The IM (Instant Messaging) conferencing workload process had the most users, with 1,587 IM users with light work out of a total of 4,800 users. All IM users are signed in after 26 min and 27 seconds. A transient interval of 30 min was therefore selected. Some of the tests could have used a shorter transient time, but standardised transient times made log parsing easier. We used a steady-state measurement period of 60 min, so that each test run took 90 minutes. The esxtool tool measured the performance of each VM in isolation from the hypervisor overhead.

5.3.3 Run Experiments

Experiments were run using the binary search protocol outlined in the Section 4. As described in Section 5.2.1, we decided to explore the system sizes 1/4, 2/4 and 4/4. Furthermore, we decided to explore both heavy and light work as described in Section 5.2.2. The complete results, which cover several pages, are presented in [16]. An overview of all the test runs is shown in Table 4 and in Table 5, for heavy and light work, respectively. In these tables, each test run has one row. The first column is the system size and the second column is the number of users. The third column is the % utilisation of the front end server divided by the allocated number of vCPUs, and the fourth column is the same for the edge server. We focus on the front end and edge server since they had the highest utilisation. The fifth column is the maximum MOS degradation.

The monitoring server measures MOS for (1) conferences, (2) peer-to-peer calls, and (3) PSTN calls. They all have the same limit of 0.5 MOS degradation. Therefore, we are only interested in the maximum of the three. From Table 4 and Table 5, we also see how MOS behaves with both lower and higher load than what gives the optimal MOS degradation.

We have only run four test runs for system size 1/4 with

Table 4: Test runs for heavy work.

Size	*# Users*	U_{front}	U_{edge}	*MOS*
1/4	200	34	28	0.10
	300	48	43	0.28
	350	58	52	0.50
	400	74	62	0.63
2/4	400	37	32	0.06
	600	53	44	0.13
	650	59	50	0.45
	700	65	56	0.64
	800	84	69	0.78
4/4	800	42	35	0.08
	1200	56	49	0.30
	1250	59	52	0.44
	1300	64	54	0.55
	1400	75	62	0.78
	1600	97	85	0.82

Table 5: Test runs for light work.

Size	*# Users*	U_{front}	U_{edge}	*MOS*
1/4	800	48	42	0.21
	1000	59	52	0.43
	1100	67	57	0.52
	1200	79	65	0.59
	1600	99	90	0.69
2/4	1200	39	32	0.07
	1800	53	46	0.24
	2000	59	51	0.47
	2200	68	56	0.59
	2400	80	63	0.78
4/4	2400	37	33	0.09
	3600	55	49	0.26
	3900	60	53	0.47
	4050	63	55	0.52
	4200	69	60	0.69
	4800	85	74	0.76

heavy work. Considering the MOS for 350 users we see that this value is exactly the limit of 0.50, and more runs are therefore not necessary.

5.3.4 Decide on Quality Metric

All quality metrics have different limits and units. In Figure 4, all parameters from the monitoring server are normalised by dividing the measured value by the limit. This means that when the curve crosses the horizontal bold line, the limit is reached. The results come from the 4/4 system size with heavy work, but the shape of the metrics curve is representative for all scenarios and system sizes.

The figure shows that the MOS metric is the first metric that crosses the limit. System administrators that monitor the system to detect and mitigate any performance issues want to have an early warning on system saturation. For this purpose, the MOS value is the best candidate to be monitored. The other metrics could also be monitored, but they would give a later warning.

Figure 4 does not include the performance counters from the LyncPerfTool clients. These counters were always 0 when the highest possible number of users were reached. Small deviations of up to 0.04 SIP error responses/sec were

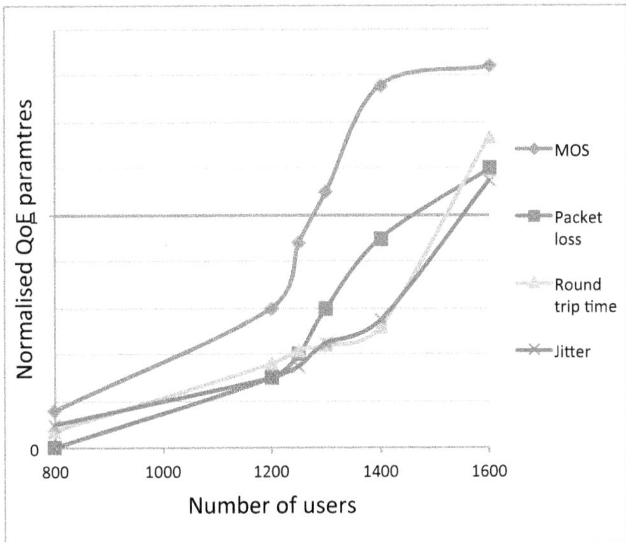

Figure 4: Normalised QoE parameters, for 4/4 system size with heavy work.

observed when the system was highly over-utilised. The SIP error messages are not considered a good metric for monitoring system saturation. The monitoring server metrics consider the real-time performance of ongoing voice sessions, etc. Real-time traffic is especially vulnerable to system contention and this may be the reason why the quality metrics gave an earlier warning compared to SIP error messages.

5.3.5 *Find Capacity per System Size*

In our case, this is already done since we know that MOS degradation is the limiting system quality, and since we then have this information from Table 4 and Table 5. We will analyse the resulting scalability more closely in Section 5.4.

5.4 Analyse Results

We are now in a position to analyse the results of the tests. We will first find the bottleneck resource and will then explore the scalability before we finally try to use the results for extrapolation.

5.4.1 *Find Bottlenecks*

Four resource types were considered in this study. Which of them was the bottleneck resource?

Memory All servers were provisioned with memory resources according to Microsoft's server recommendations. If any of the servers had too little memory, they would need to start swapping memory to disk, but memory swapping was always 0. Therefore, memory was not a bottleneck.

Network All servers had access to a 10.0 Gbit/s network interface. However, only the mediation server, edge server, and reverse proxy server used this interface. The highest network utilisation of all servers combined where 267 Mbit/s (for light work on the 4/4 system size). Not including hypervisor overhead, this gives a total utilisation of 2.67%, far below the utilisation of the processors. Network resources were therefore not a bottleneck in this Lync deployment.

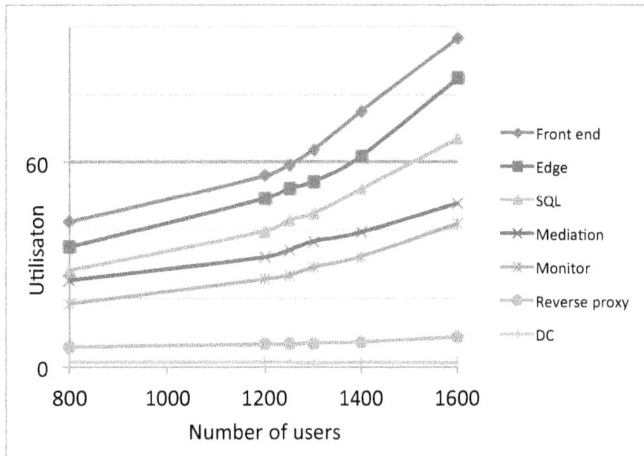

Figure 5: Normalised utilisation of all servers, for 4/4 system size with heavy work.

Disk The disk utilisation was evaluated by observing the time required to complete disk operations. As long as this time is kept below 15 ms, there is no contention on disk resources [18]. The highest observed disk service demand was 5.903 ms. This indicates that there was not a high enough utilisation of disk resources to create contention, and as a result disk resources were not a bottleneck.

CPU The utilisation of CPU resources closely correlates to system performance. Table 4 and Table 5 show that the utilisation of the front end server ranges between 58 – 60% for all maximum users measures. The processor is the highest utilised and should therefore be considered the bottleneck in this Lync deployment.

A closer look at the utilisation of each VM may give more insight into how the system behaves under different CPU loads. Figure 5 shows the normalised CPU utilisation values for all servers in the 4/4 system size with heavy work. From the figure, it is possible to observe a breakpoint in the front end server utilisation curve when the maximum number of users is met. Similar break points are also observable in other servers, such as the edge server. However, on the edge server, the break point comes after the maximum number of users is reached.

Looking at Table 4 and Table 5, we see that the utilisation of the front end server is always below 60% for all measurements where the quality metrics are within limits. For all other measurements, the utilisation is above 60%. It is therefore reasonable to say that monitoring the utilisation of the front end server would be a good indicator of system health. This measure could be used by system administrators. They could even set an alert at 50% utilisation just to get an early warning on possible performance issues and therefore have the time to mitigate the consequences.

5.4.2 *Explore Scalability*

Summarising the results from Table 4 and Table 5, gives two scalability curves for heavy and light work shown in Figure 6. From the table, we extract and plot the samples which approach the MOS degradation limit of 0.5 as close as possible without exceeding it.

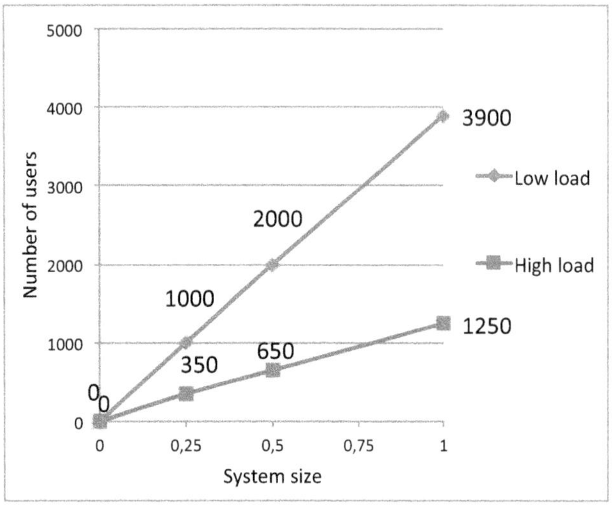

Figure 6: Measured scalability in Microsoft Lync

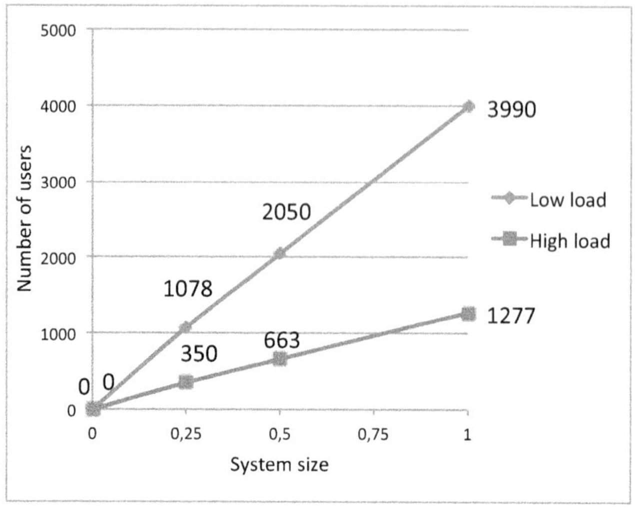

Figure 7: Interpolated scalability values

The results show a close-to-linear scalability. There are, however, some inaccuracies when measuring the maximum number of users:

- The binary search stops at a given number of iterations. This means that the unexplored search space may still be significant. Finding the "correct" value would require even more iterations, but was not feasible because of time constraints.
- Random measurement errors affect each test. This effect could be reduced by performing the same experiment several times and averaging the results. This would, however, require more time.
- Only the CPU resources are scaled, while memory, network, and storage remain the same for all system sizes. Even though the CPU has proved to be the bottleneck device, the improper scaling of other resources could create a bias towards higher number of users for 1/4 and 2/4 system size.
- The L3 cache in the 1/4 system size is scaled to 1/2 instead of 1/4. This may give a bias towards higher number of users because the processor can serve a relatively higher workload before experiencing contention because of the increased cache size. This is especially relevant since the processor is shown to be the bottleneck device of this Lync deployment.

The first inaccuracy in the above list can be reduced by using interpolation techniques. Section 4 describes why a binary search is preferred before an interpolation method when finding the maximum number of users. The quality parameters do not guarantee linearity. It is, however, safer to assume linearity over a smaller range, like the remaining search space after using the binary search for several iterations. This technique gives a more precise estimate for the maximum number of users. However, it is important to note that the MOS score is given in two digits only, so the round off error must be considered. The interpolation technique for finding the interpolated number of users ($N_{0.5}$) as well as lower and upper round off counters ($N^-_{0.5}$, $N^+_{0.5}$) is shown below

$$\begin{aligned}
N_{0.5} &= N_b + (N_a - N_b)\frac{(0.5 - MOS_b)}{(MOS_a - MOS_b)} \\
N^-_{0.5} &= N_b + (N_a - N_b)\frac{(0.5 - MOS_b - 0.005)}{(MOS_a - MOS_b + 0.01)} \\
N^+_{0.5} &= N_b + (N_a - N_b)\frac{(0.5 - MOS_b + 0.005)}{(MOS_a - MOS_b - 0.01)}
\end{aligned}$$

N_b and MOS_b denote the values measured *below* the MOS limit. N_a and MOS_a are *above* the limit. $N_{0.5}$ denotes the maximum number of users when the MOS value is interpolated to the limit value of 0.5.

The interpolation results are summarised in Table 6 and also shown in Figure 7. The figure shows the $N_{0.5}$ values only. The error bars ($N^-_{0.5}$ and $N^+_{0.5}$) are not included in the figure because they were too small to view.

The interpolated values for number of users still show close-to-linear scalability. There is, however, a slightly sub-linear tendency in the results. This tendency still exists when the round off error is considered. The improper scaling of resources (as explained in point 3 and 4 in Section5.4.2) should both give a systematic bias towards higher number of users for both 1/4 and 2/4 system size. However, it is hard to determine whether this bias alone accounts for the sub-linearity or whether the system really has a sub-linear scaling. Answering this question would require a different test setup where the bias was reduced or removed. An alternative test setup would be to run multiple identical Lync in-

Table 6: Scalability results with linear interpolation

Work	Heavy			Light		
Size	1/4	2/4	4/4	1/4	2/4	4/4
N_b	350	650	1250	1000	2000	3900
N_a	400	700	1300	1100	2200	4050
MOS_b	0.5	0.45	0.44	0.43	0.47	0.47
MOS_a	0.59	0.64	0.55	0.52	0.59	0.52
$N^-_{0.5}$	348	661	1273	1065	2038	3963
$N_{0.5}$	350	663	1277	1078	2050	3990
$N^+_{0.5}$	353	665	1283	1094	2064	4031

stances on the ESXi host instead of trying to scale down the system resources. This means that for the 1/4 system size scenario, a total of 4 identical Lync instances with the same workload would run at the same time. Then the available resources for each Lync instance would be 1/4. However, there would be more contention on resources. This kind of test setup with multiple Lync instances could not be tested because installing Lync instances was outsourced from Telenor. New Lync instances would have to be ordered and the resulting delay would postpone the master thesis work.

For practical concerns, such as when new Lync deployments are provisioned, the scalability in the 1/4 - 4/4 system size range can be considered linear. The sub-linear tendency that may exist in this system size range is too low to make an impact. This does not, however, prove that there are no scalability problems in Lync. It only shows that we have not found it in our system size range.

As shown in Table 1, the 1/4 system size was not correctly scaled. The amount of network, memory, and storage resources are not scaled at all. In addition, the level 3 cache is the same as in the 2/4 system size. A larger L3 cache will allow more instructions and data to be cached and increases performance on CPU intensive operations. As explained in Section 5.2.3, there are actually three MOS values, where the conference MOS value is highest in most scenarios, but not for some measurements for the 1/4 system size. Conferencing is a CPU intensive task (with considerable audio codec processing, in contrast to packet forwarding for peer-to-peer calls and PSTN calls). However, when the L3 cache is twice as big (compared to a "correct" scaling) the audio processing is better and other MOS values becomes the dominating factor.

5.4.3 *Extrapolation Beyond Size Range?*

Scalability testing can only be reliable within the system size range that was actually measured. In this paper the range is 1/4 - 4/4 physical servers. Using the Cisco UCS platform makes it quite straightforward to increase the number of blade servers and include them in the same virtual resource pool. For example, this would allow the system size to scale to 16/4 by including 4 blade servers. However, our standard edition license does not allow this.

To set up several servers in a pool, an enterprise edition license is needed. With the given constraint, the best configuration to achieve high performance is to put the front end server VM on a single physical server and allocate 24 vCPUs for it. Then the remaining VMs can be put on the other blade servers. However, the front end server VM requires more compute resources than the other VMs all together. Therefore, using more than two physical servers would not give significantly better performance, because the front end VM would have a much higher utilisation and become a bottleneck.

The Lync sizing study [18] shows that Lync scales linearly when the system size is varied from one to four physical front end servers. This sizing study uses the enterprise edition license and has a different user scenario than this report. However, the linear scalability conclusion from the sizing study supports the belief that the sub-linear scaling tendency observed in this report is actually caused by measurement bias alone, as discussed in section 5.4.2. Also the edge servers can be replicated on several servers.

Table 7: Approx. time per binary search iteration

Time	Task
5 min	Create user profile configuration file
5 min	Prepare parameter capture scripts on hosts
5 min	Copy configuration file to all 3 LyncPerfTool clients and start the LyncPerfTool clients
90 min	Run test (30 min transient + 60 min meas.)
10 min	Copy csv log file from hosts to home computer
5 min	Read QoE parameters from the monitoring server
10 min	Parse the results from the csv file
5 min	Compare parameters to limits and determine the correct direction for the next iteration of the binary search

5.5 Human Time Consumption

Considerable human time was required to set up the test system, tune the Lync configuration to work with LyncPerfTool, resolve issues with the performance measure tools, etc. This work would require less time for a person experienced with the specific technology who had direct access to the equipment instead of working remotely (as in our case). However, the time required to perform the actual tests would not have been affected by such concerns. The binary searches in this report were carried out in a total of 12 days. The work required for one iteration of the binary search is outlined in Table 7.

Table 7 shows that one iteration can be carried out in $\approx$ 2 hours 15 min. Configuring the tests and interpreting the results takes $\approx$ 45 min, half the time of actually running the test. Of course, this requires all steps to be carried out correctly. Since LyncPerfTool clients were situated both outside and inside the firewall (see Figure 3), an extra network interface was added to give external user simulators direct access to the internal network. This made file sharing possible without altering the standard reverse proxy configuration. This interface was for pre-test file sharing only. If the extra network interface was not disabled before the test was started, all sessions that were set up using ICE would short-cut the edge server and use the extra network interface instead. Then the performance results were biased towards better performance estimates and the whole iteration had to be run all over again with a disabled network interface. This happened twice during the test period and affected the total test time.

Automating the parameter capture could save a lot of manual work. It is possible to write scripts to start LyncPerfTool and to automate almost all parts of the capture process. However, there is one major and two minor issues that impede full automation. The major issue is that csv log files had to be copied from ESXi hosts using a Citrix Reciever GUI. Seemingly, there were no CLI options available. The minor issues were (1) creating new configuration files from UserProfileGenerator without using the GUI and (2) reading monitoring server results from a web interface. These minor issues could probably be overcome.

A person with experience in Lync setup can configure a Lync environment in 2 – 3 days. An additional 2 – 3 more days are required to install the LyncPerfTool and configure the Lync scenario accordingly. Applying the same method as described in this report, the results can be reproduced in 16 – 18 days. If a different management setup were used (excluding remote access through Citrix Receiver), the parameter capture using the binary search could probably be automated.

In our tests, we have employed binary searches to find the optimal QoE for each system size and load. Because we now know that the scalability was almost linear, we could have started this binary search algorithm with more optimal values, and not exploring very low or very high values for each system size and load. The saved time could have been used to get a better knowledge of the confidence intervals for the most important measurements.

6. CONCLUSION AND FURTHER WORK

This paper has studied the scalability of Microsoft Lync 2010 in a virtualised lab environment. The results could help to optimise the provisioning of computer resources. The lab servers where deployed on a Cisco UCS platform with one blade server. We have used both light and heavy work on the Lync Server 2010 Stress and Performance Tool. The results show that there is a close-to-linear scaling from 1/4 to 4/4 system size. A new deployment should be scaled according to the number of simultaneous end users.

When the Lync deployment is installed on a Cisco UCS platform, the CPU resources become the bottleneck. The hardware configuration of Cisco UCS is comparable to hardware from other vendors. This means that upgrading CPU resources is most important for increasing system capacity.

Among several QoE (Quality of Experience) metrics, mean opinion score (MOS) is the best metric to monitor reduction of service quality for end users. The results shows that this metric is the first to exceed limits as the system load increases. Another good performance measure is to monitor the utilisation on the front end server. When this utilisation exceeds 60%, service quality is degraded.

From the findings in this paper, there are several interesting problems that need further study:

- Test Lync with more heterogeneous work than what was possible with the LyncPerfTool, to make the conclusions more robust.
- Explore the scalability of Microsoft Lync using an enterprise edition deployment which allows several servers in the same server pool. This would allow more front end and edge servers (scaling out), instead of just improving one front end and edge server (scaling up), as shown in this paper.
- We need experience with other scalability testing scenarios to further improve and validate the method.
- The utilisation of this scalability method for optimal provisioning should be described more comprehensively.

7. ACKNOWLEDGEMENTS

This research was supported by Telenor ASA. We thank Georg Lothe and Alexander Botnen, both in Telenor Norway, for solving many problems during the measurements.

8. REFERENCES

[1] Interpreting esxtop. `http://communities.vmware.com/docs/DOC-9279`, Dec 2008. Last visited: 2013 Feb 5th.

[2] Mean opinion scores and metrics. `http://technet.microsoft.com/en-us/library/bb894481(v=office.12).aspx`, Jul 2010. Last visited: 2012 Jul 8th.

[3] Lync stress and performance tool. `http://www.microsoft.com/download/en/details.aspx?id=25005`, Mar 2011. Last visited: 2012 Jul 8th.

[4] Media quality summary report. `http://technet.microsoft.com/en-us/library/gg615012`, Jul 2011. Last visited: 2012 Jul 8th.

[5] Unified computing technology. `http://www.cisco.com/en/US/products/ps10265/technology.html`, Feb 2012. Last visited: 2012 Jul 8th.

[6] L. A. Barroso and U. Holzle. The Case for Energy-Proportional Computing. *IEEE Computer*, 40(12):33 – 37, 2007.

[7] G. Brataas, P. H. Hughes, J.-A. Fagerli, and O. C. Landmark. Exploring Architectural Scalability. *Software Engineering Notes*, 29(1):125 – 129, 2004.

[8] L. Duboc, D. Rosenblum, and T. Wicks. A Framework for Characterisation and Analysis of Software System Scalability. In *European Software Engineering Conference*, pages 375 – 384. ACM, 2007.

[9] N. Huber and et al. A Method for Experimental Analysis and Modeling of Virtualization Performance Overhead. In *Cloud Computing and Services Science*, pages 353 – 370. Springer, 2012.

[10] International Telecommunication Union. Methods for Subjective Determination of Transmission Quality. Aug 1996.

[11] P. Lewis, Abbate. *Microsoft Lync Server 2010 Unleashed*. SAMS, 1st edition, 2011.

[12] Microsoft, `http://www.microsoft.com/download/en/confirmation.aspx?id=6412`. *Quality of Experience*, Oct 2007. Last visited: 2012 Jul 8th.

[13] Microsoft, `http://www.microsoft.com/download/en/details.aspx?id=21646`. *Microsoft Lync Server 2010 Planning Guide*, Aug 2011. Last visited: 2012 Jul 8th.

[14] Microsoft, `http://technet.microsoft.com/en-us/library/gg398835`. *Server Hardware Platforms*, Mar 2011. Last visited: 2012 Jul 8th.

[15] Microsoft, `http://www.microsoft.com/download/en/details.aspx?id=12295`. *Microsoft Lync Server 2010 Capacity Calculator*, Feb 2012. Last visited: 2012 Jul 8th.

[16] K. H. Rygg. Scalability Modelling for Optimal Provisioning of Data Centres in Telenor. Master's thesis, NTNU, Trondheim, Norway, July 2012.

[17] P. Seeling. Network Performance Evaluation of Microsoft Office Communications Server 2007. In *IEEE International conference on Electro/Information Technology (EIT), 2011*, pages 1–6, May 2011.

[18] G. E. Solutions. Sizing Study: Virtualized Microsoft Lync Server 2010 Deployment. Apr 2012.

[19] VMware, `http://www.vmware.com/files/pdf/perf-vsphere-cpu\_scheduler.pdf`. *VMware vSphere 4: The CPU Scheduler in VMware ESX 4*, Des 2010. Last visited: 2012 Sept 20th.

[20] H. Winters. *Mastering Microsoft Lync Server 2010*. Sybex, 1st edition, 2012.

DataMill: Rigorous Performance Evaluation Made Easy

Augusto Born de Oliveira
Electrical and Computer Engineering
University of Waterloo
Waterloo, ON, Canada
a3oliveira@uwaterloo.ca

Jean-Christophe Petkovich
Electrical and Computer Engineering
University of Waterloo
Waterloo, ON, Canada
j2petkov@uwaterloo.ca

Thomas Reidemeister
Electrical and Computer Engineering
University of Waterloo
Waterloo, ON, Canada
treideme@uwaterloo.ca

Sebastian Fischmeister
Electrical and Computer Engineering
University of Waterloo
Waterloo, ON, Canada
sfischme@uwaterloo.ca

Abstract

Empirical systems research is facing a dilemma. Minor aspects of an experimental setup can have a significant impact on its associated performance measurements and potentially invalidate conclusions drawn from them. Examples of such influences, often called hidden factors, include binary link order, process environment size, compiler generated randomized symbol names, or group scheduler assignments. The growth in complexity and size of modern systems will further aggravate this dilemma, especially with the given time pressure of producing results. So how can one trust any reported empirical analysis of a new idea or concept in computer science?

This paper introduces DataMill, a community-based easy-to-use services-oriented open benchmarking infrastructure for performance evaluation. DataMill facilitates producing robust, reliable, and reproducible results. The infrastructure incorporates the latest results on hidden factors and automates the variation of these factors. Multiple research groups already participate in DataMill.

DataMill is also of interest for research on performance evaluation. The infrastructure supports quantifying the effect of hidden factors, disseminating the research results beyond mere reporting. It provides a platform for investigating interactions and composition of hidden factors.

Categories and Subject Descriptors

D.2.8 [**Software Engineering**]: Metrics—*performance measures*

Keywords

DataMill; performance; experimentation; infrastructure; robustness; repeatability; reproducibility

1. INTRODUCTION

Empirical computer performance evaluation is essential for computer science and industry alike. The empirical measurement of performance sees widespread use to guide the research of new ideas and the development new technologies. A performance improvement of a few percentage points may mean large savings in dollars, when applied to a large data center with billions of clients. It is also essential, then, that computer practitioners dominate the methodology necessary to evaluate computer performance correctly.

However, the research community [10, 25, 31, 32] has demonstrated that experimental evaluation in computer science is difficult. As a consequence, experiment design, setup, analysis, and data reporting are often neglected. Results obtained from experimentation are sensitive to the experiment design and setup. Conclusions drawn from performance experiments may vary across a combination of hardware and software factors. Sometimes subtle changes to an experiment's setup can have a significant impact on its outcome [21]. The lack of rigorous performance evaluation is often blamed on the fact that experimentation is difficult and costly [31], but those obstacles are not insurmountable.

To improve the current state of experimentation, we propose DataMill. DataMill is a distributed infrastructure for computer performance experimentation targeted at scientists, researchers, and aspiring researchers. DataMill aims to allow the user to *easily* produce robust, replicable, and reproducible results at low cost. To do so, DataMill executes the experiments on real hardware and incorporates the results from existing research on how to setup experiments and hidden factors. For example, the infrastructure automatically varies a selection of hardware and software factors, and therefore reduces the effort required by the user to setup the experiment while simultaneously increasing the robustness and applicability of the experimental results to a wide range of factors. The user need not know the details of the underlying mechanisms required to vary these factors and may simply take advantage of DataMill infrastructure. DataMill is mainly developed by the University of Waterloo, however, several universities (e.g., Purdue University, the University of Pennsylvania, the University of Lugano, and the Federal University of Santa Catarina) have already joined the effort and have provided benchmarking nodes.

Besides making it easy and low cost to the user, DataMill also aims to alleviate the problem of data availability and the reproduction of experimental setups. Based on our own experience, few researchers responded to inquiries for experiment setup details and raw data used in their publications. In DataMill, all experiment setup parameters and their experiment files remain stored in the infrastructure. Users can choose to make their experiments public facilitating the replication and reproduction of the their experimental setups and results. We believe DataMill has the potential to watermark performance evaluation experiments by making the data publicly and consistently available. Finally, DataMill and the public repository of repeatable experiments can serve as a valuable tool for the education of aspiring researchers.

The contributions of this paper are:

1. The design and implementation of a distributed infrastructure for computer performance evaluation;
2. The reproduction and expansion of a previously published experiment, confirming the existence of significant hidden factors;
3. A survey of current performance evaluation practices in computer science publications, demonstrating the need for more rigorous experimentation.

2. BACKGROUND

In traditional experimental design [2, 20], *factors* are properties of an experiment that affect the *response variable*; i.e., the metric of interest of that particular experiment. It is common knowledge in statistics that the one-factor-at-a-time method (OFAT), which is a sequential exploration of a *design space* (the space defined by the factor dimensions), is a poor method of experimentation. Despite its weaknesses, OFAT is the *de facto* standard in computer science. One alternative to OFAT, factorial design, attempts to ascertain the effect of all factors on the response variable at the same time, avoiding OFAT's "blind-spots".

In order to illustrate the state of experimental methodology in computer science, we conducted a survey of the recent publications to SOSP 2011, ASPLOS 2012, OSDI 2012, ATC 2012, EuroSys 2011, and PLDI 2012. The results are shown in Table 1. The columns, in order, show the total number of papers, the fraction that contains empirical performance evaluation, the fraction that describes their hardware platforms and software versions, the fraction that contains a comparison (either with a baseline or a competitor), the fraction of those that contain a formal statistical test for that comparison, the fraction that performs the experiment under different conditions as a sanity check, the fraction that contains a dispersion metric (standard deviations, confidence intervals, empirical CDFs, etc.), and the fraction that published the software being evaluated, the fraction that uses a publicly available workload (established benchmarks, input files, etc.), and, finally, the fraction that published their resulting data.

While many papers include an empirical performance evaluation, many papers do not list the versions of the software used for these experiments. Worse, even more of these publications do not contain an empirical comparison to a baseline or a competing approach. Out of those papers that include such a comparison, only a small fraction formally tests that their performance figures are different from the baseline. Only a few submissions make their experiment implementation and the results publicly available. Although the benefits for using a dispersion metric have been reiterated by many researchers [31, 10, 32] in the past, a number of papers that do a performance evaluation do not use any measure of dispersion in their performance evaluation.

We performed an experiment to demonstrate the dangers of using a single experimental setup like the majority we found in our survey. Figure 1 shows the performance measurements of Iperf [22], a network performance benchmark, repeatedly running on a loopback interface (the machine was disconnected from the network) in blocks of one hour. The x-axis shows the blocks, and the y-axis shows the bandwidth through the loopback interface. We use a standard boxplot to characterize a block, adding a $\times$ symbol to mark the mean. Each block contains 330 samples, and the machine reboots between blocks. While performance is generally stable, in two out of 72 blocks the mean performance is approximately 10% better, even though everything else *remained the same*. A "lucky" developer could mistake the rare block for the norm, and report results that are a full 10% away from the average case.

Computer science experiments are susceptible to many hidden factors such as the one seen above [34, 18, 21, 13, 19]. The systematic exploration of their effect on metrics of interest is a difficult and time-consuming task. Automated hidden factor exploration is necessary, and would best be handled by an experiment infrastructure. Many published results in computer science cannot be validated on newer platforms. Reasons include loss of data, loss of source code, loss of the original workload, and insufficiently described experimental conditions. Automated replication and publicly accessible setups are necessary.

Access to a wide variety of experimental setups is difficult to obtain, so experimental design spaces are generally small, and are often comprised of just a single machine. Automated exploration of a large design space is also necessary. Furthermore, testing an innovation on various platforms forces one to consider more corner cases. This will overall lead to more robust innovation whose performance is not only known on more platforms but also that works on more platforms. Specific exploits of particular platform features only provide limited value to the research community.

It follows, therefore, that a distributed benchmarking infrastructure with the capability of varying software and hardware factors is in order. Based on these findings, we created DataMill, an experimentation infrastructure that addresses these shortcomings automatically. DataMill enables users to make experiment packages, workloads, and results publicly available. We collect several performance counters available in the Linux environment and compute dispersion figures automatically. We support experiments that contain multiple software packages to enable a baseline comparison or a comparison with a competing approach. We believe by providing a platform that automates the factor variation and many tasks of the experiment set up, empirical performance evaluations will become more popular among researchers.

Conference	No. of Papers	Perf. Eval	HW Desc.	SW. Desc.	A/B Test	Formal Test	Sanity Check	Disp. Metric	Public SW	Public Load	Public Data
SOSP'11	27	93%	92%	60%	91%	0%	24%	44%	4%	68%	0%
ASPLOS'11	36	61%	100%	55%	54%	0%	5%	31%	4%	82%	4%
OSDI'10	30	93%	71%	61%	54%	0%	0%	32%	7%	39%	0%
ATC'12	40	95%	87%	58%	68%	0%	3%	32%	8%	26%	0%
EuroSys'11	22	86%	95%	74%	84%	12%	42%	47%	21%	58%	0%
PLDI'12	45	64%	90%	79%	93%	0%	37%	10%	24%	72%	0%

Table 1: Illustration of Experimental Rigor in Recent Scientific Publications

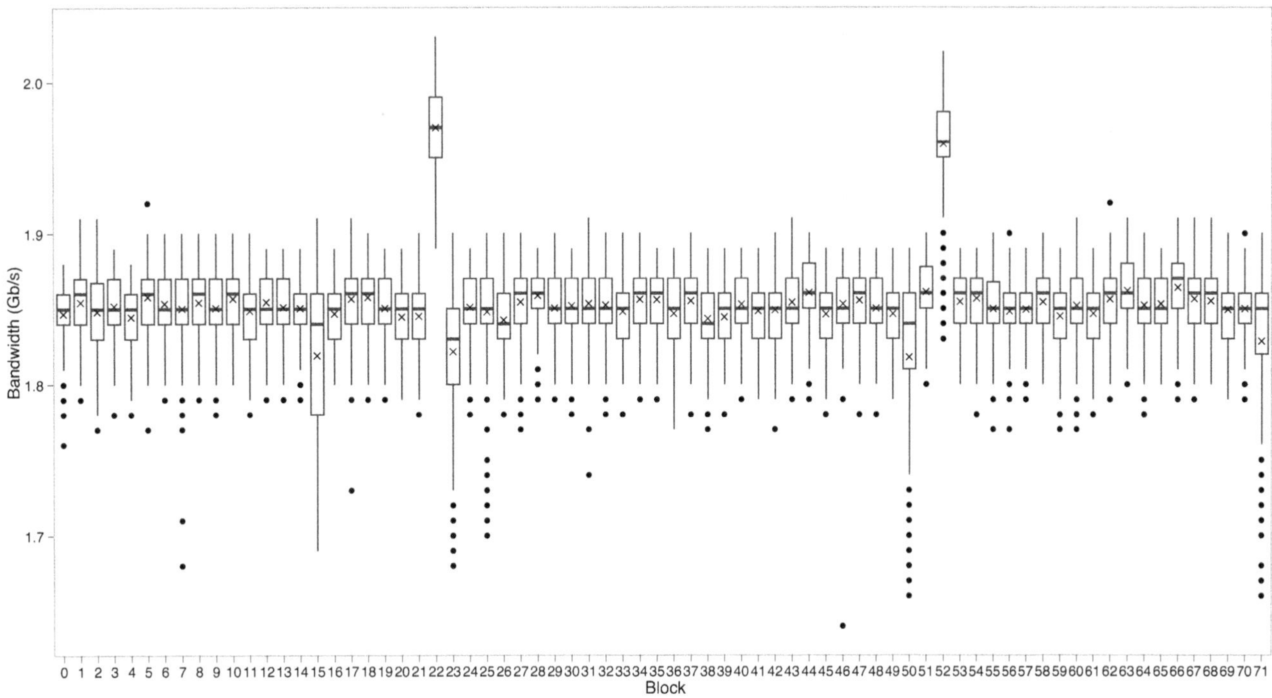

Figure 1: Iperf Performance in Blocks of 1 Hour

We now present DataMill, a world-wide open distributed performance evaluation system that aims to address the aforementioned points and implements the lower two levels of an experiment infrastructure [10].

3. DATAMILL: THE USER EXPERIENCE

DataMill is a distributed computer-performance evaluation infrastructure aimed at minimizing user effort required to get robust, reliable, and reproducible results. DataMill distributes arbitrary benchmarks and experiments to a large number of heterogeneous worker nodes, located all over the world, facilitating the creation of replicable results, and reducing the necessary effort to produce robust results.

DataMill facilitates replicable results by storing details of any particular experiment setup, allowing automated experiment reproduction. DataMill achieves robustness through replication across a wide range of experimental setups including variation of state-of-the-art known hidden factors. DataMill is a public-service infrastructure and available to scientists around world as long as they donate equipment for evaluations themselves.

In this section we describe DataMill as seen by a user, and later describe the inner workings of DataMill in Section 4. The user experience starts with packaging the experiment for DataMill to execute, then defining an experiment space from an array of hardware and software factors, and, finally, collecting and analyzing the resulting data.

3.1 Packages

Each experiment contains one or more packages. A single package experiment can quantify performance over a wide range of setups, while experiments with more packages allow performance *comparisons* over those setups.

Each package contains: (1) the source code and any input data for the experiment, and (2) auxiliary DataMill-specific scripts to set up, execute, and collect the results from the experiment. All package components are encapsulated in a compressed TAR file.

There is no restriction on experiment code, and users have administrator-level access to workers during experimentation. Users can generate free-form results data in arbitrary metrics, and then collect it via compressed packages. These features allow evaluating the performance of a wide range of software, ranging from user-space applications to kernel modules. Security concerns are minimal at this point, because participating users must contribute to the infrastructure and, therefore, are well known and trusted.

There are only two scripts that every DataMill package must contain: `run.sh` and `collect.sh`. These are the scripts that execute and collect data from the experiment, respectively. If the package requires a setup procedure (such as decompression, compilation, dependency installation, etc.), it may also contain a `setup.sh` script, which will be executed before `run.sh`. Finally, if there is the need for environment variables during execution, an `env` file may be included in the package. The contents of this environment file will be added to the experiment environment before each execution.

To exemplify the construction of a DataMill package, consider the Dhrystone [36] benchmark. It consists of a single source file, `dry.c`, which installs and runs the benchmark. To execute it, one must first set its executable flag, which is done in the setup script shown in Listing 1. Note that Dhrystone is self-compiling, so we change the file to be executable. The script assumes the source file is located in the `/dry/` directory.

```
#!/bin/sh

cd /dry/
chmod +x dry.c
```

Listing 1: Setup Script

Listing 2 shows the `run.sh` script that executes Dhrystone, capturing its output (standard out and error) to the `/dry/results` file. The output is appended to the end of the file, allowing multiple executions before collection.

```
#!/bin/sh

cd /dry/
./dry.c >> results 2>&1
```

Listing 2: Run Script

Finally, Listing 3 shows the script that packages the results file for collection. It simply compresses the result file using a unique name, and echoes the final archive's file name. This file will then be downloadable by the user once the experiment finishes executing.

```
#!/bin/sh

results=/dhrystone-`date +%Y%m%d_%H%M`.tar.gz
tar czf $results /dry/results > /dev/null 2>&1
echo $results
```

Listing 3: Collect Script

The user must package the experiment source code and DataMill scripts in a GZipped TAR file. All DataMill scripts should be in the root directory of this archive. For the Dhrystone package, the file structure of the package would be as shown in Figure 2.

```
Archive Root
├─ setup.sh
├─ run.sh
├─ collect.sh
└─ dry/
    └─ dry.c
```

Figure 2: Dhrystone Package Directory Structure

To facilitate package creation, we provide users with a virtual machine image that mimics a DataMill worker. Users can develop, test, and debug their packages in a local environment until they are sure their packages are ready for production, at which point they can submit it for execution via our experiment creation interface.

3.2 Experiment Creation

Once the packages are ready, the user submits them for execution through a file upload interface. Users must then define an experiment for DataMill to execute. Experiment definition has three steps:

1. Package selection
2. Constraint definition
3. Experiment space definition

Package selection consists of choosing which packages will be executed; one or more packages may be selected. In the constraint definition step, users inform DataMill of any limitations to their benchmark. For example, if the experiment can only execute on the ARM9 architecture and requires at least 2GB of RAM, the user expresses these constraints through an intuitive web interface.

Finally, in the experiment space definition step, users select how many experimental setups they wish to explore. The interface presents each dimension of the experiment space, divided in hardware and software categories, in a unified percentage-scale. This percentage represents what fraction of all available options should be *covered*. For example, users can define that their experiment must contain 100% of CPU available architectures and that 75% of GCC optimization flags be tested. This will cause DataMill to select machines such that every architecture in the infrastructure is represented, and to re-execute the benchmarks in a randomly selected subset of optimization flags that covers at least 75% of the total number optimization flags. Setting any dimension to 0% results in the random selection of a single level for that dimension.

After machines are selected according to the hardware dimension settings, the software dimensions are combined to generate each individual *job*. For example, if the user chose to explore all five GCC optimization flags on a single machine, there would be one job generated per optimization flag, all attributed to that machine. If, in conjunction with the optimization flags, both settings of the address randomization feature of Linux (on or off) should be tested, then the number of jobs grows to 10 (5 flags times 2 address randomization settings). Therefore, the number of jobs for each experiment scales with the experiment space defined by the user, and care must be taken to avoid combinatorial explosion and an experiment with an excessive number of jobs.

3.3 Experiment Results

After the user defines the experiment space and DataMill creates jobs, it will distribute the experiment's packages for execution, then collect the individual result files. The web interface dynamically updates experiment information as data arrives, allowing users to monitor their experiment's progress. In addition to the data collected by the collect script, DataMill collects additional metrics with minimal overhead [33], such as total execution time, and the number page faults and cache misses. This data is also made available to the user.

Finally, once all jobs associated with an experiment are finished, users can download the full experiment results file, which contains the result file from every job. Examples of how to analyse large datasets generated by DataMill are provided in Section 5.

4. DATAMILL: THE INFRASTRUCTURE

Making the user experience described in Section 3 a reality requires considerable engineering effort. The DataMill infrastructure is composed of a *master node*, responsible for the distribution of experiment trials and the collection of results, and several *worker nodes*, which execute the experiment packages provided by the users. This section describes how DataMill was implemented and how its different parts interact.

4.1 Master Node

The responsibilities of the master node are split up between four different daemons: the web front-end, the backend, the dispatcher, and the collector. The user interacts with DataMill through a web front-end. Creation of experiment configurations is based on the users' choices (e.g., experiment factors, experiment package, and desired platforms). The backend, which can consist of one or more instances, processes these experiment configurations and schedules individual instances of an experiment, which we refer to as *jobs*, on the workers. The dispatcher submits scheduled jobs to the workers. The collector accumulates results from individual jobs once they are completed on the associated worker. The web-interface provides access to these results as they arrive. In the following sections, we describe the individual components of the management layer in detail. Each of these daemons acts as a simple state-machine as shown in Figure 3. To maintain and ensure consistency of the state of the infrastructure, we have chosen to use a database for central configuration management tasks.

The Web Front-End.

The *web front-end* is the primary window through which the user interacts with DataMill. User submission of experiments triggers the backend daemon process, which processes package information and configurations and subsequently schedules jobs for worker nodes.

The web front-end also houses an XML-RPC [6] service. The service provides an API which workers call to inform the master node about state changes of the individual jobs. Workers use the same API to register with the infrastructure. To separate control and data flows, this XML-RPC interface is only used to notify the master node of state changes. The dispatcher and collector handle transferal of experiment data such as packages and results.

Backend.

The *backend* daemon, shown in Figure 3, is responsible for transforming the user-supplied specifications for a particular experiment into a set of specific trial configurations to run on the worker nodes. The creation of these configurations and the selection of the appropriate worker node to execute them is handled by an optimization solver.

The problem is then to select the minimum number of workers necessary to provide the desired factor coverage. For this, we specified an optimization problem, which the backend solves with GLPK [11] for each submitted experiment. The optimization problem is of the form:

$$w_c = \min \sum_{x_i \in \mathcal{W}} x_i \tag{1a}$$

$$\text{s. to: } \sum_{x_i \in \mathcal{P}_j} x_i \geq l_j, \qquad l_j \in \mathcal{L} \tag{1b}$$

$$l_j \geq x_i, \qquad l_j \in \mathcal{L}, x_i \in \mathcal{P}_j \tag{1c}$$

$$\sum_{l_j \in \mathcal{F}_k} l_j \geq B_k, \qquad k = 1..f \tag{1d}$$

$$x_i \in \{0,1\}, \qquad x_i \in \mathcal{W} \tag{1e}$$

$$l_j \in \{0,1\}, \qquad l_j \in \mathcal{L} \tag{1f}$$

Where $\mathcal{W}$ is the set of workers, $\mathcal{L}$ is the set of factor levels, and $\mathcal{P}_m$ is the set of machines that *provide* level m. $\mathcal{F}_k$ is the set of factor levels for factor k (a subset of $\mathcal{L}$), B_k is the user-requested minimum coverage for factor k. Equation (1b) ensures picking a factor level means at least one machine with that factor level is added to the solution, while, conversely, Equation (1c) ensures that selecting a machine adds all factor levels it provides to the tally, which is checked in Equation (1d) for every factor. This last equation guarantees that all user-provided constraints are met. After a solution is found, each worker is individually selected by their respective x_i variable.

Dispatcher.

The dispatcher daemon in Figure 3 is responsible for transmission and triggering of the execution of the packages and configurations on the appropriate worker as defined by the solution generated by the backend daemon. The dispatcher finds jobs whose associated worker is idle and transmits and calls for the execution of the associated package via secure shell.

Collector.

The collector daemon in Figure 3 is responsible for the collecting of the experiment results from the worker nodes which have completed their current job. Once the results have been collected from any particular worker the worker is marked as *idle* so that it may receive further jobs.

4.2 Worker Node

Each worker node is a separate machine running Gentoo Linux [29] with kernel version 3.3.8, and GCC 4.5.3. Gentoo Linux was chosen since it is a source-based distribution and thus supports a wide variety of architectures. Having the same distribution and tool-chain on all the machines keeps their software homogeneous.

The partition structure of a worker node is shown in Figure 4. The worker comprises two partitions, one with a minimal Gentoo installation, referred to as the *controller*

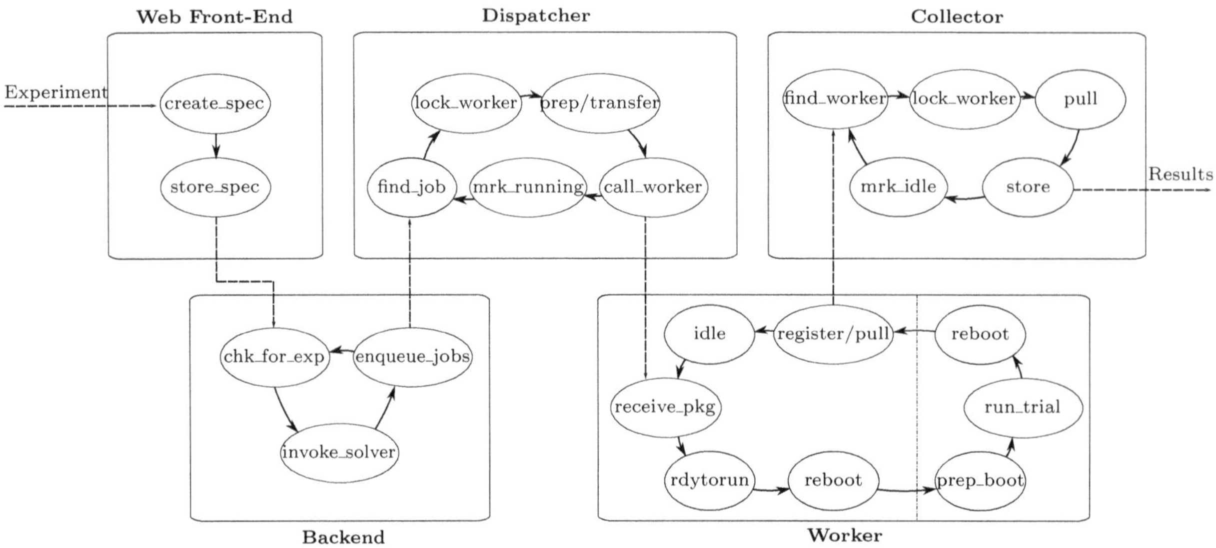

Figure 3: The DataMill Infrastructure

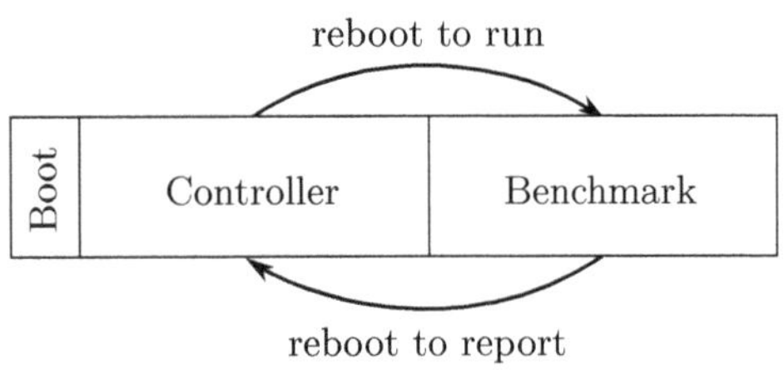

Figure 4: The Worker Node Disk Partitions

partition, which is responsible for communication with the master, and one with a more complete software environment, referred to as the *benchmark partition* on which experiments will run. The state machines for each worker partition are shown in Figure 3, left of the dotted line is the controller partition's state diagram and to the right is the benchmark partition's state diagram.

The master node communicates with workers through secure shell. Communication is kept to a one-way push architecture, from master node to worker node, where possible. The workers do not have the ability to run code on the master barring the limited XML-RPC calls that are provided to them. In this way, the communication between the master node and the worker nodes is strictly limited.

When a worker has been setup for the first time, it must notify the master of its existence. Upon its first boot, the worker collects information about itself and sends it, along with a registration request, to the master node. Once it is registered, it is marked as *idle* in the database and awaits further instruction.

When a worker receives a job from the dispatcher, the dispatcher executes a script on the worker via secure shell. The script transfers the received package to the benchmark partition and prepares it for execution. The machine subsequently reboots into the benchmark partition. The benchmark partition's init process has three tasks to perform: changing the default boot partition, executing the received package, and finally rebooting to the controller partition.

Due to the size of DataMill, worker nodes are expected to fail. The master node detects such occurrences through communication timeouts, incorrect state machine progression, and faulty result files. The master node purposefully inserts redundancies in the job scheduling minimize the impact of worker failures.

The DataMill infrastructure supports remote worker nodes. Already included in the DataMill cluster are several machines from the Purdue University, the University of Pennsylvania, the University of Lugano, and the Federal University of Santa Catarina.

4.3 Factor Variation

DataMill facilitates execution of user supplied benchmarks using a wide variety of experimental setups. Knowledge of the factors affecting performance and the details of the mechanics required to vary them are not necessary to leverage the infrastructure. Nevertheless it is important to discuss some details of DataMill's factor manipulation.

To apply software-controlled factors — compilation flags, library versions, link orders, etc. — the worker node unpacks the configuration file sent by the dispatcher, applying each configuration individually before the setup or execution. Current supported factors include various GCC flags, C and C++ object link orders, the use of address space randomization in the Linux kernel, the addition of padding bytes to the POSIX environment, reboot behavior, and dropping caches before benchmark execution.

While most factors can be easily applied (i.e., interacting with the `proc` filesystem to affect kernel options) some factors require more effort. For example, to support modifications to benchmark link order, we implemented a custom wrapper that intercepts calls to GCC. This wrapper calculates the new order of objects according to the configuration received from the master (i.e., alphabetical, or reverse alphabetical according to object file names), and then forwards the call to GCC, using the new object file order.

This method of applying software factors enables making extensive modifications to benchmarks without requiring users to implement these modifications themselves or to

Algo.	Size (kB)	Reduction (kB)	%
None	51,900	0	100.000
bzip2	13,232	38,668	25.496
XZ	12,120	39,780	23.353

Table 2: Compression Rates for Each Algorithm.

upload a series of different packages with pre-applied modifications. The list of supported factors is currently growing quickly, and we aim to support user-contributed factors in the future.

5. CASE STUDIES

This section presents two case studies: we first perform a compression algorithm performance comparison to demonstrate how easy it is to conduct a performance experiment on DataMill, then we replicate a previously published experiment that revealed performance artifacts related to the link order of a binary, demonstrating the utility of DataMill for the scientific investigation of computer performance.

5.1 XZ vs. bzip2: Best Bang for Your Buck?

XZ [30] and bzip2 [17] are widely-used compression utilities for UNIX-like operating systems. While XZ uses the LZMA2 compression algorithm, bzip2 uses the Burrows-Wheeler algorithm. These two compressors will serve as stand-ins for a "baseline-vs.-proposed-approach" performance comparison, found in the majority of computer science papers that contain empirical performance evaluations. As bzip2 is the older of the two compressors, we will treat it as the baseline.

As described in Section 3, the only preparation step required for this experiment is the creation of two DataMill packages, one for XZ 5.0.4 and another for bzip2 1.0.6. The scripts themselves are omitted for brevity, but their contents are very similar to the ones for Dhrystone and comprise a total of 32 lines. We use the system-wide *emerge* command to install both XZ and bzip2 in order to simplify installation, precluding the need to include their source code in each package.

In addition to the DataMill scripts, the packages contain the data to be compressed. For this experiment, we used the Maximum Compression [35] testset. This testset includes various kinds of files (text, executable, graphics, etc.) and has been used to compare compression algorithms since 2003.

The metrics, which the `collect.sh` script collects, were execution time and compressed file size. Since each compression algorithm leads to a different archive size, the metric used for the comparison is the byte-per-second compression rate, calculated as bytes reduced/execution time. Table 2 shows the uncompressed data size, the resulting archive size under each compressor, the absolute reduction in size, and the resulting archive size as a percentage of the original file size. Note the both compression algorithms use deterministic algorithms, so the resulting compressed files are identical between runs and machines. Machine C, an ARMv7 which uses the ext3 filesystem, reports file sizes 20kB larger than the ones reported by all other machines, which use ext4. This small 0.1% discrepancy was ignored.

If absolute compression is the only metric of interest, the XZ is clearly the winner; however, if execution time or the rate of compression are of interest, then experimentation is necessary. By using DataMill, we can easily compare the two compressors, and measure their susceptibility to different factors. The DataMill experiment space was configured to include all machines, all link orders, all optimization flags, and address randomization on and off. The number of replications was set to 15 to allow the measurement of dispersion. This led to the generation of 6300 jobs, distributed between seven machines.[1] This experiment took approximately five days to complete on the slowest machine, a 600MHz ARMv7 Beagleboard xM. All other machines completed it in less time, and were free to continue with other experiments.

Figure 5 shows an overview of the data set resulting from this experiment. This facet plot is divided by optimization flag (top header) and machine (right-hand header). Machines are indexed with a capital letter, followed by their clock speed and CPU model. Each subplot contains boxplots for each of the compressors, bzip2 and XZ, where the box denotes the median, the lower quartile, and the upper quartile, and the whisker extends to the most extreme data point that is within 150% of the interquartile range of the data set.

The first conclusion is that bzip2 has a better compression rate for all machines under all link orders and all optimization flags tested. This would suggest that, for users interested in compression speed, bzip2 is the better alternative. Also of interest is the fact that bzip2 is unaffected by the optimization flag, which suggests it is entirely I/O-bound. XZ, on the other hand, has a marked performance increase from -O0 to -O1, but negligible differences in performance after that. The omitted experimental dimensions — link order and address randomization — did not significantly affect performance, which is demonstrated by the narrow grouping of all samples in the boxplots of Figure 5 (which contain data from multiple levels in the omitted dimensions).

We use ANOVA [16] to more formally analyze the data. Table 3 shows the ANOVA table for this experiment, fitting a model with up to two-factor interactions on the execution time data for machine F, a 3.4GHz Core i7. The mean square column quantifies the change in execution time attributable to each factor, and the p-value column shows the statistical significance of each factor in respect to execution time. The model was a good fit, with $R^2 > 0.99$ (i.e., the model explains more than 99% of the variability in the data), meaning the model can be confidently used to analyze the data. The table shows that the compressor and optimization flags are significant in isolation ($P < 0.01$), but link order and address randomization are not ($P > 0.05$). Most interestingly, we can prove the interaction between the compressor and optimization flag through their interaction term, which is also significant ($P < 0.01$).

Figure 6 shows the effect of GCC optimization flags on XZ's execution time, with the x-axis ordering the different optimization flags. While optimization flags are not a continuous dimension, this ordering facilitates the visualization of the data and represents the activation of various individual GCC optimization options going from one optimization level to the next. According to GCC's manual, the -Os flag is located between -O1 and -O2 because it activates all options from -O2 that do not increase binary size, being, therefore, a middle point between the two. Data from each worker is

[1]Data for the "alphabetical" link order in the XZ was not generated, as that object order did not link successfully.

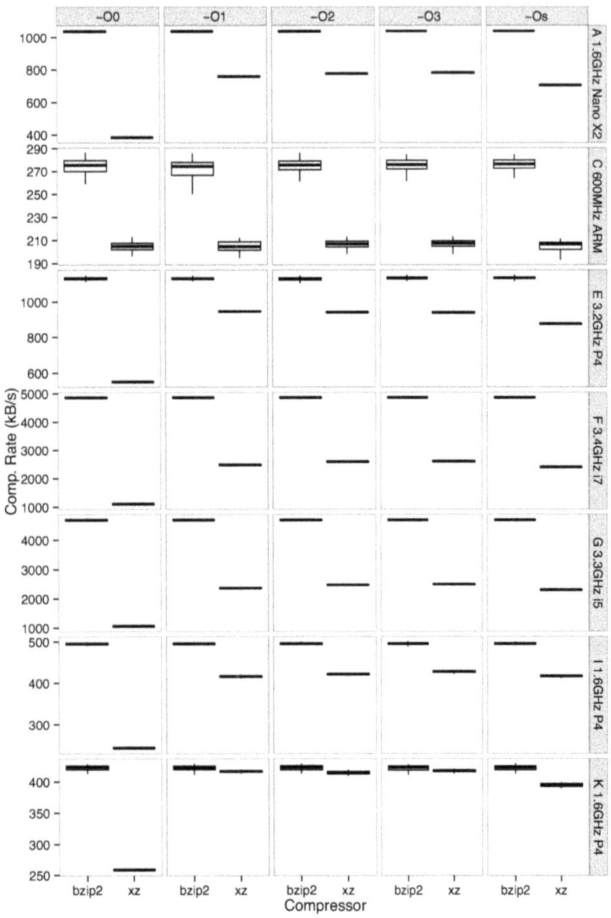

Figure 5: XZ vs. bzip2, Compression Rate

plotted along an individual line, with 95% confidence intervals shown in light gray behind each curve. The plot shows a marked and general improvement in performance going from -O0 to -O1 (as shown in Figure 5), but also reveals a more interesting point: in two of the machines, there is a *decrease* in performance going from -O1 to -O2. Despite being a small decrease, this may merit more investigation, as -O2 is the default optimization flag of several distributions.

This performance comparison demonstrates the utility of DataMill for users interested in evaluating performance: with just 32 lines of code, 6300 jobs were executed in under a week, exercising several dimensions that would normally be ignored, and leading to insight that would be unattainable through manual, one-factor-at-a-time experimentation.

5.2 Perlbench: Link Order Effect

We now demonstrate the use of DataMill for users interested in the study of computer performance evaluation. Mytkowitz et al. [21] report that the link order of a binary is correlated with runtime performance, and that the optimal link order varies from host to host. This is generally understood to be a consequence of different memory and cache layouts leading to different cache and page miss ratios. The authors showed that the performance of Perlbench — part of SPEC CPU2006 [27] — can vary by more than 8% by simply modifying the link order.

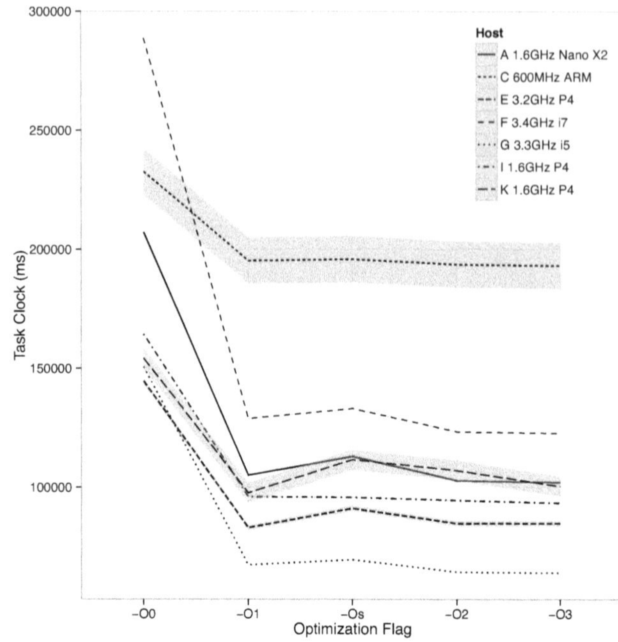

Figure 6: Effect of GCC Optimization Flags on XZ, by Host

Factor	Mean Sq.	p-value
Compressor	1.617e+12	<2e-16
Opt. Flag	1.258e+11	<2e-16
Link Order	7.397e+05	0.5807
Addr. Rand.	2.572e+04	0.8907
Compressor:Opt. Flag	1.883e+11	<2e-16
Compressor:Link Order	1.924e+06	0.2347
Compressor:Addr. Rand.	1.511e+06	0.2922
Opt. Flag:Link Order	1.312e+06	0.4624
Opt. Flag:Addr. Rand.	3.791e+06	0.0257
Link Order:Addr. Rand.	2.738e+05	0.8177
Residuals	1.360e+06	

Table 3: Reduced ANOVA Table for XZ Execution Time on Machine F

Trying to reproduce their results, we created an experiment on DataMill to explore the effect of link order and address randomization on Perlbench performance. We encapsulated Perlbench and SPEC's "train" data set in a DataMill package, with scripts and environment file totaling 33 lines. Three link orders were explored (default, alphabetical and reverse alphabetical), with Linux address randomization on and off. If address randomization is on, one would expect that the affect of link order would be neutralized, since the memory layout will be randomized. In other words, the link order and the address randomization factors should be highly correlated. We chose a number of 15 replications of each configuration to calculate dispersion, generating a total of 630 jobs over 7 machines. The DataMill took approximately 27 hours to finish the full experiment.

Figures 7 and 8 show results for the different metrics for this experiment. These facet plots are divided by address randomization (top header) and host (right header). Each subplot contains three boxplots, one for each link order ex-

Mach.	Factor	Mean Sq.	p-value
A	Link Order	135.77	<2e-16
	Addr. Rand.	0.07	0.0301
	Link Order:Addr. Rand.	0.04	0.0687
	Residuals	0.01	
K	Link Order	1856.5	<2e-16
	Addr. Rand.	1234.6	<2e-16
	Link Order:Addr. Rand.	1864.9	<2e-16
	Residuals	3.3	

Table 4: Reduced ANOVA Table for Perlbench Execution Time on Machines A and K

plored. Figure 7(a) shows execution time, and makes it clear that there is indeed a change in execution time between the different link orders on most cases. An exception to this rule is machine I, a 1.6GHz Pentium 4, where this effect is minor.

It is also clear that the link order effect does not depend on the address randomization feature of Linux being turned off; most machines show different execution times between link orders even when address randomization is turned on. However, machine K (shown in the bottom of Figure 7(a)) shows a link order effect only when address randomization is turned off, contrary to the other machines.

To help understand this effect, Figure 7(b) shows the cache misses for the experiment[2]. This plot shows that there appears to be a correlation between link order and cache misses for most machines, but they do not necessarily mirror the execution time effect seen in Figure 7(a).

Finally, Figure 8 shows that there is no apparent correlation between page faults and link order; even in the case of the Nano X2 and the Xeon machines (top two subplots), the difference in mean page fault counts between link orders is of less than 1%. A possible explanation for this is that the sum of code and data for the benchmark is small enough to fit within a page, no matter the order the object files are linked in.

Table 4 shows the reduced ANOVA table for machines A ($R^2 = 0.977$) and K ($R^2 = 0.855$). This table shows that the link order effect is significant for both of these machines ($P < 0.01$), but only in machine K do the address randomization factor ($P < 0.01$) and the interaction between address randomization and link order factors ($P < 0.01$) play a part. In machine A, both of these are not statistically significant at the 99% confidence level ($P > 0.01$ in both cases). This suggests that the effect of address randomization, which is present in machine K but not in others, is highly dependent on the machine.

Therefore, while the presence of a cache effect and the lack of a page effect would seem to explain the variations in execution time between link orders, the correlation between link order and execution time still merits more investigation. DataMill is a powerful tool for researchers in performance evaluation, since it allows the systematic variation of correlated factors, such as link order and address randomization, and the simultaneous collection of multiple relevant metrics, such as cache misses and page faults.

[2]Data is missing for the Nano X2 due to the lack of hardware performance counter support

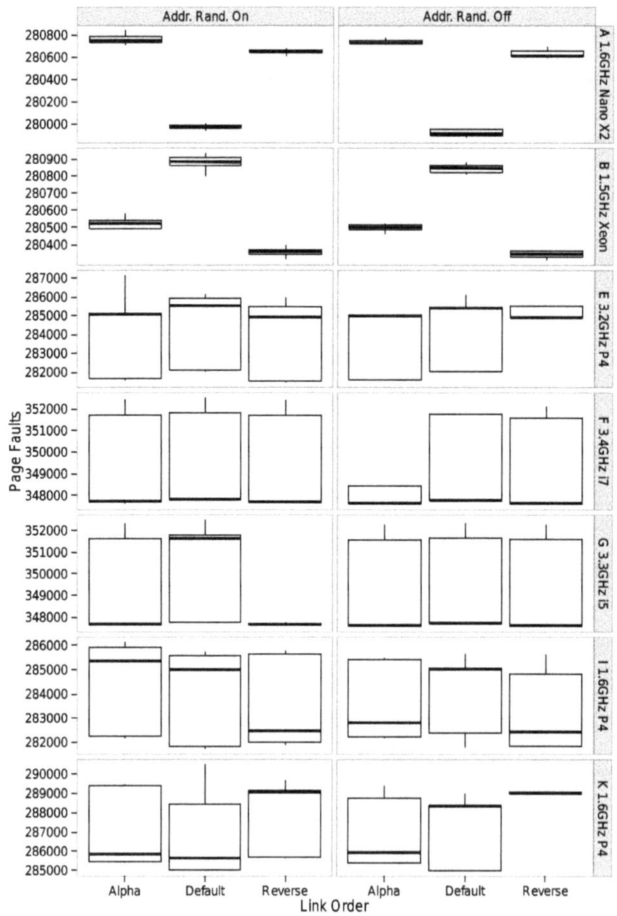

Figure 8: Effect of Link Order on Perlbench Page Faults

6. LESSONS LEARNED

The implementation of the infrastructure and the execution of the experiments detailed in Section 5 raised several interesting questions.

Debugging packages took considerably longer than expected, especially in the case of Perlbench. This was mainly due to the non-standard build system distributed by SPEC, which requires manual configuration of target architecture parameters. Our experience with it led us to create a virtual worker image on which users can debug their packages. We are currently also investigating a special "debug" experiment type which would execute packages once on each available architecture to ensure it behaves as expected.

The addition of remote nodes (located in remote universities) was a challenge, mainly because of firewalls. Our current implementation uses two-way communication, which requires special treatment for firewalls that reject all incoming connections. In the near future we plan to move to a one-way communication design where the master node is entirely passive, sidestepping this issue and minimizing the effort needed to integrate new workers.

We also realized that even though small embedded targets can run Gentoo, their performance, particularly in compile phases, is prohibitively low for large experiment design

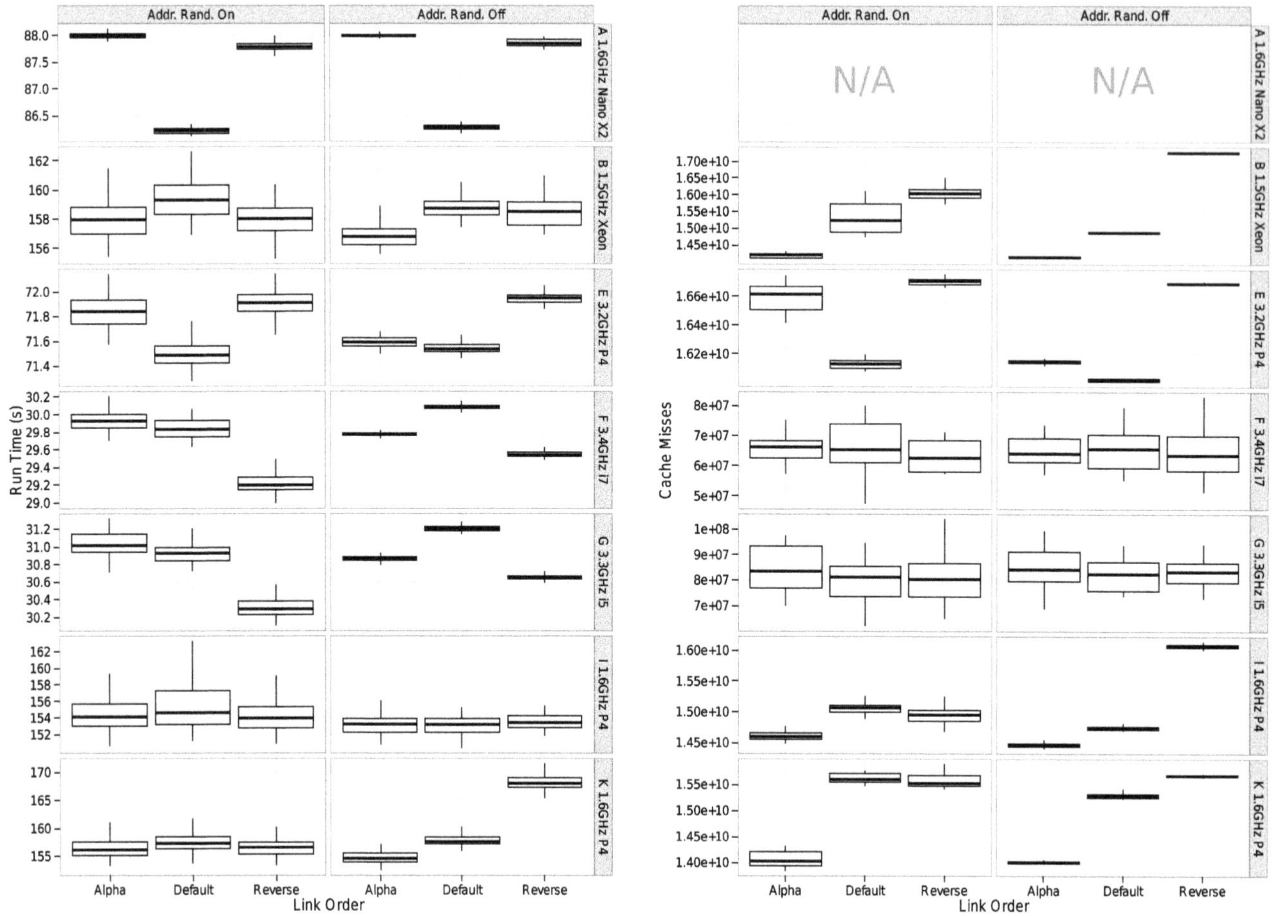

(a) Execution Time

(b) Cache Misses

Figure 7: Effect of Link Order on Perlbench Execution Time and Cache Misses

spaces. In some instances of the compression experiment, we noticed that the compile time exceeded the execution time of the experiment by several orders of magnitude. In the future, we plan to investigate providing cross-compiler support for experiments.

7. RELATED WORK

Various researchers [31, 25, 7, 8, 5, 9] argue for more statistical rigor in computer science. The paper survey by Tichy *et al.* [32] concludes that numerous publications in the field of computer performance evaluation show substantial flaws in experiment design and execution. Vitek and Kalibera [34] report that in PLDI'11, a selective conference where experimental results are commonly published, 39 of the 42 papers that published experimental results did not report a measure of uncertainty in their data, obviating the need for more rigorous statistical analysis in computer science. A survey conducted by Desprez *et al.* [10] draws similar conclusions from a paper survey they conducted. Kalibera and Jones [18] raise a similar point, presenting a random effects model tailored to computer experiments, while also noting that current textbook approaches may be insufficient in the field. Georges et al. [12] argue for the use of statistically rigorous analysis methods, however, their approach is only limited to narrow field of statistical analysis methods.

Mytkowicz *et al.* [21] demonstrate that seemingly innocuous experimental setup details, such as the UNIX environment size or the benchmark link order, can have a significant impact on performance. Harji *et al.* [13] show that the Linux kernel has had a series of performance affecting issues, and that papers that present data measured on Linux could contain incorrect results. Kalibera et al. [19] show that random symbol names generated by a compiler leads to different memory layouts at run time, and, consequently, random variations in performance.

Desprez *et al.* [10] surveys numerous large-scale computing installations that follow a similar objective to our cause; to create reproducible, extensible, applicable, and revisable experiments. The authors provide a survey of experimental methodologies and survey a selection of experimental testbeds. In contrast to our design the testbeds surveyed are often comprised of homogeneous nodes and do not exhibit a lot of variation in the hardware used. While the computing resources are vastly available and the installations provide support for complex experiments (i.e. including distributed systems), the experiment setup is described as a manual arduous process despite exposing high level interfaces.

The infrastructures have been used to implement demonstrators for federated clouds and projects in the field distributed systems. Most notably among the surveyed computing installations is OpenCirrus [4]. In addition to virtualized environments they also provide access to the physical machines. The lowest level service consists of a physical resource set (PRS). A PRS comprises a set of VLAN-isolated compute, storage and network resources. The PRSs are dynamically allocated and managed through a PRS service. Using the PRS paradigm different levels of abstractions can be configured that suit research applications reaching from low level systems research (e.g., the evaluation of OS kernel parameters) to complex distributed systems (e.g., several virtual machines that run a distributed middleware). The reader should note, while the objective of creating create reproducible, extensible, applicable, and revisable experiments are aligned with our cause, the focus of the systems surveyed by Desprez *et al.* [10] is on distributed systems. As a result the hardware infrastructure is decidedly homogeneous, which simplifies the experiment setup and configuration. While testbeds like OpenCirrus [4] support access to low-level hardware features, those features expose little variability compared to the applications we are targeting.

PlanetLab [24, 23] provides planetary-scale data services and is used by the research community to deploy, evaluate and access planetary-scale network services. Planetlab provides so called slivers to its users that consists of distributed networked virtual machines (VMs). The VMs are hosted on physical machines that are maintained in a communal fashion. In order to become a user of PlanetLab, one has to dedicate some servers to PlanetLab. PlanetLab exposes the application interface for provisioning the slivers and has facilities to isolate the network of the individual slivers. PlanetLab is used predominantly to evaluate and deploy distributed system [26], including content-distribution networks, name services, location services, file-streaming services, fault-tolerant scalable services, peer-to-peer networks, distributed anomaly detection, distributed research allocation, routing overlays, and resource discovery.

Because the experiment environment exposed to the user is a virtualized machine, PlanetLab is not an optimal choice for computer performance experiments that seek to evaluate the impact of varying hardware environment factors. This conclusion is consistent with papers listed [26]; none of the publications include computer performance experiments.

Various other experimentation infrastructures have been proposed with similar properties to PlanetLab. Jaffe *et al.* [15] describe a production platform with similar features to PlanetLab. The project links various large data centres in an effort to provide an experimentation platform for distributed systems. Unfortunately, the hardware chosen for the data centres is very homogeneous and does not aid in the exploration of large factor space.

HTCondor [28] is a distributed job scheduler designed for computation-intensive distributed workloads. HTCondor shares some of DataMill's characteristics — such as having users prepare packages and submit them for execution — but since it it geared toward maximizing a cluster's computing throughput, it is not well suited for clean-room performance evaluation.

8. FUTURE WORK

As part of our future work we consider various improvements to the existing system architecture. Huang *et al.* [14] have shown that maintaining security and accountability in a distributed experiment execution framework is challenging. Our current architecture was geared towards providing a proof-of-concept prototype. As part of our future work we want to implement a centrally managed security policy. This would add accountability for individual job executions and transparency to maintainers of remote workers such that the impact of malicious attacks on the infrastructure is minimal.

While our current prototype monitors the performance counters that are exposed by the Linux kernel, it is still up to the submitters of the experiment to analyze their experiment-specific results. In the medium-term we want to provide an application interface for distributed data-analysis tools (i.e., Hadoop [3]), and services (i.e., Amazon Web Services [1]) to facilitate the analysis of experiment data. Our plan is to provide a template engine for various standard benchmarks that can be easily customized by users.

In our current prototype, the factor variation is achieved by a loose collection of shell scripts that are distributed as part of the worker installation. In the future, we want to provide a centrally managed repository of factor variation scripts that can be distributed independently from the worker code, and provide an interface through which users can add custom factors.

9. CONCLUSION

In this paper we introduced DataMill. DataMill provides services to set up and execute robust, replicable, and reproducible experiments. DataMill enables researchers to publish their experiment software, setup, and results. That way experimental results can be reproduced easily. Many aspects of complex performance experimentation are automated by DataMill enabling users to set up performance experiments easily. Due to its support for many different hardware platforms and automated factor variation, DataMill can cover a larger experiment space than typically considered by most researchers. For example, we have shown that a complex performance experiment consisting of 6300 jobs that span various factors can be set up by configuring a package with just 32 lines of shell code. We believe that DataMill can serve as a watermark for experiments conducted for performance-oriented conferences.

10. ACKNOWLEDGMENTS

This research was supported in part by NSERC DG 357121-2008, ORF-RE03-045, ORF-RE04-036, ORF-RE04-039, APCPJ 386797-09, CFI 20314 and CMC, ISOP IS09-06-037, and the industrial partners associated with these projects. The authors would like to thank Andrew Yeung for his work on the ARM platform.

11. REFERENCES

[1] Amazon Web Services LLC. Amazon Web Services. `http://aws.amazon.com/`. Accessed Sep. 17th, 2012.

[2] J. Antony. *Design of Experiments for Engineers and Scientists*. Butterworth-Heinemann, 2003.

[3] Apache Software Foundation. Hadoop. `http://hadoop.apache.org/`. Accessed Sep. 17th, 2012.

[4] R. Campbell, I. Gupta, M. Heath, S. Y. Ko, M. Kozuch, M. Kunze, T. Kwan, K. Lai, H. Y. Lee, M. Lyons, D. Milojicic, D. O'Hallaron, and Y. C. Soh. Open CirrusTMcloud Computing Testbed: Federated Data Centers for Open Source Systems and Services Research. In *Proceedings of The 2009 Conference on Hot Topics in Cloud Computing*, HotCloud'09, Berkeley, CA, USA, 2009. USENIX Association.

[5] D. E. Comer, D. Gries, M. C. Mulder, A. Tucker, A. J. Turner, and P. R. Young. Computing as a Discipline. *ACM Communications*, 32(1):9–23, Jan. 1989.

[6] D. Winer. XML-RPC Specification. `http://www.xmlrpc.org/spec`. Nov. 1999.

[7] P. J. Denning. ACM President's Letter: What is Experimental Computer Science? *ACM Communications*, 23(10):543–544, Oct. 1980.

[8] P. J. Denning. ACM President's Letter: Performance Analysis: Experimental Computer Science as its Best. *ACM Communications*, 24(11):725–727, Nov. 1981.

[9] P. J. Denning. Is Computer Science Science? *ACM Communications*, 48(4):27–31, Apr. 2005.

[10] F. Desprez, G. Fox, E. Jeannot, K. Keahey, M. Kozuch, D. Margery, P. Neyron, L. Nussbaum, C. Perez, O. Richard, W. Smith, G. von Laszewski, and J. Voeckler. Supporting Experimental Computer Science. Technical report, Argonne National Laboratory Technical Memo, 2012.

[11] Free Software Foundation. GLPK (GNU Linear Programming Kit). `http://www.gnu.org/software/glpk/`. Accessed Sep. 17th, 2012.

[12] A. Georges, D. Buytaert, and L. Eeckhout. Statistically Rigorous Java Performance Evaluation. In *Proceedings of the 22nd Annual ACM SIGPLAN Conference on Object-Oriented Programming Systems and Applications*, OOPSLA '07, pages 57–76, New York, NY, USA, 2007. ACM.

[13] A. S. Harji, P. A. Buhr, and T. Brecht. Our Troubles With Linux and Why You Should Care. In *Proceedings of the Second Asia-Pacific Workshop on Systems*, APSys '11, pages 2:1–2:5, New York, NY, USA, 2011. ACM.

[14] M. Huang, A. Bavier, and L. Peterson. PlanetFlow: Maintaining Accountability for Network Services. *SIGOPS Oper. Syst. Rev.*, 40(1):89–94, Jan. 2006.

[15] E. Jaffe, D. Bickson, and S. Kirkpatrick. Everlab: A Production Platform for Research in Network Experimentation and Computation. In *Proceedings of the 21th Large Installation System Administration Conference*, pages 203–213, 2007.

[16] R. Jain. *The Art of Computer Systems Performance Analysis*. Wiley Professional Computing. Wiley, 1991.

[17] Julian Seward. bzip2 and libbzip2. `http://www.bzip.org/`. Accessed Sep. 17th, 2012.

[18] T. Kalibera and R. Jones. Handles Revisited: Optimising Performance and Memory Costs in a Real-Time Collector. In *Proceedings of The International Symposium on Memory Management*, ISMM '11, pages 89–98, New York, NY, USA, 2011. ACM.

[19] T. Kalibera and P. Tuma. Precise Regression Benchmarking with Random Effects: Improving Mono Benchmark Results. In *Proceedings of the Third European Conference on Formal Methods and Stochastic Models for Performance Evaluation*, EPEW'06, pages 63–77, Berlin, Heidelberg, 2006. Springer-Verlag.

[20] D. Montgomery. *Design and Analysis of Experiments*. John Wiley & Sons, 2008.

[21] T. Mytkowicz, A. Diwan, M. Hauswirth, and P. F. Sweeney. Producing Wrong Data Without Doing Anything Obviously Wrong! *SIGPLAN Notes*, 44(3):265–276, Mar. 2009.

[22] NLANR/DAST. Iperf. `http://iperf.sourceforge.net/`. Accessed Sep. 17th, 2012.

[23] L. Paterson and T. Roscoe. The Design Principles of PlanetLab. *Operating Systems Review*, 40(1):11–16, January 2006.

[24] L. Peterson, A. Bavier, M. E. Fiuczynski, and S. Muir. Experiences Building PlanetLab. In *Proceedings of The 7th Symposium on Operating Systems Design and Implementation*, OSDI '06, pages 351–366, Berkeley, CA, USA, 2006. USENIX Association.

[25] L. Peterson and V. S. Pai. Experience-Driven Experimental Systems Research. *ACM Communications*, 50(11):38–44, 2007.

[26] PlanetLab. PlanetLab Bibliography. `http://www.planet-lab.org/biblio` visited 2012-09-28.

[27] Standard Performance Evaluation Corporation. SPEC CPU2006. `http://www.spec.org/cpu2006/`. Accessed Sep. 17th, 2012.

[28] T. Tannenbaum, D. Wright, K. Miller, and M. Livny. Condor – a distributed job scheduler. In T. Sterling, editor, *Beowulf Cluster Computing with Linux*. MIT Press, October 2001.

[29] The Gentoo Foundation. Gentoo Linux. `http://www.gentoo.org/`. Accessed Oct. 5th, 2012.

[30] The Tukaani Project. XZ Utils. `http://tukaani.org/xz/`. Accessed Sep. 17th, 2012.

[31] W. F. Tichy. Should Computer Scientists Experiment More? *IEEE Computer*, 31(5):32–40, 1998.

[32] W. F. Tichy, P. Lukowicz, L. Prechelt, and E. A. Heinz. Experimental Evaluation in Computer Science: A Quantitative Study. *Systems Software*, 28:9–18, 1995.

[33] Vince Weaver. Perf Event Overhead Measurements. `http://web.eecs.utk.edu/~vweaver1/projects/perf-events/benchmarks/rdtsc_overhead/`. Accessed Sep. 17th, 2012.

[34] J. Vitek and T. Kalibera. Repeatability, Reproducibility, and Rigor in Systems Research. In *Proceedings of The Ninth ACM International Conference on Embedded Software*, EMSOFT '11, pages 33–38, New York, NY, USA, 2011. ACM.

[35] W. Bergmans. Maximum Compression. `http://www.maximumcompression.com/data/files/index.html`. Accessed Sep. 17th, 2012.

[36] R. P. Weicker. Dhrystone: A Synthetic Systems Programming Benchmark. *ACM Communications*, 27(10):1013–1030, Oct. 1984.

Workload Resampling for Performance Evaluation of Parallel Job Schedulers

Netanel Zakay Dror G. Feitelson
School of Computer Science and Engineering
The Hebrew University of Jerusalem
91904 Jerusalem, Israel

ABSTRACT

Evaluating the performance of a computer system is based on using representative workloads. Common practice is to either use real workload traces to drive simulations, or else to use statistical workload models that are based on such traces. Such models allow various workload attributes to be manipulated, thus providing desirable flexibility, but may lose details of the workload's internal structure. To overcome this, we suggest to combine the benefits of real traces and flexible modeling. Focusing on the problem of evaluating the performance of parallel job schedulers, we partition each trace into independent subtraces representing different users, and then re-combine them in various ways, while maintaining features like the daily and weekly cycles of activity. This facilitates the creation of longer workload traces that enable longer simulations, the creation of multiple statistically similar workloads that can be used to gauge confidence intervals, and the creation of workloads with different load levels.

Categories and Subject Descriptors

C.4 [**PERFORMANCE OF SYSTEMS**]: Design studies; I.6.5 [**SIMULATION AND MODELING**]: Model Development; D.4.8 [**OPERATING SYSTEMS**]: Performance—*Simulation*; D.4.1 [**OPERATING SYSTEMS**]: Process Management—*Scheduling*

General Terms

Performance

Keywords

Workload trace; Resampling; Simulation

1. INTRODUCTION

The performance of a computer system is affected by the workload it handles. Reliable performance evaluations therefore require the use of representative workloads. This means that the evaluation workload should not only have the same marginal distributions as the workloads that the system will have to handle in production use, but also the same correlations and internal structure. As a result, traces of real workloads are often used to drive simulations of new system designs, because such traces obviously contain all the structure found in real workloads.

Replaying a trace only provides a single data point of performance for one workload. But in many evaluations, several related workloads are needed. For example, in order to compute confidence intervals, one needs multiple instances of the same basic workload. The common way to satisfy this need is to create multiple synthetic workloads based on statistical workload models (which, in turn, are based on the traced data) [13, 2, 20, 24, 28]. While models provide the required variability and flexibility for evaluations, they also suffer from not necessarily including all the important features of the real workload [1] — in fact, they include only those of which the modeler was aware.

To improve the representativeness of evaluation workloads we propose to *combine the realism of real traces with the flexibility of models*. This will be done by modeling only the part of the workload that needs to be manipulated, and resampling from the real data to fill in the remaining details. Technically this is done by partitioning workload traces into their basic components and regrouping them in different ways to achieve the desired effects.

The domain of our work is parallel job scheduling. Parallel systems are increasingly relevant today, with the advent of multi-core processors (parallelism on the desktop), clusters and blade servers (parallelism at the enterprise level), and grids (parallelism across multiple locations). The jobs that run on parallel systems are composed of multiple processes that need to run on distinct processors (in large clusters and supercomputers the number of processes and processors can be in the thousands). When a job is submitted the user specifies how many processors are needed, and often also for how much time. The scheduler then determines the order in which jobs will be executed, and which processors will be allocated to each one. Accounting logs from large-scale systems are available in the Parallel Workloads Archive [11], and provide data about the workloads they served. (The logs we use in this work are listed in Table 1.) In particular, logs typically contain information about the submit time of each job, it's runtime and number of processes, the user who submitted it, and more. These logs can therefore be used to simulate the behavior of new scheduler designs and compare them with each other.

In this context, we suggest that the resampling be done at the level of users. We first partition the workload into individual subtraces for the different users, including all the jobs submitted by each user throughout the tracing period. We then sample from this pool of users to create a new workload trace. Using such resampling, we can achieve the following:

ICPE'13, April 21–24, 2013, Prague, Czech Republic.

Log	File	Period	PEs	Users	Jobs
LANL CM5	LANL-CM5-1994-3.1-cln	10/94–09/96	1024	213	201,387
CTC-SP2	CTC-SP2-1996-2.1-cln	06/96–05/97	338	679	77,222
KTH-SP2	KTH-SP2-1996-2	09/96–08/97	100	214	28,489
SDSC-SP2	SDSC-SP2-1998-3.1-cln	04/98–04/00	128	437	59,725
OSC-clust	OSC-Clust-2000-3.1-cln	01/00–11/01	178	253	36,097
SDSC-BLUE	SDSC-BLUE-2000-3.1-cln	04/00–01/03	1152	468	243,314
HPC2N	HPC2N-2002-1.1-cln	07/02–01/06	240	257	202,876
SDSC-DS	SDSC-DS-2004-1	03/04–04/05	1664	460	96,089
Intrepid	ANL-Intrepid-2009-1	01/09–09/09	163,840	236	68,936

Table 1: *Logs from the Parallel Workloads Archive (www.cs.huji.ac.il/labs/parallel/workload/) that were used in this study.*

- Create a much longer trace than the original, and use it to ensure convergence of evaluation results.
- Create multiple similar workloads, and use them to compute confidence intervals.
- Create workloads with higher or lower average loads, by using more or less concurrently active users, and use them to investigate how load affects system performance.

Importantly, while the resampled workloads differ from the original in length, statistical variation, or load, they nevertheless retain important elements of the internal structure such as sessions and the relationship between the sessions and the daily work cycle.

Workload manipulations are an important tool in the performance analyst's toolbox, that has not received its due attention in terms of methodological research. As a result, inappropriate manipulations are sometimes used, which in turn has led to some controversy regarding whether any manipulations of real workloads are legitimate. By increasing our understanding of resampling-based manipulations we hope to bolster the use of this important tool, allowing new types of manipulations to be applied to workload traces, and enabling researchers to achieve better control over their properties, as needed for different evaluation scenarios.

In the rest of this paper we describe this promising approach to using workload traces and demonstrate its effectiveness. The next section further explains the motivation for using resampling. In Section 3 we consider different resampling granularities, and justify the decision to do so at the level of all the activity of each user. Section 4 explains how the resampling is done in considerable detail, including the proposed distinction between long-term and temporary users. Section 5 then demonstrates the use of resampling to achieve the objectives listed above, and also suggests some additional potential uses. We conclude in Section 6.

2. WHY USE RESAMPLING

The Parallel Workloads Archive includes more than 20 workload traces from different systems, but this may not always suffice. Some of the traces may not be appropriate for certain system types (for example, throughput-oriented systems often allow only serial jobs). Some traces are dated and may not represent present practices. Evaluations may require certain attributes that are not available in the archive, e.g. a series of workloads whose loads differ by 5%. Even if one has access to a real system one cannot force the workload on it to conform to a desired configuration.

Resampling is a powerful technique for statistical reasoning in such situations, when not enough empirical data is available [6, 7]. The idea is to use the available data sample as an approximation of the underlying population, and resample from it. This enables multiple, quasi-independent samples to be created, which are then used to compute confidence intervals or other metrics of interest that depend on the unknown underlying distribution.

Our ideas for workload manipulation are analogous to this. We have a workload trace at our disposal. The problem is that this provides a single data point, whereas our evaluation requires the use of several (maybe many) workloads with certain variations. The proposed solution is to partition the given workload into its constituents, and re-group them in different ways to create new workloads. The simplest approach is to partition the workload into its most basic components (e.g. jobs), and resample at random. This is similar to just using the empirical distribution as a model. Our proposal is to extend this in two ways:

1. We consider different definitions of what constitutes the basic elements of the workload. For example, they could be individual jobs, batches of related jobs, complete user sessions, or even the sequence of all the sessions by each user.
2. Resampling may not be random, but guided by some specific manipulation that we want to apply to the workload, and also subject to constraints such as maintaining system stability.

The notion that this is a useful device is our working hypothesis; examples and evidence supporting this notion are given below.

We note that while we believe such resampling to be relatively novel in the context of computer workloads and performance evaluation, analogies from other fields of computer science do exist. One analogy comes from computer graphics, where texture mapping is often done by replicating a small patch of texture, with certain variations to give an impression of perspective, conform to lighting conditions, and avoid an obvious tiling effect [17]. More relevant to our work on workloads, such replication, modification, and patching together has also been done for temporal signals, such as movement specification [15] and sound [4]. Another analogy comes from the joint time-space analysis of video. Here the idea is to partition a video into patches, and then replace certain patches with others, e.g. to reconstruct missing frames or add or remove objects [30]. This technique can also be used for anomaly detection: if a piece of a new video cannot be reconstructed from snippets that exist in the system's database, then it is anomalous [3].

To the best of our knowledge resampling-based workload manipulations as we propose here have been used only in a very limited manner. The closest related work we know of is the Tmix tool, used for the generation of networking traffic. This tool extracts communication vectors describing different connections from a traffic log (sequences of ⟨req_bytes, resp_bytes, think_time⟩) and then replays them subject to feedback from the simulated system's performance [29]. A subsequent paper also mentions the possibility of resampling traces to create diverse load conditions [12], but their approach is simpler than ours as they do not use the concept of sessions nor retain phenomena like the daily cycle. A similar construction was proposed by Krishnamurthy et al. in the context of evaluating e-commerce systems [16]. In this case they reuse sequences of user operations in order to ensure that illegal sequences are not generated by mistake by the workload modeling procedure. In the domain of parallel job scheduling, Ernemann et al. resize and replicate jobs in order to make a trace suitable for simulations with a larger machine [8]. Our goal, in contrast, is to use the resampled traces to perform better and more comprehensive evaluations. Kamath et al. have suggested to merge several traces and simulate a queueing mechanism in order to increase load [14]. However, this is limited to load values that are the sums of loads from existing traces. Ebling and Satyanarayanan created micro-models of application file behaviors based on a trace, and then combined them stochastically to create test workloads [5]. Again this is similar in concept, except that we prefer to use real data directly at the bottom layer rather than to risk models that may miss important details.

3. GRANULARITY OF RESAMPLING

Resampling can be done at different levels. In many cases, the coarsest level is the activity of a user, which may be partitioned into sessions. The constituents of a session depend on what sort of work we are looking at. It can be the submittal of parallel jobs, downloads from web servers that are composed of packets being sent over the Internet, or individual accesses to file data.

Resampling at the job level is similar to resampling in statistics, e.g. as applied in the bootstrap method [6], which is similar to using the empirical distribution as a model. Note, however, that by resampling complete jobs we retain the correlations between job attributes (e.g. job size and runtime), which would be lost if we resampled from each marginal distribution independently.

Resampling is all about restructuring the workload. But at the same time, we wish to retain at least some of the local structure. Specifically, we typically want to retain the locality properties exhibited by normal work practices. Also, it may be important to retain the structure of batches of related jobs or sessions. For example, this is necessary for the evaluation of adaptive systems that learn about their workload and adapt to changing workload conditions [25, 9]; without locality and structure, such systems don't have what to exploit.

To motivate the use of resampling at the user level, we studied the correlation between each user's jobs. First we divided the work of each user into sessions [31]. Then we compared the jobs in one session with each other and with the jobs in subsequent sessions, using three attributes: the number of processors used, the jobs' runtimes, and their estimated runtimes. The metric for comparison was the ratio of the smaller value as part of the larger one: $r = \min\{j1.att, j2.att\}/\max\{j1.att, j2.att\}$. This is by definition in the range $[0, 1]$, with 0 indicating a large difference and 1 indicating identity.

Fig. 1 shows a sample of the results, using logs available from the Parallel Workloads Archive [21]. The X axis is the distance in sessions between the compared jobs, and the Y axis shows the average or median of the level of similarity. Obviously jobs in the same session tend to be more similar to each other, and the degree of similarity is reduced with distance. In effect, this testifies to the existence of locality in these workloads, which we want to retain.

To validate this result, we used bootstrapping to compare the results shown above with results that would be obtained if we sample jobs independently. To do so we retain the structure of sessions for each user, but mix the jobs randomly among the sessions. This is repeated 1000 times, and each time the degrees of similarity between jobs in the same session are computed as above. Fig. 2 shows a sample of the results. Obviously, the similarity among jobs that

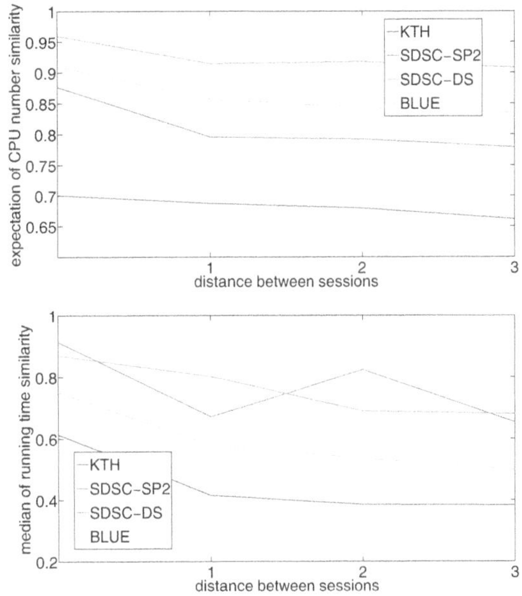

Figure 1: *Similarity between jobs as a function of the distance between them in sessions.*

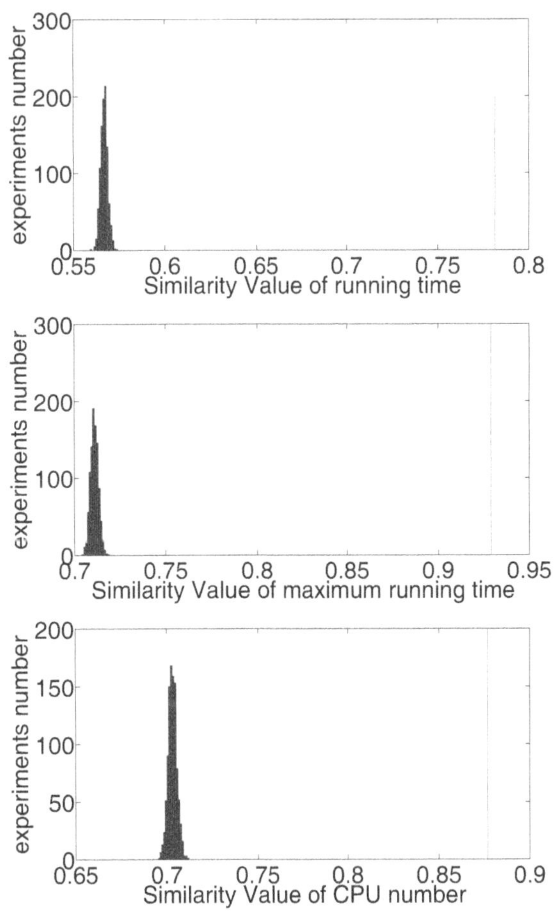

Figure 2: *Comparison of the similarity between jobs in the same session as computed from the SDSC-SP2 log (vertical line), and the distribution of similarity levels that are seen when the jobs are randomized.*

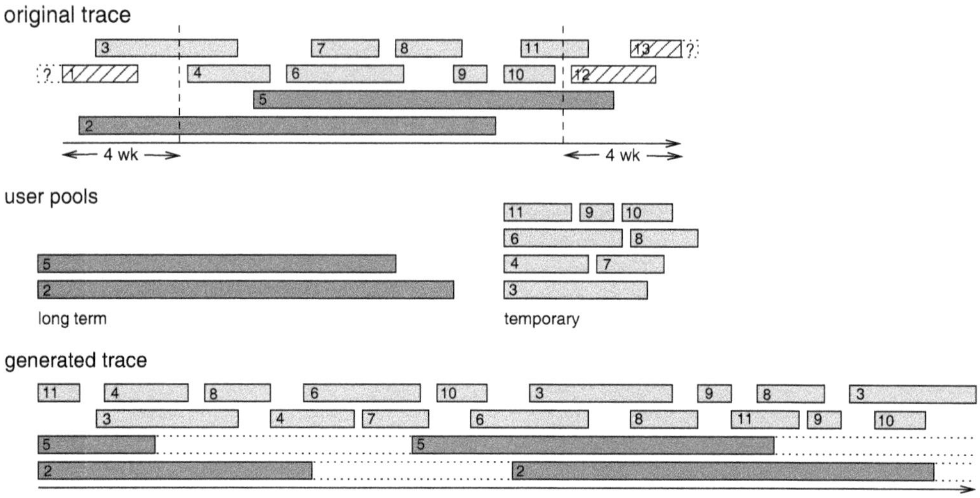

Figure 3: *Conceptual framework of dividing users into long-term and temporary, and reusing them in a generated trace.*

appear together in the original log is much higher than the similarity that would be obtained if jobs are randomized, as would happen if we resample individual jobs.

An implicit assumption in our resampling procedure is that users are independent. This is not strictly valid because users affect each other: if one user overloads the system, others may feel this and reduce their own activity. However, a large part of such interactions is due to all users operating on the same daily cycle, and we take care to retain this correlation between the resampled users. Moreover, resampling at the user level rather than at the session level allows for more sophisticated user behavior models. Specifically, we can introduce feedback effects whereby a user may decide to terminate a session because system performance is poor, and submit his subsequent jobs in a later session. Our work on incorporating such feedback effects will be reported separately; for now the main point is that if we resample at the session granularity such effects will be effectively excluded.

Based on the above considerations, we decided to perform our resampling at the user level, in order to retain the locality in the modified workloads that we produce and allow for the inclusion of feedback effects.

4. MECHANICS OF RESAMPLING

Creating a new workload by resampling users means that we dissect the given trace into sub-traces representing different users, and then recombine these sub-traces in different ways. Note that we do not manipulate each user's sub-trace. Thus the sequence of jobs representing each user will be the same as in the original trace, and the intervals between them will also be the same. This guarantees the same locality properties as in the original trace, as noted above. We also take care to synchronize the resampled users using a common timeframe, so that jobs always start on the same day of the week and the same time of the day as in the original trace. This ensures that the daily cycle of activity is retained in the produced workload, which may be important [32, 10].

An important issue in dissecting a trace into separate users is how to handle end effects. After all, there is no reason to assume that the beginning or end of the tracing period is synchronized in any way with the beginning or end of the activity of any particular user. We approach this problem by making a distinction between temporary users and long-term users (see Fig. 3).

	long-term		temporary	
Log	users	jobs	users	jobs
CTC-SP2	314	63,287	236	10,625
KTH-SP2	102	25,202	66	2,349
SDSC-SP2	173	44,251	206	8,790
SDSC-BLUE	426	221,745	31	1,435
HPC2N	178	194,429	66	7,949
SDSC-DS	230	74,764	192	9,012

Table 2: *Results of classifying users in the different logs.*

Temporary users are all the users that interact with the system for a limited time, for example while conducting a project. These users arrive to the system at a certain point, interact with it for a short while, and are expected to leave shortly after that and never return. Long-term users, in contradistinction, are the users that routinely use the system all the time. These users may be expected to have been active before logging started, and to send more jobs also after the end of the recording period.

In analyzing the log, we distinguish between temporary users and long-term users according to the interval between their first job and their last job in the log. If the interval is long enough (above 12 weeks in our implementation), the user is classified as long-term. Otherwise the classification is temporary. The threshold of 12 weeks is chosen based on observation of the distribution of periods of activity by different users. We found that for most users their period of activity was up to about 12 weeks; these are the temporary users. For the rest there was a uniform distribution from 12 weeks to the full length of the log. This is interpreted as representing long-term users whose activity was arbitrarily intersected with the logging period. The numbers of temporary and long-term users found in different logs, and the jobs that they submitted, are shown in Table 2. There tend to be somewhat more long-term users than temporary ones, but as may be expected, the long-term users submit the vast majority of the jobs.

Data about the different users is kept in separate user pools, one for temporary users and the other for long-term users. However, temporary users whose full period of activity falls within a short time (4 weeks) from the beginning or the end of the logging period are discarded. The reason for doing so is that there is a high probability that the activity of these users was truncated, but we

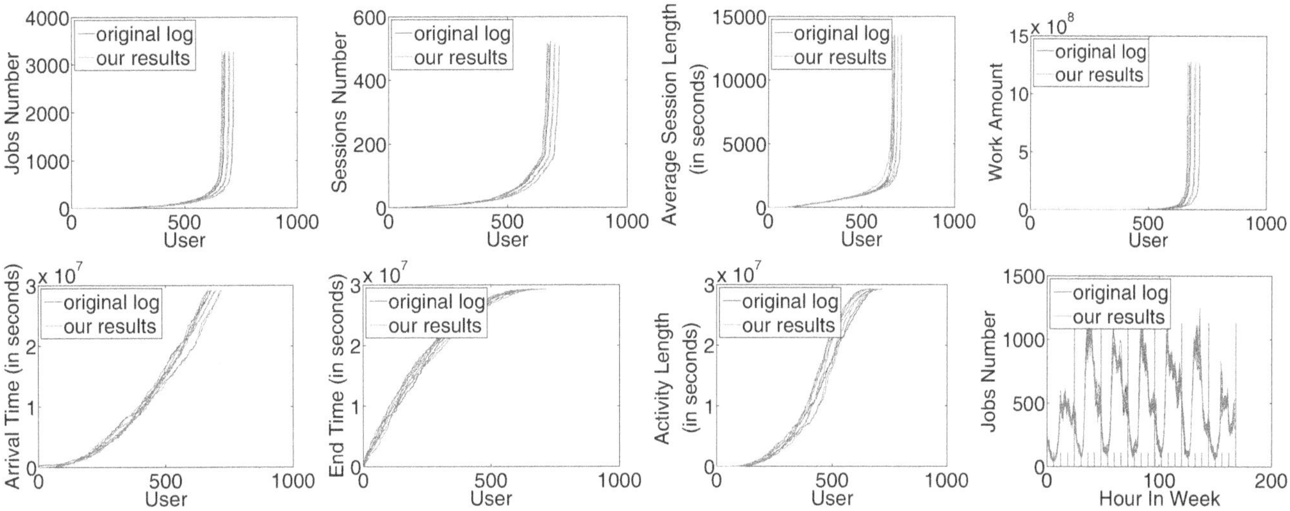

Figure 4: *Comparison between the original CTC SP2 log and 8 generated logs based on it. The last plot includes marks every 6 hours and a longer one at midnight.*

cannot know for sure. The threshold of 4 weeks is chosen because when plotting the cumulative number of users observed as a function of the number of weeks into the log, in the first few weeks the graph climbs at a higher rate. This is interpreted as being influenced by first observations of users that have already been active before. Then, when the increase settles on a lower and relatively constant average rate, this is interpreted as predominantly representing the arrivals of new users.

Given the pools of temporary and long-term users, the resampling and generation of a new trace is done as follows:

- **Initialization:** We initialize the trace with some temporary users and some long-term users. The numbers of users to use are parameters of the trace generation, and can be used to change the load or the ratio of temporary to long-term users (the defaults are the numbers of long-term users in the original log, and the average number of temporary users present in a single week of the original log). The probability to select each temporary user is proportional to the number of weeks during which the user was active in the log. Users are not started with their first job in the trace. Rather, each user is started in some arbitrary week of his traced activity. However, care is taken that jobs start on the same day of the week and time of the day in the simulation as in the original trace.
- **Temporary users:** In each new week of the simulation, a certain number of new temporary users are added. The exact number is randomized around the target number, which is a parameter of the trace generation (the default is the average rate at which temporary users arrived in the original trace). The randomization uses a binomial distribution, with a probability p equal to the fraction of temporary users expected to start every week. The selected users are started from their first traced jobs. A user can be selected from the pool multiple times, but care is taken not to select the same user twice in the same week.
- **Long-term users:** The population of long-term users is constant and consists of those chosen in the initialization. When the traced activity of a long-term user is finished, it is simply regenerated after a certain interval. While such repetitions are not very realistic, they allow us to extend the work of the long-term users as needed. We also note that repetitions only occur after rather long intervals, because logs are typically at least a year long. The interval between the regenerations corresponds to the sum of the intervals between the user's period of activity and the full logging period. Naturally the regenerations are also synchronized correctly with the time and day.

Note that this process can go on indefinitely, and indeed one of the applications of workload resampling that we describe in the next section is to extend traces and allow for longer simulations.

The exact number of users in the initialization, the week of activity from which they start, the number of temporary users added each week, and the identity of the selected users are all randomized. Therefore our simulation creates a different workload in each run. But all these workloads are based on the same sub-sequences of jobs, and are therefore all statistically similar to each other and to the original trace.

In order to perform resampling and implement the applications described in the next section it is enough to just create a new workload trace that is composed of the jobs of the different users as described above. However, we actually perform a full simulation of also scheduling these jobs. This enables us to directly use the generated workloads to evaluate various parallel job schedulers. In subsequent work we also consider adding feedback, whereby the system performance influences user behavior and may affect when subsequent jobs are submitted [23]. In any case, the simulation also creates a log file which contains the new workload. Comparing this generated workload with the original one allows us to validate the resampling process.

To validate this procedure we compare the generated workload to the original one. An example is shown in Fig. 4 based on the CTC SP2 log. The different panels show the following distributions:

- Number of jobs submitted by different users
- Number of sessions performed by users
- Average session length for different users
- Total amount of CPU time (work) used by users in all their jobs
- The users' first arrival times

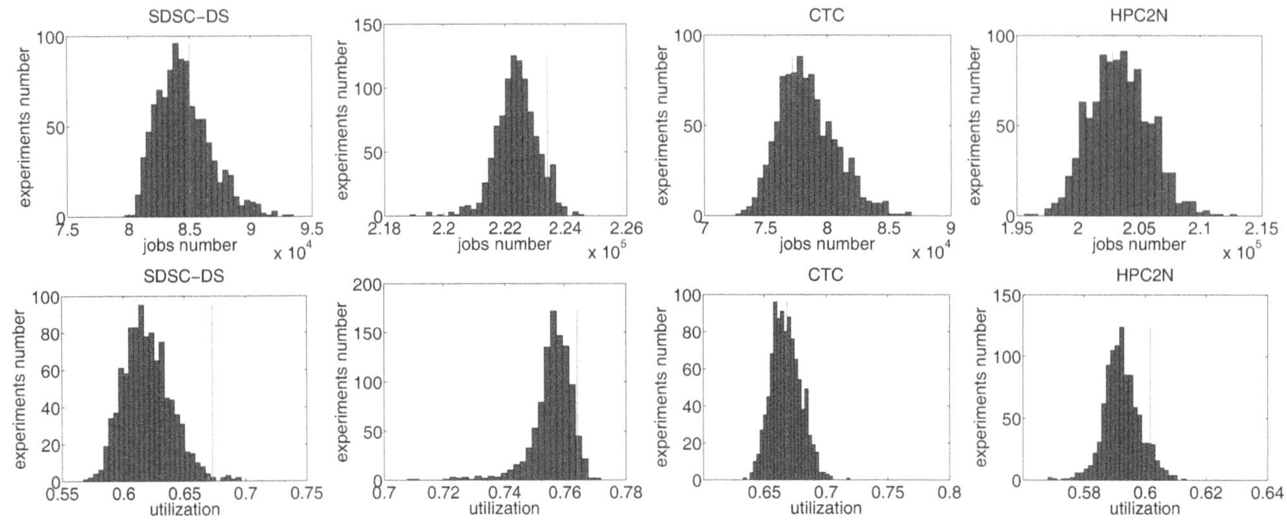

Figure 5: *Histograms of the throughput and utilization in a thousand simulations with resampled workloads compared to using the original workload (vertical red line).*

- The users' final departure times
- The users' periods of activity
- The distribution of job arrivals across days of the week, for all users

In all but the last of these, the users are first sorted according to the metric, and then the distribution is plotted. The X axis specifies the users' serial numbers after this sorting. Note that the number of users participating in each workload may be slightly different, due to the random selection of how many new users arrive each week. As we can see, all the distributions are very similar to the original one. This is attributed to the fact that despite the random mixing due to the resampling, the sequence of jobs for each user is retained.

5. APPLICATIONS OF RESAMPLING

The use of resampling is expected to lead to more reliable performance evaluations, due to being based more closely on real workload traces, and incorporating all the complexities of real workloads — including those that are unknown to the analyst. In the following we discuss some examples.

5.1 Verification of Performance Results

As noted above, one of the problems with using a workload trace directly is that it provides a single data point. This has the obvious deficiency that it is impossible to calculate any kind of confidence intervals. But with resampling we can create many resampled randomized versions of the workload, and evaluate the performance of the system with all of them, thus obtaining multiple data points that all adhere to the same underlying statistics. The distribution of these data points can then be used to compute confidence intervals for performance metrics. This is essentially an application of the well-known technique of bootstrapping used in statistical analysis [7].

Given the resampling mechanism described above, implementing this idea is trivial: simply create a large number of workloads, say 1000, based on the original log, run the scheduler simulation on all of them, and tabulate the results. But to check this we need to also examine the basic characteristics of the produced workloads, and convince ourselves that they remain representative. To do so we indeed generated 1000 resampled variants of each log, calculated various metrics on each of these 1000 variants, and created a histogram of these metric values. We also included the original values for comparison.

We performed the checks on eight different logs from the Parallel Workloads Archive, and results for four of them are shown in Fig. 5. The top row shows that the throughput (that is, average number of jobs per unit time) was typically distributed around the original value. the result for the BLUE log was the largest deviation observed; with this log 92% of the variants had a lower throughput than the original log, but the difference between the median throughput and the original was only 0.45%. For utilization (bottom row) the results were more diverse, and varied between distributions around the original value — as for CTC — and distributions that are generally below the original value — as for DS, which was the most extreme. This may indicate some systematic bias which we do not understand yet. But note that the four logs that were checked and are not shown were all less extreme than BLUE or HPC2N.

Accepting the generated workload distributions as reasonable, we turn to check the results of evaluations of the EASY scheduler, which is probably the most commonly used backfilling scheduler [18]. The results for waiting time and slowdown are shown in Fig. 6 (results for response time exhibit similar behavior to wait time). The slowdown results are the most varied. In five of the eight logs we checked, the distribution was more or less around the value obtained using the original log. This is exemplified by the BLUE and CTC logs in the figure. But in other cases the original result was at the very end of the distribution, either higher or lower than nearly all the others (as in DS or HPC2N, respectively, which were the two most extreme cases). The results for slowdown were more one-sided, being distributed either around the original values (as for DS and CTC) or largely above them (as for BLUE and HPC2N). The explanation appears to be that in some logs there are more sparse days, in which very few jobs arrive and therefore all wait times are short or nil. Our resampling tends to distribute users and jobs somewhat more evenly.

For both metrics, these results underscore the importance of using the resampling methodology to identify cases where the result

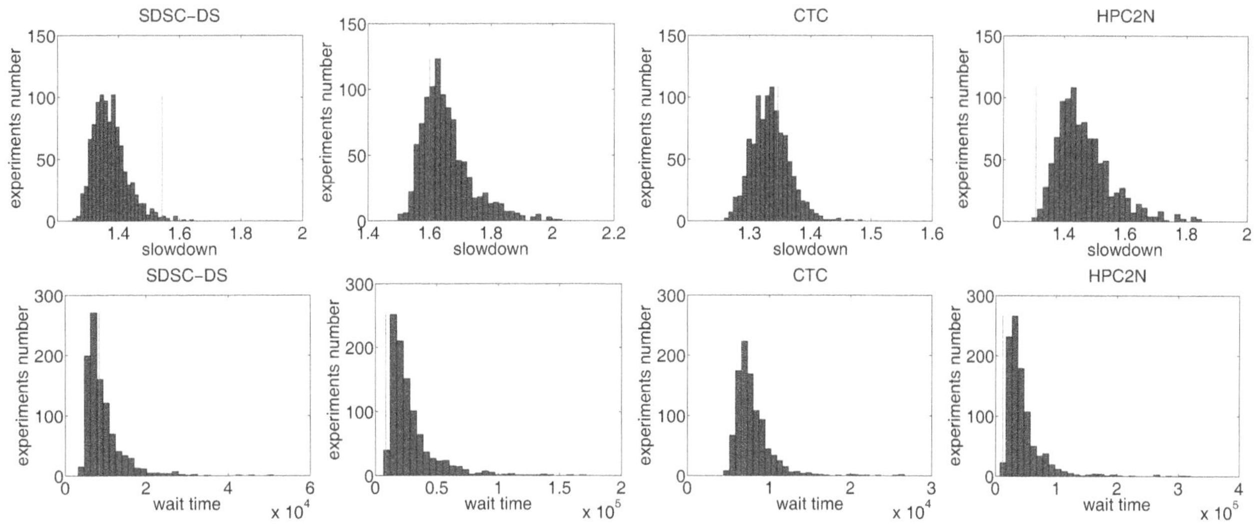

Figure 6: *Histograms of the average slowdown and waiting time in a thousand simulations of EASY on resampled workloads, compared to a simulation using the original logs (vertical red line).*

using the original log may not be truly representative. Importantly, the spread of the results indicates that the resampling indeed produces workloads that are different from each other, even though they are derived from the same source and exhibit the same statistics. On the other hand, we never saw results that were completely separated from the original result, meaning that in all cases at least some of our 1000 repetitions produced results like the original log. Also, in most cases the most extreme differences were not more than 10–20%.

Note that this application of bootstrapping serves only to provide confidence intervals for evaluations based on a single log. We consider the possibility of extending this by mixing data from multiple logs in Section 5.4.1. Such mixing will provide confidence intervals for more general evaluations that are based on all the available data.

5.2 Extending a Trace

Another simple use of workload resampling is in order to extend a trace. While some of our workload traces are pretty long, with hundreds of thousands of jobs submitted over 2 years or more, others are shorter. In addition, a significant part of the trace may be needed as a "warmup period" to ensure that the simulated system achieves its steady state [22]. Given only the raw traces, the length of the simulation may therefore be quite limited.

But with resampling we can extend the simulation to arbitrary lengths. As indicated above, this is achieved by regenerating long-term users, and randomly sampling new temporary users every week. In principle this can be continued indefinitely.

To check the resulting extended workloads, we studied three repetitions of extending given traces to five times their original length. For example, given a trace that represented one year's worth of activity, we used it to create three traces that are each five years long. We then compared the original trace with the first year, the third year, and the fifth year of each repetition. The results for the BLUE log are shown in Fig. 7, using the same distributions as in Fig. 4.

As one can see, the distributions for all three repetitions and the three periods of the extended trace all agree with each other and with the original trace data to a high degree. Note that we treat each of the three periods as a separate log, and do not carry over users that were identified in one period to another period. This causes the distributions of arrival times and end times to be separated into three, corresponding to the different periods. Remarkably, in each of these we see the same end effects as in the original shorter trace.

5.3 Changing the Load

An important aspect of systems performance evaluation is often to check the system's performance under different load conditions, and in particular, how performance degrades with increased load. Given a single trace, crude manipulations are typically used in order to change the load. These are

- Multiplying all arrival times by a constant, thus causing jobs to arrive at a faster rate and increasing the load, or causing them to arrive at a slower rate and decreasing the load. However, this also changes the daily cycle, for example causing jobs that were supposed to terminate during the night to extend into the next day. An alternative approach that has a similar effect is to multiply all runtimes by a constant. This has the deficiency of creating an artificial correlation between load and response time.
- Multiplying all job sizes (here meaning the number of processors they use) by a constant, and rounding to the nearest integer. This has two deficiencies. First, many jobs and machine sizes are powers of two. After multiplying by some constant in order to change the load, they will not be powers of two, which may have a strong effect on how they pack, and thus on the observed fragmentation. This effect can be much stronger than the performance effects we are trying to measure [19]. Second, small jobs cannot be changed with suitable fidelity as the sizes must always be integers. An alternative approach that has essentially the same effect is to modify the machine size. This at least avoids the problem presented by the small jobs.

With resampling, however, manipulating the load is relatively easy: One can simply increase or reduce the average number of active users. This changes the load while retaining all other attributes of the workload and avoiding the introduction of any artifacts. In particular, some logs have a very low utilization, in the range of 10–30%, which makes them uninteresting in terms of evaluating

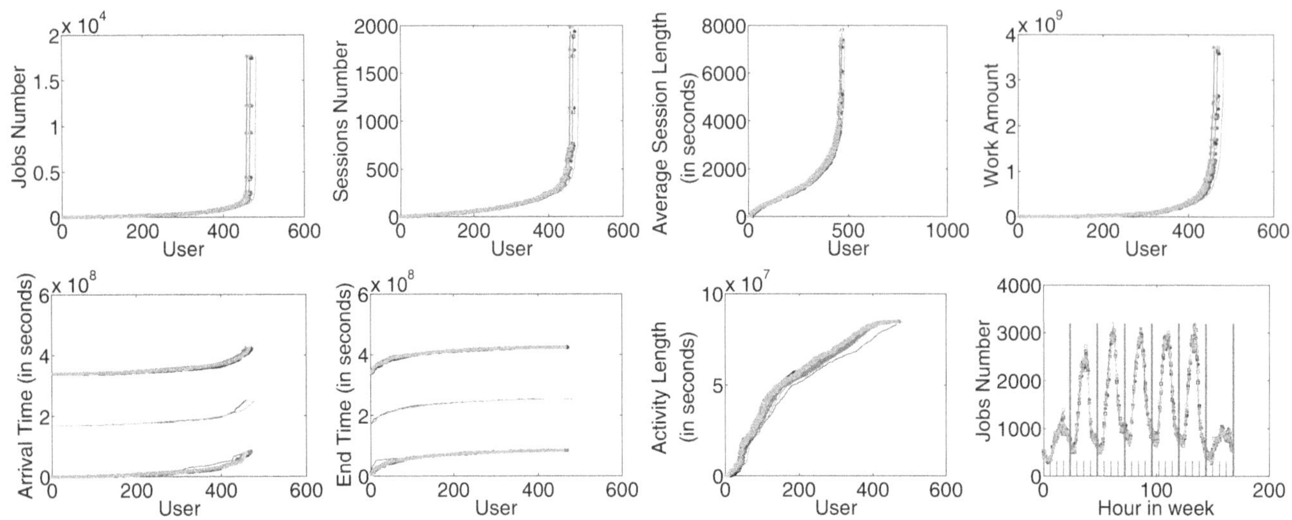

Figure 7: *Comparison between the original BLUE log to the first, third, and fifth parts of an extended resampled log that is 5 times longer.*

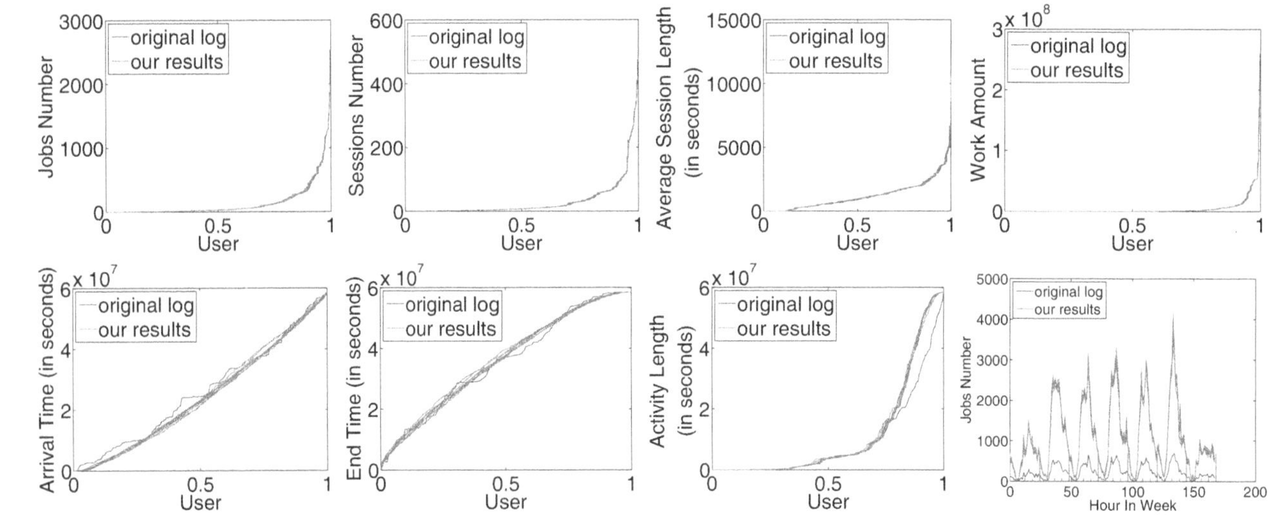

Figure 8: *Comparison between the original OSC-Cluster log to resampled workloads where the load is increased by a factor of 6.*

schedulers for parallel machines (because there are seldom enough concurrent jobs for the scheduler to have to make any decisions). Using resampling we can increase the load significantly and make these logs usable.

To implement this, three minor changes need to be made in the mechanism described above. The first is to change the number of long-term users in the initialization. Additional long-term users will be started as needed based on a random selection, taking care to use all existing long-term users before replicating one that was selected already, and also taking care that replicas of the same user will have a large difference in their start times. Likewise, we need to change the number of temporary users in the initialization. Finally, we need to change the rate at which additional temporary users arrive each week.

When users (and load) are added, the simulated system may saturate. We identify such conditions and ignore the saturated simulation results with a warning. Identifying saturation is based on noticing that the number of outstanding jobs (jobs that have arrived but not terminated yet) tends to grow. This is done as follows.

1. Tabulate the number of outstanding jobs at the beginning of each week of the simulation.
2. If the number of outstanding jobs grows due to a load fluctuation, but then decreases again, this does not indicate saturation. Therefore we replace each weekly count by the minimum count from that week to the end of the simulation, leading to a non-decreasing sequence of counts.
3. Delete the last 20% of the values, to avoid false positives based on fluctuations that occur towards the end of the simulation.
4. Use linear regression to fit a straight line to the remaining counts. If the slope is lower than 1 (meaning that on average the number of outstanding jobs grows by no more than one job per week) the simulation is declared stable. If it is higher, the simulated system is saturated.

Verifying that resampling with a modified number of users leads to reasonable workloads shows that indeed all the distributions are similar to those of the original traces (but taking into account that

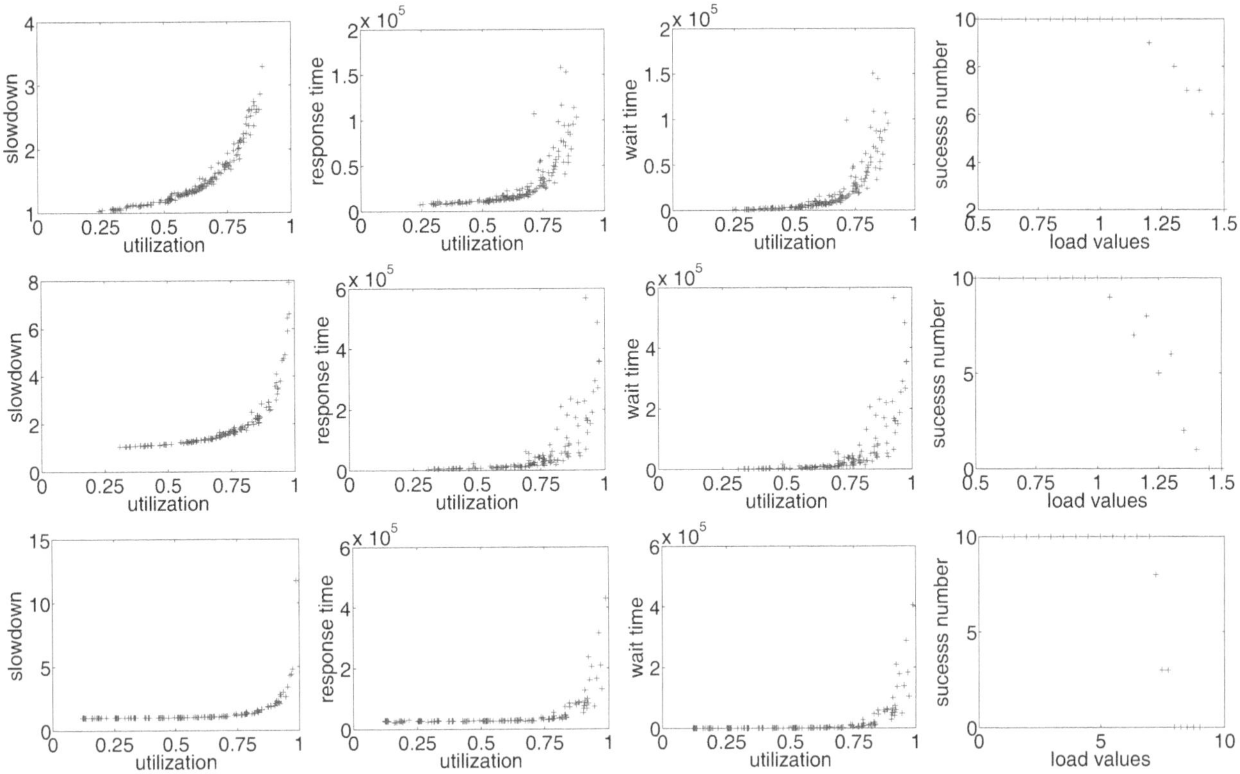

Figure 9: *The performance of EASY under different load conditions for different logs: SDSC-DS, BLUE, and OSC-Cluster respectively from top to bottom. Recall that the original OSC cluster log has very low load, so the load had to multiplied by higher factors to reach the range of interest.*

the number of users is different). Fig. 8 shows the results for one extreme case, based on the OSC cluster log. The average utilization of this log is only 12.8%, making it unusable for evaluations of parallel job schedulers. We therefore increased the number of users by a factor of 6, targeting an average utilization of approximately 76.8%. In the graphs, the user numbers on the X axis are normalized to the range $[0, 1]$ to enable comparison with the original log that has much fewer users. It is easy to see that the high-load simulations produce distributions that are very similar to the original log. The main difference is in the arrival time and end time distributions, which are smoother, because in our simulations users arrive at a constant average rate. Also, in the last graph portraying the weekly cycle of activity, one can see the big difference in the number of jobs that are being used.

The goal of all these workload manipulations is to enable the evaluation of parallel job schedulers, and in particular, their performance under different load conditions. To check this we again used simulations of the EASY backfilling scheduler [18]. For each log, we multiplied the number of users by various factors in the range 0.8 to 1.5, and performed 10 independent simulations (with different randomized resampling) for each load value. For the OSC cluster log, the range was from 1 to 9, because the original utilization of this log is very low as noted above. A sample of the results in terms of slowdown, response time, and waiting time are shown in Fig. 9.

The last panel for each log shows the fraction of simulations at each load level that were successful, meaning that our automated procedure did not conclude that the system is becoming saturated. Note that the loading factor, namely the factor by which we multiply the number of users, does not translate directly and deterministically into a commensurate change in the utilization. Due to the random selection of users there may be fluctuations in the load. Therefore we find that when the loading factor grows beyond 1, which represents the original load, the number of successful simulations begins to drop. Consequently there are fewer results for the higher loads, but all the valid results indicate a utilization of no more than 100%.

As the results in Fig. 9 show, the performance profiles are exactly as one might expect from queueing analysis. At low loads performance is good, and increasing the load has little effect. But as the system approaches saturation, the performance deteriorates precipitously. Interestingly, different systems (as represented by the logs of their workloads) have different saturation points. SDSC-DS seems to saturate at less than 90% utilization, whereas BLUE and OSC come close to 100%. This reflects the ability of the scheduler to pack jobs together and reduce fragmentation, and depends both on the scheduler and on the workload statistics.

5.4 Additional Applications

In addition to the above, we note the following ideas for using workload resampling. These have not been tested yet and are presently left for future work.

5.4.1 Mixing Traces

In many cases we have more than one trace at our disposal, typically coming from different locations or different times. To obtain generally applicable results, data from all these traces should be used. This can be done either by performing evaluations based on each trace individually, or by mixing the traces, that is by resampling from a number of traces rather than from only one trace. This mechanism has been used in the past in order to reduce the dependence of analysis results on a single trace [32], or to increase the load [14]. It was also suggested for Tmix [29].

Resampling from several traces is based on the assumption that this is the better way to achieve general results that are independent of the peculiarities of any individual trace. An interesting research question is whether this is indeed the case. And could it be that mixing and evaluations are actually transitive, and equivalent results are obtainable by averaging of results from multiple traces that are resampled independently? We intend to study this question by using both approaches and comparing the results.

5.4.2 Improving Stationarity

A special case is using resampling to create a stationary workload trace. Many of the original traces seem to be non-stationary, with different parts having different statistical properties. As a consequence performance results are then some sort of weighted average of the results under somewhat different conditions, but we don't know the details of these conditions or the weights. Resampling can be used to mix the different conditions and create a more uniform trace.

Alternatively, when examined more closely the workloads are sometimes found to be *piecewise* stationary, meaning that they are relatively stationary for some time and then change. It is therefore better to perform several stationary evaluations and combine the results, rather than using a single non-stationary model that might lose important locality properties. Resampling can then be used to create stationary segments that are long enough to be simulated reliably.

5.4.3 Improved Shaking to Reduce Sensitivity

Simulations of system performance are sometimes very sensitive to the exact value of some input parameter. For example, we have found a specific case where changing the runtime of one job from 18 hours and 30 seconds to exactly 18 hours caused the average bounded slowdown of *all* the jobs in the trace to change by about 8% [26]. We developed "shaking" as a general methodology to overcome such mishaps [27].

The idea of shaking is to cause small random perturbations to the workload and re-run the evaluation. This is repeated many times, leading to a distribution of results. This distribution is then used as the outcome of the evaluation, rather than the single point derived from the original trace. The claim is that the distribution (or its mean) more faithfully characterizes the results that would be obtained by this workload and similar ones. Our results indicate that shaking does indeed reduce instability considerably in several different cases.

The original formulation of shaking operated at the job level. Each job was moved slightly independently of the others. This could potentially cause problems if say one job was delayed and a subsequent job was moved forward and ended up before the first job. We therefore intend to now try to perform shaking at a higher level, e.g. by slightly shifting whole sessions, or even the sub-traces belonging to different users. The effect will be evaluated by comparing the original results with our current shaking results and the new shaking results. Shaking will also be compared with resampling to allow for statistical analysis as described above.

5.4.4 Selective Manipulation of the Workload

Another reason to manipulate workloads is to change their properties, so as to check the effect of these properties on system performance. In the current work we treat all users as equivalent, but this is not really so. Some users may run long jobs. Others may prefer numerous small jobs. Some use run jobs that require a lot of memory while others run more compute-intensive jobs.

The implication is that we can influence the characteristics of the workload by classifying users according to their behavior (or the behavior of their applications), and then resample with a selective bias in favor of a certain class of users. This will enable the creation of workloads that stress different parts of the system.

5.4.5 Reducing or Enhancing Locality

Finally, a special case of manipulating the workload is changing its locality properties. Locality can be very meaningful for adaptive systems that learn about their workload and adjust accordingly [9].

The mechanism for changing the locality is to introduce locality into the sampling process. Locality is typically present in user sessions (as we showed in Section 3). Therefore, to reduce locality the resampling must be done at the job level, not the session level. Regrettably, simple randomization does not work, as it violates the workload's stability properties. We will therefore need to develop a mechanism for resampling jobs subject to stability constraints. The question is how to do so and still get good randomization.

Enhancing locality can be done by introducing repetitions, i.e. specifically selecting the same jobs over and over again [9]. However, this also needs to be done subject to stability constraints, and subject to the overall statistical properties of the workload.

6. CONCLUSIONS

Workload resampling is proposed as a mechanism which allows the performance analyst to marry the realism of workload traces with the flexibility of workload models. Moreover, it combines the following attributes:

- Retaining the complex internal structure of the original trace, including features that we do not know about, and
- Allowing for manipulations that affect specific properties that we know about and want to change as part of the evaluation.

The idea is to partition a given workload trace into independent subtraces, representing the activity of individual users. These subtraces can then be re-combined in different ways in order to create new workload traces with desired attributes: they can be longer than the original, have a higher or lower load than the original, or just be different from the original so as to provide an additional data point.

A major concern in this work was how to do the resampling correctly, meaning that the created workloads will be as similar as possible to the original workload. One aspect of this was the decision to perform the resampling at the level of users, and not, say, at the level of individual sessions. This maintains the correlations between successive sessions of the same user. Another aspect was the decision to retain the time of day and day of week when each user starts (and hence also when each job starts). This leads to workloads that retain the correlations among users who all operate according to a common daily and weekly cycle.

One concern that was not handled here is the issue of workload stability constraints. Real workloads exhibit a throttling effect whereby less additional work is submitted when the system is highly loaded. Given that we use each user's sequence of jobs as-is, our generated workload traces will not display such effects. To be more realistic we therefore need to model the feedback from system performance to user behavior. Such models turn out to be rather complicated, and this work will be reported separately.

The above description focused on parallel system workloads, which are useful for evaluating the performance of parallel job schedulers. However, we believe that workload resampling has far wider applicability. Specifically, workload resampling is clearly applicable to any context in which the composition of the workload can be described as a hierarchical structure. Examples include networking, web servers, file systems, and architectural workloads. For example, it would be interesting to try to replace the large benchmark suites used in computer architecture evaluations with workload mixes based on resampling from the different applications in the suite. If successful, this may lead to an innovative fast approach for evaluating new designs.

7. REFERENCES

[1] J. Aikat, S. Hasan, K. Jeffay, and F. D. Smith, "*Towards traffic benchmarks for empirical networking research: The role of connection structure in traffic workload modeling*". In 20th *Modeling, Anal. & Simulation of Comput. & Telecomm. Syst.*, pp. 78–86, Aug 2012.

[2] P. Barford and M. Crovella, "*Generating representative web workloads for network and server performance evaluation*". In *SIGMETRICS Conf. Measurement & Modeling of Comput. Syst.*, pp. 151–160, Jun 1998.

[3] O. Boiman and M. Irani, "*Detecting irregularities in images and in video*". In 10th *IEEE Intl. Conf. Comput. Vision*, vol. 1, pp. 462–469, Oct 2005.

[4] S. Dubnov, Z. Bar-Joseph, R. El-Yaniv, D. Lischinski, and M. Werman, "*Synthesizing sound textures through wavelet tree learning*". *IEEE Comput. Graphics & Applications* **22(4)**, pp. 38–48, Jul 2002.

[5] M. R. Ebling and M. Satyanarayanan, "*SynRGen: An extensible file reference generator*". In *SIGMETRICS Conf. Measurement & Modeling of Comput. Syst.*, pp. 108–117, May 1994.

[6] B. Efron, "*Bootstrap methods: Another look at the jackknife*". *Ann. Statist.* **7(1)**, pp. 1–26, Jan 1979.

[7] B. Efron and G. Gong, "*A leisurely look at the bootstrap, the jackknife, and cross-validation*". *The American Statistician* **37(1)**, pp. 36–48, Feb 1983.

[8] C. Ernemann, B. Song, and R. Yahyapour, "*Scaling of workload traces*". In *Job Scheduling Strategies for Parallel Processing*, pp. 166–182, Springer Verlag, 2003. Lect. Notes Comput. Sci. vol. 2862.

[9] D. G. Feitelson, "*Locality of sampling and diversity in parallel system workloads*". In 21st *Intl. Conf. Supercomputing*, pp. 53–63, Jun 2007.

[10] D. G. Feitelson and E. Shmueli, "*A case for conservative workload modeling: Parallel job scheduling with daily cycles of activity*". In 17th *Modeling, Anal. & Simulation of Comput. & Telecomm. Syst.*, Sep 2009.

[11] D. G. Feitelson, D. Tsafrir, and D. Krakov, "*Experience with the parallel workloads archive*", 2012. (In preparation).

[12] F. Hernández-Campos, K. Jeffay, and F. D. Smith, "*Modeling and generating TCP application workloads*". In 4th *Broadband Comm., Netw. & Syst.*, pp. 280–289, Sep 2007.

[13] J. Jann, P. Pattnaik, H. Franke, F. Wang, J. Skovira, and J. Riodan, "*Modeling of workload in MPPs*". In *Job Scheduling Strategies for Parallel Processing*, pp. 95–116, Springer Verlag, 1997. Lect. Notes Comput. Sci. vol. 1291.

[14] P. Kamath, K.-c. Lan, J. Heidemann, J. Bannister, and J. Touch, "*Generation of high bandwidth network traffic traces*". In 10th *Modeling, Anal. & Simulation of Comput. & Telecomm. Syst.*, pp. 401–412, Oct 2002.

[15] L. Kovar, M. Gleicher, and F. Pighin, "*Motion graphs*". *ACM Trans. Graph.* **21(3)**, pp. 473–482, Jul 2002.

[16] D. Krishnamurthy, J. A. Rolia, and S. Majumdar, "*A synthetic workload generation technique for stress testing session-based systems*". *IEEE Trans. Softw. Eng.* **32(11)**, pp. 868–882, Nov 2006.

[17] V. Kwatra, A. Schödl, I. Essa, G. Turk, and A. Bobick, "*Graphcut textures: Image and video synthesis using graph cuts*". *ACM Trans. Graph.* **22(3)**, pp. 277–286, Jul 2003.

[18] D. Lifka, "*The ANL/IBM SP scheduling system*". In *Job Scheduling Strategies for Parallel Processing*, pp. 295–303, Springer-Verlag, 1995. Lect. Notes Comput. Sci. vol. 949.

[19] V. Lo, J. Mache, and K. Windisch, "*A comparative study of real workload traces and synthetic workload models for parallel job scheduling*". In *Job Scheduling Strategies for Parallel Processing*, pp. 25–46, Springer Verlag, 1998. Lect. Notes Comput. Sci. vol. 1459.

[20] U. Lublin and D. G. Feitelson, "*The workload on parallel supercomputers: Modeling the characteristics of rigid jobs*". *J. Parallel & Distributed Comput.* **63(11)**, pp. 1105–1122, Nov 2003.

[21] "*Parallel workloads archive*". URL http://www.cs.huji.ac.il/labs/parallel/workload/.

[22] K. Pawlikowski, "*Steady-state simulation of queueing processes: A survey of problems and solutions*". *ACM Comput. Surv.* **22(2)**, pp. 123–170, Jun 1990.

[23] E. Shmueli and D. G. Feitelson, "*Using site-level modeling to evaluate the performance of parallel system schedulers*". In 14th *Modeling, Anal. & Simulation of Comput. & Telecomm. Syst.*, pp. 167–176, Sep 2006.

[24] J. Sommers and P. Barford, "*Self-configuring network traffic generation*". In 4th *Internet Measurement Conf.*, pp. 68–81, Oct 2004.

[25] D. Talby and D. G. Feitelson, "*Improving and stabilizing parallel computer performance using adaptive backfilling*". In 19th *Intl. Parallel & Distrib. Processing Symp.*, Apr 2005.

[26] D. Tsafrir and D. G. Feitelson, "*Instability in parallel job scheduling simulation: The role of workload flurries*". In 20th *Intl. Parallel & Distrib. Processing Symp.*, Apr 2006.

[27] D. Tsafrir, K. Ouaknine, and D. G. Feitelson, "*Reducing performance evaluation sensitivity and variability by input shaking*". In 15th *Modeling, Anal. & Simulation of Comput. & Telecomm. Syst.*, pp. 231–237, Oct 2007.

[28] K. V. Vishwanath and A. Vahdat, "*Realistic and responsive network traffic generation*". In *ACM SIGCOMM Conf.*, pp. 111–122, Sep 2006.

[29] M. C. Weigle, P. Adurthi, F. Hernández-Campos, K. Jeffay, and F. D. Smith, "*Tmix: A tool for generating realistic TCP application workloads in ns-2*". *Comput. Commun. Rev.* **36(3)**, pp. 67–76, Jul 2006.

[30] Y. Wexler, E. Schechtman, and M. Irani, "*Space-time video completion*". In *Conf. Comput. Vision & Pattern Recog.*, vol. 1, pp. 120–127, Jun 2004.

[31] N. Zakay and D. G. Feitelson, "*On identifying user session boundaries in parallel workload logs*". In *Job Scheduling Strategies for Parallel Processing*, pp. 216–234, Springer-Verlag, 2012. Lect. Notes Comput. Sci. vol. 7698.

[32] J. Zilber, O. Amit, and D. Talby, "*What is worth learning from parallel workloads? a user and session based analysis*". In 19th *Intl. Conf. Supercomputing*, pp. 377–386, Jun 2005.

Improving the Scalability of a Multi-core Web Server

Raoufehsadat Hashemian
ECE Department, University of Calgary, Canada
rhashem@ucalgary.ca

Diwakar Krishnamurthy
ECE Department, University of Calgary, Canada
dkrishna@ucalgary.ca

Martin Arlitt
Sustainable Ecosystems Research Group
HP Labs, Palo Alto, USA
martin.arlitt@hp.com

Niklas Carlsson
CIS Department, Linköping University, Sweden
niklas.carlsson@liu.se

ABSTRACT

Improving the performance and scalability of Web servers enhances user experiences and reduces the costs of providing Web-based services. The advent of Multi-core technology motivates new studies to understand how efficiently Web servers utilize such hardware. This paper presents a detailed performance study of a Web server application deployed on a modern 2 socket, 4-cores per socket server. Our study show that default, "out-of-the-box" Web server configurations can cause the system to scale poorly with increasing core counts. We study two different types of workloads, namely a workload that imposes intense TCP/IP related OS activity and the SPECweb2009 Support workload, which incurs more application-level processing. We observe that the scaling behaviour is markedly different for these two types of workloads, mainly due to the difference in the performance characteristics of static and dynamic requests. The results of our experiments reveal that with workload-specific Web server configuration strategies a modern Multi-core server can be utilized up to 80% while still serving requests without significant queuing delays; utilizations beyond 90% are also possible, while still serving requests with acceptable response times.

Categories and Subject Descriptors

C.4 [**PERFORMANCE OF SYSTEMS**]: Performance attributes,Measurement techniques

Keywords

Performance characterization, Web server scalability, Multi-core servers

1. INTRODUCTION

As organizations increasingly use Web-based services to support their customers and employees, Quality of Service (QoS), typically measured by the response times that the users experience, becomes an important design consideration. At the same time, service providers such as Google are interested in improving the effective utilization of their infrastructure, as this improves the economic sustainability of their business. Therefore, systematic techniques are required to help organizations meet the QoS objectives of their Web customers while effectively utilizing underlying resources. One such technique is performance evaluation. While this technique was frequently used in the 1990s and 2000s [19] [3] [5] [18], such studies predate the emergence of Multi-core server architectures. Thus, revisiting Web server performance and scalability is crucial to see whether previously investigated behaviours exist on modern hardware.

In this paper, we present a detailed measurement-based study of the performance behaviour of the Lighttpd [16] Web server deployed on a 2 socket, 4-cores per socket system based on the Intel Nehalem [2] microarchitecture. We observe system behaviour under two synthetic yet realistic workloads, namely a TCP/IP intensive workload and the SPECweb2009 Support workload. The TCP/IP intensive workload emulates a server that handles a large volume of user requests to cached static content. Conversely, the SPECweb2009 Support workload contains a significant fraction of dynamic requests, which trigger application-level CPU activity. In contrast to previous work [24] [10] that consider only mean response time measures, we focus on characterizing the response time distributions for these workloads while enabling progressively more cores on the system.

Our experiments with these workloads show that the task of leveraging the performance benefits of Multi-core systems to be non-trivial. After deploying the Web server on this system and with the default configurations, the servers scaled poorly under both workloads. To eliminate bottlenecks other than the CPU and allow the Web server to scale, we first perform three initial configuration tunings related to the Web server software, network interrupt processing, and OS scheduling. Together, we find that these tunings allow the system to scale up to 69% relative to the default settings.

Tests on the tuned system reveal that the two workloads we consider scale very differently. Specifically, the TCP/IP intensive workload scales sub-linearly with increasing core count. For example, considering a 99.9^{th} percentile response time target of 4 ms, request throughput increases by a factor of 7.2 from 1 to 8 cores. For a mean response time target of 0.5 ms, throughput only increases by a factor of 6.3 from 1 to 8 cores. In contrast, throughput increases nearly linearly with core count for the SPECweb2009 Support workload for various mean and 99.9^{th} response time percentile targets.

Deeper analysis of response time distributions shows that differences in performance behaviour across the two workloads stem from the way static and dynamic requests exploit Multi-core systems. Response times of static requests are affected adversely by overheads related to migration of data belonging to the Web application's processes across cores. In particular, response time degradations are significant when data migrate across the two sockets in the system. For the more memory intensive dynamic requests, the adverse impact of process data migration is offset by an overall reduction in the amount of accesses to main memory.

Due to this workload-dependent behaviour, our results show that a Web server configuration strategy previously proposed by Gaud *et al.* [10] and Hashemian [13] to improve scalability is not likely to be effective in systems where the mix of static to dynamic requests fluctuates over time. Specifically, this strategy relies on deploying multiple Web server replicas on a host. Web server process and Network Interface Card (NIC) affinity settings are configured to distribute requests to replicas such that application-level processing of a request is carried out on the same socket that handles the network interrupt processing of that request. Our experiments show that the multiple replica approach significantly reduces inter-socket data migrations and improves scalability for the TCP/IP intensive workload. However, the new configuration decreases scalability relative to the single replica configuration for the SPECWeb Support workload. These results suggest that significant scalability improvements can be achieved by dynamically adapting Web server configuration policies at runtime based on the type of workload experienced by the system.

In summary, our paper provides insights on configuring Web servers to efficiently leverage the capacity of Multi-core hardware. With a carefully selected, workload-specific configuration strategy, our results show that a modern Multi-core server can be kept 80% busy while still serving requests without significant queuing delays. Furthermore, utilizations beyond 90% are also possible, while still serving requests with acceptable response times. A recent analysis of traces collected from an enterprise and a University environments suggests that the top 10 domains in the traces account for a vast majority of observed traffic [11]. Yet, evidence suggests that production servers in these large-scale service provider domains are intentionally very lightly utilized [6] in an attempt to ensure good response time performance. We believe such a strategy is inefficient and may need to be revisited in light of evidence presented in this paper.

The remainder of the paper is organized as follows. Section 2 reviews previous work. Section 3 describes the experimental setup, methodology, workloads and the initial Web server configuration tunings done to better exploit the parallelism of Multi-core servers. Section 4 studies the performance and scalability of the tuned server under both workloads. Section 5 explains the multiple Web replica solution and experimentally evaluates the performance gains from employing it for each of the workloads. Section 6 summarizes our work and offers directions for future research.

2. RELATED WORK

Several researchers have in recent years proposed novel OS-level enhancements to improve the performance of Multi-core systems [7] [15]. For example, Boyd *et al.* [7] modified the Linux kernel to remove resource bottlenecks that prevent applications from fully exploiting Multi-core hardware. Their study shows how the Apache Web server benefits from these kernel modifications. Kumar *et al.* [14] characterized the performance of a TCP/IP intensive workload generated on an Intel Core 2, 2-socket, 4 cores per socket system whose network stack is modified to support Direct Cache Access (DCA), which aims to improve response times by allowing a NIC to directly place data into processor caches. The authors show that a DCA-enhanced Linux kernel can perform 32% faster than a stock kernel. In contrast to these studies, we focus on simple performance improvement strategies that do not require application and kernel modifications.

Veal and Foong [24] conducted a performance evaluation of the Apache Web server deployed on a centralized memory Intel Clovertown system. Their work reports that the scalability of the Web server for a SPEC Web workload increases by only a factor of 4.8 from 1 to 8 cores. The authors establish the system's address bus as the main scalability bottleneck. We did not encounter such a bottleneck in this work as we used newer hardware that employs on-chip memory controllers, a NUMA architecture, and faster inter-socket communication mediums.

Harji *et al.* [12] compare the performance of various Web server software deployed on a quad-core socket. Using two different static workloads, the authors show that their μserver and WatPipe implementations specifically optimized for Multi-core systems outperform well-known software such as Apache and Lighttpd. They show that with careful tuning the new implementations can sustain up to 6,000 Mbps using 4 cores. In contrast to this work, our paper considers both static and dynamic workloads and investigates deployments spanning multiple sockets. Furthermore, we focus on response time measures in addition to throughput and provide a detailed characterization of low-level hardware usage triggered by Web request processing.

Gaud *et al.* [10] evaluated the performance of the Apache Web server on a 4-socket, quad core AMD Shanghai system using the SPECweb2005 benchmark. As mentioned in Section 1, the authors propose a multiple Web replica solution that minimizes migration of Web server data across sockets and show that the solution improves scalability. Our work differs from this study in many aspects. Firstly, our study is more detailed in that we compare scalability for two workloads with very different characteristics and we consider the entire response time distribution. Secondly, we present a fine-grained analysis of the response time distributions observed at various load levels for these two workloads to identify trends specific to static and dynamic requests. Thirdly, we conduct controlled experiments and report more comprehensive hardware monitoring data to identify low-level processor interactions that impact scalability. Finally, we show that for our system the multiple Web replica solution is *not* effective for workloads containing a significant fraction of dynamic requests and hence cannot be prescribed as a general scalability remedy to Web site administrators.

3. EXPERIMENTAL METHODOLOGY

3.1 System Under Study

The Multi-core server under study is a Dell PowerEdge R510 equipped with two quad-core Intel Xeon 5620 processors with the Intel Nehalem architecture. The processors operate at a frequency of 2.4 GHz. They have a three-level cache hierarchy as shown in Figure 1. Each core in a socket has private 64 KB L1 and 256 KB L2 caches. Cores in a socket share a common 12 MB L3 cache. The machine is equipped with two 8 GB memory banks, each connected to one socket through three DDR3-1333MHz channels. The server uses Intel's QuickPath Inter-connect (QPI) with a capacity of 5.6 Giga Transfers per second (GT/s) between sockets. The server has a dual-port 10 Gbps Broadcom 57711 NIC and a dual-port 1 Gbps Intel 82576 NIC. The NICs support Message Signal Interrupts-eXtended (MSI-X) and multiple Receive Side Scaling (RSS) queues that enable the distribution of network interrupts between the cores.

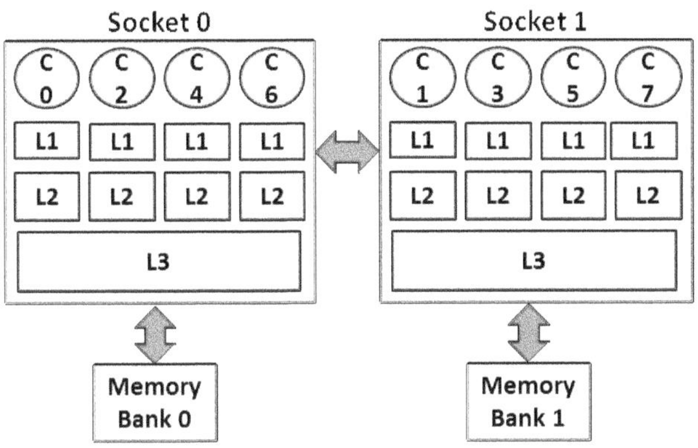

Figure 1: Server Architecture

In this study, we focus on CPU bottlenecks and their impact on scalability. Consequently, we intentionally prevent other resources (e.g., disks, network) from becoming the bottleneck. This influences our choice of the Web server and workloads used in our study. The Web server software used in our experiments is Lighttpd 1.4.28 [16]. Lighttpd was chosen as initial tests showed it scaled better than other open source Web servers such as Apache [4]. The event-driven, asynchronous architecture of Lighttpd avoids the need to manage a large number of Web server processes and decreases the size of its memory footprint. This enables the server to experience a CPU bottleneck and allowing for system performance under such a condition to be examined. Lighttpd can exploit Multi-core hardware by spawning multiple worker processes. The fork() mechanism is used to create these worker processes; therefore, they share the same address space and share some data structures in this space. As discussed in Section 3.4, our experiments used 1 worker process per core.

For the multi-tier SPECweb workload, we use the PHP language runtime as the application tier and the FastCGI [8] message exchange protocol as the means of communication between the Web and application server. In this configuration, a PHP-FastCGI process with a set of children are spawned by the Lighttpd Web server at startup. This creates a constant size pool of PHP-FastCGI processes that serve requests for dynamically generated content. We had to use 128 PHP-FastCGI processes per core in our experiments to avoid a software bottleneck.

The SPECweb benchmark emulates the behaviour of a database tier using another Web server referred to as the Backend Simulator (BeSim). We use two machines with Intel Core2 processors and 2 GB of memory for this purpose, as shown in Figure 2. Each BeSim machine executes an instance of Apache Web server with the FastCGI module. The BeSim systems were carefully monitored to confirm that they are not the bottleneck in our experiments.

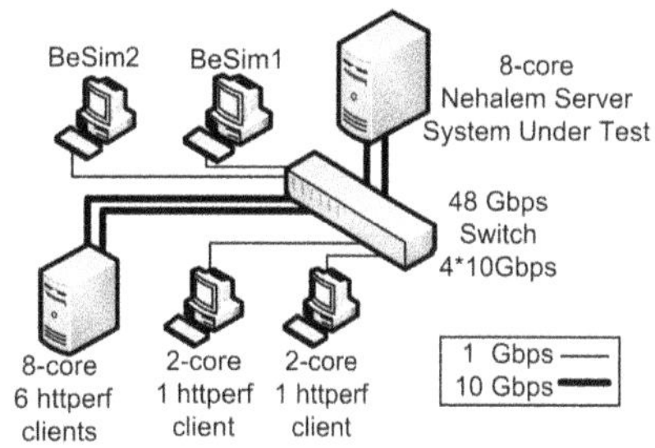

Figure 2: Testbed for Support workload

3.2 Workloads

As mentioned previously, the TCP/IP intensive workload is selected to expose CPU bottlenecks when a server is subjected to very high traffic volumes that induce significant OS activity related to processing of incoming and outgoing network packets. In each experiment, HTTP requests are sent to the server at a rate of λ requests per second. The value of λ is varied to study the behaviour of the server under a wide range of traffic intensities. To isolate the impact of TCP/IP activity, we limit the amount of application-level processing required by each request. Accordingly, all user requests are for a single static 1 KB file, which is highly likely to be served from the processors' caches. With this workload, we were able to carry out experiments up to a maximum rate of 155,000 requests per second, which resulted in a 90% average utilization of the server's cores. To generate such high request rates, we selected the *httperf* [20] Web workload generator. We used 4 instances of *httperf*, each running on a separate dual core client machine to submit this workload to the server. The client machines and the server are connected to each other through a non-blocking 10 Gbps Ethernet switch.

We also consider the SPECweb2009 Support workload. We use this workload to observe system behaviour when there is more application-level activity relative to OS level activity. The increased application-level activity is caused by a more diverse working set of static files, which may trigger processor cache misses. Furthermore, around 5% of requests are dynamically generated and rely on the services of the application tier and the BeSim tier. To prevent the server's disk from becoming a bottleneck, we change the "DIRSCALING" parameter of the Support workload from 0.25 to 0.00625. This reduces the size of the file-set to fit in the server's main memory, however, the probability distributions of file sizes and file popularity remain the same. To ensure that all files are accessed in memory and prevent disk activity, before each experiment the memory cache (including the OS file cache) is cleared and then warmed by accessing all files in a "cache filling" run.

To submit the Support workload, we use *httperf* instead of the Java-based SPECweb load generator client bundled with the benchmark suite. We previously enhanced [13] *httperf* to submit a workload into multiple server addresses simultaneously. This enhancement is used in the experiments to evaluate multiple replica configuration, that is described in Section 5.4. Eight instances of *httperf* are deployed across two dual core and one eight-core machines, as shown in Figure 2. We first instrumented the SPECweb client to record all request URLs sent to the server along with the send time stamp in an output file. A set of tests were run using multiple Java-based SPECweb clients and the lists of request URLs were recorded. In each experiment, depending on the number of concurrent users for that experiment, a subset of the recorded sessions is submitted to the server using *httperf*. We verified that the characteristics of the load generated by *httperf* is very close to that generated by the SPECweb client.

3.3 Performance Metrics

The primary scalability metric that we use in this study is the *Maximum Achievable Throughput* (MAT). We define this parameter as the maximum throughput at which a selected statistical property of the response time distribution is less than a "threshold". MAT is measured for two typical statistical metrics used in practice by performance analysts, namely the mean and the 99.9^{th} percentile of the response time distribution. The mean (average) response time is often used by queuing theorists, whereas practitioners and Internet service providers often are interested in percentiles; particularly the upper percentiles that capture the tail effects [9]. The response time thresholds differ depending on the Service Level Agreement (SLA) for a particular Web service. For instance, Amazon Dynamo Services uses an SLA where the 99.9^{th} percentile threshold is 300 msec [9]. In our experiments, the mean response time thresholds are set to about four to five times the mean response time under low load. This represents a scenario where there is contention among users for server resources yet users experience an acceptable interactive response. The threshold values are constant for all the experiments, however, the effect of selecting other thresholds is also discussed.

To quantify the server's behaviour, a variety of metrics are collected. To monitor the CPU, disk, and network utilizations of the server, the *collectl* [22] tool is used, which we confirmed incurred a negligible CPU overhead of around 1%. Low-level hardware monitoring data was collected using Performance Counter Monitor (PCM) [1] tool (specific to Intel architectures). To eliminate the effect of the hardware monitoring overhead on the scalability measurements, all the experiments were run twice, first without PCM to collect the scalability metrics, and then with PCM to collect the hardware event statistics. The experiments without PCM were repeated multiple times to achieve tight 95% confidence intervals for the reported mean response times. In all our experiments, 95% confidence intervals were less than 1% of their corresponding mean values.

3.4 Initial Configuration Tuning

In this section, we describe our initial configuration tuning exercise aimed at preventing software or OS bottlenecks from limiting the scalability of the server. Table 1 shows the quantitative values of performance improvements achieved through three configuration tunings for both TCP/IP intensive and Support workloads. The first column describes the configuration tuning while the other two columns show the improvement in the MAT for mean response time thresholds of 0.5 msec and 5.0 msec, respectively for the TCP/IP and SPECweb2009 Support workloads. We confirmed that the tuning strategy's effect on the higher percentiles is also positive. The three tuning steps are as follows:

Number of Web server processes: The Lighttpd documentation suggests a setting of two HTTP worker processes per core for CPU bound workloads. As the first configuration tuning step, we set this parameter to one. From Table 1, the MAT value was 14% and 1.5% *higher* with this setting for the TCP/IP intensive and the Support workloads, respectively, when compared to the default setting. This result indicates that for our workloads, the overhead of context switching between two worker processes is greater than the gains achieved through duplicating workers. Furthermore, the performance gain from this tuning is greater for the TCP/IP workload than for the Support workload. Careful workload-specific tuning of this parameter is therefore critical for obtaining good performance.

OS scheduling: As the next tuning strategy, we examined the Linux process affinity mechanism [17] as an alternative to the default Linux scheduler. The default scheduler dynamically migrates Web application processes among cores. We used the affinity mechanism to statically assign each of the 8 Lighttpd worker processes to its own distinct core. From Table 1, core affinity significantly outperforms the default Linux scheduler. The MAT increased 10% and 4.0% for the TCP/IP Intensive and Support workloads, respectively, compared to the MAT obtained after the previous tuning step. This result shows that for a TCP/IP intensive workload, migrating cached data pertaining to a worker process across cores can adversely impact performance. For the Support workload, we ran an additional experiment where the affinity setting was also applied to the application tier in a way that the PHP-FastCGI processes were equally divided between the cores and statically assigned to them. This further improves the MAT value by around 5% with respect to the mean response time. Therefore, the two affinity settings can improve the server's scalability up to 9% for the Support workload.

Network interrupt handling: The default NIC configuration uses a single core on the system to process interrupts from the NIC. As a result, the interrupt handling core became saturated under high loads, even though the utilization of the other cores was low. Therefore, the network interrupt handling load was distributed between the cores by creating 8 RSS queues on the NIC and affining each of the queues to a distinct core through the Linux *IRQ Affinity Setting* mechanism. From Table 1, distributing the interrupts improved the MAT value up to 45% and 20.5% for the TCP/IP Intensive and Support workloads, respectively relative to the MAT obtained after the previous tuning step.

Overall, these changes improved the MAT by 69% and 31% for the TCP/IP Intensive and Support workloads, respectively, relative to the default OS settings. The experiments presented in the next section use these tuned settings.

4. SCALABILITY EVALUATION

In this section, we describe the results of scalability evaluation experiments for the TCP/IP intensive and the SPECweb Support workloads. For each workload we change the load intensity to utilize the server to its capacity. We study the

Table 1: Effect of configuration tunings on the server's scalability

Tuning Method	Scalability Improvement	
	TCP/IP intensive workload	Support workload
Decreasing number of Lighttpd processes from 2 to 1	14.0%	1.5%
Employing core affinity	10.0%	9.0%
Balancing network interrupt handling load	45.0%	20.5%

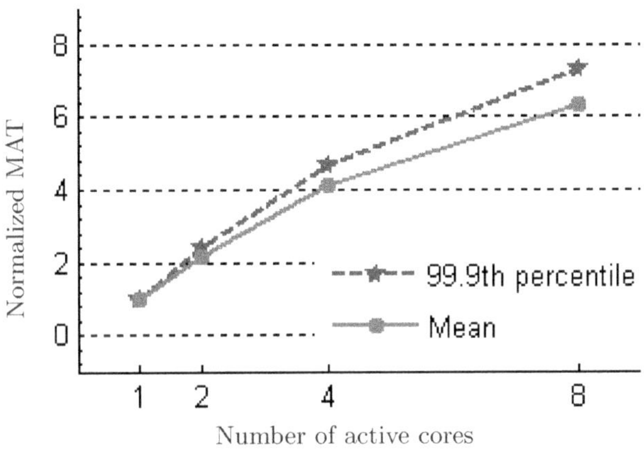

Figure 3: Normalized Maximum Achievable Throughput (MAT) with 1, 2, 4 and 8 active cores for the TCP/IP intensive workload

performance of the 8 core server when it has 1, 2, 4, or 8 active cores.

4.1 TCP/IP Intensive Workload

Figure 3 shows the normalized MAT values with respect to mean and 99.9^{th} percentile response times for 1, 2, 4 and 8 active cores. Different threshold values are used for the two different statistical properties of response time. While the acceptable mean response time is set to 0.5 msec, a 4 msec threshold is considered for 99.9^{th} percentile of response time for this workload. In the experiments with 2 and 4 active cores, all cores are selected from the same socket. We normalize MAT values with respect to the value observed with 1core. From the figures, the server scales close to proportionally from 1 to 2 cores with respect to both mean and 99.9^{th} percentile of response times. However, for more than 2 cores, the scalability factors starts to degrade. The degradation is more severe for the mean response times. The overall scalability from 1 to 8 core is 6.3 for mean and 7.2 for 99.9^{th} percentile. Selecting larger response time thresholds results in similar numbers. However, lower thresholds slightly decrease the scalability factors. These numbers suggest that there is a scalability problem in the system from 4 to 8 cores. Moreover, the problem is not confined to the tail, but observed throughout the body of distribution.

Analysis of the server's response time at different load levels reveals that scalability problems become severe when moving from 4 cores, i.e., using 1 socket in the system, to 8 cores, i.e., using both sockets on the system. The measured response time values as a function of CPU utilization are depicted in Figures 4(a) and 4(b) for the mean and 99.9^{th} percentile, respectively. Figure 4(a) show that mean response times only increase gradually till a mean per-core utilization of 80% for experiments with 1, 2 and 4 active cores. However, in the experiment with 8 active cores, the mean response times increase with a larger slope from 60% utilization. A similar trend is observable for the 99.9^{th} percentile of response times in the experiments with 2, 4 and 8 active cores. Interestingly, the 99.9^{th} percentiles are significantly higher with 1 active core. This demonstrates that when only 1 core is active in the system, a subset of requests have significantly high response rimes. We have not yet been able to determine the reason for this behaviour inspite of the large amount of hardware event data that we collected.

In summary, Figure 4(a) confirms the presence of a scalability problem in the system when all 8 cores are active while Figure 4(b) demonstrates that the problem manifests itself throughout the response time distribution. This problem prevents the Web server from fully utilizing available processor resources. In Section 5, we evaluate a Web server configuration strategy which can reduce the impact of this problem.

4.2 Support Workload

For this workload, the load on the server is changed by varying the number of concurrent users in the system. Similar to the TCP/IP intensive workload, in the experiments with 2 and 4 active cores all cores are selected from the same socket.

Figure 5 shows the MAT values with respect to mean and 99.9^{th} percentile response times for 1, 2, 4 and 8 active cores. We consider thresholds of 5 msec and 500 msec for mean and 99.9^{th} percentile of response time, respectively. From the figure, the server shows a different scalability behaviour under the Support workload. While the server scales sublinearly from 1 to 2 cores with respect to the mean response time, the scalability factors improve as we activate more cores. The overall scalability from 1 to 8 core is 7.8 and 7.9 for the mean and 99.9^{th} percentile of response time, respectively. The numbers suggest that the server scales well for the Support workload. Increasing the threshold did not have considerable effect on the scalability factors. However, selecting a lower threshold (e.g. 4 msec and 400 msec) results in a slightly better scalability from 4 to 8 cores.

Figures 6(a) and 6(b) further emphasize the differences between the TCP/IP workload and the Support workload. The figures show the mean and 99.9^{th} percentile of response times as a function of server CPU utilizations for 1, 2, 4 and 8 active cores. They demonstrate four main observations related to the server's scalability. First, from Figure 6(a), with all 8 cores enabled it is possible to utilize the cores to up to 80% without encountering significant response time increases. Second, at high loads and for a given utilization level, the experiments with 2 and 4 active cores have higher mean response times compared to the experiments with 1 and 8 active cores. A similar trend is observable for

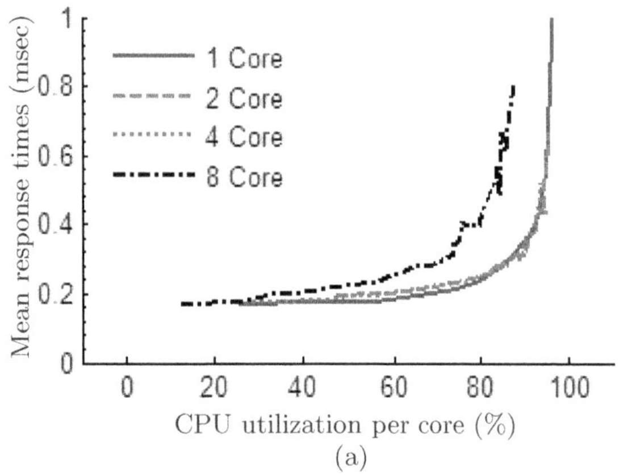

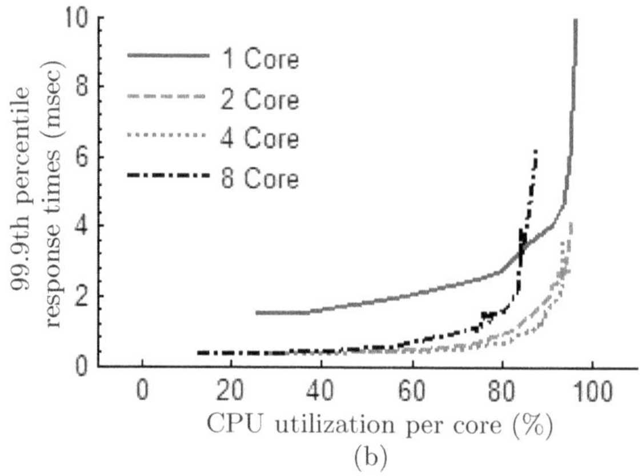

Figure 4: Web server response time vs. CPU utilization for the TCP/IP intensive workload: (a) Mean (b) 99.9th percentile

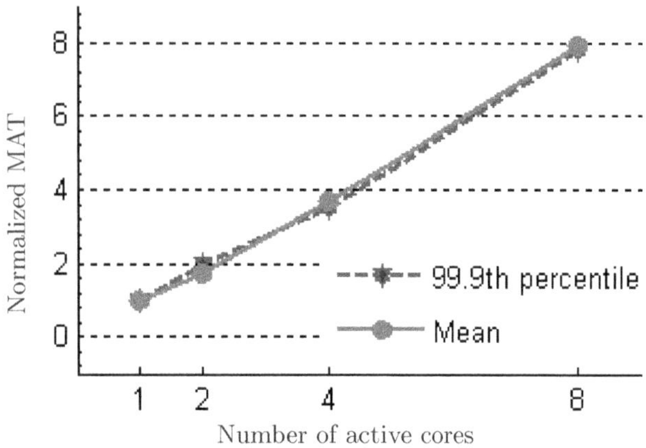

Figure 5: Normalized Maximum Achievable Throughput (MAT) with 1, 2, 4 and 8 active cores for the Support workload

the 99.9th percentile of response time as depicted in Figure 6(b). Third, the performance degradations of the 2 and 4 core curves relative to the 8 core curve are less for the 99.9th percentile of response times compared to the mean. Finally, while the experiments with 1 active core have the best performance considering the mean response time, the best 99.9th percentile response times is measured with 8 active cores, especially at high load. These observations reveal that activating more cores in a single socket has a negative effect on the server's scalability for the Support workload. This effect is more in the lower percentiles and less on the tail of the response time distribution. Moreover, activating another socket considerably improved the scalability.

We next analyze the reasons for the differences in scalability behaviour of the body and tail of the response time distribution. We plot the Cumulative Distribution Function (CDF) of response times in Figure 7 to study how the response time of different types of requests differ when we activate more cores in the server. A single test was selected from the experiments with 1, 2, 4 and 8 cores. In each experiment the selected test is the one which causes a CPU utilization of 80% per core. This way we could ensure that the comparison is done at a constant load level relative to the available resources (number of cores). From Figure 7, the response time distributions are almost identical for 1, 2, and 4 cores cases for the first 95 percentiles. Response times are significantly higher when 8 cores are active. The conditions are reversed for the last 5 percentiles. For this portion of requests, the response times in the tests with 2 and 4 core are considerably higher than the response times in the test with 1 and 8 cores. The CDF plots demonstrate that activating 8 cores causes a performance degradation for requests with smaller response times while requests with larger response time benefited from having 8 active cores.

We now focus on the identity of the requests with large response times, which benefit from having 8 active cores, i.e., both the sockets. We use the detailed response time log file output by *httperf* to plot response times as a function of the size of responses to request as shown in Figure 8. The plot was generated from the output of a test with 8 active cores at 80% per-core utilization. Plots for other numbers of cores and load levels have similar trends. Each point in the graph represents a single request submitted to the server during the test. The plot reveals that requests with medium response sizes have the highest response times. Using the *httperf* log files, these response sizes correspond to content that is dynamically generated by the application tier. Therefore, dynamic requests scale well with activating both sockets.

Considering the observations from Figures 7 and 8, we can conclude that the server has a different scalability behaviour when handling the static and dynamic requests. While the server scales well for the dynamic requests, there are some scalability problems when handling static requests, especially when moving from 4 to 8 cores, i.e., from 1 socket to 2 sockets. The fraction of dynamic requests in the Support workload is significant enough that this workload sees an overall scalability improvement with increased core count. However, the TCP/IP intensive workload contains a solitary static file and hence scales comparatively poorer with core count, especially while moving from 1 socket to both sockets.

Towards understanding the response time behaviour, we now discuss important low-level differences in system usage across the 1 socket and 2 sockets scenarios. When only one

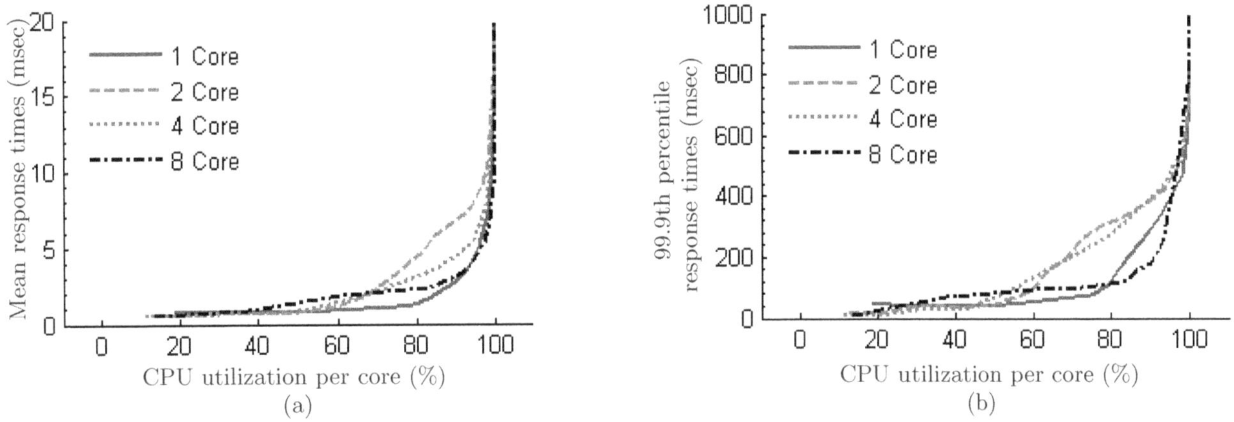

Figure 6: Web server response time vs. CPU utilization for the Support workload: (a) Mean (b) 99.9th percentile

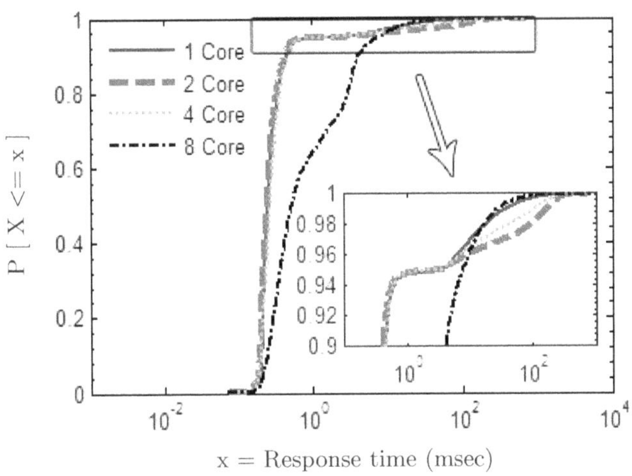

Figure 7: Cumulative Distribution Function (CDF) of response times in 80% CPU utilization for SPECweb Support workload

socket in the system is active, there is unlikely to be significant migration of data over the inter-socket QPI links since the Linux memory manager tries to allocate memory locally. In contrast, as discussed shortly, in the 2 socket scenario there can be significant data flow across the two sockets through QPI links. Contention for the QPI links can potentially hurt the performance of the 2 sockets case. However, the 2 socket case exposes twice the L3 cache capacity to the Web application processes. This can afford performance gains for the PHP-FastCGI processes, which others have shown to be memory-intensive [23][1]. In the next section, we study the relative effects of such interactions in the context of a previously proposed scalability enhancement technique that aims to reduce inter-socket communications.

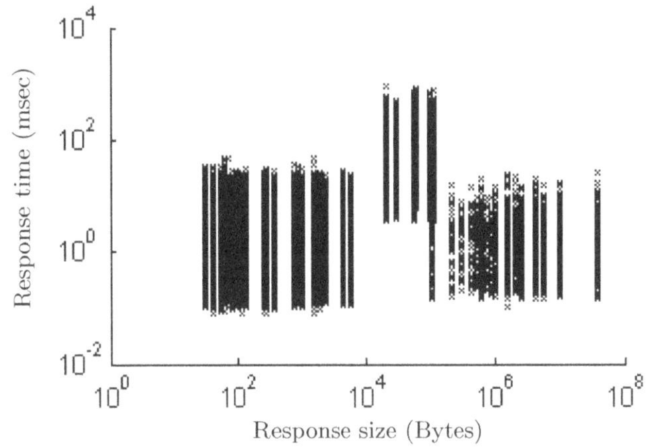

Figure 8: Response time vs. response size for a test with 8 active cores and 80% CPU utilization for the Support workload

5. SCALABILITY ENHANCEMENT

In this section, we evaluate the multiple Web site replicas approach suggested recently [10] [13] to enhance the scalability of Multi-core Web servers. To facilitate subsequent discussions, we first provide some background on how network packet processing and application data shared among Web application processes can increase the traffic in the socket inter-connect. We then describe the multiple replicas solution and present experiments to characterize its effectiveness under various circumstances.

5.1 Effect of Network Packet Processing

When the TCP packets carrying an HTTP request arrive at the server, the NIC sends an interrupt to one of the processor cores. The packet is then copied to the private cache

[1]To confirm the memory-intensiveness of PHP-FastCGI processes, we compared the memory access traffic for the Support workload with 8,000 concurrent users to a statistically similar workload where the dynamic requests were replaced by static requests of the same response size. We noticed that the workload with dynamic requests caused 3.6 times more traffic to the main memory than the workload with static requests.

of the core that is going to handle the interrupt. After the packet is processed by the core (e.g., TCP/IP protocol processing such as checksum calculation), a Web server process is chosen to serve the HTTP request. The data within the packet is then copied to the process's buffer space. There are three possibilities in this stage. First, the core on which the Web server process resides is the same core that has processed the packet interrupt. In this case, the data already exists in the private cache of that core. Second, the core on which the Web server process resides and the packet processing core are not the same, but are from the same socket. In this case, the data may exist in the shared L3 cache of that socket. Third, the Web server process residing core and the packet processing cores are from different sockets. In this case, the data should be fetched from the cache hierarchy of the other socket [21]. This causes an increase in the QPI links' traffic which can result in contention in the QPI links at high request rates.

As discussed previously, the server's MSI-X enabled NICs can divide interrupts among a set of hardware RSS queues with packet distribution among the queues performed using a hashing technique. Furthermore, interrupts assigned to each queue can then be bound to a specific core through the *IRQ Affinity Setting* supported by Linux. We show how this feature can be exploited to reduce inter-socket communications in Section 5.3.2.

5.2 Effect of Shared Application Data

Both Lighttpd and PHP-FastCGI processes can share data, e.g., kernel libraries and associated data structures, between themselves. A process residing on a given socket may request a given piece of data cached at the L3 cache of the other socket due to a previous access by another process residing on that socket. This can have a positive impact on performance since it eliminates costly accesses to main memory. However, it triggers migration of shared data across sockets that may cause heightened contention for the QPI links. Hence, the overall performance impact of shared application data depends on which of these two effects dominates.

5.3 Multiple Web site Replicas

The multiple Web site replicas solution takes advantage of the Linux *taskset* and *IRQ Affinity* mechanisms, multiple network interfaces available on modern servers, and their support for MSI-X and RSS to reduce inter-socket communications stemming from the migration of network packets and data shared among Web application processes residing on different sockets. In the following subsections, we describe how we setup the non-replica approach and two variants of the multiple replica approach on our server.

5.3.1 Original Configuration

Figure 9(a) shows the schematic of the network and Web server configuration for the server used in our setup. The server has a dual-port NIC that can operate as two individual NICs. Each NIC has its own IP address and is configured to use four queues. Queues 0 to 3 of NIC 0 are assigned to cores 0 to 3 of socket 0, and queues 0 to 3 of NIC 1 are assigned to cores 0 to 3 of socket 1[2].

The original Web server configuration used in the experiments with 8 active cores involves running a single Web server software instance with 8 Lighttpd processes. Each of these processes is bound to one of the 8 active cores as shown in Figure 9(a). When a request arrives at the server, it will first be processed by one of the queues of the NIC and next the HTTP request is handled by one of the Web server processes. The NIC and the Web server process to handle the request are each chosen with equal probabilities. Therefore, when the TCP packets of a request are processed by the cores of one socket, there is a 50% chance that the HTTP request is processed by a process residing in the other socket[3]. As mentioned previously, there are 128 PHP-FastCGI processes per core. These are not shown in the figure for the sake of clarity.

5.3.2 2-replica Direct *Mapping*

An alternative configuration is to run two instances of the Web server software, each having 4 Lighttpd processes and 4 x 128 FastCGI processes bound to the cores of one socket. Each instance is denoted as a Web site *"replica"*. The processes of replica 0 are bound to the IP address of NIC 0. Consequently, the requests coming through NIC 0 are processed by the processes of replica 0. The same setting is performed on replica 1 and NIC 1. Figure 9(b) shows the schematic of the alternative configuration with two instances of the Web server. From the figure, the processes of replica 0 reside in socket 0. The queues of NIC 0 are also assigned to the cores of socket 0. This ensures that for all the requests sent through NIC 0, both the TCP packets and the HTTP requests are processed by cores of socket 0. This means the TCP packets and HTTP request will be processed on the same socket with 100% probability. We refer to this mapping of NIC queues and replica processes as *2-replica Direct* mapping.

Employing the *2-replica Direct* mapping eliminates costly inter-socket communications involving migration of network packets. Furthermore, since both replicas are independent, a significant sharing of data among Web application processes will not trigger QPI traffic. However, as mentioned previously, the absence of sharing among sockets can represent a lost caching opportunity that might result in more main memory accesses. It follows from these discussions that the multiple replicas approach tries to leverage properties of the single socket scenarios discussed in the previous sections while still using both sockets in the system.

5.3.3 2-replica Indirect *Mapping*

To better understand the selective importance of eliminating network packet related QPI traffic and shared application data related QPI traffic, a third configuration is considered. This is done by configuring replica 1 (on socket 1) to process the requests sent through NIC 0 (on socket 0). Thus, when the TCP packets of a request are processed by the cores of one socket, the HTTP request is processed by a Web server process residing in the opposite socket. This mapping of NIC and replica shown in Figure 9(c) is referred to as *2-replica Indirect* mapping. Employing the *2-replica Indirect* mapping forces the occurrence of costly inter-socket

[2]There are other options for configuring the two NICs, such as assigning the same IP address through NIC teaming and enabling eight RSS queues for each NIC. We experimentally evaluated these configurations and did not observe a considerable performance difference for our workloads.

[3]Our workload generator sends requests to the IP addresses of both NICs of the server with equal probabilities.

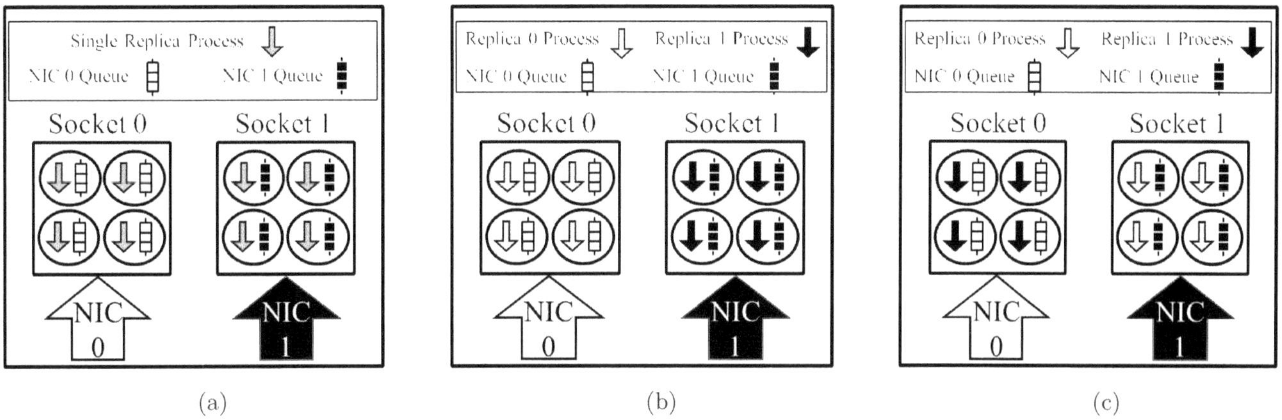

Figure 9: NIC queue and Web server process mappings: (a) Original (b) Direct mapping (c) Indirect mapping

data migrations due to the network processing. However, as Web application processes within a replica still share the same socket, this mapping is similar to the *2-replica Direct* mapping in terms of avoiding shared application data related QPI traffic.

5.4 Evaluation

This section presents experiments performed to evaluate the effectiveness of the multiple replica approach. First, the results of experiments conducted using the new configurations under the TCP/IP intensive workload are presented. We then examine the server under the Support workload. Finally, the hardware event data collected during the experiments with both workloads is presented to identify the causes for the observed performance behaviour.

5.4.1 TCP/IP Intensive Workload

Results from the original configuration with 8 active cores are considered as the baseline. Figure 10 shows the mean response times as a function of request rates for the baseline (*1 replica*), *2-replica Direct* mapping and *2-replica Indirect* mappings. From the figure, the MAT (considering a mean response time threshold of 0.5 msec) is the highest for the *2-replica Direct* mapping. Furthermore, the MAT for the *2-replica Indirect* mapping, while less than that of *2-replica Direct*, is *higher* than the baseline. Specifically, the MAT values improve around 12.3% and 6.5% with the *2-replica Direct* and *2-replica Indirect* mappings, respectively. The improvements in the MAT with respect to a 99.9th percentile threshold of 500 ms (not shown due to space constraints) are 5.2% and 3.1% for the *2-replica Direct* and *2-replica Indirect* mappings, respectively. While selecting a lower threshold reduces the scalability improvements, choosing a thresholds larger than our selected thresholds does not affect the results.

The improved scalability of the Web servers with the *2-replica Indirect* mapping compared to the baseline despite the 50% additional inter-socket migration of data due to network processing suggests that reducing QPI traffic triggered by processes accessing shared data is most responsible for observed scalability improvements. Since the TCP/IP workload has no dynamic requests and does not use the PHP-FastCGI processes, this result suggests that there is significant sharing of data among Lighttpd processes. Furthermore, since Lighttpd processes are not memory-intensive [23], this workload does not seem to be impacted by the inability of a replica to benefit from the L3 cache of the socket hosting the other replica.

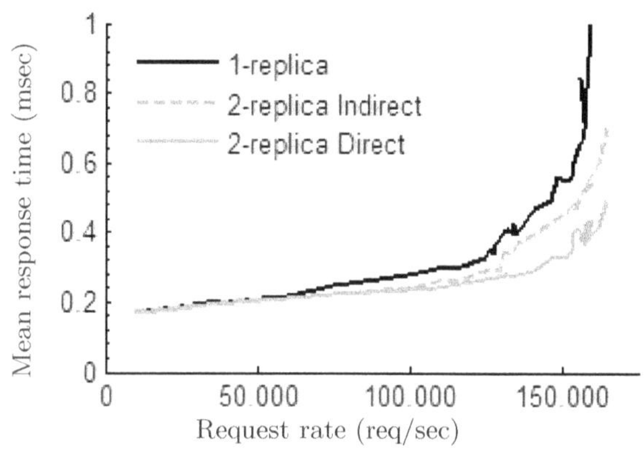

Figure 10: Web server response time vs. request rate for 8 cores with baseline and 2 replica configurations under TCP/IP intensive workload

5.4.2 Support Workload

Figure 11 compares the mean response times for the experiments for the baseline, *2-replica Direct* and *2-replica Indirect* mappings. The figure demonstrates that the server's scalability degrades significantly with the new configurations under the Support workload. From the figure, the MAT for *2-replica Direct* and *2-replica Indirect* mappings are statistically similar (considering the 95% confidence interval) and around 9.9% lower than the baseline. The graph for the 99.9th percentile of the response times (not shown) showed a similar performance degradation for the multiple replica cases. The MAT value with respect to the 99.9th percentile of the response time was 9.9% and 14.8% lower than the baseline for the *2-replica Direct* and *2-replica Indirect* mappings, respectively. This result contradicts previous studies [10] by indicating that the multiple replica configuration does not always improve a server's scalability.

We now investigate the effects of the multiple replica solution on the response time of the static and dynamic requests in the Support workload. The CDF plot of response times for the baseline, *2-replica Direct* and *2-replica Indirect* mappings for the tests with 80% CPU utilization is depicted in Figure 12. The plot was generated from the output of the tests with 7,000 concurrent users. Plots for other high load tests have similar trends. From the graph, the first 95 percentiles of response times, mainly static requests, improved significantly with 2-replica configurations compared to the baseline. This is analogous to the results obtained from the experiments under TCP/IP intensive workload. On the other hand, the last 5 percentiles of the responses times, mainly dynamic requests, are lower for tests with 2 replicas. The results reveal that the multiple replica configuration degraded performance for the dynamic requests. Since these requests consume most of the processing resources and their response times are an order of magnitude higher than the static requests, their performance decides the overall scalability of the server under the Support workload.

The results of experiments with Support workload suggest that dynamic requests do not benefit from reducing QPI traffic. Due to the memory-intensive nature of such requests, the inability of PHP-FastCGI processes of one replica to use the cache hierarchy of the socket containing the other replica seems to hurt overall performance. In the next section, we provide more evidence for these claims, using hardware event data collected during the experiments.

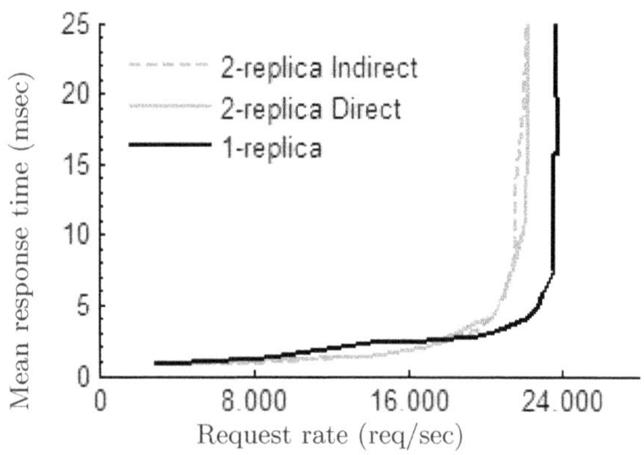

Figure 11: Web server response time vs. request rate for 8 cores with baseline and 2 replica configurations under Support workload

5.4.3 Hardware Events Data

The results presented in Sections 5.4.1 and 5.4.2 revealed that the multiple replica approach improves performance for the TCP/IP intensive workload while degrades it for the Support workload. In this section, we identify possible causes of this performance behaviour. As mentioned is Section 3.3, we collected hardware event data to analyze the root causes of the observed performance characteristics in different stages of this study. To evaluate the effect of using multiple replicas, the socket-inter-connect traffic, L3 cache hit rates, and memory access statistics were collected. Figure 13(a) shows the average QPI traffic at various load levels for the TCP/IP intensive workload. From the figure, a considerable decrease in QPI traffic is observable from the baseline to *2-replica Direct* mapping. Moreover, the QPI traffic is higher than the baseline for *2-replica Indirect* mapping (because for 50% of requests HTTP request and TCP

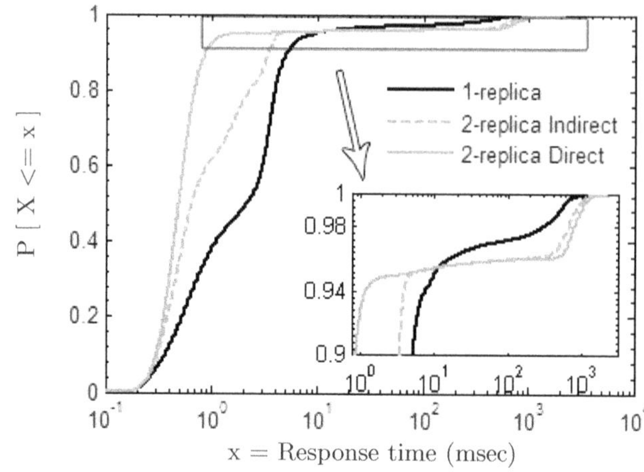

Figure 12: Cumulative Distribution Function (CDF) of response times in 80% CPU utilization with baseline and 2 replica configurations under the Support workload

packet are processed on the same socket). These measurements confirm that applying the multiple replica approach can decrease inter-socket migration under a TCP/IP intensive workload. From Figure 13(b), the average QPI traffic decreased from the baseline to *2-replica Direct* mapping for the Support workload. Moreover the QPI traffic for the *2-replica Indirect* mapping is higher than the baseline. This confirms that the multiple replica solutions were able to decrease the inter-socket migration for the Support workload as well. However, the amount of reduction is much less than the TCP/IP intensive workload. This is likely due to the fewer TCP/IP activities (lower request rate) for the Support workload which results in fewer inter-socket data migration of network processing data.

We now present measurements to discuss the reasons for the different effects of the multiple replica approach on the response times of static and dynamic requests. From Figure 14, the L3 hit ratio decreases at higher loads for the multiple replica configurations. Figure 15 shows that accesses to the main memory increase in the *2-replica Direct* and *2-replica Indirect* mappings due to a reduction the number of L3 cache hits. This confirms our previous observation that workloads with dynamic requests benefit from a PHP-FastCGI process having access to the cache hierarchies of both sockets in the single replica scenario. The results presented in this Section, confirmed that using multiple Web site replicas does not always improve the servers scalability. However, it is still effective for certain types of workloads.

6. CONCLUSIONS

In this paper, we characterized the behaviour of a Web server deployed on a Multi-core architecture under two different workloads. The non-intrusive configuration changes we explored resulted in significant scalability improvements. Investing time in server tuning allowed us to utilize all of the server's cores up to 80% without causing significant queuing delays for users; even at 90%, delays may be tolerable for peak usage. We believe that such increased utilization would enable popular service providers to deploy

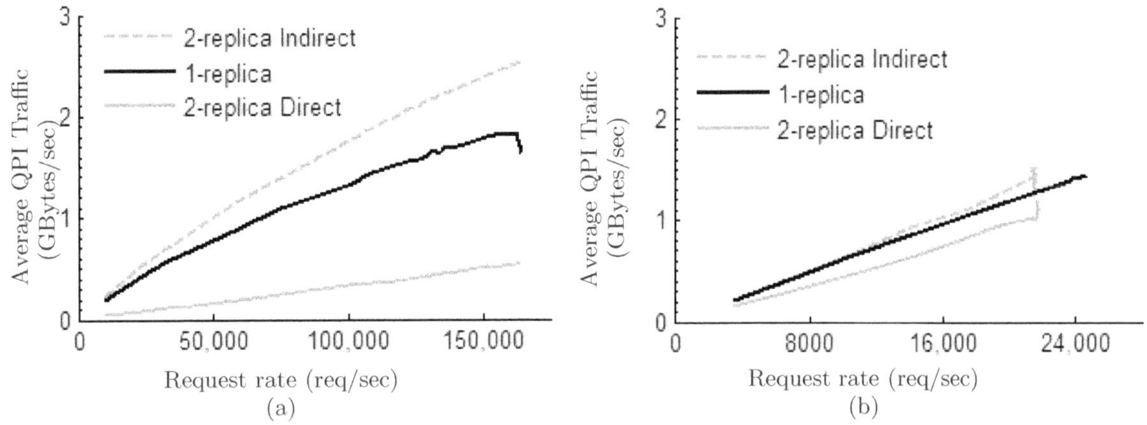

Figure 13: Average QPI traffic vs. request rate for 8 cores with baseline and 2 replica configurations: (a) TCP/IP intensive workload (b) SPECweb Support workload

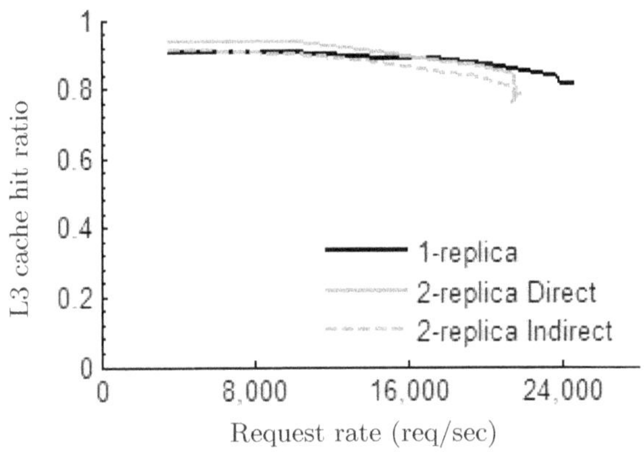

Figure 14: L3 cache hit ratio vs. request rate for 8 cores with baseline and 2 replica configurations under Support workload

fewer servers, thereby reducing both their capital and operating expenses [11]. For example, service providers such as Google and Facebook operate hundreds of thousands of servers; even a 10% reduction in the number of servers required could save these companies non-negligible amounts of money, as well as reducing their environmental footprints. Reducing the number of servers required to support a workload could also enable other savings, such as reduced cooling demand in data centers.

Our work studied in detail the relationship between workload characteristics, low level processor usage, and performance. Specifically, detailed experimentation involving two different workloads showed that the server exhibits different scaling characteristics for static and dynamic requests. The performance of static requests handled degrades as the core count increases and is especially poor when both sockets are fully utilized. In contrast, the performance of dynamic requests increases from 1 socket to 2 sockets. We showed that static requests suffer from inter-socket communication triggered while using both sockets in the system. In contrast, dynamic requests handled by the significantly more memory-intensive application tier benefit from the ability to use the cache hierarchy of both sockets. Due to these findings, our paper establishes that a previously proposed approach to enhance Web server scalability by reducing inter-socket communications [10] [13] may not be effective for workloads with dynamic requests. Our work suggests that scalability enhancement strategies need to be arrived at after careful consideration of workload characteristics. In particular, a Web server configuration can be adapted at runtime to fluctuations in the mix of static to dynamic requests. As an example, consider a popular news Web site. Most news Web sites include dynamic content. However, in the event of an important alert or breaking news that triggers a significant increase in the user traffic, the Web site administrators may disable dynamic content to reduce the load on the server. From our results, the system can in this situation benefit from having multiple Web replicas. The Web site can resume normal operation by disabling those replicas when the user traffic returns to normal.

Our study focused on a widely used processor microarchitecture (Nehalem) and a well-known combination of Web and application servers (Lighttpd, FastCGI-PHP). However, we believe that the scalability behaviour can be different for other systems with different hardware architectures or software configurations. For instance, the performance degradation of dynamic requests due to the use of multiple replicas may not appear in a system with a very large last level cache. On the other hand, a Web server with larger memory footprint, such as Apache, may even suffer performance degradation for static requests. This motivates the need for systematic experimentation to ascertain the best scalability enhancement strategy for a given system. While this time and budget investment may not be of large profit for a small organization, the achieved gain for enterprise Web service providers accumulates over their hundreds of thousands of servers. As a results, they will be willing to study the scalability behaviour of their workloads on various architectures and adapt servers' configurations accordingly.

We collected detailed hardware level statistics and used this large amount of data to analyze the performance behaviours observed in various experiments. However, in some cases, we were not able to correlate the hardware events with the measured response times. For instance, the hardware event statistics could not explain the reason for the high 99.9^{th} percentile response time in the experiments with 1 active cores and under the TCP/IP intensive workload. We believe that due to the complexity of hardware architecture of the Multi-core server, the hardware event data are not sufficient for the scalability analysis of these architectures.

Our future work will focus on other Web server software and hardware platforms and statistical analyses to identify the impact of hardware architectural differences such as cache size and socket inter-connect bandwidth on the behavior of servers. This can be of interest for researchers working on resource sharing problems, such as hosting multiple virtual machines per host. Moreover, we will work on developing an approach to dynamically enable or disable replicas based on the user traffic and the mix of requests experienced by a Web server.

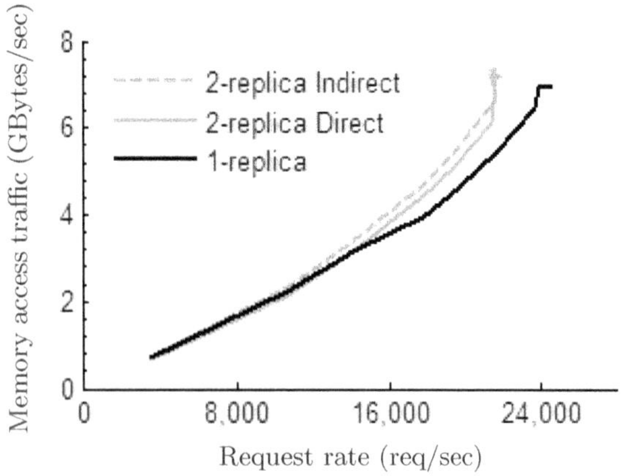

Figure 15: Total memory access vs. request rate for 8 cores with baseline and 2 replica configurations under Support workload

7. ACKNOWLEDGMENTS

This work was financially supported by the Natural Sciences and Engineering Research Council (NSERC) of Canada and HP.

8. REFERENCES

[1] Intel performance counter monitor - a better way to measure cpu utilization. http://software.intel.com/en-us/articles/intel-performance-counter-monitor/.

[2] First the tick, now the tock: Intel microarchitecture (nehalem). White Paper, 2009.

[3] J. Almeida, V. Almeida, and D. Yates. Measuring the behavior of a world-wide web server. In *Proc. HPN*, Apr. 1997.

[4] The apache http server project. http://httpd.apache.org/.

[5] M. Arlitt and C. Williamson. Internet Web servers: Workload characterization and performance implications. *IEEE/ACM Trans. Netw.*, 5:631–645, Oct. 1997.

[6] L. A. Barroso and U. Hölzle. The case for energy-proportional computing. *Computer*, 40(12):33–37, Dec. 2007.

[7] S. Boyd-Wickizer, A. T. Clements, Y. Mao, A. Pesterev, M. F. Kaashoek, R. Morris, and N. Zeldovich. An analysis of linux scalability to many cores. In *Proc. USENIX OSDI*, 2010.

[8] M. R. Brown. Fastcgi: A high-performance gateway interface. In *Proc. WWW*, May 1996.

[9] G. DeCandia, D. Hastorun, M. Jampani, G. Kakulapati, A. Lakshman, A. Pilchin, S. Sivasubramanian, P. Vosshall, and W. Vogels. Dynamo: Amazon's highly available key-value store. In *Proc. SOSP*, Oct. 2007.

[10] F. Gaud, R. Lachaize, B. Lepers, G. Muller, and V. Quema. Application-level optimizations on numa multicore architectures: the apache case study. Research Report RR-LIG-011, LIG, Grenoble, France, 2011.

[11] P. Gill, M. Arlitt, N. Carlsson, A. Mahanti, and C. Williamson. Characterizing organizational use of web-based services: Methodology, challenges, observations, and insights. *ACM Trans. Web*, 5(4):19:1–19:23, Oct. 2011.

[12] A. S. Harji, P. A. Buhr, and T. Brecht. Comparing high-performance multi-core web-server architectures. In *Proc. SYSTOR*, Jun. 2012.

[13] R. Hashemian. Revisiting web server performance and scalability. Master's thesis, University of Calgary, Nov. 2011.

[14] A. Kumar, R. Huggahalli, and S. Makineni. Characterization of direct cache access on multi-core systems and 10gbe. In *Proc. IEEE HPCA*, Feb. 2009.

[15] T. Li, D. Baumberger, D. A. Koufaty, and S. Hahn. Efficient operating system scheduling for performance-asymmetric multi-core architectures. In *Proc. ACM/IEEE SC*, Nov. 2007.

[16] Lighttpd web server. http://www.lighttpd.net/ .

[17] R. Love. Cpu affinity. http://www.linuxjournal.com/article/6799.

[18] D. A. Menasce and V. Almeida. *Capacity Planning for Web Services: metrics, models, and methods.* Prentice Hall PTR, Upper Saddle River, NJ, USA, 1st edition, 2001.

[19] J. C. Mogul. Network behavior of a busy web server and its clients. Technical Report WRL-95-5, DEC Western Research Laboratory, Palo Alto, CA, Oct. 1995.

[20] D. Mosberger and T. Jin. httperf: a tool for measuring web server performance. *SIGMETRICS Perform. Eval. Rev.*, 26:31–37, Dec. 1998.

[21] T. Scogland, P. Balaji, W. Feng, and G. Narayanaswamy. Asymmetric interactions in symmetric multi-core systems: Analysis, enhancements and evaluation. In *Proc. ACM/IEEE SC*, Nov. 2008.

[22] M. Seger. collectl. http://collectl.sourceforge.net/Documentation.html , 2008.

[23] T. Suzumura, S. Trent, M. Tatsubori, A. Tozawa, and T. Onodera. Performance comparison of web service engines in php, java and c. In *Proc. IEEE ICWS*, Sept. 2008.

[24] B. Veal and A. Foong. Performance scalability of a multi-core web server. In *Proc. ACM/IEEE ANCS*, Dec. 2007.

Modeling Performance of a Parallel Streaming Engine: Bridging Theory and Costs

Ivan Bedini Sherif Sakr Bart Theeten Alessandra Sala Peter Cogan
Alcatel-Lucent Bell Labs, Ireland and Belgium
{firstname.lastname}@alcatel-lucent.com

ABSTRACT

While data are growing at a speed never seen before, parallel computing is becoming more and more essential to process this massive volume of data in a timely manner. Therefore, recently, concurrent computations have been receiving increasing attention due to the widespread adoption of multi-core processors and the emerging advancements of cloud computing technology. The ubiquity of mobile devices, location services, and sensor pervasiveness are examples of new scenarios that have created the crucial need for building scalable computing platforms and parallel architectures to process vast amounts of generated streaming data. In practice, efficiently operating these systems is hard due to the intrinsic complexity of these architectures and the lack of a formal and in-depth knowledge of the performance models and the consequent system costs. The *Actor Model* theory has been presented as a mathematical model of concurrent computation that had enormous success in practice and inspired a number of contemporary work in this area. Recently, the *Storm* system has been presented as a realization of the principles of the Actor Model theory in the context of the large scale processing of streaming data. In this paper, we present, to the best of our knowledge, the first set of models that formalize the performance characteristics of a practical distributed, parallel and fault-tolerant stream processing system that follows the Actor Model theory. In particular, we model the characteristics of the data flow, the data processing and the system management costs at a fine granularity within the different steps of executing a distributed stream processing job. Finally, we present an experimental validation of the described performance models using the Storm system.

Categories and Subject Descriptors

C.1.2 [**Computer Systems Organization**]: Multiple Data Stream Architectures (Multiprocessors)— *Parallel processors*; C.1.4 [**Computer Systems Organization**]: Parallel Architectures—*Distributed Architecture*

ICPE'13, April 21–24, 2013, Prague, Czech Republic.

Keywords

Distributed Stream Processing Engine, Performance Model, Measurement

1. INTRODUCTION

Today's era of *Big Data* is witnessing a continuous increase of user and machine connectivity that produces an overwhelming flow of data which demands a paradigm shift in the computing architecture requirements and large-scale data processing mechanisms. Therefore, concurrent computations have been receiving increased attention due to widespread adoption of multi-core processors and the emerging advancements of cloud computing technology. For example, the MapReduce framework [?] has been introduced as a scalable and fault-tolerant data processing framework that enables the processing of a massive volume of data in parallel on clusters of horizontally scalable commodity machines. By virtue of its simplicity, scalability, and fault-tolerance, MapReduce is becoming ubiquitous, gaining significant momentum within both industry and academia. However, the MapReduce framework, open sourced by the *Hadoop*[1] Implementation, and its related large-scale data processing technologies (e.g. *Pig*[2] and *Hive*[3]) have been mainly designed for supporting *batch* processing tasks but they are not adequate for supporting *real-time stream* processing tasks [?]. The ubiquity of mobile devices, location services, sensor pervasiveness and real-time network monitoring have created the crucial need for building scalable and parallel architectures to process vast amounts of *streamed* data.

In general, distributed stream processing systems support a large class of applications in which data are generated from multiple sources and are pushed asynchronously to servers which are responsible for processing. Therefore, stream processing applications are usually deployed as continuous jobs that run from the time of their submission until their cancellation. The *Actor Model* [?] is a mathematical model of concurrent computation which has enjoyed enormous success and has inspired a number of systems in this area. Today, this field is mature and provides the proper foundation to operate a horizontally scalable system for massive parallel computations of big data [?, ?]. Recently, Twitter has open-sourced ***Storm***[4] as a parallel, distributed and fault-tolerant system which is designed to fill the gap of providing a platform that supports real-time processing of large scale

[1]http://hadoop.apache.org/
[2]http://pig.apache.org/
[3]http://hive.apache.org/
[4]https://github.com/nathanmarz/storm/wiki

streaming data on clusters of horizontally scalable commodity machines. In principle, the Storm system represents a realization of all the principles of the actor model in the context of concurrent processing on streaming data. In particular, it relies on the idea of creating and interconnecting interactive universal entities, called *Actors*, which are responsible for executing tasks concurrently and asynchronously while communicating with each other through a message-passing protocol [?, ?].

Contributions. In practice, many researchers and engineers are currently increasingly shifting to parallel and distributed systems for large-scale data precessing. However, the lack of expertise and in-depth understanding of the theoretical foundation of this area interferes with the efficient analytic design or with the proper resource allocation strategies to achieve expected performance [?, ?]. In order to tackle this challenge, we introduce, to the best of our knowledge, the first set of performance models that formalize the performance characteristics of an Actor Model based distributed stream processing engine, i.e. Storm. This performance model can play a fundamental role in guiding the user of these systems with the understanding of the system costs which are independent of the executed jobs; the level of parallelization necessary to efficiently execute a particular job; and to better predict the latency, throughput and physical resource costs. In particular, we formalize three main performance cost components as follows:

- *Data Flow Costs*: This component captures information about the amount of data flowing (transmitted) through the different processing steps of a distributed stream processing job.
- *Data Processing Costs*: This component captures information about the execution time of the different processing steps within a distributed stream processing job.
- *System Management Costs*: This component captures information about the system communication overhead between the different components of a distributed stream processing job for ensuring the reliability and fault tolerance during the execution steps.

We use these components to model the general performance characteristics of an Actor Model based implementation of a distributed stream processing system. Finally, we leverage these models to drive our experimental evaluations within the Storm framework.

RoadMap. The remainder of this paper is organized as follows. Section **??** defines the preliminary concepts used in this paper about the *Actor Model* and the *Storm* system. We present the modelling of the data flow costs in Section **??**. The modelling of the data processing costs is presented in Section **??** while the modelling of the system management costs is presented in Section **??**. The results of our experimental evaluation are presented in Section **??**. We discuss the related work in Section **??** before we finally conclude the paper in Section **??**.

2. BACKGROUND AND MOTIVATIONS

This section introduces the fundamental concepts of the Actor Model which are used in this paper and provides an overview of the Storm system which is used as our case study for implementing distributed stream processing jobs using the concurrent computations principles of the Actor Model.

2.1 Actor Model Theory

The Actor Model is one of the first computational models that proposes substantial differences from the Turing Machine Model [?]. In particular, the fundamental difference with the Turing Model is the absence of global states in the Actor Model. Indeed, the Turing Model makes use of global states that have also been described by McCarthy and Hayes in 1969, as *"the complete state of the universe at an instant of time"*. The Actor Model is defined under a fundamentally different concept of *"reacting to input events"*. These events are messages that flow through Actors and trigger responses/computations when computed in some Actor, therefore Actors are based on the concept of Partial ordering [?]. Since the arrival orderings are *indeterminate*, they cannot be deduced from prior information based on mathematical logic alone. This has been referred to as indeterminacy in concurrent computation by Hewitt [?]. Specifically, this logic evolved into the concept of unbounded nondeterminism in which the possible delay in serving a request can become unbounded given the arbitration of contention for shared resources. However, an important property is still guaranteed that *"the request will eventually be serviced"*. In principle, the fundamental concepts of the Actor Model focuses on interpreting everything as an Actor entity. The concurrency model introduced with the Actor Model is based on three fundamental concepts:

1. The *Actors* are the computational units whereby they execute a computation on some input and generate an output.
2. *Actors' input* and *output* are realized through the concept of messages that are defined as the communication unit for exchanging data among Actors.
3. *Actors' communication channels* are realized through buffers or queues from where Actors read and send the data to process. In particular, Actors communicate with each others via simple primitives such as: receipt of messages from other Actors, execution of some computation on incoming messages, generation of output messages, and finally dispatch of output messages to other Actors.

In practice, jobs usually involve multiple Actors. These Actors communicate among each other only through the concept of *addresses* and can be visualized as a *graph topology* where each node is an Actor and a direct edge from Actor A to Actor B means that there is a flow of messages from A to B. However this graph may be highly dynamic where Actors have the possibility to reconfigure their topology on the fly based on the input messages, i.e. Actors can be dynamically added and removed from the system. Therefore, any concurrent system implemented through the Actor concurrency model is theoretically perfectly parallel and scalable.

2.2 From Theory to a Real Streaming System

The Storm system has been presented as a distributed and fault-tolerant stream processing system that instantiates the fundamental principles of Actor theory. The key design principles of Storm are:

- *Horizontally scalable*: Computations and data processing are performed in parallel using multiple threads, processes and machines.
- *Guaranteed message processing*: The system guarantees that each message will be fully processed at least

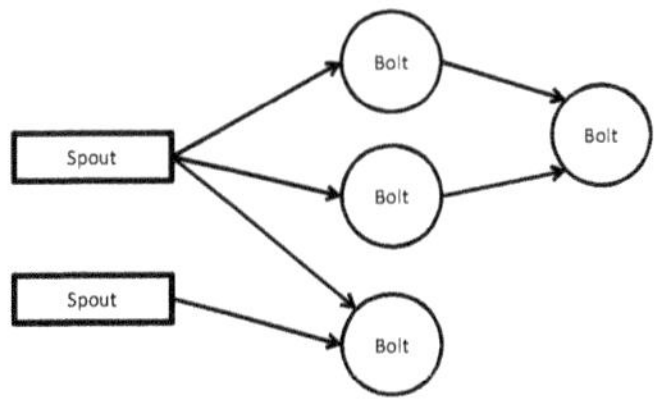

Figure 1: Sample Storm Topology

once. The system takes care of replaying messages from the source when a task fails.

- *Fault-tolerant*: If there are faults during execution of the computation, the system will reassign tasks as necessary.
- *Programming language agnostic*: Storm tasks and processing components can be defined in any language, making Storm accessible to nearly anyone. Clojure, Java, Ruby, Python are supported by default. Support for other languages can be added by implementing a simple Storm communication protocol.

2.2.1 Design Specifics in Storm

The core abstraction in Storm is the *stream*. A stream is an unbounded sequence of *tuples*. Storm provides the primitives for transforming a stream into a new stream in a distributed and reliable way.

The basic primitives Storm provides for performing stream transformations are *spouts* and *bolts*. A *spout* is a source of streams. A *bolt* consumes any number of input streams, carries out some processing, and possibly emits new streams. Complex stream transformations, such as the computation of a stream of trending topics from a stream of tweets, require multiple steps and thus multiple bolts.

A *topology* is a graph of stream transformations where each node is a spout or bolt. Edges in the graph indicate which bolts are subscribing to which streams. When a spout or bolt emits a tuple to a stream, it sends the tuple to every bolt that subscribed to that stream. Links between nodes in a topology indicate how tuples should be passed around.

Each node in a Storm topology executes in parallel. In any topology, we can specify how much parallelism is required for each node, and then Storm will spawn that number of *threads* across the cluster to perform the execution. Figure ?? depicts a sample Storm topology.

Communication Semantics. The Storm system relies on the notion of *stream grouping* to specify how tuples are sent between processing components. In other words, it defines how that stream should be partitioned among the bolt's tasks. In particular, Storm supports different types of stream groupings such as:

1. *Shuffle grouping* where stream tuples are randomly distributed such that each bolt is guaranteed to get an equal number of tuples.
2. *Fields grouping* where the tuples are partitioned by the fields specified in the grouping.
3. *All grouping* where the stream tuples are replicated across all the bolts.
4. *Global grouping* where the entire stream goes to a single bolt.

In addition to the supported built-in stream grouping mechanisms, the Storm system allows its users to define their own custom grouping mechanisms.

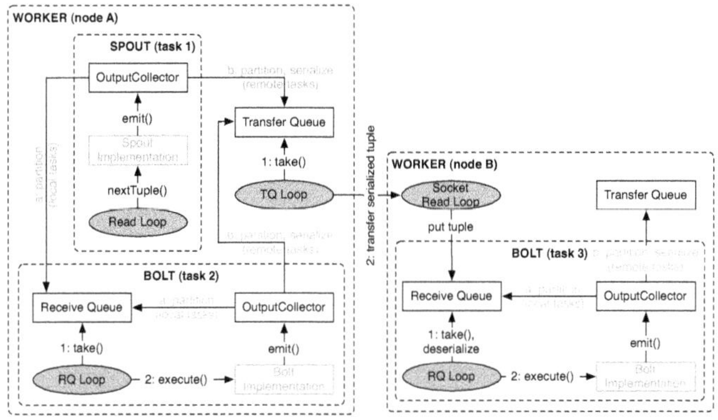

Figure 2: Execution of a Storm Topology

Storm Cluster. In general, a Storm cluster is superficially similar to a Hadoop cluster. One key difference is that a MapReduce job eventually finishes while a Storm job processes messages forever (or until the user kills it). In principle, there are two kinds of nodes on a Storm cluster:

- *The Master node* runs a daemon called ***Nimbus*** (similar to Hadoop's JobTracker) which is responsible for distributing code around the cluster, assigning tasks to machines, and handling failures.
- *The worker nodes* run a daemon called the ***Supervisor***. The supervisor listens for work assigned to its machine and starts and stops worker processes as necessary based on what Nimbus has assigned to it.

In a Storm cluster all the interactions between Nimbus and the Supervisors are done through a ***ZooKeeper***[5] cluster, an open source configuration and synchronization service for large distributed systems. Both the Nimbus daemon and Supervisor daemons are fail-fast and stateless, where all state is kept in ZooKeeper or on local disk.

Communication between workers living on the same host or on different machines is based on ZeroMQ sockets over which serialized java objects (representing tuples) are being passed.

2.2.2 The Execution Steps of Storm Topologies

In practice, a storm job is called a *topology* which continuously processes stream(s) of incoming tuples (events) within a Java Virtual Machine (JVM). Figure ?? illustrates the various phases in processing a tuple in a sample topology consisting of a single *spout* (task 1) and two *bolts* (task 2 and task 3) distributed over a 2 node cluster (Worker Node A and Worker Node B). The figure should be read starting at the *Read Loop* part of the spout. Here, tuples are continuously being read from an unspecified tuple source, which could for example be a text file or a queue. The spout's `nextTuple()` method is called continuously. The user-defined implementation of this method can for example read the next line of text or read the next tuple of the queue and transform the read data into a new tuple (i.e. a sequence of key-value pairs). Then, the implementation calls the `emit()` function on the spout's `OutputCollector` object to actually inject the tuple into the storm cluster. The `OutputCollector` applies the stream groupings onto the tuple which associates one or more *task ids* for each tuple. A task id uniquely refers to a particular task process/thread in the cluster where the tuple

[5] http://zookeeper.apache.org/

should be sent for processing. If it turns out that the task id refers to a local task, it will simply add the tuple onto the receive queue of the local task (task in the same worker). Otherwise, it needs to send the tuple to another worker (i.e. JVM), either on the same node or a different node in the cluster and therefore it is first serialized and then added to the transfer queue. The continuous loop, (*TQ Loop*), reads the serialized tuples off the transfer queue and sends them to the destination worker(s).

As shown in Figure ??, the tuples emitted by the spout are sent directly to the subsequent bolt which is executing task 2. As this is a local task which is running in the same worker as the spout, the tuple is directly placed onto the receive queue of the local bolt. The *RQ Loop* continuously reads tuples off the receive queue and calls the user-defined `execute()` method on the bolt referred to by the task id associated with that tuple. When the implementation decides that a new tuple should be triggered, it should construct a new tuple and emit it to the bolt's `OutputCollector`. In the case of task 3, which is a remote task, the new tuple is first serialized and then placed on the worker's transfer queue. The *TQ Loop* will read the tuple and send it to node B where the tuple will be placed on the receive queue of task 3. The *RQ Loop* at node B will read the tuple, deserialize it and the same process as described above continues.

3. DATA FLOW COST MODEL

The data flow cost model is designed to capture information about the size of data flowing through an active storm topology and the cost of transferring these data between the processing actors.

3.1 Data Size

The size of data flowing in a Storm job is expressed per time unit (e.g. per second) and is composed of the following two components:

- **The input data size**: It represents the total amount of tuples which are injected into the storm cluster by the various spouts in the topology.
- **The output data size**: It represents the total amount of tuples which are generated and emitted by the various processing bolts in the topology and are consumed during further processing by other processing bolts.

Input Data Size In general, an input spout can generate input tuples of different types. The data size of an input spout per time unit can therefore be modeled as:

$$S.dataSize = \sum_{t=1}^{n} S.numTuples_t * tupleSize_t$$

where S refers to a particular input spout, n refers to the number of different types of input tuples produced by that spout, $S.numTuples_t$ refers to the average number of tuples (of type t) emitted by spout S per time unit and $tupleSize_t$ represents the average byte size of tuples of type t. The total size of data input per time unit of a particular storm topology is then defined as follows:

$$T.inputSize = \sum_{i=1}^{s} S_i.dataSize$$

where T refers to a particular topology and s refers to the total number of input spouts for that topology.

Output Data Size Each processing bolt of a topology consumes at least one stream of tuples and possibly generates and emits one or more new streams of tuples. The number of output tuples of a processing bolt for a specific input tuple type t per time unit is defined as follows:

$$B.numTuples_t = \sum_{d=1}^{m} E_d.numTuples_t * B.selectivity_t$$

where E_d refers to a source executor (i.e. spout or bolt) that emits tuples for which B is a destination, m is equivalent to the number of direct edges in the topology graph from an executor E to bolt B, and $B.selectivity_t$ represents the ratio between the number of input tuples of type t and the number of new output tuples being emitted as the result of processing those input tuples by a particular bolt B. In principle, each processing bolt can have different selectivity values for the different types of tuples it can process and it can produce tuples of multiple types as output. Therefore $B.selectivity_t$ can be expanded to:

$$B.selectivy_t = \sum_{t'=1}^{n'} B.selectivity_{t,t'}$$

where n' refers to the number of different types of output tuples produced by bolt B in response to receiving tuples of type t. As a consequence, the output data size of a processing bolt per time unit is defined as follows:

$$B.dataSize = \sum_{t=1}^{n} B.numTuples_t * \overline{tupleSize}$$

where $\overline{tupleSize}$ represents the average size of output tuples, which can be further expanded as:

$$\sum_{t=1}^{n} \sum_{d=1}^{m} \left(E_d.numTuples_t * \sum_{t'=1}^{n'} (B.selectivity_{t,t'} * tupleSize_{t'}) \right)$$

where n refers to the number of different input tuple types received by bolt B and n' refers to the number of different output tuple types produced and emitted by B.

Topology Data Flow Size. The total data flow size for a storm topology can then be represented as the summation of the total data input size injected by all spouts into the topology and the total data output size emitted by all the processing bolts in the topology. This can be expressed as:

$$T.dataSize = \sum_{i=1}^{s} S_i.dataSize + \sum_{j=1}^{b} B_j.dataSize$$

where s refers to the total number of spouts and b refers to the number of bolts which are defined in the executed topology.

3.2 Data Transfer Cost

The data transfer cost is the cost of actually delivering the tuple to a destination task (i.e. bolt instance). A distinction must be made between various allocations of communicating spouts and bolts across the cluster because running tasks remotely or locally produces substantially different communication costs in terms of bandwidth consumption and communication time, i.e. latency. Therefore, we distinguish between three categories of data transfer cost (α) as follows:

1. **Local Java virtual machine (JVM) allocation** (α_{local}): where communicating actors (i.e. spout→bolt or bolt→bolt) are allocated within the same worker and thus within the same JVM. In this case, the tuple serialization/ deserialization step via the transfer queue is by-passed and tuples are immediately placed on the consuming bolt's receive queue.

2. **Local node allocation** (α_{node}): where communicating actors are allocated on different workers that are hosted on the same cluster node. In this case, a serialization/ deserialization cost to send the tuple from one JVM to another is incurred.
3. **Remote allocation** (α_{remote}): where communicating actors are allocated on different workers hosted on different cluster nodes. This is the most expensive case as both a serialization/deserialization cost and a network transfer cost are incurred.

In practice, Storm's allocation strategy *sequentially* and *iteratively* allocates all instances of each component (spout or bolt) up to the configured parallelization factor, whereby an outer loop iterates across all workers and an inner loop iterates across all nodes in the cluster. In the following subsections, we describe how the fraction of tuples that can be processed according to each category can be calculated.

Calculating the Local JVM Allocation. Let $\mathcal{P}$ be the collection of worker processes running across the cluster and let $\mathcal{D}$ be the collection of communicating actors (spout→bolt or bolt→bolt) in the topology. For each process $p \in \mathcal{P}$ and for each communicating actor $d \in \mathcal{D}$ in the sender role (i.e. emitting tuples), let $Aemitter_{p,d}$ be the number of instances of that actor that are allocated in the process p. Similarly, for each process $p \in \mathcal{P}$ and for each communicating actor $d \in \mathcal{D}$ in the receiver role (i.e. receiving tuples), let $Areceiver_{p,d}$ be the number of instances of that actor that are allocated within the same process p. Finally, for each communicating actor $d \in \mathcal{D}$, let us assume that $Aemitter_d$ and $Areceiver_d$ represent respectively the total number of instances of the sending and receiving actors across the entire cluster. It follows that the fraction of tuples which are processed within a single JVM can be computed as:

$$\alpha_{local} = \frac{\sum_{d=1}^{|\mathcal{D}|} \sum_{p=1}^{|\mathcal{P}|} \left(Aemitter_{p,d} * Areceiver_{p,d}\right)}{\sum_{d=1}^{|\mathcal{D}|} Aemitter_d * Areceiver_d}$$

Calculating the Local Node Allocation. Let $n \in \mathcal{N}$ be a node in the cluster (where $|\mathcal{N}|$ is the cluster size), and $Aemitter_{n,d}$ and $Areceiver_{n,d}$ represent the number of instances of that actor d, respectively in the sender and receiver role, that are allocated on the same node n. The fraction of tuples that are processed among tasks within a single node α_{node} can then be derived by computing the distribution of the tasks among the nodes of the cluster, which we call α_{intra_node}, given by the following formula:

$$\alpha_{intra\_node} = \frac{\sum_{d=1}^{|\mathcal{D}|} \sum_{n=1}^{|\mathcal{N}|} \left(Aemitter_{n,d} * Areceiver_{n,d}\right)}{\sum_{d=1}^{|\mathcal{D}|} Aemitter_d * Areceiver_d}$$

Thus, the fraction of tuples that are processed on the same node but on a different processes:

$$\alpha_{node} = \alpha_{intra\_node} - \alpha_{local}$$

Calculating the Remote Allocation. Given the above formulas, the fraction of tuples that are transferred across the network to another node, contributing to the overall data transfer cost can be expressed as follows:

$$\alpha_{remote} = 1 - \alpha_{node}$$

Overall Data Transfer cost. Given the information of the fraction of tuples that are processed in each node and in each JVM, we can now compute the total IO cost per time unit as follows:

Let us assume that the stream comprises of a single tuple type and let Γ_{local}, Γ_{node} and Γ_{remote} be respectively the CPU cost of sending a tuple to the local queue (i.e. within the same JVM), to the same node (but on a different JVM), and to different nodes in the cluster, the total IO cost per time unit can be represented as follows:

$$IOCost = numTuples * IOTranfer$$

where

$$IOTransfer = \alpha_{local}*\Gamma_{local}+\alpha_{node}*\Gamma_{node}+\alpha_{remote}*\Gamma_{remote}$$

In the case where the stream comprises different types of tuples, the $IOCost$ would be computed for each tuple type t and summed together which leads to:

$$IOCost = \sum_{t=1}^{n} numTuples_t * IOTransfer_t.$$

In the above formulas, $numTuples$ and $numTuples_t$ refer respectively to the total amount of tuples and those tuples of type t, flowing through the system.

4. DATA PROCESSING COST MODEL

The data processing cost model describes the execution time of the different processing steps within an active storm topology as well as the IO cost of exchanging tuples between spouts and bolts. The model is again expressed per time unit and has the following two main components:

- *Spout processing cost*: This represents the cost of reading a raw event of an unspecified source and injecting a storm tuple into the topology.
- *Bolt processing cost*: This represents the cost of processing a tuple in a bolt and possibly emitting new tuples into the topology for further processing.

In the following, we provide an in-depth analysis of the CPU processing costs for spouts and bolts, by identifying all the logical operations that are a factor of these costs.

4.1 Spout Processing Cost

Figure ?? illustrates the data processing steps in the processing component (e.g. spout or bolt) which can be split in a number of sequentially executed phases:

1. *Read*: reads a raw data item from an event source (e.g. a queue or a file).
2. *Transform* (γ): formats the raw data item into a storm tuple sequence.
3. *Emit*: adds the storm tuple sequence to the spout's output collector. There will be one destination per direct edge from the spout to a bolt in the topology.
4. *Partition*: defines where (i.e. which task) to send the tuple sequence to for each destination defined in the *emit* step. Depending on the chosen stream grouping, this could be one or more tasks (i.e. bolt instances identified by a node and port pair).
5. *Serialize*: transforms the tuple sequence in a format that can be transmitted efficiently across the network or across JVMs. This phase will only be executed when there is at least one remote destination, i.e. a bolt task that is not running within the emitting spout's JVM.

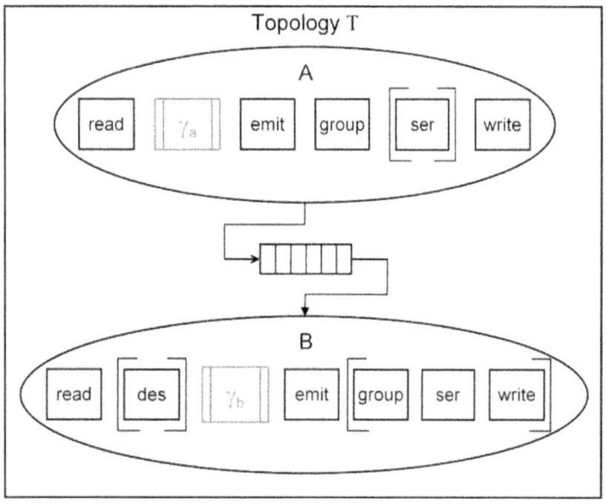

Figure 3: Data Processing Steps in a Processing Component

The *Read*, *Transform* and *Emit* phases are part of the user-defined implementation logic of the spout, while the remaining phases are part of the Storm platform.

The CPU processing cost for an input spout (S) per time unit can therefore be modeled as:

$$S.CPUCost = \sum_{t=1}^{n} S.numTuples_t * S.CPUCost_t$$

with

$$S.CPUCost_t = S.inputCost_t + S.outputCost_t$$

where

$$S.inputCost_t = S.CPURead_t + S.CPUTransform_t$$

and where

$$S.outputCost_t = \sum_{i=1}^{S.numDest_t} B_i.groupingCost_t + S.hasExternalTask_t * CPUSerialize_t$$

where

- n is the number of the different types of tuples which are emitted by the spout S.
- $S.numTuples_t$ is the number of tuples of type t, emitted by spout S per time unit.
- $CPURead_t$ is the CPU cost of reading a raw input tuple of type t.
- $CPUTransform_t$ is the CPU cost of transforming a raw input tuple of type t into a Storm tuple of type t.
- $S.numDest_t$ is the number of destination bolts that consume tuples of type t emitted by spout S.
- $B_i.groupingCost_t$ represent the cost of partitioning the input tuples of type t which varies according to the stream grouping algorithm (*Shuffle*, *Fields*, *All*, *Global*) chosen by the processing bolt B_i.
- $S.hasExternalTask_t$ is a boolean variable which stores the value 1 if the tuples of type t emitted by spout S are consumed by at least one external task. An external task is an instance of a bolt running on a different worker than the one which is hosting the instance of spout S that is emitting the tuple. Note that this could still be on the same host, but on a different JVM. Otherwise it stores the value 0.
- $CPUSerialize_t$ is the CPU cost of serializing a tuple of type t.

4.2 Bolt Processing Cost

In practice, bolt processing involves running through the same sequence of phases as spout processing with the only difference being that of instead of having a *transform* phase, there is a more general *execute* phase in which the actual bolt's processing logic is executed. It is also in this execute phase that the decision will be made as to which and how many new tuples (of type t') will be emitted by the bolt. Therefore, the CPU processing cost for a processing bolt (B) per time unit is modelled as the sum of the cost to process all input tuples of type t received per time unit and the cost of processing all resulting output tuples (of type t'), for all possible tuple types this bolt can receive. This formula is shown as:

$$B.CPUCost = \sum_{t=1}^{n} B.numInputTuples_t * B.inputCost_t + \sum_{t'=1}^{n'} (B.numOutputTuples_{t,t'} * B.outputCost_{t'})$$

with

$$B.inputCost_t = B.CPURead_t + B.CPUExecute_t + B.isExternalTask * CPUDeserialize_t$$

and

$$B.outputCost_{t'} = \sum_{i=1}^{B.numDest_{t'}} B_i.groupingCost_{t'} + B.hasExternalTask_{t'} * CPUSerialize_{t'}$$

where

- $B.numInputTuples_t$ is the number of tuples of type t received by the bolt B, which, as was shown before, can be represented as:

$$B.numInputTuples_t = \sum_{d=1}^{m} E_d.numTuples_t$$

  where E_d refers to the source executor (spout or bolt) that emits tuples for which B is a destination, m is the amount of executors that emit to this bolt and $E_d.numTuples_t$ refers to the number of tuples of type t emitted by executor E_d
- $B.numOutputTuples_{t,t'}$ is the number of new tuples of type t' generated and emitted by the bolt B in response to receiving input tuples of type t, which can be expressed as:
  $B.numOutputTuples_{t,t'} =$
  $B.numInputTuples_t * B.selectivity_{t,t'}$
- $B.CPURead_t$ is the cost of reading an input tuple of type t by bolt B, excluding potential deserialization cost.
- $B.CPUExecute_t$ is the cost of the algorithm implemented by bolt B.
- $B.isExternalTask$ is a boolean variable that stores the value 1 if the tuple being read was sent by an external task and is therefore in serialized form. Otherwise, it stores the value 0.
- $CPUDeserialize_t$ is the CPU cost of deserializing a tuple of type t.

4.3 Topology cost

The total processing cost for a storm topology represents the summation of the total CPU processing cost for all of its spouts and bolts. Therefore, the topology cost, i.e.

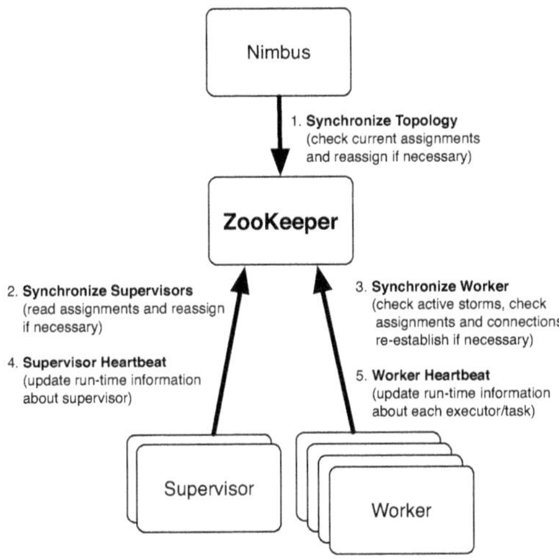

Figure 4: ZooKeeper interactions

$T.CPUCost$, for a storm topology is defined as:

$$T.CPUCost = \sum_{i=1}^{s} S_i.CPUCost + \sum_{j=1}^{b} B_j.CPUCost$$

where s represents the total number of spouts in the topology and b represents the total number of bolts in the topology.

5. SYSTEM MANAGEMENT COST MODEL

The *System Management* costs model the impact of the set of tasks, abstracted away from the user, which are required to run a Storm cluster. These tasks include provision of support for node failure/addition/removal, JVM failures, network issues etc. Storm's system management tasks are coordinated through ZooKeeper and are mainly related to the interaction between ZooKeeper and Nimbus, Supervisor and Worker as shown in Figure ??. There are 5 recurring System Management tasks that interact with ZooKeeper, and the frequency of these interactions can be configured manually. These parameters are described in the following list:

1. **Synchronize Topology**: Every configurable $\mathcal{S}_t$ seconds, Nimbus checks the active assignments and compares them to the required assignments according to the topology specification. If a difference is detected, e.g. because of node failure, Nimbus will reassign the unassigned tasks over the available worker processes in the cluster. *(period: $\mathcal{S}_t$, number of requests: $\mathcal{R}_t$, bandwidth: $\mathcal{B}_t$)*
2. **Synchronize Supervisors**: Every configurable $\mathcal{S}_s$ seconds, each supervisor reads its assignments from ZooKeeper and re-assigns them if it detects a difference between what it has currently assigned across its workers. Re-assignment takes the form of updates to ZooKeeper's assignments for the workers to query during their next poll cycle.
(period: $\mathcal{S}_s$, number of requests: $\mathcal{R}_s$, bandwidth: $\mathcal{B}_s$)
3. **Synchronize Workers**: Every configurable $\mathcal{S}_w$ seconds, each worker reads its assignments from ZooKeeper. If there is a mismatch, the missing connections are established. In addition to this, each worker also checks the active storm topologies. If the storm topology for which it is running tasks would no longer be active (because it was explicitly killed), the worker would need to stop processing.
(period: $\mathcal{S}_w$, number of requests: $\mathcal{R}_w$, bandwidth: $\mathcal{B}_w$)
4. **Supervisor Heartbeat**: Every configurable $\mathcal{S}'_s$ seconds, each supervisor will send a heartbeat to ZooKeeper. A heartbeat takes the form of storing some run-time information about the supervisor in ZooKeeper.
(period: $\mathcal{S}'_s$, number of requests: $\mathcal{R}'_s$, bandwidth: $\mathcal{B}'_s$)
5. **Worker Heartbeats**: Similarly, every configurable $\mathcal{S}'_w$ seconds, each worker sends heartbeats to ZooKeeper. A worker heartbeat includes statistics information about each task running in that worker.
(period: $\mathcal{S}'_w$, number of requests: $\mathcal{R}'_w$, bandwidth: $\mathcal{B}'_w$)

In principle, the System Management Cost has two components: a component that reflects the load on ZooKeeper, expressed in number of requests per time unit ($\mathcal{R}$) and a network load component which represents the network bandwidth consumption ($\mathcal{B}$).

5.1 Number of requests to ZooKeeper

The number of ZooKeeper requests per time unit is a function of the amount of nodes (i.e. supervisors) and the amount of workers in the cluster as reflected by the following equation:

$$\mathcal{R} = \frac{\mathcal{R}_t}{\mathcal{S}_t} + numNodes * \left(\frac{\mathcal{R}_s}{\mathcal{S}_s} + \frac{\mathcal{R}'_s}{\mathcal{S}'_s}\right)$$

$$+ numWorkers * \left(\frac{\mathcal{R}_w}{\mathcal{S}_w} + \frac{\mathcal{R}'_w}{\mathcal{S}'_w}\right)$$

where $\mathcal{R}_t$, $\mathcal{R}_s$, $\mathcal{R}_w$, $\mathcal{R}'_s$ and $\mathcal{R}'_w$ are all fixed (implementation-related) constants and $\mathcal{S}_s$, $\mathcal{S}_w$, $\mathcal{S}'_s$ and $\mathcal{S}'_w$ are configurable constants. It should be noted that the number of requests per time unit is independent of the amount of tasks/components instantiated around the cluster[6] and is linear with the size of the cluster.

5.2 Network Bandwidth Consumption

The network bandwidth consumption depends both on the number of supervisors and workers in the cluster, and also on the number of tasks running on those workers. Specifically, an increase in the number of workers leads to more worker heartbeat messages per time unit as each worker sends the heartbeats separately; an increase in number of supervisors leads to more assignment info messages being exchanged with ZooKeeper as each supervisor is in charge of managing its own assignments and finally, an increase in the number of tasks increases the size of both the worker heartbeat and assignment info messages as they contain information that are specified to each task.

In order to present a formula to calculate the bandwidth consumption of system management tasks, we will first look at two of the most important data structures used to communicate with ZooKeeper, i.e. Worker Heartbeat (WHB) and Assignment Info (AI).

Worker Heartbeats A Worker Heartbeat data structure contains information about the status of the various

[6]with the caveat as explained above that $numWorkers$ is a dynamic thing in case $numTasks$ is smaller than the sum of the configured maximum number of workers on each node.

tasks assigned to a worker. The byte size of one round of worker heartbeats can be defined as:

$$\sum_{i=1}^{numWorkers} |WHB_i| = a * numTasks + b * numWorkers$$

where a and b are constants related to the specific implementation of this data structure in Storm, which will not be further detailed to improve readability.

Assignment Information Assignment information contains a mapping from task ID to host and port and the timestamp of when the component was started.

The byte size of the assignment information data structure is given by

$$|AI| = c * numNodes + d * numTasks + e$$

where again c, d and e are constants related to the specific design of the AI data structure, which will not be further detailed. It should be noted that the overhead of the AI is independent of the amount of workers per node.

Bandwidth As was mentioned previously, an approximation of system management bandwidth consumption can be written as the sum of the bandwidth consumed by each of the recurring tasks as follows:

$$\begin{aligned} \mathcal{B} &= \frac{|\mathcal{R}_t|}{\mathcal{S}_t} + numNodes * \left(\frac{|\mathcal{R}_s|}{\mathcal{S}_s} + \frac{|\mathcal{R}'_s|}{\mathcal{S}'_s}\right) \\ &\quad + numWorkers * \left(\frac{|\mathcal{R}_w|}{\mathcal{S}_w} + \frac{|\mathcal{R}'_w|}{\mathcal{S}'_w}\right) \end{aligned}$$

where $|\mathcal{X}|$ is used to denote the size (in bytes) of $\mathcal{X}$; $numNodes$ is the number of nodes in the Storm cluster running a supervisor; $numWorkers$ is the sum of all active workers on all nodes in the Storm cluster. A worker becomes *active* if it has at least one task assigned to it.

By leveraging an in depth analysis that correlates $\mathcal{R}_t$, $\mathcal{R}_s$, $\mathcal{S}_w$, $\mathcal{S}'_s$ and $\mathcal{S}'_w$ with $numNodes$, $numWorkers$, $|AI|$ and $|WHB|$ (which we do not report here for brevity), we can express the System Management Bandwidth as:

$$\mathcal{B} = k_1 + k_2 * |AI| + numNodes * (k_3 + k_4 * |AI|)$$

$$+ \; numWorkers * (k_5 + k_6 * |AI|) + k_7 * \sum_i |WHB_i|$$

where all of k_1 through k_7 are constants.

In conclusion, with this final formulation we can see that ZooKeeper traffic bandwidth increases proportionally with the number of tasks and nodes running in the cluster, which is also validated by the experiment reported in Figure ??. In the figure we plot the measured BW (in KB) attributed to ZooKeeper communication as a function of the number of tasks running around the cluster, for various cluster sizes. Note also that ZooKeeper communication is independent of the specific algorithms implemented in spouts or bolts.

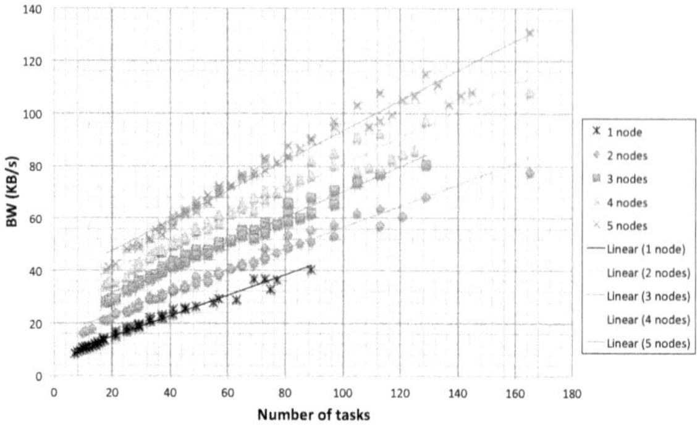

Figure 5: System management traffic as a function of the number of components

6. EXPERIMENTAL EVALUATION

This section provides a detailed experimental evaluation of the performance characteristics of the Storm system using real industrial datasets. This section aims to experimentally quantify the performance variation in running different Storm configurations by leveraging insights from the performance model. Furthermore, the intent of this experimental analysis is to detect the causes of this variability in the system performance and shed light on the precautions required in order to run the system under optimal configurations.

6.1 Dataset and Environment Setup

The experiments have been run with a real Telco dataset consisting of Performance Management (PM) observations of a large Femtocell[7] network. Femtocells were designed for use in a home or small business to improve localized cellular service and offload bandwidth usage from macrocells (i.e., traditional cell towers). During operation, Femtocells produce many low-level performance logs (e.g. number of successful handovers, number of call initiation attempts) across a range of performance categories (e.g. packet data performance, handover performance). The considered dataset is composed of hourly PM logs collected over 15 days for a network of 70k Femtocells, totalling approximately 11 million XML files. These files contain a list of 128 key-value pairs of operational and statistical counters with a mixture of integer and floating-point values. These data are usually analyzed to monitor network performance characteristics and predict traffic peaks or failures. The experiments presented in this paper deploy a simple topology composed of two actors (one spout and one bolt) whose task it is to identify the Femtocells which require the most bandwidth. The spout reads input messages from an external queue (this corresponds to the XML data) and produces a stream of tuples. The bolt receives the stream of tuples with a shuffle grouping and emits the Femtocell list.

Cluster Configuration. Each test runs on a cluster comprising 5-nodes of identical configuration. Additional machines were used to generate load into this cluster and to host the external message queue(s). Each machine is a dual 4-core Intel Xeon 3Ghz 32bit 16 GB Memory, 1 Gb/s network interface. Nodes are interconnected through a 10Gb OmniSwitch 6850 Ethernet Switch. Each node runs Linux version 2.6.32-220.4.1.el6.i686. The following software components were used: Java 1.6 OpenJDK Runtime Environment (IcedTea6 1.10.4), ZeroMQ 2.1.7, Zookeeper 3.4.2 and Kestrel 2.3.4. We used a modified version of Storm, Storm-0.8.0-SNAPSHOT (version of July 2012), to include timers

[7]Alcatel-Lucent 9360 Small Cell. http://www.alcatel-lucent.com/small-cells/

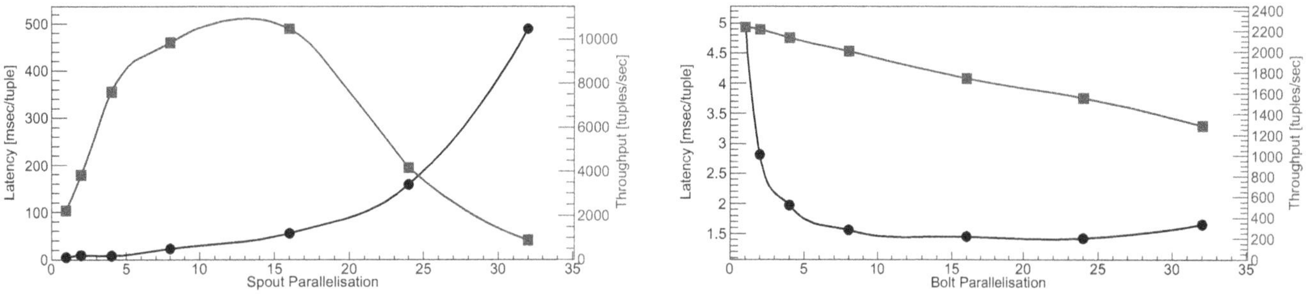

(a) Effect of Spout Parallelisation

(b) Effect of Bolt Parallelisation

Figure 6: Measuring latency and throughput on the different system configurations.

and related statistical tools for the purpose of the experiments.

6.2 Experiment Configurations

A comprehensive test automation suite was developed which is composed of various shell scripts and Java programs to automatically execute the Storm topologies and collect runtime statistics on the system performance. Storm configurations were tested while varying the number of nodes, spouts, bolts and workers. Each specific configuration was tested 10 times in order to accumulate statistical information. Careful consideration was taken to ensure that each configuration was tested under the same operational conditions.

Each individual test case is executed on a clean cluster, meaning that all processes from a previous run were killed on all cluster nodes before starting the new ones. In addition, all data generated by the previous run is erased from the filesystem. Each test consists of an initial warm-up phase of 1 minute (in which we do not collect statistics because the system may not be in its computational steady state yet) followed by a 5 minute measuring phase during which statistics are being gathered, including:

1. Throughput (μ) which represents the total number of events processed per second.
2. Latency ($\mathcal{L}$) which represents time to process a tuple both within a single actor and within the entire system.
3. External and internal queue sizes and their growth rate.
4. Network bandwidth in terms of the number of messages exchanged among workers on different nodes.
5. Various statistics on system management overhead (e.g. communication with ZooKeeper).
6. Approximate memory usage and CPU usage per thread.

Each test is run with a different Storm configuration which is determined by the following parameters:

- *Parallelization factor*: represents the number of tasks, from $\{1, 2, 4, 8, 16, 24, 32\}$, instantiated per actor.
- *Cluster size*: represents the number of nodes participating in the Storm cluster, i.e. from 1 to 4.
- *Worker pool size*: represents the number of workers (JVMs) per node, which host the tasks. In our experiment this number is selected from $\{1, 2, 4, 8, 16, 24, 32\}$.
- *Event injection rate*: represents the number of events injected per second into the queue which provides data to the spouts. In our experiments the event injection rate varies among three different rates, 5K, 10K and 50K tuples per second, to observe how Storm adjusts to different conditions.

6.3 Experimental Analysis

In order to optimize the system performance while running Storm jobs there are several variables and parameters which need manual configuration. The configuration is thus a complex task that requires a precise knowledge of the most relevant parameters and how they impact the system performance. In particular, we focus on the following metrics: the throughput (μ), the latency ($\mathcal{L}$) and the system resource utilization (CPU, memory, network bandwidth).

In the following subsections we present a set of experiments that aim to shed light on how to configure this system based on the theoretical understanding derived from the performance model presented in the previous sections.

6.3.1 *Performance Variability for Different System Configurations*

The experiments presented in this section comprise observations of the impact on latency and throughput produced by different configurations in terms of parallelization, workers and cluster size.

A configuration is expressed as the specification of the system parameters, i.e. number of nodes, spouts, bolts, and workers (hereafter simply $< n, s, b, w >$). For brevity, we only report experiments run on a single node cluster with external queue fed with 50k messages per second. However, the same experiments run on a 1 through 4 node cluster show similar results, with only a few small deviations attributed to the increased system management costs to run more nodes which will be discussed in detail in the next sections.

Spout Parallelisation In order to study the impact of the number of spouts on the overall system performance, we fixed all other parameters at 1 while varying the number of spouts in $\{1, 2, 4, 8, 16, 24, 32\}$. Figure **??** illustrates the effect upon throughput (in terms of tuples per second) and the latency (in terms of milliseconds to process a tuple) as the number of spouts is adjusted. The system throughput can be increased by increasing the number of spouts - however as the number of spouts continues to increase beyond some threshold, the throughput declines. This can be understood by observing the latency which exhibits exponential growth. Beyond some threshold (determined by the system hardware), the system is overstressed with many processes and the context switch among them kills the system performance.

Bolt Parallelisation In order to study the impact of the number of bolts on the overall system performance, we fix all other parameters at 1 while varying the number of bolts in $\{1, 2, 4, 8, 16, 24, 32\}$. Figure ?? illustrates the effect upon throughput (in terms of tuples per second) and the latency (in terms of milliseconds to process a tuple) as the number of bolts is adjusted. An increase in bolt parallelisation reduces throughput due to the extra CPU load associated with scheduling. However, a reduction in latency towards a lower limit is also observed when the number of bolt is within [8, 24] range. This limit represents the fastest possible bolt execution time, which is the cost of the system from the emission of the tuple by the spout up to the completion of the algorithm implemented by bolt B, defined in the performance model in Section ??.

6.3.2 *Costs of Distributing Computation*

This section aims to quantify the variation in performance when varying the number of workers, and the way communicating tasks are distributed across these workers. Section ?? showed that the dataflow cost depends on whether two communicating tasks are hosted (a) within the same JVM, (b) within the same node but on different JVMs, or (c) on different nodes. Specifically, cases (b) and (c) incur a serialization/deserialization cost, while case (c) additionally incurs network transfer cost. Here we demonstrate its practical impact.

Cost of Communicating JVMs Figure ?? displays the system latency $\mathcal{L}$ when some tasks are assigned to the same physical node, but to different JVMs running on that node. Each point in the plot represents a different system configuration. On the x-axis we report the specific configuration setting, i.e. the actual instantiation of $< n, s, b, w >$ and on the y-axes we plot two different results: the left-hand side axis displays the latency, in milliseconds, of the system for each configuration while the right-hand side y-axis estimates α_{node} defined in Section ??, i.e. the percentage of tuple exchanges across different JVMs on the same physical node. In this experiment, an increment of the latency corresponds to the cost of transferring the associated tuples from one JVM to another, which mainly comes from the serialization/ deserialization costs. We show that when the theoretical approximation of α_{node} increases the corresponding latency for that configuration goes up, which means that α_{node} can be used to predict the system latency of a particular configuration. Within the yellow box, starting with configuration $< 1, 8, 8, 1 >$, we observe that the system falls in an unstable configuration in the sense that the serialization/deserialization costs are not the dominant costs anymore and the latency degrades from 5ms to 28ms in the last configuration. We have increased the number of tasks and we can observe that the system is affected by extra costs such as: management overhead, Java garbage collection, scheduling and context switching among JVMs.

Impact of Increasing Workers. One of the benefits of a distributed system is its capacity to increase the amount of parallel threads of execution and reliably distribute them over the cluster nodes so as to improve processing efficiency in time and capacity. However, the *number* of allocable JVMs per single node has a direct impact on performance. This number is highly related to the number of CPUs/cores per node, therefore, the number of allocable threads per worker per CPU needs to be carefully calibrated to achieve optimal system utilization before undesirable effects on the performance show up. Figure ?? and Figure ?? show the throughput and the latency for different system configurations. In this experiment, we highlight the effect of instantiating from 1 worker with 16 threads to 16 workers each with 1 thread. The direct proportional effect of the tuple serialization/deserialization cost on the latency, which is clearly visible in Figure ??, is no longer apparent for higher thread counts. In fact, for higher thread counts, Figures ?? and ?? show that we can improve throughput and latency by instantiating multiple workers, each with a lower thread count to achieve the same overall level of parallelism.

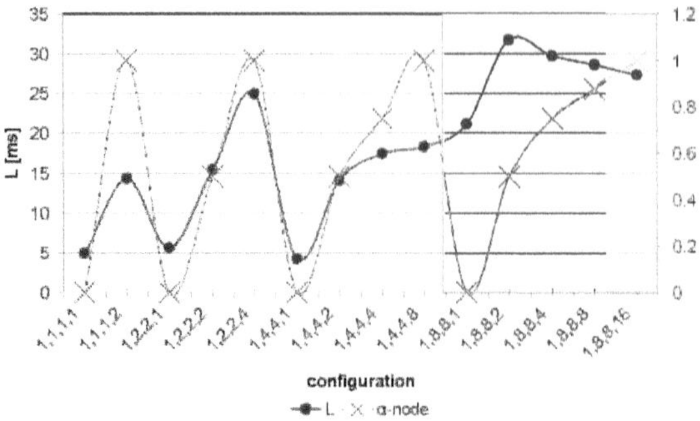

Figure 7: The latency for different system configurations.

This experiment shows that when there are more threads than cores the system performance can have unexpected behavior. For instance, if the number of threads per JVM is too high, as in configuration $< 1, 8, 8, 1 >$ with 16 threads per JVM, the throughput is lower than when there is only a single thread per JVM as in configuration $< 1, 8, 8, 16 >$. One of the factors explaining this observed behavior is the increased JVM garbage collection activity for higher thread counts. Figure ?? and ?? confirm that this trend is the same for all cluster sizes.

Bandwidth Consumption. As we have previously shown, the more cluster nodes we leverage, the better throughput we may expect to achieve. However, the participating nodes need to communicate by transferring tuples among each other, which comes at a cost that is constrained by the available network bandwidth.

In the simple topology we have used for our experiments, we can estimate total bandwidth consumption as follows:

$BW_{total} = BW_{external} + BW_{system} + BW_{dataflow}$

where

- $BW_{external}$ is the bandwidth required by the spout to read raw events from the event source (i.e. a Kestrel queue in our experiments).
- BW_{system} is the system management bandwidth we calculated in Section ??. For simplicity, we can overestimate this portion of the overall bandwidth by taking an upper bound for at least 100 active tasks as reported in Figure ??, i.e. $56K$ bytes per second for the 2-node configuration.
- $BW_{dataflow}$ is the bandwidth related to exchanging tuples between spout and bolt instances. For this portion, we can use the formulas presented in Section ??. More specifically, $BW_{dataflow} = numTuples * \alpha_{remote} * tupleSize$ where $numTuples$ represents the amount of tuples emitted by the spout per time unit.

From measurements performed on our dataset, we have determined that the size of an external event is 360 bytes

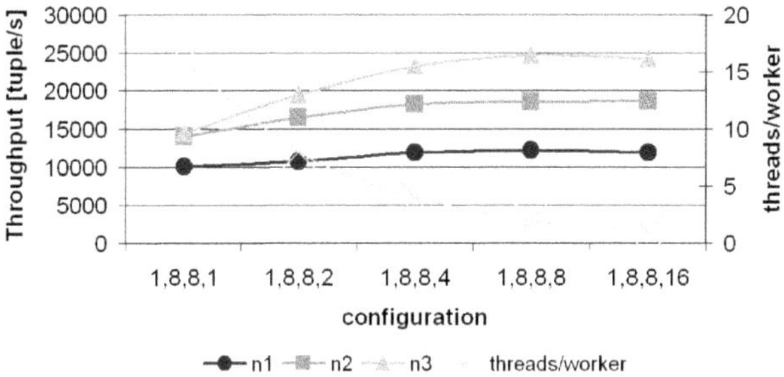

(a) Effects of the parallelism on the Throughput (μ)

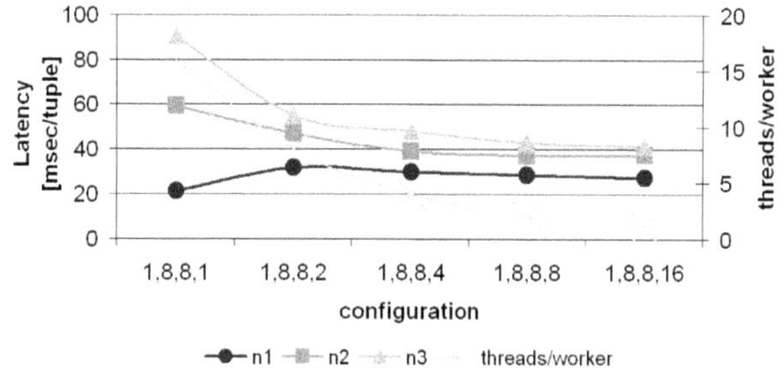

(b) Effects of the parallelism on the Latency ($\mathcal{L}$)

Figure 8: Measuring latency and throughput with 1 to 16 JVMs respectively with 1 to 16 threads per node.

and the size of a transformed event, i.e. a Storm tuple, is 44 bytes. Therefore, in this experiment, the bandwidth consumption can be computed using the following formula:

$$BW_{total} = numTuples * (360 + 44 * \alpha_{remote}) + 56000$$

Figure ?? shows the calculated bandwidth consumption as a function of system configuration for clusters consisting of 2,3 and 4 nodes. These bandwidth calculations were verified using the Unix *dstat* tool.

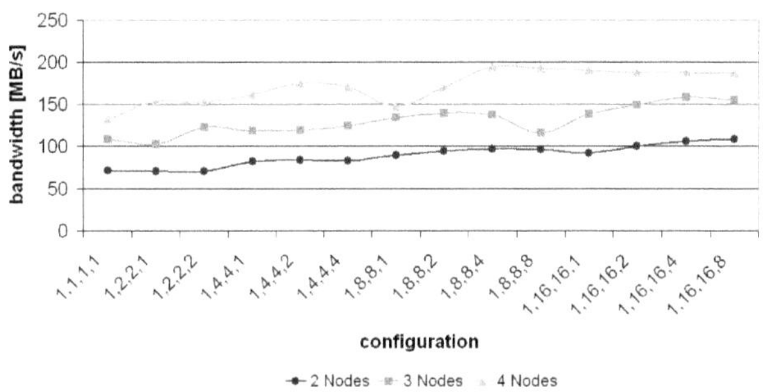

Figure 9: System Communication Cost

7. RELATED WORK

The growing demand for large-scale data processing and data analysis applications has spurred the development of novel solutions from both the industry (e.g., web-data analysis, click-stream analysis, network-monitoring log analysis) and the sciences (e.g., analysis of data produced by massive-scale simulations, sensor deployments, high-throughput lab equipment). MapReduce [?] is a framework which was introduced by Google for programming commodity computer clusters to perform large-scale data processing. The framework is designed such that a MapReduce cluster can scale to thousands of nodes in a fault-tolerant manner. However, the basic architecture of the MapReduce framework requires that the entire output of each map and reduce task be *materialized* into a local file before it can be consumed by the next stage. Therefore, it is not adequate for supporting real time processing of streaming data. Several approaches have been proposed to tackle this challenge. For example, the *MapReduce Online* approach [?] has been proposed as a modified architecture of the MapReduce framework in which intermediate data is *pipelined* between operators while preserving the programming interfaces and fault tolerance models of previous MapReduce frameworks. The *Incoop* system [?] has been introduced as a MapReduce implementation that has been adapted for incremental computations which detect the changes on the input datasets and enables the automatic update of the outputs of the MapReduce jobs by employing a fine-grained result reuse mechanism. In particular, it allows MapReduce programs which have not been designed for incremental processing to be executed transparently in an incremental manner. The M^3 system [?] has been proposed to support the answering of continuous queries over streams of data bypassing the HDFS so that data are processed only through a main-memory-only data-path and totally avoids disk access. In this approach, Mappers and Reducers never terminate where there is only one MapReduce job per query operator that is continuously executing. While these approaches could improve the performance of the MapReduce framework for incremental and streamed data processing, they are still unable to fully support the distributed *real time* processing requirements of data intensive applications.

Several distributed stream processing systems have been presented in the literature such as *Aurora* [?], *Borealis* [?], *SPADE* [?], *Stormy* [?] and *Apache S4*[8]. For example, in Borealis, the collection of continuously running queries are treated as one giant network of operators. The processing of these operators is distributed to multiple sites where each site runs an instance of the Borealis server. The query processing is controlled by an Admin component which takes care of moving query diagram fragments to and from remote Borealis nodes when instructed to do so by other components. SPADE is a declarative stream processing engine which supports a set of basic stream-relational operators with powerful windowing and punctuation semantics. The system is designed to execute a large number of long-running jobs that take the form of data flow graphs where each graph consists of a set of processing elements connected by streams and each stream carries a series of Stream Data Objects. The processing elements can communicate with each other via their input and output ports. The concepts and ideas of our proposed models can be easily adopted to be used within the context of these systems. Some studies have been presented for analyzing and modeling the service latency of event-based systems. They have been mainly focusing on different classes of implementations such as the Publish/Subscribe systems [?, ?] and message-oriented middlewares [?, ?].

When designing parallel and distributed systems, a critical problem is the one of how to tune the system configuration (e.g., in terms of number of nodes), so as to achieve certain performance goals (e.g., maximize throughput, minimize latency or meet a prefixed latency constraint). To this

[8]http://incubator.apache.org/s4/

end, various authors tried to build abstract models for the performance of such systems, starting from the well-known Amdahl law [?] and its various enhancements, some of which are summarised in [?].

Giurgiu [?] has presented an approach for estimation of performance for mobile-cloud applications. The approach tried to identify the factors which impact interaction response times, such as the application distribution schemes, workload sizes and intensities, or the resource variations of the mobile-cloud application setup. It also attempted to find correlations between these factors in order to better understand how to build a unified and generic performance estimation model. The *Starfish* system [?, ?] is the most relevant system for our work. It represents a cost-based optimizer for MapReduce programs which focuses on the optimization of configuration parameters for executing these programs on the Hadoop platform. It relies on a *Profiler* component that collects detailed statistical information from executing the programs and a *What-if Engine* for fine-grained cost estimation processing. For a given MapReduce program, the role of the cost-based optimizer component is to enumerate and search efficiently through the high dimensional space of configuration parameter settings, making appropriate calls to the What-if Engine, in order to find the optimal configuration setting. It clusters parameters into lower-dimensional subspaces such that the globally-optimal parameter setting in the high-dimensional space can be generated by composing the optimal settings found for the subspaces. Our current study represents the first step in the implementation of a similar what-if analyzer component in the more complex environment of real time distributed stream-processing engines.

8. CONCLUSIONS

The lack of expertise and in-depth understanding of the theoretical foundation for the performance modeling of large scale distributed processing system is a crucial problem. Many researchers, engineers and organizations are currently shifting their focus to exploit the advantages of the new data processing paradigm shift. In this paper, we presented a detailed set of models that formalize the performance characteristics of the Storm system as a representative of a practical distributed, parallel and fault-tolerant stream processing engine that follows the Actor Model theory. Our approach is applicable to estimate the performance characteristics of distributed stream processing jobs which follow the same principles in addition to optimizing the configuration settings of the underlying system and hardware cluster. Several directions represent good candidates for future work to further understand and optimise the complex environment of real time distributed streaming processing engines. For example, we are planning to build a What-if Engine that can help users to find the right configuration setting and parameters for their jobs. In addition, we plan to implement a more intelligent allocation strategy for the processing components across the running workers which can optimize the data transfer cost.

9. REFERENCES

[1] Abadi, D. J., et al. Aurora: A Data Stream Management System. In *SIGMOD Conference* (2003), p. 666.

[2] Abadi, D. J., et al. The Design of the Borealis Stream Processing Engine. In *CIDR* (2005), pp. 277–289.

[3] Agha, G. *Actors: a model of concurrent computation in distributed systems.* MIT Press, Cambridge, MA, USA, 1986.

[4] Aly, A. M., et al. M^3: Stream Processing on Main-Memory MapReduce. In *ICDE* (2012).

[5] Amdahl, G. M. Validity of the single processor approach to achieving large scale computing capabilities. In *Proceedings of AFIPS '67, Spring Joint Computer Conference* (1967), pp. 483–485.

[6] Bhatotia, P., et al. Incoop: MapReduce for Incremental Computations. In *SOCC* (2011).

[7] Cherkasova, L. Performance modeling in mapreduce environments: challenges and opportunities. In *ICPE* (2011), pp. 5–6.

[8] Clinger, W. D. Foundations of Actor Semantics. Tech. rep., Cambridge, MA, USA, 1981.

[9] Condie, T., et al. MapReduce Online. In *NSDI* (2010), pp. 313–328.

[10] Cucinotta, T. Optimum scalability point for parallelisable real-time components. In *Proceedings of SOMRES* (2011).

[11] de Gooijer, T., et al. An industrial case study of performance and cost design space exploration. In *ICPE* (2012), pp. 205–216.

[12] Dean, J., and Ghemawat, S. MapReduce: Simplified Data Processing on Large Clusters. In *OSDI* (2004), pp. 137–150.

[13] Gedik, B., et al. Spade: the system s declarative stream processing engine. In *SIGMOD Conference* (2008), pp. 1123–1134.

[14] Giurgiu, I. Understanding performance modeling for modular mobile-cloud applications. In *ICPE* (2012), pp. 259–262.

[15] Herodotou, H., et al. MapReduce Programming and Cost-based Optimization? Crossing this Chasm with Starfish. *PVLDB 4*, 12 (2011), 1446–1449.

[16] Herodotou, H., et al. Starfish: A Self-tuning System for Big Data Analytics. In *CIDR* (2011), pp. 261–272.

[17] Hewitt, C. ORGs for Scalable, Robust, Privacy-Friendly Client Cloud Computing. *IEEE Internet Computing 12*, 5 (2008), 96–99.

[18] Hewitt, C. ActorScript(TM): Industrial strength integration of local and nonlocal concurrency for Client-cloud Computing. *CoRR abs/0907.3330* (2009).

[19] Hewitt, C. Actor Model for Discretionary, Adaptive Concurrency. *CoRR abs/1008.1459* (2010).

[20] Hewitt, C., et al. A Universal Modular ACTOR Formalism for Artificial Intelligence. In *IJCAI* (1973), pp. 235–245.

[21] Loesing, S., et al. Stormy: an elastic and highly available streaming service in the cloud. In *EDBT/ICDT Workshops* (2012), pp. 55–60.

[22] Meseguer, J., and Talcott, C. L. A partial order event model for concurrent objects. In *Proceedings of CONCUR* (1999).

[23] Mühl, G., et al. Stochastic Analysis of Hierarchical Publish/Subscribe Systems. In *Euro-Par* (2009).

[24] Sachs, K., et al. Benchmarking of message-oriented middleware. In *DEBS* (2009).

[25] Sachs, K., et al. Performance evaluation of message-oriented middleware using the SPECjms2007 benchmark. *Perform. Eval. 66*, 8 (2009).

[26] Sakr, S., et al. A Survey of Large Scale Data Management Approaches in Cloud Environments. *IEEE Communications Surveys and Tutorials 13*, 3 (2011), 311–336.

[27] Schröter, A., et al. Stochastic performance analysis and capacity planning of publish/subscribe systems. In *DEBS* (2010).

Supporting Operations Personnel through Performance Engineering

Len Bass
NICTA
Sydney, Australia
len.bass@nicta.com.au

ABSTRACT

Operations costs are around 50% of the worldwide costs of Information Technology. Yet operations problems are mainly viewed as the problem of IT and not of the problem of product developers or performance engineers. I will discuss some areas where performance engineers could be of great assistance to operations and where an understanding of operations process could inform the models created by performance engineers.

Some of these areas are:

- Upgrades from version N to version N+1 are a common occurrence. Some systems are updated as frequently as 10 times a day. Approximately 10% of these updates fail and one common source of this failure is incorrect configurations. Configuration settings that adversely affect performance are among the most difficult to find and correct. For example, a mistaken configuration setting caused a routing of messages unnecessarily through a different region in the cloud. Inconsistent specification of thread pools between two different libraries used by a product is another example.
- Operations personnel monitor system wide performance to determine appropriate resource assignments. This monitoring could be much more effective if applications exported relevant information about their particular performance characteristics and needs.
- Operations processes dictate particular steps when managing applications. Again, considering the upgrade problem, a rolling upgrade causes servers to be systematically taken out of service, upgraded, and restored to service. The workload assumptions of performance models are, typically, not automatically connected to process descriptions. Why could an embedded performance model not be aware of the operations process currently being executed and could adjust the behavior of an application in response.

In general, my argument is that operations personnel are important stakeholders for systems. Performance engineers could make a significant contribution to smoother operations activities and operations processes provide an important source of problems for performance engineers.

ICPE'13, April 21–24, 2013, Prague, Czech Republic.
ACM 978-1-4503-1636-1/13/04.

Categories and Subject Descriptors

D.2.1 [**Requirements/Specifications**]: Elicitation methods

K.6.2 [Installation Management] Performance and usage measurement

Keywords

Systems administration, devops, environmental change

1. CONENTS

The motivation for this work is characterized by the following quote [1]:

"Through 2015, 80% of outages impacting mission-critical services will be caused by people and process issues, and more than 50% of those outages will be caused by change/configuration/release integration and hand-off issues."

When designing a system, we must attend to the problems of all of the stakeholders including operations staff. The problem becomes – how can our systems be designed so that people and process issues do not cause so many outages?

In order to design systems to support the operations staff, we should know what the operations staffs' duties are and the causes of their problems. In general, operators monitor and control the operations of computer systems, install new or upgraded applications/middleware/hardware, and support business continuity through back-ups and disaster recovery. Problems typically arise in association with a change in the environment – upgrading to a new version of a software system or library; changing configuration parameters to reflect changing loads, hardware, or network configurations; or during shift handoffs.

Performance problems that occur during environmental changes are not the most numerous but they are the most subtle and difficult to find.

I advocate a threefold attack on the detection of performance problems during environmental change:

1. Make systems operations aware. This includes version aware, shift hand off aware, and configuration change aware.
2. Develop performance models that enable detection of subtle changes in performance of individual applications.
3. Have systems expose application specific information important for the models developed in 2.

4. Use either external tools or internal monitors to detect and report and correct performance problems caused by environmental changes.

2. CONCLUSIONS

Reducing the outage rate caused by operations problems will require modifications both to process and product. Product modifications should involve making the product operations aware in a variety of dimensions and building models and tools that take advantage of information gathered from the applications.

3. REFERENCE

[1] Colville, R. and Spafford, G. http://www.rbiassets.com/getfile.ashx/42112626510

Self-Adaptive Workload Classification and Forecasting for Proactive Resource Provisioning

Nikolas Roman Herbst
Karlsruhe Institute of Technology
Am Fasanengarten 5
76131 Karlsruhe, Germany
herbst@kit.edu

Nikolaus Huber
Karlsruhe Institute of Technology
Am Fasanengarten 5
76131 Karlsruhe, Germany
nikolaus.huber@kit.edu

Samuel Kounev
Karlsruhe Institute of Technology
Am Fasanengarten 5
76131 Karlsruhe, Germany
kounev@kit.edu

Erich Amrehn
IBM Research & Development
Schoenaicher Str. 220
71032 Boeblingen, Germany
amrehn@de.ibm.com

ABSTRACT

As modern enterprise software systems become increasingly dynamic, workload forecasting techniques are gaining in importance as a foundation for online capacity planning and resource management. Time series analysis offers a broad spectrum of methods to calculate workload forecasts based on history monitoring data. Related work in the field of workload forecasting mostly concentrates on evaluating specific methods and their individual optimisation potential or on predicting Quality-of-Service (QoS) metrics directly. As a basis, we present a survey on established forecasting methods of the time series analysis concerning their benefits and drawbacks and group them according to their computational overheads. In this paper, we propose a novel self-adaptive approach that selects suitable forecasting methods for a given context based on a decision tree and direct feedback cycles together with a corresponding implementation. The user needs to provide only his general forecasting objectives. In several experiments and case studies based on real-world workload traces, we show that our implementation of the approach provides continuous and reliable forecast results at run-time. The results of this extensive evaluation show that the relative error of the individual forecast points is significantly reduced compared to statically applied forecasting methods, e.g. in an exemplary scenario on average by 37%. In a case study, between 55% and 75% of the violations of a given service level agreement can be prevented by applying proactive resource provisioning based on the forecast results of our implementation.

ICPE'13, April 21–24, 2013, Prague, Czech Republic.

Categories and Subject Descriptors

C.4 [**Computer Systems Organization**]: Performance of Systems

Keywords

workload forecasting, arrival rate, time series analysis, proactive resource provisioning

1. INTRODUCTION

Virtualization allows to dynamically assign and release resources to and from hosted applications at run-time. The amount of resources consumed by the executed software services are mainly determined by the workload intensity. Therefore, they both typically vary over time according to the user's behavior. The flexibility that virtualization enables in resource provisioning and the variation of the resource demands over time raise a dynamic optimisation - problem. Mechanisms that continuously provide appropriate solutions to this optimisation problem could exploit the potential to use physical resources more efficiently resulting in cost and energy savings as discussed for example in [1, 23].

Commonly, mechanisms that try to continuously match the amounts of demanded to provisioned resources are reactive as they use threshold based rules to detect and react on resource shortages. However, such reactive mechanisms can be combined with proactive ones that anticipate changes in resource demands and proactively reconfigure the system to avoid resource shortages. For building such proactive mechanisms, the time series analysis offers a spectrum of sophisticated forecasting methods. But as none of these methods is the best performing in general, we work on the idea to intelligently combine the benefits of the individual forecasting methods to achieve higher forecast accuracy independent of the forecasting context.

In this paper, we present a survey of forecasting methods offered by the time series analysis that base their computations on periodically monitored arrival rate statistics. We group the individual forecasting methods according to their computational overhead, their benefits and drawbacks as well as their underlying assumptions. Based on this analysis, we design a novel forecasting methodology that dynamically selects at run-time a suitable forecasting method for a given context. This selection is based on a decision tree that captures the users' forecasting objectives, requirements of individual forecasting methods and integrates direct feedback cycles on the forecast accuracy as well as heuristics for further optimisations using data analysis. Users need to provide only their general forecast objectives concerning forecasting frequency, horizon, accuracy and overhead and are not responsible to dynamically select appropriate forecasting methods. Our implementation of the proposed Workload Classification and Forecasting (WCF) approach is able to continuously provide time series of point forecasts of the workload intensity with confidence intervals and forecast accuracy metrics in configurable intervals and with controllable computational overhead during run-time.

In summary, the contributions of this paper are: (i) We identify the characteristics and components of an observed workload intensity behavior and define metrics to quantify these characteristics for the purpose of automatic classification and selection of a forecasting method. (ii) We provide a survey of state-of-the-art forecasting approaches based on time series analysis (including interpolation, decomposition and pattern identification techniques) focusing on their benefits, drawbacks and underlying assumptions as well as their computational overheads. (iii) We propose an novel forecasting methodology that self-adaptively correlates workload intensity behavior classes and existing time series based forecasting approaches. (iv) We provide an implementation of the proposed WCF approach as part of a new `WorkloadClassificationAndForecasting` (WCF) system[1] specifically designed for online usage. (v) We evaluate our WCF approach using our implementation in the context of multiple different experiments and case studies based on real-world workload intensity traces.

The results of our extensive experimental evaluation considering multiple different scenarios show that the dynamic selection of a suitable forecast method significantly reduces the relative error of the workload forecasts compared to the results of a statically selected fixed forecasting method, e.g. in the presented exemplary scenario on average by 37%. Even when limiting the set of forecasting methods considered in the dynamic selection, our self-adaptive approach achieves a higher forecast accuracy with less outliers than individual methods achieve in isolation. In the presented case study, we apply proactive resource provisioning based on the forecast results of the introduced WCF approach and achieve to prevent between 55% and 75% of the violations of a given service level agreement (SLA).

The major benefit of our self-adaptive WCF approach is an enhancement in foresight for any system that is capable of dynamically provision its resources. In other words, our WCF approach is able to extend the reconfiguration times a system needs for its optimisation. Additionally, the forecast results that are provided by the WCF system can be used for anomaly detection.

The remainder of the paper is structured as follows: In Section 2, we define several concepts and terms that are crucial for understanding the presented approach. In the next step, we analyze characteristics of a workload intensity behavior and give a compact survey on existing workload forecasting methods based on time series analysis. We present our self-adaptive approach for workload classification and forecasting in Section 3. In Section 4, we evaluate the WCF approach by presenting and discussing an exemplary experiment and a case study. We review related work in Section 5, and present some concluding remarks in Section 6.

[1]http://www.descartes-research.net/tools/

2. FOUNDATIONS

In this section, we start by defining crucial terms in the context of software services, workload characterization and performance analysis that are helpful to build a precise understanding of the presented concepts and ideas and are used in the rest of the paper:

We assume that *Software services* are offered by a computing system to a set of users which can be humans or other systems. In our context, a software service can be seen as a deployed software component.

Each *Request* submitted to the software service by a user encapsulates an individual usage of the service.

A *request class* is a category of requests that are characterized by statistically indistinguishable resource demands.

A *resource demand* in units of time or capacity is a measure of the consumption of physical or virtual resources incurred for processing an individual request.

The term *workload* refers to the physical usage of the system over time comprising requests of one or more request classes. This definition deviates from the definition in [29], where *workload* is defined as a more general term capturing in addition to the above the respective applications and their SLAs.

A *workload category* is a coarse-grained classification of a workload type with respect to four basic application and technology domains as defined in [29]: (i) Database and Transaction Processing, (ii) Business Process Applications, (iii) Analytics and High Performance Computing, (iv) Web Collaboration and Infrastructures.

A *time series* X is a discrete function that represents real-valued measurements $x_i \in R$ for every time point t_i in a set of n equidistant time points $t = t_1, t_2, ..., t_n$: $X = x_1, x_2, ..., x_n$ as described in [26]. The elapsed time between two points in the time series is defined by a value and a time unit.

The number of time series points that add up to the next upper time unit or another time period of interest is called the *frequency* of a time series. The frequency attribute of a time series is an important starting point for the search of seasonal patterns.

A *time series of request arrival rates* is a time series whose values represent $n_i \in N$ unique request arrivals during the corresponding time intervals $[t_i, t_{i+1})$.

A *workload intensity behavior* (WIB) is a description of a workload's characteristic changes in intensity over time like the shape of seasonal patterns and trends as well as the level of noise and bursts as further described in the following section. The WIB can be extracted from a corresponding time series of request arrival rates.

2.1 Workload Intensity Behavior

According the theory of the time series analysis [8, 16, 28], a time series can be decomposed into the following three components, whose relative weights and shapes characterise the respective WIB:

The *trend* component can be described by a monotonically increasing or decreasing function (in most cases a linear function) that can be approximated using common regression techniques. A break within the trend component is caused by system extrinsic events and therefore cannot be forecast based on historic observations but detected in retrospect. It is possible to estimate the likelihood of a change in the trend component by analysing the durations of historic trends.

The *season* component captures recurring patterns that are composed of at least one or more frequencies, e.g. daily, weekly or monthly patterns (but they do not need to be integer valued). These frequencies can be identified by using a Fast Fourier Transformation (FFT) or by auto-correlation techniques.

The *noise* component is an unpredictable overlay of various frequencies with different amplitudes changing quickly due to random influences on the time series. The noise can be reduced by applying smoothing techniques like weighted moving averages (WMA), by using lower sampling frequency, or by a low-pass filter that eliminates high frequencies. Finding a suitable trade-off between the amount of noise reduction and the respective potential loss of information can enhance forecast accuracy.

A time series decomposition into the above mentioned components is illustrated in Figure 1. This decomposition of a time series has been presented in [31]. The authors offer an implementation of their approach for time series decomposition and detection of breaks in trends or seasonal components (BFAST)[2]. In the first row, the time series input data is plotted. The second row contains detected (yearly) seasonal patterns, whereas the third row shows estimated trends and several breaks within these trends. The remainder in the undermost row is the non-deterministic noise component computed by the difference between the original time series data and the sum of the trend and the seasonal components.

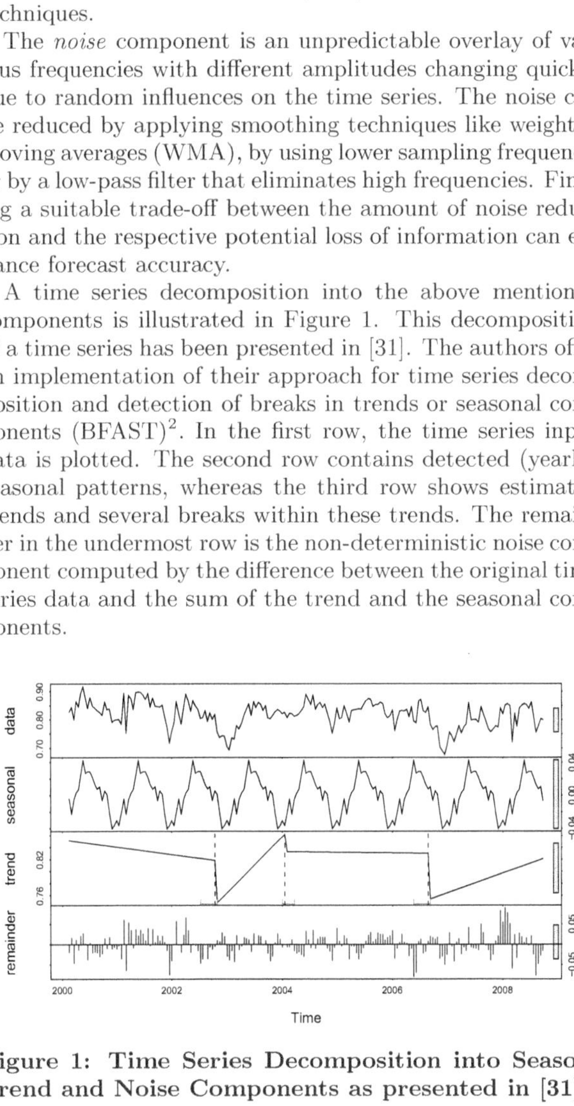

Figure 1: Time Series Decomposition into Season, Trend and Noise Components as presented in [31]

The theory of time series analysis differentiates between static and dynamic stochastic processes. In static process models, it is assumed that the trend and season component remain constant, whereas in dynamic process models, these components change or develop over time and therefore have to be approximated periodically to achieve good quality forecasts. Still, the trend and season components are considered to be deterministic and the quality of their approximation is important for the workload forecasting techniques. Depending on the applied stochastic model, the season, trend and noise components can be considered either as multiplicative or additive.

A nearly complete decomposition of a time series into the three components described above using for example the BFAST approach [31] induces a high computational overhead. Hence, more easily computable characteristics of a time series are crucial for efficient workload classification at run-time as described in the following:

The *burstiness* index is a measure for the spontaneity of fluctuations within the time series and calculated by the ratio of the maximum observed value to the median within a sliding window.

The *length* of the time series mainly influences the accuracy of approximations for the above mentioned components and limits the space of applicable forecasting methods.

The number of *consecutive monotonic values* either upwards or downwards within a sliding window indirectly characterises the influence of the noise and seasonal components. A respective small value can be seen as a sign of a high noise level and a hint to apply a time series smoothing technique.

The *maximum*, the *median* and the *quartiles* are important indicators of the distribution of the time series data and can be unified in the *quartile dispersion coefficient* (QDC) defined as the distance of the quartiles divided by the median value.

The standard deviation and the mean value are combined in the *coefficient of variation* (COV) which characterizes the dispersion of the distribution as a dimensionless quantity.

Absolute *positivity* of a time series is an important characteristic because intervals containing negative or zero values can influence the forecast accuracy and even the applicability of certain forecasting methods. As arrival rates cannot be negative by nature, a time series that is not absolutely positive should be subjected to a simple filter eliminating the negative values or analyzed using specialized forecasting method.

We define a *relative gradient* as the absolute gradient of the latest quarter of a time series period in relation to the median of this quarter. It captures the steepness of the latest quarter period as a dimensionless value. A positive relative gradient shows that the last quarter period changed less than the median value, a negative value indicates a steep section within the time series (e.g., the limb of a seasonal pattern).

The *frequency* of a time series represents the number of time series values that form a *period* of interest (in most cases simply the next bigger time-unit). These values are an important input as they are used as starting points for the search for seasonal patterns.

The presented metrics are sufficient to capture the most important characteristics of a WIB, though they have low computational complexity. Hence, they are of major importance for our online workload classification process.

2.2 Survey of Forecasting Methods

In this section, we compare the most common forecasting approaches based on the time series analysis and highlight

[2]http://bfast.r-forge.r-project.org/

Table 1: Forecasting Methods based in Time Series Analysis

Name	Operating Mode	Horizon	Requirements	Overhead	Strengths	Weaknesses	Optimal Scenario
Naïve Forecast ≘ ARIMA 100	The naïve forecast considers only the value of the most recent observation assuming that this value has the highest probability for the next forecast point.	**very short** term forecast (1-2 points)	single observation	nearly none O(1)	no historic data required, reference method	naïve -> no value for proactive provisioning	constant arrival rates
Moving Average (MA) ≘ ARIMA 001	Calculation of an arithmetic mean within a sliding window of the x most recent observations.	**very short** term forecast (1-2 points)	two observations	very low O(log(n))	simplicity	sensitive to trends, seasonal component	constant arrival rates with low white noise
Simple Exponential Smoothing (SES) ≘ ARIMA 011	Generalization of MA by using weights according to the exponential function to give higher weight to more recent values. 1st step: estimation of parameters for weights/exp. function 2nd step: calculation of weighted averages as point forecasts	**short** term forecast (< 5 points)	small number of historic observations (> 5)	experiment: below **80ms** for less than **100 values**	more flexible reaction on trends or other developments than MA	no season component modeled, only damping, no interpolation	time series with some noise and changes within trend, but no seasonal behavior
Cubic Smoothing Splines (CS) [18] ≘ ARIMA 022	Cubic splines are fitted to the one-dimensional time series data to obtain a trend estimate and linear forecast function. Prediction intervals are constructed by use of a likelihood approach for estimation of smoothing parameters. The cubic splines method can be mapped to an ARIMA 022 stochastic process model with a restricted parameter space.	**short** term forecast (< 5 points)	small number of historic observations (> 5)	experiment: below **100ms** for less than **30 values** (no improved accuracy)	extrapolation of the trend by a linear forecast function	sensitive seasonal patterns (step edges), negative forecast values possible	strong trends, but minor seasonal behavior, low noise level
ARIMA 101 auto-regressive integrated moving averages	ARIMA 101 is an ARIMA stochastic process model instance parameterized with p = 1 as order of AR(p) process, d = 0 as order of integration, q = 1 as order of MA(q) process. In this case a stationary stochastic process is assumed (no integration) and no seasonality considered.	**short** term forecast (< 5 points)	small number of historic observations (> 5)	experiment: below **70ms** for less than **100 values**	trend estimation, (more careful than CS), fast	no season component in modeled, only positive time series	time series with some noise and changes within trend, but no seasonal behavior
Croston's Method intermittent demand forecasting using SES [27]	Decomposition of the time series that contains zero values into two separate sequences: a non-zero valued time series and a second that contains the time intervals of zero values. Independent forecast using SES and combination of the two independent forecasts. No confidence intervals are computed due to no consistent underlying stochastic model.	**short** term forecast (< 5 points)	small number of historic observations, containing zero values (> 10)	experiment: below **avg. 100 ms** for less than **100 values**	decomposition of zero-valued time series	no season component, no confidence intervals (no stochastic model)	zero valued periods, active periods with trends, no strong seasonal comp., low noise
Extended Exponential Smoothing (ETS) innovation state space stochastic model framework	1st step: model estimation: noise, trend and season components are either additive (A), or multiplicative (M) or not modeled (N) 2nd step: estimation of parameters for an explicit noise, trend and seasonal components 3rd step: calculation of point forecasts for level, trend and season components independently using SES and combination of results	**medium to long** term forecast (> 30 points)	at least 3 periods in time series data with an adequate frequency of [10;65]	experiment: up to **15 seconds** for less than **200 values,** high variability in computation time	capturing explicit noise, trend and season component in a multiplicative or additive model	only positive time series	times series with clear and simple trend and seasonal component, moderate noise level
tBATS trigonometric seasonal model, Box-Cox transformation, ARMA errors, trend + seasonal components [10]	The tBATS stochastic process modeling framework of innovations state space approach focuses modeling of complex seasonal time series (multiple/high frequency/non-integer seasonality) and uses Box-Cox transformation, Fourier representations with time varying coefficients and ARMA error correction. Trigonometric formulation enables identification of complex seasonal time series patterns by FFT for time series decomposition. Improved computational overhead using a new method for maximum-likelihood estimations.	**medium to long** term forecast (> 5 points)	at least 3 periods in time series data with an adequate frequency of [10;65]	experiment: up to **18 seconds** less than **200 values,** high variability in computation time	modeling capability of complex, non-integer and overlaying seasonal patterns and trends	only positive time series, complex process for time series modeling and decomposition	times series with one or more clear but possibly complex trend and seasonal components, only moderate noise level
ARIMA auto-regressive integrated moving averages stochastic process model framework with seasonality	The automated ARIMA model selection process of the R forecasting package starts with a complex estimation of an appropriate ARIMA(p,d,q)(P,D,Q)$_m$ model by using unit-root tests and an information criterions (like the AIC) in combination with a step-wise procedure for traversing a relevant model space. The selected ARIMA model is then fitted to the data to provide point forecasts and confidence intervals.	**medium to long** term forecast (> 5 points)	at least 3 periods in time series data with an adequate frequency of [10;65]	experiment: up to **50 seconds** for less than **200 values,** high variability in computation time	capturing noise, trend and season component in (multiplicative or additive) model, Achieves close confidence intervals	only positive time series, complex model estimation	times series with clear seasonal component (constant frequency), moderate noise level

their requirements, advantages and disadvantages presenting a short summary based on [8, 16, 17, 28, 10, 18, 27]. All considered forecasting methods have been implemented by Hyndman in the R forecast package[3] and documented in [17]. The implemented forecasting methods are based either on the state-space approach [16] or the ARIMA (auto-regressive integrated moving averages) approach for stochastic process modelling [8]. These two general approaches for stochastic process modeling have common aspects, but are not identical as in both cases there exist model instances that have no counterpart in the other approach and both have different strength and weaknesses as discussed, e.g., in [17]. In Table 1, we briefly describe the considered forecasting approaches which are ordered according to their computational complexity. Besides from [17], detailed information on the implementation, parameters and model estimation of the individual methods can be taken from cited sources within Table 1. Capturing the complexity in common O-notation in most cases is infeasible except for the two first simple cases. As the shape of seasonal patterns contained in the data as well as the used optimisation thresholds during a model fitting procedure strongly influence the computational costs, the length of time series size represents only one part of the problem description. Therefore, the computational complexity of the individual methods has been evaluated based on experiments with a representative amount of forecasting method executions on a machine with a Intel Core i7 CPU (2.7 GHz). The forecasting methods make use of only a single core as multi-threading is not yet fully supported by the existing implementations.

[3]http://robjhyndman.com/software/forecast/

2.3 Forecast Accuracy Metrics

A number of error metrics have been proposed to capture the differences between point forecasts and corresponding observations. They are summarized, explained and compared in [19]. The authors propose to use the *Mean Absolute Scaled Error* (MASE) as forecast accuracy metric to enable consistent comparisons of forecasting methods across different data samples. The MASE metric for an interval $[m, n]$ is defined as follows with $b_{[m,n]}$ as the average change within the observation:

(I) $error_t = forecast_t - observation_t, t \in [m, n]$

(II) $b_{[m,n]} = \frac{1}{n-m} \cdot \sum_{i=m+1}^{n} |observation_i - observation_{i-1}|$

(III) $MASE[m, n] = mean_{t=[m+1,n]}(\left|\frac{error_t}{b_{[m,n]}}\right|)$

Computing the MASE metric enables direct interpretation whether the forecast accuracy is higher compared to the naive forecasting approach which typically serves as a reference. Therefore, we use the MASE metric as the central element to build a reliable and direct feedback mechanism for our WCF approach as presented in the following section.

3. APPROACH

As stated in the Introduction, the major contribution of this paper is the construction of an online workload classification mechanism for optimized forecasting method selection. In this section, we present the individual concepts that are incorporated into our WCF approach as sketched in Figure 2.

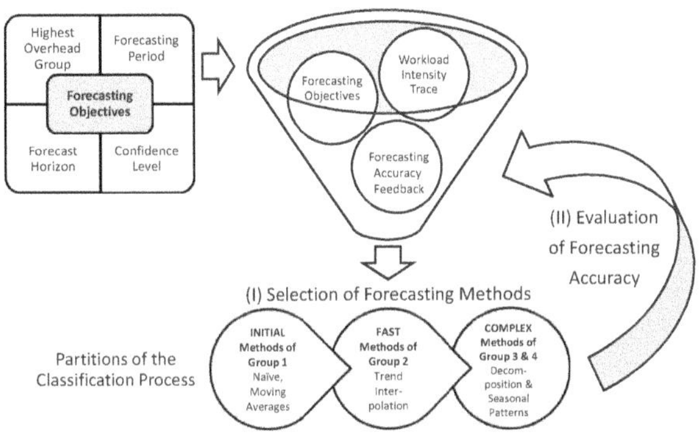

Figure 2: Workload Classification Process

We will use the term *WIB class* to refer to the current choice of the most appropriate forecasting method. Each class identifier is equal to the name of the corresponding forecasting method. A WIB may change and develop over time in a way that the classification is not stationary and therefore it needs to be periodically updated. The WIB class at a time point t corresponds to the currently selected forecasting method A providing the best accuracy compared to the other available forecasting methods. At a later point in time $t+i$, the WIB may be reclassified and may now correspond to another forecasting method B. As in some cases different forecasting methods may possibly deliver identical results, the class of a WIB is not necessarily unique at each point in time.

3.1 Overhead of Forecasting Methods

The forecasting methods as presented in Table 1 are grouped into four subsets according to their computational overhead:

Group 1 stands for nearly no overhead and contains the *Moving Average (MA)* and the *Naive* forecasting methods.

Group 2 stands for low overhead and contains the fast forecasting methods *Simple Exponential Smoothing (SES)*, *Cubic Spline Interpolation (CS)*, the predefined *ARIMA101* model and the specialized *Croston's Method* for intermittent time series. The processing times of forecasting methods in this group are below $100ms$ for a maximum of 100 time series points.

Group 3 stands for medium overheads and contains the forecasting methods *Extended Exponential Smoothing (ETS)* and *tBATS*. The processing times are below 30 seconds for less than 200 time series points.

Group 4 stands for high overheads and contains again *tBATS* and additionally the *ARIMA* forecasting framework with automatic selection of an optimal *ARIMA* model. The processing times remain below 60 seconds for less than 200 time series points.

3.2 Forecasting Objectives

Given that forecast results can be used for a variety of purposes like manual long term capacity planning or short term proactive resource provisioning, a set of forecasting objectives is introduced to enable tailored forecast result processing and overhead control. The following parameters characterizing the forecasting objectives are defined:

(i) The *Highest Overhead Group* parameter is a value of the interval $[1,4]$ specifying the highest overhead group from which the forecasting methods may be chosen in the classification process.

(ii) The *Forecast Horizon* parameters are defined as a tuple of two positive integer values quantifying the number of time series points to forecast. The first value defines the start value, the second value accordingly the maximum forecast horizon setting. Due to the significant differences of the forecasting methods in term of their processing times and capabilities for long term forecasts, the start value need to be increased up to the given maximum value during the learning phase (until the periods are contained in the time series data) using multipliers that are defined as part of the configuration of the classification process.

(iii) The *Confidence Level* parameter can be a value in the interval $[0,100)$ and sets the confidence level *alpha* as percentage for the confidence intervals surrounding the forecast mean values.

(iv) *Forecasting Period* is a positive integer parameter i specifying how often in terms of the number of time series points a forecasting method execution is triggered. The configuration of the presented parameters allows to customize the forecasting strategy according to the characteristics of the considered scenario.

3.3 Partitions of the Classification Process

We partition the classification process according to the amount of historic data available in the time series to support an initial partition, a fast partition and a complex partition. These partitions are connected to the overhead groups of the forecasting methods. Having a short time series, only forecasting methods in the *overhead group 1* can be applied. A medium length may allow application of methods contained in the *overhead group 2* and a long time series enables the use of the methods in *overhead group 3 and 4*. The two thresholds that define when a time series is short, medium or long can be set as parameters in the classification setting. Based on experience gained from experiments, it is our recommendation to set the initial-fast threshold to a value between five (as this is the minimal amount of point needed for Cubic Spline Interpolation) and the value corresponding to half a period. The fast-complex threshold should be set to a value as high as covering three complete time series periods for the simple reason that most methods in the respective overhead group need at least three pattern occurrences to identify them.

3.4 Optional Smoothing

A high noise level within the *time series of request arrival rates* complicates every forecasting process. To improve the forecast accuracy it is conducive to observe metrics that cap-

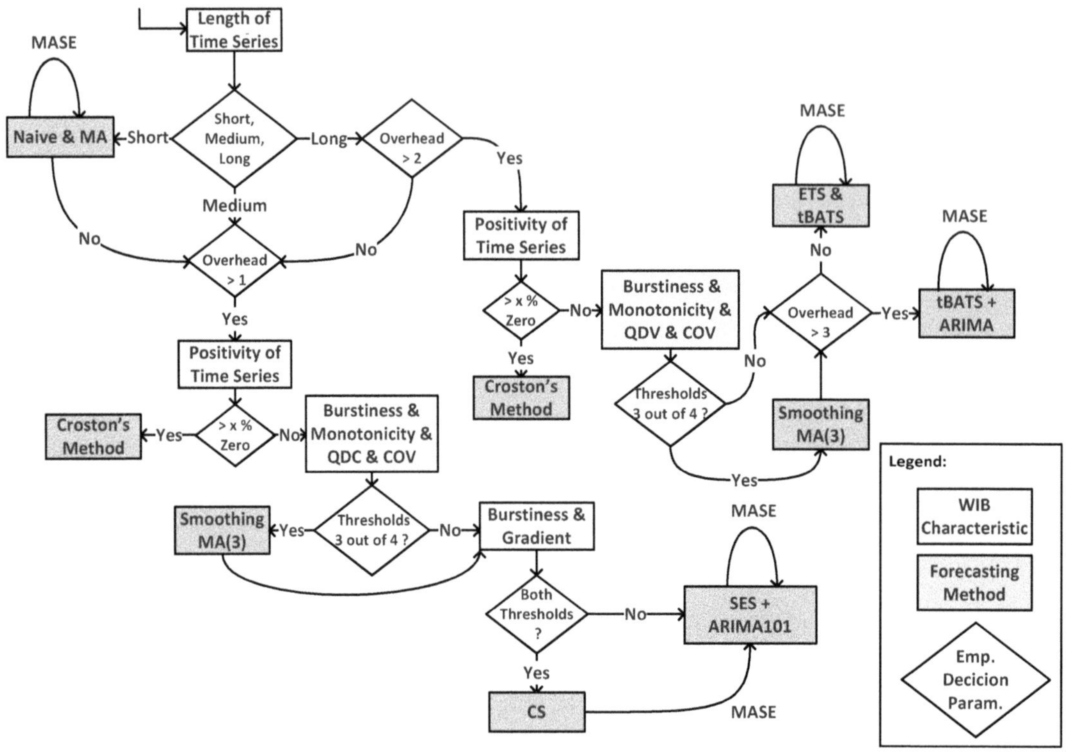

Figure 3: Decision Tree for Online Classification and Selection of Appropriate Forecasting Methods

ture the noise level within the time series and define thresholds to trigger a smoothing method before the time series is processed by the forecasting method. However, any smoothing method has to be applied carefully not to eliminate valuable information. The four metrics (burstiness index, relative monotonicity, COV and QDC) introduced in Section 2.1 are suitable and sufficient to quantify noise related characteristics within a time series and therefore are repeatedly applied to the set of the most recent time series points of configurable length n. Smoothing of the time series can be achieved for example using the elementary method Weighted Moving Averages (WMA) where the level of smoothing can easily be controlled by the window size and the weights. In our WCF approach, we observe the four mentioned noise-related metrics and apply WMA when three of them cross configurable thresholds. We use a window size of three, a weight vector $w = (0.25, 0.5, 0.25)$ and a control variable to assure that no time series value is smoothed more than one time. With this setting we are carefully smooth out a high noise level, but avoid the loss of valuable information. Time series smoothing can also be achieved by far more complex methods like using a Fast Fourier Transformation to build a low-pass filter eliminating high frequencies. However, this would introduce additional overheads and would require an inadequately high amount of time series data.

3.5 Decisions based on Forecast Accuracy

As mentioned in Section 2.3 and extensively discussed in [19], the MASE metric is the metric of choice for quantifying the forecast accuracy. We describe in this Section how the MASE metric is used in our adaptive classification process to identify and validate the current WIB class.

Given that the MASE metric compares the forecast accuracy against the accuracy of the naive forecasting method, it is directly visible when a method performs poorly. If the MASE value is close to one or even bigger, the computed forecast results are of no value, as their accuracy is equal or even worse than directly using monitoring values. The closer the metric value is to zero, the better the forecasting method delivers trustworthy forecast results. The forecasting method having the smallest MASE value defines the WIB class of our adaptive classification process.

In our approach, two or more forecasting methods are executed in parallel and compared in configurable intervals to check whether the recent WIB class is still valid. This WIB class validation is also triggered when the forecast result seems implausible or shows low accuracy (MASE value > 1). The MASE metric is estimated during the forecast execution itself as at this point in time observation values are not yet available. This estimation needed when two or more forecasting methods are executed in parallel to output the more promising result. Just before the next forecast execution is triggered, the MASE metric is then precisely calculated as observation values and forecast results are available at this point in time and used to validate the WIB classification.

3.6 Decision Tree with Feedback Cycles

The presented concepts and mechanisms were combined to build our WCF approach that automatically adapts to changing WIBs with the result of optimized and more constantly adequate forecasting accuracy. The combination of a classification process with a set of forecasting methods enables a configuration according to given forecasting objectives. Figure 3 illustrates the classification process of the WCF approach. The classification process can be seen as independent from the forecasting method executions and is triggered according to the configuration of the *classification period* parameter every n time series points, just before a forecasting method execution is requested. The classification process uses the forecasting objectives specified by the user and the WIB characteristics to reduce the space of possible forecasting methods. The suitable forecasting methods are then processed together and their estimated forecast accuracy is evaluated to output the more promising result. Before the next forecasting method execution, the classification (which is the selection of the better performing method) is validated by the observed MASE metric and used until the next classification process execution is triggered. The WCF parameter set we propose in this section follows rules of thumb from time series analysis literature and has been experimentally calibrated using several independent real-world workload traces provided by IBM. These traces are not used for the evaluation of the WCF approach. In the presented exemplary scenario and the case study in Section 4 the recommended parameter set is applied without any changes. For further case specific optimisations, the WCF approach can be controlled in detail by adjusting the parameter settings and thresholds.

4. EVALUATION

In this section, we present results of an extensive evaluation of our WCF approach based on real-world scenarios. Due to the space constraints, we focus on presenting a selected experiment focusing a forecast accuracy comparison and an exemplary case study where the forecast results are interpreted for proactive resource provisioning. In general, real-world WIB traces are likely to exhibit strong seasonal patterns with daily periods due to the fact that the users of many software services are humans. The daily seasonal patterns are possibly overlaid by patterns of a longer periods such as weeks or months. Depending on the monitoring precision, a certain noise level is normally observed, but can be reduced by aggregation of monitoring intervals or smoothing techniques as discussed in Section 3.4. Deterministic bursts within a WIB trace are often induced by planned batch tasks (for example in transaction processing systems). In addition, a WIB trace can exhibit non-deterministic bursts that cannot be foreseen by any time series analysis technique due to system extrinsic influences. For the evaluation of the WCF approach we used multiple real-world WIB traces with different characteristics representing different types of workloads. The results presented in this paper are based on the following two WIB traces:

(i) *Wikipedia Germany Page Requests*: The hourly number of page requests at the webpage of Wikipedia Germany has been extracted from the publicly available server logs[4] for October 2011.

(ii) *CICS* transaction processing monitoring data was provided by IBM of an real-world deployment of an IBM z10 mainframe server. The data reports the number of started transactions during one week from Monday to Sunday in 30 minute windows.

[4] http://dumps.wikimedia.org/other/pagecounts-raw/

4.1 Exemplary Scenario

In the presented exemplary scenario, the WCF system uses forecasting methods from all four overhead groups and is compared to the fixed use of the Extended Exponential Smoothing (ETS) method. Additionally, the Naive forecasting method (which is equivalent to system monitoring without forecasting) is compared to the other forecasting methods to quantify and illustrate the benefit of applying forecasting methods against just monitoring the request arrival rates. The individual forecasting methods have been executed with identical exemplary forecast objectives on the same input data and are therefore reproducible. The forecast results are provided continuously over the experiment duration, which means that at the end of an experiment there is for every observed request arrival rate a corresponding forecast mean value with a confidence interval. This allows to evaluate whether the WCF system successfully classifies the WIB. A successful classification would mean that the forecasting method that delivers the highest accuracy for a particular forecasting interval is selected by the WCF system. To quantify the forecast result accuracy, a relative error is calculated for every individual forecast mean value: $relativeError_t = \frac{|forecastValue_t - observedArrivalRate_t|}{observedArrivalRate_t}$. The distributions of these relative errors are illustrated using cumulative histograms which have inclusive error classes on the x-axis and the corresponding percentage of all forecast points on the y-axis. In other words, an $[x, y]$ tuple expresses that y percent of the forecast points have a relative error between 0% and x%. Accordingly, given that the constant line $y = 100\%$ represents the hypothetical optimal error distribution, in each case the topmost line represents the best case of the compared alternatives with the lowest error. We chose to illustrate the error distribution using cumulative histograms to obtain monotone discrete functions of the error distribution resulting in less intersections and therefore in a clearer illustration of the data. To improve readability, the histograms are not drawn using bars but simple lines connecting the individual $[x, y]$ tuples. In addition, statistical key indices like the arithmetic mean, the median and the quartiles as well as the maximum were computed to enable direct comparison of the relative error distributions and rank the forecasting methods according to their achieved forecast accuracy. Finally, directed paired t-tests from common statistics were applied to ensure the statistical significance of the observed differences in the forecast accuracy of the different forecasting methods.

This experiment compares the forecast accuracy of the WCF approach with the ETS method and the Naive method during a learning process which means that at the beginning there is no historical data available to the individual forecasting methods that are compared. Concerning the forecasting objectives, the forecast horizon (in this scenario identical to the forecasting frequency) for all forecasting methods is configured to increase stepwise as shown in Table 2. The maximum overhead group is set to 4 for an unrestricted set of forecasting methods. The confidence level is set to 95%, but not further interpreted in this scenario.

The cumulative error distribution for each of the methods are shown in Figure 5 as well as in Figure 6 which demonstrate that the WCF method achieves significantly better forecast accuracy compared to ETS and Naive. ETS can only partially achieve slightly better forecast accuracy than the Naive method, however, it induces processing over-

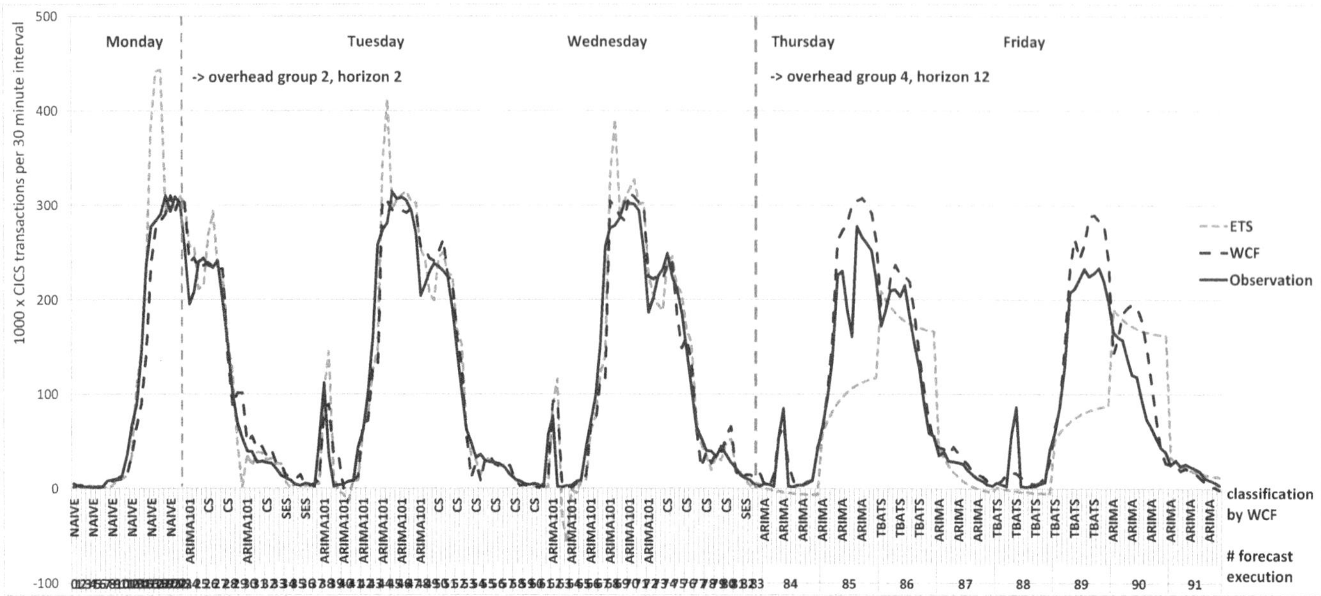

Figure 4: Exemplary Scenario: Comparison Chart of WCF and ETS

Table 2: Exemplary Scenario

Experiment Focus	Comparison of unrestricted WCF to static ETS and Naive
Forecasting Methods (overhead group)	WCF(1-4), ETS(3), Naive(1)
Input Data WIB trace	CICS transactions, Monday to Friday, 240 values in transactions per 30 minutes, frequency = 48, 5 periods as days
Forecast Horizon (h) (= Forecasting Frequency)	h = 1 for 1st half period, h = 2 until 3rd period complete, h = 12 for 4th and 5th period

heads of 715 ms per forecasting method execution compared to 45 ms for the Naive method (computation of the confidence intervals). The WCF method has an average processing overhead of 61 ms during the first three periods of the WIB trace. For the fourth and fifth period (Thursday and Friday), when forecasting methods of the overhead group 4 are selected, WCF's processing overhead per forecasting method execution is on average 13.1 seconds. The forecast mean values of the individual methods and the respective actual observed request arrival rates in the course of time are plotted in Figure 4. In this chart it is visible that the ETS method forecast values have several bursts during the first three periods and therefore do not remain as close to the actual observed values as the WCF forecast values do. During the last eight forecast executions the WCF approach successfully detects the daily pattern early enough to achieve better results than ETS.

The WCF approach shows the lowest median error value of 20.7% and the lowest mean error value of 47.4%. In addition, WCF has a significantly smaller maximum error value. Though the ETS method is a sophisticated procedure on its own, the results of the presented exemplary scenario demonstrate that the application of ETS on its own leads to poor

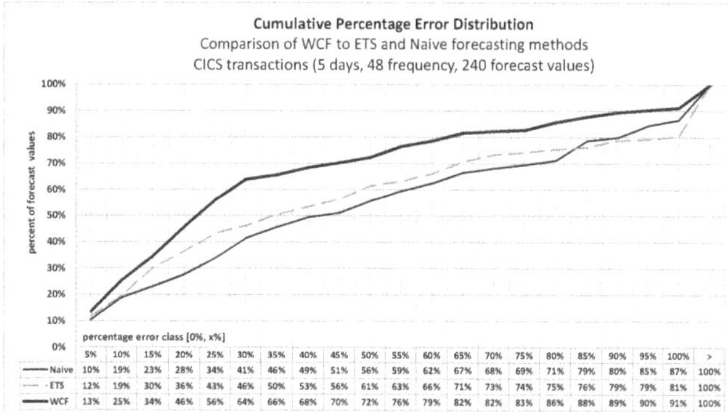

Figure 5: Exemplary Results: Cumulative Error Distribution of WCF, ETS and Naive

forecast accuracy with average errors almost as high as for the Naive method and even higher maximal errors.

4.2 Case Study

In this section, we present a case study demonstrating how the proposed WCF method can be successfully used for proactive online resource provisioning. WCF offers a higher degree of flexibility due to the spectrum of integrated fore-

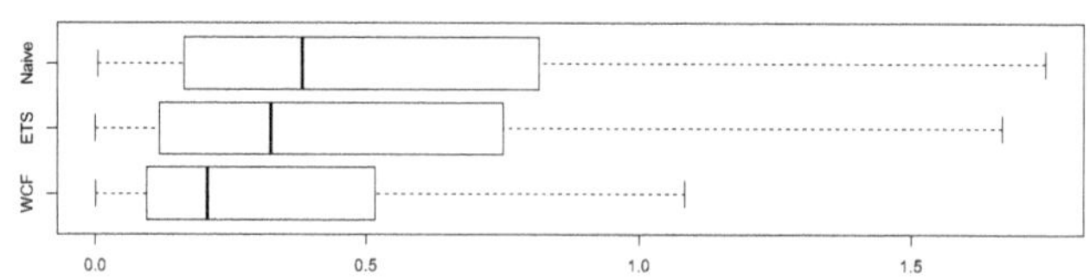

Figure 6: Exemplary Results: Box & Whisker Plots of the Error Distributions

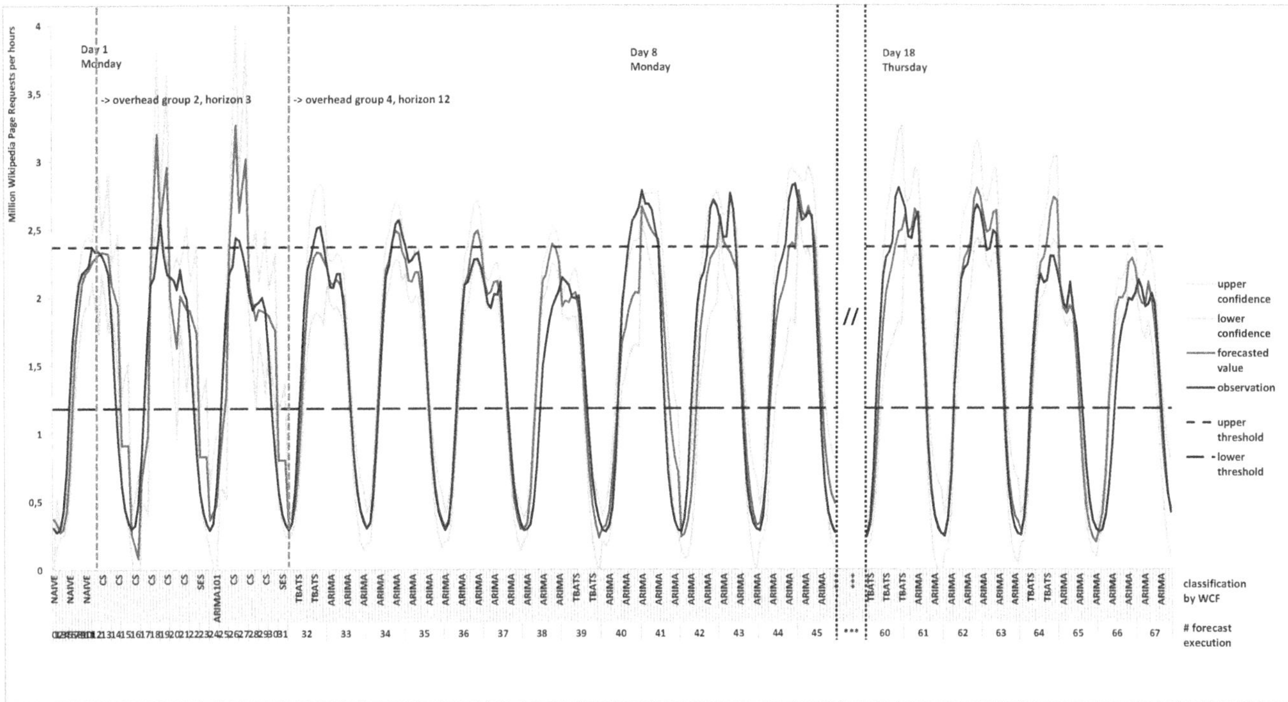

Figure 7: Case Study Chart: 21 Days Wikipedia Page Requests, WCF4 approach

cast methods. None of these forecasting methods can offer this degrees of flexibility on their own. For this case study scenario, there is no historic knowledge available at the beginning and the WCF has to adapt to this and wait for the first three periods (Monday, Tuesday, Wednesday) until forecasting methods of the overhead group 4 can be applied. For an arrival rate higher than a given threshold, the SLA of the average response time is violated. We assume that the underlying system of the analyzed workload intensity behavior is linearly scalable from zero to three server instances. This assumption implies a constant average resource demand per request for all system configurations. Furthermore we assume the presence of a resource provisioning system that reacts on observed SLA violations as well as on forecast results. For an arrival rate lower than a given threshold, the running server instances are not efficiently used and therefore one of them is shut down or put in stand by mode. In this scenario, the WCF system provides forecast mean values and confidence intervals to the resource provisioning system which then can proactively add or remove server instances at that point in time, when the resource or server is needed, in addition to solely reacting on SLA violations. Details on the WCF configuration for this case study scenario are given in Table 3.

In Figure 7, the WCF forecast values are plotted together with the corresponding confidence intervals and the actual observed workload intensities. The two dotted lines represent the given thresholds that define when a server instance needs to be started or shut down. The upper threshold is placed in a way that it is not reached constantly in every daily seasonal period (for example not on the weekends). The additions or removals of server instances are triggered either in a reactive manner after an SLA violation or in a

Table 3: Case Study Scenario

Forecast Strategy (overhead group)	WCF(1-4)
Input Data WIB trace	Wikipedia Germany, 3 weeks, 504 values in page requests per hour, frequency = 24, 21 periods as days
Forecast Horizon (= Forecasting Frequency) (number of forecast points (h))	h = 1 for 1st half period h = 3 until 3rd period is complete h = 12 from 4th on
Confidence Level	80 %

proactive fashion anticipating the points in time when SLA violations would occur. It can be seen that SLA violations cannot be forecast reliably in the first three daily periods. For the following daily periods, SLA violations can be correctly anticipated in the majority of cases. Only when the amplitude of the daily pattern changes, for example before and after weekends, the forecast mean values deliver false positives or do not anticipate the need for additional computing resources on time. To obtain the results summarized

Table 4: Case Study: SLA Violations Summary

Comparison	Reactive vs. proactive WCF based provisioning of server instances
Reactive	76 SLA violations
Proactive (WCF based)	42 SLA violations correctly anticipated and 15 more nearly correctly 6 cases of false positives 13 cases of false negatives (not detected)

in Table 4, we counted correct, incorrect and nearly correct forecasts for all threshold crossings of the real workload. In the worst case 34 and in the best case 19 SLA violations occur when using the proactive approach based on WCF compared to 76 SLA violations in the reactive case. 17% of SLA violations were not anticipated by the proactive WCF-based approach in addition to 8% reported false positives.

4.3 Result Interpretation

As part of our comprehensive evaluation, we considered a number of different scenarios under different workloads with different forecasting objectives and different respective WCF configurations. The experiment results show that the WCF system is able to sensitively select appropriate forecasting methods for particular contexts thereby improving the overall forecast accuracy significantly and reducing the number of outliers as compared to individual forecasting methods applied on their own. In the presented exemplary scenario, WCF achieved an improvement in forecast accuracy on average by 37%. As demonstrated in the case study scenario, the interpretation of WCF forecast results by proactive resource provisioning reduces the number of SLA violations by between 55% to 75%. In addition, our proposed self-adaptive WCF approach supports the system users to select a forecasting method according to their forecasting objectives. Especially at the beginning of a WIB's lifetime, when no or few historical data is available, a static decision made by a user would not fit for the WIB's lifetime. With its dynamic design and the flexibility to react on changes in the WIB, the WCF system is able to adapt to such changes, thereby increasing the overall accuracy of the forecast results. Our WCF system enables online and continuous forecast processing with controllable computational overheads. We achieve this by scheduling forecasting method executions and WIB classifications in configurable periods. In all experiments, the processing times of all forecasting methods remained within the boundaries specified by their corresponding overhead group in such a way that the forecast results are available before their corresponding request arrival rates are monitored. The latter is crucial especially for the execution of forecasting methods at a high frequency with short horizons.

For a WIB with strong daily seasonal pattern and high amplitude variance due to known calendar effects, the forecast accuracy might be strongly improved by splitting this WIB into two separate time series: regular working days in the first and weekends and public holidays in the second. This can reduce the impact of the possibly strong overlay of weekly patterns. As part of our future work, we plan to introduce support for such time series splittings as well as for the selection of the data aggregation level for varying forecasting objectives. This can be achieved by a realisation of an intelligent filter applied to the monitoring data before it is provided to the WCF system in form of time series.

5. RELATED WORK

We divide related work into the following three groups: (i) The first group of related work has its focus on evaluating forecasting methods applied to either workload related data or performance monitoring data. (ii) In the second group of related work we consider approaches for workload forecasting that are not based on the methods of the time series analysis, as opposed to our proposed WCF approach. (iii) In the third group of related work we cover research that has its focus on approaches for proactive resource provisioning using tailored forecasting methods.

(i) In 2004, Bennani and Menascé have published their research results of an robustness assessment on self-managing computer systems under highly variable workloads [4]. They come to the conclusion that proactive resource management improves a system's robustness under highly variable workloads. Furthermore, the authors compare three different trend interpolating forecasting methods (polynomial interpolation, weighted moving averages and exponential smoothing) that have been introduced as means for workload forecasting in [24]. Bennani and Menascé propose to select the forecasting methods according to the lowest R^2 error as accuracy feedback. Still, the focus of this work is more on the potentials of proactive resource provisioning and the evaluation of forecasting methods is limited to basic trend interpolation methods without any pattern recognition for seasonal time series components. In [11], Frotscher assesses the capabilities to predict response times by evaluating the forecast results provided by two concrete ARIMA models without seasonality, simple exponential smoothing (SES) and the Holt-Winters approach as a special case of the extended exponential smoothing (ETS). His evaluation is based on generated and therefore possibly unrealistic times series data. The author admits that the spectrum of forecasting methods offered by time series analysis is not covered and he is critical about the capability to predict response times of the evaluated methods as their strengths in trend extrapolation does not suit to typically quickly alternating response times.

(ii) In [21] and a related research paper [5], the authors propose to use neuronal nets and machine learning approaches for demand prediction. This demand predictions are meant to be used by an operating system's resource manager. Goldszmidt, Cohen and Powers use an approach based on Bayesian learning mechanisms for feature selection and short term performance forecasts described in [13]. Shortcoming of these approaches is that the training of the applied neuronal nets on a certain pattern need to be completed before the nets can provide pattern based forecasts. This stepwise procedure limits the flexibility and implies the availability of high amounts of monitoring data for the mining of possibly observable patterns.

(iii) The authors of [12] use a tailored method to decompose a time series into its dominating frequencies using a Fourier transformation and then apply trend interpolation techniques to finally generate a synthetic workload as forecast. Similarly in [14], Fourier analysis techniques are applied to predict future arrival rates. In [9], the authors base their forecast technique on pattern matching methods to detect non-periodic repetitive behavior of cloud clients. The research of Bobroff, Kochut and Beaty presented in [7] has its focus on the dynamic placement of virtual machines, but workload forecasting is covered by the application of static ARMA processes for demand prediction. In [20], an approach is proposed and evaluated that classifies the gradients of a sliding window as a trend estimation on which the resource provisioning decision are then based. The authors Grunske, Aymin and Colman of [3, 2] focus on QoS forecasting such as response time and propose an automated and scenario specific enhanced forecast approach that uses a combination of ARMA and GARCH stochastic process modeling frameworks for frequency based representation of time series data. This way, the authors achieve improved forecast accuracy for QoS attributes that typically have quickly alternating values. In [25], different methods of the time series analysis and Bayesian learning approaches are applied. The authors propose to periodically selected a forecasting method using forecast accuracy metrics. But it is not further evaluated how significantly this feedback improves the overall forecast accuracy. No information is given on how the user's forecast objectives are captured and on how the computational overheads can be controlled. In contrast to our research, the authors concentrate on the prediction of resource consumptions.

However, besides of [25], a common limitation of the above mentioned approaches is their focus on individual methods that are optimised or designed to cover a subset of typical situations. Therefore, they are not able to cover all possible situations adequately as it could be achieved by an intelligent combination of the innovation state space frameworks (ETS and tBATS) and the auto-regressive integrated moving averages (ARIMA) framework for stochastic process modeling. Furthermore, in all mentioned approaches besides in [14], the resource utilisation or average response times are monitored and taken as an indirect metric for the recent workload intensity. But these QoS metrics are indirectly influenced by the changing amounts of provisioned resources and several other factors. The forecasting computations are then based on these values that bypass resource demand estimations per request and interleave by principle performance relevant characteristics of the software system as they can be captured in a system performance model.

6. CONCLUSION

Today's resource managing systems of virtualized computing environments often work solely reactive using threshold-based rules, not leveraging the potential of proactive resource provisioning to improve a system performance while maintaining resource efficiency. The results of our research is a step towards exploiting this potential by computing continuous and reliable forecasts of request arrival rates with appropriate accuracy.

The basis of our approach was the identification of workload intensity behavior specific characteristics and metrics that quantify them. Moreover, we presented a survey on the strengths, weaknesses and requirements of existing forecast methods from the time series analysis and ways to estimate and evaluate the forecast accuracy of individual forecast executions. As the major contribution of this paper, we presented an approach classifying workload intensity behaviors to dynamically select appropriate forecasting methods. This has been achieved by using direct feedback mechanisms that evaluate and compare the recent accuracy of different forecasting methods. They have been incorporated into a decision tree that considers user specified forecasting objectives. This enables online application and processing of continuous forecast results for a variable number of different WIBs. Finally, we evaluated an implementation of our proposed approach in different experiments and an extensive case study based on different real-world workload traces. Our experiments demonstrate that the relative error of the forecast points in relation to the monitored request arrival rates is significantly reduced as in the exemplary scenario by 37% on average compared to the results of a static application of the Extended Exponential Smoothing (ETS) method. More importantly, the major benefit of our approach has been demonstrated in an extensive case study, showing how it is possible to prevent between 55% and 75% of SLA violations.

6.1 Future Work

We plan to extend the functionality of the WCF system to apply it in further areas. One extension is the combination of the WCF system with an intelligent filter that helps a system user to fit the aggregation level of the monitored arrival rates for specific forecasting objectives, e.g, continuous short term forecasts for proactive resource allocation or manually triggered long term forecasts for server capacity planning. For divergent forecasting objectives, such a filter would multiplex an input stream of monitoring data and provide it as individual WIBs in different aggregation levels at the data input interface of the WCF system. A second WCF system application area will be a combination with an anomaly detection system like ΘPAD as outlined in [6] at the data output interface of the WCF system. Such a system can compute continuous anomaly ratings by comparing the WCF system's forecast results with the monitored data. The anomaly rating can serve to analyze the workload intensity behavior for sudden and unforeseen changes and in addition as an reliability indicator of the WCF system's recent forecast results.

Ultimately, we will connect the WCF system's interface (data input) to a monitoring framework like Kieker [30] to continuously provide online forecast results. These forecast results can then be used as input for self-adaptive resource management approaches like [15], giving them the ability to proactively adapt the system to changes in the workload intensity. This combination of workload forecasting and model-based system reconfiguration is an important step towards the realization of self-aware systems [22].

7. REFERENCES

[1] H. Abdelsalam, K. Maly, R. Mukkamala, M. Zubair, and D. Kaminsky. Analysis of energy efficiency in clouds. In *Future Computing, Service Computation, Cognitive, Adaptive, Content, Patterns, 2009. COMPUTATIONWORLD '09. Computation World:*, pages 416–421, nov. 2009.

[2] A. Amin, A. Colman, and L. Grunske. An Approach to Forecasting QoS Attributes of Web Services Based on ARIMA and GARCH Models. In *Proceedings of the 19th IEEE International Conference on Web Services (ICWS)*. IEEE, 2012.

[3] A. Amin, L. Grunske, and A. Colman. An Automated Approach to Forecasting QoS Attributes Based on Linear and Non-linear Time Series Modeling. In *(Accepted for publication to appear) Proceedings of the 27th IEEE/ACM International Conference on Automated Software Engineering (ASE)*. IEEE/ACM, 2012.

[4] M. N. Bennani and D. A. Menasce. Assessing the robustness of self-managing computer systems under highly variable workloads. In *Proceedings of the First International Conference on Autonomic Computing*, pages 62–69, Washington, DC, USA, 2004. IEEE Computer Society.

[5] M. Bensch, D. Brugger, W. Rosenstiel, M. Bogdan, W. G. Spruth, and P. Baeuerle. Self-learning

prediction system for optimisation of workload management in a mainframe operating system. In *ICEIS*, pages 212–218, 2007.

[6] T. C. Bielefeld. Online Performance Anomaly Detection for Large-Scale Software Systems, March 2012. Diploma Thesis, University of Kiel.

[7] N. Bobroff, A. Kochut, and K. Beaty. Dynamic placement of virtual machines for managing sla violations. In *Integrated Network Management, 2007. IM '07. 10th IFIP/IEEE International Symposium on*, pages 119 -128, 21 2007-yearly 25 2007.

[8] G. Box, G. Jenkins, and G. Reinsel. *Time series analysis : forecasting and control.* Wiley series in probability and statistics. Wiley, Hoboken, NJ, 4. ed. edition, 2008. Includes index.

[9] E. Caron, F. Desprez, and A. Muresan. Pattern matching based forecast of non-periodic repetitive behavior for cloud clients. *J. Grid Comput.*, 9:49–64, March 2011.

[10] A. M. De Livera, R. J. Hyndman, and R. D. Snyder. Forecasting time series with complex seasonal patterns using exponential smoothing. *Journal of the American Statistical Association*, 106(496):1513–1527, 2011.

[11] T. Frotscher. Prognoseverfahren für das Antwortzeitverhalten von Software-Komponenten, March 2011. Christian-Albrechts-Universität zu Kiel, Bachelor's Thesis.

[12] D. Gmach, J. Rolia, L. Cherkasova, and A. Kemper. Workload analysis and demand prediction of enterprise data center applications. In *Proceedings of the 2007 IEEE 10th International Symposium on Workload Characterization*, IISWC '07, pages 171–180, Washington, DC, USA, 2007. IEEE Computer Society.

[13] M. Goldszmidt, I. Cohen, and R. Powers. Short term performance forecasting in enterprise systems. In *In ACM SIGKDD Intl. Conf. on Knowledge Discovery and Data Mining*, pages 801–807. ACM Press, 2005.

[14] M. Hedwig, S. Malkowski, C. Bodenstein, and D. Neumann. Towards autonomic cost-aware allocation of cloud resources. In *Proceedings of the International Conference on Information Systems ICIS 2010*, 2010.

[15] N. Huber, F. Brosig, and S. Kounev. Model-based Self-Adaptive Resource Allocation in Virtualized Environments. In *6th International Symposium on Software Engineering for Adaptive and Self-Managing Systems (SEAMS 2011), Honolulu, HI, USA*, May 23-24 2011.

[16] R. Hyndman, A. Köhler, K. Ord, and R. Snyder, editors. *Forecasting with Exponential Smoothing : The State Space Approach.* Springer Series in Statistics. Springer-Verlag Berlin Heidelberg, Berlin, Heidelberg, 2008.

[17] R. J. Hyndman and Y. Khandakar. Automatic time series forecasting: The forecast package for R, 7 2008.

[18] R. J. Hyndman, M. L. King, I. Pitrun, and B. Billah. Local linear forecasts using cubic smoothing splines. Monash Econometrics and Business Statistics Working Papers 10/02, Monash University, Department of Econometrics and Business Statistics, 2002.

[19] R. J. Hyndman and A. B. Koehler. Another look at measures of forecast accuracy. *International Journal of Forecasting*, pages 679–688, 2006.

[20] H. Kim, W. Kim, and Y. Kim. Predictable cloud provisioning using analysis of user resource usage patterns in virtualized environment. In T.-h. Kim, S. S. Yau, O. Gervasi, B.-H. Kang, A. Stoica, and D. Slezak, editors, *Grid and Distributed Computing, Control and Automation*, volume 121 of *Communications in Computer and Information Science*, pages 84–94. Springer Berlin Heidelberg, 2010.

[21] S. D. Kleeberg. Neuronale Netze und Maschinelles Lernen zur Lastvorhersage in z/OS. Universität Tübingen, Diploma Thesis.

[22] S. Kounev, F. Brosig, N. Huber, and R. Reussner. Towards self-aware performance and resource management in modern service-oriented systems. In *Proceedings of the 2010 IEEE International Conference on Services Computing*, SCC '10, pages 621–624, Washington, DC, USA, 2010. IEEE Computer Society.

[23] L. Lefèvre and A.-C. Orgerie. Designing and evaluating an energy efficient cloud. *The Journal of Supercomputing*, 51:352–373, 2010. 10.1007/s11227-010-0414-2.

[24] D. A. Menascé and V. A. F. Almeida. *Capacity planning for Web performance: metrics, models, and methods.* Prentice-Hall, Inc., Upper Saddle River, NJ, USA, 1998.

[25] X. Meng, C. Isci, J. Kephart, L. Zhang, E. Bouillet, and D. Pendarakis. Efficient resource provisioning in compute clouds via vm multiplexing. In *Proceedings of the 7th international conference on Autonomic computing*, ICAC '10, pages 11–20, New York, NY, USA, 2010. ACM.

[26] T. Mitsa. *Temporal Data Mining.* Chapman & Hall/CRC data mining and knowledge discovery series. Chapman & Hall/CRC, 2009.

[27] L. Shenstone and R. J. Hyndman. Stochastic models underlying croston's method for intermittent demand forecasting. *Journal of Forecasting*, 24(6):389–402, 2005.

[28] R. H. Shumway. *Time Series Analysis and Its Applications : With R Examples.* Springer Texts in StatisticsSpringerLink : Bücher. Springer Science+Business Media, LLC, New York, NY, 2011.

[29] J. Temple and R. Lebsack. Fit for Purpose: Workload Based Platform Selection. Technical report, IBM Corporation, 2010.

[30] A. van Hoorn, J. Waller, and W. Hasselbring. Kieker: A framework for application performance monitoring and dynamic software analysis. In *Proceedings of the 3rd ACM/SPEC International Conference on Performance Engineering (ICPE 2012).* ACM, Apr. 2012. Invited tool demo paper. To appear.

[31] J. Verbesselt, R. Hyndman, A. Zeileis, and D. Culvenor. Phenological change detection while accounting for abrupt and gradual trends in satellite image time series. *Remote Sensing of Environment*, 114(12):2970 – 2980, 2010.

COSBench: Cloud Object Storage Benchmark

Qing Zheng* , Haopeng Chen
School of Software, Shanghai Jiao Tong University, Shanghai, China
{ mark, chen-hp } @ sjtu.edu.cn

Yaguang Wang, Jian Zhang, Jiangang Duan
Intel Asia-Pacific R&D Ltd., Shanghai, China
{ yaguang.wang, jian.zhang, jiangang.duan } @ intel.com

ABSTRACT

With object storage systems being increasingly recognized as a preferred way to expose one's storage infrastructure to the web, the past few years have witnessed an explosion in the acceptance of these systems. Unfortunately, the proliferation of available solutions and the complexity of each individual one, coupled with a lack of dedicated workload, makes it very challenging for one to evaluate and tune the performance of different systems. To help address this problem, we present the *Cloud Object Storage enchmark* (COSBench). It is a benchmark tool that we have developed at Intel with the goal of facilitating both performance comparison and system optimization of these systems. In this paper, we describe the design and implementation of this tool, focusing on its extensibility and scalability. In addition, we discuss how people can use this tool to perform system characterization and how the latter can facilitate system comparison and optimization. To demonstrate the value of our tool, we report the results of our experiments conducted on two Swift setups we built in our lab. We also share some of our experiences in turning our setups to achieve higher performance.

Categories and Subject Descriptors

H.3.4 [**Systems and Software**]: Performance evaluation

Keywords

Object Storage, Benchmark Tool

1. INTRODUCTION

With our world being increasingly connected and enriched by newly emerged computing technologies, the data produced is gigantic [31]. To combat such data tsunami, object storage has been introduced to fulfill the common need of a scale-out storage infrastructure, with acceptance of these systems growing every year [1]. However, the proliferation of object storage systems and the complexity of each individual one makes it very difficult for one to choose the best system. In order to better understand existing options, careful performance characterization of each system is typically needed, so one can quantitatively compare different candidates and then pick out a system that can best accommodate its specific requirements. For example, people dealing with large objects may confidently steer their attention away from a system that only excels in handling small objects.

Tuning systems to achieve maximum performance is also challenging. It entails not only choosing the optimal values in a variety of software configurations, but also assigning the correct hardware to different components constituting the system. In order to approach the best combination, numerous experiments have to be conducted across the large spectrums of possibilities at different dimensions, which can be quite time-consuming.

To alleviate these problems, we present the *Cloud Object Storage enchmark* (COSBench). It is a benchmark tool that we have developed to help people better compare different hardware and software stacks as well as better identify bottlenecks to drive system tuning and optimization. Capable of generating workloads that mimic diverse applications, this tool allows people to characterize systems under their own usage patterns. In addition, a lot of attention has been paid to ensuring the tool's scalability and extensibility, so that people can easily simulate a large number of concurrent clients and adapts these clients to a myriad of storage systems. In order also to secure ease-of-use, we have designed our tool to be operated with either web browsers or command-line utilities, making it friendly not only to human users, but automation tools as well.

Although file systems and block storage can also be employed to manage data, we focus on object storage systems with the following reasons. Firstly, there are many existing tools [26, 17, 25, 16] available for measuring other types of storage systems. Besides, recent years have seen a trend in building file systems and relational databases upon object storage infrastructures [19, 27, 28, 29], which compels our attention to the performance of these underlying services. Finally, the lack of a nontrivial workload for novel object storage systems best warrants our development of the tool.

In this paper, we describe the design and implementation of the COSBench, with discussions on how people can use it to perform system characterization and how the latter can facilitate system comparison and optimization. We

*This paper was performed in Intel where the author is an intern.

ICPE'13, April 21–24, 2013, Prague, Czech Republic.

Figure 1: Object storage interface

also share the results of our experiments conducted on two different Swift setups we deployed in our lab.

The rest of this paper is organized as follows. Section 2 gives an overview of object storage systems while Section 3 discusses system characterization. Section 4 introduces the workload model used in the COSBench with details of the tool itself covered in Section 5. Section 6 and 7 share the results and the experiences we have obtained from our experiments, followed by Section 8 which reports the related works. The paper is concluded in Section 9 where future works are also described.

2. OBJECT STORAGE OVERVIEW

In this section we present an overview of the object storage by discussing the common motivation, storage interface, and system architecture shared among different implementations.

2.1 Motivation

Object storage is largely motivated by the gap between file systems and block storage. On one hand, many applications nowadays find themselves less dependent on traditional POSIX file systems. For instance, applications dealing with images can sit comfortably upon a file system that only supports a flat namespace and weak consistency. Content sharing applications can also painlessly devise their own ways to handle concurrent modifications without resorting to file system locks. On the other hand, although a reduction of file system features frequently fuels scalability, developers still need storage services to free them from the burden of managing blocks and various other storage metadata. To this end, object storage systems are often implemented with eclectic designs: they are less sophisticated than file systems but still more intelligent than block devices.

The rise of cloud computing has also assisted the emergence of object storage. Object storage echoes the very spirit of cloud computing. It provides people with an unlimited and on-demand data depository accessible from anywhere in the world, and it helps to achieve data consolidation and economy of scale, both facilitating cost-effectiveness. With cloud computing getting increasingly adopted, so are object storage systems.

2.2 Storage Interface

A careful design of the storage interface is essential for object storage to become a building block in web computing: (i) the interface should ensure ubiquitous access on the part of millions of applications running all over the Internet; (ii) appropriate security mechanisms have to be properly incorporated into the interface to facilitate user authentication and data encryption; (iii) it is also necessary to keep the interface stateless as session management notoriously hinders the scalability of systems; (iv) object storage must eschew binding its interface to any specific programming languages in favor of a more interoperable approach. With all above criteria honored, object storage systems are designed as RESTful web services and are accessed via HTTP requests, as is exemplified in Figure 1.

To use object storage systems, people create *containers* and put *objects* into these containers for storage. Objects are just like files in regular file systems, though they cannot be locked or updated partially. Containers are identical to directories except that they cannot be nested. Both objects and containers are identified by unique names distinguishing one from each other. Notwithstanding the resemblance among the interfaces of various object storage systems, details do differ. While efforts [3] have been made towards a standard protocol, time is yet needed before an extensive acceptance is eventually reached.

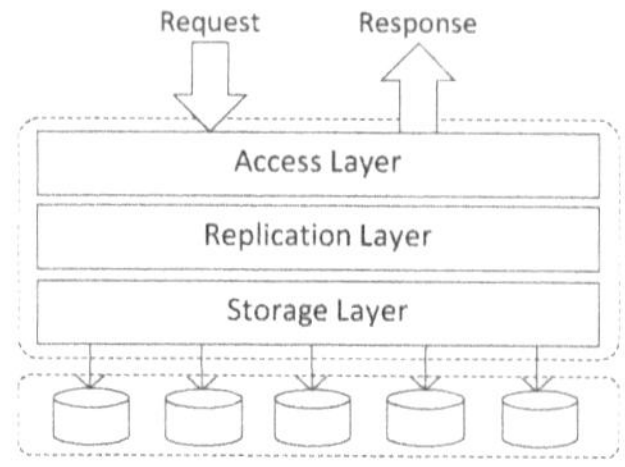

Figure 2: Architecture of object storage systems

2.3 Architecture

The techniques related with implementing object storage systems can be stratified into three distinct layers: storage layer, replication layer, and access layer, as illustrated in Figure 2. The storage layer is responsible for storing objects to physical devices. The key issue the storage layer has to deal with is performance: it has its own discretion in optimizing either for the read performance or the write. As most storage layers use file systems to handle their workloads, it is then important to choose the correct file system and to mount it with the proper options. A benchmark tool can help people compare different file systems and options, and tune the storage layer to its best capacity.

The replication layer sitting upon the storage layer holds all the secrets of transforming a pool of storage devices to a web-scale data serving platform. As a first step, the replication layer should keep an eye on the durability of each object stored in the system. After all, hardware faults inevitably occur and high quality devices often incur prohibitive expenses. To this regard, some systems create multiple copies to offset disk failures, others employ erasure codes [24] to break each object into redundant fragments. However, in both cases, it follows that the replication layer should start to take care of the consistency of the copies or fragments associated with each object. This is often resolved by rebuilding missing copies and quarantining corrupted ones. One key issue here is to shorten the inconsistency window as much as possible while also averting impact on the main activities of the system. With a benchmark tool in hand, people can better evaluate and study these impacts, obtaining insights to guide system optimization.

Flexibility and scalability are also addressed in the replication layer, with this layer bearing the responsibility of scattering the incoming workload to all storage servers. Load

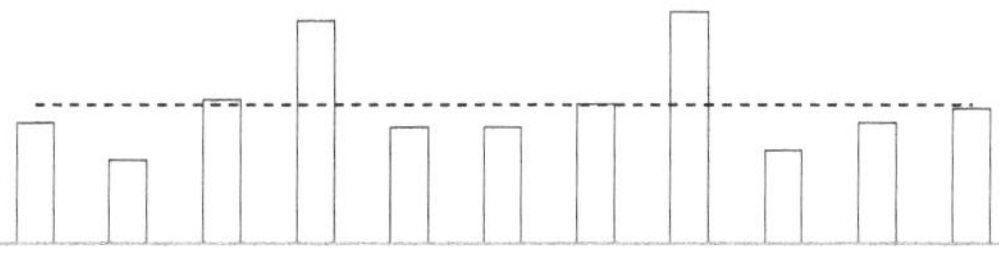

(a) a system exhibiting poor predictability but lower average response time.

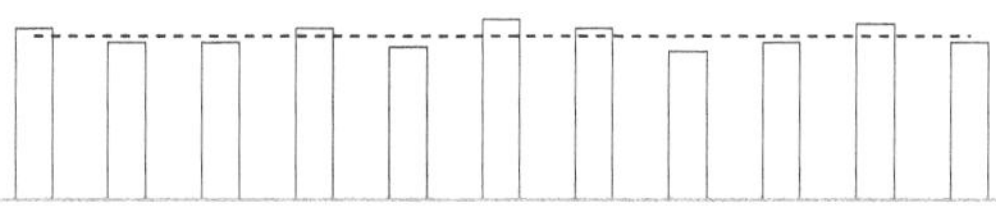

(b) a system exhibiting great predictability though higher average response time.

Figure 3: Characterizing the performance

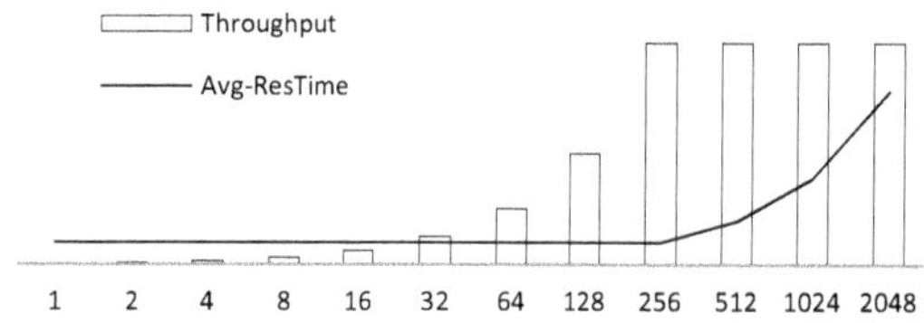

(a) an ideal system that keeps scale until its capacity is reached. The horizontal line represents numbers of clients.

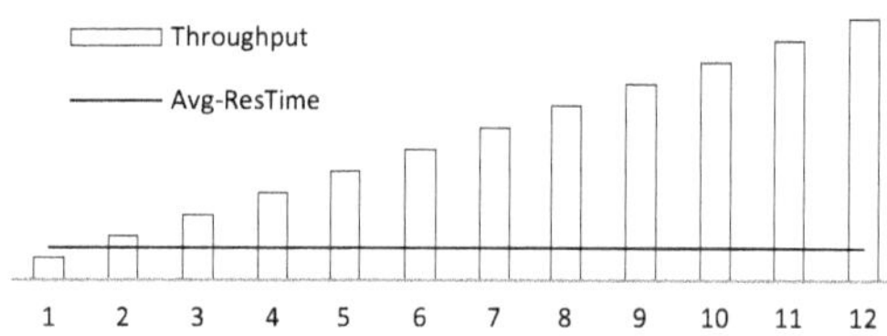

(b) an ideal system that scales in proportion to its hardware resources. The horizontal line represents resource units.

Figure 4: Characterizing the scalability

balance in fact also encompasses assigning part of the workload to newly joined servers and revoking from overloaded or faulty ones. More judicious algorithms may also take hot spots into consideration by adaptively rebalancing the workload accordingly. However, the key issue here is to keep the system efficient, available, and manageable. To achieve these targets, a benchmark tool can be employed to drive and verify better system designs.

Finally, the access layer on top of the system serves as the interface to external applications. One crucial duty of the access layer is to deal with all kinds of failures, thus to ensure high availability. In most cases, it is important to keep a system both writable and readable in spite of the presence of errors. The access layer is also regarded as the last resort for achieving high performance. For example, some people may employ concurrent reads to reduce latency at the expense of throughput; others may introduce asynchronization to boost write performance with both durability and consistency compromised. One important requirement of the access layer is to delicately prioritize and balance different system properties, with the goal of best satisfying the need of their applications. To this regard, a benchmark tool capable of generating customizable workloads will be extremely valuable to people exploring these tradeoffs.

3. SYSTEM CHARACTERIZATION

In this paper, we focus on system characterization. As will be discussed in this section, data obtained from characterization can be quite useful in many possible ways: (i) it provides tangible insights for one to compare different object storage offerings; (ii) it brings knowledge and experience on better system deployment that are important for tuning performance; (iii) it establishes a baseline performance by which further research or engineering efforts can be envisaged and channelled; and finally, (iv) the profiling data acquired during system characterization can help inspire new ideas of improvements and drive optimization.

In this section, we introduce three different aspects that one should bear in mind during system characterization.

3.1 Performance

The first aspect people need to focus on is undoubtedly the performance. One interesting issue regarding performance is its involvement of two related but competing properties displayed by a system: (i) the efficiency of a system which is expressed by throughput or average response time, and (ii) the predictability of a system which is associated with the distribution of its response times. To achieve both is difficult; one has to meticulously design the system to reconcile the inherent conflict between the above two properties. Since most SLAs stipulate bounds for latencies, it is then important to ensure the predictability of a system in order to conform to these agreements. Ideals of such enforcement often turn to amortization: the time spent on certain unforeseeable and high-latency events is split among all requests, thus diminishing the variance of their response times. Unfortunately, the ensuing extra efforts taken by the system to perform such actions inevitably dilute resource utilization, contracting system efficiency with an increased average response time and a decreased system throughput.

Examples of such efforts include the concurrent reads and the asynchronous writes we mentioned in Section 2, which are also illustrated in Figure 3-a and 3-b. Note that the key point behind these tradeoffs is that we must take both efficiency and predictability into consideration when performing system characterization. Therefore, a legitimate benchmark tool should be able to measure both properties, facilitating unbiased analyses against these systems. A system with better performance is one that exhibits higher efficiency while maintaining sufficient predictability. While results of such an analysis can help people better evaluate and compare different object storage systems, they are equally helpful in forming baseline performance for future optimization. The core ideal here is to decide on an acceptable latency, and then to maximize the throughput as much as possible.

3.2 Scalability

Different from the previous aspect which focuses on the balance between efficiency and predictability, scalability seeks to explore the performance under a dynamic context marked by either an increasing number of clients or hardware resources, as is shown in Figure 4-a and 4-b. In the first case, the hardware is fixed but the number of clients is raised exponentially. While an ideal system can answer this situation with a constant average response time and an increasing throughput proportionate to the growing number of clients, real systems are often subject to the contentions of various resources such as CPU, disks, networks, or software locks, thus cannot exhibit such perfect trends. The core ideal here is to improve the parallelism of a system to sustain higher performance on a given hardware configuration.

Unfortunately, every system has its maximum capacity: once it is saturated, more clients will have to compete with each other for resources, resulting in unacceptable response times; the best possible case is shown in Figure 4-a.

Ideally, after a system is saturated, in order to sustain more clients, only commensurate amount of hardware will have to be provisioned to stabilize the response time and to keep the throughput increasing accordingly, as is shown in Figure 4-b. However, in real situations, adding hardware resources does not necessarily earn proportionate performance enhancement, if at all. Many factors can be held culpable for impeding such improvements: metadata updating, load imbalance, or hardware limits. The core ideal here is to identify and eliminate these bottlenecks so as to better harness the underlying hardware resources, allowing the system to grow into thousands of nodes or even beyond.

By projecting performance to the dimensions of clients and resources, scalability plays a significant role in comparing different object storage systems. A better system is one that can engage more clients with fewer servers. Besides, results of scalability tests can also be studied to uncover system bottlenecks, which provides insights to guide system tuning. Note that it is crucial for a benchmark tool to be scalable as well, so that a large number of clients can be simulated to stress these storage systems in the first place.

3.3 Hardware Profiles

Finally, our last aspect of system characterization lies in the status of various hardware resources including CPU, memory, disks, and networks. To start with, one can always use these data to identify pathological resource utilizations and then take corresponding actions to get rid of them. For example, high CPU utilization is often associated with inadequate CPU power and may be resolved by either switching to more powerful nodes, or provisioning more nodes to share workloads. Profiling data can also be used to examine the balance among different servers or different CPU cores or disks within a single server. Any blatant imbalance exposed here is either a potential path for future optimization, or a symbol suggesting hardware faults. For example, it is not uncommon for a small group of disks to hold the "hot spot" data and hence to be more frequently accessed. To tackle this, an extra layer of cache could be deployed to level off the exorbitant burden of these disks, thus restoring balance. Moreover, profiling data can assist people to better invest money on hardware resources by revealing the resource usage patterns of a given system in the face of a certain workload. For example, if it is the CPU power that matters to the overall performance, it might not be that sensible for one to still spent money on high-end disks.

4. WORKLOAD MODEL

In this section, we present the workload model that underlies the COSBench system. A workload model abstracts the common patterns shared among a variety of object storage applications; it is crucial to make the model flexible and representative so that diverse usage patterns can be readily simulated to yield informative observations.

Table 1: Storage Interface

Operation	Description
`LOGIN`	retrieve a token representing an identity
`READ`	download an object
`WRITE`	upload a new object
`REMOVE`	delete an existing object
`INIT`	create a new container
`DISPOSE`	delete an empty container

4.1 Storage Interface

Storage interfaces are protocols defined by object storage systems to expose service. As there is no widespread standard available at this moment, different systems assume different protocols. In order for our workload model to work with most object storage systems, we have defined our storage interface as a small bundle of common operations seen in various specific protocols. So far, we have six core operations defined in our interface, as summarized in Table 1. These operations are sufficient for one to take on tasks such as bottleneck locating and capacity measuring, making this benchmark tool quite practical. That said, we are still actively working on adding more operations, but only in a way that will keep the tool simple, practical, and universal.

4.2 Usage Patterns

To simulate diverse usage patterns, our workload model can be flexibly configured in terms of *concurrency patterns*, *access patterns*, and *usage constraints*. In concurrency patterns, one specifies the number of clients that should be simulated by the workload generator. This directly enables people to evaluate the scalability of a system against rising workloads. In access pattern, one defines a set of operations that each client should issue to the target system. According to our design, each operation can be customized in terms of container path, object path, object size[1], and operation ratio. As can been seen, the access pattern lends people a lot of possibilities for designing their workloads: it can be read-intensive, write-intensive, or hybrid; it can deal with small objects, large objects, or both; and these objects can also be spread among many or only a few containers. It is conceivable that such richness of the workload design space can benefit not only people who want to evaluate and compare different object storage systems, but also those who want to probe specific systems for an in-depth analysis. The last aspect of the usage patterns involves various usage constraints. They can be expressed as total running time of a workload, total number of operations to be submitted, or total number of bytes to be transferred. If more than one such limit is presented, each of them can individually lead to the exit of a running workload.

4.3 Workload Model

In order to make the workload model even more flexi-

[1]Note that not every operation requires the size of an object to be specified, such as read or remove operations.

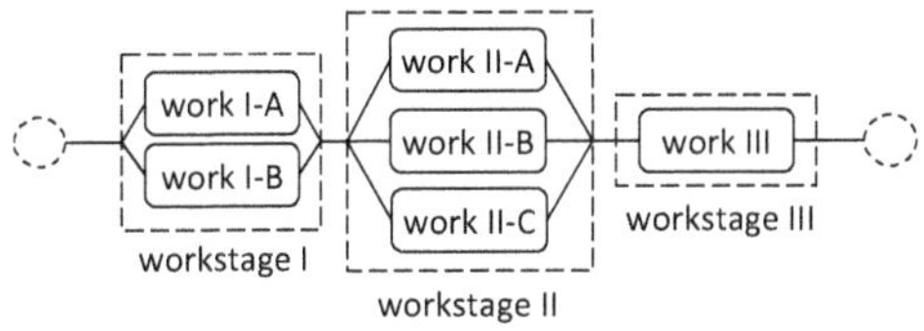

Figure 5: Workload modeled as a workflow

ble, we have employed workflow. In our model, a workload is modeled as a *workflow* consisting of multiple *workstages*, with each workstage representing a step in the workflow and focusing on a particular activity. For example, a workflow can have three different workstages: an initial stage, a main stage, and a cleanup stage. The initial stage can be configured to setup the benchmark environment (e.g., by uploading certain objects and creating dedicated containers). The main stage is responsible for launching the real workload. Finally, the cleanup stage removes all the data involved in this workload, restoring the target system to its original state. When a workflow is executed, each workstage will be executed in sequence. A workload is marked finished only after all its workstages have been completed. Failure of any workstage in the middle will terminate the overall workload execution immediately.

In order for a workstage in our model to perform multiple independent tasks simultaneously, a workstage is further regarded as a group of *works*, with each work representing a special job within a workstage. For example, a workstage can be defined to comprise two different works: one mimics a web application, frequently issuing read-intensive operations on small objects, while the other pretends to be an archive program constantly streaming up large objects. When a workstage is executed, all of its internal works are executed at the same time. A workstage is considered completed only after all its works have been completed. If an error occurs in any one of these works, the remaining works will be automatically aborted and the overall workstage will be deemed to have failed.

Works are atomic and are decoupled from each other. Each work can enjoy its own configuration context including its usage patterns, user credentials, and even the target system. As a result, it is possible for one to simulate multiple users at the same time, as well as to statically perform load balancing among different system portals. At runtime, each work is executed by a certain number of *workers*. A worker is a logic concept of a client; the number of workers is derived from the concurrency patterns associated with the work. The overall workflow model is illustrated in Figure 5.

5. BENCHMARK IMPLEMENTATION

We have developed a tool named COSBench to execute the workflows described in the previous section. In order to make sure that our tool can work with various object storage systems, we have cautiously designed this software so that people can easily adapt our storage interface to their specific implementations. Attention has also been paid to ensuring the scalability of the tool to sustain large numbers of clients. In this section, we describe the system architecture of our tool and highlight its extensibility and scalability.

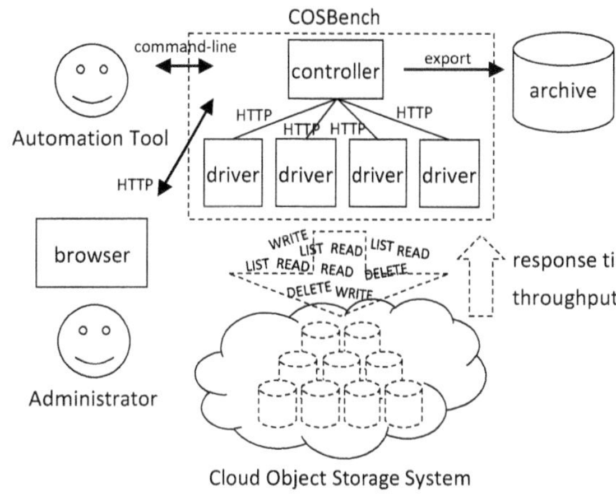

Figure 6: COSBench system overview

5.1 System Architecture

The COSBench is a web based Java application capable of generating workloads to a given object storage system and measuring the performance this specific system exhibits. As is illustrated in figure 6, our tool consists of two kinds of components: the *controller* and the *driver*. The controller serves as the facade of COSBench: it provides interfaces through which users can operate and control the benchmark tool as a whole. When users submit their workloads to the controller, internally, the controller divides these workloads into multiple tasks and schedules these tasks to different driver instances as *missions* for execution. As such, the controller communicates with drivers through the web interfaces they expose to assign missions to them or to retrieve execution status and results from these drivers. At the driver side, in order to execute a mission, an *auth* agent is first created to perform the required user login. After that, a certain number of *worker* agents are spawned to stress the target storage system in three phases: the *ramp-up* phase, the *transaction* phase, and the *ramp-down* phase. Drivers are independent from each other as they receive commands only from the controller, not from any other driver instances. The controller is responsible for synchronizing the execution of different missions associated with different drivers, ensuring the integrity of a workload. During the execution of a workload, the controller constantly polls related drivers to gather runtime status of its missions. When a workload is finished, the controller aggregates the performance data retrieved from those outsourced missions and archives the final report into a persistent place.

In order to facilitate comprehensive performance examinations, our final report comprises average, 90th, 95th, 99th, and 100th response times, throughput, bandwidth, success ratio, and a detailed latency histogram for each operation in each stage of a workload. In addition, a time series plot of the performance readings from ramp-up phase to ramp-down phase is also provided for each operation in each stage. To secure ease-of-use, we have made our tool accessible from both web browsers and command-line utilities, making it friendly not only to human users, but automation tools as well. Users can use these interfaces to submit or cancel workloads, check runtime status, exam or export final reports.

5.2 Extensibility

Several efforts have been made to ensure the extensibility of our tool. Firstly, we have made user authentication a dedicated interface, independent from the storage. According to our experience, it is possible for an object storage system to support multiple user authentication protocols (e.g., both Swauth [14] and Keystone [7] can serve as the authentication mechanism for Swift). One the other hand, it is also possible for multiple object storage systems to share the same authentication protocol (such as OAuth [9] and SAML [13]). By recognizing separate interfaces for both authentication and storage, our tool not only supports convenient switch among different authentication alternatives, but encourages software reuse as well.

Another effort lies in the workload configuration. In our tool, a workload is configured in an XML file prepared by a user when submitting a workflow. Unfortunately, difficulties arise when we try to design the schema of such XML. Since different object storage systems have varying definitions of their operation parameters, it is therefore impractical and infeasible for us to accommodate all possibilities in our workload configuration. As an example, people can specify certain "policies" when creating containers. In some systems, people must set a strategy that controls erasure code. For others, distinct options should be supplied to control various features provided by the system, such as container-to-container synchronization, object versioning, and others. To assure that our configuration can adapt to these manifold parameters, we regard all parameters as key-value pairs and merge all of these pairs into a single field which we call `config`. For example, a write operation might be configured as follows.

<operation type="writ", ratio="30",
config="containers=u(1,100); objects=u(1,100);
sizes=u(64,128)KB; object-expiration=10" />

As can be seen, there are four configuration pairs defined: `containers`, `objects`, `sizes`, and `object-expiration`. While the former three (which define the ranges and sizes of the containers and objects involved) are core to all implementations and are thus required by the COSBench, the last one is of a foreign nature and is forwarded by the COSBench to the adaptor for further interpretation and handling. With different adaptors responsible for defining and checking their own parameters within this unified framework, we successfully decoupled our configuration schema from various specific protocols, therefore achieving not only extensibility but flexibility as well.

Finally, the modular design of our system also considerably facilitates the overall extensibility of the tool. To be more specific, the COSBench is built upon Eclipse Equinox [4], an OSGi [11] implementation that supports dynamic module management. Layered above this infrastructure, our system consists of multiple modules known to the infrastructure as bundles. Bundles can expose services to a central service registry or import services from it. At runtime, when bundles are launched, the service registry will automatically map service providers (bundles that expose services) to their consumers (bundles that import services), allowing service consumers to use services without the knowledge of the providers. To add new storage types into the COSBench, people create new bundles, implement the storage interface inside their bundles, and configure their bundles to expose

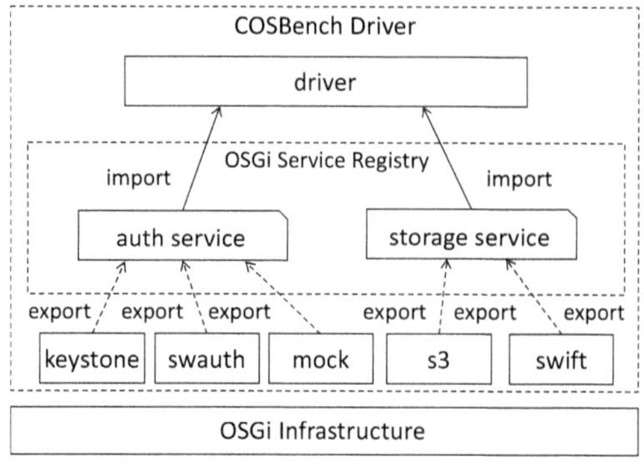

Figure 7: Modular design of the COSBench

their implementation as services. At runtime, once system is booted, all exposed storage services will be automatically bound to the driver. (The driver bundle is configured to import all available storage services). When a mission (Section 5.1) is assigned, the driver will iterate over all storage services bound to it to see if any of them is capable of handling the incoming mission. If such service is found, this service will be invoked, creating a new adaptor instance, which will then be attached to this particular mission for execution. Otherwise, the mission will be rejected. The relationship among the driver, adaptors, and service registry is depicted in Figure 7. As adaptors are decoupled from the driver and are dynamically plugged into the system at runtime, new adaptors can be independently developed, configured, and integrated without modifications (or awareness) of the core system.

5.3 Scalability

One important issue regarding characterizing web-scale storage systems is that the benchmark tool should be highly scalability so that a large number of clients can be simulated to place a substantial pressure on the target system. In our tool, as workloads are executed at the driver side, the overall throughput of the tool is therefore proportional to the number of driver instances being provisioned: more drivers instances means more CPU cores and more bandwidth, and thus more clients the tool is able to sustain. Besides, within each driver, workers are also carefully kept isolated from one another. Each worker has a dedicated execution context including a connection to the target system, a random number generator, various counters for performance measurements, as well as other status variables required. As such, workers can be executed in a fully parallel manner, which leads to the high scalability of COSBench.

6. EXPERIMENTS

The latest stable release of the COSBench is 2.0.0-GA. Internally, we have been using this tool to support our characterization on multiple object storage setups we deployed in our lab. While some of these systems are built with open source technologies, others are obtained directly from the commercial worlds. So far, we have found it quite smooth for us to write adaptors for different object storage systems. Besides our internal uses, the COSBench is also shared among different parties with various backgrounds both inside and outside Intel: some are object storage engineers

Table 2: Workloads used in our experiments

Workload	Read/Write Pattern	Object Size
Workload A	Read	128 KB
Workload B	Write	128 KB
Workload C	Read	10 MB
Workload D	Write	10 MB

who employ workloads to better verify their system designs; some are decision makers who are responsible for gauging and weighing different object storage offerings; and some are cloud builders who need data to promote their solutions. Of course, there are also pure researchers just like us.

In this section, we report the results of the experiments we have performed on two Swift setups under four distinct workloads. As our intention is to obtain a deep understanding of different object storage systems and insights to guide optimization as well, more have to be conducted before we can be rewarded with such knowledge. For this, we are still actively performing various experiments in order to reach our targets. We plan to share our further findings in a separate report where more data will be collected and more systems compared.

6.1 Workloads

We pay attention to the performance, scalability, and resource usage patterns exhibited by different systems under distinct read-write patterns and object sizes. The workloads are illustrated in Table 2. Although in the real world no application ever displays such simple and naive patterns, these workloads in fact suffice to serve as the starting points towards understanding a storage system. Besides, the baseline performance revealed under these workloads can help channel our future research efforts to the most compelling parts. As our ultimate goal is to disclose various factors throttling the performance of a system and to propose ideals for resolving such bottlenecks, simple and pure workloads can actually better help us to achieve this specific goal.

In our experiments, for read-only workloads, the target system were prepared with 1 user account, 100 containers, and 10,000 objects. For write-only ones, the systems were prepared similarly except that no objects are pre-inserted. To examine scalability, we varied the worker number from 5, 10, 20 ... to 2,560 (increased exponentially). We repeated each test case with three separate runs (each consisted of a 300s ramp-up phase and a 300s transaction phase) and took the average as the final result[2].

Besides the results reported by the COSBench, we also collected various hardware status monitored by utilities such as `iostat`, `vmstat`, and `sar`. This was performed by an automation tool we had separated developed with which the COSBench was integrated. The integration in fact took trivial efforts, thanks to the command-line interface exposed by the COSBench (Section 5.1).

6.2 Experimental Setup

For our experiments, we built two independent Swift setups on 22 server-class machines with each cluster enjoying five client nodes, five storage nodes and one proxy node.

[2]In the beginning, it was five runs and 500s for both phases. However, we later found that three runs and 300s were fine for both setups (saving us 64% of the total execution time).

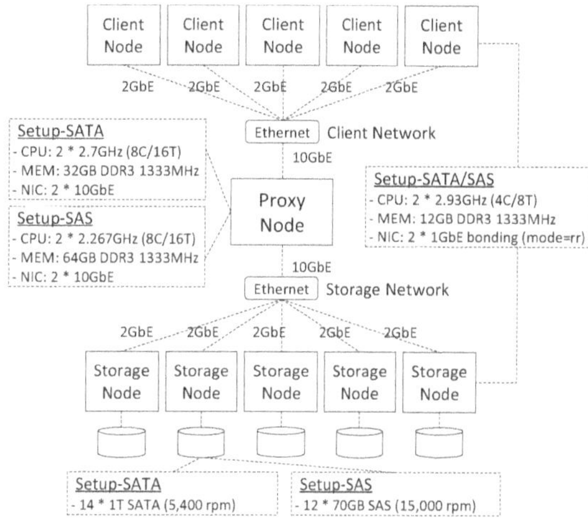

Figure 8: Hardware configurations

The client nodes ran the COSBench, with each running a `driver` instance and a special one running an additional `controller` instance. The storage nodes and the proxy node were all used to run Swift, with the proxy node serving as a front-end controller forwarding requests to backend storage nodes where specific operations were carried out. As such, we ran a certain number of `proxy-server` instances on the proxy node, and a certain number of `object-server`, `container-server`, and `account-server` instances on each storage node. Although in production environments other services, such as `object-replicator`, `object-updater`, `object-auditor`, to list just a few, should also be launched on storage nodes to keep data consistent and durable, we turned these services off to render less variance to system's performance[3]. When our experiments were performed, the latest release of Swift was 1.4.8. As Swift is still under heady development, increases or changes are both likely to be seen in the performance of its future releases. However, as we share our data mainly to demonstrate the value of our tool in facilitating performance comparison and system optimization, our data are not that meaningless.

As can been seen from Figure 8, there were a few differences between the above two setups: Setup-SATA possessed a more powerful proxy node whereas Setup-SAS was attached with more sophisticated disks. We will review the impacts of such hardware discrepancies when we discuss the results of our experiments. Note that in both setups, the aggregate bandwidth from client nodes to the proxy node was made equal to that from the proxy node to storage nodes, suggesting that our clients were able to consume as much throughput as the storage system could produce. We also allocated separate disks to run operating systems and a separate network for system control, protecting our environment from unintended noises.

For both setups, we tried our best to tune our systems to reach maximum performance. Since our setups were built upon different hardware infrastructures, each setup required different configurations to achieve its optimal performance. We share some of our efforts in Section 7.

[3]According to our experience, these "background" services will produce haphazard impacts on the overall system, resulting in as much as 18% of variance of performance among multiple runs.

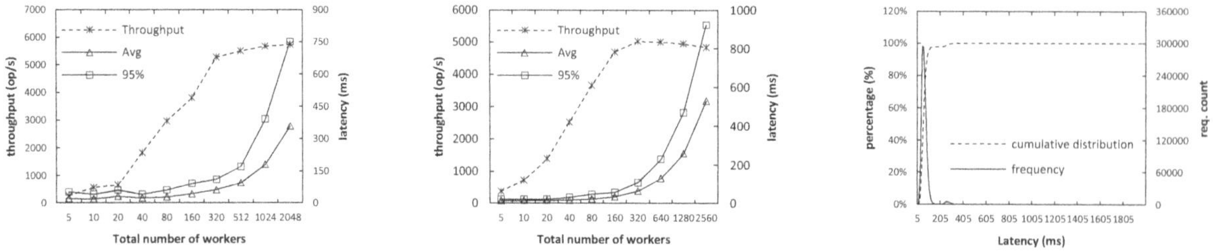

(a) Setup-SATA (b) Setup-SAS (c) latency histogram of Setup-SAS under its peak performance

Figure 9: Results of workload A. In Setup-SATA, the performance peaked at 320 workers (5,266 op/s in throughput, 110 ms in 95% latency); in Setup-SAS, 320 workers (5,019 op/s in throughput, 106 ms in 95% latency).

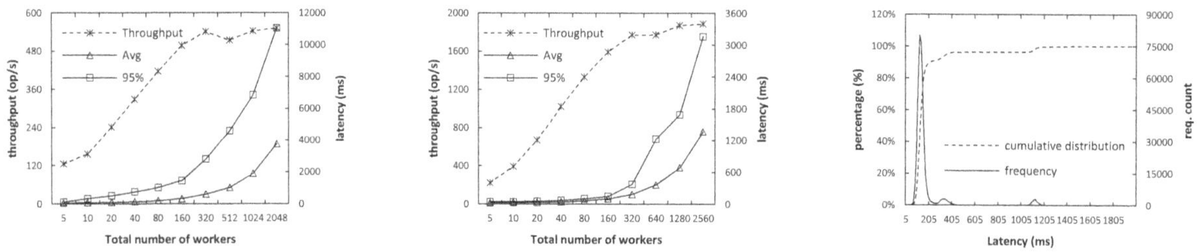

(a) Setup-SATA (b) Setup-SAS (c) latency histogram of Setup-SAS under its peak performance

Figure 10: Results of workload B. In Setup-SATA, the performance peaked at 320 workers (540 op/s in throughput, 2,810 ms in 95% latency); in Setup-SAS, 320 workers (1,769 op/s in throughput, 370 ms in 95% latency).

6.3 Workload A: Small Object Reads

The results of workload A are illustrated in Figure 9, which shows the throughput, average, and 95% response time observed under mounting number of workers. For both setups, with more workers stressing the system, response time increased as throughput increased. Besides, both setups exhibited good predictability in their response times. As can be seen from Figure 9-c, most requests were completed somewhere between 35 ms and 75 ms in Setup-SAS, suggesting that the variance among different requests was quite low.

Although Setup-SATA ran on slower disks, it actually exhibited 5% more throughput than Setup-SAS, which led us to conjecture that it was the CPU at the proxy node that mattered most to the performance of these systems under this workload. This was later ascertained by the exorbitant CPU utilization found in both setups in particular under their peak performance. After all, sustaining over 5,000 requests per seconds does require a nontrivial amount of computation power.

To further investigate scalability, we varied the number of storage nodes from one to five in Setup-SAS and obtained the throughput versus worker curve for each different setting. The results are showed in Figure 11. Note that when one single storage node was provisioned, we configured Swift to maintain only one copy for each object uploaded. If there were two storage nodes, the number of copies maintained was increased to two. Otherwise, the number was set to three, the default one. Note also that in normal cases Swift only takes one copy to serve a read request (more copies will be consulted only if all previous attempts have failed). As can be seen from the results, the throughput increased as the number of storage nodes increased until the proxy node became the bottleneck and prevented the system from gaining any further capacity. This could have be alleviated by provisioning more proxy nodes to lend more CPU power to the proxy tier. In our specific case, it is quite clear that every three storage nodes should be matched with one proxy node, which indicates that we should have provisioned two proxy nodes rather than a single one.

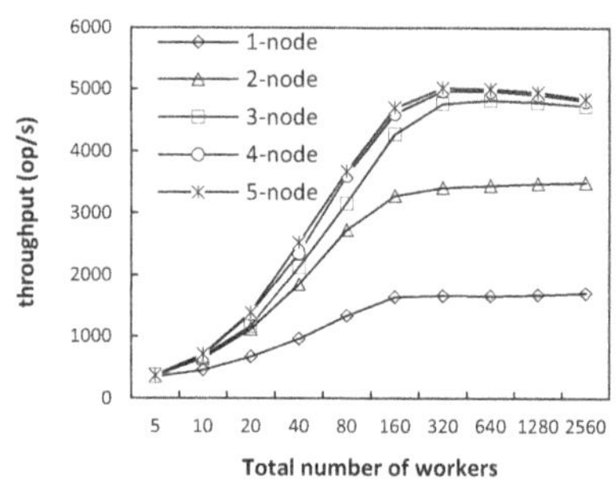

Figure 11: Results of the scalability test in Setup-SAS under workload A.

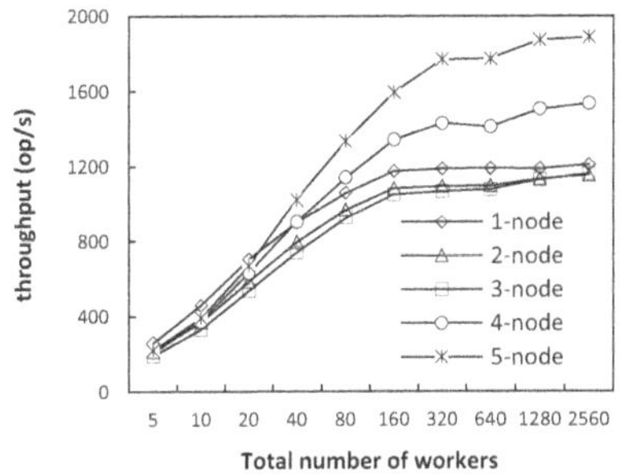

Figure 12: Results of the scalability test in Setup-SAS under workload B.

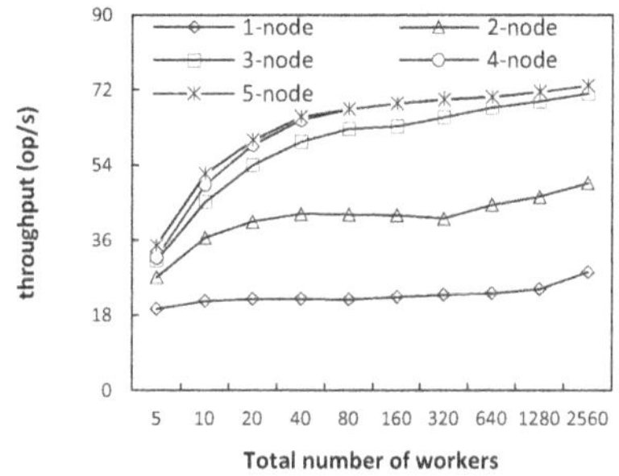

Figure 13: Results of the scalability test in Setup-SAS under workload C.

6.4 Workload B: Small Object Writes

Workload B shows a different picture from workload A by focusing on the write performance instead. The results are provided in Figure 10. This time, Setup-SATA presented a much worse performance than Setup-SAS in that the 95% response time curve in the former was noticeably steeper and that the throughput of it was also terribly lower. As Setup-SATA was not so fortunate to be blessed with such nice disks as the ones given to Setup-SAS, it followed that the poor performance exhibited by Setup-SATA was due to its limited disk power. This is within one's expectation since each object uploaded to Swift will cause three copies of data to be created and three replicas of container metadata to be updated as well. As a consequence, there are more works left for disks than for the CPU. To convince ourselves, we replaced all storage disks used in Setup-SATA with SSD disks. Results showed that the throughput increased significantly from, for example, the original 550 op/s to 1,800 op/s under 2,048 workers, which confirmed our above presumption that the root cause of the poor performance in Setup-SATA was its slow disks.

As can be seen from Figure 10-c, Setup-SAS achieved a slightly lower predictability in its response times than it did in workload A, with most requests completed in 85 ms to 175 ms under its peak performance this time. After all, each write operation has to undergo six internal round trips with three for storing object copies and the other three for updating container, which inescapably introduces more variance to the overall latency.

A similar scalability test for Setup-SAS was performed, with the results plotted in Figure 12. The throughput initially went down slightly as the number of storage nodes rose from one to three. This however was understandable since the number of copies that must be created for each object uploaded increased as well. As the number of storage nodes kept rising from three to five, a proportionate increase of the throughput was observed, indicating that the workload was well balanced among the backend storage nodes.

6.5 Workload C: Large Object Reads

Workload C intends to evaluate the ability of a system to send out large objects. The results are illustrated in Figure 14. Sadly, neither setup exhibited as good predictability in response times as they had achieved in previous workloads. Figure 14-c shows that individual response times observed in Setup-SAS under its peak performance spanned a long range from 650 ms to 1,900 ms, indicating that the bandwidth witnessed by different workers varied a lof from one request to another.

Since the performance of our two setups was similar to each other in this workload, we assumed that the bottleneck should be neither the disk nor the CPU. After a careful scrutiny, we realized that the key to further optimization actually resided in the client network. Although our aggregate bandwidth between storage nodes and client nodes should be as large as 1,250 MB/s (10Gb/s) in theory, we only got 700 MB/s at most in practice. We later found that this was due to the overhead incurred by bonding network links together. In our case, after two 1Gb links were bonded on each storage node and client node, the new link can only receive data at a low speed of 140 MB/s (56% of the theoretical maximum). As a consequence, our five client nodes could only read data at a maximum speed of 700 MB/s, thus rendering throughput an upper bound of 70 op/s. To resolve this issue, two different approaches were taken. In the first one, we provisioned five more client nodes to the setup with each node adding 140 MB/s more bandwidth to the aggregate bandwidth. In the second one, we set up four virtual machines on each client node using SR-IOV instead of bonding (with each VM assigned a 1Gb SR-IOV NIC). The results showed that both approaches were able to improve network bandwidth to its theoretical capacity; the throughput increased from 700 MB/s to over 1,100 MB/s.

Figure 13 shows the results of the scalability test under this workload in Setup-SAS. The throughput increased as the number of storage nodes increased, until the network became the bottleneck. If the network issue had been resolved, we would have seen further increases in throughput proportionate to the number of storage nodes beyond three. Another thing worth mentioning here is that our proxy node was capable of sustaining more storage nodes (at least five nodes) in this workload than it was in workload A (three nodes at most), which constitutes a concrete example of how workload characteristics affect system configuration and performance tuning.

6.6 Workload D: Large Object Writes

Finally, the results of workload D are illustrated in Figure 15. Unfortunately, both setups exhibited very poor predictability in their response times. As can been seen from Figure 15-c, which shows the latency histogram of Setup-SAS under its peak performance, the response times almost

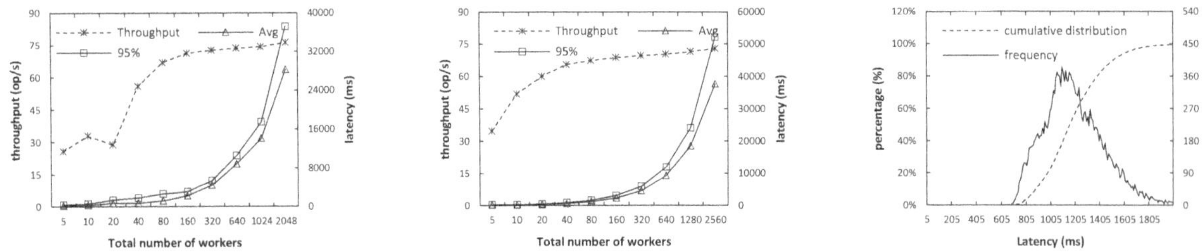

(a) Setup-SATA

(b) Setup-SAS

(c) latency histogram of Setup-SAS under its peak performance

Figure 14: Results of workload C. In Setup-SATA, the performance peaked at 160 workers (71 op/s in throughput, 3,060 ms in 95% latency); in Setup-SAS, 80 workers (67 op/s in throughput, 1,636 ms in 95% latency).

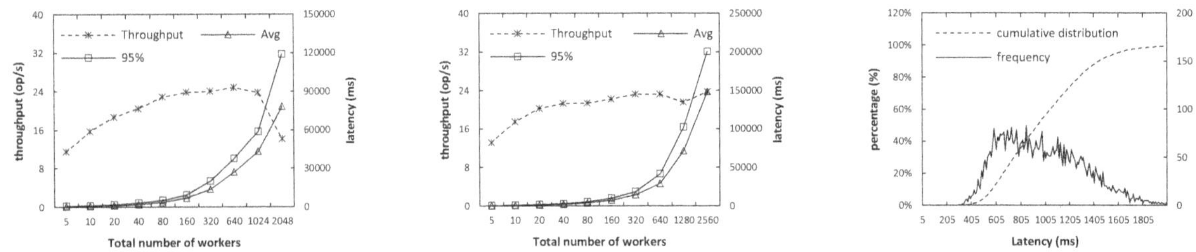

(a) Setup-SATA

(b) Setup-SAS

(c) latency histogram of Setup-SAS under its peak performance

Figure 15: Results of workload D. In Setup-SATA, the performance peaked at 80 workers (23 op/s in throughput, 5,100 ms in 95% latency); in Setup-SAS, 20 workers (20 op/s in throughput, 1,569 ms in 95% latency).

evenly distributed all over the time line from 500 ms to 1,600 ms.

Since the throughput exhibited by both setups in this workload happened to be a third of the throughput they exhibited in the previous one, this led us to believe that the key performance issue for this workload was still at the network layer, and that the high contention of the network bandwidth should take the blame for the high variance in the response times we observed. Note that in this workload, the data actually flew from the clients to the storage nodes, so it was the storage nodes that were reading the data and thus suffering from the poor receiving performance caused by network bonding. As such, it was legitimate to say that the throughput would be improved if more storage nodes had been provisioned, and worsen if less. In fact, the scalability test we performed later yielded evidences which could support our above assumption. As can be seen from Figure 16, the throughput increased linearly to the number of storage nodes as the latter rose from three to five. We actually also checked the status of both the CPU and the disks on storage nodes; however, no symptom of bottleneck had been found in these hardware resources. As a result, it is safe to take the storage network as the culprit for inhibiting scalability.

7. EXPERIENCES

As we mentioned early, we tried our best to tune both our setups to achieve their maximum performance. Since our two setups were associated with different hardware configurations, each of them had to be considered and optimized individually. For example, compared with Setup-SAS, we found that more server instances (`proxy-server`, `object-server`, and the like) have to be launched in Setup-SATA in order to match and compensate its slow disks. However, attempting to launch an equal number of instances in Setup-SAS as in Setup-SATA would only do a disservice to the performance of the former. As a lot of efforts have been taken to better configure our systems, in this section, we share some of our experiences by discussing two factors that, we think, people should take into consideration when tuning their systems.

Firstly, one should make sure that the hardware has been appropriately configured. Ideally, each element of the system should be kept balance with the rest to preclude any explicit bottleneck. For example, in our early experiments, the proxy node in Setup-SATA only had one single 10Gb NIC port configured, which was connected both to the client network and the storage network. We later realized that the system was ill balanced since this specific port turned out to be a bottleneck. Once a separate port was enabled, the performance immediately improved from 3,893 op/s to 4,726 op/s in the case of workload A. Besides watching the hardware, it is also important to be familiar with the internal logic of a system. We find that this kind of knowledge can help people better understand the behavior of a system and better read test results as well. For example, in our experiments, we initially configured all our workloads to use just one single container. The performance we observed in work-

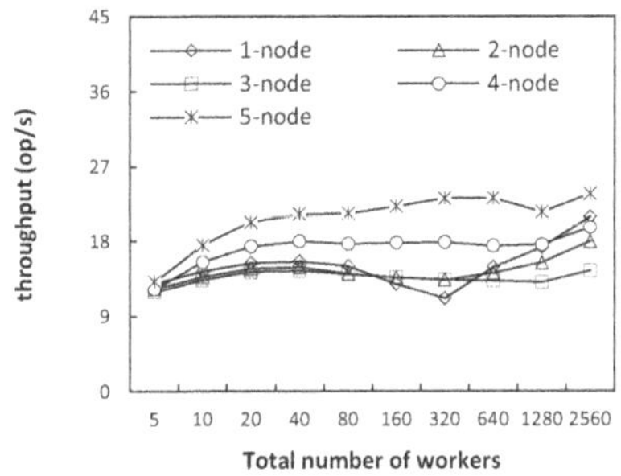

Figure 16: Results of the scalability test in Setup-SAS under workload D.

load B under this setting was only 197 op/s for Setup-SATA. Such low performance happened to remind us of one important detail we had learned from the source code: when each object is uploaded, besides creating three copies, Swift will additionally update the corresponding container to register this newly inserted object. Since the status of each container is persisted in a dedicated sqlite DB file with three replicas, the poor performance was very likely to be caused by the lock contention often associated with concurrent database modifications. To resolve this issue, we tuned our systems as follows: (i) we configured our workloads to use 100 containers instead of a single one to alleviate lock contention; then (ii) we increased the number of `container-server` instances launched in both setups to improve responsiveness; and (iii) we allocated dedicated SSD disks to hold all container DB files so as to work around the limitation of the HDD disks (this was for Setup-SATA only). Results showed that each of these actions was able to individually enhance performance. They collectively achieved a 280% increase in throughput in the case of workload B for Setup-SATA, from the previous 197 op/s to what we listed in Section 6.4.

8. RELATED WORK

Object storage systems are no longer strangers to developers, with the proliferation of these systems observed during recent years. Some object storage systems (such as Walrus in Eucalyptus [5] and Cumulus in Nimbus [8]) are implemented as immediate adaptor over traditional storage technologies. Some (such as Swift in Openstack [10]) are designed from bottom up with efforts devoted to ensure scalability, durability, and cost-effectiveness. Besides, there are also ones (such as RADOS [30] and HYDRAstor [23]) that are intended mainly to support file systems, and ones (such as Windows Azure Storage [20] and Yahoo! Walnut [21]) that are built with the goal of creating a unified enterprise storage infrastructure which can sustain better flexibility, manageability and application performance. Last but not least, one can also adopt object storage by using cloud services offered by Amazon [2], Rackspace [12] or Google [6].

The ideal of creating a benchmark tool to evaluate the performance of a system is not unusual. Efforts of implementing such a tool have already been seen in the newly-emerged area of object storage. For example, there is a benchmark utility included in the distribution of Swift, which is called SwiftBench [15]. Our tool is similar to it as both can be used to characterize Swift setups. However, while the former is dedicated for Swift to facilitate its system development, our tool is designed to be a general benchmark tool and is intended for a variety of purposes including system comparison, tuning, and optimization. Another example of such efforts is Haystress [18], which is provided to support the performance testing of Haystack, which is in turn a distributed photo repository used exclusively inside Facebook. Our tool is similar to Haystress since both can simulate different kinds of object storage applications. However, our tool is built with a more sophisticated workload model and is therefore capable of generating more diverse usage patterns. Besides these serious programs, scripts written by perl or curl have also been witnessed in the characterization of object storage systems. Although scripts are easier to implement and are not bound to pre-designed workload models, they lack the extensibility to work with different systems and various features to ensure ease-of-use. To this regard, we have designed our tool with attention paid on both issues, thus allowing our users to focus more on their innovation rather than benchmark details. Besides, our work also resembles YCSB [22] though we focus on object storage systems while they "nosql" data stores.

One prototype version of our tool was presented in a previous short report [32]. Compared with this release, our current version has been fully revised with a stronger system architecture which gives birth to the high extensibility and scalability of the tool. In addition, the workload model as well as the user-interface accompanying the tool has also been considerably improved in our current distribution, making the tool more practical and easier to use.

9. CONCLUSION

We have presented a benchmark tool we named *Cloud Object Storage enchmark*. This tool is developed with the main intention of helping people better evaluate and compare different object storage systems. It is also designed to facilitate system tuning and optimization. To demonstrate the effectiveness of our tool, we have used it to characterize the performance of two Swift setups under four distinct workloads, and discussed the impacts of different hardware configurations on system performance. We also showed how hardware profiling data can be utilized to identify performance bottlenecks and guide system optimization. As our tool can be flexibly adapted to other object storage systems, people can easily use this tool to characterize their own systems, either for facilitating performance comparison or for driving system optimization.

COSBench is still under our active development and is being enhanced in many aspects. One important part of our future work is to make the tool open source, so that it will be easier for people to use this tool and to extend it with their own storage adaptors. Besides, we are also working on adding additional features into this software. For example, one feature that we are implementing can empower people to detect possible data corruptions: it can be enabled by configuring the tool to generate and check the MD5 value of each object it handles. Another track of improvements concerns ease-of-use. For example, we are implementing a web interface which enables our users to create workload configurations directly in their browsers. As a result, this "online web editor" can free our users from the burden of expressing their usage patterns in raw XML.

10. REFERENCES

[1] 905 billion objects and 650,000 requests/second. http://aws.typepad.com/aws.

[2] Amazon s3. http://aws.amazon.com/s3.

[3] Cdmi. http://www.snia.org/cdmi.

[4] Equinox. http://www.eclipse.org/equinox.

[5] Eucalyptus. http://www.eucalyptus.com.

[6] Google cloud storage. http://www.google.com/enterprise/cloud.

[7] Keystone. http://docs.openstack.org/developer/keystone.

[8] Nimbus. http://www.nimbusproject.org.

[9] Oauth. http://oauth.net.

[10] Openstack. www.openstack.org.

[11] Osgi. http://www.osgi.org/Main/HomePage.

[12] Rackspace cloud files. http://www.rackspace.com/cloud.

[13] Saml. https://www.oasis-open.org/committees/security.

[14] Swauth. https://github.com/gholt/swauth.

[15] Swift bench. https://github.com/openstack/swift.

[16] N. Agrawal, A. C. Arpaci-Dusseau, and R. H. Arpaci-Dusseau. Generating realistic impressions for file-system benchmarking. In *AST*, 2009.

[17] E. Anderson, M. Kallahalla, M. Uysal, and R. Swaminathan. Buttress: A toolkit for flexible and high fidelity i/o benchmarking. In *AST*, 2004.

[18] D. Beaver, S. Kumar, H. C. Li, J. Sobel, and P. Vajgel. Finding a needle in haystack: facebook's photo storage. In *OSDI*, 2010.

[19] M. Brantner, D. Florescu, D. Graf, D. Kossmann, and T. Kraska. Building a database on s3. In *SIGMOD*, 2008.

[20] B. Calder, J. Wang, A. Ogus, N. Nilakantan, A. Skjolsvold, S. McKelvie, Y. Xu, S. Srivastav, J. Wu, H. Simitci, J. Haridas, C. Uddaraju, H. Khatri, A. Edwards, V. Bedekar, S. Mainali, R. Abbasi, A. Agarwal, M. F. u. Haq, M. I. u. Haq, D. Bhardwaj, S. Dayanand, A. Adusumilli, M. McNett, S. Sankaran, K. Manivannan, and L. Rigas. Windows azure storage: a highly available cloud storage service with strong consistency. In *SOSP*, 2011.

[21] J. Chen, C. Douglas, M. Mutsuzaki, P. Quaid, R. Ramakrishnan, S. Rao, and R. Sears. Walnut: a unified cloud object store. In *SIGMOD*, 2012.

[22] B. F. Cooper, A. Silberstein, E. Tam, R. Ramakrishnan, and R. Sears. Benchmarking cloud serving systems with ycsb. In *SoCC*, 2010.

[23] C. Dubnicki, L. Gryz, L. Heldt, M. Kaczmarczyk, W. Kilian, P. Strzelczak, J. Szczepkowski, C. Ungureanu, and M. Welnicki. Hydrastor: a scalable secondary storage. In *AST*, 2009.

[24] C. Huang, H. Simitci, Y. Xu, A. Ogus, B. Calder, P. Gopalan, J. Li, and S. Yekhanin. Erasure coding in windows azure storage. In *USENIX ATC*, 2012.

[25] A. W. Leung, S. Pasupathy, G. Goodson, and E. L. Miller. Measurement and analysis of large-scale network file system workloads. In *USENIX ATC*, 2008.

[26] D. Roselli, J. R. Lorch, and T. E. Anderson. A comparison of file system workloads. In *USENIX ATC*, 2000.

[27] C. Ungureanu, B. Atkin, A. Aranya, S. Gokhale, S. Rago, G. Calkowski, C. Dubnicki, and A. Bohra. Hydrafs: a high-throughput file system for the hydrastor content-addressable storage system. In *AST*, 2010.

[28] M. Vrable, S. Savage, and G. M. Voelker. Bluesky: a cloud-backed file system for the enterprise. In *AST*, 2012.

[29] S. A. Weil, S. A. Brandt, E. L. Miller, D. D. E. Long, and C. Maltzahn. Ceph: a scalable, high-performance distributed file system. In *OSDI*, 2006.

[30] S. A. Weil, A. W. Leung, S. A. Brandt, and C. Maltzahn. Rados: a scalable, reliable storage service for petabyte-scale storage clusters. In *PDSW*, 2007.

[31] N. Yezhkova, R. L. Villars, L. Conner, and B. Woo. Worldwide enterprise storage systems 20102014 forecast: Recovery, efficiency, and digitization shaping customer requirements for storage systems. *IDC*, 2010.

[32] Q. Zheng, H. Chen, Y. Wang, J. Duan, and Z. Huang. Cosbench: A benchmark tool for cloud object storage services. In *C OUD*, 2012.

Parallelism Profiling and Wall-time Prediction for Multi-threaded Applications

Achille Peternier
Institute of Computational Science (ICS)
University of Lugano (USI)
Lugano, Switzerland
achille.peternier@usi.ch

Walter Binder
Akira Yokokawa
Dynamic Analysis Group (DAG)
University of Lugano (USI)
Lugano, Switzerland
{first.last}@usi.ch

Lydia Chen
IBM Research Lab
Zurich, Switzerland
YIC@zurich.ibm.com

ABSTRACT

A detailed and accurate characterization of the parallelism of applications is essential for predicting their wall-time on different platforms, both for an application running in isolation and for a set of consolidated applications executing on the same platform. However, prevailing profilers are often based on sampling and do not provide exact information on the parallelism of the profiled application. In this paper we present a novel profiler that logs all thread scheduling activities within the operating system kernel. These logs enable us to accurately characterize applications' parallelism on a given platform by computing the number of threads that are active at each moment. We also present a simple mathematical prediction model to estimate wall-time for program execution on a k_2-core machine using profiles collected using a k_1-core machine (of the same architecture and running at the same clock speed). We use our profiler to assess the parallelism of several CPU-bound DaCapo benchmarks and evaluate the accuracy of our prediction model.

Categories and Subject Descriptors

D.2.8 [**Software Engineering**]: Metrics—*performance measures*

Keywords

Profiling, workload characterization, performance prediction, multicores

1. INTRODUCTION

The level of parallelism is a crucial parameter in determining applications' performance, especially in the multicore era. An application with high level of parallelism is able to leverage the abundant hardware resource available in today's computing system, particulary the CPU cores. However, many applications have limited parallelism, which can potentially result in system under-utilization. The straightforward solution to this problem is to consolidate such applications on a single system. The challenge here is how to allocate resources to those consolidated applications based on their level of parallelism, such that the overall system efficiency is maximized.

As today's applications are built on a complex software stack, it is not a trivial task to know the parallelism level of applications, especially when runtime systems of managed languages—which often feature concurrent dynamic code optimization and parallelized Garbage Collection (GC)—are part of the stack. To predict the wall-time of an application via its parallelism level, profiling methods usually require the execution of applications on all system configurations of interest. The complexity of profiling thus grows in the number of applications as well as system configurations. In this paper, we seek a light-weight profiler, which only collects per-application parallelism on a base system with k_1 CPU cores and enables the wall-time prediction on different systems with k_2 cores. We will assume that the k_1-core and k_2-core systems only differ in their number of cores, but otherwise feature the same architecture and have the same hardware and software components. That is, our parallelism characterization and wall-time predictions are platform-specific.

In this work we first present a novel profiler which operates at the Linux-kernel level and offers detailed and configurable traces of scheduler behavior. Based on these traces, a detailed characterization of the parallelism of applications is possible. Second, we provide a simple wall-time prediction model for (multi-threaded) applications on k-core machines using our parallelism characterization. Third, we present a detailed parallelism characterization of selected CPU-bound benchmarks of the DaCapo suite [3]. Fourth, we predict the wall-time of these benchmarks on a multicore machine with a varying number of cores and assess the relative error of our predictions.

2. KERNEL-LEVEL PARALLELISM PROFILER

Our approach is based on empirical analysis and measurements. We aim at minimizing the perturbations introduced by monitoring the applications we are profiling. To this end, we implemented a kernel-level module based on [12] that keeps track of all the state changes happening within a specified list of threads defined by their process and thread identifiers (Linux PIDs and TIDs). Each time a state change happens, the modified kernel scheduler logs the correspond-

ICPE'13, March 21–24, 2013, Prague, Czech Republic.

ing event into the system log directory. This information enables *post-mortem* analysis of the parallelism of an application during its runtime.

The parallelism profiler is composed of three main elements: (1) the kernel module, (2) the command-line configurator, and (3) the trace generator.

The kernel module patches the Linux kernel by adding per-thread state logging to the methods that are invoked by the OS scheduler upon each thread state change. In this way, internal kernel-space events are logged in user-space and become accessible through log files. Each time such an event is generated, timestamp, TID, PID, and the new state are recorded.

In Linux-based systems, the thread state is encoded as R when the thread is running or runnable, as S when it is waiting for an event to complete, as T when it is stopped, etc. In our work, we are mainly interested in distinguishing the R state from all the other states to identify when and how long a thread has been *active* (i.e., running or runnable).

The command-line configurator is used to intercept all the threads that are spawned within a specified process. The configurator is built on top of *libmonitor*[1] and logs the timestamps, TIDs, and PIDs of the various threads and subprocesses starting and terminating.

The trace generator gathers the log files produced by both the kernel module and the command-line configurator to combine the information using the timestamps and to generate a trace of the state changes that happened during the observed time for each thread, including thread creation and termination events.

Thanks to the traces generated by our parallelism profiler, we can evaluate the maximum degree of parallelism that an application is using (i.e., identify how many concurrent threads are being used), and for each level of parallelism $j \geq 0$, we can compute the wall-time spent when j threads were active. Based on the information provided by the traces, parallelism profiles can be created to point out an application's parallelism changes over time. Metrics such as the level of parallelism reached within a specific time interval, or for how long a specific level of parallelism is available can then be computed to produce charts such as the ones used in our evaluation.

The total overhead added by our monitoring infrastructure depends on the state change frequency. For the DaCapo benchmarks we used in our evaluation, the overhead is 2.16% on average, with a maximum of 7.98%.

3. WALL-TIME PREDICTION MODEL

In this section, we present a predictive model for multicore systems using our profiling tool. The idea is based on Amdahl's Law [9], which states that the execution time can be improved by speeding up the parallel part of the program on additional hardware. The challenge of applying Amdahl's Law lies in identifying and quantifying the parallel part of a program. In a multi-threaded program, the parallel part can be viewed as the fraction of the program where there are more than one active threads. As the number of active threads is not constant, the degree of parallelism in a program also fluctuates. One needs to incorporate variability of parallelism in applying Amdahl's Law to predict the program execution time on a k-core system. Furthermore, depending on the number of available cores, those fractions of time can be accelerated accordingly.

To such an end, we first leverage the timing information about the number of active threads provided by our profiling tool. Let $T_j(1)$ denote the execution time when there are j active threads executing the program on a single core. The number of active threads can range from zero to J, that is, $j \in \{0, \ldots J\}$. Zero active thread denotes that no thread is active (e.g., all threads are blocked on I/O). Also let k denote the number of cores available for execution. On the one hand, when the number of cores is greater or equal to the number of threads (i.e., $k \geq j$), all j threads can run on available cores without any competition, assuming no dependencies among those threads. On the other hand, when there is not a sufficient number of cores for j threads (i.e., $j > k$), the maximum speedup is restricted by the number of available cores. As such, when executing on a k-core system, the execution under j active threads, $T_j(k)$, can be further improved by the minimum value between j threads and k cores:

$$T_0(k) = T_0(1)$$

$$T_j(k) = \frac{T_j(1)}{\min\{j,k\}}, j > 0.$$

Note that when $j = 0$ or $j = 1$, no acceleration is possibile. The overall execution time of a program on a k-core system, $E(k)$, is thus the summation of fractions of time in all possible numbers of active threads:

$$\begin{aligned} E(k) &= \sum_{j=0}^{J} T_j(k) \\ &= T_0(1) + \sum_{j=1}^{J} \frac{T_j(1)}{\min\{j,k\}} \end{aligned}$$

To accommodate wider profiling and prediction scenarios, that is, on any different number of cores, we further generalize the aforementioned analysis. The profiling of active threading time is executed on a system with k_1 cores and we aim at obtaining the prediction of execution time on a system with k_2 cores. To derive $T_j(k_2)$, we first classify two cases, (1) $k_1 \leq k_2$ and (2) $k_1 > k_2$, and then discuss the performance improvement (resp. slowdown) due to the parallelism.

In case one, various parts of execution time of a program can be improved, depending on the relationship between the degree of parallelism, $T_j(k_1)$, k_1, and k_2. When $j \leq k_1$, it implies this part of the program is at its best speed and cannot be further accelerated by being executing on k_2-cores. We thus know $T_j(k_1) = T_j(k_2)$, $j \leq k_1 \leq k_2$. As for the times when $j > k_1$, $T_j(k_1)$ can be improved by leveraging the extra cores on a k_2-core system, given the available threads and cores. Essentially, the execution time of j threads on k_2 cores is speed by a factor of $\frac{\min\{j,k_2\}}{k_1}$, from $T_j(k_1)$. The rational behind is that as $T_j(k_1)$ is already fully acerbated by k_1 cores, the improvement of $T_j(k_2)$ needs to be first normalized by k_1 and then multiplied by the maximum parallelism, that is, $\min\{j, k_2\}$. As such, one can derive $T_j(k_2) = T_j(k_1)/(\frac{\min\{j,k_2\}}{k_1})$. Similar analysis can be applied for case two. The only difference is that some part of the program can be scaled down, due to the decrease of parallelism when moving from k_1 cores down to k_2 cores.

[1] https://outreach.scidac.gov/projects/libmonitor/

In summary, the generalization of our predictive model is the following:

$$E(k_2) = \sum_{j=0}^{J} T_j(k_2),\ where$$

$$T_j(k_2) = \begin{cases} T_j(k_1), & j \leq \min\{k_1, k_2\}; \\ T_j(k_1)\frac{k_1}{\min\{j, k_2\}}, & k_1 \leq k_2, k_1 \leq j. \\ T_j(k_1)\frac{\min\{j, k_1\}}{k_2}, & k_2 \leq k_1, k_2 \leq j. \end{cases}$$

4. EVALUATION

In this section we first explore the parallelism profiles of a series of DaCapo benchmarks using a single core ($k_1 = 1$). We then use these profiles to predict wall-time using more cores (e.g., $k_2 = 2$, $k_2 = 3$, and $k_2 = 4$).

4.1 Testing Environment and Settings

Experiments have been conducted on a desktop Dell Optiflex 760 PC with 8 GB of RAM equiped with an Intel Core2Quad Q9650 3.0 GHz CPU with 4 cores, referred to as Intel-4. To obtain more stable results, we disabled frequency scaling.

As software environment we use Ubuntu Linux Server 64bit version 11.04 with a custom modified version of kernel 2.6.39.4 to log thread state changes. The Java Virtual Machine (JVM) used is Oracle's Hotspot 1.6.0_23 64bit Server VM. We focus on a selection of nine DaCapo benchmarks, version 9.12 Bach (with default workload size) [3]. We based our selection on benchmarks that are CPU-bound. When necessary, we restrict the scheduling of threads to selected cores to simulate machines with a configuration different than $k = 4$ cores. To this end, we use the *taskset* command.

Since the DaCapo benchmarks have very different execution times (by default, some of them are very short), we perform a series of iterations over each test to have longer runs within the same JVM process. In more detail, a single benchmark run executes 15 iterations for *batik*, 40 for *fop*, 5 for *h2*, 5 for *jython*, 30 for *luindex*, 10 for *lusearch*, 10 for *pmd*, 10 for *sunflow*, and 10 for *xalan*. For each experiment, we measure the wall-time of the whole execution, including the JVM bootstrapping time and all the iterations. Before each measured experiment, we run the same experiment once (without measurement) to ensure that disk caches already hold the required data (e.g., the Java class files linked during benchmark execution); this helps keep the JVM startup phase (which is less interesting for us) short.

4.2 Parallelism Characterization of DaCapo Benchmarks

We used our parallellism profiler to characterize the selected DaCapo benchmarks. In this phase, we produce the input data for our prediction model. We generate a parallelism profile for each benchmark by restricting the scheduling of the whole application (including all the threads belonging to the JVM) to a single core (i.e., $k_1 = 1$).

The results are reported in Figure 1, showing the levels of parallelism reached by each benchmark and for how long during its execution time. These results represent the values observed in a single run. Several independent runs of the same benchmark show very similar execution times. The non-determinism observed during repeated runs (in separate JVM processes) is due to application-level non-determinism, differences in the thread scheduling, different identity hashcodes for objects [2], and other phenomena studied in the literature [7, 11].

The DaCapo benchmark suite is based on Java and it is an interesting evaluation environment since several system threads are automatically generated by the JVM for performing tasks such as GC, thus making the system more complex. The JVM inspects the hardware configuration to determine how many system threads to create upon startup.

Since our single-core scheduling constraint (to enforce $k_1 = 1$) is invisibile for the JVM, the total number of system threads is related to the total number of CPU cores available on the machine (Intel-4). For example, by default, the Hotspot VM used in our evaluation tends to allocate one GC thread per core.

For example, this behavior is clearly pointed out by the information gathered through our parallelism profiler for the *xalan* benchmark. According to the DaCapo documentation[2], *xalan* allocates a thread pool with a number of worker threads corresponding to the number of available CPU cores. By looking at the bars in Figure 1, *xalan* has two parallelism-level peaks at 4 and 6 threads. This means that for the majority of its execution time, *xalan* has 4 respectively 6 active threads. A comparison of the wall-time recorded for each parallelism level with the GC and just-in-time compilation times monitored through the Java ThreadMXBean API is left for future work.

The *sunflow* parallelism profile shows that the benchmark is designed to use four threads for its execution: in fact, this benchmark allocates one thread per core available to perform parallel ray-tracing. As ray-tracing is commonly known as an efficient case taking advantage from parallel execution, most of the application execution time is spent running 4 threads.

Both the *batik* and *luindex* benchmarks are mainly single-threaded (using some helper threads for a few tasks). Their parallelism profiles confirm this characteristic, showing that most of the execution time is consumed at a parallelism level of 1, while during GC phases there are several active GC threads.

The *pmd* benchmark uses one thread per hardware core to analyze a set of Java classes for a range of source code problems. Since many objects are allocated and released during the application life-time, the GC is particularly stressed. According to the documentation, this benchmark is "driven by a single client thread that is internally multi-threaded using one worker thread per hardware thread". This behavior is well visible in the parallelism profile, through the first parallelism peak reached at 1 (for the single thread client), and between 3 and 6 (when both the client and worker threads, as well as additional internal JVM threads are used).

4.3 Wall-time Prediction for DaCapo Benchmarks

Figure 2 shows the measured ( ) and predicted ( ) wall-time of the DaCapo benchmarks considered in our evaluation. The measurements are obtained by running the benchmarks on a varying number of cores $k_2 \in \{1, \ldots 4\}$. All predictions are based on the parallelism characterization pre-

[2]http://dacapobench.org/threads.html

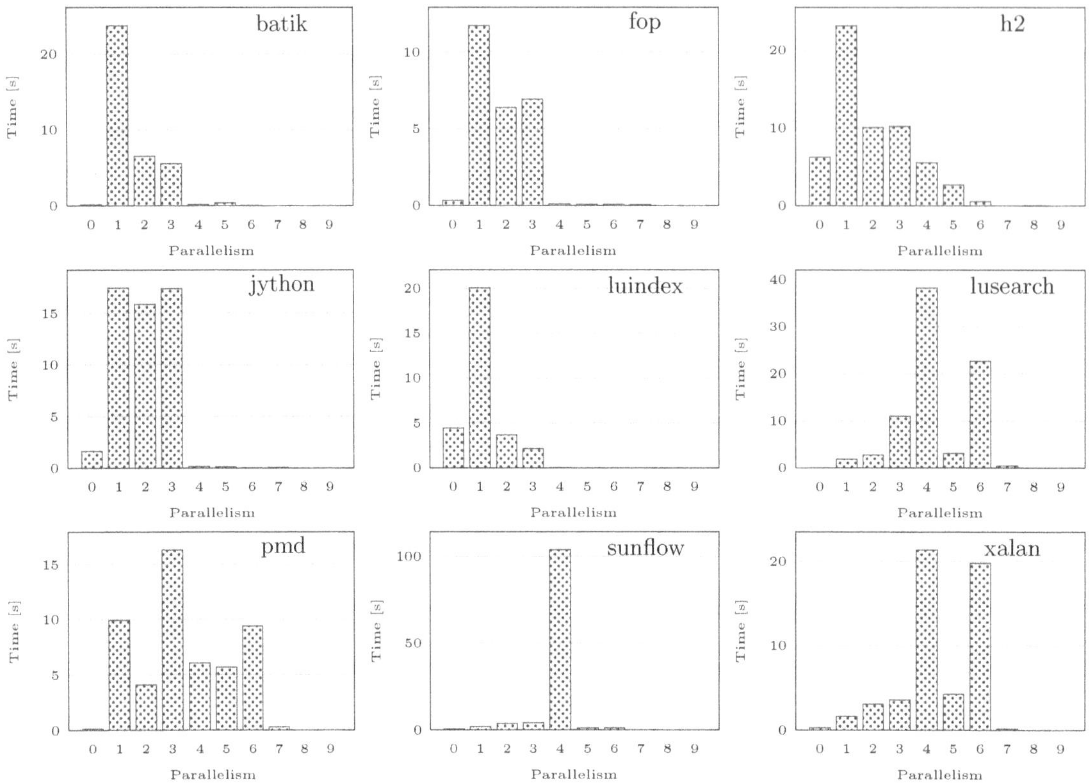

par.	batik	fop	h2	jython	luindex	lusearch	pmd	sunflow	xalan
0	0.1135	0.3272	6.2622	1.6383	4.4334	0.1052	0.1286	0.6091	0.2665
1	23.7125	11.7187	23.1446	17.4342	20.0666	1.9308	9.9018	1.9215	1.6435
2	6.5175	6.3705	10.0869	15.8579	3.6819	2.7649	4.0905	3.6845	3.0751
3	5.5487	6.9116	10.2254	17.3543	2.1347	11.0191	16.2830	4.1474	3.5842
4	0.1705	0.0939	5.5579	0.1589	0.0280	38.1997	6.0497	103.8943	21.3744
5	0.3767	0.0812	2.7235	0.1283	0.0178	3.1146	5.6721	1.1907	4.2616
6	0.0333	0.0804	0.5365	0.0247	0.0001	22.6459	9.3826	1.1717	19.8270
7	-	0.0707	0.0194	0.0560	0.0106	0.4857	0.2956	0.0347	0.1675
8	-	2.40×10^{-5}	0.0260	1.40×10^{-5}	-	0.0006	0.0001	1.20×10^{-5}	0.0004
9	-	-	-	-	-	1.00×10^{-5}	-	-	1.70×10^{-5}
tot.	36.473	25.654	58.582	52.653	30.373	80.266	51.804	116.654	54.200

Figure 1: Workload parallelism characterization on Intel-4 using a single core. These measurements refer to the wall-time (in second) spent by each benchmark when a specific number of threads were active. The total time refers to the wall-time computed from the profiles.

sented in Figure 1 using the model described in Section 3. Figure 2 also presents the relative error of each prediction.

Despite of our model's simplicity, the average of the absolute relative errors is only 4.11% for $k_2 = 2$, 6.39% for $k_2 = 3$, and 6.43% for $k_2 = 4$, which can be sufficiently accurate for use in practice. Note that exact reproducibility of measurement results is typically impossible in such complex systems with many sources of non-determinism [7, 11]. Hence, also subsequent measurements usually yield different execution times.

To some extent, the prediction errors can be attributed to the perturbations introduced by our profiler. Interestingly, the wall-time prediction is particularly inaccurate for *h2* (with an absolute relative error of 15.64% for $k_2 = 4$), for which also the profiling overhead (7.98%) is four times higher than for all other benchmarks.

In general, prediction errors can be attributed to many factors that are not captured by our simple model. Our model does not consider contention (e.g., on the memory bus) that may occur when executing the application threads on a machine with more cores (i.e., $k_2 > k_1$). Furthermore, due to contention on shared data structures, compare-and-swap instructions may fail more often if more active threads execute in parallel. In addition, if active threads are spinning, executing them in parallel on additonal cores does not reduce application execution time, as it would be computed by our model.

There are several essential preconditions for the use of our prediction model. First, the application must not change its threading behavior according to the available number of CPU cores. The number of threads created by the application must be the same on the profiling machine and on the

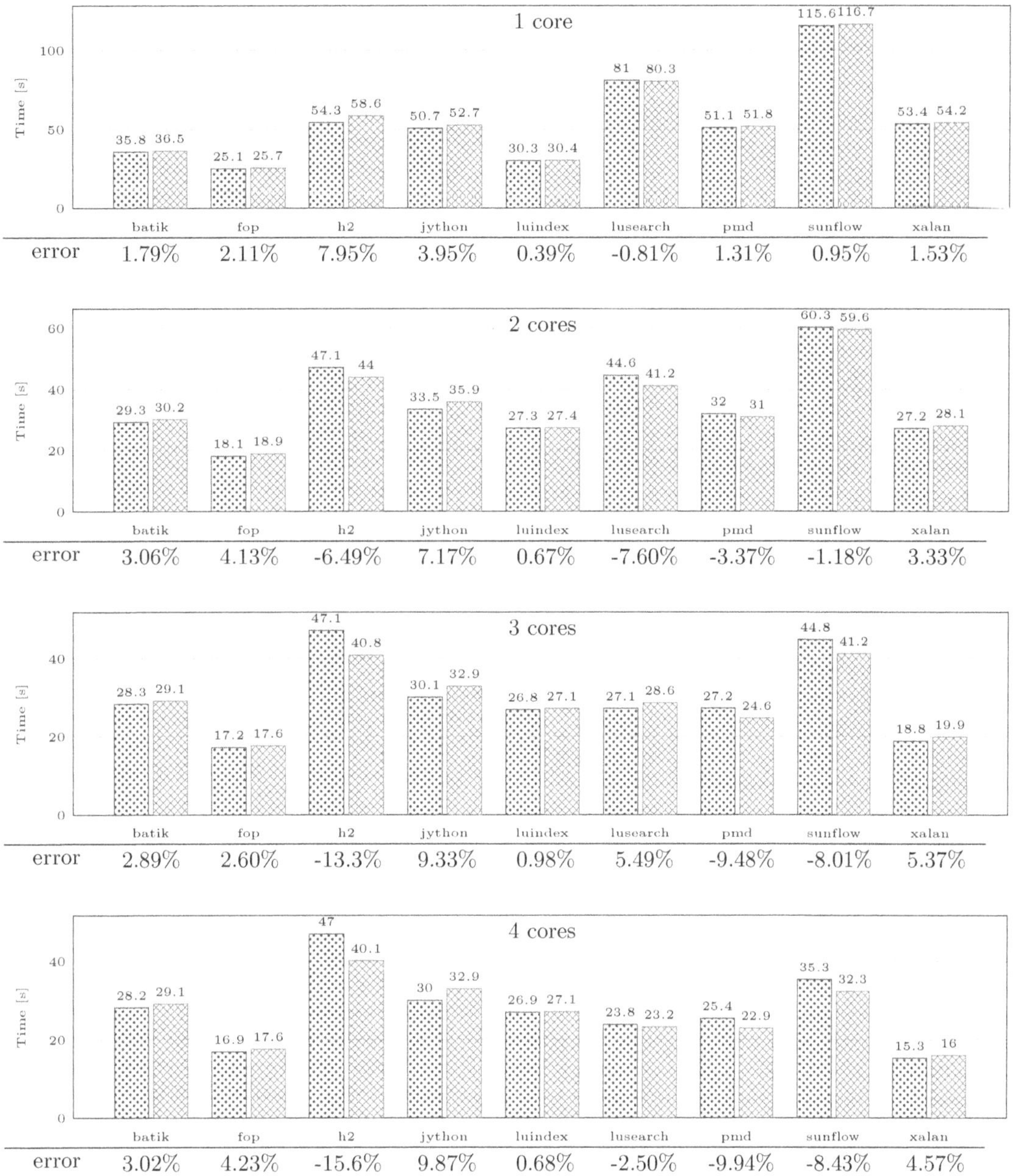

Figure 2: Performance measurement (▒) and prediction (▒) on Intel-4 running DaCapo benchmarks on 1–4 cores. Error denotes the relative error of the predicted wall-time.

target machine for prediction. In fact, both the HotSpot VM and some DaCapo benchmark adapt the number of created threads to the number of CPU cores. Hence, in our evaluation we had to ensure that Intel-4 was always seen as a quad-core machine, and we resorted to scheduler constraints to keep certain cores idle for $k_2 \in \{1, \dots 3\}$. Second, the profiling machine and the target machine for prediction must be equivalent apart from the number of CPU cores, and the cores must operate at a constant clock speed independently of the number of active threads executed in parallel. Our measurement machine was configured to meet these constraints.

5. RELATED WORK

In this section we discuss relevant related work in the field of profiling and parallel performance prediction.

Wong et al. developed PAS2P to predict parallel application performance with an accuracy of 97% in [14]. Their approach is based on signatures created by monitoring message-passing activities, while we rely on a more generic observation of the parallelism levels and per-thread execution times.

Cook et al. used a complex statistical approach to model and predict performance of various modern architectures in [5], providing good approximations but requiring a significant effort for model calibration.

Kismet [10] is a tool that predicts parallel performance of serial (unparallelized) programs. Like our approach, Kismet does not require any modification at the program level and is based on empirical measurements, thus quickly providing results without additional efforts. Our profiler deals with multi-threaded applications, by analyizing the parallelism level of any concurrent program.

There is a significant body of research [4, 1, 6, 13, 15] addressing the performance analysis of multi-threaded applications on multicore systems, especially from the single resource's perspective, such as cache. Dey et al. [6] propose a general methodology to characterize any multi-threaded application for its last-level cache contention and private cache contention. Chen et al. [4, 1] applied queueing network models to predict application response times, relying on a profiling approach on measure resource demands, that is, required execution time on CPU and disk. Tallent and Mellor-Crummey [13] focus on Cilk programs and built a runtime performance analysis, which aims at pinpointing and quantifying serialization, using attributing work, parallel idleness, and parallel overhead to logical calling contexts. Yang et al. [15] presented a simple performance model of on-chip interconnect and intra-core bandwidth contention, for single- and multi-threaded Gaussian 03 computational chemistry code.

Our performance model is applicable to predict the overall wall-time of generic applications, instead of performance measures for single resources. Moreover, our model relies on a small set of parameters, which is readily provided by our profiling tools with moderate overheads.

Amdahl's Law [8] is widely used to predict the performance of parallel programs, ranging from the field of high performance computing to the multicore era. Hill and Marty [9] took a hardware perspective and developed a speed model for a multicore chip, applying the concept of Amdahl's Law. However, the parallelism of software aspects, that is, multi-threaded applications, is not considered.

6. CONCLUSIONS

In this short paper we presented a new profiler for characterizing the parallelism level within applications. Our profiler generates traces based on kernel scheduler events related to thread state changes. We present a prediction model using these traces that can forecast the wall-time for executing multi-threaded applications, given their parallelism profile and a target number of cores. However, the parallelism profile and the resulting wall-time predictions are platform-specific.

Using our profiler, we determined the parallelism profiles of a selection of CPU-bound DaCapo benchmarks and predicted their wall-time using a varying number of cores. Our predictions show a relative error in the range of 4.11% to 6.43% on average.

As future work, we plan to take additional factors into account, such as dynamic frequency scaling and simultaneous multi-threading to allow our model to perform predictions on hardware platforms featuring such technologies.

7. ACKNOWLEDGMENTS

The research presented in this paper has been supported by the Swiss National Science Foundation (Sinergia project CRSI22_127386 and project 200021_141002) and by the European Commission (Seventh Framework Programme grant 287746).

8. REFERENCES

[1] D. Ansaloni, L. Y. Chen, E. Smirni, and W. Binder. Model-driven consolidation of Java workloads on multicores. In *Proc. of DSN*, pages 229–234, 2012.

[2] W. Binder, J. Hulaas, P. Moret, and A. Villazón. Platform-independent profiling in a virtual execution environment. *Softw., Pract. Exper.*, 39(1):47–79, 2009.

[3] S. M. Blackburn, R. Garner, C. Hoffmann, A. M. Khang, K. S. McKinley, R. Bentzur, A. Diwan, D. Feinberg, D. Frampton, S. Z. Guyer, M. Hirzel, A. Hosking, M. Jump, H. Lee, J. E. B. Moss, A. Phansalkar, D. Stefanović, T. VanDrunen, D. von Dincklage, and B. Wiedermann. The DaCapo benchmarks: Java benchmarking development and analysis. *SIGPLAN Not.*, 41(10):169–190, Oct. 2006.

[4] L. Y. Chen, D. Ansaloni, E. Smirni, A. Yokokawa, and W. Binder. Achieving application-centric performance targets via consolidation on multicores: Myth or reality? In *Proc. of HPDC*, pages 229–234, 2012.

[5] J. Cook, J. Cook, and W. Alkohlani. A statistical performance model of the Opteron processor. *SIGMETRICS Perform. Eval. Rev.*, 38(4):75–80, Mar. 2011.

[6] T. Dey, W. Wang, J. Davidson, and M. Soffa. Characterizing multi-threaded applications based on shared-resource contention. In *Proc. of ISPASS*, pages 76–86, 2011.

[7] A. Georges, D. Buytaert, and L. Eeckhout. Statistically rigorous Java performance evaluation. *SIGPLAN Not.*, 42(10):57–76, Oct. 2007.

[8] A. Grama, G. Karypis, V. Kumar, and A. Gupta. *Introduction to Parallel Computing*. Pearson Education, 2003.

[9] M. D. Hill and M. R. Marty. Amdahl's law in the multicore era. *IEEE COMPUTER*, 2008.

[10] D. Jeon, S. Garcia, C. Louie, and M. B. Taylor. Kismet: parallel speedup estimates for serial programs. *SIGPLAN Not.*, 46(10):519–536, Oct. 2011.

[11] T. Mytkowicz, A. Diwan, M. Hauswirth, and P. F. Sweeney. Producing wrong data without doing anything obviously wrong! *SIGPLAN Not.*, 44(3):265–276, Mar. 2009.

[12] A. Peternier, D. Bonetta, W. Binder, and C. Pautasso. Overseer: Low-level hardware monitoring and management for Java. In *Proc. of PPPJ*, pages 143–146, Denmark, 2011.

[13] N. Tallent and J. Mellor-Crummey. Effective performance measurement and analysis of multithreaded applications. In *Proc. of PPoPP*, pages 229–240, 2009.

[14] A. Wong, D. Rexachs, and E. Luque. Pas2p tool, parallel application signature for performance prediction. In *Proc. of PARA*, pages 293–302, 2012.

[15] R. Yang, J. Antony, and A. P. Rendell. A simple performance model for multithreaded applications executing on non-uniform memory access computers. In *Proc. of HPCC*, pages 79–86, 2009.

Stream-Based Event Prediction Using Bayesian and Bloom Filters

Miao Wang
Performance Engineering Lab
School of Computer Science and Informatics
University College Dublin, Ireland
miao.wang@ucdconnect.ie

Viliam Holub
Performance Engineering Lab
School of Computer Science and Informatics
University College Dublin, Ireland
viliam.holub@ucd.ie

John Murphy
Performance Engineering Lab
School of Computer Science and Informatics
University College Dublin, Ireland
j.murphy@ucd.ie

Patrick O'Sullivan
Collaboration Solution System
IBM Software Group
Dublin, Ireland
patosullivan@ie.ibm.com

ABSTRACT

Nowadays, enterprise software systems store a large amount of operational information in logs. Manually analysing these data can be time-consuming and error-prone. Although a static knowledge database eases the task to capture recurring problems, maintaining such a knowledge repository requires periodic knowledge updates by domain experts. Moreover, as the repository grows, the problem of memory efficiency will also arise.

Our goal is to enable administrators to efficiently capture interesting data in a high volume stream of events in real-time. We are proposing a statistical approach for software applications to be automatically trained with a smaller dataset to efficiently predict interesting data under such conditions. The proposed solution maintains a stable memory usage by migrating keywords from a dynamic data structure to fixed sized data structures (Bloom Filter). In particular, the solution has achieved better prediction results by enhancing the Bayesian theory with belief modifiers.

Categories and Subject Descriptors

H.3.3 [**Information Storage and Retrieval**]: Information filtering; H.3.4 [**Systems and Software**]: Performance evaluation (efficiency and effectiveness)

General Terms

Algorithms, Design, Performance

Keywords

Stream Event Prediction, Statistical, High Performance

1. INTRODUCTION

Over the last decade, computers have been fast developed with many advances in hardware technologies from powerful servers to lightweight handsets. The highly developed technologies have also caused the amount of computer data to rapidly grow such that gigabytes of data are now very common. In particular, data streams have gained a great popularity for many real-life use cases such as video/voice streaming, distributed file sharing, and sensor networks. However, as the information contained in a data stream would continuously flow through the system, the data processing algorithms need to be designed as efficiently as possible to avoid data to be queued in the network. As a result, when dealing with a large volume of stream data, processing the same piece of information multiple times is not feasible, instead, single pass techniques are often preferred. Furthermore, data streams can also lead to other problems in relation to limited CPU and memory resources [1].

A common approach to recognise recurring problems and interesting data is to use a set of patterns and rules maintained in a static knowledge database, which must be defined prior to the data matching process [12, 14, 31]. However, most of pattern-based approaches require domain expert knowledge and extra maintenance costs for periodically updates [4], of which tasks are often manually carried out [26]. In summary, when dealing with a high volume of stream events, which are single raw text messages such as log records, a pattern-based approach would suffer following problems: a) Can not easily be extended to recognise new patterns; b) If the size of patterns keep growing, there can be too much information saved in the memory repository; c) A disk-based pattern repository can cause significant performance overhead in the pattern detection process due to much slower data access speed caused by disk I/O operations. As a result, an automatic way to detect problematic and interesting events has been strongly required by system administrators [10].

The statistical approach [10, 19, 26, 27] is dynamic and enables a software application to be automatically trained with a fraction of the dataset to capture interesting data, which may not even have been seen before. The focus of this paper is to discuss the statistical techniques to automatically predict interesting events through a raw event stream by assuming no dependency exists between each event and also no prior knowledge on incoming events either.

The bag-of-words is a traditional strategy to save training keywords and corresponding statistics in the memory [28]. In particular, the tree data structure allows keywords with the same prefixes to share a single branch, which is more memory efficient than saving each individual keyword separately. However, a tree has to be expanded for each distinct keyword, in which case the size of the tree can easily grow up depending on the number of distinct keywords and their length. As a result, using the tree data structure can potentially carry a risk of the memory-intensive problem, especially in a large-scale heterogenous environment such as Cloud [15], where a variety of software applications can work together and produce a high volume of dynamic data contents such as transaction IDs and session IDs.

Furthermore, we are proposing a statistical approach to predict interesting events through a dynamic stream events. Our approach has been built top of Bayesian and Bloom filter techniques with the ability to predict the interesting events through a large volume of stream data produced by enterprise software applications. In particular, this paper contains following contributions:

1. Using normalised probabilities to avoid the lowest and highest probabilities especially for the zero probability to cause inaccurate prediction results.

2. Allowing the memory to be effectively maintained by periodically migrating keywords from a dynamic data structure (Tree) to fixed sized data structures (Bloom Filter), which are created based on minimal and maximal probabilities. In particular, such a keyword migration does not require the quality of prediction results to be sacrificed.

3. Proposing the belief modifier to improve the accuracy of the Bayesian filtering technique.

4. The proposed solution enables both event training and prediction processes to deliver a high system throughput to cope with a large volume of stream events.

5. In addition, we are able to show that the proposed event prediction mechanism can also be used to discover interesting data through static documents.

2. BACKGROUND AND MOTIVATIONS

Log data analysis is an important task for administrators to understand the behaviour of software applications. However, due to the amount of information, manually performing such a task is time-consuming and thus ineffective. To ease the log analysis tasks, Run Time Correlation Engine (RTCE) [11] has been developed as a framework to automatically collect large amounts of log data from different application logs, and correlate them together to provide a single view to monitor all application instances. RTCE was designed to handle generic data types and developed in conjunction with our industrial partner (IBM Dublin Software Group), where RTCE has been successfully deployed by their testing teams. In particular, many teams have achieved significant time saving by using RTCE to automate their log analysis tasks [30] for variants of real industrial applications. In this paper, we are using the RTCE framework, which contains various log adapters to automate event conversion tasks for enterprise application logs.

The current success of RTCE has been mainly built based on the efficient run-time data analysis tasks such as the symptom matching algorithm [31], which enables the XPath-based evaluation engine to efficiently process a large set of XPath queries to match against stream events. However, maintaining a large symptom database, which contains hundreds of thousands of entries is not an easy task, which requires domain expert knowledge for each enterprise application and frequent data updates. To ensure our research work is able to solve more real-life problems for real industries, we have also carried our regular weekly meetings with IBM System Verification Test (SVT) team managers and experts to discuss further improvements to ease their testing experiences. As a result, we have decided to discover a new approach to enable their daily tasks to maintain the knowledge database to be automated as well as automating the problem detection process. The goal is to enhance the current RTCE framework with additional automated functionality for testing teams to gain more time saving on system monitoring related tasks.

3. EVENT PREDICTION SYSTEM

In order to support efficient event training and prediction processes for timely data analysis without sacrificing the quality of result, we have developed a new stream processing engine consisting of following components:

- **Event Trainer**: continuously breaks incoming events into a set of keywords and sends these keywords with statistics to appropriate data structures.

- **Tree**: saves keywords and statistics, which have not rarely or popularly occurred.

- **Bloom Filter (BlmF)** [6]: maintains two bit arrays separately such that one for rare keywords and the other one for popular keywords.

- **Event Predictor**: continuously examines the probability of the belief for each incoming event to find interesting ones with strong believes.

- **Bayesian Filter (BayeF)** [24]: calculate the overall probability of the belief for a set of keywords using the tree and BlmFs.

The purpose of distinguishing rare and popular keywords from the normal ones is to reduce the memory overhead caused by the tree. Typically a rare keyword is a set of distinct characters from other keywords. To save such a keyword, the tree data structure has to be expanded with a new branch and leaf, which will cause the size of tree grows up. On the other hand, the popular keywords are the ones that can be migrated to reduce the performance overhead for frequent calculations, which are mostly repeated. Furthermore in addition to the core system components, we have also used following thresholds to ease the event training and prediction process:

- **Minimal probability** (P_{min}): The minimal frequency of a keyword that could possibly occur in interesting events. This value has to be normalised to be greater than zero.

- **Maximal probability** (P_{max}): The maximal frequency of a keyword that could possibly occur in interesting events. This value has to be normalised to be less than one.

- **Strong threshold** (ST): A probability threshold to decide if an event is interesting such that the overall probability of the event has to be greater than or equal to this value.

- **Strong belief modifier** (T_s): The probability of how likely an event can be believed interesting by a given keyword, which has occurred in interesting events.

- **Weak belief modifier** (T_w): The probability of how likely an event can be believed interesting by a given keyword, which has not occurred in interesting events.

3.1 Event Trainer Overview

The event training engine is responsible for constructing a repository of keywords and corresponding statistics of their occurrences. The aim to allow a high volume of stream events to be efficiently trained without sacrificing the quality of the repository for event prediction during run-time.

As shown in Figure 1, when an event arrives in the system, the event training will use the *'Keyword Extractor'* to convert the message body into a list of keywords. Then the extracted keywords will be added to the *'Tree'* data structure and the statistics of each keyword will also be updated on associated leaf nodes. Once the tree is fully updated, all the updated and new leaf nodes will be sent to the *'Threshold Examiner'*, which periodically checks if any keyword has occurred with a rare or popular frequency, which has been calculated as:

$$p = Freq_{(interesting)} = \frac{occurrence\ in\ interesting\ events}{the\ total\ occurrence}$$

(In this example, we use P_{min} as the rare frequency and P_{max} as the popular frequency for rare and popular keywords, which have 1% and 99% probabilities to occur in interesting events respectively while both values can be re-adjusted depending on the actual use cases.)

If any leaf node with a rare or popular frequency is found, the *'TreeNode Extractor'* will find the corresponding keyword from the *'Tree'* using the parent reference contained in each tree node. During the keyword extraction, all the standalone tree nodes will be discarded to reduce the memory usage of the tree. Thereafter, the *'TreeNode Extractor'* will send rare and popular keywords to the rare and popular BlmF processors respectively.

During the training phase, a user has to define the pattern to select interesting events depending on the actual datasets. For example, the keyword *'spam'* would allow the *Event Trainer* to select all spam messages as interesting events. The purpose is to allow our system to be trained with a little user input (common keywords) to automatically predict future events.

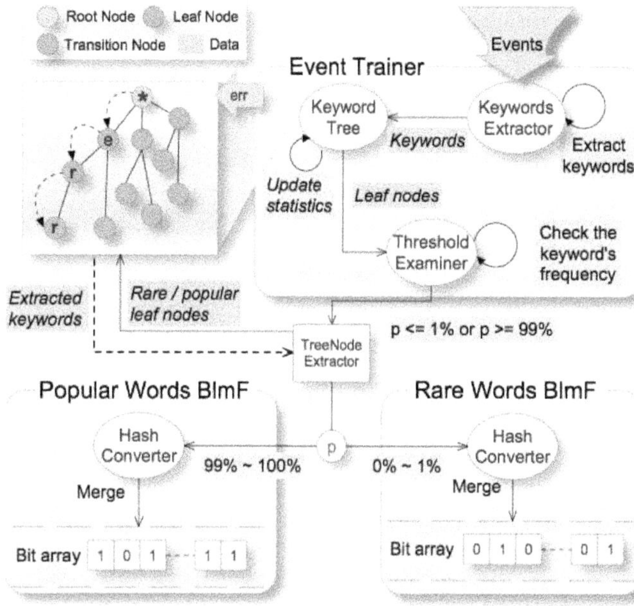

Figure 1: Stream-Based Event Training System.

3.2 Tree Structure

The tree structure (Figure 1) consists of a static root node and a list of leaf and transition nodes. New transition nodes are created when none of the existing nodes can form an incoming keyword. The last character of a keyword always settles on a leaf node. In the case of common strings being shared by multiple keywords, we allow all existing nodes to be reusable for them. As a result, a transition node can also become a leaf node if one keyword string is a subset of another keyword. As shown in the graph, the leaf node *'r'* will be reused as a transition node, if the keyword *'error'* comes. Furthermore, a path traversal between a leaf node and the root node can always precisely form a previously processed keyword. Consequently, each keyword's statistics such as the frequency occurring in interesting events and overall occurrence only need to be saved on their corresponding leaf nodes. For our current implementation, the statistics are collected as integers with a maximal limit of $2^{63} - 1$, which we expect to work for most situations.

During the tree node extraction phase, the engine will iterate all leaf nodes that have rare and popular occurrences in interesting events. From each leaf node the engine will traverse back to the root node in order to form the corresponding keyword for BlmFs. During the tree traversal the engine will also discard the current leaf node and all other transition nodes, unless any of them have been either linked by other descendant nodes or used as the leaf nodes by other keywords. By performing such a keyword migration process, the memory usage for the tree data structure can be greatly saved and the effectiveness is shown in the experiment section 4.2.

Furthermore, the rare and popular thresholds are defined as the lower and upper bounds of keywords' occurrence that core engine calculates. In this paper, we mainly use percentages for statistical calculations thus 1% and 99% have been used as the thresholds for determining rare and popular keywords. By defining the lower and upper bounds of keywords' occurrence, we are also able to assign fixed probabilities in the BlmFs during keyword lookups to avoid unnecessary calculations for a better performance.

From the memory consumption perspective, there are also other competitive techniques available. For example, keywords can be converted into hash values and saved in a hash table; A list of keywords can also be compressed using various data compression techniques. Although both approaches would allow keywords to be stored with a reduced memory consumption than the tree data structure, during the keyword lookup process the tree data structure would become more optimal. Because a hash table has to iterate through a list of records to search for a given keyword and results in longer response time if the number of records is large. By using data compression techniques, keywords need to do the data decompression, which will introduces extra performance overhead. Comparing other techniques, a key feature of using the tree data structure as the main knowledge repository to save keywords along with their occurrences is because of its efficient keyword lookups, which allows a given keyword to be searched through a large set of keywords via a single pass.

3.3 Bloom Filter

The BlmF technique has the advantage of space efficiency by encoding and merging each keyword into a fixed sized bit array. When a BlmF stores a keyword, it encodes the keyword into hash values (a bit array with 0 and 1 values) using a hash function and then merges the result into a predefined bit array. Then when a BlmF checks a keyword's existence, it has to convert the keyword into a hash value and check if all 1's in the keyword's hash value can be matched to the 1's in the BlmF's bit array. If all 1's are matched, then it means the BF has previously stored the same keyword in the bit array. Otherwise, the keyword has not been stored. As all keywords are saved together, it is impossible to assign or fetch statistics for individual keywords from the BlmF's bit array. Instead when a rare or popular keyword is found, the BlmF processor will simply return a fixed frequency of P_{min} or P_{max} accordingly.

We use the following function to calculate the size of the bit array:

$$m = -\frac{n \cdot \log f}{(\log 2)^2}$$

Where *'m'* is the number of bits required in the array, *'n'* is the total number of keywords going to be inserted into the bit array (the maximal capacity of the BlmF), and *'f'* is the false positive rate, which can be tolerated by this BlmF. (During the implementation, we have defined each BlmF, which consumes less than 1 MB memory space to allow maximal 100,000 keywords to be saved by tolerating 0.01 false positive.) To efficiently convert each keyword into hash values, the MurmurHash2 [1] algorithm has been used.

Furthermore rare and popular keywords are migrated from the tree data structure based on their extreme occurrences, during the keyword lookup process the fixed occurrence values for BlmFs' keywords are immediately returned as the minimal and maximal likelihoods to believe an event to be interesting. By defining such normalised thresholds to represent the lower and upper bounds of likelihoods, we assume that the prediction process can tolerant little precision loss, in which case it might result in mismatching a small fraction of interesting events that have their overall likelihoods slightly below the predefined (ST). However, depending on the actual deployment environment, the optimal value of ST can be re-adjusted to either capture more mismatched interesting events by reducing ST or less non-interesting events by increasing ST.

[1] http://code.google.com/p/smhasher/wiki/MurmurHash2

The BlmF-based solution has been proposed as a space efficient technique for examining the keywords' existence. However, a BlmF has to risk certain chances of false positive, which causes a keyword to be falsely reported as being contained in the bit array. The false positive problem occurs because the 1's bits of a keyword are fully satisfied by the BlmF's bit array but in fact some 1's bits in the BlmF are set by storing other keywords' hash values rather than the current keyword. An example of such a problem is illustrated in Figure 2. As assuming the original bit array has not stored the hash value for neither *'Keyword A'*, *'Keyword B'*, nor *'Keyword C'*, when *'Keyword A'* arrives for the training purpose, all the corresponding 1's bits must be turned on within the original bit array. However, the new bit array produced after such a merging process might also contain all necessary 1's bits for other keywords, which have never been stored before. For example, during the querying phase, although both *'Keyword B'* and *'Keyword C'* have never been merged into the bit array, the hash value of *'Keyword C'* has just become fully satisfied after the new bit array has the 1's bits turned for both *'Keyword A'* and previous keywords. Consequently, the *'Keyword C'* would be falsely reported as a keyword, which has been theoretically stored in the bit array previously.

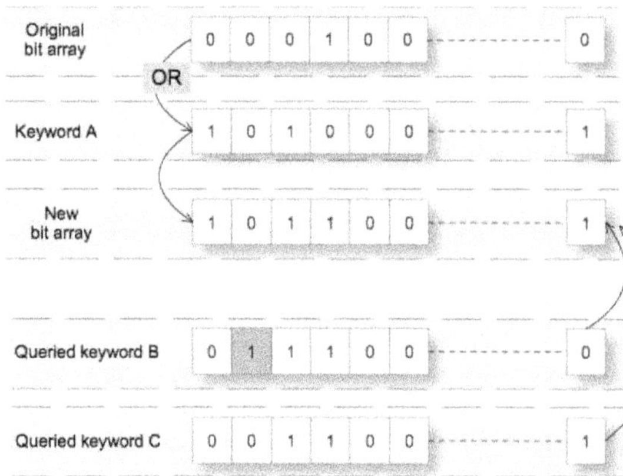

Figure 2: The False Positive Problem in Bloom Filter.

3.4 Event Predictor

The event prediction engine is responsible for evaluating the probability of each arrival event to be believed interesting based on previous keywords and statistics, which have been saved in the *'Tree'* and rare and popular BlmFs. The aim is to develop an efficient event prediction algorithm to capture interesting events through a high volume of stream events in real-time.

During the event prediction process, once a stream event arrives, the event prediction engine uses the Algorithm EP to calculate the overall probability of the event and determine how likely the event is believed to be interesting. Before an

Algorithm EP BayeF-based Event Prediction (*event*)

1: $T_s = 90\%$ {Strong belief modifier}
2: $T_w = 10\%$ {Weak belief modifier}
3: $ST = 90\%$ {Strong Threshold}
4: $P_{min} = 1\%$ {Minimal probability}
5: $P_{max} = 99\%$ {Maximal probability}
6: $k = 0$ {Keyword Counter}
7: **for all** keywords in event **do**
8:   **if** keyword exists on the tree **then**
9:     $p_k^t = \dfrac{occurrence\ in\ interesting\ events}{the\ total\ occurrence}$
10:   **else if** keyword exists in rare BlmF **then**
11:     $p_k^t = P_{min}$
12:   **else if** keyword exists in popular BlmF **then**
13:     $p_k^t = P_{max}$
14:   **else** {keyword is unknown}
15:     skip and continue with the next keyword
16:   **end if**
17:   $p_k = p_k^t \times T_s$
18:   $p'_k = (1 - p_k^t) \times T_w$
19:   $k = k + 1$
20: **end for**
21: **if** k is equal to 0 **then**
22:   **return** No keyword found
23: **else**
24:   $P = \dfrac{\prod_{i=1}^{k}(p_i)}{\prod_{i=1}^{k}(p_i) + \prod_{i=1}^{k}(p'_i)}$
25:   **if** P is greater than or equal to ST **then**
26:     **return** Event is interesting
27:   **else**
28:     **return** Event is not interesting
29:   **end if**
30: **end if**

event is processed, the engine will initialise a variable *'k'* as the counter for the number of processed keywords, which are recognisable within the training repository (See line: 6). Then for each arrival event, the engine will extract all the keywords from the message body (See line: 7) and collect the associated probability for each keyword (See line: 8 – 16).

In order to reduce the chances of having false positive problems in both BlmFs, the execution order of the probability collection process needs to be properly arranged such that the engine always tries to find an extracted keyword from the tree first before using BlmFs. As a result, the keywords saved on the tree will not be falsely found from BlmFs. Thereafter, if a keyword is found on the tree, the corresponding probability for the current event to be believed interesting is going to be calculated based on the previous formula '$Freq_{(interesting)}$'. If the keyword can be found in the rare BlmF, a static probability will be assigned to the keyword using the value of P_{min}. Consequently, if a keyword can only be found in the popular BlmF, the P_{max} will be assigned to the keyword. However, if the keyword has not been found in neither *'Tree'* nor the two BlmFs, the keyword is considered neutral such that its probability will be skipped for the current event's probability calculation.

Furthermore for the best practice, the execution order between rare and popular BlmFs should be adjusted based on the number of merged keywords in their bit arrays such that the one with a smaller number is always executed before the other. As more keywords are merged into the bit array, more 1's bits will be turned on, in which case the chances of having false positive problems will be increased. Therefore when collecting the probability for each keyword, executing the BlmF with less merged keywords prior to the other can practically reduce the risk of false positive problems. Such an optimisation can be implemented before the *'for loop'* (line: 7) by assigning BlmF to two separate variables and then replacing line10, 12 with the BlmF variables containing less and more keywords respectively, while assigning the corresponding probability to p_k^t (line: 11, 13).

Once a probability is successfully calculated for a keyword, the engine will use the Bayesian theory to calculate the likelihood statistics of the current event based on the processed keywords. The basic formula is defined as:

$$P = \frac{p_1 \cdot p_2 \cdot p_3 \ldots \cdot p_k}{p_1 \cdot p_2 \cdot p_3 \ldots \cdot p_k + p'_1 \cdot p'_2 \cdot p'_3 \ldots \cdot p'_k}$$

P is the probability of how likely the event is believed to be interesting, p_k is the probability of the keyword, which has occurred in interesting events to allow an event to be believed interesting, and p'_k is the probability of the keyword, which has not occurred in interesting events to allow an event to be believed interesting. If p_k^t is the probability of the keyword occurring in interesting events ($Freq_{(interesting)}$), then:

- $p_k = p_k^t \times T_s$: By applying the strong belief modifier to the frequency of the keywords occurring in interesting events, the probability of of an event believed to be interesting can be more realistically reflected, because a single interesting keyword should not make the event believed to be interesting with 100% certainty. Instead, an interesting keyword should only give a strong belief (90%) to the event for being interesting.

- $p'_k = (1 - p_k^t) \times T_w$: As opposite to p_k, p'_k is calculated based on a weak belief modifier to the frequency of the keyword occurring in non-interesting events. Although in this case the probability represents the chance of a keyword to occur in non-interesting events, the belief of such events to be interesting should not be absolutely impossible. Instead, a non-interesting keyword should give a weak belief (10%) to the event for being interesting.

As shown in the Algorithm EP, the engine updates the product over the overall probability from the $1st$ to current keywords with two realistic probabilities of p_k and p'_k (See line: 17 – 19). When the engine finishes collecting all probabilities with the keyword counter *'k'*, the overall probability of the belief for the current event being interesting is calculated using the Bayesian theory as line 24. In this case, $\prod_{i=1}^{k}(p_i)$ is the probability of the event to be interesting based on keywords occurred in interesting events and $\prod_{i=1}^{k}(p'_i)$ represents how likely the event is interesting based on keywords occurred in non-interesting events. An the end of the algorithm, the engine will compare the overall probability *'P'* to the predefined threshold *'ST'* to determine whether the event is interesting (See line: 25 – 29).

4. EXPERIMENT

The implementation of the proposed solution has been developed in Java and evaluated using a standard desktop machine[2]. During all experiment tests, the implemented application has been running with a default memory setting, which allows maximal 123MB to be consumed and using 20% $\sim$ 31% of the quad-core CPU on average. To clearly demonstrate the performance of our solution, the experiment has been broken down into 7 sub-sections.

In Section 4.1, we will explain the memory-intensive problem caused by the tree data structure. In Section 4.2, we will show the effectiveness of the keyword migration mechanism for stream-based event training. In Section 4.3, we will evaluate the quality of our solution from the following aspects:

- **False positive**: representing the percentage of the non-interesting events, which should not be captured out of the entire set of captured events. The value gives an impact on the amount of non-interesting events that an administrator needs to filter out from the predicted events.

- **False negative**: representing the percentage of the missing events out of the interesting events, all of which should be captured. The value gives an impact on the amount of interesting events that an administrator needs to filter out from the entire event stream.

As a high quality result, the value of both evaluation criteria should be kept low. In the case that only one of evaluation criteria can be maintained low, the false positive is always sacrificed to achieve a low false negative value. Because finding interesting data through a high volume of stream events is much more difficult and time-consuming than finding non-interesting data through a small fraction of the stream dataset.

In particular, Section 4.3 will also demonstrate the system throughput for stream-based event training and prediction processes in addition to the accuracy assessment. The aim is to show the applicability of applying our solution for real-time event prediction use cases.

In Section 4.4, we will show the effectiveness of the belief modifiers by comparing to the BayeF without belief modifiers. In Section 4.5, we will demonstrate the performance of RTCE before and after being integrated with our solution. The aim is to show the performance overhead introduced by our solution to the original RTCE framework. In Section 4.6, we will evaluate the quality of our solution by processing static data using two open source datasets. The aim is to show the applicability of our solution for various data types. In Section 4.7, we will compare three other classifiers against our approach.

4.1 Memory-Intensive Problem for Tree

During our experimental test, we use the B-Tree data structure to store indexing keywords. Each tree node consists of three possible fields: a word character in all tree nodes, a word frequency value on the leaf node, and a list of node transitions on the transition nodes. Such a tree data structure can achieve space efficiency for storing keywords with the same prefix on a single tree branch. In order to simulate unique keywords, our test data consists of a list of random UUIDs[3] (Universal Unique IDentifier [20]).

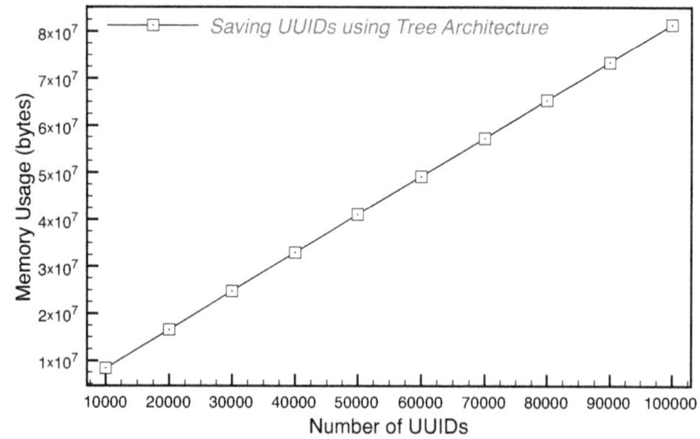

Figure 3: Memory consumptions for saving UUIDs in a tree architecture

Figure 3 illustrates the memory consumptions for storing different amounts of UUIDs in the tree data structure. As the number of stored UUIDs increases, the more memory are consumed by the tree. To be precise, the size of the tree is increased by 8 megabytes for every 10,000 UUIDs under our Java implementation. In the case that tens of thousands of UUIDs need to be processed, the memory exhaustion would become significant. In addition, the memory consumption shown in the graph is only for storing a series of UUIDs. If storing real log events, which may contain the UUID, timestamp, and message body for each log entry, then the size of the tree will consume even more memory space. Therefore, we would conclude that purely using the tree data structure to store the training keywords and statistics is not a feasible solution to deal with a large volume of stream events.

4.2 Keyword Migration for Event Training

As explained previously, processing a dynamic stream data can potentially cause the memory-intensive problem. Beyond the solution of upgrading machines with more physical memory chips, we have proposed a new approach to migrate rare and popular keywords into two fixed sized BlmFs. The effectiveness of our approach is shown in Figure 4.

During the experiment, the implemented application has been constantly trained with a set of log events collected from the IBM SVT lab environment. The overall dataset consists of $27m$ lines of events and $2.5g$ in size. The number of processed keywords is $62k$, which consequently require $563k$ tree nodes to be created in total. The number of tree nodes is periodically examined against the pre-defined threshold of $100k$ to trigger the keyword migration process, which has caused the number of keywords to be dramatically reduced from the tree as shown in Figure 4.

Moreover, the set of log files used in this experiment is only a snapshot of a typical enterprise application in software industries. In the case of handling a real production application running for weeks or even months, the training dataset can be much bigger. As a result, purely saving all keywords and statistical data on a tree data structure would require much more memory space than our approach.

[2]Ubuntu 8.10, x86_64 GNU/Linux, J2RE 1.6.0 J9 2.4 Linux amd64-64, Intel Q9400 2.66GHz, 8GB RAM, Western Digital 7200rpm SATA-II

[3]http://docs.oracle.com/javase/1.5.0/docs/api/java/util/UUID.html

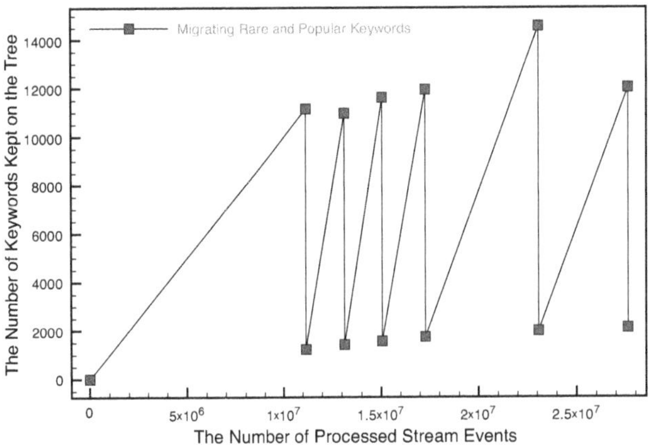

Figure 4: Keyword Migration

Table 1: Performance of Stream Event Prediction

Stream Source:	Application Server	UI Server
Training Throughput	118k events/sec 10.2m bytes/sec	115k events/sec 10.7m bytes/sec
Training Dataset	1.4m events (First 25% of stream events)	2.7m events
Interesting Events	20,826	42,184
Rare Words	8,937	2,170
Popular Words	792	40
Training Words	10,394	3,134
Prediction Throughput	119k events/sec 10.3m bytes/sec	112k events/sec 10.4m bytes/sec
Prediction Dataset	5.9m events	10.8m events
Neutral Words	137,667	3,936
Interesting Events (Predicted)	80,898	173,753
Total Events (Predicted)	93,092	192,725
False Positive	13%	9.8%
False Negative	0.3%	0%

4.3 Stream-Based Event Prediction

For the stream-based event prediction, we have used another desktop machine (Pentium4) as the load generator connected to the implemented application via a 100Mbps local network. The data source consists of two sets of log files produced by various system components of an enterprise social application [13]. During the experiment, the load generator has been constantly sending each line of log events to the implemented application until the last event is reached. (The load generator is capable of sending up to 130k events per second on average.) Based on the experiment result, we have seen that using 90% as the threshold value for ST and T_s and 10% as T_w has been able to predict interesting events with a reasonable accuracy, which is expected to tolerate 10% false positives and 1% false negatives.

As shown in Table 1, the two stream sources are collected from an application server and UI server. From the system throughput perspective, the UI Server events have been processed with a slightly lower event processing rate (events/sec) but a higher data processing rate (bytes/sec). The main reason is because each UI Server event is longer than the application server. More importantly, the implemented application is able to deliver a high system throughput for both applications during event training and prediction phases. As a result, the proposed solution should be efficient enough to handle real-time event prediction use cases.

During the event training phase, the implemented application has only been trained with a 25% of the overall stream data source produced by each enterprise application. For the best practise, the training dataset for the implemented application should be made as small as possible to reduce the performance overhead introduced by the training process. From the QoS perspective, the more data is trained, the more accurate results can be achieved because less neutral words will be experienced during the prediction phase. However, sometimes the increased training dataset can not significantly improve the quality of the result, instead a significant performance overhead can be introduced. As a result, after trying different sizes of training datasets, the percentage 25% has been found as the most optimal training threshold for all use cases presented in this paper.

In order to maintain a stable heap usage, the keyword migration mechanism has been used to move the rare and popular keywords from the tree to BlmF. As shown in the table, the majority of processed event keywords have rarely occurred, while a small set of keywords have popularly occurred in both application servers. Because the BlmFs are fixed sized data structures, the significant amount of migrated keywords removed from the tree can dramatically reduce the overall heap usage.

From the performance perspective, RTCE has been able to process over a hundred thousands of events per second for both training and prediction processes. The data processing throughput has almost reached the maximal theoretical limitation on the network bandwidth (12.5m bytes/sec). However due to time constraints, we have been able to carry out additional tests in a more powerful hardware environment, which would be done in our future work. In this paper, the main purpose of demonstrating the result of system performance for the current experiment is to show the prospective efficiency of our event training and prediction processes to handle a high volume of stream data. In particular, all the event training and prediction results are almost processed in real-time with no sophisticated buffering techniques being enabled.

Furthermore, due to a small dataset being trained for the implemented application, there has been many neutral keywords found during the event prediction phase. For the application server, 57% of the neutral keywords belong to interesting events and for the UI server 19% of neutral keywords belong to interesting events. Although providing a complete training with the entire stream would make more realistic statistics available for the probability calculation, such an experiment has not been able to significantly improve the quality of the result. In both stream sources, the non-interesting keywords have got low occurrences, which can only give little impact on the probabilities ('P') of interesting events. As a result, the false negative values for both servers have remained the same for a complete training.

From the false positive perspective, by training the full dataset the application server has achieved a lower value of

12%, while the UI Server has got a slightly higher value of 10.3% comparing to the 25% of training data. In the application server, most of the interesting keywords have been seen in the interesting events but only a small number of interesting keywords exist in the non-interesting events. When the number of non-interesting keywords increases with a complete training, the probabilities (*'P'*) on many of the non-interesting events have been dramatically reduced such that thousands of such events have no longer believed to be interesting. As a result, the value of false positive has been reduced to 12% from 13%.

However, with a complete training the UI server has achieved a slightly different result, which has increased the value of false positive to 10.3% from 9.8%. For a small fraction of the stream events, the probability of the belief (*'P'*) has surpassed the threshold of interesting events because some of the newly recognised interesting keywords have gained higher occurrences than the non-interesting keywords. But as the probabilities of the majority of the non-interesting events have still been dominated by the non-interesting keywords, most of the non-interesting events were still classified in the result. Thus, a small variance of the false positive has been seen.

More importantly, we have seen that a complete training has only been giving small impact on the quality of result. As the training workload and duration have been significantly increased for training, the trade-off between accuracy and performance has not been equalised. Therefore, a 25% of training dataset is optimal for event training and sufficient for prediction. In real use cases, such a configuration can be predefined based on the requirement of the monitored application. For example, if a testing team is going to evaluate the stability of an enterprise application based on a 4-days run, our solution can be trained with stream data produced by the application during the 1*st* day.

4.4 Belief Modifier Effectiveness

In order to examine how effectively the strong and weak belief modifiers can impact on the event prediction results. Two experiment tests have been carried out for the same stream dataset, which consists of log events produced by the Application and UI servers. The experiment results have shown that applying the belief modifiers the average probability of each stream event has been increased and enabled the engine to make more accurate decisions.

As shown in Figure 5, without using the belief modifiers the event probabilities calculated by the BayeF has varied between 1.0^{-77} and 0.99. By using the belief modifiers, the probabilities varied between 1.0^{-33} and 0.99. In addition, without belief modifiers the most of the events have achieved the probabilities between 1.0^{-34} and 1.0^{-14} while using belief modifiers most events have had probabilities between 1.0^{-15} and 1.0^{-7}. Although most of the events have gained the probabilities below the strong threshold, the belief modifiers have enabled the engine to capture interesting events with a relatively high accuracy as shown in the previous Table 1.

Figure 6 illustrates a similar event distribution to what the application server has achieved. The average event probabilities with the belief modifiers are higher than the ones without using belief modifiers. While most of the stream events, which contained a small of interesting keywords have gained low probabilities (lower than the strong threshold), by using belief modifiers the proposed solution can still capture all the interesting events in the prediction result dataset, which also contains 9.8% of non-interesting events. (See Table 1)

Table 2 gives a detailed breakdown of the event prediction result without using the belief modifiers. During this test, we have trained the engine with the same dataset (25%) for each event stream. However, the experiment has shown a bad quality for the event prediction result, which have gained 62% and 3.4% of false negative rates for the application and UI servers respectively. As discussed previously, the false negative is a worse quality indicator than false positive. Because to deal with a false positive problem, administrators only need to go through a small dataset containing all the predicted events. In the case that a high false negative is achieved, the administrator has to go through the entire stream dataset to find interesting events, which were missing. However, finding the other 62% of interesting events for the application server through millions of events can be an extremely time-consuming task for administrators to carry out.

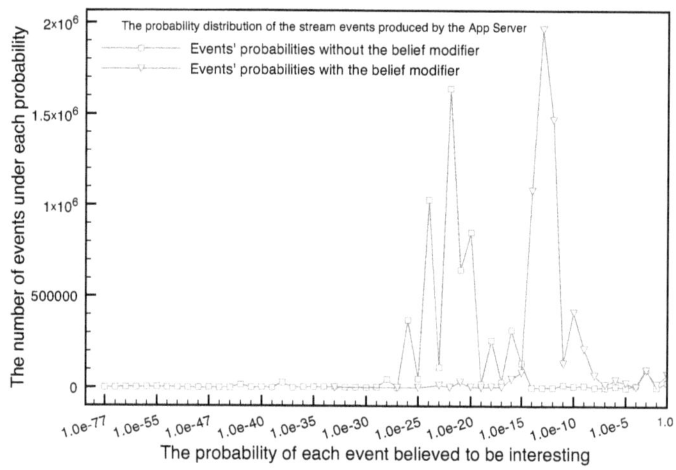

Figure 5: The Event Distribution of Bayesian Probabilities across Application Server's Event Stream.

Table 2: The Prediction Results Without Belief Modifiers

Stream Source:	Application Server	UI Server
Interesting Events (Predicted)	30,778	167,844
Total Events (Predicted)	30,778	167,867
False Positive	0%	0.014%
False Negative	62%	3.4%

4.5 RTCE Integration

In addition to the experiment tests for the standalone application, we have also integrated the solution into the RTCE framework. The goal is to extend the tool with more useful functionality to enable testing teams to easily perform system monitoring related tasks.

During the experiment, we have increased the maximal heap usage to 256 MB for running the RTCE framework

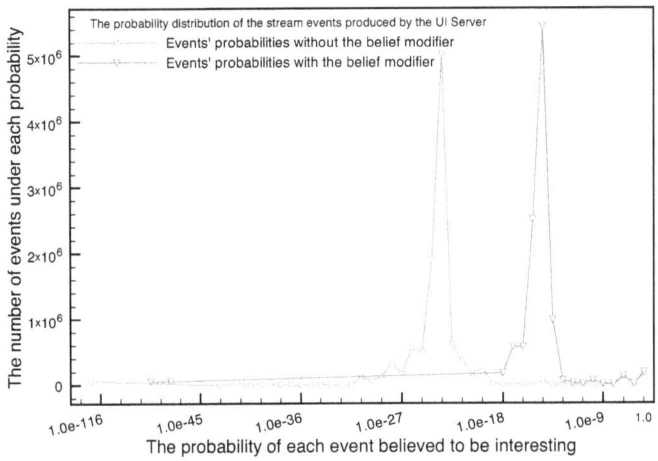

Figure 6: The Event Distribution of Bayesian Probabilities across UI Server's Event Stream.

due to more system components being loaded. Because the proposed solution has mainly focused to enhance the runtime data analysis tasks, the data persistence functionality of RTCE, which is the *'Run'* system component described in the previous work [11] has been disabled during each test. Also, the same stream dataset has been used for all the tests.

As shown in Table 3, the first experiment test has been carried out to show the baseline for the RTCE performance assessment. The result has shown that the original RTCE has achieved a high system throughput of 108.4k events per second, which equally gives the data processing rate of 10.06 megabytes per second. Thereafter, the result of the second test has shown that after integrating our solution to the RTCE framework, the system throughput has slightly decreased to 102.2k events per second for the event training process. Finally, when the integrated RTCE framework performs the event prediction tasks, the system throughput has slightly increased to 106.2k events per second but still lower than the original RTCE performance. From the CPU usage perspective, the integrated RTCE framework has shown a significant increase than the original RTCE due to extra data mining tasks being performed.

Table 3: The Performance of RTCE with Event Prediction Integrated

RTCE Performance	CPU Usage	System Throughput (events/sec)	System Throughput (bytes/sec)
Original	27.5% $\sim$ 30%	108.4k	10.06m
Integrated (Training)	41% $\sim$ 48.5%	102.2k	9.48m
Integrated (Prediction)	40% $\sim$ 46%	106.2k	9.86m

4.6 SMS Spam and Farm Ad Prediction

Beyond the stream-based event prediction experiment, we have also assessed our solution against static documents: SMS messages [2] and farm ads [23], which have been used by other research works of [3] and [22] respectively.

As shown in Table 4, both documents have been trained to predict interesting records, which are spam SMS messages and accepted farm ads. Because the size of each document is small such that our solution can almost process both documents instantly, the training and prediction throughputs are excluded in this table. By training first 25% of each document, the SMS and farm-ad documents have been trained with 202 spams and 568 accepted-ads respectively.

Table 4: SMS Spams & Farm Ads Prediction

Data Source:	SMS Messages	Farm Ads
Training Dataset	1,393 Messages	1,035 Ads
Interesting Records	Spam: 202	Accepted: 568
Training Words	4,240	17,971
Prediction Dataset	5,574 Messages	4,143 Ads
Neutral Words	10,627	68,001
Interesting Events (Predicted)	741	1,708
Total Events (Predicted)	865	3,027
False Positive	14.9%	43.5%
False Negative	0.8%	22.7%

During the event prediction phase, a significant amount of neutral keywords have been seen for both SMS and farm-ad documents. As discussed previously, depending on the ratio of interesting to non-interesting keywords in the new messages, the values of false positive and false negative can vary a lot.

For the SMS messages, our solution has achieved a 14.9% of false positive and 0.8% of false negative. For the farm ads, 43.5% of false positive and 22.7% of false negative have been achieved. Comparing to the SMS document and other stream-based log events, the farm-ads document has been predicted with much higher false positives due to stronger believes built based on interesting keywords in the new events. Because in many of the new non-interesting events, the interesting keywords have occurred more frequently than non-interesting keywords. As the neutral keywords can not give any impact on the probabilities, such non-interesting events have been falsely included in the interesting event category.

Furthermore during the prediction phase, 99.2% of new spam messages in the SMS dataset have higher recurrence of interesting keywords than non-interesting keywords. As a result, there has been only 0.8% of spam messages missing. In the farm-ad dataset, 24% of accepted-ads have been missing due to many non-interesting keywords occurred in the new accepted-ads, which probabilities have been reduced below the strong threshold due to the lack of unrecognised interesting keywords contained in the neutral keyword set.

4.7 Comparing to Other Classifiers

In order to further assess the quality of our solution, we have compared our approach with three other popular classifiers: Naive-Bayes Tree (NBTree) [18], $C4.5$ [25], and Sequential Minimisation Optimisation (SMO) [17]. $C4.5$ is a decision tree based algorithm featuring the branch pruning ability. NBTree is a hybrid algorithm consisting of both Naive-Bayes and decision tree classifiers. SMO is the successor to the Support Vector Machine (SVM) and optimises the data processing speed. To compare the performance of each algorithm against our approach, we have obtained the implementation for each algorithm from Weka [4], which is an open source project with a wide range of machine learning algorithms implemented in Java.

During the first experiment, we have used the farm-ads dataset to assess the performance of each algorithm. As shown in Table 5, we have trained all four algorithms with 25% of farm-ads for prediction. As a result, both $C4.5$ and NBTree have achieved the highest number of correctly classified farm-ads, while RTCE and SMO being the second and third accurate approaches. From the memory efficiency perspective, the $C4.5$ and RTCE have both been able to successfully run with the default JVM setting, which is 123MB. Furthermore, the time taken to build and predict farm-ads has also been measured for each algorithm to assess their capabilities to fulfil the real-time event prediction requirement. As shown in the table, RTCE has been the least time consuming approach to process farm-ads, while NBTree being the slowest approach.

Table 5: Classifier Comparison Using Farm Ads

Classifiers:	C4.5	SMO	NBTree	RTCE
Training Data	1,035 Farm Ads			
Prediction Data	4,143 Farm Ads			
Correctly Classified	2,361 56.98%	2,210 53.34%	2,361 56.98%	2,322 56.04%
Incorrectly Classified	1,782 43.01%	1,933 46.65%	1,782 43.01%	1,821 43.95%
Memory Setting	default	512MB	512MB	default
Time to Build	$10.64s$	$2.11s$	$45.44s$	$0.077s$
Time to Predict	$2.58s$	$2.79s$	$5.56s$	$0.698s$

During the second experiment, we have used the SMS message dataset to assess the performance. The same as farm-ads experiment each algorithm has been trained with 25% of SMS messages. As shown in Table 6, RTCE has achieved the highest number of correctly classified SMS messages. Furthermore while $C4.5$ and RTCE are still the most memory efficient approaches, RTCE has been found to be the fastest approach to build and predict SMS messages.

Table 6: Classifier Comparison Using SMS Spams

Classifiers:	C4.5	SMO	NBTree	RTCE
Training Data	1,393 SMS Messages			
Prediction Data	5,574 SMS Messages			
Correctly Classified	4,827 86.59%	4,827 86.59%	4827 86.59%	5,438 97.56%
Incorrectly Classified	747 13.40%	747 13.40%	747 13.40%	136 2.43%
Memory Setting	default	256MB	256MB	default
Time to Build	$3.64s$	$0.64s$	$7.6s$	$0.035s$
Time to Predict	$0.8s$	$1.09s$	$1.47s$	$0.09s$

Additionally we have also carried out a training test using the full dataset of farm-ads for $C4.5$, which has been the most memory efficient approach beyond RTCE. However, the experiment has shown that $C4.5$ requires a minimum $1.5GB$ memory setting to successfully process the full farm-ads dataset while taking $354.38s$ and $34.41s$ to build and predict data respectively. Comparing to the our approach, which has been able to process the full dataset almost instantly ($0.109s$ to build and $0.792s$ to predict) with the default JVM memory setting, the $C4.5$ is still much less memory efficient and slower.

[4]Weka: Machine Learning Software in Java, http://sourceforge.net/projects/weka/files/weka-3-6/3.6.8

5. RELATED WORK

Naive Bayes Classifier (NBC) [24] has been proposed as an efficient solution to classify raw text documents with multiple attributes. The probabilities of various terms for certain attributes are calculated based on their frequencies in the training dataset. Comparing to other learning methods, NBC is more efficient and suitable for processing a large set of data due to the avoidance of the searching process for possible hypothesis. In addition, NBC also classifies documents when the contents of data are independent. In the case that one of the terms has gained a 0 probability, an estimating probability has to be calculated by adding *'mp'* to the *'numerator'* and *'m'* to the *'dominator'*, where *'m'* is the equivalent sample size and *'p'* is the estimated prior probability. Such a 0 probability problem is avoided in our solution by assigning the minimal probability (1%). Furthermore, a following research work [5] has found that NBC scores can unrealistically move close to 0 or 1 when the number of input words increases.

Iterative Bayes Classifier (IBC) [7] and Adaptive Bayes Classifier (ABC) [8] are two of extended approaches to the core NBC technique. The IBC can further improve the prediction probability by iterating the training dataset. After a normal NBC training process, the IBC needs to cycle through the same dataset again to allow the desired classes to be predicted more confidently, in which case IBC would increase the probability of predicted classes and decrease the probabilities of other classes that are not selected. As a result, the confidence of selecting the predicted classes will become stronger and give more significant impact on the believes of further predictions. For the ABC approach, the confidence of predicted classes is improved by learning from new data contents instead of iterating the old training dataset. In the case of predicting stream events, the ABC approach is more suitable for handling dynamic data contents, which can not be accurately predicted by IBC if there is any changes occurring in the stream. However, neither NBC, IBC, nor ABC has considered the memory-intensive problem to fulfill the requirement of processing a large volume of stream data.

The Conceptual Clustering and Prediction (CCP) framework [16] has been proposed as a stream-based email filtering system, which assumes there is only a limited memory resource available. In their related work section, a number of previous works have been discussed to emphasis the memory constraint problem for stream processing. In order to solve such a memory issue, CCP uses a mapping function to transform every batch of stream data into conceptual vectors and groups multiple vectors into clusters. For each cluster, CCP needs to assign a corresponding classifier, which is also trained and updated by the batch of data. By only keeping the cluster centroid and corresponding classifiers in the main memory, the CCP framework can easily cope with a stream data while there is a constraint on the memory usage. However, in order to achieve an optimal performance, the proper batch size can only be selected after the preliminary experiment results are gained.

Context-aware Ensemble (CAE) [9] is one of recently proposed stream processing engines, which use the ensemble approach [29] to train and update classifiers. CAE maintains a pool of classifiers and based on their weights, an appropriate classifier is used and trained for recurring concepts processing. CAE can improve the adaptation to recurring concepts by exploiting the relationship between contexts and concepts. To deal with memory constraints, CAE can periodically prune away classifiers, which have the lowest utilisation when the maximal number of classifiers is reached. However, the experiment result shows that the accuracy of the CAE approach would be significantly affected when there is some noise data occurring in the underlying stream.

Classifier4j [21] is an open source tool supporting text classification based on the Bayesian theorem. Classifier4j uses the relational database (RDBMS) technology to save the processed keywords and corresponding statistics. Because RDBMS persists data directly onto the hard disk, the memory-intensive problem can be easily avoided by Classifier4j due to a much smaller set of data being kept in the memory. However, as each RDBMS data access requires one disk I/O operation, which is much slower than the RAM data access, the overall system throughput can be limited by a large amount of I/O waiting time while dealing with a high volume of stream data. Moreover, Classifier4j only uses the keywords' frequency as the probability of the belief by assuming a 100% belief for keywords existing in both interesting and non-interesting events. As discussed previously, this mechanism is likely to produce inaccurate probability results when dealing with a high volume of dynamic stream data, where many interesting keywords in predicting events have only been trained with low probabilities.

From the log analysis perspective, Xu.W and et al. [32] have proposed an interesting four-step approach to detect problems through system console logs. Their first step enables unstructured log messages to be parsed based on the logging points of source codes. The second step allows predefined functions to be executed to provide fine-grained information for the third step, which detects the anomalies through variants of variables and identifiers. The last visualisation step shows a decision tree graph to allow users to gain a detailed problem description. By making use of several open source libraries such as Apache Lucene and Hadoop, they have been able to process millions of messages per minute with dozens of nodes while achieving a high result accuracy. Comparing to their approach, which mainly focuses anomaly detection based on related numerical variables and identifiers, our solution is a more light-weighted approach to only predict the existence of an anomaly message rather than message correlation. However, during the experiment, their approach has failed to parse some of the "message type" attributes containing long strings, in which case a complete understanding of all message types, which are not required by our approach is required by their approach to properly process various logs. Furthermore to ease troubleshooting tasks, LogEnhancer [33] has been proposed to enhance logging points in the source code. Comparing to these two approaches, both of which would require the access to the source code and domain expert knowledge to understand it, our solution has been mainly focusing on anomaly detection with non-expert knowledge and assuming source code access is not always possible.

6. CONCLUSIONS

In this paper, we have explained the advantage of using a statistical approach to predict interesting data based on limited training resources. Thereafter, the memory-intensive problem has been detailed under the context of coping with a large volume of stream events. The proposed solution tackles such a problem by using two BlmFs, which are both fixed sized and efficiently save periodically migrated keywords from the tree data structures. To avoid inaccurate prediction results, the lowest and highest probabilities have been normalised using minimal and maximal probabilities respectively. By assigning the minimal and maximal probabilities to each BlmF, the migrated keywords have avoided the sacrifice to the quality of prediction results. The proposed solution has also enhanced the Bayesian theory with strong and weak belief modifiers to achieve better prediction results. Furthermore during the experiment, the implemented event prediction system has been able to predict interesting data through a high system throughput for log-based stream events produced by enterprise software applications, while only being trained with a fraction of the entire stream. As a result, we would conclude that the goal of the research, which develops an effective and efficient event prediction technique to allow stream processing engines to capture interesting data with a high throughput in real-time has been achieved. In the future work, we would like to find a way to automatically discover the optimal belief modifiers without relying on any experiment experience and also try our solution in a much larger environment with much higher workloads over a long period of time.

7. ACKNOWLEDGMENTS

This research was funded by Science Foundation Ireland (SFI) via grant 08/SRC/11403 FAME SRC (Federated, Autonomic Management of End-to-End Communications Services – Scientific Research Cluster).

8. REFERENCES

[1] C. C. Aggarwal. Chapter 1: Introduction to data streams. In *Data Streams - Models and Algorithms*, volume 31 of *Advances in Database Systems*, pages 1–6. Springer, Heidelberg, 2007.

[2] T. A. Almeida and J. M. G. Hidalgo. Sms spam collection, Sept. 2012. Available [online]: http://www.dt.fee.unicamp.br/~tiago/smsspamcollection/.

[3] T. A. Almeida, J. M. G. Hidalgo, and A. Yamakami. Contributions to the study of sms spam filtering: new collection and results. In *Proceedings of the 11th ACM symposium on Document engineering*, DocEng '11, pages 259–262, New York, NY, USA, 2011. ACM.

[4] I. Androutsopoulos, J. Koutsias, K. V. Chandrinos, and C. D. Spyropoulos. An experimental comparison of naive bayesian and keyword-based anti-spam filtering with personal e-mail messages. In *Proceedings of the 23rd annual international ACM SIGIR conference on Research and development in information retrieval*, SIGIR '00, pages 160–167, New York, NY, USA, 2000. ACM.

[5] P. Bennet. *Assessing the Calibration of Naive Bayes' Posterior Estimates*. In Technical Report CMU-CS-00-155. School of Computer Science, Carnegie Mellon University, 2000.

[6] B. H. Bloom. Space/time trade-offs in hash coding with allowable errors. *Commun. ACM*, 13:422–426, July 1970.

[7] J. Gama. Iterative naive bayes. In *Proceedings of the Second International Conference on Discovery Science*, DS '99, pages 80–91, London, UK, UK, 1999. Springer-Verlag.

[8] J. Gama and G. Castillo. Adaptive bayes. In F. Garijo, J. Riquelme, and M. Toro, editors, *Advances in Artificial Intelligence, IBERAMIA 2002*, volume 2527 of *Lecture Notes in Computer Science*, pages 765–774. Springer Berlin / Heidelberg, 2002.

[9] J. a. B. Gomes, E. Menasalvas, and P. A. C. Sousa. Learning recurring concepts from data streams with a context-aware ensemble. In *Proceedings of the 2011 ACM Symposium on Applied Computing*, SAC '11, pages 994–999, New York, NY, USA, 2011. ACM.

[10] C. Green and P. Edwards. Using machine learning to enhance software tools for internet information management. In *Proceedings of the AAAI Workshop on Internet-based Information Systems*, pages 48–55, 1996.

[11] V. Holub, T. Parsons, P. O'Sullivan, and J. Murphy. Run-time correlation engine for system monitoring and testing. In *ICAC-INDST '09: Proceedings of the 6th international conference industry session on Autonomic computing and communications industry session*, pages 9–18, New York, NY, USA, 2009. ACM.

[12] IBM. Websphere application server version 6.0.2, Sept. 2010. Available [online]: http://publib.boulder.ibm.com/infocenter/wasinfo/v6r0/index.jsp?topic=/org.eclipse.hyades.log.ui.doc.user/concepts/cesdb.htm.

[13] IBM. Ibm collaboration software, Sept. 2012. Available [online]: http://www-01.ibm.com/software/lotus.

[14] B. Jacob, R. Lanyon-Hogg, D. K. Nadgir, and A. F. Yassin. Ibm redbooks, 2004. Available [online]: www.redbooks.ibm.com/redbooks/pdfs/sg246635.pdf.

[15] S. J.Carolan. Introduction to cloud computing, June 2009. Available [online]: http://www.scribd.com/doc/17274860/Introduction-to-Cloud-Computing-Architecture.

[16] I. Katakis, G. Tsoumakas, and I. Vlahavas. Tracking recurring contexts using ensemble classifiers: an application to email filtering. *Knowl. Inf. Syst.*, 22(3):371–391, Mar. 2010.

[17] S. S. Keerthi, S. K. Shevade, C. Bhattacharyya, and K. R. K. Murthy. Improvements to platt's smo algorithm for svm classifier design. *Neural Comput.*, 13(3):637–649, Mar. 2001.

[18] R. Kohavi. Scaling up the accuracy of naive-bayes classifiers: a decision-tree hybrid. In *proceedings of the second international conference on knowledge discovery and data mining*, pages 202–207. AAAI Press, 1996.

[19] K. Lang. Newsweeder: Learning to filter netnews. In *in Proceedings of the 12th International Machine Learning Conference*, 1995.

[20] P. Leach, M. Mealling, and R. Salz. A Universally Unique IDentifier (UUID) URN Namespace, July 2005.

[21] N. Lothian. Classifier4j, Sept. 2012. Available [online]: http://classifier4j.sourceforge.net.

[22] C. Mesterharm and M. J. Pazzani. Active learning using on-line algorithms. In *Proceedings of the 17th ACM SIGKDD international conference on Knowledge discovery and data mining*, KDD '11, pages 850–858, New York, NY, USA, 2011. ACM.

[23] C. Mesterharm and M. J. Pazzani. UCI machine learning repository: Farm ads data set, Sept. 2012. Available [online]: http://archive.ics.uci.edu/ml/datasets/Farm+Ads.

[24] T. M. Mitchell. Chapter 6: Bayesian learning. In *Machine Learning*, pages 156–158, 177–179. McGraw-Hill, Inc., New York, USA, 1st edition, 1997.

[25] J. R. Quinlan. *C4.5: programs for machine learning.* Morgan Kaufmann Publishers Inc., San Francisco, CA, USA, 1993.

[26] M. Sahami, S. Dumais, D. Heckerman, and E. Horvitz. A bayesian approach to filtering junk E-mail. In *Learning for Text Categorization: Papers from the 1998 Workshop*, Madison, Wisconsin, 1998. AAAI Technical Report WS-98-05.

[27] F. Sebastiani. Machine learning in automated text categorization. *ACM Comput. Surv.*, 34(1):1–47, Mar. 2002.

[28] B. Sriram, D. Fuhry, E. Demir, H. Ferhatosmanoglu, and M. Demirbas. Short text classification in twitter to improve information filtering. In *Proceedings of the 33rd international ACM SIGIR conference on Research and development in information retrieval*, SIGIR '10, pages 841–842, New York, NY, USA, 2010. ACM.

[29] W. N. Street and Y. Kim. A streaming ensemble algorithm (sea) for large-scale classification. In *Proceedings of the seventh ACM SIGKDD international conference on Knowledge discovery and data mining*, KDD '01, pages 377–382, New York, NY, USA, 2001. ACM.

[30] M. Wang, V. Holub, T. Parsons, J. Murphy, and P. O'Sullivan. Scalable run-time correlation engine for monitoring in a cloud computing environment. In *Proceedings of the 2010 17th IEEE International Conference and Workshops on the Engineering of Computer-Based Systems*, ECBS '10, pages 29–38, Washington, DC, USA, 2010. IEEE Computer Society.

[31] M. Wang, V. Holub, T. Parsons, P. O'Sullivan, and J. Murphy. Symptom matching for event streams. *IET Software*, 6(4):296–306, 2012.

[32] W. Xu, L. Huang, A. Fox, D. Patterson, and M. I. Jordan. Detecting large-scale system problems by mining console logs. In *Proceedings of the ACM SIGOPS 22nd symposium on Operating systems principles*, SOSP '09, pages 117–132, New York, NY, USA, 2009. ACM.

[33] D. Yuan, J. Zheng, S. Park, Y. Zhou, and S. Savage. Improving software diagnosability via log enhancement. In *Proceedings of the sixteenth international conference on Architectural support for programming languages and operating systems*, ASPLOS XVI, pages 3–14, New York, NY, USA, 2011. ACM.

On-Line Fair Allocations Based on Bottlenecks and Global Priorities

Yoel Zeldes
Department of Computer Science
The Hebrew University
91904 Jerusalem, Israel
yhotdog@gmail.com

Dror G. Feitelson
Department of Computer Science
The Hebrew University
91904 Jerusalem, Israel
feit@cs.huji.ac.il

ABSTRACT

System bottlenecks, namely those resources which are subjected to high contention, constrain system performance. Hence effective resource management should be done by focusing on the bottleneck resources and allocating them to the most deserving clients. It has been shown that for any combination of entitlements and requests a fair allocation of bottleneck resources can be found, using an off-line algorithm that is given full information in advance regarding the needs of each client. We extend this result to the on-line case with no prior information. To this end we introduce a simple greedy algorithm. In essence, when a scheduling decision needs to be made, this algorithm selects the client that has the largest minimal gap between its entitlement and its current allocation among all the bottleneck resources. Importantly, this algorithm takes a global view of the system, and assigns each client *a single priority* based on his usage of *all* the resources; this single priority is then used to make coordinated scheduling decisions on all the resources. Extensive simulations show that this algorithm achieves fair allocations according to the desired entitlements for a wide range of conditions, without using any prior information regarding resource requirements. It also follows shifting usage patterns, including situations where the bottlenecks change with time.

Categories and Subject Descriptors

C.2.3 [**COMPUTER-COMMUNICATION NETWORKS**]: Network Operations—*Network management*; D.4.1 [**OPERATING SYSTEMS**]: Process Management—*Scheduling*; K.6.2 [**MANAGEMENT OF COMPUTING AND INFORMATION SYSTEMS**]: Installation Management—*Pricing and resource allocation*

General Terms

Design, Management, Performance

Keywords

Fairness; Bottlenecks; Resource allocation; Entitlements; Online algorithm; Priority inversion

ICPE'13, April 21–24, 2013, Prague, Czech Republic.

1. INTRODUCTION

Fairness is a basic issue in scheduling and resource management, and there are many different definitions of what "fairness" means. We use the definition that a fair system is one which allocates resources according to each client's entitlement: a client that is entitled to 30% of the system should get 30% of the resources[1]. However, this definition is problematic, because clients may have uneven requirements. For example, if a client is entitled to 30% of the system, and he would like to use 70% of the network bandwidth but only 10% of the CPU cycles, what would be a fair allocation?

A system in which clients alternate between using multiple resources, e.g. the CPU, network, and disk, can be modeled as a queueing system. In queueing systems one finds that clients spend the majority of their time waiting in queue for a specific subset of the resources, namely the *bottleneck* resources. The degree to which clients make progress is constrained by the service they receive on these bottlenecks [14]. It has therefore been suggested that bottleneck resources are the most important resources in terms of scheduling and resource allocation [7, 5]. By controlling allocations of the bottlenecks, one controls allocations across the whole system.

Given the insight about focusing on bottlenecks, one can define a fair allocation as follows: *each client should be given his entitlement of the system bottleneck* [7, 5]. Thus if the network is the bottleneck, our example client will be cut beck and given only 30% of the bandwidth. But if the CPU turns out to be the bottleneck, and the network is not, then he can get all he wants, because his requirements on the bottleneck are lower than his entitlement.

This definition can be extend to the case where the system suffers from multiple bottlenecks by guaranteeing that each client receives his entitlement on *at least one* bottleneck. It has been shown that, for any combination of entitlements and requests, a fair allocation according to this definition exists [4]. However, the algorithm to find this allocation works off-line and assumes full information is available.

Our main contribution is to extend this result with an on-line algorithm, that *achieves a fair allocation without knowing each client's requirements in advance*. In fact, it doesn't even have to know which clients are going to be active and when. All it needs to know is their relative entitlements. It then serves the requests they make in an order that is based on their entitlements and the allocations they have received so far. Essentially, this is a simple greedy algorithm.

In the next section we provide more details about previous work and the required background. Section 3 then explains the details of the on-line algorithm. An evaluation using extensive simulations to

[1]Thus entitlements are similar to capacity or throughput guarantees in SLAs.

show that this algorithm indeed achieves the desired allocations is presented in Section 4. We conclude in Section 5.

2. BACKGROUND AND RELATED WORK

The two main considerations for scheduling algorithms are fairness and quality of service. In the context of interactive and real-time applications, quality of service (as reflected in response times and meeting deadlines) is paramount [8, 21]. We focus on fairness, which is more relevant to servers, especially in the context of clouds and virtualization where several servers are consolidated onto a common physical platform.

The notion of fairness is somewhat tricky to nail down. At its core lies a conflict between accounting, where we just count how much each client receives, and utilitarianism, where we consider the benefit derived from the allocated resources [20]. In the context of computing systems we typically do not know the benefits, so we are left with different forms of accounting. In particular, striving to satisfy given entitlements is known as "fair-share" scheduling. In this context, a major aspect of fairness is isolation between competing clients: guaranteeing a certain share of the resources to one client implies that this share will be received irrespective of the actions of other clients. For example, this may be important in real-time settings and to prevent certain denial-of-service attacks [15].

An example of a large-scale system based on such fairness considerations is the Intel NetBatch platform [23]. This is essentially a global enterprise grid system with hundreds of clusters and tens of thousands of machines in dozens of locations around the world. Intel business and engineering groups purchase computing resources according to their needs and make them available to other groups as part of the grid. However, each group retains ownership of their resources. This means that each group's jobs receive priority for running on their own machines: if no alternative resources are available, whatever is running on these machines will be suspended and the owner's jobs will run instead. Thus allocations reflect ownership, and by implication allocations reflect the entitlements as represented by the relative fraction of the resources that have been purchased by each group.

Many schemes have been devised to achieve fair-share scheduling. These include simple accounting where allocations are simply not made to clients who have achieved their entitlement [13], lottery scheduling where lottery tickets are distributed according to entitlements and the actual allocations are randomized [19], manipulation of a process's priority so as to nudge it towards achieving the desired allocation [6], and using the min-max principle to prioritize those clients who are lagging behind [17].

A whole class of fair allocation schemes is based on the notion of virtual time [22, 16]. This essentially means that time is counted at different rates for different clients, thereby leading to different but controlled allocations. A more sophisticated approach is to define fairness according to the divergence between what a client has received and what he was entitled to receive up to now [18]. This can also be used directly to prioritize scheduling decisions [8, 3].

All the above schemes are oblivious to the system state. There has been very little work considering scheduling decisions that are driven by resource contention. Unlike bottleneck-based scheduling (to be defined more fully below), this work is usually concerned more with preventing bottlenecks than with focusing on them. The idea is to prefer clients that make little use of contended resources, so as to reduce the danger of congestion [1, 11].

The schemes described above were typically applied to one specific resource, most commonly the CPU or the network. This leaves the question of how to control the joint allocation of multiple resources. The simplest approach is asset fairness. This means that we sum up the usage of the different resources, and ensure that this sum is proportional to the entitlement. However, this is inefficient in the sense that someone who is the sole client of an unpopular resource will be held back due to contention by others for a resource in which he is not interested.

Another approach is dominant-resource fairness [10, 9]. In this scheme each client is represented by the resource he wants the most of, which is called his dominant resource. The allocations of the dominant resources are then made proportional to the entitlements. A simple algorithm that achieves this has been shown, which is also "strategy proof". This means that a client cannot increase his allocation and his throughput by increasing his demands. However, note that as long as fairness (however one may decide to define it) is maintained, there is no way that the system can distinguish between a client that genuinely requires more resources and one that is trying to game the system. In particular, if a client can cause the system to give him a larger allocation, but this larger allocation is still considered fair, this is not necessarily a problem. It is therefore not clear that the property of strategy proofness as defined above is indeed appropriate.

Asset fairness and dominant-resource fairness are oblivious to the system state. The alternative is to monitor the system to identify the current bottlenecks, which may change from time to time [7, 5]. In bottleneck-based fairness an allocation is considered fair if all clients do not have justified grounds for complaining. This means that either they *receive all they want*, or else they *receive their entitlement on at least one bottleneck*. The claim that they cannot complain about this is based on the fact that bottlenecks are by definition contended resources, so giving more would have to come at the expense of other clients who also have their entitlements[2]. It has been shown that an allocation based on this principle is guaranteed to exist [4, 12]. Note that asset fairness and dominant resource fairness do not possess the "no justified complaints" property: in those schemes, a client may be prevented from receiving resources that could have been allocated without hurting anyone else.

In order to sharpen the differences between the above approaches, we suggest the fruit salad metaphor. Assume a fruit salad buffet with a bowl of diced apples, a bowl of diced oranges, a bowl of cherries, etc. Each diner is invited to create his or her own mix. Asset fairness then means that each diner gets a small bowl of the same size and can fill it in whatever way he chooses. This is intuitively fair but inefficient: if ten people like cherries, but only one each like apples, oranges, and bananas, why not let these three take everything that is there? Dominant resource fairness means each diner gets the same amount of his favorite fruit. This suffers similar inefficiency as asset fairness. The bottleneck-based approach means that each diner gets the same amount from *one of the fruits that ran out*, thus emphasizing the importance of those fruit which are in high demand, while explicitly allowing you more of fruit that nobody else wants. This analogy also shows the problem in finding a fair solution according to this definition: when you are making the allocations, how do you know which fruit are going to run out?

Given a scheduling scheme, the remaining question is how to apply it. The simplest approach is to schedule each resource individually. However, this may lead to undesirable results. For example, a client may use less than his entitlement of one resource, and more than his entitlement of another. Assume the first resource is a bottleneck, while the second is not. This client should therefore be prioritized in order to provide it with its due entitlement of the

[2]Note that a resource may become a bottleneck even if it is used by only one client. But in that case this client is fully utilizing the resource, so there is no more to allocate.

bottleneck resource. But when it is waiting for the second, non-bottleneck resource, it will receive a low priority. This will cause delays that affect the fairness as reflected by using the bottleneck resource. The solution to such mixed priorities is to use a global priority across the whole system. This global priority is *computed* based on the bottleneck resources, but it is *applied* to *all* the resources.

A building block that can be used to implement the idea of bottleneck-based fairness is the RSVT scheduler, which is a modular scheduler that can be "glued" to different resources to provide virtual-time-based prioritization [2]. This needs to be combined with a centralized monitoring facility that identifies bottleneck resources, and sets the global priorities. In this paper we fill in the algorithmic details and simulate how such an implementation may be expected to operate when faced with various scenarios of contending clients with different requirements. The actual implementation, and an evaluation of the overheads and performance characteristics, are left for future work.

A basic assumption of all the above works dealing with the allocation of multiple resources is that clients have constant resource usage profiles. For example, a client may always use twice as much disk as CPU, so if we allocate 30% of the CPU he will also use 60% of the disk bandwidth, and if we limit his disk usage to 20% we are effectively also limiting his CPU usage to 10%. This assumption is reasonable in cases where clients are performing essentially the same actions all the time, for example serving incoming requests that all have the same nature. However, our on-line algorithm does not rely on such persistence. Rather, it just tracks the cumulative usage of different resources by each client, regardless of the precise patterns in which requests are made. In particular, clients are free to change their behavior, and the algorithm will follow them and even identify changes in the set of bottlenecks.

3. THE ON-LINE ALGORITHM

Consider a set of clients with different entitlements, each with his unique resource usage profile. We know that a bottleneck-based fair allocation exists for any combination of entitlements and requirements. But the proofs given in [4, 12] are non-trivial, use complete information about the requirements in order to find the solution, and require the resource usage profile to be constant. The difficulty stems from the fact that we do not know in advance which resources will turn out to be bottlenecks. Moreover, it turns out that there may be multiple solutions, and even different solutions in which different resources become the bottlenecks.

An interesting question is therefore whether the principles of bottleneck-based fairness can be distilled into an on-line algorithm that will maintain a fair allocation despite not knowing what the requests are in advance. Obviously such an algorithm also does not know the bottlenecks in advance, but we assume it can discover them when they materialize, by monitoring the utilization of each resource. Thus it can also adjust if the bottlenecks change with time.

Our algorithm is based on the following two principles:

1. The "no justified complaints" condition requires that clients receive their entitlement on at least one bottleneck resource. This leads to the idea that a client's priority should be proportional to the divergence between his entitlement and his actual consumption, on the bottleneck resource where this divergence is *minimal*. The reason is that as we allocate more resources and enable him to run, this is where the gap will be closed first. A process that suffers from a larger minimal gap should be given priority in order to enable it to catch up and achieve its entitlement.
2. Given that there may be several bottlenecks and any one of them may be used to satisfy a client, a global viewpoint is required. Thus each client will be assigned a global, system-wide priority value, and scheduling decisions on all the different resources will be made based on these system-wide priorities.

While the algorithm based on these principles is simple (as shown below), it is not self evident. In particular, basing priorities on the minimal *gap* contradicts the commonly used max-min approach, which bases priorities on the minimal *consumption*, which is equivalent to using the *maximal* gap. But in the multi-resource scenario with fixed usage profiles this is problematic, because boosting the usage of the least-used resource also increases the usage of the heavily used resources, which may be impossible (or at least undesirable) if they too are bottlenecks.

Importantly, using global prioritization implies that scheduling on the different resources will be coordinated. This is in stark contrast with the prevailing methodology used today, where each resource has its own scheduler, and the scheduling is done based on a myopic view of each individual resource. As a result, we prevent situations in which a scheduler responsible for some resource that is not contended decides to run client A and let client B wait in the queue, due to some local efficiency consideration, when A is actually receiving all he wants while B is not receiving his due allocation on the system bottlenecks.

To formalize these ideas we need some notation. Let there be N clients and M resources. Client i has entitlement e_i, where $\sum_i e_i = 1$. Ideally, the fraction of the system that client i was entitled to receive up to time t is simply $a_i(t) = e_i t$. For each resource j, his consumption of resource j up to time t will be denoted $c_i(j,t)$. Both allocation and consumption are expressed in the same units, namely time using the resource, and these units apply equally to all resources (as opposed to units like bandwidth which are specific to a certain resource). Let $J(t)$ denote the set of resources that are bottlenecks at time t. Based on the above considerations, we calculate the global priority of client i as the minimal difference between the ideal allocation and the consumption, across bottleneck resources:

$$pri_i(t) = \min_{j \in J(t)} \{a_i(t) - c_i(j,t)\}$$

The queue of each resource will be sorted according to these global priorities, and the highest priority client will be selected. If several clients have the same global priority, the local scheduler may use its local considerations to break the tie. In either case, the selection is always for a limited time: a quantum of CPU time, the sending of a packet, the reading of a disk block, etc. This causes $c_i(j,t)$ to grow for the selected client, while at the same time $a_i(t)$ grows for all clients (but by less). As a result another client will most probably be selected the next time around, and if all the clients have the same entitlements our algorithm will produce simple round-robin scheduling.

But what happens if there are no bottlenecks? In this case we prioritize the different processes according to the difference between their allocation and their total consumption on all resources, namely we revert to asset fairness:

$$pri_i(t) = a_i(t) - \sum_j c_i(j,t)$$

Note, however, that this only affects the scheduling order and does not affect the allocations. When there are no bottlenecks there is no real contention and therefore all requests will be granted.

The above formalization cannot be used as described because it assumes that all clients arrived at the system at time 0 and have been active continuously ever since. In general, of course, this is not the case. Clients may come and leave at different times, and may suspend waiting on some external event. Therefore the following adjustments must be made.

First, at each time t only active clients are considered: clients who have terminated or have not arrived yet (or are suspended) will be excluded. Thus the entitlements need to be renormalized. Denoting the set of active clients by A, the entitlements used to calculate the priorities will be $e'_i = e_i / \sum_{i \in A} e_i$.

Second, if a new client enters the system at time t_0, his entitlement will be computed as $a_i(t) = e'_i(t - t_0)$. As e'_i may change with time, this is actually computed piecemeal by summing over intervals where e'_i is constant.

Third, handling clients who become suspended or do not use their full allocation is done similarly to RSVT [2]. The problem is that clients may divide their time across the resources in very different ways. Thus we may encounter situations where a client predominantly uses a resource that nobody else is interested in. As a result, his consumption may outstrip his entitlement (and there is no reason to prevent this, as otherwise the resource would stay idle). But if later some other client starts using this resource, and the resource becomes a bottleneck, then the first client will have a hugely negative priority and will be locked out for a long time. This should be prevented.

Conversely, consider a resource that nobody uses to a great degree. This resource is mostly idle, and for all clients their consumption will lag way below their allocation. Then, if the client with the highest entitlement starts to use this resource extensively and turns it into a bottleneck, and assuming this is the only bottleneck, he will lock out all others, because his allocation had outgrown his consumption more than for all others. This too should be prevented.

The above scenarios can be summarized as follows. Ideally, we want to make allocations based on entitlements, and have the consumption follow these allocations. But if the consumption does not follow the allocation, and becomes either too big or too small relative to the allocation, we need to bound this difference [22]. The way to do this is simple: instead of accumulating allocations and consumptions since the client arrived, we only consider a window of the last T seconds. In such a limited window allocations and consumptions reflect the most recent entitlements and actual use, but they cannot diverge by more than T.

An important concern for online algorithms is their overhead. Our algorithm is centralized in the sense that it utilizes global information for its prioritization. Hence it faces the danger of not scaling for large systems. However, we note that only the accounting needs to be performed on each and every dispatch decision, and this can be incorporated into the mechanism performing the dispatch — for example, the CPU scheduler or the mechanisms controlling packet sending and disk I/O. The prioritization which uses this information can be limited to a desired granularity, e.g. once every so many seconds, as we indeed do in the simulations.

4. SIMULATION RESULTS

In this section we describe some of our simulation results, focusing on those that best illustrate various features of the scheduling algorithm.

4.1 The Simulator

To assess the behavior of the algorithm presented above we wrote an event-based simulator. This simulates the evolution of a system supporting several processes (or a server with several virtual machines). The processes represent the clients. Each process runs a "program", which specifies its resource usage. The programs are composed of multiple iterative phases, where each iteration consists of using different resources in the system one after the other. Use of a resource is described by a tuple of the form $\langle res,\ mean,\ width,\ dist \rangle$. res is the resource being used. The other three parameters describe the distribution of service times when the resource is used ($dist$ can be UNIForm, EXPonential, or NORMal; $mean$ is obviously the mean, and $width$ is half the range for UNIF or the standard deviation for NORM). As each use of each resource is simulated, the service time for this instance is selected at random from the specified distribution. We typically use the normal distribution, because the service times are expected to be influenced by many independent factors.

```
# Priority:
34
# Requests:
loop 1
CPU_CS 2 0 NORM
loop 40
DISK_CS 0.1 0.05 NORM
CPU_CS 0.1 0.05 NORM
```

Figure 1: *Example process program in the simulation.*

An example program is shown in Fig. 1. This process has a relative priority (entitlement) of 34. Its program consists of using the CPU for precisely 2 seconds, and then performing 40 iterations of using the disk for 0.1 seconds and then the CPU for an additional 0.1 seconds, both with a standard deviation of 0.05 seconds.

Each process that wishes to use a certain resource is passed on to that resource's scheduler. Each scheduler keeps track of every process's usage of its resource in the last 3 seconds (This is the T mentioned above). A global prioritization agent collects the prioritization data from the schedulers of bottleneck resources so as to create the global priorities. Bottleneck resources are those that have been active for more than 90% of the time in the last 3 seconds (these are all configurable parameters). Thus the identification of a bottleneck will lag 3 seconds behind the time it actually became a bottleneck.

Allocations can either be continuous, or, if the resource name ends with _CS (allows context switching), they are done in quanta of 0.1 seconds. This may represent a scheduling quantum, the sending of a packet, or the reading of a disk block. Note, however, that the resource names and the service times should not be taken too literally. Obviously in many real systems CPU quanta are much shorter than disk access times. We use "CPU", "disk", and "network" to easily refer to the resources, but they could just as well be abstracted as "R1", "R2", etc. Our goal in the simulations is to probe the behavior of the algorithm under diverse conditions, not to approximate specific benchmarks.

The simulator as described has the limitation that each process uses one resource at a time. This is not completely realistic, as a process may spawn multiple parallel threads on a multiprocessor system, and asynchronous I/O can be done in parallel to computation. Note, however, that this is a limitation of only the simulator and not of the algorithm itself.

4.2 Simple Experiments

The first experiments are meant to demonstrate that the simulator works as expected on simple scenarios with a single resource, and to explain the way we present the simulation results.

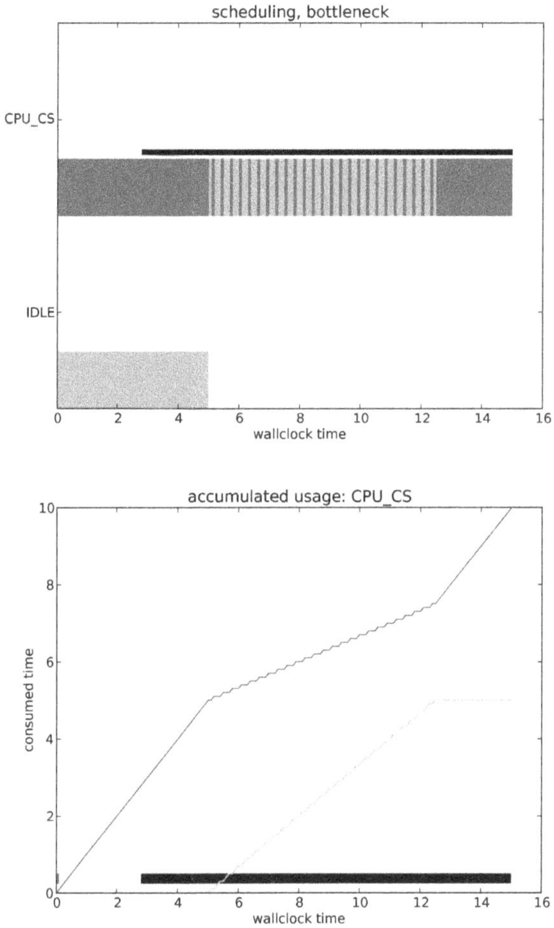

Figure 2: *An example of two processes with different priorities contending for a resource. Service times are deterministic.*

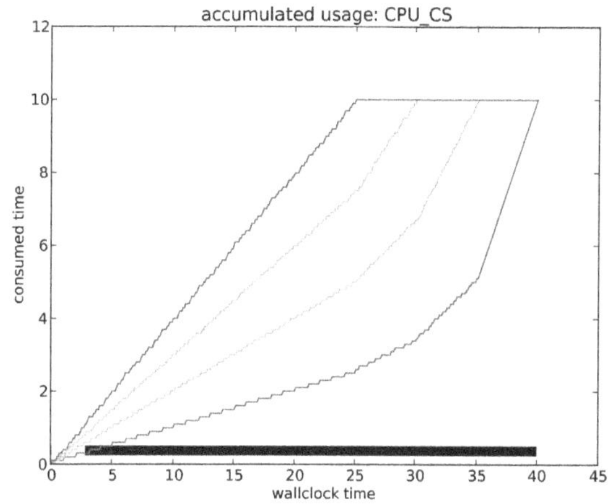

Figure 3: *An example of four processes with different priorities contending for a resource. Service times are deterministic.*

Each experiment involves one or more processes, using one or more resources. In the graphs, the horizontal axis represents wall-clock time. The first graph for each experiment is actually a Gnatt chart, with a lane for each resource. This lane is color-coded according to which process is using the resource at each instant; no coloring indicates that the resource is idle. A black line segment above the lane identifies those intervals in which the resource is considered to be a bottleneck. The second graph shows the cumulative resource usage by each process. Each process is represented by a line, with the same color as in the Gnatt chart. If a process is continuously active, on whatever resource, this will be a straight line with slope 1. When a process waits for a resource, this is represented by a horizontal line segment. When two or more processes share a resource the effect is to produce a slope smaller than 1, but this is actually a sequence of small steps. We occasionally also look at the cumulative usage of a select resource, rather than all resources together.

A simple example is shown in Fig. 2, involving two processes. The red process has a relative priority of 33% and wants to use the CPU for 10 seconds, while the blue one has 67% and wants 5 seconds. Initially only the red process runs, and gets full use of the CPU (note that we use the "idle" resource to delay the blue process). 5 seconds later, when the blue process starts, their relative shares become governed by their priorities. This is achieved by giving the blue process two quanta each time, while the red process only receives one. As the blue process has a higher priority it terminates first, again leaving the CPU exclusively to the red process. A generalization to four processes that start together with priorities of 10%, 20%, 30%, and 40% is shown in Fig. 3. As they all want to run for 10 seconds, the one with the highest allocation terminates first, and then the others in decreasing order. As each one terminates, the CPU is divided among the remaining processes according to their designated shares.

Note the smooth transition as the blue process starts running in Fig. 2. This reflects two features of the scheduling algorithm. First, when the red process runs alone, its effective relative priority is 100% because there is no other process in the system. As it indeed receives full use of the CPU, it does not accumulate any lag between its entitlement and its consumption. Second, when the blue process starts, it is initialized with zero entitlement and consumption. As a result both processes start from an equal footing, and immediately receive allocations according to their relative entitlements.

4.3 Properties of the Scheduling Algorithm

Fig. 4 shows how the algorithm adjusts when different processes use different resources. Here we have 3 processes, that repeatedly (5 times) use the CPU for 0.1 second and then some other resource for 10 seconds. The blue process has an entitlement of 70%, and uses the network. The green and red processes have entitlements of 20% and 10%, respectively, and use the disk. As we see, the blue process actually receives nearly 100% of the network, despite having an entitlement of only 70%, because no other process requests it. Green and red receive approximately 67% and 33% of the disk, which are also much higher than their entitlements, but maintain the correct ratio. The CPU is hardly used, and therefore does not affect these allocations.

Note that due to the randomization used in the simulations the total time in any specific run does not necessarily sum to exactly 50 seconds of resource use. However, as shown in the bottom plot, the average of multiple runs does come out right. The error bars show the distribution of values observed in the individual runs. Importantly, in the initial part of the simulation, when all processes are active, there are no such variations. In the sequel, we will typi-

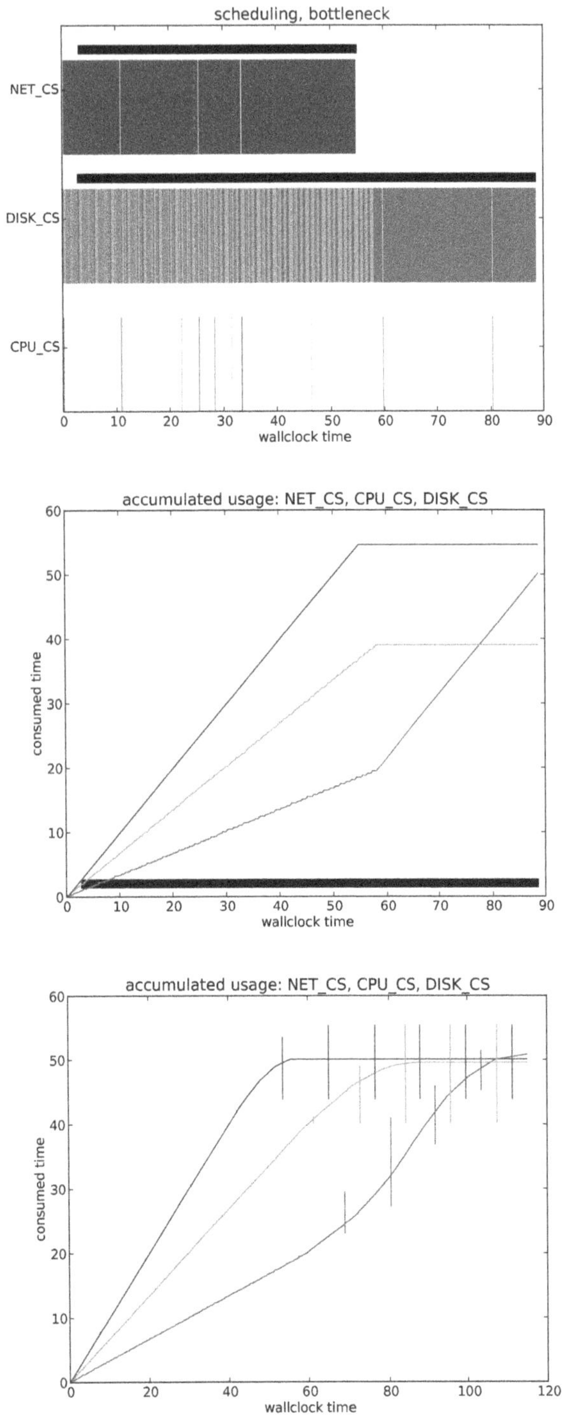

Figure 4: *An example of three processes using different resources, and thus enjoying a higher effective entitlement. Service times are randomized, with the top two graphs showing a specific run while the bottom one shows an average of many runs.*

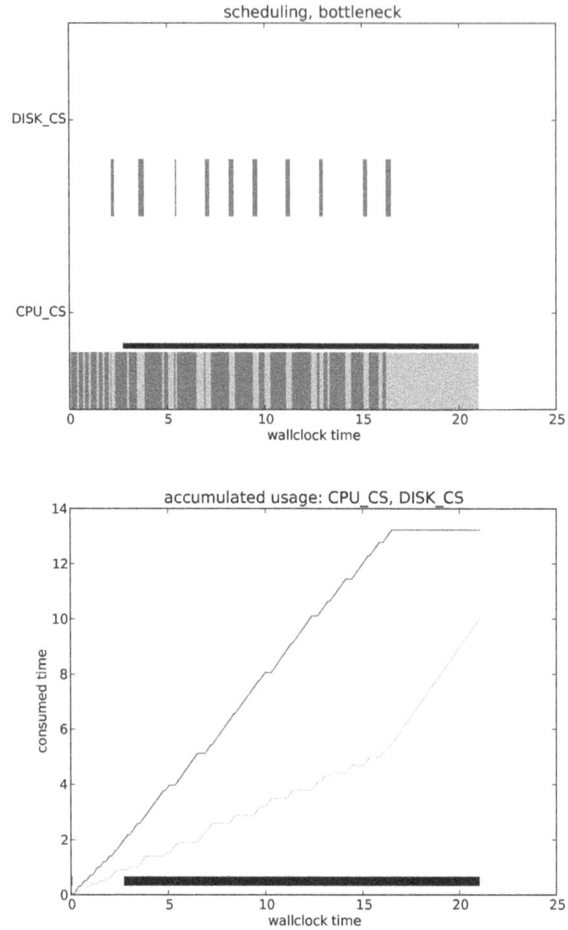

Figure 5: *Example where the grace period increases the ideal allocation also when a resource is not used. Service times are randomized.*

cally use averages and error bars to show the results of randomized experiments.

An important feature of the algorithm is the way it handles processes that skip from one resource to another, or become suspended. On one hand, we want continuity: if a process does not use a resource for a short time, its allocation should nevertheless continue to grow, so as to be available once it requests to use this resource again. On the other hand, we do not want the allocation to grow excessively relative to the consumption, so as to avoid situations that give the process an unbeatable priority that will allow it to lock out all other processes.

Following the RSVT scheduler [2] we define a grace period of 1 second during which the allocation continues to grow. The effect of this grace period is demonstrated in the following two experiments. In both experiments, the red process represents the user and has an entitlement of 70%, and the blue process represents some background activity with an entitlement of only 30%. In the first experiment (Fig. 5) the red process alternates between using the CPU for 2 seconds and then accessing the disk for 0.5 second. This is shorter than the grace period, so the allocation of the CPU continues to rise, and when he returns he gets exclusive access for a short while in order to make up for the deficit in consumption. In the other experiment (Fig. 6) the red process uses the CPU for 2

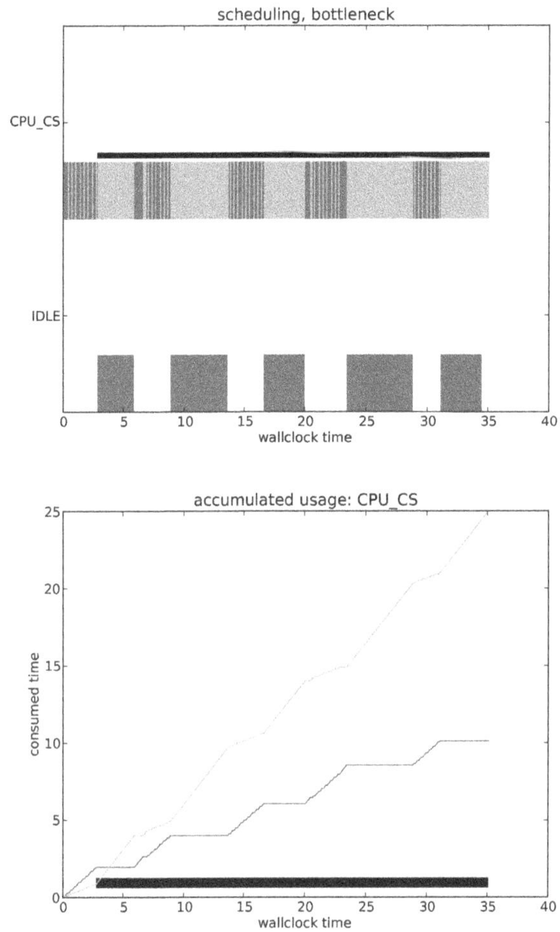

Figure 6: *Example where the grace period is not long enough to increase the ideal allocation also when a resource is not used. Service times are randomized.*

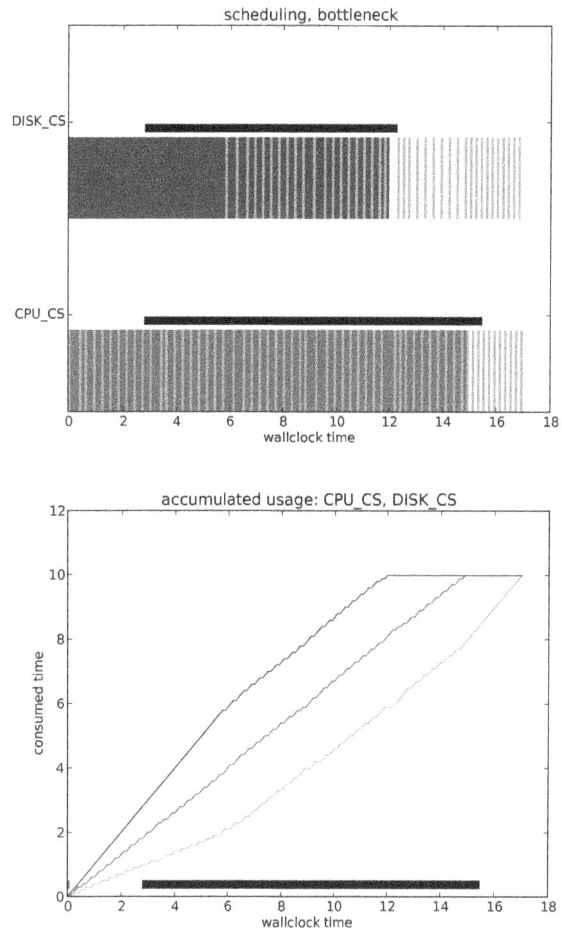

Figure 7: *An example of three processes with different priorities contending for two resources. Service times are deterministic.*

seconds and then suspends waiting for user input for 4 seconds on average. This is longer than the grace period, so when he returns he has to share the CPU with the blue process. To enable this to be seen clearly, we do not average multiple random runs but rather show a single run.

4.4 Further Examples of Controlled Allocations

A slightly more involved example is shown in Fig. 7. Here we have 3 processes with different priorities contending for the use of two resources:

- Red, with priority 40%, wants 10 seconds on the CPU
- Blue, with priority 40%, wants 10 seconds on the disk
- Green, with priority 20%, wants 2 seconds on the CPU and then 8 seconds on the CPU and disk alternately

The results of the run are as follows. Initially blue uses the disk with no contention, while red and green share the CPU according to their relative priorities. Then green starts using the disk. As a result we find that now all three processes make progress at the same overall rate. However, on each resource, green gets half what the other process gets. Thus red and blue receive their relative entitlements of 67% on their resources, and green receives his relative entitlement of 33% on each of the resources, for a total of 67%.

Fig. 8 shows what happens when the priorities are changed. In the first plot, all three processes have equal priorities of 33%. As a result the green process receives the same allocation as the others on each resource. But due to the fact that it uses both resources, its total rate of progress becomes double that of the others. This highlights the difference between asset fairness and bottleneck-based fairness: In bottleneck-based fairness, we are not bothered by the total allocations, but only by the allocations on each bottleneck.

The second plot shows what happens when the green process has a priority of 50, and the other two 25. We would then expect the green process to make progress at twice the rate on each resource, and at four times the rate in total because it uses both resources. However, this does not happen. The explanation is that when green alternates between the two resources, it leaves each of them idle half of the time. The other processes pick up the slack rather than leaving them idle. This illustrates the increased efficiency relative to dominant resource fairness.

Interestingly, these effects are observed only when the service times are all deterministic and equal, i.e. when the standard deviations in the distributions are 0 (and indeed this is the configuration shown in the graphs). When this is the case, the processes proceed

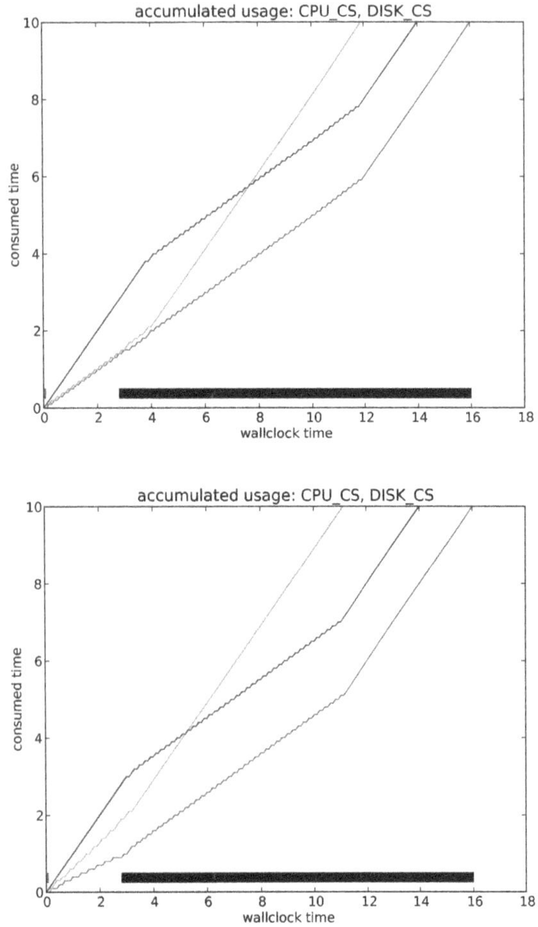

Figure 8: *The same scenario as in Fig. 7 but with different relative priorities. Service times are deterministic.*

in lockstep and the different allocations match perfectly. But if randomization is introduced, the synchronization is broken. The green is then not always available at the right instant to receive his share of each resource, but blue and red, which only use one resource each, are always there waiting. As a result green is cut back, and makes progress at about the same rate as the other two.

We now turn to a couple of more complex scenarios. A rather extreme example is shown in Fig. 9. Here four processes execute the same type of iterations 3 times each. The iterations are composed of 0.5 second network, 1 second CPU, and 2 seconds disk (as may happen when serving requests). The differences are in their priorities: purple has a priority of 97%, and the other three 1% each. Obviously purple's priority is much higher than the others, and indeed we find that it never has to wait for a resource (except to wait for the end of the current quantum) and makes the most progress. As a result it manages to finish its 3 iterations way ahead of the others. They are then left to contend with each other, and enjoy similar performance on average. However, each run may actually be somewhat different. Specifically, the in run depicted in the top two graphs of Fig. 9, blue was lucky and required slightly less CPU than green and red in the first iteration. As a result it managed to get to the disk first, and continued to run out-of-phase with the other two and to make better progress. Nevertheless, each process still

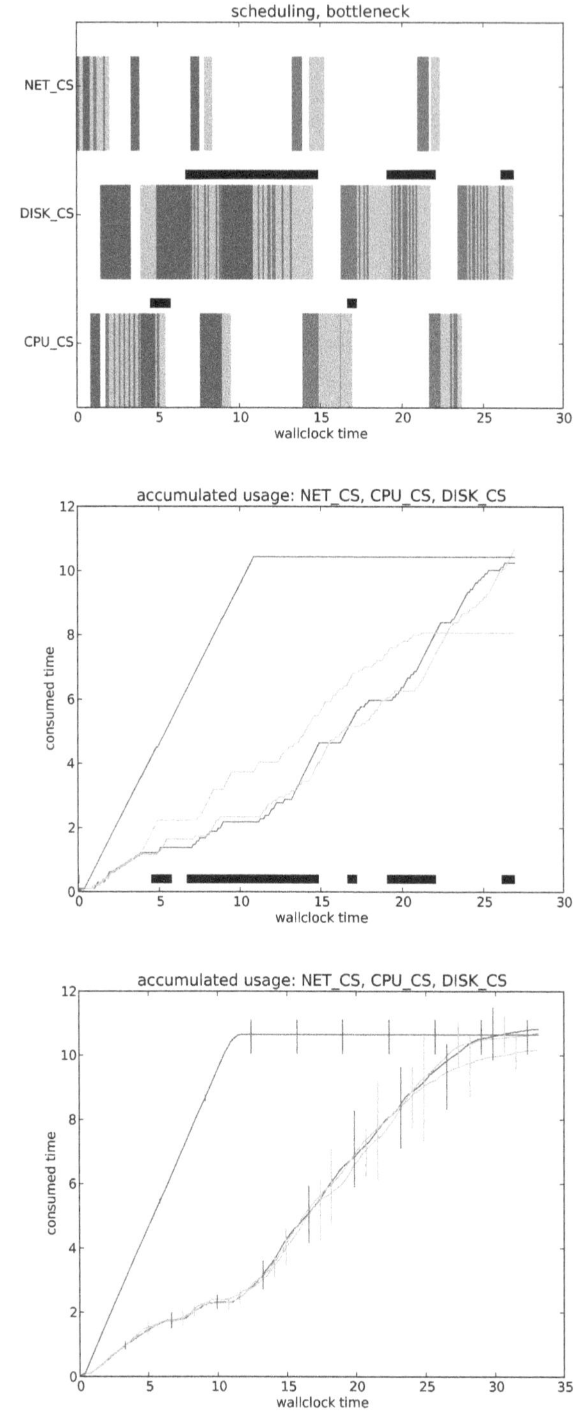

Figure 9: *An example of four processes with widely different priorities. Service times are randomized, with the top two graphs showing a specific run while the bottom one shows an average of many runs.*

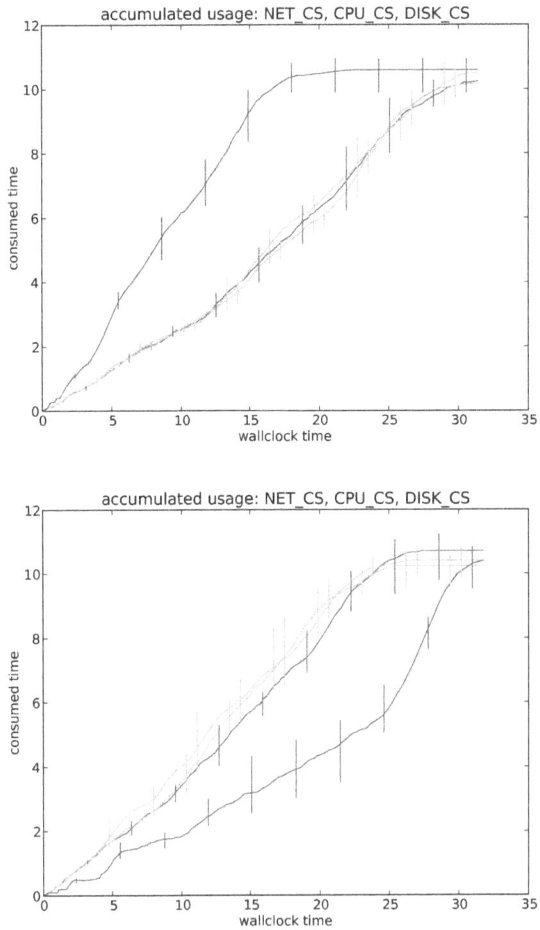

Figure 10: *Variations on the experiment shown in Fig. 9. Service times are randomized.*

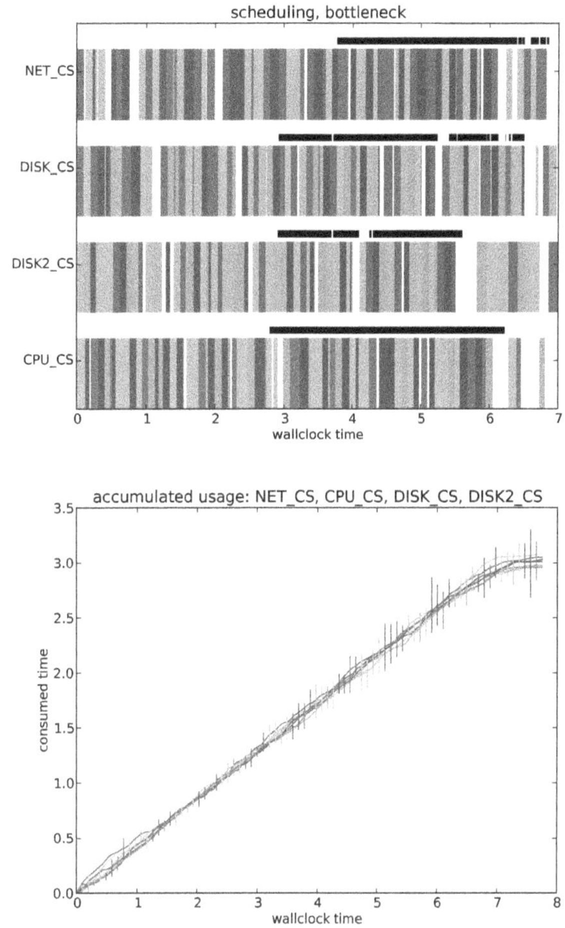

Figure 11: *An example where several different solutions are possible, and the algorithm gravitates towards the symmetrical solution. Service times are randomized.*

receives at least his entitlement on the bottleneck, in this case the disk.

Fig. 10 shows two variations on this experiment. In the first the priorities are not so extreme: purple is only 40%, and the other three are 20% each. In the second variation the priorities are inverted: purple is down to 10%, and the other three are 30% each. In these variations individual runs may be even more noisy than in the original experiment, but on average they are pretty clear. In the second variation purple appears to achieve more than $\frac{1}{3}$ of the others, because they sometimes need to wait for each other to release some resource, and thus leave another resource idle. This can be seen more clearly in longer runs, where the noise becomes smaller in relative terms, and therefore individual runs are very close to the average of many runs.

As mentioned in Section 3, there exist configurations in which several off-line solutions are in principle possible. Moreover, different solutions may involve different sets of bottleneck resources. One example of such a situation is the following. We have 8 processes and 4 resources. All the processes have equal entitlements, and they are arranged in pairs. Each pair uses 3 resources iteratively, for 0.1 seconds each time. The pattern is completely symmetric: if we arrange the resources in a circle, then each pair uses a different set of 3 consecutive resources on this circle. Possible solutions that abide by the no justified complaints criterion are as follows:

- All resources become bottlenecks, and all processes continuously use their 3 resources for $\frac{1}{6}$ of the time each.
- Two of the pairs run for 25% of the time on each of their resources, while the other two run for 37.5% of the time (this is why we need pairs: a single process cannot utilize 3 resources at 37.5% each). As a result two resources become bottlenecks, while the other two are only utilized 87.5% of the time. There are six such solutions with different sets of bottleneck resources.

Simulating this configuration with deterministic service times leads to the symmetric solution, with all resources utilized 100% of the time. The result of a randomized simulation is shown in Fig. 11. This is slightly less clear-cut than the deterministic version, and none of the resources maintain their bottleneck status without some gaps and idle periods. However, when observing the progress made by the 8 processes, they are found to achieve largely the same rate.

4.5 Allocations Based on Bottleneck Usage

In the previous examples the bottlenecks did not play a dominant role in the scheduling. However, there are situations where the

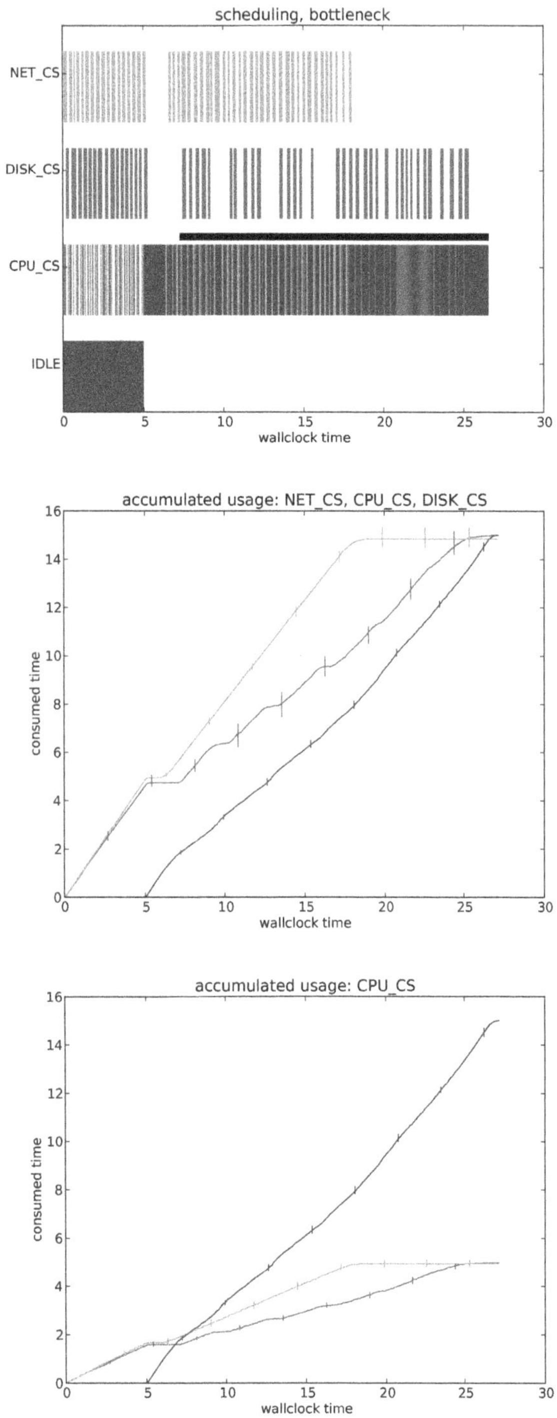

Figure 12: *An example where allocations change once a resource becomes a bottleneck. Service times are randomized.*

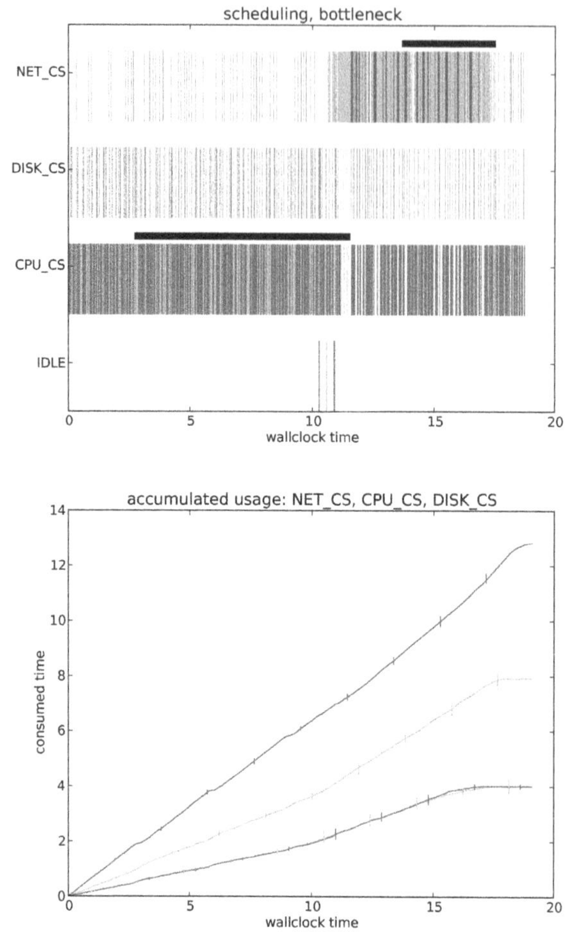

Figure 13: *An example where the focus shifts from one bottleneck to another. Service times are randomized.*

bottleneck resource actually has a decisive effect. An example is shown in Fig. 12. We start with 2 processes, with entitlements 33 and 67. The red process uses 1 second of CPU per 2 seconds of disk. The green process uses 1 second of CPU per 2 seconds of network. These requirements to not stress the system, so each runs continuously, and all 3 resources are 67% utilized. As there is no contention, the entitlement do not influence the allocations. After 5 seconds a third process (blue) becomes active. This process has entitlement 100, and only uses the CPU. The CPU then becomes a bottleneck. As a result we allocate 50% of the CPU (on average) to the blue process, 17% to the red process, and 33% to the green process. For the red process this is half what he got before, and as a result the usage of the disk also drops to half what it was. For the green process it happens to be exactly what he received before the blue process arrived, so nothing much changes.

Another example shows how the algorithm manages to track a shift from one bottleneck to another (Fig. 13). There are 4 processes, with entitlements of 50, 25, 12.5, and 12.5. The first uses the CPU 80% of its time, and the disk and network 10% each. The other three initially iterate between the CPU and disk, but then add the network too, for 70% of each iteration. As a result the CPU is the bottleneck in the first part of the simulation, but the network becomes the bottleneck in the later part, leading to a change in relative allocations.

Our last example is derived from the recent paper by Ghodsi et al. about using dominant resource fairness in network settings [9]. In that paper they suggest a test case where two processes each predominantly uses a different resource, and speculate that bottleneck-based scheduling would lead to strong oscillations as the system tries to satisfy them in turn. They further conjecture that a third pro-

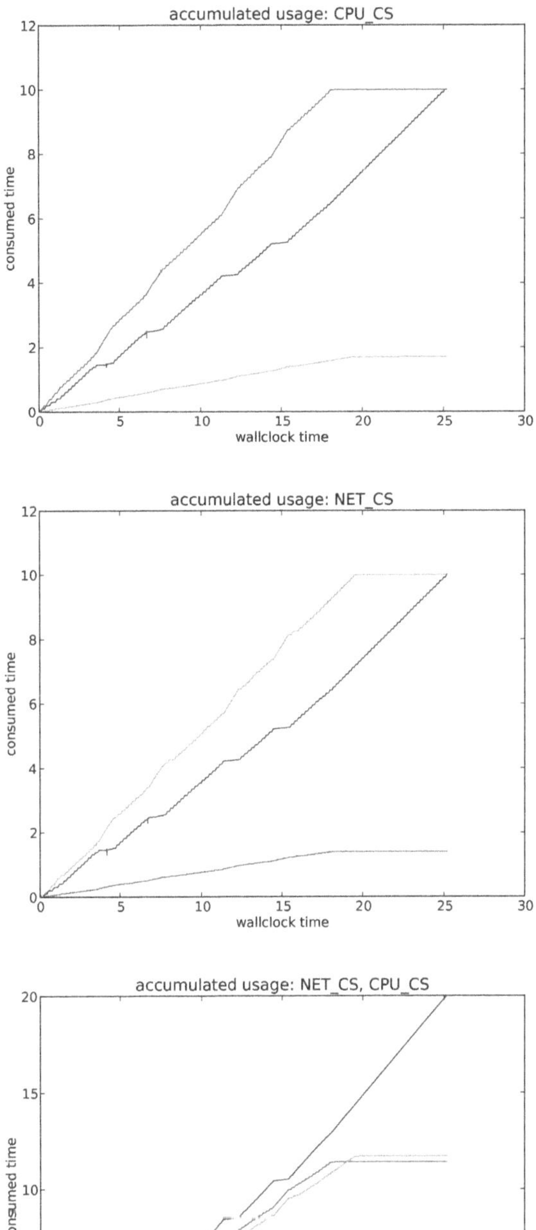

Figure 14: *Results for processes with complementary requirements. Service times are randomized.*

cess that wants both resources would not receive its due allocation at all. We implemented this scenario, with one process requesting 0.014 seconds on the network for each 0.1 second quantum on the CPU, a second requesting 0.1 seconds of the network for 0.017 of the CPU, the a third requesting equal use (in 0.1 second quanta) of both. All three processes have equal entitlements.

The simulation results show that the concerns were unfounded (Fig. 14). As expected, both the CPU and network become bottlenecks, and retain this status until one of the processes terminates. While some oscillations do occur, they are small. Moreover, these oscillations involve the third process, which tends to get ahead of the other two rather then not receiving its due entitlement. The scheduler responds by occasionally stopping it momentarily to allow the other two to catch up. The erroneous prediction seems to follow from a mindset where at each instant only one of the first two processes is active. But in reality their usage profiles are actually complementary, and the system quickly falls into a pattern where this is exploited.

5. CONCLUSIONS

We define fair allocations of resources based on bottlenecks, and specifically, require that each process receive at least its entitlement on at least one bottleneck resource. Previous work showed that for any combination of entitlements and requirements it is possible to find a fair allocation according to this definition. However, the proof was based on an off-line procedure that assumed full knowledge and a static configuration. We now augment this with a dynamic on-line algorithm that achieves a fair allocation without prior knowledge, and adjusts to changing conditions.

The algorithm itself is essentially a greedy algorithm. In a nutshell it can be described as follows:

1. Define a global (system-wide) priority order, and schedule processes according to this order on all resources.
2. The priority of a process is the minimum of its priorities on the different bottleneck resources.
3. The priority on each bottleneck resource is calculated based on the lag between what the process is entitled to receive and what it had actually received so far.

The main innovations in this algorithm are that it uses a global view, and that it focuses on the system bottlenecks. The prioritization of processes is system-wide: each process has a *single* priority, which is calculated based on *all* the bottlenecks, and is valid for *all* the resources. This produces coordinated scheduling decisions, thus preventing resource-specific schedulers from counteracting each other and causing priority inversions.

The algorithm provides the conceptual framework for an online bottleneck-based scheduling and allocation mechanism, that accommodates shifting usage patterns and provides allocations according to pre-defined entitlements. In future work we intend to incorporate these ideas into a working system, by using an RSVT scheduler [2] to control each resource, in conjunction with a monitoring facility that will identify the system bottlenecks. The monitoring will involve a recording of the periods during which resources are busy, and identifying those that are busy a large fraction of the time (e.g. above 90%). To reduce overheads, the calculation of priorities will be done at a fixed granularity, e.g. once a second. These priorities will then be used by all dispatch decisions until new priorities are computed.

Apart from the implementation, there is also more to be done regarding the algorithm itself. Group accounting (e.g. multiple threads in a process that share their entitlement and allocations, or asynchronous I/O leading to concurrent use of multiple resources by the same process) is a challenging and interesting issue. Further on, we would also like to investigate additional resources that might become bottlenecks, such as cache space, bus bandwidth, accelerators such as GPGPUs, and memory.

6. ACKNOWLEDGMENTS

This research was supported by the Israel Science Foundation (grant no. 28/09) and by an IBM faculty award.

7. REFERENCES

[1] Y. Amir, B. Awerbuch, A. Barak, R. S. Borgstrom, and A. Keren, "*An opportunity cost approach for job assignment in a scalable computing cluster*". *IEEE Trans. Parallel & Distributed Syst.* **11(7)**, pp. 760–768, Jul 2000.

[2] T. Ben-Nun, Y. Etsion, and D. G. Feitelson, "*Design and implementation of a generic resource sharing virtual time dispatcher*". In 3rd *Ann. Haifa Experimental Syst. Conf.*, May 2010.

[3] A. Chandra, M. Adler, P. Goyal, and P. Shenoy, "*Surplus fair scheduling: A proportional-share CPU scheduling algorithm for symmetric multiprocessors*". In 4th *Symp. Operating Systems Design & Implementation*, pp. 45–58, Oct 2000.

[4] D. Dolev, D. G. Feitelson, J. Y. Halpern, R. Kupferman, and N. Linial, "*No justified complaints: On fair sharing of multiple resources*". In 3rd *Innov. Theor. Comput. Sci.*, pp. 68–75, Jan 2012.

[5] N. Egi, A. Greenhalgh, M. Handley, G. Iannaccone, M. Manesh, L. Mathy, and S. Ratnasamy, "*Improved forwarding architecture and resource management for multi-core software routers*". In 6th *IFIP Intl. Conf. Network & Parallel Comput.*, pp. 117–124, Oct 2009.

[6] D. H. J. Epema, "*Decay-usage scheduling in multiprocessors*". *ACM Trans. Comput. Syst.* **16(4)**, pp. 367–415, Nov 1998.

[7] Y. Etsion, T. Ben-Nun, and D. G. Feitelson, "*A global scheduling framework for virtualization environments*". In 5th *Intl. Workshop System Management Techniques, Processes, and Services*, May 2009.

[8] Y. Etsion, D. Tsafrir, and D. G. Feitelson, "*Process prioritization using output production: scheduling for multimedia*". *ACM Trans. Multimedia Comput., Commun. & App.* **2(4)**, pp. 318–342, Nov 2006.

[9] A. Ghodsi, V. Sekar, M. Zaharia, and I. Stoica, "*Multi-resource fair queueing for packet processing*". In *ACM SIGCOMM Conf.*, pp. 1–12, Aug 2012.

[10] A. Ghodsi, M. Zaharia, B. Hindman, A. Konwinski, S. Shenker, and I. Stoica, "*Dominant resource fairness: Fair allocation of multiple resource types*". In 8th *Networked Systems Design & Implementation*, pp. 323–336, Mar 2011.

[11] F. Guim, I. Rodero, and J. Corbalan, "*The resource usage aware backfilling*". In *Job Scheduling Strategies for Parallel Processing*, E. Frachtenberg and U. Schwiegelshohn (eds.), pp. 59–79, Springer Verlag, 2009. Lect. Notes Comput. Sci. vol. 5798.

[12] A. Gutman and N. Nisan, "*Fair allocation without trade*". In 11th *Autonomous Agents & Multiagent Syst.*, Jun 2012.

[13] G. J. Henry, "*The fair share scheduler*". *AT&T Bell Labs Tech. J.* **63(8, part 2)**, pp. 1845–1857, Oct 1984.

[14] E. D. Lazowska, J. Zahorjan, G. S. Graham, and K. C. Sevcik, *Quantitative System Performance: Computer System Analysis Using Queueing Network Models*. Prentice-Hall, Inc., 1984.

[15] A. Mancina, D. Faggioli, G. Lipari, J. N. Herder, B. Gras, and A. S. Tanenbaum, "*Enhancing a dependable multiserver operating system with temporal protection via resource reservations*". *Real-Time Syst.* **43(2)**, pp. 177–210, Oct 2009.

[16] J. Nieh, C. Vaill, and H. Zhong, "*Virtual-Time Round Robin: An O(1) proportional share scheduler*". In *USENIX Ann. Technical Conf.*, pp. 245–259, Jun 2001.

[17] B. Radunović and J.-Y. Le Boudec, "*A unified framework for max-min and min-max fairness with applications*". *IEEE/ACM Trans. Networking* **15(5)**, pp. 1073–1083, Oct 2007.

[18] D. Raz, H. Levy, and B. Avi-Itzhak, "*A resource-allocation queueing fairness measure*". In *SIGMETRICS Conf. Measurement & Modeling of Comput. Syst.*, pp. 130–141, Jun 2004.

[19] C. A. Waldspurger and W. E. Weihl, "*Lottery scheduling: Flexible proportional-share resource management*". In 1st *Symp. Operating Systems Design & Implementation*, pp. 1–11, USENIX, Nov 1994.

[20] M. E. Yaari and M. Bar-Hillel, "*On dividing justly*". *Social Choice and Welfare* **1(1)**, pp. 1–24, May 1984.

[21] T. Yang, T. Liu, E. D. Berger, S. F. Kaplan, and J. E. B. Moss, "*Redline: First class support for interactivity in commodity operating systems*". In 8th *Symp. Operating Systems Design & Implementation*, pp. 73–86, Dec 2008.

[22] L. Zhang, "*Virtual clock: A new traffic control algorithm for packet switching networks*". In *ACM SIGCOMM Conf.*, pp. 19–29, Sep 1990.

[23] Z. Zhang, L. T. X. Phan, G. Tan, S. Jain, H. Duong, B. T. Loo, and I. Lee, "*On the feasibility of dynamic rescheduling on the Intel distributed computing platform*". In 11th *Intl. Middleware Conf. (Industrial Track)*, pp. 4–10, Nov 2010.

Survivability Models for the Assessment of Smart Grid Distribution Automation Network Designs

Alberto Avritzer
Siemens Corporation
Princeton, NJ 08540

Sindhu Suresh
Siemens Corporation
Princeton, NJ 08540

Daniel Sadoc Menasché
Federal University of Rio de Janeiro, Brazil

Rosa Maria Meri Leão
Federal University of Rio de Janeiro, Brazil

Edmundo de Souza e Silva
Federal University of Rio de Janeiro, Brazil

Morganna Carmem Diniz
Federal University of the State of Rio de Janeiro, Brazil

Kishor Trivedi
Duke University
Durham, NC 27706

Lucia Happe
Karlsruhe Institute of Technology, Germany

Anne Koziolek
University of Zurich
Zurich, Switzerland

ABSTRACT

Smart grids are fostering a paradigm shift in the realm of power distribution systems. Whereas traditionally different components of the power distribution system have been provided and analyzed by different teams through different lenses, smart grids require a unified and holistic approach that takes into consideration the interplay of communication reliability, energy backup, distribution automation topology, energy storage and intelligent features such as automated failure detection, isolation and restoration (FDIR) and demand response.

In this paper, we present an analytical model and metrics for the survivability assessment of the distribution power grid network. The proposed metrics extend the system average interruption duration index (SAIDI), accounting for the fact that after a failure the energy demand and supply will vary over time during a multi-step recovery process. The analytical model used to compute the proposed metrics is built on top of three design principles: state space factorization, state aggregation and initial state conditioning. Using these principles, we reduce a Markov chain model with large state space cardinality to a set of much simpler models that are amenable to analytical treatment and efficient numerical solution. In the special case where demand response is not integrated with FDIR, we provide closed form solutions to the metrics of interest, such as the mean time to repair a given set of sections.

We have evaluated the presented model using data from a real power distribution grid and we have found that survivability of distribution power grids can be improved by the integration of the demand response feature with automated FDIR approaches. Our empirical results indicate the importance of quantifying survivability to support investment decisions at different parts of the power grid distribution network.

ICPE'13, April 21–24, 2013, Prague, Czech Republic.

Categories and Subject Descriptors

C.4 [**Performance of Systems**]: Modelling Techniques

Keywords

Survivability; Transient Analysis; Smart Grid; Fault Tolerance

1. INTRODUCTION

Information and communication technologies (ICT) are being deployed to the distribution power grid to facilitate the management of energy demand and supply. The automatic management of customer consumption in response to variations in supply or as a result of power failures is referred to as *demand response*. The automatic detection, isolation and restoration of failures is known as *distribution automation* (DA). The automation of the smart grid brings novel challenges to the power grid engineers, such as the assessment of the tradeoffs involved to accurately engineer the power distribution reliability.

The introduction of automation (intelligence) to power distribution networks has created a need for the holistic assessment of the distribution network. The distribution network reliability is a function of the correct operation of several architecture artifacts such as electrical power components, telecommunications, distribution network topology, failure detection isolation and restoration, demand response and distributed generation and storage. The automation of power distribution requires a more integrated perspective across these domains. However, in the current mode of operation they are still being engineered separately.

Traditionally, the reliability of power systems has been quantified using average metrics, such as the system average interruption index (SAIDI). SAIDI is used by public service commissions in the United States to assess utilities' compliance with the commission rules. It was developed to track manual restoration times, and according to Standard 166-1998, the median value for North American utilities is roughly one and a half hours. In smart grid networks, power failure and restoration events will have a finer level of granularity, due to the deployment of reclosers, which isolate faulty sections, and demand side management system activities, such as distributed generators and demand response application systems. Therefore, there is a need to extend the SAIDI metric, and to develop new models and tools for the accurate computation of customer interruption indexes *after power failure events occur*, even if the occurrence of such events is rare. The *survivability* of a mission-critical application is the ability of the system to continue functioning during and after a failure or disturbance [17].

In [14] we presented a proposal for a common analysis framework to support the survivability analysis of distribution automation using extensions of the IEC-standardized common information model. The paper presented a case study of the application of the proposed method to the survivability analysis of a simple distribution automation network that was derived from a real power distribution network. In [19] we have evaluated the impact of available active and reactive power supply after a section failure on the distributed automation survivability metric and we derived closed-form expressions for certain survivability related metrics.

In this paper we present an analytical model to assess the survivability of distributed automation power grids and to predict SAIDI and related metrics as a function of different system parameters related to communications, distributed generation, demand response and other smart grid features. We use a performability model to capture how the system recovers from a failure. Our model accounts for the fact that the topology is sectionalized. Given a failure in section i, our key insight is to aggregate the sections of the network that may be fed by backup sources into a single node, denoted by $i+$. This aggregation allows us to efficiently quantify *transient metrics* of the network after a failure, also referred to as *survivability metrics*. For example, our model allows us to compute how the *energy not supplied* (ENS) after a failure varies over time as a function of the available backup power, the demand response application and of the state of the information and communication network.

After a power failure event, some power grid areas of the network may experience restoration times of the order of magnitude of minutes, while other power grid areas may require hours for the manual repair events to take place. Our model allows for the accurate assessment of the power grid network survivability by tracking the time-dependent state of the system under study.

The main contributions of this paper are the following.

Survivability model: We present a Markov chain model that supports the survivability assessment of power grid metrics accounting for the sectionalizing of distribution automation topology, the available excess power, the unreliability of the telecommunications network and the interaction with the demand response application. Our model can be generated and solved in a cost-efficient manner.

Implications of system integration: We bring awareness to the importance of accurate holistic power engineering that considers the interactions between telecommunications reliability and the reliability benefits of integration with other distribution automation features, such as the integration of failure recovery with demand response. In particular, we show that if demand response can be activated after a failure occurs, the reliability of the system significantly increases.

Extension of the SAIDI metric to support distributed automation: We present an extension of the SAIDI metric that captures the dynamic nature of the smart-grid by taking into account the number of customers impacted by the service interruption, the service impact of the interruption (*e.g.*, Energy not Supplied) and the duration of the recovery period. We use the analytical solution of the survivability model to capture the time spent in each state during the recovery period and the reward associated with each state to capture the service impact of the interruption.

The outline of this paper is as follows. In Section 2 we present a survey of the related literature. In Section 3 we present an overview of demand response and failure detection isolation and restoration applications. In Section 4 we introduce the survivability metrics that can be derived from our model. We present the model used in this paper in Section 5. The analysis of our empirical results is presented in Section 6. Section 7 presents our conclusions and suggestions for future research.

2. LITERATURE REVIEW

The available literature on power systems reliability is extensive [4, 8, 1]. Recently, researchers have studied how to improve power systems reliability with smart grid techniques [27, 22]. To our knowledge, our work is the first to assess survivability metrics of power systems accounting for the implications of electromechanical and computer-based strategies to address failures in an integrated manner.

Elmakias [8] presents a review of computational methods in power system reliability. The focus of the review is on the application of Markov models to reliability assessment. To address failures in the distribution system, the author studied a number of approaches such as the reduction of main feeder line length, the introduction of sectionalizer switches, and the automatic connection of backup power supply to sections isolated by a failure. The analysis focuses on steady state metrics, whereas in this paper our focus is on studying the system after a failure occurs. Conditioning the initial state to be a failure state is important in order to evaluate metrics such as the mean energy not supplied until recovery.

The impact of adding Distributed Generation (DG) as a backup source in a power system has been studied in [27, 29, 26, 28]. Wassem [27] and Zou *et al.* [29] analyzed the impact of DG placement on SAIDI, comparing several main feeder topologies. They concluded that placing the DG source at the end of the main feeder line provided the best improvement in power reliability. Wang *et al.* [26] propose an analytical model to evaluate system reliability. They use the model to obtain the placement and sizing of DG's that maximizes power reliability. An optimization approach for the optimal sizing of DG's is also presented by Zhang *et al.* [28]. These works are related to ours, as these results can be used to obtain the probability that a backup source can supply energy to the affected sections.

Janev [11] presents the implementation and evaluation of a power flow algorithm for power distribution grids with distributed generation. Specifically, the power flow approach presented in [11] was evaluated using an adaptation of a power distribution benchmark [23]. This benchmark employs the following types of renewable power generation equipment: PV, Wind, Small Hydro and Biomass.

Martins [18] presents a model for active distribution systems expansion planning that considers distributed generation together with traditional alternatives for distribution expansion such as rewiring, network reconfiguration and installation of protection devices. The authors evaluate different alternatives for distribution automation using average reliability metrics (SAIDI, SAIFI) and cost.

Brown [4, Section 2.2.5] presents a detailed discussion of the shortcomings of existing indices to assess reliability of power systems. In addition, the author proposes a novel metric, the system average interruption duration exceeding threshold, or SAIDET. In this paper, we argue that survivability metrics also play a key role in the assessment of smart grid networks.

Heegaard and Trivedi [10] studied the survivability of telecommunication systems. They presented a phase recovery model to capture the transient properties of the system after a failure. In this paper, we account for features that are specific to the smart grid domain and leverage such features for the efficient solution of the proposed model (see Section 5.1).

Performability metrics have been defined to measure the ability of a system to continue to operate after a failure but at (possibly) different performance levels [20, 21]. Performability is usually concerned with the quality of service provided that the system is operational. The initial system state is chosen accordingly. In this paper, in turn, our focus is on survivability metrics. In this case, the initial state of the system is set to a failure state, so survivability is "conditional performability" [16].

Keshav and Rosenberg [13] argued that concepts pioneered by the Internet are applicable to the design of smart grids, and suggest the initiation of a dialogue between the Internet community and the electrical grid research community. Our work is a product of such a dialogue [10, 25].

3. DISTRIBUTION SYSTEMS BACKGROUND

In this section we introduce some background on distribution systems, focusing on the aspects relevant to our model, namely 1) the demand response application and 2) fault detection, isolation and restoration. Currently, the two features are implemented by separate distribution automation systems, so we discuss the potential benefits of the integration of demand response and fault detection, isolation and restoration features. We start with a brief primer on distribution automation.

3.1 Distribution Automation Primer

The smartening of distribution networks can bring significant benefits to operators and customers, but will require considerably more effort than the smartening of transmission networks. Distribution networks have many more nodes to be instrumented and managed, and there is a need to meet stringent requirements for communication reliability. Distribution systems connect to nearly all electricity customers (excluding some large industrial customers that are connected directly to the transmission system). In addition, future distribution networks will become very complex with the introduction of new technologies, such as distributed generation and variable/dispatchable resources, and new load types, such as electric vehicles. Therefore, there is a need to quantitatively engineer and manage the distribution automation technology complexity and the associated costs, with the goal of optimizing the power grid to the benefit of all the stakeholders: power utilities, regulatory entities and customers.

The integration of smart grid related technology into the distribution side will lead to significant changes in the power system configuration. The current power grid distribution system is designed to meet the expected power load requirements. In the future, the distribution automation (DA) feature will be responsible for maintaining power reliability. Specifically, several alternatives will be available to reconfigure the power system distribution topology after a power event such as a failure. Two alternatives are: (1) depending on the load that is required to be met to recover from a power failure, demand response applications or distributed generators might be initiated, and (2) if several power distribution areas are interconnected through the use of tie line switches, spare power from one area can be used to meet the power demand from the failed area.

Therefore, the interconnection of several distribution areas and the introduction of distribution side energy management schemes like demand response, electric vehicle, energy storage and distributed generation can help manage power reliability by decreasing the mean fault clearing times and reducing the burden to be carried by the power protection devices.

3.2 Demand Response Application

Demand response are a set of the incentive payments designed to induce lower electricity use at times of high wholesale market prices or when system reliability is jeopardized. Demand response can be defined as the action taken by consumers to reduce electricity demand in response to price, monetary incentives, or utility directives so as to maintain reliable electric service or avoid high electricity prices.

A decade ago distribution side networks were considered mostly demand nodes, while now they can act both as a power generation source or a power demand sink. Bidirectional power flow can have a significant impact on the protection and reliability of the system. In addition, evolution of the nodal market creates a location based pricing mechanism, where the price is not only a function of the available energy but also depends on the congestion.

The introduction of demand response applications and the emergence of wholesale energy and reliability markets create new opportunities for demand-side resources by enabling customer loads to participate in the wholesale energy market. In addition, the application of demand response has the potential of enhancing power reliability and helping operators manage peak demand by reducing congestion on critical transmission lines.

3.3 Fault Detection, Isolation and Restoration

Fault detection, isolation and restoration is concerned with the detection of faults on the feeder line, determining the location of the fault as defined by the two feeder switches that determine the fault boundary, isolation of the faulty feeder section, and automated restoration of power to the feeder sections located outside the fault boundary, *i.e.,* the non-faulty feeder sections.

The granularity of fault detection, isolation and restoration depends on the type of switch/recloser used for dividing the feeder line into sections, and the availability of backup power to feed the healthy sections of the feeder line.

The time required for fault detection isolation and restoration depends on the level of automation implemented in the infrastructure deployed by the Utility to support the *FDIR feature*. Customers usually report outages 5-10 minutes after a fault occurrence. Power can be automatically restored to the healthy sections of the feeder line in about 2 minutes, when automated fault detection, isolation and restoration is implemented. When automated fault detection and isolation is not implemented, it may take up to 1 hour for power to be restored to the healthy sections of the feeder, because manual fault location and manual switching has to be performed. Repair of the faulty section of the feeder line may take 1 to 4 hours. When the feeder line is connected to a tie switch, the healthy parts of the feeder line can be powered by a secondary substation, after the faulty section is isolated. Table 1, which will be further discussed with the model presented in Section 5, summarizes the above numbers. The numbers in Table 1 reflect the multiple time scales at which repairs occur. In this paper, these numbers are set based on expert knowledge, but they could as well be adjusted based on time series or event logs.

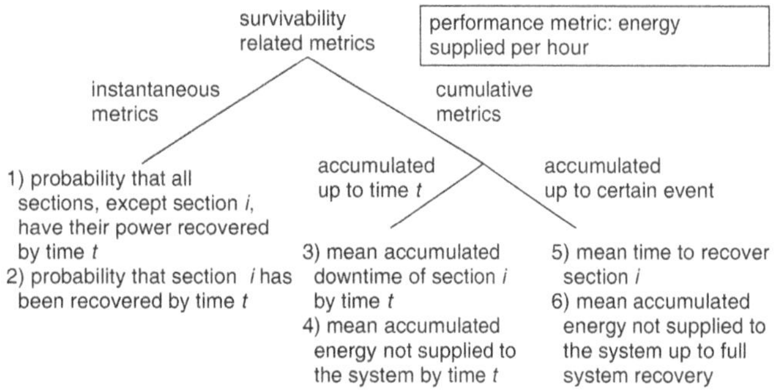

Figure 1: A taxonomy of survivability related metrics

3.4 Integration of Demand Response and Failure Recovery

In current distribution systems, demand response services are not integrated with failure recovery services. Usually, these two services are designed independently. Nonetheless, in this paper we consider the possibility of demand response being activated in response to a failure event with the aim of improving the power distribution survivability.

4. SURVIVABILITY METRICS

Survivability has been defined by ANSI as the transient performance of a system after an undesirable event [2]. The metrics used to quantify survivability vary according to applications, and depend on a number of factors such as the minimum level of performance necessary for the system to be considered functional, and the maximum acceptable outage duration of a system. Survivability metrics are transient metrics computed after the occurrence of a failure. In the remainder of this paper, time t refers to the time since a failure occurred and is measured in hours.

Survivability metrics are computed with respect to a measure of interest $\mathcal{M}$, also referred to as the *performance metric* [10]. In the realm of power systems, the performance metric $\mathcal{M}$ is the energy supplied per hour, measured in kilowatts. Assuming that $\mathcal{M}$ has value μ just before a failure occurs, the survivability behavior is quantified by attributes such as the relaxation time for the system to restore the value of $\mathcal{M}$ to μ. In this paper we compute metrics related to the relaxation time, focusing on the mean energy not supplied per hour after a failure occurs.

4.1 Metrics Taxonomy

Figure 1 shows the taxonomy of the survivability related metrics considered in this paper. We classify the metrics into two broad categories. Instantaneous metrics are transient metrics that capture the state of the system at time t. An example of an instantaneous metric is the probability that a given section i has been recovered by time t.

Cumulative metrics are obtained in our model by assigning *reward rates* to system states. A reward is gained per time unit in a state, as determined by the reward rate assigned to that state. The *accumulated reward* is the result of the accumulation of rewards since the failure up to time t or up to a certain event. The mean accumulated downtime of a given section by time t and the mean accumulated energy not supplied by time t are examples of cumulative metrics computed up to time t. The mean accumulated energy not supplied up to the full recovery of the system is an example of a cumulative metric computed up to a certain event occurs. The mean time to recover a given section is also an example of the latter class of metrics, where the accumulated reward in this case is the time itself, obtained by assigning a reward of one per time unit at every state. Other definitions of transient metrics can be found in [6]. In Section 6 we present the evaluation of the metrics described above as applied to the case study presented in this paper.

4.2 From SAIDI to Survivability Related Metrics

We now define and extend one of the key metrics of interest in the realm of power systems, the System Average Interruption Duration Index (SAIDI). SAIDI is an important measure of the power utility's ability to cope with recovery from failures. It is a measure of average customer impact of system interruptions as it computes the sum of customer interruption durations over the total number of customers [4].

Given a topology with C sections, let N be the total number of customers, let $N_{j,k}$ be the number of customers in the system impacted by the k-th failure at section j and let K_j be the number of failures at section j during a pre-established large observation period, $j = 1, \ldots, C$, $k = 1, \ldots, K_j$. Let $\varphi_{j,k}$ be the outage duration due to the k-th failure that occurred at section j, measured in hours. Let φ_j be the average outage duration due to all failures at section j. Let ϕ_j be the average number of failures at section j, during the same pre-established observation period. The observation period is usually assumed to be one year so that φ_j and ϕ_j are the annual average outage duration and number of failures, respectively.

The System Average Interruption Duration Index (SAIDI) is an important measure of the power utility's ability to cope with recovery from failures. It is a measure of average customer impact of system interruptions as it computes the sum of customer interruption durations over the total number of customers [24],

DEFINITION 4.1. *The SAIDI index is the average outage duration for each customer served,*

$$\mathsf{SAIDI} = \sum_{j=1}^{C}\sum_{k=1}^{K_j} \varphi_{j,k}\frac{N_{j,k}}{N} \qquad (1)$$

After a failure, the energy not supplied will vary over time during a multi-step recovery process. Let $\{m_j(t), t \geq 0\}$ be a stochastic process in which the random variable $m_j(t)$ characterizes the energy not supplied per unit time, after a failure in section j, $j = 1, \ldots, C$, t units of time after the failure; $m_j(t)$ accounts for the effect of one single failure in section j. If a full system recovery occurs at time T, we set $m_j(t) = 0$ for $t \geq T$. Let $\overline{m}_j(t)$ be the mean value of $m_j(t)$. In the remainder of this paper, given a random variable Z we denote its mean by $\overline{Z}$.

Let $M_j(\tau)$ be the accumulated energy not supplied by time τ after a failure in section j, $j = 1, \ldots, C$,

$$\overline{M}_j(\tau) \quad = \quad \int_{t=0}^{\tau} \overline{m}_j(t)dt, \quad j = 1, \ldots, C \qquad (2)$$

Note that the total energy demanded per unit time can also vary during recovery. This occurs, for instance, if demand response is integrated with failure recovery. Let $\{d_j(t), t \geq 0\}$ be a stochastic process in which the random variable $d_j(t)$ characterizes the total energy demanded per unit time at time t during the recovery from a failure in section j. Let $D_j(\tau)$ be the energy demanded over the first τ time units during the recovery from a failure in section j,

$$\overline{D}_j(\tau) \quad = \quad \int_{t=0}^{\tau} \overline{d}_j(t)dt, \quad j = 1, \ldots, C \qquad (3)$$

Let ϕ_j be the expected number of failures at section j during a pre-established large observation period (typically one year). We define the extended SAIDI index (ESAIDI) as the outage duration accounting for the energy demanded and not supplied during the first τ units of time after a failure at a section, averaged over all sections,

DEFINITION 4.2. *The extended SAIDI index is given by*

$$\mathsf{ESAIDI}(\tau) = \sum_{j=1}^{C} \phi_j \tau \left(\frac{\overline{M}_j(\tau)}{\overline{D}_j(\tau)} \right) \quad (4)$$

The term inside parentheses in (4) is the fraction of the mean energy not supplied over the mean energy demanded by time τ after a failure. Note that we assumed that τ is a scalar value. Alternatively, let X_j be a random variable characterizing the time to full system recovery after a failure at section j, $j = 1, \ldots, C$, and $\mathbf{X} = (X_1, \ldots, X_C)$. Replacing τ in (4) by the corresponding mean recovery times yields

$$\mathsf{ESAIDI}(\mathbf{X}) = \sum_{j=1}^{C} \phi_j E[X_j] \left(\frac{E[\overline{M}_j(X_j)]}{E[\overline{D}_j(X_j)]} \right) \quad (5)$$

where

$$E[\overline{M}_j(X_j)] = \lim_{\tau \to \infty} \overline{M}_j(\tau) \quad (6)$$

Let $\overline{M}_j = E[\overline{M}_j(X_j)]$ and let N_j be the average number of customers affected by a failure at section j. The equality in (6) follows from the fact that if a full system recovery occurs at time T, $\overline{m}(t) = 0$ for $t \geq T$. Quantity $\overline{m}_j(t)$ is precisely that defined by ANSI [2] as *survivability*. Cumulative quantities such as $\overline{M}_j$ are defined as extension to the basic survivability measure and called Excess Loss due to Failures (ELF) [15].

Next, we show conditions according to which Definition 4.1 follows as a special case of Definition 4.2. To this goal, assume that the energy demanded per user per unit time is constant and equal to E, and that the number of customers affected by a failure at section j is also constant and equal to N_j. Therefore, $\overline{m}_j(t) = N_j E$ and $\overline{d}_j(t) = NE$. In addition, we also assume that the mean outage duration due to one failure at section j is τ, and the number of failures at section j is independent of the outage duration due to a failure at that section, $\varphi_j = E[X_j]\phi_j = \tau\phi_j$. Let $\widehat{\varphi}_j$ be the mean outage duration due to one failure at section j, $\widehat{\varphi}_j = \varphi_j/\phi_j$. Then, $\mathsf{ESAIDI}(\widehat{\varphi}) = \mathsf{SAIDI}$, where $\widehat{\varphi} = (\widehat{\varphi}_1, \ldots, \widehat{\varphi}_C)$.

$\mathsf{ESAIDI}(\tau)$ is a function of ϕ_j, $\overline{D}_j(\tau)$ and $\overline{M}_j(\tau)$, $j = 1, \ldots, C$. ϕ_j is computed from an availability model while $\overline{D}_j(\tau)$ and $\overline{M}_j(\tau)$ are computed from a survivability model. In the remainder of this paper our focus will be on the survivability model, which allows us to compute the terms in parentheses in equations (4) and (5), as described in Section 5 and illustrated in a case study in Section 6.

5. SURVIVABILITY MODEL

In this section we present the model used to compute survivability metrics of power distribution systems. We describe the modeling challenges and design principles, followed by the model overview and the specific model instantiation used throughout the remainder of the paper.

5.1 Challenges and Design Principles

Survivability assessment of the power grid distribution topology is implemented by taking advantage of the power line design that uses sections for failure isolation.

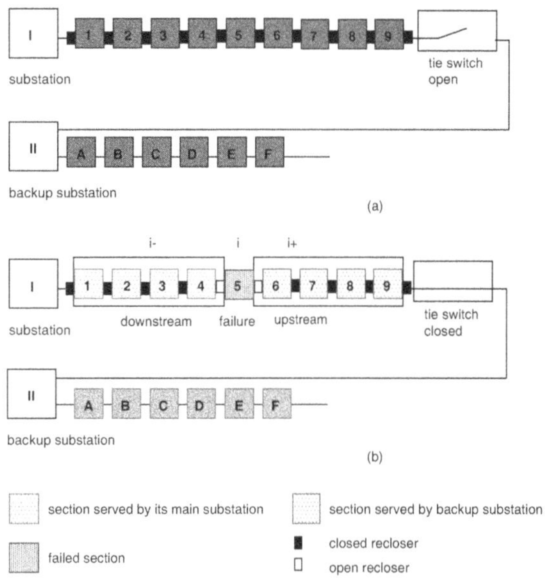

Figure 2: Failed section and its upstream and downstream

Initially, we attempted to characterize the individual behavior of each of the sections by conducting an exhaustive quantitative analysis accounting for the failure rates of each of the sections and their multiple possible states. During this initial model design phase we encountered the following challenges:

- capturing the state of each section individually leads to a large state space, as the number of states grows exponentially as a function of the number of sections;
- generating, storing, and solving a model with a very large state space is computationally challenging;
- accounting for component failures that occur at a much coarser level of granularity than the failure repair yields a model that is numerically hard to solve.

The methodology presented in this paper addressed the above challenges by relying on three key principles: (1) state space factorization; (2) state aggregation and (3) initial state conditioning.

5.1.1 State Space Factorization

Our methodology encompasses a set of models, where each model characterizes the system evolution after the failure of a given section. Given a topology with C sections, our methodology yields C models, where each model is tailored to the characteristics of the failed section. The advantages of such a state space factorization are:

i) **flexibility:** having a model tailored to a given section enables us to capture specific details about the impacts of failures on that particular section;

ii) **reduced complexity:** the computational complexity to compute the metrics of interest is reduced by considering a set of models as opposed to a single model with cardinality C times larger. Consider a distribution automation topology that has C sections, and let K be the number of states at which a section can be found. The computational complexity to solve the non-factorized model is $O(C^3K^3)$, using, for instance, the GTH solution method [9] to compute the steady

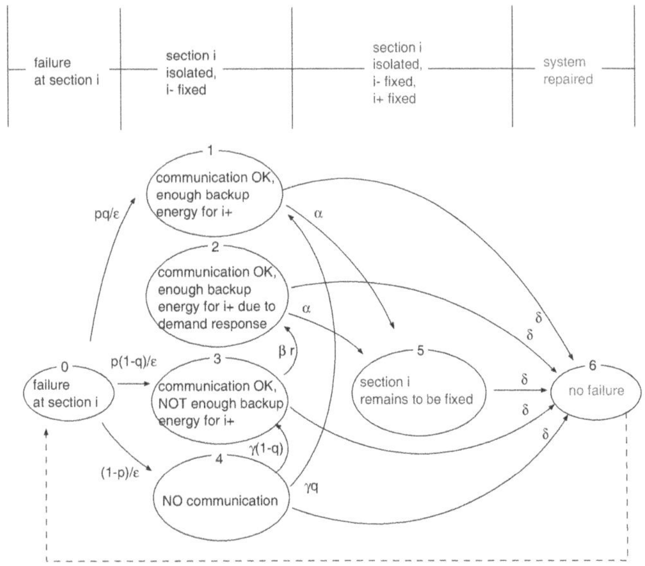

Figure 3: Phased recovery model

state probabilities of an associated ergodic model (described in the Appendix), without taking advantage of any possible special structure in the model. The decomposition approach, in contrast, requires the solution of C models, each one requiring $O(K^3)$ steps to be solved, which results in a complexity of $O(CK^3)$.

5.1.2 State Aggregation

One of the insights of this paper is the observation that after a failure of a given section the remaining sections of the distribution automation topology can be aggregated into groups of affected and non-affected sections. In the scenario considered in the remainder of this paper after the failure of section i, section i is isolated and the non-failed sections can be aggregated into two groups: the downstream sections that are aggregated into a set of sections $i-$ and are served by their original substation and the upstream sections that are aggregated into into a set of sections $i+$ and might be served by a backup substation, if enough backup power is available (see Figure 2). State aggregation yields significant reduction in the computational complexity required to obtain the desired metrics, since the system state space can be described in terms of the aggregated section states.

5.1.3 Initial State Conditioning

The computations of the metrics of interest are performed by assuming that the initial state is a failure state. Our models do not capture the failure *rates* of different components. Instead, the models are parameterized by using the *conditional probability* that specific system components are still operational after a specific section failure. In the remainder of this paper we will consider conditional probabilities to account for the probability that a substation backup power is able to supply isolated sections (q), the reliability of the telecommunications network (p) and the effectiveness of the demand response application (r).

5.2 Model Overview

Automatic and manual restoration events are initiated after a section failure event. The restoration process is a combination of electro-mechanical and computer-based events. In what follows, we describe the sequence of events initiated after the failure of section i.

The isolation of the failed section is automatically performed by *reclosers*, within 10-50 ms after the failure, and power is instantaneously restored to the *downstream* sections ($i-$). The *upstream* sections ($i+$) have their power restored depending on the following factors:

- **communication:** communication is needed for all failure detection, isolation and recovery operations. In particular, communication is used by the *supervisory control and data acquisition* (SCADA) system at a substation to detect failure location, recalculate flow and close the tie switch to feed the upstream sections ($i+$);
- **backup power:** sufficient spare backup power must be available at a backup substation;
- **demand response:** demand response applications can reduce the load in the system after a failure, increasing the probability that the available backup power is able to supply energy to the upstream sections.

Recall from Section 3.3 that, after a section failure, if the communication system is available and the backup power is able to restore energy to the upstream sections, it takes an average of 1-2 minutes to execute the automated restoration feature (see Table 1). If there is not enough available backup power for the restoration of the upstream sections, but communication is available, the demand response feature might be used to adjust the demand accordingly. When the demand response is effective, demand of sections $i+$ can be lowered to the target values within 15 minutes on average. Note that we do not explicitly model demand that is shifted to a later point in time (i.e. load shedding [3]). If the communication system is not available after the section failure, a 1 hour repair time is required for manual restoration of the communication system. This time is dominated by the time it takes for a truck to arrive at the failure site.

Finally, section i may require manual repair, *e.g.*, to remove weather related damage and restore the damaged components to their original condition. After section i is repaired, if the upstream sections are still not recovered, these sections will be connected to the main substation through section i. The average time to manually repair a section is 4 hours.

5.3 Model Description

A Markov chain with rewards is used to model the phased recovery of the distribution automation network. The states of the model correspond to the different recovery phases at which the system might be found as shown in Figure 3. Each state is associated with a reward rate that corresponds, for instance, to the energy not supplied per hour or the number of customers not served per hour in that state. In this paper we assume that state residence times are exponentially distributed, which serves to illustrate our methodology in a simple setting. Future work consists of extending the model to allow for general distributions for the state residence times. The system states and the state rewards are described in the following subsections.

5.3.1 Phased Recovery Model

The phase recovery model is characterized by the following states and events. After a section failure the model is initialized in state 0. The residence time at state 0 corresponds to the time required for

Parameter	Description	Value
ϵ	mean time for recloser to isolate failed section	≈ 0
α	automatic restoration rate	30
β	demand response rate	4
γ	communication repair rate	1
δ	manual repair rate	1/4

Table 1: Model Parameters (rates are given in units of events/hour)

Variable	Description
C	number of sections
i	failed section
$i+$	upstream of section i (sections $\{i+1,\ldots,C\}$)
$i-$	downstream of section i (sections $\{1,\ldots,i-1\}$)
p	probability that communication works after failure
q	probability that backup power suffices to supply isolated sections
r	probability that demand response is effective after failure

Table 2: Table of Notation

the recloser to isolate the section, which takes an average of ϵ. As mentioned in Section 5.2, a recloser isolates a section within 10-50 ms after a failure, so in the remainder of this paper we assume $\epsilon = 0$. Let p be the probability that the communication network is still operational after a section failure and q be the probability that there is sufficient backup power to supply energy for sections $i+$. After the isolation of section i is completed the model transitions to one of three states:

1. with probability pq the model transitions to state 1, where the distribution network is amenable to automatic restoration,

2. with probability $1-p$, the model transitions to state 4, where the communication system requires manual repair, which occurs at rate γ,

3. with probability $p(1-q)$ the model transitions to state 3, where the effectiveness of demand response will determine if the system is amenable to automatic restoration.

At state 3, demand response takes place after a period of time with average duration $1/\beta$. Let r be the probability that demand response effectively reduces the load of the system to a level that is supported by the backup substation. In this case, the model transitions from state 3 to state 2 with rate βr. When the model is in states 1 or 2 the distribution network is amenable to automatic restoration, which occurs after a period of time with average duration $1/\alpha$. What distinguishes states 1 from state 2 is the fact that state 1 can be reached in one step transition after a failure, whereas state 2 is reached only after the successful activation of the demand response feature. Therefore, the state reward rates associated to states 1 and 2, such as the energy not supplied per hour at those states, are usually different. A manual repair of section i takes on average $1/\delta$ hours (and can occur while the system is in states 1-5). After a manual repair, the model transitions to state 6, which corresponds to a fully repaired system.

We now describe the computation of the survivability metric (Energy Not Supplied) by using the phased recovery model described in Figure 3. In each state of the model of Figure 3 we associate the energy not supplied per hour at that state, the state *reward rate*. Let $\pi_k(t)$ be the transient probability associated with state k and σ_k be the reward rate (*e.g.*, mean energy not supplied per hour) associated with state k, $k = 0, \ldots, 6$. Let $L(t)$ be a random variable characterizing the reward accumulated by time t after a failure (*e.g.*, accumulated energy not supplied by time t). The mean reward accumulated by time t is

$$\overline{L}(t) = \sum_{k=0}^{6} \int_{y=0}^{t} \sigma_k \pi_k(y) dy \qquad (7)$$

Let s_k be the residence time at state k before reaching state 6 (*i.e.*, up to full system recovery), $k = 0, \ldots, 5$. Let L be a random variable characterizing the accumulated energy not supplied up to full system recovery. The mean reward accumulated up to full system recovery is

$$\overline{L} = \lim_{t\to\infty} \overline{L}(t) = \sum_{k=0}^{5} \sigma_k \overline{s}_k \qquad (8)$$

Note that (7) is the mean energy not supplied in the interval $[0,t]$ after a failure, defined in (2), and (8) is the ELF measure defined in (6), with subscript j dropped since we consider a single failure. In the Appendix we show how to compute $\overline{s}_k$ and $\pi_k(t)$, and present their closed form solutions when $r = 0$ and $\epsilon = 0$.

5.4 Model Solutions

To compute the metrics of interest presented in Section 4 we used standard techniques for the solution of Markov chains. In the Appendix we show that if $r = 0$, that is, demand response is not enabled, closed form solutions can be derived for the probability distributions of states 1-6 of the Markov model shown in Figure 3. If $r > 0$ we can still find closed form solutions, although they cannot be written in a compact form. Therefore, when $r > 0$ we use standard Markov chain numerical methods to solve the model. For the computation of the average accumulated reward at a given point in time, we use the techniques based on uniformization implemented at the Tangram-II tool [5, 7].

6. ANALYSIS

In this section we present the analysis of the empirical results obtained using the analytical modeling approach introduced in Section 5.

6.1 Setup Description

Our experimental setup is based on an adaptation of the data reported in [27], about the energy load in a number of sections in the US state of Virginia. Figure 2 illustrates the topology considered in our experiments. The topology consists of nine sections. The average number of customers and load per section, in KW, are shown in Table 3.

section	users	load	load amenable to demand response	net load
1	21	49.50	2.03	47.47
2	25	54.80	**0.00**	54.80
3	9	12.00	**4.07**	7.93
4	15	23.22	**11.58**	11.63
5	28	47.80	**0.00**	47.80
6	111	142.25	**15.28**	126.97
7	12	27.40	**0.00**	27.40
8	50	178.40	**0.00**	178.40
9	4	6.90	**1.40**	5.50
total	275	542.27	34.36	507.80

Table 3: Load per section (in KW)

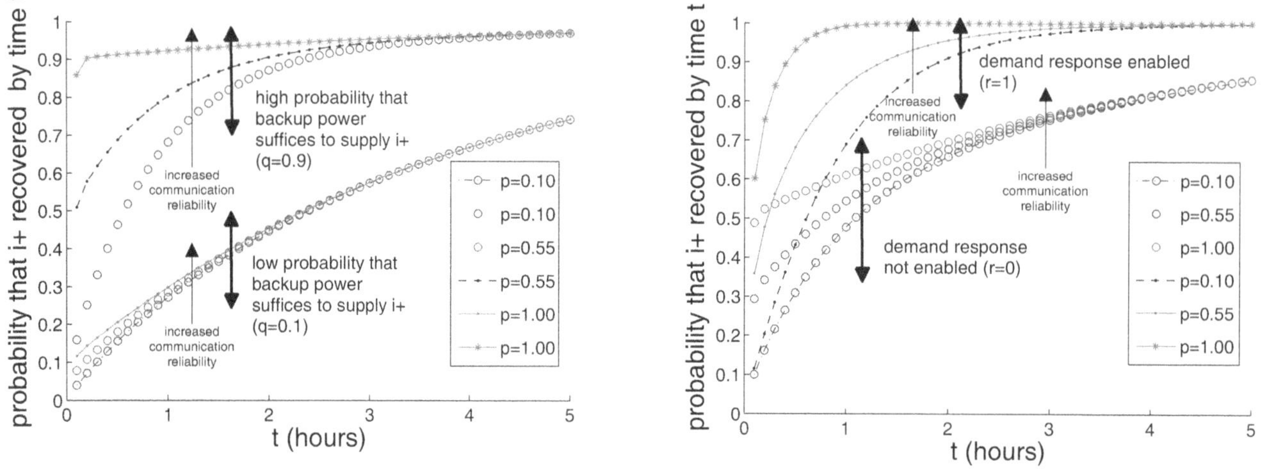

(a) (b)

Figure 4: Probability of $i+$ being recovered by time t, varying a) p and q (no demand response enabled, $r = 0$); b) p and r (backup power suffices to supply $i+$ with probability 0.5, $q = 0.5$).

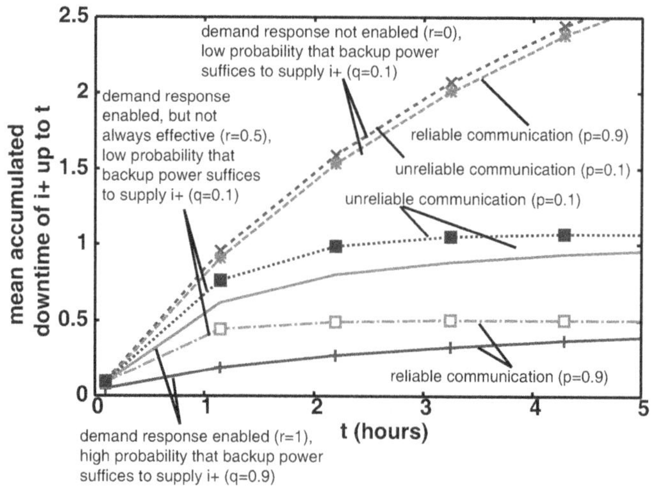

Figure 5: Mean accumulated downtime of $i+$.

The energy supplied and not supplied per hour at each of the model states can be obtained from Table 3. We consider the worst case scenario in which section 1 fails, $i = 1$, maximizing the demand placed on the backup substation to supply the $i+$ sections. Table 4 shows the values of the rewards at the different model states. In states 0, 1, 3 and 4, the energy not supplied per hour is 542.27 KW, which corresponds to the average power demand placed on the distribution network. In state 2, the energy not supplied per hour decreases by 32.33 (sum of elements in bold in Table 3), to 509.94 KW, due to the activation of demand response. In state 5, the energy not supplied per hour is 49.5 KW as section 1 remains to be fixed. The energy supplied in state 5, in turn, depends on whether demand response was enabled. To simplify presentation, we set the energy supplied per hour in state 5 to its lower bound, 460.38 KW, assuming that demand response is always enabled at that state. Since the approximation has minimal impact in the results that follow, we proceed with our analysis under such simplification, noting that a straightforward extension of the model consists of splitting state 5 into two states, to account for whether demand response is enabled or not at state 5. Finally, we assume that the failed section 1 is not affected by demand response.

state	0 - 1	2	3 - 4	5	6
ES/h	0.00	0.00	0.00	460.38	542.27
ENS/h	542.27	509.94	542.27	49.50	0.00

Table 4: Rewards: lower bound on energy supplied per hour (ES/h) and energy not supplied per hour (ENS/h).

In the next subsections we evaluate how the metrics introduced in Section 4 vary as a function of the following parameters: the probability that the substation backup power is able to supply the isolated sections $i+$ (q), the reliability of the telecommunications network (p) and the effectiveness of the demand response application (r).

In what follows, we present experimental results for the time to recover (Section 6.2) and for the energy not supplied (Section 6.3). The results presented in Section 6.2 are independent of the rewards at different states and of the index of the failed section because in this experiment we did not include the capacity of the backup stations as an exogenous parameter. Instead, we captured the substation backup capacity indirectly through the parameter q. In contrast, the results presented in Section 6.3 depend on the rewards and the index of the failed section as presented above.

6.2 Upstream Recovery Time

We consider the probability that the upstream $i+$ sections have recovered by time t. When $r = 0$, the expression of the probability that $i+$ has been recovered by t is given by (11) in the Appendix. Figure 4(a) shows how the probability that $i+$ has recovered increases over time. We observe that if q, the probability of the backup energy being sufficient to supply $i+$, is small, the communication infrastructure is ineffective to recover the $i+$ stations. In contrast, Figure 4(b) shows that if demand response is integrated with failure recovery ($r = 1$), the mean time for the system to recover decreases. Furthermore, with demand response, communication significantly impacts the probability that $i+$ has been recovered by time t. We observe that with demand response the probability that $i+$ has been recovered by time t is roughly 1 for t greater than 3. Otherwise, this probability does not reach 0.9 even after five hours.

Figure 5 shows the mean accumulated downtime of the upstream $i+$ sections as a function of the parameters evaluated in this study. If the evaluated parameter settings don't allow for the automated recovery of the $i+$ sections, the accumulated downtime will grow until a manual repair event occurs.

In contrast, if the evaluated parameter settings allow for automated recovery of the $i+$ sections the mean accumulated downtime will level off after a shorter period of time. For example, for the parameter values of $p = 0.9$, $q = 0.1$ and $r = 0.5$, Figure 5 shows that the mean accumulated downtime of $i+$ levels off after one hour, and never surpasses 0.5 hours.

Closely related to the mean accumulated downtime is the mean time to recover $i+$, shown in Figure 8. The mean time to recover decreases as p or q increase. As r increases from 0 to 0.5, the impact of q decreases because if demand response is integrated with failure recovery, demand response can reduce the load demand to the power grid after a section failure.

6.3 Quantifying Energy Not Supplied

In this subsection we present the computation of the accumulated energy not supplied, from the time of occurrence of a failure event up to time t. We use the rate rewards associated with each of the model states as described in Section 6.1. Figures 6(a) and 6(b) show the mean accumulated energy not supplied by time t, and the fraction of mean energy not supplied over mean energy demanded by time t, as a function of time. The former corresponds to $\overline{M}(t)$ (eq. (2)), and the latter corresponds to the term inside parentheses in the definition of $\mathsf{ESAIDI}(t)$, $\overline{M}(t)/\overline{D}(t)$ (eq. (4)). $\overline{M}(t)$ and $\overline{D}(t)$ are computed using (7), setting $reward_k$ to the mean energy not supplied per hour and the mean energy demanded per hour, respectively. If $q = 0.9$, that is, there is a high probability that the backup power suffices for sections $i+$, demand response does not have a significant impact on the energy not supplied, since the backup station is likely to support the additional load demand even in the absence of demand response. In contrast, if $q = 0.1$ demand response plays a key role, because sections $i+$ can be automatically restored when demand response is effective. The plots in Figures 6(a) and 6(b) also demonstrate the significant impact of integrated demand response on the mean accumulated energy not supplied when the probability that backup power suffices to supply $i+$ is low ($q = 0.1$).

The curves corresponding to $q = 0.9$, $r = 0$ and $q = 0.1$, $r = 1$ cross each other in Figures 6(a) and 6(b) because during the first moments after a failure, it is beneficial to have a high value of q independently of the value of r, as demand response takes an average of 15 minutes to become operational. After two hours the mean accumulated energy energy not supplied is smaller when $q = 0.1$, $r = 1$ as opposed to $q = 0.9$, $r = 0$.

Figure 7 shows $\overline{M}/E[\overline{D}(X)]$, the mean fraction of mean energy not supplied over mean energy demanded up to full system recovery (term inside parentheses in the definition of $\mathsf{ESAIDI}(X)$, eq. (5), with subscript j removed since we consider a single failure). $\overline{M}$ is computed using (8), for different values of p and r, with $q = 0.1$ ($E[\overline{D}(X)]$ is computed similarly). The figure indicates that even a slight increase in r yields a substantial decrease in the mean accumulated energy not supplied up to full system recovery. It also shows that communication reliability, p, becomes more relevant as r increases, because communication is needed for all automatic failure recovery operations, including the activation of demand response for failure recovery.

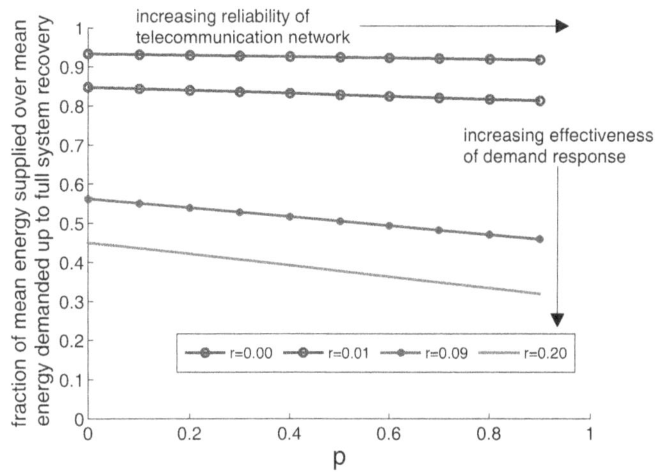

Figure 7: Fraction of mean energy not supplied over mean energy demanded up to full system recovery, $\overline{M}/E[\overline{D}(X)]$ ($q = 0.1$).

6.4 Discussion

With the presented results, investment decisions for the studied circuit can be based on survivability considerations. With the extended SAIDI as presented in (4), different options of how to improve reliability and survivability can be evaluated and compared to each other. For example, the effect of investing in the reliability of the communication infrastructure (*e.g.*, by adding redundancy) can be compared to investing in demand response mechanisms (*e.g.*, by subsidizing smart meters at the consumer).

7. CONCLUSION

In this paper we have argued, using analytical models, that new metrics are important to assess the quality of future intelligent power grids. Specifically, it was shown that survivability metrics can be used to drive a holistic engineering approach and are able to capture dynamic behavior of the smart grid after the occurrence of events of interest, *e.g.*, section failures.

We have proposed a methodology to assess the impact of different system parameters on the survivability of distribution automation power grids. Specifically, we have shown the interactions between three key parameters evaluated in this study and their impact on six survivability metrics. The empirical results obtained indicate that the integration of demand response with failure recovery can yield significant reductions in the amount of energy not supplied after a failure. In addition, the communications reliability parameter is most important when demand response is integrated. These results illustrate the need for holistic and accurate approaches to guide investment decisions on different parts of the network.

We believe that our work opens up several avenues for future exploration. Our models can serve to quantify the tradeoffs between investment cost and reliability gains. In face of such tradeoffs, one can devise algorithms to issue recommendations on how to invest on smart grids in light of prospective survivability gains. Specifically, the analytical solution of the survivability model enables the analytical optimization of the distribution automation investment given the costs of achieving a certain communication reliability (p), available backup power (q) or demand response effectiveness (r).

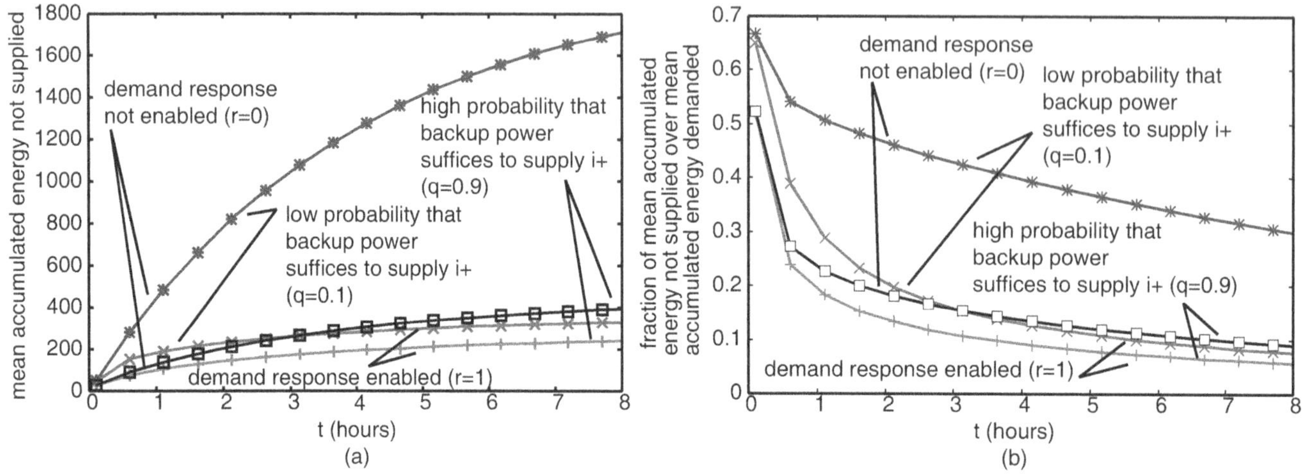

Figure 6: (a) Mean accumulated energy not supplied, by time t, $\overline{M}(t)$, and (b) fraction of mean accumulated energy not supplied over mean accumulated energy demanded, by time t, $\overline{M}(t)/\overline{D}(t)$ ($p = 0.9$).

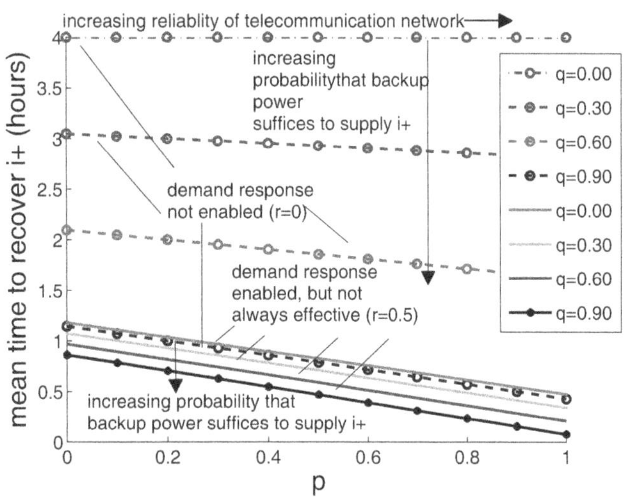

Figure 8: Mean time to recover $i+$, varying p, q and r. Dotted and solid lines correspond to demand response not enabled ($r = 0$) and enabled but not always effective ($r = 0.5$), respectively.

Detailed models for communications, back-power and demand response are needed to validate that the desired values of p, q and r are achieved for the distribution automation network design.

Other future research topics include the extension of the presented methodology to account for more complex topologies, more complex failure trees (cascading failures) and other features that are required to be included in the modeling of industrial distribution networks.

8. ACKNOWLEDGEMENT

This work was supported in part by grants from CNPq, FAPERJ/Brazil and Siemens Corporate Research, Inc and by the Swiss Research Foundation (SNF) as a Forschungskredit of the University of Zurich, grant no. 53510401.

9. REFERENCES

[1] A.A.Chowdhury, S. Agarwal, and D. Koval. Reliability modeling of distributed generation in conventional distribution systems planning and analysis. *IEEE Transactions on Industry Applications*, 39(5):1493–1498, 2003.

[2] ANSI T1A1.2 Working Group on Network Survivability Performance. *Technical report on enhanced network survivability performance*. TR No 68, ANSI, February 2001.

[3] J.-Y. L. Boudec and D. Tomozei. Satisfiability of elastic demand in the smart grid. In *Energy*, 2011.

[4] E. R. Brown. *Electric Power Distribution Reliability*. CRC Press, 2nd edition, 2002.

[5] E. de Souza e Silva, D. R. Figueiredo, and R. M. M. Leão. The TANGRAM-II integrated modeling environment for computer systems and networks. *SIGMETRICS Perform. Eval. Rev.*, 36(4):64–69, 2009.

[6] E. de Souza e Silva and R. Gail. *Computational Probability*, chapter Transient solution for Markov chains. Kluwer, 2000.

[7] E. de Souza e Silva and H.R. Gail. The Uniformization Method in Performability Analysis. In *Performability Modeling: Techniques and Tools*. John Wiley & Sons, 2001.

[8] D. E. Elmakias. *New Computational Methods in Power System Reliability*. Springer, 2010.

[9] W. Grassmann, M. I. Taksar, and D. Heyman. Regenerative analysis and steady state distributions for markov chains. *Operations Research*, 33(5), 1985.

[10] P. Heegaard and K. Trivedi. Network survivability modeling. *Computer Networks*, pages 1215–1234, june 2009.

[11] V. Janev. *Implementation and evaluation of a distribution load flow algorithm for networks with distributed generation.* M.Sc. Thesis, Swiss Federal Institute of Technology, Zurich, 2009.

[12] J. Kemeny and J. Snell. *Finite Markov Chains*. Springer, 1983.

[13] S. Keshav and C. Rosenberg. How internet concepts and technologies can help green and smarten the electrical grid. *ACM SIGCOMM Computer Communication Review*, 2011.

[14] A. Koziolek, L. Happe, A. Avritzer, and S. Suresh. A common analysis framework for smart distribution networks applied to survivability analysis of distribution automation. In *Software Engineering for the Smart Grid (SE4SG), 2012 International Workshop on*, pages 23 –29, june 2012.

[15] Y. Liu, V. Mendiratta, and K. Trivedi. Survivability analysis of telephone access network. In *ISSRE*, pages 367 – 377, Nov. 2004.

[16] Y. Liu and K. Trivedi. Survivability quantification: The analytical modeling approach. *International Journal of Performability Engineering*, 2(1):29–44, 2006.

[17] Z. Ma. Towards a unified definition for reliability, survivability and resilience (I): the conceptual framework inspired by the handicap principle and ecological stability. In *Aerospace Conference, 2010 IEEE*, pages 1–12, march 2010.

[18] V. Martins and C. Borges. Active distribution network integrated planning incorporating distributed generation and load response uncertainties. *Power Systems, IEEE Transactions on*, 26(4):2164 –2172, nov. 2011.

[19] D. Menasché, R. M. M. Leão, E. de Souza e Silva, A. Avritzer, S. Suresh, K. Trivedi, R. A. Marie, L. Happe, and A. Koziolek. Survivability analysis of power distribution in smart grids with active and reactive power modeling. In *GreenMetrics Workshop (to appear in special issue of Performance Evaluation Review)*, 2012. `http://www.sigmetrics.org/greenmetrics/2012/papers/Menasche.pdf`.

[20] J. Meyer. On evaluating the performability of degradable computing systems. *IEEE Transactions on Computers*, 29:720–731, 1980.

[21] J. F. Meyer. Performability: a retrospective and some pointers to the future. *Performance evaluation*, 14:139–156, 1992.

[22] K. Moslehi and R. Kumar. A reliability perspective of the smart grid. *IEEE Transactions on Smart Grid*, pages 57–64, june 2010.

[23] K. Rudion, A. Orths, Z. Styczynski, and K. Strunz. Design of benchmark of medium voltage distribution network for investigation of dg integration. In *Power Engineering Society General Meeting, 2006. IEEE*, page 6 pp., 0-0 2006.

[24] T. Short. *Electric Power Distribution Handbook*. CRC, 2003.

[25] K. S. Trivedi, D. S. Kim, and R. Ghosh. Resilience in computer systems and networks. In *Proc. ICCAD*, pages 74–77. IEEE/ACM, 2009.

[26] S. Wang, W. Zhao, and Y. Chen. Distribution system reliability evaluation considering DG impacts. In *Int. Conf. Electric Utility Deregulation and Restructuring and Power Technologies*, pages 2603–2607, April 2008.

[27] I. Waseem. *Impacts of Distributed Generation on the Residential Distributed Network Operation*. M.Sc. Thesis, Virginia Polytechnic Institute, 2008.

[28] J. Zhang and Z. Bo. Research of the impact of distribution generation on distribution network loss. In *Universities Power Engineering Conference 2010*, pages 1–4, Sep. 2010.

[29] K. Zou, W. Keerthipala, and S. Perera. SAIDI minimization of a remote distribution feeder. In *AUPEC 2007*, pages 1–5, dec. 2007.

Appendix - Closed Form Solutions

Next, we present closed form solutions to the proposed model. We consider the case in which the time it takes for a recloser to isolate a section is zero, $\epsilon = 0$ (in practice, it is much smaller than all the other time intervals considered in the system) and demand response is not integrated with the failure recovery system, $r = 0$.

Let Y_k be the k-th visited state after a failure, $Y_0 = 0$. Let $P(Y_1 = j)$ be the system *initial condition* after the failure, which is determined by p and q, $P(Y_1 = 1) = pq$, $P(Y_1 = 3) = p(1-q)$ and $P(Y_1 = 4) = 1 - p$. Let $\pi_k(t)$ be the probability that the system is at state k at time t, $t \geq 0$. Then,

$$\pi_k(t) = \sum_{l=1}^{6} \pi_k(t|Y_1 = l)P(Y_1 = l), \quad k = 1, \ldots, 6 \qquad (9)$$

To compute $\pi_k(t|Y_1 = j)$ we note that there is at most one sample path from each state to every other. Let $\Upsilon_{i,j}$ be the sample path from state i to state j. Let $\Upsilon_{j,i}(k)$ be the k-th state in the path $\Upsilon_{j,i}$, $1 \leq k \leq |\Upsilon_{j,i}|$. Let $P(\Upsilon_{j,i})$ be the probability that $Y_k = \Upsilon_{j,i}(k)$, $k = 1, \ldots, |\Upsilon_{j,i}|$, conditioned on $Y_1 = j$. Then,

$$\pi_k(t|Y_1 = j) = \pi_k(t|Y_1 = j, \Upsilon_{j,i})P(\Upsilon_{j,i}) \qquad (10)$$

Substituting (10) into (9) yields closed form expressions for $\pi_k(t)$, $k = 0, \ldots, 6$,

$$\begin{aligned}
\pi_0(t) &= \pi_2(t) = 0 \\
\pi_1(t) &= e^{-(\alpha+\delta)t}pq + \left(e^{-(\gamma+\delta)t} - e^{-(\alpha+\delta)t}\right)\left(\frac{\gamma q(1-p)}{\alpha-\gamma}\right) \\
\pi_3(t) &= e^{-\delta t}p(1-q) + (1-q)(e^{\gamma t}-1)e^{-(\delta+\gamma)t}(1-p) \\
\pi_4(t) &= e^{-(\delta+\gamma)t}(1-p) \\
\pi_5(t) &= (1-e^{\alpha t})e^{-\delta t}pq + \\
&\quad + \left(e^{-\delta t} - \frac{\alpha e^{-(\delta+\gamma)t} - \gamma e^{-(\delta+\alpha)t}}{\alpha-\gamma}\right) q(1-p) \\
\pi_6(t) &= 1 - e^{-t\delta}
\end{aligned}$$

The probability that $i+$ has been recovered by time t is $\pi_{5\cup 6}(t)$ and is given by

$$\pi_{5\cup 6}(t) = \pi_5(t) + \pi_6(t) \qquad (11)$$

We now compute the mean time to fix $i+$, *i.e.*, to reach states 5 or 6 ($\overline{M}$, the mean time to reach state 6, defined by (6), follows similarly). To this goal, we consider a Markov chain with state space Ω' and infinitesimal generator Q'_ψ, where ψ is a a real valued parameter. The states in Ω' are indexed by 1, 3, 4 and $\star$ and the infinitesimal generator Q' is obtained from Figure 3 after 1) replacing the transitions to states 5 and 6 by transitions to state $\star$ and 2) adding transitions from state $\star$ to states 1, 3 and 4 with rates ψpq, $\psi p(1-q)$ and $\psi(1-p)$, respectively,

$$Q'_\psi = \begin{bmatrix} -(\alpha+\delta) & 0 & 0 & \alpha+\delta \\ 0 & -\delta & 0 & \delta \\ \gamma q & \gamma(1-q) & -(\delta+\gamma) & \delta \\ \psi p q & \psi p(1-q) & \psi(1-p) & -\psi \end{bmatrix} \begin{matrix} 1 \\ 3 \\ 4 \\ \star \end{matrix} \quad (12)$$

Let $\boldsymbol{\pi}'$ be the steady state solution of the MC above, satisfying $\boldsymbol{\pi}' Q'_\psi = 0$ and $\pi'_1 + \pi'_3 + \pi'_4 + \pi'_\star = 1$. Then,

$$\pi'_1 = \psi\delta q(\gamma + \delta p)/\Delta \quad (13)$$

$$\pi'_3 = \psi(\alpha+\delta)(1-q)(\gamma+\delta p)/\Delta \quad (14)$$

$$\pi'_4 = \psi\delta(\alpha+\delta)(1-p)/\Delta \quad (15)$$

$$\pi'_\star = \delta(\alpha+\delta)(\delta+\gamma)/\Delta \quad (16)$$

where Δ is a normalization constant to ensure $\pi'_1+\pi'_3+\pi'_4+\pi'_\star = 1$.

Let C be a random variable characterizing the time to reach state $\star$ immediately after leaving $\star$. This corresponds to the mean time to reach states 5 or 6 in the original Markov chain, starting from state 0. The mean time to repair $i+$ is given by $E[C]$. It follows from (13)-(16) and [12, Chapter III] that

$$E[C] = \psi^{-1}(1/\pi'_\star - 1) \quad (17)$$

(17) is used to obtain the curves in Figure 4.

Let $s_k(\{5,6\})$ be the mean time spent at state k before reaching states 5 or 6 from state 0 in the original Markov chain, $k = 1, 3, 4$. It follows from (13)–(17) that

$$\overline{s}_k(\{5,6\}) = \frac{\pi'_k}{\psi(\pi'_1 + \pi'_3 + \pi'_4)} \left(\frac{1}{\pi'_\star} - 1 \right), \quad k = 1, 3, 4$$

Let $L(\{5,6\})$ be the mean accumulated reward up to reaching states 5 or 6 of the original Markov chain,

$$\overline{L}(\{5,6\}) = \sum_{i=1,3,4} \sigma_k \overline{s}_k(\{5,6\}) \quad (18)$$

In this appendix we have shown how to compute $\overline{s}_k(\{5,6\})$, $k = 1, 3, 4$ and $\overline{L}(\{5,6\})$. The mean time spent at state k before reaching state 6, $\overline{s}_k$, $k = 1, 3, 4, 5$, and the mean accumulated reward up to reaching state 6, $\overline{L}$, defined by (8), follow similarly. An adaptation of (18) is used to obtain the curves in Figure 7.

Although in this appendix we assumed that $r = 0$ and $\epsilon = 0$, a similar approach is applicable when $r > 0$ or $\epsilon > 0$.

Benchmarking Approach for Designing a MapReduce Performance Model

Zhuoyao Zhang
University of Pennsylvania
zhuoyao@seas.upenn.edu

Ludmila Cherkasova
Hewlett-Packard Labs
lucy.cherkasova@hp.com

Boon Thau Loo
University of Pennsylvania
boonloo@seas.upenn.edu

ABSTRACT

In MapReduce environments, many of the programs are reused for processing a regularly incoming new data. A typical user question is how to estimate the completion time of these programs as a function of a new dataset and the cluster resources. In this work[1], we offer a novel performance evaluation framework for answering this question. We observe that the execution of each map (reduce) tasks consists of specific, well-defined data processing phases. Only map and reduce functions are custom and their executions are user-defined for different MapReduce jobs. The executions of the remaining phases are *generic* and depend on the amount of data processed by the phase and the performance of underlying Hadoop cluster. First, we design *a set of parameterizable microbenchmarks* to measure generic phases and to derive *a platform performance model* of a given Hadoop cluster. Then using the job past executions, we summarize job's properties and performance of its custom map/reduce functions in a compact job profile. Finally, by combining the knowledge of the job profile and the derived platform performance model, we offer *a MapReduce performance model* that estimates the program completion time for processing a new dataset. The evaluation study justifies our approach and the proposed framework: we are able to accurately predict performance of the diverse set of twelve MapReduce applications. The predicted completion times for most experiments are within 10% of the measured ones (with a worst case resulting in 17% of error) on our 66-node Hadoop cluster.

Categories and Subject Descriptors: C.4 [Computer System Organization] Performance of Systems, D.2.6.[Software] Programming Environments.

General Terms: Measurement, Performance, Design.

Keywords: MapReduce, benchmarking, job profiling, performance modeling

1. INTRODUCTION

MapReduce and Hadoop represent an economically compelling alternative for efficient large scale data processing and cost-effective analytics over "Big Data" in the enterprise. There is a slew of interesting applications associated with live business intelligence that require completion time guarantees. While there were a few research efforts to design different models and approaches for predicting performance of MapReduce applications [6, 5, 8, 9, 10], this question still remains a challenging research problem. Some of the past modeling efforts aim to predict the job completion time by analyzing the execution times' distribution of map and reduce tasks [9], and deriving some scaling factors for these execution times when the original MapReduce application is applied for processing a larger dataset [10, 8]. Some other efforts [6, 5, 8] aim to perform a more detailed (and more expensive) job profiling at a level of phases that comprise the execution of map and reduce tasks.

In this work, we offer a new approach for designing a MapReduce performance model that can efficiently predict the completion time of a MapReduce application for processing a new dataset as a function of allocated resources. In some sense, it combines the useful rationale of the detailed *phase* profiling method [6] in order to more accurately estimate durations of map and reduce tasks, and then apply fast and efficient analytical models designed in [9]. However, our new framework proposes a very *different approach* to estimate the execution times of these job phases. We observe that the executions of map and reduce tasks consist of specific, well-defined data processing phases. Only map and reduce functions are custom and their computations are user-defined for different MapReduce jobs. The executions of the remaining phases are *generic*, i.e., strictly regulated and defined by the Hadoop processing framework. The execution time of each generic step depends on the amount of data processed by the phase and the performance of underlying Hadoop cluster. In the earlier papers [6, 5, 8], profiling is done for all the phases (including the generic ones) for each application separately. Then these measurements are used for predicting a job completion time. In our work, we design *a set of parameterizable microbenchmarks* to measure generic phases and to derive *a platform performance model* of a given Hadoop cluster. The new framework consists of the following key components:

- *A set of parameterizable microbenchmarks* to measure and profile different phases of the MapReduce processing pipeline of a given Hadoop cluster. These microbenchmarks might be executed in a small cluster deployment, e.g., having only 5 nodes. This test cluster employs a similar hardware and Hadoop configuration as a production cluster, and its advantage is that the benchmarking process does not interfere with the production jobs, and therefore, can be performed much faster while testing a large set of diverse processing patterns. The input parameters of microbenchmarks im-

[1]This work was completed during Z. Zhang's internship at HP Labs. Prof. B. T. Loo and Z. Zhang are supported in part by NSF grants CNS-1117185 and CNS-0845552.

ICPE'13, April 21-24, 2013, Prague, Czech Republic.

pact the amount of data processed by different phases of map and reduce tasks. We concentrate on profiling of general (non-customized) phases of MapReduce processing pipeline. Intuitively, the executions of these phases on a given Hadoop cluster are similar for different MapReduce jobs, and the phase execution time mostly depends on the amount of processed data. By running a set of diverse benchmarks on a given Hadoop cluster we collect a useful training set that characterizes the execution time of different phases while processing different amounts of data.

- *A platform profile* and *a platform performance model* that characterize the execution time of each *generic phase* during MapReduce processing. Using the created training set and a robust linear regression we derive a platform performance model that estimates each phase duration as a function of processed data.
- *A compact job profile* for each MapReduce application of interest that is extracted from the past job executions. It summarizes the job's properties and performance of its custom map and reduce functions. This job profile captures the inherent application properties such as the job's map (reduce) selectivity, i.e., the ratio of the map (reduce) output to the map (reduce) input. This parameter describes the amount of data produced as the output of the user-defined map (reduce) function, and therefore it helps in predicting the amount of data processed by the remaining generic phases.
- *A MapReduce performance model* that combines the knowledge of the extracted job profile and the derived platform performance model to estimate the completion time of the programs for processing a new dataset.

The proposed evaluation framework aims to **divide** *i)* the performance characterization of the underlying Hadoop cluster and *ii)* the extraction of specific performance properties of different MapReduce applications. It aims to derive **once** an accurate performance model of Hadoop's generic execution phases as a function of processed data, and then **reuse** this model for characterizing performance of generic phases across different applications (with different job profiles).

We validate our approach using a set of 12 realistic applications executed on a 66-node Hadoop cluster. We derive the platform profile and platform performance model in a small 5-node test cluster, and then use this model together with extracted job profiles to predict the job completion time on the 66-node Hadoop cluster. Our evaluation shows that the designed platform model accurately characterizes different execution phases of MapReduce processing pipeline of a given cluster. The predicted completion times of most considered applications are within 10% of the measured ones.

The rest of this paper is organized as follows. Section 2 provides MapReduce background and presents our phase profiling approach. Section 3 describes microbenchmarks and objectives for their selection. Sections 4 , 5 introduce main performance models that form a core of the proposed framework. Section 6 evaluates the accuracy and effectiveness of our approach. Section 7 outlines related work. Section 8 presents conclusion and future work directions.

2. MAPREDUCE PROCESSING PIPELINE

In the MapReduce model [4], the main computation is expressed as two user-defined functions: *map* and *reduce*. The map function takes an input pair and produces a list of intermediate key/value pairs. The reduce function then merges or aggregates all the values associated with the same key. MapReduce jobs are executed across multiple machines: the map stage is partitioned into *map tasks* and the reduce stage is partitioned into *reduce tasks*.

The execution of each map (reduce) task is comprised of a specific, well-defined sequence of processing phases (see Fig. 1). Note, that only map and reduce phases with customized map and reduce functions execute the user-defined pieces of code. The execution of the remaining phases are *generic* (i.e., defined by Hadoop code), and mostly depend on the amount of data flowing through a phase. Our goal is to derive *a platform performance model* that predicts a duration of each generic phase on a given Hadoop cluster platform as a function of processed data. In order to accomplish this, we plan to run a set of microbenchmarks that create different amounts of data for processing per map (reduce) tasks and for processing by their phases. We need to profile a duration of each generic phase during the task execution and derive a function that defines a phase performance as a function of the processed data from the collected measurements.

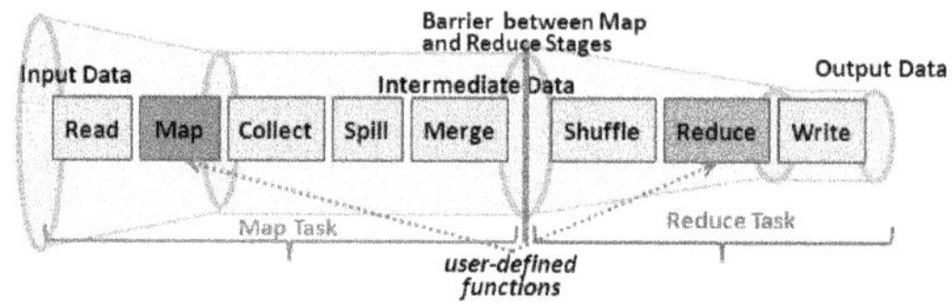

Figure 1: MapReduce Processing Pipeline.

Map task consists of the following generic phases:

1. *Read* phase – a map task typically reads a block (e.g., 64 MB) from the Hadoop distributed file system (HDFS). However, written data files might be of arbitrary size, e.g., 70 MB. In this case, there will be two blocks: one of 64 MB and the second of 6 MB, and therefore, map tasks might read files of varying sizes. We measure the duration of the read phase as well as the amount of data read by the map task.

2. *Collect* phase – this generic phase follows the execution of the map phase with a user-defined map function. We measure the time it takes to buffer map phase outputs into memory and the amount of generated intermediate data.

3. *Spill* phase – we measure the time taken to locally sort the intermediate data and partition them for the different reduce tasks, applying the combiner if available, and then writing the intermediate data to local disk.

4. *Merge* phase – we measure the time for merging different spill files into a single spill file for each destined reduce task.

Reduce task consists of the following generic phases:

1. *Shuffle* phase – we measure the time taken to transfer intermediate data from map tasks to the reduce tasks and merge-sort them together. We combine the shuffle and sort phases because in the Hadoop implementation, these two sub-phases are interleaved. The processing time of this phase depends on the amount of intermediate data destined for each reduce task and the Hadoop configuration parameters. In our testbed, each JVM (i.e., a map/reduce slot) is allocated 700 MB of RAM. Hadoop sets a limit (~46% of the allocated memory) for in-memory sort buffer. The portions of shuffled data are merge-sorted in memory, and a spill file (*~320 MB*) is written to disk. After all the data is shuffled, Hadoop merge-sorts first 10 spilled files and writes them in the new sorted file. Then it merge-sorts next 10 files and writes them in the next new sorted file. At the end, it merge-sorts these new sorted files. Thus, we can expect that the duration of the shuffle phase might be approximated with a different linear function when the intermediate dataset per reduce task is larger than *3.2 GB* in our Hadoop cluster. For a differently configured Hadoop cluster, this threshold can be similarly determined from the cluster configuration parameters.

2. *Write* phase – this phase follows the execution of the reduce phase that executes a custom reduce function. We measure the amount of time taken to write the reduce output to HDFS.

Note, that in platform profiling we do not include phases with user-defined map and reduce functions. However, we do need to profile these custom map and reduce phases for modeling the execution of given MapReduce applications:

- *Map (Reduce)* phase – we measure a duration of the entire map (reduce) function and the number of processed records. We normalize this execution time to estimate a processing time per record.

Apart from the phases described above, each executed task has a constant overhead for setting and cleaning up. We account for these overheads separately for each task.

For accurate performance modeling it is desirable to minimize the overheads introduced by the additional monitoring and profiling technique. There are **two** different approaches for implementing phase profiling.

1. Currently, Hadoop already includes several *counters* such as the number of bytes read and written. These counters are sent by the worker nodes to the master periodically with each heartbeat. We *modified the Hadoop code* by adding counters that measure durations of the six generic phases to the existing counter reporting mechanism. We can activate the subset of desirable counters in the Hadoop configuration for collecting the set of required measurements.

2. We also implemented the alternative profiling tool inspired by Starfish [6] approach based on *BTrace* – a dynamic instrumentation tool for Java [1]. This approach does have a special appeal for production Hadoop clusters because it has a zero overhead when monitoring is turned off. However, in general, the dynamic instrumentation overhead is much higher compared to adding new Hadoop counters directly in the source code. We instrumented selected Java classes and functions internal to Hadoop using *BTrace* in order to measure the time taken for executing different phases.

In this work, first, we use profiling to create *a platform performance model*. We execute a set of microbenchmarks (described in the next Section 3) and measure the durations of six *generic* execution phases for processing different amount of data: *read, collect, spill,* and *merge* phases for the map task execution, and *shuffle* and *write* phases in the reduce task processing. This profiling is done on a small test cluster (5-nodes in our experiments) with the same hardware and configuration as the production cluster. While for these experiments both profiling approaches can be used, the Hadoop counter-based approach is preferable due to its simplicity and low overhead, and that the modified Hadoop version can be easily deployed in this test environment.

The *MapReduce performance model* needs additional measurements that characterize the execution of user-defined map and reduce functions of a given job. For profiling the map and reduce phases of the given MapReduce jobs in the production cluster we apply our alternative profiling tool that is based on *BTrace* approach. Remember, this approach does not require Hadoop or application changes, and can be switched on for profiling a targeted MapReduce job of interest. Since we only profile map and reduce phase executions the extra overhead is relatively small.

3. MICROBENCHMARKS AND PLATFORM PROFILE

We generate and perform a set of parameterizable microbenchmarks to characterize execution times of generic phases for processing different data amounts on a given Hadoop cluster by varying the following parameters:

1. *Input size per map task* (M^{inp}): This parameter controls the size of the input read by each map task. Therefore, it helps to profile the *Read* phase durations for processing different amount of data.
2. *Map selectivity* (M^{sel}): this parameter defines the ratio of the map output to the map input. It controls the amount of data produced as the output of the map function, and therefore directly affects the *Collect, Spill* and *Merge* phase durations in the map task. Map output determines the overall amount of data produced for processing by the reduce tasks, and therefore impacting the amount of data processeded by *Shuffle* and *Reduce* phases and their durations.
3. *A number of map tasks* N^{map}: increasing this parameter helps to expedite generating the large amount of intermediate data per reduce task.
4. *A number of reduce tasks* N^{red}: decreasing this parameter helps to control the number of reduce tasks to expedite the training set generation with the large amount of intermediate data per reduce task.

Thus, each microbenchmark MB_i is parameterized as

$$MB_i = (M_i^{inp}, M_i^{sel}, N_i^{map}, N_i^{red}).$$

Each created benchmark uses input data consisting of 100 byte key/value pairs generated with *TeraGen* [3], a Hadoop utility for generating synthetic data. The map function simply emits the input records according to the specified map selectivity for this benchmark. The reduce function is defined as the identity function. Most of our benchmarks consist of a specified (fixed) number of map and reduce tasks. For example, we generate benchmarks with 40 map and 40 reduce tasks each for execution in our small cluster deployments with 5 worker nodes (see setup details in Section 6). We run benchmarks with the following parameters: M^{inp}={2MB, 4MB, 8MB, 16MB, 32MB, 64MB}; M^{sel}={0.2, 0.6, 1.0, 1.4, 1.8}. For each value of M^{inp} and M^{sel}, a new benchmark is executed. We also use benchmarks that generate special ranges of intermediate data per reduce task for accurate characterization of the shuffle phase. These benchmarks are defined by N^{map}={20,30,...,150,160}; $M^{inp} = 64MB$, $M^{sel} = 5.0$ and $N^{red} = 5$ which result in different intermediate data size per reduce tasks ranging from 1 GB to 12 GB.

We generate the *platform profile* by running a set of our microbenchmarks on the small 5-node test cluster that is similar to a given production Hadoop cluster. We gather durations of generic phases and the amount of processed data for all executed map and reduce tasks. A set of these measurements defines the *platform profile* that is later used as the training data for a *platform performance model*:

- *Map task processing:* in the collected platform profiles, we denote the measurements for phase durations and the amount of processed data for *read, collect, spill,* and *merge* phases as $(Dur_1, Data_1)$, $(Dur_2, Data_2)$, $(Dur_3, Data_3)$, and $(Dur_4, Data_4)$ respectively.
- *Reduce task processing:* in the collected platform profiles, we denote phase durations and the amount of processed data for *shuffle* and *write* as $(Dur_5, Data_5)$ and $(Dur_6, Data_6)$ respectively.

Figure 2 shows a small fragment of a collected platform profile as a result of executing the microbenchmarking set. There are six tables in the platform profile, one for each phase. Figure 2 shows fragments for *read* and *collect* phases. There are multiple map and reduce tasks that process the same amount of data in each microbenchmark. This is why there are multiple measurements in the profile for processing the same data amount.

Row number j	Data MB $Data_1$	*Read* msec Dur_1
1	16	2010
2	16	2020
...	...	...

Row number j	Data MB $Data_2$	*Collect* msec Dur_2
1	8	1210
2	8	1350
...	...	...

Figure 2: A fragment of a platform profile for *read* and *collect* phases.

4. PLATFORM PERFORMANCE MODEL

Now, we describe how to create a *platform performance model* M_{Phases} which characterizes the phase execution as a function of processed data. To accomplish this goal, we need to find the relationships between the amount of processed data and durations of different execution phases using the set of collected measurements. Therefore, we build *six* submodels $M_1, M_2, ..., M_5$, and M_6 that define the relationships for *read, collect, spill, merge, shuffle,* and *write* respectively of a given Hadoop cluster. To derive these submodels, we use the collected platform profile (see Figure 2).

Below, we explain how to build a submodel M_i, where $1 \leq i \leq 6$. By using measurements from the collected platform profiles, we form a set of equations which express a phase duration as a linear function of processed data. Let $Data_i^j$ be the amount of processed data in the row j of platform profile with K rows. Let Dur_i^j be the duration of the corresponding phase in the same row j. Then, using linear regression, we solve the following sets of equations (for each $i = 1, 2, \cdots, 6$):

$$A_i + B_i \cdot Data_i^j = Dur_i^j, \quad \text{where } j = 1, 2, \cdots, K \qquad (1)$$

To solve for (A_i, B_i), one can choose a regression method from a variety of known methods in the literature (a popular method for solving such a set of equations is a non-negative Least Squares Regression). To decrease the impact of occasional bad measurements and to improve the overall model accuracy, we employ robust linear regression [7].

Let $(\hat{A}_i, \hat{B}_i)$ denote a solution for the equation set (1). Then $M_i = (\hat{A}_i, \hat{B}_i)$ is the submodel that defines the duration of execution phase i as a function of processed data. The *platform performance model* is $M_{Phases} = (M_1, M_2, ..., M_5, M_6)$.

In addition, we have implemented a test to verify whether two linear functions may provide a better fit for approximating different segments of training data (ordered by the increasing data amount) instead of a single linear function derived on all data. As we will see in Section 6, the shuffle phase is better approximated by a combination of two linear functions over two data ranges: less than 3.2 GB and larger than 3.2 GB (confirming the conjecture that was discussed in Section 2).

5. MAPREDUCE PERFORMANCE MODEL

In this section, we describe *a MapReduce performance model* that is used for predicting a completion time of a given MapReduce job. We do it by applying the *analytical model* designed and validated in ARIA [9]. The proposed performance model utilizes the knowledge about average and maximum map (reduce) task durations for computing the lower and upper bounds on the job completion time as a function of allocated resources (map and reduce slots). Typically, the estimated completion time based on the average of lower and upper bounds serves as a good prediction: it is within 10% of the measured completion time as shown in [9].

To apply the analytical model, we need to estimate the map and reduce tasks distributions to approximate the average and maximum completion time of the map and reduce tasks. To achieve this, for a given MapReduce job, a special compact job profile is extracted automatically from the previous run of this job. It includes the following metrics:

- Map/reduce selectivity that reflects the ratio of the map/reduce output size to the map/reduce input size;
- Processing time per record of map/reduce function.

We also collect characteristics of the input dataset such as 1) the number of data blocks, and 2) the average/maximum data block size (both in bytes and in the number of records). Such information defines the number of map tasks and the average/maximum input data per map task.

For map tasks, the task completion time is estimated as a sum of durations of all the map stage phases. The generic phase durations for *read, collect, spill* and *merge* are estimated according to the platform model $M_{Phases} = (M_1, M_2, ..., M_5, M_6)$ by applying a correspoding function to the data amount processed by the phase. Note, that the data size for *collect, spill,* and *merge* phases is estimated by applying the map selectivity to the map task input data size (this information is available in the extracted job profile). The *map* phase duration depends on the user-defined map functions and is estimated according to the number of input records and the map function processing time per record (again available from the extracted job profile). Depending on the average and maximum input data size, we estimate the average and maximum map task durations respectively.

For reduce tasks, the task completion time is estimated as a sum of durations of *shuffle, reduce,* and *write* phases. Similarly to the map task computation described above, the *shuffle* and *write* phase durations are estimated according to the platform model and the phase input data size. The *reduce* function duration is estimated according to the number of reduce records and the reduce function processing time per record available from the job profile.

The input size for the *shuffle* phase depends on the overall data amount of the map outputs and the number of reduce tasks. Thus the input size of the reduce task can be estimated as:

$$Data_{shuffle} = (M^{inp} \times M^{sel} \times N^{map})/N^{red}, \qquad (2)$$

where we assume that each map output is uniformly distributed across the reduce tasks.

6. EVALUATION

All experiments are performed on 66 HP DL145 G3 machines. Each machine has four AMD 2.39GHz cores, 8 GB RAM and two 160 GB 7.2K rpm SATA hard disks. The machines are set up in two racks and interconnected with gigabit Ethernet. We used Hadoop 0.20.2 with two machines dedicated as the JobTracker and the NameNode, and remaining 64 machines as workers. Each worker is configured with 2 map and 2 reduce slots. The file system blocksize is set to 64MB. The replication level is set to 3. We disabled speculative execution since it did not lead to significant improvements in our experiments.

To profile the *generic* phases in the MapReduce processing pipeline of a given production cluster, we execute the designed set of microbenchmarks on the small 5-node test cluster that uses the same hardware and configuration as the large 66-node production cluster. Figure 3 shows the relationships between the amount of processed data and the execution durations of different phases for a given Hadoop cluster. Figures 3 (a)-(f) reflect the platform profile for six generic execution phases: *read, collect, spill,* and *merge* phases of the map task execution, and *shuffle* and *write* phases in the reduce task. Each graph has a collection of dots that represent phase duration measurements (Y-axes) of the profiled map (reduce) tasks as a function of processed data (X-axes). The red line

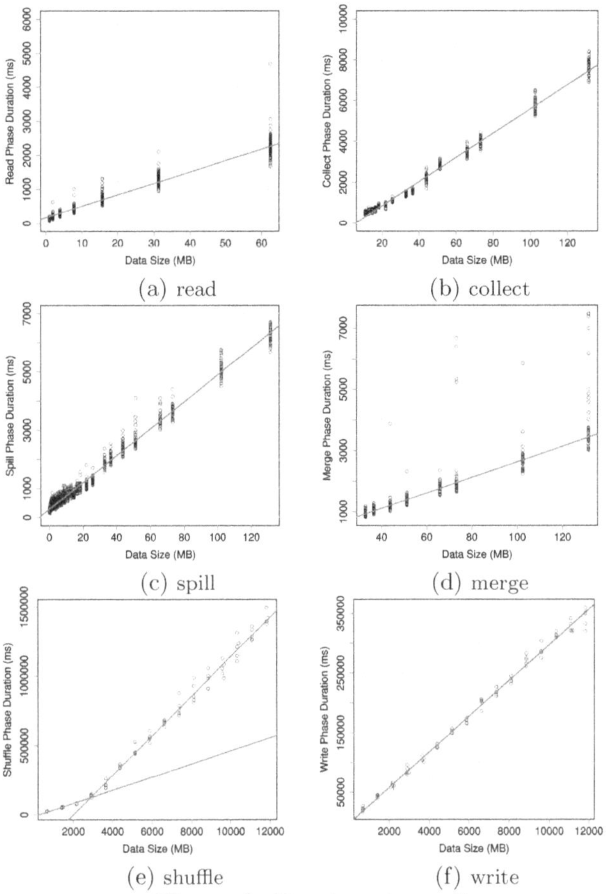

Figure 3: Benchmark results.

on the graph shows the linear regression solution that serves as a model for the phase. As we can see (visually) the linear regression provides a good solution for five out of six phases. As we expected, the shuffle phase is better approximated by a linear piece-wise function comprised of two linear functions (see a discussion about the shuffle phase in Section 2): one is for processing up to 3.2 GB of intermediate data per reduce task, and the second segment is for processing the datasets larger than 3.2 GB.

In order to formally evaluate the accuracy and fit of the generated model M_{Phases} we compute for each data point in our training dataset a *prediction error*. That is, for each row j in the platform profile we compute the duration dur_i^{pred} of the corresponding phase i by using the derived model M_i as a function of data $Data^j$. Then we compare the predicted value dur_i^{pred} against the measured duration d_i^{measrd}. The relative error is defined as follows:

$$error_i = \frac{|d_i^{measrd} - d_i^{pred}|}{d_i^{measrd}}$$

We calculate the relative error for all the data points in the platform profile. Table 1 summarizes the relative errors for derived models of six generic processing phases. For example, for the *read* phase, 66% of map tasks have the relative error less than 10%, and 92% of map tasks have the relative error less that 20%. For the *shuffle* phase, 76% of reduce tasks have the relative error less that 10%, and 96% of reduce tasks have the relative error less that 20%.

In summary, almost 80% of all the predicted values are within 15% of the corresponding measurements. Thus the

Table 1: Relative error distribution

phase	error ≤ 10%	error ≤ 15%	error ≤ 20%
read	66%	83%	92%
collect	56%	73%	84%
spill	61%	76%	83%
merge	58%	84%	94%
shuffle	76%	85%	96%
write	93%	97%	98%

derived platform performance model fits well the collected experimental data.

Next, we validate the accuracy of the proposed MapReduce performance model by predicting completion times of 12 applications made available by the Tarazu project [2].

Table 2 gives a high-level description of these 12 applications with the job settings (e.g, number of map and reduce tasks). Applications 1, 8, and 9 process synthetically generated data, applications 2 to 7 use the Wikipedia articles dataset as input, while applications 10 to 13 use the Netflix movie ratings dataset. We will present results of running these applications with: *i) small input datasets* defined by parameters shown in columns 3-4, and *ii) large input datasets* defined by parameters shown in columns 5-6 respectively.

Table 2: Application characteristics.

Application	*Input data (type)*	*Input (GB) small*	*#Map, Reduce tasks*	*Input (GB) large*	*#Map, Reduce tasks*
TeraSort	Synthetic	2.8	44, 20	31	495, 240
WordCount	Wikipedia	2.8	44, 20	50	788, 240
Grep	Wikipedia	2.8	44, 1	50	788, 1
InvIndex	Wikipedia	2.8	44, 20	50	788, 240
RankInvIndex	Wikipedia	2.5	40, 20	46	745, 240
TermVector	Wikipedia	2.8	44, 20	50	788, 240
SeqCount	Wikipedia	2.8	44, 20	50	788, 240
SelfJoin	Synthetic	2.1	32, 20	28	448, 240
AdjList	Synthetic	2.4	44, 20	28	508, 240
HistMovies	Netflix	3.5	56, 1	27	428, 1
HistRatings	Netflix	3.5	56, 1	27	428, 1
KMeans	Netflix	3.5	56, 16	27	428, 50

Figure 4 shows the comparison of the measured and predicted job completion times [2] for 12 applications (with a small input dataset) executed using 5-node test cluster. The graphs reflect that the designed MapReduce performance model closely predicts the job completion times. The measured and predicted durations are less than 10% for most cases (with 17% error being a worst case for *WordCount* and *HistRatings*). Note the split at Y-axes in order to accommodate a much larger scale for a completion time of the *KMeans* application.

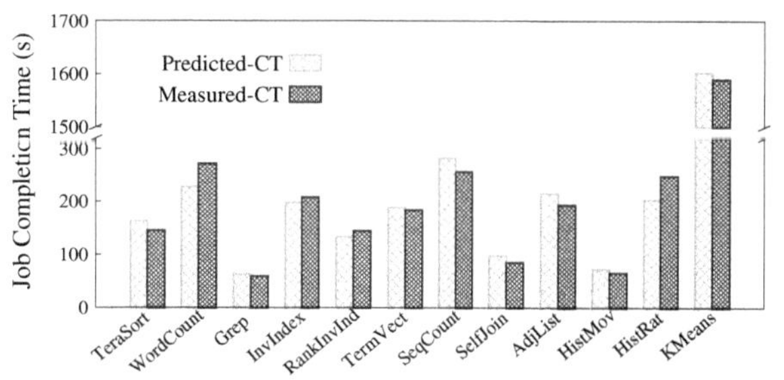

Figure 4: Predicted vs. measured completion times of 12 applications on the *small 5-node test cluster*.

The next question to answer is whether the *platform performance model* constructed using a small 5-node test cluster can be effectively applied for modeling the application performance in the larger production clusters. To answer this question we execute the same 12 applications (with a large input dataset) on the 66-node production cluster. Figure 5 shows the comparison of the measured and predicted job completion

[2]All the experiments are performed five times, and the measurement results are averaged. This comment applies to the results in Figure 4, 5.

times for 12 applications executed on the 66-node Hadoop cluster. The predicted completion times closely approximate the measured ones: for 11 applications they are less than 10% of the measured ones (a worst case is *WordCount* that exhibits 17% of error). Note the split at Y-axes for accomodating the *KMeans* completion time in the same figure.

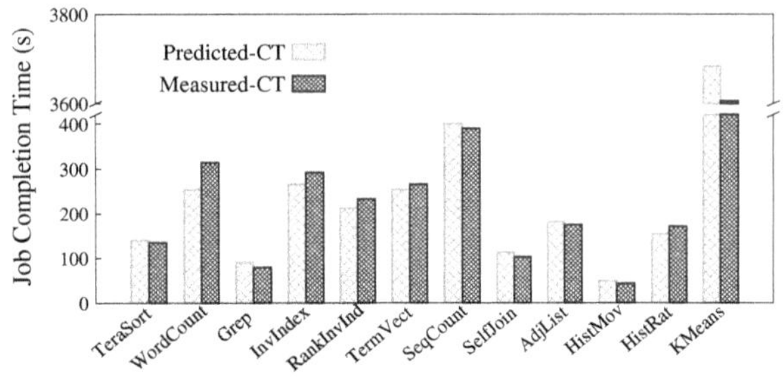

Figure 5: Predicted vs. measured completion times of 12 applications (with a large input dataset) on the ***large 66-node production cluster.***

These results justify our approach for building the platform performance model by using a small test cluster. Running benchmarks on the small cluster significantly simplifies the approach applicability, since these measurements do not interfere with production workloads while the collected platform profile leads to a good quality platform performance model that can be efficiently used for modeling production jobs in the larger enterprise cluster.

7. RELATED WORK

In past few years, performance modeling and workload management in MapReduce environments have received much attention, and different approaches [6, 5, 8, 9, 10] were offered for predicting performance of MapReduce applications.

Starfish [6] applies dynamic Java instrumentation to collect a run-time monitoring information about job execution at a fine granularity and by extracting a diverse variety of metrics. Such a detailed job profiling enables the authors to predict job execution under different Hadoop configuration parameters, automatically derive an optimized cluster configuration, and solve cluster sizing problem [5]. However, collecting a large set of metrics comes at a cost, and to avoid significant overhead profiling should be applied to a small fraction of tasks. Another main challenge outlined by the authors is a design of an efficient searching strategy through the high-dimensional space of parameter values. Our phase profiling approach is inspired by *Starfish* [6]. We build a light-weight profiling tool that only collects selected phase durations and therefore, it can profile each task at a minimal cost. Moreover, we apply counter-based profiling in a small test cluster to avoid impacting the production jobs.

Tian and Chen [8] propose an approach to predict MapReduce program performance from a set of test runs on small input datasets and small number of nodes. By executing 25-60 diverse test runs the authors create a training set for building a regression-based model of a given application. The derived model is able to predict the application performance on a larger input and a different size Hadoop cluster. It is an interesting approach but it cannot be directly applied for job performance optimization and parameter tuning problems.

ARIA [9] proposes a framework that automatically extracts compact job profiles from the past application run(s). These job profiles form the basis of a *MapReduce analytic performance model* that computes the lower and upper bounds on the job completion time. ARIA provides a fast and efficient capacity planning model for a MapReduce job with timing requirements. The later work [10] enhances and extends this approach by running the application on the smaller data sample and deriving the scaling factors for these execution times to predict the job completion time when the original MapReduce application is applied for processing a larger dataset.

In our current work, we consider a more detailed profiling of MapReduce jobs via six generic and two customized execution phases. The proposed models aim to estimate durations of these phases and accordingly the durations of map and reduce tasks when the application is executed on the new dataset. Once, we obtain predicted map and reduce task durations (by applying the proposed platform performance model and using the extracted job profile), we can utilize performance models designed in [9] for predicting the job completion time as a function of allocated resources.

8. CONCLUSION

Hadoop is increasingly being deployed in enterprise private clouds and also offered as a service by public cloud providers (e.g. Amazon's Elastic Map-Reduce). Many companies are embracing Hadoop for advanced data analytics over large datasets that require completion time guarantees.

In this work, we offer a new benchmarking approach for building a MapReduce performance model that can efficiently predict the completion time of a MapReduce application. We use a set of microbenchmarks to profile generic phases of the MapReduce processing pipeline of a given Hadoop cluster. We derive an accurate platform performance model of a given cluster **once**, and then **reuse** this model for characterizing performance of generic phases of different applications. The introduced MapReduce performance model combines the knowledge of the extracted job profile and the derived platform performance model to predict the program completion time on a new dataset. In the future work, we intend to apply the proposed approach and models for optimizing program performance, e.g., by tuning the number of reduce tasks and tailoring the resource allocation to meet the required completion time guarantees.

9. REFERENCES

[1] BTrace: A Dynamic Instrumentation Tool for Java. http://kenai.com/projects/btrace.

[2] F. Ahmad, S. Chakradhar, A. Raghunathan, and T. Vijaykumar. Tarazu: Optimizing MapReduce on heterogeneous clusters. In *Proc. of ASPLOS*, 2012.

[3] Apache. Hadoop: TeraGen Class. http://hadoop.apache.org/common/docs/r0.20.2/api/org/apache/hadoop/examples/terasort/TeraGen.html.

[4] J. Dean, and S. Ghemawat. MapReduce: Simplified Data Processing on Large Clusters. *Communications of the ACM*,51(1), 2008.

[5] H. Herodotou, F. Dong, and S. Babu. No one (cluster) size fits all: Automatic cluster sizing for data-intensive analytics. In *Proc. of ACM Symposium on Cloud Computing*, 2011.

[6] H. Herodotou, H. Lim, G. Luo, N. Borisov, L. Dong, F. Cetin, and S. Babu. Starfish: A Self-tuning System for Big Data Analytics. In *Proc. of 5th Conf. on Innovative Data Systems Research (CIDR)*, 2011.

[7] P. Holland and R. Welsch. Robust regression using iteratively reweighted least-squares. *Communications in Statistics-Theory and Methods*, 6(9):813–827, 1977.

[8] F. Tian and K. Chen. Towards Optimal Resource Provisioning for Running MapReduce Programs in Public Clouds. In *Proc. of IEEE Conference on Cloud Computing (CLOUD 2011)*.

[9] A. Verma, L. Cherkasova, and R. H. Campbell. ARIA: Automatic Resource Inference and Allocation for MapReduce Environments. *Proc. of the 8th ACM International Conference on Autonomic Computing (ICAC)*, 2011.

[10] A. Verma, L. Cherkasova, and R. H. Campbell. Resource Provisioning Framework for MapReduce Jobs with Performance Goals. *Proc. of the 12th ACM/IFIP/USENIX Middleware Conference*, 2011.

Towards Building Performance Models for Data-intensive Workloads in Public Clouds

Rizwan Mian, Patrick Martin,
Farhana Zulkernine
School of Computing
Queen's University
Kingston, Ontario, Canada, K7L3N6
{mian, martin, farhana}@cs.queensu.ca

Jose Luis Vazquez-Poletti
Departamento de Arquitectura de Computadores y Automatica
Universidad Complutense de Madrid
28040. Madrid, Spain
jlvazquez@fdi.ucm

ABSTRACT

The cloud computing paradigm provides the "illusion" of infinite resources and, therefore, becomes a promising candidate for large-scale data-intensive computing. In this paper, we explore experiment-driven performance models for data-intensive workloads executing in an infrastructure-as-a-service (IaaS) public cloud. The performance models help in predicting the workload behaviour, and serve as a key component of a larger framework for resource provisioning in the cloud. We determine a suitable prediction technique after comparing popular regression methods. We also enumerate the variables that impact variance in the workload performance in a public cloud. Finally, we build a performance model for a multi-tenant data service in the Amazon cloud. We find that a linear classifier is sufficient in most cases. On a few occasions, a linear classifier is unsuitable and non-linear modeling is required, which is time consuming. Consequently, we recommend that a linear classifier be used in training the performance model in the first instance. If the resulting model is unsatisfactory, then non-linear modeling can be carried out in the next step.

Categories and Subject Descriptors

C.4 [**Performance of Systems**]: Modeling techniques; H.2.4 [**Database Management**]: Systems – *query processing, transaction processing, concurrency.*

Keywords

Performance Model, Performance Variables, Prediction Techniques, Multi-Tenancy.

1. INTRODUCTION

The pay-as-you-go flexibility and the absence of an up-front commitment in clouds are attractive to new businesses or companies seeking to lower their operational costs or wishing to experiment with new applications. Force.com, for example, currently supports over 55,000 organizations by hosting their applications and data on shared resources as tenants [35]. Tenants operate in virtual isolation from one another. The organizations are allowed controlled access to their objects. Further, they can use and customize their tenants, while the data and customizations remain secure and insulated from other tenants. Each tenant can service multiple workloads, and a workload can access multiple tenants with appropriate permissions. In our case, the tenants exist in a public cloud, which offers its infrastructure and computing resources to the general public over the internet.

Predicting the workload behaviour sets the expectations for execution time and cost. In particular, this enables more reliable guarantees towards service level agreements (SLA) prior to any workload execution. Knowing the workload behaviour a priori is also very useful for many administrative tasks such as capacity planning, admission control and effective scheduling of workloads.

We see a performance model as a key component when predicting the cost of workload execution in the clouds. We propose a framework [18] that uses search algorithms, and associated performance and cost models to minimize the cost of executing workloads in a cloud. A search algorithm in our framework hunts for a suitable resource configuration given a workload and an objective such as minimal dollar-cost. It employs a cost model to estimate the expense of a resource configuration, which in turn uses a performance model to predict the workload behaviour on the resource configuration.

In our previous work, we used a Queuing Network Model (QNM) as a performance model [19]. We found that the response times for queries on a Virtual Machine (VM) as predicted by simple single server centre models, varied by as much as 70% from the measured response times. A simple model does not capture the impact on the workload performance of the interactions among different query types. Developing a more detailed QNM for a VM is not feasible because of the difficulties in acquiring detailed performance parameters in a public cloud environment. Further, high performance variability in clouds poses a great challenge for database applications in guaranteeing SLAs [25].

In this paper, we explore an experiment-driven approach for creating performance models for data-centric workloads on Infrastructure-as-a-Service (IaaS) public clouds such as the Amazon EC2 [7], which appeared around 2006. These clouds pose unique challenges due to multi-tenancy, heterogeneity of the virtual machine types, non-linear changes in resources and the interplay among mixed workloads. Serving multiple tenants increases: (a) competition among different tenants over the shared resources (like memory), and (b) interference amongst the concurrently executing requests. We believe that our experimental-based approach is particularly suited to the variability of cloud environments. Our performance models predict throughputs for transactions, and response times for queries. We enumerate the variables that impact variance, but limit the scope of our performance models by accounting for a subset of these variables. Furthermore, we identify different data patterns in the measurements.

We also compare different prediction techniques for their suitability as a base classifier, and choose the most promising one. We find that *linear classifiers* are suitable for most request types and are fast to build and validate. They require less involvement on a developer's part and can often be employed straight out-of-the-box with default parameters in a commonly used machine learning toolkit such as Weka [13]. However, the results are unsatisfactory where there are non-linear trends in the performance data. In such cases, we explore *non-linear modeling* methods, which require choosing a suitable kernel and searching for appropriate parameter values. The search time can be in the order of hours, which becomes prohibitive for large number of request types. As a result, we recommend that a linear classifier be used in training the performance model in the first instance. If the resulting model is unsatisfactory, then the non-linear modeling can be carried out as the next step. In both cases, the models obtained can be *serialized* into java *bytestream*, and used by other applications [1].

The paper is structured as follows. We survey related work in Section 2. The performance variables are discussed in Section 3. We customize the typical experiment-driven model building approach, determine the suitable number of samples to build a representative performance model, and compare different prediction techniques in Section 4. Section 5 contains evaluation of the performance model. We choose diverse request types from three tenant databases, and justify our choice; identify different data patterns observed during the construction of the performance models; and present the validation results with a linear classifier. Section 6 discusses non-linear modeling. We discuss the evaluation results in Section 7. Finally, we present the summary and conclusions in Section 8.

2. RELATED WORK

Analytical performance models have enjoyed great popularity in the database management systems (DBMSs) area. Weikum et al. [34] provide a survey of the advances in autonomic tuning in database technology. They conclude that self-tuning should be based on a feedback control loop and should use mathematical models with proper deployment into the system components. Abouzour et al. [2] use analytical modeling to set the multi-programming level (mpl) of a DBMS for improved throughput. Analytical models have also been used for answering "what-if" questions to study the effects of system changes, such as system upgrades and service migrations [27]. Analytical models, however, are hard to evolve with the underlying system and make simplifying assumptions that make them oblivious to the interactions of the dynamically changing workloads and their effects [26]. These effects are amplified by the variance in the cloud [19]. Therefore, there is increasing interest in experiment-driven machine learning and statistical modeling.

Ganapathi et al. [11] predict multiple performance metrics for individual query types with less than 20% error for 85% of the test cases. Their work however, focuses on single query types and ignore interactions and query mixes. Gupta et al. [12] study the problem of predicting the execution time of a query on a data warehouse with a dynamically changing workload. They use a machine learning approach that takes the query plan, combines it with the observed load vector of the system and uses the new vector to predict the execution time of the query.

Courtious et al. [10] propose a prediction technique, called *regression splines*, that builds a non-linear regression function piecewise using a set of linear functions. In addition, they automate building of a regression function up to a desired accuracy by performing progressive sampling and experimentation. They evaluate their work by predicting CPU demands of an event based server. We find their idea of automation useful but consider standard predictive models available in Weka in our case.

Some recent papers view the *transaction mix* as the combination of different transactions that execute during a time interval window without considering which of these transactions execute simultaneously. This is fundamentally different to our notion of a *concurrent* request mix, where request instances execute simultaneously at any given time. Transaction mix models have been used for capacity planning [40], workload management [39], preempting congestion [38] and detecting anomalies in performance [17].

Much of the above work does not consider interactions between the concurrently executing requests, which can have a significant impact on database system performance [3]. Ahmad et al. [4] develop an interaction-aware query scheduler that targets report-generation workloads in Business Intelligence (BI) settings. Under certain assumptions, the schedule found by this scheduler is within a constant factor of optimal, and consistently outperforms conventional schedulers that do not account for query interactions.

Ahmad et al. [5] use a combination of an offline statistical model trained on sample query mixes and an online interaction-aware simulator to estimate workload completion times. No prior assumptions are made about the internal workings of the database system or the cause of query interactions, making the models robust and portable.

Tozer et al. [28] use a linear regression response time model for throttling long running queries. A performance model built using linear regression is unable to model non-linear trends in the response times of a query. Sheikh et al. [26] propose performance modeling based on Gaussian Processes, which can model non-linear trends, update online and reuse prior knowledge.

The performance models used in the above literature are typically built for workloads accessing a single data tenant. Further, the performance models usually provide predictions for response time only, and are validated on a local server or a local VM. In contrast, our performance model predicts both throughput and response times for transactional and analytical workloads, and operates over a multi-tenant data-service. We propose usage of different classifiers that vary in their modeling scope and development effort. We believe that this is the first attempt to build such models in a public cloud.

3. VARIABLES IN BUILDING A PERFORMANCE MODEL

Cloud components such as CPU, memory and I/O suffer from high performance unpredictability, especially when compared to a physical machine in a local network [25]. This is a major problem in building a performance model for workload execution, which is used for providing SLAs [19].

Therefore, we discuss some variables that play an important role in performance variance and their possible values in building a performance model. As a rule of thumb, the wider the scope of the performance model, the greater the variance it has to capture, and the lower is its prediction accuracy. The possible combinations of these variables in our experimental environment are large so, for purposes of the presentation, we choose to use combinations that provide us with modestly generic and accurate performance models. Hence, we acknowledge the variables' presence but explore a subset due to practical reasons.

We view a workload as a set of request types, where each request type has zero or more instances. All request instances are executed concurrently. Therefore, a workload is synonymous with a request mix that executes at a data service at any particular time, and we use them interchangeably. A performance model is trained and validated against a number of samples, where each sample is a different workload or request mix.

We consider the following variables when building a performance model for workloads executing at a data service in a cloud:

1. **Workload and tenant diversity:** The workload for a tenant may be entirely Online Transaction Processing (OLTP), entirely Online Analytical Processing (OLAP) or a mix of the two. The OLAP queries and OLTP transactions place significantly different requirements on a data service. The OLAP queries may take hours to execute, while OLTP transactions usually complete in a sub-second timeframe. In reality, a data service is rarely an analytical data source. Further, multi-tenancy increases the variance in the behaviour of a data service, since the tenants compete for shared resources such as memory. In this paper, we build performance models for *any type* of workload executing against a *multi*-tenant data service.

2. **Execution Platform:** The system attributes, such as number of cores, memory and I/O performance, of the cloud VMs do not vary linearly [6]. Even the servers hosting the same VM type do not necessarily have the same processor type. For example, the *xlarge* VM type in the Amazon cloud is powered by either a Xeon or an Optron processor. This amplifies the variance many folds, which can be about 35% compared to the 0.1% on a local physical machine for the CPU [25].

   Given the non-linearity of the VM types and the heterogeneity of the processors, the workload behaviour is likely to be non-linear across VM types and their host servers. In order to limit the variance, we decide to build models for specific VM and processor types in this paper.

3. **Day of the week:** Schad et al. [25] note that the CPU performance of a VM type also varies by the day of a week. This increases the variance in the training data for the performance model. Daily variance can be explicitly modelled using some additional attributes, say day of the week, or using techniques like time series. Either approach comes at the cost of additional complexity. Further, taking daily measurements is possible but labor-intensive. We want our performance model to be time independent. However, we leave the modelling of daily variance for future work. For now, we overlook the daily variance in building and validating our performance models.

4. **Model specificity:** Building a performance model specific to a subset of workloads reduces the variance in the training data. Consequently, a workload specific performance model is more accurate compared to a model built for any type of workload. However, the building effort is exponential in the number of workloads in the worst-case scenario. This becomes excessively brute force and highly synonymous to a lookup table. Instead, we use stratified sampling over the workload space to give us a modest coverage. This is discussed further in Section 4.1.

5. **Prediction technique:** The performance model can employ a number of regression methods or base classifiers such as linear regression or multi-layer perceptron. These classifiers vary in their ability to capture variance at the cost of training and runtime complexity. The training complexity places a requirement on the number of samples required for a *representative* model. Meanwhile, the runtime complexity can undermine the performance model if it is embarrassingly large. We compare different classifiers in Section 4.5.

4. BUILDING THE PERFORMANCE MODEL

Our approach is a typical experiment-driven performance modeling method customized to clouds. It consists of three stages: (a) sampling the space of possible request types and their instances for a request mix, (b) collecting data by executing possible request mixes or samples, and (c) pre-processing data and building performance models. In addition, we empirically determine a suitable number of samples for building a representative performance model, and we compare different prediction techniques to determine their suitability in our performance model. We believe that the latter two exercises need not be repeated every time a similar performance model is built.

4.1 Sampling the space of request mixes

The possible combinations of request mixes are exponential, so an effective sampling approach is essential. Similar to Tozer et al. [28], we randomly sample the N-dimensional space, where N is the number of request types, using a Latin Hypercube Sampling (LHS) protocol [15]. This protocol significantly reduces the number of experiments needed while providing a *normal* coverage of the possible request mixes. This is because the distribution of all the request instances, R, considering all the samples approximates a normal distribution around the mean value of R. This results in a somewhat narrow distribution of load, which is not desirable for building performance models for widely varying loads. That is why Sheikh et al. [26] perform uniform sampling across two dimensions: 1) total number of queries, and 2) the number of different types of concurrent queries. Nonetheless, we still settle for the LHS protocol to control the diversity in the samples.

4.2 Experiment-driven data collection

Once the samples are obtained, we execute them in a public cloud for each VM type. Both the client and the data-service exist in a public cloud to avoid communication delay over a WAN. We wrap up the tenant databases with the MySQL dbms and the ubuntu linux, and store that as an *image*.[1] This greatly simplifies the engineering process, and the workloads can start execution as soon as the compute and storage resources are available, i.e. when the image is instantiated on a VM. On instantiation, the buffer pool occupies 80% of the total memory of a VM instance, and is partitioned in proportion to the number of tenants. Each sample is executed for some time (say around 10m). The request mix remains constant throughout the execution of the sample. The

[1] Our image (ami-7bc16e12) is publicly available at: http://thecloudmarket.com/owner/966178113014. Once the image is instantiated, the clients can connect (ssh in) to the instance and access the MySQL dbms as root user with wlmgmt password.

client collects run time statistics such as throughput and response times for each request type in a sample. This is the most time-consuming part of the model building process and takes tens of hours to complete. Fortunately, this process need not be repeated for each request type unlike suitable parameter value search for non-linear modeling.

4.3 Constructing the request mix model

After all the samples are executed, we collect the sample execution results from the data repository of the client. We pre-process the raw data before training a performance model. The pre-processing involves (a) adjusting the scale of the units e.g. converting response time from milliseconds to seconds, (b) analyzing the data to identify any data patterns such as non-linear trends, and (c) cleaning the data e.g. removing outliers. Understanding and treating data patterns improves the quality of the performance model (discussed later in detail in Section 5.3). Then we can train a performance model on the pre-processed data. We can use any regression method such as linear regression or multi-layer perceptron (mlp). We compare different regression techniques on the basis of correlation as discussed in Section 4.5. Finally, we validate the performance model against new data (presented in Section 5.4).

4.4 Determining a suitable number of samples

Traditional wisdom says that more samples produce a more *representative* prediction model. Executing samples is an expensive exercise, and therefore, we need some way of determining an appropriate sample size that gives reasonable accuracy and confidence in the performance model. We see two approaches of determining the appropriate number of samples: (a) a theoretical approach such as estimating the number of samples based on the confidence level and interval, or (b) an empirical approach such as using experimentation to determine the appropriate number of samples).

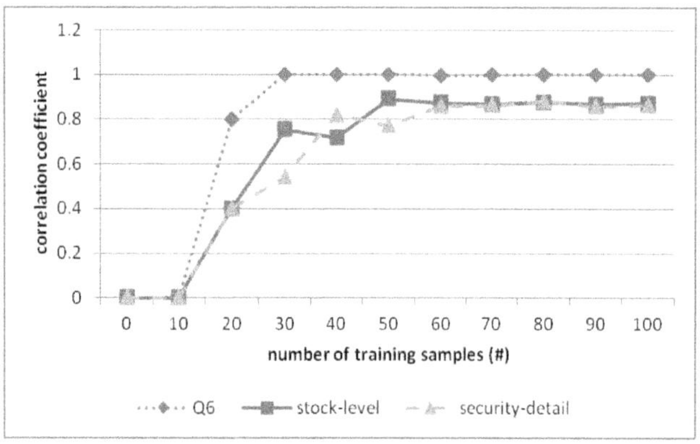

Figure 1: Observing changes in correlation coefficients for multi-variate linear regression (LR) on a hp-xl VM instance against number of training samples (up to 100).

In this paper, we explore the empirical approach further. We choose the High-CPU Extra Large (*hp-xl*) VM type in the Amazon cloud for analyzing the change in the quality of the performance model as the number of samples and their observed metrics (collectively called *training samples*) are increased. For example, we compare the correlation coefficient for three request types (Q6, stock-level and security-detail), each from a different database tenant (TPC-H [32], TPC-C [29] and TPC-E [31] respectively). Q6 is an analytical query, while the other two request types are transactions. The intention is to use diverse request types in our analysis.

We execute about 100 samples obtained using the LHS protocol (experimental setup similar to Section 5.1). We divide the training samples into 10 intervals. Each interval also contains the training samples of its predecessor. We choose multi-variate linear regression (LR) as the base classifier for the performance model. We build and validate the LR models on each interval using 10 folds cross-validation. We plot the correlation values of the LR models against the number of training samples as shown in Fig. 1. We see the greatest gains in accuracy when the numbers of training samples are in the first half of the plot. After that, there are diminishing returns, and the correlation coefficients seem to stabilize.

We observe a similar pattern for a number of other request types. We analyze the effect of the sample size (up to 620) on the correlation values of the same request types as shown in Fig 2. We see the greatest gains in correlation value for about the first 150 samples and limiting gains afterwards. This is similar to the observation of Sheikh et al. [26], who see greatest gains in accuracy for the first 100 samples and diminishing returns subsequently. Based on this analysis, the sample size of a few hundred should provide us with a representative prediction model.

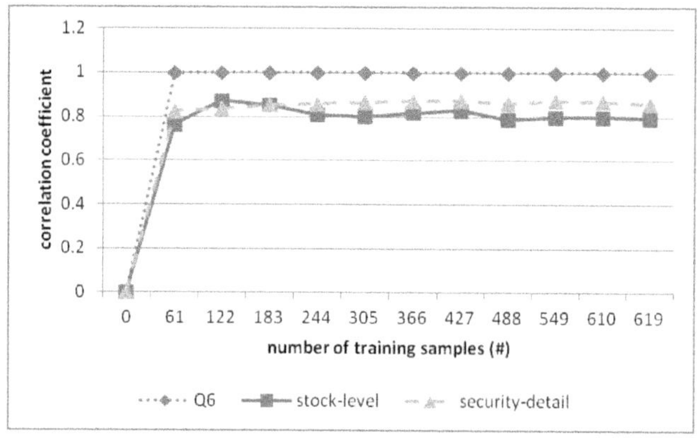

Figure 2: Observing changes in correlation coefficient for multi-variate linear regression (LR) on a hp-xl VM instance against the number of training samples (up to 620).

4.5 Comparison of prediction techniques

Cautious of high variance in the clouds, we consider a number of base classifiers for usage in our performance model. We compare four classification techniques: multi-variate linear-regression (LR), Gaussian Processes (GP), Multi-Layer Perceptrons (mlp) and Support Vector Machine (SVM). We consider two implementations of SVM (nuR [9] and SMO [22]). Tozer et al. [28] find LR to be sufficient as a base classifier for their performance model. LR is a simple regression model, and serves as a baseline in our case. Sheikh et al. find GP to be particularly accurate and adaptive to unseen request types, so we include that in our analysis. SVM has seen major development and fame in the last few years due to its robustness and transparency. Compared to other classifiers, mlp is a "black-box" type model and serves as an alternate comparison point.

GP and SVM can employ various kernel functions. The kernel functions used in SVM rearrange the original training samples into a high dimensional space using a set of mathematical functions. The motivation is to better identify the boundaries between the training samples. The trade-off is higher prediction accuracy at the cost of increased computational complexity and the risk of over-fitting.

We consider two kernel functions: (a) linear, and (b) Radial Basis Function (RBF) for both GP and SVM. Linear kernel is similar to an identify function, while Radial Basis Function is a popular kernel. We use both linear and RBF kernels for GP and SMO with their *default parameter values* in Weka. We consider the default parameter values for each classifier to be a good starting point. More importantly, Witten et al. [36] suggest that the Weka Explorer generally chooses sensible default values for parameters. We consider 150 training samples for comparison. We compare the correlation values of the classifiers obtained after 10-folds cross-validation, in the bar chart (as shown in Fig. 3). We group variations of the same technique. Each group is distinguished by a different shade of grey.

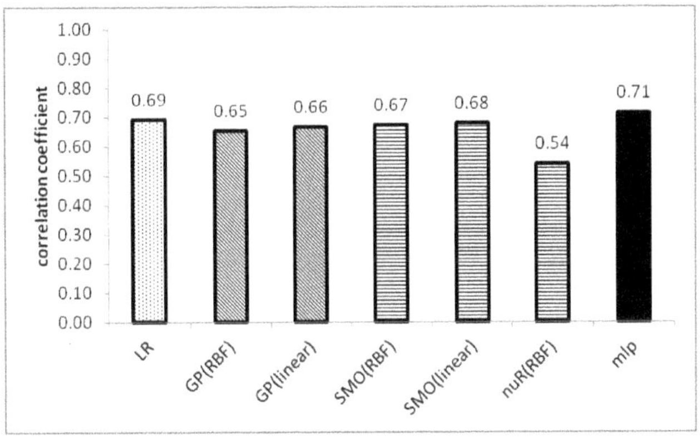

Figure 3: Comparing correlation values for trade-update. About 150 samples have been executed on a *large* VM instance. Variants of the same technique are grouped and distinguished by a different shade of grey.

The high coefficient value of mlp is attractive but possibly misleading. A major drawback of a mlp consisting of any hidden layers is that the hidden neurons are essentially opaque. Secondly, a mlp is prone to over-training. The next candidate is LR, which tends to exaggerate the errors in the general case since it minimizes the predictions' squared errors instead of absolute errors, which is the case with SMO with a linear kernel i.e. SMO (linear). Like LR, the basic idea of SMO (linear) is to find a function that approximates the training points well by minimizing the predictions' absolute errors [36]. The crucial difference is that all deviations up to a user-specified parameter are ignored. We feel that SMO (linear) is an upgrade of LR. It uses SVM constructs while minimizing over-fitting right from the outset. Linear classifiers are preferable over non-linear classifiers, because the latter usually require specification of additional parameter values and are prone to over-fitting. Incorrect parameter values lead to poor correlation values, and we see that in the case of nuR(RBF).

GP and its variants have been used for performance modelling [5; 26], and the accuracy of a GP can be further improved by using an appropriate kernel and parameter values. A GP is defined by a mean function and a co-variance function [24]. The co-variance function itself can have some parameters called hyper-parameters. Sheikh et al. [26] develop a configuration model to generate hyper-parameter values, which enables fast learning of unknown configurations. This is relevant for various types of unseen VMs and/or workloads, where prediction models have not been trained previously. Unfortunately, the GP implementation in Weka does not allow tuning of hyper-parameter values.

As we show in Section 5.4, SMO (linear) suffices as the base classifier in modeling performance for many request and VM types. However, it is unsuitable for modeling non-linear trends in data, and we use non-linear modeling in such cases. Standard SVM training has $O(n^3)$ time and $O(n^2)$ space complexities, where n is the training set size [33]. Platt's original Sequential Minimal Optimization (SMO) is linear in the amount of memory required for the training set size, and the training time of SMO empirically scales between O(n) and $O(n^{2.2})$ on various test cases [22].

5. EVALUATION

5.1 Experimental setup and validation method

The LHS protocol draws a specified number of samples from a multi-dimensional space given the lower and the upper bounds of each dimension (i.e. a request type). The minimum number of instances of a request type in a sample mix is 0. Meanwhile, we set the upper bound to be the optimal mpl value on a VM type (e.g. 14 for *small* VM type for the request type considered). We determine the optimal mpl value for each VM type experimentally.

We use the LHS protocol to generate two sets of samples with different random seeds. We consider a larger set (150 samples) for training and a smaller set (100 samples) for validation to be appropriate. We execute both sets in the Amazon cloud using separate VMs and clients. We execute each sample at the data service twice. The first round of sample execution is for warm-up, and the second round is for taking throughput and response time measurements. We employ SMO (linear) for learning. Then, we validate the performance model against the test set.

We use popular metrics from the literature for comparison, namely correlation coefficient [28] and mean prediction %errors [26][1]. Correlation quantifies the similarity between the actual and modeled trends – they may be far apart, and yet we can have excellent correlation. Meanwhile, prediction errors quantify the gap between the predicted and the measured values. Correlation coefficient and prediction accuracy are complementary, and we use both.

Table 1: Specifications of the VM types considered in the Amazon cloud

VM Type	Cores (#)	Memory (gb)	Cost/hr($)	Optimal mpl
Small	1	1.7	0.065	14
Large	2	7.5	0.260	75
Xlarge	4	15	0.520	115

High correlation coefficients (around 0.80 or above) and low prediction errors (around 20% or below) indicate the success of our performance model. We set these boundaries based on the existing literature [11; 26; 28]. The ideal value of a correlation coefficient is one, while the ideal value of mean-%error is zero.

We consider three heterogeneous VM types for evaluation: (a) *small*, (b) *large*, and (c) *xlarge*. They vary in price, processing power and capacity to hold data in memory. Further, xlarge VM type has either Xeon or Optron processor, and we build our

[1] Percentage-error (%error) = |measured value – predicted value|/measured value

models for Xeon processor only. The data service is configured to occupy most of the available memory on a VM type. The specifications of the VM types we used are stated in Table 1.

5.2 Tenant databases and selection of request types

We use databases in well-known benchmarks as tenant databases in evaluating our work. We consider databases of two transactional benchmarks (TPC-C [29] and TPC-E [31]), and the database of an analytical benchmark (TPC-H [32]). Our workloads consist of a mix of queries and transactions from the stated benchmarks (as shown in Table 2).

Table 2. Selected queries and transactions from the standard benchmarks

Benchmark	Request types
TPC-H (OLAP)	Q1, Q6, Q12, Q21
TPC-C (OLTP)	new-order, payment
TPC-E (OLTP)	trade-order, trade-update

A request type in a workload may have multiple request instances that execute concurrently. Our workloads are bound by time. Until then, a request instance is continuously re-submitted if it finishes early. This ensures that the request mix remains constant at a data service throughout the time bound or an experimental run.

We need *data-intensive* request types, which spend *significant* part of their execution time accessing (reading and/or writing) data. We also want diverse request types that place different requirements on the data service. Hence, we use both queries and (read/write) transactions in our workloads. Many request types cannot execute independently. For example, a payment transaction in the TPC-C benchmark assumes the presence of an unpaid order. In the absence of the unpaid order, the payment transaction fails and no change is made to the database. The payment failure is the exception rather than the norm, which also makes intuitive sense since the customer is not charged twice for the same order.

We execute the samples over tens of hours, which results in millions of transactions being executed. We want to avoid millions of failed transactions, or worse, a mix of successful and failed transactions. The latter skew the execution results, such as throughput. Avoiding such transaction failures places a constraint on the selection of our workload. That is, the request type is either independent, or at most dependent on the other transactions in the workload.

We select a subset of queries and transactions from the transactional and analytical benchmarks. We briefly describe each benchmark, and the role of the requests chosen from them. TPC-C models the principal activities (transactions) of an order-entry environment. These transactions include entering and delivering orders, recording payments, checking the status of orders, and monitoring the level of stocks at the warehouses. The payment transactions depend on the results of the new-order transactions, while the new-order transactions can execute independently. The pair of a new-order and a payment transaction can execute independently of the other transactions in the TPC-C benchmark. Therefore, we choose both of them for our workloads.

TPC-E models the activity of a brokerage firm that must manage customer accounts, execute customer trade-orders, and manage interactions of customers with financial markets. The TPC-E benchmark has many "read-only" transactions, but only four "read/write" transactions. Amongst the read/write transactions, the trade-order can execute independently, while the trade-update depends on the results of the trade-order transaction. The trade-order transactions represent buying or selling a security, and the trade-update transactions enable minor corrections or updates to a set of trades. We select them as candidate request types for our workloads.

TPC-H is a decision support benchmark. It consists of a suite of business oriented ad-hoc queries and concurrent data modifications. We experimentally profile data access of all the TPC-H queries and calculate the ratios of *data access* to *execution time* for all the queries. We choose the top four queries (Q1, Q6, Q12, Q21) according to the values of ratios ordered decreasingly. Q1 reports the amount of business that was billed, shipped, and returned. Q6 quantifies the increase in revenue because of the elimination of some discounts. Q12 determines whether cheaper transportation is adversely affecting the priority of orders. Q21 identifies suppliers whose shipments are late.

5.3 Data patterns: Identification and treatment

Our initial attempts at building a performance model with SMO (linear) were met with poor results. We discovered different patterns in the data upon investigation (as shown in Fig. 4).

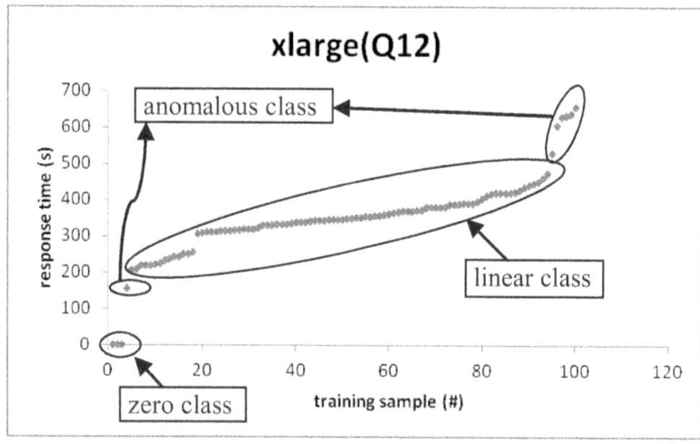

Figure 4: Different data classes in the response time measurements of Q12 on xlarge VM type.

Understanding patterns helps us to improve the quality of the performance models. For example, it allows us to distinguish between outliers and non-linear trends. Removing outliers can sometime improve the mean-%errors significantly. Non-linear trends may require data transformations and place advanced modeling requirements on the performance model.

We identify patterns in the execution results or the data, and the reasons for their existence. In hindsight, they uncover interactions among request types and their impacts on the performance metrics. For example, a large number of concurrent trade-update instances in a request mix reduce the trade-order throughput significantly due to locks on the shared tables and frequent accesses to disks. We describe the data pattern classes below, and their possible treatments before training a SVM.

5.3.1 Data classes

We categorize the data into four general classes: (a) zero, (b) linear, (c) anomalous, and (d) alternate. We also describe the heuristics for identifying each class. In this process, we use the Inter-Quartile Range (IQR) filter to identify any obvious oddities.

Zero class: The zero class represents the samples that have no instances of a request type (r) in the request mix, hence zero metric value for r (as shown in Fig. 4). This is normal behaviour, and we leave them in the data set. Unfortunately, we see some non-zero prediction values for the samples in the zero class. A simple adjustment to the performance model can fix this glitch. In this fix (called zero-fix), the performance model provides zero metric values for the samples where the request instance value is zero in the request mix. This fix considerably improves the correlation relationship between the predicted and measured metric values.

Linear class: The members of the linear class represent a near-linear change in the throughput or response time when sorted by the measured metric value in ascending order (as shown in Fig. 4). Naturally, the linear classifier performs best when the data set is mostly comprised of training samples in the linear class.

Anomalous class: The anomalous class represents training samples with an unusually high or unusually low value of a metric (as shown in Fig. 4). They are *few* in number, say around 2% or less of the entire data set. For example, the training samples in the anomalous class for Q1 exist with unusually low response times due to the smaller number of concurrent Q1 query instances in the request mix in the case of xlarge VM type (discussed in Section 5.4.3). We leave them in the data set unless they skew the validation results *significantly*. The reason for the divergence is sometimes unknown.

Alternate class: If there are *many* training samples with unusually high or low metric values not following the linear trend, then they belong to the alternate class. This is the normal behaviour and the alternate class training samples are left in the data set. The unusual values for the metrics stand out from the values in the linear class due to non-linear trends. For example, about 23% of the training data for the trade-order transaction on the large VM instance consists of unusually high but legitimate metric values. The throughput for trade-order decreases exponentially (non-linearly) with increasing number of concurrent trade-update instances in the request mix. Catering to the combination of alternate and linear classes requires non-linear modeling.

We collectively call all classes, except linear, the *irregular class*. We explored if existing clustering methods can give us the above (or different) classes. Wu et al. [37] put k-means in the second place in the top 10 algorithms in data mining. Unfortunately, we find that k-means and its extension x-means [21], are unable to differentiate the subtle boundaries between the classes. Raatikainen et al. [23] also find the workload classes obtained by k-means to be unsuitable. We want to classify the patterns based on the reasons for their existence, which is something not obvious to a clustering method. For example, members of *zero* and *alternate* classes can have zero and unusually small values respectively but exist due to different reasons. X-means, however, treats them as a single cluster. Therefore, we leave the exploration of other clustering methods for another venue, and currently identify classes manually. Fortunately, the IQR filter helps us *partially* in this job. We use an offset (of 1.5) on IQR over the entire data set to find the training samples that standout from the rest of the training samples.

5.4 Validation results

As mentioned in Section 4.5, we use SMO with a linear kernel as our base classifier. We build our performance model with the training set, and validate against the test set. We see that the linear classifier meets the validation criteria for most request types but performs poorly with non-linear trends. We deal with non-linear trends in Section 6.

We analyze the validation results[1] of the VM types in the following order: *large*, *small* and *xlarge*. This is because only two tenants fit in the memory of the large VM type, and it represents a "middle" case. All tenants fit in the memory of the xlarge VM type, and none fit in the memory of the small VM type. We discussed our choice for different request types in Section 5.2. Every request type needs a separate SVM. We build response time SVMs for the queries Q1, Q6, Q12, and Q21, and throughput SVMs for the transactions i.e., new-order, payment, trade-order, and trade-update.

5.4.1 Large VM type (optimal mpl=75)

We evaluate the SVMs of the queries first. We plot the frequency histogram of observed response times of Q1 after excluding members of the zero class (as shown in Fig. 5). We see that the distribution is normal-like. The frequency histograms for the remaining queries are similar to that in Fig. 5.

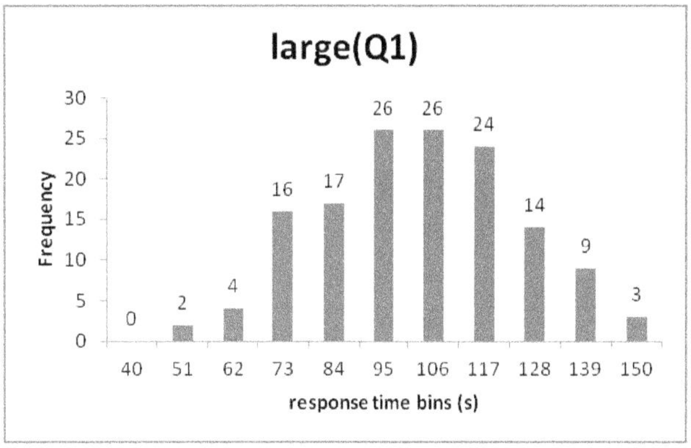

Figure 5: Frequency histogram for observed response times of Q1 on the large VM type instance.

The evaluation metrics are reported in Table 3. The SVMs for the queries have near-ideal correlation coefficients and single digit mean-%errors. We attribute the errors to a few anomalies which exist with unusually low response times due to the smaller number of concurrent queries in the request mix.

Table 3: Evaluation metrics for the response time SVMs built for the large VM type

	Q1	Q6	Q12	Q21
Correlation coefficient	0.99	1	0.96	0.97
mean-%error	3.86	2.71	9.28	5.11

For example, unusually low response times are observed for Q12 as there is a smaller number of concurrent Q12 queries in the request mix compared to the request mixes belonging to other classes. This situation apparently leads to less load on the data service and hence lower response times for Q12.

[1] The raw data used to build the performance model is present at: http://research.cs.queensu.ca/home/mian/index_files/Page485.htm

We plot the correlation relationship between the predicted and actual response times for Q1 and Q12 in Fig. 6 and Fig. 7, respectively.

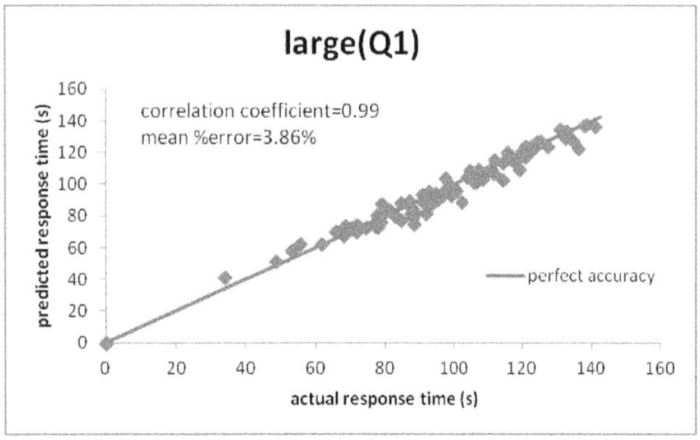

Figure 6: Predicted vs. measured response times for Q1 on a large VM type.

The correlation plots for the remaining queries are similar. The SVMs for all queries have excellent correlation coefficients (close to 1), and low mean-%errors. Therefore, they all meet the validation requirement.

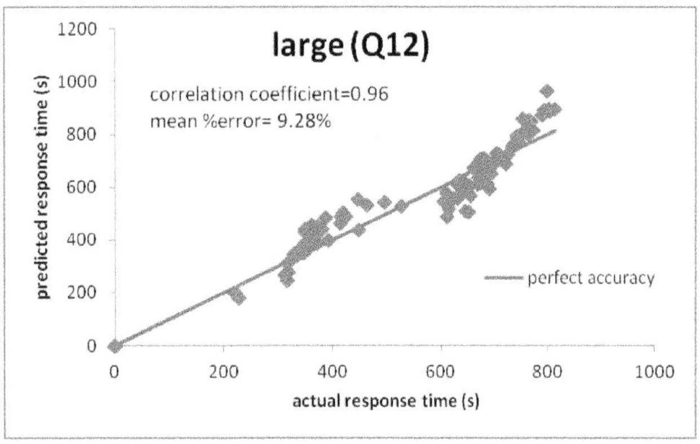

Figure 7: Predicted vs. measured response time for Q12 on a large VM type.

Next, we evaluate the SVMs for transactions. We state the evaluation metrics for the throughput SVMs in Table 4. The SVMs for the new-order and payment transactions have excellent correlation coefficients and acceptable mean-%errors. Therefore, they pass the validation test. The mean-%errors are generally higher than their query counter-part. The irregular class for new-order and payment mostly contains members of zero class and a few anomalies. The anomalies consist of unusually high throughputs, which are caused by a high number of concurrent new-order and payment transactions in a request mix.

Table 4: Evaluation metrics for the throughput SVMs built for the large VM type

	New-order	Payment	Trade-order	Trade-update
Correlation coefficient	0.97	0.97	***0.58***	0.80
mean-%error	14.79	14.65	***111.69***	13.61

The irregular class for trade-update also contains mostly members of zero class and a few anomalies. The anomalies have lower numbers of query instances compared to the training samples in the linear class. This suggests that the data service is under-loaded, and is able to execute more transactions. The trade-update has acceptable mean-%error but the correlation coefficient is on the borderline. We plot the frequency histogram of observed throughput for trade-update after excluding members of zero class in Fig. 8. We see that the distribution is positively skewed.

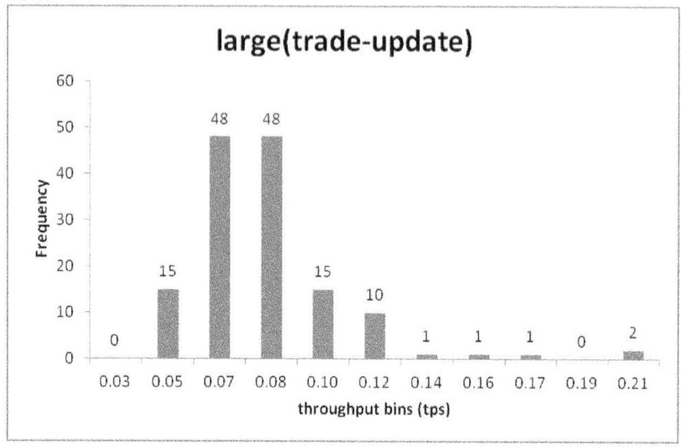

Figure 8: Frequency histogram for observed throughput for trade-update on the large VM type instance.

Regression assumes that variables have normal distributions, and not non-normally distributed variables (highly skewed or kurtotic variables, or variables with substantial outliers) can distort relationships and significance tests [20]. It is possible to employ transformations (e.g. square root, log, or inverse), to the improve normality, but this complicates the interpretation of the results, and should be used in an informed manner. Due to an acceptable mean-%error, we do not perform any data transformation and consider that the trade-update SVM passes the validation test on the border.

Meanwhile, trade-order has a large irregular class, which is mostly comprised of members from the alternate class. For example, we see 34 irregularities in the training set. This is about 23% of the training data, and represents an unusually large proportion. Inspecting the training data for the respective samples, we see that the number of concurrent request instances of trade-update is low in the alternate class. The trade-order and trade-update share some tables in their operations [31]. In addition, the trade-update transaction generates a high disk I/O because it looks for older records that are usually not in the buffer pool due to their age and frequency of access [30].

We suspect that the low trade-order throughput (in the linear class) is due to the high lock contention over the shared tables and frequent access to the disk by trade-update. We perform a simple experiment to confirm this suspicion, in which we set the trade-update instances to zero in all the samples. We see very high throughput values for trade-order, and this confirms our suspicion. The throughput for trade-order decreases exponentially (non-linearly) with increasing number of concurrent trade-update instances in the request mix.

We plot the frequency histogram of observed throughput for trade-order after excluding members of zero class in Fig. 9. We see that the distribution is heavily skewed with a long tail. We find that applying common transformations such as square root, log and inverse do not improve the normality of the measured throughput. In this case, the non-linearity in the data must be explicitly modeled using a non-linear kernel for SVM, for example. We perform non-linear modeling in Section 6. The trade-order SVM fails the validation test with SMO (linear).

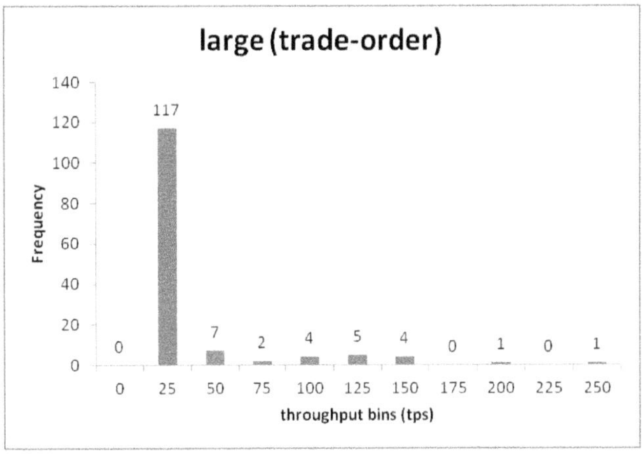

Figure 9: Frequency histogram for observed throughputs for trade-order on the large VM type instance.

5.4.2 Small VM type (optimal mpl=14)

First, we evaluate the SVMs of the queries (see Table 5). The SVMs for Q1, Q12 and Q21 have excellent correlation values and acceptable mean-%errors, consequently, passing the validation test.

Table 5: Evaluation metrics for the response time SVMs built for the small VM type

	Q1	Q6	Q12	Q21
Correlation coefficient	0.90	***0.86***	0.90	0.90
Mean-%error	15.65	***25.43***	13.07	17.56

As highlighted in Table 5, Q6 has a relatively poor correlation value and high mean-%error. We observe three classes in the Q6 data set: (a) zero (28 members), (b) linear (65 members), and (c) alternate (7 members). On average, the alternate class has a higher number of concurrent request instances compared to the linear class. We believe this situation leads to a greater load on the data service and hence higher response times for Q6. The alternate and linear classes represent non-linear trends, and the Q6 SVM does not meet validation requirement due to poor mean-%error.

Next, we evaluate the throughput SVMs for the transactions as shown in Table 6. The SVMs for new-order and payment have excellent correlation coefficients and acceptable mean-%errors. Therefore, they pass the validation test.

Table 6: Evaluation metrics for the throughput performance model built for the small VM type

	New-order	Payment	Trade-order	Trade-update
Correlation coefficient	0.96	0.97	***0.90***	***0.63***
Mean-%error	13.52	9.72	***1465.84***	***34.80***

Interestingly, trade-order's SVM has a high correlation value and an extremely poor mean-%error. Examining the samples (request mixes) in the test set, we see that the resulting throughput values belong to three classes: (a) linear (62 members), (b) zero (29 members), and (c) alternate (10 members). We find that the number of concurrent trade-update transactions is at most two in any request mix. Consequently, the trade-order has high throughput (in tens) for most samples in the test set (i.e. linear class). As noted earlier, providing a performance model for both linear and alternate classes with acceptable evaluation metric values requires modeling of non-linear trends. Therefore, the SVM for trade-order fails validation.

Similarly, trade-update consists of three classes in the test set : (a) linear (56 members), (b) zero (29 members), and (c) alternate (15 members). The respective SVM performs poorly and fails validation.

5.4.3 Xlarge VM type (optimal mpl=115)

First, we evaluate the SVMs of the queries as shown in Table 7. The SVMs for all the queries have excellent correlation values and mean-%errors within the set boundary. Therefore, they all pass the validation test. All SVMs except one (i.e. Q21) have single digit mean-%errors. The training samples for Q21 fall in three classes: (a) linear, (b) zero, and (c) anomalous. The anomalies exist with unusually low response times due to the smaller number of concurrent Q21 queries in the request mix.

Table 7: Evaluation metrics for the response time SVMs built for the xlarge VM type

	Q1	Q6	Q12	Q21
Correlation coefficient	0.99	0.99	0.93	0.93
mean-%error	5.46	3.21	5.86	11.72

Table 8: Evaluation metrics for the throughput SVMs built for the xlarge VM type

	New-order	Payment	Trade-order	Trade-update
Correlation coefficient	0.97	0.97	***0.46***	0.87
mean-%error	18.26	16.68	***94.77***	14.03

Next, we evaluate the throughput SVMs for the transactions as shown in Table 8. The SVMs for new-order and payment have excellent correlation coefficients and acceptable, though high, mean-%errors. They pass the validation requirements. We attribute the high mean-%errors for the new-order SVM to a few anomalies in the new-order test set. The anomalies exist due to a high number of concurrent new-order transactions and a low number of other concurrent request types. The same is true for the payment transaction.

We find a poor correlation coefficient for the SVM of trade-order. Like the data sets of trade-order in the case of large VM type, we see that many training samples belong to the alternate class. The throughput for trade-order decreases exponentially (non-linearly) with an increasing the number of concurrent trade-update instances in the request mix. The trade-order SVM fails the validation test with SMO (linear).

The irregular class for trade-update contains members of two classes: (a) zero, and (b) anomalous. The numbers of trade-order and trade-update request instances in the request mix are similar across anomalous and linear classes. However, the anomalous class has a lower number of query instances compared to the samples in the linear class. This suggests that the data service is under-loaded, and is able to execute more transactions. This cannot be said conclusively given the small size of the anomalous class (only 3 members in the test set), and we leave it as an observation. We also leave all anomalies in the data sets for trade-update. We still get good correlation value with acceptable mean-%error. Therefore, the trade-update SVM passes the validation test.

6. MODELING NON-LINEAR BEHAVIOUR

We have seen that SMO (linear) is sufficient for many request types, particularly where an alternate trend in the performance is non-existent. We consider a popular (non-linear) RBF Kernel to cater to the combination of linear and alternate trends. We further explore request types that failed their validation test with SMO (linear). We demonstrate the use of the RBK Kernel with the small VM type, but the process is the same for all the VM types. We choose the small VM type for demonstration because the SMO (linear) for the small VM type fails for the highest number of requests amongst all the VM types considered.

Appropriate kernel and parameter settings can greatly improve the SVM classification accuracy. Suitable values for the penalty parameter, C, and the kernel function parameter, γ, are unknown beforehand. For medium-sized problems, the *grid search* approach is an efficient way to find the best C and γ [16]. In grid search, pairs of (C, γ) are tried and the one with the best cross-validation accuracy is chosen.

Before the grid search, we revise our data set to give us more training data out of the existing data set. This is possible using the *0.632 bootstrap* sampling method [14]. We aggregate the training and test data sets to provide us with a combined data set. We remove members of the zero class from the aggregated data set. This is because we can augment the performance model to provide zero values for samples belonging to the zero class, and therefore we do not need to train the performance model to cater for zero class. The revised aggregated dataset of n training samples is randomly sampled n times, with replacement, to give a *learning* dataset of n training samples. Because some elements in the learning dataset are (almost certainly) repeated, there must be some instances in the aggregated dataset that have not been picked by the random selection, which become part of the *validation* set. The size of the validation set is approximately 1/3rd of n. In this way, we retain the property of unseen training samples for validation similar to the test set. As stated in Section 5.1, training and test samples were generated randomly using different seeds.

Then, we perform a grid search on the learning set using 10-folds cross-validation. In this search, we explore exponentially growing sequences of C ($2^{-5}, 2^{-3}, \ldots, 2^{15}$) and γ ($2^{-15}, 2^{-11}, \ldots, 2^{5}$), since Hsu et al. [16] find this to be a practical method to identify good parameter values. They further recommend a coarse grid search first, and then a finer grid search on a "promising" region. Once the search identifies *good* kernel parameters, we train a SMO (RBF) with these parameters using the entire learning set and validate against the validation set. We use this search, train and validate method for Q6.

We find further data transformations for trade-order and trade-update throughput to be appropriate. We discuss the transformation and the justification below. We sort the training samples of the trade-order transaction in the increasing order of throughput value. We plot throughput values against the instance numbers. Despite removing the zero class, the trade-order trend is still fairly non-uniform (as shown in Fig 10). While it is possible to model the trend, there is a large chance of over-fitting to obtain high accuracy given the number of bends and turns required for the curve fitting this trend.

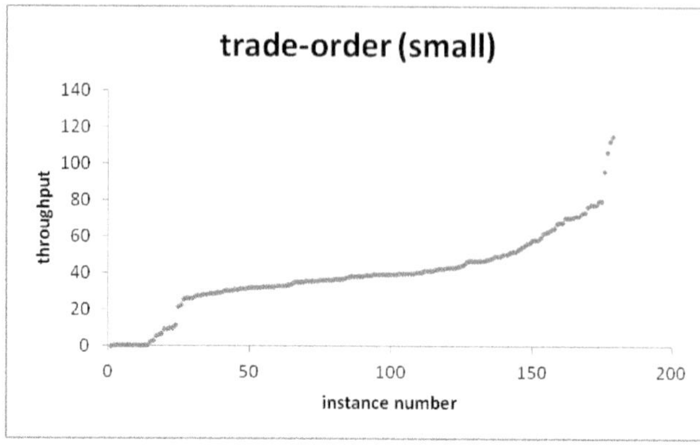

Figure 10: Trade-order throughput values sorted ascendingly.

Instead, the level of symmetry increases significantly if we apply a logarithmic function (as shown in Fig. 11). The bends are near-linear, and there is only one turn. We can train a SVM on the transformed trend without being overly concerned about over-fitting. Therefore, we find the above data transformations for trade-order and trade-update throughput to be appropriate and apply them to the aggregated data set prior to bootstrap sampling.

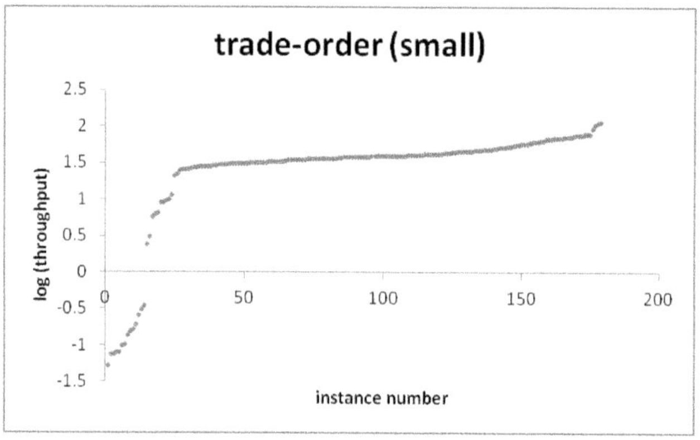

Figure 11: Ascendingly sorted logarithmic trade-order throughput values.

We state the evaluation metrics for the non-linear SVMs trained using the revised training and validation schemes in Table 9. We find that the mean-%errors greatly improve, and the correlation coefficients are excellent for the two request types when we explicitly model the non-linear trends.

Table 9: Evaluation metrics for the non-linear SVMs built for the small VM type

	Q6	Trade-order	Trade-update
Correlation coefficient	0.97	0.98	0.79
mean-%error	8.09	3.43	3.92

We can further improve the *representation* of the real trend in the model for the above request types. Presently, linear and alternate class sizes are not in balance. A good strategy for producing a high accuracy classifier on imbalanced data is to classify any example as belonging to the majority class, which is called the majority-class classifier [8]. This is what we have done although it may not be very useful in practice. The problem with the success rate is that it assigns equal importance to errors made on examples belonging to the majority class and the minority class. To address the imbalance in the data, we need to assign different penalty parameter values for misclassification to each class.

We further observe that the workload behaviour changes dramatically depending on the amount of trade-update transactions in the request mix. For example, we see different phases in the trade-order throughput (as shown in Fig. 10). We can model each phase separately. This will likely: (a) improve the representation of the real trend in the model, and (b) avoid over-fitting. However, this comes at the cost of identifying phases and boundaries between them. The process of identification may require validation. Presently, we leave class size and phase aware modeling as part of the future work.

7. DISCUSSION

We validate the performance models for the VM types that vary in their system capacity, and in particular their physical memory. As a result, the optimal mpl values as well as the range on the number of request instances vary on these VM types. The behaviour of a request is affected by other concurrently executing requests both in terms of the request types and their number of instances. For example, a smaller number of query instances in the request mix results in less load, and consequently an overall lower response time and for queries high throughput for transactions. We also observed that lock contentions and interactions between concurrently executing requests can have a significant impact on the performance of a database system. This supports the claim by Ahmed et al. [3], that interactions between concurrently executing requests can have a significant impact on the performance of a database system.

We showed that the linear classifier is suitable for 19 out of 24 request types, and can be modeled using an out-of-the-box tool such as Weka. However, it fails where there are non-linear trends in the performance data. In such cases, we explore non-linear modeling methods that require choosing a suitable kernel and a search for appropriate parameter values. Efficient search approaches such as the grid search can take several hours. As a result, we suggest that the linear classifier be used first to train the performance model, and in the case of unsatisfactory results, non-linear modeling be used as the next step.

Osborne et al. [20] state a number of assumptions for multiple regression that the researchers should test. We see that their assumptions (e.g. variables are normally distributed) do not always hold in our case. For example, the throughput distribution for trade-order in case of large VM type is highly skewed. Simple transformations, such as inverse, do not improve the normality of the distribution. Instead, explicit modeling of non-linear trends is required. We find grid search over RBF kernel promising, and see significant improvements in the evaluation metrics.

Our workloads contain at most eight request types each with a different number of instances. This is reasonable since TPC-C and TPC-E benchmarks have five and ten transactions, respectively, although TPC-H has 22 queries. We believe a realistic data service is rarely a read-only or a write-only service. It usually serves a combination of transactional and analytical workloads.

8. SUMMARY

We employ an experiment-driven approach for building a performance model applicable in a cloud environment. The samples are generated using stratified sampling, and measurements are collected by executing these samples in a public cloud. We provide a comparison of different underlying prediction techniques based on accuracy, and justify our choice. Some data patterns are identified and their possible treatments are suggested. Then, we train our performance models using the measured and treated data. The performance models are judged against multiple evaluation metrics, and validated against fresh data. Finally, we analyze the performance models built for different types of Amazon VMs accessible on their public IaaS EC2 cloud.

Recent literature typically builds performance models for a single database tenant, provides response time predictions only, and validates the models on a local server or a local VM. Our performance model predicts throughput for transactions, and response times for queries. The performance model is built for workloads executing at a multi-tenant data-service hosted in a cloud.

Presently, the performance model provides *raw* predictions without expressing any confidence in them. This is an important issue since the errors are cumulative in our framework, and we need some method of managing the errors across the framework components. Therefore, we intend to look at mechanisms which manage the error levels across the framework.

We also intend to make the performance model adapt online, which is particularly relevant for provide predictions for unseen request types and the cloud environment. The magnitude of the errors may be large in the beginning but would reduce over time as the models learns the new setting.

9. ACKNOWLEDGMENTS

The authors acknowledge research support from National Science and Engineering Research Council of Canada (NSERC), MEDIANET (Comunidad de Madrid S2009/TIC-1468), HPCcloud (MICINN TIN2009-07146) and 4CaaSt (European Commission's IST priority of the 7th Framework Programme under contract number 258862). The authors also thank Dr. Skillicorn and Dr. Hussain for their input on non-linear modeling and class identification and validation.

10. REFERENCES

[1] *Use WEKA in your Java code.* http://weka.wikispaces.com/Use+WEKA+in+your+Java+code.

[2] Abouzour, M., Salem, K., and Bumbulis, P., 2010. Automatic tuning of the multiprogramming level in Sybase SQL Anywhere. In *2010 IEEE 26th International Conference on Data Engineering Workshops (ICDEW)*, Long Beach, California, USA, 99-104.

[3] Ahmad, M., Aboulnaga, A., and Babu, S., 2009. Query interactions in database workloads. In *Proceedings of the Second International Workshop on Testing Database Systems* ACM, Providence, Rhode Island, US, 1-6.

[4] Ahmad, M., Aboulnaga, A., Babu, S., and Munagala, K., 2008. Modeling and exploiting query interactions in database systems. In *Proceedings of the 17th ACM conference on Information and knowledge management* ACM, Napa Valley, California, USA, 183-192.

[5] Ahmad, M., Duan, S., Aboulnaga, A., and Babu, S., 2011. Predicting completion times of batch query workloads using interaction-aware models and simulation. In *Proceedings of the*

14th International Conference on Extending Database Technology (EDBT'11) ACM, Uppsala, Sweden, 449-460.

[6] Amazon, *EC2 Instance Types*. http://aws.amazon.com/ec2/instance-types/.

[7] Amazon, *Elastic Compute Cloud (EC2)*. http://aws.amazon.com/ec2/.

[8] Ben-Hur, A. and Weston, J., 2010. A user's guide to support vector machines. *Methods in Molecular Biology 609*, 2, 223-239.

[9] Chang, C.-C. and Lin, C.-J., 2011. LIBSVM: A library for support vector machines. *ACM Transactions on Intelligent Systems and Technology (TIST) 2*, 3, 1-27.

[10] Courtois, M. and Woodside, M., 2000. Using regression splines for software performance analysis. In *Proceedings of the 2nd international workshop on Software and performance* ACM, Ottawa, Ontario, Canada, 105-114.

[11] Ganapathi, A., Kuno, H., Dayal, U., Wiener, J.L., Fox, A., Jordan, M., and Patterson, D., 2009. Predicting Multiple Metrics for Queries: Better Decisions Enabled by Machine Learning. In *IEEE 25th International Conference on Data Engineering, 2009. (ICDE '09)*. IEEE, Shanghai, China, 592-603. http://ieeexplore.ieee.org/xpls/abs_all.jsp?arnumber=4812438&tag=1.

[12] Gupta, C., Mehta, A., and Dayal, U., 2008. PQR: Predicting Query Execution Times for Autonomous Workload Management. In *International Conference on Autonomic Computing, 2008. (ICAC '08)*. IEEE, Chicago, IL 13-22.

[13] Hall, M., Frank, E., Holmes, G., Pfahringer, B., Reutemann, P., and Witten, I.H., 2009. The WEKA data mining software: An update. *ACM SIGKDD Explorations Newsletter 11*, 1, 10-18.

[14] Han, J., Kamber, M., and Pei, J., 2012. *Data mining: concepts and techniques (Third Edition)*. Morgan Kaufmann.

[15] Hicks, C.R. and Turner Jr, K., 1999. *Fundamental concepts in the design of experiments*. Oxford University Press, New York.

[16] Hsu, C.W., Chang, C.C., and Lin, C.J., 2003. *A practical guide to support vector classification*. National Taiwan University. http://www.csie.ntu.edu.tw/~cjlin/papers/guide/guide.pdf.

[17] Kelly, T., 2005. Detecting performance anomalies in global applications. In *Proceedings of the 2nd conference on Real, Large Distributed Systems - Volume 2* USENIX Association, San Francisco, CA, 42-47.

[18] Mian, R. and Martin, P., 2012. Executing data-intensive workloads in a Cloud. In *CCGrid Doctoral Symposium 2012 in conjuction with 12th IEEE/ACM International Symposium on Cluster, Cloud and Grid Computing (CCGrid)*, Ottawa, Canada, 758-763.

[19] Mian, R., Martin, P., and Vazquez-Poletti, J.L., 2012. Provisioning data analytic workloads in a cloud. *Future Generation Computer Systems (FGCS)*, in press http://dx.doi.org/10.1016/j.future.2012.1001.1008.

[20] Osborne, J.W. and Waters, E., 2002. Four assumptions of multiple regression that researchers should always test. *Practical Assessment, Research & Evaluation 8*, 2, 1-9.

[21] Pelleg, D. and Moore, A.W., 2000. X-means: Extending K-means with Efficient Estimation of the Number of Clusters. In *Proceedings of the Seventeenth International Conference on Machine Learning* Morgan Kaufmann Publishers Inc., 727-734.

[22] Platt, J., 1998. *Sequential Minimal Optimization (SMO): A fast algorithm for training support vector machines.* Microsoft Research. http://www.bradblock.com/Sequential_Minimal_Optimization_A_Fast_Algorithm_for_Training_Support_Vector_Machine.pdf.

[23] Raatikainen, K.E.E., 1993. Cluster analysis and workload classification. *SIGMETRICS Perform. Eval. Rev. 20*, 4, 24-30.

[24] Rasmussen, C.E. and Williams, C.K.I., 2006. *Gaussian Processes for Machine Learning*. The MIT Press.

[25] Schad, J., Dittrich, J., and Quiane-Ruiz, J.-A., 2010. Runtime measurements in the cloud: observing, analyzing, and reducing variance. *Proc. VLDB Endow. 3*, 1-2, 460-471.

[26] Sheikh, M.B., Minhas, U.F., Khan, O.Z., Aboulnaga, A., Poupart, P., and Taylor, D.J., 2011. A bayesian approach to online performance modeling for database appliances using gaussian models. In *8th ACM international conference on Autonomic computing (ICAC)* ACM, Karlsruhe, Germany, 121-130.

[27] Thereska, E., Narayanan, D., and Ganger, G.R., 2006. Towards self-predicting systems: What if you could ask 'what-if'? *The Knowledge Engineering Review 21*, 03, 261-267.

[28] Tozer, S., Brecht, T., and Aboulnaga, A., 2010. Q-Cop: Avoiding bad query mixes to minimize client timeouts under heavy loads. In *IEEE 26th International Conference on Data Engineering (ICDE)*, Long Beach, CA, USA, 397-408.

[29] TPC-C, *Order Processing Benchmark*. http://www.tpc.org/tpcc/.

[30] TPC-E, *Detailed description*. http://www.tpc.org/tpce/.

[31] TPC-E, *Trading Benchmark*. http://www.tpc.org/tpce/.

[32] TPC-H, *Decision Support Benchmark*. http://www.tpc.org/tpch/.

[33] Tsang, I.W., Kwok, J.T., and Cheung, P.-M., 2005. Core vector machines: Fast SVM training on very large data sets. *Journal of Machine Learning Research 6*, 363-392.

[34] Weikum, G., Moenkeberg, A., Hasse, C., and Zabback, P., 2002. Self-tuning database technology and information services: from wishful thinking to viable engineering. In *Proceedings of the 28th international conference on Very Large Data Bases* VLDB Endowment, Hong Kong, China, 20-31.

[35] Weissman, C.D. and Bobrowski, S., 2009. The design of the force.com multitenant internet application development platform. In *Proceedings of the 35th SIGMOD international conference on Management of data* ACM, Providence, Rhode Island, USA. http://dl.acm.org/citation.cfm?id=1559942.

[36] Witten, I.H., Frank, E., and Hall, M.A., 2011. *Data Mining: Practical machine learning tools and techniques (3rd edition)*. Morgan Kaufmann Pub.

[37] Wu, X., Kumar, V., Ross Quinlan, J., Ghosh, J., Yang, Q., Motoda, H., McLachlan, G., Ng, A., Liu, B., Yu, P., Zhou, Z.-H., Steinbach, M., Hand, D., and Steinberg, D., 2008. Top 10 algorithms in data mining. *Knowledge and Information Systems 14*, 1, 1-37.

[38] Zhang, M., Martin, P., Powley, W., Bird, P., and McDonald, K., 2012. Discovering Indicators for Congestion in DBMSs. In *Proceedings of the International Workshop on Self-Managing Database Systems (SMDB'12) in Conjunction with the International Conference on Data Engineering (ICDE'12)*, Washington, DC, USA, in press.

[39] Zhang, M., Niu, B., Martin, P., Powley, W., Bird, P., and McDonald, K., 2011. Utility Function-based Workload Management for DBMSs. In *Proceedings of the 7th International Conference on Autonomic and Autonomous Systems (ICAS 2011)*, Mestre, Italy, 116-121.

[40] Zhang, Q., Cherkasova, L., Mathews, G., Greene, W., and Smirni, E., 2007. R-Capriccio: A Capacity Planning and Anomaly Detection Tool for Enterprise Services with Live Workloads Middleware 2007. *Lecture Notes in Computer Science 4834*, 244-265.

vPerfGuard: an Automated Model-Driven Framework for Application Performance Diagnosis in Consolidated Cloud Environments

Pengcheng Xiong and Calton Pu
Georgia Institute of Technology
{pxiong3,calton}@cc.gatech.edu

Xiaoyun Zhu and Rean Griffith
VMware Inc.
{xzhu,rean}@vmware.com

ABSTRACT

Many business customers hesitate to move all their applications to the cloud due to performance concerns. White-box diagnosis relies on human expert experience or performance troubleshooting "cookbooks" to find potential performance bottlenecks. Despite wide adoption, the *scalability* and *adaptivity* of such approaches remain severely constrained, especially in a highly-dynamic, consolidated cloud environment. Leveraging the rich telemetry collected from applications and systems in the cloud, and the power of statistical learning, *vPerfGuard* complements the existing approaches with a model-driven framework by: (1) *automatically* identifying system metrics that are most predictive of application performance, and (2) *adaptively* detecting changes in the performance and potential shifts in the predictive metrics that may accompany such a change. Although correlation does not imply causation, the predictive system metrics point to potential causes that can guide a cloud service provider to zero in on the root cause.

We have implemented vPerfGuard as a combination of three modules: a sensor module, a model building module, and a model updating module. We evaluate its effectiveness using different benchmarks and different workload types, specifically focusing on various resource (CPU, memory, disk I/O) contention scenarios that are caused by workload surges or "noisy neighbors". The results show that vPerfGuard automatically points to the correct performance bottleneck in each scenario, including the type of the contended resource and the host where the contention occurred.

Categories and Subject Descriptors

D.4.8 [**OPERATING SYSTEMS**]: Performance—*Modeling and prediction; Measurements; Monitors*

Keywords

Automated framework, Cloud computing, Model-Driven, Performance analysis, Performance diagnosis, Statistical methods

1. INTRODUCTION

Diagnosing and resolving application performance problems in consolidated cloud environments is hard. For example, consider the following scenario. Alice, a cloud service provider, hosts Bob's *vSlashdot* news aggregation site on her virtualized infrastructure as shown in Fig. 1. Since Bob's site composes of Web, App and DB servers, Alice deploys each server in a *virtual machine* (VM) because such a distributed deployment enables better consolidation and more flexible resource sharing.

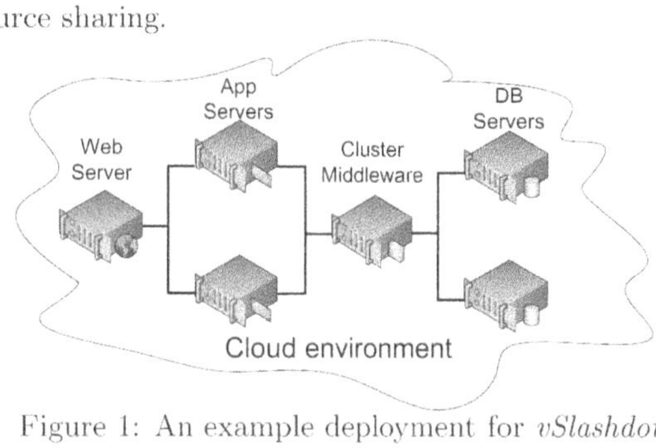

Figure 1: An example deployment for *vSlashdot*

After the *vSlashdot* service goes online, Bob notices that the performance of the web site degrades over time – its throughput drops and the response times increase. Bob immediately issues a support call with Alice. Traditionally, an IT service provider either relies on human experts with deep technical knowledge to identify performance bottlenecks, with the help of performance monitoring tools and system logs, or follows standard procedures in performance troubleshooting "cookbooks" [26] for problem localization and root cause analysis.

However, such *white-box* performance analysis methods [24] can no longer meet the demand for fast and accurate diagnosis in a highly-dynamic, large-scale, complex cloud environment, for the following reasons: (1) these methods have highly variable resolution times (from minutes to weeks); (2) they are not easily *scalable* to analyzing the behavior of many hosts and VMs in consolidated environments and many heterogeneous and distributed applications; (3) performance "cookbooks" only provide guidelines for problems that were seen before, whereas a dynamic cloud environment is likely to see emergent behavior or new interactions. For example, the performance of one application may suffer due to demand spikes in other applications (i.e., *noisy neighbors*) sharing the same physical infrastructure.

To overcome these limitations, we believe a data-driven approach [17] to performance diagnosis should be adopted to leverage the rich telemetry collected from applications and systems in the cloud. However, any performance analysis tool may be overwhelmed by the sheer amount of performance data and statistics collected at different levels of the stack. A recent report by TRAC Research [7], cited in [27], showed that 42% of the surveyed IT organizations reported challenges regarding usability of performance data. These include amount of time spent correlating performance data (63%), amount of performance data that is irrelevant (61%), issues they are not able to see (false negatives) (42%), and getting invalid alerts (false positives) (32%).

In order to address the above challenges, we propose *vPerfGuard*[1], a model-driven framework for achieving *automated*, *scalable*, and *adaptive* performance diagnosis in consolidated cloud environments. vPerfGuard accomplishes this by learning a performance model for an application using the system metrics that are most predictive of the application performance, and by adapting the model online by automatically detecting changes in the performance and the potential shifts in the predictive metrics that may accompany such a change. More specifically, the vPerfGuard architecture, as shown in Figure 2, consists of three modules - a sensor module, a model building module, and a model updating module.

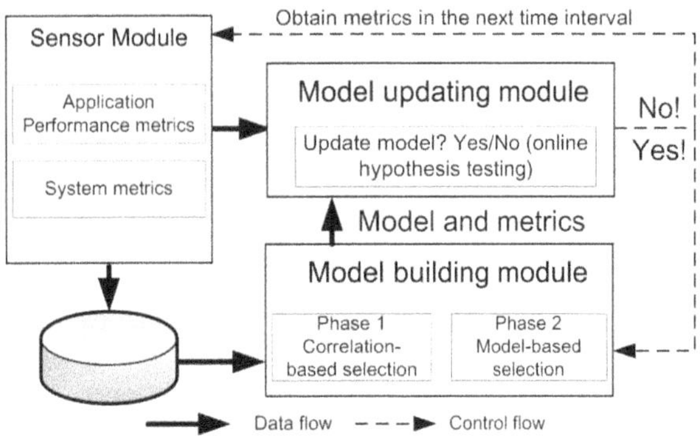

Figure 2: vPerfGuard framework

Whenever a performance degradation is observed and a performance troubleshooting request is received, vPerfGuard presents the top predictive system metrics in the current performance model to the cloud service provider such as Alice. These metrics can provide hints to Alice regarding potential causes for the observed performance problem, including the critical component within a complex, distributed application as well as the suspected resource bottleneck in the associated host or VM. Alice can then use this information to determine the real root cause and take remediation steps such as VM re-sizing or migration. Moreover, the ability of these models to predict application performance using system metrics can enable the development of performance control systems that further automate the process of remediation. The last goal is the focus of our ongoing work and will not be discussed further in this paper.

This paper makes two main contributions, using both theoretical reasoning and experimental validation:

1. We apply appropriate statistical learning techniques to construct performance models that capture the relationship between application performance and system resources. This is implemented in the model building module of vPerfGuard. The statistical learning techniques (1) filter thousands of system metrics and select those that are most strongly correlated with observed application performance, eliminating a large number of irrelevant metrics, and (2) further reduce redundancy in the selected metrics and build a performance model using a small set of metrics that give the best prediction accuracy. The automatic model generation process successfully overcomes the *scalability* challenge.

2. We apply appropriate statistical hypothesis tests to detect the need to update the performance model when it no longer accurately captures the relationship between performance and system resources. This is implemented in the model updating module of vPerfGuard. The statistical hypothesis tests (1) detect the change-point due to variations in workloads (such as demand spikes) or system conditions (such as resource contention), and (2) trigger the model building module to update the set of predictive metrics and rebuild the model at runtime. The automatic model updating process effectively overcomes the *adaptivity* challenge.

[1]vPerfGuard stands for: virtual Performance Guard.

The remainder of the paper is organized as follows. We describe the design of vPerfGuard in Section 2. In Section 3, we introduce our testbed setup. We then present the experimental results in Sections 4 and 5, and describe results visualization in Section 6. In Section 7, we review related work. Finally, we offer concluding remarks in Section 8.

2. SYSTEM DESIGN

In this section, we introduce the design of the three modules - the sensor module, the model building module, and the model updating module in vPerfGuard.

2.1 Sensor module

The objective of this module is to continuously collect system metrics and application performance metrics. More specifically, it collects two categories of system metrics - *VM metrics* from the operating systems within individual VMs and *host metrics* from the physical hosts running the hypervisors. In our experimental evaluations, the sensor can collect thousands of system metrics, which we refer to as *raw metrics*. It also collects the application performance metrics of interest, e.g., throughput, response times, etc. Workload metrics such as offered load and transaction mix have been required by others for performance predictions [33]. Our framework does not preclude these metrics but we do not *require* their inclusion because: (1) we prefer vPerfGuard to be application-agnostic to free cloud service providers from needing detailed knowledge about the inner operations of their customers' applications, and (2) results from our experimental studies show that they are filtered out and do not appear in any of our final performance models.

2.2 Model building module

The objective of this module is to automatically utilize the thousands of *raw metrics* from the sensor module to derive a performance model that captures the relationship between application performance and system resources.

However, a performance model that is built using all the raw metrics can be computationally expensive to construct and can lead to model over-fitting. First, since the size of the search space with *thousands* of raw metrics is huge, machine learning algorithms operate slowly. Second, many raw metrics are irrelevant or redundant, e.g., a VM's CPU utilization observed from its host is closely related to the CPU utilization observed from within the VM itself. Such dependencies among metrics increase the amount of redundant information in the model and can degrade model quality.

This necessitates the selection of a small number of highly predictive metrics. After removing many of the irrelevant and redundant metrics, the model accuracy can be improved in some cases while the model can be more easily interpreted in other cases. We leverage two categories of algorithms for feature selection [23] to our metric selection: *filters* [25] evaluate features according to heuristics based on general characteristics of the data while *wrappers* [25] use the learning algorithm itself to evaluate the usefulness of features.

We achieve the objectives of metric selection and model building using a two-phase algorithm which is a combination of filters and wrappers: first (in phase 1) selecting a small number of *candidate metrics* that are most strongly correlated with the application performance from among the raw system metrics, and then (in phase 2) identifying even fewer *predictor metrics* that can give the best prediction accuracy for a specific model from among the candidate metrics.

A. Phase 1: Correlation-based selection.

In phase 1 (see Algorithm 1), we aggressively reduce the number of raw metrics considered by filtering out the raw metrics that are not highly correlated with the observed application performance. We denote the application performance metric (e.g., mean response time) as $perf$, and the time series of the $perf$ metric ending at time interval t as a vector $\overrightarrow{perf(t)} = [perf(t), perf(t-1), ...]$. We denote a raw system metric (e.g., CPU consumption of a VM) as m, and the set of all the raw metrics as M. We then denote the time series of each metric ending at time interval t as a vector $\overrightarrow{m(t)} = [m(t), m(t-1), ...]$. For each metric $m \in M$, we use r_m to denote the absolute value of the correlation coefficient between $perf$ and m, and p_m to denote the associated p-value for testing the hypothesis of no correlation. Each p-value is the probability of getting a correlation coefficient as large as the observed value by random chance, when the true correlation is zero. If the p-value is small, say less than 0.05, then the observed correlation is significant.

Algorithm 1: Phase 1

procedure Metrics selection by correlation coefficient
Input: performance metric $perf$ and a set of raw metrics M;
Output: a set of candidate metrics $M_{can} \subset M$;
Tunable Parameter: number of candidate metrics N_{can};
$\forall m \in M$, $r_m = |\text{corrcoef}(\overrightarrow{perf}, \overrightarrow{m})|$,
$p_m =$p-value$(\overrightarrow{perf}, \overrightarrow{m})$;
Select top N_{can} metrics with highest r_m and lowest p_m;
Sort N_{can} metrics in descending r_m and ascending p_m;
Return the set of N_{can} metrics, denoted as M_{can}.

To limit the number of candidate metrics for our model, we select N_{can} top metrics as the candidate metrics for the phase 2 algorithm from all the N_{raw} raw metrics based on the absolute correlation coefficient value and the p-value. We also sort N_{can} metrics for visualization purpose. The time complexity is $O(N_{raw}) + O(N_{can} \log N_{can})$ when the BFPRT algorithm [5] is used for selection and the quicksort algorithm is used for sorting, respectively. N_{can} is a configurable parameter for managing the tradeoff between better model accuracy and lower overhead in the second phase.

B. Phase 2: Model-based selection.

In phase 2 (see Algorithm 2), we explore various combinations of the candidate metrics generated in phase 1, and choose a combination that gives the best prediction accuracy measured by the average R^2 (coefficient of determination) value [19] of the performance model using a 10-fold cross validation [29]. We evaluate and compare the predictive capability of the following four specific types of performance models [20] — linear regression model, k-nearest neighbor (k-NN), regression tree, and boosting approach.

Although the number of metric combinations has been reduced from $2^{N_{raw}}$ to $2^{N_{can}}$ after phase 1, the exploration process is still clearly prohibitive for all but a small number of metrics. We use a heuristic, hill climbing [30] search strategy, i.e., given a set of selected metrics, we choose the additional metric from the remaining set that can give the best improvement in the R^2 value. The algorithm ends when the improvement is smaller than a given threshold. For N_{can} candidate metrics, the computation complexity of the phase 2 model-based selection is $O(N_{pred} * N_{can})$, where N_{pred} is the final number of predictor metrics selected.

Algorithm 2: Phase 2

procedure Metrics selection by a specific model
Input: a performance metric $perf$ and M_{can} from phase 1;
Output: a set of predictor metrics $M_{pred} \subset M_{can}$ and the associated model $F(M_{pred})$ with learned parameter values;
Tunable Parameter: type of model F (e.g., "linear"), R^2_{inc} for the minimum incremental R^2 improvement;
$selected = \emptyset$, $left = M_{can}$, $R^2_{old} = 0$, $R^2_{best} = 0$;
while *true* **do**
    for $m \in left$ **do**
        $metrics = selected \cup \{m\}$;
        Use $\overrightarrow{perf}$ and all $\overrightarrow{m}$ in $metrics$, obtain R^2_{new} following a 10-fold cross validation;
        if $R^2_{new} > R^2_{best}$ **then**
            $R^2_{best} = R^2_{new}$;
        end
    end
    if $R^2_{best} - R^2_{old} > R^2_{inc}$ **then**
        move m from $left$ to $selected$;
        $R^2_{old} = R^2_{best}$;
    else
        break;
    end
end
Build the final model F using the metric set $selected$;
Return $M_{pred} = selected$ and the model $F(M_{pred})$.

Hall [23] proposes a method to select a subset of metrics based on a heuristic "merit" of the subset. The motivation is that phase 1 (i.e., filters [25]) may pick many metrics which individually have high correlation with the output metric, but that when combined together in a model do not provide much additional useful information. We have implemented his method and compared it with our two-phase algorithm. We find that (1) his method has comparable overhead with ours; (2) our phase 2 algorithm can also overcome the limitation of phase 1; (3) the final metrics and models are very similar if the final number of predictor metrics (N_{pred}) is a small number.

2.3 Model updating module

The objective of this module is to automatically detect the change-point when the performance model derived from the model building module no longer accurately captures the relationship between application performance and system resources.

In a highly-dynamic, consolidated cloud environment, the relationship between application performance and system resources could be altered due to time-varying workload patterns, aggravated resource contention, different VM-to-host mappings, or other changes. We define such a relationship change as a *change-point*. This is different from detecting changes in performance alone. For example, if a 20% increase in the workload leads to degraded performance, our module should not flag this as a change-point if the relationship between performance and system resources still holds.

We assume that the distribution of the model's prediction errors (residuals) is stationary across adjacent time intervals when there is no change. Motivated by this, we use an online change-point detection technique [14] to determine whether a change occurs by performing hypothesis testing on the model's prediction errors across adjacent time intervals.

More specifically, given an existing performance model constructed in time window $W(t')$, its prediction errors in $W(t')$ and those in an adjacent time window $W(t)$ as shown in Fig. 3, we adopt an unpaired 2-sample t-test [19] to determine whether the prediction errors observed in $W(t')$ and $W(t)$ come from the same distribution, (could have the same statistical mean), i.e., the null hypothesis is that "there is no significant difference in the statistical mean between $W(t')$ and $W(t)$." If the result of the hypothesis test suggests that to be the case, then our performance model is likely as (in)accurate as when we accepted it for use in production and, absent other information, we have no reason to discontinue using the model. A significant difference in the statistical mean is a sufficient but not necessary condition for a significant difference in the distribution and our t-testing does not assume that the variances of the prediction errors in $W(t')$ and $W(t)$ are equal.

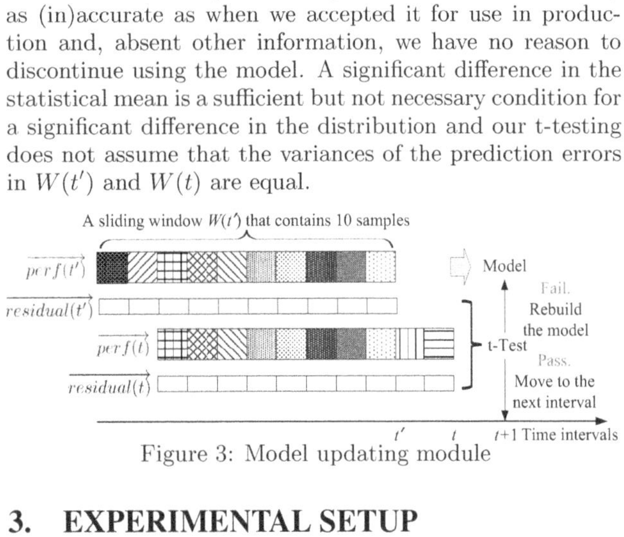

Figure 3: Model updating module

3. EXPERIMENTAL SETUP

3.1 Hypervisor and sensor module

The vPerfGuard framework is generic and can work with different virtualized platforms and monitoring tools.

For the system metrics, we run VMware ESX 4.1 [9] as the hypervisor on each host. The sensor module of vPerfGuard can collect the host metrics (~1800 metrics per host) from the "esxtop" [2] interface on ESX systems. Whereas our evaluation is done using VMware's hypervisor and tools, our framework generalizes to other virtualized platforms where similar tools exist to gather system-level metrics, e.g., "xentop" for Xen-based systems [13] and "Hyper-V Monitor Gadget" for Hyper-V-based systems [10]. For the VM metrics, we run "dstat" [1] and "iostat" [3] tools within the guest VMs so that the sensor module of vPerfGuard can collect the VM metrics (48 metrics per VM) from them.

For the application performance metrics, we collect the throughput and response times per sampling interval directly from the workload generator. In future work, we plan to leverage monitoring tools that can measure application metrics from the hosting platform. One such tool is the VMware vFabric Hyperic [8], which offers out-of-the-box performance monitoring for a suite of Web applications.

3.2 Benchmarks and workloads

Although we run various benchmarks [2] on our virtualized testbed, due to space limitations, we focus on the results from the RUBBoS [4] and the TPC-H [6] benchmarks.

In our experiments, we deploy the RUBBoS application with the browsing-only transaction mix in a 4-tier setup, including one Apache server, two Tomcat severs, one CJDBC cluster server and two MySQL servers as shown in Figure 4(a). The sampling interval is 1 minute. We deploy the TPC-H benchmark with a scaling factor 3 using PostgresSQL. The total database size is 4571MB including all the data files and index files. The original benchmark contains 22 queries, i.e., Q1 to Q22. We choose Q6, Q7, Q12 and Q14 for our experiments because these are IO-intensive queries and they can be completed within a sampling interval of 6 minutes. For both benchmarks, we modify the original workload generator to dynamically vary the number of concurrent users in the system.

3.3 Testbed setup and configurations

We run the RUBBoS benchmark on eight hosts, as shown in Fig. 4(a). Four hosts, *ESX1* through *ESX4*, are used to run the six VMs hosting the individual application tiers, labeled as *Web, App1, App2, CJDBC, DB1*, and *DB2*, respectively. We also run some *co-hosted* VMs on *ESX1* and *ESX4* to induce resource contention on the respective host. The four client VMs run on the other four hosts.

We run the TPC-H benchmark on three hosts shown in Fig. 4(b). Two virtual machines, $TPCH_F$ (foreground) and $TPCH_B$ (background), are deployed on one host, *ESX5*, running two instances of the TPC-H DB server. The two client VMs run on the other two ESX hosts.

The host and VM configurations are shown in Tables 1 and 2, respectively. All the VMs run Linux kernel 2.6.32. vPerfGuard runs on a separate host.

Table 1: Configuration of hosts

Testbed	RUBBoS	TPC-H, vPerfGuard
Model	Dell Power Edge 1950	Dell OptiPlex 780
CPU	2 Intel Xeon E5420 2.5 GHz Quad-Core	1 Intel Core2 Q9650 3.0 GHz Quad-Core
Memory	32 GB	16 GB
Storage	Clariion CX-40 SAN	7200 RPM local disk

[2]RUBiS, RUBBoS with the browsing-only and the read-write transaction mixes, TPC-W and TPC-H.

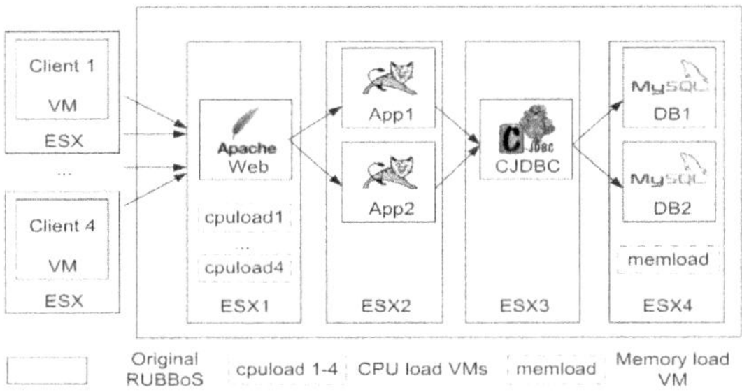

(a) Test bed for RUBBoS benchmark

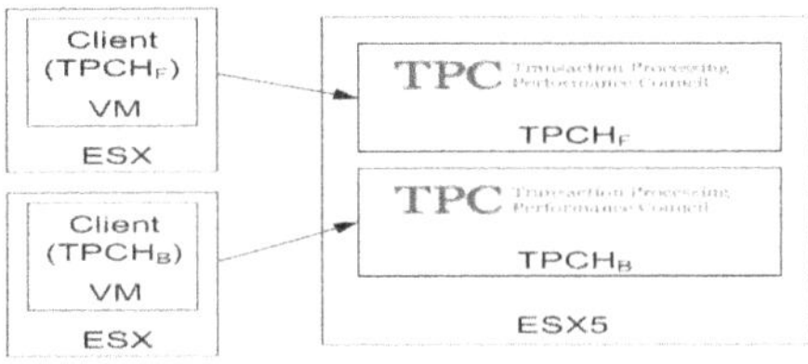

(b) Test bed for TPC-H benchmark

Figure 4: Setup of two experimental testbeds

Table 2: Configuration of VMs

Testbed	RUBBoS	TPC-H
Application VM vCPU	2	4
Application VM vRAM	1 GB	2 GB
Client VM vCPU	2	4
Client VM vRAM	8 GB	4 GB

Table 3: Metrics naming convention

Application perf. metrics	$THR, MRT, RT_{std}, RT_{50p}, RT_{75p}, RT_{90p}, RT_{95p}, RT_{99p}$
Host metrics	$H\_\{ESX\}\_\{VM\}\_\{Metric\}\_\{Details\}$
VM metrics	$V\_\{VM\}\_\{Metric\}\_\{Details\}$

3.4 Naming convention

We describe the naming convention for all the metrics we collect in Table 3. For example, THR, MRT and RT_{95p} are application performance metrics, denoting throughput, mean response time and 95th percentile response time, respectively. The other metrics that begin with H and V are host and VM metrics, respectively. For instance, metric $H\_ESX1\_Web\_CPU\_System$ represents the CPU "System" counter for the "Web" VM running on the "ESX1" host, and metric V_CJDBC_Int represents the "Interrupt" counter for the "CJDBC" VM.

4. EVALUATION OF MODEL BUILDING MODULE

In order to evaluate the model building module and compare the predictive capabilities of different performance models, we use the RUBBoS benchmark in the setup shown in Fig. 4(a), without the co-hosted VMs. We first run a calibration experiment where we vary the number of users from 400 to 4000 with a step size of 400, and observe that the application reaches a performance bottleneck at around 3000 concurrent users. This can be seen in Fig. 5(a) in the saturation of the application throughput and the near 100% CPU utilization of the "Web" VM. We then run a "random" workload for 400 minutes, where the number of users changes randomly between 400 to 4000. A sampling interval of 1 minute is used in the sensor module, resulting in 400 measurement samples that are used for the evaluation in this section. Each sample is a high-dimensional vector, consisting of the following 7522 metrics: 8 application performance metrics as shown in Table 3, 7226 host metrics from the four ESX hosts, and 288 VM metrics from the six VMs.

4.1 Evaluation of phase 1

For evaluation purposes, instead of limiting the number of candidate metrics from phase 1 as described in Algorithm 1, we report the number of candidate metrics selected by the phase 1 algorithm as a function of two threshold values — a lower bound, r_{LB}, on the absolute value of the correlation coefficient, and an upper bound, p_{UB}, on the p-value of the observed correlation.

We use throughput as the $perf$ metric and the 7226 ESX host metrics as the raw metrics. The number of selected host metrics is shown in Fig. 5(b). For example, for $r_{LB} = 0.8$ and $p_{UB} = 0.1$, a total of 132 metrics are selected out of the 7226 raw metrics. That means these 132 metrics (or 2% of the raw metrics) are correlated with the observed throughput with $r_m \geq 0.8$ and $p_m \leq 0.1$. We can also infer that 98% of the raw metrics are not highly correlated with the application throughput. The number of selected metrics is reduced as the minimum correlation level increases or as the maximum p-value decreases. The latter indicates an increased level of confidence in the observed correlation.

We observe similar trends when MRT or RT_{95p} is chosen as the $perf$ metric. We also observe similar trends when we use throughput or MRT as the $perf$ metric and the 288 VM metrics as the raw metrics.

4.2 Evaluation of phase 2

To evaluate the phase 2 algorithm, we set the number of candidate metrics from phase 1 to be 100 and the minimum incremental R^2 improvement in phase 2 to be 0.01. For illustration, we provide an example of building a linear regression model for MRT in Figure 5(c), which shows the R^2 value for the model when one, two and three predictor metrics are selected sequentially. The phase 2 algorithm first chooses the network UDP active status of the Web VM on the $ESX1$ host ($V\_Web\_UDP\_Act$), resulting in an R^2 value of 0.668 for the single-metric linear MRT model. The algorithm then adds the second metric, the percentage of CPU Used of the Web VM on the $ESX1$ host ($H\_ESX1\_Web\_vCPU\_Used$), increasing the R^2 value of the model to 0.731. After adding the third metric, the total CPU Used on the $ESX1$ host ($H\_ESX1\_CPU\_TotalUtil$), the algorithm stops searching because the model quality improvement falls below the minimum incremental R^2 improvement threshold (0.01) when a 4th metric is added.

To evaluate the impact of the phase 2 metric selection algorithm, Fig. 5(d) reports the R^2 values for different model types in two scenarios: (1) using the 100 candidate metrics from phase 1 directly as the predictor metrics (without phase 2 selection), and (2) using the smaller number of predictor metrics selected from phase 2. For the first three model types (linear regression, k-NN, regression tree), the additional metric selection in phase 2 helps improve the accuracy of the final model.

4.3 Sensitivity analysis

The effectiveness of the two-phase algorithm depends on the values of the tunable parameters, including (1) the num-

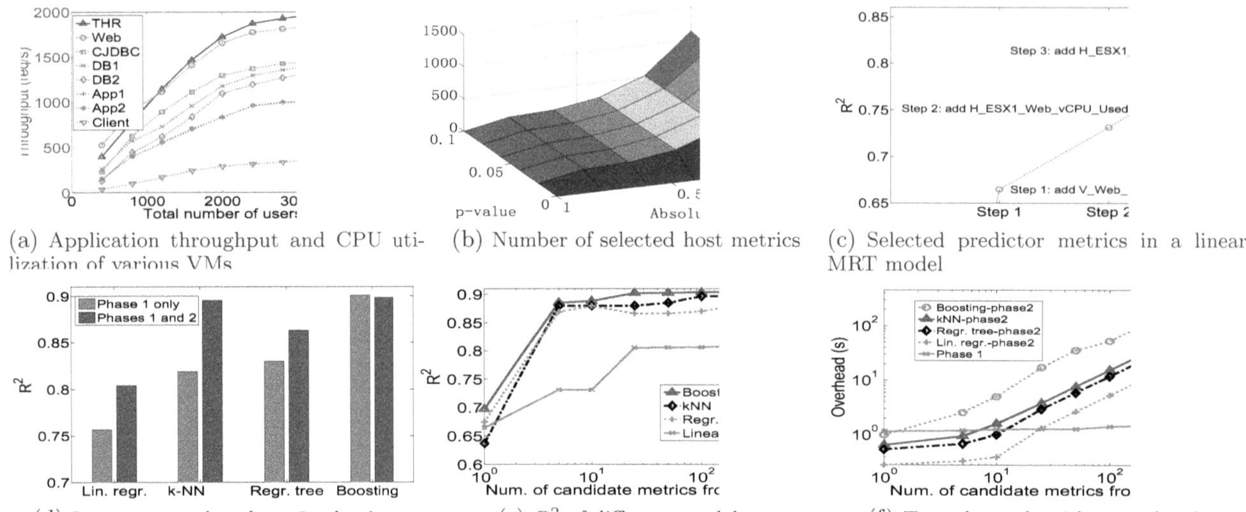

(a) Application throughput and CPU utilization of various VMs

(b) Number of selected host metrics

(c) Selected predictor metrics in a linear MRT model

(d) Improvement by phase 2 selection

(e) R^2 of different models

(f) Two phase algorithm overhead

Figure 5: Evaluation results for 2-phase metric selection and model building algorithm

ber of selected candidate metrics from phase 1 (N_{can}), (2) the type of model F chosen in phase 2, and (3) the minimum incremental R^2 improvement in phase 2 (R^2_{inc}). We evaluate the impact of these parameters in two aspects — model accuracy and computation overhead. We assume that MRT is chosen as the *perf* metric.

Firstly, we fix the minimum incremental R^2 improvement in phase 2 at 0.01, and vary the other two tunable parameters. More specifically, for each of the four types of models, we limit the number of candidate metrics from phase 1 at [5, 10, 25, 50, 100, 200, 400, 800], and run the phase 2 algorithm to build a model for the MRT. In both experiments, a 10-fold cross validation [29] is used to compute the R^2 value for each model type.

Figure 5(e) shows the R^2 value of the final model as a function of the number of candidate metrics from phase 1 (in *log* scale). Different lines represent different model types. We make the following observations: (1) As more metrics are selected in phase 1, the model accuracy from phase 2 is generally improved for all four model types; (2) All the models achieve reasonably good accuracy ($R^2 > 0.8$) with 25 or more candidate metrics from phase 1, although the linear model's R^2 value is slightly lower than those from the nonlinear models.

Figure 5(f) shows the computation time of both phase 1 and phase 2 as a function of the number of candidate metrics from phase 1 (in *log* scale). Different lines for phase 2 represent different model types. We make the following observations: (1) The phase 1 overhead increases slowly with the number of candidate metrics; (2) For all four model types, the overhead in the phase 2 algorithm grows as we increase the number of candidate metrics from phase 1; (3) The boosting model has the most overhead, and the linear regression model has the least.

For demonstration purposes, we also run the phase 2 algorithm directly on all the raw metrics, without initial metric selection in phase 1. The result shows that, for all the model types, the metric selection in phase 1 helps achieve better accuracy in the final model as well as reducing the overhead in model building in phase 2.

Secondly, we run similar experiments to evaluate the effect of the minimum incremental R^2 improvement in phase 2 (R^2_{inc}). We omit detailed results due to space constraints. As the threshold value becomes smaller, we have better model accuracy and more computational overhead in phase 2. We find that the threshold value of 0.01 for R^2 improvement strikes a good balance between better accuracy and lower overhead. Larger values require larger model-accuracy improvements to add metrics to the model, pre-empting the consideration of additional metrics.

Finally, we choose the linear regression model as our default model because it has the best *human-interpretability* and lowest overhead with only slightly lower accuracy relative to the nonlinear models. In spite of better accuracy, the regression tree is not a good candidate because (1) if we use shallow trees, the marginal ratio between performance and predictor metric is zero at most points, making it unable to infer which system metric is the bottleneck, and (2) if we use deep trees, the over-fitting issue prevents the model from generalizing faithfully. The k-NN and boosting approaches are also not preferred because they are harder to interpret directly due to model complexity.

5. EVALUATION OF MODEL UPDATING MODULE

To evaluate the model updating module of vPerfGuard, we run the tool against four typical, dynamic workload scenarios a cloud service provider such as Alice may experience, including an application performance bottleneck caused by a surge in the workload intensity, and performance degradation in one application due to the CPU, memory, or disk I/O contention from co-hosted VMs (*aka. noisy neighbors*).

Our evaluation criteria focus on three aspects of the performance models generated: (1) **prediction:** whether a model provides an accurate prediction of the application performance using the selected system metrics; (2) **diagnosis:** whether the selected system metrics point to the correct performance bottlenecks, including the critical application component, the resource under contention, or the host where the

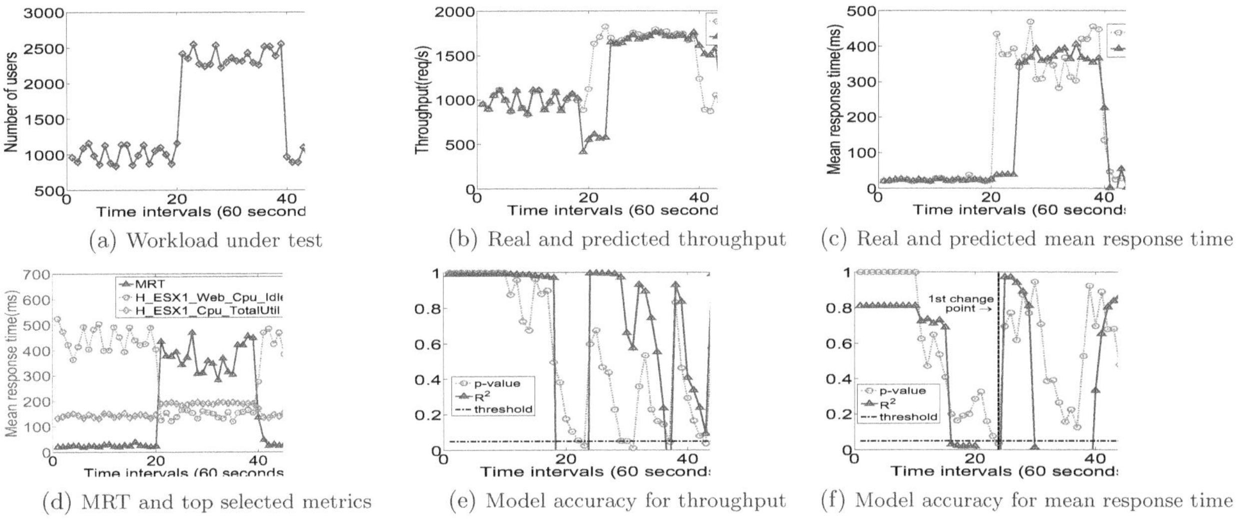

(a) Workload under test (b) Real and predicted throughput (c) Real and predicted mean response time

(d) MRT and top selected metrics (e) Model accuracy for throughput (f) Model accuracy for mean response time

Figure 6: Experimental results for the workload surge scenario

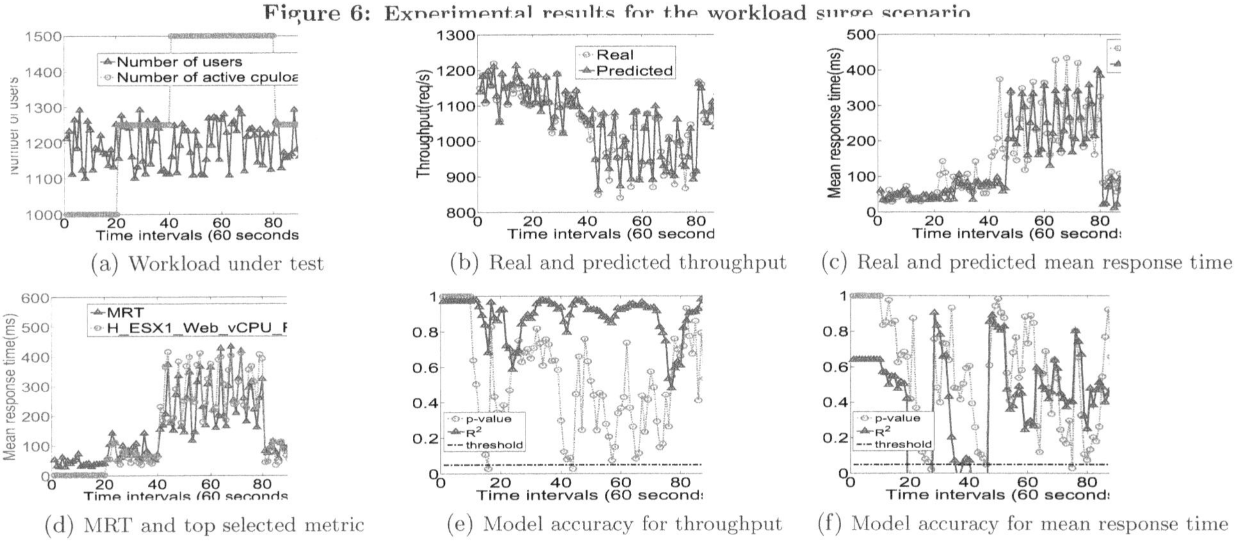

(a) Workload under test (b) Real and predicted throughput (c) Real and predicted mean response time

(d) MRT and top selected metric (e) Model accuracy for throughput (f) Model accuracy for mean response time

Figure 7: Experimental results for the CPU contention scenario

contention occurred; (3) **adaptivity:** whether the model is adaptive to the changes in relationship between application performance and system resources. Note that we do not expect the human analyst to interpret the models directly. We will show a graphical user interface in the next section to illustrate how the top suspicious metrics and the associated coefficients can be presented to the user for inspection.

The following subsections describe the experimental settings of the four scenarios, present the detailed results in Figures 6-9, and summarize the evaluation in Tables 4 and 5. In particular, each figure is organized as follows. Fig. (a) shows the workload(s) used, Fig. (b) and Fig. (c) compare the real and the model-predicted throughput and mean response time (MRT), respectively, Fig. (d) shows the MRT and the top selected metrics in the MRT model, and finally, the model accuracy measures including the p-value and the R^2 value for the throughput and the MRT models are shown in Fig. (e) and Fig. (f), respectively.

5.1 Workload surge

A. Experimental settings.

In this scenario, we use the testbed in Fig. 4(a) with the RUBBoS application only (i.e., no co-hosted VMs). The browsing-only workload mix is run for an hour (60 time intervals), with a surge in the workload intensity that goes from 1000 to 2300 users (with small, random variation) and lasts from the 21st to the 40th intervals (Fig. 6(a)).

B. Evaluation.

According to the experimental setting and our previous knowledge from Fig. 5(a), the mean response time increases during the workload surge period due to the CPU resource bottleneck in the Web tier of the RUBBoS application. As shown in Fig. 6(e) and Fig. 6(f), the online module detects a change-point multiple times, resulting in 6 throughput models and 3 MRT models for the duration of the experiment, such that we maintain a high confidence in the learned models (p-value ≥ 0.05). The throughput models starting from the 1st, 24th, 32nd, 38th, 44th and 51st intervals are:

$$THR = 0.19 \times V\_CJDBC\_Int - 519.60,$$

$$THR = 0.05 \times H\_ESX1\_Web\_vSwitch\_PcksTrans/s + 36.25$$

$$THR = 28.53 \times H\_ESX2\_App2\_vSwitch\_MBitsRec/s + 22.53 \times V\_DB2\_CPU\_Sys + 358.79$$

$$THR = 16.16 \times H\_ESX2\_CPU\_Idle\_Overlap + 1277$$
$$THR = 0.08 \times V\_Web\_ContextSwitch - 202.37$$
$$THR = 28.53 \times H\_ESX2\_App2\_vSwitch\_MBitsRec/s + 113.75$$

The MRT models for the mean response time starting from the 1st, 25th and 54th intervals are:

$$MRT = 2.17 \times H\_ESX1\_CPU\_TotalUtil - 20.91$$
$$MRT = -8.10 \times H\_\mathbf{ESX1\_Web\_CPU\_Idle} + 545.10$$
$$MRT = 0.49 \times V\_CJDBC\_UDP\_Act - 162.75$$

We make the following observations. (1) The models are predictive of the application performance most of the time according to Figs. 6(b), 6(c) and Table 4. The signs of the coefficients in each THR or MRT model make sense, e.g., when throughput increases, interrupts, CPU utilization, context switches and network packets transmitted or received also increase. (2) The selected system metrics in all the THR models do not point to the correct cause of the performance degradation. Three of the six models choose network attributes as the top metrics, and two other models choose system interrupts or context switches as the top metrics. The selected system metric $H\_ESX1\_Web\_CPU\_Idle$ in the second MRT model as shown in Fig. 6(d) directly points to not only the bottleneck host (ESX1), but also the bottleneck VM (Web) and the critical resource (CPU).

Following the above observations, we conclude that, (1) both THR and MRT models have good *performance-prediction* capability, and THR models have better prediction accuracy due to its linear relationship with many system metrics, and (2) THR models are not suitable for diagnosis, and MRT models have good *diagnosis* capability during the periods of performance bottlenecks. This observation is also validated in the next three scenarios and in [22]. When the application experiences a performance bottleneck, the THR remains almost constant, making it harder for our correlation-based selection to identify critical system metrics. At the same time, a small change in a critical system metric may lead to a large change in the MRT, making it easier for our algorithm to identify the correlation. For conciseness, we do not show the throughput models in the following scenarios, and we show only the MRT models during the periods of performance degradation to illustrate their diagnosis capability.

The MRT models are also *adaptive* to the surge in the workload, with only 3 intervals of delay in response. As Fig. 6(f) shows, after the workload surge at the 21th interval, the application MRT increases dramatically. It takes 3 intervals for the p-value of the first MRT model to drop below the threshold value of 0.05. The vertical line in the figure (24th interval) indicates where the first change-point is detected, after which the Web VM's CPU idle time is chosen as the key metric for learning a new MRT model starting from the next interval.

5.2 CPU contention

A. Experimental settings.

In this experiment, we run the RUBBoS benchmark with a workload intensity randomly varying between 1100 and 1300 users, as shown in Fig. 7(a). To create a CPU contention scenario, we use the four co-hosted VMs, *cpuload*1 to *cpuload*4, as noisy neighbors, to share the CPUs on the ESX1 host with the RUBBoS *Web* VM (see Fig. 4(a)). Each *cpuload* VM is configured with 2vCPUs, 1GB vRAM, and runs a "CPU eater" workload that consumes the vCPUs at a specified utilization level for a specified time period. We let the CPU utilization of each VM vary periodically between 10% and 40% with a period of 2 minutes. All the four co-hosted VMs are idle for the initial 20 time intervals. We then start the workload in *cpuload*1 and *cpuload*2 at the 21st interval, and start *cpuload*3 and *cpuload*4 at the 41st interval. The workloads in these VMs are idle again at the 81st and the 101st intervals (Fig. 7(a)).

B. Evaluation.

According to the experimental setting, the application's mean response time starts to increase after the 21st interval due to the CPU resource bottleneck on the ESX1 host, caused by the active workloads in the four *cpuload* VMs. The MRT models starting from the 28th, 47th, 76th and 91st intervals are:

$$MRT = 1.97 \times H\_ESX4\_DB1\_Mem\_Active + 1.13 \times H\_\mathbf{ESX1\_CPU\_Util} - 89.7$$
$$MRT = 752.76 \times H\_\mathbf{ESX1\_CPULoad\_1}MinuteAvg - 562.87$$
$$MRT = 12.48 \times H\_\mathbf{ESX1\_Web\_vCPU\_Ready} - 25.01$$
$$MRT = 6.38 \times H\_\mathbf{ESX1\_Web\_vCPU\_Ready} + 48.72$$

We make the following observations. (1) These MRT models have good prediction capability as shown in Figs. 7(b), 7(c) and Table 4. (2) These models have good diagnosis capability because they all point to the correct performance bottleneck. For example, one of the top system metrics, $H\_ESX1\_Web\_vCPU\_Ready$, as shown in Fig. 7(d), specifies not only the bottleneck host (ESX1), but also the bottleneck VM (Web) and the critical resource (CPU). (3) The models are adaptive to the increased CPU load from the noisy neighbors as shown in Fig. 7(f). After two of the cpuload VMs become active in the 21th interval, it takes the model updating module 6 intervals to detect the change and build a new model.

5.3 Memory contention

A. Experimental settings.

In this experiment, we run the RUBBoS benchmark with a workload intensity randomly varying between 900 and 1100 users, as shown in Fig. 8(a). Because the size of the MySQL database is 498.88MB, the total size of database files on the two database VMs, i.e., $DB1$ and $DB2$, are approximately 1GB. We use a co-hosted VM, *memload* (configured with 4vCPUs, 1GB vRAM), as the noisy neighbor, to run on the $ESX4$ host along with $DB1$ and $DB2$ (see Fig. 4(a)). To create memory contention, we configure the $ESX4$ host with 4GB of physical memory. Since about 3GB of memory is reserved by the hypervisor, only 1GB of memory is available for the three VMs ($DB1$, $DB2$, and *memload*) to share. As a result, the total memory commitment during these 40 intervals is much more than the shared 1GB memory.

For the initial 40 intervals, the *memload* VM remained idle, so the total memory commitment on $ESX4$ is close to 1GB. Between the 41st and the 80th intervals, a four-thread "memory eater" application is started inside the *memload*

VM. Each thread in the application allocates 120-180MB memory and randomly touches the allocated pages to keep them actively used. As a result, the total memory commitment during these 40 intervals is much more than the shared 1GB memory. When the RUBBoS DB servers cannot access enough physical memory, more requests require disk accesses, reducing throughput (Fig. 8(b)) and increasing response times (Fig. 8(c)).

B. Evaluation.
The MRT models starting from the 47th and 52th intervals are:

$$MRT = -0.75 \times H\_ESX4\_Network\_PksReceived/sec + 3072.86$$

$$MRT = -25.38 \times H\_\mathbf{ESX4\_DB1\_Mem\_GrantedMB} + 4876.55$$

We make the following observations. (1) The models have reasonably good prediction capability as shown in Figs. 8(b), 8(c) and Table 4. (2) The second model for mean response time has good diagnosis capability, since the top system metric, $H\_ESX4\_DB1\_Mem\_GrantedMB$ as shown in Fig. 8(d), indicates not only the bottleneck host (ESX4), but also to the critical resource (Memory). (3) The models are adaptive to the increased memory load in the system with 10 intervals of delay as shown in Fig. 8(f).

5.4 Disk I/O contention

A. Experimental settings.
In this experiment, we run two instances of the TPC-H benchmark in parallel. A foreground database VM ($TPCH_F$) and a background database VM ($TPCH_B$) are co-located on the $ESX5$ host, as shown in Fig. 4(b). Since the total database size is 4571MB, which cannot fit into the VM's 2GB vRAM, some of the queries much involve disk I/O. Fig. 9(a) shows the workloads used in both VMs with a 6 minute sampling interval. The workload for $TPCH_F$ has the number of users randomly varying between 3 and 4 throughout the experiment. The background VM, $TPCH_B$, is idle for the initial 20 intervals. Between the 21st and the 40th intervals, the TPC-H workload is activated inside the $TPCH_B$ VM, a noisy neighbor, with the number of users randomly varying between 1 and 2.

B. Evaluation.
The MRT model between the 25th and 49th intervals for the $TPCH_F$ application is:

$$MRT = 180.62 \times H\_\mathbf{ESX5\_PhysicalDisk\_Writes/s} + 29.24$$

We make the following observations. (1) The models can capture the main trend in the application MRT as shown in Figs. 9(b), 9(c) and Table 4. (2) The top system metric $H\_ESX5\_PhysicalDisk\_Writes/s$ selected, as shown in Fig. 9(d), points not only to the bottleneck host (ESX5), but also the critical resource (disk I/O). Note that database servers often write temporary files (e.g., sorting files) to the disk when the physical memory is scarce. (3) The models are adaptive to the occurrance of the disk I/O bottleneck with only 3 intervals of delay as shown in Fig. 9(f).

5.5 Evaluation summary

For the four dynamic workload scenarios that we test, we summarize the evaluation results for the model prediction and the model diagnosis accuracies of vPerfGuard.

A. Prediction.
Besides showing the model prediction accuracy in R^2 over a sliding window of samples in the previous figures, we also report in Tab. 4 the statistics (mean, standard deviation, 50th percentile, 90th percentile) of the relative error for the individual THR and MRT samples, i.e., $(|(predicted - real)/real|)$. In the last column, we also show the *overall* relative error as (sum of $|predicted - real|$)/(sum of $real$). We can see that vPerfGuard achieves reasonably good prediction accuracy. We also notice that the relative error for THR is much smaller than that for MRT in all four scenarios except the disk I/O contention scenario. This validates our earlier observation in Sec. 5.1 that linear models capture the relationship between application throughput and system metrics well in most cases.

Table 4: Relative error for THR and MRT

Scenarios(perf)	Mean(std)	50p	90p	overall
Workload(THR)	0.10(0.20)	0.01	0.53	0.10
Workload(MRT)	0.41(0.64)	0.12	0.96	0.35
CPU(THR)	0.01(0.01)	0.01	0.03	0.01
CPU(MRT)	0.29(0.26)	0.23	0.64	0.27
Memory (THR)	0.12(0.16)	0.06	0.33	0.10
Memory (MRT)	0.51(0.45)	0.25	0.95	0.37
Disk I/O(THR)	0.95(3.32)	0.11	1.58	0.24
Disk I/O(MRT)	0.29(0.39)	0.16	0.70	0.39

B. Diagnosis and adaptivity.
Table 5 shows the diagnosis accuracy of the MRT models using *precision* and *recall* measures from pattern recognition literature, computed only for the performance bottleneck period. In our context, we define precision to be the fraction of all the selected metrics that are relevant (i.e., point to the correct bottleneck); and if a metric appears in n intervals, it's counted n times. We define recall to be the fraction of all the intervals in which the selected metrics are relevant; and for intervals with multiple selected metrics, that interval is counted using only the fraction of the relevant metrics. We report precision and recall for the detection of bottleneck resource and bottleneck host, separately. We use the CPU contention scenario as an example where the length of the bottleneck period is 80 intervals In intervals 21-27, the model contains an irrelevant metric; in intervals 28-46, the models contains two metrics with one being relevant; in the remaining intervals the model contains one relevant metric. Hence, precision$=(19+54)/(7+19\times2+54) = 74\%$, and recall$=(19/2+54)/80 = 79\%$. In the last column, we also report the delay in change-point detection in number of intervals. We can see that models built by vPerfGuard achieve good diagnosis accuracy in terms of precision and recall, with short delays in model updates.

Table 5: Diagnosis summary

Scenarios	Precision		Recall		Delay
	resource	host	resource	host	
Workload	100%	100%	100%	100%	3
CPU	74%	74%	79%	79%	6
Memory	73%	85%	73%	85%	10
Disk I/O	80%	100%	80%	100%	3

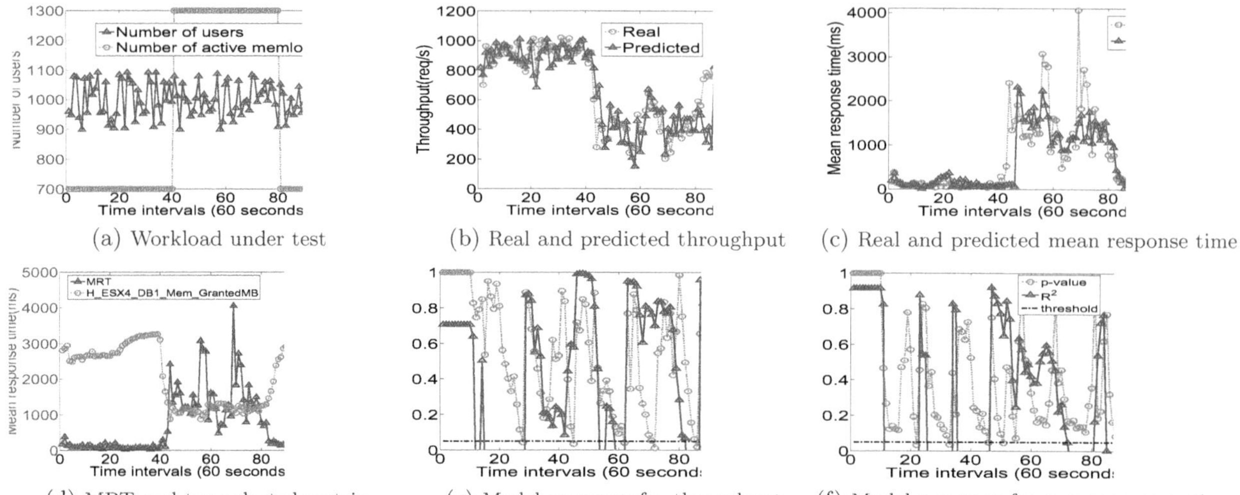

(a) Workload under test (b) Real and predicted throughput (c) Real and predicted mean response time

(d) MRT and top selected metric (e) Model accuracy for throughput (f) Model accuracy for mean response time

Figure 8: s for the memory contention scenario

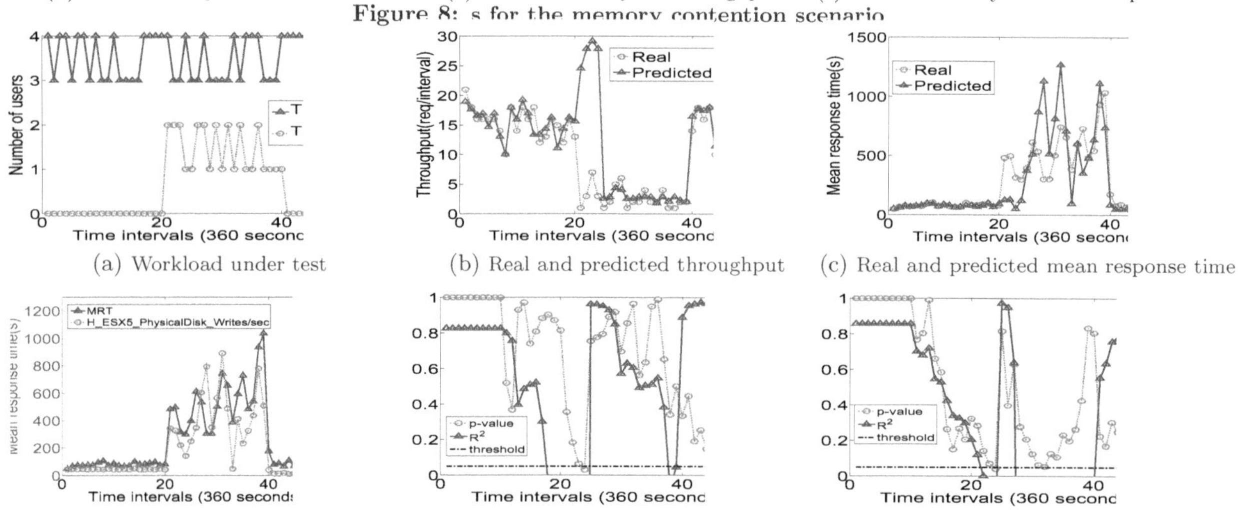

(a) Workload under test (b) Real and predicted throughput (c) Real and predicted mean response time

(d) MRT and top selected metric (e) Model accuracy for throughput (f) Model accuracy for mean response time

Figure 9: Experimental results for the disk I/O contention scenario

5.6 Discussion

A. Adaptation overhead.

In Table 6, we show the mean and standard deviation of the overhead in running vPerfGuard online for four dynamic scenarios, using 10 consecutive samples for both model building and change-point detection. The average overhead is 67ms (standard deviation = 37ms) for the sensor module to pull the application performance metrics and the system metrics from the hosts and the VMs. The metric selection and model building module takes an average of 221ms, the longest among all. The average model testing time is 54ms, and the average hypothesis testing time is 5ms.

Table 6: Online model adaptation overhead(mean(std))

Metrics collection(ms)	Building time(ms)	Testing time(ms)	Hypothesis testing(ms)
67(37)	221(148)	54(23)	5(18)

B. Sensitivity analysis.

We perform a sensitivity analysis using different change-point detection criteria (p-value < 0.05 or $R^2 < 0.8$) or a different number of samples to explore the tradeoff between prediction accuracy and diagnosis accuracy. In Table 7, we summarize the average positive R^2 value, the number of models generated, the number of models indicating the correct bottleneck, the total number of selected metrics in all the models, and the number of metrics identifying the correct bottleneck for the CPU contention scenario, during the contention period. As we increase the number of samples, the average R^2 increases (as one would expect), but the percentage of correct models or metrics decreases. For change-point detection, if a criterion of $R^2 < 0.8$ is used instead of using hypothesis testing with p-value < 0.05, we may generate too many models (due to over-fitting) for the human analyst to reason about. This result indicates that using hypothesis testing is a more robust method for change-point detection than using the R^2 value directly.

Table 7: Sensitivity analysis

Method (# of samples)	average R^2	# of models (correct)	# of metrics (correct)
p-value (10)	0.62	4 (4)	5(4)
p-value (20)	0.62	3 (2)	4 (2)
p-value (30)	0.76	3 (0)	3 (0)
R^2 (10)	0.74	49 (32)	56 (32)
R^2 (20)	0.73	40 (18)	45 (18)
R^2 (30)	0.80	38 (12)	38 (12)

6. VISUALIZATION OF RESULTS

We introduce a primitive graphic user interface (GUI) for visualization of the results from vPerfGuard. The GUI consists of four panels: a configuration tab, a real-time tracking tab, a real-time analysis tab, and a real-time diagnosis tab. The configuration tab is used to configure the necessary parameters for vPerfGuard to operate. The real-time tracking tab composes of two windows: one shows the real and the predicted performance metric values for comparison; the other shows the corresponding p-value and R^2 value. Next we describe the other two tabs using the workload surge scenario as an example.

The real-time analysis tab (shown in Figure 10) presents the selected metrics and models in real time. The top-left window shows the selected metrics from phase 1 with absolute correlation coefficient in descending order. If a highlighted metric is double-clicked, the time series of the metric and the application performance metric will be shown in the top-right window. In the same tab, the bottom-left window shows the series of models built in phase 2 with the top system metrics, Because an abstract model may be hard to interpret by a cloud service provider, when a highlighted metric is double-clicked, the GUI translates the abstract metric name into human readable description at the top of the window. At the same time, the time series of the metric and the application performance metric are displayed in the bottom-right window.

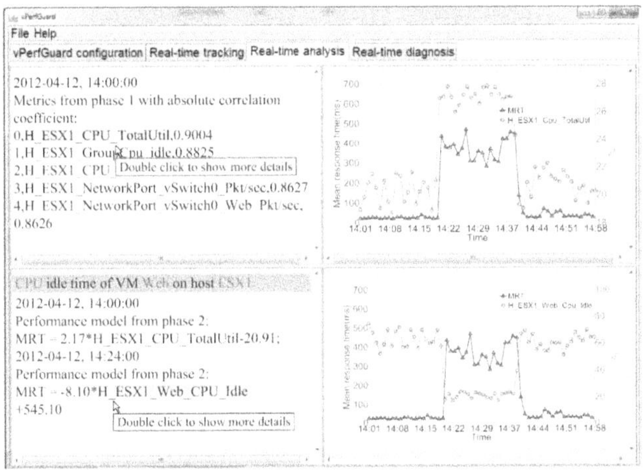

Figure 10: GUI for real-time analysis

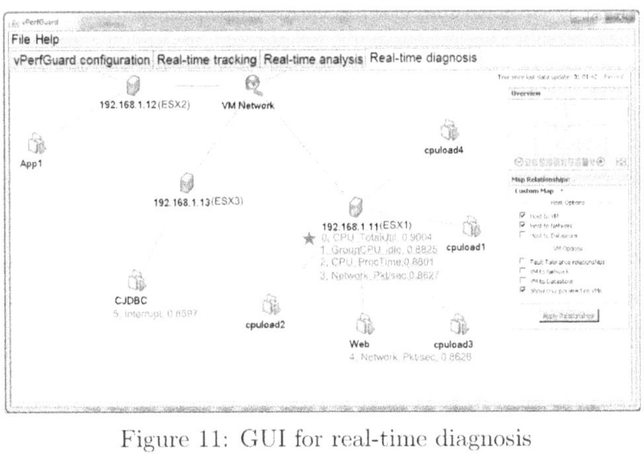

Figure 11: GUI for real-time diagnosis

The real-time diagnosis tab (shown in Figure 11) points to possible performance bottleneck locations. We utilize vCenter map [9], a visual representation of the vCenter Server topology that captures the relationships between the virtual and physical resources managed by the vCenter Server. The selected system metrics are overlayed on top of the associated components in the system. We use red fonts to locate the metrics from phase 1 and red star icon to locate the metrics from phase 2, respectively. As a result, this tab offers a cloud service provider better visibility into the potential bottlenecks in a complex and distributed environment.

One useful feature we'd like to implement is allowing a user to manually add a metric to the model, by right clicking on a specific metric. This offers a cloud service provider an interface to provide inputs to the diagnosis process by applying domain knowledge. While model-driven performance diagnosis is useful in highlighting potential bottlenecks, incorporating domain knowledge from an experienced human analyst may lead to even better results in the timely determination of the real root causes of the performance problems.

7. RELATED WORK

In this section we survey prior work in performance diagnosis in two categories: knowledge-lean analysis of passive measurements and model-driven analysis.

Several projects build on low-overhead end-to-end tracing (e.g., [11, 12, 16, 31, 21]), which captures the flow (i.e., path and timing) of individual requests within and across the components of a distributed system. For example, Aguilera *et al.* [11] develop two different algorithms, i.e., RPC messages based and signal-processing based ones for inferring the dominant causal paths through a distributed system. Magpie [12] extracts the resource usage and control path of individual requests in a distributed system and tags incoming requests with a unique identifier and associating resource usage throughout the system with that identifier. Chen *et al.* [16] describe Pinpoint, a system for locating the components in a distributed system most likely to be the cause of a fault. vPerfGuard is similar to these systems in that it relates application performance to VMs and hypervisors that host application components. The key difference is that we consider metrics collected within VMs and hypervisors rather than communication patterns among components. Therefore, our approach is complementary to theirs.

Other recent research seeks to adopt a performance model that connects application performance with system metrics. A performance model may be constructed using a *white-box* approach based on human expert experience and domain knowledge [28, 18, 15, 32] or a *data-driven* approach [17] where a model is built automatically using statistical learning techniques. Compared with a *a white-box* performance model, a *data-driven* model assumes little or no domain knowledge; it is therefore generic and has potential to apply to a wide range of systems and to adapt to changes in the system and the workload. Shen *et al.* [32] present a performance debugging approach based on a whole-system performance model according to the design protocol/algorithms of the target system. Bodík *et al.* [15] present a methodology for automating the identification of performance crises using a fingerprint to represent the state of a datacenter. Tan *et al.* [34] present *PREPARE* which can predict recurrent performance anomalies by combining attribute value prediction with supervised anomaly classification methods. However, the approach in [32] requires deep understanding of I/O systems while the approach in [15] and [34] currently only work with recurrent anomalies. Compared with these prior approaches, vPerfGuard is more flexible to capture emergent behavior or new relationships inside a dynamic cloud environment that have not been seen before due to its data-driven nature. Our work complements theirs by helping identify performance bottlenecks using models. Probably the work closest to ours is [17] where Cohen *et al.* also use a data-driven approach to build a tree-augmented naive (TAN) Bayesian network model to learn the probabilistic relationship between the SLO state and system metrics. However, their models are built offline after an SLO violation has occurred to identify performance bottlenecks. Instead, we use an online approach for continuous adaption of our performance models.

8. CONCLUSION AND FUTURE WORK

In this paper, we propose vPerfGuard, an automated model-driven framework for application performance diagnosis in consolidated cloud environments. The key difference of vPerfGuard from existing approaches is the application of a model-driven approach. First, compared with existing *white-box* performance diagnosis [28, 32, 15, 24] that relies heavily on human expert experience and domain knowledge, vPerfGuard automatically construct models to bridge the application performance and system resources. Second, compared with the offline diagnosis approach in [17], vPerfGuard adaptively updates the models to accommodate changes in highly dynamic consolidated cloud environments. The model-driven approach not only helps a cloud service provider track the application performance using models, but also presents her with suspicious system metrics that can lead to the root cause of the performance problem.

Recall Bob's performance issue (Section 1). Running vPerfGuard with his application constructs a model mapping system metrics to application performance. Alice can track Bob's application performance with models and differentiate between performance problems that may be resolved by the cloud user and those that may have to be resolved by the cloud provider (e.g., contention due to noisy neighbors). The models also allow for richer automated diagnostics to be built on top of them and provide a solid foundation for constructing (and understanding) automatic performance control systems, two active areas of research for our future work.

9. REFERENCES

[1] dstat. `http://dag.wieers.com/home-made/dstat`.

[2] Interpreting esxtop Statistics. `http://communities.vmware.com/docs/DOC-9279`.

[3] iostat. `http://linux.die.net/man/1/iostat`.

[4] RUBBoS. `http://jmob.ow2.org/rubbos.html`.

[5] Selection algorithm: en.wikipedia.org/wiki/Selection_algorithm.

[6] TPC-H. `http://www.tpc.org/tpch`.

[7] TRAC Research. `www.trac-research.com/application-performance-management`.

[8] VMware vFabric Hyperic. `http://www.vmware.com/products/application-platform/vfabric-hyperic/`.

[9] VMware vSphere `www.vmware.com/products/vsphere/overview.html`.

[10] Windows Hyper-V Server. `www.microsoft.com/hyper-v-server/`.

[11] Aguilera, M. K., Mogul, J. C., Wiener, J. L., Reynolds, P., and Muthitacharoen, A. Performance debugging for distributed systems of black boxes. In *Proc. of SOSP* (2003).

[12] Barham, P., Donnelly, A., Isaacs, R., and Mortier, R. Using Magpie for request extraction and workload modelling. In *Proc. of OSDI* (2004).

[13] Barham, P., Dragovic, B., Fraser, K., H, S., Harris, T., Ho, A., Neugebauer, R., Pratt, I., and Warfield, A. Xen and the art of virtualization. In *Proc. of SOSP* (2003).

[14] Basseville, M., and Nikiforov, I. V. *Detection of Abrupt Changes*. Prentice Hall, 1993.

[15] Bodik, P., Goldszmidt, M., Fox, A., Woodard, D. B., and Andersen, H. Fingerprinting the datacenter: automated classification of performance crises. In *Proc. of EuroSys* (2010).

[16] Chen, M. Y., and et al. Path-based faliure and evolution management. In *Proc. of NSDI* (2004).

[17] Cohen, I., Goldszmidt, M., Kelly, T., Symons, J., and Chase, J. S. Correlating instrumentation data to system states: A building block for automated diagnosis and control. In *Proc. of OSDI* (2004).

[18] Doyle, R. P., Chase, J. S., Asad, O. M., Jin, W., and Vahdat, A. M. Model-based resource provisioning in a web service utility. In *Proc. of USITS* (2003).

[19] Draper, N. R., and Smith, H. *Applied regression analysis*. Wiley-Interscience, 1998.

[20] Duda, R. O., Hart, P. E., and Stork, D. G. *Pattern Classification*, 2nd ed. Wiley, New York, 2001.

[21] Erlingsson, U., Peinado, M., Peter, S., and Budiu, M. Fay: extensible distributed tracing from kernels to clusters. In *Proc. of SOSP* (2011).

[22] Georges, A., and Eeckhout, L. Performance metrics for consolidated servers. In *Proc. of HPCVirt* (2010).

[23] Hall, M. A. Feature selection for discrete and numeric class machine learning. Morgan Kaufmann, pp. 359–366.

[24] Jain, R. *The Art of Computer Systems Performance Analysis*. Wiley-Interscience, New York, 1991.

[25] John, G. H., Kohavi, R., and Pfleger, K. Irrelevant features and the subset selection problem. In *Proc. of ICML* (1994).

[26] Kumar, C., and Rosenberg, H. Troubleshooting Performance Related Problems in vSphere 4.1 Environments. Feb. 2011.

[27] Livni, N. Blog: Six Application Performance Challenges facing IT. `www.correlsense.com/blog/six-application-performance-challenges-facing-it/`.

[28] Padala, P., Hou, K., Zhu, X., Uysal, M., Wang, Z., Singhal, S., Merchant, A., and Shin, K. Adaptive control of virtualized resources in utility computing environments. In *Proc. of EuroSys* (2007).

[29] Refaeilzadeh, P., Tang, L., and Liun, H. Cross validation. *Encyclopedia of Database Systems* (2009).

[30] Rich, E., and Knight, K. *Artificial intelligence*. McGraw-Hill, Inc., New York, NY, USA, 1991.

[31] Sambasivan, R. R., and et al. Diagnosing performance changes by comparing request flows. In *Proc. of NSDI* (2011).

[32] Shen, K., Zhong, M., and Li, C. I/O system performance debugging using model-driven anomaly characterization. In *Proc. of FAST* (2005).

[33] Stewart, C., Kelly, T., and Zhang, A. Exploiting nonstationarity for performance prediction. In *Proc. of EuroSys* (2007).

[34] Tan, Y., Nguyen, H., Shen, Z., Gu, X., Venkatramani, C., and Rajan, D. Prepare: Predictive performance anomaly prevention for virtualized cloud systems. In *Proc. of ICDCS* (2012).

Predictive Performance Modeling of Virtualized Storage Systems using Optimized Statistical Regression Techniques

Qais Noorshams, Dominik Bruhn, Samuel Kounev, Ralf Reussner
Chair for Software Design and Quality
Karlsruhe Institute of Technology
Karlsruhe, Germany
{noorshams, kounev, reussner}@kit.edu,
dominik.bruhn@student.kit.edu

ABSTRACT

Modern virtualized environments are key for reducing the operating costs of data centers. By enabling the sharing of physical resources, virtualization promises increased resource efficiency with decreased administration costs. With the increasing popularity of I/O-intensive applications, however, the virtualized storage used in such environments can quickly become a bottleneck and lead to performance and scalability issues. Performance modeling and evaluation techniques applied prior to system deployment help to avoid such issues. In current practice, however, virtualized storage and its performance-influencing factors are often neglected or treated as a black-box. In this paper, we present a measurement-based performance prediction approach for virtualized storage systems based on optimized statistical regression techniques. We first propose a general heuristic search algorithm to optimize the parameters of regression techniques. Then, we apply our optimization approach and create performance models using four regression techniques. Finally, we present an in-depth evaluation of our approach in a real-world representative environment based on IBM System z and IBM DS8700 server hardware. Using our optimized techniques, we effectively create performance models with less than 7% prediction error in the most typical scenario. Furthermore, our optimization approach reduces the prediction error by up to 74%.

Categories and Subject Descriptors

C.4 [**Performance of Systems**]: Modeling techniques, Performance attributes; G.1.6 [**Numerical Analysis**]: Optimization; I.2.8 [**Artificial Intelligence**]: Problem Solving, Control Methods, and Search—*Heuristic methods*

Keywords

I/O, Storage, Performance, Prediction, Virtualization

ICPE'13, March 21–24, 2013, Prague, Czech Republic.

1. INTRODUCTION

The high growth rate of modern IT systems and services demands a powerful yet flexible and cost-efficient data center landscape. Server virtualization is a key technology used in modern data centers to address this challenge. By enabling the shared resource usage by multiple virtual systems, virtualization promises increased resource efficiency and flexibility with decreased administration costs. Furthermore, modern cloud environments enable new pay-per-use cost models coupled with flexible on-demand resource provisioning.

Modern cloud applications increasingly have I/O-intensive workload profiles (cf. [2]), e.g., mail or file server applications are often deployed in virtualized environments. With the increasing popularity of such applications, however, the virtualized storage of shared environments can quickly become a bottleneck and lead to unforeseen performance and scalability issues. Performance modeling and evaluation techniques applied prior to system deployment help to avoid such issues and ensure that systems meet their quality-of-service requirements.

In current practice, however, virtualized storage and its performance-influencing factors are often neglected or treated as a black-box due to their complexity. Several modeling approaches considering I/O-intensive applications in virtualized environments have been proposed, e.g., [1, 17], however, without explicitly considering storage configuration aspects and their influences on the overall system performance. Moreover, the increasing complexity of modern virtualized storage systems poses challenges for classical performance modeling approaches.

In this paper, we present a measurement-based performance prediction approach for virtualized storage systems. We derive optimized performance models based on statistical regression techniques capturing the complex behaviour of the virtualized storage system. Furthermore, the models explicitly consider the influences of workload profiles and storage configuration parameters. More specifically, we first propose a general heuristic search algorithm to optimize the parameters of regression techniques. This algorithm is not limited to a certain domain and can be used as a general regression optimization method. Then, we apply our optimization approach and create performance models based on systematic measurements using four regression techniques. Finally, we evaluate the models in different scenarios to assess their prediction accuracy and the improvement achieved by our optimization approach. The scenarios comprise interpolation and extrapolation scenarios as well as scenarios when the workload is distributed on multiple virtual machines. We evaluate our approach in a real-world environment based on IBM System z and IBM DS8700 server hardware. Using our optimized techniques, we effectively create performance models with less than 7% prediction error in the most typical interpolation scenario. Furthermore, our optimization approach reduces the prediction error by up to 74% with statistical significance.

In summary, the contribution of this paper is multifold: i) We present a measurement-based performance modeling approach for virtualized storage systems based on four statistical regression techniques. ii) We propose a general heuristic optimization method to maximize the predictive power of statistical regression techniques. To the best of our knowledge, this is the first automated regression optimization method applied for performance modeling of virtualized storage systems. iii) We validate our approach in a real-world environment based on the state-of-the-art server technology of the IBM System z and IBM DS8700. iv) We comprehensively evaluate our approach in multiple aspects including interpolation and extrapolation scenarios, scenarios with multiple virtual machines, and the improvement achieved by our optimization approach.

The remainder of this paper is organized as follows: Section 2 further motivates our approach. Section 3 discusses the regression techniques we considered for performance modeling. In Section 4, we introduce our system environment and methodology. Section 5 presents our heuristic optimization algorithm. In Section 6, we extensively evaluate our performance models and optimization approach. Finally, Section 7 presents related work, while Section 8 summarizes and concludes with the lessons learned.

2. MOTIVATION

The use of virtualization technology has significant implications on the landscape of modern data centers. By consolidating multiple systems to increase resource efficiency, virtualization presents today's systems with increasing challenges to meet the non-functional requirements. Especially for virtualized storage systems, this trend is evident. At the cost of highly increased complexity, storage systems have evolved significantly from simple disks to sophisticated systems with multiple caching and virtualization layers. Furthermore, modern virtualized storage systems feature intricate scheduling and optimization strategies.

Without any doubt, there are a variety of performance influences of virtualized storage systems (cf. [23]), e.g., the architecture of the virtualization platform and the implications of workload consolidation. Moreover, the many logical layers between the application and the physical storage lead to complex effects and interdependencies of influencing factors, which are more and more difficult to quantify. The effect of a few simple influencing factors alone leads to a wide range of performance characteristics. Illustrated in Figure 1a for instance, the response time of an application spreads by more than the factor of 35 when varying five factors, e.g., the request size or the I/O scheduler. Moreover, Figure 1b exemplifies a shift in the influencing factors. The CFQ scheduler, which is the standard I/O scheduler since many years, impairs performance drastically for certain configurations. Consequently, the performance traits of an application de-

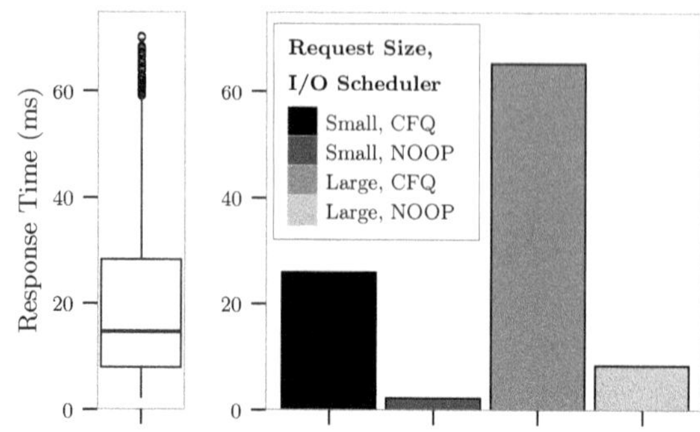

(a) Distribution (b) Examples

Figure 1: Response Time Depending on Influencing Factors

ployed in such an environment are highly unclear and might be also subject to change.

This development is the motivation for our approach. The increasing complexity of modern virtualized storage systems poses challenges for classical performance modeling approaches to manually create, calibrate, and validate explicit performance models within a limited time period. The effects and interdependencies of the influencing factors are more and more complex, which needs to be accounted for and quantified. Furthermore, the effects are subject to change and need to be captured flexibly as the systems keep evolving. Therefore, we target at an automated measurement-based approach to virtualized storage performance modeling based on regression techniques. Our goal is to create practical performance models while lifting the burden for performance engineers to manually develop them. Building a regression-based performance model, however, is also not straightforward due to the many techniques and their parameters. To briefly introduce the techniques we considered, the following section analyzes multiple regression techniques and their parameters.

3. REGRESSION TECHNIQUES

Regression techniques are used to model the effect of *independent variables* on *dependent variables*, e.g., the effect of I/O request sizes on the request response time. In this section, we introduce and analyze the regression techniques we applied for modeling the performance of virtualized storage systems and discuss their parameters. Furthermore, we highlight the relationships between the considered techniques.

3.1 Linear Regression Models

One of the most popular techniques is based on *linear regression models (LRM)* [12] which assume a linear relationship between the dependent variable and the independent variables. Thus, the dependent variable is represented as a linear combination of the independent variables with an added constant intercept. More formally, for a vector of independent variables $\vec{x} := (x_1, \ldots, x_n)$ a model f with coefficients $\beta_0, \ldots, \beta_n$ is created of the form

$$f(\vec{x}) = \beta_0 + \sum_{i=1}^{n} \beta_i x_i.$$

The model f, however, does not necessarily have to be linear in the independent variables x_i. The independent variables can be derived, e.g., $x_2 = x_1^2$ or $x_3 = x_1 x_2$. For the sake of clarity, we explicitly exclude such definitions and specifically consider two forms: The *linear regression models* are explicitly linear in the independent variables as well. The *linear regression models with interaction* can include terms expressing the interaction between variables, e.g., x_1x_2. This model is of maximum degree n in the independent variables x_i, however, the effects of the independent variables remain linear in the model. When creating a model based on a set of training data, the coefficients of the model are determined to minimize the error between the model and the training data.

Model Derivation. The most popular approach to create a model is the *method of least squares.* For a set of training data $\{(\vec{x}_1, y_1), \ldots, (\vec{x}_N, y_N)\}$ where $\forall i$: $\vec{x}_i := (x_{i1}, \ldots, x_{in})$, this method finds the coefficients $\vec{\beta} := (\beta_0, \ldots, \beta_n)$ such that the *residual sum of squares* defined as $RSS(\vec{\beta}) := \sum_{i=1}^{N}(y_i - f(\vec{x}_i))^2$ is minimized. To find the minimum, the derivative of RSS is set to zero and solved for $\vec{\beta}$. As RSS is a quadratic function, the minimum is guaranteed to always exist.

Parameters.
- None.

3.2 Multivariate Adaptive Regression Splines

Multivariate Adaptive Regressions Splines (MARS) [9] consist of piecewise linear functions (so-called *hinge functions*). Formally, these functions with so-called *knot* t are defined as $(x-t)_+ := \max\{0, x-t\}$ and $(t-x)_+ := \max\{0, t-x\}$. Let $\vec{x} = (x_1, \ldots, x_n)$ be the vector of independent variables as above, $H := \{(x_i - t)_+, (t - x_i)_+\}_{i=1,\ldots,n}$ and $t \in T$, where T is the set of observed values of each independent variable. Then, MARS constructs a model f of the form

$$f(\vec{x}) = \beta_0 + \sum_{i=1}^{n} \beta_i \, h_i(\vec{x})$$

with $h_i \in H$ and coefficients $\beta_0, \ldots, \beta_n$. Similar to the linear regression models, we explicitly distinguish between *MARS* and *MARS with interactions*. The latter also includes hinge functions that are a product of one or more functions in H.

Model Derivation. The parameters β_i are estimated by the method of least squares as for the linear regression model. The model is constructed iteratively in a forward step. Starting from a constant function, the algorithm adds the hinge functions in H that result in the maximum reduction of the residual squared error. If interactions are permitted, the algorithm can also choose a hinge function as a factor. However, each variable can only appear once in a term, i.e., higher-order powers of a variable are excluded. This step stops if a predefined number of maximum terms is reached or if the residual error falls below a predefined threshold. Finally, the model is pruned to avoid overfitting. The terms that produce the smallest increase in residual squared error are removed iteratively until a predefined number of terms is reached. Frequently, the pruning step also considers the model complexity combined with the residual squared error. This results in a trade-off between accuracy and complexity of the model.

Parameters.
- `nnodes`: Maximum number of terms in the forward step.
- `threshold`: Maximum acceptable residual error.
- `nprune`: Maximum number of terms after pruning.

3.3 Classification and Regression Trees

Classification and Regression Trees (CART) [5] are a group of algorithms that model data in a tree-based representation. CART models are binary trees with conditions in their non-leaf nodes and constant values in their leaf nodes. To determine the value of the dependant variable corresponding to a set of values of the independent variables, the evaluation starts at the root and the condition in this node is checked. If the condition is true, the left edge is followed, otherwise the right edge. This is repeated until a leaf is reached.

Model Derivation. Similar to MARS, the algorithm comprises a forward and a pruning step. In the forward step, an initial tree with a single node is created. The leafs in the tree are split iteratively at a certain splitting variable and a splitting point. The splitting point is determined such that the residual squared error is minimized. The algorithm splits a leaf until it contains less samples than a predefined value. The algorithm stops if no leaf can be split any further. In the pruning step, the tree is reduced using *cost-complexity pruning* (cf. [12]). Let m be the m-th leaf, R_m be the region specified by the conditions to the m-th leaf, and $l(T)$ be the number of leaves in the tree T. Furthermore, for the training data $\{(\vec{x}_i, y_i)\}_i$, let the number of observations in a region $n_m := |\{\vec{x}_i \,|\, \vec{x}_i \in R_m\}|$, the mean of the observations values $\hat{c}_m := n_m^{-1} \sum_{\vec{x}_i \in R_m} y_i$, and the mean squared difference between each observed value and the mean of the observations values $q_m(T) := n_m^{-1} \sum_{x_i \in R_m} (y_i - \hat{c}_m)^2$. Then, the cost complexity criterion with parameter α is defined as

$$c_\alpha(T) := \sum_{m=1}^{l(T)} n_m q_m(T) + \alpha \, l(T)$$

that is to be minimized. The parameter α is a trade-off between complexity and goodness-of-fit to the training data. It is determined according to the residual sum of squares with cross-validation. This splits the training data into two partitions consisting of model creation and model testing data. For a given α, the tree is greedily pruned using *weakest link pruning.* Iteratively, the node with the smallest per-node increase in $\sum_{m=1}^{l(T)} n_m q_m(T)$ is collapsed.

Parameters.
- `minsplit`: Minimum number of samples to split a leaf.
- `cp`: Maximum acceptable tree complexity after pruning.

3.4 M5 Trees

M5 trees [25] are – figuratively speaking – a combination of linear regression models and CART trees. Similar to CART trees, M5 models are binary decision trees with conditions in their non-leaf nodes. The leaf nodes, however, contain linear regression models instead of the constant values as in CART trees. For the prediction of a value, starting in the root node, the conditions are evaluated until a linear regression model in a leaf is reached. Then, the model is used as the predictor.

Model Derivation. Similar to MARS and CART, the tree is first built in a forward step and pruned in the next step. To build a M5 tree, the initial model consists of a one-node tree T. The tree is split iteratively into subtrees T_i until a predefined maximum number of splits is reached. M5 splits the tree at the condition that maximizes the expected reduction in error Δe with

$$\Delta e := \sigma(T) - \sum_i \frac{|T_i|}{|T|} \sigma(T_i),$$

where σ is the standard deviation. For each leaf, a linear regression model is created. Finally, the M5 model f is greedily simplified to decrease the complexity c with

$$c(T) := n^{-1} \sum_i |y_i - f(\vec{x}_i)| \cdot \frac{n+p}{n-p},$$

where $\{(\vec{x}_i, y_i)\}_i$ is the training data, n is the number of training data and p is the number of parameters in the model. The term $\frac{n+p}{n-p}$ allows an increase in error if the complexity is decreased. For each leaf node, parameters of the linear model are removed, whereas for every non-leaf node, the node is collapsed if this reduces $c(T)$.

Parameters.
- `nsplits`: Maximum number of splits in the forward step.

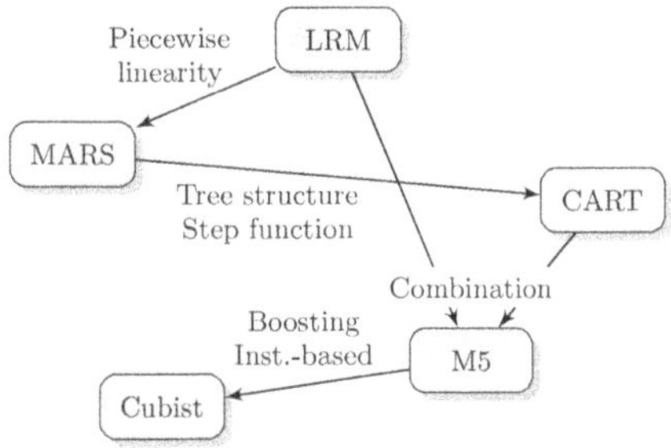

Figure 2: Relation between the Regression Techniques

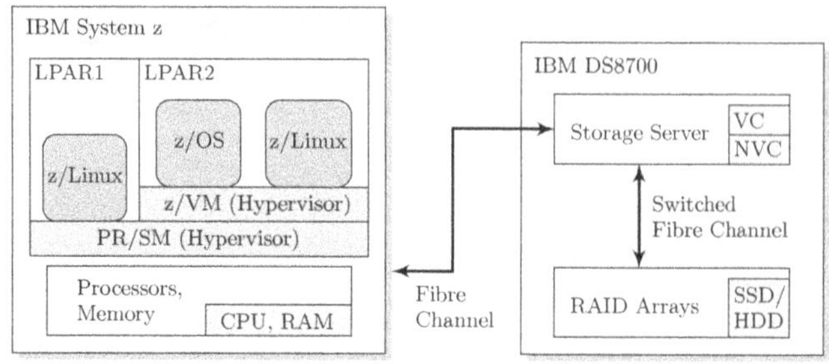

Figure 3: IBM System z and IBM DS8700

3.5 Cubist Forests

Cubist forests [26, 18] are an extension of M5 trees. Thus, Cubist forests are rule-based model trees with linear models in the leaves. Compared to M5, Cubist introduces two extensions. First, it follows a boosting-like approach, i.e., it creates a set of trees instead of a single tree. To obtain a single value, the tree predictions are aggregated using their arithmetic mean. Second, it combines model-based and instance-based learning (cf. [24]), i.e., it can adjust the prediction of unseen points by the values of their nearest training points.

Model Derivation. Initially, the maximum number of trees in Cubist is defined to construct a forest. The first tree is created using the M5 algorithm. The following trees are created to correct the predictions of the training points by the previous tree $f_t(\vec{x})$. Each value of a training point y_i is modified by $y_i' := 2\,y_i - f_t(\vec{x})$ to compensate for over- and under-estimations. Then, the tree creation is repeated. In contrast to, e.g., Random Forests [4] combining the prediction trees with the mode operator, Cubist aggregates the values predicted by each tree using arithmetic mean. Finally, the prediction of unseen points can be adjusted by the values of a possibly dynamically determined number of training points (so-called instance-based correction), cf. [24]. The prediction of a new point $\vec{x}$ is adjusted by the weighted mean of the nearest training points (so-called neighbors) with weight $w_n := 1/(m(\vec{x}, \vec{n}) + 0.5)$ for every neighbor $\vec{n}$, where $m(\vec{x}, \vec{n})$ is the Manhatten distance of $\vec{x}$ and $\vec{n}$. M5 is a special case of Cubist with one tree and no instance-based correction.

Parameters.

- `nsplits`: Maximum number of splits in the forward step.
- `ntrees` : Number of trees.
- `ninstances` : Size of instance-based correction.

3.6 Summary

Figure 2 shows an overview of the considered regression techniques illustrating the relationships among them. As described above, the MARS algorithm can be seen as an extension of LRM by allowing piecewise linear models. However, MARS regulates the number of linear terms. While MARS and CART seem relatively different, the forward step of MARS can be transformed into the one of CART by using a tree-based structure with step functions, cf. [12]. M5 in turn can be seen as a combination of LRM and CART. However, M5 differs from CART in the complexity criterion and the pruning procedure. Finally, Cubist extends M5 by introducing a boosting like scheme creating several trees that are aggregated by their mean. Furthermore, Cubist introduces an instance-based correction to include the training data in the prediction of unseen data.

4. METHODOLOGY

In our approach, we apply statistical regression techniques to create performance models based on systematic measurements. In this section, we present our experimental environment and setup as well as our measurement methodology and performance modeling approach.

4.1 System Under Study

A typical virtualized environment in a data center consists of servers providing computational resources connected to a set of storage systems. Such storage systems typically differ significantly from traditional hard disks and native storage systems due to the complexity of modern storage virtualization platforms.

In this paper, we consider a representative virtualized environment based on the IBM mainframe *System z* and the storage system *DS8700*. They are state-of-the-art high-performance virtualized systems with redundant or hot swappable resources for high availability. The System z combined with the DS8700 represent a typical virtualized environment that can be used as a building block of cloud computing infrastructures. It supports on-demand elasticity of pooled resources with a pay-per-use accounting system (cf. [22]). The System z provides processors and memory, whereas the DS8700 provides storage space. The structure of this environment is illustrated in Figure 3.

The *Processor Resource/System Manager (PR/SM)* is a hypervisor managing logical partitions (*LPARs*) of the machine and enabling CPU and storage virtualization. For memory virtualization and administration purposes, IBM introduces another hypervisor, *z/VM*. The System z supports the classical mainframe operating system *z/OS* and special Linux ports for System z commonly denoted as *z/Linux*. The System z is connected to the DS8700 via fibre channel. Storage requests are handled by a storage server having a volatile cache (VC) and a non-volatile cache (NVC). The storage server is connected via switched fibre channel with SSD or HDD RAID arrays. As explained in [8], the storage server applies several pre-fetching and destaging algorithms for optimal performance. When possible, read-requests are served from the volatile cache, otherwise they are served from the RAID arrays and stored in the volatile cache for future requests. Write-requests are written to the volatile as well as the non-volatile cache, but they are destaged to the RAID arrays asynchronously.

In such a virtualized storage environment, a wide variety of heterogeneous performance-influencing factors exists, cf. [23]. In many cases, the factors have a significantly different effect compared to traditional native storage systems. As Figure 1b illustrates, e.g., the standard Linux I/O scheduler CFQ performs significantly worse than the NOOP scheduler

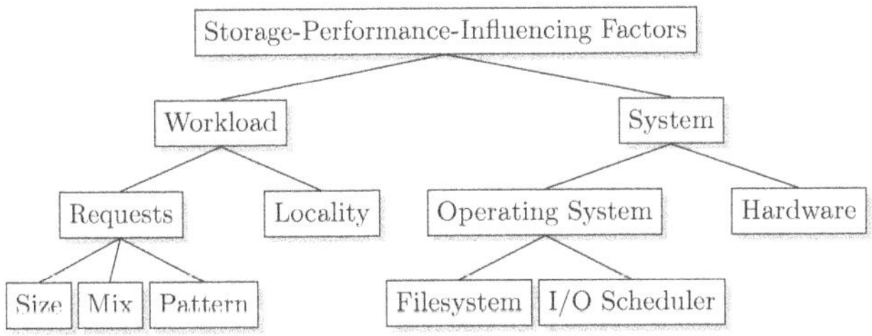

Figure 4: Performance Influences (derived from [23])

for certain configurations. Figure 4 gives an overview of the major storage-performance-influencing factors used as basis for our analysis. In general, we distinguish between workload and system factors. Workload factors comprise average request size, read/write mix and random or sequential access pattern. Furthermore, the locality of requests affects the storage cache effectiveness. System factors comprise operating system and hardware configurations. The major factors of the operating system are the file system and the I/O scheduler. Hardware factors can be analyzed on different abstraction levels, but are often system specific.

4.2 Experimental Setup

In our experimental environment, the DS8700 contains 2 GB NVC and 50 GB VC with a RAID5 array containing seven HDDs. Measurements are obtained in a z/Linux virtual machine (VM) with 2 IFLs (cores) and 4 GB of memory. Using the `O_DIRECT` POSIX flag, we isolate the effects of operating system caching in order to focus our measurements on the storage performance. However, we explicitly take into account the cache of the storage system by varying the overall size of data accessed in our workloads. As a basis for our experimental analysis, we used the open source *Flexible File System Benchmark*[1] (FFSB) due to its fine-grained configuration possibilities. FFSB runs at the application layer and measures the end-to-end response time covering all system layers from the application all the way down to the physical storage. The system and workload parameters for the measurements match the factors described in Section 4.1 and illustrated in Figure 4. The locality of the workload can be deduced from the overall size of the set of files accessed in the workload since FFSB requests are randomly distributed among the files. The considered parameter values are summarized in Table 1. The parameter space is fully explored leading to a total of 1120 measurement configurations.

4.3 Experimental Methodology

For a given configuration of a benchmark run, the measurements run in three phases: First, a set of 16 MB files is created. Second, the target number of workload threads are started and they start reading from and writing into the initial file set. After an initial warm up phase, the measurement phase starts and the benchmark begins obtaining read and write performance data. For each workload thread, the read and write operations consist of 256 sub-requests of the specified size directed to a randomly chosen file from the file set. If the access pattern is sequential, the sub-requests access subsequent blocks within the file. Each thread issues a request as soon as the previous one is completed.

For each of the 1120 configurations, we configured a one minute warm up phase and a five minute measurement phase.

[1] `http://github.com/FFSB-prime` (extending `http://ffsb.sf.net`)

Table 1: Experimental Setup Configuration

System Parameters	
File system	ext4
I/O scheduler	CFQ, NOOP
Workload Parameters	
Threads	100
File set size	1 GB, 25 GB, 50 GB, 75 GB, 100 GB
Request size	4 KB, 8 KB, 12 KB, 16 KB, 20 KB, 24 KB, 28 KB, 32 KB
Access pattern	random, sequential
Read percentage	0%, 25%, 30%, 50%, 70%, 75%, 100%

The measurement phase consists of five intervals of one minute length each. In one minute, the benchmark obtains between approximately 90 000 and 2 800 000 measurement samples depending on the configuration and approximately 575 000 measurement samples on average. Furthermore, we analyzed the measurement intervals to ensure stable and meaningful results. Illustrated in Figure 5, we evaluated the relative standard error (RSE) of the response time means for each configuration over the five intervals. For read requests, the 90th percentile of the RSE is 8.45% and the mean RSE is 3.35%. For write requests, the 90th percentile of the RSE is 5.35% and the mean RSE is 2.10%. Thus, we conclude that the measurement samples are sufficiently large and the measurements are sufficiently stable. This is important since our goal was to keep the measurement length limited.

Since a manual measurement and evaluation approach is highly tedious and error prone, we automated the process as part of a new tool we developed called *SPA (Storage Performance Analyzer)*[2]. Illustrated in Figure 6, our tool basically consists of a *benchmark harness* that coordinates and controls the execution of the FFSB benchmark and a tailored *analysis library* used to process and evaluate the collected measurements. The benchmark harness component runs on a controller machine managing the measurement process. Using SSH connections, the benchmark controller first configures the benchmark, then it executes the target workload, and it finally collects the results into an SQLite database. Furthermore, the benchmark controller guarantees a synchronized execution of experiments on multiple targets, i.e., on multiple VMs that can be deployed on the same system. The evaluation is automated using tailored analysis functions implemented using the open source statistics tool *R* [27]. The analysis library comprises the analysis, optimization and regression functions we created and applied for regression optimization and performance model creation. For more information on the technical realization, we refer the reader to [6].

4.4 Performance Modeling Methodology

We employ the regression techniques presented in Section 3. As independent variables, we used the system and the workload parameters listed in Table 1. As dependent variables, we used the system response time and throughput. For each regression technique, we first optimize the parameters to create effective models as explained in detail in the next section. We then create dedicated models for read and write requests as well as for response time and throughput predictions. Due to the structure of the independent variables, the LRM and MARS models *without* interaction perform *significantly worse* than all the other techniques. Therefore, we explicitly

[2] `http://sdqweb.ipd.kit.edu/wiki/SPA`

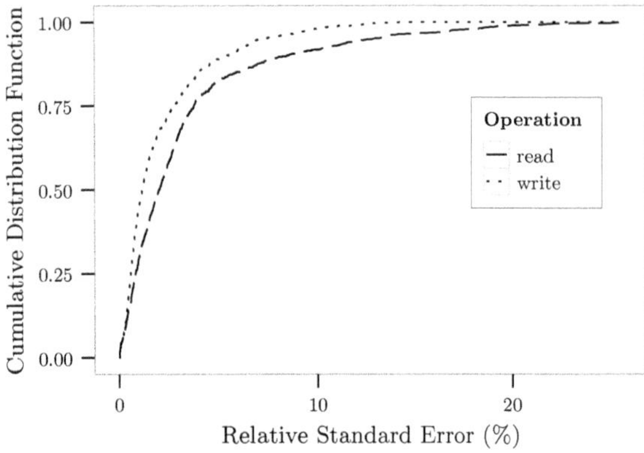

Figure 5: Distribution of the Relative Standard Error

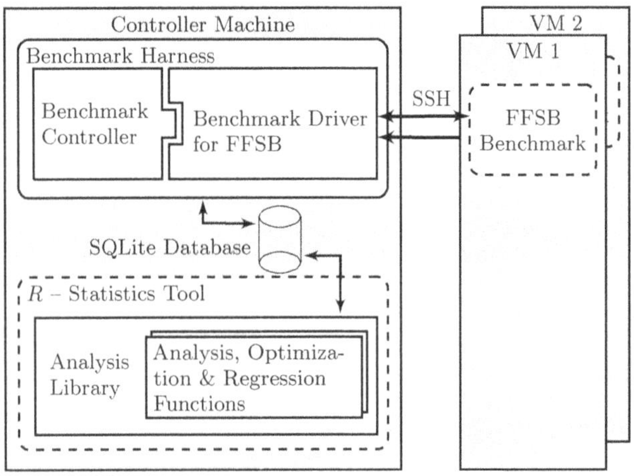

Figure 6: Overview of our Automated Methodology

exclude LRM and MARS models without interactions from consideration. Furthermore, Cubist stems from M5 both algorithmically and in terms of implementation and they are identical for Cubist's standard parameter values. Therefore, in the following, we do not explicitly distinguish between the two. Thus, having four models for each regression technique, we create a total of 16 performance models evaluated in detail in Section 6.

5. REGRESSION OPTIMIZATION

Regression techniques often have parameters, which influence the effectiveness of the technique in a certain application scenario. Usually, the parameters are left to their standard values or chosen based on an educated guess. However, both approaches are insufficient to systematically create powerful models.

To create effective performance models, we search for optimal parameters for each of our considered techniques[3], cf. Table 2. A full factorial search would take *weeks* due to the high computational costs of creating millions of regression models. Therefore, we propose a heuristic iterative search algorithm shown in Algorithm 1 and illustrated in Figure 7. In a nutshell, we split the parameter space into multiple subspaces and iteratively refine the search in the most promising regions.

[3]Note that there can be more input parameters than the ones considered here depending on the implementation of the technique. In this paper, we use the implementation in the statistics tool *R* [27].

Algorithm 1 Iterative Parameter Optimization

`Configuration:`
$k \leftarrow$ Number of splits
$n \leftarrow$ Number of explorations
$\mathcal{S}$ *// Stopping criterion*
`Definitions:`
$\vec{p} := (p_1, \ldots, p_l), L := \{1, \ldots, l\}$ *// l parameters*
$p_i \in [a_i, b_i], \forall i \in L$ *// Ranges of parameters*
$c(\vec{p}) :=$ Mean RMSE in cross-validation of parameters
`Init:`
$E \leftarrow \{(\vec{a} := (a_1, \ldots, a_l), c(\vec{a}))\}$
// Evaluate first border parameters
$M \leftarrow E$ *// M: Set of best parameters*
`Algorithm:`
while $\mathcal{S}$ does not apply **do**
  `j-th Iteration:`
  for all $(\vec{\nu}^{j}_{(h)}, \cdot) \in M$ **do** *// h-th pivot*
    for all $i \in L$ **do**
      $A_i \leftarrow \{p_i \mid (p_i, \cdot) \in E \wedge p_i < \nu^{j}_{(h)i}\}$
      if $A_i \neq \emptyset$ **then** $a_i^* \leftarrow \max A_i$
        else $a_i^* \leftarrow a_i$
      $B_i \leftarrow \{p_i \mid (p_i, \cdot) \in E \wedge p_i > \nu^{j}_{(h)i}\}$
      if $B_i \neq \emptyset$ **then** $b_i^* \leftarrow \min B_i$
        else $b_i^* \leftarrow b_i$
      $s_i^* \leftarrow \frac{b_i^* - a_i^*}{k+1}, \forall i \in L$ *// Step width*
      $S_i^* \leftarrow \{a_i^*, a_i^* + s_i^*, \ldots, a_i^* + k\, s_i^*, b_i^*\}$
      for all $s \in S_i^*$ **do**
        if s invalid **then** round s to next valid value
      end for
    end for
    $E^{j}_{(h)} \leftarrow \{(\vec{x}, c(\vec{x})) \mid \vec{x} \in S_1^* \times \ldots \times S_l^*\}$
    // Evaluate parameters if not evaluated yet
  end for
  $E \leftarrow E \cup \bigcup_h E^{j}_{(h)}$ *// Save all evaluated parameters*
  $M \leftarrow$ Find n best tuples in E (w.r.t. c)
end while

Algorithm. The algorithm's configuration parameters are the number of splits (i.e., splitting points) in each iteration and the number of subspace explorations in the next iteration. The number of splits configures the sampling frequency of the search space and should be higher if multiple high but narrow optima can be assumed. The number of explorations configures how many local optima are analyzed in their adjacent area and should be higher if many local optima can be assumed. The algorithm can be configured with an arbitrary stopping criterion, e.g., when the improvement between iterations falls below a given threshold or when a specified maximum number of iterations is reached. For each of the given parameters $\vec{p} := (p_1, ..., p_l)$, a reasonable range $p_i \in [a_i, b_i]$ needs to be defined to limit the search space. The lower bound usually exists for most parameters and the upper bound can be derived based on the used training data. For a given set of parameter values, we evaluate the model using the mean *root mean square error* (RMSE) of a 10-fold cross-validation, denoted as $c(\vec{p})$. This means that the measurement data is separated into 10 groups and each group is used once as a prediction set while the rest of the data is used for model building. This evaluation approach is key to prevent overfitting of the prediction models.

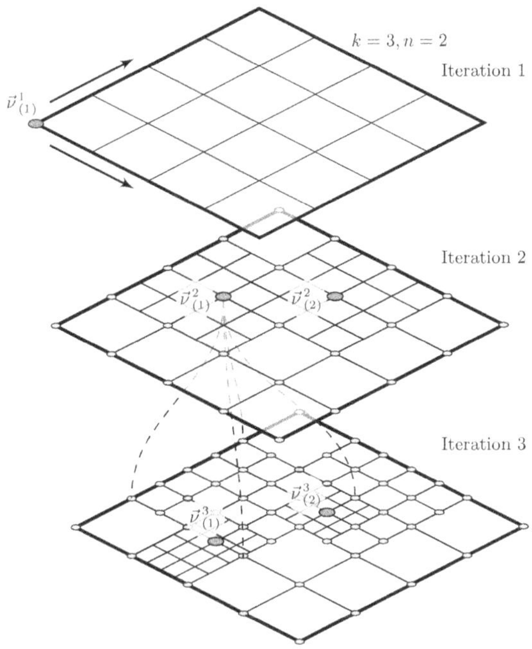

Figure 7: First Iterations of the Parameter Optimization

The algorithm begins with the evaluation of one corner point in the parameter space, cf. Figure 7. As initially this is the only evaluated point, it is the only point saved in a set M for exploration in the next iteration.

In each iteration, we explore every point $\vec{x}$ in the set M containing the points corresponding to the best parameter value assignments. The unevaluated space around $\vec{x}$ is split into $(k+1)^l$ subspaces. Initially, this is the whole parameter space. Each corner point of the subspaces is then evaluated and the best n found so far are saved in M for the next iteration. To save computation time, we maintain a set E of already evaluated parameter values. The algorithm repeats each iteration until the stopping criterion is fulfilled.

Example. In Figure 7, the first three iterations of the algorithm with $k = 3$ and $n = 2$ are illustrated. Initially, one corner point is evaluated. This pivot $\vec{\nu}^1_{(1)}$ is explored in the first iteration. Since there are no other evaluated points, the whole parameter space is cut with $k = 3$ splits for each parameter indicated by the grid lines (at the Iteration 1 level). Each of the intersection points depicted as small gray points (at the Iteration 2 level) is evaluated. After the best $n = 2$ evaluated points are found, they become the next pivots $\vec{\nu}^2_{(1)}$ and $\vec{\nu}^2_{(2)}$. Now, when $\vec{\nu}^2_{(1)}$ is explored, the space around it limited by the neighbouring evaluated points is explored, cf. dashed lines between Iteration 2 and Iteration 3. The same is done for the second pivot $\vec{\nu}^2_{(2)}$. This process is repeated in the next iterations and the best parameters are used for model building. Note that the space does not necessarily have to be split equidistantly for all parameters, cf. $\vec{\nu}^3_{(1)}$. That is why the complexity evaluation is independent from the exact values the algorithm explores.

Complexity. In our implementation of the algorithm, we used indexed data structures to realize the logic. Thus, the regression model building by means of the cross-validation evaluation represents by far the most computationally expensive part of the algorithm. Therefore, we evaluate the complexity of the cross-validation evaluation. In the worst case, in every iteration we evaluate every new set of parameter values except the initial corner one in the first iteration and the four corner ones in every other iteration. Thus, the number of evaluations in our search in the worst case is limited by

$$\mathcal{O}((k+2)^l - 1 + (\max j - 1) \cdot n\,((k+2)^l - 4)) = \mathcal{O}(\max j \cdot n\,k^l),$$

where $\max j$ is the maximum number of iterations. By configuring $\max j$, n and k, the overhead of the algorithm can be adjusted and kept within bounds. As discussed in the next section, our evaluation showed that even moderately low values of these parameters lead to substantial improvements in the model accuracy.

Optimality. Our main goal is to achieve such model accuracy improvements efficiently. However, if we do not limit the number of iterations and allow to search long enough, our approach is able to find locally optimal solutions. Thus, for convex problems our approach also guarantees globally optimal solutions. This is since, let $k > 1$ and $n > 0$, $\forall i \in L$ one has if $j \to \infty$, then $s^*_i \to 0$ for continuous parameters and $s^*_i \to 1$ for discrete parameters. In other words, as we iteratively explore the current best solutions of the iteration, at least one solution will be fully explored.

Optimization Results. Based on the parameter optimization algorithm, for each regression technique we create four different models as described in Section 4: A read response time model (RT_r), a write response time model (RT_w), a read throughput model (TP_r), and a write throughput model (TP_w). We configured the number of splits k to four and the number of explorations n to five. We stop after 10 iterations. Table 2 summarizes the parameters of the regression techniques and lists their standard values as well as their valid ranges. For unbounded parameters, we chose a generous upper limit. For each model, the recommended parameter value assignments for our performance models determined by the optimization algorithm are shown. In the next section, we evaluate the accuracy of the resulting optimized regression models and demonstrate the benefits of our modeling approach.

Table 2: Parameters of the Regression Techniques

	Parameter	Standard	Valid/Used Range	Best $\vec{p}$ (RT_r, RT_w, TP_r, TP_w)
MARS[4]	`nnodes`	21 (der.)[5]	$[3,\infty)/[3,800]$	255, 353, 131, 523
	`threshold`	0.001	$[0,1)/[0,0.999]$	$5.01e^{-6}$, $3.07e^{-7}$, $5.05e^{-5}$, $2.66e^{-5}$
	`nprune`	`nnodes`	$[2,\infty)/[2,800]$	98, 134, 99, 141
CART	`minsplit`	20	$[2,\infty)/[2,400]$	3, 2, 2, 2
	`cp`	0.01	$[0,1)/[0,0.999]$	$3.07e^{-7}$, $4.09e^{-7}$, 0, 0
Cubist	`nsplits`	100	$[2,\infty)/[2,400]$	173, 217, 173, 217
	`ntrees`	1	$[1,100]$	77, 41, 11, 26
	`ninstances`	0	$[0,9]$	2, 4, 2, 1

6. EVALUATION

In this section, we analyze and evaluate our approach in three steps. First, we evaluate the performance models w.r.t. their goodness-of-fit as well as their predictive power for interpolation and extrapolation scenarios. Then, we analyze scenarios where the workload is distributed on multiple VMs. Finally, we evaluate the improvements in model accuracy achieved through our heuristic optimization. In each of the

[4] The standard MARS algorithm does *not* include interactions. However, we only consider MARS *with* interactions, cf. Section 4.4.

[5] Derived value ($\min(200, \max(20, 2 \cdot \#IndependentVariables)) + 1$).

Table 3: Goodness-of-Fit of Performance Models

	Coefficient of determination R^2				
Model	RT_r	RT_w	TP_r	TP_w	$\frac{1}{4}\sum$
LRM	0.9720	0.9703	0.9699	0.9546	0.9667
MARS	0.9955	0.9991	0.9951	0.9973	0.9968
CART	0.9983	1	1	1	0.9996
Cubist	0.9990	0.9994	0.9995	1	0.9995

three steps, we use the optimized models generated by our SPA tool as a basis for the evaluation.

6.1 Model Accuracy

Goodness-of-Fit. For each of the optimized performance models, we evaluate how well it reflects the observed system behavior. To evaluate this goodness-of-fit, we analyze the coefficient of determination R^2 defined as

$$R^2 := 1 - \frac{SSE}{SST} = 1 - \frac{\sum_i (y_i - f(\vec{x_i}))^2}{\sum_i (y_i - \frac{1}{N}\sum_j y_j)^2} \leq 1,$$

where SSE and SST are the residual and the total sum of squares, respectively, $\{(\vec{x_i}, y_i)\}_{i=1,\ldots,N}$ represent the measurement data and f is the respective regression model. R^2 values close to 1 indicate a good fit. The results of our evaluation for the different models are shown in Table 3. Overall, most of the models fit very well to the measurements. Especially MARS performs particularly well even with a comparably low number of terms (cf. `nprune` parameter values). CART, however, has an expectedly good fit due to the low `minsplit` and `cp` parameters allowing for a highly splitted tree. Furthermore, as expected Cubist also fits very well due to the large number of trees and especially the instance-based correction. Finally, LRM exhibits a fairly well fit, however, exhibiting the lowest fit of all the models.

Predictive Power for Interpolation Scenarios. We evaluate 100 configuration scenarios with parameter values randomly chosen within the ranges shown in Table 4. For each scenario, we compare the model predictions with measurements on the real system.

Figure 9 shows the mean relative error for the various models. Overall, the models perform very well and especially MARS and Cubist exhibit excellent performance prediction accuracy with less than 7% and 8% error, respectively. The CART trees are highly splitted, yet with approximately 10% error their accuracy is acceptable. Finally, the LRM models exhibit the highest error with approximately 13%.

Figure 8 depicts the empirical cumulative distribution functions of the prediction errors. In all cases, the majority of the error is less than 10%. For MARS and Cubist, the error is less than 10% for more than 75% of the samples. Most notably, the 90th percentile of the error of both MARS and Cubist is less than 25%. Moreover, even the 99th percentile of the error is less than 50% for almost all models.

Predictive Power for Extrapolation Scenarios. We now consider configuration scenarios with parameter values outside of the ranges used to build the models. More specifically, we extrapolate the request size, the file set size, and the read percentage in two steps in each direction. We evaluate 256 scenarios by considering all combinations of the parameter values shown in Table 5. We explicitly distinguish between small (i.e., 1 KB and 2 KB) and large request sizes (i.e., 36 KB and 40 KB).

As shown in Figure 10, the predictive power of the models for extrapolation scenarios differs significantly for smaller

Table 4: Interpolation Setup Configuration

Parameter	Range
I/O scheduler	{CFQ, NOOP}
File set size	[1 GB, 100 GB] rounded to multiples of 16 MB
Request size	[4 KB, 32 KB] rounded to multiples of 512 bytes
Access pattern	{random, sequential}
Read percentage	[25%, 100%] for read requests [0%, 75%] for write requests

Table 5: Extrapolation Setup Configuration

Parameter	Values
I/O scheduler	{CFQ, NOOP}
File set size	{512 MB, 768 MB, 110 GB, 120 GB}
Request size	{1 KB, 2 KB, 36 KB, 40 KB}
Access pattern	{random, sequential}
Read percentage	{15%, 20%, 80%, 85%}

and larger request sizes especially for throughput predictions. Scenarios with smaller request sizes are predicted fairly well for response time. The error for the best performing model Cubist is less than 15%. Throughput predictions, however, exhibit much higher errors. This is due to the large performance overhead of serving many small requests simultaneously. The file system introduces significant overhead since the standard extent sizes (in which data is stored) are 4 KB. Furthermore, the optimization strategies in the various layers (especially the I/O scheduler) are very eager to optimize the requests. The latter is difficult since multiple threads issue requests simultaneously. This effect is multiplied by the penalty introduced by frequent backend RAID array accesses if the file set size is larger than the NVC and the VC, respectively. On the other hand, scenarios with larger request sizes are predicted very well. This is notable since the models have to cope with unexplored regions in the parameter space. Especially, Cubist predicts response times with less than 9% error and throughput with approximately 16% error. Also, MARS performs impressively well with approximately 11% error for response time and approximately 21% error for throughput.

Summary. Overall, the performance models exhibited good accuracy and predictive power. To highlight the key aspects, the models managed to effectively capture the performance influences with very high goodness-of-fit characteristics. In interpolation scenarios, the models exhibited excellent performance prediction accuracy for arbitrarily chosen configuration scenarios. Here, MARS and Cubist performed best with approximately 7%-8% prediction error. Also, in extrapolation scenarios, the models showed promising results. Except for throughput predictions for smaller request sizes, the best performing model Cubist had a prediction error of only 9%-16%.

6.2 Workload Distribution

We now evaluate scenarios where the workload is distributed on multiple co-located VMs as done for example in the context of server consolidation. We distribute the workload evenly across two and three VMs with shared cores, i.e., both the number of threads and the file set size are scaled inversely by the number of VMs. We explicitly distinguish between the different I/O scheduler implementations as well as between random and sequential workloads. Thus, having 100 randomly chosen configurations based on the parameter ranges in Table 4, we conduct 25 measurements for each

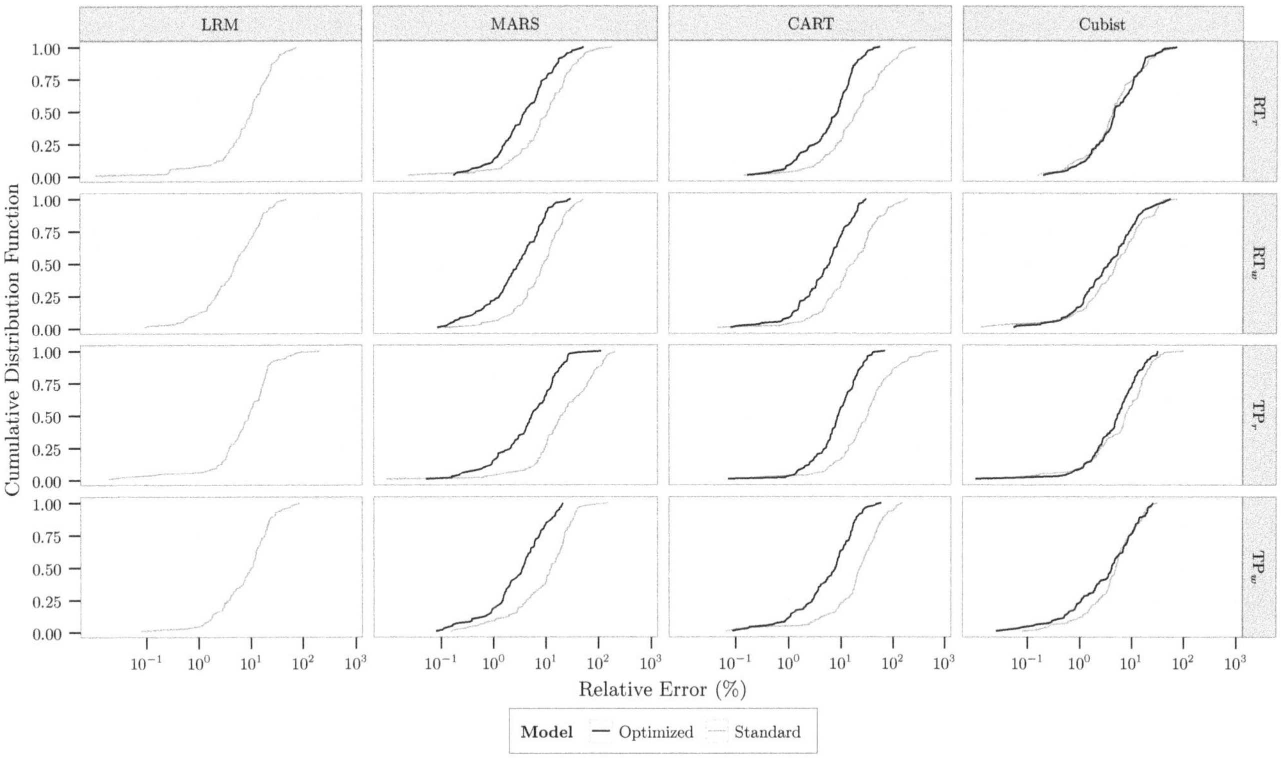

Figure 8: Cumulative Distribution Function of the Relative Error of the Performance Models (Note the Logarithmic x-Axis)

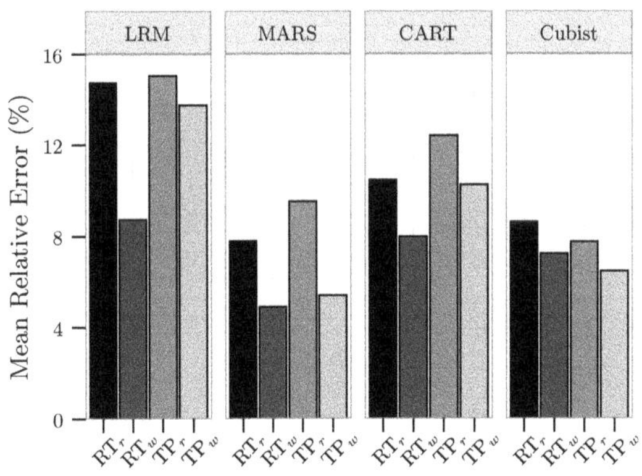

	Mean Relative Error (%)				
Model	RT_r	RT_w	TP_r	TP_w	$\frac{1}{4}\sum$
LRM	14.73	8.72	15.05	13.74	13.06
MARS	7.80	4.90	9.52	5.39	6.90
CART	10.47	7.98	12.41	10.26	10.28
Cubist	8.63	7.23	7.75	6.44	7.51

Figure 9: Prediction Accuracy of Interpolation

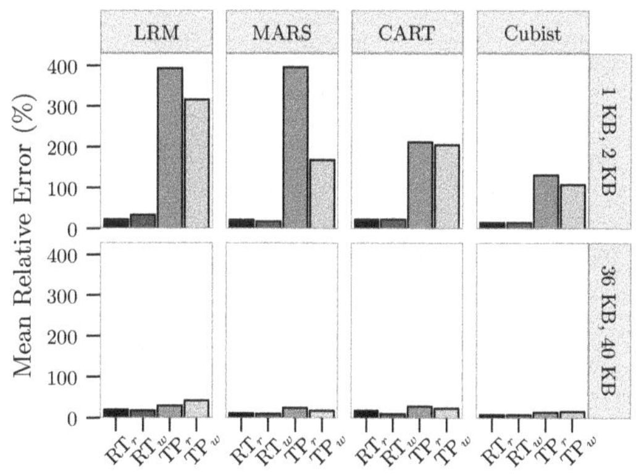

	Mean Relative Error (%)				
Model	RT_r	RT_w	TP_r	TP_w	$\frac{1}{4}\sum$
Request Size: 1 KB, 2 KB					
LRM	22.36	33.65	393.19	315.81	191.25
MARS	21.79	17.54	395.66	168.26	150.81
CART	22.18	22.59	211.69	204.60	115.27
Cubist	14.47	14.96	130.76	108.38	67.14
Request Size: 36 KB, 40 KB					
LRM	19.90	18.34	29.77	42.47	27.62
MARS	11.66	10.00	24.86	17.98	16.13
CART	18.10	9.96	28.85	24.25	20.29
Cubist	8.70	8.52	14.12	15.74	11.77

Figure 10: Prediction Accuracy of Extrapolation

CFQ & NOOP scheduler and random & sequential workload combination. These measurements are used to evaluate the prediction error for each model. Note that the models are still extracted from measurements obtained in one VM to show the effectiveness of our approach.

For two and three VMs, respectively, Figure 11 and Figure 12 show the mean relative error for all models structured according to the chosen I/O scheduler and workload pattern. Since CFQ performs significant optimizations that depend on the current load, the model extracted from measurements in one VM does not fit well with the measurements conducted in multiple VMs for sequential workload. For random workload, however, this effect is less significant. Response time is predicted fairly well and especially throughput is predicted very well. The NOOP scheduler performs only optimizations based on request splitting and merging. Thus, the model extracted from measurements in one VM fits much better for multiple VMs in most cases. Only for write response times,

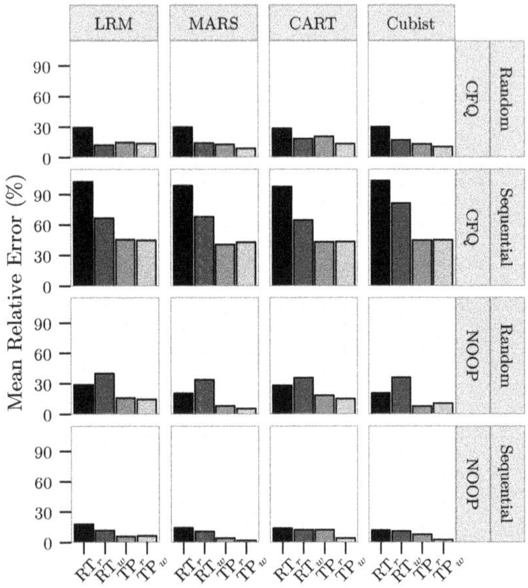

Model	Mean Relative Error (%) RT_r	RT_w	TP_r	TP_w	$\frac{1}{4}\sum$
Scheduler: CFQ, Random Requests					
LRM	29.42	12.24	14.86	13.88	17.60
MARS	29.62	13.96	12.80	9.01	16.35
CART	28.27	18.39	20.51	13.44	20.15
Cubist	30.17	17.03	13.24	10.47	17.73
Scheduler: CFQ, Sequential Requests					
LRM	102.32	66.32	45.42	44.50	64.64
MARS	98.55	67.73	40.58	42.84	62.43
CART	97.45	64.34	43.08	43.20	62.02
Cubist	103.40	81.14	44.81	45.10	68.61
Scheduler: NOOP, Random Requests					
LRM	28.85	39.83	15.79	14.58	24.76
MARS	20.31	33.82	7.97	5.48	16.90
CART	28.11	35.70	18.30	15.00	24.28
Cubist	20.61	36.07	7.78	10.46	18.73
Scheduler: NOOP, Sequential Requests					
LRM	17.71	11.47	5.82	6.43	10.36
MARS	14.35	10.57	4.00	1.82	7.69
CART	14.05	12.13	12.19	4.22	10.65
Cubist	12.08	10.95	7.81	2.49	8.33

Figure 11: Prediction Accuracy of the One-VM-Models when the Workload is Distributed on Two VMs

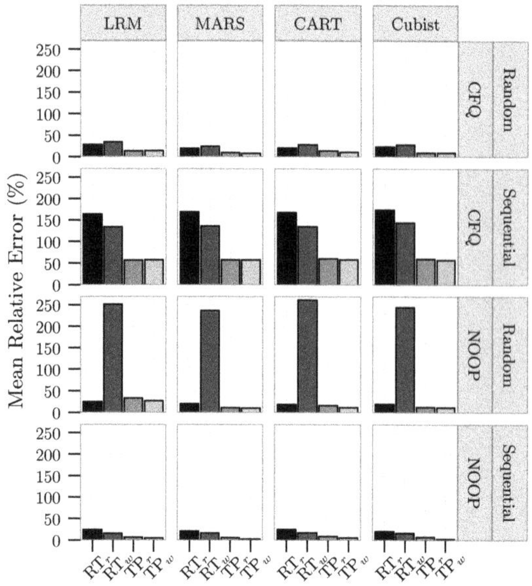

Model	Mean Relative Error (%) RT_r	RT_w	TP_r	TP_w	$\frac{1}{4}\sum$
Scheduler: CFQ, Random Requests					
LRM	28.43	34.32	13.74	14.66	22.79
MARS	20.37	24.44	10.08	8.40	15.82
CART	21.12	28.34	14.97	11.88	19.08
Cubist	23.88	28.60	9.89	10.13	18.12
Scheduler: CFQ, Sequential Requests					
LRM	163.93	133.93	57.13	57.37	103.09
MARS	169.06	136.32	57.91	57.85	105.29
CART	167.72	134.81	60.63	57.98	105.29
Cubist	173.29	143.90	59.63	57.51	108.58
Scheduler: NOOP, Random Requests					
LRM	25.21	251.33	33.34	27.04	84.23
MARS	20.14	237.03	11.02	10.64	69.71
CART	18.65	261.30	16.55	12.19	77.17
Cubist	19.85	244.44	12.32	11.58	72.05
Scheduler: NOOP, Sequential Requests					
LRM	24.01	16.13	7.04	5.17	13.09
MARS	21.16	16.49	5.83	2.59	11.52
CART	24.95	17.23	8.99	5.87	14.26
Cubist	20.30	16.89	7.35	2.70	11.81

Figure 12: Prediction Accuracy of the One-VM-Models when the Workload is Distributed on Three VMs

the model exhibits a high error particularly for larger file set sizes exceeding the storage cache. We conclude that the frequent cache misses and write request optimizations in the storage system lead to higher prediction errors. Nonetheless, in most cases the performance model still provides a valid approximation of the system performance.

Summary. Distributing the workload among multiple VMs leads to complex optimization and virtualization effects. While our models were not tuned for these effects, they still provided promising prediction results in most scenarios. In some scenarios, the error was evident and expected as, e.g., the CFQ scheduler is known for very active and eager request optimization. Still, the mean error remains less than 10%-20% for the MARS and Cubist models in most cases. While further in-depth analysis of the virtualization and scheduler influences is out of scope of this paper, the results are very encouraging.

6.3 Regression Optimization

To evaluate the improvements in model accuracy achieved through our regression optimization algorithm, we compare the accuracy of the models when using the optimized regression parameters vs. the standard parameters, respectively. We evaluate the performance prediction error for each model with 100 random configurations based on the ranges shown in Table 4, cf. Section 6.1.

In Figure 13, we show the reduction in error for each model. Overall, the results show significant improvements of the model accuracy, i.e., reduced modeling error. Especially MARS and CART benefit from the parameter optimization exhibiting an error reduction of 66.30% and 74.08%, respectively. The improvement of Cubist is less, however, an error reduction of approximately 16% is still observed. Interestingly, the highest error reduction was achieved for read throughput predictions across all models. The MARS models showed most improvements for throughput metrics, while the CART models were almost evenly improved across all metrics. Surprisingly, only a small error reduction was achieved for Cubist's read response time predictions, whereas significant improvements were seen for write response time and read throughput predictions.

In Figure 8, we evaluate the error distribution of the different techniques in the form of empirical cumulative distribution functions. The results show that our optimization especially eliminated large prediction errors. Furthermore, the results confirm that the optimization especially improved CART and MARS with a decisive reduction in large errors. For Cubist, the improvement was less than for the other techniques, however, the error reduction is still evident. This is due to the fact that the technique performs competitively well even with standard parameters and the optimization is primarily able to decrease the number of extreme errors. Still, the optimization process was key for achieving the high quality results of the MARS and the Cubist models presented in Section 6.1 and Section 6.2. These two models exhibited the highest accuracy in almost all of the considered evaluation scenarios

Finally, we evaluate the statistical significance of the optimization results. In a *paired t-test*, we evaluate the mean of each difference of the prediction error of the respective regression technique between the optimized and the standard parameter values. Thus, for each regression technique the null hypothesis H_0 is that the true mean of differences of the prediction error is equal to zero. For the t-test, the p-value of both MARS and CART is less than $2.2e^{-16}$ and the p-value of Cubist is 0.0003394. Thus, H_0 is rejected confirming that the optimization is statistically significant for every considered regression technique.

Summary. Overall, the parameter optimization was crucial for creating effective performance models. The mean error reductions of approximately 16%, 66%, and 74% for Cubist, MARS, and CART, respectively, lead to performance models with high accuracy and predictive power. Especially the improvements for MARS lead to accurate models that, together with the Cubist models, exhibited the best prediction accuracy in our scenarios. Furthermore, the improvement achieved by the parameter optimization was statistically significant for every considered regression technique.

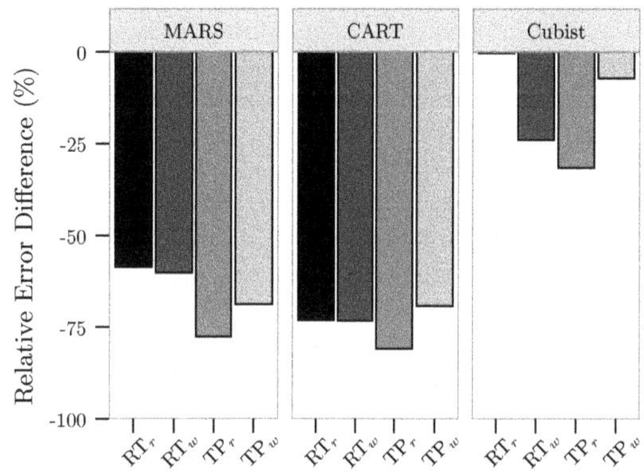

Model	RT_r	RT_w	TP_r	TP_w	$\frac{1}{4}\sum$
	Relative Error Difference (%)				
MARS	−58.64	−60.20	−77.59	−68.78	−66.30
CART	−73.14	−73.24	−80.78	−69.15	−74.08
Cubist	−0.25	−24.01	−31.52	−7.23	−15.75

Figure 13: Error Reduction by Parameter Optimization

7. RELATED WORK

Many general modeling techniques for storage systems exist, e.g., [7, 11, 20, 21, 28], but they are only shortly mentioned here as our work is focused on virtualized environments.

The work closely related to the approach presented in this paper can be classified into two groups. The first group is focused on modeling storage performance in virtualized environments. Here, Kraft et al. [17] present two approaches based on queueing theory to predict the I/O performance of consolidated virtual machines. Their first, trace-based approach simulates the consolidation of homogeneous workloads. The environment is modeled as a single queue with multiple servers having service times fitted to a Markovian Arrival Process (MAP). In their second approach, they predict storage performance in consolidation of heterogeneous workloads. They create linear estimators based on mean value analysis (MVA). Furthermore, they create a closed queueing network model, also with service times fitted to a MAP. Both methods use monitored measurements on the block layer that is lower than typical applications run. In [1], Ahmad et al. analyze the I/O performance in VMware's ESX Server virtualization. They compare virtual to native performance using benchmarks. They further create mathematical models for the virtualization overhead. The models are used for I/O throughput degradation predictions. To analyze performance interference in a virtualized environment, Koh et al. [16] manually run CPU bound and I/O bound benchmarks. While they develop mathematical models for prediction, they explicitly focus on the consolidation of different types of workloads, i.e., CPU and I/O bound. By applying different machine learning techniques, Kundu et al. [19] use artificial neural networks and support vector machines for dynamic capacity planning in virtualized environments. Further, Gulati et al. [10] present a study on storage workload characterization in virtualized environments, but perform no performance analysis.

The second group of related work deals with benchmarking and performance analysis of virtualized environments not specifically targeted at storage systems. Hauck et al. [13] propose a goal-oriented measurement approach to determine performance-relevant infrastructure properties. They examine OS scheduler properties and CPU virtualization overhead. Huber et al. [14] examine performance overhead in VMware ESX and Citrix XenServer virtualized environments. They create regression-based models for virtualized CPU and memory performance. In [3], Barham et al. introduce the Xen hypervisor comparing it to a native system as well as other virtualization plattforms. They use a variety of benchmarks for their analysis to quantify the overall Xen hypervisor overhead. Iyer et al. [15] analyze resource contention when sharing resources in virtualized environments. They focus on analyzing cache and core effects.

8. CONCLUSION

Summary. We presented a measurement-based performance prediction approach for virtualized storage systems. We created optimized performance models based on statistical regression techniques capturing the complex behaviour of the virtualized storage system. We proposed a general heuristic search algorithm to optimize the parameters of regression techniques. This algorithm is not limited to a certain domain and can be used as a general regression optimization method. We applied our optimization approach and created performance models based on systematic measurements using four regression techniques: LRM, MARS, CART, and Cubist. We evaluated the models in different scenarios to assess their prediction accuracy and the improvement achieved by our optimization approach. The scenarios comprised interpolation and extrapolation scenarios as well as scenarios when the workload is distributed on multiple virtual machines. We evaluated the models in a real-world environment based on IBM System z and IBM DS8700 server hardware. In the most typical scenario, the interpolation ability of the models were excellent with our best models having less than 7% prediction error for MARS and less than 8% for Cubist. The extrapolation accuracy was promising. Except for throughput predictions of 1 KB and 2 KB requests, the prediction error of our best model Cubist was always less than 16%. Furthermore, using our models we were able to successfully approximate the performance if the workload is distributed on two and three virtual machines. The average prediction error of both MARS and Cubist was less than 20% in most scenarios. Finally, our optimization process showed to be crucial and reduced the prediction error by 16%, 66%, and 74% for Cubist, MARS, and CART, respectively, with statistical significance.

Lessons Learned. i) Our performance modeling approach was able to extract powerful prediction models. For interpolation and extrapolation scenarios, the models showed to have excellent prediction accuracy. ii) Even if the workload is distributed on several virtual machines, the performance models are able to approximate the expected performance in most cases. iii) Our optimization process showed to be key for the prediction accuracy of the models and achieved statistically significant improvements for every considered regression technique. iv) While LRM and CART showed to perform well, MARS and Cubist consistently showed the best fit and the best prediction accuracy.

Application Scenarios. Generally, our approach is targeted at the analysis and evaluation of virtualized storage systems. It is especially beneficial for complex systems and in cases that prohibit explicit fine-grained performance models due to, e.g., time constraints for the manual performance model creation, calibration, and validation. Our automated performance modeling approach can aid (e.g., the system developer) to measure the system once and extract a performance model that represents the environment. This model can be used in different scenarios (e.g., by customers) to assess the system characteristics and evaluate deployment decisions.

9. ACKNOWLEDGMENTS

This work was funded by the German Research Foundation (DFG) under grant No. RE 1674/5-1 and KO 3445/6-1. We especially thank the Informatics Innovation Center (IIC)[6] for providing the system environment of the IBM System z and the IBM DS8700.

10. REFERENCES

[1] I. Ahmad, J. Anderson, A. Holler, R. Kambo, and V. Makhija. An analysis of disk performance in VMware ESX server virtual machines. In *WWC-6*, 2003.

[2] M. Armbrust, A. Fox, R. Griffith, A. D. Joseph, R. Katz, A. Konwinski, G. Lee, D. Patterson, A. Rabkin, I. Stoica, and M. Zaharia. A view of cloud computing. *Commun. ACM*, 53(4):50–58, 2010.

[3] P. Barham, B. Dragovic, K. Fraser, S. Hand, T. Harris, A. Ho, R. Neugebauer, I. Pratt, and A. Warfield. Xen and the art of virtualization. *SIGOPS Oper. Syst. Rev.*, 37:164–177, 2003.

[4] L. Breiman. Random Forests. *Machine Learning*, 45, 2001.

[5] L. Breiman, J. Friedman, C. J. Stone, and R. Olshen. *Classification and Regression Trees*. The Wadsworth and Brooks-Cole statistics-probability series. Chapman & Hall, 1984.

[6] D. Bruhn. Modeling and Experimental Analysis of Virtualized Storage Performance using IBM System z as Example. Master's thesis, Karlsruhe Institute of Technology, Karlsruhe, Germany, 2012.

[7] J. S. Bucy, J. Schindler, S. W. Schlosser, G. R. Ganger, and Contributors. *The DiskSim Simulation Environment - Version 4.0 Reference Manual*. Carnegie Mellon University, Pittsburgh, PA, 2008.

[8] B. Dufrasne, W. Bauer, B. Careaga, J. Myyrrylainen, A. Rainero, and P. Usong. IBM System Storage DS8700 Architecture and Implementation. `http://www.redbooks.ibm.com/abstracts/sg248786.html`, 2010.

[9] J. H. Friedman. Multivariate Adaptive Regression Splines. *Annals of Statistics*, 19(1):1–141, 1991.

[10] A. Gulati, C. Kumar, and I. Ahmad. Storage workload characterization and consolidation in virtualized environments. In *VPACT '09*.

[11] P. Harrison and S. Zertal. Queueing models of RAID systems with maxima of waiting times. *Performance Evaluation*, 64:664–689, 2007.

[12] T. Hastie, R. Tibshirani, and J. Friedman. *The Elements of Statistical Learning: Data Mining, Inference, and Prediction*. Springer Series in Statistics. Springer, 2nd edition, 2011.

[13] M. Hauck, M. Kuperberg, N. Huber, and R. Reussner. Ginpex: deriving performance-relevant infrastructure properties through goal-oriented experiments. In *QoSA-ISARCS '11*.

[14] N. Huber, M. von Quast, M. Hauck, and S. Kounev. Evaluating and Modeling Virtualization Performance Overhead for Cloud Environments. In *CLOSER '11*.

[15] R. Iyer, R. Illikkal, O. Tickoo, L. Zhao, P. Apparao, and D. Newell. VM3: Measuring, modeling and managing VM shared resources. *Computer Networks*, 53:2873–2887, 2009.

[16] Y. Koh, R. Knauerhase, P. Brett, M. Bowman, Z. Wen, and C. Pu. An Analysis of Performance Interference Effects in Virtual Environments. In *ISPASS '07*.

[17] S. Kraft, G. Casale, D. Krishnamurthy, D. Greer, and P. Kilpatrick. Performance Models of Storage Contention in Cloud Environments. *SoSyM*, 2012.

[18] M. Kuhn, S. Witson, C. Keefer, and N. Coulter. Cubist Models for Regression. `http://cran.r-project.org/web/packages/Cubist/vignettes/cubist.pdf`, 2012. Last accessed: Oct 2012.

[19] S. Kundu, R. Rangaswami, A. Gulati, M. Zhao, and K. Dutta. Modeling Virtualized Applications using Machine Learning Techniques. In *VEE '12*.

[20] A. S. Lebrecht, N. J. Dingle, and W. J. Knottenbelt. Analytical and Simulation Modelling of Zoned RAID Systems. *The Computer Journal*, 54:691–707, 2011.

[21] E. K. Lee and R. H. Katz. An analytic performance model of disk arrays. *SIGMETRICS Perform. Eval. Rev.*, 21(1), 1993.

[22] P. Mell and T. Grance. The NIST definition of cloud computing. *National Institute of Standards and Technology*, 53(6):50, 2009.

[23] Q. Noorshams, S. Kounev, and R. Reussner. Experimental Evaluation of the Performance-Influencing Factors of Virtualized Storage Systems. In *EPEW '12*, volume 7587 of *LNCS*. Springer, 2012.

[24] J. R. Quinlan. Combining Instance-Based and Model-Based Learning. In *ICML '93*.

[25] J. R. Quinlan. Learning with Continuous Classes. In *AI '92*. World Scientific.

[26] RuleQuest Research Pty Ltd. Data Mining with Cubist. `http://rulequest.com/cubist-info.html`, 2012. Last accessed: Oct 2012.

[27] The R Project for Statistical Computing. `http://www.r-project.org/`, 2012. Last accessed: Oct 2012.

[28] M. Wang, K. Au, A. Ailamaki, A. Brockwell, C. Faloutsos, and G. R. Ganger. Storage Device Performance Prediction with CART Models. In *MASCOTS '04*.

[6] `http://www.iic.kit.edu/`

Experimental Analysis of Task-based Energy Consumption in Cloud Computing Systems

Feifei Chen, John Grundy, Yun Yang, Jean-Guy Schneider and Qiang He
Faculty of Information and Communication Technologies
Swinburne University of Technology
Melbourne, Australia 3122
{feifeichen, jgrundy, yyang, jschneider, qhe}@swin.edu.au

ABSTRACT

Cloud computing delivers IT solutions as a utility to users. One consequence of this model is that large cloud data centres consume large amounts of energy and produce significant carbon footprints. A common objective of cloud providers is to develop resource provisioning and management solutions that minimise energy consumption while guaranteeing Service Level Agreements (SLAs). In order to achieve this objective, a thorough understanding of energy consumption patterns in complex cloud systems is imperative. We have developed an energy consumption model for cloud computing systems. To operationalise this model, we have conducted extensive experiments to profile the energy consumption in cloud computing systems based on three types of tasks: computation-intensive, data-intensive and communication-intensive tasks. We collected fine-grained energy consumption and performance data with varying system configurations and workloads. Our experimental results show the correlation coefficients of energy consumption, system configuration and workload, as well as system performance in cloud systems. These results can be used for designing energy consumption monitors, and static or dynamic system-level energy consumption optimisation strategies for green cloud computing systems.

Categories and Subject Descriptors

C.4 [**Computer System Organization**]: Performance of Systems; K.4.1 [**Public Policy Issues**]: Use/abuse of power; B.8.2 [**Performance Analysis and Design Aids**]

General Terms

Measurement, Performance, Experimentation

Keywords

Cloud computing, green cloud, energy consumption, performance analysis, energy efficiency.

1. INTRODUCTION

Cloud computing is a new and promising computing paradigm which delivers computing as a utility [1]. It provides rented services for computation, application software, and data storage via the Internet. Key advantages for consumers include flexible scaling on demand to their computing and data storage needs without the traditional and potentially large upfront investment in computing infrastructure. Over the last few years many large-scale data centres have been built due to the massive growth in demand for high performance cloud data and computational services. As cloud computing becomes more widespread, increasing data storage and computation needs raise the energy consumed by large cloud infrastructures. Thus, energy consumption has become a critical concern in designing modern cloud systems.

Firstly, a common economic objective of cloud providers is to minimise their total operational costs. High energy consumption directly contributes to operational costs, especially as energy unit costs continue to rise significantly. A utility provider in Virgina estimates that in 2012, 10% of all the energy it supplies to northern Virgina will be consumed by data centres [2]. Power consumption currently contributes up to 42% of a data centre's monthly expenses [3]. Secondly, due to the need to respond on-demand to customer load, many data centres consume electricity produced by "brown" generation facilities, such as coal. Thirdly, cloud system performance must not be jeopardised while minimising cloud system energy consumption. Therefore, energy consumption, as well as its impact on cloud system performance, operational cost and the environment, have become critical issues in green cloud computing systems [4].

Many efforts have been made to improve the energy efficiency of cloud systems. Some simple techniques provide basic energy management for servers in cloud systems, for example, turning on and off servers, putting them to sleep or using Dynamic Voltage/Frequency Scaling (DVFS) [5] to adjust servers' power states. DVFS adjusts CPU power (consequently the performance level) according to the workload. However, the scope of DVFS optimisation is limited to CPUs. Another approach for improving energy efficiency is to adopt virtualisation techniques to obtain better resource isolation and reduce infrastructure energy consumption through resource consolidation and live migration [6]. Based on virtualisation techniques, several energy-aware resource allocation policies and scheduling algorithms have been proposed to optimise energy consumption in cloud systems [7]. However, energy consumption and performance of cloud systems vary drastically with different system configuration and resource allocation strategies, as well as the workloads and types of runtime tasks in cloud systems [8]. By nature, the workloads of cloud systems are highly variable, application-specific and can often not be predicted in advance. Ideally, system performance should not be adversely impacted while energy consumption is being minimised. We thus require a thorough understanding of

ICPE'13, April 21–24, 2013, Prague, Czech Republic.

energy consumption patterns of different types of tasks in cloud systems, for example, data retrieval and data processing tasks. We also need to understand how energy consumption of such tasks and cloud system performance are affected by different workloads and system configurations, respectively.

We proposed an energy consumption model to calculate the energy consumption of specific types of tasks in cloud systems, and to use this model to statically and/or dynamically organise cloud application load [9]. In our model, runtime cloud tasks are divided into three types: computation-intensive, data-intensive and communication intensive. The model identifies what factors determine the energy consumption of a specific task. However, the following major issues have yet to be addressed:

- How do we characterise and profile the energy consumption of different types of tasks?
- How do system configuration and resource allocation strategies affect the energy consumption?
- What is the relationship between energy consumption and workload of tasks?
- What is the relationship between energy consumption and cloud system performance?

In order to address these issues, we have conducted a series of experiments to empirically analyse the energy consumption in cloud systems. Based on our experimental results, we have identified the correlation coefficients of energy consumption, system configuration, workload, and system performance in cloud systems. Our analytical results provide guidelines for deriving energy consumption models based on different types of runtime cloud computing tasks. In addition, our analytical results also provide guidelines for statically planning tasks and scheduling on available cloud platforms, and dynamically monitoring energy consumption. They may be used to support system- or application-level energy consumption and performance optimisation (or both).

We briefly summarise the state-of-the-art of energy consumption models, energy-saving policies and analysis approaches in Section 2. In Section 3, we describe our energy consumption model to profile energy consumption. Our experimental energy consumption profiling setup and method are described in Section 4. In Section 5, we present our experimental profiling results and detailed analysis. We discuss observations derived from our analysis in Section 6. Finally, we summarise key findings and discuss directions for future work in Section 7.

2. RELATED WORK

Energy consumption of cloud systems has become a popular research topic in recent years. Several efforts have been made to build energy consumption models for cloud systems. Jung et al. [10] focused on the energy consumed by physical servers. Their energy consumption model does not take into account the impact of specific workloads running in cloud systems. Mach and Schikuta [11] proposed a method for energy consumption calculation based on the number of Java Virtual Machine (JVM) instances. However, it is difficult to measure the actual numbers of JVMs because of the dynamic nature of the JVM life cycle. Lee and Zomaya [12] proposed an energy model of cloud tasks for developing energy-conscious task consolidation algorithms that reduce energy consumption of cloud systems. However, in their model it is assumed that the relation between CPU utilisation and energy consumption is linear. This is a huge limitation because the utilisation of other resources such as memory and hard disk can also greatly influence the energy consumption.

Energy-saving policies of cloud systems have also been investigated in the past few years. Liu et al. [13] described a new cloud infrastructure which can dynamically consolidate Virtual Machines (VMs) based on CPU utilisation of servers in order to identify idle physical servers. Identified idle physical servers can be turned off to save energy. Verma et al. [14] used the characteristics of VMs, such as cache footprint and the set of applications running on the VMs, to drive power-aware placement of VMs. VirtualPower [15] was proposed to exploit power management decisions of guest VMs on virtual power states. The virtual power states of guest VMs were considered as preconditions to run local and global energy management policies across the computation. These energy saving policies do not take the workload in cloud systems into consideration.

Research efforts have also focused on profiling and analysing the energy consumption of cloud systems. Most existing profiling efforts were conducted by using energy benchmarks or closely monitoring the energy profiles of individual system components at runtime, such as CPU, cache, hard disk and memory. A framework has been proposed by Stoess et al [16] for energy optimisation and development of energy-aware operation systems based on the availability of energy models for each hardware component. Chen et al. [17] developed a linear power model that presents the behaviour and power consumption of individual hardware components of a single physical server. Joulemeter, a power meter for VMs [18], makes use of software components to monitor the resource usage of VMs and then converts the resource usage to energy consumption based on the power model of each individual hardware component. Although some of the profiling and analysis are conducted based on specific applications in cloud systems, the evaluation only includes a single type of cloud application. For instance, Lefèvre and Orgerie [19] evaluated energy efficient cloud systems on a multicore platform. They focused only on CPU cores and conducted their evaluation of the energy consumption during migration of VMs only with computation-intensive cloud applications.

Some of the above work have made initial efforts in profiling the energy consumption and performance of cloud systems. However, none of them has identified the relationship between energy consumption and runtime tasks under different cloud system configurations and the correlation with cloud system performance. In order to address these issues, in this paper we present our energy consumption profiling results obtained from extensive experiments conducted on a cloud test-bed. Our goal is to investigate and characterise the impact of both system configurations and workload on the energy consumption and system performance.

3. ENERGY CONSUMPTION MODEL

Our energy consumption profiling and analysis is motivated by our existing energy consumption model proposed for cloud-wide energy analysis [9]. It provides a basis for characterising the energy consumption in cloud systems under different system configurations.

In this model, the total energy consumption of a cloud workload is divided into a fixed part and a dynamic part. The fixed part of energy consumption includes the energy consumption during idle time and energy consumption of the cooling system, defined as

EC_{Fix}. The dynamic energy consumption is the energy consumed by running tasks in the cloud system, defined as EC_{Var}. Thus, the total energy consumption defined as EC_{Total} is formulated as:

$$EC_{Total} = EC_{Fix} + EC_{Var} \quad (1)$$

Instead of individual hardware components, we set single task running in a cloud system as the fundamental unit for energy consumption calculation. All runtime tasks in cloud systems utilise computation, storage and communication resources. However, the percentage of each resource used by different types of tasks is significantly different. Based on the major type of resource consumed by the task, we characterise all cloud runtime tasks three types: *computation-intensive*, *data-intensive* and *communication-intensive*. For a computation-intensive task t^i_{comp}, $1 \leq i \leq n$, where n is the total number of computation-intensive tasks, the energy consumption of t^i_{comp} is defined as $EC(t^i_{comp})$. For a data-intensive task t^i_{data}, $1 \leq i \leq m$, where m is the total number of data-intensive tasks, the energy consumption of t^i_{data} is defined as $EC(t^i_{data})$. For a communication-intensive task t^i_{com}, $1 \leq i \leq k$, where k is the total number of communication-intensive tasks, the energy consumption of t^i_{com} is defined as $EC(t^i_{com})$. Due to scheduling and other overheads, which is denoted as EC_{sche}, the total energy consumption of two tasks is usually not the sum of the energy individually consumed by the tasks. In our model, the total energy consumption is defined as:

$$EC_{Total} = EC_{Fix} + \sum_{i=1}^{n} EC(t^i_{comp}) + \sum_{i=1}^{m} EC(t^i_{data}) + \sum_{i=1}^{k} EC(t^i_{com}) + EC_{sche} \quad (2)$$

For each task, its energy consumption is dependent on the task workload. The factors related to workload influence energy consumption of the task directly and largely. In our model, the parameters of a task taken into account in the calculation of the energy consumption include: number of processes for the task, defined as *PT*; the size of data to be processed, defined as *DS*; and the size of data to be transmitted, defined as *DT*. In addition, system configurations, such as hardware of the physical server and the scale of the configured VMs, have significant impact on energy consumption. Hence, the energy consumed by each task is determined by the task parameters and the system configuration, denoted as *C*. Thus, the energy consumption of each type of task are defined as:

$$EC(t^i_{comp}) = f_{comp}\left(PT^i_{comp}, DS^i_{comp}, DT^i_{comp}, C^i_{comp}\right) \quad (3)$$

$$EC(t^i_{data}) = f_{data}\left(PT^i_{data}, DS^i_{data}, DT^i_{data}, C^i_{data}\right) \quad (4)$$

$$EC(t^i_{com}) = f_{com}\left(PT^i_{com}, DS^i_{com}, DT^i_{com}, C^i_{com}\right) \quad (5)$$

In order to determine the energy consumption model of a cloud system, it is essential to specify the factors that have impact on the energy consumption of each type of tasks.

4. EXPERIMENTAL SETUP

Our energy consumption profiling and analysis were performed based using the energy consumption model described in Section 3. Based on the energy consumption profiling results, the impact of the factors indicated in equations (3)-(5) on the energy consumption of different tasks will be verified. Our objective was to collect data and analyse the correlation coefficients of energy consumption, system configuration, workload, and system performance. The analysis result will be adopted to refine and formalise our energy consumption model, used to support predictive analysis of workload energy consumption, and used to support static and dynamic optimisation of energy usage.

We profiled the energy consumption by creating heterogeneous Virtual Machines in a cloud system and running computation-intensive tasks, data-intensive tasks and communication-intensive tasks, respectively. For each type of task, we also profiled the energy consumption and system performance with various task parameters, that is, the number of processes, the size of data to be processed, the size of data to be transmitted, and task workload. We introduce the experimental setup in this section.

4.1 Testbed

SwinCloud is a private cloud that provides a common computational infrastructure to researchers at Swinburne University of Technology. Our experiments were conducted in SwinCloud as we have detailed knowledge of the cloud hardware, networking, operating system versions, and other application software. SwinCloud was experimented in the Energy Research Lab (ERL) at Swinburne University of Technology. This lab focuses on energy-related research and development. Using the extensive and sensitive power monitoring facilities provided by this lab, we could precisely monitor the power consumption of the SwinCloud server and network devices, including network cards, switches and routers. The power consumption measurement was realised and managed using PowerNode, a power usage profiling equipment developed by GreenWave Reality [1]. It supports measurement of both immediate and average power consumption. Collected power readings were reported to the GreenWave Gateway, which is used for creating a mesh-based Home Area Network (HAN). The GreenWave Gateway then sent the data to the GreenWave Reality data centre, where the information could be viewed and analysed.

Table 1. Specifications of HP Z400

Basic Specification		Notes
Number of Cores	4	
Number of Threads	2	Intel Hyper-Threading Technology
CPU Frequency	2.8GHz	Fixed CPU Frequency
Memory	10GB	Memory Speed 1333 MHz
Hard Drive	1TB 7200 RTM SATA	
Network Interface	Intel e1000 Gb	

A cloud computing system is composed of multiple servers. The energy consumption of the cloud system includes the energy consumed by individual servers and the scheduling and communication overhead across different servers. In this paper, we focus on the energy consumption of individual servers as it is the predominant part [20]. Furthermore, the cross-server scheduling and communication overhead of one cloud system can be significantly different from the other, depending on the scheduling mechanism adopted by the cloud systems and the distribution of the constituent servers. Thus, we conducted our experiments by measuring energy consumption of tasks running

[1] http://www.greenwavereality.com/

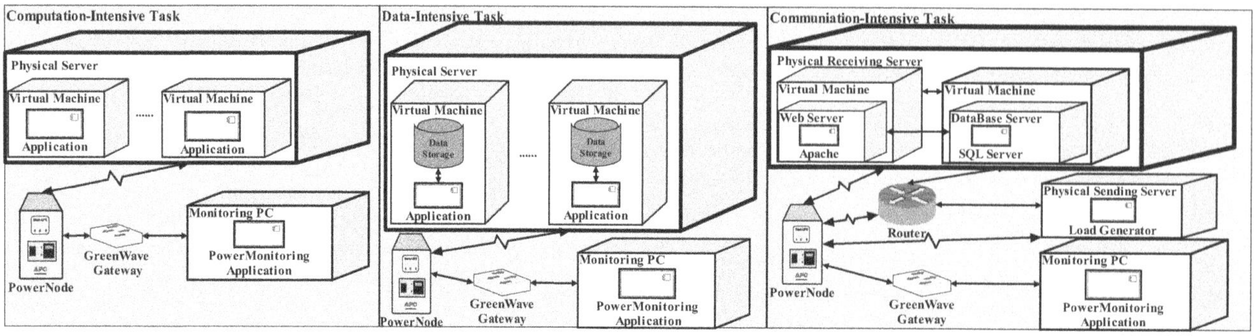

Figure 1. Energy profiling Framework

on a single discrete server. We left research to determine the energy consumption incurred by cross-server scheduling for future work. Moreover, we did not adopt any existing energy-saving policies in our experiments such as DVFS [5] or VirtualPower [15], in order to be able to isolate the factors causing unexpected energy consumption. Our experimental set up can however be reused to include these options enabled.

A type of the servers currently deployed in SwinCloud is HP Z400. Table 1 lists the specifications of the HP Z400. The Virtual Machine Manager (VMM) utilised for VM management is VMware ESX 4.1 and the operating systems running on the virtual machines are either Window Server 2008 or Windows XP Professional.

Figure 1 presents the energy profiling framework for different tasks used in our experiments. For computation-intensive tasks and data-intensive tasks, stand-alone applications were deployed. For communication-intensive tasks, a Web Application was deployed. The user requests were sent to the Web server through the router from a client PC. The major energy consumption of computation-intensive tasks was introduced by the computation of the deployed application. For data-intensive tasks and communication-intensive tasks, the energy consumption of the deployed application caused by computation is minimal. The major energy consumption of data-intensive tasks is introduced by reading/writing data in data storage. For communication-intensive tasks, the major energy consumption includes the energy consumed to send user request, the energy consumed to serialise/deserialise data and the energy consumed to process user requests. In order to focus on the communication aspects, the workload applied for communication-intensive tasks in our experiments was arranged to minimize load on sender/receiver so almost all energy is the encoding and decoding overhead.

Table 2. Type of VM

Virtual Machine	Number of Cores	RAM	Hard Disk
Small	1	2GB	80GB
Medium	2	4GB	80GB
Large	3	6GB	80GB
XLarge	4	8GB	80GB

In the experiments, all VMs were assigned with 2GB, 4GB, 6GB or 8GB memory. The number of virtual CPUs (vCPUs) of each VM varied from 1 to 4 in steps of 1. The number of vCPUs corresponded to the number of physical cores assigned to the VM. The configuration of VMs in different scales is presented in Table 2. System configurations related to specific task types will be introduced in Sections 5.1, 5.2 and 5.3, respectively.

4.2 Profiling Method

The energy consumption result of a task equals the difference of average power consumption between the server with and without workload multiplied by the execution time of the task. We firstly retrieved the average power consumption measured by PowerNode with no hosted application workload in the cloud system as our idle state benchmark. Then, we retrieved the average power consumption measured by PowerNode during the running time of each task. Finally, we multiplied the average power consumption by the execution time of the task to obtain the energy consumption of the task.

As discussed in Section 3, the task parameters, including the number of processes, the size of data to be processed, the size of data to be transmitted, have impact on the task energy consumption and system performance based on the energy consumption model. These task parameters determine the workload of a single task. In addition, the number of running tasks determines the overall workload in the cloud system. Energy consumption is highly influenced by the workloads in the cloud system as higher workloads consume more resources that leads to higher energy consumption. Moreover, system configurations, such as the number and scale of the configured VMs, also impact energy consumption significantly. As such, we took the cloud system workload and system configuration as inputs of our experiments. We set energy consumption and system performance as outputs of our experiments. We selected the throughput of the system as the profiled system performance measure. This is because throughput is often a key performance parameter monitored in cloud systems. It has the advantage of reflecting resource usage accurately [21]. For computation-intensive tasks, throughput is defined as the number of computation tasks completed per hour. For data-intensive tasks, throughput is defined as the amount of data transferred per second. For communication-intensive tasks, throughput is defined as the total number of user interactions requested and completed successfully per second.

4.3 Test Case Design

Again, we used three types of tasks: computation-intensive, data-intensive and communication-intensive tasks. We designed and conducted three series of experiments, described as follows:

1. **Computation-intensive tasks**: The major resources consumed by computation-intensive tasks are the CPU cores. In order to make sure the computation workload would be distributed to all available CPU cores, an application which calculates Fibonacci sequence has been developed and parallelism has been applied to the application. We deployed multiple

processes to calculate Fibonacci sequences to implement computation-intensive tasks. Each calculation was considered as a computation-intensive task. As the largest number of the Fibonacci sequence determines the duration of each calculation task, we mapped this number to the workload of each computation-intensive task – defined as *LN*. As this application purely consumes CPU resources, the energy consumption caused by other resources such as memory can be eliminated. In order to analyse the impact of workload and cloud system configurations on the total energy consumption and throughput, we designed three sets of test cases in this series of experiments. We first ran a single task to calculate Fibonacci sequence and increased the *LN* of the task gradually with fixed system configuration in test set 5.1.1 as described in Section 5.1. Then, we increased the resource allocated to the task while keeping *LN* constant. Computation-intensive tasks mainly consume CPU resources and Hyper-Threading (HT) technology is usually applied to improve the overall performance of the CPU. Thus, we measured the energy consumption and system performance with HT enabled and disabled in test set 5.1.2 as described in Section 5.1, to analyse the impact of HT technology on energy consumption and system performance. In these tests we kept the resources allocated constant and increased the *LN* of the tasks gradually and turned on and off HT. In test set 5.1.3 as described in Section 5.1, we ran multiple tasks with the same *LN* simultaneously and kept the system configuration and resource allocation constant to analyse the scheduling overhead.

2. **Data-intensive tasks:** As a data-intensive task mainly consumes the storage resources in cloud systems, we ran IOzone benchmark[2] in our tests, a disk and file system benchmarking application. Using IOzone, we generated a large number of I/O operations and stressed the disk I/O as required. Each run of IOzone benchmark was considered as a data-intensive task. The total amount of data read/writer determined the workload of a task. In each task, we set the read and write ratio to 50% - 50%. The parameters of a data-intensive task to be changed include the process number and the size of the data record. We set the size of the data record to be transferred at 4K, 8K and 64K bytes as our basic test suite. 4K is the memory page size, 8K is what Microsoft Windows uses for network transfers, and 64K is the typical record size that Windows uses when applications try to transfer blocks of data bigger than 64K [22]. We also tested a few very large data records, including 8MB and 16MB data records. A total of five sets of tests were designed and run in this series of experiments. Test set 5.2.1 as described in Section 5.2 aimed to reveal the impact of workload on energy consumption and system throughput while keeping the task parameters and system configuration constant. The purpose of test set 5.2.2 as described in Section 5.2 was to analyse the impact of system configuration on energy consumption and system throughput. So we kept workload and task parameters constant and changed the number and type of virtual machines. In test set 5.2.3 as described in Section 5.2, we aimed to evaluate the impact of task parameters on energy consumption and system throughput by keeping workload and system configuration constant. Furthermore, we evaluated the impact of scheduling overhead on energy consumption and system throughput in test set 5.2.4 as described in Section 5.2. Similar to computation-intensive tasks, resource allocated to data-intensive tasks would affect the energy consumption and system performance. Therefore, we measured the energy consumption and system throughput by changing the total resource allocation in test set 5.2.5 as described in Section 5.2.

3. **Communication-intensive tasks:** We deployed a Java Web Application named JPetstore [3] in SwinCloud as the communication-intensive application. It is a Web e-commerce application which has been used as a representative application for a transactional workload that stresses the servers and network devices as required in our tests. JPetstore uses a Web server to handle the user requests and a database server to process the database queries in response to the user requests. Therefore, it is a composite task which is composed of communication-intensive task, data-intensive task and a small level of computation-intensive task. For our experiment, we used its communication aspects only. As mentioned in Section 4.1, we arranged the workload of communication-intensive task to minimize load on sender/receiver in order to measure the energy consumption of the communication aspects only of this reference application. All network traffic generated by customers was emulated using JMeter[4]. JMeter is a load testing tool for analysing and measuring the performance of variety of Web applications. Each user session generated by JMeter consisted of a series of sequential user requests, such as browsing the list of products, adding items to the shopping cart, checking out, and so on. Each run of JMeter was considered as a communication-intensive task and the number of user sessions generated by JMeter determined the workload of the task. All user sessions were started with fixed time interval in each test set. The time interval decreased when the number of user sessions increased. The parameters of a communication-intensive task to be changed included the user request type, the number of requests in each user session and the packet size of each user request. We developed three test sets for communication-intensive tasks. We applied different workloads for each test set by mixing browsing requests or shopping requests in different percentages. In test set 5.3.1 as described in Section 5.3, we fixed the task parameter and the workload and changed virtual machine types. In test set 5.3.2 as described in Section 5.3, we changed the workload of the task while fixing the task parameter, the system configuration and the resource allocation. We only changed the task parameter and kept all other factors constant to evaluate the impact of task parameters on energy consumption and system performance in test set 5.3.3 as described in Section 5.3.

5. EXPERIMENTAL RESULTS

We conducted three major sets of tests to analyse the energy consumption of three types of cloud-hosted application tasks in order to analyse the relationship between the system energy consumption and system performance. For each test set, we took cloud system workload and system configuration as inputs. Energy consumption of each task and the system throughput were the outputs of our experiments. We modified the configuration of SwinCloud, including the number of VMs, the hardware and software resources allocated to the VMs. We then measured the energy consumption of SwinCloud running different types of tasks. Each set of tests was repeated ten times to reduce measurement error. We analysed the impact of system

[2] http://www.iozone.org

[3] http://java.sun.com/developer/releases/petstore/

[4] http://jmeter.apache.org/

configuration, task parameters and workload on the energy consumption and performance of the cloud system based on the experimental results.

5.1 Computation-intensive Tasks Results

A computation-intensive task usually requires a number of isolated processes to perform the computation in a cloud system. The total energy consumption might increase with the number of processes since the increased overhead of scheduling will cause extra energy consumption. Moreover, the energy consumption of the cloud system is subject to change under different system configurations.

We deployed multiple processes that calculate Fibonacci sequences to implement computation-intensive tasks. Parallelism was applied to make sure the computation workload would be distributed to all available vCPUs. As the largest number of the Fibonacci sequence determines the duration of each calculation task, we mapped this number to the workload of each computation-intensive task – defined as *LN* (see Section 4.3). In order to control the execution time of each task within a reasonable value, we set *LN* from 52 to 56 in the following tests.

Test Set 5.1.1: Keeping the number of tasks constant, while gradually increasing the number of active CPU cores allocated to the task, and the workload of the task. The total number of tasks was wet to one. This set of tests was run on an XLarge virtual machine (see Table 2 for specification details). The server power usage is presented in Figure 2. We observed increasing power consumption caused by increasing CPU usage. The power consumption was linear to CPU usage. The energy consumption per task is displayed in Figure 3. We noticed that the energy consumption of the task was impacted by the number of CPU cores allocated to the task. Moreover, the largest *LN* of the Fibonacci sequence also affected the energy consumption. As shown in Figure 3, the energy consumption of each task increased with the workload of the task. Moreover, the energy consumption of each task decreased dramatically as the number of cores allocated to the task increased. This is because the execution time of a task will decrease as more computation resources are allocated to the task. However, the increase in average energy usage rate caused by extra CPU cores is not as much as the execution time of the task. Therefore, the energy consumption will decrease accordingly. In addition, we observed a slight turning point of the energy consumption when the number of cores allocated to the task reaches three. For instance, when we set the largest number of the Fibonacci sequence *LN* to 56, the energy consumption with four cores allocated increased 4% compared to energy consumption with three cores. This shows that the overhead of scheduling an extra core can cancel out the task running time saved and will also cause more energy consumption. Therefore, the energy consumption as a function of number of cores is highly nonlinear, with a minimum at three. In addition, we observe that most of the reduction of energy consumption is in the changing from one core to two cores, and then it does not change much any more. The system throughput is presented in Figure 4. As expected, the more resources allocated to the task the better the system throughput obtained. This result shows that, for computation-intensive tasks, the system throughput rises with the number of allocated cores and the increase of system throughput is nonlinear. Together with the results of Figure 3, we can conclude that energy consumption of computation-intensive tasks is a more complex nonlinear function of allocated cores and execution time.

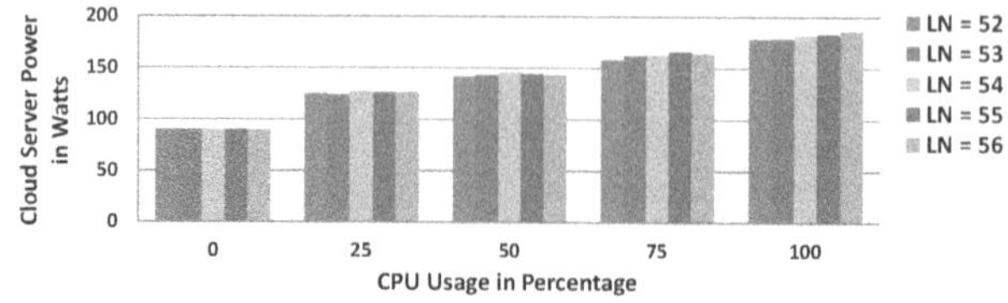

Figure 2. Server Power Consumption with Different Workload.

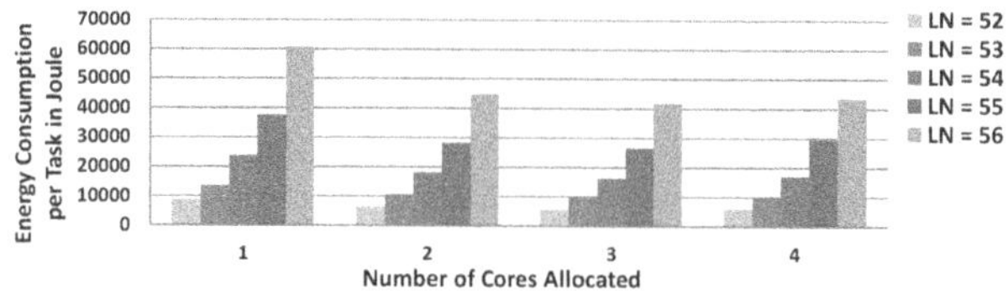

Figure 3. Energy Consumption per Task with Different Workload.

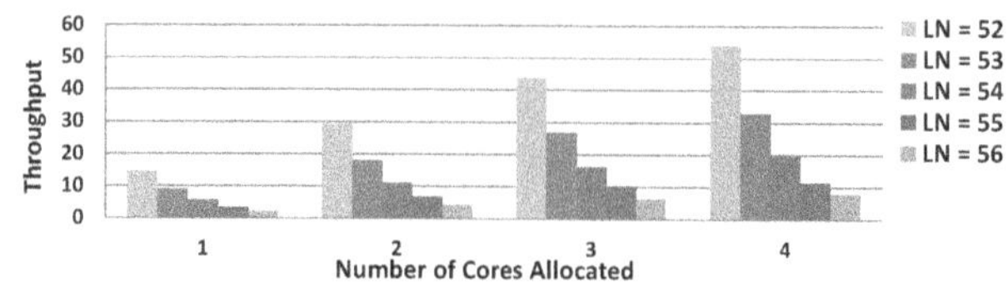

Figure 4. System Throughput with Different Workload.

Test Set 5.1.2: Keeping the number of tasks and resources allocated constant, gradually increasing the workload of the task, and turning Hyper Threading on and off. Only one task was running on the test-bed at one time. This set of tests was run on an XLarge virtual machine. The results are presented in Figures 5 and 6. As showed by Figure 5, with Hyper-Threading enabled, there is a significant increase in the energy consumption of the task – the Hyper-Threading causing extra power consumption. In the case of "*LN* = 56", the extra energy consumption caused by Hyper-Threading is 8%. However, as shown in Figure 6, the extra system throughput obtained by enabling Hyper-Threading is not as much as expected. In cases of "*LN* = 53", "*LN* = 54"and "*LN* = 55", Hyper-Threading even led to decrease in system throughput. This suggests that Hyper-Threading be disabled for computation-intensive tasks since there is little to gain form Hyper-Threading if a core's execution resources are already well utilised.

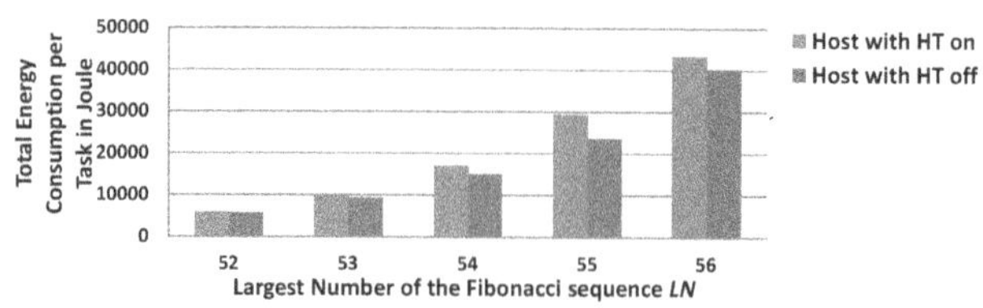

Figure 5. Energy Consumption per Task with HT on and off.

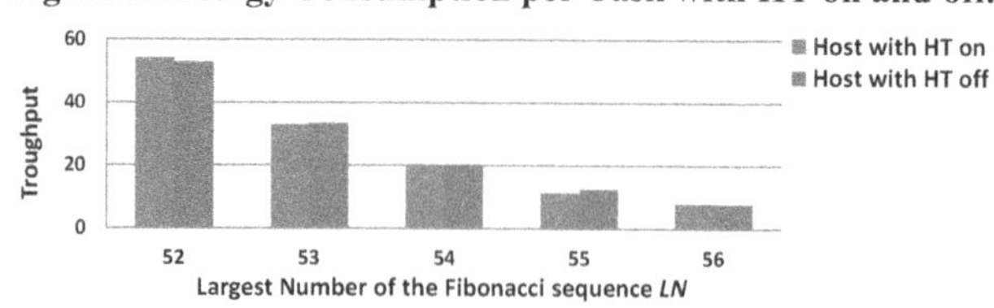

Figure 6. System Throughput with HT on and off.

Test Set 5.1.3: Keeping the workload of each task constant, and increasing the total number of tasks and number of VMs allocated to tasks. In this set of tests, the type of each VM was set to Small. Figure 7 shows the results. The number of running tasks can influence the average energy consumption of a single task: a larger number of tasks lead to higher average energy consumption of individual tasks. When the total number of tasks increased, more consequential scheduling overhead caused extra energy consumption. Moreover, it is clear that the increase in number of VMs significantly resulted in increase in per-task energy consumption. For instance, in the case of 6 tasks, the per-task energy consumption increased by 5% when we increased the number of VMs from one to two. As more VMs were deployed, the VM maintenance overhead increased as well. This caused more memory consumption, resulting in more power consumption. On the other hand, as depicted in Figure 8, the throughput was minimal when only one VM was configured. The throughput of two VMs and three VMs were at the same level. However, three VMs consumed 3% more energy than two VMs on average, as presented in Figure 7. Thus the two VMs configuration demonstrates higher energy efficiency than other configurations when there are multiple computation-intensive tasks running simultaneously.

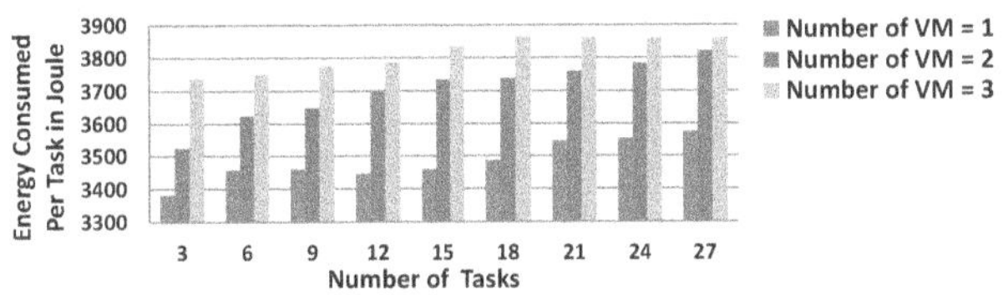

Figure 7. Energy Consumption per Task with Different VM Numbers.

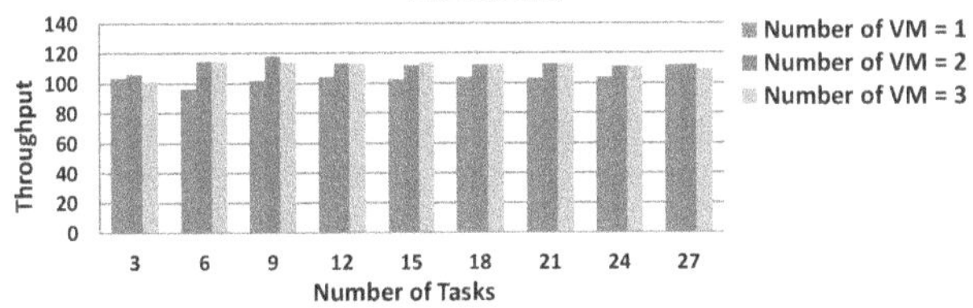

Figure 8. Throughput with Different VM Numbers.

5.2 Data-intensive Tasks Results

A data-intensive task needs to retrieve or store a large amount of data stored in one or more data storage servers. It requires high local disk I/O bandwidth in order to meet performance requirements. In reality the storage servers can be deployed in different data centres located in different geographic locations. However, we only consider the energy consumption in one data centre for the purpose of simplicity as the characteristics of data centres of the same service provider are often very similar.

We focused on the correlation between the energy consumption and the amount of data transferred in and out of the storage server. To profile and analyse the energy consumption caused by different data sizes and system configurations, we stress the cloud test-bed with data-intensive tasks generated using IOzone.

In every set of tests, we observed major spikes of energy consumption at the beginning and at the end of each task. They corresponded to the moments when disk I/O were stressed. We conducted five sets of tests in this series of experiments.

Test Set 5.2.1: Keeping the total number of tasks, process number of each task, resources allocated and VM configuration constant, while increasing the total amount of data transferred and record size. This set of tests was performed on an XLarge virtual machine. Only one task was running on the cloud test-bed at one time and the number of processes was fixed at one. Therefore, the impact of scheduling overhead caused by multiple tasks and processes on energy consumption were avoided. As presented in Figure 9, server power consumption increased as the data record size decreased. However, the effect of record sizes on power consumption is nonlinear as displayed in Figure 9, with larger record sizes leading to convergence to a certain value. For instance, server power consumption did not change too much when we increased the record size from 8MB to 16MB. Smaller data record sizes led to generation of more overhead information for keeping track of where the data were located on the storage media. The overhead information consisted of the directory information, the space allocation and any other data that was not a part of the data to be transferred. As a result, the tasks used more computational resources, consuming more energy.

The results of system energy consumption and system throughput are displayed in Figure 10 and Figure 11, respectively. From Figure 10, we see that the energy consumption increases proportionally to the total amount of transferred data. The per-task energy consumption increases in a linear manner. As the total amount of data transferred increased, smaller record size would cause higher gradient when the total energy consumption increased because the execution time and the energy consumption increased simultaneously. As the record size decreased, more overhead information needed to be processed and stored. In summary, if only one process is configured to run the data-intensive task, bigger record size is more energy-efficient. As shown in Figure 11, the system throughput was higher when we set the record size to 64K, compared to other record sizes. As 64K is the typical record size Windows uses when applications try to transfer blocks of data bigger than 64K, data is transferred most efficiently in this size.

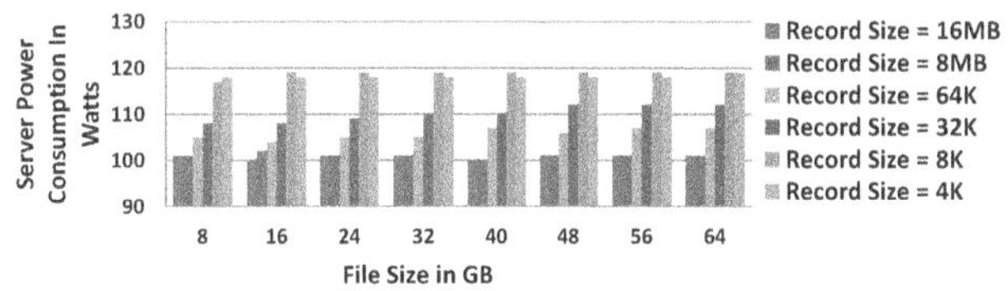

Figure 9. Server Power Consumption with Different File Size.

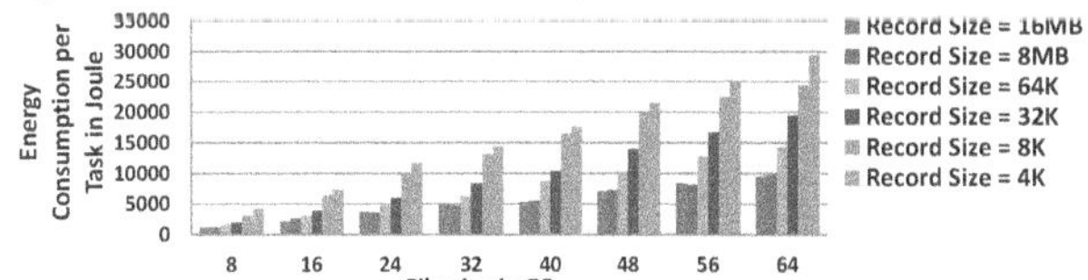

Figure 10. Energy Consumption per Task with Different File Size.

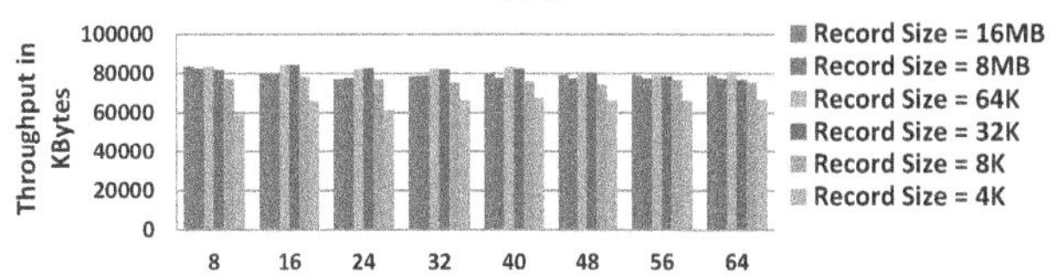

Figure 11. Throughput with Different File Size.

Test Set 5.2.2: Keeping total number of tasks, process number per task, total amount of data transferred and resource allocated constant, while changing VM configurations and record size of the task. This set of experiments was running on one XLarge VM, two Medium VMs and four Small VMs respectively so as to fix the size of the total virtual memory and the total number of virtual cores allocated to the VMs. The total number of tasks was set to four and the number of processes instantiated for each task was set to one. All the tasks were distributed on the VMs evenly. The total amount of data to be transferred was 64GB.

The system energy consumption and system throughput are presented in Figure 12 and Figure 13. We observed that the energy consumption increased and the throughput decreased as we changed the VM deployment from one XLarge VMs to four Small VMs. On one hand, VMs compete with each other for I/O bandwidth. On the other hand, more VMs will cause more read/write latency that lead to longer execution time of each task. As shown in Figure 12, the per-task energy consumption increased in a sub-linear manner. As shown in Figure 13, the best system throughput presented when we ran the workload on one XLarge VM. From these tests we conclude that running one VM for the entire I/O operation is most energy-efficient.

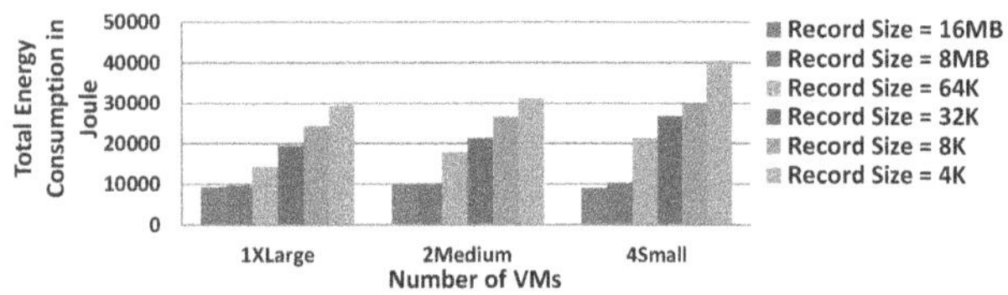

Figure 12. Energy Consumption with Different VM Configurations.

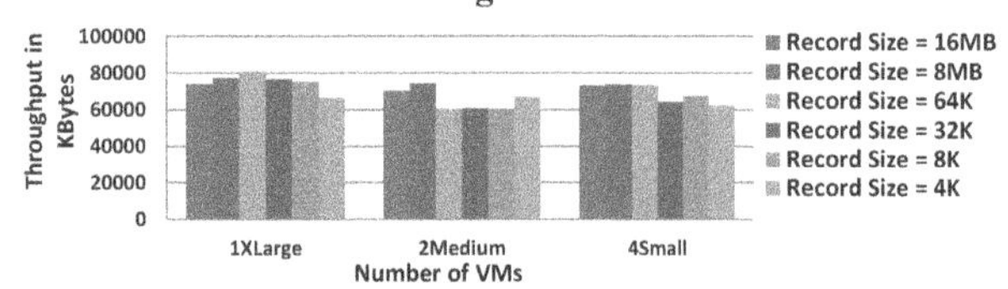

Figure 13. Throughput with Different VM Configurations.

Test Set 5.2.3: With single task, keeping the total amount of data transferred stable, resource allocation and VM configuration constant while increasing the process numbers for task and the record size of the task. An XLarge VM was deployed to run this set of tests. Only one task was running on the VM and it was assigned to transfer a 64GB file. As depicted in Figure 14, the per-task energy consumption increased with the number of processes. Although the power consumption of the server decreased slightly as the number of processes increased, as presented by Figure 15, multiple processes caused scheduling and synchronisation overhead. Thus, the total execution time of one data-intensive task increased accordingly. As a result, the system throughput decreased, as shown in Figure 16. Our conclusion is that running data-intensive tasks with one sequential I/O operation is the most energy-efficient when the total amount of data transferred is fixed.

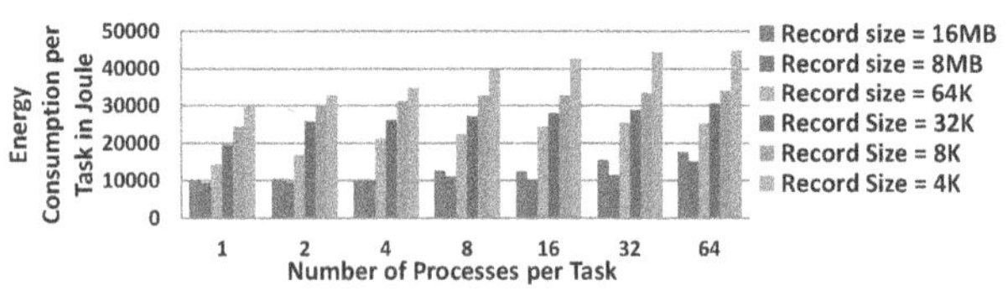

Figure 14. Energy Consumption per Task with Multiple Processes and Single VM.

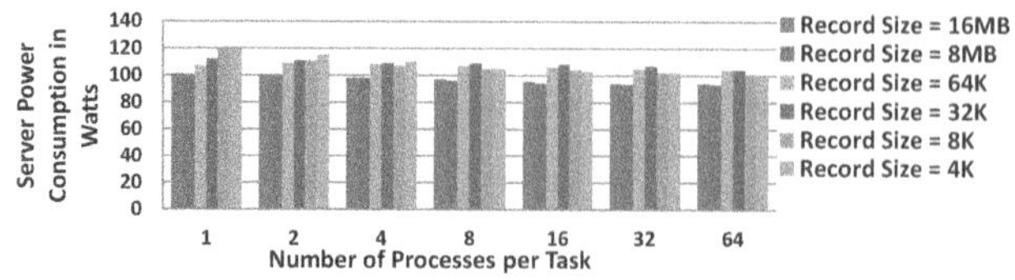

Figure 15. Server Power Consumption with Multiple Processes and Single VM.

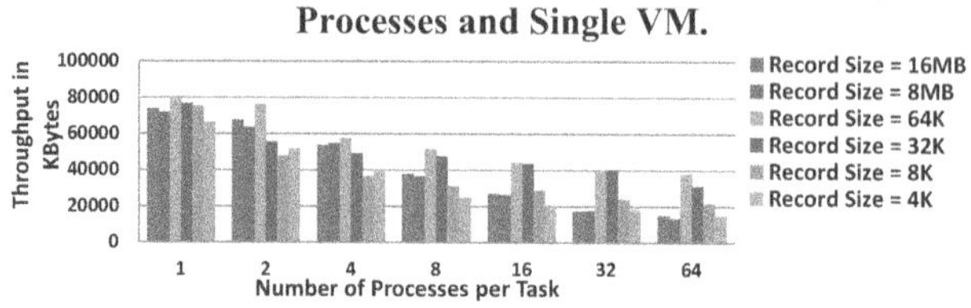

Figure 16. Throughput with Multiple Processes and Single VM.

Test Set 5.2.4: With multiple tasks, fix the total amount of data transferred, the resource allocation and number of processes for each task, changing VM configuration and record size for each task. The total number of tasks was set to four and the tasks were distributed to all the VMs evenly. We fixed the number of processes for each task running on each virtual machine at four and the total amount of data to be transferred of all the tasks at 64GB. The system energy consumption and system throughput are shown in Figure 17 and Figure 18, respectively. As we already observed in previous tests, energy consumption is greatly impacted by data record size. Smaller record size results in more energy consumption. The highest throughput was approximately 30% less than the highest throughput we obtained in test set 5.3.2. The numbers of process in both test sets were the same. However, there were multiple tasks running simultaneously this test set while only one task was running in test set 5.3.2. Therefore, multiple task scheduling causes extra scheduling overhead.

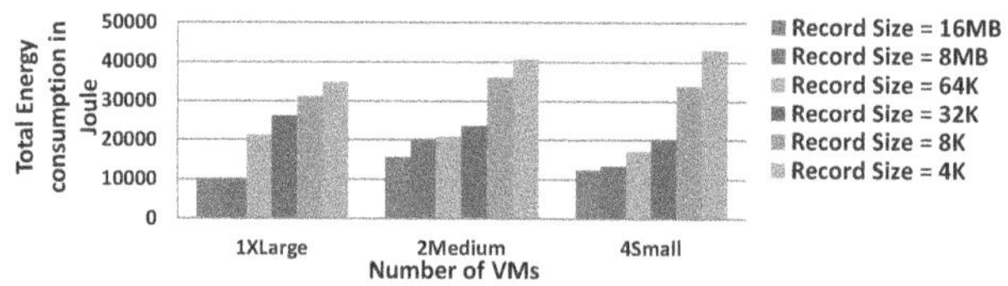

Figure 17. Energy Consumption with Multiple Processes and Multiple VMs.

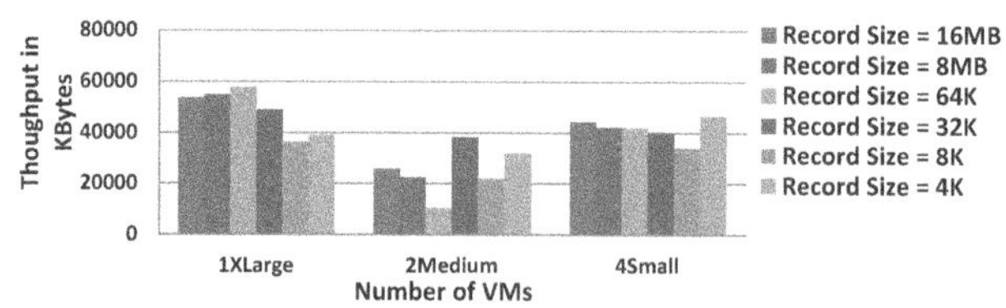

Figure 18. Throughput with Multiple Processes and Multiple VMs.

Test Set 5.2.5: Keeping the total number of tasks, the total amount of data transferred and the number of processes for each task constant, while changing the resources allocated, VM configuration and record size of each task. Only one task was running on the cloud test-bed at one time. We fixed the number of processes for the data-intensive task at one and the total amount of data transferred at 64GB and scaled up the size of VM from Small to XLarge. These results are shown in Figure 19 and Figure 20. Compared to computation-intensive tasks, the VM type did not impact the energy consumption and throughput significantly when the record size was fixed. Running on an XLarge virtual machine was most energy-efficient when the record size was smaller than 64K. When the record sizes were 8M and 16M, running the data-intensive task on a Small virtual machine was most energy-efficient.

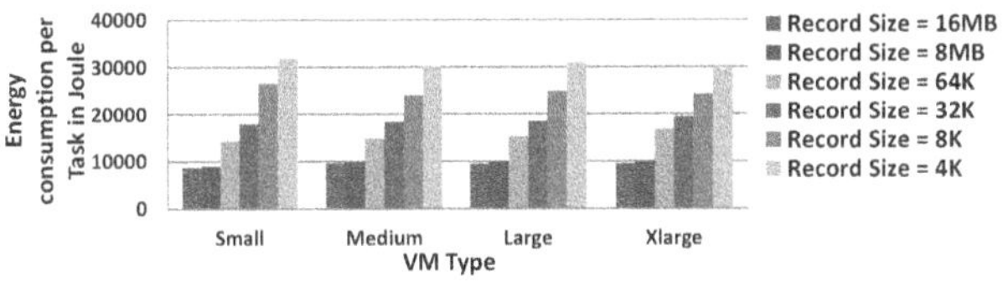

Figure 19. Energy Consumption per Task with Different

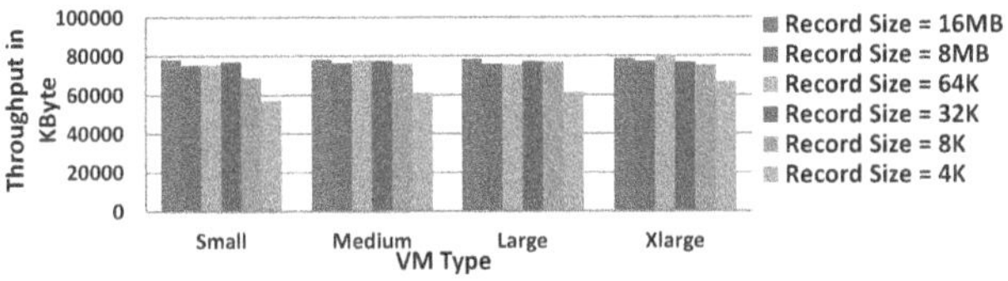

Figure 20. Throughput with Different VM.

5.3 Communication-intensive Tasks Results

A communication-intensive task requires a lot of network resources to transmit large amounts of data. Switches form the basis of the interconnection fabric of a private cloud network. Thus, switches, plus network cards, are the main energy consumers of the network resources. Traditionally, energy consumption of a switch depends on hardware parameters, such as the type of switch, the number of ports and the port transmission rates. However, the energy consumption may increase with the amount of data transferred over the network because of the data processing overhead. In addition, the total energy consumption might be impacted by network congestion because of the imbalance between the computation speed and the communication speed. The computation speed of the switch may be slower than the data transmission speed. We investigated this issue by applying different network workloads. A Cisco 2960 switch connected the cloud servers and client PCs. All parameters of the switch were set as default to avoid potential impact caused by switch configuration. We deployed the Web server and database server of JPetstore onto two different VMs to isolate the impact of different user requests.

In order to reveal the impact of the network packet size, the types of user requests and the system configuration on the energy consumption of communication-intensive tasks, we conducted three sets of tests in this series of experiments, described as follows:

Test Set 5.3.1: Keeping the total number of user requests each user session, the resources allocated and the VM configuration of the Web server constant, while changing the total number of user sessions and workload type. In this set of tests, each user session consisted of mixed browsing requests or shopping requests in different percentages. There are five types of mixed workload, as presented in Table 4. Browsing requests include checking home page, viewing catalogue, viewing products, searching products and so on. Shopping requests include checking out, updating shopping cart, filling order forms, ordering inquiry, and so on. For each type of the mixed workload, we increased the number of concurrent user sessions from 100 to 700 in steps of 100. The resulting system energy consumption and system throughput are shown in Figures 21 and 22.

Table 4. Mixed Workload

	Mixed 1	Mixed 2	Mixed 3	Mixed 4	Mixed 5
Browsing Request	100%	80%	50%	20%	0%
Shopping Request	0%	20%	50%	80%	100%

As the number of concurrent user sessions increased, the energy consumption increased but the throughput decreased when we fixed the type of VM as presented in Figures 21 and 22. This was due to the increase in the number of concurrent user sessions causing extra scheduling overhead as each user request was processed. However, the decrease in throughput was not proportional to the increase in energy consumption. For instance, when running Mixed 1 on an XLarge virtual machine and after we changed the number of concurrent user sessions from 600 to 700, the system energy consumption increased by 24.7% and the system throughput decreased by 3.7%. In addition, we noticed that Mixed 1 consumed the most energy while Mixed 5 consumed the least energy among all five workload types, despite the type of VM. Meanwhile, we observed that Mixed 1 resulted in the lowest throughput compared to the other four types of mixed workload. Since Mixed 1 contained the most browsing requests and Mixed 5 contained the most shopping requests, Mixed 1 had the most disk and memory access of all as opposed to Mixed 5. Thus, it took longer to complete requests of type Mixed 1 compared to the other four workload types. As a result, the energy efficiency of Mixed 1 was lower than the other four workload types.

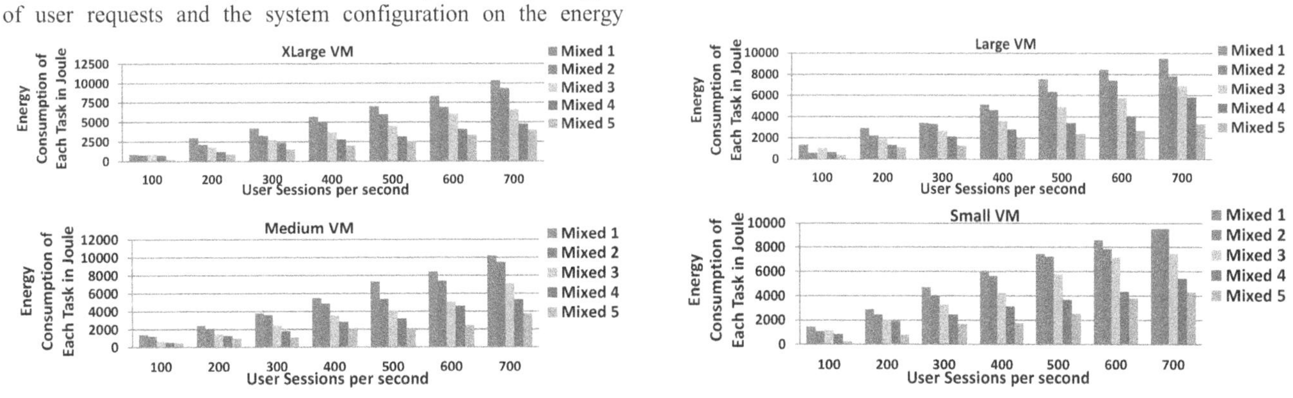

Figure 21. Energy Consumption per Task with Different VMs.

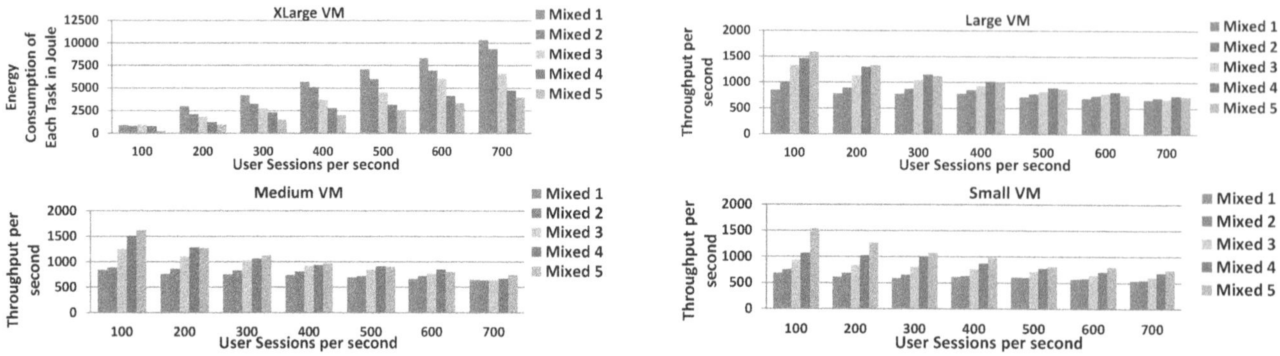

Figure 22. Throughput per Task with Different VMs.

Test Set 5.3.2: Keeping the total request number of each user session and the workload type constant, while changing the number of user sessions, the resources allocated and VM deployment on the Web server. The energy consumption and system throughput are presented in Figures 23 - 27. We observed that when the size of the VM deployed for the Web server increased from Small to Large, the system throughput increased while the energy consumption decreased in general. Intuitively, the more resources used the greater the energy consumption. However in this case, the smaller the instance the higher the disk accesses due to the thrashing of the cache, which leads to increase in energy consumption. Even for the Medium VM, which has two cores, the extra energy consumption by the additional core was cancelled out by the much larger added memory, reducing the number of accesses to the database server. Noticeably, when the size of the VM deployed for the Web server changed from Large to XLarge, the system throughput decreased and the system energy consumption increased in general.

When we set the type of the VM to Large, the total capacity of the Web server and the database server reached the full capacity of the physical server. However, when we set the size of the VM for the Web server at XLarge, the total capacity of the Web server and the database server exceeded the full capacity of the host. In this situation, extra overhead occurred because of the scheduling of VMs, which caused extra energy consumption.

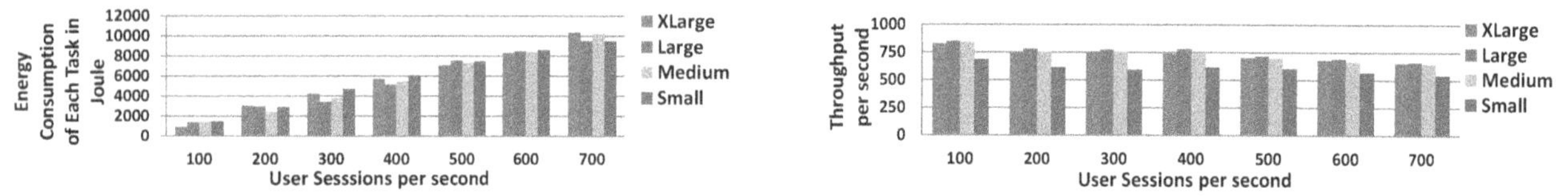

Figure 23. Energy Consumption and Throughput per Task with Mixed 1 Workload Type.

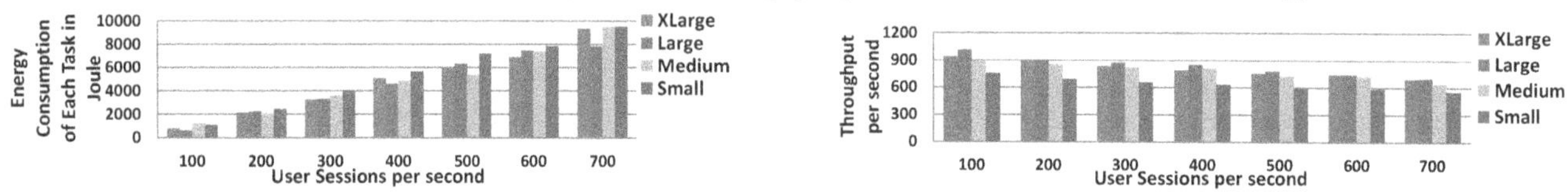

Figure 24. Energy Consumption and Throughput per Task with Mixed 2 Workload Type.

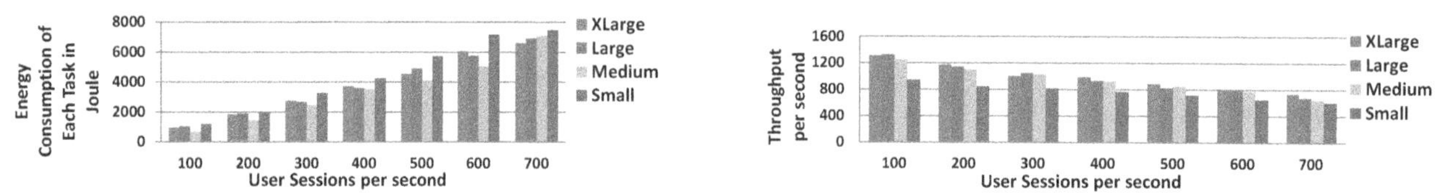

Figure 25. Energy Consumption and Throughput per Task with Mixed 3 Workload Type.

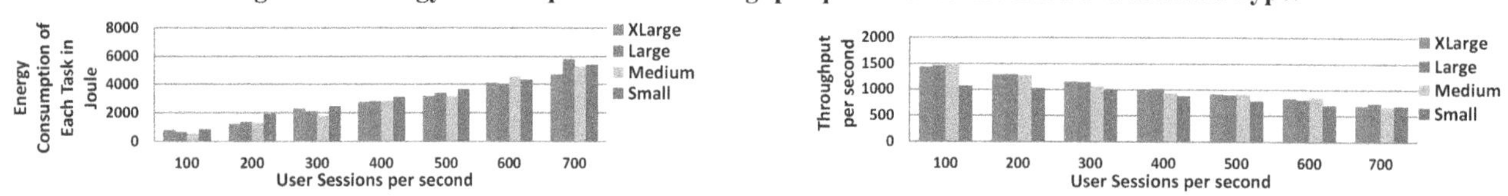

Figure 26. Energy Consumption and Throughput per Task with Mixed 4 Workload Type.

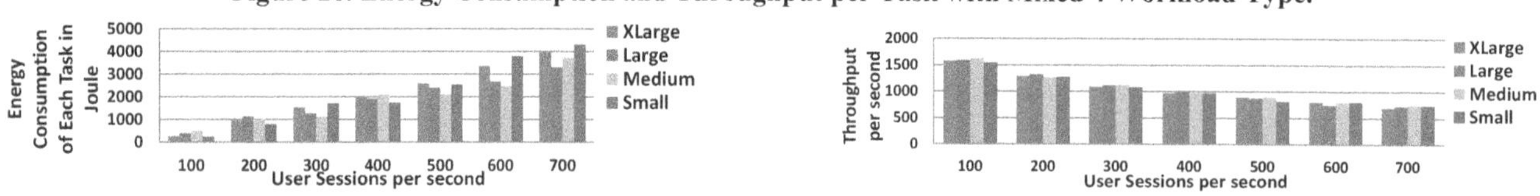

Figure 27. Energy Consumption and Throughput per Task with Mixed 5 Workload Type.

Test Set 5.3.3: Keeping the workload type, the number of user sessions constant, while changing the packet size, the resource allocation and the VM configuration. We fixed the number of concurrent user sessions of the workload at 300. First, we ran the test with only browsing requests and gradually increased the packet size of the browsing requests on different VM of different types. The packet size of each browsing request is listed in Table 5. These packet sizes represent the majority of all the browsing requests of JPetstore. The packet size of request "Browsing 1" is the smallest while packet size of request "Browsing 3" is the largest. The corresponding system energy consumption and system throughput are presented in Figures 28 and 29. As demonstrated, there was a slight increase in the energy consumption of each task when we increased the packet size of the requests. Furthermore, the throughput decreased as the packet size increased. Bigger packet size usually leads to more transmission time over the network and more processing time on both the servers and switches. Accordingly, throughput decreases and energy consumption increases for each communication-intensive task.

Table 5. Browsing Request Packet Size

	Browsing 1	Browsing 2	Browsing 3
Packet Size(KB)	3758	4190	4853

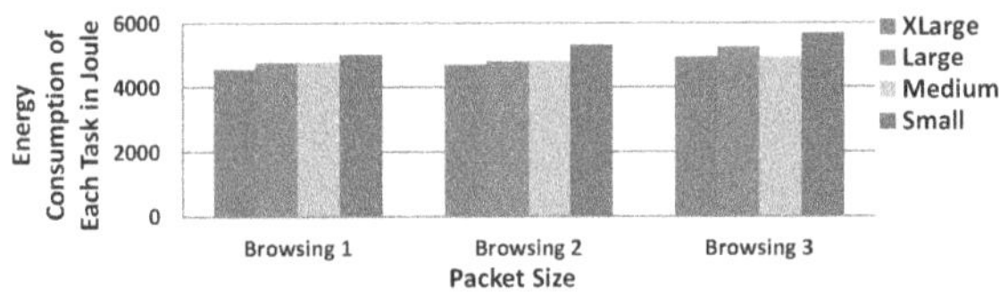

Figure 28. Energy Consumption per Task with Increased Packet Size.

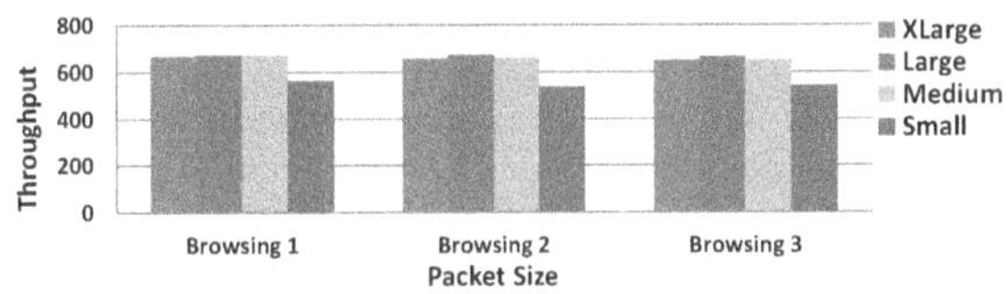

Figure 29. Throughput per Task with Increased Packet Size.

6. DISCUSSION

Based on our observations, we have derived a set of guidelines which can be adopted to achieve energy efficient resource provisioning and management for green cloud computing.

The more resources used by a single task, the more energy it consumes, and the better the system performance tends to be. However, based on the type of the runtime task, the declining throughput results show that overcommitted resource allocation would result in significant increase in energy consumption and decrease in overall system performance. For instance, energy consumption increased by 11% on average and the overall system performance decreased by 4% when the resource allocation exceeded the full capacity of the host during the tests of communication-intensive tasks. In cloud systems, the resource allocation should dynamically adapt to the customers' needs, taking both performance (and other) SLAs and energy efficiency needs into account. Dynamic scaling-up is needed when the peak workload is likely to exceed the capacity of the cloud system. Based on our results, it is worth finding the trade-off between the energy cost caused by the overcommitted resource allocation and the cost introduced by adding new resources in cloud systems.

The types and workload of runtime tasks impact energy consumption significantly. The energy consumption of each task is highly coupled with the resource utilisation in cloud systems. Thus, it is important to predict the required resources accurately based on the types and workloads of runtime tasks. For some applications, their workload is either known or can be empirically determined and is relatively constant. However, due to the dynamic nature of many cloud applications and the demand of different hosting platforms, the workload of different runtime tasks in cloud systems can drastically change over time. The need to find out the workload patterns for different runtime tasks in cloud systems, in order to schedule them for optimal performance and energy consumption, has emerged.

For a specific type of task, the various configuration parameters associated with the task, that is, the number of processes, the size of data to be processed, and the size of data to be transmitted, can greatly affect the task's energy consumption. These task parameters, that may originally come from application requirements, are closely linked to system configurations. Even with the same resource allocation, different system configurations can result in different energy consumption based on our observation. Therefore, dynamically changing cloud systems configurations is needed to adapt to different tasks based on their various configuration parameters.

Differing task types, task workload, task parameters, and cloud platform configurations can dramatically affect task throughput performance and energy consumption. As discussed, some of these effects are predictable, while others are counter-intuitive. Certain configurations give optimal balance of maximising workload vs minimising energy consumption. However, there may still be times that customers and/or cloud providers choose to prioritise one over the other, that is, to sacrifice performance to maximise energy efficiency further, or to sacrifice energy efficiency for improved performance.

7. CONCLUSIONS AND FUTURE WORK

Understanding cloud system energy consumption dynamics is valuable for developing efficient energy-aware resource management techniques for green cloud computing. We conducted a number of empirical experiments and profiled the energy consumption and performance with different types of runtime tasks on a controlled, representative cloud system. We treated a single task as a unit and measured the energy consumed by the task under various system configurations and task workload. The correlation of system throughput and energy consumption were analysed based on our experimental results. These results provide guidelines for developing energy management techniques for cloud systems that aim to reduce the energy consumption while achieving sufficient system performance to meet customer Service Level Agreements (SLAs).

Currently, we are determining the functional shape of our energy consumption model based on the experimental results analysis. Moreover, we are running experiments including large scale composite workload on larger cloud platform. We compare the energy consumption of individual task against the energy consumption of composite load predicted by our model. For computation-intensive tasks, we are selecting CPU-intensive and memory-intensive applications which inherently have single threaded and multi threaded algorithms instead of developing a new application. The experiment results will be together to help us improve our energy consumption model. In addition, we will integrate an energy cost rate and an "energy dirtiness rate" into our energy consumption model to factor in the costs – monetary and environmental – of cloud energy generated by different resources. This enhanced energy cost model will be investigated in order to minimise total energy costs while meeting system performance needs.

8. ACKNOWLEDGMENTS

We thank Professor Ryszard Kowalczyk for providing the facilities of Swinburne Energy Research Lab. This research is supported by the Australian Research Council under Discovery Project DP110101340.

9. REFERENCES

[1] Armbrust, M., Fox, A., Griffith, R., Joseph, A. D., Katz, R. H., Konwinski, A., Lee, G., Patterson, D., Rabkin, A., Stoica, I. and Zaharia, M. Above the clouds: a Berkeley view of cloud computing. Technical Report UCB/EECS-2009-28, UC Berkeley Reliable Adaptive Distributed Systems Laboratory, USA, 2009.

[2] Fanara, A., Haines, E. and Howard, A. The State of Energy and Performance Benchmarking for Enterprise Servers. In *Proceedings of the 1st TPC Technology Conference(TPCTC2009)*, pages 52-56, Lyon, France, 2009.

[3] Hamilto, J. Cooperative expendable micro-slice servers (CEMS): low cost, low power servers for internet-scale services. In *Proceedings of the 4th Biennial Conference on Innovative Data Systems Research(CIDR2009)*, pages 1-8, Asilomar, California, USA, 2009.

[4] Baliga, J., Ayre, R. W. A., Hinton, K. and Tucker, R. S. Green cloud computing: balancing energy in processing, storage, and transport. *Proceedings of the IEEE*, 99 (1): 149-167, 2011.

[5] Shang, L., Peh, L.-S. and Jha, N. K. Dynamic voltage scaling with links for power optimization of interconnection networks. In *Proceedings of the 9th International Symposium on High-Performance Computer Architecture(HPCA2003)*, pages 91-102, Anaheim, California, USA, 2003.

[6] Clark, C., Fraser, K., Hand, S., Hansen, J. G., Jul, E., Limpach, C., Pratt, I. and Warfield, A. Live migration of virtual machines. In *Proceedings of the 2nd Symposium on Networked Systems Design and Implementation(NSDI2005)*, pages 273-286, Boston, Massachusetts, USA, 2005.

[7] Raghavendra, R., Ranganathan, P., Talwar, V., Wang, Z. and Zhu, X. No "power" struggles: coordinated multi-level power management for the data center. In *Proceedings of the 13th International Conference on Architectural Support for Programming Languages and Operating Systems(ASPLOS2008)*, pages 48-59, Seattle, WA, USA, 2008.

[8] Zhang, Z. and Fu, S. Characterizing power and energy usage in cloud computing systems. In *Proceedings of the 3rd IEEE International Conference on Cloud Computing Technology and Science(CloudCom2011)*, pages 146-153, Athens, Greece, 2011.

[9] Chen, F., Schneider, J.-G., Yang, Y., Grundy, J. and He, Q. An energy consumption model and analysis tool for Cloud computing environments. In *Proceedings of the 1st International Workshop on Green and Sustainable Software(GREENS2012)*, pages 45-50, Zurich, Switzerland, 2012.

[10] Jung, G., Hiltunen, M. A. and Joshi, K. R. Mistral: dynamically managing power, performance, and adaptation cost in cloud infrastructures. In *Proceedings of the International Conference on Distributed Computing Systems(ICDCS2010)*, pages 62-73, Genova, Italy, 2010.

[11] Mach, W. and Schikuta, E. A consumer-provider cloud cost model considering variable cost. In *Proceedings of the 9th IEEE International Conference on Dependable, Autonomic and Secure Computing(DASC2011)*, pages 628-635, Sydney, Australia, 2011.

[12] Lee, Y. C. and Zomaya, A. Y. Energy efficient utilization of resources in cloud computing systems. *The Journal of Supercomputing*, 60 (2): 268-280, 2012.

[13] Liu, L., Wang, H., Liu, X., Jin, X., He, W., Wang, Q. and Chen, Y. GreenCloud: a new architecture for green data center. In *Proceedings of IEEE Conference on Autonomic Computing(ICAC2009)*, pages 29-38, Barcelona, Spain, 2009.

[14] Verma, A., Ahuja, P. and Neogi, A. Power-aware dynamic placement of hpc applications. In *Proceedings of the 22nd Annual International Conference on Supercomputing(ICS2008)*, pages 175-184, Island of Kos, Greece, 2008.

[15] Nathuji, R. and Schwan, K. VirtualPower: coordinated power management in virtualized enterprise systems. In *Proceedings of the 21st ACM Symposium on Operating Systems Principles(SOSP2007)*, pages 265-278, Stevenson, Washington, USA, 2007.

[16] Stoess, J., Lang, C. and Bellosa, F. Energy management for hypervisor-based virtual machines. In *Proceedings of the 2007 USENIX Annual Technical Conference(USENIX2007)*, pages 1-14, Santa Clara, CA, USA, 2007.

[17] Chen, Q., Grosso, P., van der Veldt, K., de Laat, C., Hofman, R. and Bal, H. Profiling energy consumption of VMs for green cloud computing. In *Proceedings of the 9th IEEE International Conference on Dependable, Autonomic and Secure Computing(DASC2011)*, pages 768-775, Sydney, Australia, 2011.

[18] Kansal, A., Zhao, F., Kothari, N. and Bhattacharya, A. A. Virtual machine power metering and provisioning. In *Proceedings of the 1st ACM Symposium on Cloud Computing(SoCC2010)*, pages 39-50, Indianapolis, Indiana, USA, 2010.

[19] Lefèvre, L. and Orgerie, A.-C. Designing and evaluating an energy efficient Cloud. *Journal of Supercomputing*, 51 (3): 352-373, 2010.

[20] Zhang, Q., Cheng, L. and Boutaba, R. Cloud computing: state-of-the-art and research challenges. *Journal of Internet Services and Applications*, 1 (1): 7-18, 2010.

[21] Koller, R., Verma, A. and Neogi, A. WattApp: An Application Aware Power Meter for Shared Data Centers. In *Proceedings of the 7th International Conference on Autonomic Computing(ICAC2010)*, pages 31-40, Reston, VA, USA, 2010.

[22] Schuster, A. Searching for processes and threads in Microsoft Windows memory dumps. *Digital Investigation*, 3 (1): 10-16, 2006.

Towards a Standard Event Processing Benchmark

Marcelo R. N. Mendes
CISUC, University of Coimbra
Dep. Eng. Informática – Pólo II
Coimbra, Portugal
mnunes@dei.uc.pt

Pedro Bizarro
CISUC, University of Coimbra
Dep. Eng. Informática – Pólo II
Coimbra, Portugal
bizarro@dei.uc.pt

Paulo Marques
CISUC, University of Coimbra
Dep. Eng. Informática – Pólo II
Coimbra, Portugal
pmarques@dei.uc.pt

ABSTRACT
There has been an increasing interest both in academia and industry for systematic methods for evaluating the performance and scalability of event processing systems. A number of performance results have been disclosed over the last years, but there is still a lack of standardized benchmarks that allow an objective comparison of the different systems. In this paper, we present our work in progress: the BiCEP benchmark suite, a set of workloads, datasets and tools for evaluating different performance aspects of event processing platforms. In particular, we introduce *Pairs*, the first of the BiCEP benchmarks, aimed at assessing the ability of CEP engines in processing progressively larger volumes of events and simultaneous queries while providing quick answers.

Categories and Subject Descriptors
C.4 [Performance of Systems]: Measurement techniques

General Terms
Design, Experimentation, Measurement, Performance.

Keywords
Benchmark, CEP, Event Processing, Performance, Scalability.

1. INTRODUCTION
Even-driven applications are becoming increasingly prevalent in the most diverse domains of industry, such as capital markets, telecom, healthcare, sensor networks, and many others [5]. Most of these applications, however, are mission-critical or performance-sensitive, generating huge amounts of data and requiring very short response times. In order to guarantee that these applications have their requirements met, systematic methods for evaluating the performance of their enabling technologies are required. Aware of this need, SPEC has released the first industry standard benchmark for evaluating message-oriented middlewares (MOM) – SPECjms2007 [7].

MOM platforms play a fundamental role in event-driven applications, by ensuring that events are reliably disseminated to the appropriate destinations. However, exchanging messages that carry the information about events occurrence is only one of the features required by event-driven applications. After message delivery, they still need to filter the incoming events, aggregate their data, correlate seemingly unrelated events, detect situations of interest and react to them. These operations are typically performed with the aid of another class of systems, the so-called *Complex Event Processing* (CEP) engines, for which there is currently no standard benchmarks. In fact, a great part of the known performance numbers come from tests and studies conceived or sponsored by vendors (e.g., [8], [9]), many of which disclosed without the necessary details for replicating the results. A few others (e.g., [3], [6]), though neutral and detailed, used simple workloads and datasets that do not fully represent the typical usage of CEP engines. In order to address this lack of systematic evaluation methods, novel benchmarks are required. These benchmarks need to be clearly specified and easily understood. More importantly, they must exercise the entire spectrum of features offered by event processing platforms in a realistic manner.

In this paper we propose a first step towards filling this gap. We introduce our work-in-progress, the BiCEP benchmark suite, and the first of its domain-specific benchmarks: *Pairs*. The *Pairs* benchmark is set on the capital markets environment and exercises a wide range of features commonly found in most event processing applications, including:

- Filtering, aggregation, and correlation of events;
- Detection of event patterns and trends;
- State maintenance;
- Large number of simultaneous queries (increasing with the system scale);
- Changing load conditions.

Pairs was designed to assess the ability of the CEP systems in processing increasingly larger number of continuous queries and event arrival rates while providing quick answers – three quality attributes equally important for an event processing engine. The benchmark was also designed to be fully customizable, so that users can carry out performance studies that resemble more closely their own environments.

2. DESIGN PRINCIPLES
Even though most event processing applications share some common characteristics (e.g., the need for automated and timely answers), the several CEP domains differ significantly in their functional and performance requirements. For example, users in the capital markets are generally very concerned about processing latency, as short response times represent competitive advantage. Thus, sub-millisecond latencies are typically expected

ICPE'13, April 21–24, 2013, Prague, Czech Republic.

in the algorithmic trading domain and the shorter the system takes to react, the better. However, other applications, like fraud detection and traffic monitoring, are generally not so latency-sensitive, and tolerate response times in the order of few seconds or even minutes. Performance concerns in these cases might revolve around volume of data or queries state size.

These significant differences in the performance requirements of the application domains makes virtually impossible for a single benchmark, with a single metric, to be representative of the entire spectrum of applications and provide all the information required by its heterogeneous target audience. For this reason, we opted to devise BiCEP as a set of smaller, domain-specific benchmarks, each with its own workload, dataset and metrics. Our goal is that each benchmark will allow evaluating one or more aspects of event processing systems (e.g. latency, scalability with respect to number of queries and rules, storage efficiency, etc.). Besides measuring CEP engines in multiple ways, a benchmark suite has a number of other advantages:

- *Extensibility*: as more information about CEP use-cases become available, new tests can be added to the suite, making it more comprehensive.
- *Understandability*: users can more easily relate the individual domain-specific benchmarks to their real applications.
- *Configurability*: the focus of the evaluation can be controlled by selecting whether all tests will be executed or only a subset of them will be considered.
- *Fairness/Portability*: CEP engines differ considerably in their focus, capabilities and query languages. Having multiple tests allows a system that does not perform well on a given benchmark (or is not able to implement it at all), to showcase its capabilities on a different scenario.

The BiCEP benchmark suite is the ongoing result of years of research and analysis of real event processing use-cases and platforms. Throughout the rest of this paper we outline its first benchmark, *Pairs*, and discuss our plans for extending the suite.

3. THE *Pairs* BENCHMARK

In this section we provide a brief overview of the *Pairs* benchmark. Note that due to space limitations, only the main aspects of the benchmark are discussed. For a complete specification please refer to [2].

3.1 Application Scenario

The scenario for *Pairs* is an investment firm where a number of analysts interact with an enterprise trading system responsible for automating and optimizing the execution of orders in stock markets. Users of the system pose trading strategies which are continuously matched against live stock market data. The exercised trading strategies belong to a category broadly known in the financial domain as *statistical arbitrage* and consist in monitoring the prices of two historically correlated securities, looking for temporary digressions that indicate an opportunity to capitalize on market inefficiencies.

The general structure of the benchmark scenario, including the main entities and the corresponding cardinalities, is depicted in Figure 1. Per every stock market M, a number of symbols (100) are monitored by the system, from which half are known to be mutually correlated. Each of the users of the system manages exactly five strategies. The number of users per market ranges from five up to fifty, depending on the scale factor. In the simplest case (5 users), there will be 25 strategies, each defined over a unique pair of correlated symbols. On the limit, each pair of correlated securities is monitored by ten strategies of different users, each with its own parameters.

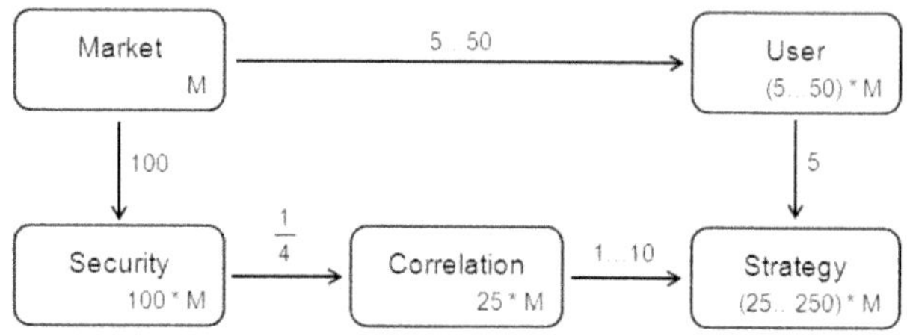

Figure 1: Entities and relationships in the *Pairs* benchmark

This structure allows evaluating not only if the tested system performs well on a multi-query scenario, but also its ability in sharing resources among similar queries.

3.2 Input Data

Input data for the *Pairs* benchmark consists in a stream of simulated stock market data with the following schema:

(*symbol*: **string**, *price*: **int**, *size*: **int**, *tickTS*: **long**, *TS*: **long**)

Each incoming tuple represents a trade operation executed in the stock market, such that *symbol* identifies the security being traded, *price* indicates the value in cents of the transaction, *size* represents the amount of shares negotiated, *tickTS* is the time, in milliseconds, at which the trade has been executed (i.e., simulation clock time) and *TS* is the actual time the record was sent to the CEP engine (i.e., wall clock time, added by the benchmark driver).

In the standard configuration, 2 hours of simulated market data are produced by a data generator application and submitted afterwards to the *system under test* (SUT). Tick arrivals follow a Poisson process [1], with its λ parameter – which represents the average arrival rate – varying over time. The reason for having a varying input rate is to simulate more realistically what happens in most real event processing applications, where new data arrives at different rates depending on the period of the day.

3.3 Workload

The benchmark workload consists in processing simultaneously a number of *Pairs* strategies. All strategies perform the same set of operations, described below, although using different parameters:

1. *Compute indicators*: the prices of a pair of securities are aggregated over a given time interval and then correlated to produce a *ratio*. The values of this ratio are then aggregated again, producing a moving average and upper and lower bands (these are usually referred as "Bollinger Bands").
2. *Signal opportunities*: the indicators produced in the previous step are used to determine possible opportunities to capitalize. This happens when the line formed by the values of the ratio crosses one of the bands.
3. *Position*: once a possible opportunity has been spotted, the system checks if it must change its current market position.

4. *Place orders*: if a change in market positioning is indeed required, the system must emit orders. This step involves identifying the appropriate values for the parameters of each order (i.e., size and price).
5. *Manage risk*: once a market position has been assumed, the system must detect if the prices keep drifting apart, countering the expected reversal trend, to prevent losses.

All the operations above are performed by each running strategy. Ideally the systems under test will be able to identify similarities among them and share resources during strategies execution.

3.4 Output

The output of the *Pairs* benchmark consists in two event streams: *Indicator* and *MarketOrder*. The former represents the output of the first step in the strategy execution process and is used in the benchmark scenario for visualization and auditing purposes (*the stream serves to produce a chart like Figure 2 that allows users to better understand the decisions taken by each strategy*).

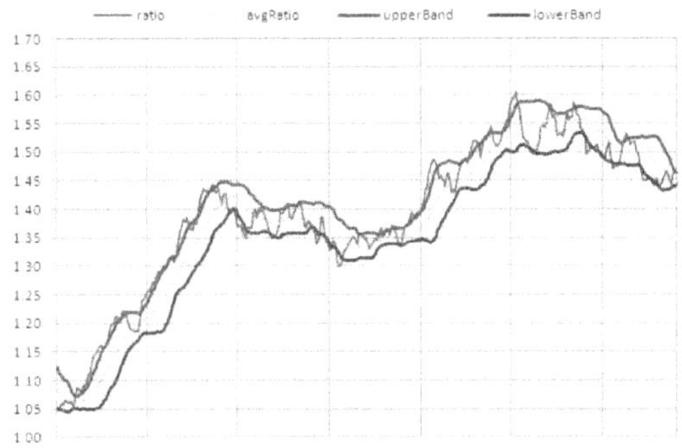

Figure 2: Chart showing the values of the *Indicator* stream

The second stream represents the orders that were issued as a result of the execution of the strategies. The schemas of the two output streams are shown below:

Indicator (		***MarketOrder*** (	
strategy	: **string**,	*strategy*	: **string**,
ratio	: **double**,	*type*	: **string**,
avgRatio	: **double**,	*symbol*	: **string**,
upperBand	: **double**,	*price*	: **int**,
lowerBand	: **double**,	*size*	: **int**,
inputTickTS	: **long**,	*inputTickTS*	: **long**,
inputTS	: **long**	*inputTS*	: **long**
)		)	

Tuples of the *Indicator* stream consist in a field *strategy*, indicating which strategy generated the result, and the fields *ratio*, *avgRatio*, *upperBand* and *lowerBand*, containing the values of the indicators. The *MarketOrder* stream consists in the fields *strategy*, again identifying the strategy that triggered the output, *type*, identifying the order as '*BUY*' or '*SELL*', and the fields *symbol*, *price* and *size*, which have the same meaning as in the input stream *StockTick*. Besides the payload, tuples from both streams include two timestamps: *inputTickTS* and *inputTS*. Both are derived from the input event that triggered the emission of the output tuple and represent respectively the tick occurrence time (simulation clock) and its arrival time (wall clock). The former is used for checking the correctness of the results while the latter is used for response time computation purposes.

3.5 Scaling

The scale factor (SF) in *Pairs* affects the number of users, and consequently the number of strategies executed in parallel as follows:

- Number of users: 5 SF
- Total number of strategies: 25 SF

Additionally, per every increment of ten in the scale factor, the input rate is incremented by 5,000 and the number of symbols is increased by 100 (this is to avoid too many similar strategies over the same symbols and to allow to observe how the system scales with changes in input rate and cardinality). The effect is as if a whole new market were now being monitored by a new team of analysts. For instance, for a scale factor of 15, there will be 75 users, each managing 5 strategies, on a total of 375 strategies running in parallel on the trading system, from which 250 are over the first set of 100 symbols and 125 are over the second set of 100 symbols.

3.6 Measures

As mentioned earlier, the purpose of the *Pairs* benchmark is to evaluate the ability of CEP systems in *processing increasing loads* while *providing quick answers*. Naturally, different users have different perceptions on the value of each dimension depending on their requirements (e.g. for some, the best system is simply the one that replies faster, while for others it is the one that handles more load). Nonetheless, in order to facilitate comparison among the several platforms and benchmark runs, we have defined a p_{score} metric to represent overall system performance[1]:

$$p_{score} = \frac{load}{99^{th}\, latency}$$

When defining the metric above, we tried to benefit systems that are able not only to process high volumes of events, but also react quickly and scale well with respect to the number of concurrent queries. Note, though, that the p_{score} exists essentially for comparison purposes, and that a *Pairs* report should always include a number of other measures (e.g., throughput, average and maximum latency, latency histogram, etc.) to help users to better understand the performance of the system under test and judge whether it fits their needs or not.

3.7 Is *Pairs* a Good Workload Scenario?

There are a number of reasons why we believe the *Pairs* benchmark represents a good test case for CEP platforms. First, the workload exercises several common features that appear repeatedly in most event processing applications: it *filters* out ticks from securities which are not of interest, *aggregates* events data over *temporal* and *count-based windows*, *correlates* price data for interrelated securities, *detects patterns* from price movements, keeps track and updates strategies' *state* upon the occurrence of certain events, and performs *lookups* to determine orders price and size. In addition, different from most benchmarks, which have a fixed set of queries, the number of queries in *Pairs* increases with the system size. This is in

[1] The term "*load*" in the formula is a function of the number of strategies and the input rate. Further information on how the p_{score} is computed can be found in the *Pairs* specification [2].

conformance with what happens in many real event processing applications and allows evaluating important aspects like query scalability and resource sharing.

Other key benefits of *Pairs* are understandability and representativeness. The benchmark mimics a niche of application where event processing platforms have perhaps been most successful – capital markets. In fact, most products use simple financial use-cases to exemplify the usage of their features and languages in their documentation, so in principle it should be easy for anyone reasonably familiar with CEP to understand *Pairs*. In addition, *Pairs* is loosely based on a real use-case, and as such has a good chance to be representative of its application domain.

Finally, *Pairs* allows a great deal of customization. Users can control load intensity by setting high-level workload parameters like input rate and number of simultaneous strategies, or by altering scenario characteristics such as number of securities and configuration of the strategies. While the results obtained from these "customized" runs cannot be compared to standard runs, the ability to customize the workload enables users to exercise the systems in a manner closer to their own real environment.

3.8 Implementation and Preliminary Results

The *Pairs* benchmark should be implemented and executed as illustrated in Figure 3 below:

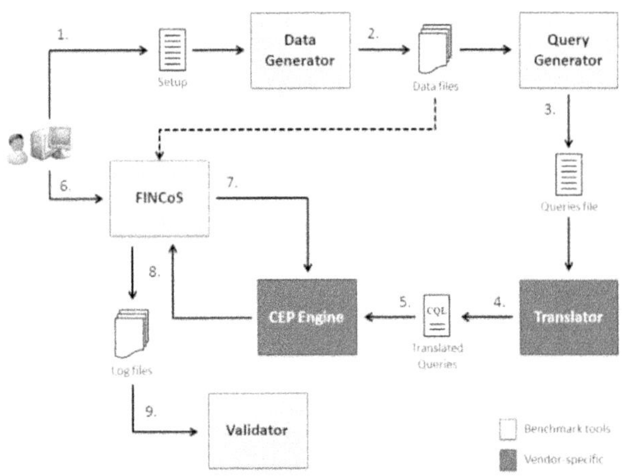

Figure 3 The execution flow of the *Pairs* benchmark

Initially, the user specifies a couple of workload parameters or, alternatively, uses the standard benchmark configuration to create a test setup (1). Then, a *data generator* application generates data and auxiliary files (2), which are used afterwards by a *query generator* to produce the strategies that compose the benchmark workload (3). The output of the query generator is then parsed by a vendor-specific *translator*, which converts the workload, initially represented in a neutral format (e.g., xml file), into the query language used by the SUT (4). After loading the query/rule set into the SUT (5), the user starts a performance run (6). During the run, the benchmark driver (*FINCoS*) loads the generated data file and submits the events on it to the SUT (7), which in turn returns the corresponding results to the framework (8). After test completion, a *validator* verifies the correctness of the answers produced by the SUT (9).

All the aforementioned tools are written in Java and require very little effort to be executed. The *data generator*, *query generator* and *validator* applications are specific to the *Pairs* benchmark and can be downloaded from [2]. The *FINCoS framework*, on the other hand, is benchmark-independent and can be found at [4].

We have recently implemented *Pairs* in a widely used open-source CEP engine and run some preliminary performance tests. In our experiments, the SUT managed to reach a maximum scale factor of 3, with processing latencies ranging from 5 up to 2,810 milliseconds (average: 33 ms; 99th-perc.: 106 ms), obtaining a p_{score} of 0.085. The tests also revealed some interesting facts about the system performance, like a regular increase of response time when faced with larger load levels and an odd variation on CPU utilization across the measurement interval. Further experiments and analysis are still required to fully understand these results, though. At the present moment, we are implementing the benchmark on another CEP engine to establish a basis for comparison.

4. SUMMARY

So far very little information about the performance of CEP engines has been made available, in spite of those systems being used in mission-critical or performance-sensitive scenarios. In this paper we delineate the path for filling this gap by introducing the BiCEP suite, and the first of its domain-specific benchmarks – *Pairs*. Both BiCEP and *Pairs* are work in progress, which we intend to further develop over the next months. In particular, we plan to extend the suite with new benchmarks as well as enhance *Pairs* and its tools. We promptly invite the community to contribute to this discussion.

5. ACKNOWLEDGMENTS

This research has been supported in part by the Portuguese Science and Technology Foundation (FCT), under grant N° 45121/2008.

6. REFERENCES

[1] Aldridge, Irene, High-frequency trading: a practical guide to algorithmic strategies and trading. Wiley trading series. ISBN 978-0-470-56376-2.

[2] BiCEP project web site: http://bicep.dei.uc.pt

[3] Dekkers, P. Master Thesis Computer Science. Complex Event Processing. Radboud University Nijmegen, Thesis number 574, October 2007.

[4] FINCoS Framework: https://code.google.com/p/fincos/

[5] Hinze, A., Sachs, K., Buchmann, A.P. Event-based applications and enabling technologies. In Proceedings of DEBS 2009.

[6] Mendes, M.R.N., Bizarro, P., Marques, P. 2009. A Performance Study of Event Processing Systems. In Proceedings of the 1st TPC Technology Conference (Lyon, France, August 2009), 221-236.

[7] SPECjsm2007 benchmark: http://www.spec.org/jms2007/

[8] STAC Report: Aleri Order Book Consolidation on Intel Tigertown and Solaris 10. Available at: http://www.stacresearch.com/node/3844

[9] White, S., Alves, A., Rorke, D. 2008. WebLogic event server: a lightweight, modular application server for event processing. In Proceedings of DEBS 2008, 193-200.

Towards a Methodology Driven by Relationships of Quality Attributes for QoS-based Analysis*

Steffen Becker
University of Paderborn
Paderborn, Germany
steffen.becker@upb.de

Lucia Happe (Kapova)
Karlsruhe Insitute of Technology
Karlsruhe, Germany
kapova@ipd.uka.de

Raffaela Mirandola
Politecnico di Milano
Milano, Italy
mirandola@elet.polimi.it

Catia Trubiani
University of L'Aquila
L'Aquila, Italy
catia.trubiani@univaq.it

ABSTRACT

Engineering high quality software is a tough task. In order to know whether a certain quality attribute has been achieved or degraded, it has to be quantified by analysis or measured. However, determining what to quantify and how these quantities are related to each other is the difficult part. Early analysis of the quality attributes of a software system on the basis of the system's planned architecture allows informed decisions on design trade-offs. Such decisions can be later validated by measurements on the running system.

In this paper, we revisit software quality attributes. In particular, we introduce a generic taxonomy of quality attributes, the relationship between the attributes is argued, and finally we devise future work leading to an attribute-based methodology for evaluating software architectures. The goal is reasoning about multiple quality attributes of software systems to achieve the ability to quantitatively evaluate and trade-off them.

Categories and Subject Descriptors

D.2.8 [**Software Engineering**]: Metrics—*complexity measures, performance measures*

Keywords

QoS-based analysis, quality attributes, relationships.

1. INTRODUCTION

The quality attributes of a software system and their relationships must be carefully understood early in the development process, thus the architect can design an accurate architecture that satisfies them. However, dealing with quality attributes is not a straightforward task. They are usually complex with several relationships and interactions among them, therefore, it is even difficult to make a systematic list of the concerns to be managed. Additionally, some quality attributes have major implications on the core functionality of the system applications and priorities can be useful to roughly fulfill the more relevant end user expectations.

The definition of quality attributes is the first step for defining and enforcing system quality, and the more difficult step is to associate the quality attributes with each other, thus to model their relationships.

In this work, we introduce a generic taxonomy of quality attributes and define relationships between the specified attributes, thus providing easy accessible, early guidance for software architects. The ultimate goal of our work is to enable reasoning about multiple software quality attributes. This would allow to quantitatively evaluate and trade-off quality attributes on the basis of pre-defined relationships.

The paper is organized as follows. Section 2 presents the contribution of this paper, i.e. a graph of relationships to deal with the quality attributes of software architectures and their relationships. Section 3 discusses the benefits of using the graph of relationships as a support to reason and analyze different and conflicting design decisions. In Section 4 we briefly survey the state-of-the-art on the field, and Section 5 concludes the paper and outlines future research directions.

2. GRAPH OF RELATIONSHIPS

Figure 1 illustrates our current proposal of the graph of relationships for quality attributes: nodes (see Section 2.1) represent the quality attributes that we consider, such as *performance*, *reliability*, etc., whereas arcs (see Section 2.2) represent their underlying relationships, such as *trade-off*, *include*, etc.

2.1 Definition of nodes

In this section we provide a brief overview of the quality attributes our graph considers. For sake of space we do not provide a complete description of each quality attribute since it is out of interest of this paper and can be easily retrieved from the international standard for the evaluation of software quality, *ISO/IEC 25010* [1].

*This work has been partially supported by VISION ERC project (ERC-240555).

ICPE'13, April 21–24, 2013, Prague, Czech Republic.

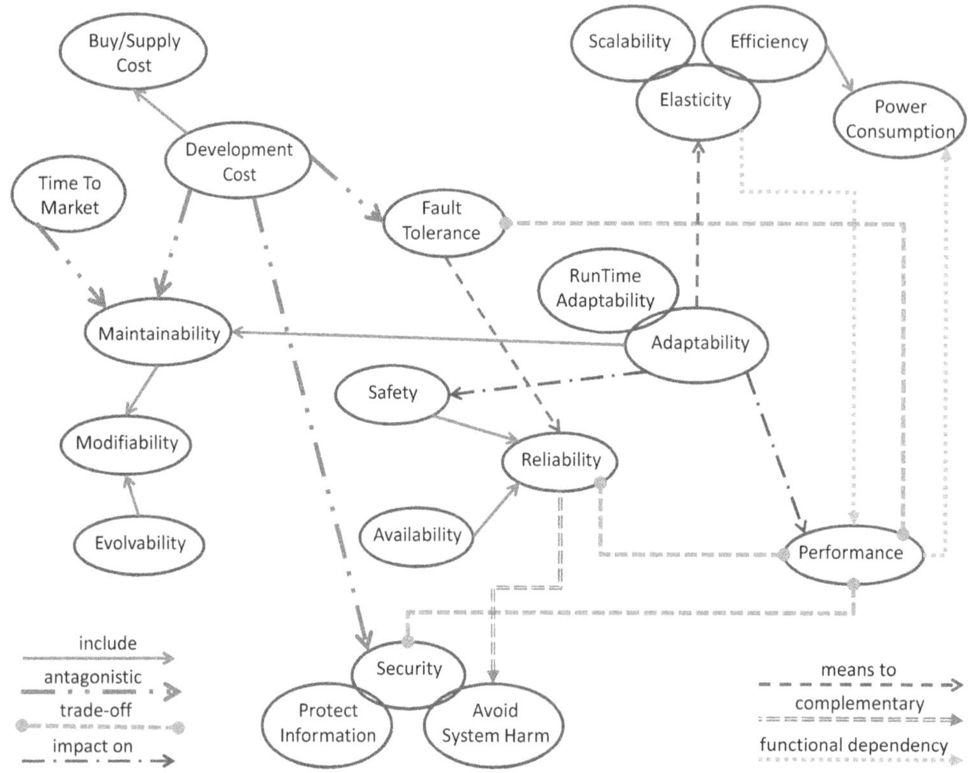

Figure 1: Graph of relationships between quality attributes.

Our graph does not claim to be exhaustive, however, we think it represents a good approximation to conduct a QoS-based analysis. The most common quality attributes are: *performance*, *reliability*, *fault tolerance*, *safety*, *availability*, and *power consumption*. We additionally select other quality attributes that aim at measuring the capacity: (i) to adapt to system changes (*maintainability*, *modifiability*, *evolvability*, *adaptability*, and *run time adaptability*); (ii) to comply with the market constraints (*time to market*, *development costs*, *buy/supply costs*); (iii) to cope with an increasing and/or decreasing number of requests (*scalability*, *elasticity*, *efficiency*); (iv) to protect the system against malicious users (*security*, *protect information*, *avoid system harm*).

2.2 Definition of arcs

In this section we provide a brief overview of the relationships between the quality attributes our graph considers:

include: a quality attribute QA_x includes a quality attribute QA_y if the latter attribute is part of the former one. For example, the *Development Cost includes Buy/Supply Cost* since the latter attribute (i.e. buy software by third party organizations, supply software by implementing system functionalities) is part of the *Development Cost* attribute whose meaning is broader since it contains additional costs (i.e. integration costs, testing costs, etc.).

antagonistic: a quality attribute QA_x is antagonistic with a quality attribute QA_y whether QA_y is negatively influenced by QA_x. For example, the *Time To Market* is *antagonistic* with the *Maintainability* since the property of maintaining a software system is negatively affected by the time it needs to be delivered.

trade-off: it is a bidirectional relationship and it means that two quality attributes (e.g. QA_x and QA_y) depend on each other, in particular the raising of one quality attribute positively or negatively affects the other one. For example, the *Security* and the *Performance* are in a *trade-off* relationship since raising the level of security inevitably affects the performance of the software system.

impact on: a quality attribute QA_x *impacts on* a quality attribute QA_y if the latter attribute is affected by the former one. For example, *Adaptability impacts Performance* since the capacity of the system to adapt on the basis of changes implies that the performance of the system is affected by the actual modifications.

means to: a quality attribute QA_x is a *mean to* a quality attribute QA_y if the latter attribute is achieved through the former one. For example, *Fault Tolerance* is a *mean to Reliability* since fault tolerance mechanisms are aimed at strengthening the system reliability. Several fault tolerance mechanisms can be devised to this scope such as redundancy (i.e. providing multiple instances of the same system to switch in case of failure) or diversity (i.e. providing multiple implementations of the same specification to reduce errors).

complementary: a quality attribute QA_x is *complementary* to a quality attribute QA_y if the latter attribute is indirectly supported by the former one, even if the quality attributes apparently do not seem to be related each other. For example, *Reliability* is *complementary* to *Avoid System Harm* since reliability mechanisms support the avoidance of malicious users by decreasing their probability of intrusion.

functional dependency: a quality attribute QA_x has a *functional dependency* with an attribute QA_y if the latter attribute is used from a functional point of view to cope with the former one. For example, *Performance* has a *functional dependency* to *Power Consumption* since the use of more computing power also requires more electrical power.

3. HOW TO USE THE GRAPH

In this section we discuss the benefits of using the graph of relationships for QoS-based evaluation of software architectures. We demonstrate that the graph can be used as a reasoning (see Section 3.1) and analysis (see Section 3.2) support to guide design decisions of software architects.

3.1 Reasoning on selected quality attribute relationships

In this section we explain the adopted criteria to define the relationships between quality attributes, using our previous experience gained in the field.

Performance vs Security: ***trade-off.*** The critical aspect that we found between performance and security is the need to quantify the performance degradation incurred to achieve the security requirements. From our previous work [8, 9, 17] we experimented that the solution of a performance model that embeds security aspects allows to quantify the *trade-off* between security and performance in software architectures. In particular, the values of indices coming from the solution of the performance model (i.e. the one that includes security aspects) can be compared to the ones obtained for the same model (i) without security solutions, (ii) with different security mechanisms and (iii) with different implementations of the same security mechanism. Such comparisons help software architects to decide whether it is feasible to introduce/modify/remove security strategies on the basis of (possibly new) performance requirements.

Performance vs Power Consumption: ***functional dependency.*** The *green computing* discipline addresses the problem of building and managing computing infrastructures that provide the same quality of service with lower energy consumption [16]. Rudimentary techniques for power management, e.g., shutting down idle servers, can impact the ability of the hosting center to meet the Service Level Agreements (SLAs) implicitly or explicitly stipulated with clients. Shutting down servers ensures the maximum power saving; however, bringing up a machine when needed can require a significant start-up time (up to several minutes, depending also on the time needed to start the applications). Start-up delays can severely impact the ability of the system to promptly handle workload fluctuations. Besides, repeated on-off cycles can stress hardware components, increasing the probability of failures, which add further costs for repair or replacement of broken devices.

A different approach suggests to reduce the power consumption by making the system *power proportional* [4], meaning that the power consumption of devices is kept proportional to their utilization. Ideally, a device should consume zero power when its utilization is zero, and full power when its utilization is one. Unfortunately, such "perfect" power proportionality has not been achieved yet.

Fault-Tolerance vs Reliability: ***means to.*** Reliability can be defined as the probability that a given component or system will perform its required functions without failure, for a given period of time, in a specified environment. When a reliability problem is encountered it means that the service delivered deviates from the correct service, and this is called *service failure.* An *error* is the part of the system state that leads to the occurrence of a failure, and is caused by a *fault.* Therefore, if the system includes mechanisms able to detect and repair errors and faults, the failure probability decreases [2, 18, 14, 15].

There are two well-known strategies for software fault tolerance: error processing and fault treatment [2, 18]. *Error processing* aims to remove errors from the software state and can be implemented by substituting an error-free state in place of the erroneous state, called error recovery, or by compensating for the error by providing redundancy, called error compensation. The second strategy, fault treatment, aims to prevent activation of faults and so action is taken before the error occurs. The two steps in this strategy are fault diagnosis and fault passivation. The nature of faults which typically occur in software has to be thoroughly understood in order to apply these strategies effectively. Techniques for tolerating faults in software have been divided into three classes - design diversity, data diversity, and environment diversity (see [2, 18, 15] for details).

Specifically, these techniques can be integrated in a rule-based approach like proposed in [15]. Once the source of the failure has been identified, through a *sensitivity* analysis for example, a set of rules for the mitigation of the reliability problem, based on fault-tolerance techniques has been defined.

3.2 Further analysis

In this section we discuss the main fields that may benefit from our graph of relationships as a more sophisticated instrument to quickly highlight the quality attributes relationships.

Requirements Analysis. The analysis of the requirements is fundamental since they are aimed at determining the goals, the functions, and the constraints of the software systems. Additionally, multiple stakeholders may define a different priority for the requirements, hence the quality attributes can be associated to stakeholders' utilities/weights. Such analysis can be supported by our graph, in fact it is possible to determine if system requirements are unfeasible depending on the stated relationships between the quality attributes, and the challenge becomes to slightly modify them in order to find the better feasible solution.

Attribute-based methodology for evaluating software architectures. The QoS-based evaluation of software architectures can be performed by looking at their required quality attributes and associating appropriate architectural styles [13], since they represent engineering artifacts defining classes of designs along with their known properties. Architectural styles can be explicitly associated to our graph of relationships that is meant to give suggestions about the most appropriate styles.

Service Level Agreement (SLA) negotiation. A service-level agreement is a part of a service contract where the level of service is negotiated between two parties, i.e. the customer and the service provider. We are interested in SLAs related to the QoS properties of the software systems. For example, these SLAs have SLOs such as the response time, throughput, or utilization for performance, or mean time between failures, mean time to repair for reliability. Our graph of relationships may drive the SLA negotiation by suggesting to the customers the feasible agreements and the ones that may not be fulfilled due to the defined relationships while analyzing the different quality attributes.

4. RELATED WORK

In recent decades, several efforts have been made in the literature for dealing with quantifying software qualities at an appropriate abstraction level. To this end, often software architecture has emerged as a good starting point [7, 19] and several results have been obtained, concerning the definition of methods and tools able to evaluate quality at the software architecture level (see, for example, [3, 10, 19]).

In general, there are *single attribute approaches* that aim at evaluating one quality attribute at time. On the contrary, similarly to the analytic hierarchy process (AHP), there are *multiple attributes approaches* that consider more than quality attribute and mostly aim at comparing the attributes while minimizing and/or maximizing them while looking at optimization techniques (e.g. the Pareto front) that compare two quality attributes at time.

More common today are still methods which address single quality attribute (e.g., performance or availability). However, a major challenge in system development is finding the best balance between different (possibly) conflicting quality requirements that a system has to meet (e.g., maximize performance, maximize availability and minimize cost).

There are multi-attribute approaches where the definition of a software architecture model can embody not only the software qualities of the resulting system, but also the trade-offs decisions taken by designers [5, 6, 20]. Efforts to explore such trade-offs have produced the so-called *scenario-based* analysis methods, such as SAAM and ATAM [11, 12] and others reviewed in [10]. However, these methods provide qualitative results and are mainly based on the experience and the skill of designers and on the collaboration with different stakeholders.

With respect to the state of the art, in this paper we propose a high-level, easy to use, simultaneous analysis of multiple quality attributes with the aim of pointing out their relationships, thus to anticipate negative and/or positive consequences between them. In this way, the software architect has a deeper knowledge of the system under study and architectural decisions can be driven by our graph of relationships.

5. CONCLUSION

This paper presents work in progress towards a systematic approach to do quantitative trade-offs of a whole bunch of quantifiable quality attributes. As a first step, a graph of relationships highlighting the dependencies among quality attributes is presented. This early graph already helps requirements engineers and software architects to check, that not quality attribute has been forgotten. Additionally, trade-offs among conflicting attributes can be identified easily.

In future work, the graph needs to be tested and extended according to experiences. Ideally, quantitative analysis methods are extended to take these relationships into account in order to find systematic and rational trade-offs.

6. REFERENCES

[1] ISO/IEC 25010 - Systems and software Quality Requirements and Evaluation (SQuaRE) - System and software quality models, 2011.

[2] A. Avizienis. Fault-tolerant computing-progress, problems and prospects. In *IFIP Congress*, pages 405–420, 1977.

[3] S. Balsamo, A. D. Marco, P. Inverardi, and M. Simeoni. Model-based performance prediction in software development: A survey. *IEEE Trans. Software Eng.*, 30(5):295–310, 2004.

[4] L. A. Barroso and U. Hölzle. The case for energy-proportional computing. *IEEE Computer*, 40(12):33–37, 2007.

[5] L. J. Bass, F. Bachmann, and M. Klein. Making variability decisions during architecture design. In *Software Product-Family Engineering (PFE)*, pages 454–465, 2003.

[6] P. C. Clements. On the importance of product line scope. In *Software Product-Family Engineering(PFE)*, pages 70–78, 2001.

[7] P. C. Clements. Process validation, session report. In *Software Product-Family Engineering(PFE)*, pages 388–389, 2001.

[8] V. Cortellessa and C. Trubiani. Towards a library of composable models to estimate the performance of security solutions. In *WOSP*, pages 145–156, 2008.

[9] V. Cortellessa, C. Trubiani, L. Mostarda, and N. Dulay. An Architectural Framework for Analyzing Tradeoffs between Software Security and Performance. In *ISARCS*, pages 1–18, 2010.

[10] L. Dobrica and E. Niemelä. A survey on software architecture analysis methods. *IEEE Trans. Software Eng.*, 28(7):638–653, 2002.

[11] R. Kazman, L. J. Bass, M. Webb, and G. D. Abowd. Saam: A method for analyzing the properties of software architectures. In *International Conference on Software Engineering (ICSE)*, pages 81–90, 1994.

[12] R. Kazman, M. H. Klein, M. Barbacci, T. A. Longstaff, H. F. Lipson, and S. J. Carrière. The architecture tradeoff analysis method. In *ICECCS*, pages 68–78, 1998.

[13] M. H. Klein, R. Kazman, L. J. Bass, S. J. Carrière, M. Barbacci, and H. F. Lipson. Attribute-based architecture styles. In *WICSA*, pages 225–244, 1999.

[14] J.-C. Laprie. Dependability modelling and evaluation of software and hardware systems. In *Fehlertolerierende Rechensysteme*, pages 202–215. Springer, 1984.

[15] Q-ImPrESS Consortium. The Q-ImPrESS project. Project website: http://www.q-impress.eu, 2010.

[16] P. Ranganathan. Recipe for efficiency: principles of power-aware computing. *Commun. ACM*, 53:60–67, April 2010.

[17] R. J. Rodríguez, C. Trubiani, and J. Merseguer. Fault-Tolerant Techniques and Security Mechanisms for Model-based Performance Prediction of Critical Systems. In *ISARCS*, 2012.

[18] K. S. Trivedi. Reliability evaluation for fault-tolerant systems. In *Computer Performance and Reliability*, pages 403–416, 1983.

[19] L. G. Williams and C. U. Smith. Pasa^{sm}: a method for the performance assessment of software architectures. In *WOSP*, pages 179–188, 2002.

[20] J. Yang, G. Huang, W. Zhu, X. Cui, and H. Mei. Quality attribute tradeoff through adaptive architectures at runtime. *Journal of Systems and Software*, 82(2):319–332, 2009.

Assessing Computer Performance with SToCS

Leonardo Piga, Gabriel F. T. Gomes, Rafael Auler, Bruno Rosa, Sandro Rigo, Edson Borin
Institute of Computing, University of Campinas, Brazil
{leonardo.piga,gabriel.ferreira,rafael.auler,bruno.rosa}@lsc.ic.unicamp.br, {sandro,edson}@ic.unicamp.br

ABSTRACT

Several aspects of a computer system cause performance measurements to include random errors. Moreover, these systems are typically composed of a non-trivial combination of individual components that may cause one system to perform better or worse than another depending on the workload. Hence, properly measuring and comparing computer systems performance are non-trivial tasks.

The majority of work published on recent major computer architecture conferences do not report the random errors measured on their experiments. The few remaining authors have been using only confidence intervals or standard deviations to quantify and factor out random errors. Recent publications claim that this approach could still lead to misleading conclusions.

In this work, we reproduce and discuss the results obtained in a previous study. Finally, we propose SToCS, a tool that integrates several statistical frameworks and facilitates the analysis of computer science experiments.

Categories and Subject Descriptors

C.4 [**Performance of Systems**]: Measurement techniques; G.3 [**Probability and Statistics**]: Experimental design, Nonparametric Statistics

Keywords

Performance Analysis, Statistics, Hypothesis Tests

1. INTRODUCTION

Computer performance measurement is subject to random and systematic errors. Mytkowicz *et al.* [15] show that the code layout produced by assemblers and linkers may impact hardware mechanisms, such as cache memory and branch predictors, in non-deterministic ways. Lilja [11] states that several other factors influence the measurement of computer performance, for example the precision of the measuring tools. In order to assist experimenters in the analysis of random errors in the measured data, we propose SToCS (Statistical Tool for Computer Science) [13], a tool that integrates several statistical frameworks and facilitates the analysis of computer science experiments.

Another problem faced when assessing computer performance is the summarization of performance metrics collected for individual benchmarks in a suite (e.g. SPEC CPU 2006). Several metrics have been proposed, leading to decades of debates. This became known as the "War of the benchmark means" [14], where several authors discussed about which summarization technique better describes the overall performance of a computer across benchmarks [8, 9, 18]. We evaluate these metrics with data collected from the execution of benchmarks on real machines and with synthetic data and observe that these metrics provide similar results in our experiments.

More recently, Chen *et al.* [7] argued that parametric statistical techniques should not be used when assessing computer systems performance data and introduced a hierarchical performance test framework (HPT) that applies consecutive statistical hypothesis tests on the performance scores of two computing systems. The framework reports how much a system is faster than the other and with which confidence level. Our results indicate that HPT is conservative when compared with other summarization techniques.

Last, Lilja states that the analysis of computer systems performance should be thought of as a combination of measurement, interpretation, and communication [11]. According to the American Physics Society, the success and credibility of science rely on the exposition of ideas and results to independent testing [1]. We suggest that providing the maximum amount of information in computer science research contributes to the advance of the area.

The main motivation for the development of SToCS are the intriguing questions raised by the discrepancies between HPT and other, more conventional, statistical frameworks, such as the geometric mean of ratios used by SPEC.

The contributions of this work can be summarized as follows: 1) We conduct experiments similar to those performed by the authors of the HPT test and verify that the distribution of the samples is indeed not normal. 2) We also experiment with the test proposed by the authors and notice that the conclusions that can be inferred from our experimental results are conservative, when compared with the traditional means. 3) We propose a tool that implements all these data analysis and helps experimenters to report their results.

The remaining of the text is organized as follows: Sec-

ICPE'13, April 21–24, 2013, Prague, Czech Republic.

tion 2 motivates the creation of SToCS by analyzing the overhead of a binary translator. Section 3 presents issues that are raised when summarizing performance scores. Section 4 evaluates the HPT framework on a synthetic experiment where the true means are known. Finally Section 5 presents our conclusions.

2. MOTIVATION

To motivate the discussion, we analyze the overhead caused by emulation in a dynamic binary translator (DynamoRIO [4]) when running SPECint 2006 using the Hierarchical Performance Test (HPT) [7].

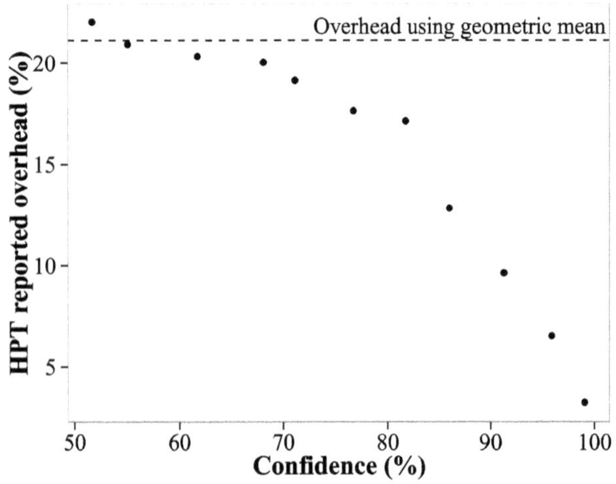

Figure 1: HPT reported overhead for a varying confidence interval caused by the DynamoRIO binary translator. Note that HPT reports a lower bound for the overhead.

The quality of a DBT (Dynamic Bynary Translator) is measured by its fitness to translate computer programs efficiently and transparently [3, 10]. Figure 1 shows the emulation overhead (slowdown) reported by HPT given different confidence levels. The graph shows that the higher the confidence level, the lower is what can be asserted about the overhead. HPT uses Wilcoxon tests to determine lower bounds on the differences between two systems. For example, with a confidence level of 61%, the overhead is *no less* than 20.3%. Moreover, with a confidence level of 99%, we can only assert that the delay added by emulation is *no less* than 3.2%.

The outputs of the HPT framework are always accompanied by confidence levels, on the other hand they report more modest speedups in comparison with the geometric mean of ratios (dashed line in Figure 1). We refer to this behavior as "conservative". The difference between the methods raised the question about which of them we should use and motivated the creation of a tool that aggregates several statistical frameworks.

To help experimenters to report their performance number we developed SToCS, a command line tool implemented in Python that calculates several statistics of an experiment and compares measurements between two computers using different statistical frameworks. It processes CSV files to calculate statistics such as arithmetic, geometric, harmonic means, standard deviation, minimum and maximum values of a given experiment. Furthermore, when two data files are provided, the tool is able to compare the measurements of the given files using different metrics such as the HPT-speedup and the geometric mean of ratios (e.g. SPECrate) [7, 11]. The tool implements the HPT framework, which is not available in conventional tools such as R [16]. R does have the Wilcoxon tests that are used by HPT, but it does not include the HPT framework itself.

3. SUMMARIZING COMPUTER PERFORMANCE SCORES

This section describes issues that are raised when summarizing performance scores, the consequences of making wrong assumptions about data distribution, and the applicability of the Central Limit Theorem (CLT) in computer science experiments.

Confidence intervals can be used to represent the index of dispersion of individual benchmarks provided that the experimenter guarantees that the actual data distribution is normal and techniques such as resampling are not used. But even when this is the case, summarizing the overall performance score of a computer across several benchmark applications with a single number can be misleading.

Figure 2 shows the individual execution time ratios in Actina Solar 220 X3 system compared with the base system defined by SPEC – data collected from SPEC.org. The discrepancy observed in *libquantum* influences indexes of central tendency in different ways. For example, in the ratio of the arithmetic means, a single value has more influence than in the geometric mean of ratios, which amortizes the discrepancies in the summarized data.

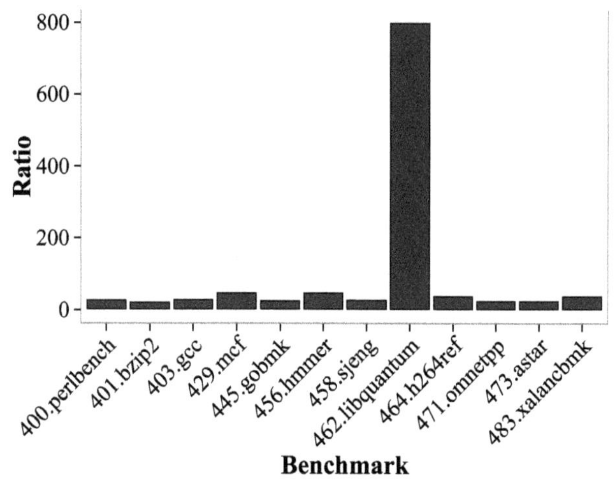

Figure 2: Speedups for all SPECint 2006 applications for the Actina Solar 220 X3 system

On the other hand, Lilja [11] advocates that the geometric mean of ratios does not represent the performance metric that we are ultimately interested in (i.e. shortest execution time). He suggests using ratio of arithmetic means. Calculating the speedup of the Actina system relative to the base system defined by SPEC (i.e. geometric mean of ratios) results in a value of 35.87 while using the ratio of arithmetic means results in a value of 31.50.

3.1 Checking variability

We verify the hazard of assuming that computer data follow a normal distribution in an experiment of 10000 executions of the benchmark libquantum from the SPEC CPU 2006 suite and empirically show that the data distribution may not be normal. Due to time constraints, we select the *test* input subset of the libquantum and run the application on an Intel Core 2 Quad at 2.4GHz with 4GB RAM.

Figure 3 shows the probability density function generated by Naive Normality Fitting (NNF) – bell curve line – and Kernel Parzen Window (KPW) – line above the gray area. The NNF technique assumes a normal distribution and plots the conventional bell curve, while KPW displays a smoothed picture of the actual distribution. Following the same methodology presented by Chen *et al.* [7], we visually compare the plots against each other and determine that the data do not follow a normal distribution. We also apply Lilliefors test (Kolmogorov-Smirnov test) [12] on the measurements and confirm that they were not taken from a population that follows a normal distribution.

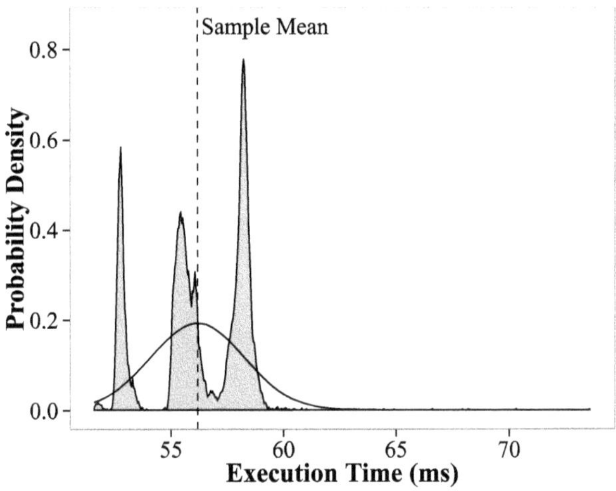

Figure 3: Probability density function for the test input of libquantum running on an Intel Core 2 Quad at 2.4GHz with 4GB RAM

Our result complies with those obtained by Chen *et al.* [7], which has shown that computer science experiments can easily happen to be non-normal. In these cases, parametric techniques that assume a normal distribution of the actual data (e.g. confidence intervals) should not be used. We advise that experimenters check if the sample distribution conforms with the normal before using any statistical framework. In this sense, SToCS implements a statistical framework to check the normality of the distribution by applying Lilliefors (Kolmogorov-Smirnov) test on a sample.

3.2 Applicability of the Central Limit Theorem

The Central-Limit Theorem (CLT) states that, even if the data are not normally distributed, the distribution of the sample means with a sufficiently large number of elements can be safely approximated by a Gaussian [6]. However, checking this approximation is usually neglected in computer science experiments [7]. This section checks the applicability of the CLT using the execution times of 1000 runs of the TeraGen Hadoop workload [5] configured to create a 8 GB HDFS file. This benchmark is one of the key workloads that are used in cloud deployments characterization [2, 17]. Due to its distributed nature, this environment presents large variability.

We characterize the I/O of TeraGen on a cluster composed of 10 AMD A8-3850 APUs which contains 4 CPU cores at 2.9 GHz, each with a 1MB L2 cache. Each node is configured with 16GB of DDR3-1333 memory, 2 SATA HD (1 TB each @ 7200 rpm) configured in RAID 0 mode, where each cluster node is connected using a 1Gbit Ethernet switch running Red Hat Enterprise Linux 6.2 and Hadoop 0.20.2.

SToCS splits the data set of 1000 points into m samples of size n so that $m \times n = 1000$. Then, it applies Lilliefors test on the m sample means to check whether they were taken from a normal distribution. The test shows that the distribution of sample means follows a normal distribution when n is greater than 60. Therefore, to provide a confidence interval for the mean, the experimenter must have at least 180 measurements to have 3 samples of 60 measurements. Such large amount of measurements may require a large amount of time to perform the experiments. This result corroborates the finds in Chen *et al.* [7], which reports that computer science experiments may require large number of measurements to use CLT.

We emphasize that this result is specific for this benchmark. This large amount of measurements may not be required for other applications in order to use CLT.

4. SYNTHETIC DATA COMPARISON

So far, we have discussed different metrics to compare computer performance. We have analyzed real computer performance using ratios of arithmetic means, geometric means of ratios, and the HPT framework. In this section, we evaluate the behavior of these metrics using synthetic data that comes from populations which have normal distribution with known mean and standard deviation.

To create the synthetic data, we create two hypothetical computers A and B, running at 2.4GHz and 2.9GHz, respectively. We generate 10 benchmarks with a varying number of instructions and define an IPC (instructions per cycle) for each benchmark, b, on each computer, c. Using the IPCs and the number of instructions of the benchmarks, we calculated the execution time, $t_{b,c}$ of each application on each machine and generate 1000 samples with a normal probability density function with mean $t_{b,c}$ and standard deviation $t_{b,c}/100$. Table 1 presents these values.

We define, for this experiment, that the total execution time of all applications is the metric that interest us, as suggested by Lilja [11]. Since we generate the samples of execution time with a normal probability density function, we know the population mean, which is the true execution time of each benchmark; hence, we also know the true ratios between the total execution time on each computer. By using SToCS to calculate the geometric mean of ratios, the ratios of arithmetic means, and the HPT-speedup, we evaluate which computer is faster when running the whole benchmark suite. Then, we compare these metrics, shown in Figure 4, with the expected values.

As expected the sample geometric mean of the ratios and the sample ratio of arithmetic means for the total execution time are very close to the expected result (i.e. Computer B

Table 1: **Benchmark characteristics and IPCs for hypothetical computers A (2.4GHz) and B (2.9GHz)**

Bmk.	Number of Instructions	IPC		Time (s)	
		A	B	A	B
a	$9.770 \cdot 10^{12}$	2.86	2.64	1423	1276
b	$9.650 \cdot 10^{12}$	3.04	0.69	1322	4822
c	$8.050 \cdot 10^{12}$	0.82	2.66	4090	1043
d	$9.120 \cdot 10^{12}$	1.46	2.44	2602	1288
e	$10.490 \cdot 10^{12}$	0.48	1.28	9105	2825
f	$9.330 \cdot 10^{12}$	1.74	1.60	2234	2010
g	$12.100 \cdot 10^{12}$	1.23	1.00	4098	4172
h	$20.720 \cdot 10^{12}$	3.42	2.37	2524	3014
i	$22.130 \cdot 10^{12}$	2.32	1.12	3974	6813
j	$6.250 \cdot 10^{12}$	1.58	1.15	1648	1874
Total	$117.61 \cdot 10^{12}$			33025	29142

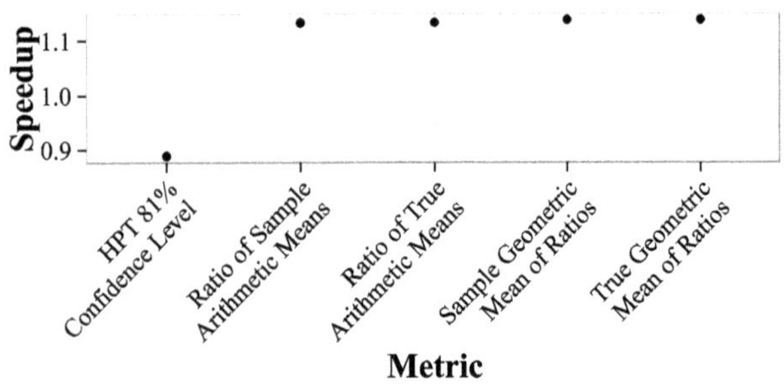

Figure 4: Performance comparison using different metrics

is 1.13 times faster then Computer A). However, we point that HPT might not be comparable to geometric mean of ratios or to ratio of arithmetic means. HPT uses hypothesis tests and, in this particular scenario, it is not able to infer which computer is faster rendering the test inconclusive.

5. CONCLUSION

In this paper we propose SToCS, a tool that assists in the analysis of experimental data by centralizing several statistical frameworks. We used our tool to evaluate these techniques and concluded that they present similar results.

Confidence intervals give an estimate of the precision of measurements, but they require knowledge about the actual distribution of these errors. This requirement does not apply to HPT. On the other hand, the results obtained by this tool indicate that it is conservative when compared to other summarization techniques.

We conclude that publications in computer science should include the maximum amount of information available and that statements about the comparison of computers should be as clear as possible, making it easier for readers to understand the choices and decisions of the authors.

6. ACKNOWLEDGMENTS

The authors would like to thank the anonymous reviewers for their insightful comments and suggestions to improve this paper. They are also grateful to FAPESP, CNPq and CAPES for their financial support.

7. REFERENCES

[1] American Physics Society. What is science? http://www.aps.org/policy/statements/99_6.cfm.

[2] C. Bennett, R. L. Grossman, D. Locke, J. Seidman, and S. Vejcik. Malstone: towards a benchmark for analytics on large data clouds. In *16th ACM SIGKDD International Conference on Knowledge Discovery and Data Mining*, 2010.

[3] E. Borin and Y. Wu. Characterization of DBT overhead. In *Proceedings of the IEEE International Symposium on Workload Characterization*, 2009.

[4] D. Bruening, T. Garnett, and S. Amarasinghe. An infrastructure for adaptive dynamic optimization. In *Proceedings of the International Symposium on Code Generation and Optimization*, 2003.

[5] J. Bughin, M. Chui, and J. Manyika. Clouds, big data, and smart assets: Ten tech-enabled business trends to watch, 2010.

[6] L. L. Cam. The central limit theorem around 1935. *Statistical Science*, 1986.

[7] T. Chen, Y. Chen, Q. Guo, O. Temam, Y. Wu, and W. Hu. Statistical performance comparisons of computers. In *Proceedings of HPCA-18*, 2012.

[8] D. Citron, A. Hurani, and A. Gnadrey. The harmonic or geometric mean: does it really matter? *ACM SIGARCH Computer Architecture News*, 2006.

[9] P. J. Fleming and J. J. Wallace. How not to lie with statistics: the correct way to summarize benchmark results. *Commun. ACM*, 1986.

[10] J. D. Hiser, D. W. Williams, W. Hu, J. W. Davidson, J. Mars, and B. R. Childers. Evaluating indirect branch handling mechanisms in software dynamic translation systems. *ACM Transactions on Architecture and Code Optimization*, 2011.

[11] D. J. Lilja. *Measuring computer performance: a practiotioner's guide.* Press Syndicate of the University of Cambridge, 2000.

[12] H. W. Lilliefors. On the kolmogorov-smirnov test for normality with mean and variance unknown. *Journal of the American Statistical Association*, 1967.

[13] Statistical Tool for Computer Science (SToCS). http://lampiao.lsc.ic.unicamp.br/stocs/.

[14] J. R. Mashey. War of the benchmark means: time for a truce. *SIGARCH Comput. Archit. News*, 2004.

[15] T. Mytkowicz, A. Diwan, M. Hauswirth, and P. F. Sweeney. Producing Wrong Data Without Doing Anything Obviously Wrong! In *Proceedings of the International Conference on architectural support for programming languages and operating systems*, 2009.

[16] R Development Core Team. *R: A Language and Environment for Statistical Computing.* R Foundation for Statistical Computing, Vienna, Austria, 2008. ISBN 3-900051-07-0.

[17] N. B. Rizvandi, J. Taheri, and A. Zomaya. On using pattern matching algorithms in mapreduce applications. In *2011 IEEE Ninth International Symposium on Parallel and Distributed Processing with Applications Workshops*, 2011.

[18] J. E. Smith. Characterizing computer performance with a single number. *Communications of the ACM*, 1988.

Towards a Workload Model for Online Social Applications

[ICPE 2013 Work-in-Progress Paper]

Alexandru-Corneliu Olteanu
University Politehnica of Bucharest, Romania
alexandru.olteanu@cs.pub.ro

Alexandru Iosup
Delft University of Technology, the Netherlands
a.iosup@tudelft.nl

Nicolae Ţăpuş
University Politehnica of Bucharest, Romania
nicolae.tapus@cs.pub.ro

ABSTRACT

Popular online social applications hosted by social platforms serve, each, millions of interconnected users. Understanding the workloads of these applications is key in improving the management of their performance and costs. In this work, we analyse traces gathered over a period of thirty-one months for hundreds of Facebook applications. We characterize the popularity of applications, which describes how applications attract users, and the evolution pattern, which describes how the number of users changes over the lifetime of an application. We further model both application popularity and evolution, and validate our model statistically, by fitting five probability distributions to empirical data for each of the model variables. Among the results, we find that most applications reach their maximum number of users within a third of their lifetime, and that the lognormal distribution provides the best fit for the popularity distribution.

Categories and Subject Descriptors

I.6.5 [**Simulation and Modeling**]: Model Development; K.8 [**Computing Milieux**]: Personal Computing—*Games*

Keywords

workload model, workload characterization, online social applications, social gaming, statistical modeling

1. INTRODUCTION

Online social applications hosted by social platforms, such as Facebook, gather an increasing number of interconnected users. With a total of more than 1 billion Facebook users, there are applications that reach tens of millions of daily active users. Moreover, the hosting platform itself is a heterogeneous system, where many third-party developers bring into the platform tens of thousands of applications—for gaming, multimedia streaming, travel, communication, etc.—and their own infrastructure to host them. Given the diversity in both infrastructure and user coverage, understanding usage patterns and modeling the workload is crucial in providing means to optimize the performance and the cost of hosting such applications.

We analyse traces gathered over a period of thirty-one months for 630 of the most popular Facebook applications, with the purpose of modeling their workload. To identify usage patterns and to understand the workload of social applications, we investigate two research questions: *What is the number of users per application?* and *How does the number of active users evolve over time for online social applications?*.

Although tens of recent studies have focused on the workloads of web applications [1] and gaming workloads [8], relatively few studies focus on online social applications. Existing models for online social applications focus on social interactions and use a limited number of applications over a limited period of time [6]. Our effort is conducted on large datasets with the objective of proposing a workload model. We will use this model to generate test workloads and we will further investigate the possibility of insightful predictions based on it. In our future efforts, we will use the model to provide resource offloading and to fine-tune provisioning and allocation for online social applications. Our main contribution is three-fold:

1. We collect data describing the usage of a number of top applications over a multi-year period (Section 2);
2. We propose a model with two components: the popularity distribution, and the evolution of the number of users during the lifetime of an application (Section 3);
3. We use a data-driven empirical method to provide a characterization and validate our model statistically, for each of the two components (Section 4).

2. DATASETS

For this study, we have collected several large-scale datasets. We describe in the following the steps of the data collection process: crawling, parsing, storing, and sanitizing.

Our analysis is based on data collected from Facebook and from various third-party websites, which report data about Facebook applications. Facebook reports daily, through its APIs, application usage data, such as daily active users (*DAU*) and monthly active users (*MAU*). In our analysis we use DAU and MAU to investigate both popularity and the evolution of the number of users. However, we are also interested in other data that might provide insights, such as application developer, category and subcategory. We use *app* as an abbreviation for application.

To gather information, we have been *crawling* two websites, namely *appdata.com* and *developeranalytics.com*. We have chosen these websites because they report data about

ICPE'13, April 21–24, 2013, Prague, Czech Republic.

Table 1: Volume of crawled, parsed, and stored information.

Source	Files	Apps	Samples
appdata.com	47,056	16,664	1,864,812
developeranalytics.com	49,177	630	133,594
graph.facebook.com	16,901	16,901	16,901

a large number of Facebook applications, in a paged-tabular form. From such data, it is easy to obtain information about the applications that are top ranking at the time of the crawling, but also to follow several individual applications for a time. When choosing our data sources, we have considered *threats to validity*, as described in several works focused on workloads [2][10]. We have cross-checked the values reported by our two sources. Moreover, we have not used data such as ranks, which could be skewed by the sampling that the websites are using.

In this paper, we present data collected between January 1st, 2010 and July 31st, 2012. From *appdata.com* we started to extract MAU and rank for the top 6,000 apps, but, due to a change in the query parameters in March 2010, our crawlers retrieved only information regarding top 40 apps. From *developeranalytics.com* we extracted DAU, MAU, rank, and developer for apps while they were in top 100 during the thirty-one months. We have also followed a sample of 50 apps, that were in the top in our first day of crawling, during the whole period. Moreover, *developeranalytics.com* displays charts with historical data, going back to the beginning of 2008, which gave an increased coverage for around 50 apps. Overall, we have gathered information about around 150 apps for more than 12 consecutive months.

At the end of the crawling period, we developed Python parsers that we used to *parse* and *store* useful data in an unified SQLite database. We completed the datasets with information regarding developer, category and subcategory by crawling *graph.facebook.com* for all the apps obtained from the other two sources. A quantification of the volume of crawled, parsed, and stored data can be found in Table 1.

To *sanitize* the data, we tried to identify and remove any reporting and reading errors. For example, we consider a reporting error when several applications appear to have no users at a given date, although they have users both before and after that date.

3. MODEL OVERVIEW

We propose a model with two components: the popularity distribution, which describes how users are spread throughout the population of applications, and the evolution pattern, which describes how the number of users varies during the lifetime of an application. We detail our model in the remainder of this section.

3.1 Popularity

To quantify application popularity, we investigate the relation between maximum number of DAU and the rank of the application. We expect this relation to conform with a Pareto-like principle, with top ranking applications gathering many more users even than the ones ranked immediatly below them.

We investigate how the popularity distribution varies over time. Given the increasing number of Facebook users, it is interesting to analyse how new users distribute along popular apps. We also investigate whether the popularity distribution is affected or not by seasonality.

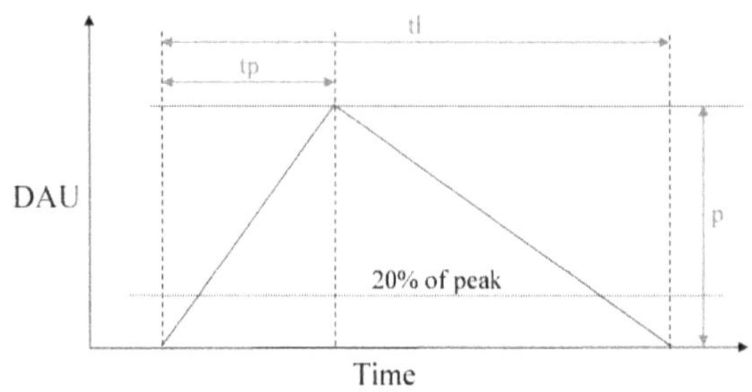

Figure 1: A simplified model to approximate the growth and decay of the number of users for social apps. Labels: p, peak value; t_p, time to peak; t_l, total length of life.

To model popularity, we conduct curve fitting on the empirical distribution for various reference distributions, namely Exponential, Weibull, Pareto, Log-normal, and Gamma. We use maximum likelihood estimation method (MLE) to estimate the distribution parameters, and the Kolmogorov-Smirnov (KS) test to evaluate goodness of fit (GOF). In this paper, we only report the parameters that provide the best fit for each of the reference distributions. For results in terms of the p-values and D-values, we refer to our technical report [7].

3.2 Evolution

In previous studies [9], the number of users over time seems to first grow quickly, then reach a peak, then decrease. Such an evolution can span over months, and even years for popular applications, but people eventually quit using them as new versions or competing applications appear. Thus, we propose in this section a model that captures the growth and decay processes observed for online social applications.

Specifically, we propose a linear model consisting of two line segments. The first segment is on a line with a positive slope, depicting the growth of the number of users, and the second segment is on a line with a negative slope, depicting the decay of the number of users (Figure 1).

We parameterize this model with: peak value (p), growth rate (m), and decay rate (n). Peak value is measured as the maximum DAU throughout the lifespan of the app. Growth rate is given by $m = \frac{p}{t_p}$ and decay rate is given by $n = \frac{-p}{t_l - t_p}$.

We further investigate the underlying relation between growth rate and decay rate by studying the maturity of the app at the peak (r), expressed as the ratio between the time it took the app to reach the peak (t_p) and its total lifetime (t_l). Dividing the equations for growth rate and decay rate, we obtain: $m = (1 - \frac{1}{r})n$

For each of the four parameters described in this section, we plot the cumulative distribution function (CDF), and we conduct curve fitting following the same procedure that we use for fitting the popularity distribution.

4. EXPERIMENTAL RESULTS

In this section we present experimental results related to the two aspects that constitute the focus of our study, app popularity and evolution of the number of app users. For each, we first empirically characterize the aspect using the datasets introduced in Section 2, then validate empirically the models proposed in Section 3. We only present selected results; for more details, we refer to our technical report [7].

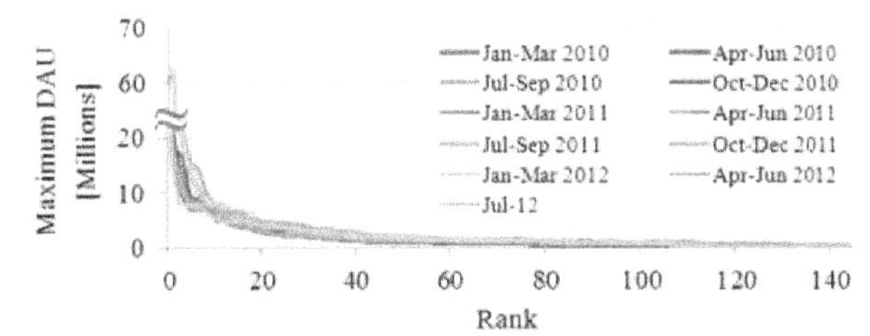

Figure 2: Popularity distribution, depicted as the distribution of maximum DAU over rank. One curve per trimester.

Table 2: Parameters for the reference distributions (Exponential, Weibull, Log-normal, and Gamma) that provide best fit for the popularity distribution. Values for μ_e, λ and θ are expressed in multiples of 10^6. One trimester per year.

Series	Exp(μ_e)	Wbl(k_w, λ)	LogN(μ_l, σ)	Gam(k_g, θ)
Jan-Mar'10	1.76	0.82 1.53	13.69 1.03	0.84 2.09
Jan-Mar'11	2.31	0.84 2.03	14.02 0.90	0.92 2.52
Jan-Mar'12	3.43	0.80 2.84	14.31 0.99	0.81 4.26

4.1 Popularity

We analyse the maximum observed DAU and MAU as metrics of app popularity. To account for seasonal transitions yet still be able to follow the results, we analyse the information we have for these metrics per trimester. We characterize and model, in the remainder of this section, the trimestrial datasets resulting from our 31 months of data.

Characterization: For each trimester in our datasets, we select the maximum DAU for each of the apps, then sort the maxima in descending order to rank the apps. Figure 2 depicts the maximum app DAU over app rank. The shape of the curves for different trimesters are very similar, which indicates that seasonal effects do not influence the relative order of maxima; a more detailed investigation resulting in the same conclusion is included in our technical report [7]. Each trimestrial curve exhibits the same phenomenon: **the best-ranked 5-10% apps** (apps rank 1 through 11-21) **attract many more users than all the remaining apps taken together**. Specifically, the first 20 apps have each over 5,000,000 DAUs; many of them also have over 20,000,000 MAUs (not depicted here). This phenomenon and the observed values indicate that there may be two types of app operators: operators of the first 20 apps, who due to their app sizes could largely own the physical infrastructure on which their apps operate, and the other operators, who could lease infrastructure from a cloud, only when needed.

Modeling through Curve Fitting: We conduct curve fitting for all the eleven datasets. The p-values and D-values obtained through curve fitting, whose exact values are included in our technical report [7], show that **the lognormal distribution provides the best fit with the data we analysed, from the five distributions we have tried.** The parameters obtained for each distribution are summarized in Table 2.

4.2 Evolution

We analyse in this section the evolution of the values observed for app DAU and MAU. Similarly to the previous section, we present characterization and modeling results.

Characterization: To compare apps while accounting for the high variability among them—both lifespan, and evolution of DAU and MAU—, we first normalize both the lifespan and the number of users of each app. In the characterization plots presented in this section, we express for each app the number of users at a moment in time as a percentage of the app's peak number of users, and the moment in time as a percentage of the app's total lifetime. We further present results only for a selection of apps with at least 5,000,000 DAUs or 20,000,000 MAUs, which are thresholds inspired by the results obtained while characterizing app popularity (Section 4.1); these apps account together for more than 80% of the total DAUs and MAUs observed in our datasets.

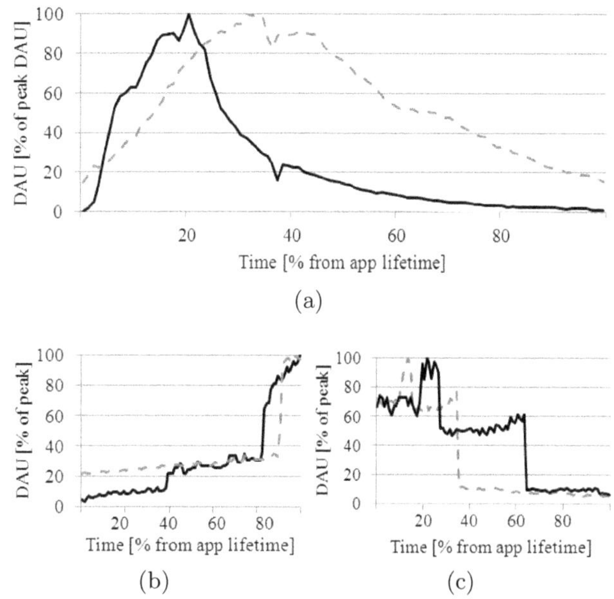

Figure 3: Examples of DAU evolution for several applications: (a) typical single-peaked pattern, (b) suddenly ascending, (c) suddenly descending.

We divide the selected apps into three classes. Class-1 apps (the **common pattern**) grow steadily to a peak, then decay steadily (Figure 3a). The growth and decay can occur with various velocities, but the decay occurs slower than the growth. The growth and decay periods may be interrupted by limited periods (maximum 2 weeks) of opposite effect. Apps in this class are favorable for predictive provisioning of resources for the infrastructure on which they operate. However, the periods of opposite effect may lead to inefficient resource provisioning which, especially when these periods occur close to the peak, may lead to significant unnecessary costs.

Class-2 apps (the **bursty pattern**) exhibit bursts, that is, sharply ascending or descending periods (Figure 3b,c). These apps represent a difficult case for resource provisioning, requiring not only adequate predictions of when the burst will occur, but also finding a provider that can provide the large amount of needed resources.

For Class-3 apps (the **inconclusive pattern**), the evolution cannot be characterized, as for these apps the metrics may be continuously ascending or descending (not depicted here); these apps take a long time to reach their peak or we track only a part of their lifetime in our datasets. These apps are less difficult to provision for than class-1 apps, but require constant managerial attention to avoid missing the peak and thus cause unnecessary costs, and may require a special commercial agreement between the app operator and the resource provider.

Table 3: Percentage of apps in each type of evolution pattern, for two samples of apps: with peak DAU larger than 5,000,000 (column "DAU") and with peak MAU larger than 20,000,000 ("MAU").

	Parameter	
Trend of covered lifespan	DAU	MAU
Single peak	61.0	64.4
Suddenly ascending	7.3	6.7
Suddenly descending	9.9	8.9
Continuously ascending	17.0	13.3
Continuously descending	4.8	6.7
All	100	100

Table 4: Parameters for the reference distributions that provide best fit for each of the parameters of the evolution model (see Section 3.2).

Series	Exp(μ_e)	Wbl(k_w, λ)	LogN(μ_l, σ)	Gam(k_g, θ)
p	4.36×10^6	0.87 4.04×10^6	14.58 1.33	0.83 5.26×10^6
m	50,792	0.66 35,256	9.68 1.62	0.54 93,242
n	8,409	0.68 6,382	7.96 1.66	0.58 14,619
r	19	1.37 21	2.64 0.85	1.76 10

We summarize the percentage of each type of pattern among our selected apps in Table 3. **The common pattern occurs in over 60% of all cases, which leaves a significant percentage of apps exhibiting un-common patterns** and, thus, are difficult cases for provisioning.

In further analysis, with results presented only in our technical report [7], we have found that **over 50% of the apps reach their peak DAU within 15% of their lifetime, and that 85% of the apps reach their peak within their first third of their lifetime**.

Modeling through Linear Fitting: We validate the linear model proposed in Section 3.2 empirically, using our datasets. However, we can only use data coming from apps for which we have a full lifespan coverage, which would greatly reduce the size of our sample. Instead, we find a compromise in using, from our datasets, all the apps having their first and their last DAU reading below 20% of the peak DAU value. With this compromise, we can use 53 apps for modeling, from the 630 apps with at least one DAU reading.

We report from our modeling results the p-values and D-values in our technical report [7], and the distribution parameters in Table 4. **The Weibull and Gamma distributions provide best-fits for both peak value and maturity at the peak. The growth rate can be best approximated with the Log-normal distribution and for the decay rate Weibull, Log-normal, and Gamma are close approximations**.

5. RELATED WORK

Our work is closely related to studies of Internet workloads, including web applications [1, 3, 4], peer-to-peer file sharing [9], and online gaming [8]. In contrast, our study focuses on online social applications, which represent a new class of applications.

Closest to this work, several research efforts focus on online social applications. Nazir et al. [6] study the workload of three Facebook applications, with a focus on social interactions. Kirman et al. [5] study two Facebook games in comparison with a stand-alone game, and find that the social networks show a sharp cut-off, in comparison with the scale-free nature of the social network of the stand-alone game. In contrast to this body of work, ours is conducted on a much larger data set and over a much longer period of time, and the focus of our investigation provides new characterization and modeling insights.

6. CONCLUSION

Understanding the workload of online social applications is an important step in optimizing the performance and reducing the operational costs of these applications, with impact on millions of customers. In this work, we have collected data for hundreds of popular Facebook applications over a period of thirty-one months. Then, we have conducted a data-driven study of the popularity and evolution of online social applications, for which we have provided characterization and modeling results.

We are currently extending our model with a micro-model of user sessions arrival and in-session behavior, which will describe the workload with a finer granularity. We plan to use the extended model to investigate prediction-based resource provisioning and allocation.

Acknowledgments

We thank Boxun Zhang (TUD) for his help. This work is funded by the Sectoral Operational Programme Human Resources Development 2007-2013 of the Romanian Ministry of Labour through POSDRU/107/1.5/S/76909 and by the STW/NWO Veni grant 11881.

7. REFERENCES

[1] M. Arlitt and C. Williamson. Web server workload characterization: The search for invariants. In *ACM SIGMETRICS*, volume 24, pages 126–137, 1996.

[2] Y. Borghol, S. Mitra, S. Ardon, N. Carlsson, D. Eager, and A. Mahanti. Characterizing and modelling popularity of user-generated videos. *Performance Evaluation*, 68(11):1037–1055, 2011.

[3] L. Cherkasova, Y. Fu, W. Tang, and A. Vahdat. Measuring and characterizing end-to-end internet service performance. *ACM TOIT*, 3(4):347–391, 2003.

[4] A. Iyengar, M. Squillante, and L. Zhang. Analysis and characterization of large-scale web server access patterns and performance. *WWW*, 2(1):85–100, 1999.

[5] B. Kirman, S. Lawson, and C. Linehan. Gaming on and off the social graph: The social structure of facebook games. In *CSE*, pages 627–632, 2009.

[6] A. Nazir, S. Raza, D. Gupta, C. Chuah, and B. Krishnamurthy. Network level footprints of facebook applications. In *IMC*, pages 63–75, 2009.

[7] A. C. Olteanu, A. Iosup, and N. Ţăpuş. Towards a workload model for online social applications: Extended report. Tech.Rep. PDS-2013-003, TU Delft, January 2013. `http://www.pds.ewi.tudelft.nl/fileadmin/pds/reports/2013/PDS-2013-003.pdf`.

[8] M. Suznjevic and M. Matijasevic. Player behavior and traffic characterization for mmorpgs: a survey. *Multimedia Systems*, pages 1–22, 2012.

[9] B. Zhang, A. Iosup, J. Pouwelse, and D. Epema. Identifying, analyzing, and modeling flashcrowds in bittorrent. In *P2P*, pages 240–249, 2011.

[10] B. Zhang, A. Iosup, J. Pouwelse, D. Epema, and H. Sips. Sampling bias in bittorrent measurements. *Euro-Par*, pages 484–496, 2010.

A Robust Optimization for Proactive Energy Management in Virtualized Data Centers

Ibrahim Takouna* Wesam Dawoud* Kai Sachs± Christoph Meinel*

* Hasso Plattner Institute
University of Potsdam
Potsdam, Germany
firstname.lastname@hpi.uni-potsdam.de

±SAP AG
Walldorf, Germany
kai.sachs@sap.com

ABSTRACT

Energy management has become a significant concern in data centers to reduce operational costs and maintain systems' reliability. Using virtualization allows server consolidation, which increases server utilization and reduces energy consumption by turning off unused servers. However, server consolidation and turning off servers can cause also consequences if they are not exploited efficiently. For instance, many researchers consider a deterministic demand for capacity planning, but the demand is always subject to uncertainty. This uncertainty is an outcome of the workload prediction and the workload fluctuation. This paper presents a robust optimization for proactive capacity planning. We do not presume that the demand of VMs is deterministic. Thus, we implement a range prediction approach instead of a single point prediction. Then, we implement a robust optimization model exploiting the range-based prediction to determine the number of active servers for each capacity planning period. The results of the simulation show that our approach can mitigate undesirable changes in the power-state of the servers. Additionally, the results indicate an increase in the servers' availability for hosting new VMs and reliability against a system failure during power-state changes. As future work, we intend to apply our approach to dynamic workload such as a web application. We plan to investigate applying our approach to other resources, where we consider only the CPU demand of VMs. Finally, we compare our approach against the approaches using stochastic optimization.

Categories and Subject Descriptors

k.6 [**Management of computing and information systems**]: General; k.6.2 [**Installation Management**]: Performance and Usage Measurement

General Terms

Management, Performance, Experimentation

Keywords

Energy-aware; Robust Optimization; Prediction; Virtualization

1. INTRODUCTION

Energy efficient resource management has become a significant concern in virtualized data centers to reduce operational costs. As an idle server consumes approximately 70% of the power consumed by the server running at full capacity [1], turning off idle servers to reduce energy consumption has been widely proposed by many researchers [1][2][3]. However, these approaches considered the number of VMs and their capacity demands are deterministic. Hence, they built deterministic models that did not take into account the uncertainty of the demand. Furthermore, a typical server has multiple power states including on, off, sleep, and hibernated. Some related work has ignored the energy consumption during the change of the server's power-state, which we consider as a wasted energy. During a power-state change, a server consumes energy without performing any useful work. For instance, a normal PC takes around 25 seconds to switch from one state to off state and vice versa [4]. Furthermore, we found that a server with 1TB memory requires 5 minutes for a clean boot, which includes the hardware check stage. On the other hand, Mao et al. [5] have observed that the start-up time of a VM is proportional to its image size. The start-up time of a VM is very crucial for online applications such as web applications. Thus, implementing a proactive optimization solution can assist to avoid SLA's violations. By predicting the number of the required VMs in the next planning period, we can prepare these VMs images and the physical server in advance. Many researchers consider a deterministic demand for capacity planning, but the demand is always subject to uncertain. This uncertainty is an outcome of the workload prediction and the workload fluctuation. Ignoring the uncertainty in real world applications can make the usual optimal solution infeasible [7]. Unlike deterministic models, Dance et al. [8] have used the stochastic optimization for considering uncertainty. However, the stochastic optimization requires to know the probability distributions of the demand. In our approach, we used the robust optimization that addresses data uncertainty. The robust optimization model assumes that uncertain parameters belong to a bounded range. Importantly, the robust optimization can outperform the stochastic optimization when selecting an appropriate robustness level, and it is less computationally intensive

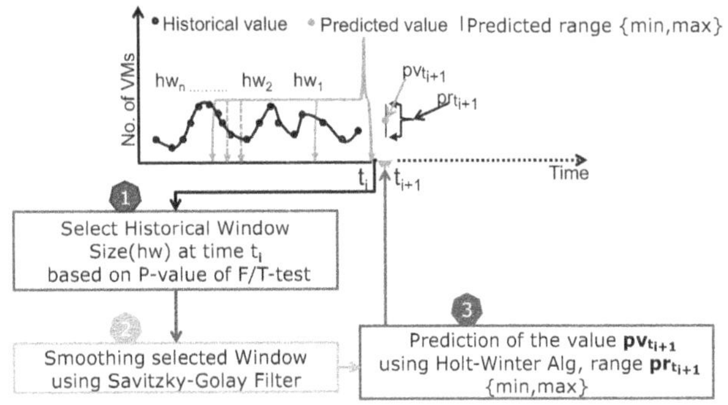

Figure 1: The stages of the range-based prediction approach

when the distribution of uncertain parameters is complicated [9]. To build the robust optimization model for proactive capacity planning, we implement an adaptive range-based workload prediction instead single point prediction for predicting the number of the requested VMs. The results of the simulation show that our approach can mitigate undesirable changes in the power-state of the servers. Additionally, the results indicate an increase in the servers' availability for hosting new VMs and reliability against a system failure during power-state changes.

2. RANGE-BASED PREDICTION

Typically, point value prediction approaches might not cover the workload fluctuation. The approaches solve the problem as a deterministic optimization, which assume the precise knowledge of the workload demand. Furthermore, optimization based on the mean-value or the max-value of the workload can produce low provision or high provision, respectively. This is costly in both cases. In this section, we present a range-based prediction approach with an adaptive window-size algorithm to predict the number of demanding VMs in a data center. Our approach consists of three stages as shown in Figure 1: (1) selecting the historical window size based on the statistical test (F/T-test); (2) smoothing the values of the selected historical window; and (3) predicting the next number of active VMs and its minimum and maximum range. Most of the related work in the context have been done for grid computing [10][11]. For example, Wu et al. [10] have proposed an adaptive prediction of grid performance with a confidence window for the historical values. They used an auto-regression to find a model for the historical interval by which predicts the future workload. However, we consider the historical data variation to enhance the prediction accuracy and bound the predicted range. As shown in Figure 2, the number of VMs shows a random behavior, but it also depicts a certain pattern. For instance, it starts low at morning then increases reaching the peak at around the midday. At evening, it starts to go down again. Our approach uses an adaptive window-size of historical values to provide a high accurate prediction range. In Figure 1, the measured workload values are shown by a series of line-dots up to time t. On the other hand, the gray dot represents the predicted workload value.

- Window selection: our interest is to predict the number of VMs for the next 5 minutes from the historical window hw. The historical window-size is determined based on the P-value of both F-test and T-test to filter out the values that are very unlikely to be in the same window. We used F-test and T-test to probe the significance of the change in variance and mean between two samples of populations, respectively. Using the P-value of F-test and T-test, we can decide whether the two samples have almost the same variance and the same mean. For example, after performing F-test, if we find out that the P-value is less than $\alpha = 0.05$, we reject the null hypothesis. This means that these values do not belong to the same historical window. Thus, the algorithm stops going back to take more historical values and moves to the next stage, which is window smoothing.
- Window smoothing: using prediction algorithms with the historical values causes errors. Thus, we used a smoothing filter to remove noise and prevent its influence on the prediction algorithm. There are many smoothing filters, but we selected Savitzky-Golay filter due its effectiveness in keeping the peak values and removing the spikes, which can be considered as noise. The filter has two significant parameters that guide the smoothing process: the frame size and polynomial degree. In our approach, the frame size is not constant, and it equals to the selected historical window-size. In contrast, Wu et al. [10] fixed the frame size.
- Range prediction: we used Holt-Winter implemented in the R-tool, because it dynamically optimizes and determines the level and the trend of the time series. Importantly, we determine the predicted range based on the single predicted point value pv and the standard deviation of the selected window σ_{hw}. The predicated range pr $\{r_l\ ,\ r_h\}$ where r_l and r_h equal $(pv - \sigma_{hw})$ and $(pv + \sigma_{hw}.)$

2.1 Planet-lab Workload Traces

The monitoring infrastructure project of Planet-lab [6] provides traces of historical data for CPU utilization, which measured every 5 minutes. These traces are for more than a thousand machines running in more than 500 locations around the world. Here, we present data for two days that have different workload fluctuations. We assume that these machines are hosted as VMs.

To simulate the on-demand concept of the cloud computing environment (i.e., the open system behavior), we terminate the VMs with less than 5% CPU utilization. In other words, we considered it as being destroyed and exited the system. Then, when the trace shows a VM with a CPU

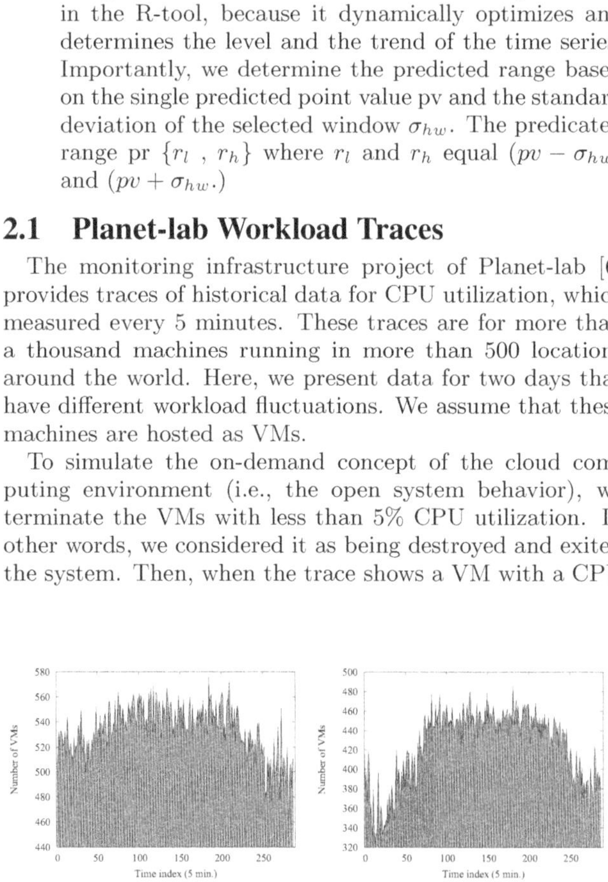

(a) Day03 (b) Day06

Figure 2: Planet-lab workload traces

utilization higher than 5%, we consider a new request for provisioning a VM.

2.2 Implementation and Evaluation

We implemented the proposed approach using Java with integration of the R-tool, which consists of many statistical functions and the required filters. Here, we present the results of our approach. Figure 3 shows the predicted range for each value of workload (i.e., number of VMs). The low predicted r_l is shown by a red dashed-line meanwhile the high predicted r_h is represented by a blue dashed-line. The purple sold-line represents the single point predicted value. The predicted range is propositional to the workload fluctuation. For instance, due to the low fluctuation of the workload around the time index 58, the predicted range is small. Contrarily, a large range is predicted around the time index 225.

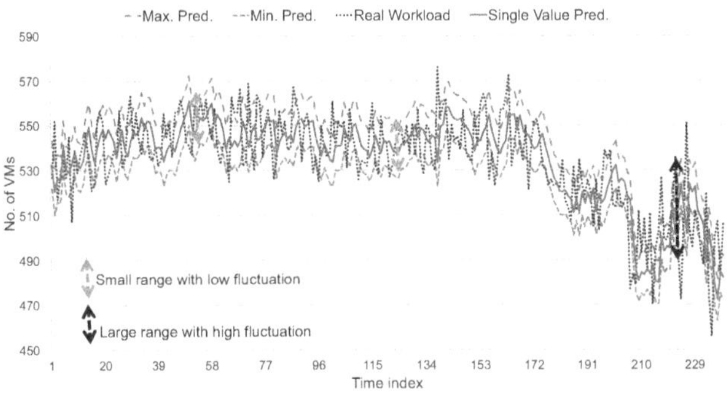

Figure 3: The results of the range-based prediction approach

3. ROBUST OPTIMIZATION

As we implemented the prediction approach based on a range not a single value, we will study the implementation of robust optimization for capacity planning. Robust optimization deals with optimization problems whereas the robustness is sought against uncertainty or deterministic variability in the value of a parameter of the problem (i.e., the workload). The principle of robust optimization considers point prediction meaningless and it is replaced by range prediction. Thus, robust optimization addresses data uncertainty by assuming that uncertain parameters belong to a bounded range. In our approach, we avoid the assumption that considers the precise knowledge of the workload demand in the planning horizon where many proposed solutions have solved the problem as a deterministic optimization [1][2][3]. We have a predicted range, centered at the nominal prediction $\bar{d}$, for the demand at each time period. The robust optimization approach replaces each deterministic demand $\bar{d}$ by an uncertain parameter $\tilde{d} = \bar{d} + \hat{d}$*z, where $|z| \leq 1$. Furthermore, it guarantees that the constraints hold for a given uncertainty set. cient of its constraint matrix. to solve this problem

3.1 Robust Problem Formulation

In this section, we present a robust formulation of the capacity planning problem. We use the following notation: e_{idle} is the energy consumption of a server during idle state (i.e., $p_{idle}*t_{idle}$). The parameter e_{swt} is the energy consumption of a server to change from power-state to another, which consumes p_{swt} and takes t_{swt} seconds. So, e_{swt} the power-state change energy wastage equals $p_{swt}*t_{swt}$. The binary variables n and b represent currently active servers and previously active servers, respectively. The parameters $\tilde{nv}$, $\bar{nv}$, and $\hat{nv}$ represent the uncertain number of VMs, the mean of the predicted number of VMs, and the standard deviation of the predicted number of VMs, respectively. The parameters $\tilde{vc}$, $\bar{vc}$, and $\hat{vc}$ represent the uncertain utilization of a VM, the mean of the VM's utilization, and the standard deviation of the VM's utilization, respectively. Finally, the parameter $\tilde{d}$ is the total uncertain demand of the number of VMs and their utilization. The objective function in Equation 1 is to minimize the wastage energy that might result from keeping the server idle and switching the power-state of a server. Equation 2 guarantees that the constraints hold for a given uncertainty set of the demand in Equation 3. Equation 4 and Equation 5 represent uncertain demand of a number of VMs and uncertain utilization of VMs, receptively. We formulate this problem based on the following assumption. When a VM is requested, it occupies a certain portion of a server capacity. For instance, a small-instance with 1 vCPU might take 1/4 of a server has 4 logical cores. Then, after running the VM for while the real resource consumption of this VM can be revealed and will be taken into consideration for the next planning period. The constant parameters t_{swt}, t_{idle}, p_{swt} and p_{idle} are 150s, 300s, 120watts, 100watts, respectively. Observably, the switching power is slightly greater than the idle power due to the CPU utilization. The power constants were set based on SPECpower [12] results of HP ProLiant ML110 G3 server [13]. A scalar variable z models the demand uncertainty. We do not presume exact knowledge of the actual distribution of demand, but instead we assume that the distribution is characterized by the mean of the number of VMs $\bar{nv}$ and their capacity $\bar{vc}$ and standard deviation of the demand of the number of VMs $\hat{nv}$ and their capacity $\hat{vc}$. Moreover, we have an accurate estimations of the most optimistic uncertainty z_{min} and the most pessimistic uncertainty z_{max}. These parameters form lower and upper bounds of z, respectively.

Minimize:

$$\sum_{i=1}^{ns} e_{idle} + \sum_{i=1}^{ns} e_{swt} * (n * (1-b) + b(1-n)) \qquad (1)$$

Subject to:

$$\tilde{d} \leq \sum_{i=1}^{ns} \tilde{sc_i} * n \qquad (2)$$

$$\tilde{d} = \sum_{i=1}^{nv} \tilde{vc_i} \qquad (3)$$

$$\tilde{nv} = \bar{nv} + \hat{nv} * z_{nv} \qquad (4)$$

$$\tilde{vc} = \bar{vc} + \hat{vu} z_{vc} \qquad (5)$$

$$|z| \leq 1, \quad z = [z_{min}, z_{max}], \quad and \quad n, b \in \{0, 1\}$$

3.2 Implementation and Evaluation

To solve a robust optimization problem, we used Robust Optimization Made Easy (ROME), which is an algebraic modeling tool implemented in the MATLAB environment [14]. ROME operates as an intermediate layer between the modeler and optimization solver engines. It helps converting the original uncertain optimization problem into its robust counterparts. Its core functionality consists of translating modeling code into an internal structure in ROME. Then, it translated into a solver-specific input format for solving by linear optimization solvers. This can be done manually, but it is tedious and error-prone [14]. We used IBM ILOG CPLEX as optimization solver.

Figure 4 depicts simulation results of a deterministic optimization and different values of uncertainty scalar z. The lift axis represents the amount of energy and the right axis represents the percentage of the energy consumption by the power-state switching and the idle state. The deterministic result means that prediction of the VMs demand is 100% accurate. So, we used the real traces to perform capacity planning using deterministic optimization. Then, we compared the results with the range-based prediction taking into account uncertainty of demand. This shows the results with different range of uncertainty scalar z. Figure 4 depicts the following observations. In deterministic, there is no idle energy consumption, because we assumed that the demand is known, and the total wastage energy results from changing a server's power-state. On the other hand, the robust optimization considering the uncertainty of the demand, we had different results by changing the uncertainty scalar z. First, the range of uncertainty is very wide, z{-1,1}. The total of the objective function and the total of idle energy are the highest, and the total of switching energy is the lowest compared to the other results. Second, when changing the uncertainty scalar z from z{-1,1} to z{-0.5,0.5}, we decreased the uncertainty range. Thus, we could save energy by reducing an unbeneficial power-state switching and keeping some server idle. This also can increase the system availability and reliability. Finally, when z was set {-0.25,0.25}, we could achieve more energy savings from both power-state switching and idle servers. However, this can cause some under provisioning of capacity and violation of allocation some VMs. The calculated mean and standard deviation of under provisioning VMs are 3VMs and 5VMs, respectively. Importantly, the execution time of our proposed approach is less than 1 second while using a computer with Pentium 2.6GHz processor.

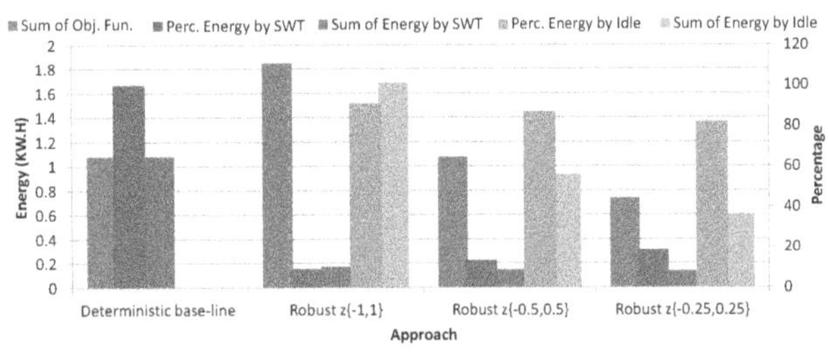

Figure 4: The result of the proposed robust optimization approach

4. CONCLUSIONS AND FUTURE WORK

In this paper, we presented a proactive robust optimization approach for capacity planning in virtualized data centers. To achieve this, we implemented a range-based prediction algorithm, which allows formulating the problem using the robust optimization. The robust optimization model takes into account the prediction uncertainty. We compared the results of deterministic and robust of capacity planning, and we found that the robust optimization more realistic to be used in data centers where VMs demand and their utilization are uncertain. Importantly, by using our approach, we could achieve energy saving and provide high availability and reliability for the system. As future work, we will consider heterogeneous servers and VMs size. Furthermore, we intend to extend our approach for a dynamic provision of web applications. Thus, we intend to compare our approach against the other approaches that use stochastic and deterministic optimization for dynamic provision of web applications. Furthermore, we intend to investigate applying our approach to other resources, where we just consider only the CPU demand of VMs.

5. REFERENCES

[1] R. Nathuji, K. Schwan, "Virtualpower: Coordinated power management in virtualized enterprise systems," *ACM SIGOPS Operating Systems Review*, 41(6) pp. 265-278, 2007

[2] A. Verma, P. Ahuja, A. Neogi, "PMAPPER: Power and Migration Cost Aware Application Placement in Virtualized Systems," *proceedings of the 9th ACM/IFIP/USENIX Middleware*,pp. 243-264,Belgium.

[3] A. Gandhi, M. Harchol-balter, R. Das, C. Lefurgy,"Optimal Power Allocation in Server Farms," *proceedings of the 11th International Joint Conference on MMCS*, ACM New York, USA, 2009; 157-168.

[4] Booting-time.[online].available:http://www.toms hardware.co.uk/ubuntu-oneiric-ocelot-benchmark-review,review-32377-15.html.[accessed: 1-jun-2012].

[5] M. Mao and M. Humphrey," A Performance Study on the VM Startup Time in the Cloud," *proc. of fifth IEEE Conference on Cloud Computing*, 2012.

[6] Planet-lab. available :http://www.planet-lab.org/

[7] A. Ben-Tal and A. Nemirovski, "Robust Solutions of Linear Programming Problems Contaminated with Uncertain Data," mathematical programming v. 88 (2000), 411-424.

[8] C. Dance and A. Gaivoronski, "Stochastic Optimization for Real Time Service Capacity Allocation under Random Service Demand," annals of operations research, vol. 193, no. 1, pp. 221-253, feb. 2011.

[9] W. Yanbo, "Empirical Comparison of Robust, Data Driven and Stochastic Optimization," MIT 2008.

[10] Y. Wu, K. Hwang, Y. Yuan , and W. Zheng, "Adaptive Workload Prediction of Grid Performance in Confidence Windows," *IEEE trans. on PDS 2009*

[11] M. J. Clement and M.J. Quinn, "Analytical Performance Prediction on Multicomputers," J. Supercomputing, pp. 886-894, 1993.

[12] K. D. Lange,"Identifying Shades of Green: The SPECpower Benchmarks," IEEE Computer 42(3), pp. 95-97, 2009.

[13] SPEC-results.[online].available. http://www.spec.org/power$_s$sj2008/results/res2011q1 /power$_s$sj2008-20110127-00342.html

[14] J. Goh and M. Sim, "Robust Optimization Made Easy with Rome," Operations Research, 2011, pp.973-985.

A Meta-Model for Performance Modeling of Dynamic Virtualized Network Infrastructures*

[Work-in-Progress Paper]

Piotr Rygielski
Karlsruhe Institute of Technology
Am Fasanengarten 5
Karlsruhe, Germany
piotr.rygielski@kit.edu

Steffen Zschaler
King's College London
Department of Informatics
London, UK
szschaler@acm.org

Samuel Kounev
Karlsruhe Institute of Technology
Am Fasanengarten 5
Karlsruhe, Germany
kounev@kit.edu

ABSTRACT

In this work-in-progress paper, we present a new meta-model designed for the performance modeling of dynamic data center network infrastructures. Our approach models characteristic aspects of Cloud data centers which were not crucial in classical data centers. We present our meta-model and demonstrate its use for performance modeling and analysis through an example, including a transformation into OMNeT++ for performance simulation.

Categories and Subject Descriptors

C.4 [**Performance of Systems**]: Modeling techniques

Keywords

performance modeling, data center networks, meta-modeling

1. INTRODUCTION

The increasing popularity of Cloud Computing is leading to the emergence of large virtualized data centers hosting increasingly complex and dynamic IT systems and services. The performance of applications running in such data centers depends on the performance of three main subsystems: computing, storage, and networking. Different applications utilize these subsystems in different ways. The amount of computational, storage and communication resources assigned to applications influences significantly the quality-of-service (QoS) delivered to end users.

Due to the common adoption of virtualization technologies, Cloud data centers are becoming increasingly dynamic. Virtual machines, data, and services can be migrated on demand between physical hosts to optimize resource utilization while enforcing service-level agreements (SLAs) [2]. This dynamism of data center infrastructures turns an accurate and timely performance analysis into a real challenge [7].

In our research, we focus on network infrastructures of Cloud data centers. The main incentive for this is the fact that network infrastructures in virtualized and dynamic environments are introducing several new challenges for performance analysis. First, the growing density of Cloud data centers due to resource sharing leads to drastic increase in the number of network end points deployed on top of the same physical infrastructure. This situation makes performance analysis difficult due to, for example, high traffic volume from various sources traversing over a single physical link. Second, the volume of inner-data center traffic (having its source and destination inside the same data center) is much higher in Cloud environments than in classical data centers. Third, new traffic sources in the management layer of Cloud environments emerge (e.g., traffic caused by the migration of virtual machines). Finally, there is a lack of standardized and well established technology for network virtualization supporting performance isolation and QoS assurance. This requires performance models that are largely technology independent.

In this paper, we propose a new performance meta-model aiming for modeling virtualized network infrastructures addressing the above mentioned challenges. The meta-model is designed as a part of the Descartes Meta-Model (DMM) [10], a new meta-model for run time QoS and resource management in virtualized service infrastructures. It covers, among others, computing infrastructures, storage, middleware, software and services. DMM is designed to serve as a basis for autonomic resource management during operation ensuring that system QoS requirements are continuously satisfied while infrastructure resources are utilized efficiently.

Some approaches for modeling and analyzing network performance in data centers already exist in the literature. We discuss a selection below. Most of the existing approaches (e.g., [3, 13]) are based either on black-box models or on highly-detailed protocol-level simulation models (e.g., [11, 4, 13]). The black-box models do not consider the internal network structure and topology while the protocol-level simulation models focus only on selected parts of the network infrastructure and do not capture the link to the running applications and services which are sources of the network traffic. Furthermore, existing modeling approaches do not

*This work is a part of RELATE project supported by the European Commission (Grant no. 264840ITN). We thank Jörg Henß for providing the meta-model of the OMNeT++.

ICPE'13, April 21–24, 2013, Prague, Czech Republic.

consider network virtualization techniques and their performance influences [14].

In [5], the authors proposed a modeling approach named Syntony. They use Syntony to model the Ad-hoc On-Demand Distance Vector protocol and compare the model-based analysis to the native OMNeT++ implementation. In [4], the authors use proSPEX (an approach based on the ITU-T Z.109 [1] profile) to evaluate the performance of the ESRO (Efficient Short Remote Operations) transport protocol. Both Syntony and proSPEX focus only on modeling specific network protocols. Approaches based on ITU-T SDL/MSC[1] (Specification and Description Language/Message Sequence Charts) have general focus and mainly model communication protocols [11]. The Palladio Component Model (PCM) [3] treats network infrastructures as a black-box and does not support any detailed network abstractions. GANA [12] (Generic Autonomic Network Architecture) models only autonomic components of networks on the protocol level. Additionally, in [13], the classical black-box and low-level simulation based performance modeling approaches in the area of computer networks are described.

We stress that none of these works explicitly take into consideration the influences of network virtualization techniques on the system performance and the mentioned specific aspects of Cloud data center networks.

2. ENVISIONED APPROACH

We propose a model-based approach to performance analysis. The networking domain of data centers is described by a new meta-model presented in this paper. The DNI Meta-Model (*Descartes Network Infrastructure*) is the base for DNI models instantiation. A DNI model can utilize external information about the deployed software and the traffic it produces. The general approach is depicted in Figure 1.

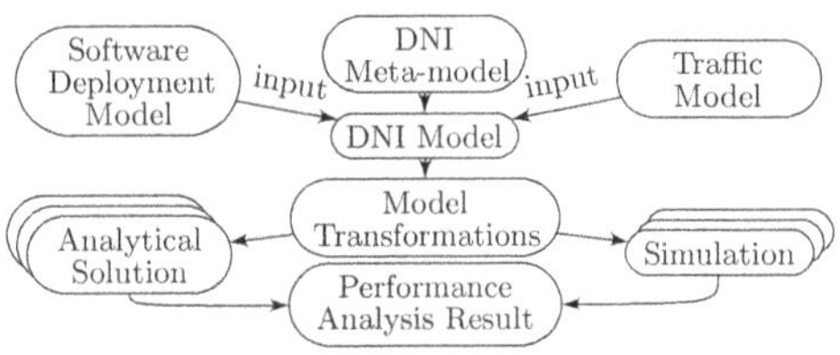

Figure 1: The approach diagram with model inputs.

The obtained DNI model is transformed into known performance models (e.g., queueing networks, Petri nets) and analyzed using existing performance analysis methods. As the research develops, DNI models can be transformed into multiple performance models and deliver multiple performance analysis results, e.g., with different accuracy.

As our work on the meta-model is still in progress, the current version of the DNI Meta-Model does not cover all important, performance-influencing aspects of data center network infrastructures. Our development roadmap includes expansion of the meta-model by adding entities covering, e.g., QoS assurance, routing, network virtualization by tunneling and overlaying. In the next section, we present the DNI Meta-Model in its current state of development.

3. META-MODEL

The DNI Meta-model[1] covers four parts of data center network infrastructures: network protocols, structure of the physical and virtual network infrastructure (nodes, interfaces, links), deployed virtual machines and applications (traffic sources), and a part that describes the network traffic in the data center. The meta-model is implemented in Ecore using the Eclipse Modeling Framework (EMF).

[1]Available online: `http://bit.ly/DNI-model-ICPE2013`

Network protocols control the process of any data exchange between communicating nodes. They define means of identification of communicating parties; that is, addresses of the source and destination. Additionally, the implementation details of a given protocol have strong influence on the performance of data transmission. In the DNI Meta-Model, the `NetworkProtocols` are organized in `NetworkProtocolStacks` that allows the application of a complete stack to a given part of the network. Each protocol can provide a `ProtocolAddress`—for example, to a `NetworkInterface`. This part of the DNI Meta-Model is presented on the class diagram depicted in Figure 2.

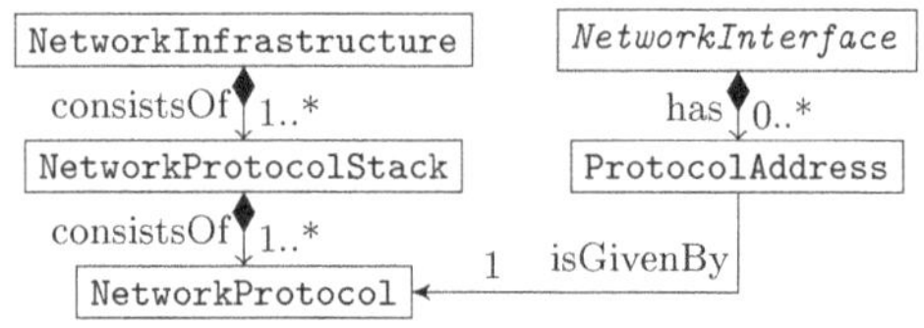

Figure 2: Network protocols and protocol stacks.

The protocols are described in the meta-model with the information about introduced overheads. Due to the wide variety of protocols used nowadays, we have captured only common, generic features describing a protocol, e.g., data unit header size, and whether communication is connection-oriented.

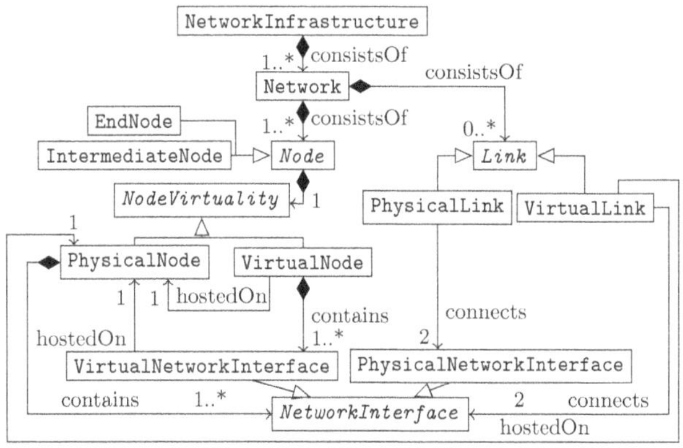

Figure 3: Network's physical and virtual structure.

The part of the meta-model representing the physical and virtual structure of a network is depicted in Figure 3. The `NetworkInfrastructure`, apart from the `NetworkProtocolStack`, consists of `Networks`. Each `Network` is built from `Links` and `Nodes` which are connected by `NetworkInterfaces`. As their impact on network performance differs, `Nodes` are categorized as `EndNode` (e.g., server, VM, gateway to an external network) or `IntermediateNode` (e.g., switch, router, firewall). Each of the `Network` components can be further characterized as physical or virtual where each virtual entity must be hosted on a physical one.

The information about physical to virtual resource allocations of `EndNodes` is included in the meta-model in a simplified way—a virtual entity gets a percentage of the physical resources (removed from Fig. 3 for clarity)—although the data is assumed to be imported from other parts of DMM.

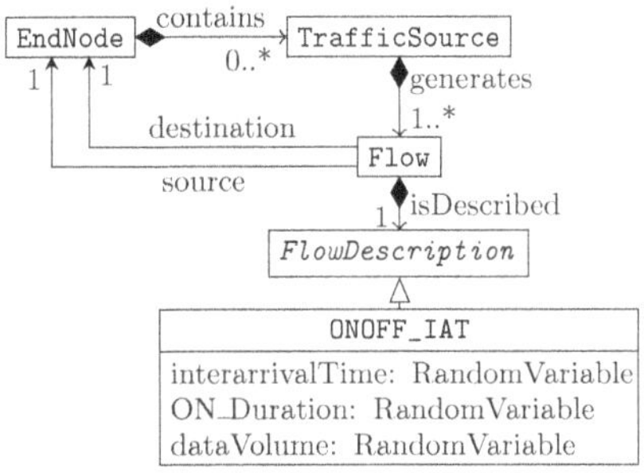

Figure 4: Network traffic and software deployment.

In a data center, most of the network traffic is generated by deployed applications. As depicted in Figure 4, network traffic is generated by `TrafficSource`s that are deployed on `EndNode`s. Each `TrafficSource` generates `Flow`s which have exactly one source and destination located in the `EndNodes` and can be uniquely identified by the set of protocol-level addresses. For now, we assume that there is exactly one route from the source to destination. Finally, each `Flow` can be described using various models. In this paper, we provide the classical description of an ON-OFF source with inter-arrival times specified by a probability distribution. However, a variety of other traffic models can be found in the literature (e.g., [8]) which can be integrated into our meta-model.

4. EXAMPLE

In this section, we first present a selected fragment of an exemplary data center network infrastructure which is modeled with DNI. Next, to obtain performance analysis results, we apply a model-to-model transformation of the DNI instance into an instance of OMNeT++ simulation [6, 15]. In this example, selected parts of the transformation have manually hard-coded performance impacts, however there exist systematic relations between the source and destination model and the hard-coded values will be removed in a future version of the transformation.

In the considered scenario (cf. Fig. 5a), the network consists of three servers and one switch. On two servers two virtual machines (VMs) are deployed. Each of the VMs is connected internally to a software hypervisor switch. The servers are connected to the physical switch using an IP network. Fig. 5b shows part of the model of this scenario (focusing on only one server). The complete model, the meta-model, the transformation and its result can be found online (see the URL in Section 3). The network infrastructure depicted in Figure 5 lacks the information about any software deployed in the machines as this information can be provided manually (e.g., from measurements) or imported from other state-of-the-art models (e.g., PCM [3] or DMM [10]) using model transformations.

The entities from the DNI model are transformed into entities recognized by OMNeT++ using the transformation rules described in Table 1. In the destination model, we use the `SimpleHypervisor`[2] module that is a modification of the `StandardHost` in order to enable sharing of a physical network interface among all hosted VMs. The transformation has been implemented using the Epsilon Transformation Language [9]. Below we discuss two excerpts from the transformation code in more detail.

In Listing 1, we show a fragment of the transformation code describing how the generic features of network protocols are translated into overheads in data transmission. We assume that overheads of protocols of lower layers (i.e., these addressing a *`NetworkInterface`*) are decreasing the capacity of a link transmitting data. The overhead of higher layer protocols (e.g., TCP that addresses an application) increase the amount of data produced by a `TrafficSource`. In the future, we plan to abstract from the classical ISO/OSI layers and support layering in a generic fashion.

Listing 1: Utilizing network protocols information in the model transformation (language: ETL).

```
operation CreateChannels
var p=stack.getProtocols();
var headLen=p.select(pr|pr.layer=2).headerLenght;
headLen=headLen+p.select(pr|pr.layer=3).headerLenght;
var l2UnitSize=p.select(pr|pr.layer=2).dataUnitSize;
var overhead = (100*headLen)/l2UnitSize;
var channel = new OMNET!Channel;
var par = new OMNET!DoubleParameter;
par.name = "datarate";
var realThr=link.maxThroughput*(1.0-(overhead/100.0));
par.value = realThr.equivalent();
par.unit = link.unit.equivalent();
channel.parameters.add(par);
```

In Listing 2, we demonstrate how the physical resources are shared among virtual `EndNodes` bound by *hostedOn* relation. Using the simplified resource allocation information, the total bandwidth available to a physical node is divided among virtual nodes using the percentage value and a new transmission delay is calculated.

Listing 2: Sharing physical resources among virtual entities.

```
operation networkAddConnections
for(link in DNI!VirtLink.allInstances){
 var physRes=link.hostedOn.getTotalBandwidth();
 var conn=new OMNET!ChannelConnectionItem;
 var ch=OMNET!Channel.allInstances.selectOne(
    c|c.name==link.name);
 var delay=ch.parameters.selectOne(p|p.name=="delay");
 var rate=ch.parameters.selectOne(p|p.name=="datarate");
 var l2=DNI!NetworkProtocol.allInstances.selectOne(
    p|p.layer=2);
 var resAlloc=link.hostedOn.hasResAlloc;
 var ra=resAlloc.selectOne(r|r.virtNode==conn.leftNode);
 var nDelay=l2.dataUnitSize/(rate.value*ra.resShare);
 ch.setDelay(delay.value + nDelay);
 ch.setDatarate(0.0);
 conn.channel=ch; }
```

5. CONCLUSIONS

In this work-in-progress paper, we have presented a new meta-model aiming for performance modeling of virtualized data center network infrastructures. Our meta-model addresses the challenges of performance analysis in Cloud environments by abstracting too detailed and technology-specific information, while covering important performance-relevant aspects of network infrastructures. We have presented an example of model transformation to a popular simulation framework where performance analysis can be run. Our future research steps concern primarily further development of the meta-model to support resource allocation, QoS aspects, and network virtualization techniques. Furthermore,

[2]Available online; See the URL in Section 3.

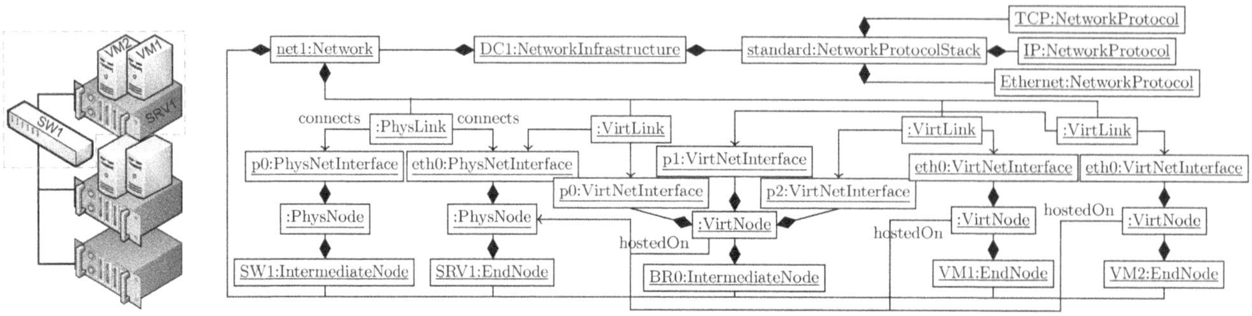

(a) (b)

Figure 5: Selected fragment of an exemplary data center network (a) presented on the object diagram (b).

Table 1: Selected transformation rules from DNI to OMNeT++

Source Entity	Destination Entity	Comments
`NetworkProtocolStack`	a parameter of a `DatarateChannel`	Protocols used in the stack introduce transmission overheads. An excerpt of the transformation code is presented in Listing 1.
`NetworkInfrastructure`	`network`	A `network` in OMNeT++ represents the whole network infrastructure. DNI `Networks` are distinguished by protocol-level addresses.
`EndNode`	`SimpleHypervisor`	Applies to both `PhysicalNode` and `VirtualNode`.
`IntermediateNode`	e.g., `Switch` or `Router`	Depends on the value of `IntermadiateNode.type` enumeration.
`Link`	`DatarateChannel`	Applies to both `PhysicalLink` and `VirtualLink`.
`NetworkInterface`	`inout` gate in e.g., `Router`.	Applies to the `gates` section of a *`Node`* the gate is attached to. Applies to both `Physical-` and `VirtualNetworkInterface`.
`TrafficSource`	e.g., `TcpApp`, `UdpApp`	Choice of application type depends on the `NetworkProtocol` in the L4 (e.g., for TCP we use `TcpApp`).
`FlowDescription`, `ONOFF_IAT`	configuration of e.g., `TcpApp`	The choice of concrete *`FlowDescription`* depends on traffic generation implementation and configuration of `TrafficSources`.
mapping of `Virtual-` on `Physical-` entities	a parameter of a `DatarateChannel`	Sharing of physical resources causes modification of virtual links throughput. See transformation code in Listing 2.

we plan to provide more ways of traffic modeling and integrate the approach with DMM.

6. REFERENCES

[1] SDL combined with UML. ITU-T Z.109, 2000.

[2] M. Armbrust, A. Fox, R. Griffith, A. D. Joseph, R. Katz, A. Konwinski, G. Lee, D. Patterson, A. Rabkin, I. Stoica, and M. Zaharia. Above the Clouds: A Berkeley View of Cloud Computing. Technical report, University of California, 2009.

[3] S. Becker, H. Koziolek, and R. Reussner. The Palladio component model for model-driven performance prediction. *Journal of Systems and Software*, 82(1):3–22, 2009.

[4] N. de Wet and P. Kritzinger. Using UML models for the performance analysis of network systems. *Comput. Netw.*, 49(5):627–642, 2005.

[5] I. Dietrich, F. Dressler, V. Schmitt, and R. German. SYNTONY: network protocol simulation based on standard-conform UML2 models. In *Proc. of the ValueTools '07*, pages 21:1–21:11, 2007.

[6] J. Henss. OMPCM – An OMNeT++ simulator for Palladio. In *Palladio Days 2012*, Paderborn, Germany, 2012. Special Focus Talk.

[7] N. Huber, M. von Quast, M. Hauck, and S. Kounev. Evaluating and Modeling Virtualization Performance Overhead for Cloud Environments. In *Proc. of the 1st Int. Conf. on Cloud Computing and Services Science*, pages 563–573, 2011.

[8] T. Karagiannis, M. Molle, and M. Faloutsos. Long-range dependence: Ten years of internet traffic modeling. *IEEE Internet Computing*, 8(5):57–64, 2004.

[9] D. Kolovos, R. Paige, and F. A. Polack. The Epsilon Transformation Language. In *Theory and Practice of Model Transformations, vol. 5063 of LNCS*, pages 46–60. Springer, 2008.

[10] S. Kounev, F. Brosig, and N. Huber. Descartes Meta-Model (DMM). Technical report, Karlsruhe Institute of Technology, 2013.

[11] A. Mitschele-Thiel and B. Müller-Clostermann. Performance engineering of SDL/MSC systems. *Comput. Netw.*, 31(17):1801–1815, 1999.

[12] A. Prakash, Z. Theisz, and R. Chaparadza. Formal methods for modeling, refining and verifying autonomic components of computer networks. In *Transactions on Computational Science XV*, pages 1–48. Springer, 2012.

[13] R. Puigjaner. Performance modelling of computer networks. In *Proc. of the 2003 IFIP/ACM Latin America conf. on Towards a Latin American agenda for network research*, LANC '03, pages 106–123, New York, NY, USA, 2003. ACM.

[14] P. Rygielski and S. Kounev. Network Virtualization for QoS-Aware Resource Management in Cloud Data Centers: A Survey. *PIK — Praxis der Informationsverarbeitung und Kommunikation*, 36(1), 2013. In print.

[15] A. Varga. The OMNeT++ discrete event simulation system. In *Proc. of the European Simulation Multi-conference (ESM)*, pages 319–324, 2001.

Performance Modelling of Database Contention using Queueing Petri Nets

David Coulden Rasha Osman William J. Knottenbelt
Department of Computing
Imperial College London
London SW7 2AZ, UK
{drc09, rosman, wjk}@imperial.ac.uk

ABSTRACT

Most performance evaluation studies of database systems are high level studies limited by the expressiveness of their modelling formalisms. In this paper, we illustrate the potential of Queueing Petri Nets as a successor of traditionally-adopted modelling formalisms in evaluating the complexities of database systems. This is demonstrated through the construction and analysis of a Queueing Petri Net model of table-level database locking. We show that this model predicts mean response times better than a corresponding Petri net model.

Categories and Subject Descriptors

C.4 [Performance of Systems]: *Modelling techniques.*

General Terms

Performance

Keywords

Queueing Petri nets, performance modelling, database locking.

1. INTRODUCTION

The data landscape has changed dramatically in size and complexity in the past decade. The Internet, ubiquitous communication, cloud services and e-Science applications have led to an explosion in data generation and storage. A large proportion of this data is stored and managed in databases. Market growth for relational database management systems is expected to double by 2016 [11], making performance of these large DBMSs a critical issue for users and vendors alike.

Database system performance is influenced by complex and interdependent functionalities (e.g. transaction usage scenarios, database cache management, disk contention, concurrency and lock contention and the implementation of logical and physical structures in DBMS). In the performance evaluation literature there have been many performance studies of different components of database systems and many methodologies developed for their performance evaluation [14]. However, the impact of these studies on industry has been limited. One of the reasons is the lack of detailed modelling needed to represent production-grade database systems. A main cause is the interaction of physical and logical resources within database systems, which is difficult to represent using traditional modelling formalisms. In this work, we return to the issue of modelling database systems by modelling table level Two Phase Locking using queueing Petri nets and illustrate the potential that this new approach has in performance modelling of database systems.

Queueing Petri Nets (QPNs) [1] extend coloured stochastic Petri nets by incorporating queues and scheduling strategies into places forming *queueing places*, thus producing a very powerful modelling formalism that has the synchronization capabilities of Petri nets (PNs) while also being capable of modelling queueing behaviours. These queueing places consist of two components: the *queue*, and the *depository* where serviced tokens (customers) are placed. Tokens enter the queueing place through the firing of input transitions, as in other Petri nets; however, as the entry place is a queue they are placed in the queue according to the scheduling strategy of the queue's server. Once a token has been serviced it is deposited in the depository where it can be used in further transitions. Queueing places can have variable scheduling strategies and service distributions; these are known as *timed queueing places*. *Immediate queueing places* impose a scheduling discipline on arriving tokens without a delay.

Queueing Petri nets have been recently applied in the performance evaluation of component-based distributed systems [4, 5] and grid environments [10]. In this paper, we apply QPN in modelling locking contention in database systems. We have chosen QPN as the modelling formalism over other variations of Petri nets, as the queueing places allow for the representation of lock scheduling in database systems, while the places and transitions naturally represent the flow of execution of a transaction in the system. Even though queueing network models (QNMs) are currently the prevailing formalism for performance modelling of database systems, QPNs are more expressive when representing simultaneous resource possession and blocking.

The rest of this paper is organized as follows. Section 2 overviews related work. Section 3 details the QPN model for a particular measured system. Section 4 analyzes the results and Section 5 concludes the paper.

2. RELATED WORK

2.1 Queueing Network Models

Osman and Knottenbelt [14] surveyed queueing network performance models of database systems. They found that the majority of studies that model concurrency control in database systems assume a uniform distribution of the locks over the total number of data items, in addition to representing update only transactions in the models. While these assumptions produce tractable models, they neglect hot-spots and do not represent the

ICPE'13, April 21–24, 2013, Prague, Czech Republic.

effect of read transactions holding shared locks on the execution of update transactions that hold conflicting locks. Moreover, these models assess the performance of the transaction at the physical hardware level; therefore they are incapable of representing the effect of lock conflicts at the table or row level, which is more beneficial to database performance tuning.

2.2 Petri Nets

There is a paucity of studies that apply Petri nets to database systems in comparison to the available research that applies queueing network models. Of the studies that do exist, Chen [2] analyzes deadlock detection scheduling in centralized databases using stochastic Petri nets. The places in the model represent transaction execution states, i.e., waiting for a lock, locking an item, CPU processing, etc. Firing of the timed transitions represents the delay in moving from one state to the next.

For parallel databases, Mikkilineni et al. [9] use a Petri net to model concurrent parallel query execution plans in a distributed database. In the PN model, the transitions represent query operations and the places represent data blocks. The firing of a transition represents data communications. Jenq et al. [3] analyze two-phase locking in a parallel database machine using a two-layered model. The higher-layer model is a Petri net representing the parallel and synchronized execution of the relational operations of a transaction. The lower-level model is a QNM that represents the hardware resources and lock waiting queues.

The modelling approach presented in this paper differs from that of previous work in that we do not model the synchronization of query execution plans. Instead, we borrow the concept of modelling the execution of the transaction at the database table level from previous work in modelling database systems using queuing networks [13]. This work is an improvement over previous models of database systems in that we are able to represent locking contention between read and write transactions in a way similar to that of actual systems. Moreover, our QPN models have a more intuitive structure that maps the transaction and database design to the QPN model. This makes the model easier to comprehend by database developers and administrators.

3. Queueing Petri Net Model

3.1 Measured System

DBMSs implement concurrency control through locking protocols. The most widely used protocol is Strict Two Phase Locking (Strict 2PL) [15]. Strict 2PL forces transactions to hold *exclusive locks* to modify data and *shared locks* to read data. For a transaction to acquire an exclusive lock on a data object, no other transaction should hold a shared or exclusive lock on that object. Transactions can acquire shared locks on data objects only if no transaction has an exclusive lock on the objects. In our case study, we model Strict 2PL at the table level. A version of table level Strict 2PL is the default locking method implemented in the MySQL default storage engine [12]. In our experiments we use the PostgreSQL 9.1 [16] DBMS. In order to mimic table level Strict 2PL we explicitly lock the table within the transactions.

The measured system has two types of transactions: shared and exclusive that compete to access *Table A* (100,000 rows). Both transactions explicitly lock the table in the appropriate lock mode (shared or exclusive) and read or modify 1% of the table rows. Access to the table is achieved through a full table scan, i.e. no index is utilized by either transaction. The structure of the transactions is shown in Figure 1. In order to simulate a TPC-W-like workload in which update transactions are longer than read transactions [17], the execution time of the exclusive transaction is artificially lengthened by 40ms. Clients submit transactions to the DB server with exponentially distributed think times with mean 500ms. The mean response time of each transaction type is measured and compared to that emerging from the QPN model for different transaction mixes. The measured system was run on an Intel$^{(R)}$ Core$^{(TM)}$ i7-2600 CPU@3.40GHz box running Ubuntu 12.10 64-bit and PostgreSQL 9.1.

```
shared        BEGIN;
transaction     LOCK TABLE Table A in
                                ACCESS SHARE MODE;
                SELECT count(*) FROM Table A
                                WHERE id > value;
              END;

exclusive     BEGIN;
transaction     LOCK TABLE Table A in
                                ACCESS EXCLUSIVE MODE;
                UPDATE Table A SET other-id = other-value
                                WHERE id > value;
                 SELECT pg_sleep(0.04);
              END;
```

Figure 1. Structure of shared and exclusive transactions.

3.2 QPN Model of Table Level Locking

The QPN model for the measured system was developed using QPME2.0 [6, 7]; it is detailed in Figure 2. The clients are represented by a timed queueing place with an infinite-server queue. The tokens in the *client* place have two colors; each color represents a client of one transaction type. Clients submit transaction jobs to the database server after an exponentially distributed think time. Then, transactions enter the *lock waiting* place where they wait for the lock on the table to be free. The lock waiting place is an immediate queueing place with FIFO departure discipline which ensures that the transactions are serviced in order of arrival.

The table-level locking mechanism is represented using a *lock repository* place, which is an ordinary place containing *lock* tokens. A share transaction will require one lock token and an exclusive transaction will require the maximum number of tokens defined for the lock repository place. By setting the number of lock tokens within the lock repository place to be equal to the maximum number of share transactions, all share transactions will be able to run simultaneously and an exclusive transaction will be forced to wait if there is a least one share transaction accessing the table. Any transaction entering the database queues behind any waiting transaction. Once a transaction has acquired a lock it will access *Table A*. Table A is represented by a timed queuing place with an infinite server queue which models transaction execution. Each type of transaction is treated as having an exponentially distributed service time that models the entire execution of the transaction. When a transaction has been serviced it will be passed back into the client place to repeat the process.

In our QPN model, we are assuming logical resources are the bottleneck, not physical resources. Therefore, the model does not directly capture disk and CPU contention and performance. However, the effects of processing are partially reflected in the

G/M/∞-IS queueing place representing *Table A*. Infinite server scheduling models the forking of PostgreSQL processes for each database connection. To minimize the effect of DBMS automated disk access, the default PostgreSQL configuration has been modified[1]. This modified configuration will not eliminate disk access but configures the DBMS for performance instead of durability [16].

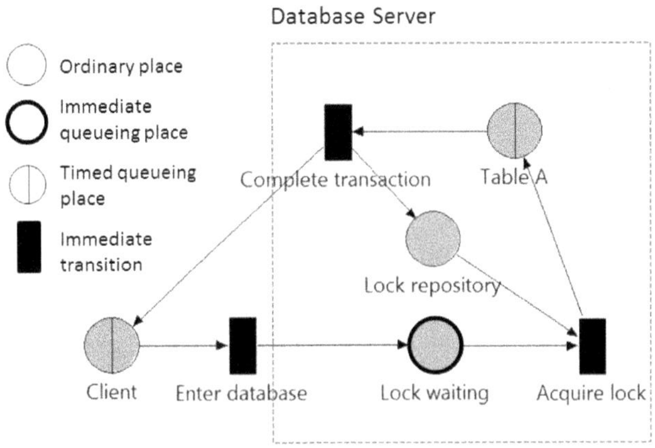

Figure 2. Queueing Petri net model of table level locking.

4. RESULTS

The experiment was based on the workloads of the TPC-W benchmark. The TPC-W benchmark is an e-commerce benchmark implementing an on-line bookstore. It has three workload mixes [8]: the *browsing mix* has 95% reads and 5% updates, the *shopping mix* has 80% reads and 20% updates, and the *ordering mix* has 50% reads and 50% updates.

Each transaction type was executed in isolation, i.e. without any locking contention, and the mean response time calculated. The measured mean response time for the shared transaction was 18.8ms and for the exclusive transaction 60.4ms.

We compare the measured system with our QPN model and an equivalent PN model. The PN model is the same as the QPN model in Figure 2 except that the *lock waiting* place is an ordinary place[2]. We have not compared to a QNM, as this type of simultaneous resource possession and blocking is difficult to express using QNMs.

The results for the browsing, shopping and ordering workloads are presented in Figure 3. First, we will discuss the performance of the actual system. For the browsing workload (Figure 3(a)) the share transactions dominate the traffic and their performance is minimally affected by the increase in exclusive transactions. The *step-like* trend for both transactions is caused by the constant number of exclusive transactions for the corresponding number of clients. For the shopping mix (Figure 3(b)) the increased number of exclusive transactions has affected the performance of both types of transactions, i.e. lock waiting time has increased in comparison to transaction execution time. This is especially evident for the shared transactions where we notice a sharp increase in the mean response time for large number of clients. For the ordering workload (Figure 3(c)) in which the number of shared and exclusive transactions are equal, lock waiting time dominants transaction execution which is evident in the performance degradation of the shared transactions, leading to approximately the same response times for both transaction types at high client numbers.

The PN model severely underestimates the performance of the exclusive transactions for the browsing workload with an error of 97% at 60 clients. However, it overestimates the performance of the shared transactions, with an error of 13% at 60 clients. This is due to the fact that in the PN model the transactions do not queue for locking, as in the real system; and therefore the exclusive transactions are starved. This trend continues as the percentage of exclusive transactions increases in the shopping and ordering workloads. From Figures 3(b) and 3(c), the PN model overestimates the performance of the shared transaction by 83% and 94%, and underestimates the performance of the exclusive transactions by 43% and 17% at 60 clients for the shopping and ordering workloads respectively. The increase in exclusive transactions in the workloads increases their probability of holding the lock token, thus the accuracy of the PN increases as the number of exclusive transactions increase. However, the opposite effect is seen on the shared transactions as without a scheduling discipline they are able to skip ahead of the exclusive transactions whenever a lock token is held by at least one share transaction.

The QPN model underestimates the performance of both transactions for the browsing workload with an error of 32% at 60 clients. The accuracy of the QPN model increases as the number of exclusive transactions increase in the system, i.e. when the lock waiting times dominate the response times. For the shopping workload the QPN underestimates the performance of both transactions with an error of 10% for the shared transaction and 13% for the exclusive transaction at 60 clients. The QPN model overestimation is possibly due to the unaccounted multi-core processing. For the ordering workload, the QPN correctly predicts that both share and exclusive transactions have approximately the same response times. Unlike the previous workloads, the QPN model underestimates the performance of both transaction types when the number of clients is less than 20, with an average error of 6%. When the number of clients is 20 or more, the QPN model overestimates the performance of both transaction types with an average error of 8% at 60 clients. The overestimate is due to the high updates that cause more disk access which, in turn, affects the response times of both transaction types. This is not accounted for in the QPN model. Nonetheless, the QPN model is able to follow the performance trend for both transactions for all workloads, especially the share transaction in comparison to the PN model.

[1] The modified server configuration parameters are: fsync=off, synchronous_commit=off and checkpoint_segments =600. The reader is referred to [16] for a definition of these parameters.

[2] Access to *Table A* should have been modelled as a timed transition with infinite service. However, QMPE2.0 does not support timed transitions [6], so this was approximated by a serial network consisting of an immediate transition, a timed queueing place and a second immediate transition similar to the QPN model.

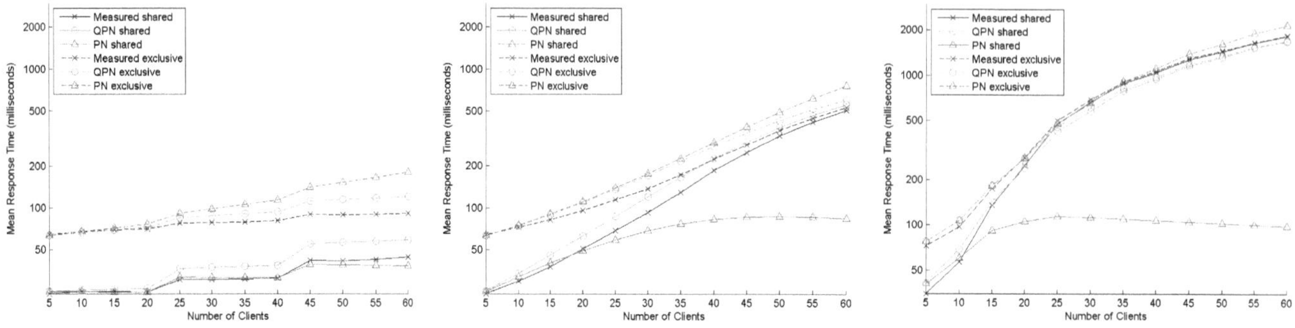

(a) browsing (95% shares and 5% exclusives) **(b)** shopping (80% shares and 20% exclusive) **(c)** ordering (50% shares and 50% exclusive)

Figure 3. Mean response time for TPC-W workload.

5. CONCLUSIONS AND FUTURE WORK

In this paper, we have demonstrated the potential in modelling relational database systems using Queueing Petri Nets, thus overcoming some of the limitations of queueing network models and Petri nets which are currently the main modelling formalisms used to represent database systems.

We have presented a QPN model of two-phase table-level locking. The QPN model was able to approximate the queueing of the lock requests to the table for varying workloads, in contrast to an equivalent Petri net model.

This paper is a starting point for further investigations, which will include the extension of the QPN model to incorporate the effect of hardware contention on the system. This will lead to modelling of more representative database systems with more realistic workloads. Furthermore, for this approach to be feasible and applicable, an automated mapping tool will be developed. We will also investigate emerging paradigms, e.g. NoSQL databases, although the diversity of their data models and implementations will likely mean a generic modelling framework will be infeasible.

6. REFERENCES

[1] Bause, F. 1993. Queueing Petri Nets—A Formalism for the Combined Qualitative and Quantitative Analysis of Systems. In *Proc. 5th Int'l Workshop Petri Nets and Performance Models* (Oct 19-22, 1993), 14-23.

[2] Chen, I.-R. 1995. Stochastic Petri Net Analysis of Deadlock Detection Algorithms in Transaction Database Systems with Dynamic Locking. *The Computer Journal*, 38, 9 (1995), 717-733. DOI: 10.1093/comjnl/38.9.717

[3] Jenq, B.-C., Twichell, B.C.; Keller, T.W. 1989. Locking performance in a shared nothing parallel database machine. *IEEE Transactions on Knowledge and Data Engineering*, 1, 4 (Dec 1989), 530-543.DOI: 10.1109/69.43427

[4] Kounev, S. 2006. Performance modeling and evaluation of distributed component-based systems using queueing Petri nets. *IEEE Trans. Software Engineering*, 32, 7 (July 2006), 486-502.

[5] Kounev, S. and Buchmann, A. 2003. Performance modelling of distributed e-business applications using queuing petri nets. In *Proc. IEEE International Symposium on Performance Analysis of Systems and Software* (Mar 2003), 143-155.

[6] Kounev, S. and Spinner, S. 2011. *QPME 2.0 User's Guide*. (May 2011) Karlsruhe Institute of Technology, Germany. http://descartes.ipd.kit.edu/fileadmin/user_upload/descartes/QPME/QPME-UsersGuide.pdf

[7] Kounev, S., Spinner, S. and Meier, P. 2010. QPME 2.0-A Tool for Stochastic Modeling and Analysis Using Queueing Petri Nets. In *From Active Data Management to Event-Based Systems and More*, LNCS vol 6462, 293-311.

[8] Menasce, D. A. 2002. TPC-W: a benchmark for e-commerce. *IEEE Internet Computing*, 6, 3 (May/June 2002), 83-87. DOI: 10.1109/MIC.2002.1003136

[9] Mikkilineni, K. P., Chow, Y.-C. and Su, S. Y. W. 1988. Petri-net-based modeling and evaluation of pipelined processing of concurrent database queries. *IEEE Trans. Software Engineering*, 14, 11 (Nov 1988), 1656-1667. DOI: 10.1109/32.9053

[10] Nou, R., Kounev, S., Julia, F. and Torres, J. 2009. Autonomic QoS control in enterprise Grid environments using online simulation. *Journal of Systems and Software*, 82, 3 (March 2009), 486-502.

[11] Olofson, C. W. 2012. *Worldwide Relational Database Management Systems 2012–2016 Forecast*. International Data Corporation (August 2012), Doc # 236273. www.idc.com

[12] Oracle Corporation 2013. *MySQL 5.6 Reference Manual. Internal Locking Methods*, http://dev.mysql.com/doc/refman/5.6/en/internal-locking.html

[13] Osman, R., Awan, I. and Woodward, M. E. 2011. QuePED: Revisiting Queueing Networks for the Performance Evaluation of Database Designs. *Simulation Modelling Practice and Theory*, 19, 1 (Jan 2011), 251-270.

[14] Osman, R. and Knottenbelt, W. J. 2012. Database System Performance Evaluation Models: A Survey. *Performance Evaluation*, 69, 10 (Oct 2012), 471–493. DOI=10.1016/j.peva.2012.05.006

[15] Ramakrishnan, R. and Gehrke, J. 2003. *Database management systems*. McGraw-Hill, Boston, Mass.

[16] The PostgreSQL Global Development Group 2012. *PostgreSQL 9.1.7 Documentation*. http://www.postgresql.org/docs/9.1/static/index.html.

[17] The Transaction Processing Performance Council 2003. *TPC-W Benchmark version 2*. http://www.tpc.org/tpcw/

CloudScale: Scalability Management for Cloud Systems

Gunnar Brataas[*,**]
Erlend Stav[*]
* SINTEF ICT, ** IDI, NTNU
Trondheim, Norway
{Gunnar.Brataas,
Erlend.Stav}@sintef.no

Sebastian Lehrig
Steffen Becker
Universität Paderborn
Paderborn, Germany
{Sebastian.Lehrig,
Steffen.Becker}
@uni-paderborn.de

Goran Kopčak
Darko Huljenic
Ericsson Nikola Tesla
Zagreb, Croatia
{Goran.Kopcak,
Darko.Huljenic}@ericsson.com

ABSTRACT

This work-in-progress paper introduces the EU FP7 STREP CloudScale. The contribution of this paper is an overall description of CloudScale's engineering approach for the design and evolution of scalable cloud applications and services. An Electronic Health Record (EHR) system serves as a motivation scenario. The overall CloudScale method describes how CloudScale will identify and gradually solve scalability problems in this existing applications. CloudScale will also enable the modelling of design alternatives and the analysis of their effect on scalability and cost. Best practices for scalability will further guide the design process. The CloudScale method is supported by three integrated tools and a scalability description modelling language. CloudScale will be validated by two case studies.

Categories and Subject Descriptors

D.2.2 [**Software**]: Programmer workbench; H.1.0 [**Information Systems**]: Models and Principles, General; D.2.11 [**Software**]: Patterns

Keywords

Scalability, Cloud, Software architecture, Provisioning, Domain specific modelling language, Tools

1. INTRODUCTION

Cloud providers theoretically offer their customers virtually unlimited infrastructure resources for their applications on an on-demand basis. However, scalability is not only determined by the available resources, but also by how the control and data flow of the application or service is designed and implemented. Implementations that do not consider their effects can either lead to low performance (under-provisioning, resulting in high response times or low throughput) or high costs (over-provisioning, caused by low utilisation of infrastructure resources).

ICPE'13, April 21–24, 2013, Prague, Czech Republic.

CloudScale [2] aims at providing an engineering approach for building scalable cloud applications and services with the following objectives:

1. Make cloud systems scalable by design so that they can optimally exploit the elasticity of the cloud, as well as maintaining and also improving scalability during system evolution. At the same time, a minimum amount of computational resources shall be used.
2. Enable model-driven analysis of scalability of basic and composed services in the cloud.
3. Ensure industrial relevance and uptake of the CloudScale results so that design for scalability becomes easier for cloud systems.

CloudScale's engineering approach for scalable applications and services will enable small and medium enterprises as well as large players to fully benefit from the cloud paradigm by building scalable and cost-efficient applications and services based on state-of-the-art cloud technology and software know-how. Furthermore, the engineering approach reduces risks and costs for companies entering the cloud market.

The consortium for the EU FP7 STREP CloudScale consists of five partners and is lead by the independent research institute SINTEF. SAP and Ericsson Nikola Tesla (ENT) are industrial partners, while the University of Paderborn (UPB) is an academic partner. XLAB is responsible for tool implementation and hosting activities.

The contribution of this work-in-progress paper is an overall description of the CloudScale approach to scalability management. The paper is structured as follows: First, we describe a motivating scenario in Section 2. CloudScale's Method, i.e., its engineering approach, is explored in Section 3. Validation and demonstrators are discussed in Section 4. Finally, Section 5 offers conclusions and an outlook on future activities.

2. SCENARIO

Ericsson Nikola Tesla (ENT) provides the Ericsson Healthcare Exchange (EHE) platform as part of its healthcare portfolio. The EHE platform is deployed on a national level in Croatia and several platform services are integrated in many healthcare provider institutions as shown in Figure 1. For example, the e-Prescription service is integrated in 2300+ general practitioner offices, 2500+ dentist offices, 192 paediatrician offices, 180 gynaecologist offices, and 1100+ pharmacies.

One of the main components of EHE is Electronic Health Records (EHR), which provide digitally stored health information for patients supporting care, education and research [7]. EHR stores information on medications, past medical history, immunisations, laboratory data, radiology reports etc [3]. The EHR system, as part of the EHE platform, needs to share data with healthcare providers, insurance institutions, government agencies, and patients.

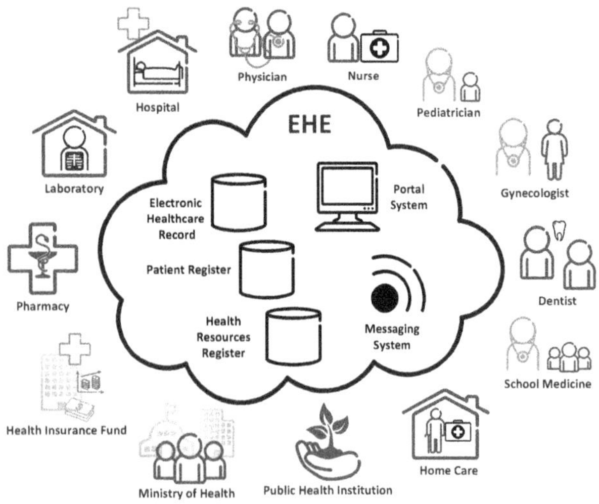

Figure 1: Ericsson Healthcare Exchange Platform.

The current non-cloud version of the EHR system is centralised and consists of the EHR database and a service layer developed above this database. Users can access EHR data either via direct database access or via the service layer. However, the main resource is always the database. Database mechanisms such as database clusters, connection pools, and database replication ensure system scalability and responsiveness up to some level that can be guaranteed by a large, but limited, system size. The resource demand per database user is non-linear.

A cloud-based solution is desired because of cost advantages, but also because of EHR requirements for scalability and elasticity. A cloud solution would also facilitate the provisioning of healthcare products and services to patients located in remote areas and others with limited access to quality medical services. EHR has the following characteristics which are particularly relevant for CloudScale's setting:

Variable load Since this solution is developed on a national level, the number of patients for which the healthcare records will be generated is known. Although the number of patients is predictable, the total number of EHR users can grow because of new services and integration of new institutions. Ericsson may also want to deploy the EHR technology in larger countries.

For specific annual events like regular flu or even major (not so frequent) events like bird flu, large earthquakes, or tsunami, there is a need for elastic system scale up. However, in holiday periods the system needs to scale down to reduce operational costs.

Constant system monitoring is necessary to be able to determine when and how to adjust the system size. Hence, a major advantage of a cloud based EHR solution, compared to the current non-cloud based solution, is system size elasticity.

Increasing work The size of the records will grow unpredictably, and some of EHR data records can be very large, e.g., treatment images, video files, sensor data (for remote patient monitoring).

Service Level Agreements (SLAs) Patient records may be required immediately. Users can be segmented in different hierarchy groups with different access priority, e.g., users from emergency healthcare institutions could have high priority.

Simple switching to a cloud infrastructure does not solve these problems without additional analysis and refactoring. Using CloudScale's approach, it should be possible to validate new requirements, evaluate system scalability, and provision an optimal amount of infrastructure resources. Now, it is very hard to predict the appropriate amount of system resources in a cost effective way, especially for custom defined user requirements.

In the process of switching from non-cloud EHR solution to the cloud infrastructure, Ericsson can use the CloudScale method and tools to monitor and optimise system scalability and also for predicting system size for a given workload. CloudScale can offer techniques analysing the existing solution, detecting architectural problems, and propose good architectural patterns for system evolution. When developing new services and identifying new customer requirements, design-time predictions could be made and this would also enable tuning of system specifications. In that way, EHR provider can forecast scalability properties and necessary system resources and have early operation cost predictions. During runtime, a constant monitoring processes can provide system elasticity to changes in workload.

3. METHOD

To illustrate how CloudScale aims at achieving the solution to the issues explained in Section 2, this section presents CloudScale's Method, its engineering approach for scalable cloud applications. We describe (1) the process steps as dictated by the Method, (2) main artefacts used in the Method (the scalability description language ScaleDL as well as (scalability) patterns and anti-patterns), and (3) the three tools Analyser, Extractor, and Spotter.

The proposed Method builds on an overall system lifecycle process illustrated in Figure 2 that focuses on scalability. As any other development process, it starts with requirements identification (Process (1)). During requirements identification, the main focus is on describing the load and work scaling path, i.e., the anticipated evolution of load and work on the system. The resource scaling path describes how our cloud platform can increase its amount of infrastructure resources, and a quality metric describe what is acceptable system quality to the users, e.g., a particular response time. For example, an emergency room using the EHR system could require to get one particular health record in less than one second. This process results in a requirement specification document.

Based on this requirement specification, the process of system construction and analysis starts (Process (2)) where we will use a model for specifying the system. We can do this in two ways:

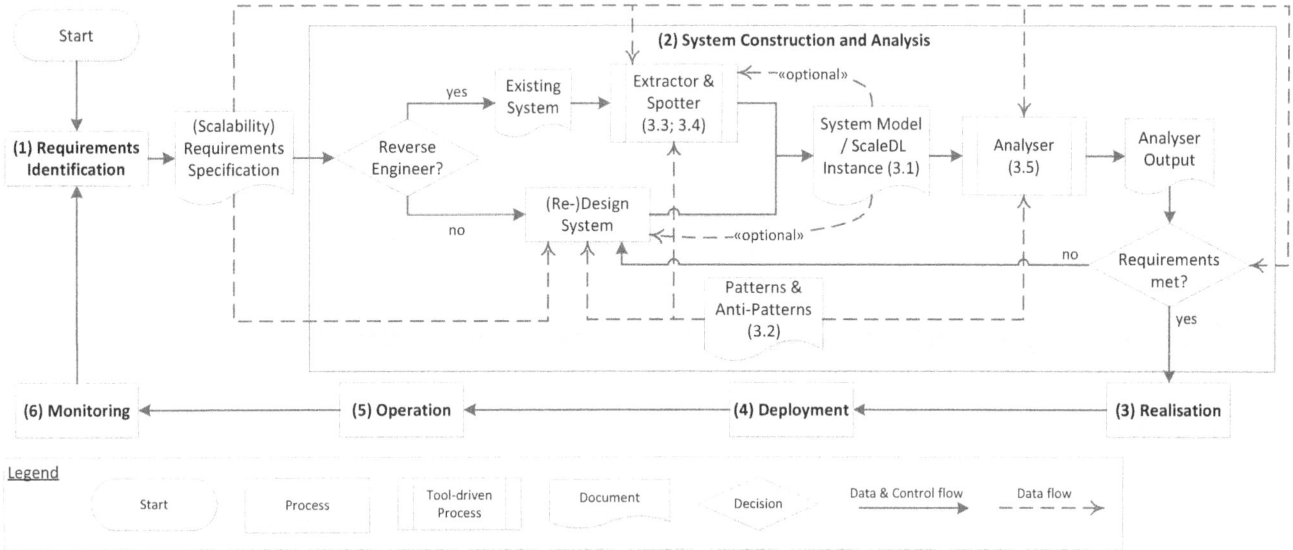

Figure 2: The overall CloudScale Method.

- *Reverse engineer* using an existing code base for creating an initial or adapting an existing system model. This process is driven by our tools Extractor and Spotter as described in Section 3.3 3.4, respectively.

- *(Re-)Designing* a system on the model level using our Analyser tool as described in Section 3.5. For specifying this model, we will use the ScaleDL language in Section 3.1.

We guide our design decisions along the requirements specification and support it with known patterns for good architectures regarding scalability in cloud environments. Section 3.2 provides details on patterns and anti-patterns.

The next step involves the analysis of the modelled system to check whether it meets the identified requirements. This process is driven by our Analyser tool as described in Section 3.5 and is repeated with different system alternatives until the requirements are met. The system construction and analysis process finally ends by reaching satisfactory results. We may also find that our requirements are infeasible and, therefore, have to relax them by reducing the complexity (and consequently work) of the services offered during high load.

After that, the system will be implemented (Process (3)), deployed (Process (4)), and put in operation in a cloud environment (Process (5)). For operation, it has well-defined resource requirements that enable cloud infrastructure provider to provision resources and fulfil load and work requirements of the deployed application reflecting the application evolution path.

Monitoring is another process (Process (6)) that is active during the system operation and enables control of system behaviour. Collecting measurements for performance parameters also belongs in this step. Based on the operational parameters and system quality metrics, monitoring control can require some changing system requirements and triggering the need to rerun our process cycle (evolution).

3.1 ScaleDL

ScaleDL will be a language for making precise models of the scalability properties of basic and composite cloud services. ScaleDL models will include structural elements of the system, parameters affecting scalability such as their resource usage, and elasticity information. Infrastructure resources will be described in a uniform manner, so that different configurations can be compared with each other. Among the resources to include in the model are processors, storage and network connections.

ScaleDL models can be created manually by the modeller based on knowledge and educated guessing, automatically by code analysis, measurements and predictions, or a combination of these. Ultimately, the scalability models can be used to determine resource scalability and cost scalability of the system, and help service providers to determine, e.g., what capacity the system will provide for a given cost. One of the design criteria for ScaleDL is that it must be useable in combination with scalability requirements specifications for the system, allowing the analysis of different scenarios and of fulfilment and consequences of the requirements.

ScaleDL will build on and extend existing languages and concepts where practical and appropriate. This includes the performance prediction language PCM of Palladio [5], the concept of system size from the scalability framework described in [6], and languages such as MARTE [4]. The design of ScaleDL will also be aligned with the ongoing development of the CloudML modelling language [1] which focuses on deployment models for cloud services.

3.2 Patterns and Anti-Patterns

Best practice scalability patterns show service providers how to build cloud systems with good scalability, while anti-patterns describe typical ways of making systems with poor scalability. For each anti-pattern, a recommended pattern is provided as well as a process for the transformation. Our EHR system can, for example, be suggested to be restructured according to a pattern that removes scalability issues with its database connection pool.

3.3 The Extractor

In case there is already existing code, the service provider can use the Extractor to extract an initial ScaleDL model to start from. This tool will enhance the reverse engineering tool Archimetrix [8] to take scalability parameters into account. The Extractor is also used during system evolution to keep the code and the scalability model in sync. In the EHR system, the Extractor would, e.g., find the data store component and its properties.

3.4 The Spotter

CloudScale's Spotter supports the service provider in spotting the places in a system that may cause scalability problems. It uses a ScaleDL model, AST's of code, monitoring data, and performance predictions to detect scalability anti-patterns. The Spotter can, e.g., be used to identify what needs to be changed to comply with new requirements and load/work plans. In the EHR system, the Spotter could, identify any non-linearites in the connection pooling that cause too many required connections when new users arrive.

3.5 The Analyser

Sooner or later, changing requirements or a changing work or load plan will trigger system redesign to still operate cost efficiently. The Analyser tool will make scalability predictions both of completely new designs or of redesigns based on existing systems. It will also analyse composed services. Using the Analyser, design alternatives can be analysed for their effect on scalability and cost. Internally, model transformations are employed to create scalability prediction models that build on established performance analysis approaches. We will base ourself on the open source tool set Palladio [5], which supports simulation as well as analytical solving.

Regarding our requirement that the EHR system should respond within one second, the Analyser may predict that it is capable of handling the increased load in case a bird flue occurs. Particularly, the Analyser takes the cloud systems' elasticity into account for its analysis.

4. VALIDATION AND DEMONSTRATORS

Demonstrators will show all the aspects of CloudScale in the form of prediction, analysis, metrics, measurements, anti-patterns, etc. They will provide a story line to support dissemination and make a platform for exploitation. The demonstrators will be completely open and realistic and will overcome any confidentiality issues of the industrial partners SAP and ENT in presentation of their internal systems.

The validation of CloudScale's showcases will collect evidence that (a) the CloudScale methods and tools spot the right bottlenecks, (b) our scalability predictions are sound, and (c) near optimal scalability improvements of a system are suggested. We also validate the feasibility of the CloudScale Method and report on the efforts needed to use it.

5. CONCLUSIONS

This paper has presented the CloudScale project at an overall level. The planned results of CloudScale are aimed at different types of people, organisations and roles, offering benefits to each.

- *End users* being satisfied also during peak loads because of improved scalability of the systems.
- For *developers* of software services, improved scalability management becomes a selling point. CloudScale tools help developers to make sensible decisions about which parts of the system which require extra care.
- *System architects* are able to understand and predict the scalability of services resulting from compositions.
- *Service providers* are able to make timely decisions about purchase or deployment of more hardware in order to prevent scalability bottlenecks before they show up. They are also able to plan reduction in non-essential features to retain core functionality during periods of extreme demand.
- *IaaS (Infrastructure as a Service)* providers will optimise their business due to more effective use of resources by their customers, thanks to improved scalability. Moreover, they are able to serve more customers with the same hardware through better management of scalability of their own systems. Thereby, they can operate with a smaller safety margin and greater profit.

The CloudScale project started in October 2012 and will continue until September 2015. The first versions of the results are planned in October 2013. This includes a refined Method, ScaleDL, the three tools Analyser, Spotter, and Extractor, as well as patterns and anti-patterns.

6. ACKNOWLEDGEMENT

The research leading to these results has received funding from the European Union Seventh Framework Programme (FP7/2007-2013) under grant no 317704 (CloudScale).

7. REFERENCES

[1] Cloud modelling language (CloudML). www.cloudml.org.

[2] CloudScale Project. www.cloudscale-project.eu.

[3] Healthcare Information and Management Systems. www.himss.org.

[4] Modelling and Analysis of Real-Time and Embedded Systems (MARTE). www.omgmarte.org.

[5] S. Becker, H. Koziolek, and R. Reussner. The palladio component model for model-driven performance prediction. *Journal of Systems and Software*, 82(1):3 – 22, 2009.

[6] G. Brataas, P. H. Hughes, J.-A. Fagerli, and O. C. Landmark. Exploring Architectural Scalability. *Software Engineering Notes*, 29(1):125 – 129, 2004.

[7] I. Iakovidis. Towards personal health record: current situation, obstacles and trends in implementation of electronic healthcare record in Europe. *International Journal of Medical Informatics*, 52(1-3):105 – 115, 1998.

[8] M. Platenius, M. von Detten, and S. Becker. Archimetrix: Improved software architecture recovery in the presence of design deficiencies. In *Software Maintenance and Reengineering (CSMR), 2012 16th European Conference on*, pages 255 –264, march 2012.

A Generic Approach for Architecture-Level Performance Modeling and Prediction of Virtualized Storage Systems

Qais Noorshams, Andreas Rentschler, Samuel Kounev, Ralf Reussner
Chair for Software Design and Quality
Karlsruhe Institute of Technology
Karlsruhe, Germany
{noorshams, rentschler, kounev, reussner}@kit.edu

ABSTRACT

Virtualized environments introduce an additional abstraction layer on top of physical resources to enable the collective resource usage by multiple systems. With the rise of I/O-intensive applications, however, the virtualized storage of such shared environments can quickly become a bottleneck and lead to performance and scalability issues. The latter can be avoided through careful design of the application architecture and systematic capacity planning throughout the system life cycle. In current practice, however, virtualized storage and its performance-influencing design decisions are often neglected or treated as a black-box. In this work-in-progress paper, we propose a generic approach for performance modeling and prediction of virtualized storage systems at the software architecture level. More specifically, we propose two performance modeling approaches of virtualized systems. Furthermore, we propose two approaches how the performance models can be combined with architecture-level performance models. The goal is to cope with the increasing complexity of virtualized storage systems with the benefit of intuitive software architecture-level models.

Categories and Subject Descriptors

C.4 [**Performance of Systems**]: Modeling techniques;
D.2.11 [**Software Engineering**]: Software Architectures

Keywords

I/O, Storage, Performance, Virtualization

1. INTRODUCTION

In recent years, virtualization technology has been widely adopted. According to analysts, this development has yet to reach its peak. Over the period 2012–2016, the server virtualization market is expected to grow annually by more than 31% [18]. In a parallel development, the I/O resource demands of modern IT systems have grown exponentially over the past couple of decades [17]. Until the year 2020, the amount of digital data is expected to double every year, 40% of which is expected to be either processed or stored in the cloud [5]. The danger of this trend is, however, that the virtualized storage of shared environments becomes a bottleneck, thus leading to significant performance and scalability issues as there are many performance-influencing factors. To avoid such issues, performance modeling and evaluation techniques are needed for capacity planning at deployment time as well as to guarantee performance requirements during operation upon changes in the workload.

In current practice, however, virtualized storage and its performances influences on the overall system performance are often neglected or treated as a black-box due to their complexity. Since traditional storage systems were not designed for the intricate data consumption in virtualized environments [7], virtualized storage systems have evolved significantly to sophisticated systems with multiple caching and virtualization layers.

Existing approaches model the storage performance of I/O-intensive applications in virtualized environments, e.g., [1, 10], however, with validation limited to basic environments or with focus on consolidation scenarios only. Moreover, it is difficult to include such models into software architecture models, because the required information between the two modeling abstraction levels needs to be synchronized. Software architecture models are on a higher abstraction level making them simpler and more intuitive to use. However, they may have several access points to the storage resources, thus, having mutual performance effects with low-level storage models. A media streaming portal, for instance, can have multiple different requests at the software architecture level, e.g., watermarking and media downloading. Those requests translate to intertwined read and write accesses at the storage system.

In this work-in-progress paper, we propose a generic approach for performance modeling and prediction of virtualized storage systems and combining high-level software architecture models with low-level storage models. More specifically, we propose a black- and a white-box performance modeling approach for virtualized storage systems. Furthermore, we propose two approaches how the performance models can be combined with architecture-level performance models. The goal is to cope with the increasing complexity of virtualized storage systems with the benefit of intuitive software architecture models. The overall methodology, however, is generic and can be applied for enhancing any software architecture models by more complex resource or subsystem models.

ICPE'13, April 21–24, 2013, Prague, Czech Republic.

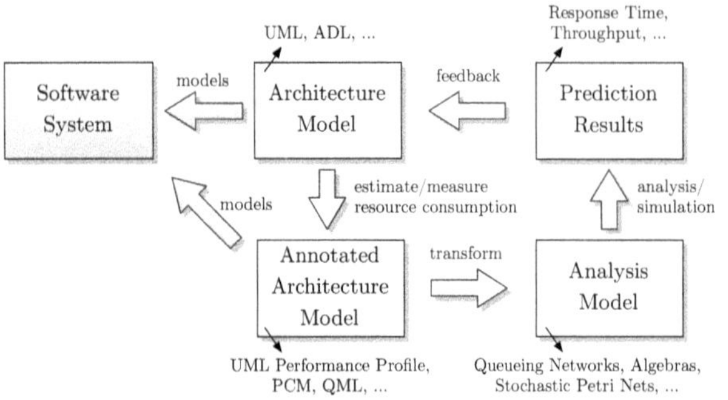

Figure 1: Model-based Performance Prediction (Source: [2])

2. PROPOSED APPROACH

In this section, we first present the background of our work originated from model-based performance prediction. Then, we present our performance modeling methodology as well as our combination approach integrating performance models of virtualized storage systems into software architecture models.

2.1 Background

In general, our approach can be applied to any performance-aware software architecture models. We apply our proposed approach to the Palladio Component Model (PCM) [3], a modeling language for component-based software architectures realizing a model-based performance prediction approach. The PCM is aligned with the component-based software engineering (CBSE) development process and provides modeling constructs to describe the i) software components, ii) system architecture, iii) resource allocations, and iv) usage profile of a component-based software system.

Using model transformations, e.g., to queueing networks or performance prototypes, a PCM model instance can be analytically solved or simulated to predict the system performance. The performance prediction serves as feedback and enables a model-based quality assessment of software systems. This process is illustrated in Figure 1.

2.2 Performance Modeling Methodology

Illustrated in Figure 2, in this section we show how our approach fits into the model-based performance prediction approach by first analyzing performance-influencing factors and then creating black- and white-box storage analysis models, i.e., performance prediction models. The storage analysis models are combined with the software architecture level model in the next section.

2.2.1 System Under Study

As a proof-of-concept of our approach, we study a representative virtualized environment based on the IBM mainframe *System z* and the storage system *DS8700*. They are state-of-the-art high-performance virtualized systems with redundant or hot swappable resources for high availability. The System z combined with the DS8700 represents a typical virtualized environment that can be used as a building block of cloud computing infrastructures. It supports on-demand elasticity of pooled resources with a pay-per-use accounting system (cf. [12]). The System z provides processors and memory, whereas the DS8700 provides storage space.

2.2.2 Performance-Influencing Factors

Aligned with our modeling methodology, in [16] we identified and evaluated the performance-influencing factors of applications deployed in the considered environment. The factors comprise both workload-specific characteristics and system-specific configuration parameters. An important aspect is to find an appropriate abstraction level for the factors. On the one hand, too coarse-grained factors lead to inaccuracies in the performance models. On the other hand, too fine-grained factors cannot be specified by software developers and deployers. Furthermore, we develop a fully automated approach to extract the workload characteristics for a given application. Tools like *blktrace*[1] are able to provide detailed workload-level information, e.g., the read/write ratio and the requested block addresses. Based on this information, we calculate important request metrics, e.g., the average size. Moreover, we develop a heuristic classification strategy to identify multiple, possibly intertwined, sequential accessing clients, since this has a significant performance impact [16].

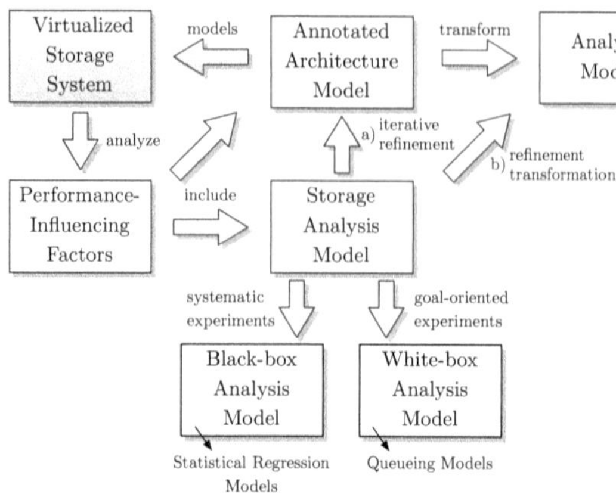

Figure 2: Proposed Approach

2.2.3 Black-box Analysis Model

In [15] we created black-box analysis models by applying multiple statistical regression techniques on systematic measurements. Regression techniques are very powerful in terms of prediction accuracy for interpolation scenarios. Furthermore, we proposed a general heuristic optimization algorithm to optimize the parameters of the regression techniques. The main advantage of our black-box modeling approach is that it is fully automated as part of our tool *SPA* [4], thus, lifting the burden for performance engineers to manually develop the models. We are working on extending our automated methodology to further evaluate the approach with another load-driver and in another system environment. Moreover, we plan to use the automatically extracted workload characterization, which will be integrated into SPA, as input for the regression-based models to be able to predict arbitrary workloads, cf. Figure 3. This approach aims at creating the black-box models once and map different applications to the required black-box model input parameters.

2.2.4 White-box Analysis Model

The black-box modeling approach requires systematic experiments that cover the whole parameter space. To create performance models with a set of goal-oriented experiments and gain further understanding of the system environment, we develop an iterative white-box performance modeling methodology and model our virtualized storage system in a queueing network, cf. Figure 4. Our approach follows the

[1] http://www.cse.unsw.edu.au/~aaronc/iosched/doc/blktrace.html

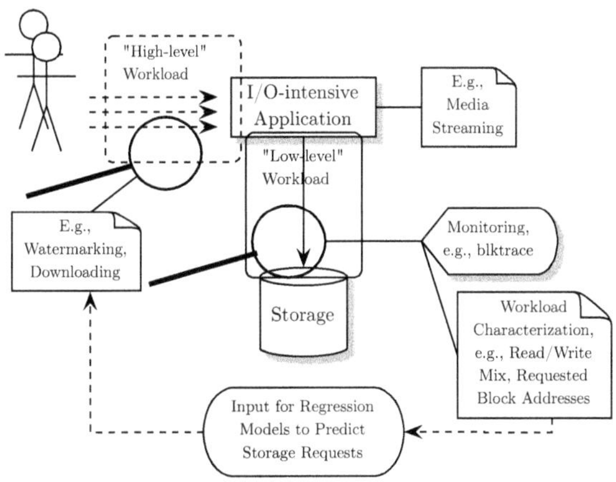

Figure 3: Analysis Model Generalization Approach

common generic steps in classical performance engineering comprised of performance model *creation*, *calibration*, and *validation* established by existing approaches [13, 14]. We first analyze the system environment identifying the system-specific performance-relevant aspects including the generic performance-influencing factors. Furthermore, we identify the workload scenarios we want capture as part of the model initially. We then create the model calibrated with experiments that we iteratively refine to account for more complex scenarios. Our preliminary results model the caching and the RAID array resource of our environment in homogeneous queueing models. The mean service times μ of the service stations are parameterized and calibrated based on response time measurements, such that

$$\mu(t, s, d) = c_1\, ts \cdot \ln(d) + c_2\, t \cdot \ln(d) + c_3\, s \cdot \ln(d) + c_4\, ts + c_5\, t + c_6\, s + c_7 \ln(d) + c_8; \quad c_i \in \mathbb{R},$$

where t is the number of simultaneous requests (clients), s is the request sizes, and d is the overall amount of data the workload operates on. Shown in Figure 5, the results demonstrate the high prediction accuracy of our white-box models for both interpolation and extrapolation (or consolidation) scenarios. Figure 5a shows the relative error for 1200 completely random configurations. The mean error (depicted as cross) is 7.22% and 2.39% for read and write requests, respectively. Figure 5b shows the relative error for 480 workload consolidation scenarios with 2 and 3 VMs. Here, the mean error is 8.22% and 2.74% for read and write requests, respectively. Next, we plan to extend the queueing models to account for mixed workloads. Furthermore, we plan to generalize the white-box model similar to the black-box model generalization approach.

2.3 Combination Approach

We propose two approaches to combine the low-level storage models with high-level software architecture models, *iterative refinement* and *refinement transformations*. In both approaches, it is important to design the *storage interface* between the two modeling abstraction levels and integrate the interface into the architecture model.

2.3.1 Storage Interface Design

The storage interface includes the required dynamic information (i.e., the workload-specific performance-influencing factors) in the software architecture model that is passed as input to the storage analysis model (e.g., the distinction between read and write requests and their access pattern). Furthermore, the set of static storage-relevant information (i.e., the system-specific performance-influencing factors) is used to configure or initialize the storage analysis model (e.g., the chosen I/O scheduler). At this point, the main research question is what to include in the storage interface and at which abstraction level.

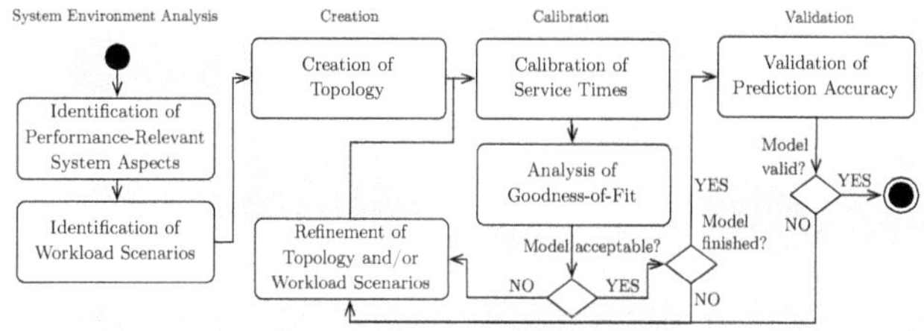

Figure 4: Iterative White-box Modeling Methodology

2.3.2 Iterative Refinement

An iterative refinement approach allows for combining a specific software architecture model instance with the storage analysis model. The software architecture model instance contains multiple accesses to different active and passive resources (e.g., thread pools) and, more specifically, contains storage resource consumptions that are represented by a parametric delay, e.g., when simulating the architecture model instance. The architecture model instance is coupled with the storage analysis model as follows (cf. Figure 6):

1. The storage delays are initialized with a small value.
2. The architecture model instance is simulated to obtain the performance characteristics of the model instance. During the simulation, the storage accesses of the model instance over the storage interface are monitored to extract the storage workload profile.
3. The workload profile and the static storage-relevant information is used as input for the storage analysis model to obtain specific storage delays for the architecture model.
4. The storage delays in the simulation are updated with the delays obtained by the storage analysis model.
5. Step 2–4 is repeated until the performance characteristics of the model instance converge.

The caveat is, however, that the performance characteristics are not guaranteed to converge. Moreover, it is an open question to what extent this approach can be automated.

2.3.3 Refinement Transformations

A second approach is to include the storage analysis model with refinement transformations. Illustrated in Figure 7, the transformations include both i) a *completion transformation* [19] as well as ii) an *analysis transformation*. The

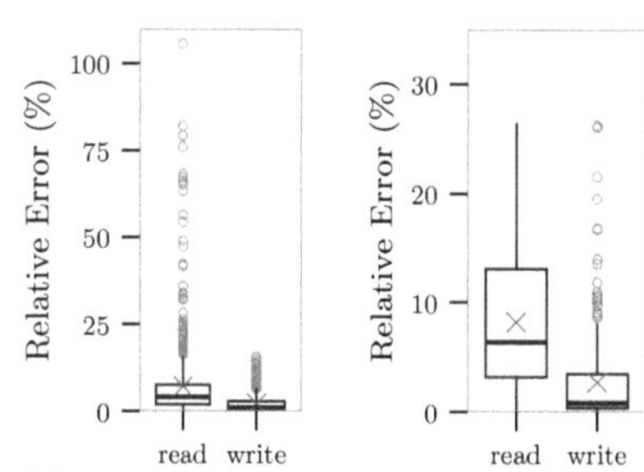

(a) Interpolation (b) Consolidation

Figure 5: Prediction Accuracy of Queueing Models

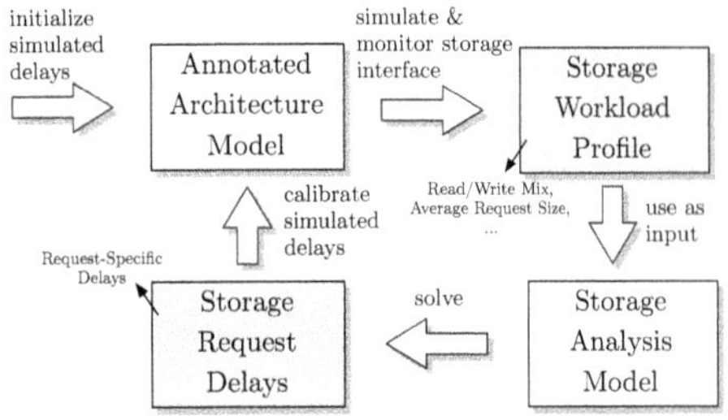

Figure 6: Iterative Refinement Approach

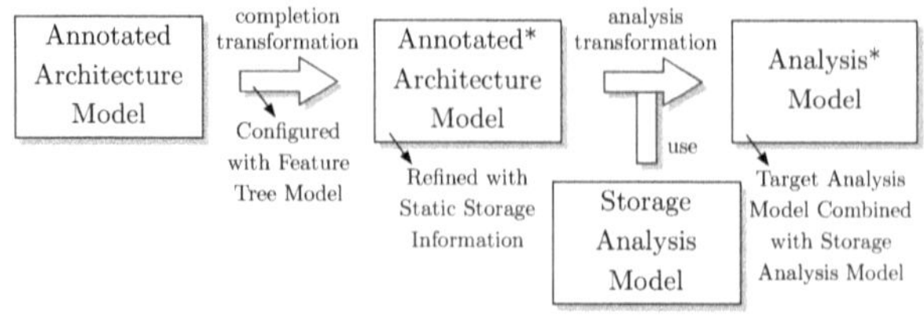

Figure 7: Refinement Transformation Approach

completion transformation refines the architecture model to fill in the static storage-relevant information and is configured, e.g., using feature tree models, cf. [8]. The analysis transformation uses the architecture model and creates the target analysis model to predict the system performance, cf. Figure 1. This transformation is extended to combine the target analysis model with the storage analysis model over the storage interface using a bridge or adapter. Generally, this approach allows for an arbitrary target analysis model. However, the conceptual interoperability of different modeling formalisms is still subject to research. Thus, if different formalisms are to be combined, their interoperability is required to be well-defined.

3. RELATED WORK

Several existing approaches have been proposed for performance analysis of I/O-intensive applications in virtualized environments. However, none of these approaches considers the system performance at the software architecture level as realized in the approach presented in this paper. Most related, Kraft et al. [10] propose a queueing network-based approach to predict the I/O performance of consolidated virtual machines. They monitor and model low-, block-level I/O requests. Their approach is focused on predicting consolidation scenarios. Further, Ahmad et al. [1] analyze the I/O performance in VMware's ESX Server virtualization. They compare virtual against native performance using benchmarks. They create mathematical models scaling the native performance down for I/O throughput predictions in virtualization. To analyze performance interference in a virtualized environment, Koh et al. [9] manually run CPU bound and I/O bound benchmarks. While they develop mathematical models for prediction, they explicitly focus on the consolidation of different types of workloads, i.e., CPU and I/O bound. By applying machine learning techniques, Kundu et al. [11] use artificial neural networks and support vector machines for dynamic capacity planning in virtualized environments. In [6], Gulati et al. present a study on storage workload characterization in virtualized environments, however, they do not propose any performance prediction techniques.

4. CONCLUSION

In this work-in-progress paper, we proposed a methodology for performance modeling and analysis of a) I/O-intensive applications in b) virtualized environments at the c) software architecture level. To the best of our knowledge, there are no other approaches that combine a), b), and c) to our extent. For this approach, we will extend the Palladio Component Model, a model-based performance prediction approach for component-based software architectures. For performance modeling, we create both black-box models based on statistical regression techniques and white-box models based on queueing networks. We apply our approach in real world environments as the IBM System z and the IBM DS8700. Our main contribution is that we provide a generic approach for combining software architecture level performance models with resource-specific storage models. More specifically, our contributions are the following:

– We analyze virtualized storage systems and identify the performance-influencing factors.
– We create both black- and white-box analysis models of virtualized storage systems.
– We combine high-level software architecture models and low-level storage performance models.

Acknowledgments This work was funded by the German Research Foundation (DFG) under grant No. RE 1674/5-1 and KO 3445/6-1. We especially thank the Informatics Innovation Center (IIC) – `http://www.iic.kit.edu/` – for providing the system environment of the IBM System z and the IBM DS8700.

5. REFERENCES

[1] I. Ahmad, J. Anderson, A. Holler, R. Kambo, and V. Makhija. An analysis of disk performance in vmware esx server virtual machines. In *WWC-6*, 2003.
[2] S. Becker. *Coupled model transformations for QoS enabled component-based software design*. PhD thesis, Universität Oldenburg, 2008.
[3] S. Becker, H. Koziolek, and R. Reussner. The palladio component model for model-driven performance prediction. *J. of Systems and Software*, 82(1), 2009.
[4] D. Bruhn, Q. Noorshams, S. Kounev, and R. Reussner. Storage Performance Analyzer (SPA). http://sdqweb.ipd.kit.edu/wiki/SPA, 2012.
[5] J. Gantz and D. Reinsel (IDC). THE DIGITAL UNIVERSE IN 2020: Big Data, Bigger Digital Shadows, and Biggest Growth in the Far East. http://idcdocserv.com/1414, 2012. Last accessed: Jan 2013.
[6] A. Gulati, C. Kumar, and I. Ahmad. Storage workload characterization and consolidation in virtualized environments. In *VPACT '09*.
[7] InformationAge. The year of virtual storage. http://www.information-age.com/channels/the-cloud-and-virtualization/perspectives-and-trends/1596523/the-year-of-virtual-storage.thtml, 2011. Last accessed: Jan 2013.
[8] L. Kapova and T. Goldschmidt. Automated feature model-based generation of refinement transformations. In *SEAA '09*.
[9] Y. Koh, R. Knauerhase, P. Brett, M. Bowman, Z. Wen, and C. Pu. An analysis of performance interference effects in virtual environments. In *ISPASS '07*.
[10] S. Kraft, G. Casale, D. Krishnamurthy, D. Greer, and P. Kilpatrick. Performance models of storage contention in cloud environments. *Springer Journal of Software and Systems Modeling*, 2012.
[11] S. Kundu, R. Rangaswami, A. Gulati, M. Zhao, and K. Dutta. Modeling Virtualized Applications using Machine Learning Techniques. In *VEE '12*.
[12] P. Mell and T. Grance. The nist definition of cloud computing. *National Institute of Standards and Technology*, 53(6), 2009.
[13] D. Menascé and V. Almeida. *Scaling for E-Business: Technologies, Models, Performance, and Capacity Planning*. Prentice Hall, 2000.
[14] D. Menascé, V. Almeida, L. Dowdy, and L. Dowdy. *Performance by Design: Computer Capacity Planning by Example*. Prentice Hall science explorer. Prentice Hall, 2004.
[15] Q. Noorshams, D. Bruhn, S. Kounev, and R. Reussner. Predictive Performance Modeling of Virtualized Storage Systems using Optimized Statistical Regression Techniques. In *ICPE '13*.
[16] Q. Noorshams, S. Kounev, and R. Reussner. Experimental Evaluation of the Performance-Influencing Factors of Virtualized Storage Systems. In *EPEW '12*, volume 7587 of *LNCS*. Springer.
[17] S. Oliveira, K. Furlinger, and D. Kranzlmuller. Trends in computation, communication and storage and the consequences for data-intensive science. In *IEEE HPCC-ICESS'12*, pages 572–579.
[18] TechNavio. Global Server Virtualization Market 2012-2016. http://www.technavio.com/content/global-server-virtualization-market-2012-2016, 2013. Last accessed: Jan 2013.
[19] M. Woodside, D. Petriu, and K. Siddiqui. Performance-related completions for software specifications. In *ICSE '02*.

Adaptive Deployment in Ad-Hoc Systems Using Emergent Component Ensembles: Vision Paper

Lubomír Bulej[1,2] Tomáš Bureš[1,2] Vojtěch Horký[1] Jaroslav Keznikl[1,2]

[1]Charles University in Prague
Faculty of Mathematics and Physics
Malostranské náměstí 25
118 00 Prague 1, Czech Republic

[2]Academy of Sciences of the Czech Republic
Institute of Computer Science
Pod Vodárenskou věží 2
182 07 Prague 8, Czech Republic

{bulej,bures,horky,keznikl}@d3s.mff.cuni.cz

ABSTRACT

Mobile cloud computing in the context of ad-hoc clouds brings new challenges when offloading computation from mobile devices. The management of application deployment needs to ensure that the offloading provides users with the expected benefits, but it suddenly needs to cope with a highly dynamic environment which lacks a central authority and in which computational nodes appear and disappear.

We propose an approach to the management of ad-hoc systems in such dynamic environment using component ensembles that connect mobile devices with more powerful computation nodes. Our approach aims to address the challenges of scalability and robustness of such systems without the need for central authority, relying instead on simple patterns that lead to reasonable adaptation decisions based on limited and imprecise information.

Categories and Subject Descriptors

[**Computer systems organization**]: Other architectures — Self-organizing autonomic computing, Distributed architectures — Cloud computing; [**Software and its engineering**]: Extra-functional properties — Software performance

Keywords

ad-hoc cloud, ensembles, adaptive deployment

1. INTRODUCTION

Increasing capabilities of handheld devices and improvements in mobile network infrastructures pave the way for mobile cloud computing [1], an architectural solution where mobile devices offload computation to the cloud to gain advantage for example in increased computing power or reduced battery usage. Another motivation is the emergence of ad-hoc clouds [2], whose computing power comes from pooled resources of nearby general-purpose computing devices rather than from dedicated servers. Our work, carried out in the scope of the ASCENS project [3], aims to combine and extend the two trends by blurring the traditionally strict [4] boundary between the client devices and the cloud infrastructure. We envision an ad-hoc cloud formed as a multitude of dynamically emerging groups of computational devices that share their computing power. The groups will typically involve a number of mobile devices along with locally situated general-purpose computers, potentially connected to a remotely situated dedicated cloud infrastructure.

Important specifics of our ad-hoc cloud concept are that it has a dynamic, mostly uncontrollable architecture with fluctuating computing power and partially limited resources, e.g. the available battery charge for mobile devices. The promise of the ad-hoc cloud is in maintaining the usual benefits associated with offloading computation to a cheap, flexible and resilient environment—however, the ad-hoc cloud applications must dynamically adapt their deployment to deliver the expected user experience in such specific conditions.

The challenge in application adaptation is related to the open character of the ad-hoc cloud. Where a common cloud application can react to increased utilization by requesting additional computing resources, an application in an ad-hoc cloud must act in presence of other adapting applications that share the same resources. Even the scheduling solutions for computational grids, which do cope with shared resources, assume a degree of centralization knowledge and control over the grid that is not available in the ad-hoc cloud [5]. We therefore believe that the application of adaptation solutions in ad-hoc clouds will not assume the shape of complex algorithms that compute close-to-optimal deployment under dynamic conditions, but rather the shape of relatively simple patterns that lead to reasonable adaptation decisions relying on limited and imprecise information.

In this vision paper, we focus on the problem of adaptive deployment planning in ad-hoc clouds and outline an adaptation approach based on a component system with emergent component ensembles [6, 3]. We assume an external mechanism would be responsible for the actual migration [7]. We discuss the potential benefits and scalability of this approach.

2. MOTIVATING EXAMPLE

Besides smart phones, we consider tablet computers to be a perfect target for mobile computing in ad-hoc clouds—modern applications take advantage of their relatively high computational power and users tend to use them both for

ICPE'13, April 21–24, 2013, Prague, Czech Republic.

work and entertainment. However, they are still constrained by the limited battery life.

We start our vision with an example, where we consider a user travelling in a train or a bus, who wants to do productive work using her tablet computer or review travel plans and accommodation. Her tablet registers the presence of an offload server machine located in the bus itself, and to save battery, it offloads most computationally intensive tasks to that machine. Later, when the bus approaches the destination, the offload server notifies her tablet that its service will soon become unavailable and tasks will start moving back to the tablet. When the bus enters the terminal, the tablet will discover another offload server, provided by the terminal authority, and move some of its tasks to the newly found machine.

Many similar examples can be found, and they would follow similar pattern. Abstracting away from the details, we can try to capture such examples formally under one general umbrella. Assume a mobile device M (tablet in our example) and two stationary devices S and T (offload servers in our example). M executes application A, which is internally split into two parts: a frontend Af, responsible for the interaction with a user, and a backend Ab, responsible for the computationally intensive tasks.

In our scenario, M discovers S and assesses that offloading the computationally intensive Ab to S could save M's battery. After some time, S signals that it is going to be unavailable, but M discovers that there are other devices available. Ab is thus migrated to the one that appears most suitable for running Ab—device T in our scenario.

The challenge is in predicting which deployment scenario will—in the context of ad-hoc cloud—deliver the expected user experience. We assume that each application will have a simple performance model that, given specifics of the execution environment and other constraints, will provide a rough estimate of the expected user experience (e.g. what frame rate could be achieved given CPU and GPU budget). The application deployment would be then planned dynamically, taking into account the expected user experience estimated by the model, possibly corrected for measured accuracy of the model from past deployments.

3. COMPONENT ENSEMBLES

Although the scenario of the running example is relatively straightforward, it is relatively difficult to realize due to the inherent dynamicity of the whole ecosystem (i.e. all applications and devices). Further, the combination of the dynamicity and the autonomy of the applications and devices imply the absence of the notion of a global state. In fact, every information about the ecosystem has the form of a "belief" – i.e. an information valid to only a certain extent.

To cope with these issues, we suggest in this paper to take advantage of a component system based on emergent component ensembles [6, 3] and use it to represent situation in such a dynamic ecosystem and to manage the belief about it. To this end, we outline in the rest of the section the basic principles of component ensembles and explain their use in addressing the adaptation in Section 4.

Emergent component ensembles are based on the idea of implicit communication via implicit bindings. Specifically, an ensemble is a dynamically formed group of components, where a component constitutes knowledge (i.e., data) and processes (i.e., active threads operating upon the knowledge). The membership of a component in an ensemble is determined dynamically (the task of the component system runtime framework) according to the membership condition of the ensemble specified upon the knowledge of the components. In an ensemble one component plays the role of the ensemble's coordinator while others play the role of members. A single component can be member and/or coordinator of multiple ensembles at the same time; thus an ensemble forms an independent logical overlay over components.

The interaction among the components forming an ensemble takes the form of knowledge exchange, carried out implicitly (by the runtime framework); i.e., the runtime framework transfers knowledge from one component of the ensemble to another independently on the components' execution.

The benefit of such ensembles is that they allow for capturing communication (i.e., exchange of knowledge) among a (potentially) large, declaratively defined set of components in a concise way.

4. ADAPTATION ARCHITECTURE

To address the aforementioned challenges, we present a generic architecture of adaptation logic, based on emergent component ensembles. The proposed *adaptation architecture* follows the following basic principles.

To ensure separation of concerns, the adaptation logic forms a separate overlay architecture mirroring the architecture of the adapted application.

Additionally, the entities important for deciding adaptation, i.e., (a) computation nodes (and their NFPs), (b) individual adapted applications (and their NFP preferences), and (c) the applications' components (and their NFPs), are explicitly reflected in the adaptation architecture.

4.1 Adaptation architecture components

The adaptation architecture (Figure 1) is formed by following components:

Planner. Each adapted application, and particularly its NFP preferences, are represented by the Planner component. Specifically, the Planner selects a (potentially optimal) deployment of the application, given the alternatives for deploying each of the application's components. We assume an external mechanism [7] to interpret the deployment plan provided by the Planner and perform the adaptation (e.g., by migrating a component). The alternatives comprise important NFP-related data (NFPData) indicating the (potential) performance of the corresponding application component in that particular deployment (e.g., FPS, energy, etc.). The Planner also advertises definitions of Monitors for individual application components (MonitorDef); see Device.

Monitor. Each application component, particularly each of its deployment alternatives, is reflected by the Monitor component, which is responsible for obtaining the NFPData for that particular alternative. Monitor operates in one of the two modes, depending on the actual deployment of the corresponding application component.

- Monitor is in the *running mode* if it resides on the same computation node as the corresponding application component, i.e. it reflects the actual deployment. NFPData is obtained by performance measurement and analysis of the running application component.
- Monitor is in the *mock mode* if it resides on a different computation node, i.e. it represents a potential deployment alternative. NFPData is obtained from

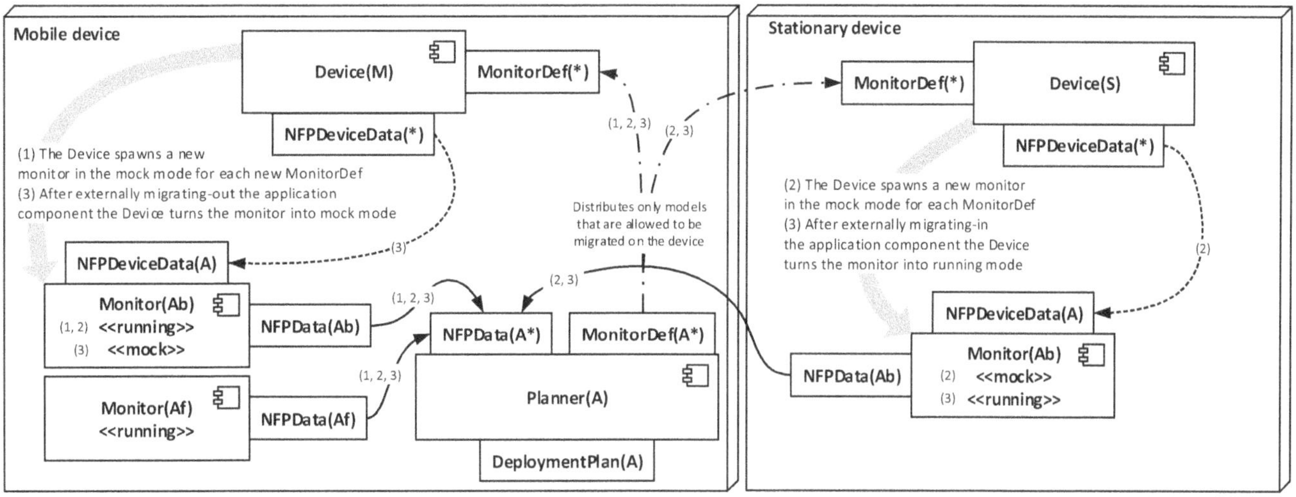

Figure 1: Adaptation architecture of the running example: phases 1 (M isolated), 2 (S discovered), and 3 (Ab migrated to S). Phases 1,2,3 are in the figure denoted by (1), (2), (3).

the included performance dependency model of the corresponding application component (e.g., the function $CPU \times GPU \rightarrow FPS$). In other words, Monitor predicts – based on the model – the performance of the application component if it would be deployed on that computation node. The model might depend on particular machine-specific performance data (NFPDeviceData, e.g., available CPU speed, etc.); see Device.

Device. Each computation node is reflected by the Device component. Specifically, a Device component ensures management of the Monitors (e.g., it instantiates Monitors advertised by newly discovered Planners) and it provides NFPDeviceData for Monitors in the mock mode.

4.2 Adaptation architecture ensembles

The expectation is that the number of available computation nodes, as well as the number of Monitors, changes dynamically. Therefore, the communication among the components exploits the concept of emergent component ensembles. The architecture involves the following ensembles (Figure 1):

Planner and Device(s). Each Planner is a coordinator of an ensemble that distributes MonitorDefs (including the performance dependency model) of application components to Devices representing currently available computation nodes (including the one the Planner is running on). The Planner is able to constraint which MonitorDefs should be distributed to which Devices (effectively constraining the potential migration destinations for a particular application component). A simplified example of a definition of this ensemble is in Figure 2. It specifies that only reachable devices within 2 network hops are to be considered and that this check is to be performed every 15 seconds. The distribution of the MonitorDefs is performed by adding the MonitorDef to the target component's knowledge.

Planner and Monitor(s). Each Planner is a coordinator of an ensemble that aggregates NFPData from all Monitors corresponding to the components of the application reflected by the Planner. Thus, this ensemble aggregates all the deployment alternatives for the application.

```
ensemble PlannerToDevice:
  coordinator: Planner
  member: Device
  membership: HopDistance(Planner.device, Device) ≤ 2
  knowledge exchange:
    Device.monitorDef[Planner.app] := Planner.monitorDef
  scheduling: periodic( 15s )
```

Figure 2: Example of an ensemble definition.

Device and Monitor(s). Each Device component is a coordinator of an ensemble that distributes NFPDeviceData to the Monitors in the mock mode residing on the corresponding computation node.

4.3 Adaptation architecture in action

In this section, we illustrate on the motivation example the adaptation architecture interaction at runtime.

At first (phase 1, Figure 1), the ensemble distributes the MonitorDefs of both Af and Ab from Planner of A to the Device component of the mobile device (M), which subsequently spawns Monitors for both components and sets them to the running mode. The Monitors start measuring NFPData of the running components which are then aggregated back to the Planner. So far no deployment alternatives are discovered.

After the stationary device (S) is discovered (phase 2, Figure 1), the ensemble propagates MonitorDefs of the components that could be (potentially) migrated (i.e., Ab) to its Device component, which spawns a new Monitor. Since Ab is deployed on a different Device this Monitor runs in the mock mode. Thus, the Device component of the stationary device feeds the Monitor with NFPDeviceData allocated for A. Based on this NFPDeviceData and the performance dependency model of Ab the Monitor produces NFPData reflecting the expected performance of Ab on S. Consequently, another ensemble aggregates all the currently produced NFPData for Af and Ab to the Planner. The Planner thus eventually discovers that there are two deployment alternatives for Ab (i.e., one actually running on M and one modeled on S) and finally decides to deploy Ab on the stationary device.

After Ab is migrated to the stationary device (phase 3, Figure 1), the Monitor on S is set to the running mode, while the Monitor on M is set to the mock mode and the whole monitoring and planning process repeats.

In the case of discovering further stationary devices, new Monitors in the mock mode are spawned which eventually results in new deployment alternatives aggregated in the Planner (similarly, if devices disappear).

5. BENEFITS

Scalability and robustness. By exploiting the features of the ensembles, the adaptation architecture scales well with the number of computation nodes, applications, and components per application. In fact, the adaptation architecture does not require any changes when increasing the number of nodes/applications/components. Furthermore, it is very robust with respect to emergence of computation nodes.

Transparent trade-off management. Due to the declarative nature of ensembles, it is possible to easily manage the trade-offs between the benefit of migration and the effort necessary for monitoring and planning. For instance, Monitors do not have to be spawned on all available nodes but only on a subset; e.g., only the nodes in the same subnet.

Respecting interests of all involved parties without central authority. Although each application is planned autonomously, it is possible (without any centralized authority) to take into account the interests of the other applications and of host devices by regulation of the NFPDeviceData and management of the application's Monitors — e.g., the NFPDeviceData may reflect only a portion of device's resources.

Flexible NFP data acquisition. The NFPData produced by Monitors may contain any information important for deciding adaptation as along as it is obtainable via measurements and/or performance dependency model, e.g., latency between Ab and Af, expected up-time of the computation node (for detecting shutdowns), etc. Moreover, a Monitor can decide between accepting NFPDeviceData given by Device and measuring its own, e.g., Monitor(Ab) can either individually measure latency to Af or rely on the network latency information given by Device(S). Although the performance dependency model employed by a Monitor will usually provide only a rough approximation of the expected performance, it can be potentially improved by actual measurements.

Scalable extensions. Being declarative, the ensembles allow the design to scale with respect to potential extensions of the basic architecture. For instance the Planner itself can be subject to migration in case the application does not have any frontend. Additionally, when understanding the Planner as an entity controlling the NFPs of the application, it is possible to foresee the existence of multiple Planners per application, thus hierarchically decomposing the adaptation.

6. RELATED WORK

In our previous work [8] we proposed to use Stochastic Performance Logic (SPL) [9] to express rules for adaptation in component systems based on real and predicted performance of individual components. The rules controlling the adaptation are similar to the decision logic of the Planner that compares deployment alternatives for Ab.

The issue and challenges of dynamic deployment adaptation has been formulated [4] and addressed [10] previously. However, in spite of the variety of solutions to the individual challenges (e.g., parameters of decision, migration to stationary only or also to mobile devices, acquisition of NFP-related data, etc.), in the majority of the approaches a predetermined solution is used. On the other hand, our adaptation architecture is dynamic enough to allow for combining multiple solutions simultaneously and selecting among them dynamically. It also provides general means to address the remaining challenges (e.g., scheduling of NFP data acquisition or estimation of cost for running before real execution).

A significant body of work has been devised in the related area of Mobile Cloud Computing (MCC) [1]. Although many of the challenges and solutions can be adopted in ad-hoc clouds, there is a significant difference in perceiving the role of the mobile device, i.e., MCC considers the mobile device as separate from the cloud while ad-hoc clouds do not distinguish among the role of the devices.

We assume an external mechanism responsible for the deployment/migration-related aspects of our approach since it has been intensively researched separately, e.g., in [7].

7. CONCLUSION

In this paper, we have presented our vision on addressing the problem of planning deployment adaptation in ad-hoc clouds. In particular, we have described a generic architecture for deployment adaptation logic that is based on the concept of emergent component ensembles. We have also discussed the potential benefits and scalability of this architecture.

8. ACKNOWLEDGEMENTS

This work has been supported by EU project 257414 ASCENS and GACR project P202/10/J042 FERDINAND.

9. REFERENCES

[1] L. Guan *et al.*, "A survey of research on mobile cloud computing," in *Proc. ICIS'11*, pp. 387–392, IEEE CS, 2011.

[2] G. N. C. Kirby *et al.*, "An approach to ad hoc cloud computing," *CoRR*, vol. abs/1002.4738, 2010.

[3] M. Hölzl *et al.*, "Engineering Ensembles: A White Paper of the ASCENS Project." ASCENS Deliverable JD1.1, 2011. Online: `http://www.ascens-ist.eu`.

[4] B.-G. Chun and P. Maniatis, "Dynamically partitioning applications between weak devices and clouds," in *Proc. MCS'10*, pp. 1–5, ACM, 2010.

[5] C. Jiang *et al.*, "A survey of job scheduling in grids," in *Advances in Data and Web Management*, vol. 4505 of *LNCS*, pp. 419–427, Springer, 2007.

[6] J. Keznikl *et al.*, "Towards Dependable Emergent Ensembles of Components: The DEECo Component Model," in *Proc. WICSA/ECSA'12*, IEEE, 2012.

[7] S.-H. Hung *et al.*, "Executing mobile applications on the cloud: Framework and issues," *Computers & Mathematics with Applications*, vol. 63, no. 2, 2012.

[8] L. Bulej *et al.*, "Performance awareness in component systems: Vision paper," in *Proc. COMPSAC'12 Workshops*, pp. 514–519, IEEE CS, 2012.

[9] L. Bulej *et al.*, "Capturing Performance Assumptions using Stochastic Performance Logic," in *Proc. ICPE'12*, pp. 311–322, ACM, 2012.

[10] M. Sharifi, S. Kafaie, and O. Kashefi, "A survey and taxonomy of cyber foraging of mobile devices," *IEEE Commun. Surveys Tuts.*, vol. 14, 2012.

Keynote Talk

Data Centric Computing for Internet Scale Enterprises

Yuqing Gao
IBM T. J. Watson Research Center
Yorktown Heights, NY

ABSTRACT

In the era of exploding internet usage, social and mobile, enterprises are facing both the challenges and business opportunities that are introduced by Big Data, which has the characteristics of high volume, high velocity, and high variety. Big Data and the emergence of Internet-facing workloads will blur the separation between traditional transactional and analytics workloads. To extract business value and make actionable insight from the unprecedented volume of the data with the agility required from the business, it requires transformational innovations from many fronts. For example, in data management layer, how unstructured data is stored and retrieved efficiently, how data-intensive analytic computation can be done on commercial systems effectively, how the distributed cache should be designed to make use of the latest network protocols so the network-connected memory data can be accessed remotely and seamlessly. Moreover, the trend also motivates many architectural and technological advancement, such as moving from a transaction-centric to a data-centric architecture that supports extreme low and predicable latency, massive scale-out, high concurrency, and real-time situational awareness and analytics, and that requires orders of magnitude improvement over existing systems across each of these characteristics. At the same time, new applications in the Mobile and social space leverage new open source software stacks written in multiple programming languages, e.g., Java, JavaScript, Ruby, PHP, where the developer chooses the best tool for the job. How a polyglot runtime platform can be built that serves as a best practice platform for the programmers' community and in the meantime, optimized for enterprises with elastic, lightweight, resilient, agile runtime for business computing. Last, but not least, how the benchmarks should be enriched to measure the new runtimes, new data-centric systems and architectures.

In this talk, I will talk about some of the research activities at IBM Research that addresses these challenges. We examined several enterprise-grade java workloads running on commercial multicore systems for massive parallelization, identified lock contentions and worked towards a streamlined methodology for lock-contention analysis of Java workloads. I will use this to describe the excitement around node.js framework. I will also talk about our design of data centric computing systems, particularly, in the area of data access latency, data ingestion, and massive scale-out distributed caching in the exemplary context of an eCommerce application. I will describe the architecture of a global secondary index to greatly improve data access latency of Hadoop Database (HBase), an open-source key-value distributed datastore. I will describe an innovative distributed caching system that exploits low latency interconnects to utilize hash maps of data keys on each server for local lookup while data resides and are accessed across clustered systems. The distributed cache can achieves 100 to 1000-fold performance gain over many caching methods. Last, I will talk about our early activities in developing technologies for an elastic, scalable, resilient polyglot runtime system. I will conclude with my views on the challenges for benchmarking community for next decade.

Categories and Subject Descriptors

H.3.4 [**Information Storage and Retrieval**]: Systems and Software

General Terms

Performance

Keywords

Big data, data access, workload characterization

ICPE'13, April 21–24, 2013, Prague, Czech Republic.
ACM 978-1-4503-1636-1/13/04.

Further Implementation Aspects of the Server Efficiency Rating Tool (SERT)

Klaus-Dieter Lange
Hewlett-Packard Company
klaus.lange@hp.com

Jeremy A. Arnold
IBM Corporation
arnoldje@us.ibm.com

Hansfried Block
Fujitsu Technology Solutions GmbH
hansfried.block@ts.fujitsu.com

Nathan Totura
Intel Corporation
nathan.totura@intel.com

John Beckett
Dell Inc.
john_beckett@dell.com

Mike G. Tricker
Microsoft Corporation
mike.tricker@microsoft.com

ABSTRACT

The Server Efficiency Rating Tool (SERT) has been developed by the Standard Performance Evaluation Corporation (SPEC) at the request of the US Environmental Protection Agency (EPA). Almost 3% of all electricity consumed within the US in 2010 went to running datacenters. With this in mind, the EPA released Version 2.0 of the ENERGY STAR for Computer Servers program in early 2013 to include the mandatory use of the SERT. Other governments world-wide that are also concerned with growing power consumption of servers and datacenters are considering the adoption of the SERT.

Categories and Subject Descriptors

H.3.4 [**Systems and Software**]: Performance evaluation (efficiency and effectiveness)

General Terms

Design, Experimentation, Measurement, Performance, Reliability, Standardization.

Keywords

SPEC, Benchmark, Energy Efficiency, Server, System Performance, Performance Engineering, Memory, System Discovery, Affinitization, Framework, Reporting, Energy Star, Environment Protection Agency (EPA).

1. INTRODUCTION

This paper builds on those published in 2011 [2] and 2012 [4], which described the initial design and development phases of the SERT [5].

As the SERT is now released, this paper first provides an overview of the SERT explaining its different components, a description of the Chauffeur framework, the graphical user interface (GUI), and the automated discovery of hardware and software information. Next, this paper states the challenges of automation of processor affinitization. The SERT is an efficiency evaluation tool that does not offer a single scoring model or metric (unlike a benchmark [8]); therefore, in a section of this paper, the memory subsystem is taken as an example in order to provide a detailed description of how the scores are derived.

A future aspect of the SERT is addressed with initial test results and comparisons between DC and AC powered servers.

ICPE'13, April 21–24, 2013, Prague, Czech Republic.

2. SERT OVERVIEW

The use of multiple power analyzers and temperature sensors is supported by the SERT in order to measure a large scope of system configurations. The most basic SERT measurement configuration requires one **power analyzer**, one **temperature sensor**, a system under test **(SUT)**, and a **Controller** system.

The SERT's test harness, named **Chauffeur**, controls the software installed on the SUT and **Controller**. Chauffeur also handles the logistical side of measuring and recording the power consumption and inlet temperature of the SUT.

The SUT gets instructed by the **Director** (Chauffeur instance) to execute the suite, which is comprised of a set of **workloads**. The workload consists of a set of **Worklets**, which exercise the SUT while Chauffeur collects the power and temperature data. The Worklets are the actual code designed to stress a specific system resource or resources, such as the CPU, memory, or storage IO.

Each power analyzer and temperature sensor interacts with its dedicated instance of the **SPEC PTDaemon**, which gathers their readings while the Worklets are executed.

The **Reporter**, executed after all measurements phases are completed, compiles all of the environmental, power, and performance data for a complete test run into an easy-to-read HTML report as well as an extensible markup language (XML) report; the HTML report includes a graphical visualization of the results.

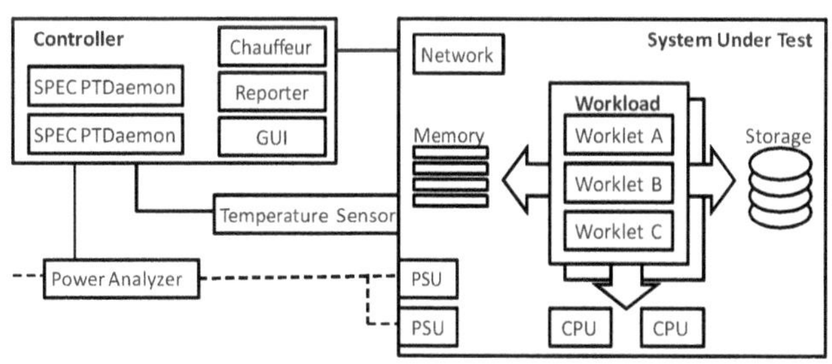

Figure 1 - Discovery Workflow

3. CHAUFFEUR FRAMEWORK

SPEC implemented the Chauffeur framework to support most of the following common functionality, and built the SERT on top of this framework. Future tools and benchmarks could be developed more quickly by taking advantage of Chauffeur.

Energy Measurements

One of the key features of Chauffeur compared to other benchmark frameworks is support for measuring power consumption under a variety of loads. The SPEC PTDaemon interfaces to power analyzers and temperature sensors, and Chauffeur makes the necessary calls to PTDaemon in order to collect the data for the appropriate time intervals. The resulting data is stored together with the runtime performance information so that the data does not have to be correlated after the test run is completed.

The ability to run at multiple utilizations is also an inherent property of Chauffeur. This support is based on the same principles used in SPECpower_ssj2008 to vary the load by first determining the maximum transaction throughput during a calibration process, and then scheduling transactions with appropriate delays to drive the system at lower levels of utilization. This support is implemented generically in Chauffeur and can be applied to any workload that is composed of a series of short-running transactions.

Scalable

The Chauffeur framework is inherently multi-process and multi-threaded, providing scalability across a wide range of servers. Multiple-node runs are also supported, enabling Chauffeur-based workloads to run across multiple blade servers. This support also could be extended to do runs that span multiple virtual servers on one or more physical hosts, although the SERT does not currently allow this.

Chauffeur's Director (the component that instructs the Host JVM to start executing the workload) normally runs on a system other than the SUT, and is usually collocated with the SERT GUI and SPEC PTDaemon. It communicates with a host process that runs on each system that is being measured. The Chauffeur Host will automatically launch one or more Client JVMs as needed, using platform-specific affinity commands as described in Section 4.

Chauffeur can adjust heap settings for each Client JVM based on the amount of memory in the system and the number of Clients being used. The algorithms for calculating heap sizes can be adjusted for individual Worklets; for example, the SERT uses nearly all available memory when running the memory Worklets, but normally uses only 256 MB per processor for CPU Worklets.

Ease of Use

Benchmark configurations can be complex, particularly when power measurements are involved. Chauffeur includes a number of features intended to simplify testing as much as possible.

A particular challenge for industry standard benchmarks and tools like SERT is reporting a complete and accurate description of the system configuration. To assist users in this process, Chauffeur supports automatic collection of system configuration data. This data is included in the results file, and also made available to the SERT GUI to allow users to review and edit the information. This process is described in more detail in Section 4.

Chauffeur also supports automatic validation of results, both at runtime and for completed results. These validation checks can confirm that the configuration is valid, that transactions did not fail, that power analyzer data met the requirements specified by the run rules, and various other requirements. The current validation checks in Chauffeur are defined for the SERT, but the framework is generic to allow checks to be added, changed, or removed for other workloads. This validation gives users confidence that the workload is configured and running properly and that the results are accurate.

Portable

Chauffeur is implemented primarily in Java to simplify portability across different systems. The SERT currently supports 64-bit Windows and Linux on x86 processors, and AIX on the Power architecture. Limited testing has also been performed on other platforms with minimal difficulty.

Platform-specific code is included in two areas of Chauffeur: System Configuration Discovery (see Section 4) and Affinity (see Section 5). In both cases Chauffeur will continue to run on platforms without explicit support. Future versions of Chauffeur can be extended easily by adding support for these features on other platforms.

Although the framework is written in Java, Worklets can make use of other programming languages, as the Storage Worklets in the SERT do. Communication between the Chauffeur Host and Clients is intentionally language-neutral, enabling a possible future native implementation of the Chauffeur Client.

Flexible

Chauffeur implements the core functionality required for the SERT. It also offers flexibility for changing the runtime behavior, either for research purposes or for future Chauffeur-based benchmarks. Many aspects of Chauffeur behavior can be changed via configuration files, without modifications to Chauffeur itself. For example, a virtualization benchmark may require the ability to run different virtual servers at different utilization levels in order to mimic dynamic heterogeneous usage patterns. Chauffeur does not currently support this usage model, but a custom implementation of a "Sequence" [6] could be implemented and plugged in through the configuration file.

It is often desirable to collect a variety of different types of data during a benchmark run, particularly for research and development purposes. Chauffeur provides a Listener interface that allows custom data collection to be performed without modifications to Chauffeur. These listeners are notified at various stages of the run, such as at the beginning and at the end of each measurement interval. Listeners can launch platform-specific tools or collect data that is not directly supported by Chauffeur. Data obtained by listeners can be included in the main Chauffeur results file so it does not need to be correlated after the run completes.

The Chauffeur Reporter is also designed for flexibility. It reads the XML-based results file from a Chauffeur run and produces HTML or plain-text reports. The contents of these reports are defined using an XSL transform, which allows changes to be made to the reports without code changes to Chauffeur. Advanced users can create their own reports, or generate output in comma-separated values (CSV) format for easy import into spreadsheets or statistical packages.

4. SYSTEM CONFIGURATION DISCOVERY

One of the largest challenges in benchmark and rating tool submissions is correctly identifying and capturing all of the characteristics of the SUT. Unintentional errors readily occur when collecting identification details of the various hardware components and recording the details into the benchmark report. Formatting errors overlooked while entering data may only become obvious once the final report is ready for submission. Correcting such errors and oversights in the final report can be cumbersome. The SERT addresses these issues with an automated hardware discovery process and an easy-to-use GUI workflow that assists the user in generating high quality accurate reports. The GUI therefore reduces the burden of test configuration, execution and report editing so that the user can focus on obtaining results.

4.1 Workflow

The relationship between the main components (Hardware Discovery, Test Environment Editor, and Preview Report) of the SERT discovery workflow is shown in Figure 2.

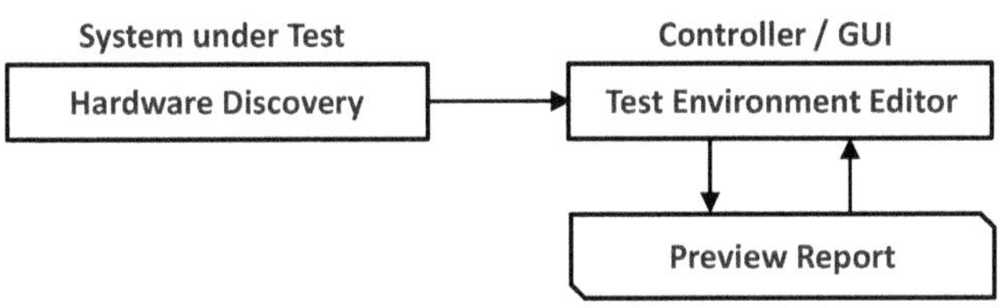

Figure 2 - Discovery Workflow

The initial phase of discovery occurs immediately after the SERT host is started on the SUT. On Windows systems, the hardware discovery scripts execute on the SUT using Windows Management Instrumentation (WMI). The solution on Linux distributions is a combination of scripts that parse the data found in the `/proc` and `/sys` file systems. The burden added to startup is small, and hardware discovery on the Windows platform completes within approximately 15 seconds. The process is efficiently executed once for the lifetime of the SUT host. Once the Controller connects to the SUT host, the discovered data is transferred and is made available for multiple purposes. The data is first used to attach directly to the final report. This ensures that the report reviewer always has an unmodified reference of the discovered data for any discrepancies encountered while reviewing. The discovered data is then sent to the GUI for assisting the user in updating the test environment configuration.

The GUI makes a request to the Controller to retrieve the discovery data as part of one of the first workflow panels the user encounters. The discovered name value pairs are presented in a list and the end-user has an opportunity to review the individual discovered items.

Once the user reviews the data, the GUI workflow is advanced and the discovered data is inserted into the final test configuration file. A mapping is contained within the raw discovered data that assists the GUI in aligning discovered values with the corresponding test environment configuration values. One of the GUI's main responsibilities is to provide an interface for the end-user to access the test environment configuration values. The Test Environment Panel of the GUI, as seen in Figure 3, serves this purpose and is used for verifying, editing, adding, removing, and viewing test environment configuration values.

The panel in Figure 3 illustrates some of the main features of the GUI that assist the user with system reporting. Color coding and status of the individual sections draw the user's attention to the components that still need manual intervention. "Incomplete" items are colored red and indicate that a value within the collection is completely missing.

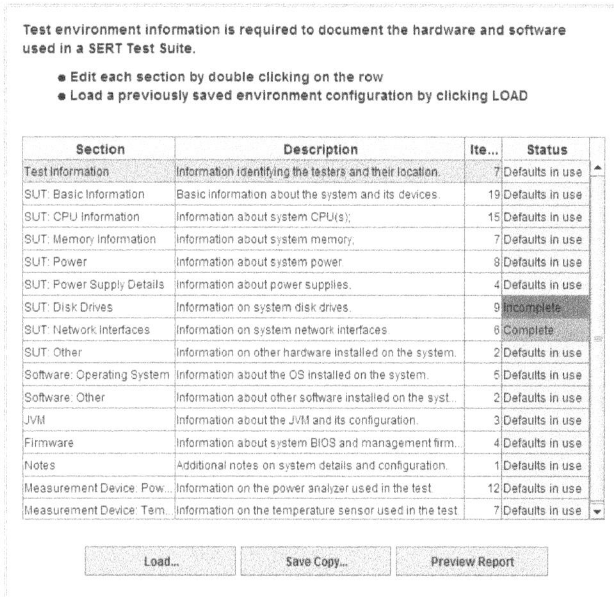

Figure 3 - Screenshot: SERT Test Environment Panel

Sections that contain default values are color-coded yellow and have a status of "Defaults in use". The default values are used to provide a hint to the user as to what a report value should contain. These default values are demarked with a leading underscore.

The section editor dialog is visible in Figure 4 and is opened when editing a specific section.

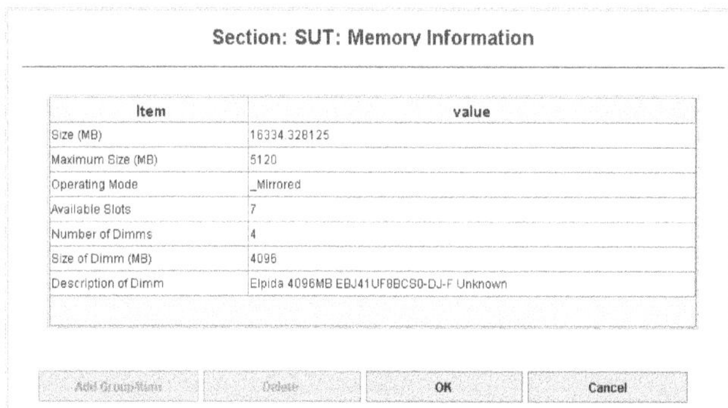

Figure 4 - Screenshot: SERT Section Editor Dialog

The example Memory Information section contains several values and demonstrates a couple of key GUI features. Visible is the value for "Operating Mode" set to "_Mirrored"; the leading underscore indicates a parameter is currently at a default value and needs updating. The default value is there to give an example of what type of information is expected in this field.

The remaining Memory Information values were automatically populated by the hardware discovery process and a user simply needs to verify that the values represent the system correctly. In the example, the Size (MB) field is not formatted in the best possible human readable format. The user now has the opportunity to edit each line and fix any other issues. The benefit of hardware discovery is evident in the "Description of DIMM" item value. Determining the dual in-line memory module (DIMM) description without the assistance of hardware discovery would have been tedious and error prone if a lengthy description such as this was entered by hand.

Once all items have been reviewed and updated, the GUI contains a very useful "Preview Report" button that launches a browser with a SERT Report populated with the current test environment values. Figure 5 is an example of a SERT Report preview. A similar status color coding is used to draw attention to values that still need to be updated.

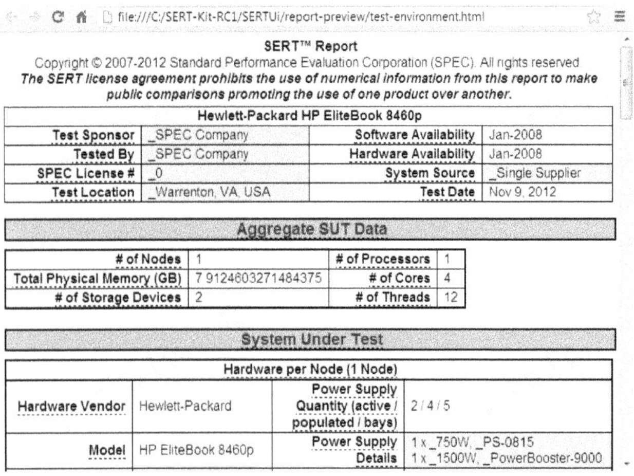
file:///C:/SERT-Kit-RC1/SERTUi/report-preview/test-environment.html

SERT™ Report

Copyright © 2007-2012 Standard Performance Evaluation Corporation (SPEC). All rights reserved

The SERT license agreement prohibits the use of numerical information from this report to make public comparisons promoting the use of one product over another.

Hewlett-Packard HP EliteBook 8460p			
Test Sponsor	_SPEC Company	Software Availability	Jan-2008
Tested By	_SPEC Company	Hardware Availability	Jan-2008
SPEC License #	_0	System Source	_Single Supplier
Test Location	_Warrenton, VA, USA	Test Date	Nov 9, 2012

Aggregate SUT Data

# of Nodes	1	# of Processors	1
Total Physical Memory (GB)	7.9124603271484375	# of Cores	4
# of Storage Devices	2	# of Threads	12

System Under Test

Hardware per Node (1 Node)			
Hardware Vendor	Hewlett-Packard	Power Supply Quantity (active / populated / bays)	2 / 4 / 5
Model	HP EliteBook 8460p	Power Supply Details	1 x _750W, _PS-0815 1 x _1500W, _PowerBooster-9000

Figure 5 - Screenshot: SERT Preview Report

Another key feature of a SERT Report is the hyperlink found on each field name. Following the link brings the user to a help page with a full, detailed description and definition of the field.

The GUI additionally provides "Save" and "Load" buttons to organize test configurations, a useful feature if one Controller system is used for multiple SUTs. Each new configuration can be saved under a descriptive unique name and the applicable configuration file can be reloaded easily when needed. Configurations can also be moved to different machines to serve as a starting point or reference for the next system to test.

4.2 Implementation Challenges

Several major challenges presented themselves when scoping the effort of automated Hardware Discovery. Perhaps one of the greatest challenges among these was the realization that, depending on the system where Discovery was run, different vendor implementations meant that hardware information was reported in varying ways. To better understand how to filter and format the discovery results, a survey was undertaken by the SPECpower Committee. This survey characterized an early version of the discovery script in order to analyze which fields were consistent across vendor platforms. It also identified which fields could be salvaged with additional post-discovery script logic, for inclusion in the final formatted SERT result output.

This survey sampled 21 separate platform types across five vendors. It enabled the researchers to characterize the suitability of each discovery field in determining which fields were consistent across vendors, and which fields needed additional logic to transform varied output strings into data that could be useful for the final report. The survey found that many details required interpretation and transformation across platforms. Vendor-specific fields such as Vendor and Model number were generally consistent, but other elements such as CPU details, Memory DIMM population, and Disk information were less easily interpreted. In addition other elements, such as Network Device information, were hampered by large numbers of pseudo network interface controller (NIC) devices, depending on the software installed on the target SUT.

In the analysis of CPU details, many separate elements were characterized such as the name, characteristics, frequency, core and thread counts, and the total number of CPUs populated on the system. The survey showed that while many of these elements were correct, others needed considerable rework in order to return correct and relevant information. In particular, the CPU cache fields presented problems. Reporting L1 and L2 cache in a per physical processor combined manner made it difficult to parse and almost impossible to transform into per-core information that would match with standard processor vendor specifications. The number of cores and threads per core were also initially problematic, but allowed for the insertion of additional logic to transform these values into data that could be readily transformed to a per-core/per-processor/per-system format.

Memory population details were another challenging area of the discovery process implementation. All the servers tested were correctly reporting the total installed memory and individual DIMM capacity in a consistent manner. However, elements such as individual slot population and memory manufacturer information could differ significantly depending on the platform. Another focal point of the discovery process was reporting the number and type of storage devices on the SUT, and the Host Bus Adapter (HBA) to which these devices were connected. It was also observed that almost all RAID HBAs analyzed would hide detailed information about the storage device vendor and any other element beyond simple capacity information. However, if the HBA was a simple SATA controller, this information was typically not obscured and additional detail could be derived from the underlying storage devices.

One of the persistent challenges in the development of the discovery process was filtering out extraneous NIC information that is common in some operating systems (OS) and driver models. In some cases, many pseudo-NIC devices were being discovered along with the actual network adapters that were the target. In such circumstances, it was determined that additional logic needed to be added to the discovery script to avoid reporting additional OS pseudo-NIC devices and other interface entries created by VPN and similar drivers. Significant effort was made to correctly determine which interface was connected and to report the connection speed accurately. As it is common on many servers for multiple NIC devices of the same type to be present but only one connected for the SERT run, the discovery process needed to be able to account for this in a fashion that could be correctly interpreted for the purposes of creating a final report.

Another major challenge for the implementation of the discovery mechanism was the fact that some fields were found to be duplicates of one another in cases where multiple instances of the same type of device were discovered. This applies not only to multiple processors, but also to DIMMs, NICs, and storage devices. To allow latter portions of the discovery process to operate correctly on this information after the discovery script itself was executed, an additional key/value pair was added for discovery elements where this behavior was present. The "_Uniquefields" tag was added to the discovery script output for subsets of elements that could be identified as multiples of the same device type, which enabled the Reporter to automatically generate device counts and correctly represent device elements.

The SERT Automated Discovery process supports two major operating systems: Microsoft Windows Server and multiple Linux distributions. The challenges presented in implementing the Windows discovery process were somewhat more straightforward than Linux due to the common Windows Management Instrumentation (WMI) query structure provided by Microsoft. For Linux, there was no unified query mechanism that could be counted on being installed by default on any major distribution. This meant that all the major discovery elements that were successfully discovered in Windows needed to be compiled into a list, and a determination made as to what Linux resources could be queried in Linux to report the same information. System files in the `/proc` and `/sys` filesystems that could be queried either directly or through common commands made the writing of the Linux Perl discovery script less challenging than originally envisioned. After refinement, both the Windows and Linux discovery scripts reliably report the same information in similar formats.

4.3 Additional Advantages

The SERT Automated Discovery process significantly eases the user burden in documenting and tracking different SUT configuration details. In addition, the discovery data can be used to cross-check the validity of user edits made to the included data after the fact during the submission checking process. Discovery is executed independently of the GUI for any SERT run, and that information is encoded into the results.xml file. This gives the result reviewer the ability to detect whether the final report system description details vary significantly from what was actually found during initial discovery. Almost all the system description fields that show up in the final text and HTML reports can be user-modified, either through the GUI or after the run is complete and the HTML report is regenerated. However, this could lead to erroneous or intentional obfuscation of important system hardware configuration details, misrepresenting the state of the platform that was characterized by a particular SERT run. Since the raw discovery data is present in a form that prevents tampering the result reviewer has greater confidence that the result was actually executed on the same type of hardware as described by the submitter in the HTML report.

5. PROCESSOR AFFINITIZATION

A major challenge in the design of SERT was the need to remove the burden of manually setting Java client (JVM) core affinity. This is required to take the best advantage of processor cache sharing and non-uniform memory access (NUMA) nodes present on multi-socket systems. One of the major design goals was to automate this sometimes complex and burdensome process with an automated affinity generator. This was required to operate across three separate operating systems, one- to eight-processor sockets, up to two NUMA nodes per processor socket, and with different platform ACPI presentations to the OS. The SERT affinity generator also needed to generate the correct affinity masks regardless of how many JVMs needed to be affinitized. This was due to the JVM count being expected to vary depending on the processor type, populated socket count, and workload choice.

Proper NUMA node affinity for multi-socket systems is important for optimal performance. This may be impacted by the additional latency costs incurred when a thread requests memory non-local to the core to which that worker thread is bound. It is therefore critical to be able to identify the NUMA layout of a particular system. This information will be used to group worker threads of a JVM instance to a single physical CPU for maximal resource sharing through explicit binding, and to ensure that memory allocation occurs only on the local NUMA node. Failure to adhere to these best practices will cause performance degradation and high run-to-run variability.

All three OS families supported by the SERT, including Microsoft Windows Server, 64-bit Linux Server distributions, and IBM AIX, were enabled with automatic affinity generator support. Some assumptions were made regarding the topology of the SUT to reduce complexity. Among these assumptions are that the total number of logical processors on the system is evenly divisible by the number of JVMs. It is also assumed that the total JVM count will be evenly spread across all NUMA nodes of a given system. This assumption should hold true if each NUMA node has the same number of logical processors. There are potential situations where these assumptions may not hold true, in which case the SERT will still operate correctly but the affinity may not be completely optimal. The SERT developers expect that configurations with which automatic affinity generators perform sub-optimally will be very rare.

5.1 Linux Affinity

In Linux operating systems, the cores enumerated during kernel boot time are based on the presentation by the ACPI table. This ACPI presentation differs considerably between separate vendors and even between platforms and BIOS releases in some cases. This means that two platforms from different vendors populated with the same processor type and socket count can order cores very differently due to the ACPI table presentation. It was therefore recognized during the SERT development process that the Linux affinity generator needed to be able to handle any type of core enumeration strategy.

Figure 6 and Figure 7 display two different core enumeration strategies utilizing the same 8-core Intel Xeon processor type and socket count with Hyper-Threading enabled. The NUMA Node notations show which set of memory is local to which physical processor and associated cores. Note that the numbers separated by the "/" character indicate the real core and Hyper-Threaded sibling from an OS logical processor enumeration standpoint. As an example, if one were to compare the actual location of OS CPU 1 between the two illustrations, one would see that the first example is located as the first core on the second socket with memory local to NUMA Node 1. The second illustration shows OS CPU 1 as the second core on the first socket, with memory local to NUMA Node 0. These illustrations show two of several enumeration strategies known to exist. It is clear, given the disparate enumeration strategies encountered, that the SERT developers were unable to count on consistency from vendor to vendor, so all affinity strategies needed to be calculated on a per-system basis.

Modern server-oriented 64-bit Linux distributions come with tools such as `numactl` to determine the NUMA node count as well as displaying which cores are local to any given NUMA node. However, the challenge of Hyper-Threaded platforms meant that it was also necessary to be able to pair individual cores with their Hyper-Threaded siblings. This meant that all logical processor data from the Linux `/proc/cpuinfo` file needed to be parsed to locate this information in order to ensure optimal affinity mask generation.

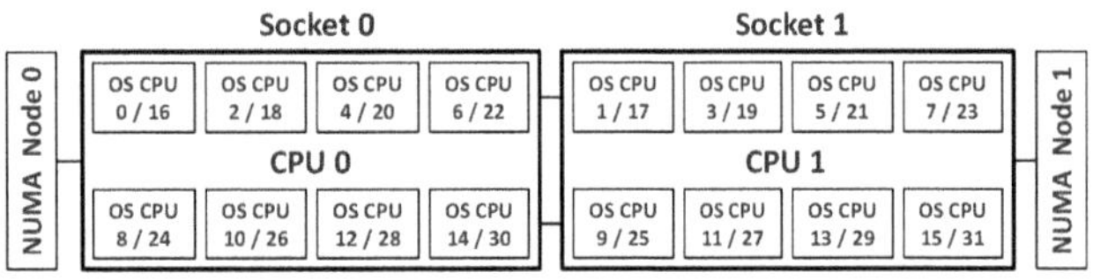

Figure 6 – Core Enumerations – Example 1

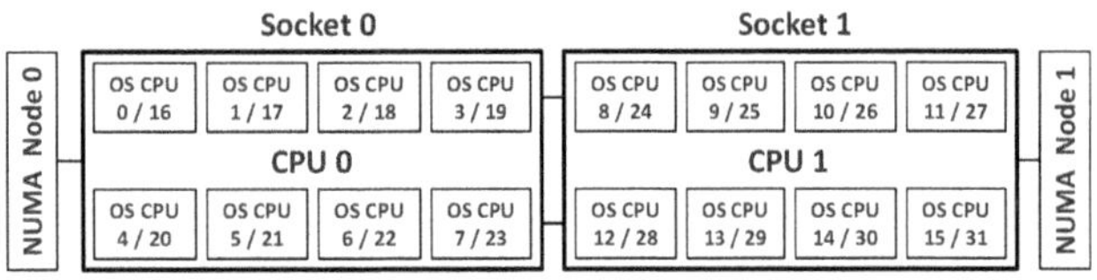

Figure 7 – Core Enumerations – Example 2

Each JVM needs a different affinitization command, which is provided as a set of OS CPUs in a comma-separated list, and added to a `numactl` command which will pin the JVM instance to the logical processor list and closest NUMA node.

5.2 Windows Affinity

Unlike Linux operating systems, the Microsoft Windows Server OS variants enforce their own core enumeration strategy regardless of ACPI presentation. This means that for two platforms with the same processor type and socket count, the location of each core or Hyper-Threaded sibling will be the same regardless of the vendor. This advantage was balanced with the difficulty of easily detecting the number of NUMA nodes on a given system, where no default system tools exist to discover this information.

The lack of the ability to easily determine the NUMA node count meant that the SERT developers needed to design a method for determining this information as part of the SERT package. This was accomplished with .dll helper files called through Java JNI that call specific Windows APIs to provide the NUMA node count and total number of logical processors.

Given the discovered JVM and NUMA node counts the WindowsAffinityProvider Java class generates individual hexadecimal affinity masks for each JVM as it is being spawned. These individual hexadecimal masks are appended to a Windows `start /affinity` command. This ensures that each JVM being spawned is bound to a particular set of logical processors that are on the same physical CPU and NUMA node, and are capable of sharing cache resources for optimal performance. In addition, the `/NODE` switch is used in cases where there is more than one NUMA node on a particular platform. This ensures correct NUMA node locality and makes hexadecimal masks considerably shorter for systems with a high logical processor count. In addition, the use of the `/NODE` switch allows correct affinity masks to be generated for systems with greater than 64 logical processors for Windows Server 2008 R2 and newer operating systems.

5.3 AIX Affinity

The IBM AIX automatic affinity generator was developed after the Linux implementation was complete, and shares some of the same general structure. However, AIX system commands for topology discovery and affinitization are completely different, and the IBM Power Processor Architecture shares little similarity with x86 implementations. Much like the Linux affinity generator, the AIX affinity generator relies on system commands to discover NUMA topology and core/thread enumeration and to bind JVM instances to specific processor resources. The `lssrad` command is used to determine the logical CPU set associated with each NUMA node. The `smtctl` command is utilized to identify which logical processors are actually hardware threads on the same physical processor core. This is used because, although OS logical processors are typically grouped in a predictable fashion, Symmetrical Multi-Threading (SMT) levels can change post-boot in AIX environments. Using this command allows the SERT to handle processors in ST (single-thread), SMT-2 (two threads per core) and SMT-4 (four threads per core) modes. The `execrset` command is used to launch JVM instances and bind them to a list of appropriate hardware threads. To ensure correct memory allocation on local NUMA nodes, `MEMORY_AFFINITY=MCM` environmental variable is set.

6. MEMORY WORKLETS

The Memory Worklets included in the SERT [1] RC1 release have been used for an extensive series of experiments on various system configurations, including different numbers and models of CPUs and DIMMs. The tests were executed on a Fujitsu PRIMERGY computer server model with 24 DIMM slots under both Microsoft Windows Server 2008 R2 and Red Hat Enterprise Linux (RHEL) 6.2. SPEC PTDaemon was employed for the power measurements of total system power.

The reporting capabilities of the SERT were used to document the performance and power results of these experiments. The scaling properties of the two memory Worklets, comparing performance and power consumption of the tested configurations at different load levels, are shown.

The server side Java (SSJ) workload of SPECpower_ssj2008 [7] was the first industry-standard benchmark for measuring computer server efficiency. It primarily stresses the CPU and a limited amount of main memory of computer servers. Higher memory capacities do not increase SSJ performance but do increase power consumption, thus worsening the efficiency result. This deficit restricts its usability for governmental regulatory programs; e.g. the EPA ENERGY STAR [1].

The Memory Worklets developed for SERT generate synthetic loads on server storage devices mimicking basic access patterns from real world usage models. The design goals specified in the SERT Design Document [6] state that the Memory Worklets should give credit to higher bandwidth, lower latency, and increased total size of computer server main memory. The tests described in this paper were performed to check the suitability of the implementation for the designed purpose, especially testing whether these design goals are met.

6.1 Test Configurations

The experiments described in this paper are based on the RC1 version of SERT. The results are representative for this specific SERT release and may differ for the final release.

In order to show the scaling capabilities of the Memory Worklets, a series of tests was executed on a two-socket Fujitsu PRIMERGY RX300 S7 rack server with 24 DIMM slots. Throughput and power consumption have been measured running the SERT Memory Worklets under different CPU and Memory configurations.

6.1.1 Power Measurement Setup

For the experiments described in this paper the SERT standard power measurement setup was used, only measuring the total system power at the Power Supply Unit (PSU) input. Because no adapters for measuring memory power consumption on the system main board have been available for our tests, only overall system power can be provided.

A temperature sensor was used in all test scenarios to measure the ambient temperature and ensure that it always stays above the required minimum of 20°C.

6.1.2 The SERT Memory Workload

The SERT memory workload includes two memory Worklets named Flood and Capacity, briefly introduced below. A detailed description of these Worklets is given in the SERT Design Document [6].

Memory Flood Worklet

The Flood Worklet is based upon STREAM, a popular benchmark that measures memory bandwidth across four common and important array operations. For the *long* (64-bit) integer arrays used in Flood, the following amounts of memory are involved per assignment:

1. **COPY:** *a(i) = b(i)* -- 8 bytes read + 8 bytes write
2. **SCALE:** *a(i) = k * b(i)* -- 8 bytes read + 8 bytes write
3. **ADD:** *a(i) = b(i) + c(i)* -- 16 bytes read + 8 bytes write
4. **TRIAD:** *a(i) = b(i) + k * c(i)* -- 16 bytes read + 8 bytes write

The Flood score is based upon the aggregate system memory bandwidth calculated from the average of these four tests multiplied by the amount of physical memory installed in the SUT. While Flood is based upon STREAM, it uses no STREAM code and is implemented entirely in Java.

Memory Capacity Worklet

The Capacity Worklet uses modified code from the SERT XMLvalidate Worklet, which exercises Java's XML validation package.

Memory scaling in Capacity is done through a scheme known as input data caching (IDC). In IDC, the universe of possible input data (here, randomized XML file data) is pre-computed and then cached within memory before the start of the workload. During workload execution, the input data for a particular transaction instance is then chosen randomly and retrieved from this cache rather than computed on the fly.

The data store size is increased incrementally with each interval. If the data store size is less than the amount of physical memory available to the Worklet, data is retrieved from the cache. Once the data store size is larger than the maximum size of the data cache, a 'cache miss' penalty is incurred when the transaction randomly chooses a data store element that is not currently in the cache. When this occurs, multiple iterations of re-generating a cache element are performed to apply a cache miss penalty and the transaction rate decreases. The more memory the system has, the larger a data store access can be executed before the transaction rate begins to lower as a result of cache misses.

In addition to the transaction characteristics, the maximum cache size is applied to the scoring algorithm. Cache size is computed as: Physical Memory * data-cache-to-heap-ratio (currently 0.6)

While this Worklet does contain transactions that are memory-oriented, there is still a component that is influenced by CPU performance.

Memory Workload Execution

The Flood Worklet uses the SERT fixed iteration execution model (i.e., it always deploys a fixed number of iterations). The amount of memory under test will automatically adjust to fully utilize installed DRAM, so runtime will vary depending upon system configuration.

The Capacity Worklet measurement intervals run for a fixed amount of time, 2 minutes plus pre- and post-measurement phases of 15 seconds Each interval consists of a No Delay Series (i.e., the code runs unrestricted at highest possible speed), where the parameter data-store-size changes with each interval. Table 1 describes the load levels of the two Memory Worklets.

6.1.3 The Tested Configurations

The basic configuration of the SUT used for the test series is described below. The CPU and memory configuration as well as the OS, have been varied to measure how these configuration changes influence Worklet performance and power consumption.

- CPU: 2 x Intel Xeon E5-2690, E5-2620, E5-2603
- RAM: 8, 16, 24 x 16GB PC3L-12800R DIMMs
  8, 16, 24 x 8GB PC3L-12800R DIMMs
- OS: Microsoft Windows Server 2008 R2
  Red Hat Enterprise Linux 6.2
- JVM: Oracle HotSpot 1.7.0_09-b05
- RAID Controller: 1 x LSI 2108 SAS
- Storage Device: 1 x 146GB SAS 2.5" 10k rpm (boot dev.)
- PSU: 1 x 800W/230V AC CSCI Platinum Standard
  1 x 800W/48V DC CSCI Gold Standard

Table 1. Memory Workload Load Levels

Worklet	Load Levels	Description
Flood	Full, Half	2 load levels using full and half of the available memory capacity
Capacity	4, 8, 16, 32, 64, 128, 256, 512, 1024	9 load levels with increasing IDC data store sizes specified in GB

Due to time constraints, each of these configurations was tested with a single SERT run only. The SERT test configuration file (config-all.xml) was modified to execute the Flood, Capacity, and Idle Worklets only, resulting in a reduced execution time of about 40 - 50 minutes per test run depending on the configuration.

6.2 Memory Worklet Test Results

This section presents the results of the experiments executed on the different configurations. Specifically, it shows the scaling capabilities of the memory Worklets and compares their power consumption. The different configurations tested are described in Table 2. Pairs of three Intel Xeon CPU models with differing number of cores/threads and clock frequencies were used. All tested CPUs include an on-chip memory controller providing four memory channels (i.e. a total of eight memory channels). DIMMs of size 16GB and 8GB have been tested in configurations of 8, 16 and 24 [i.e., 1, 2, and 3 DIMMs Per Channel (DPC)]. Table 2 shows the effective memory frequencies which are defined by the capabilities of DIMMs and CPUs (e.g., the E5-2603 CPU restricts memory frequency to 1066MHz although the tested DIMMs support up to 1600MHz). For all CPU models the memory frequency is reduced to 1066MHz for configurations with 24 DIMMs or 3 DPC.

Table 2. Memory Scaling Configurations

#	Configuration Description
1	E5-2690 8C/16T 2.90G.Hz, 8x16=128GB@1600MHz
2	E5-2690 8C/16T 2.90GHz, 16x16=256GB@1600MHz
3	E5-2690 8C/16T 2.90GHz, 24x16=388GB@1066MHz
4	E5-2690 8C/16T 2.90GHz, 8x8=64GB@1600MHz
5	E5-2690 8C/16T 2.90GHz, 16x8=128GB@1600MHz
6	E5-2690 8C/16T 2.90GHz, 24x8=192GB@1066MHz
7	E5-2620 6C/12T 2.00GHz, 8x8=64GB@1600MHz
8	E5-2620 6C/12T 2.00GHz, 16x8=128GB@1600MHz
9	E5-2620 6C/12T 2.00GHz, 24x8=192GB@1066MHz
10	E5-2603 4C/4T 1.80GHz, 8x8=64GB@1066MHz
11	E5-2603 4C/4T 1.80GHz, 16x8=128GB@1066MHz
12	E5-2603 4C/4T 1.80GHz, 24x8=192GB@1066MHz

In the following description we will reference the tested configurations using the numbers defined in Table 2.

6.2.1 Memory Flood Results

Figure 8 shows the performance results of the Flood Worklet for both load levels, **Full** (solid lines) and **Half** (dotted lines), on all tested hardware and software configurations. The corresponding values are printed in Table 3.

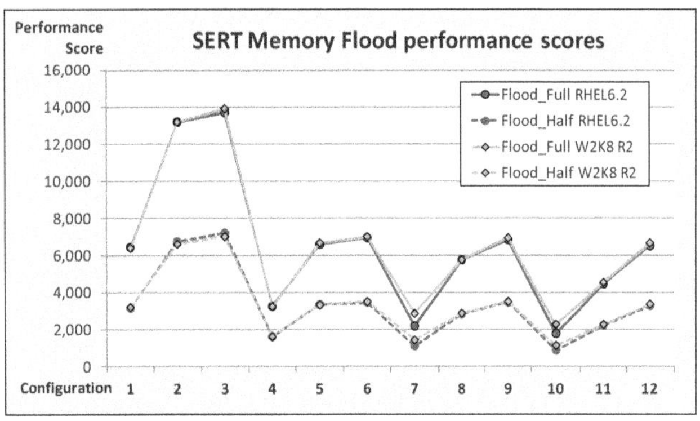

Figure 8 – Flood Worklet Performance Scores

Remarks:

- The performance scores presented here are calculated by the Flood Worklet code from measured aggregate system memory bandwidth multiplied by the amount of physical memory.
- The scores increase with capacity for configurations with the same bandwidth [e.g., (1, 2), (4, 5), (7, 8) and (10, 11, 12)].
- Configurations 3, 6, and 9 show a smaller performance increase with capacity because of lower bandwidth due to reduced memory frequency.
- The scores from the load level **Half** are lower because tested physical memory is cut by half for both measurement and score calculation (i.e., although the bandwidth is about the same the score is halved). This load level was included in the Flood Worklet to facilitate higher efficiency scores with memory power management potentially powering off unused DIMMs.
- The scores are almost identical under both OSs.

Generally, the measured results reflect the desired behavior.

Table 3. Flood Worklet Performance Scores

	RHEL6.2		W2K8 R2	
#	Full	Half	Full	Half
1	6,434	3,163	6,396	3,200
2	13,185	6,743	13,189	6,591
3	13,698	7,216	13,901	7,005
4	3,257	1,622	3,238	1,620
5	6,586	3,345	6,659	3,331
6	6,940	3,451	6,991	3,502
7	2,192	1,117	2,860	1,433
8	5,747	2,836	5,752	2,875
9	6,826	3,460	6,926	3,501
10	1,790	897	2,283	1,143
11	4,454	2,231	4,538	2,274
12	6,493	3,274	6,651	3,365

In order to get performance values in the same order of magnitude from all Worklets the individual performance scores of each Worklet are divided by a fixed reference score. The reference score for each Worklet was determined by averaging the performance scores across several SERT test runs on a well-defined reference configuration under different operating systems.

Figure 9 shows the normalized performance results of the Flood Worklet for the peak load level (i.e., **Full**, on all tested hardware and software configurations together with the corresponding power readings and Idle power).

The bars represent the normalized performance score (left y-axis) and the lines show the corresponding power consumption in watt (right y-axis).

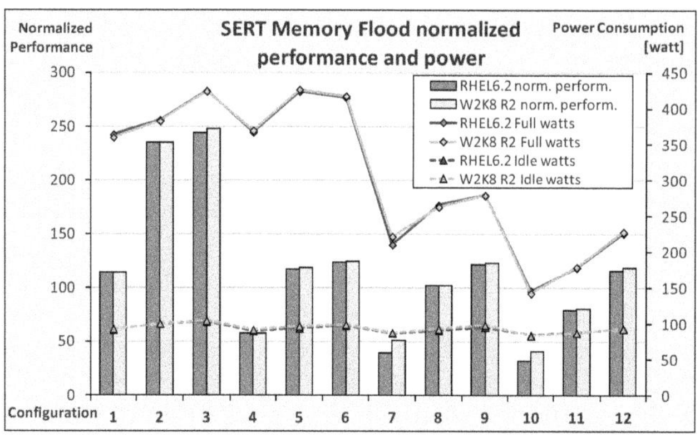

Figure 9 – Flood Worklet Normalized Performance and Power

Normalized performance follows the basic performance score described previously. Idle power for all configurations is almost the same and is only marginally influenced by the number and the capacity of the DIMMs. Peak load power is dominated by CPU power as can be seen from the three power levels in Figure 9 corresponding to the 3 CPU models. Within each group power increases with the number of DIMMs. Configurations 6 and 9 show a smaller power increase or even decrease due to lower bandwidth caused by reduced memory frequency. This effect is countered by higher DIMM power consumption in configuration 3. Power consumption is the same under both operating systems.

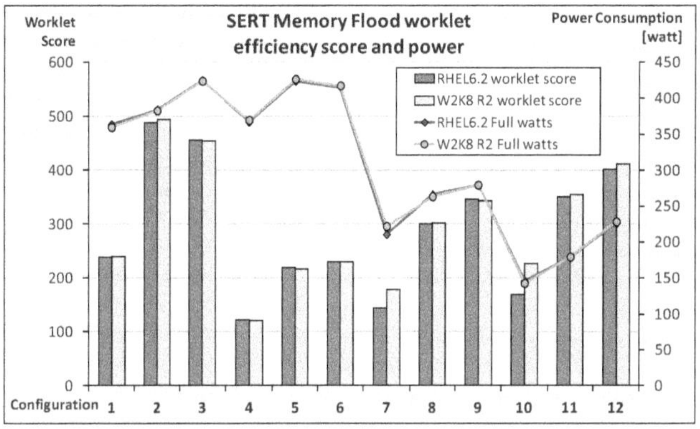

Figure 10 – Flood Worklet Efficiency Score and Power

Figure 10 finally shows the Flood efficiency scores which are calculated as the sum of the normalized performance for each interval divided by the sum of the average-watts for each interval. Efficiency for the Idle Worklet is undefined as the performance part is missing by definition. Please note that Idle power is **not** included in the per Worklet efficiency score calculation.

Efficiency is best for the high end configuration and it increases for most configurations with memory capacity, which is due mainly to the high weight of capacity in the performance calculation. Only configuration 3 has a lower efficiency than the smaller configuration 2, caused by the big increase of power consumption.

6.2.2 Memory Capacity Results

Figure 11 shows the performance results of the Capacity Worklet for selected load levels, 4GB, 64GB, 128GB, and 1024GB data store size, on all configurations with 8GB DIMMs under RHEL6.2 OS. The corresponding values are printed in Table 4.

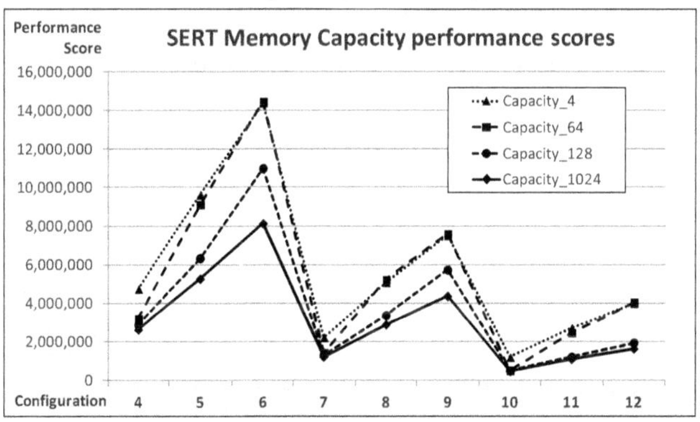

Figure 11 – Capacity Worklet Performance Scores

Remarks:

- The performance scores presented here are calculated by the Capacity Worklet code as explained in chapter 6.1.2.
- Performance is dominated by CPU capabilities as can be seen from the three performance levels in Figure 11 corresponding to the three CPU models.
- In the peak load level the complete data store of 4GB can always be cached completely in available physical memory (i.e., there are no cache misses and the XML transformation rate only depends on CPU capabilities). This load level defines the upper performance limit.
- In the lowest load level, the 1024GB data store never fits into available physical memory resulting in high cache miss rates causing high CPU load for regenerating cache elements. It defines the lower performance boundary.
- Because of additional memory overhead needed for XML translation work, the 64GB physical memory in configurations 4, 7, and 10 is not sufficient for fully caching the 64GB data store. For the configurations with 128GB and 192GB the performance is close to the 4GB results because now the data store is totally cached.
- Cache hit rates increase significantly with additional physical memory for 128GB data store, but it cannot be cached fully in the 192GB configurations due to the overhead explained above.
- The remaining load levels are not shown in this chart for better readability. They reach scores between the upper and lower limit defined by the 4GB and 1024GB data store sizes depending on how much data can be cached.

Generally, the measured results reflect the desired behavior.

Table 4. Capacity Worklet Performance Scores

	Capacity Performance Scores RHEL6.2			
Load	**4**	**64**	**128**	**1024**
4	4,752,281	3,192,736	2,864,201	2,621,257
5	9,629,606	9,116,005	6,334,995	5,264,811
6	14,395,461	14,464,419	10,978,881	8,130,920
7	2,219,895	1,437,212	1,302,491	1,213,172
8	5,085,719	5,221,489	3,369,048	2,886,206
9	7,508,036	7,602,138	5,730,170	4,362,105
10	1,192,931	516,047	490,574	467,977
11	2,712,958	2,469,347	1,203,025	1,084,737
12	4,011,082	4,053,182	1,936,115	1,621,643

Figure 12 shows the normalized performance results of the Capacity Worklet for the peak load level on all tested hardware and software configurations together with the corresponding power readings and Idle power.

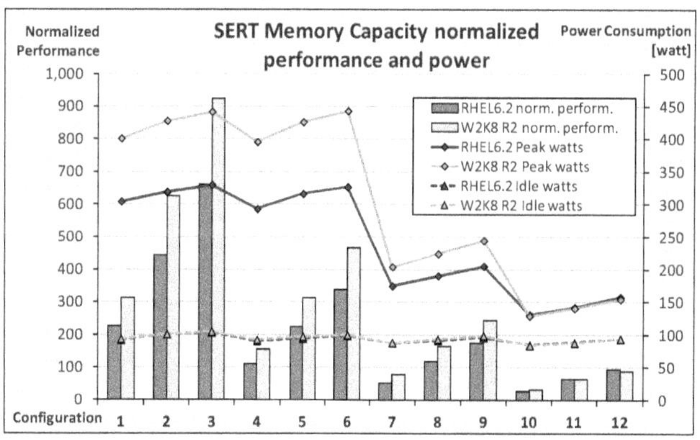

Figure 12 – Capacity Worklet Normalized Performance and Power

Normalized performance follows the basic performance score. Peak load power is dominated by CPU power as can be seen from the three power levels in Figure 12 corresponding to the three

CPU models. Within each group both power and performance increase with the number of DIMMs.

Performance is significantly higher under the Windows Server 2008 R2 OS compared to Red Hat Enterprise Linux 6.2. Power increases proportionally with the performance. Currently, there is no explanation for these differences. Additional experiments are required to analyze this anomaly.

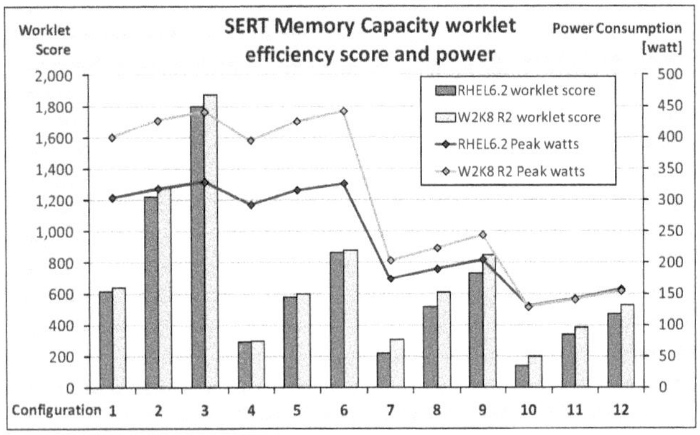

Figure 13 – Capacity Worklet Efficiency Score and Power

The Capacity efficiency scores are shown in Figure 13. Again, efficiency is best for the high end configurations and it increases for all configurations alongside memory capacity, which is mainly due to the high weighting of capacity in the performance calculation. Although performance was higher under Windows, the efficiency scores are close to each other or even equal for some configurations. This is due to the higher power consumption under Windows which compensates the performance advantage.

7. AC-DC COMPARISON

The EPA has received requests from stakeholders to support DC-powered servers with Version 2.0 of the ENERGY STAR Enterprise Servers Specification [1]. Currently, this cannot be achieved because the SERT does not support DC loads, as stated in the SERT Design Document [6]. The general SERT design allows for such measurements, once a DC-capable version of the SPEC PTDaemon becomes available. In order to evaluate the DC capabilities of the SERT and to compare the characteristics of AC and DC power consumption a modified version of SPEC PTDaemon has been implemented, which supports DC measurements for a test series. Typically the power analyzer uncertainties of DC measurements are significantly higher than those of AC measurements, specifically with lower voltages. In order to stay below the 1% uncertainty threshold required for SERT measurements, a high precision power analyzer had to be used. This special beta version of the SPEC PTDaemon is for internal use only. Currently there are no plans for releasing this version with the final SERT kit.

Three consecutive full SERT runs have been executed on the test system described in chapter 6.1.3 with 8 x 8GB DIMMs using an 800W/230V AC PSU. A second series of full SERT tests was performed on the same configuration but exchanging the PSU against an 800W/48V DC model.

Figure 14 shows the normalized peak performance for all SERT Worklets in both configurations and the corresponding power consumption values.

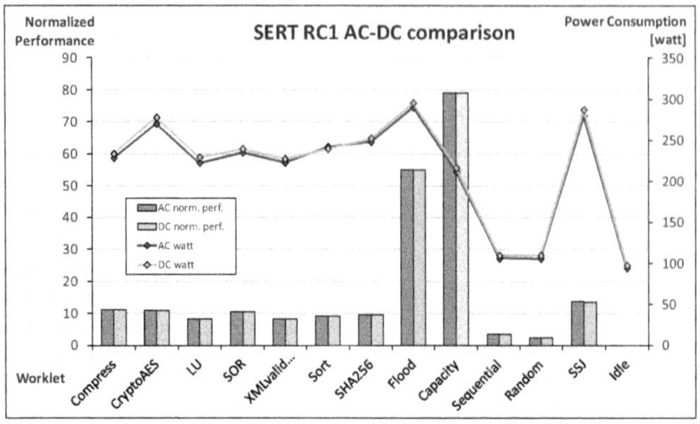

Figure 14 – SERT RC1 AC – DC Comparison

The detailed result values are provided in Table 5, which also includes a column showing the power consumption delta.

Table 5. SERT AC – DC comparison test results

PSU	800W AC CSCI Platinum		800W DC CSCI Gold		Delta
Worklet	Norm. Perf.@ 100%	Avg. Watts @ 100%	Norm. Perf. @ 100%	Avg. Watts @ 100%	DC - AC watts
Compress	11.1	228	11.1	233	2.2%
CryptoAES	10.9	270	10.9	277	2.8%
LU	8.3	222	8.3	229	3.1%
SOR	10.5	235	10.5	240	2.0%
XMLvalidate	8.3	222	8.2	227	2.1%
Sort	9.1	242	9.1	240	-1.0%
SHA256	9.6	248	9.6	253	1.9%
Flood	54.9	290	54.9	296	2.0%
Capacity	79.0	212	79.0	216	2.1%
Sequential	3.5	106	3.5	110	3.4%
Random	2.4	106	2.4	110	3.9%
SSJ	13.7	280	13.7	287	2.6%
Idle		94		97	3.6%

7.1 AC-DC Comparison Results

The performance scores are almost identical for all Worklets. Power consumption is however typically 2 – 4% higher for the DC PSU configuration. Only the Sort Worklet consumes less power in the DC configuration. This is probably due to high run-to-run variations seen during our test, which may be caused by the aggressive JVM tuning flags currently used by default. These flags will be revised for the final SERT release with the goal of minimizing run-to-run variations.

The higher power draw of the DC configuration is partly caused by the lower efficiency standard of the DC PSU, which is CSCI Gold compared to CSCI Platinum for the AC PSU. Based on the efficiency curves for both PSUs, it is estimated that about half of the power delta is due to higher power losses in the DC model.

Generally it is assumed that DC PSUs would show the higher efficiency because AC-DC conversion losses will not occur. These tests have shown the opposite behavior. This is most probably due to the lower voltage of the DC PSU (48V) compared to the AC PSU (230V), which results in much higher currents having to be handled by the PSU for voltage conversion. Higher currents cause increased power loss and this effect dominates the missing AC-DC conversion loss.

Repeating this comparison with a 400V DC PSU would most probably end in favor of the DC configuration, because of the absence of AC-DC conversion losses and lower currents resulting in reduced voltage conversion losses.

8. CONCLUSION

The SERT was released in February 2013, for use in Version 2.0 of the EPA ENERGY STAR for Computer Servers program. While previous papers described the initial design and implementation of the SERT, this one has focused on the design decisions, implementation trade-offs and validation performed to actually deliver the tool.

This paper discusses in detail the design decisions taken to simplify configuring and running the SERT, with customer feedback during the beta phases as a key input. Experience gained with customers of SPECpower_ssj2008, and the results reviewed by the SPECpower Committee also inspired simplification of use. Five years of ssj2008 submission highlighted some of the complexities of configuring test hardware and power analyzers to perform measurements. This directly drove the development of the affinity mask generator, and most significantly, the GUI and hardware discovery components.

It has been stressed throughout the development of the SERT that it is not a benchmark. However there continue to be requests for scores and metrics that can be used to help differentiate between similar servers from different vendors.

The SERT implementation also shows that the underlining Chauffeur framework provides the features needed for future performance and energy efficiency benchmark implementations.

Finally, this paper describes some of the future evolution of the SERT that is under active development or consideration. Adding support for DC power is in the experimental phase. The SERT is currently providing results that realistically represent what might be observed with production workloads on servers running on DC power. The intent is for a future release to add this support, broadening the appeal and usability of the SERT to non-traditional server industries, such as telecommunications.

Considerable interest has been shown in the SERT by several countries. It is therefore reasonable to hope for widespread adoption in the next few years. This should enable consistency across markets, which will benefit computer manufacturers and users, as well as the environment as a whole.

The SERT offers ease of use, cross platform support, a strong range of synthetic Worklets, and a highly modular and extensible architecture. These features are intended to ensure its ability to evolve along with the computer industry, leading to a common baseline for server power measurements across geographies.

Work is already underway to further improve and extend the SERT. It is hoped that this leads to even broader adoption across other industries, and for other types of workloads.

9. ACKNOWLEDGMENTS

The authors would like to acknowledge Karin Wulf for providing a modified version of SPEC PTDaemon supporting DC power measurements, as well as Thomas Brand and Charlie Cha for executing numerous measurements for this paper.

The authors also wish to acknowledge current and past members of the SPECpower Committee who have contributed to the design, development, testing, and overall success of the SERT: Sanjay Sharma, Greg Darnell, Karl Huppler, Van Smith, Paul Muehr, David Ott, Cathy Sandifer, Jason Glick, and Dianne Rice, as well as the late Alan Adamson and Larry Gray.

10. REFERENCES

[1] ENERGY STAR Enterprise Servers Specification Version 2.0: http://www.energystar.gov/products/specs/node/142

[2] Lange, K. D., and Tricker, M. G. 2011. The Design and Development of the Server Efficiency Rating Tool (SERT). In *Proceedings of the second joint WOSP/SIPEW international conference on Performance engineering* (Karlsruhe, Germany, March 14 - 16, 2011). DOI= http://dx.doi.org/10.1145/1958746.1958769

[3] Lange. K. D., Identifying Shades of Green: The SPECpower Benchmarks, IEEE Computer, V42 #3 2009, 95-97, DOI= http://dx.doi.org/10.1109/MC.2009.84

[4] Lange, K. D., Tricker, M. G., Arnold, A. A., Block, H., and Koopmann, C. 2012. The Implementation of the Server Efficiency Rating Tool (SERT). In *ICPE '12 Proceedings of the third joint WOSP/SIPEW international conference on Performance Engineering* (Boston, USA, April 22 - 25, 2012). DOI= http://dx.doi.org/10.1145/2188286.2188307

[5] Server Efficiency Rating Tool - Home Page: http://www.spec.org/sert/

[6] Server Efficiency Rating Tool - Design Document: http://www.spec.org/sert/docs/SERT-Design_Document.pdf

[7] SPECpower_ssj2008 - Home Page: http://www.spec.org/power_ssj2008/

[8] SPEC Power and Performance Benchmark Methodology: http://www.spec.org/power/docs/SPEC-Power_and_Performance_Methodology.pdf

Variations of the Star Schema Benchmark to Test the Effects of Data Skew on Query Performance

Tilmann Rabl
Middleware Systems
Reseach Group
University of Toronto
Ontario, Canada
tilmann@msrg.utoronto.ca

Meikel Poess
Oracle Corporation
Califonia, USA
meikel.poess@oracle.com

Hans-Arno Jacobsen
Middleware Systems
Research Group
University of Toronto
Ontario, Canada
jacobsen@eecg.utoronto.ca

Patrick O'Neil
Department of Math and C.S.
University of Massachusetts
Boston
Massachusetts, USA
poneil@cs.umb.edu

Elizabeth O'Neil
Department of Math and C.S.
University of Massachusetts
Boston
Massachusetts, USA
eoneil@cs.umb.edu

ABSTRACT

The Star Schema Benchmark (SSB), has been widely used to evaluate the performance of database management systems when executing star schema queries. SSB, based on the well known industry standard benchmark TPC-H, shares some of its drawbacks, most notably, its uniform data distributions. Today's systems rely heavily on sophisticated cost-based query optimizers to generate the most efficient query execution plans. A benchmark that evaluates optimizer's capability to generate optimal execution plans under all circumstances must provide the rich data set details on which optimizers rely (uniform and non-uniform distributions, data sparsity, etc.). This is also true for other database system parts, such as indices and operators, and ultimately holds for an end-to-end benchmark as well. SSB's data generator, based on TPC-H's dbgen, is not easy to adapt to different data distributions as its meta data and actual data generation implementations are not separated. In this paper, we motivate the need for a new revision of SSB that includes non-uniform data distributions. We list what specific modifications are required to SSB to implement non-uniform data sets and we demonstrate how to implement these modifications in the Parallel Data Generator Framework to generate both the data and query sets.

Categories and Subject Descriptors

K.6.2 [**Management of Computing and Information Systems**]: Installation Management—*benchmark, performance and usage measurement*

Keywords

Star Schema Benchmark; PDGF; data skew; data generation; TPC-H

1. INTRODUCTION

Traditionally, relational schemas were designed using database normalization techniques pioneered by Edgar F. Codd [4]. Database normalization minimizes data redundancy and data dependencies. However, Kimball argues that this traditional approach to designing schemas is not suitable for decision support systems because of poor query performance and usability [7]. He further argues that the design of a decision support system should follow a star schema, which features a fact table containing real live events, e.g., sales, returns etc., and dimensions further describing these events by supplying attributes such as sales price, item descriptions etc. While traditional normalized schemas are designed to reduce the overhead of updates, star schemas are designed to increase the ease-of-use and speed of retrieval. These fundamental design goals are achieved by minimizing the number of tables that need to be joined by queries.

Having realized the need to measure the performance of decision support systems in the early 1990s, the Transaction Processing Performance Council (TPC) released its first decision support benchmark, TPC-D, in April 1994. Because at that time the majority of systems followed Codd's normalization approach, TPC-D is based on a 3rd normal form schema consisting of 8 tables. Even TPC-H, TPC-D's successor, which was released in April 1999 continued using the same 3rd normal form schema.

While for the technology available at that time, TPC-D and TPC-H imposed many challenges, both on hardware and on database management systems (DBMS). However, their 3rd normal form schema and uniform data distributions are not representative any more of current decision support systems. As a consequence, O'Neil et al. proposed a star schema version of TPC-H, the Star Schema Benchmark (SSB) in [11]. SSB denormalizes the data that TPC-H's data generator, dbgen, generates. However, its underlying

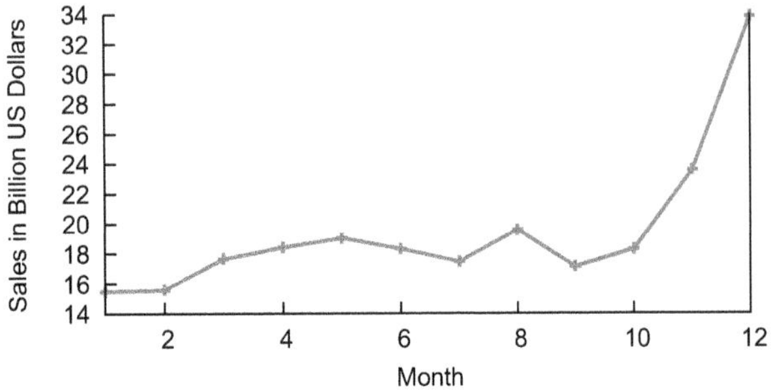

Figure 1: US Retail Sales in Billion US Dollars as Reported by the US Cenus

data generation principles were not changed from TPC-H, especially its uniform data distributions.

We argue that uniform data distributions are not representative for real life systems because of naturally occurring data skew. For instance, retail sales tend to be highly skewed towards the end of the year, customers are clustered in densely populated areas and popular items are sold more often than unpopular items.

Non-uniform data distributions impose great challenges to database management systems. For instance, *data placement algorithms*, especially those on large scale shared nothing systems, are sensitive to data skew, but also common operations such as *join*, *group-by* and *sort* operations can be affected by data skew, most notably the *query optimizer*, is highly challenged to generate the optimal execution plan for queries in the presence of data skew. This is supported in a SIGMOD 1997 presentation by Levine and Stephens in which they noted that a successor of TPC-D should include data skew [2].

Consider the retail sales numbers per month as reported by the US census in Figure 1. The graph shows that US retail sales vary between 15 Billion in January and 35 Billion in December. Sales stay under 20 Billion per month until November when they shoot to 25 Billion and then to 35 Billion in December. It is not uncommon that during the year-end holiday season, often as much as 30% of annual sales are made within the last two months of the year (and much of that in the last half of December) [14].

TPC-DS, published in April 2012, includes some data skew [10], implemented using "comparability zones", which are ranges in the data domain of columns with uniform distributions. The number of comparability zones for a particular column is limited only by the number of unique values in this column. However, because of the underlying design principles of TPC-DS, execution of individual queries is restricted to one comparability zone. While TPC-DS's approach for data skew is a step in the right direction, it limits the effect of data skew greatly, because individual queries still operate on uniform data.

So far there has not been any industry standard benchmark that tests query performance in the presence of highly skewed data. This paper addresses the need for a modern decision support benchmark that includes highly skewed data. The resulting new version of SSB includes data skew in single fact table measures, single dimension hierarchies and multiple dimension hierarchies. To demonstrate the effects of data skew on query performance, several experiments were run on a publicly available DBMS. The results of these experiments underline the need for the introduction of data skew in SSB.

Because SSB's current data generator is an adapted version of TPC-H's data generator, which hard codes most of the data characteristics, a new data generator was implemented using the Parallel Data Generation Framework (PDGF). It's design allows for quickly modifying data distributions as well as making structural changes to the existing tables. In addition it allows the generation of valid SQL from query templates similar to qgen of SSB. This is particularly important as PDGF allows for the generation of queries in the presence of data skew. While this paper demonstrate the key aspects of the data generator, the fully functional data generator is available at `http://www.paralledatageneration.org/`.

In summary, our contributions in this paper are:

- We design and implement a complete data generator for SSB based on the Parallel Data Generation Framework (PDGF) that can be easily adapted and extended.
- We design and implement a query generator for SSB based on PDGF, which enables the generation of more realistic queries with parameters that originate from the data set rather than randomly chosen values.
- We introduce skew into SSB's data model in a consistent way, which makes it possible to reason about the selectivity of queries.
- We implement the proposed data skew in our data generator and demonstrate the effects of the skew on query processing times.

The remainder of this paper is organized as follows. Section 2 reviews those parts of SSB that are relevant to understanding the concepts discussed in this paper. Section 3 presents our data and query generator implementations in PDGF. Section 4 suggests and discusses variations of the star schema benchmark that include various degrees of data skew including experimental results. Before concluding our paper, we discuss related work in Section 5.

2. STAR SCHEMA BENCHMARK

The Star Schema Benchmark is a variation of the well studied TPC-H benchmark [12], which models the data warehouse of a whole sale supplier. TPC-H models the data in 3rd normal form, while SSB implements the same logical data in a traditional star schema, where the central fact table Lineorder contains the "sales transaction information" of the modeled retailer in form of different types of measures as described in [7]. SSB's queries are simplified versions of the queries in TPC-H. They are organized in four flights of three to four queries each.

This paper focuses on data skew in columns that are used for selectivity predicates because data skew in those columns exhibit the largest impact on query performance. Hence, the data set and query set need to be considered when introducing skew into SSB. Therefore, the following two sections first recap the main characteristics of SSB's schema and data set and then analyze the four query flights to ultimately suggest a representative set of columns for which data skew could be implemented.

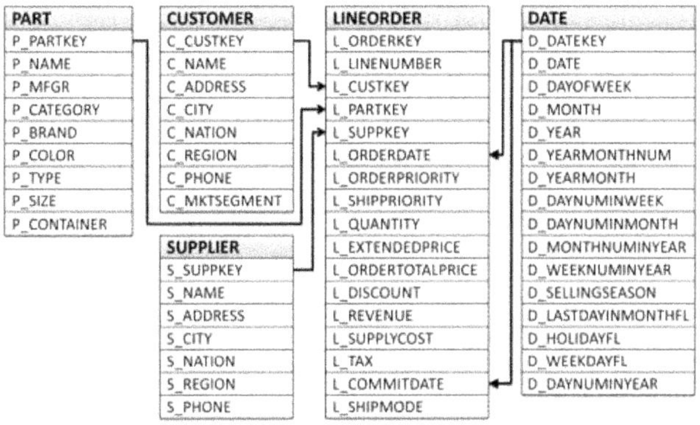

Figure 2: SSB Schema

The uniform characteristics of the TPC-H [1] data set have been studied extensively in [13]. According to the TPC-H specification the term "random" means "independently selected and uniformly distributed over the specified range of values" . That is, n unique values V of a column are uniformly distributed if $P(V = v) = \frac{1}{n}$. Because TPC-H and SSB's data generators use pseudo random number generators and data is generated independently in parallel on multiple computers, perfectly uniform data distributions are impossible to guarantee. Hence, we follow the definition of uniform as presented in [13].

2.1 SSB Schema and Data Population

In order to create a star schema of TPC-H, SSB denormalizes several tables: (i) the Order and Lineitem tables are denormalized into a single Lineorder fact table and (ii) the Nation and Region tables are denormalized into the Customer and Supplier tables and a city column is added. This simplifies the schema considerably, both for writing queries and computing queries as the two largest tables of TPC-H are pre-joined. Queries do not have to perform the join and users writing queries against the schema do not have to express the join in their queries. In SSB, the Partsupp table is removed, because this table has a different temporal granularity than the Lineorder table. This is an inconsistency in the original TPC-D and TPC-H schemas as described in [11]. Finally, an additional Date table is added, which is commonly used in star schemas. To summarize, SSB consists of one large fact table (Lineorder) and four dimensions (Customer, Supplier, Part, Date). The complete schema is depicted in Figure 2.

Like in TPC-H, all data in SSB is uniformly distributed. To be consistent with the general star schema approach the definition of the SSB tables is slightly different from the definition of TPC-H. The most notable change is the introduction of selectivity hierarchies in all dimension tables. They are similar to the manufacturer/brand hierarchy in TPC-H, where the manufacturer (P_Mfgr) value is a prefix of the brand (P_Brand) value. We denote this dependency with P_Brand → P_Mfgr, meaning that for each P_Brand value there is exactly one P_Mfgr value. In the Part table, SSB adds a category, which extends the TPC-H hierarchy to P_Brand1 → P_Category → P_Mfgr. Because all data is uniformly distributed this hierarchy serves as a simple way to control row selectivity of queries. There are five different values for P_Mfgr, resulting in selectivity of $\frac{1}{5}$ each. For each of those there exist five different values in P_Category with a selectivity of $\frac{1}{25}$ each. Finally, the P_Brand1 field has 40 variations per P_Category value for a selectivity of $\frac{1}{1000}$ each. The hierarchies in Customer and Supplier are C_City → C_Nation → C_Region and S_City → S_Nation → S_Region, respectively. Date features two natural hierarchies D_DayNumInMonth → D_Month → D_Year and D_DayNumInWeek → D_Week → D_Year .

2.2 SSB Queries

Instead of TPC-H's 22 queries, O'Neil et al. propose four flights of queries that each consist of three to four queries with varying selectivity [11]. Each flight consists of a sequence of queries that someone working with data warehouse system would ask, e.g., for a drill down.

Query Flight 1 (Q1.1, Q1.2 and Q1.3) is based on TPC-H's Query 6 (see Listing 1 for the SQL code of Q1.1). It performs a single join between the Lineorder table and the small Date dimension table. Selectivity predicates are set on D_Year, Lo_Discount and Lo_Quantity. Flight 1 simulates a drill down into the Date dimension. From one query of the flight to the next join selectivity is reduced by refining the predicate on the Date component. Q1.1's Date range is one year, Q1.2 selects data within a range of one month and Q1.3's refines this further by selecting data of one week. Q1.3 further changes the predicate on Lo_Quantity from `less than n` to `between n and m`.

```
SELECT SUM(lo_extendedprice*lo_discount) AS revenue
  FROM lineorder, date
 WHERE lo_orderdate = d_datekey
   AND d_year = 1993
   AND lo_discount BETWEEN 1 AND 3
   AND lo_quantity < 25;
```

Listing 1: Query Q1.1 of SSB

In order to measure the effect of data skew on query execution times of queries in Flight 1, we propose data skew in (i) Lo_Orderdate, (ii) Lo_Discount and (iii) Lo_Quantity. Data skew in the Lo_Orderdate column is realistic as retail sales are low and flat in the summer, but pick up steeply after Thanksgiving to top in December (see Figure 1). Data Skew in Lo_Discount and Lo_Quantity are also realistic, which can be categorized as single column skew of single tables because these columns are statistically independent.

Queries in Flight 2 join the Lineorder, Date, Part and Supplier tables (see Listing 2 for Q2.1). They aggregate and sort data on Year and Brand. Selectivity is modified using predicates on P_Category, S_Region and P_Brand. That is, they use the P_Brand1 → P_Category → P_Mfgr hierarchy in the Part table.

```
SELECT SUM(lo_revenue), d_year, p_brand1
  FROM lineorder, date, part, supplier
 WHERE lo_orderdate = d_datekey
   AND lo_partkey = p_partkey
   AND lo_suppkey = s_suppkey
   AND p_category = 'MFGR#12'
   AND s_region = 'AMERICA'
 GROUP BY d_year, p_brand1
 ORDER BY d_year, p_brand1;
```

Listing 2: Query Q2.1 of SSB

In addition to data skew in Lo_Orderdate, as proposed for Flight 1, for Flight 2 we propose data skew in P_Category and S_Region, which is realistic as not every manufacturer produces the same number of parts and not all suppliers are distributed uniformly in all regions.

Flight 3 aggregates revenue data by Customer and Supplier nations as well as the years in which orders were placed. Starting from a specific customer, supplier, region and date, Flight 3 drills down into specific cities and years. It uses the City → Nation → Region hierarchies in Customer and Supplier. Listing 3 shows the third query in Flight 3. Only flights 3 and 4 add disjunctive predicates.

```
SELECT c_city, s_city, d_year,
       SUM(lo_revenue) AS revenue
  FROM customer, lineorder, supplier, date
 WHERE lo_custkey = c_custkey
   AND lo_suppkey = s_suppkey
   AND lo_orderdate = d_datekey
   AND (c_city_skewed='City1'
    OR c_city_skewed='City2')
   AND (s_city_skewed='City2'
    OR s_city_skewed='City3')
   AND d_year >= 1992
   AND d_year <= 1997
 GROUP BY c_city, s_city, d_year
 ORDER BY d_year ASC, revenue DESC;
```

Listing 3: Query Q3.3 of SSB

As this flight eventually drills down into one month, data skew in the Lo_Orderdate and the region columns makes sense for this query as well.

Flight 4 is the most complex one, as it joins all tables. Query Q4.1 can be seen in Listing 4. It drills down into region and manufacturer, using both the P_Brand → P_Category → P_Mfgr and the C_City → C_Nation → C_Region hierarchies in Supplier. Hence, data skew in the Customer, Part and Supplier hierarchies makes sense.

```
SELECT d_year, c_nation,
       SUM(lo_revenue - lo_supplycost) AS profit
  FROM date, customer, supplier, part, lineorder
 WHERE lo_custkey = c_custkey
   AND lo_suppkey = s_suppkey
   AND lo_partkey = p_partkey
   AND lo_orderdate = d_datekey
   AND c_region = 'AMERICA'
   AND s_region = 'AMERICA'
   AND (p_mfgr = 'MFGR#1' OR p_mfgr = 'MFGR#2')
 GROUP BY d_year, c_nation
 ORDER BY d_year, c_nation;
```

Listing 4: Query Q4.1 of SSB

The above query analysis shows that implementing data skew in the single columns Lo_Discount and Lo_Quantity, as well as in the dimension hierarchies P_Brand → P_Category → of Part, C_City → C_Nation → C_Region of Customer, and C_City → C_Nation → C_Region of Supplier is sufficient to vary the selectivity of all queries and demonstrate the affects of data skew to query performance.

3. SSB IMPLEMENTATION IN PDGF

The Parallel Data Generation Framework (PDGF) is a generic data generator that was implemented at the University of Passau [15]. It was mainly built to generate relational data for database benchmarking purposes. However, it is able to create any kind of data that can be expressed in a relational model. PDGF exploits the parallelism in *xorshift* random number generators (PRNG) to generate complex dependencies by re-calculating dependent row data (a.k.a. field values) rather than storing them, which would greatly limit the maximum data size and parallelism that can be generated on a given system (for details on xorshift generators see, e.g., [9]). The underlying approach is straight forward: the random number generator is a hash function, which can generate any random number out of a sequence in O(1). With this approach every random number in the sequence of the PRNG can be computed independently. Using the random numbers, generated by the PRNG, arbitrary values are generated using mapping functions, dictionary lookups and such. Quickly finding the right random numbers is possible by using a hierarchical seeding strategy. Furthermore, by using bijective permutations it is possible to generate consistent updates of the data where values may be inserted, changed and deleted [6].

3.1 Data Generation in PDGF

Although pseudo random number generation is in general deterministic, i.e., a repeated execution of the data generator produces the identical results, most data generation approaches do not allow a repeated generation of an individual field value without generating data for all preceding rows. This is due to dependencies within the generation. For example, if the generation of a timestamp in a certain row of a table is dependent on the timestamp in the previous row (e.g., $t_n = t_{n-1}+$ random number of seconds) all previous timestamps have to be computed recursively in order to re-calculate each timestamp. With this approach it is impossible to generate data that references such timestamps without reading the data after its generation. For large data sets and distributed data generation, re-reading data is not feasible. To circumvent these kinds of problems, PDGF uses a repeatable data generation methodology. All field values are generated using functions that map random numbers (input) to row data, the *field value generators*. They have to follow the requirement, that the same input number always creates the same output value and that the value generator is stateless. That is, each field value generator needs to take a seed value as input parameter. If field value generators follow this requirement the challenge of repeatable data generation can be reduced to repeatable random number generation. We achieve this by using the seeding hierarchy depicted in Figure 3. The hierarchy assigns a virtual identifier to each field in the database. The identifier is mapped to a random number by iteratively reseeding random number generators (schema → table → column → update → row). It has to be noted that most of these reseeding operations only have to be done during initialization. Output of the seeding hierarchy is a deterministically seeded random number generator that is used to generate the required input numbers for the generation of the field value. This is especially important if complex field values are generated, such as Markov chain generated pseudo text, which does not consume the same number of random values for each field value.

PDGF can be configured using two XML files: the *Schema Configuration File* describes the underlying relational schema of the data, the *Generation Configuration File* defines

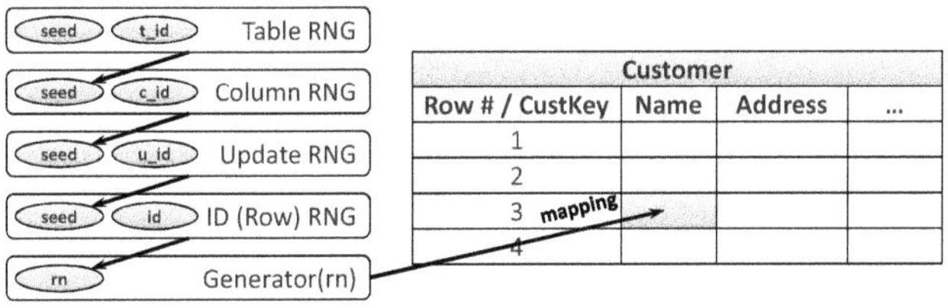

Figure 3: PDGF's Seeding Strategy

how the data is formatted. The schema configuration has a similar structure as the SQL definition of a relational schema, it defines a set of tables which consist of fields. Each field has name, size, data type and generation specification. Listing 5 shows the first two fields of the Supplier table and the definition of dependencies in PDGF. For instance, the S_Name field is defined as the string "Supplier", concatenated with the value of the field S_Suppkey and padded to 9 places. PDGF automatically resolves this dependency if the *OtherFieldValueGenerator* is used. The additional transformations can be specified with meta-generators that change the values generated by subsequent generators. In the example, the padding is done by a *PaddingGenerator* and the prefix is added with another generator.

```
<table name="SUPPLIER">
  <size>${S}</size>

  <field name="S_SUPPKEY" size="" type="NUMERIC"
         primary="true" unique="true">
    <gen_IdGenerator />
  </field>

  <field name="S_NAME" size="25" type="VARCHAR">
    <gen_PrePostfixGenerator>
      <gen_PaddingGenerator>
        <gen_OtherFieldValueGenerator>
          <reference field="S_SUPPKEY" />
        </gen_OtherFieldValueGenerator>
        <character>0</character>
        <padToLeft>true</padToLeft>
        <size>9</size>
      </gen_PaddingGenerator>
      <prefix>Supplier</prefix>
    </gen_PrePostfixGenerator>
  </field>
[..]
```

Listing 5: Excerpt of the Supplier Table Definition

One peculiarity of the SSB schema is the definition of the Lineorder table. It resembles the denormalized Lineitem $\leftrightarrow$ Order relationship of TPC-H. One row in the Order table corresponds to $n \in [1,7]$ rows in the Lineitem table. In its denormalized version each row of the Lineorder table contains the individual price of the item as well as the total price of the order. Hence, there is a linear dependency between Lineitem and Orders. This is resolved in PDGF in the following way: Instead of treating each line item as a single row we generate all line items of a particular order in one row. This is similar to the generation of Lineorder in the original generator of SSB and the generation of Lineitem and Order in TPC-H, respectively. The PDGF implementation of Lineorder ist shown in Listing 6. The field Lo_Number_Of_Lineitems determines the number of lineitems within a certain order. This field is not printed. All order related values, e.g., Lo_Orderkey, are specified once, while values that are specific to a lineitem, e.g., Lo_Linenumber are specified for each lineitem.

```
    <table name="LINEORDER">
        <size>${L}</size>

        <field name="LO_ORDERKEY" size=""
            type="NUMERIC">
            <gen_FormulaGenerator>
                <formula>
                  (gc.getID() / 8 * 32) +
                      (gc.getID() % 8)
                </formula>
                <decimalPlaces>0</decimalPlaces>
            </gen_FormulaGenerator>
        </field>

        <field name="LO_NUMBER_OF_LINEITEMS" size=""
            type="NUMERIC">
            <gen_LongGenerator>
                <min>1</min>
                <max>7</max>
            </gen_LongGenerator>
        </field>

        <field name="LO_LINENUMBER_1" size=""
            type="NUMERIC">
            <gen_StaticValueGenerator>
                <value>1</value>
            </gen_StaticValueGenerator>
        </field>
[..]
```

Listing 6: Excerpt of the Lineorder Table Definition

The output of Lineorder data as multiple rows is specified in the generation configuration file, which describes the output location and format of the output data. Furthermore, the generation configuration file allows for specifying arbitrary post-processing operations. In Listing 7, an excerpt of the generation specification for Lineorder is shown. The value of the field in Column 2, Lo_Number_Of_Lineitems, determines the number of rows that have to be generated.

```
<table name="LINEORDER" exclude="false">
  <output name="CompiledTemplateOutput">
  <fileTemplate>outputDir + table.getName()
               + fileEnding</fileTemplate>
  <outputDir>output/</outputDir>
  <fileEnding>.tbl</fileEnding>
  <charset>UTF-8</charset>
  <sortByRowID>true</sortByRowID>
  <template><!--
    int noRows
        =(fields[1].getPlainValue()).intValue();
    for (int i = 0; i < noRows; i++) {
      buffer.append( fields[0] ); // LO_ORDERKEY
      buffer.append('|').append( fields[2+i] );
[..]
```

Listing 7: Excerpt of the Generation Definition for Lineorder

Post-processing in the generation configuration, as shown in Listing 7, is only necessary if the relational definition in the schema definition file is not identical to the format on disk.

Another example in the SSB schema is the address information. Because the dimension tables are denormalized the Region and Nation fields have to be generated consistently (e.g., France should always be in Europe). We implemented this by specifying a virtual table Region_Nation, which is referenced by other tables and is not actually generated. This way, we ensure consistency in the data generation because the virtual table Region_Nation contains only valid Region-Nation combinations.

3.2 Query Generation in PDGF

In addition to the data generation, we also designed and implemented a query generator in PDGF, which converts the SSB query templates into valid SQL. In order to generate a large query set, many benchmarks including TPC-H, TPC-DS and SSB define query templates that can be converted into valid SQL queries. For TPC-H and TPC-DS this is done by tools that substitute tagged portions of queries with valid values, such as selectivity predicates, aggregation functions or columns (qgen, DSQgen) [14].In order for such tools to generate a valid and predictable workload they need to have knowledge about the schema and data set. Otherwise the chosen values, e.g., selectivity predicates might not qualify the desired number of rows. TPC-H's qgen and its counterpart for SSB generate queries by selecting random values within the possible ranges of the data. The following shows the definition of the supplier name ranges as specified in the TPC-H specification: `S_NAME text appended with digit ["Supplier", S_SUPPKEY]`

TPC-H's and SSB's qgen implementations have major drawbacks: (i) the parameter ranges for each substitution parameter are hard-coded into a c-program, (ii) scale factor dependent substitution parameter handing is difficult, and (iii) this approach only works on uniformly distributed data.

Instead of hard-coding valid parameter ranges for each query, we use PDGF's post-processing functionality to implement our query generator for SSB. For each query we define one table that is populated with all valid parameters of that query. We refer to these tables as *parameter tables*. The key benefit of our approach is that query parameters are generated from existing values in the data set. Using our approach we generate queries by selecting random field values from the data set, i.e., field values that actually exist in the generated data. This is a more flexible approach as it does not statically define the data range.

In Listing 8, an excerpt for the table specifying the parameters for Query Q1.1 can be seen.

```
<table name="QUERYPARAMETERS_Q1.1">
  <size>13 * ${QUERY_ROUNDS}</size>
  <field name="YEAR" size="4" type="NUMERIC">
    <gen_LongGenerator>
      <min>1993</min>
      <max>1997</max>
    </gen_LongGenerator>
  </field>
[..]
</table>
```

Listing 8: Schema Definition of the Parameter Table for Query Q1.1 (excerpt)

The query template itself is specified in the generation configuration. For Query Q1.1 this can be seen in Listing 9. As can be seen in the listing, the template is embedded in a XML comment in Java-style String format. This format will be compiled into a regular Java binary at runtime. Future versions of PDGF will include a new output plugin that is able to read the template from files and directly generate this Java representation.

```
<table name="QUERY_PARAMETERS" exclude="false" >
  <output name="CompiledTemplateOutput" >
    [..]
    <template><!--
int y = (fields[0].getPlainValue()).intValue();
int d = (fields[1].getPlainValue()).intValue();
int q = (fields[2].getPlainValue()).intValue();
String n = pdgf.util.Constants.DEFAULT_LINESEPARATOR;
buffer.append("-- Q1.1" + n);
buffer.append("select sum(lo_extendedprice *");
buffer.append("         lo_discount) as revenue" + n);
buffer.append("from lineorder, date" + n);
buffer.append("where lo_orderdate = d_datekey" + n);
buffer.append("and d_year = " + y + n);
buffer.append("and lo_disc between " + (d - 1));
buffer.append("                  and " + (d + 1) + n);
buffer.append("and lo_quantity < " + q + ";" + n);
 --></template>
  </output>
</table>
```

Listing 9: Excerpt of the Template Specification for Query Q1.1

4. STAR SCHEMA BENCHMARK VARIATIONS

Our PDGF implementation of the original Star Schema Benchmark enables easy alterations to the schema and the data generation as well as the query generation. Columns can be easily added or data distributions changed and queries modified and added. This section discusses different approaches of introducing data skew into the original SSB and the reasoning why some of these changes work and some do not. The variations are based on the findings of Section 2.2 and can be subdivided into four categories:

- Skew in foreign key relations
- Skew in a fact table measures
- Skew in a single dimension hierarchies
- Skew in multiple dimension hierarchies

We discuss each of the proposed data skews below and conduct experiments to show their effects on query elapsed times. Each experiment focuses on one of the above proposed data skews by running queries on the original uniform and on the proposed skewed data sets. For each experiment we choose a representative query template, generate queries for all possible substitution parameters and run them in three modes: (i) Index forced: The query optimizer is forced to use indexes; (ii) Index disabled: The query optimizer is not allowed to use indexes, forcing the system to perform full table scans; and (iii) Index automatic: The query optimizer is free to utilize indexes.

We do not claim that the above three modes are an exhaustive list to analyze the effects of data skew on query elapsed time. However, indexes are widely used to improve query performance and index technology is available on most relational database systems (RDBMS) today. With the three modes, we demonstrate the ability/inability of the query optimizer to choose the best performing plan for each of the substitution parameters and any related side effects, such as large scans and large intermediate result sets. We further demonstrate the difference in query processing time for uniform and skewed data under equal conditions.

Our experiments were conducted on a publicly available RDBMS as a proof of concept. The findings are RDBMS agnostic as the technology used is available in all of today's RDBMS implementations. Thus, the following experiments are not conducting a performance analysis of a specific RDBMS. They are included to show the effects of the proposed distribution changes, thereby underlining the importance of data skew in database benchmarks. The displayed elapsed times are averages from five consecutive runs of the same query.

4.1 Skew in Foreign Key Relations

The characteristics of real life data when loaded into a star schema result in skew of foreign key relations, i.e., fact table foreign keys. For instance, popular items are sold more often than unpopular, some customer purchase more often than others and sales do not occur uniform throughout the year, but tend to be highly skewed towards the holiday season (see also Figure 1).

These characteristics of real life data result in potentially highly skewed foreign key distributions, which impact the way joins are executed, data structures are allocated and plan decisions are made by query optimizers. Implementing skew in foreign keys is easy in PDGF. However, there are drawbacks to adding this kind of skew. Due to surrogate keys being used in star schemas, distributions in foreign keys cannot be correlated to the values in the dimensions that they reference, e.g., values in hierarchy fields. As a result skew in foreign keys does not necessarily translate into skew in the join to dimensions. The high selectivity in dimension fields evens out the skew in the foreign keys.

As an example consider Query Q2.1 in Listing 2. Even if we introduce skew into the foreign keys to the Part table, i.e., Lo_Partkey, many Lineorder tuples will still match the same P_Mfgr value. Since there are only 5 different P_Mfgr values, the resulting selectivities per P_Mfgr value will be almost uniformly distributed. The same is true for the P_Category and P_Brand1 selectivities. For this reason, we do not introduce skew in foreign key references.

In the original SSB the foreign keys, Lo_Custkey, Lo_Partkey, Lo_Suppkey and Lo_Orderdate are distributed uniformly. E.g., the keys in Lo_Orderdate have a coefficient of variation of 0.0044. This allows us to introduce skew into the values of the dimensions themselves. Due to the uniform distributions of the foreign keys the skew of dimension values directly translates into skew of the foreign keys. This is a better way to control the skew in the cardinality of joins between the fact table and dimension tables.

4.2 Skew in Fact Table Measures

Lo_Quantity is an additive measure of the Lineorder fact table. It contains natural numbers $x \in [1, 50]$. In the original uniform distribution values of Lo_Quantity occur with a likelihood of about 0.02 with a coefficient of variation of 0.000284. We modify the Lo_Quantity distribution to skew quantities towards smaller values in the following way: Let $x \in [1, 50]$ be the values for Lo_Quantity, then the likelihood of x is $p(x) = P(X = x) = \frac{0.3}{1.3^x}$. In order to calculate the number of rows with value x, i.e., $r(x)$, we multiply n by the table cardinality, $n = |\text{Lineorder}| : r(x) = n * p(x)$. This results in a exponential probability distribution.

In PDGF this kind of distribution can be implemented very easily by adding a distribution node to the data generator specification. This can be seen in Listing 10. The exponential distribution is parameterized with $\lambda = 0.26235$, which results in the desired distribution.

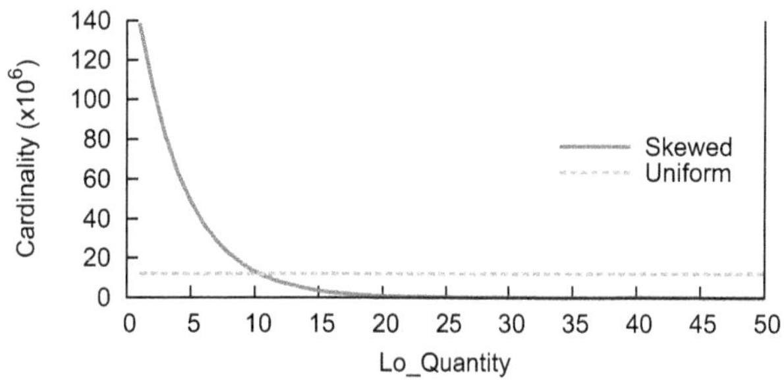

Figure 4: Lo_Quantity Distribution

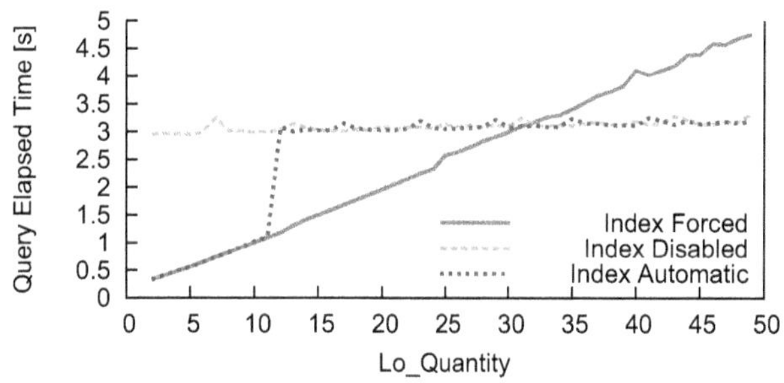

Figure 5: Elapse Times of Query Q1.1 with Uniformly Distributed Lo_Quantity

```
<field name="LO_QUANTITY" size="2" type="NUMERIC">
  <gen_LongGenerator>
    <min>1</min>
    <max>50</max>
    <distribution name="Exponential"
         lambda="0.26235" />
  </gen_LongGenerator>
</field>
```

Listing 10: Schema Definition of Lo_Quantity

Figure 4 shows the uniform (original) and skewed distributions of Lo_Quantity. The x-axis plots the values for Lo_Quantity ranging from 1 to 50 and the y-axis plots the number of rows with those values. The cardinalities in the uniform case for scale factor 100 are constant at around 12 Million with a coefficient of variation of 0.00000557, while the cardinalities in the skewed case start at 140 Million and decrease steeply to 336 with a coefficient of variation of 0.047465541.

Using the skewed and uniform distributions for Lo_Quantity we run Q1.1 and Q1.2 of Flight 1, which contain selectivity predicates on Lo_Quantity. Query Q1.1 contains the "less than" predicate `lo_quantity < x`, while Query Q1.2 contains the "between" predicate `lo_quantity BETWEEN x AND x+9`. We vary the selectivity predicates on Lo_Quantity in the following way: for Query Q1.1 $x \in [2, 51]$ and Query Q1.2 $x \in [1, 41]$. To take advantage of the extreme distribution of Lo_Quantity, we create an index on Lo_Quantity.

Figures 5 and 6 show the elapsed times of Query Q1.1 us-

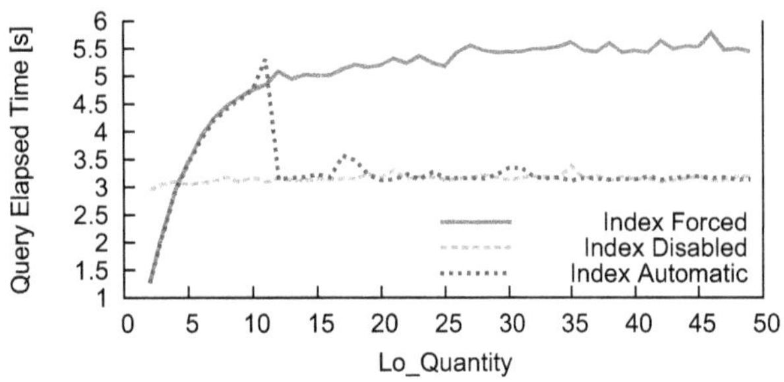

Figure 6: Elapse Times of Query Q1.1 with Skewed Distributed Lo_Quantity

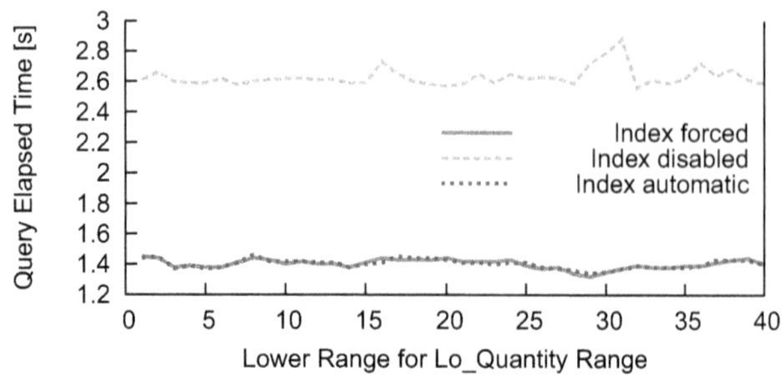

Figure 7: Effect of skewed Lo_Quantity on Query Q1.2 elapsed times

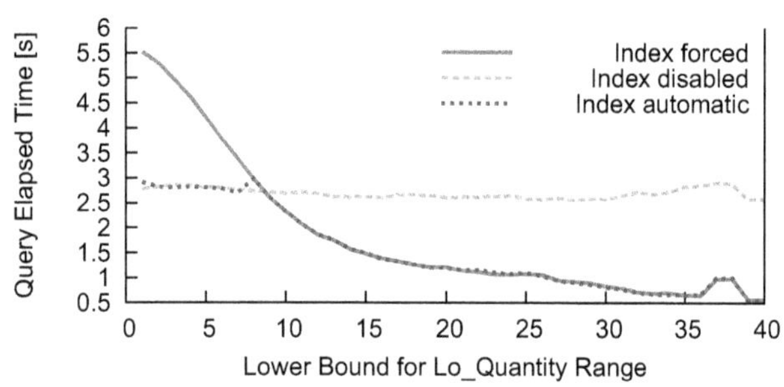

Figure 8: Effect of skewed Lo_Quantity on Query Q1.2 elapsed times

ing the three modes explained above and varying the predicate `lo_quantity < x`, $x \in [2, 51]$ on both the uniform and skewed data sets. The solid line in Figure 5 shows the elapsed times of index driven queries. The elapsed times increase linearly with increased values of Lo_quantity. It starts with about 0.4s at $x = 2$ and increases linearly to 4.8s at $x = 51$. This is not surprising as the advantage of the index diminishes with increased values for x, i.e., the number of rows qualifying for the index driven join increases-linearly with increased values of x. The dashed line shows the elapsed time of the non-index driven join queries. This line is flat at about 3s. This is also not surprising as the non-index driven join queries have to scan the entire Lineorder table regardless of substitution parameters. That is, the index driven queries outperform the non-index driven queries until $x = 31$. The dotted line shows the elapsed time of queries when the query optimizer is free to choose the best execution plan. This line shows that the query optimizer switches from an index driven join to an non-index driven join at $x = 12$. This is suboptimal as the real cutoff of index driven joins is at about $x = 31$. That is the query optimizer switches too early from an index driven join to a non-index driven join.

Figure 6 repeats the same experiments on skewed data. As in the uniform case the elapsed times of the index driven queries, as indicated by the solid line, and the elapsed times of the non-index driven queries, as indicated by the dashed line follow the skewed Lo_Quantity distribution. Due to the extreme skew in Lo_Quantity, the index driven queries outperform the non-index driven queries only for values $x < 5$. The dotted line shows the elapsed time of queries when the query optimizer is free to choose the best execution plan. It shows that the optimizer switches from an index driven plan to a non-index driven plan at the same value for x as in the uniform experiment, namely at $x = 11$. However, this switch is suboptimal. It should have occurred earlier, i.e., at $x = 5$.

These experiments show that the optimizer factors the selectivity of certain predicates in the query planning. However, the fact that the switch in the query plan from index driven to non-index driven occurs at the same predicate parameter for skewed and uniform data suggests that the optimizer does not consider the data distribution for this decision.

Figures 7 and 8 show the elapsed times of Query Q1.2 on uniform and skewed data in the above described three optimizer modes. Similarly, as in the experiments for Query Q1.1, we vary $x \in [1, 41]$ in `lo_quantity between x and x+9`.

Figure 7 shows that the elapsed times of Query Q1.2 stay constant for each of the three modes regardless of the substitution parameter x. On average the index driven queries execute in 2.63s with a coefficient of variation of 0.024, while the non-index driven queries and the queries chosen in the automatic mode execute on average in 1.42s with a coefficient of variation of 0.021. The index driven queries outperform the non-index driven queries by 46% and the lines for index forced and index automatic overlap.

Figure 8 shows that in the skewed case Query Q1.2 benefits from non-index driven joins in cases of high selectivity. The elapsed times of the non-index driven queries are agnostic to the substitution parameter x. With an average of 2.69s at a coefficient of variation of 0.038, which is similar to the uniform case, they outperform the index driven queries for $x < 10$. The index driven queries follow the distribution for Lo_quantity (see Figure 4), i.e., they monotonously decrease with increasing x. For $x >= 10$ the index driven queries outperform the non-index driven queries. The line showing the elapsed times of queries in the automatic mode indicates that the system makes the right choice switching from non-index driven queries to index driven queries at $x = 10$.

4.3 Skew in Single Dimension Hierarchies

Selectivity in many of the original SSB queries is controlled by predicates on hierarchy columns. Each dimension table defines one three-column hierarchy. Each column of a hierarchy defines a different range of values. Column cardinality increases as individual queries of flights

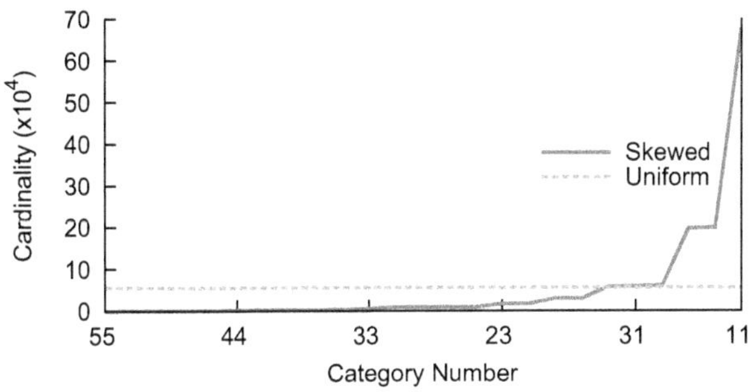

Figure 9: P_Category Distribution Uniform and Skewed

drill down into hierarchies. For example, the Part table defines the following hierarchy: P_Brand → P_Category → P_Mfgr. Values for these hierarchy columns are generated in the following way: Values in P_Mfgr are used as prefixes in P_Category whose values again serve as prefixes in P_Brand. While P_Mfgr contains five different values $m \in \{MFGR\#1, ..., MFGR\#5\}$, P_Category contains 25 different values $c \in \{MFGR\#11, ..., MFGR\#55\}$ and P_Brand contains 1000 values $b \in \{MFGR\#1101, ..., MFGR\#5540\}$. By varying the projection predicates in queries for these three columns, the join selectivity can be varied as well. Consider Query Q2.1 in Listing 2; due to the hierarchy and the uniform distributions in all values and references one can easily see that the selectivity on Part and thus also on Lineorder is $\frac{1}{5}$ of the size of the tables. By changing the restriction to Category, e.g., `p_category = 'MFGR#22'`, the join selectivity can be reduced to $\frac{1}{25}$. The set ratio between the number of rows qualifying a specific substitution parameter is exploited in the query flights.

By introducing skew in these columns the above property may be lost. However, if we control the skew and are able to estimate the frequency of each value, the selectivity calculations are still possible. Therefore, we set a predefined probability for each value in the part hierarchy. For the numbers in Mfgr and Category we use the following probabilities: 1 70%, 2 20%, 3 6%, 4 3%, and 5 1%. For instance, the value "Mfgr#1" appears with a probability of 70% and the category "Mfgr#11" occurs with a probability of $70\% * 70\% = 49\%$. The values where chosen to have one range that has the same probability as in the uniform case (i.e., "MFGR#2" and "MFGR#22"). This makes it possible to specify queries with the same selectivity for the skewed and uniform data. SSB Query Q1.3 includes a range selection on Brand1. Therefore, we define ranges with the same probability in column Brand1. The selectivities in Brand1 are: 1-10 70%, 11-20 25%, 21-30 4.5%, and 31-40 0.5%. Thus brands in the range of "MFRG#2211"-"MFRG#2220" have the same probability as in the uniform case.

Figure 9 displays the uniform and skewed distribution of values in the P_Category column. The cardinalities of the uniform distribution stay at around 4% with a coefficient of variation of 0.0044. The cardinalities in the skewed distribution vary between 0.01% and 48.36% with a coefficient of variation of 2.5028.

Skew in P_Category can be implemented in PDGF as described in Listing 11. It shows the definition of the P_Mfgr field. Each value consists of two parts, the prefix "MFGR#"

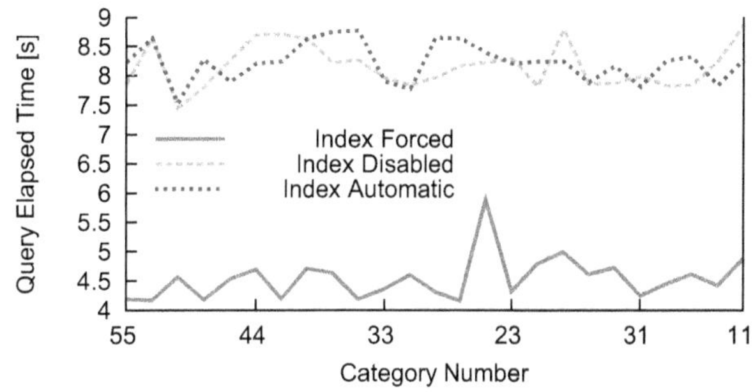

Figure 10: Elapsed Times of Query Q2.1 with Uniformly Distributed P_Category

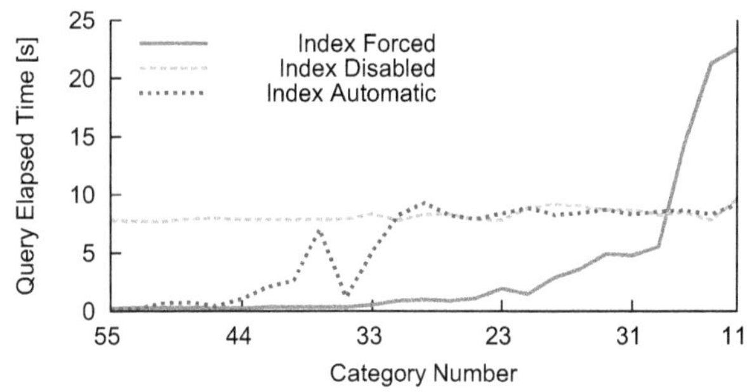

Figure 11: Elapsed Times of Query Q2.1 with Skewed Distributed P_Category

and a number between one and five. The probability for each number is explicitly defined as specified above. The definition of the fields P_Category and P_Brand1 is implemented in the same way, by explicitly giving the probabilities.

```
<field name="P_MFGR" size="6" type="VARCHAR">
 <gen_PrePostfixGenerator>
  <gen_ProbabilityGenerator>
   <probability value="0.70">
    <gen_StaticValueGenerator>
     <value>1</value>
    </gen_StaticValueGenerator>
   </probability>
   <probability value="0.20">
    <gen_StaticValueGenerator>
     <value>2</value>
    </gen_StaticValueGenerator>
   </probability>
   [..]
  </gen_ProbabilityGenerator>
  <prefix>MFGR#</prefix>
 </gen_PrePostfixGenerator>
</field>
```

Listing 11: Definition of Skew in P_Mfgr Field

Query Q2.1, as defined in Listing 2, can be used to demonstrate the effects of the skewed distribution on query execution times. In Figure 10, the execution times of all possible values of P_Category are shown for the uniform, i.e., original distribution, and in Figure 11, they are shown for the skewed distribution. As can be seen in the graph, the execution times for the queries in the uniform case are almost constant as expected, because the query predicate is an "equal" predicate in Q2.1. Interestingly, the query optimizer consistently chooses the wrong query plan, i.e., an index-driven plan. In the skewed case, the execution times are increasing with the selectivity for the index driven query plan and constant for the plan without index. As in previous experiments, the query optimizer is aware of the skew in the selectivity of P_Category as can be seen at the automatic plan in the skewed case. However, the query optimizer switches from an index to non-index plan too early, thus significantly increasing the query elapsed time. This behavior can only be observed on the skewed data set. With the regular data set the source of the less efficient query plan cannot be determined.

4.4 Skew in Multiple Dimension Hierarchies

This section discusses the introduction of data skew in multiple dimensions and its effect on query elapsed times. Similar to the data skew in the P_Brand1 → P_Category → P_Mfgr hierarchy of the Part table, data skew can be implemented in the City → Nation → Region hierarchies of the Supplier and Customer tables. Having five regions in each table (Supplier, Customer) results in a selectivity of $\frac{1}{5}$ in the uniform case. Each region contains 5 nations, each with a selectivity of $\frac{1}{25}$. Each nation contains 10 cities for a total selectivity of $\frac{1}{250}$. Queries in flights 3 and 4 compare customer with supplier using various levels in the above hierarchies.

For our experiments, we modify S_City and C_City to follow an exponential distribution. For simplicity, we assign each city of the Supplier and Customer tables a unique number: $c_s \in [1, 250]$ for Supplier and $c_c \in [1, 250]$ for Customer. The likelihoods of Supplier and Customer cities are then defined as $p(c_s) = P(C = c_s) = \frac{0.0309}{1.0309^{c_s}}$ and $p(c_c) = P(C = c_c) = \frac{0.04}{1.04^{c_c}}$. In order to calculate the number of rows $r(c_s)$ with value c_s and the number of rows $r(c_c)$ with values c_c, we multiply the likelihoods by the table cardinality $n_s = |\text{Supplier}| : r(c_s) = n_s * p(c_s)$ and $n_c = |\text{Customer}| : r(c_c) = n_c * p(c_c)$. This results in exponential probability distributions for Supplier and Customer cities as depicted in Figure 12. The cardinality of the Customer table is 15 times larger than the cardinality of the Supplier table. In order to plot graphs for S_City and C_City in one graph, we normalized both by dividing the cardinality of each city by the cardinality of the largest city: $max|\text{Supplier}_{\text{City}}| = 6799$, $max|\text{Customer}_{\text{City}}| = 121135$. As depicted in Figure 12 both distributions are similar.

Query Q3.3 (see Listing 3) contains selectivity predicates on Date(D_Year), Supplier(S_City) and Customer(C_City). The predicates on S_City and C_City are disjunctions. We use the same city for each part of the disjunction. Hence, we have two substitution parameters (city1, city2) for Q3.3, where city1 is used as a projection of Supplier and city2 is used as a projection of Customer. Because there is no correlation between the city distributions in Supplier and Customer and the Date table, not all city pairs are guaranteed to occur for every year. Hence, before running our experiments we compute the valid Supplier and Customer city pairs. It turns out that out of the total combination of $250^2 = 62500$ city pairs only 1569 occur within `D_Year=1997` (default for Query Q3.3) in the original SSB data. To reduce the number of experiments we select a subset of 156, i.e., every 10th, city pair. Figure 13 shows the join cardinalities $|C_{c\_city=city1} \bowtie L \bowtie S_{s\_city=city2} \bowtie D_{d\_year=1997}|$ of the subset of valid Supplier and Customer city pairs that we use in subsequent experiments. The graph shows that

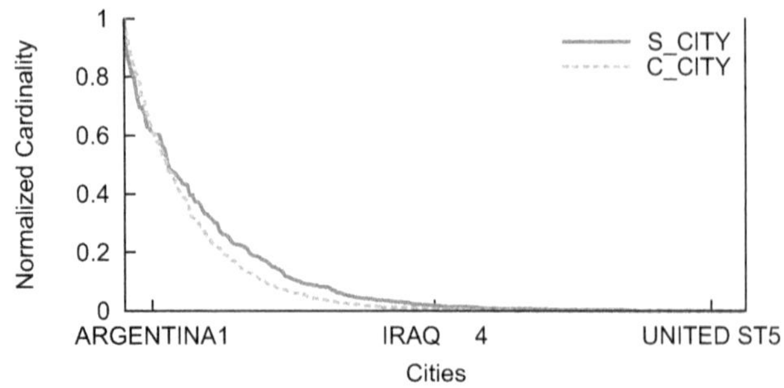

Figure 12: C_City and S_City Skewed Distributions

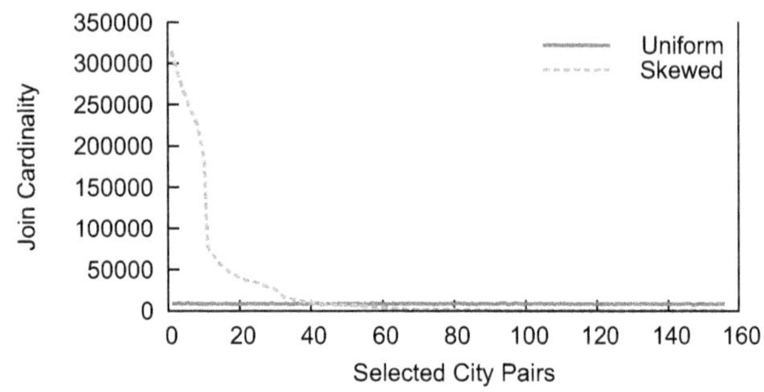

Figure 13: Query Q3.3 Join Cardinalities of the Skewed Distributions

the skewed distributions of cities in Supplier and Customer translate into a similar distribution of the join cardinalities. They start very high at around 300k and decrease negative exponentially to single digit values for high city pairs. At city pair 41 the two lines cross, i.e., before city pair 41 the join cardinalities of the skewed case are much higher compared to the uniform case. After that skewed cardinalities are much lower compared to the uniform cardinalities.

In PDGF the above distribution can be implemented very easily by adding a distribution node to the data generator specification. For the city fields the exponential distribution is implemented as in the Lo_Quantity field as shown above in Listing 10. In this experiment, we are mainly interested in the effect of multiple skewed dimensions and, therefore, we only skew the most selective attribute, which is S_City and C_City, respectively. As discussed in Section 3.1, in our PDGF implementation nation and region are references to a virtual table, as is the nation prefix of the City attribute. The definition of the S_City attribute can be can be seen in Listing 12. A value of S_City is made from a prefix that is a padded or truncated version of a Nation value (strings longer than 9 characters are truncated, shorter strings are padded) and a number between 0 and 9. In order to generate a skewed output, the reference to Nation_Region can be skewed, as can the generation of the number that is concatenated with the prefix. In our PDGF implementation for Supplier we only insert a skew in the reference to Nation_Region, while for Customer, we additionally add a skew in the number generation.

Figures 14 and 15 show the elapsed times of Q3.3 on uniform and skewed distributions of S_City and C_City. Query Q3.3 is more complicated compared to the queries used in previous experiments as it joins four tables: Lineorder, Sup-

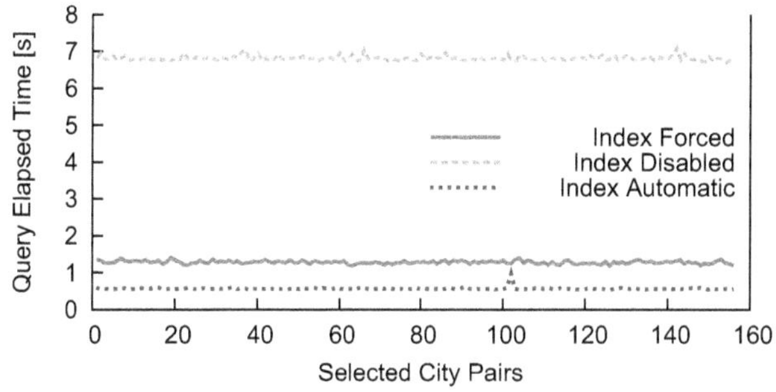

Figure 14: Elapsed Times of Query Q3.3 with Uniform C_City and S_City

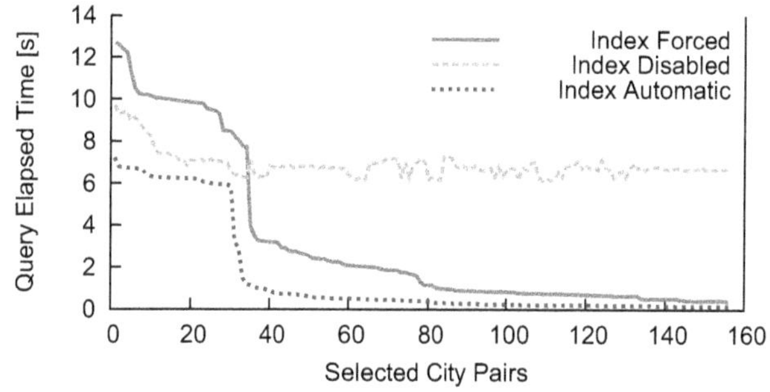

Figure 15: Elapsed Times of Query Q3.3 with Skewed C_City and S_City

```
<field name="S_CITY" size="10" type="VARCHAR">
  <gen_SequentialGenerator concatenateResults="true">
    <gen_PaddingGenerator>
      <gen_DefaultReferenceGenerator id="S_CITY_id">
        <reference table="NATION-REGION"
            field="NATION" />
      </gen_DefaultReferenceGenerator>
      <character> </character>
      <padToLeft>false</padToLeft>
      <size>9</size>
    </gen_PaddingGenerator>
    <gen_LongGenerator>
      <min>0</min>
      <max>9</max>
    </gen_LongGenerator>
  </gen_SequentialGenerator>
</field>
```

Listing 12: Definition of S_City in the Supplier table

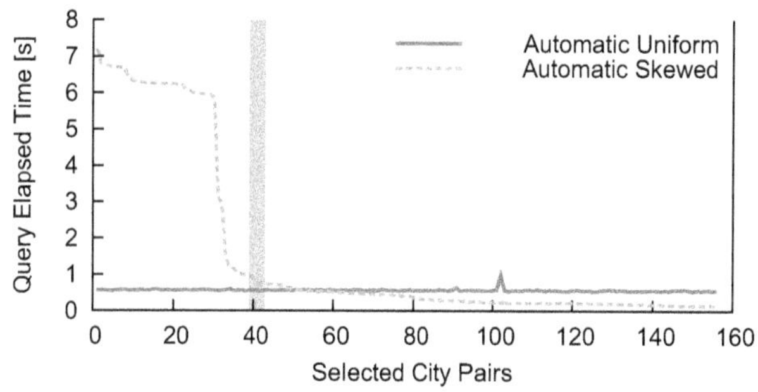

Figure 16: Elapsed Times of Query Q3.3 with Uniform and Skewed C_City and S_City using Automatic Plans

plier, Customer and Date. Due to the increased complexity, the choices for indexes and query plans for Query Q3.3. are much larger compared to those in previous experiments. Hence, the index driven join forces the system to use a query plan that does not necessarily coincides with the query plan chosen by the optimizer in automatic mode.

Figure 14 shows the elapsed times for all three modes on a uniform data set. Each line shows a constant query elapsed time regardless of the city pair chosen. This is not surprising as the cardinalities of at which each city pair occurs in the database is constant (see Figure 12. The non-index driven queries have the longest elapsed times while the automatic queries have the shortest elapsed times. The index forced queries execute slightly longer than the index driven queries.

Figure 15 shows the elapsed times for all three modes on the skewed data set. The graph reveals a couple of very interesting characteristics of the system. The elapsed times of the non-index driven queries, dashed line, is more or less constant at about 7s. Queries using the first 20 city pairs, which result in very high join cardinalities (50k to 300k) show slightly higher elapsed times (up to 9s). This is due to the larger intermediate result set at these high join cardinalities. Contrary, the lines for index forced and index automatic plans show a steep decrease in elapsed time starting at city pair 30 for the automatic plan and 34 for the index forced plan. The steep decrease is due to the steep decrease in the join cardinality. That is the index driven queries are able to take advantage of the index to filter out most of the rows to beat the non-index driven queries. Again, this is due to a sub-optimal query plan that was forced in the index-forced mode. Also, like in the uniform case, the automatic query outperforms both the index driven and non-index driven queries for all city pairs.

Figure 16 compares the automatic case for the uniform and skewed distributions. The queries run on uniform data execute constantly under 1s, while the queries on skewed data follow the cardinality distribution as described above. With a decrease in join cardinality the queries on skewed data outperform those on uniform data, which is not surprising. However, the cross over point seems to be at city pair 52. We would have expected the cross over to be closer to city pair 41, where the join cardinalities of the skewed and uniform data sets are identical. This shows that the system has a consistently increased latency for skewed data. This trend also holds for the index-forced and no-index plans.

5. RELATED WORK

While the importance of data skew in database processing has been identified and, to some degree, quantified in various publications, especially in the field of parallel systems [8, 16], there is no industry standard benchmark that measures the effect of data skew on query processing. In many cases, performance of parallel DBMS is adversely affected by data skew, because it introduces load imbalances, stresses the creation of internal data structures, and it reveals poor design decisions in algorithms and query optimizers. Having realized the above, Crolotte and Ghazal proposed the introduction of data skew into TPC-H similarly to TPC-DS, i.e., using comparability zones [5]. As a consequence, queries are still run with substitution parameters that are chosen from within a comparability zone with uniform distribution.

[5] discusses two approached how skew can be introduced into the nation keys of the Customer and Supplier tables. The first approach scales the numbers of customers and suppliers who reside in a nation proportional to the actual population of this nation (e.g., obtained from the US census). Although being realistic this approach is not feasible in the context of TPC-H as the queries would impose different workloads for different customers. Their second approach, which is feasible for TPC-H is creating skew with a step function. The first 13 nations have a small population, while the remaining 12 nations have a large population. They implement the above described skew using Teradata SQL syntax with a random function, i.e., they read the original TPC-H data and apply a random function for nation keys.

Closer to our approach for introducing data skew in standard benchmarks is Chaudhuri and Narayasa modified version of TPC-D [3]. TPC-D was the precursor of TPC-H. The two benchmarks share the same schema and data model, but vary in their workload slightly. Hence, the proposed changes also apply to TPC-H. Chaudhuri and Narayasa modified TPC-D/H's data generator dbgen to support Zipfian distributed data in all columns. The parameter to the distribution (ρ), which controls the degree of skew in the data, is given as a parameter to the modified version of dbgen. ρ can be set to any value larger or equal to 0; $\rho = 0$ generates a uniform data whereas $\rho = 4$ generates a highly-skewed distribution. In addition they also implemented functionality that allows dbgen to randomly chose ρ values for each column, thereby creating a "mixed" data distribution.

Their approach differs from Crolotte and Ghazal's approach as it does not guarantee workload predictability, which is a major factor in industry standard benchmarks and in particular in TPC-H. However, Chaudhuri and Narayasa approach can be easier implemented as the data generator is freely available, but it is limited to the Zipf distribution and cannot be customized on a column per column basis.

Data skew has been addressed to some extend in TPC-DS. The TPC after realizing that the uniform data distributions in TPC-D, TPC-H and TPC-R were not challenging today's DBMS vouched to implement data skew in its next decision support benchmark TPC-DS. The TPC opted for a solution that introduced zones of comparability – essentially flat spots in the data distribution – that can be used to provide both the variability and the comparability that the eventual user of the generated data requires. These comparability zones differ in size within the column domain and in number between column domains.

6. CONCLUSION

In this paper, we presented an extension of the Star Schema Benchmark that introduces skew in single table columns and single table hierarchies. Due to the limited extensibility of the original data generator, we implemented a new data generator and query generator based on the Parallel Data Generation Framework. Our extensive experimental analysis shows that the introduction of skew in the data sets can uncover unpredicted behavior in query processing.

PDGF is continuously extended and improved. A recent extension was presented in [6]. It allows for the efficient generation of consistent update data. This enables PDGF to generate more realistic updates on the data compared to TPC-H and SSB. However, we did not explore this option in the presented work. Information on the current status of PDGF as well as downloadable versions can be found on `www.paralleldatageneration.org`.

For future work, we will explore the possibilities of using references to the generated data set in query generation. This will make it possible to introduce skew in the query workload and thus make it possible to examine the influence of query caching and other optimizations in a meaningful way. In combination with the deterministic data generation the reference-based query generation will enable precomputation of query results for certain kinds of queries and thus provide means for verification beyond the capabilities of current benchmarks. We will combine these and the above presented parts into a complete benchmark suite that will make it possible to test the different influences of skew in separated and combined workloads.

7. REFERENCES

[1] TPC Benchmark H. `http://www.tpc.org/tpch/spec/tpch2.15.0.pdf`, 2012.

[2] J. S. Charles Levine. Standard Benchmarks for Database Systems. `http://www.tpc.org/information/sessions/sigmod/sigmod97.ppt`, 1997.

[3] S. Chaudhuri and V. Narasayya. Program for Generating Skewed Data Distributions for TPC-D. `ftp://ftp.research.microsoft.com/users/viveknar/tpcdskew`, 1997.

[4] E. F. Codd. A Relational Model of Data for Large Shared Data Banks. *Communications of the ACM*, 13(1):377–387, 1970.

[5] A. Crolotte and A. Ghazal. Introducing Skew into the TPC-H Benchmark. In *TPCTC '11*, pages 137–145, 2011.

[6] M. Frank, M. Poess, and T. Rabl. Efficient Update Data Generation for DBMS Benchmark. In *ICPE '12*, 2012.

[7] R. Kimball and M. Ross. *The Data Warehouse Toolkit: The Complete Guide to Dimensional Modeling*. John Wiley and Sons, Inc., 2002.

[8] M. S. Lakshmi and P. S. Yu. Effect of Skew on Join Performance in Parallel Architectures. In *DPDS*, pages 107–120, 1988.

[9] G. Marsaglia. Xorshift RNGs. *Journal Of Statistical Software*, 8(14):1–6, 2003.

[10] R. O. Nambiar and M. Poess. The Making of TPC-DS. In *VLDB '06*, pages 1049–1058, 2006.

[11] P. O'Neil, E. O'Neil, X. Chen, and S. Revilak. The Star Schema Benchmark and Augmented Fact Table Indexing. In *TPCTC '09*, pages 237–252, 2009.

[12] M. Poess and C. Floyd. New TPC Benchmarks for Decision Support and Web Commerce. *SIGMOD Record*, 29(2000):64–71, 2000.

[13] M. Poess, T. Rabl, M. Frank, and M. Danisch. A PDGF Implementation for TPC-H. In *TPCTC*, pages 196–212, 2011.

[14] M. Poess and J. M. Stephens. Generating Thousand Benchmark Queries in Seconds. In *VLDB*, pages 1045–1053, 2004.

[15] T. Rabl, M. Frank, H. M. Sergieh, and H. Kosch. A Data Generator for Cloud-Scale Benchmarking. In *TPCTC '10*, pages 41–56, 2010.

[16] C. B. Walton, A. G. Dale, and R. M. Jenevein. A Taxonomy and Performance Model of Data Skew Effects in Parallel Joins. In G. M. Lohman, A. Sernadas, and R. Camps, editors, *VLDB '91*, pages 537–548. Morgan Kaufmann, 1991.

Monitoring-as-a-Service in The Cloud

SPEC PhD Award (Invited Abstract)

Shicong Meng
Georgia Institute of Technology
smeng@cc.gatech.edu

Thesis Supervisor: Ling Liu
Georgia Institute of Technology
lingliu@cc.gatech.edu

ABSTRACT

State monitoring is a fundamental building block for Cloud services. The demand for providing state monitoring as services (MaaS) continues to grow and is evidenced by CloudWatch from Amazon EC2, which allows Cloud consumers to pay for monitoring a selection of performance metrics with coarse-grained periodical sampling of runtime states. This dissertation research [1], awarded with the SPEC Distinguished Dissertation Award 2012, is dedicated to innovative research and development of an elastic framework for providing state monitoring as a service. We analyze limitations of existing techniques, systematically identify the need and the challenges at different layers of a Cloud monitoring service platform, and develop a suite of distributed monitoring techniques to support flexible monitoring infrastructure, cost-effective state monitoring and monitoring-enhanced Cloud management.

Categories and Subject Descriptors

D.4.7 [**Distributed Systems**]: Scalable Monitoring; C.2.4 [**Distributed Systems**]: Distributed applications

General Terms

Algorithms, Design, Performance, Reliability

Keywords

Cloud Monitoring, State Monitoring, Multi-Tenancy

1. INTRODUCTION

Today's Cloud datacenters are complex composition of large-scale servers, virtual machines, physical and virtual networks, middleware, applications, and services. Their growing scale and complexity challenge our ability to closely monitor the state of various entities, and to utilize voluminous monitoring data for better operation. Providing Monitoring-as-a-Service (MaaS)[5] brings a number of benefits to both Cloud providers and consumers.

First, MaaS minimizes the cost of ownership by leveraging state of the art monitoring tools and functionalities. MaaS makes it easier for users to deploy state monitoring at different levels of Cloud services compared with developing ad-hoc monitoring tools or setting up dedicated monitoring hardware/software. Second, MaaS enables the pay-as-you-go utility model for state monitoring. This is especially important for users to enjoy full-featured monitoring services based on their monitoring needs and available budget. MaaS also brings Cloud service providers the opportunity to consolidate monitoring demands at different levels (infrastructure, platform, and application) to achieve efficient and scalable monitoring. Finally, MaaS pushes Cloud service providers to invest in state of the art monitoring technology and deliver continuous improvements on both monitoring service quality and performance. With the consolidated services and monitoring data, Cloud service providers can also develop value-add services for better Cloud environments and creating new revenue sources.

ICPE'13, April 21–24, 2013, Prague, Czech Republic.
ACM 978-1-4503-1636-1/13/04.

2. CONTRIBUTIONS

This dissertation research tackles critical challenges in providing MaaS with a layered approach that systematically addresses monitoring efficiency, scalability, reliability and utility at the monitoring infrastructure level, the monitoring functionality level and the monitoring data utility level. We analyze key limitations of existing techniques, and develop new techniques to offer more effective Cloud monitoring capabilities in this layered design. In addition, we built systems that help Cloud developers and users to access, process and utilize Cloud monitoring data. Specifically, this dissertation makes the following contributions.

2.1 Monitoring Infrastructure

At the monitoring infrastructure level, we propose REMO [3, 4] and Tide [6] which contribute to a Cloud-scale monitoring infrastructure that ensures the efficiency, scalability and multi-tenancy support of Cloud monitoring.

Monitoring Topology Planning [3, 4]. Large-scale monitoring can incur significant overhead on distributed nodes participating in collection and processing of monitoring data. Existing techniques that focus on monitoring task level efficiency often introduce heavily skewed workload distributions on monitoring nodes and cause excessive resource usage on certain nodes. We developed REMO, a resource-aware monitoring system that considers node-level resource constraints, e.g., monitoring-related CPU utilization should less than 5%, as the first-class factor for scheduling multiple monitoring tasks collectively. REMO optimizes the throughput of the entire monitoring network without causing excessive resource consumption on any participating node, which ensures performance isolation in multi-tenent monitoring environments. It also explores cost sharing opportunities among tasks to optimize monitoring efficiency. We deployed REMO on Sysem S, a large-scale distributed stream processing system built at IBM TJ Watson Research Center. Through resource-aware planning, REMO achieves 35%-45% error reduction compared to existing techniques.

Self-Scaling Monitoring Infrastructure [6]. From traces collected in production datacenters, we found that monitoring and management workloads in Cloud datacenters tend to be highly volatile due to their on-demand usage model. Such workloads often make the management server a performance bottleneck. To address this problem, we developed Tide, a self-scaling management system which automatically scales up or down its capacity according to the observed workloads. We built the prototype of Tide by modifying VMware's vSphere management server and leveraging non-SQL Hadoop based HBase for scalable state persistence. The experimental results show that Tide provides consistent performance even with extreme volatile management workloads through self-scaling.

2.2 Monitoring Functionalities For Cloud

At the monitoring functionality level, we aim at providing new monitoring techniques to meet the unique and diverse Cloud monitoring needs, and we propose WISE [8, 7], Volley and CrystalBall [2] which deliver accurate, cost-effective and reliable monitoring results by employing novel distributed monitoring algorithms for error-prone Cloud environments.

Continuous State Violation Detection [7][8]. Most existing works on distributed state monitoring employ an instantaneous monitoring model, where the state is evaluated based on the most recent collected results, to simplify algorithm design. Such a model, however, tends to introduce false state alerts due to noises and outliers in monitoring data. To address this issue, we proposed WISE, window based state monitoring which utilizes temporal windows to capture continuous state violation in a distributed setting. WISE not only delivers the same results as those of a centralized monitoring system with a distributed implementation, but also decouples a global monitoring task into distributed local ones in a way that minimizes the overall communication cost.

Violation-Likelihood based Monitoring. Asynchronized monitoring techniques such as periodical sampling often introduce cost-accuracy dilemma, e.g., frequent polling state information may produce fine-grained monitoring data but may also introduce high sampling cost for tasks such as deep packet inspection based network monitoring. To address this issue, we proposed Volley, a violation likelihood based approach which dynamically tunes monitoring intensity based on the likelihood of detecting important results. More importantly, it always safeguards a user-specified accuracy goal while minimizing monitoring cost. Volley also coordinates sampling over distributed nodes to maintain the task-level accuracy, and leverages inter-task state correlation to optimize multi-task sampling scheduling. When deployed in a testbed datacenter environment with 800 virtual machines, Volley reduces monitoring overhead up to 90% with negligible accuracy loss.

Fault-Tolerant State Monitoring [2]. While we often assume monitoring results are trustworthy and monitoring services are reliable, such assumptions do not always hold, especially in large scale distributed environments such as datacenters where transient device/network failures are the norm rather than the exception. As a result, distributed state monitoring approaches that depend on reliable communication may produce inaccurate results with the presence of failures. We developed CrystalBall, a robust distributed state monitoring approach that produces reliable monitoring results by continuously updating the accuracy estimation of the current results based on observed failures. It also adapts to long-term failures by coordinating distributed monitoring tasks to minimize accuracy loss. Experimental results show that CrystalBall consistently improves monitoring accuracy even under severe message loss and delay.

2.3 Monitoring Enhanced Cloud Management

At the monitoring data utility level, we study intelligent techniques that utilize monitoring data to offer advanced monitoring management capabilities. As an initial attempt, we propose Prism which offers an innovative application provisioning functionality based on knowledge learned from cumulative monitoring data. We aim at utilizing multi-tier Cloud application performance data to guide application provisioning. Prism is a prediction-based provisioning framework that simplifies application provisioning by using performance prediction to find a proper provisioning plan for a performance goal in a huge space of candidate plans. As its unique feature, Prism isolates and captures the performance impact of different provisioning options, e.g., virtual machine types and cluster configurations, from performance monitoring data with off-the-shelf machine learning techniques. This technique avoids exploring the huge space of candidate provisioning plans with experiments. As a result, Prism can quickly find the most cost-effective plan with little cost for training performance prediction models.

Acknowledgement

This dissertation research was partially sponsored by grants from NSF NetSE and SaTC program, Intel ISTC on Cloud computing and an IBM PhD Fellowship.

3. REFERENCES

[1] S. Meng. Monitoring-as-a-service in the cloud. *PhD thesis, Georgia Institute of Technology*, 2012.

[2] S. Meng, A. Iyengar, I. Rouvellou, L. Liu, K. Lee, B. Palanisamy, and Y. Tang. Reliable state monitoring in cloud datacenters. In *IEEE CLOUD*, pages 951–958, 2012.

[3] S. Meng, S. R. Kashyap, C. Venkatramani, and L. Liu. Remo: Resource-aware application state monitoring for large-scale distributed systems. In *ICDCS*, pages 248–255, 2009.

[4] S. Meng, S. R. Kashyap, C. Venkatramani, and L. Liu. Resource-aware application state monitoring. *IEEE Trans. Parallel Distrib. Syst. (TPDS)*, 23(12):2315–2329, 2012.

[5] S. Meng and L. Liu. Enhanced monitoring-as-a-service for effective cloud management. *IEEE Transactions on Computers (TC)*, to appear.

[6] S. Meng, L. Liu, and V. Soundararajan. Tide: Achieving self-scaling in virtualized datacenter management middleware. In *Middleware'10.*

[7] S. Meng, L. Liu, and T. Wang. State monitoring in cloud datacenters. *IEEE Trans. Knowl. Data Eng. (TKDE)*, 23(9):1328–1344, 2011.

[8] S. Meng, T. Wang, and L. Liu. Monitoring continuous state violation in datacenters: Exploring the time dimension. In *ICDE*, pages 968–979, 2010.

Deriving Coloured Generalised Stochastic Petri Net Performance Models from High-Precision Location Tracking Data

Nikolas Anastasiou
Department of Computing
Imperial College London
South Kensington Campus
London SW7 2AZ
na405@doc.ic.ac.uk

William Knottenbelt
Department of Computing
Imperial College London
South Kensington Campus
London SW7 2AZ
wjk@doc.ic.ac.uk

ABSTRACT

Stochastic performance models are widely used to analyse systems that involve the flow and processing of customers and resources. However, model formulation and parameterisation are traditionally manual and thus expensive, intrusive and error-prone. Our earlier work has demonstrated the feasibility of automated performance model construction from location tracking data. In particular, we presented a methodology based on a four-stage data processing pipeline, which automatically constructs *Generalised Stochastic Petri Net* (GSPN) performance models from an input dataset of raw location tracking traces. This pipeline was enhanced with a presence-based synchronisation detection mechanism.

In this paper we introduce *Coloured Generalised Stochastic Petri Nets* (CGSPNs) into our methodology to provide support for multiple customer classes and service cycles. Distinct token types are used to model customers of different classes, while Johnson's algorithm for enumerating elementary cycles in a directed graph is employed to detect service cycles. Coloured tokens are also used to enforce accurate customer routing after the completion of a service cycle. We evaluate these extensions and their integration into the methodology via a case study of a simplified model of an Accident and Emergency (A&E) department. The case study is based on synthetic location tracking data, generated using an extended version of the LocTrackJINQS location-aware queueing network simulator.

Categories and Subject Descriptors

C.4 [**Performance of Systems**]: Modelling Techniques; I.6.5. [**Model Development**]: Modelling Methodologies

Keywords

Location Tracking, Performance Modelling, Data Mining, Coloured Generalised Stochastic Petri Nets

1. INTRODUCTION

Complex customer-processing systems are frequently encountered in everyday life. Such systems, e.g. airports, hospitals and factory assembly lines, are subject to stringent Quality of Service (QoS) requirements. Failure to comply with these requirements can have severe financial implications and may lead to high profile fiascos, such as the opening of Heathrow Terminal 5[1] and the unacceptably long waiting times of patients in Accident and Emergency departments in the UK[2]. Accurate performance models of these systems can be used as a predictive tool to ensure the provision of adequate QoS under foreseeable workloads. Unfortunately, the availability and quality of such models are currently limited due to the associated construction cost, both in terms of time and money, and the high degree of manual intervention required, in terms of data collection, model construction and model parameterisation.

The gradual realisation of the Internet of Things and the increasing adoption of Real Time Location Systems (RTLSs) have led to the availability of low-level data which describes the movements of customers and/or resources in a given system. In this context, our earlier work demonstrated the feasibility of automated performance model construction from location tracking data [2]. In particular, we presented a methodology, based on a four-stage data processing pipeline, which automatically constructs *Generalised Stochastic Petri Net* (GSPN) [15, 4] performance models from raw location tracking traces. The developed methodology proved its ability to capture the stochastic features and abstract structure of customer-processing systems which satisfy the following assumptions: static service centres, random service discipline, single customer class, single-server service semantics and probabilistic customer routing. The resulting models can be visualised using PIPE2, the Platform Independent Petri net Editor [5, 7].

In subsequent research, we developed a presence-based synchronisation detection mechanism [3]. This determines whether the processing of customers at each service area depends on the presence of a certain number of customers at other service areas. If synchronisation is detected, then the extracted service time samples are adjusted accordingly.

[1]`http://www.computerweekly.com/news/2240086013/British-Airways-reveals-what-went-wrong-with-Terminal-5`

[2]`http://www.guardian.co.uk/society/2012/may/31/patients-waiting-four-hours-2004`

We now introduce *Coloured Generalised Stochastic Petri Nets* (CGSPNs) [14] into our methodology in order to address some of its limitations. In particular we provide support for multiple customer classes and allow customers to be routed according to *service cycles*. Different token types (represented by distinct colours) are used to model customers of different classes. Johnson's algorithm for enumerating elementary cycles in a directed graph [11] is used to identify service cycles. Coloured tokens are also used to enforce accurate customer routing after the completion of a service cycle. Together these extensions allow for more realistic modelling of complex customer-processing systems.

The rest of this paper is organised as follows. Section 2 presents previous research related to our work. Section 3 provides an overview of our existing and modified methodologies. Section 4 presents a case study to examine the accuracy and effectiveness of the modified approach. This uses synthetic location tracking data generated using an extended version of LocTrackJINQS [9]. The paper concludes with a summary of the results and a discussion on future work.

2. RELATED WORK

Previous research conducted by Kounev et *al.* examined the automatic construction of performance models in a different context [12]. Their work presents an approach to automatically extract the performance model of an *Enterprise Data Fabric* (EDF) in the form of a *Queueing Petri Net* (QPN) [4]. An EDF is a distributed enterprise middleware, located between the application and the host network, and it is used to allocate and manage data and resources across multiple, physically separate, hardware nodes. The authors parameterise the model using readily available monitoring data which are provided by the EDF. This approach is implemented as part of the simulation-based tool Jewel whose purpose is the automated performance prediction and capacity planning for EDFs.

Another research endeavour in the field of automatic construction of GSPN models has been made by Xue et *al.* [17]. Their methodology is applicable to *Flexible Manufacturing Systems* (FMSs) and it is implemented by the FMSPet program. An FMS consists of two components: the actual manufacturing system, i.e. an assembly line, and a controller which allows the system to react to various changes. FMSPet requires as input the description of the FMS, specified by an input language called FMSDL (presented in [17]), and produces a GSPN model for the underlying system. However, the methodology does not extract the model from data; the FMS must be explicitly defined in the input file.

Earlier work conducted by Horng et *al.* [10] is the most closely related to ours. The authors present a methodology designed to infer simple Queueing Network performance models from high-precision location tracking data. The constructed models capture the structure of certain restricted classes of underlying system – specified in terms of the routing probabilities of the customer flow – and the service time distributions of its service areas. The inter-arrival time distribution of customers at service centres is also inferred. Unlike in our present approach, however, there is no ability to infer automatically the locations and radii of service areas, the existence of presence-based synchronisation between service areas or the presence of service cycles.

3. INFERRING PNPMS FROM HIGH-PRECISION LOCATION TRACKING DATA

3.1 Existing Data Processing Pipeline

Here, we provide an overview of our prior work [2, 3] since some basic understanding of the existing methodology is required in order to introduce new features. A high-level description of the original data processing pipeline is shown in Figure 1. The pipeline takes as input a set of raw location tracking data, which describes the customer flow in a customer-processing system, and outputs a Petri Net performance model (PNPM) of the underlying system.

The first stage of the pipeline prepares the raw location traces[3] for processing by the subsequent stages. In particular, it converts the input data into a standardised format and separates it into customer paths, one for each recorded customer. A customer path consists of the location updates associated with a particular customer's movements throughout its stay in the system. A speed-based filter is then applied on each customer path in order to remove location updates that imply infeasibly rapid movement.

The second stage consists of three phases which infer the locations and radii of stationary service areas in the system. It operates under the assumption that customers stop or slow down while receiving service. The three phases are: speed filtering, density filtering and the application of the DBSCAN clustering algorithm [8]. The speed and density filters are applied on each customer path and aim to identify the multiset of positions (two-dimensional Cartesian coordinates) where the customer was stationary or moving at 'low' speed. These multisets – one for each customer path – are aggregated into a single dataset on which the DBSCAN algorithm is applied. The centroids of the produced clusters are used to approximate the locations of the system's service areas. The radius of a service area is conservatively approximated as 110% of the 95th percentile of the distance between the corresponding cluster's centroid and each of its contained points.

Stage three constructs the basic structure of the derived PNPM, beginning with the places and transitions required to represent the flow of customers in the system. We distinguish between two types of places: places associated with service areas (inferred from stage two) and places associated with customer movement between service areas. We call these server and travel places respectively. Similarly, we distinguish between two types of transitions: those which connect server to travel places and those connecting travel to server places. The latter are called travelling time transitions and the former service area service time transitions. These transitions are not currently parameterised; they are replaced during the next stage by a GSPN subnet that accurately reflects the distribution of the relevant time delays. In preparation for this, we compute response time samples – for each customer – inside a service area broken into waiting time and service time. Samples of the time required by each customer to transit between two service areas are also computed (travelling time samples). Then, the presence-based synchronisation detection mechanism [3] is applied. Whenever synchronisation involving the processing of customers

[3] A typical location update reading from an RTLS is a tuple of the form (`tagName, type, time, x, y, z, stderr`).

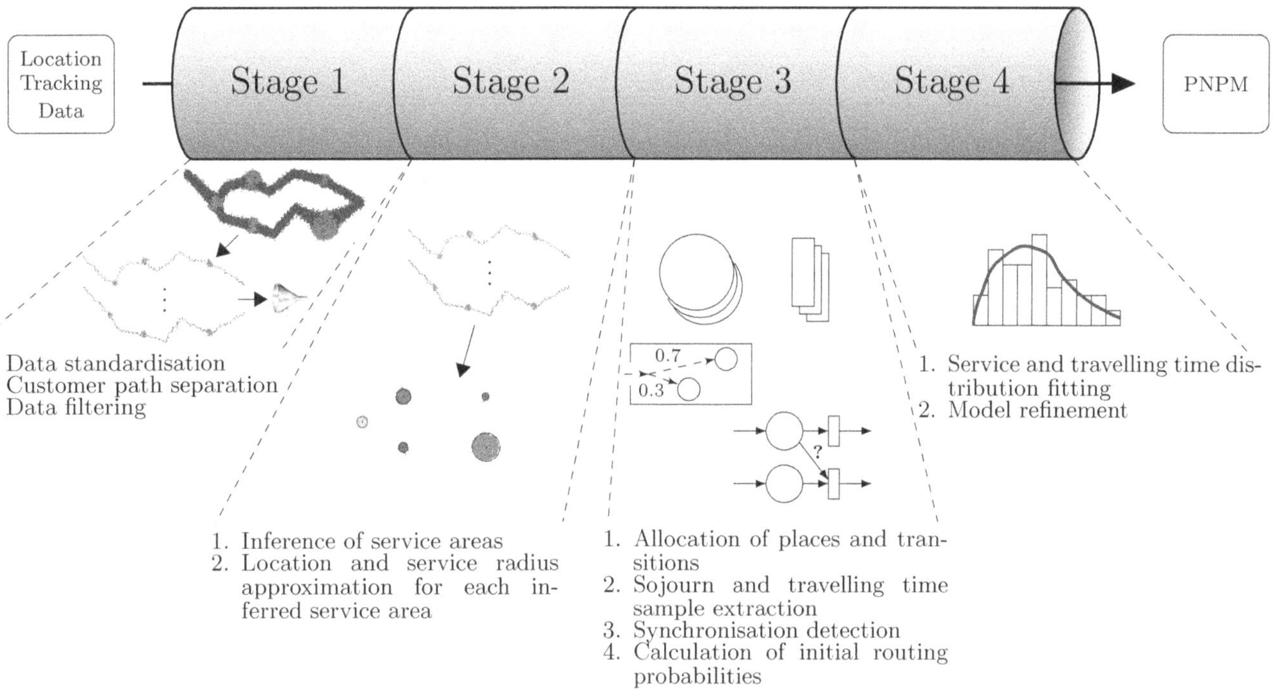

Figure 1: The four-stage data processing pipeline as in [3].

at a particular service area is detected, the corresponding service time samples of those customers are adjusted to take into account the proportion of time during which the synchronisation condition(s) are satisfied[4]. Finally, a counting mechanism calculates the initial and inter-routing probabilities of the customer flow.

The fourth stage of the pipeline replaces the service area service time transitions and travelling time transitions with GSPN subnets that reflect the distributions of the corresponding service and travelling time samples computed in stage three. For each set of these time samples we fit several hyper-Erlang distributions (HErDs) using the G-FIT tool [16]. Each GSPN subnet is constructed in such a way that it reflects the best-fit HErD which is selected using the Akaike Information Criterion (AIC) [1].

3.2 Modelling Multiple Customer Classes

The first extension introduces CGSPNs into the data processing pipeline in order to support multiple customer classes. The main assumption here is that the class of each customer in the system is known and provided through the customer's associated location updates, e.g. the `type` field of a typical location update contains the category the monitoring tag belongs to, or by a static mapping of each customer's unique identifier (`tagName`) to the class it belongs to. Here we focus on the third stage of the pipeline, especially on the way we form the initial structure of the PNPM.

In a CGSPN, each transition can support many firing modes, i.e. a transition can be enabled under several different markings. For example, consider the CGSPN depicted in Figure 2 where p_1 is marked with three token types: two blue, two red and one black. The transition t_1 may require two blue tokens to be enabled, or one red, or one black, or even a combination of distinct types, e.g. one blue, one red and one black. The set of markings under which t_1 is

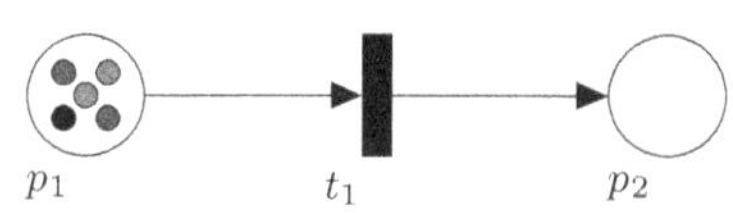

Figure 2: A simple CGSPN with three token classes.

enabled must be specified so that the dynamic behaviour of the model accurately reflects the underlying system's behaviour. We note that a different firing rate or weight can be associated with each supported firing mode of a timed or immediate transition respectively.

To facilitate the unambiguous visualisation of a transition's firing modes we use a different transition for each firing mode that we wish to support. That is, if we consider the latter example, assuming that t_1 has three firing modes (see Table 1) we use three transitions; this is shown in Figure 4.

Section 3.1 presented an overview of the existing pipeline, which included the various tasks performed during the third stage. When dealing with multiple customer classes the process of place construction remains the same. Service area service time transitions are created in a similar manner as before but instead of having one such transition we now have multiple transitions: one for each class of customers that was processed by the corresponding service area. The same principle applies for the construction of travelling time transitions. The total number of customer classes is obtained during the computation of the response and travelling time samples by simply counting the distinct classes observed. The operation of this process requires no modifications; how-

[4]We assume that service progresses only when the synchronisation condition(s) are met.

Mode	$I^-(p_1,t_1)$(red)	$I^-(p_1,t_1)$(blue)	$I^-(p_1,t_1)$(black)	$I^+(p_2,t_1)$(red)	$I^+(p_2,t_1)$(blue)	$I^+(p_2,t_1)$(black)
1	1	0	0	1	0	0
2	0	1	0	0	1	0
3	1	1	1	1	1	1

Table 1: The supported (three) firing modes of transition t_1 shown in Figure 2. $I^-(p_i,t_j)(c_k)$ denotes the number of tokens of colour c_k that must be present in p_i in order for t_j to be enabled. $I^+(p_i,t_j)(c_k)$ denotes the number of tokens of colour c_k that are created in p_i when t_j fires.

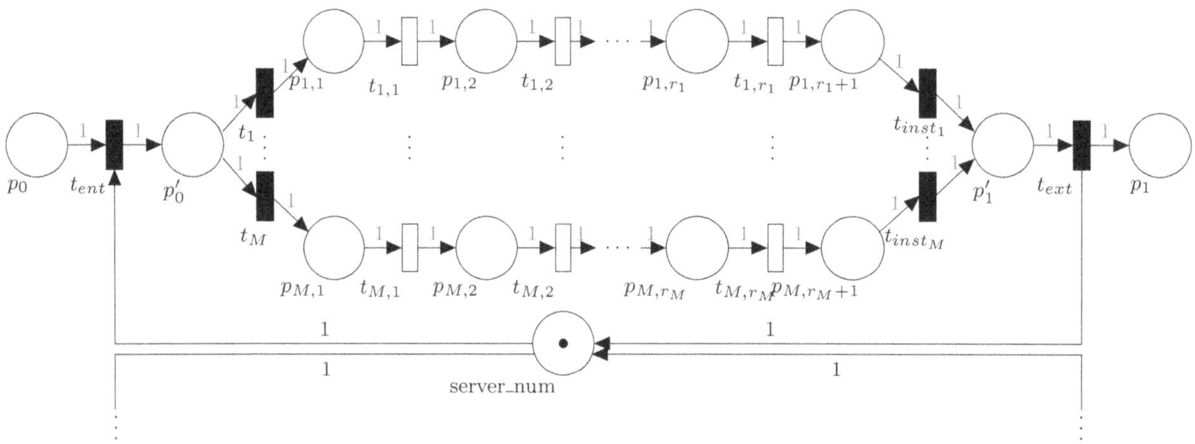

Figure 3: CGSPN subnet representation for service area service time transitions. The colour of tokens supported varies according to the customer class to which the net applies. The complementary place labelled as server_num is shared among subnets associated with the same service area. This allows only one token to be in a subnet at a time and thus prohibits parallel service of distinct customer classes (we assume single-server semantics). p_0 represents a service area and p_1 is associated with the movement of customers between p_0 and some other service area.

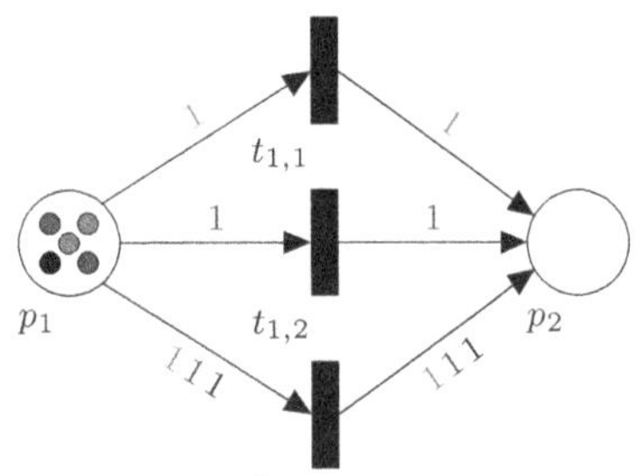

Figure 4: The CGSPN shown in Figure 2 with one transition for each different firing mode of t_1 (cf. Table 1).

ever, the computed service and travelling time samples are now separated into groups, one for each customer class.

The GSPN subnet reflecting the best-fit HErD of a particular group of time samples now becomes a CGSPN. The firing mode of each transition in the subnet is set according to the class of customers the group of time samples corresponds to. The general form of a CGSPN subnet used to replace service area service time transitions is shown in Figure 3. All such subnets that are associated with a particular service area share the same complementary place server_num. This place contains one token and ensures that no more than one token is allowed in a subnet simultaneously (thus prohibiting parallel service of customers of different classes). Subnets used to replace travelling time transitions are similarly structured but without the complementary place server_num.

3.3 Service Cycles and Customer Routing

The approach presented here assumes one class of customers. The extension to multiple customer classes is discussed in the next section.

Often, in customer-processing systems, routing is not probabilistic but deterministic with several phases of processing. The journey through these processing phases may involve repeated visits to the same service centres. In order to be able to model this kind of customer behaviour we introduce the concept of a *service cycle*. An example of such a system is shown in Figure 5 where the service cycle consists of Service Area A, B and C. Formally,

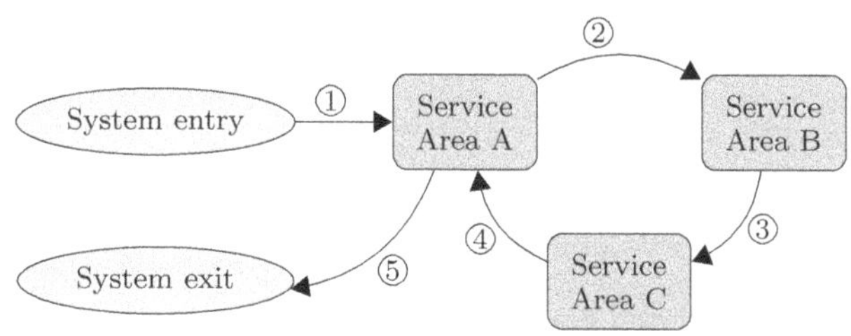

Figure 5: A customer-processing system which provides services to customers in a cyclic fashion. Arrows indicate the flow of customers in the system.

Definition 1. Consider a physical customer-processing system consisting of the set of service areas $S = \{S_i \mid i = 1,\ldots,N\}$, for some finite $N \in \mathbb{N}$. A finite sequence of *distinct* service areas from which the customers of the system obtained some service, is said to be a service cycle given that the following conditions hold:

1. If the customers initiate their service sequence at S_i, $S_i \in S$, they must also terminate it at S_i after service completion at S_k, $S_k \in S$ and $i \neq k$.
2. When the customers complete their service at S_i for the second time, i.e. after obtaining service at S_k, they do not repeat the same service sequence.

The service area which marks the initiation and termination of the service cycle is called the *head* of the service cycle.

If the existing methodology is applied to a set of location tracking data retrieved from the system depicted in Figure 5, under the assumption of one customer class, then the inferred GSPN model (non-parameterised version) is shown in Figure 6. Although the structure of the model resembles

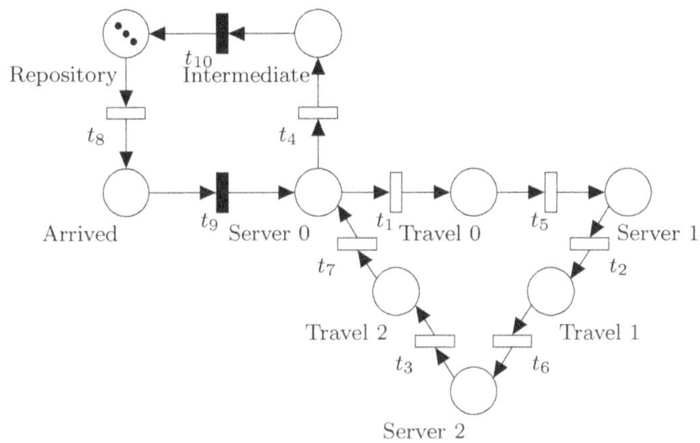

Figure 6: The GSPN model (non-parameterised version) for the system depicted in Figure 5 using our existing methodology. Places labelled as Server 0, Server 1 and Server 2 represent the service areas A, B and C respectively.

the underlying system, the model fails to capture the desired behaviour and routing of customers in the system. In particular, the second condition of Definition 1 can be possibly violated because of the race condition between transitions t_1 and t_4; when the place Server 0 is marked with at least one token t_1 and t_4 are simultaneously enabled. The transition to fire is selected probabilistically and the probability of each transition depends on their firing rates.

The key idea here is to identify the transitions involved in such race conditions and adjust their firing mode, i.e. change the supported token type, so that the race condition ceases to exist. The different token types are used to express the customers' "phases of service". That is, we distinguish between customers who have obtained service from each service area contained within the service cycle and customers who are about to enter it, i.e. obtain service from the head of the service cycle. We classify the former and latter groups of customers as serviced and unserviced respectively. The CGSPN model (non-parameterised version) that is obtained for the previous example when this modelling approach is applied, is shown in Figure 7.

To apply this approach, the system being modelled must first be examined for the presence of service cycles. To perform this task we obtain a directed graph G, $G = (V, E)$,

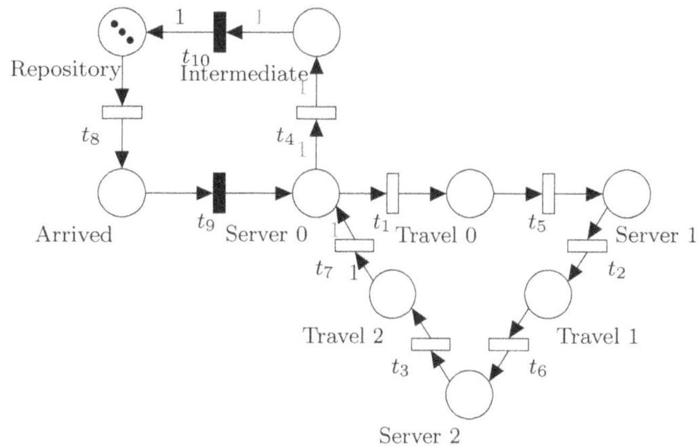

Figure 7: The CGSPN model (non-parameterised version) for the system depicted in Figure 5, which enables the distinction between serviced (red tokens) and unserviced (black tokens) customers. Where no arc inscription is explicitly shown one black (default) token is assumed.

which represents the flow of customers in the system. V denotes the set vertices of the graph and E the set of its edges. This graph is derived from the non-parameterised model by mapping each place in the model to a vertex and each transition – along with its incident arcs – to a directed edge (cf. Figure 8). We then use Johnson's algorithm [11]

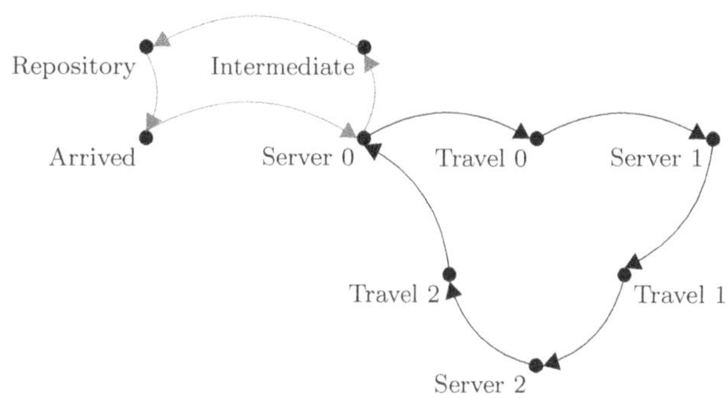

Figure 8: The directed graph produced for the Generalised Stochastic PNPM depicted in Figure 6. The paths consisting of black and red edges are two elementary cycles beginning at vertex Server 0.

to ascertain whether the derived graph contains elementary cycles. An elementary cycle is a sequence of vertices $c_v = (v_1 = v, v_2, \ldots, v_k = v)$ such that $(v_i, v_{i+1}) \in E$ for $1 \leq i < k$, where no vertex, apart from the first and last, appears twice. For more information we refer the reader to a standard graph theory textbook, e.g. [6]. This algorithm requires as input the directed graph and outputs the sequence of vertices that form each identified elementary cycle. The direct application of the algorithm on G would always return at least one elementary cycle which will include the Repository place, e.g. see Figure 8. This place of the PNPM represents the entry/exit point of the underlying system and thus, is actually not part of any service cycle. To avoid such unwanted cycles to be detected by the algorithm we examine the subgraph of G induced by the set of vertices

Transition	Original Firing Mode		Modified Firing Mode	
t_4	I^-(Server 0, t_4)(black) = 1	I^+(Interm., t_4)(black) = 1	I^-(Server 0, t_4)(red) = 1	I^+(Interm., t_4)(red) = 1
t_7	I^-(Travel 2, t_7)(black) = 1	I^+(Server 0, t_7)(black) = 1	I^-(Travel 2, t_7)(black) = 1	I^+(Server 0, t_7)(red) = 1
t_{10}	I^-(Interm., t_{10})(black) = 1	I^+(Rep., t_{10})(black) = 1	I^-(Interm., t_{10})(red) = 1	I^+(Rep., t_{10})(black) = 1

Table 2: The required firing mode modifications, if any, for each transition of the PNPM shown in Figure 6 to enable the accurate representation of the underlying system's customer flow (Figure 7). Interm. and Rep. denote the places labelled as Intermediate and Repository respectively.

V − {Repository} instead of G. Also, according to the following theorem we restrict the algorithm's search to begin only with vertices which have at least two *predecessors*[5] and at least two *successors*.

THEOREM 1. *Given a customer-processing system and its corresponding directed graph, a service cycle exists only if the vertex v_i which corresponds to the head of the service cycle S_i has at least two predecessors and at least two successors.*

The proof of Theorem 1 is shown in the Appendix. When the algorithm completes its search, it returns the ordered list of vertices V_i which constitute the elementary cycle – one list for each cycle i. For each such list we obtain the (ordered) list of the corresponding places P_i; these places are obtained by reversing the mapping used earlier to derive the directed graph from the PNPM. Consider our initial example of the customer-processing system shown in Figure 5 and its corresponding PNPM (cf. Figure 6). In this example we have only one service cycle: Server 0, Travel 0, Server 1, Travel 1, Server 2, Travel 2, Server 0.

Now, using the latter information, we wish to accurately represent the flow of customers in the system. We begin by retrieving the sets of input and output transitions of the place $p_H^{(i)}$ that corresponds to the head of the ith service cycle. For each transition t_{in} contained in $\bullet p_H^{(i)}$ we obtain its input place[6] and similarly, for each transition t_{out} contained in $p_H^{(i)}\bullet$ its output place. Here $\bullet p$ and $p\bullet$ denote the sets of input and output transitions of a place p. Similarly, $\bullet t$ and $t\bullet$ denote the sets of input and output places of a transition t. In our example (cf. Figure 6) we have: $p_H^{(i)}$ = Server 0, $\bullet$Server 0 = $\{t_7, t_9\}$, Server 0$\bullet$ = $\{t_1, t_4\}$, $\bullet t_7$ = {Travel 2}, $\bullet t_9$ = {Arrived}, $t_1\bullet$ = {Travel 0}, and $t_4\bullet$ = {Intermediate}.

The next step is to determine the transition t_{in}, $t_{in} \in \bullet p_H^{(i)}$, whose input place is in cycle i, and change its firing mode. In particular, we only need to modify its forward incidence function to support another token type. That is,

$$I^+(p_H^{(i)}, t_{in})(black) = 1 \rightarrow I^+(p_H^{(i)}, t_{in})(c) = 1$$

for some (hitherto unused) colour c other than black. We then identify the transition t_{out}, $t_{out} \in p_H^{(i)}\bullet$, whose output place p_k is not in cycle i, and change both its backward and forward incidence functions accordingly, i.e.

$$I^-(p_H^{(i)}, t_{out})(black) = 1 \rightarrow I^-(p_H^{(i)}, t_{out})(c) = 1$$

$$I^+(p_k, t_{out})(black) = 1 \rightarrow I^+(p_k, t_{out})(c) = 1.$$

[5]When two vertices $v_i, v_j \in V$ are adjacent through the edge $(v_i, v_j) \in E$, then v_j is called a successor of v_i and v_i is called a predecessor of v_j [6].

[6]At this point, each transition in the model has only one input and only one output place.

We proceed by recursively changing the backward and forward incidence functions of all subsequent connected transitions in the non-parameterised model; we consider two transitions t_i, t_j to be connected if $t_i \bullet \cap \bullet\, t_j \neq \emptyset$. This process continues until we reach the transition whose output place is the Repository. For this transition we only change its backward incidence function since we wish the original token type to be restored in the Repository. Table 2 lists the required changes of the transitions' firing mode for our example. The resulting model is shown in Figure 7.

3.4 Service Cycles and Multiple Customer Classes

When multiple customer classes exist and in particular, when at least two classes of customers perform the same service cycle, the modelling approach presented in Section 3.3 is not applicable. Here, we demonstrate the reason the latter approach fails and propose a solution.

Consider the customer-processing system shown in Figure 5 and assume the existence of two customer classes. Following the methodology presented in Section 3.2 we obtain the CGSPN model depicted in Figure 9. If we were to map the places and transitions of that model to vertices and edges to form the corresponding directed graph (as described in Section 3.3), we would obtain a directed multigraph, also known as p-graph with $p = 2$ (see Figure 10). In a p-graph, p denotes the maximum number of arcs having the same initial and terminal vertices. If we have a system with three customer classes where each class performs the same service cycle, the corresponding graph of the PNPM for that system would be a 3-graph. While Theorem 1 is still applicable for this graph, Johnson's algorithm considers the three elementary cycles as being the same and thus outputs only one. The original approach fails since we can neither generalise the detected service cycle to all customer classes, nor deduce the class which performs the service cycle.

To resolve this issue we obtain the directed graph that represents the flow of customers of each class – one graph for each customer class – and then iteratively apply the same approach as before, on each graph. To find these graphs we proceed as follows. Assuming that we have N customer classes in the system, we decompose the set of the model's transitions T into disjoint subsets T_{c_i}, where c_i, $i = 1, \ldots, N$, denotes the token type that a transition's firing mode supports – this is unique since at this stage of model construction there is a one-to-one correspondence between each customer class and each token type. Of course when the modelling approach presented in Section 3.3 is applied, a one-to-many, in particular $1 : k+1$, correspondence will exist between each customer class and each token type. Here k denotes the number of service cycles performed by a particular customer class. Then, for each T_{c_i} – along with the connected set of places – we apply the same mapping as before to derive the directed graph that corresponds to the jour-

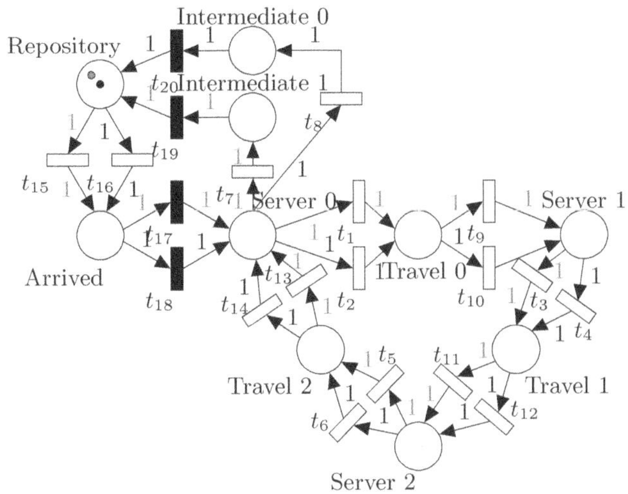

Figure 9: The CGSPN model (non-parameterised version) for the system depicted in Figure 5 when two customer classes exist. Its corresponding 2-graph is shown in Figure 10.

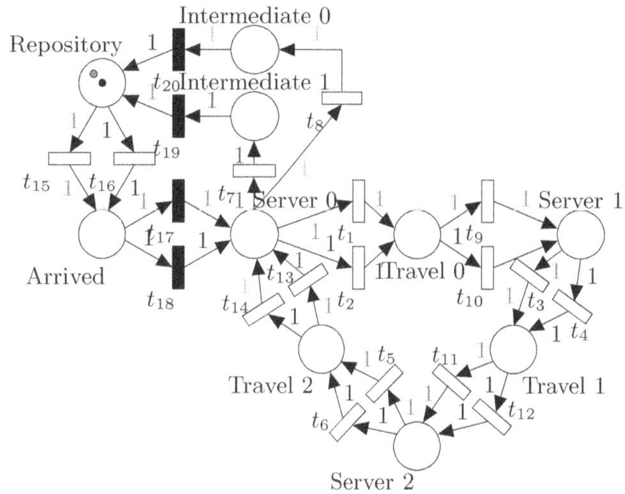

Figure 11: The CGSPN model (non-parameterised version) for system depicted in Figure 5 when two customer classes exist. Black and green tokens are used to distinguish between unserviced and serviced customers of class one. Red and blue tokens are used to distinguish between unserviced and serviced customers of class two.

ney undertaken by each customer class. In our example, we assumed two classes of customers represented by black and red tokens respectively, i.e. $N = 2$, $c_1 :=$ black, $c_2 :=$ red. If we denote the set of edges that were formed from the mapping of T_{black} as E_{black}, and similarly those formed from the mapping of T_{red} as E_{red}, then the directed graphs that correspond to the first and second customer classes are, respectively, $G_1 = (V, E_{\text{black}})$ and $G_2 = (V, E_{\text{red}})$ (see Figure 10).

Figure 11 shows the CGSPN model obtained via this extended version of our initial approach for the latter example.

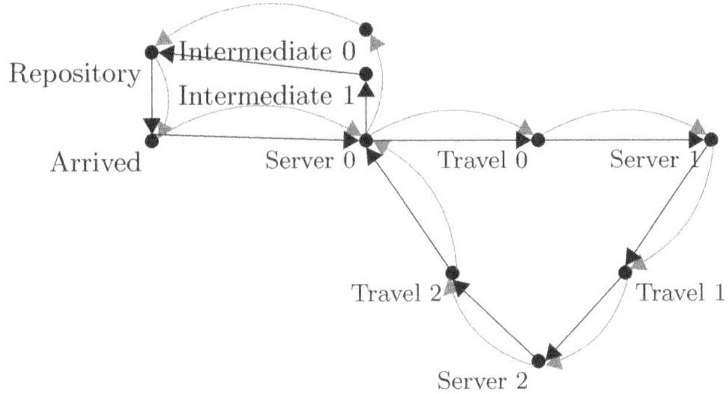

Figure 10: The directed 2-graph produced for the CGSPN model depicted in Figure 9. If E_{black} and E_{red} denote the set of black and red edges then the 2-graph can be defined as $G = (V, E_{\text{black}} \cup E_{\text{red}})$. The directed graphs for the first and second customer classes are $G_1 = (V, E_{\text{black}})$ and $G_2 = (V, E_{\text{red}})$ respectively.

3.5 Calculation and Representation of Inter-routing Probabilities

This section presents the calculation and representation of inter-routing probabilities of the customer flow within the system, when multiple classes of customers exist.

During the response and travelling time extraction process (third stage of the pipeline) we count the total number of customers of each class who received service at each service area. For each service area S_i we maintain tables – one for each customer class – whose tuples consist of two fields: the identity of another service area S_k, $k \neq i$, and the number of customers who visited S_k immediately after they completed their service at S_i. To calculate the routing probability from S_i to S_k for some k, $k \neq i$, we divide the number of customers of class j who visited S_k immediately after S_i, say $n_{i,k}^{(j)}$, by the total number of customers of the same class who were serviced by S_i, $n_i^{(j)}$. The routing of each customer class j is represented in the model as follows:

1. For each service area S_i and for each customer class j, an intermediate place is created, where tokens are placed after the associated service time transition fires.
2. For each destination service area S_k of each origin service area S_i, immediate transitions are created from the intermediate places of S_i to a so-called "travel" place. Their weights are set equal to $n_{i,k}^{(j)}/n_i^{(j)}$.
3. Travelling time transitions are created, connecting each constructed travel place to the corresponding destination service area S_k.

4. CASE STUDY

We present a case study to evaluate the pipeline's newly added features and their integration in the existing methodology. For this case study we have generated location tracking data using an extended version of LocTrackJINQS [9]. Synthetic data provides two advantages over real traces: it allows us to characterise the degree of accuracy of the inferred distributions and their parameters – as the exact model parameters and processes are known – and its generation is performed in a time efficient manner rather than engaging in long experimental procedures.

Here, a simplified model of an A&E department is employed. It comprises five service areas: a reception (S1), an examination room (S2), an x-ray operation room (S3), an x-ray room (S4) and a treatment room (S5). The simulation takes place in a 40 m × 28 m virtual environment. Figure 12 depicts the experimental setup. Each service area consists

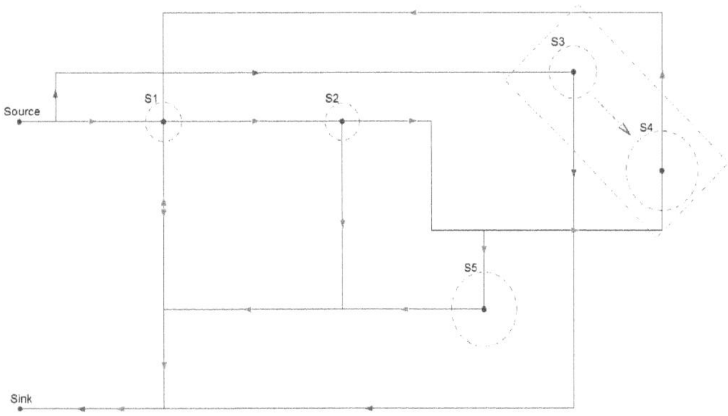

Figure 12: The experimental setup in terms of abstract system structure for this case study. The red arrows represent the flow of customer classes 0 and 1. Blue arrows are used to indicate the flow of customer classes 2 and 3. The blue dotted arrow (contained in the blue dotted rectangle) indicates the synchronisation conditions imposed by S3 on S4: one nurse (customer class 2) and one radiologist (customer class 3) must be present in the x-ray operation room (S3) so that the screening process of patients in the x-ray room (S4) can be initiated. After service completion at S4 and S5, customers of classes 0 and 1 obtain service from S1 (for the second time) and then exit the system, thus defining two service cycles.

of a single customer-processing server with a random service discipline. The location update error is normally distributed with mean 0.15 m and standard deviation 0.2 m.

We assume minor and major patients classes which correspond to customer classes zero and one respectively. Furthermore, two more classes of customers exist: nurses (customer class two) and radiologists (customer class three). The arrival process for each customer class is shown in Table 3. Patients entering the system are routed to the reception in order to register and then proceed to the examination room. From there they get discharged, or are sent to the x-ray room (S4) or to the treatment room (S5). This routing occurs with probability 0.3, 0.4 and 0.3 respectively. Patients routed to the x-ray or treatment room must then revisit the reception to schedule a followup examination before exiting the system, i.e. we have two service cycles: S1, S2, S4, S1, and S1, S2, S5, S1. Nurses and radiologists are immediately routed to the x-ray operation room (S3) and when their service is terminated they exit the system. In order for the scanning process of patients to initiate and progress, at least one nurse and one radiologist must be present in S3. The customers are assumed to travel between service areas with speed drawn from a normal distribution whose parameters – for each class – are shown in Table 3. The service time for each server and each customer class may follow a different density function. Table 4 shows each service area's actual location and service radius, as well as its service time density for each customer class.

Customer Class	Arrival Distribution	Speed Distribution
0	Exp(0.01)	Normal(0.38, 0.1)
1	Exp(0.008)	Normal(0.25, 0.1)
2	Exp(0.004)	Normal(0.4, 0.2)
3	Exp(0.004)	Normal(0.5, 0.15)

Table 3: The customer arrival process and speed distribution for each customer class.

	Server Location	Service Radius	Service Time Density	
S1	(10.0, 12.5)	1.0	Customer Class 0	Exp(0.033)
			Customer Class 1	Exp(0.05)
S2	(20.0,12.5)	0.95	Customer Class 0	Erlang(2, 0.035)
			Customer Class 1	Normal(40, 10)
S3	(33.0,10.0)	1.33	Customer Class 2	Erlang(6, 0.05)
			Customer Class 3	Erlang(5, 0.04)
S4	(38.0,15.0)	1.8	Customer Class 0	Exp(0.033)
			Customer Class 1	Exp(0.033)
S5	(28.0,22.0)	1.5	Customer Class 0	Exp(0.013)
			Customer Class 1	Erlang(2, 0.013)

Table 4: The parameters for each service area in the system, for each customer class.

4.1 Results

The locations and service radii of the service areas are in general approximated well (see Table 6). The largest error for the service area location inference – specified in terms of the distance between the real and inferred points – is 0.320 metres. The maximum service radius approximation error is 0.277 metres.

Service Area	Customer Class	Simulated Probability	Inferred Probability (3 d.p.)
S4	0	0.4	0.429
	1	0.4	0.397
S5	0	0.3	0.280
	1	0.3	0.298
Sink	0	0.3	0.291
	1	0.3	0.305

Table 5: The simulated and inferred routing probability (to three decimal places) of the customer flow from service area S2.

The quality of the extracted service time samples and the best-fit HErD are assessed by Kolmogorov-Smirnov tests which examine their compatibility (see Table 7). To examine the degree of accuracy of the approximation of the theoretical (simulated) service time distribution by the best-fit HErD we compute the relative entropy [13] of the two distributions. As shown in Table 8, good level of agreement is observed between the parameters of the best-fit HErD and the parameters of the corresponding theoretical service time distribution (relative entropy for all fitted HErDs is bounded by 0.119 nat). Figure 14 shows the cumulative histogram of the extracted service time samples and its best-fit HErD compared with the corresponding theoretical distribution for customer class 0.

The inferred CGSPN performance model is shown in Figure 13. Customer classes 0, 1, 2 and 3 are represented by

	Server Location			Service Radius		
	Real	Inferred	Error	Real	Inferred	Error
S1	(10.0, 12.5)	(9.955, 12.511)	0.046	1.0	1.095	0.095
S2	(20.0, 12.5)	(20.014, 12.546)	0.048	0.95	1.025	0.075
S3	(33.0, 10.0)	(33.017, 10.014)	0.022	1.33	1.454	0.124
S4	(38.0, 15.0)	(37.814, 14.976)	0.188	1.8	2.077	0.277
S5	(28.0, 22.0)	(28.136, 22.290)	0.320	1.5	1.609	0.109

Table 6: The inferred location and service radius for each service area in the system accompanied with the absolute error. These values are given to three decimal places.

red, blue, green and black coloured tokens. Places labelled as Server 0, Server 1, Server 3, Server 2 and Server 4 correspond to service areas S1, S2, S3, S4 and S5. The two service cycles involving service areas S1, S2, S4 and S1, S2, S5, have been correctly identified and modelled. In particular, the CGSPN subnets with labels HErD21 and HErD15 connect travel places, associated with customer movement after S4 and S5, to S1 thus completing the service cycles for customer class 0 (subnets HErD19 and HErD13 for customer class 1). We note the token colour change for the first customer class (class 0) from red to brown (by HErD21 and HErD15) and for the second customer class (class 1) from blue to yellow (by HErD19 and HErD13). Table 5 shows the inferred routing probability from S2 to S4, S5 and Sink (represented by the Repository place). Also, the synchronisation conditions between S3 (synchronising service area) and S4 (synchronised service area) have been modelled correctly; one nurse (green colour) and one radiologist (black colour) are required to be present in S3 so that patients in S4 can be scanned.

5. CONCLUSION AND SUMMARY

This paper has introduced extensions to an existing data processing pipeline for the automated derivation of performance models from location tracking traces. These increase its domain of applicability to customer-processing systems supporting multiple customer classes and service cycles. The calculation and representation of inter-routing probabilities of the customer flow between the system's service areas, when multiple customer classes exist, has also been presented. All extensions are incorporated within the third stage of the pipeline.

The case study results indicate that the extended methodology can infer the stochastic features of certain multi-class systems accurately, even in the presence of synchronisation and service cycles, at least when synthetically-generated location tracking data is used.

Our current work assumes single-server semantics, random service discipline and static customer classes. In future work we intend to relax these assumptions. It should be straightforward to modify the subnets used to model the service times of service areas (cf. Figure 3) to support multiple servers by changing the number of tokens contained in the complementary place which controls the number of tokens (customers) which can be simultaneously in the subnet[7]. More challenging tasks include devising a means to infer the number of servers present in each service area from location tracking data, and supporting dynamic customer classes (where class changes may occur after completion of service at a service area).

Ultimately, we would like to examine the applicability and performance of our pipeline in the context of more complicated scenarios using location tracking data gathered by a real RTLS. In fact, experiments on a real A&E department would be of particular interest given the favourable outcome of the synthetic A&E case study presented in this paper.

		Class 0		Class 1	
S1 - 1st visit	Test Statistic	0.0741		0.0638	
	α	0.1	0.05	0.1	0.05
	Critical Values	0.0807	0.0920	0.0877	0.0999
	Compatible ?	Yes	Yes	Yes	Yes
S1 - 2nd visit	Test Statistic	0.0798		0.0422	
	α	0.1	0.05	0.1	0.05
	Critical Values	0.0973	0.1109	0.1110	0.1265
	Compatible ?	Yes	Yes	Yes	Yes
S2	Test Statistic	0.0479		0.0455	
	α	0.1	0.05	0.1	0.05
	Critical Values	0.0809	0.0922	0.0901	0.1027
	Compatible ?	Yes	Yes	Yes	Yes
S4	Test Statistic	0.1095		0.1293	
	α	0.1	0.05	0.1	0.05
	Critical Values	0.1244	0.1418	0.1443	0.1645
	Compatible ?	Yes	Yes	Yes	Yes
S5	Test Statistic	0.0928		0.1011	
	α	0.1	0.05	0.1	0.05
	Critical Values	0.1529	0.1743	0.1671	0.1905
	Compatible ?	Yes	Yes	Yes	Yes
		Class 2		Class 3	
S3	Test Statistic	0.0708		0.1374	
	α	0.1	0.05	0.1	0.05
	Critical Values	0.1370	0.1562	0.1430	0.1630
	Compatible ?	Yes	Yes	Yes	Yes

Table 7: Kolmogorov-Smirnov test at significance levels 0.1 and 0.05 applied to the extracted service time samples for each service area in this case study. The null hypothesis is that each extracted sample belongs to the corresponding best-fitted HErD.

6. REFERENCES

[1] H. Akaike. A New Look at the Statistical Model Identification. *Automatic Control, IEEE Transactions on*, 19(6):716 – 723, December 1974.

[2] N. Anastasiou, T.-C. Horng, and W. Knottenbelt. Deriving Generalised Stochastic Petri Net performance models from High-Precision Location Tracking Data. In *Proc. 5th Intl. Conference on Performance Evaluation Methodologies and Tools (VALUETOOLS 2011)*, Paris, France, May 2011.

[3] N. Anastasiou, W. Knottenbelt, and A. Marin. Automatic Synchronisation Detection in Petri Net Performance Models Derived from Location Tracking Data. In *Proc. 8th European Performance Engineering Workshop*, pages 29–41, 2011.

[4] F. Bause and P. Kritzinger. *Stochastic Petri Nets.* Friedrich Vieweg & Sohn Verlag, 2002.

[7]For timed transitions contained in these subnets we assume infinite-server semantics.

		Service Time Density	Fitted HErD Parameters			Relative Entropy
			Phase Lengths	Rate (3 d.p.)	Weights (3 d.p.)	(3 d.p.)
S1 - 1st visit	Class 0	Exp(0.033)	1	0.034	1.000	0.001
	Class 1	Exp(0.05)	2, 2	0.084, 1.184	0.899, 0.101	0.054
S1 - 2nd visit	Class 0	Exp(0.033)	1	0.034	1.000	0.001
	Class 1	Exp(0.05)	1, 2	0.070, 0.088	0.325, 0.675	0.045
S2	Class 0	Erlang(2, 0.035)	2, 8	0.032, 5.686	0.985, 0.015	0.017
	Class 1	Normal(40, 10)	17	0.385	1.000	0.119
S3	Class 2	Erlang(6, 0.05)	5	0.037	1.000	0.049
	Class 3	Erlang(5, 0.04)	5	0.036	1.000	0.030
S4	Class 0	Exp(0.033)	1	0.027	1.000	0.021
	Class 1	Exp(0.033)	1	0.027	1.000	0.016
S5	Class 0	Exp(0.013)	1	0.013	1.000	0.000
	Class 1	Erlang(2, 0.013)	2	0.013	1.000	0.001

Table 8: The HErD parameters fitted by G-FIT for each server's service time density and each customer class, along with the relative entropy (in nat) between the theoretical and fitted probability density function. The rates and weights of each fitted HErD, as well as the relative entropy values are given to three decimal places.

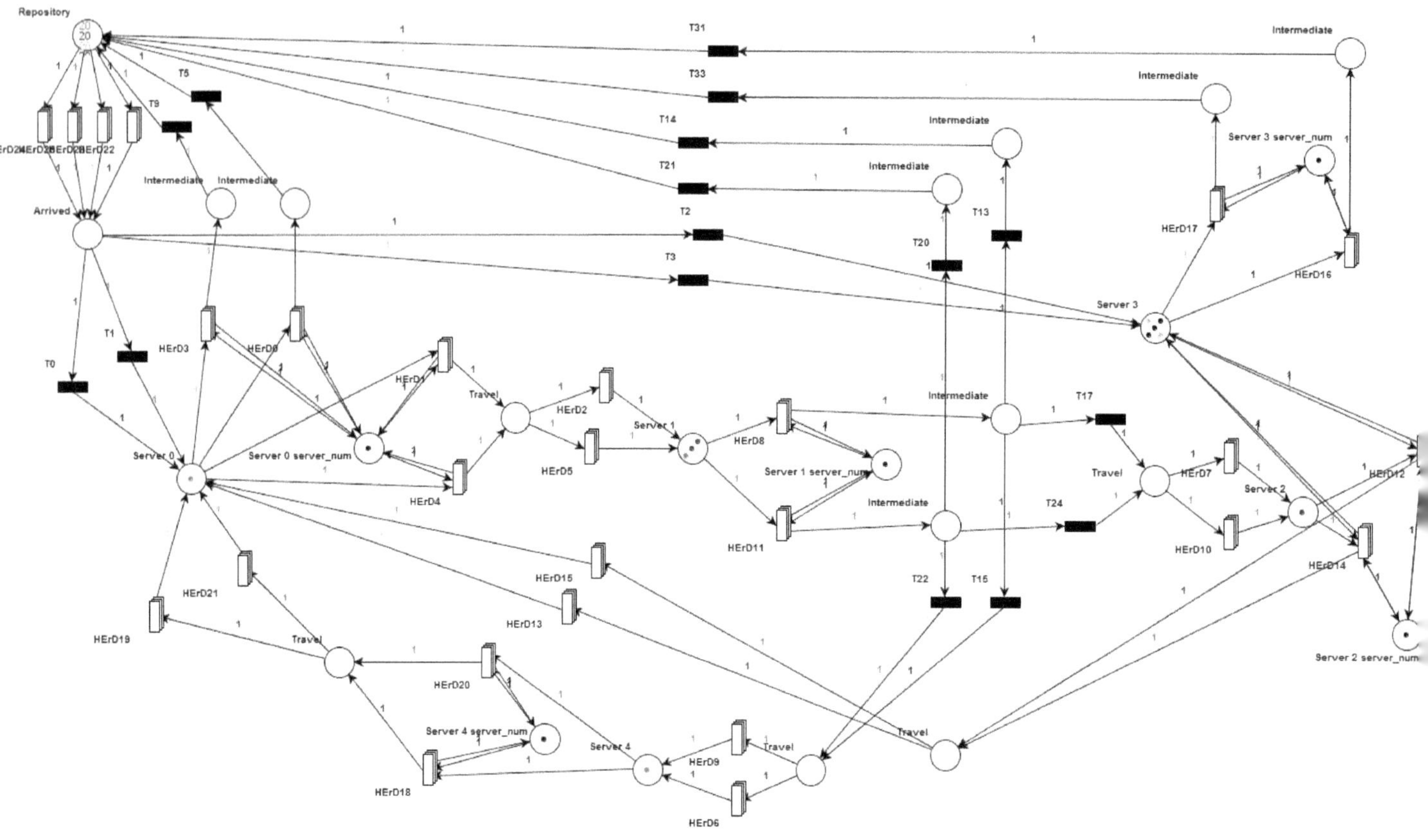

Figure 13: The inferred CGSPN performance model for this case study, visualised in PIPE2, (in compact transition form). Red, blue, green and black tokens represent customer class 0, 1, 2 and 3 in respective order.

[5] P. Bonet, C. Llado, R. Puijaner, and W. Knottenbelt. PIPE v2.5: A Petri Net Tool for Performance Modelling. In *Proc. 23rd Latin American Conference on Informatics (CLEI 2007)*, San Jose, Costa Rica, October 2007.

[6] B. Carré. *Graphs and networks.* Clarendon Press, 1979.

[7] N. Dingle, W. Knottenbelt, and T. Suto. PIPE2: A tool for the performance evaluation of Generalised Stochastic Petri Nets. *ACM SIGMETRICS Performance Evaluation Review*, 36(4):34–39, 2009.

[8] M. Ester, H.-P. Kriegel, J. Sander, and X. Xu. A density-based algorithm for discovering clusters in large spatial databases with noise. In *Proc. 2nd Intl. Conf. on Knowledge Discovery and Data Mining*, 1996.

[9] T.-C. Horng, N. Anastasiou, and W. Knottenbelt. LocTrackJINQS: An Extensible Location-aware Simulation Tool for Multiclass Queueing Networks. In *Proc. 5th Intl. Workshop on Practical Applications of*

Stochastic Modelling (PASM 2011), Karlsruhe, Germany, March 2011.

[10] T.-C. Horng, N. Dingle, A. Jackson, and W. Knottenbelt. Towards the Automated Inference of Queueing Network Models from High-Precision Location Tracking Data. In *Proc. 23rd European Conference on Modelling and Simulation (ECMS 2009)*, pages 664–674, May 2009.

[11] D. Johnson. Finding All the Elementary Circuits of a Directed Graph. *SIAM Journal on Computing*, 4(1):77–84, 1975.

[12] S. Kounev et al. Automated Simulation-Based Capacity Planning for Enterprise Data Fabrics. In *Proc. 4th Intl. Conference on Simulation Tools and Techniques (SIMULTOOLS 2011)*, pages 27–36, 2011.

[13] S. Kullback and R. A. Leibler. On Information and Sufficiency. *The Annals of Mathematical Statistics*, 22(1):79–86, 1951.

[14] M. Li and N. Georganas. Colored Generalized Stochastic Petri Nets for Integrated System Protocol Performance Modelling. In *Proc. GLOBECOM 1988*, volume 3, pages 1798–1802, 1988.

[15] M. Marsan, G. Conte, and G. Balbo. A Class of Generalized Stochastic Petri Nets for Performance Evaluation of Multiprocessor Systems. *ACM Transactions on Computer Systems*, 2(2):93–122, 1984.

[16] A. Thümmler, P. Buchholz, and M. Telek. A Novel Approach for Phase-Type Fitting with the EM Algorithm. *IEEE Transactions on Dependable and Secure Computing*, 3:245–258, 2005.

[17] Y. Xue, R. Kieckhafer, and F. Choobineh. Automated construction of GSPN models for flexible manufacturing systems. *Computers in Industry*, 37:17–25, 1998.

APPENDIX

Proof of Theorem 1.

PROOF. We consider a physical customer-processing system consisting of the set of service areas $S = \{S_i \mid i = 1, \ldots, N\}$, for some finite $N \in \mathbb{N}$. Assume that a service cycle exists, and consider the directed graph which corresponds to the PNPM (non-parameterised version) of the underlying system.

We define:

1. $S_c \subseteq S$ to be the subset of service areas that make up the service cycle,
2. $P_c = \{v_j, v_{j+1}, \ldots, v_{j+k}\}$ to be the set of vertices which correspond either to the places representing service areas $S_i \in S_c$ or to the travel places that exist between pairs of service areas S_i, S_k, $i \neq k$, $S_i, S_k \in S_c$.

We examine the two conditions of the theorem individually:

At least two predecessors are required.
If no predecessor of the vertex v_j which corresponds to S_j, exists, then it is trivial that no service cycle exists. Now let us assume that only one predecessor of the vertex v_j exists, say v_p with $p \neq j$. We need to consider two cases. If $v_p \in P_c$, it corresponds to the travel place which connects the last service area of the service cycle, say S_l, $l \neq j$, $S_l \in S_c$, to the head of the service cycle S_j. This means that the customers initiated their service sequence at the service area $S_l \in S_c$. But since S_j is the head of the service cycle and $j \neq l$, we reach a contradiction. If $v_p \notin P_c$, then v_j is not accessible from any vertex $v \in P_c$, and consequently customers cannot reach S_j from the last service area of the service cycle, meaning that no service cycle exists. Thus, we obtain a contradiction.

At least two successors are required.
If no successor of the vertex v_j which corresponds to S_j exists, then it is trivial that no service cycle exists. Now assume that only one successor of the vertex v_j exists, say v_p, $p \neq j$. If $v_p \in P_c$, it corresponds to the travel place which connects S_j to the next service area in the service cycle, say S_{j+1}. Since no other vertex is a successor of v_j this means that all customers completing their service cycle will be re-routed to S_{j+1}, thus repeating the same service cycle. This violates the second condition of definition 1 and therefore, a contradiction is obtained. If $v_p \notin P_c$, then none of the vertices corresponding to service areas S_l, $l = j+1, \ldots, j+k$, are accessible from S_j, hence no service cycle exists. Again, we reach a contradiction. □

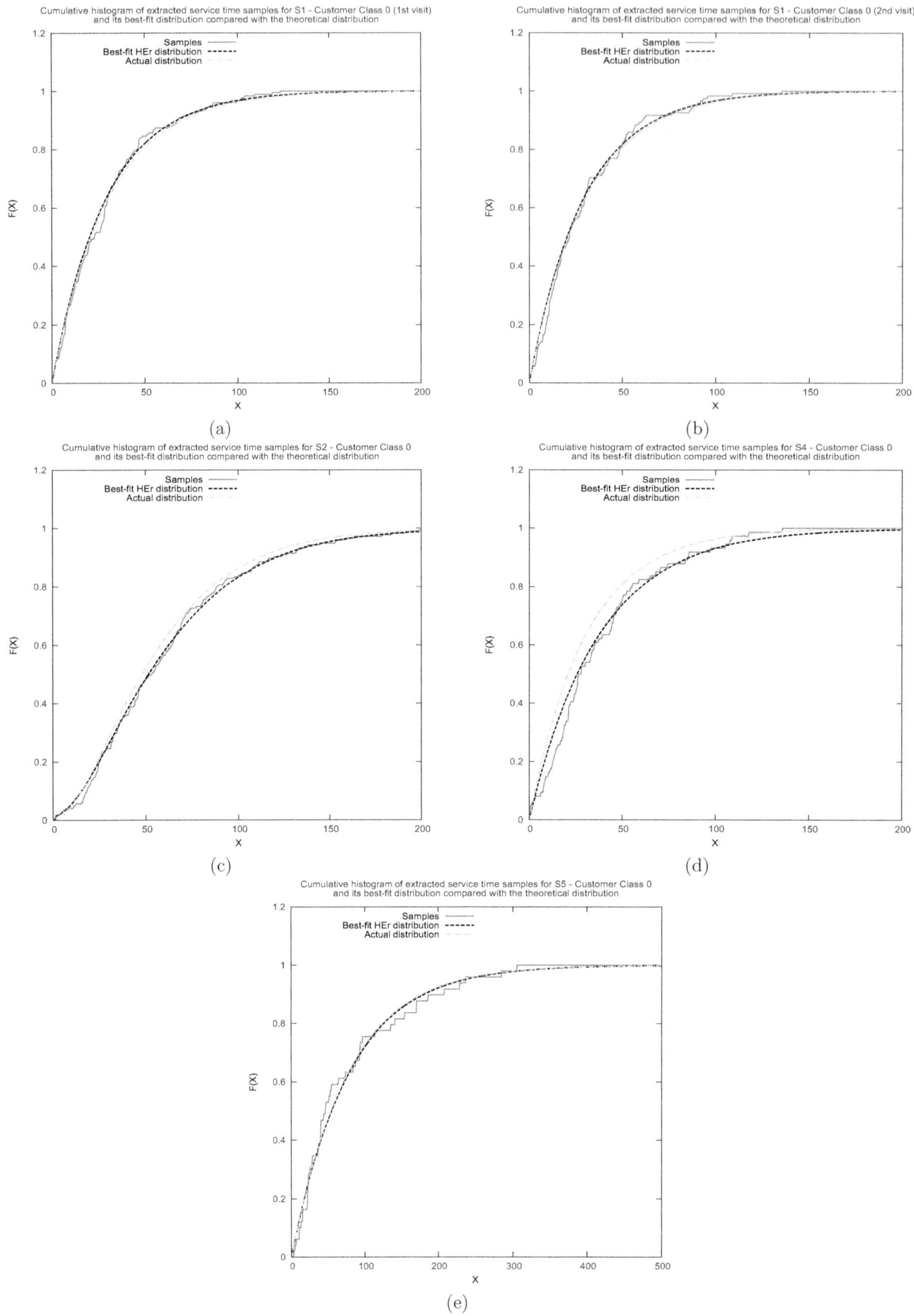

Figure 14: Case Study: Graphs 14(a), 14(b), 14(c), 14(d) and 14(e) show the cumulative histogram of the extracted service time samples (adjusted for synchronisation in 14(d)) for customer class 0 and its best-fit hyper-Erlang distribution compared with the corresponding theoretical distribution for S1 (entry to service cycle), S1 (exit from service cycle), S2, S4 and S5 respectively.

Resource Availability Based Performance Benchmarking of Virtual Machine Migrations

Senthil Nathan, Purushottam Kulkarni and Umesh Bellur
Department of Computer Science and Engineering
Indian Institute of Technology, Bombay
Mumbai, India
{cendhu,puru,umesh}@cse.iitb.ac.in

ABSTRACT

Virtual machine migration enables load balancing, hot spot mitigation and server consolidation in virtualized environments. Live VM migration can be of two types - adaptive, in which the rate of page transfer adapts to virtual machine behavior (mainly page dirty rate), and non-adaptive, in which the VM pages are transferred at a maximum possible network rate. In either method, migration requires a significant amount of CPU and network resources, which can seriously impact the performance of both the VM being migrated as well as other VMs. This calls for building a good understanding of the performance of migration itself and the resource needs of migration. Such an understanding can help select the appropriate VMs for migration while at the same time allocating the appropriate amount of resources for migration. While several empirical studies exist, a comprehensive evaluation of migration techniques with resource availability constraints is missing. As a result, it is not clear as to which migration technique to employ under a given set of conditions. In this work, we conduct a comprehensive empirical study to understand the sensitivity of migration performance to resource availability and other system parameters (like page dirty rate and VM size). The empirical study (with the Xen Hypervisor) reveals several shortcomings of the migration process. We propose several fixes and develop the Improved Live Migration technique (ILM) to overcome these shortcomings. Over a set of workloads used to evaluate ILM, the network traffic for migration was reduced by 14-93% and the migration time was reduced by 34-87% compared to the vanilla live migration technique. We also quantified the impact of migration on the performance of applications running on the migrating VM and other co-located VMs.

Categories and Subject Descriptors

C.4 [**PERFORMANCE OF SYSTEMS**]: [Performance attributes]; D.4.8 [**OPERATING SYSTEMS**]: Performance—*Measurements*

ICPE'13, April 21–24, 2013, Prague, Czech Republic.

Keywords

Virtualization; Live Migration

1. INTRODUCTION

Advances in virtualization technologies have enabled hosting of applications and services in isolated and resource-guaranteed virtual machines. Typically, a single physical machine (PM) hosts multiple applications, each executing in it's own virtual machine (VM). In addition, today's virtualization technologies [4, 12, 6] allow VMs to be migrated from one physical machine to another during execution (VM live migration). This feature along with dynamic resource provisioning for a virtual machine can be used to either balance load [15, 11] across hosts (physical machines) in a data center or to consolidate virtual machines [5, 14, 7] onto fewer physical machines for power savings and reducing operational costs (i.e., server consolidation).

Live VM migration entails an iterative process which transfers memory contents of a VM from one host to another. Since the VM is executing, memory pages are dirtied while migration happens. Thus, multiple iterations are required to transfer the state. Further, the migration can be of two types– a non-adaptive process in which a VM's memory pages are transferred at the maximum possible rate, and an adaptive process in which the rate of page transfers is tied to the page dirty rate. Heuristics for migration coupled with memory usage patterns of the VM impact the performance of live migration. The important performance/cost metrics of migration are: *migration time*—the total duration for the migration to complete; *down time*—the duration of time for which the virtual machine is not executing (during switchover); CPU and network resources consumed for migration; and performance implications on migrating VM and other co-hosted VMs.

The resource management heuristics [15, 11, 5, 16] employed in today's virtualized data centers to mitigate overload situations are migration-aware and base their migration decisions on factors such as resource utilization levels across machines, the "size" of the virtual machine to be migrated, the availability of resources at target machines, the number of migrations to achieve dynamic provisioning goal, etc. Accurate characterization of the cost-benefit trade-off between performance/cost metrics and configurations of the migration process is vital for migration-based resource management. Consider the following scenario, involving a cluster of PMs, each hosting several VMs (a typical hosting situation). A dynamic resource management problem is to modify resource allocations for the VMs to ensure SLA guarantees of services/applications executing in the VMs. In such a scenario, suppose the resources allocated to a VM need to be increased but its host PM has no spare capacity (an overload condition). A migration based reconfiguration of VMs migrate the highly loaded VM to a lightly loaded PM or migrate other VMs to create additional local capacity. For example, some existing approaches select VMs for migration based on the size (e.g., memory size) of the VMs. Verifying the efficacy and applicability of this basis is important. Selection of input parameters for migration and characteristics of the VM (e.g., memory size, memory usage dynamics etc.) impacts the cost-benefit trade-off of migration-enabled resource management. This manifests into several interesting questions, (i) given a cost metric (e.g., minimize interference to other VMs due to migration) which set of VMs should be migrated?, (ii) under which conditions would adaptive migration be preferred over non-adaptive migration?, (iii) which destination host should be selected, to minimize migration costs?, (iv) how much quantity of the CPU and network resources should be allocated to the migration process itself, to minimize migration costs?

While many studies related to migration performance exist in current literature, none of them answer all the questions mentioned above. A comprehensive empirical study of migration performance and related cost-benefit analysis will enable improved decision making for managing resources. Our contributions are as follows:

- We performed a comprehensive empirical study of VM live migration and compared the performance of adaptive live migration and non-adaptive live migration with respect to parameters which affect migration performance, such as VM size, page dirty rate, resource availability.
- We identified and implemented optimizations to the migration technique based on the above studies. In particular, we propose an enhanced migration process—the Improved Live Migration (ILM) technique.
- We quantified the performance degradation in the migrating VM and in other VMs during adaptive live migration, non-adaptive live migration and ILM technique.

In the rest of the paper, we present a short background of live migration and motivate the need for a migration performance study in Section 2. In Section 3, we describe the experimental setup and methodology. An empirical comparison of *non-adaptive* live migration and *adaptive* live migration is presented in Section 4. The Improved Live Migration techniques and its evaluation are presented in Section 5. In Section 6, we present the impact of migration on application's performance. Related work is presented in Section 7, while Section 8 concludes the paper.

2. BACKGROUND

Virtual machine live migration [10] is a process of copying memory pages of the virtual machine to the destination machine in rounds—known as as the *iterative pre-copy* phase. Iteration #1 transfers all memory pages allocated to a VM. Subsequent iterations transfer only pages dirtied during the previous iteration. However, pages dirtied before their transmission are **skipped** from transmission in the current iteration, based on the expectation that these pages may get dirtied again. The Xen live migration process [4] transfers pages in batches, i.e., a set of pages at a time instead of one page at a time. After a batch is transferred, the migration process locates pages that were dirtied (with the help of Xen shadow page tables) during the last batch transfer time. The yet-to-be-transferred dirty pages are added to a **skip list**. The rate at which pages are transferred is based on two approaches—**non-adaptive** and **adaptive**. Non-adaptive live migration transfers pages at the maximum available network capacity, whereas the adaptive technique adjusts the rate of page transfers to the rate at which pages are dirtied by the virtual machine. Iteration #1 transfers pages at the rate of r_{min} Mbps, and subsequent iterations (i>1) transfer pages at a rate r_i Mbps, which is equal to

$$r_i = \min\{\max(d_{i-1} + \delta, r_{min}), r_{max}\} \quad (1)$$

where r_{min}, r_{max} and δ are configurable variables. d_{i-1} is the **page dirty rate** (unit in Mbps) during iteration #(i-1) and is equal to,

$$d_{i-1} = \frac{\text{\#pages dirtied in iter. \#(i-1)} \times \text{page\_size}}{\text{time taken by iter. \#(i-1)}} \quad (2)$$

Based on the page dirty rate dynamics, the *iterative pre-copy* may never complete. To avoid such situations, a *stop-and-copy* condition is used. The switch from *iterative pre-copy* to *stop-and-copy* occurs when either one of the following four conditions occur:

1. a fixed number n of *pre-copy* iterations are completed.
2. the total network traffic generated is greater than a threshold, i.e., m times greater than the VM size.
3. the number of pages dirtied during the current iteration is less than a threshold, i.e., less than p pages.
4. the page dirty rate of the last iteration is greater than a threshold, i.e., greater than r_{max} Mbps **and** the number of pages sent in the current iteration is greater than the previous iteration (imply a decrease in the page dirty rate).

With *non-adaptive* live migration, only the first three stopping conditions are applicable, while all four conditions are applicable for *adaptive* live migration. In the *stop-and-copy* phase, the virtual machine being migrated is suspended and pages dirtied during the last *iterative pre-copy* round are transferred at a rate equal to the available network capacity (with both *adaptive* and *non-adaptive* live migration techniques). The performance of live migration process can be evaluated in terms of:
(i) **migration time**—the time between the initiation of migration and the successful completion, and
(ii) **downtime**—the time for which the service provided by the VM is unavailable, i.e., the time taken by the *stop-and-copy* phase.

The *cost* of the live migration process can be evaluated in terms of: (i) total **CPU utilization** at the source and destination PMs to transfer and receive pages, respectively, (ii) the total **network traffic** generated during migration—i.e., total number of pages transferred (some possibly multiple times) during migration, (iii) **migration rate**—the rate at which pages are transferred. Cost of migration is a function of these parameters as each of them can impact the performance of applications running in virtual machines at the source and destination PMs as well as other machines which share the network link.

Since the migration process consumes memory, network and CPU resources, the following parameters affect the *performance* and *cost* of the migration process:

1. Memory size allocated to the virtual machine.

2. Available CPU resource levels at the source and destination physical machines.
3. Available network capacity between source and destination physical machines.
4. The rate at which the application dirties memory pages.

2.1 Need for Migration Performance Study

Virtual machine migration based dynamic resource management opens up several questions:

• What amount of CPU and network resources should be allocated for the migration process? Migration performance itself is proportional to the resources allocated to it. However, resource allocation for the migration process usually comes at a cost of degraded performance in other VMs, from which resources are taken away to enable migration. Should we then treat the migration activity as secondary? Or should we speed it up in order to free resources on the PM from which the VM is migrating? These questions can only be answered by a thorough understanding of the dependencies between migration performance and resource allocation.

• Given an operating state of the VMs, which migration technique is best suited to minimize migration cost?

• What parameters should be considered while selecting a VM for migration? Existing solutions either select a VM randomly or consider only size of the VM to reduce migration time, downtime and migration cost. Is considering only VM size adequate?.

• What are the factors need to be considered while selecting a new destination PM for a VM? Existing approaches select a PM which is least loaded. Should the network bandwidth availability between source and destination PM, and CPU availability at destination PM for migration process be considered?

• How do the four stop-and-copy conditions impact the performance of live migration?

1. It is not clear how the number of iterations affects migration time and downtime.
2. What should be the threshold for iterative pre-copy condition 2?
3. The iterative pre-copy condition 4 is applicable for adaptive live migration but not for non-adaptive live migration. It is not clear what will be the impact of condition 4 on adaptive live migration.

While many studies of migration performance exist in the literature, we are not aware of any that have explored all of these questions. A comprehensive empirical study of migration performance can answer these and related questions. This will help to make effective resource-availability based decisions for virtual machine based load balancing and server consolidation.

3. METHODOLOGY AND SETUP

In order to compare the performance of adaptive live migration and non-adaptive live migration and answer other questions, migrations of VMs hosting four different workloads are performed by (i) varying the VM size and by keeping the load level of each workload constant, (ii) varying the load level of each workload (to create different page dirty rates) and by keeping the VM size and available resources constant, and (iii) varying the available resources and by keeping the load level of each workload and virtual machine size constant. We measured migration time, downtime, total network traffic generated and CPU resource utilization. We performed experiments on the Xen virtualization platform.

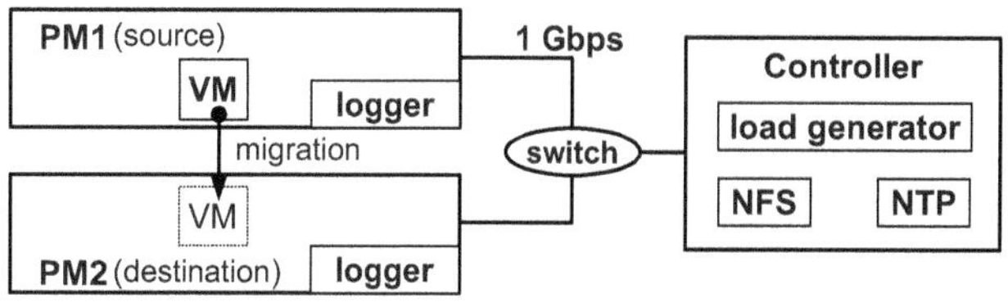

Figure 1: Experimentation setup for performance benchmarking.

We used the following workloads as application running within VMs (to be migrated). We picked these workloads as each one has a different page dirty pattern.

1) **Linux idle**: an idle Linux installation that does not host any application. This workload is mainly used as a baseline for comparison.

2) **RUBiS** [2]: an auction site prototype modeled after eBay.com. Clients perform browse and bid on existing items, register and sell items. We used the *PHP* implementation of RUBiS v1.4.3 with *MySQL*. We hosted the RUBiS web server (Apache/PHP) and database server (MySQL) on two different virtual machines. This workload results in high CPU usage and low I/O usage. The load was varied by varying the number of clients in the RUBiS workload generator.

3) **File Server**: an HTTP file server that hosts several files of different sizes. We used four clients to download four different files of size ranging from 1 GB to 4 GB using *wget*. This is mainly an I/O intensive workload. Unless otherwise specified, the default transfer rate is 240 Mbps. We controlled the transfer rate with the help of *wget --rate-limit* parameter.

4) **Kernel Compile**: a CPU intensive workload that compiles the Linux kernel v2.6.39 with the default configuration. Unless otherwise specified, the default number of threads used for compilation was 2.

Migrations of these workloads were performed using the setup shown in Figure 1. Our setup consisted of five physical machines, each with a 2.8 GHz Intel Core i5 760 CPU (4 cores), and 4GB of main memory. Out of the five physical machines, three acted as a *controller* whereas the other two physical machines (PM1 and PM2) were installed with *Xen hypervisor v4.0.1*. All five machines were connected through a 1 Gbps D-Link DGS-1008D switch. All physical machines and virtual machines were installed with the *Ubuntu Server 10.04 64-bit* Linux distribution.

We used three *controllers* to make sure that it does not become a bottleneck during the experiments. Each *controller* acted as a Network File System (NFS) server (i.e., shared storage) providing a specific set of virtual machine's disk and swap images. It also acted as Network Time Protocol (NTP) server (for time synchronization). Each *controller* executed a *load generator* daemon which issued commands to (i) create or destroy a virtual machine, (ii) specify the size of a VM and allocate CPU for the migration process, (iii) migrate the VM from PM1 to PM2, and (iv) generate load for the virtual machines.

The *logger* daemon executed in the management domain (Dom0) of PM1 and PM2. The Dom0 runs the Xen man-

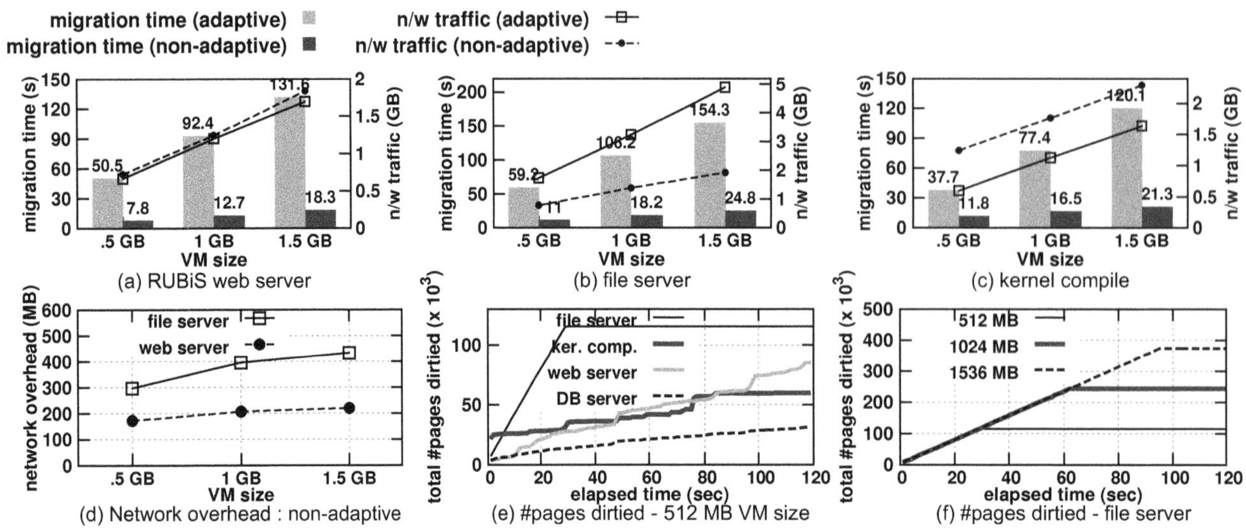

Figure 2: Effect of VM size on the performance of live migration

Table 1: Application's downtime during live migration

	VM size (GB)								
	0.5 GB	1 GB	**1.5** GB	**0.5** GB	1 GB	**1.5** GB	**0.5** GB	1 GB	**1.5** GB
Benchmark	**adaptive migration rate** (Mbps)			**adaptive downtime** (ms)			**non-adaptive downtime** (ms)		
RUBiS web server	122	108	106	33	37	43	33	35	41
RUBiS db server	108	107	106	27	32	34	26	31	33
kernel compile	132	120	112	541	640	696	362	377	475
file server	243	252	263	822	988	1149	23	28	28

agement tool stack that performs the migration of virtual machine. The *logger* daemon recorded the CPU utilization of Dom0 (using *Xentop*[1]) and network utilization (using */proc/net/dev* file) during migration. The duration of each iteration, number of pages dirtied, the number of pages transmitted, the migration rate, etc. during each *pre-copy* iteration were logged in */var/log/xen/* by the Xen hypervisor. The Dom0 of PM1 and PM2 were configured with 512 MB of main memory and 1 CPU core for all the experiments. The *logger* daemon utilized less than 2% of Dom0 CPU per second and was considered as negligible. In all experiments, a virtual machine was migrated from PM1 to PM2. Each migration experiment was repeated 10 times and results reported are averaged over 10 trails.

In Xen Hypervisor, by default, the following values were assigned to variables in iterative pre-copy termination conditions, i.e., n = 29 iterations, m = 3, p = 50 pages and r_{max} = 500 Mbps. In equation (1), the variables δ and r_{min} were assigned with 50 Mbps and 100 Mbps, respectively.

4. COMPARISON BETWEEN ADAPTIVE AND NON-ADAPTIVE LIVE MIGRATION

In this section, we compare the performance of two variants of the live migration process and try to answer the questions presented in Section 2.1.

4.1 Change in VM size

In a virtualized server environment, each virtual machine is allocated some amount of memory along with other resources to handle its peak workload. To evaluate the performance of adaptive live migration and non-adaptive live migration, we migrated VMs of different memory sizes hosting the four workloads described in Section 3.

Figures 2(a), (b) and (c) plot the migration time and the total volume of network traffic generated over different memory sizes for both *adaptive* and *non-adaptive* live migration.

Observation 1: *The time taken by adaptive live migration was three to eight times* ***higher*** *than that of non-adaptive live migration.* This is because migration rates with adaptive live migration were three to eight times lower compared to non-adaptive live migration. Table 1 shows the migration rate of adaptive live migration. The network bandwidth utilized by non-adaptive live migration for the file server workload was 640 Mbps (which was the available capacity), whereas for the other workloads it was observed to be 830 Mbps to 900 Mbps (network bandwidth requirement of other workloads was lower as compared to the file server workload).

Observation 2(a): *The total volume of network traffic generated (i.e., cost of migration) was on average 2.45 times* ***higher*** *with adaptive live migration for the file server workload as compared to non-adaptive live migration.* The reasons for this result are: (i) *Adaptive* live migration spent more time in each iteration as compared to *non-adaptive* live migration as the migration rate was low. Further, the number of unique pages dirtied increased significantly over time for the file server workload – refer Figure 2(e). Hence, the number of unique pages dirtied and the number of pages transferred during an iterative pre-copy iteration were always higher with *adaptive* live migration compared to the corresponding iteration in non-adaptive live migration. (ii) Both *adaptive* live migration and *non-adaptive* live migration with the file server workload executed more or less same number of iterations ($\approx$10 iterations).

Observation 2(b): *For the kernel compile workload, the total network traffic generated was on average 1.75 times* ***lower*** *with adaptive live migration compared to non-adaptive live migration. Similarly, for the RUBiS workload it was*

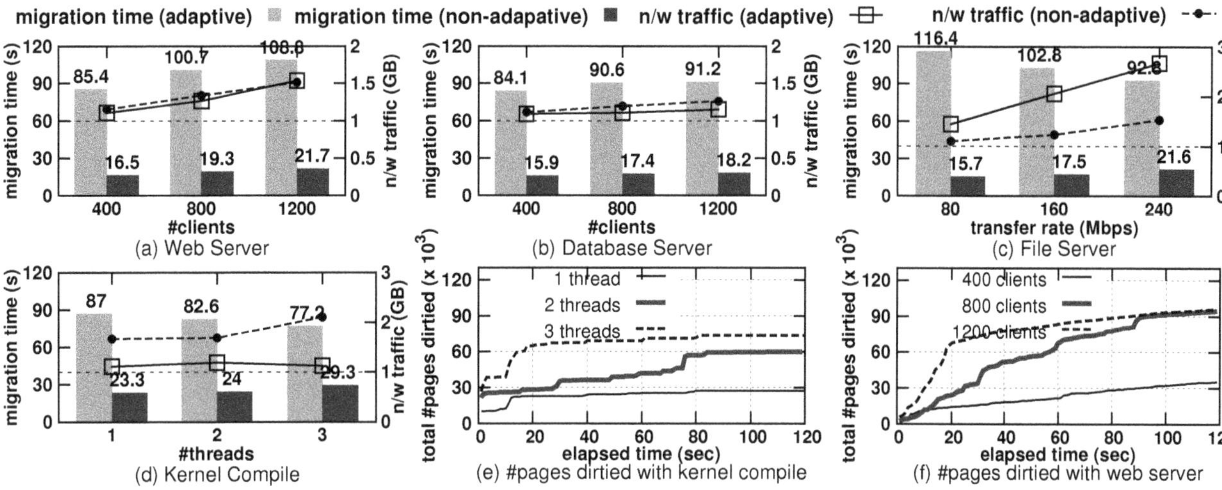

Figure 3: Effect of page dirtying rate on the performance of *non-adaptive* live migration

1.08 times lower. The reasons for this behavior are: (i) Though the time for each iteration was higher with adaptive live migration as compared to the corresponding iteration with non-adaptive live migration, the number of unique pages dirtied and the number of pages transferred during an iteration increased only marginally. Figure 2(e) shows the slowly increasing page dirty rate of the RUBiS and the kernel compile workloads. (ii) The *non-adaptive* live migration and *adaptive* live migration of these two workloads executed 29 iterations and 12 iterations, respectively. The adaptive live migration aggressively terminated iterative pre-copy due to the "page dirty rate is greater than a threshold" condition which is not used as a stopping condition for non-adaptive live migration. With increase in the number of iterations executed, the network traffic generated increased.
Takeaway: *Irrespective of the migration technique, the network traffic generated during migration depends on the page dirtying characteristics of the VM and the migration rate.*

Observation 3: *With non-adaptive live migration, migration time for RUBiS web server workload with 1 GB of memory and kernel compile workload with .5 GB of memory were* ***almost the same****. Similarly, the total network traffic generated during the migration of these two workloads were observed to be same.* The reason is that 27 out of 29 iterations spent less than 20 seconds for both of the workloads. The number of unique pages dirtied (refer Figure 2(e)) during those 27 iterations was higher with kernel compile workload as compared to RUBiS web-server workload.
Takeaway: *Considering only VM size to reduce the migration cost (i.e., migration time and resource utilization) is not adequate. It is necessary to account for page dirty rate of VM as well.* Though number of pages dirtied per second with kernel compile workload (25,000 pages/sec) was higher than file server workload (8,000 pages/sec), number of unique pages dirtied over time was higher with file server. Hence, only the number of unique pages dirtied over time should be considered.

Observation 4: *The downtime with adaptive live migration was observed to be* ***higher*** *as compared to non-adaptive live migration.* The reasons are as follows: (i) For an iteration #i, the number of unique pages dirtied with adaptive live migration was always higher than non-adaptive live migration (as mentioned in point (i) of observation 2(a) & 2(b)). (ii) Number of pages dirtied in iteration #i was always lesser than iteration #(i-1) for both adaptive live migration and non-adaptive live migration. This is because the iteration time decreased with increase in iteration number as the number of pages to be transferred per iteration decreased. As number of iterations with adaptive live migration was always lower as compared to non-adaptive live migration and also due to the reason (i) mentioned above, the number of pages to be transferred during stop-and-copy phase was higher.

Observation 5: *With increased VM size, the network overhead, i.e., the difference between total network traffic generated and size of the VM, increased.* This is plotted in Figure 2(d). As the size of the VM increased, the time taken by iteration #1 increased. As a result, the total number of unique pages dirtied during iteration #1 increased. Thus, the time taken by later iterations and the unique number of pages dirtied per iteration also increased. Further, with the file server workload, increase in VM size resulted in increase in size of the page caches. This resulted in higher number of pages dirtied per second. Figure 2(f) shows the increasing application page dirty rate of the file server workload for different VM sizes.

Increase in time per iteration and pages dirtied per iteration resulted in larger number of pages to be copied in the stop-and-copy phase. As a result, the downtime of applications increased with increase in VM size, as reported in Table 1.

Observation 6: *The migration rate of adaptive live migration decreased as the VM size increased except for the file server workload.* Table 1 presents the migration rate of adaptive live migration. This is because increase in VM size increased the duration of each iteration significantly but increment in the number of unique pages dirtied per iteration was less significant. As a result, the migration rate calculated using equation (1) and (2) for each iteration decreased with increase in the VM size. With the file server workload, the number of unique pages dirtied per iteration also increased significantly. As a result, the migration rate increased with increase in the virtual machine size of the file server workload.

Observation 7: *CPU utilization at the destination PM was observed to be* ***higher*** *than at the source PM.* For exam-

Table 2: Migration rate with adaptive live migration.

Benchmark	migration rate (Mbps)		
	load L_1	load L_2	load L_3
RUBiS web server	104	106	117
RUBiS db server	105	107	109
kernel compile	105	119	121
file server	102	165	237

ple, with non-adaptive live migration, with a migration rate of 640 Mbps, Dom0 CPU utilization at the source and destination PMs was observed to be 24% and 48%, respectively. The reasons for this behavior are: (i) At the source PM, the TCP segmentation offload feature was enabled to reduce the CPU overhead of TCP/IP. (ii) As decoding of packets at the destination PM tend to be more expensive than encoding at the source, CPU utilization was higher. (iii) In addition, memory write operation (i.e., storing memory pages of VM) at the destination PM is higher in cost compared to page read at the source.

Through these experiments, we observe that page dirty rate and migration rate play an important role in deciding the performance and cost of the live migration process. In the following sections, we study the impact of page dirty rate and resource availability (indirectly the migration rate) on the performance and cost of virtual machine migration.

4.2 Impact of Page Dirty Rate

In this section, we analyze and quantify the impact of the page dirty rate on the performance of the migration process. In order to generate different page dirty rates, we executed three workloads each with three different configurations (load levels). As expected, the number of unique pages dirtied increased as the load levels increased. All VMs were allocated memory of 1 GB each and the available network capacity for the migration process was 600 Mbps.

Figures 3(a), (b), (c), and (d) plot migration time and total network traffic over different load levels for both adaptive live migration and non-adaptive live migration. The number of unique pages dirtied over time at different load levels for kernel compile and RUBiS web server workloads are plotted in Figures 3(e), and 3(f), respectively.

Observation 8: *With adaptive live migration, only the migration time of the RUBiS workload increased as the load level increased, whereas the migration time of the file server and the kernel compile workloads decreased.* The reason is that, with the file server and kernel compile workloads, the migration rate increased significantly as the page dirty rate increased. Remember that page dirty rate is calculated using equation (2). With the RUBiS workload, the increase in migration rate was not significant. Table 2 presents the migration rate for adaptive live migration. This is because most of the iterations with adaptive live migration of all three workloads spent less than 10 seconds. The unique number of pages dirtied during that period was larger for the file server and kernel compile workloads (Figures 3(e) and 3(f))) compared to the RUBiS workload.

Observation 9: *The total volume of network traffic increased as the load level increased, except with adaptive live migration of VM hosting kernel compile workload.* This is because the total number of pages dirtied during any particular iteration increased as the page dirty rate increased. As a result, the number of pages transmitted during any particular iteration also increased. However, when three threads were used in the kernel compile workload, the total network traffic generated was lower (due to fewer iterations) than the two-threaded kernel compile workload. As the page dirty rate of three threads kernel compile workload was higher, the iterative pre-copy phase terminated aggressively with fewer iterations due to the "page dirty rate of previous iteration is greater than a threshold and #pages sent in current iteration is greater than previous iteration" condition as compared to the two threads kernel compile workload.

In order to understand the impact of page dirty rate further on network traffic, we performed migration of a VM hosting a micro-benchmark, where we can configure the number of pages dirtied per second (n), and a *hot-page* factor (h), which denotes the percentage of "n" pages that are dirtied every second (i.e., same pages). The remaining $100 - h$ percentage of "n" pages are selected randomly. The size of the VM was 1 GB and the available network capacity for the migration process was 1 Gbps.

Takeaway: *Allocation of network resources to non-adaptive live migration process should be much higher than the page dirty rate of the VM that is being migrated to reduce total network traffic. When the number of pages dirtied per second is greater than the network capacity, it is better to reduce the threshold value in iterative pre-copy condition 2 to reduce migration cost.*

The reasons are as follows: Figure 4(a) plots total network traffic generated with non-adaptive live migration over different number of pages dirtied per second. When the number of pages dirtied per second was less than the network capacity (i.e., < 32768 pages), the iterative pre-copy phase terminated due to the "number of pages dirtied in current iteration is less than a threshold" condition. This is because the rate at which pages dirtied during every iteration was less than the migration rate. As a result, the number of unique pages dirtied decreased in each iteration, as the number of iteration increased. When the number of pages dirtied per second was greater than the network capacity, the iterative pre-copy phase terminated due to the "total network traffic generated is greater than a threshold" condition. This is because the rate at which pages dirtied during every iteration was greater than the page transfer rate. With a decrease in the hot-page factor (h), the total network traffic increased as more pages were dirtied.

Takeaway: *To reduce the total network traffic generated with adaptive live migration, it is better to remove part (ii) from iterative pre-copy condition 4. This reduces the total resource utilization significantly but with little increase in the downtime.*

The reasons are as follows: Figure 4(b) plots the total network traffic generated with adaptive live migration over different number of pages dirtied per second. With a hot-page factor of 100%, the total network traffic generated is almost same for different number of pages dirtied per second. This is because the iterative pre-copy phase terminated at iteration #3 due to the "(i) page dirty rate of the last iteration is greater than a threshold and (ii) #pages sent in current iteration is greater than the last iteration" condition (the dirty rate is calculated using equation (2)). When the hot-page factor was less than 100% (implying increase in the page dirty rate), we expected the iterative pre-copy to terminate at iteration #3 or earlier due to the above mentioned iterative pre-copy condition. However, a larger number of

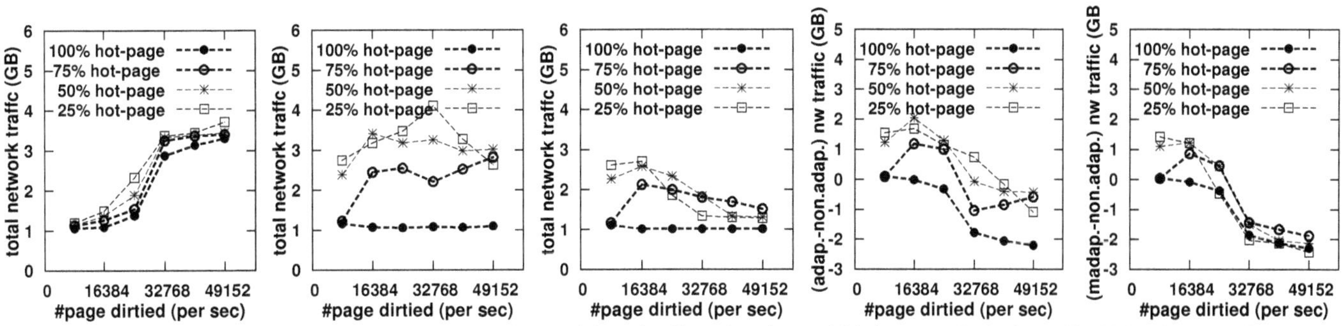

(a) Non-adaptive live migration (b) Adaptive live migration (c) madap live migration (d)(adap-non.adap) n/w traffic (e)(madap-non.adap) nw traffic

Figure 4: Impact of page dirtying rate on the migration generated network traffic

Table 3: Downtime with adaptive and *modified-adaptive* live migration (madap) techniques.

#pages dirtied n	adaptive(secs)		madap(secs)	
	h=25	h=75	h=25	h=75
16384	0.22	0.44	0.5	0.67
32768	1.2	3.4	1.52	5.09
49152	2.2	5.3	3.2	5.47

iterations were required to meet "# pages sent in current iteration is greater than the last iteration" condition. As a result, the network traffic was higher.

To speed up termination of the migration process and hence reduce the network overhead, we eliminate the stopping condition #(ii) and study its impact on the performance and cost of migration. We call this updated technique the *modified-adaptive* (madap) migration technique. Figure 4(c) plots the total network traffic generated with *modified-adaptive* live migration over different number of pages dirtied every second. The total network traffic observed was quite low compared to adaptive live migration. However, the downtime of the VM increased and is presented in Table 3. The reason is that, the number of iterative pre-copy iterations executed with *modified-adaptive* live migration was less than adaptive live migration. However, with the four workloads listed in Section 2, we did not observe a significant difference (< 100 MB) in network traffic with adaptive live migration and *modified-adaptive* live migration and we observed that the downtime increased on average by 500 milliseconds.

Observation 10: *When the number of pages dirtied per second is greater than the network capacity, the total network traffic generated with adaptive live migration and* modified-adaptive *live migration was less than non-adaptive live migration.* This behavior occurs because adaptive live migration and *modified-adaptive* live migration executed fewer iterations when the number of pages dirtied per second was greater than the network capacity. The iteration terminated due to iterative pre-copy termination condition #4 (refer Section 2). Figures 4(d) and 4(e) plot the difference in network traffic generated by adaptive and modified-adaptive live migration (madap) with respect to non-adaptive live migration.

Takeaway: *If we enable iterative pre-copy termination condition #4 for non-adaptive live migration, the total network traffic generated with non-adaptive will become less than adaptive.* We present the impact of this modification on non-adaptive live migration in Section 5.1.

In these experiments, we do not restrict the resources to be utilized by the migration process. However, dynamic resource management techniques may constrain the availability of resources for migration—finite resources are dynamically distributed to balance performance and resources utilized over all VMs on a host. In the following section, we study the performance of the migration processes with resource constraints.

4.3 Impact of Resource Availability

In this section, we quantify the impact of resource availability, (i) network capacity, and (ii) Dom0 CPU, on the performance of VM migration process. First, we performed adaptive and non-adaptive live migration of VMs by restricting the network bandwidth utilization. To restrict the bandwidth utilization of a migration process, we introduce a new variable named *max_bw_limit* in the source code. As a result, the migration process's transfer rate does not exceed *max_bw_limit* Mbps. To achieve this, we re-used some lines of code from the adaptive live migration. The RUBiS workload was configured with 1,200 clients and all VMs were allocated 1 GB of memory.

Figures 5(a) and 5(b) plot the migration time of the four workloads over different available network capacities for non-adaptive live migration and adaptive live migration, respectively. Increase in the available network capacity results in a non-linear decrease in the migration time. The time to migrate the file server workload was higher than the other workloads, as the unique number of pages dirtied over a given period was higher. Also (referring to Figure 2(e)), migration time is directly proportional to the unique number of pages dirtied over a period of time. The downtime of the workloads are presented in Table 4 for different available network capacities.The downtime decreased as the available network capacity increased.

Takeaway: *Allocating low resource level for the VM migration process results in disproportionately higher migration time and downtime. Hence, we should not allocate too low resources for migration process.*

Observation 11: *Adaptive live migration under-utilizes available resources, which results in higher migration times.* The reasons are as follows: Migration time decreased till the available network capacity reached the page dirty rate, and then it became constant. This is because adaptive live migration fixed the migration rate to the rate at which pages were dirtied. The page dirty rate of the workloads was lower than the available network capacity.

Observation 12: *With increase in the available network capacity, the network traffic for migration either decreased non-linearly or remained more or less constant.* Figures 5(d) and 5(e) plot the total network traffic generated during migration over different available network capacities for the file

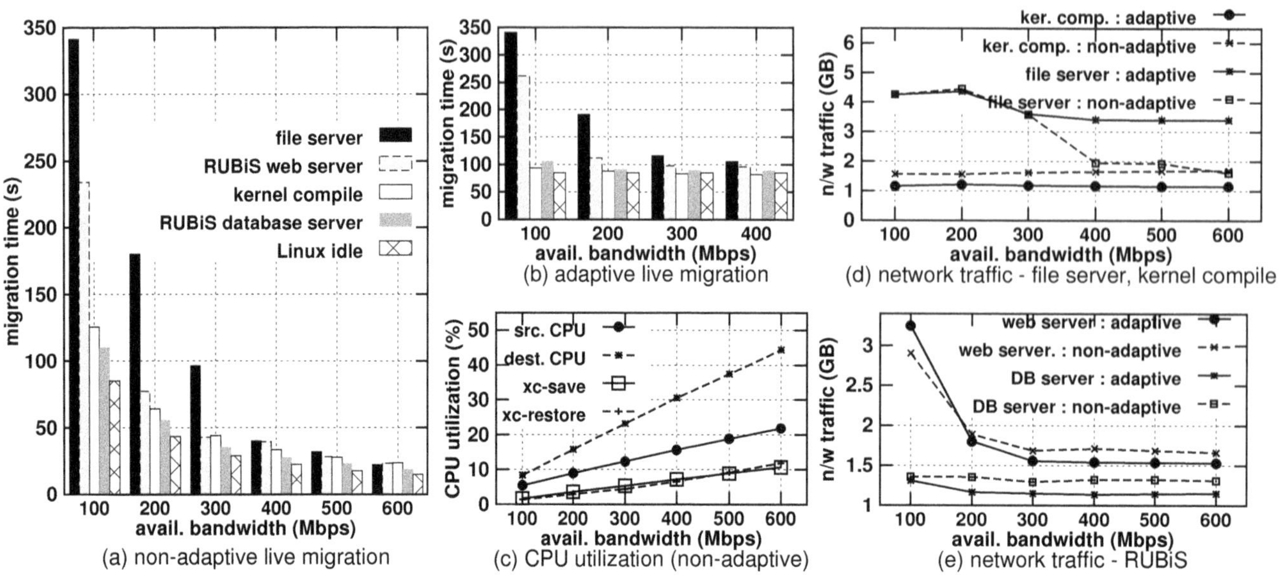

Figure 5: Impact of available network capacity on migration time, network traffic and Dom0 CPU utilization

Table 4: Impact of available network capacity on downtime of the application

	migration rate (Mbps)											
	100	200	300	400	500	600	100	200	300	400	500	600
Benchmark	**downtime** : adaptive (sec)						**downtime** : non-adaptive (sec)					
RUBiS web server	15.1	2.54	2.2	0.65	0.46	0.36	8.78	0.63	0.41	0.61	0.32	0.3
RUBiS database server	2	0.57	0.4	0.26	0.22	0.17	1.38	0.63	0.43	0.33	0.29	0.25
kernel compile	9.6	4.2	2.96	1.86	1.45	1.21	7.19	3.41	1.73	1.32	1	.778
file server	74.6	33.5	6.9	3	2.41	2.01	74.6	36.7	6.62	0.12	0.09	0.05

server, kernel compile, and RUBiS workloads, respectively. This is because with a decrease in the migration rate, the time spent on each iteration increased. As a result, the total number of unique pages dirtied per iteration increased. Hence, the network traffic generated also increased. This is because higher network traffic may result in higher overall performance degradation. However, with the kernel compile workload, the total network traffic increased. The reasons are as follows: As the migration rate increased, efficiency of the skip list decreased. Figure 6 plots the percentage of dirtied pages skipped during the migration of the kernel compile and RUBiS database server workloads over different available network capacities for non-adaptive live migration. As the available network capacity increased, total number of pages dirtied decreased for both the kernel compile and RUBiS database server workloads (for other workloads as well) because of decrease in time taken by each iteration. Also, the percentage of the dirty pages skipped decreased as the migration rate increased. The reason is that, with increase in the migration rate, the batch-transfer time (refer Section 2.1) decreased. As a result, only a few pages were added to the **skip list** and more pages were dirtied after their transmission. When the migration rate was low, more pages were added to the **skip list** and fewer pages were dirtied after their transmission due to the high batch-transfer time. However, the total number of pages dirtied increased with decrease in migration rate. With the kernel compile workload, the percentage of the dirty pages skipped decreased significantly, whereas with RUBiS database server workload, the decrease in percentage was not significant. Also, the total number of pages dirtied was higher with the kernel compile workload (implies decrease in the skip-list effectiveness with increase in migration rate).. As a result, the

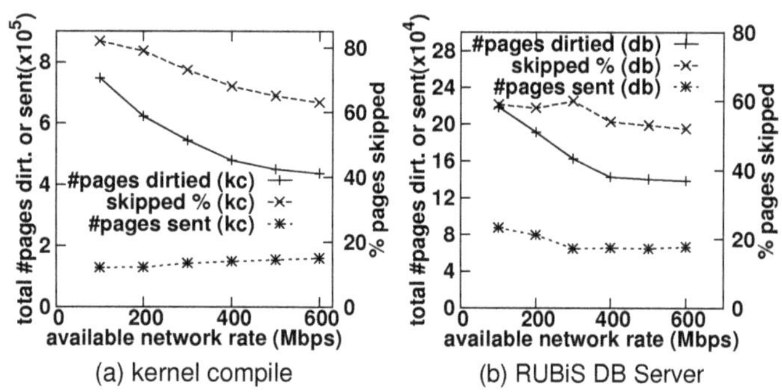

Figure 6: Efficiency of skip list over different available network capacity for non-adaptive live migration

amount of network traffic generated increased with increase in the available network capacity for non-adaptive live migration of kernel compile workload.

Takeaway: *Non-adaptive live migration with finite resources allocated to the migration process provides a tuning knob to adjust the performance and cost of migration. With higher resources allocated for migration, migration time, down time and network traffic decrease monotonically. Quantum of resources allocated can be determined based on minimizing the impact of migration on application performance of other co-located VMs.*

Skip-list effectiveness: To understand the effectiveness of the skip list, we migrated the workloads by disabling the skip list functionality (in source code) for the adaptive and non-adaptive live migration. Table 5 presents the performance of non-adaptive live migration with and without the skip list functionality for the RUBiS web server and the kernel compile workloads. Without the skip list functionality, migration time, network traffic and downtime were higher (almost double) as compared to the migration with the skip list functionality.

Observation 13: *As the available network capacity increased, the migration rate increased linearly and so did the CPU utilization at the source and destination physical machines.* The process *xc_save* that implements the live migration algorithm transfers memory pages from the source PM to process *xc_restore* which receives memory pages and starts the virtual machine at the destination PM. The CPU utilization (excluding network activity) of these two processes is reported in Figure 5(c). The CPU utilization of *xc_save* and *xc_restore* were similar and increased linearly.

Further, we migrated virtual machines by restricting the Dom0 (the management domain in Xen) CPU utilization levels of *xc_save* and *xc_restore* processes with the help of the *cpulimit* tool. We performed three sets of experiments by restricting Dom0 CPU utilization level of (i) the *xc_save* process, (ii) the *xc_restore* process, and (iii) both *xc_save* and *xc_restore* processes. The CPU levels allocated to *xc_save* and *xc_restore* were the same as the values plotted in Figure 5(c). As the Dom0 CPU availability decreased, the migration rate decreased linearly and the migration time and network traffic were similar to the experiment which restricted available network capacity. We observed that, the migration rate was proportional to the minimum of (i) available Dom0 CPU for *xc_save* process and (ii) available Dom0 CPU for *xc_restore* process.

As a result, each allocated CPU level has a corresponding network rate which can be supported, and either of which can be used to allocate network resources to the migration process.

5. IMPROVED LIVE MIGRATION

The two main qualitative inferences from the empirical evaluation (presented in the previous section) are: (i) Cost of migration—migration time, downtime, network traffic for migration, is lower for non-adaptive live migration as compared to adaptive live migration. The exception to this being workloads with high page dirty rate. (ii) The adaptive migration technique seldom uses all the available resources for migration, whereas the non-adaptive technique does not account for "available" resources.

We propose to combine the benefits of both the adaptive and non-adaptive techniques by simultaneously being sensitive to both page dirty rate of applications and available resources for migration. Towards this we propose the **Improved Live Migration** (ILM) technique. The spirit of the Improved Live Migration technique is to use non-adaptive live migration but with finite resources for migration (to reduce interference towards other applications) and to be sensitive to an application's page dirty behavior (to reduce the cost of migration). We propose three main improvements as part of ILM:

(i) Stopping criteria tuned to page dirty rate: The "page-dirty-rate-greater-than-threshold" condition in the adaptive migration technique aggressively terminates the migration process if the dirty rates are very high (see results in Section 4.2). The exact condition being, *page dirty rate of previous iteration is greater than a threshold and number of pages sent in current iteration is greater than previous iteration.* Since no such condition exists in the non-adaptive technique we often observed that the procedure terminated only after higher number of iterations (see results in Section 4). The first component of ILM is to add this condition as a stopping criteria to the non-adaptive live migration procedure.

(ii) Finite resources for migration: The non-adaptive migration technique attempts to use as many resources as possible to reduce the migration time and downtime metrics. This is oblivious to the fact that other applications may need to pay the "cost" during migration. To balance the benefits of improved migration performance and minimize the impact on other applications, we allot deterministic resources to the non-adaptive migration procedure.

The two main resources consumed by the migration procedure are network bandwidth and Dom0 CPU (for the Xen architecture). Restricting usage of these resources will minimize interference with other applications. For the purpose of this work we assume that the "available" levels of resources for migration are known. This level of resource availability depends on the left-over (unallocated) resources after accounting for resources needed to meet the SLA requirements of other VMs. We assume that this estimation happens through a higher-level migration decision process and is orthogonal to the migration procedure optimization.

Further, as seen in Section 4.3, the network utilization of the migration process and the Dom0 CPU utilization are directly (and linearly) related. Hence, as the second component of ILM, we propose to restrict the maximum network bandwidth available to then migration process.

(iii) Only transfer required pages: Conventional migration procedures transfer free-pages and page-cache pages as part of the migration process. Transfer of free-pages and page-cache pages is not required for the correctness of the migration process—free-pages have no relevant content and page-cache misses result in additional disk reads. The usefulness of the page-cache depends on the I/O access patterns and the associated hit rates. As part of ILM, we trade-off the potential I/O access benefit to reduce in the migration cost. Reduction in migration cost is more deterministic, as it is based on reducing the number of pages transferred as compared to the utility of the page cache pages. The third component of ILM is to ignore free-pages and page-cache pages from the set of pages to be migrated to the destination machine.

In the next section, we evaluate the performance of the Improved Live Migration technique (ILM) and compare it with the conventional adaptive and non-adaptive live migration techniques.

5.1 ILM Evaluation

To quantify the improvement in *performance* and *cost* of live migration techniques resulting from the proposed modifications, we performed migrations of a VM (of size 1 GB) hosting four different workloads with adaptive live migration, non-adaptive live, Improved Live Migration technique without dropping free-pages and page-cache pages (we call this ILM-fc) and Improved Live Migration technique (ILM)–with all three modifications.

Table 6 presents the network traffic (NT), migration time (MT) and Downtime (DT) with conventional live migration techniques and Improved Live Migration techniques. The network traffic generated with ILM-fc was 1 to 2.24 times lower as compared to *adaptive* live migration and 1.21 to 1.41 times lower as compared to *non-adaptive* live migration.

Table 5: Effectiveness of skip list functionality.

Benchmark	with skip-list			without skip-list		
	migration time (s)	network traffic (GB)	downtime (ms)	migration time (s)	network traffic (GB)	downtime (ms)
RUBiS web server	28.1	1.68	325	40.3	2.43	1257
kernel compile	27.7	1.67	1000	53.8	3.26	1707

Table 6: Available Network Capacity = 400 Mbps

Benchmark	adaptive			non-adaptive			ILM-fc			ILM		
	NT (GB)	MT (sec)	DT (sec)	NT (GB)	MT (sec)	DT (sec)	NT (GB)	MT (sec)	DT (sec)	NT (GB)	MT (sec)	DT (sec)
RUBiS web server	1.46	100.1	0.69	1.71	39.5	0.61	1.34	27.8	0.65	1.25	26	0.98
RUBiS db server	1.13	89.2	0.39	1.32	27.5	0.33	1.09	23	0.35	0.14	6.5	0.2
kernel compile	1.16	81.9	1.86	1.64	33.62	1.32	1.16	24.48	2.2	0.21	4.7	1.38
file server	3.39	105.8	3	1.94	40	0.12	1.51	35.8	0.2	0.23	5.2	0.24

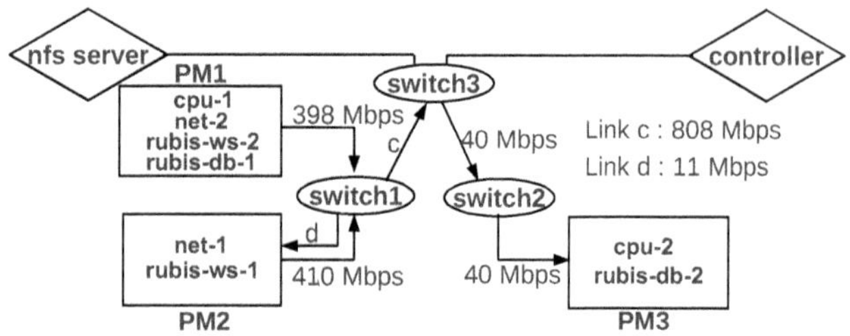

Figure 7: Experimentation setup used for measuring performance degradation.

The reason is that, ILM-fc and adaptive live migration executed more or less the same number of iterations and lower than non-adaptive live migration. We already observed that the higher the number of iterations, the higher the network traffic and shorted the downtime. As a result, downtime with ILM-fc was 1.06 to 1.66 times higher as compared to *non-adaptive* live migration and 1.06 to 15 times lower as compared to *adaptive* live migration.

The network traffic generated with ILM was 1.16 to 14 times lower as compared to adaptive live migration and 1.3 to 8.4 times lower as compared to non-adaptive live migration. The RUBiS web server workload did not have more free-pages and page-cache pages. As a result, the improvement in performance and cost of migration was low. Whereas, in case of other workloads, the memory was mainly occupied with page-cache pages.

In the following section, we study the impact of migration with and without resource restrictions on an application's performance

6. IMPACT OF MIGRATION ON APPLICATION PERFORMANCE

To quantify the impact of migration techniques on the performance of applications running in the VMs, we performed experiments with the setup shown in Figure 7. We used three machines (PM1 to PM3, whose configurations were same as the one reported in Section 3). The set of workloads collectively hosted on the three PMs are: two web servers (rubis-ws-1 and rubis-ws-2), two RUBis database servers (rubis-db-1 and rubis-db-2), two file server workloads (net-1 and net-2) and two CPU intensice PHP scripts (cpu-1 and cpu-2). RUBiS-1 (rubis-ws-1+rubis-db-1) and RUBiS-2 (rubis-ws-2+rubis-db-2) represent the two RUBiS applications. An initial mapping of workloads (running in VMs) to PMs is show in Figure 7. The average network utilization of each link is also marked in Figure 7.

We allocated resources for each workload to handle a specific load level. We used Apache JMeter to generate requests for each workload. The throughput (τ) of each workload is presented in Table 7 (refer column $\tau_{without-migration}$). The throughput of the VM hosting CPU intensive PHP script (cpu-1) was 8 requests/sec. We introduce an overload situation in physical machine PM1 by increasing the request arrival rate of cpu-1 from 8 requests/sec to 23 requests/sec, which requires an additional 50% of CPU. However, there is not enough CPU resource available at physical machine PM1 to allocate for the cpu-1. Hence, this overload situation can be solved only through VM migration. In order to mitigate the hot-spot, we migrated rubis-ws-2 (allocated with 100% of CPU) of size 1 GB from PM1 to PM3 (we assume that we cannot migrate cpu-1) to free CPU resources. The average network utilization of the link connecting PM1 to PM3 was 808 Mbps (it varied from 700 Mbps to 940 Mbps). Hence, network link was the bottleneck during VM migration and observations made below are considering this bottleneck scenario. The available Dom0 CPU at PM1 and PM3 were adequate to migrate a VM at a rate of 1 Gbps. We measured the performance degradation in terms of throughput (reqs/sec) of the workloads with adaptive live migration, non-adaptive live migration, and ILM-fc with four different migration rates. Table 7 presents migration time (MT), downtime (DT), migration rate (MR), network traffic generated (NT), throughput of each workload without migration ($\tau_{without-migration}$), and throughput drop (τ_{drop}) during migration.

Observation 14: *The throughput drop per second (τ_{drop}) for RUBiS-1, RUBiS-2, net-1 and net-2 workloads with non-adaptive live migration was observed to be 1.6 to 2.4 times higher than with adaptive live migration. However, the total #requests dropped for each workload was lower with non-adaptive live migration* and are presented in Table 8. This is because the total amount of resources utilized was higher with adaptive live migration. The network traffic generated with adaptive live migration was 1.23 times higher than non-adaptive live migration. Also, the downtime of adaptive live migration was three orders of magnitude higher than non-adaptive live migration.

Observation 15: *As the migration rate of the ILM-fc technique increased, the total number of requests dropped during migration decreased.* This is because with increase in migration rate, the total network traffic generated decreased (refer Section 4.3). As a result, the impact of migration on other applications also decreased.

As the migration rate of ILM-fc decreased from 400 Mbps to 100 Mbps, the percentage of the throughput drop was de-

Table 7: Impact of migration on the performance of application

Technique	MT sec	DT msec	MR Mbps	NT MB	$\tau_{without-migration}$ req/s				τ_{drop} req/s			
					rubis-1	rubis-2	net-1	net-2	rubis-1	rubis-2	net-1	net-2
adaptive	79	1,251	135	1,293	412	591	3.25	2.7	117	223	0	0.35
non-adaptive	26	3	350	1,051	448	545			287	363	0.07	0.81
ILM-fc (100 Mbps)	96	2,972	108	1,405	410	596			53	262	0	0.12
ILM-fc (200 Mbps)	52	2,632	205	1,283	419	592			157	273	0	0.46
ILM-fc (300 Mbps)	34	2,700	298	1,205	433	563			253	293	0.03	0.76
ILM-fc (400 Mbps)	26	4	351	1041	448	545			288	360	0.07	0.82

Table 8: Total number of requests drop per workload during migration

Technique	rubis-1 (#reqs)	rubis-2 (#reqs)	net-1 (#reqs)	net-2 (#reqs)	cpu-1 (#reqs)	total #reqs
adaptive	9,243	17,617	0	27	1,185	28,072
non-adaptive	7,462	9,438	2	21	390	17,313
ILM-fc (100 Mbps)	5,088	25,152	0	12	1,440	31,692
ILM-fc (200 Mbps)	8,164	14,196	0	22	780	23,162
ILM-fc (300 Mbps)	8,602	9,962	1	26	510	19,101
ILM-fc (400 Mbps)	7,488	9,360	2	22	390	17,262

creased disproportionately from 66% to 43% for RUBiS-2 workload. Whereas with RUBiS-1 workload, it decreased from 64% to 12%. This is because of the impact of enabling the shadow page table and retrieving the dirty bitmap after every batch transfer to construct skip list during migration. These two operations increased the CPU utilization of rubis-ws-2 during migration, making the CPU a bottleneck even at lower migration rate (i.e., 100 Mbps). The CPU utilization of rubis-ws-2 was always between 50% and 70% before the migration. However, during the migration, the CPU utilization was often at 100%. As a result, even with lower migration rate, the throughput drop was higher.

Takeaway: *In a network bottleneck scenario, if we migrate VMs at low network rate, the throughput drop (requests per seconds) of applications will be low but the total #requests dropped (during migration) will be high.*

In this section, we studied the impact of migration on the performance of web applications by making network link as bottleneck during migration. As a part of future work, we would like to quantify the performance degradation by making Dom0 CPU as the bottleneck. Also, we need to study the impact of dropping page cache pages on the application performance.

7. SUMMARY

- We conducted a comprehensive empirical study of migration techniques along the axes of migration time, downtime, and network traffic generated during migration. We showed that non-adaptive live migration performs better (implies lower migration time, downtime and network traffic) than adaptive live migration in most of the cases.
- We showed that only considering VM size as a parameter to reduce migration cost is inadequate. The page dirty rate of the VM along with the migration rate (indirectly governed by the available network capacity) are vital parameters deciding the cost of migration.
- Adaptive live migration under utilizes available CPU and network resources and cannot minimize the migration cost. We proposed ILM, an improved migration technique, which allocated finite resources to the non-adaptive live migration technique, and delivers the least migration cost as compared to the other techniques.
- Resource allocation levels for the migration process affects the throughput drop (requests per second) and the total #requests dropped (during migration) in opposite proportions. Allocation levels should follow the user's requirements in such cases.

8. RELATED WORK

The parameters of migration and related performance metrics considered in studies so far are summarized in Table 9.

Huang et al. [8] studied the performance of *non-adaptive* live migration with both the Xen [4] and KVM [12] virtualization technologies. Their study was restricted to two workloads—a Java virtual machine benchmark and a CPU-intensive benchmark. Xen and KVM implementations of *non-adaptive* live migration were compared based on downtime, migration time, total volume of network generated, and the performance degradation of the benchmark in the VM which was migrated. This work did not consider the different system parameters that can affect migration performance.

Akoush et al. [3] presented a simulation model to predict migration time and downtime of *non-adaptive* live migration. The migration technique was simulated using the page dirty trace collected from Xen shadow page tables. The model assumed that the recorded set of pages dirtied will be approximately representative of the application and will repeat under similar environments. Experiments were performed with 100 Mbps, 1 Gbps and 10 Gbps link capacity, and different VM sizes. However, the impact of available resources on the performance of non-adaptive live migration was not studied.

Hai et al. [9] evaluated the performance of *non-adaptive* live migration against page dirty rate using a micro-benchmark. Only downtime of the VM was measured and discussed. They also performed migration of idle VMs by reserving Dom0 CPU for migration and reported migration time. However, the impact of resource availability on downtime, total network traffic and CPU utilization were not discussed. Also, the impact of page dirty rate of the workloads was not explicitly considered or discussed.

Clark et al. [10] studied the performance of *adaptive* live migration for various workloads. However, only the page dirty rate was considered as a parameter which affected the performance. How each iteration's migration rate and network traffic in each iteration varied were shown for different

Table 9: Parameters and Performance metrics accounted in current literature

	Huang et. al. [8]	Akoush et. al. [3]	Liu et. al. [13]	Clark et. al. [10]	Hai et. al. [9]	Kejiang et. al. [17]	Our work
Live Migration Algorithms Evaluated							
Non-adaptive	√	√			√		√
Adaptive			√	√		√	√
Parameters Considered							
VM Size						√	√
Page dirtying rate		√	√	√	√		√
Network bandwidth				√			√
Dom0 CPU						√	√
Performance Metrics Accounted							
Migration time, downtime	√	√	√	√	√	√	√
Network traffic	√		√		√		√
CPU utilization							√

workloads. The effect of migration on web server transmission rates was also studied. However, this work did not study the impact of page dirty rate on the CPU resources required for migration and performance of non-adaptive migration.

Liu et al. [13] presented mathematical models and techniques to predict the energy consumption, migration time, downtime, and total volume of network traffic for an *adaptive* live migration. The model used resource-usage and migration-related parameters from past VM migrations to predict the migration performance parameters. However, this work covered only the page dirty rate parameter while evaluating the migration technique.

Ye et al. [17] evaluated the performance of adaptive live migration (only migration time and downtime) against available Dom0 CPU level at the source and destination machines. As adaptive live migration utilizes lower resources, they could not find any significant impact of available Dom0 CPU level. Migration rate, CPU utilization of migration process and page dirty rate of the workloads were not reported.

As can be seen, no prior work benchmarks the performance of VM migration comprehensively— against all parameters that affect migration and measure all performance metrics. Our work fills this gap and comprehensively evaluates the impact of all systems parameters on migration performance. Also, this is the first study to compare both *non-adaptive* live migration and *adaptive* live migration techniques.

9. CONCLUSION AND FUTURE WORK

In this paper, we presented a comprehensive empirical study of the Xen live migration techniques—adaptive live migration and non-adaptive live migration. We showed how the four parameters—VM size, Dom0 CPU, network capacity and application page dirty rate affect the performance of live migration in terms of migration time and downtime, and migration cost in terms of total CPU utilization and total network traffic. We also compared adaptive live migration and non-adaptive live migration, identified shortcomings and proposed three improvements. For our experimental setup and workloads, the Improved Live Migration technique (ILM) reduced network traffic for migration reduced by 14-93% and migration time reduced by 34-87% as compared to the vanilla live migration techniques.

As a part of future work, we intend to create a mathematical model to predict migration performance. The aim of model would be to predict migration time, downtime, total network traffic and CPU utilization by taking VM size, resource availability and application page dirty rate as inputs. Such prior predictions can be utilized for effective migration decisions for purposes of load balancing, consolidation and hot-spot mitigation within a virtualized environment.

10. REFERENCES

[1] http://linux.die.net/man/1/xentop.

[2] RUBiS: Rice University Bidding System. http://rubis.ow2.org/.

[3] S. Akoush, R. Sohan, A. Rice, A. W. Moore, and A. Hopper. Predicting the Performance of Virtual Machine Migration. In MASCOTS, Aug. 2010.

[4] P. Barham, B. Dragovic, K. Fraser, S. Hand, T. Harris, A. Ho, R. Neugebauer, I. Pratt, and A. Warfield. Xen and the Art of Virtualization. In SOSP, Oct. 2003.

[5] T. Das, P. Padala, V. N. Padmanabhan, R. Ramjee, and K. G. Shin. LiteGreen: Saving Energy in Networked Desktops using Virtualization. In USENIX ATC, June 2010.

[6] E. L. Haletky. *VMware ESX Server in the Enterprise: Planning and Securing Virtualization Servers*. 1 edition.

[7] F. Hermenier, X. Lorca, J.-M. Menaud, G. Muller, and J. Lawall. Entropy: a Consolidation Manager for Clusters. In VEE, March 2009.

[8] D. Huang, D. Ye, Q. He, J. Chen, and K. Ye. Virt-LM: a Benchmark for Live Migration of Virtual Machine. In ICPE, March 2011.

[9] H. Jin, L. Deng, S. Wu, X. Shi, and X. Pan. Live Virtual Machine Migration with Adaptive, Memory Compression. In IEEE CLUSTER, Aug. 2009.

[10] C. C. Keir, C. Clark, K. Fraser, S. H, J. G. Hansen, E. Jul, C. Limpach, I. Pratt, and A. Warfield. Live Migration of Virtual Machines. In NSDI, May 2005.

[11] G. Khanna, K. Beaty, G. Kar, and A. Kochut. Application Performance Management in Virtualized Server Environments. IEEE/IFIP NOMS, April 2006.

[12] A. Kivity, Y. Kamay, D. Laor, U. Lublin, and A. Liguori. KVM: The Linux Virtual Machine Monitor. In OLS, June 2007.

[13] H. Liu, C.-Z. Xu, H. Jin, J. Gong, and X. Liao. Performance and Energy Modeling for Live Migration of Virtual Machines. In HPDC, June 2011.

[14] A. Verma, P. Ahuja, and A. Neogi. pMapper: Power and Migration Cost Aware Application Placement in Virtualized Systems. In Middleware, Dec. 2008.

[15] T. Wood, P. Shenoy, A. Venkataramani, and M. Yousif. Black-box and Gray-box Strategies for Virtual Machine Migration. In NSDI, April 2007.

[16] J. Xu and J. Fortes. A Multi-objective Approach to Virtual Machine Management in Datacenters. In ICAC, June 2011.

[17] K. Ye, X. Jiang, D. Huang, J. Chen, and B. Wang. Live Migration of Multiple Virtual Machines with Resource Reservation in Cloud Computing Environments. In IEEE CLOUD, July 2011.

Overcoming Memory Limitations in High-Throughput Event-Based Applications

Marcelo R. N. Mendes†
CISUC, University of Coimbra
Dep. Eng. Informática – Pólo II
Coimbra, Portugal
mnunes@dei.uc.pt

Pedro Bizarro
CISUC, University of Coimbra
Dep. Eng. Informática – Pólo II
Coimbra, Portugal
bizarro@dei.uc.pt

Paulo Marques
CISUC, University of Coimbra
Dep. Eng. Informática – Pólo II
Coimbra, Portugal
pmarques@dei.uc.pt

ABSTRACT

The last decade has witnessed the emergence of business critical applications processing streaming data for domains as diverse as credit card fraud detection, real-time recommendation systems, call-center monitoring, ad selection, network monitoring, and more. Most of those applications need to compute hundreds or thousands of metrics continuously while coping with very high event input rates. As a consequence, large amounts of state (i.e., moving windows) need to be maintained, very often exceeding the available memory resources. Nonetheless, current event processing platforms have little or no memory management capabilities, hanging or simply crashing when memory is exhausted. In this paper we report our experience in using secondary storage for solving the performance problems of memory-constrained event processing applications. For that, we propose *SlideM*, a novel buffer management algorithm that exploits the access pattern of sliding windows in order to efficiently handle memory shortages. The proposed algorithm was implemented in a real stream processing engine and validated through an extensive experimental performance evaluation. Results corroborate the efficacy of the approach: the system was able to sustain very high input rates (up to 300,000 events per second) for very large windows (about 30GB) while consuming small amounts of main memory (few kilobytes).

Categories and Subject Descriptors

C.4 [**Performance of Systems**]: Design studies

Keywords

Disk, Event Stream Processing, Memory Management, Query Plan Sharing, Sliding Window.

1. INTRODUCTION

Event stream processing has recently passed from a pure academic subject to a well-established field in industry, with many of the original research projects [1][7][17] turning into fully-functional products (e.g., [9][18][23]). As the technology matures, these *stream processing engines* (SPEs) start to gain increased popularity among real-world users, finding application in the most diverse domains such as financial, telecom and sensor networks [12]. Many of these applications involve the computation of *aggregates over sliding windows*, which allow users to better measure the quality-level of their businesses and systems in real-time. For example, consider a call-center monitoring application where information about customers' calls is constantly analyzed by a stream processing engine. A typical query executing at the SPE in such scenario is:

Query 1: "*Report, every second, the average time during which customers are waiting for service, across the last hour.*"

```
SELECT AVG(waitTime)
FROM   calls [RANGE 1 HOUR SLIDE 1 SECOND]
```

The query above, expressed using the CQL language [2] syntax, specifies an AVG aggregation query over a sliding window with two parameters: RANGE, which defines the span of the window (*i.e., for how long tuples of the event stream "calls" are considered in query answer computation*); and SLIDE, which defines when tuples are expired out of the window and controls the frequency in which updated results are produced.

For a number of reasons (e.g., stringent latency requirements, transient data items, etc.), SPEs use main memory for processing their continuous queries. Unfortunately, many queries have an unbounded space cost, which cannot be determined a priori. For instance, the memory consumption of Query 1 typically depends on the event arrival rate, which might vary significantly during query execution. In those circumstances, one would expect SPEs to be prepared to deal with memory shortages. Interestingly, previous work [16] has shown that many commercial SPEs deal badly with this situation – some suffer from thrashing due to OS paging or excessive garbage collection activity while others simply crash with out-of-memory errors.

In this paper we report our experience in addressing this problem for analytic workloads composed by a large number of continuous aggregation queries. We introduce *SlideM*, a buffer management algorithm that selectively offloads sliding windows state to secondary storage when main memory becomes insufficient. Our approach is based on the observation that for most aggregation queries the memory consumption and access pattern is dictated by the sliding window, and that this access pattern can be exploited to achieve excellent performance while using small amounts of main memory.

This work was motivated by a series of real event processing applications we have been working with recently. Most of them

† Work partially carried out while at industrial partner FeedZai.

ICPE'13, April 21–24, 2013, Prague, Czech Republic.

involve the computation of several metrics (aggregates) over very large sliding windows, under stringent memory requirements. Throughout the rest of the paper, we use one of such use-cases, the call-center monitoring application introduced earlier and described in detail in Appendix A, to illustrate the problem. The requirements of this application include the computation of about 150 KPIs over 24-hour sliding windows, under an input rate of around 1,000 events per second, running on a machine with 2 gigabytes of main memory. Note that, given the moderate input rate, relative simplicity of the KPIs being computed (SUM, COUNT and AVG aggregates) and the large size of the window, the application tends to be memory-bound rather than CPU-bound. A simplistic calculation gives an idea on the dimension of the problem: considering that the average event size is 94 bytes, and that all tuples need to be maintained until they are expired out of the window, a single aggregation query will require at least: 1000 x 24 x 3600 x 94 bytes = 7.6 gigabytes (*we show in Section 2 that the commonly used technique of pre-aggregating data instead of storing tuples may require even more space*). Our goal is to allow such memory-intensive applications to run smoothly on an SPE whether they fit on available memory or not. In summary, we make the following contributions:

1. We analyze current proposals for executing sliding-window aggregates and show that frequently-used techniques, designed to reduce memory consumption and create opportunity for resource sharing, in many cases do not produce the desired effects (Section 2).
2. We propose an optimal buffer management algorithm to deal with memory shortages during execution of sliding-window aggregation queries. We demonstrate that, contrary to common sense, storing windows data on disk can be appropriate even for applications with very high event arrival rates (Section 3).
3. We build upon the proposed buffer management algorithm and develop a strategy to share computational resources when processing multiple aggregation queries over overlapping sliding windows (Section 4).
4. We implement our proposed techniques in a real SPE [19] and validate their effectiveness through an extensive experimental evaluation (Section 5).

2. SLIDING-WINDOW AGGREGATES

Continuous queries in SPEs are computed over infinite event streams rather than bounded datasets. However, many operations cannot be executed over infinite inputs in bounded memory, and some *blocking* operations require seeing the entire input before producing any result (e.g., join) [11]. Traditionally these two issues have been addressed in SPEs by limiting the amount of data over which operations take place through the use of *sliding windows*. A sliding window is a construct that retains only the most-recently arrived tuples of an event stream. The *size* of a sliding window determines the amount of data to be retained, and can be specified in number of tuples (*count-based* windows) or through an interval (*time-based* windows). Stale tuples are purged when window *slides*, due to arrival of new event or time passing.

A *sliding-window aggregate* (SWA) computes an aggregation function over a sliding window content and produces an updated result every time the window slides. For example, consider our motivating scenario where a stream of statistics continuously generates new information about call-center interactions with its customers. Query 1 defines a sliding window that retains the data items that arrived in the last hour, computes the average "waiting time" from this set of elements, and reports a new result every second. A common variation of this query structure is to have a *grouped aggregation*, where the input stream is logically partitioned into sub-streams, based on a grouping key, and one aggregate is produced for each partition. For instance, a GROUP-BY clause could be added to Query 1 in order to produce the average waiting time per customer region or per employee.

SWA is recognized as a fundamental operation of SPEs and has been extensively studied in previous work [3][4][12][14]. In the rest of this section we discuss how SWAs have been traditionally implemented. We present the two most frequently-used approaches, and compare how well they utilize memory resources in different workload scenarios.

2.1 SWA Implementation

Many important aggregates such as AVG, SUM, COUNT, MIN and MAX can be computed incrementally, in a single-pass over data items. This, in principle, allows an aggregation operator to discard events right after they have been processed. For instance, an AVG aggregate can be computed in O(1) space by simply keeping two variables – *sum* and *count* – and updating them upon event arrivals. However, when the aggregate is applied over a sliding window, tuples are eventually expired and this tuple removal has to be reflected into the query answer – in the AVG example, this means subtracting from the *sum* variable the value of the aggregated field in the expired tuples and decrementing the *count* variable by the number of expired tuples. Therefore, an SWA operator needs to maintain information about events that arrived previously so that the result can be updated properly when they eventually leave the window.

Traditionally, two approaches have been used to keep this information about past events in an SWA. The first, simply keeps all tuples in the window until it is time to expire them [2]. Normally, the sliding window is an operator by itself which forwards incoming and expired tuples to subsequent, aggregate operators (Σ), as depicted in Figure 1. The second approach, first introduced in [14], and adopted in subsequent proposals (e.g., [12]) sub-aggregates the incoming stream using smaller windows and then aggregates these sub-aggregates into a window of the original size in order to produce the final query result (*see Figure 2*). Taking Query 1 as example, SUM and COUNT sub-aggregates are computed over a 1-second window and then aggregated over a 1-hour window, thus producing the final result. The main advantage of this *two-level aggregation* (2LA) scheme over the former *one-window* (1W) approach is that the space cost of the query no longer

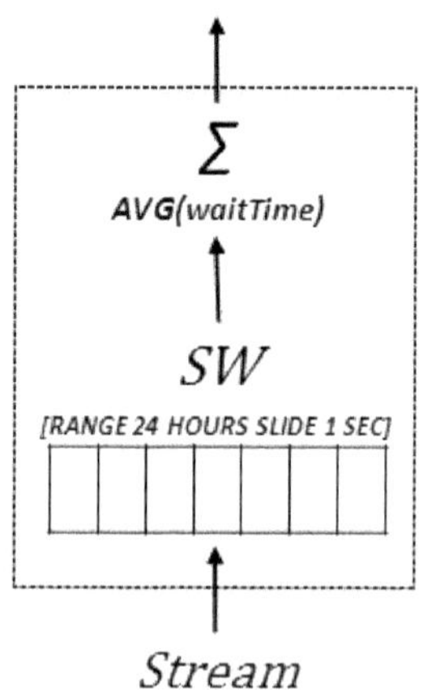

Figure 1: Plan for Query 1 using the single-window (*1W*) scheme.

depends on the input rate. However, as we are going to discuss next, it does not guarantee a bounded space cost, and for some workloads results in increased memory consumption.

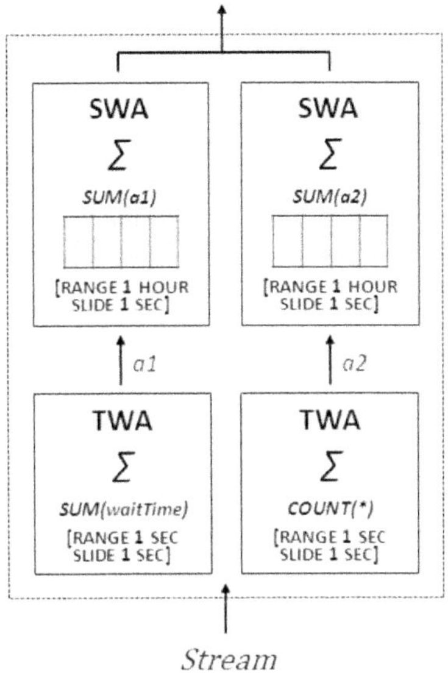

Figure 2: Execution plan for Query 1 using the *2LA* scheme

2.2 Space Cost Analysis

In this section we examine the space cost of the two widely-used SWA implementations. In particular, we demonstrate that either approach can incur in considerable memory costs, eventually bringing event processing applications to run out of memory. We also show that the *2LA* technique, originally designed to reduce space cost of SWAs, might in many cases aggravate the problem.

2.2.1 Two-Level Aggregation (2LA)

The *2LA* technique can be very useful for reducing space and computation cost of periodic sliding-window aggregates, particularly when input rates are high and/or the answer does not need to be updated often. However, there are a couple of issues that limit its effectiveness in many important scenarios. Probably the most relevant of them is that its space cost grows linearly with the number of aggregates being computed. This is particularly critical since most monitoring applications compute not one, but several aggregates – either because different aggregation functions are needed, distinct sets of attributes of the input streams are aggregated, or different grouping criteria are used. For instance, the users of the call-center monitoring application are interested not only in the *average waiting time*, but also in the *total* waiting time, the average *call time*, and the average waiting time *per region*. With the *2LA* scheme, each such aggregate results in a pair of operators, a *tumbling-window aggregate* (TWA) and a subsequent *sliding window aggregate* (SWA), as illustrated in Figure 2.

Another issue is that aggregations with a GROUP-BY clause implemented using the *2LA* scheme have their space and computation cost directly affected by the number of groups. This is because the TWA operator will produce as much aggregates as the number of distinct groups seen during the lifetime of its window. Each of these aggregates consumes space and computation in the subsequent SWA operator. Moreover, since the number of groups seen during the interval of the TWA window cannot be determined a priori, the *2LA* scheme does not guarantee a bounded space cost for grouped sliding-window aggregates. Taking into account the aforementioned factors, the total space cost when computing a set of aggregates using the *2LA* scheme is determined as follows[1].

Consider that N sliding-window aggregates, $\Sigma_1,\ldots,\Sigma_N$, are to be computed – *each representing a unique combination (Φ_i, F_i, P_i) of aggregation function (Φ), aggregated fields (F) and grouping criteria (P)* – over a common sliding window with size W and update interval U. The space cost of each aggregate Σ_i is given by the sum of the costs of its inner *tumbling-window* and *sliding-window* aggregates:

$$Space_{2LA} = \mathrm{N}\cdot(Space_{TWA} + Space_{SWA})$$

The TWA operator does not keep tuples in a window and only consumes the space required to maintain an aggregation state s' for each of the g groups seen during period U as shown below:

$$Space_{TWA} = \mathrm{g}\cdot \mathrm{s'}$$

The SWA operator, on the other hand, keeps both the tuples produced by TWA and a per-group aggregation state:

$$Space_{SWA} = \frac{g}{U}\cdot W\cdot \mathrm{t}_{agg} + G\cdot \mathrm{s}$$

In the formula above, g/U is the rate at which tuples arrive at SWA from TWA, t_{agg} is the size of the tuples produced by TWA, G is the total number of groups seen during W, and s is the size of the aggregation state per-group. The final space cost of the *2LA* scheme for A aggregates is then given by the formula below:

$$Space_{2LA} = N\cdot\left[g\cdot\left(\mathrm{s'} + \frac{W\cdot t_{agg}}{U}\right) + G\cdot \mathrm{s}\right] \qquad (2.1)$$

Note that depending on the function being computed, the aggregation state sizes s and s' can be constant or grow with the number of tuples in the corresponding window. As discussed elsewhere ([3] and [14]), *subtractable* aggregates like SUM, COUNT, AVG and VARIANCE can be computed with constant storage, but *distributive* (e.g., MIN and MAX) and *holistic* (e.g., QUANTILE) functions require O(N) space.

2.2.2 To Sub-Aggregate or not to Sub-Aggregate

We now examine the space cost of the *one-window* approach. As it can be seen from Figure 1, the memory consumption for the *1W* scheme corresponds to the sum of the space required to maintain the tuples in the main sliding window, and the state of each aggregate Σ_i. The former is obtained by the product of the window size W, the input rate λ, and the tuple size t. The latter is given by the product of the number of aggregates N, the total number of groups G seen during W, and the state size of each aggregate operator s. Or algebraically:

$$Space_{1W} = \lambda\cdot W\cdot t + N\cdot G\cdot s \qquad (2.2)$$

We can see from formulas 2.1 and 2.2 that each approach is more sensitive to a given factor than the other. With the *1W*

[1] For the sake of brevity, we limit our discussion to time-based windows, but similar analysis applies to count-based windows.

implementation, the space cost will be substantial for large windows if the input rate is high. On the other hand, the *2LA* scheme is immune to the input rate[2], independently on how large the window is, but can be severely penalized if the workload has many aggregates or groups.

As an example, we compare the memory consumption of the two different approaches using parameters taken from the call-center monitoring use-case. Let W=24 hours, U=10 seconds, λ=1000 events/sec, N=104 aggregates, G=10000 groups, g=1000 groups, s=s'=16 bytes[3], t=94 bytes, and t_{agg}=20bytes. The space cost of each scheme is in this case:

$$Space_{2LA} = 104 \cdot \left[1000 \cdot \left(16 + \frac{86400 \cdot 20}{10} \right) + 10000 \cdot 16 \right] \cong 17GB$$

$$Space_{1W} = 1000 \cdot 86400 \cdot 94 + 144 \cdot 10000 \cdot 16 \cong 8GB$$

As we can see, for this particular use case, performing a *two-level aggregation* in the end results in less efficient usage of memory resources than when computing the aggregates with a single window. More importantly, considering that in this application the available memory is limited to less than 2GB, neither of the two approaches allows the workload to run entirely at RAM. In this situation, it is necessary to selectively spill part of the queries state to disk so that the application does not run out of memory. Note that in either scheme the queries space cost is largely dominated by the state of the sliding window(s). For this reason, we address the problem of insufficient memory resources during computation of SWAs with an algorithm to manage the content of sliding windows.

3. *SlideM*: A BUFFER MANAGEMENT ALGORITHM FOR SLIDING WINDOWS

In this section we introduce *SlideM*, an algorithm for managing the working set of sliding windows. The proposed algorithm exploits the fact that sliding-window operators are most of the time manipulating only a small fraction of their data set and are doing so in a very predictable pattern – once a tuple is stored on the window it is not going to be accessed by the sliding-window operator until it is time to expire it, which may take long (e.g., consider a 6-hour time-based window).

SlideM is employed on a per-operator basis, that is, each window physical operator in the query plan is given a *repository* to hold its tuples. The actual location of tuples (either RAM or disk) is encapsulated by this repository, which internally implements a buffer management strategy based on *SlideM*. The repository includes a buffer pool (BP) for holding the memory-resident part of the window and a handle for accessing tuples at secondary media. Both the buffer pool and the data file at disk are divided in non-spanned blocks with a fixed block factor. These blocks are the unit of transfer between main memory and disks.

[2] Assuming that the number of groups *g* seen during period U is not affected by the input rate, which typically is not the case.

[3] Only *subtractable* aggregates are computed. We simplify discussion using a single value to represent the average state size of the different aggregation functions in the application.

The algorithm operates as illustrated in Figure 3: when the buffer pool gets full, it first sends to disk the block containing the most recently arrived tuples because these are the ones that are not going to be needed for the longest time. Similarly, when the oldest block at RAM is expired and hence the BP has space left once more, *SlideM* brings back from disk the least recently written (LRW) block, because it contains the tuples which are going to be needed next among the ones currently at disk. This behavior ensures that memory will always contains the tuples that are going to be needed by the sliding-window operator in the shortest time. The algorithm operation is described in details in Figure 4 and Procedures 1 and 2.

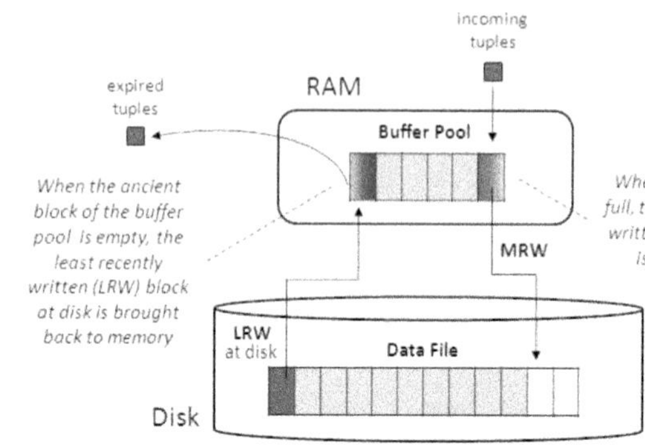

Figure 3: Overview of *SlideM* operation.

Procedure 1 add(Tuple t)

let *recent_block*: block at RAM holding recently arrived tuples

```
if recent_block is full then
  if buffer pool is full then
    if there are free blocks at disk then
      diskPos ← address of least recently released block
    else
      diskPos ← end_of_file
    end if
    WRITETODISK(diskPos, recent_block);
  end if
  recent_block ← ALLOCATENEWBLOCK();
end if
recent_block.APPENDTUPLE(t);
```

Procedure 1 describes event arrivals: every time a new tuple needs to be stored in the window, the algorithm checks whether the block at the tail of the window still has space left. If so, it stores the tuple normally at the end of the block; otherwise it allocates a new block at the buffer pool (*as shown in step 2 of Figure 4*). If the buffer pool is full (*3*), the algorithm first spills the most recently written (MRW) block to disk (*4*) to free space for the new block that will be soon allocated. The MRW block is written at a free position on disk (*9*) or at the end of the data file, if there are no free blocks (*5*).

Tuple expiration happens as in Procedure 2: when the repository receives a request to remove a tuple at the beginning of the window it checks whether the oldest block is now empty. If so, the block is removed from the buffer pool (*as shown in step 6 of*

Figure 4); if there is data at disk, the LRW disk block is brought to the buffer pool, at the position of the just-dismissed block (*7*). Then, the tuple is finally removed from the (newly arrived to memory) ancient block.

Procedure 2 expireOldest()

```
let ancient_block: block at RAM holding soon-to-expire tuples

if ancient_block is empty then
    buffer_Pool.REMOVE(ancient_block);
    if there is data at disk then
        // position at disk of least recently written block.
        lrw ← GETLRWDISKBLOCKADDRESS();
        lrwDiskBlock ← READFROMDISK(lrw);
        buffer_pool.ADD(lrwDiskBlock);
    end if
    ancient_block ← buffer_pool.GETLRWBLOCK();
end if
ancient_block.REMOVEOLDESTTUPLE();
```

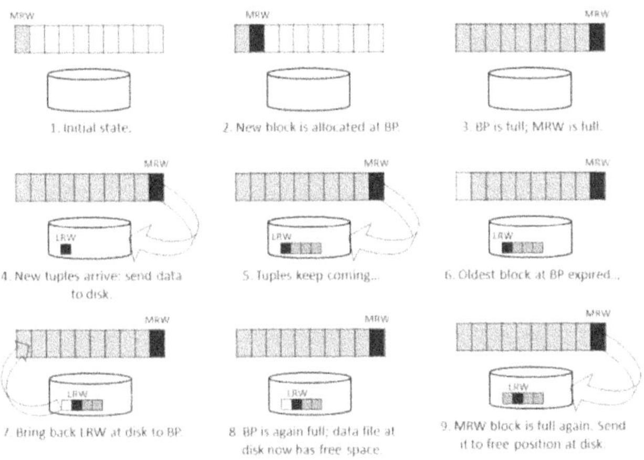

Figure 4: *SlideM* algorithm in several phases of its execution.

It should be clear that *SlideM* operation results in a very small memory consumption. In its strict sense, the algorithm needs only the equivalent to two blocks, one for holding the oldest part of the window (*ancient block*) and another for accommodating the newly arriving events (*recent block*). In fact, the performance of *SlideM* should not be affected by buffer pool size, unless BP becomes large enough to hold the entire working set of the window – if the window does not fit at RAM the algorithm will necessarily be swapping data to/from disk. Additionally, *SlideM* is optimal in terms of the amount of generated I/O as it always evicts to disk the block that is not going to be referenced for the longest time (*the optimality of clairvoyant page replacement policies has been first established at [5]; a short proof can be found at [20]*).

3.1 Discussion: I/O Load

We now examine the I/O demand of the *SlideM* buffer management algorithm. As explained before, *SlideM* issues disk *read* requests every time blocks at RAM are expired and performs disk *write* operations whenever a new block needs to be created at buffer pool but there is no space left. Thus, the number of I/O operations requested per second (IOPS) by *SlideM* for a single sliding window operator is given by:

$$IOPS = Expired\_Blocks/sec + New\_Blocks/sec$$

Assuming that events are fixed-size and there is a balance between event arrival and expiration rates – which is always true for count-based sliding windows and is also frequently the case across a period [τ, τ+WINDOW_RANGE] of a time-based sliding window – the following property holds:

$$Expired\_Blocks/sec = New\_Blocks/sec$$

From which we derive:

$$IOPS = 2 \cdot New\_Blocks/sec$$

Now, the rate at which new blocks are produced is a function of the event arrival rate, λ, as follows:

$$New\_Blocks/sec = \lambda \,/\, block\_factor$$

Where *block_factor* represents the number of tuples stored inside a block. This relation gives us the final I/O demand:

$$IOPS = 2 \cdot \lambda \,/\, \lfloor block\_size \,/\, tuple\_size \rfloor \qquad (3.1)$$

$$IO_{bandwidth} = IOPS \cdot block\_size \qquad (3.2)$$

EXAMPLE: For an input rate of λ=1000 events/sec, 94-bytes tuples, as found in the call-center use-case, and a block size of 64KB, the IO demand of *SlideM* will be (*assuming the 1W scheme is used*):

$$IOPS = 2 \cdot 1000 \,/\, \lfloor 64 \cdot 1024 \,/\, 94 \rfloor = 2.9\ iops$$

$$IO_{bandwidth} = 2.9 \cdot 64 \,/\, 1024 = 0.18\ MB/sec$$

Note that these numbers are far less than the theoretical transfer rate of modern hard drives (e.g., up to 204 MB/sec [21]), or the maximum measured disk bandwidth achieved under workload conditions similar to the modeled application (around 25MB/sec). Therefore, *SlideM* is capable of handling much larger input rates than the ones mentioned so far or to process a much larger number of simultaneous sliding window operators before the I/O subsystem starts to become a bottleneck. Nevertheless, the scalability of the algorithm can still be greatly improved by sharing the content of overlapping windows as we discuss next.

4. SHARING STATE OF OVERLAPPING SLIDING WINDOWS

The previous section introduced *SlideM*, an efficient algorithm to manage the state of a *single* sliding window operator. Now we extend the discussion to a multi-query scenario, where *multiple overlapping* sliding windows are defined over a common event stream. The problem is of foremost importance as large-scale monitoring applications usually process several aggregation queries over different time granularities – e.g., average price of a stock in the last hour, 12 hours, last day and so on. In a naïve approach, each of these overlapping windows would be mapped into an operator inside the query execution plan. Obviously, this limits system scalability and performance since having one operator per window implies that tuples (*or pointers to tuples*) are stored multiple times at different places, thus wasting memory space. Using the algorithm of the last section only address partially this issue as the bottleneck is eventually moved from the memory system to the I/O subsystem. We then build upon the *SlideM* algorithm and propose a shared execution scheme we call *Shared SlideM* (*SSM*) to improve the usage of computational resources when processing multiple overlapping sliding windows.

4.1 Shared *SlideM* (SSM)

We consider the problem of processing a set of N aggregation queries over N sliding windows of different sizes, defined over a common event stream S. For example, assume that three SWA queries are defined over a stream "*calls*" as follows:

```
Q1:  SELECT AVG(waitTime)
     FROM calls [RANGE 1 HOUR]

Q2:  SELECT AVG(waitTime)
     FROM calls [RANGE 6 HOURS]

Q3:  SELECT AVG(waitTime)
     FROM calls [RANGE 12 HOURS]
```

A direct translation of this set of queries would result in an execution plan like the one shown in Figure 5, with aggregation (Σ) and sliding window (ω) operators being replicated for every query in the set. This naïve approach simplifies query plan generation, but wastes memory during query execution, since the tuples stored in the smaller windows are also, by definition, present in the larger windows, and could be instead maintained in a single, shared, location. This is illustrated in Figure 6: the tuples that arrived in the interval [τ_{now} - *1hour*, τ_{now}] are part of all three windows; similarly, tuples belonging to the interval [τ_{now} - *6hours*, τ_{now} - *1hour*] are shared by both the 6-hour and 12-hour windows. In this case, having one operator per window implies that roughly half of the tuples in the query set are stored more than once.

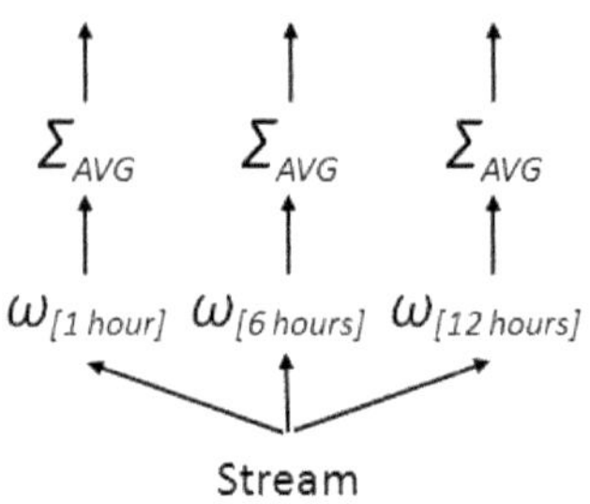

Figure 5: Unshared execution plan for three SWA queries

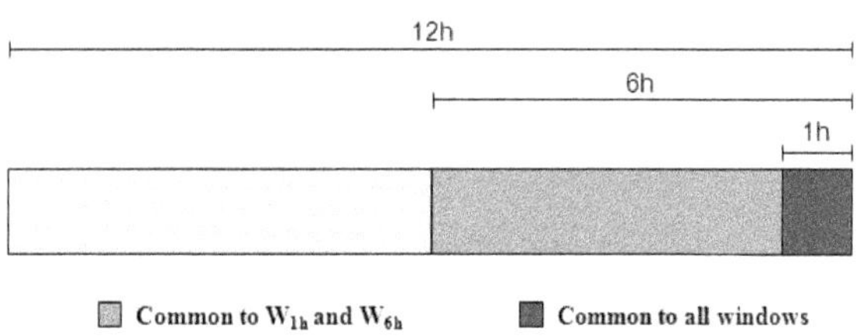

Figure 6: Three overlapping sliding windows

Using the *SlideM* algorithm reduces the pressure over main memory since portions of the windows can be offloaded to disk, but does not solve the problem of unnecessary data redundancy. Moreover, assuming that the windows do not fit in main memory, each query will produce a pair of IO operations from time to time (*read* for the ancient part of the windows and *write* for the recent segment). Eventually, as the number of queries increases, the I/O subsystem will become saturated.

To overcome these issues, we propose *SSM*, an adaptation of the *SlideM* algorithm in which multiple overlapping sliding windows are processed in a shared way. *SSM* works much like *SlideM*, in the sense that it sends parts of the window to disk when main memory is insufficient and brings data back from disk when it is time to expire them. However, contrary to *SlideM*, *SSM* manages a tuple repository that serves multiple *logical* window operators. We use the term 'logical' here because the several windows are in fact implemented by a single operator (Ω) as illustrated in Figure 7. This allows sharing computation of tuple arrivals as we explain next.

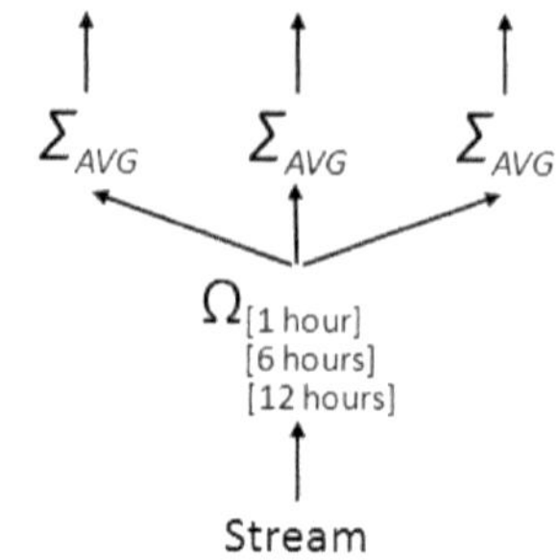

Figure 7: Shared execution plan for three SWAs, with *SSM*

A *SSM* tuple repository shared by multiple overlapping windows looks like the structure shown in Figure 8. The recent block (MRW), which stores the newly-arriving tuples, is common to all windows, but each window has its own ancient block (LRW), containing the tuples which are about to expire.

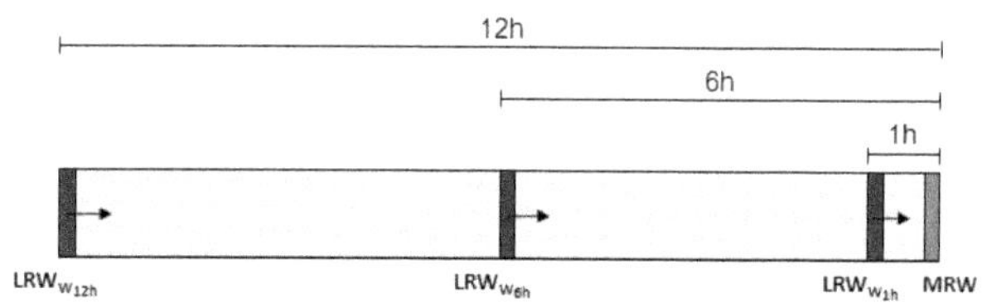

Figure 8: Shared tuple repository serving multiple windows

As the recent block is shared by all windows in the set, incoming tuples are processed only once by the shared operator Ω. Tuple arrivals in *SSM* occur essentially in the same way as in *SlideM* (see Procedure 1), with the exception that the request for adding tuples in the repository is now shared by multiple windows and the block sent to disk when the buffer pool is full is not necessarily the recent block (line 9 in Procedure 1) – a victim block must be selected instead. The major differences are, though, on the way tuple expirations are handled: first, tuples are not purged out of the repository unless the window which requested expiration is the largest one in the set. Intuitively, a tuple can only be discarded when it no longer belongs to any window, which happens when the largest window in the set requests its expiration – the same holds in a, coarser, block-level granularity. Another difference is that *SSM* does not prefetch data from disk when a block at RAM gets empty as *SlideM* does. This is because the LRW block at disk is not necessarily the block which is going to be needed next by the set of windows – with multiple windows the disk access pattern is no longer strictly sequential. Since determining which block will be required next is a potentially expensive operation, *SSM* skips prefetching and only brings data from disk when a request to a non-memory-resident block is issued. As a consequence, when the ancient block of a window gets empty it is no longer guaranteed that its "new" ancient block will be already at RAM, and as such it might be necessary to bring data from disk. Additionally, if the buffer pool is full, it will be also necessary to send a block to disk in order to open room for the upcoming block. The expiration process under the *SSM* scheme is described in detail in Procedure 3 (note that the procedure has a parameter to indicate which window the request comes from).

Procedure 3 expireOldestShared(*window_rank*)

let *ancient_block*: block at RAM holding soon-to-expire tuples of the window passed as argument
let *valid_index*: index of the oldest, non-expired tuple in the ancient block of the window passed as argument

```
1:  ancient_block ← GETANCIENTBLOCK(window_rank);
2:  valid_index ← GETVALIDINDEX(window_rank);
3:  if ancient_block has only expired tuples then
4:    if window_rank is the largest then
5:      buffer_pool.REMOVE(ancient_block);
6:    end if
7:    new_AB ← GETNEXT(window_rank, ancient_block);
8:    if new_AB is at buffer pool then
9:      new_AB_Addr ← GETBPBLOCKADDRESS(new_AB);
10:     ancient_block ← buffer_pool.GET(new_AB_Addr);
11:   else
12:     new_AB_Addr ← GETDISKBLOCKADDRESS(new_AB);
13:     ancient_block ← READFROMDISK(new_AB_Addr);
14:     if buffer pool is full then
15:       victim_block ← GETVICTIMBLOCK();
16:       buffer_pool.SWAP(victim_block, ancient_block);
17:       WRITETODISK(new_AB_Addr, victim_block);
18:     else
19:       buffer_pool.ADD(ancient_block);
20:     end if
21:   end if
22:   SETANCIENTBLOCK(window_rank, ancient_block);
23:   valid_index ← 0;
24: end if
25: valid_index ← valid_index + 1;
26: SETVALIDINDEX(window_rank, valid_index);
```

4.2 Discussion: I/O Load and Eviction Policy

The major advantage of *SSM* lies in a better use of memory space by avoiding that tuples are stored multiple times in the several window operators. This guarantees that no matter how many windows are defined over a given stream, the space cost will never exceed the size of the largest window in the set. As a consequence, *SSM* can handle a much greater number of queries than an unshared approach with the same amount of available memory before having to resort to secondary storage.

Now another important aspect is once the memory has been exhausted and access to disk is required, how much load *SSM* puts into the I/O subsystem. In order to determine that, consider that there are N windows of different sizes: $w_1 < w_2 < \ldots < w_N$. Let ρ be the likelihood of the next oldest block of a window being already at RAM after the current ancient block gets empty (see line 8 in Procedure 3)[4]. Assuming the buffer pool is full, the I/O pattern will be as follows: i) a write request will be issued every time a new recent block is created and ii) expiration of the ancient block of window w_i will incur, with likelihood $(1-\rho)$: one read request, and, if $i < N$, one additional write request (for $i = N$, the ancient block is effectively removed from the buffer pool, and as such, there is no need to send data to disk to open room for the new ancient block). Algebraically:

$$\#IO = W + (1-\rho) \cdot [(N-1) \cdot (R+W) + R] \qquad (4.1)$$

Note that in the limit, for $\rho=0$, the amount of I/O generated when processing the query set using *SSM* will be exactly the same as in *SlideM* (one pair of read and write request per window):

$$\#IO = W + (N-1) \cdot (R+W) + R = N \cdot (R+W)$$

This means that the shared execution mechanism will never perform more I/O than the unshared approach, and in the worst case the I/O pressure of the two schemes will be equivalent. For any $\rho>0$, *SSM* will reduce the amount of I/O, and the reduction will be as large as ρ. As discussed in previous section, the optimal eviction policy that maximizes ρ is the one that sends to disk the block that is not going to be needed for the longest time. For the single-window case, the choice is straightforward: the optimal victim block is always the most recently written block. This does not hold for multiple windows though, as the MRW block might be needed earlier by a small window than an intermediate block by larger windows. Instead, the optimal victim block in a multi-window scenario can be determined by computing the distance – in number of blocks or time units – of the candidate blocks at the buffer pool to the ancient block of each window as follows: let d_{ki} be the distance of block k at BP to the ancient block of window w_i. For any block k, ref_k denotes the next time the block will be referenced by any window, and corresponds to the minimum value in the set of distances: $ref_k = \min\{d_{ki} \mid d_{ki} > 0\}$. The victim block v is the one with the maximum value for ref_k among the B candidates at buffer pool: $v = (k \mid ref_k = \max\{ref_1, \ldots, ref_B\})$.

This *distance-based* replacement policy creates clusters of blocks in the BP, immediately after the ancient block of each window as illustrated in Figure 9 below:

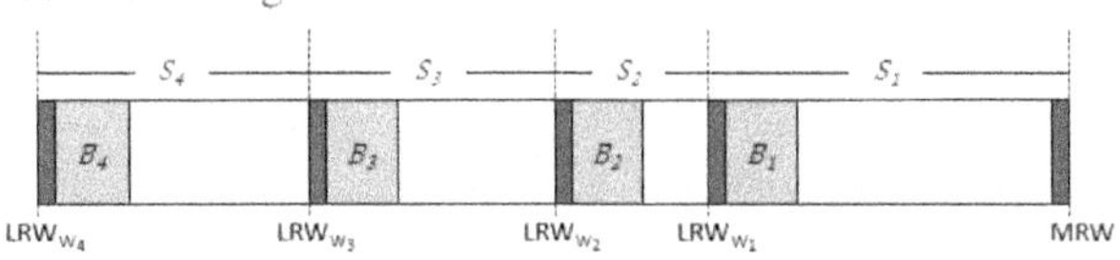

Figure 9: Typical arrangement of blocks at the buffer pool when using *SSM* block replacement policy.

Since each block is referenced only once by each window, the buffer pool hit rate of the scheme is given by the average percentage of blocks residing at memory of each segment:

$$\rho_{SSM} = \frac{1}{N}\sum_{i=1}^{N}(B_i / S_i) \qquad (4.2)$$

where B_i is the number of memory-resident blocks of each segment and S_i is the corresponding total number of blocks.

Clearly, the more memory is available (larger B_i) and the more overlapping the windows are (smaller S_i), the higher the buffer pool hit rate ρ_{SSM} will be.

[4] In fact, ρ represents the hit rate of the buffer pool.

5. PERFORMANCE STUDY

In this section we present an extensive experimental evaluation of the *SlideM* algorithm and the sharing scheme *SSM*. We have performed a wide variety of experiments with the objective of assessing:

1. *Effectiveness of SlideM in a real-world use-case*: we demonstrate the ability of the proposed algorithm in addressing memory shortages in a real scenario and compare its performance against the conventional memory-only implementations discussed earlier in this paper (Section 5.2).
2. *High-performance nature of SlideM*: we examine *SlideM* performance under heavy load conditions. Results reveal that the algorithm was capable of handling very high input rates for multi-gigabyte windows while keeping latency under desirable levels (Section 5.3).
3. *Performance and scalability of SSM*: we show that the sharing mechanism *SSM* scales significantly better than an unshared approach (Section 5.4).

For the first set of experiments, we used queries and stream definitions taken from the real use-case, as described in Appendix A. For the other two sets, we used synthetic queries and datasets. Tests setup and methodology are described next.

5.1 Experimental Setup and Methodology

We implemented our proposed techniques in Pulse [19], a Java-based stream processing engine from industrial partner FeedZai. Experiments were conducted on a server with two Intel Xeon E5420 2.50 GHz Quad-Core processors, 4 GB of RAM, and 4 SATA-300 disks distributed in two RAID-0 arrays, running Windows Server 2008 x64 and Hotspot x64 JVM (*configured with a 1 GB heap size*). One RAID array was used to host the OS while the other was used to hold window data during tests. Measurements were taken as follows:

- A single Java application was responsible for generating, submitting and consuming tuples during the performance runs. Input data was submitted to the SPE through local method calls using its API.
- Tests consisted in a *warmup phase*, during which the SPE was brought to a steady state, and a subsequent *measurement interval* (MI), when the performance of the system was measured. The duration of both warmup and MI was set to the time necessary for traversing 1.5 times the window – *e.g., an experiment with a 6-hour window ran for 18 hours (9h of warmup plus 9h for MI)*.
- We collected both application-level and system-level metrics. Average throughput was computed as the ratio between processed tuple count and elapsed time. Latency was computed through the `nanoTime()` method of the Java runtime, called immediately before and after sending tuples to the SPE. Memory consumption was computed by the end of tests using standard Java SDK methods. CPU, disk, and process metrics were collected using the *System Monitor* tool of MS-Windows.

All experiments with *SlideM* and *SSM* used a fixed block size (64 kilobytes).

5.2 Call-Center Use-Case Results

Our first set of experiments mimics the workload conditions of a real event processing application, and consists in processing a number of aggregations like Query 2 below:

Query 2: *Compute call-center statistics in the last day*

```
SELECT    COUNT(*),
          SUM(busyTime),
          AVG(busyTime)
FROM      calls [RANGE 24 HOURS SLIDE 10 SECONDS]
GROUP BY  serviceId
```

On total, the application computes 144 aggregates, resulting in 104 distinct SWA operators – *the query above is replicated for 6 different fields and 8 distinct grouping criteria, with the* COUNT *aggregate being shared by the queries with different fields; a more detailed description of the use-case can be found in Appendix A*. We then compare the performance of memory-only SWA implementations against application performance when paging sliding window content to disk through the *SlideM* algorithm. Both the *1W* and the *2LA* SWA approaches were tested in each case. Results are presented in Figure 10.

The three uppermost graphs show the performance of the two non-managed implementations, and illustrate what typically occurs with memory-constrained event processing applications in most Java-based SPEs: as the application working set approaches the available memory threshold, the system spends progressively more time with garbage collection, increasing the tuple processing latency and preventing the SPE to cope with the data input rate. Since there is no data to be purged until the sliding window closes, the application eventually crashes with an out-of-memory error. In our tests this happened before 3 hours for the *1W* implementation and before 1 hour when using the *2LA* scheme, as signalized in Figure 10.

The results above contrast with application behavior when the *SlideM* algorithm is employed to manage the state of the sliding windows, as illustrated in the bottom part of Figure 10. Using our proposed algorithm allowed the experiments to complete without errors, while keeping performance metrics in desirable levels. As expected, memory consumption and tuple processing latency was larger when using the *2LA* scheme than with the *1W* approach in this use-case.

5.3 Performance of *SlideM*

Many event processing applications, like those found in the financial trading environment, require that SPEs be able to process a considerable volume of data within very short periods of time. Using disks in these cases might be inadequate if they are not able to cope with the very high arrival rates and stringent latency requirements. In this section we examine how *SlideM* performs in such critical scenarios. For that, we employ a simple microbenchmark, which consists in computing one or more aggregations over a stock market data stream. Each input tuple has 4 attributes: *Timestamp*, *Symbol*, *Price* and *Volume* (about 28 bytes). We fill tuples by repeatedly cycling through a list of 100 stock symbols and assigning the tuple creation time to the timestamp field and random values to the other two. The workload consists in computing the volume-weighted average price (VWAP) of each stock over the last hour, as shown in Query 3. Six runs of this experiment are performed,

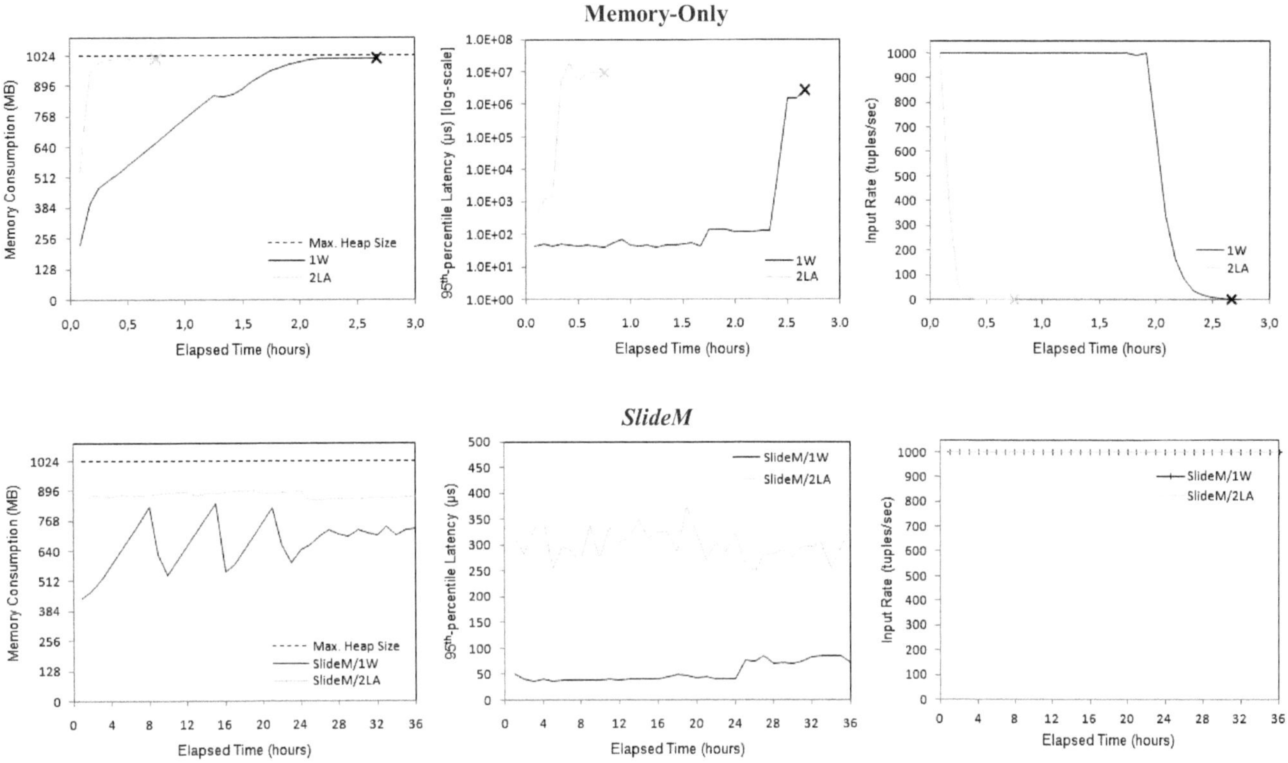

Figure 10: Comparison: performance of *SlideM* vs. memory-only implementations in real-world-based workload conditions

progressively scaling the injection rate from 50,000 up to 300,000 tuples per second.

Query 3: *Compute the VWAP of each stock over the last hour*

```
SELECT   Symbol, SUM(Volume*Price)/SUM(Volume)
FROM     Stock [RANGE 1 HOUR]
GROUP BY Symbol
```

Note that the window definition in the query above does not include a SLIDE, which means that the result must be updated whenever a new tuple arrives at the *Stock* stream[5]. In all experiments, the buffer size was set to the minimum, 2 blocks (128KB), so that system performance is measured under maximum I/O pressure. Results are shown in Table 1.

As we can see, the system was able to handle up to 300,000 tuples per second, with the CPU being the limiting factor at that point. Average processing latency was fairly low in all experiments (a few microseconds), and even the absolute maximum latency remained under acceptable levels as the load was increased. Disk utilization was also quite low in all runs, as it can be seen from the average disk queue length (ADQL) metric in Table 1. The reason is that the disk bandwidth required by *SlideM* at the maximum load of 300,000 tuples per second in this benchmark is only 16 MB/sec, which is still far from the maximum measured disk transfer rate, as discussed in Section 3.1. This moderate load posed by *SlideM* into the I/O subsystem was crucial to remove a bottleneck (memory) without creating a new one, thus allowing the SPE to fully exploit the available CPU power.

Table 1: *SlideM* Performance, scaling injection rate

Injection Rate (tuples/sec)	Space Cost (GB)	Avg. Latency (ms)	Max. Latency (ms)	% CPU	ADQL
50,000	4.7	0.009	2.9	5%	0.016
100,000	9.4	0.012	3.6	13%	0.035
150,000	14.1	0.007	3.9	23%	0.056
200,000	18.7	0.012	16.7	37%	0.071
250,000	23.5	0.008	14.4	69%	0.103
300,000	28.2	0.008	18.9	100%	0.145

5.4 Performance of *SSM*

We now examine the performance of the shared execution scheme *SSM* and quantify to which extent it scales better than an unshared approach. Experiments here consist in processing N instances of Query 3, using either the unshared *SlideM* algorithm or its shared counterpart, *SSM*, under an input rate of 5,000 tuples per second. The number of queries in each experiment, N, took the following values: N = {2, 4, 8, 16, 32}. All N queries in the set have different window sizes, uniformly distributed in the interval [3600, 7200] seconds. Available memory was set to

[5] An aggregation query without a SLIDE clause would probably make little sense if its result were to be output (i.e., used for monitoring purposes). In many cases, however, the result of an aggregation is used as input for further processing (e.g., pattern detection), and updated results must be produced as soon as new data is available.

512MB in all experiments (for the non-shared version, this amount was equally divided among the N buffer pools). We then measured for each algorithm the total space cost and the amount of pressure put onto the I/O subsystem. Results are depicted in Figure 11:

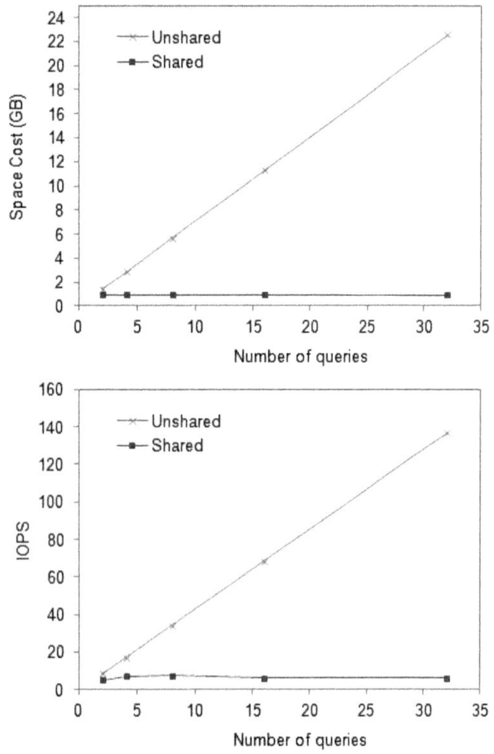

Figure 11: Performance of *SlideM* (unshared) and *SSM* (shared) in a multi-query scenario.

As expected, the total space cost (*memory and disk*) of the unshared approach grows linearly with the number of windows while with the shared strategy the space cost remains constant (*it is bounded to the size of the largest window*). *SSM* was also significantly more I/O-efficient, issuing up to 22 times less disk requests than the unshared implementation. The explanation for this remarkable difference in the number of I/O requests between the two algorithms is that *SSM* benefits from the fact that adding more queries to the set reduces the distance between the overlapping windows, thus increasing the hit rate of the buffer pool as expressed in formula 4.2. This way, while the I/O pressure of the unshared implementation consistently grows as more windows are used, with *SSM* it tends to stabilize since the increase on buffer hit rate compensates for the increased number of simultaneous queries. The result is a much better scalability as we can see in Figure 11.

6. RELATED WORK

There has been considerable work on resource management in stream processing systems [4] [11] [17]. For dealing with memory shortages, two approaches have been widely employed. The first consists in providing approximate answers by shedding load [8] [22] [24], with research on this area focusing essentially on minimizing error of approximations. However, many event processing applications rely on exact answers to perform complex data analysis and support real-time decision making. In these cases, techniques such as load shedding or approximation are not applicable. The alternative, then, is to use secondary storage as an extension of main memory. Indeed, such disk-based approach has been adopted in a number of proposals such as [6], [10] and [15]. The focus of those works, though, is on processing of *join queries*. Liu et al [15] consider queries with multiple operators and propose strategies to choose which part of the operator states to spill during query execution in order to maximize the overall throughput. As discussed before, this is not an issue for sliding-window aggregates since the data access pattern can be accurately predicted. Farag and Hamad [10] propose a two-phase external-memory algorithm that joins the arriving tuples of one stream with the memory-resident data of the other streams, and postpones matching with the disk-resident portion until the stream runs out-of-space or arrival of new tuples stalls. Chakraborty and Singh [6] propose an *Exact Window Join* algorithm that deals with memory shortages by deferring the load during high workload, and processing the deferred load during the period of low workload. This strategy, however, results in high delays (> 5 seconds) even for moderate data input rates (450 tuples per second). To the best of our knowledge, our work is the first to address the problem of exact answer computation of aggregations in memory-limited, high-throughput, environments.

Shared processing of sliding-window aggregates has been previously explored in a couple of proposals. Arasu and Widom [3] devise two algorithms for sharing execution of multiple sliding-window aggregates, where a common aggregation function is computed over different window sizes. These algorithms assume an *aperiodic* scenario, where results are produced on-demand (when user polls a query). Our proposed strategy, on the other hand, is for periodic aggregates and applies even when different aggregation functions are used. In [12] Krishnamurthy proposes a strategy for sharing the execution of multiple periodic sliding-window aggregates implemented under the *2LA* scheme. The strategy focuses on *computation sharing*, and consists in computing the partial aggregates with only one shared operator, rather than using one operator per query. It does not address, however, the *space sharing* problem introduced in this paper, as the partial aggregates are still stored several times at the main window.

7. CONCLUSIONS

In this work we introduce techniques for overcoming the traditional memory limitations faced by stream processing engines when processing aggregation queries over sliding windows. We address the problem by proposing a novel buffer management algorithm, *SlideM*, which offloads sliding window state to disk during memory shortages. In order to further increase algorithm scalability, we also proposed *SSM*, a query sharing strategy that prevents explosion of space cost by storing the state of multiple overlapping sliding windows in a single, shared, repository. Experimental results demonstrated that the two techniques together provide significant performance and scalability benefits. With *SlideM* the system was able to handle up to 300,000 events per second for multi-gigabyte windows while consuming only 128 kilobytes of main memory. In a scenario with multiple simultaneous queries, *SSM* reduced space cost by a factor of up to 24, issuing up to 22 less disk requests.

There are a couple of issues we did not explore in this paper and constitute interesting avenues for future work. First, the techniques proposed here are designed to exploit the access pattern of sliding window operators during event arrivals and expirations, which allows excellent performance for SWA queries. Our implementation currently supports random access to tuples inside the window, but probably in a way far from optimal for operations such as joins. Therefore, we intend to investigate state-spilling mechanisms that work well for a more diversified gamma of queries. A promising direction is to combine per-operator *SlideM* repositories with a global buffer manager responsible for serving queries with less predictable access patterns. Another possible direction for future work is to extend the proposed algorithms and develop new disk-based techniques in order to enable SPEs to recover their state in the advent of system failures. Finally, the results presented in this work were obtained with conventional hard drives. It shall be interesting to observe the behavior of the proposed techniques in conjunction with faster storage technologies like solid-state disks (SSDs) or phase-change memories (PCMs).

Acknowledgements

We would like to thank the development team at FeedZai for all the feedback and valuable discussions during the implementation of the query execution engine of the Pulse SPE. This research has been supported in part by the Portuguese Science and Technology Foundation (FCT), under grant Nº 45121/2008, and by the industrial partner FeedZai.

8. REFERENCES

[1] Abadi, D.J., Carney, D., Çetintemel, U., Cherniack, M., Convey, C., Lee, S., Stonebraker, M., Tatbul, N., and Zdonik, S.B. Aurora: a new model and architecture for data stream management. In *Proceedings of VLDB Journal*. 2003, 120-139.

[2] Arasu, A., Babu, S., and Widom, J. The CQL continuous query language: semantic foundations and query execution. In *Proceedings of VLDB Journal* Vol. 15 Issue 2, 2006, 121-142.

[3] Arasu, A. and Widom, J.: Resource Sharing in Continuous Sliding-Window Aggregates. In *Proc. of the 30th VLDB Conference* (Toronto, Canada, September 2004), 336-347.

[4] Babcock, B., Babu, S., Datar, M., Motwani, R., and Widom, J.: Models and Issues in Data Stream Systems. In *Proceedings of the 21st Symposium on Principles of Database Systems*, pages 1–16, June 2002.

[5] Belady, L.A.: A Study of Replacement Algorithms for Virtual-Storage Computer. In *Proceedings of IBM Systems Journal*. 1966, 78-101.

[6] Chakraborty, A., and Singh, A.: Processing Exact Results for Sliding Window Joins over Time-Sequence, Streaming Data Using a Disk Archive. In *Proceedings of the 1st Asian Conference on Intelligent Information and Database Systems* (Vietnam 2009), 196-201.

[7] Chandrasekaran, S., et al. TelegraphCQ: Continuous Dataflow Processing for an Uncertain World. In *Proceedings of CIDR 2003* (Asilomar, California, USA).

[8] Dobra, A., Garofalakis, M., Gehrke, J., Rastogi, R.: Processing Complex Aggregate Queries over Data Streams. In *Proceedings of the 2002 ACM SIGMOD*, Madison, Wisconsin, USA.

[9] Esper: http://esper.codehaus.org/

[10] Farag, F., and Hamad M.A: Adaptive Execution of Stream Window Joins in a Limited Memory Environment. In *Proc. of the 11th International Database Engineering and Applications Symposium* (Banff, Canada, 2007), 12-20.

[11] Golab, L., and Ozsu, M.: Issues in Data Stream Management. SIGMOD Record, Vol. 32, Issue 2, 5–14.

[12] Hinze, A., Sachs, K., Buchmann, A.P. Event-based applications and enabling technologies. In *Proceedings of DEBS* 2009 (New York, USA), Art. 1, 15 pages.

[13] Krishnamurthy, S. *Shared Query Processing in Data Streaming Systems*. Ph.D. Thesis, University of California, Berkeley, 2006.

[14] Li, J., Maier, D., Tufte, K., Papadimos, V., and Tucker, P.A. No Pane, no Gain: Efficient Evaluation of Sliding-Window Aggregates over Data Streams. SIGMOD Record (2005), Vol. 34, Issue 1, 39-44.

[15] Liu, B., Zhu, Y., and Rundensteiner, E.A.: Run-Time Operator State Spilling for Memory Intensive Long-Running Queries. In *Proceedings of the 2006 ACM SIGMOD*, (Chicago, Illinois, USA), 347-358.

[16] Mendes, M.R.N., Bizarro, P., and Marques, P.: A Performance Study of Event Processing Systems. In *Proceedings of TPCTC 2009* (Lyon, France), 221-236.

[17] Motwani, R., and et al.: Query Processing, Approximation, and Resource Management in a Data Stream Management System. In *Proceedings of CIDR 2003* (Asilomar, California, USA).

[18] Oracle CEP: http://docs.oracle.com/cd/E13157_01/wlevs/docs30/

[19] Pulse: http://www.feedzai.com/products/pulse

[20] Roy, B.V. A Short Proof of Optimality for the MIN Cache Replacement Algorithm. In *Proceedings of Information Processing Letters*, 2007, 72-73.

[21] Seagate Cheetah hard disk Data Sheet: http://www.seagate.com/files/docs/pdf/datasheet/disc/cheetah-15k.7-ds1677.3-1007us.pdf

[22] Srivastava, U., and Widom, J.: Memory-Limited Execution of Windowed Stream Joins. In *Proc. of the 30th VLDB Conference* (Toronto, Canada, September 2004), 324-335.

[23] StreamBase: http://www.streambase.com/

[24] Tatbul, N., Etintemel, U., Zdonik, S. B., Cherniack, M., and Stonebraker M.: Load shedding in a Data Stream Manager. In *Proceedings of the 29th VLDB Conference* (Berlin, Germany, September 2003), 309-320.

APPENDIX

A. USE-CASE DETAILS

In this section we present some of the details of the call-center monitoring application that we referred to throughout this paper. The description provided here is for a proof-of-concept prototype, which corresponds to a simplified version of the full application, still under development.

As mentioned before, the purpose of the application is to provide a real-time view of the operation of a large call center chain. The company is spread over 20 geographical sites and has around 12,000 agents serving more than 3 million customer requests per day. A statistical module collects information about the calls and produces a stream of data items describing each step of the interactions between the call center and its customers. This data stream, whose schema is shown in Figure 12, is then fed into the stream processing engine, where 144 KPIS are continuously computed. Each KPI corresponds to a 24-hour sliding-window aggregate over a given field of the stream. Specifically, three functions (SUM, COUNT and AVG) are applied over 6 attributes of the stream (`alertingTime`, `busyTime`, `wrapUpTime`, `waitTime`, `helpTime`, and `availableTime`), using 8 different grouping keys (`instance`, `serviceId`, `agentId`, `mediaId`, `interactionLegId`, `agentSite`, `callSite`, and `direction`).

In our experimental evaluation we filled the tuples of the stream *AgentInteractions* with synthetic data since real datasets were not available due to confidentiality issues. The generated data, however, respected the critical properties of the original input stream, such as the cardinality of the attributes used as grouping key in the queries and the distribution of these groups over time. We did not replicate eventual oscillations on tuple arrival rate though, keeping the injection rate fixed in 1,000 tuples per second. All the tests were performed in a virtual machine with 8 cores, 2GB of RAM, and running Window Server 2008, as found in the production environment.

```
AgentInteractions (
   timestamp                 long,
   instance                  int,
   start                     long,
   sessionId                 int,
   serviceId                 int,
   agentId                   int,
   interactionLegId          int,
   alertingTime              int,
   busyTime                  int,
   wrapUpTime                int,
   waitTime                  int,
   direction                 int,
   mediaId                   int,
   helpTime                  int,
   agentReleased             bool,
   mediaOutcome              int,
   finalSegment              bool,
   agentSite                 int,
   callSite                  int,
   availableTime             int,
   availableTimeByService    int,
   held                      int,
   help                      int
)
```

Figure 12: Schema of the input data stream in the call-center monitoring application

Systematic Performance Evaluation based on Tailored Benchmark Applications

Christian Weiss
SAP Research
Vincenz-Priessnitz-Strasse 1
76131 Karlsruhe, Germany
christian.florian.weiss@sap.com

Dennis Westermann
SAP Research
Vincenz-Priessnitz-Strasse 1
76131 Karlsruhe, Germany
dennis.westermann@sap.com

Christoph Heger
Karlsruhe Institute of Technology
Am Fasanengarten 5
76131 Karlsruhe, Germany
christoph.heger@kit.edu

Martin Moser
SAP AG
Dietmar-Hopp-Allee 16
69190 Walldorf, Germany
martin.moser@sap.com

ABSTRACT

Performance (i.e., response time, throughput, resource consumption) is a key quality metric of today's applications as it heavily affects customer satisfaction. SAP strives to identify and fix performance problems before customers face them. Therefore, performance engineering methods are applied in all stages of the software lifecycle. However, especially in the development phase continuous performance evaluations can introduce a lot of overhead for developers which hinders their broad application in practice. In order to evaluate the performance of a certain software artefact (e.g. comparing two design alternatives), a developer has to run measurements that are tailored to the software artefact under test. The use of standard benchmarks would create less overhead, but the information gain is often not sufficient to answer the specific questions of developers. In this industrial paper, we present an approach that enables exhaustive, tailored performance testing with minimal effort for developers. The approach allows to define benchmark applications through a domain-specific model and realizes the transformation of those models to benchmark applications via a generic Benchmark Framework. The application of the approach in the context of the SAP Netweaver Cloud development environment demonstrated that we can efficiently identify performance problems that would not have been detected by our existing performance test infrastructure.

ICPE'13, April 21–24, 2013, Prague, Czech Republic.

Categories and Subject Descriptors

C.4 [**Performance of Systems**]: [measurement techniques, design studies, performance attributes]; D.2.5 [**Software Engineering**]: Testing and debugging—*testing tools*; D.2.5 [**Database Management**]: Logical Designs—*data models*

General Terms

Java Persistence API (JPA), Benchmarking, Measurelet, Automated Performance Testing, Model-based Testing, Performance Regression Testing

Keywords

Performance, Benchmarking, Database, JPA, Feedback

1. INTRODUCTION

Identifying performance problems before customers face them is highly relevant for web-based software solutions such as websites or Software- and Platform-as-a-Service (SaaS/-PaaS) products. Recent industrial studies and field reports have shown that performance (response time, throughput, resource usage) often directly correlates with revenue [16]. Engineering methods (e.g., monitoring, testing, modelling) can help software providers to understand performance behaviour and identify performance problems early.

However, evaluating performance in a realistic environment involving large subsystems (e.g., databases, middleware, legacy) is still a challenging task. Especially in the development phase, continuous performance evaluations can introduce a lot of overhead for developers which hinders their broad application in practice.

In this industrial paper, we report on our efforts towards a systematic and lean approach for evaluating the performance of the persistence layer of a platform solution. The persistence layer of an application is often one of the most performance-critical components as it is quite likely that it becomes a bottleneck. In a PaaS solution, platform providers offer a ready-to-use persistence layer to developers that create applications on top of the platform. This implies the following challenges.

The platform service provider needs to support many different application types with different usage profiles. Standard benchmarks such as provided by SPEC [15] and TPC [18] do not test broad enough in order to enable a detailed understanding of performance characteristics. In order to simulate a comprehensive set of workloads at development time, it requires tailored performance tests that simulate many different usage patterns and specifically focus on the concrete system under test [7]. However, creating and running these detailed tests is very time-consuming and requires detailed expert knowledge.

Developers that consume the platform service are mainly interested in understanding how the performance of the platform service affects the performance of their application. The platform service provider could support the consumers by providing a prediction that maps the service usage to the expected performance. However, in order to provide such a prediction, the platform service provider needs an even more detailed understanding of how the service behaves under different usage profiles. In our scenario, classical model-driven prediction approaches (such as surveyed in [10]) fail due to the complexity of the service and its continuous modification which makes it hard to create and maintain the models. Hence, exhaustive, systematic measurements are the means that we choose to derive predictions. This underlines our requirement for an efficient approach to set up and run tailored performance tests.

In this paper, we present our approach for creating and running performance tests tailored to the persistence service of a platform solution. In our scenario, the service interface is the Java Persistence API (JPA) [3]. Based on a JPA data model provided by the platform consumer, an object-relational mapping (ORM) component persists the application data in the underlying database. We introduce a measurement API that separates benchmark applications from their runtime environment. Moreover, we developed a Benchmark Framework that generates all necessary project and source files of a benchmark application based on information provided in a benchmark model. Furthermore, the Benchmark Framework compiles the generated code, deploys it on the test environment, executes the measurements and returns the results.

The presented approach simplifies the creation and execution of performance tests for developers by hiding complexity and automating many tasks. This simplification makes it easier to introduce performance tests which results in a larger set of tests. Having more tailored tests increases the probability of uncovering a performance problem. Moreover, it allows us to quickly reproduce a performance issue on the test environment that we observed for a consumer application on the real environment. In the paper, we demonstrate how we use the approach for performance regression testing and performance prediction function construction. Moreover, we report on the first results that we achieved by applying the approach in the development of the SAP Netweaver Cloud platform [14]. Based on the initial experience with the approach, we discuss open issues and future research directions that we see in the performance engineering field.

The remainder of this paper is organized as follows. Section 2 gives an overview on the approach and introduces the key components. Sections 3 and 4 describe the measurement API and our approach to automate the construction, execution and analysis of benchmark applications, respectively. In Section 5, we demonstrate the application of the approach in our industrial setting. In Section 6, we discuss potential research directions based on our experience with the presented approach. Related research is introduced in Section 7. Finally, Section 8 concludes the paper.

2. OVERVIEW

The main goal of the work presented in this paper is to support developers as well as consumers of a persistence service in understanding the performance impact of design and implementation decisions. Figure 1 illustrates the basic scenario.

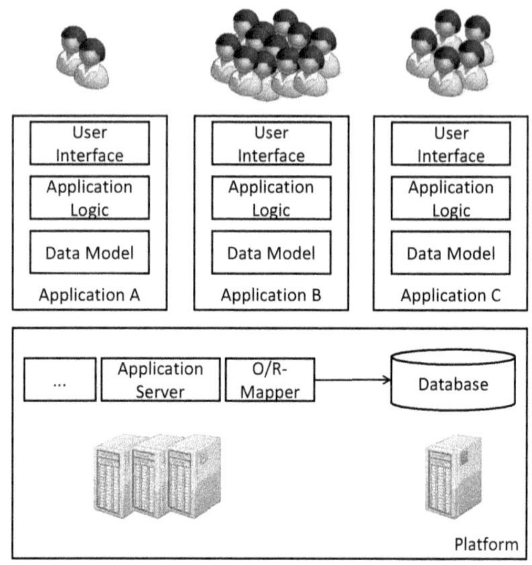

Figure 1: Platform Scenario

The platform hosts a set of different applications with different usage profiles. In addition to hosting the applications, the platform provides further services to the application developers. In this paper, we focus on the persistence service of the platform which allows application developers to store and load data based on a simple interface. Hence, application developers do not have to deal with setting up, configuring, or maintaining their own persistence solution. The results presented in this paper are derived for a persistence service that uses JPA [3] as the interface to the application developer. In that scenario, the main components of the persistence service are the database and the object-relational mapping component (such as Hibernate [6] or EclipseLink [23]). The application developer designs the entities that make up the data model of the application and bundles it together with the application logic and the user interface to a deployable unit that runs on the the platform.

When it comes to quality assurance in terms of performance, the persistence service provider as well as the application developers need to answer different questions. Examples from a persistence provider view are:

- Does my planned database update affect performance?
- Which test cases reflect the actual usage profile best?
- Should I use O/R-Mapper A or B?

Application developers face questions like:

- Is variant A or variant B of my data model better with respect to performance?
- What is the performance of my application on platform X?
- Did my latest change affect the performance of my application?

The approach presented in this paper provides a means to support persistence service providers and application developers in answering these questions efficiently. In Section 5, we provide some examples on how we used the approach for performance regression testing and providing performance feedback. Figure 2 illustrates the basic idea of the approach based on a simple example.

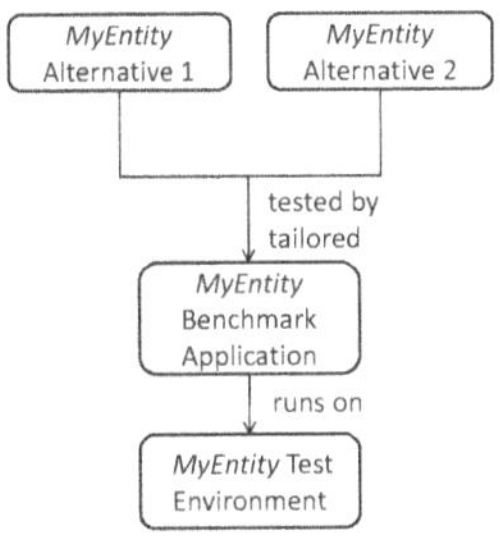

Figure 2: Basic Idea

If the developer of a data model wants to make a decision between two design alternatives for the entity *MyEntity*, a tailored benchmark application is constructed that tests the performance of the two alternatives of this specific entity on a test environment. Thereby, the design of the benchmark application as well as the test environment depend on the goal of the test (e.g. comparing two design alternatives with respect to an achievable throughput for a certain workload). This procedure requires to create and run a large set of tailored measurements. In the following, we give an overview on how we manage this requirement.

Creating and running a single performance measurement in the described context requires the following steps:

1. Developing a JPA data model.
2. Developing a test application logic that uses the data model.
3. Developing a load driver that simulates a usage profile.
4. Adding instrumentation that enables performance monitoring.
5. Deploying the test application on a performance test environment.
6. Executing the performance test.
7. Analysing and interpreting the measured data.

In the following, we refer to the software that is created during steps 1 to 4 as benchmark application. With our approach, we aim at reducing the manual effort in creating and executing benchmark applications while retaining the flexibility in creating these individual benchmark applications. Therefore, we facilitate the definition of such benchmark applications through a generic domain-specific model and provide a Benchmark Framework to realize the transformation of those models to benchmark applications (see Section 4) which allows us to efficiently create and run benchmark applications. Furthermore, we adopt a concept known from the web development domain where developers build *Servlets* that are executed in a *Servlet Container* [12]. The *Servlet Container* provides the infrastructure to run the *Servlets* and takes over management and controlling of *Servlets*. In our approach, the benchmark applications are called *Measurelets* and the mechanisms to control such *Measurelets* are provided by a *Measurelet Container*. A *Measurelet Container* executes *Measurelets* by providing an appropriate infrastructure and can provide additional services concerning the processing of measurements (e.g., recovery of failed runs or repeating runs until a certain confidence level has been reached). The benefit of this approach is that the developer of a benchmark application can focus on the design of the benchmark application (i.e., the *Measurelet*) and does not have to deal with setting up the execution environment and controlling the measurements. In order to ensure the compatibility between the *Measurelets* and the *Measurelet Container* we introduce a *Measurelet API* as part of the approach (see Section 3). The *Measurelet API* supports developers in building and running a single benchmark application. However, in order to answer some of the questions stated above, the construction and execution of a multitude of benchmark applications is required (e.g. for exhaustive performance regression testing such as described in Section 5).

Developing all these benchmark applications manually would be too time-consuming and thus impracticable. Therefore, we introduce a Benchmark Framework that allows us to automatically create, run and analyse *Measurelets*. Thereby, the construction and execution of benchmark applications can be parametrisable. A *Measurelet* can, for example, contain the code of the application logic that accesses the data model via a certain interface. The actual implementation of the data model can then be varied which allows comparing different design alternatives. In order to support the efficient parametrised construction and execution of a multitude of *Measurelets*, we introduce a model-based representation of *Measurelets*, called JPA Benchmark Model (JBM). Having the information about the content of a *Measurelet* as well as a set of variation points specified in a model allows us to leverage model-driven development advantages such as code generation and consistent design documentation. The model is interpreted by the Benchmark Framework that automates the execution of the three steps illustrated in Figure 3.

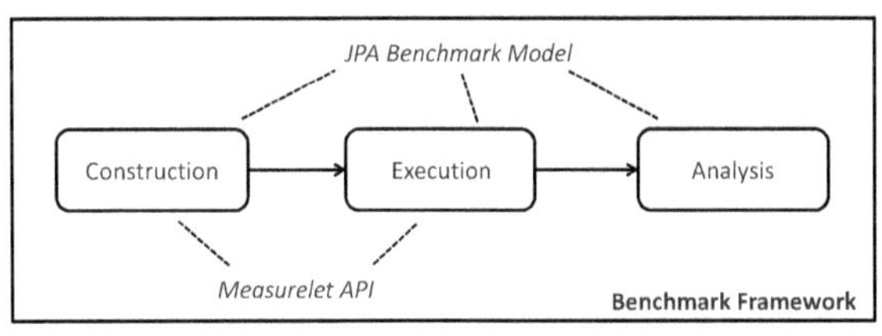

Figure 3: Benchmark Framework

The Benchmark Framework automates construction, execution, and analysis of benchmark applications. It is parametrisable via the information specified in the JPA Benchmark Model. In the construction step the Benchmark Model triggers the generation of code and configuration files and packages the benchmark application (i.e., the *Measurelet*) to a deployable unit. In the execution phase the *Measurelet* is deployed to the *Measurelet Container* which runs the application. Finally, the analysis step analyses the measured results and correlates the measurements with the information specified in the JPA Benchmark Model. In Section 4, we introduce this process in more detail.

Both, the *Measurelet* concept as well as the automation capabilities of the Benchmark Framework combined with the JPA Benchmark Model helped us to reduce the effort for detailed performance testing. In Section 5, we report on the first results that we achieved by applying exhaustive, tailored performance measurements in the development process of a platform persistence service.

3. MEASURELET API

The *Measurelet API* is an approach to reduce the efforts for creating and executing tailored performance measurements. We apply the idea of the *Servlet API* [12] to the performance measurement domain. Benchmark applications are called *Measurelets* and created based on a common abstract interface. The execution of Measurelets is controlled by so-called *Measurelet Containers*. Hence, a Measurelet is focused on benchmark structure, workloads, and monitoring probes. The management and execution of Measurelets is conducted by Measurelet Containers based on the abstract Measurelet API. The Measurelet Containers adapt to the specific platform on which the Measurelets (i.e., the benchmark applications) should be executed (i.e. the SAP Netweaver Cloud platform in our scenario).

Figure 4 shows the common interface for the implementation of Measurelets.

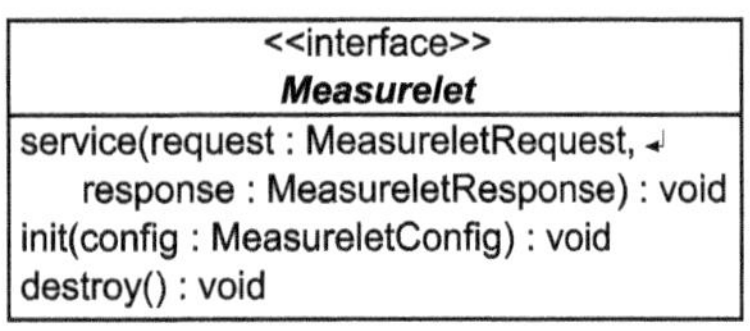

Figure 4: Measurelet Interface

The method `service` realizes the actual measurement. This method expects two input parameters which represent the interface to the caller of the benchmark application. The parameters provide access to the caller input and return the response values. The values for those parameters are provided by the Measurelet Container, which abstract, for example, from the concrete communication protocol or deployment format of the execution platform. Examples for different communication protocols are HTTP and Java Remote Method Invocation (RMI) for synchronous invocations and messaging protocols like the Advanced Message Queuing Protocol (AMQP) for asynchronous invocations.

To create a Measurelet, a developer has to create a class which implements the Measurelet Interface shown in Figure 4. The developer does not need to provide extensive, platform-specific execution and communication logic as this is realized in the Measurelet Containers. Rather, he can focus on the core benchmark application design and rely on the common interfaces to access the inputs and provide the outputs. Using a common, abstract interface for the implementation of Measurelets and their execution through Measurelet Containers makes the implementation of these components independent from each other. Thus, Measurelets are not bound to a specific Measurelet Container, which makes them executable on different platforms which cover, for example, different JPA-based persistence services. Listing 1 shows an example Measurelet implementation.

```
1  public class MyMeasurelet
2      implements Measurelet {
3    public void service(
4        MeasureletRequest request,
5        MeasureletResponse response) {
6      int count = request.getParam("count");
7      long time = System.nanoTime();
8      for (int i = 0; i < count; ++i) {
9        persist(createRandomEntity());
10     }
11     time = System.nanoTime() - time;
12     response.setValue("persist.time", time);
13   }
14   // ...
15 }
```

Listing 1: Sample Measurelet Implementation

The benchmark application is realized within the `service` method and uses the input parameters provided by the caller to control a measurement and evaluate its result. Line 6 shows how input parameters are accessed. In line 12 the measured value is stored in the result object. The implementation artefacts of such a Measurelet code can be deployed and executed on Measurelet Containers using its integration tools. For example, such a tool could package the artefacts in a JAR file to deploy and invoke it on a remote application server.

In the following sections, we present an approach that aims at the automated construction and execution of Measurelets using a model-based description. This allows us to create and run an exhaustive set of benchmark applications for a certain target platform.

4. AUTOMATED, TAILORED BENCHMARKING

Exhaustive, tailored benchmarking can be used in multiple scenarios (see also Section 5). Examples are (1) performance regression testing, (2) comparing performance characteristics of different platforms, or (3) comparing design alternatives. In all use cases, the following basic actions are to be conducted (cf. Figure 3):

- create a benchmark application,
- execute the benchmark application on the target platform, and
- analyse the results.

While all use cases have to execute the same actions, they differ in how the actions are included in a process flow. For example, in the performance regression testing use case, the same benchmark application is executed multiple times while the execution platform evolves. Whereas, in the design alternative comparison use case, we compare the results of two different benchmark applications that are executed on the same platform.

The effort to implement the use cases mentioned above can be very high, especially if the basic actions have to be implemented manually for each use case. Hence, our approach extracts the common actions into components that can be reused in different use cases. This allows us to simply compose the basic actions in order to implement the process flow of a specific use case. Moreover, we use a model-based description of benchmark applications that allows us to leverage advantages of model-driven development such as code generation and consistent design documentation.

In the following sections, we introduce our JPA Benchmark Model as well as the component-based Benchmark Framework that we developed to automate the creation, execution and analysis of benchmark applications.

4.1 JPA Benchmark Model

A *JPA Benchmark Model* (JBM) is a means to describe benchmark applications that are supposed to run on a platform that provides a JPA-based persistence service. Using a model, performance analysts can provide a declarative description of benchmark applications. This description is then used to automatically construct, execute and analyse the benchmark application (see Section 4). A JPA Benchmark Model is constructed based on a metamodel that allows a complete specification of JPA entity structures as it is aligned to the elements of the *Metamodel API* described in the JPA Specification [3]. Moreover, the metamodel allows for the registration of additional information in a key-value fashion. This additional information is required to describe the other parts (i.e., other than the data model) of the benchmark application (e.g. information about test data or test workload). Listing 2 illustrates an example for the information described in a JPA Benchmark Model.

```
1  JPABenchmarkModel:
2    entities:
3      - name: Customer
4        operationsToTest: [insert, update]
5        attributes:
6          - name: id
7            type: Integer
8            id: true
9          - name: primaryAddress
10           type: String
11           length: 128
12         - name: secondaryAddresses
13           type: List<String>
14           size: 4
15           stringItemLength: 128
```

Listing 2: Custom JPA Benchmark Model

The model consists of a single entity class called `Customer`, which has 3 persistent attributes called `id`, `primaryAddress` and `secondaryAddresses`. Line 4 represents the additional information which states that for this entity class the performance of the persistence operations `insert` and `update` should be tested. In line 11, additional information is added to the attribute `primaryAddress` which defines to use string values with a length of 128 characters. Lines 14 and 15 specify that the list of `secondaryAddresses` should contain 4 items and the length of those string items should be 128 characters.

A JPA Benchmark Model can be used to automatically create executable applications through model to code transformation which has several benefits. Experts like benchmark designers can create the necessary transformations whereas application developers and performance analysts create the model. Moreover, it is possible to create model instances via an API which allows to combine the approach with different applications and tools such as systematic experimentation [20] or code extraction [5]. In the following section, we provide a detailed description on how the models are used to automate the benchmarking process.

4.2 Benchmark Framework

We developed a *Benchmark Framework (BF)* which implements this component-based approach by providing a generic component model for the three basic actions outlined above. For each action, we define a specific component type. The *BF* manages the components and enables their usage in multiple use cases. The generic component model of the framework is based on OSGi [17]. Components are created and managed as OSGi services and are available through the service registry. Each action type specifies the interface of the implementing service component leading to the specification of an interface for constructing, executing, and analysing benchmark applications. The interfaces form the basis for component implementations and ensure their type-safe usage. In the following subsections, we describe the different component types and their interfaces in more detail. Moreover, we describe how we implemented the components to apply them in the use cases introduced in Section 5.

4.2.1 Construction Component Type

The construction component is responsible for creating benchmark applications. Therefore, it uses the information specified in a *JPA Benchmark Model (JBM)* to build an executable application via a model-to-code transformation. The resulting benchmark application realises the measurements of interest. The construction component accepts an instance of the *JBM* as input and provides an executable JAR file including the benchmark application as output (see Figure 5).

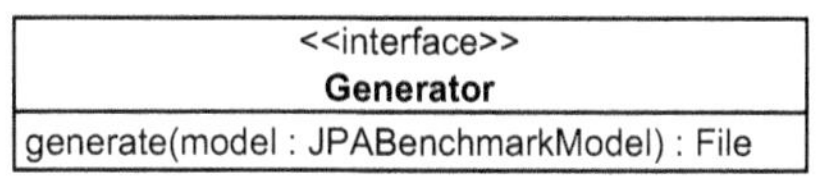

Figure 5: Construction Component Interface

The benchmark applications are implemented as *Measurelets* in order to increase the reusability of the component (see also Section 3). The construction component applied by the use cases described in Section 5 uses Apache Velocity Templates [1] to generate Java source code. The place holders and control structure within the templates are aligned to the elements of the *JBM* which is why an JBM is the only input parameter required by the templates. Listing 3 outlines an example Velocity template used to create the source code of an entity class.

```
@Entity
public class ${entity.name} {
  #foreach ( $att in $entity.attributes )
    #if ( $att.id == true )
      @Id
    #end
    private ${att.type} ${att.name};
    // ...
  #end
  // ...
```

```
11    public static class Factory {
12      public ${entity.name} randomInstance() {
13        ${entity.name} ri =
14          new ${entity.name}();
15        #foreach ($att in $entity.attributes)
16          #if ( $att.type == "String" )
17            ri.${att.name} =
18              randomString(${att.length});
19          #end
20        #end
21        return ri;
22  }}}
```

Listing 3: Sample Code Template

Lines 3-9 iterate over the attributes of an entity defined in the *JBM* and create the attribute signatures in the entity source code. In line 15, the attribute information defined in the JBM is used to generate the code for the factory method that is responsible for creating instances of the entity. Here, the template also considers additional information like the length of a String attribute (cf. line 18 in Listing 3 and line 11 in Listing 2).

The implementation uses the visitor pattern to create the different artefacts specified in the JBM. The visitor traverses the model and calls the template engine appropriately. The artefacts are created within a folder according to the Java conventions. The artefacts include the source code of JPA entities, the JPA persistence configuration file, as well as the Measurelet interface and the corresponding configuration as an OSGi declarative service. The resulting artefacts are compiled by the Java Compiler API and bundled in a JAR file using the JVM libraries of the `java.util.zip` package.

4.2.2 Execution Component Type

The execution component is responsible for running the benchmark applications on the target platform and gathering the metrics of interest. The execution can be influenced by parameters specified in the *JBM* (e.g. by specifying the database instance that should be used for the measurements). The input of the execution component is (i) a JAR file that contains an executable benchmark application and (ii) the corresponding *JBM* (see Figure 6). The output are the results of the execution (e.g. measured response times for a certain service call).

<<interface>> **Runner**
run(model : JPABenchmarkModel, benchmark : File) : Future<MeasurementResult>

Figure 6: Execution Component Interface

The execution is based on the Measurelet API which increases the reusability of the component. The execution component used by the use cases described in Section 5 implements a Measurelet Container that adapts to the *SAP Netweaver Cloud* platform. The container registers a Servlet according to the *OSGi Web Application Specification* [17]. The Servlet allows to load JAR files into the container that are then installed as OSGi bundles. The bundles can register Measurelets as OSGi services and registered Measurelets can be executed via the Servlet. Thus, the deployed benchmark applications can be triggered by other applications using the HTTP protocol (e.g. by a nightly-build job for performance regression testing).

4.2.3 Analysis Component Type

The results derived by running one or many benchmark applications are processed by analysis components. Analysis components can perform a variety of analysis and data processing methods. For example, results can be illustrated in graphs or processed for textual representation. Moreover, it is possible to trigger follow-up actions based on the measurement results. When analysing the measured data, the component implementation can use the information specified in the *JBM*. Hence, the input of the analysis component is the execution result of a benchmark application and the corresponding *JBM* describing the application and the execution context (see Figure 7).

<<interface>> **Processor**
process(model : JPABenchmarkModel, result : MeasurementResult) : void handleError(model : JPABenchmarkModel, error : Throwable) : void

Figure 7: Analysis Component Interface

Despite the fact that the interface only accepts the result of a single execution of a benchmark application as input, a complementary component can enhance the capabilities by storing and providing results of previous benchmark application runs. This allows to implement more sophisticated analyses such as the performance regression analysis applied in one of the use cases described in Section 5.

5. USE CASES

In this section, we introduce two use cases that leverage from the approach presented in this paper. In the performance regression testing use case, we apply the approach to run an exhaustive set of tailored performance test for the persistence service of the SAP Netweaver Coud platform. The test are triggered by a nightly job on the build server and thus ensure that the development teams are timely informed on introduced performance problems. The second use case describes how we use the approach to provide performance feedback in the development environment of application developers that use the persistence service within an application that runs on the Netweaver Cloud platform.

5.1 Performance Regression Testing

Performance regression testing is a commonly used means to monitor whether a new software version still adheres to a specified quality goal or agreement. Performance tests are (in most cases) automatically executed on a regular basis (e.g., nightly). If a performance regression is observed, the development team can immediately dig into the issue. However, the probability of actually observing an issue as well as the effort for identifying its root cause is highly dependent on the number and quality of performance tests executed on a regular basis. Using the approach presented in this paper, we have been able to increase the number of performance regression test significantly. Moreover, it allows us to create performance tests that are tailored to the specifics of the persistence service which improves the value of the information provided by the tests. We developed a set of benchmark applications that include different data models, different application logic methods accessing a data model as well as different types of usage profiles. As a result, we have an exhaustive set of benchmark applications all testing a different aspect of the persistence service. This set of benchmark applications is executed on a nightly basis. Moreover, we can easily add additional tests as soon as we identify a performance problem in an application that has not been detected by the existing tests. The concept is similar to the unit testing approach where specific tests are created for each concrete feature and tests are added after fixing a problem in order to prevent from introducing the problem again. Figure 8 shows a performance regression that we observed after having the tailored benchmark set in place.

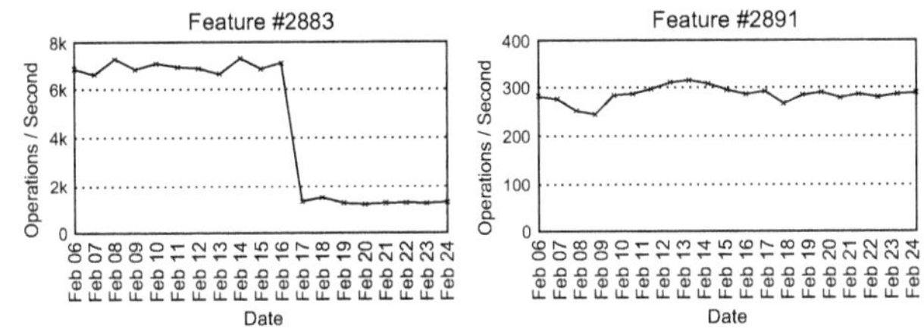

Figure 8: Identified Regression

The graphs show the measured throughput for two benchmark applications over a certain period of time. The benchmark application on the left side executes a named query that retrieves all instances of an entity in a certain data model. The benchmark application on the right side executes a query that stores a number of instances of the same entity to the database. The graph on the left side of Figure 8 shows that for this benchmark application a performance regression of factor 4 has been introduced. As the tests run on a nightly basis, we have been able to identify the root cause for the issue very quickly which happened to be an update of the database version that has been conducted at that day. An interesting observation is that the regression has only been observed in one test out of the set of benchmark applications. The test shown on the right side of Figure 8 does, for example, not show a performance regression. This observation underlines the assumption that more and tailored performance tests increase the probability of detecting a performance issue. Moreover, knowing the exact conditions under which a problem occurs and under which not can be very helpful in fixing a performance issue.

5.2 Performance Feedback

In data-centric applications, the data model can limit performance and scalability of the overall application. When developing the data model and the application logic that uses the data model, it is often unclear to developers how different design decisions or usage profiles affect the application's performance. With our ongoing research [19], we aim at increasing the performance awareness of developers by providing immediate feedback about the expected performance of the artefact under development. We integrate the performance feedback into the development environment (e.g. eclipse) to lower the burden for developers (cf. Figure 9). For the persistence scenario described in this paper, we identified two feedback usage types.

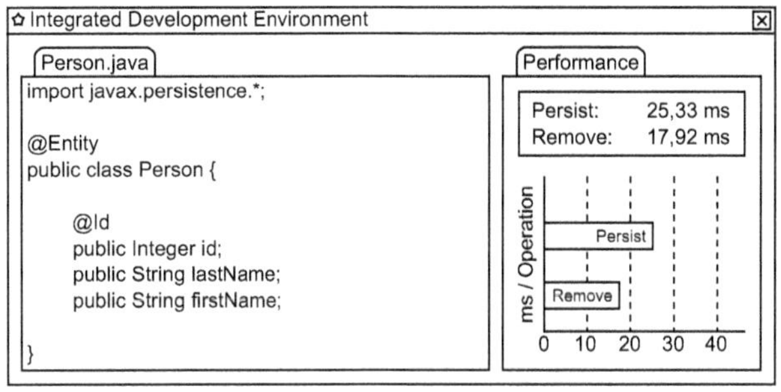

Figure 9: Immediate Feedback in the IDE.

1. Developers can use the feedback to continuously track the performance impact of changes applied to the artefact.

2. Developers can use the feedback to evaluate design alternatives with respect to data model entities (e.g. distribution of attributes across entity classes) and entity usage (e.g. number of parallel reads).

For the first usage type, the performance feedback relates to the currently focused software artefact (e.g. a data model entity or a method using persistence operations) and the performance values are updated when changes are applied to the software artefact. Figure 9 sketches how this could look like in the IDE of the developer.

While the developer is developing the entity *Person* and adding additional attributes, the performance feedback view on the right side of the figure shows how the changes affect the average response time for persist and remove operations on this entity for a predefined test workload.

In the second scenario, developers directly compare different implementation alternatives against each other in order to understand the performance characteristics of each alternative. Figure 10 shows an example for this kind of feedback.

Part (a) of Figure 10 sketches the two implementation alternatives. The functional requirement for the developer is to store 32 numbers in a *Container* entity. *Alternative 1* implements this requirement by adding 32 fields of type *Long* to the entity. *Alternative 2* uses a list field that can hold values of type Long. Part (b) of Figure 10 illustrates the performance feedback view for that example. It shows the throughput that can be achieved for the insert, update, remove, and persist operations using the respective alternative. In the example, the throughput that can be achieved with *Alternative 1* is 4 times higher than with *Alternative 2*. We derived the feedback for this example by integrating the two entities in two benchmark applications that differ only in the described point. The workload as well as the test platform have been identical.

Basically, the feedback can be based on two sources. Either by directly measuring the performance of the artefact of interest (as we did it for the example above) or by predicting the performance using performance prediction models. Measured performance is more accurate but is not immediately available to developers. Performance prediction is immediately available but requires the prior derivation of performance prediction models. In the remainder of this section, we describe how we apply the approach presented in this paper to implement the two feedback sources.

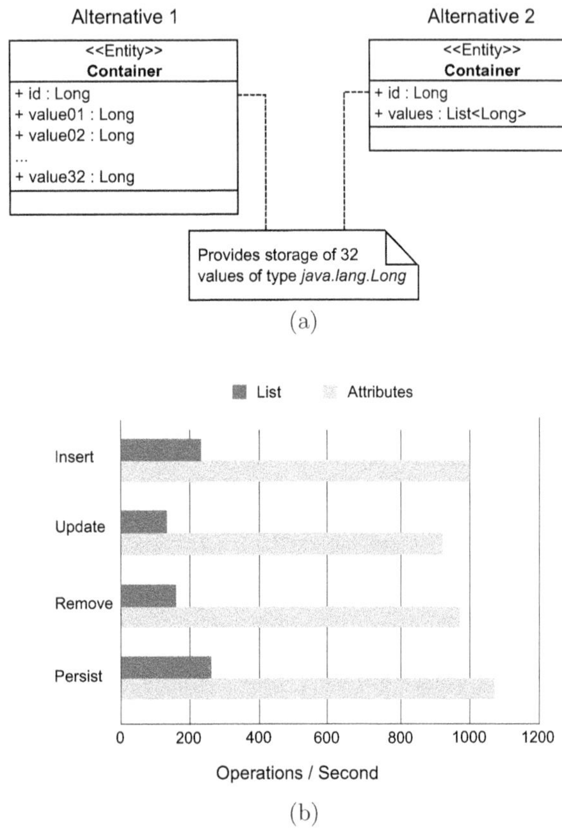

Figure 10: Evaluation of Design Alternatives.

5.2.1 *Measurement-based Feedback*

In case of measurement-based feedback, we execute a benchmark application on a test environment and measure the performance of different persistence service operations. Therefore, a JPA Benchmark Model is created by the IDE plugin. The plugin uses static code analysis to derive information about the data model that is to be tested and adds information about the test data and test workload that is to be used for the test. Then, the Benchmark Framework is applied to create a benchmark application based on the model, run it on the test platform and return the measurement result to the IDE plugin which channels the feedback to the developer.

5.2.2 *Prediction-based Feedback*

For the prediction scenario, we do not have real-world results yet. However, we give a short outline on the overall idea in order to illustrate how the approach presented in this paper can be used for our ongoing research. Our approach to derive performance prediction models is based on systematic experimentation by applying measurements and sophisticated statistical analyses (see also [21]). To define and control the experiments required to derive a prediction function, we apply our Software Performance Cockpit tool [20] in combination with the approach presented in this paper. The Software Performance Cockpit executes experiments by systematic variation of parameter values (e.g. the number of the entity's Long attributes). The benchmark framework creates and executes benchmark applications with the parameters specified by the Software Performance Cockpit and returns the performance measurement results. Based on the results provided by multiple benchmark applications SoPeCo derives a functional relationship between a set of input parameters and a performance metric of interest. For the example shown in Figure 10, the function that evaluates the achievable throughput for the persist operation could be as follows:

$$TP_{persist}(n_{LongAtt}) = 2300 + (-12 * numLongFields)$$

Deriving a complete prediction function for different data models and applications is a challenging task and requires detailed future research (see also Section 6). However, the approach presented in this paper forms the basis for this research as it provides an efficient means to run an exhaustive set of benchmark applications. Our next steps towards the automated construction of performance prediction functions for JPA-based applications are setting up systematic experiments to identify performance-relevant aspects of JPA-based data models and to identify a set of synthetic workload classes that are representative for applications of a certain domain.

6. RESEARCH DIRECTIONS

The approach of exhaustive, tailored performance testing has immediately shown its benefits when we applied it to the development of the persistence service. The proof of concept implementations of the *Measurelet* idea and the Benchmark Framework have demonstrated that it is possible to create and run a multitude of benchmark applications with limited effort. Based on the experience we gained through applying the presented approach, we discuss different research directions that we see in this area.

How to design the actual benchmark applications?
While the approach presented in this paper simplifies the construction and execution of benchmark applications, it is out of the scope of this work to determine how the design of the benchmarks should look like. In order to answer questions like „What are the usage profiles that I should test in my benchmark applications?", „What are performance-critical data model designs?", or „How to deal with caching effects that can bias the results?", it requires sophisticated engineering methods. An example for such a method could be the automated derivation of benchmark applications based on real user monitoring data.

How to deal with the large parameter space?
The approach allows developers to create and run large sets of benchmark applications based on parametrisable *Measurelet* construction. However, the potential degrees of freedom for a developer of a benchmark application are huge. There are, for example, a vast amount of different potential data models. The same is true for the application logic and the usage profiles. Moreover, the number of test environments as well as the time available to execute the tests are limited. Thus, smart approaches are required that provide maximum information gain with a limited number of benchmark applications.

How to derive prediction models to provide performance feedback to application developers?
In Section 5 we introduced a scenario that aims at providing feedback to application developers with respect to the

performance of the persistence service for their specific application. However, deriving a prediction model that provides accurate predictions for different real-world applications is a challenging task. Besides the two questions stated above, one has to deal with finding an appropriate abstraction level that on the one hand limits the complexity of applications and platform service, but on the other hand includes enough detail to enable accurate predictions. A systematic search for proper heuristics in combination with sophisticated statistical analysis and machine learning techniques could be a potential direction towards answering this question.

How to apply the approach to other components? The idea of having a Measurelet API and a parametrisable framework that creates and runs benchmark applications is independent of JPA and the persistence service. Except for the JPA Benchmark Model Model introduced in Section 4.1 all the components presented in this paper are reusable for other performance measurements. However, so far we do not have any information on the effort for applying the approach to a different component (e.g., a messaging or identity management service). Moreover, other components can introduce additional challenges with respect to automated performance testing.

7. RELATED WORK

The approach presented in this paper combines automated model-based benchmarking and benchmark application generation with systematic measurements. In the following, we discuss existing work that addresses these aspects.

Benchmarking is a common and widely spread task in industry. Standard benchmarks, such as those offered by the Standard Performance Evaluation Corporation (SPEC) [15], the Transaction Processing Council (TPC) [18] (e.g. TPC-H and SPECjEnterprise2010), or software vendors like SAP [13], provide standardized performance measurement for comparing and ranking of platform configurations or platforms of different vendors. The JPA Performance Benchmark [9] is the largest benchmark that deals with JPA-based applications and allows the comparison of different JPA providers (e.g. EclipseLink, Hibernate) and Database Management Systems (e.g. Derby, MySQL). However, standard benchmarks do not satisfy individual information needs and are often difficult to setup. Therefore, specifically tailored and easily employable benchmark applications are needed to satisfy a certain information need.

Wright et al. [22] present Auto-Pilot a tool for the automation of performance measurements. Benchmark scripts execute performance measurements described using a procedural language. Auto-Pilot provides the ability to run analysis on performance measurements but does not use descriptive models for performance measurements in general. Kalibera [8] et al. present an approach for automatic performance measurements in a distributed and heterogeneous environment. They use an execution framework to take performance measurements and a benchmarking framework to control the execution. However, the execution framework requires the implementation of a certain run time environment and communication protocol. Moreover, Kalibera et al. do not use models to describe performance measurements and benchmark applications.

Apfelbaum and Doyle [2] as well as Malik et al. [11] present approaches for generating functional tests from models that describe the system under test. However, our JPA Benchmark Models describe performance measurements and not functional tests. In [24], Zhu et al. present a model-based approach for automated usage of performance measurements. The benchmark software is described in UML based models that are enhanced with annotations and elements describing the usage of the benchmark in order to conduct performance measurements. The model of the benchmark software can be derived automatically from UML models of the system under test with a model to model transformation. The model-based approach using UML for model description focuses on the integration of performance measurements in model-based software development. In contrast, our JPA Benchmark Model is domain specific and targets tailored measurements by describing test-relevant parts of the benchmark application.

In contrast to a model-based approach, Denaro [4] describes an approach for benchmark application construction based on existing software artefacts. Unimplemented parts are replaced by stubs which provide only minimal requirements (e.g. just an interface). The assumption is that the results delivered by the benchmark applications approximate the final performance characteristics of the software system. However, benchmark applications and required stubs have to be developed like common software system artefacts. Our benchmark framework complements the approach by automatically generating benchmark applications. The benefit of our model-based approach is to replace stubs with the concrete implementation of the artefact as development evolves.

8. CONCLUSIONS

In this paper, we introduced an approach that allows for exhaustive, tailored performance testing. The two main components of the approach are the *Measurelet* concept as well as the model-based construction and execution of benchmark applications. Using the presented approach, developers can create and run a large set of tailored benchmark applications with minimal effort. We have illustrated how we use the capabilities of the approach to ensure the quality of a JPA-based persistence service via detailed performance regression testing. Besides this platform service provider view, we also reported on how we plan to use the approach to provide valuable performance feedback to the consumer of the platform service (i.e., the application developers) in order to support them in developing high-quality applications. In our future work, we focus on implementing the performance feedback for the platform service consumer. Therefore, we are investigating sophisticated methods to derive performance prediction functions based on the systematic execution of parametrised benchmark applications. Moreover, we plan to adopt the exhaustive performance regression testing approach to other platform services.

Acknowledgment

This work is supported by the German Research Foundation (DFG), grant RE 1674/6-1 (Transfer project KIT-SAP).

9. REFERENCES

[1] Apache Software Foundation. The Apache Velocity Project. http://velocity.apache.org/, 2012.

[2] L. Apfelbaum and J. Doyle. Model based testing. *Software Quality Week Conference*, pages 1–14, 1997.

[3] L. DeMichiel and J. P. E. Group. JSR 317: Java Persistence API, Version 2.0. http://jcp.org/en/jsr/detail?id=317, 2009.

[4] G. Denaro, A. Polini, and W. Emmerich. Early performance testing of distributed software applications. In *Proceedings of the 4th international workshop on Software and performance*, pages 94 – 103. ACM, 2004.

[5] Forschungszentrum Informatik (FZI). SQools: Software Analysis Tools. http://www.sqools.org/, 2012.

[6] Hibernate. Relational Persistence for Java and .NET. http://www.hibernate.org, 2012.

[7] R. Jain. *The Art of Computer Systems Performance Analysis: Techniques for Experimental Design, Measurement, Simulation, and Modeling.* Wiley professional computing. Wiley, 1991.

[8] T. Kalibera. Regression benchmarking environment. *Proceedings of WDS*, 2004.

[9] I. Kirsh. JPA Performance Benchmark (JPAB). http://www.jpab.org/, 2012.

[10] H. Koziolek. Performance evaluation of component-based software systems: A survey. *Performance Evaluation Journal*, 2009.

[11] Q. A. Malik, J. Lilius, and L. Laibinis. Model-based testing using scenarios and event-b refinements. *Methods, Models and Tools for Fault Tolerance*, pages 177–195, 2009.

[12] R. Mordani. JSR 315: Java Servlet Specification, Version 3.0 Rev a. http://jcp.org/en/jsr/detail?id=315, 2010.

[13] SAP. Standard Application Benchmarks. http://www.sap.com/campaigns/benchmark, November 2011.

[14] SAP AG. Sap netweaver cloud. http://www.sap.com/solutions/technology/cloud/netweaver/index.epx, 2012.

[15] SPEC. Standard Performance Evaluation Corporation. http://www.spec.org/, 2012.

[16] S. Stefanov. Book of speed: The business, psychology and technology of high-performance web apps. http://www.bookofspeed.com, 2012.

[17] The OSGi Alliance. OSGi service platform core specification. http://www.osgi.org/Specifications, 2012.

[18] TPC. Transaction Processing Performance Council. http://www.tpc.org/, 2012.

[19] D. Westermann. A generic methodology to derive domain-specific performance feedback for developers. In *Proc. of the 34th International Conference on Software Engineering (ICSE 2012), Doctoral Symposium*, 2012.

[20] D. Westermann, J. Happe, M. Hauck, and C. Heupel. The Performance Cockpit Approach: A Framework for Systematic Performance Evaluations. In *Proceedings of the 36th EUROMICRO Conference on Software Engineering and Advanced Applications (SEAA 2010)*, pages 31–38. IEEE Computer Society, 2010.

[21] D. Westermann, J. Happe, R. Krebs, and R. Farahbod. Automated inference of goal-oriented performance prediction functions. In *Proceedings of the 27th IEEE/ACM International Conference on Automated Software Engineering*, ASE 2012, pages 190–199, New York, NY, USA, 2012. ACM.

[22] C. P. Wright, N. Joukov, D. Kulkarni, Y. Miretskiy, and E. Zadok. Auto-pilot: A platform for system software benchmarking. *ATEC '05 Proceedings of the annual conference on USENIX Annual Technical Conference*, pages 1–19, 2005.

[23] G. York. EclipseLink JPA: An Advanced ORM Persistence Framework. Dzone Refcardz, http://refcardz.dzone.com/refcardz/eclipselink-jpa, 2011.

[24] L. Zhu, N. B. Bui, Y. Liu, and I. Gorton. MDABench: Customized benchmark generation using MDA. *Journal of Systems and Software*, 80(2):265–282, Feb. 2007.

On Time-Average Limits in Deterministic and Stochastic Petri Nets*

Tomáš Brázdil
brazdil@fi.muni.cz

Ľuboš Korenčiak
korenciak@fi.muni.cz

Jan Krčál
krcal@fi.muni.cz

Jan Křetínský
jan.kretinsky@fi.muni.cz

Vojtěch Řehák
rehak@fi.muni.cz

Faculty of Informatics, Masaryk University, Botanická 68a, 60200 Brno, Czech Republic

ABSTRACT

In this poster paper, we study performance of systems modeled by deterministic and stochastic Petri nets (DSPN). As a performance measure, we consider long-run average time spent in a set of markings. Even though this measure often appears in DSPN literature, its existence has never been considered. We provide a DSPN model of a simple communication protocol in which the long-run average time spent in a fixed marking is *not* well-defined due to a highly unstable behavior of the model. Further, we introduce a syntactical restriction on DSPN which preserves most of the modeling power yet guarantees existence of the long-run average.

Categories and Subject Descriptors

G.3 [**Mathematics of Computing**]: Probability and statistics—*Stochastic processes*

General Terms

Performance, Theory

Keywords

deterministic and stochastic Petri nets, performance analysis, steady-state distribution, time-average limits

1. INTRODUCTION

Petri nets are widely used in modelling of concurrent systems, such as network components. A Petri net is specified by a finite set of *places*, a finite set of *transitions*, and non-negative numbers $i_{t,p}$ and $o_{t,p}$ for each transition t and place p. Configuration of a Petri net is called *marking* and assigns to each place a non-negative number of tokens. A transition t is *enabled* if every place p contains at least $i_{t,p}$ tokens. *Firing* an enabled transition t consist of removing $i_{t,p}$ tokens from each place p, and adding $o_{t,p}$ tokens to each place p. The net starts in an initial marking and changes its marking by subsequent firing of enabled transitions.

For the sake of performance evaluation, Petri nets are turned into a continuous-time model as follows. Whenever a transition t becomes enabled, its *time to fire* is set to a real number. As long as t is enabled, its time to fire then decreases with the time flow until it reaches zero and the transition fires. If there are more transitions with zero time to fire, the non-determinism is resolved according to fixed priorities of transitions[1]. In deterministic and stochastic Petri nets (DSPN) [3], there are transitions of three types with time to fire set as follows. For an *immediate* transition, it is set to 0; for a *deterministic* transition t, it is set to a constant $d_t > 0$; and for an *exponential* transition t, it is set randomly according to the exponential distribution with rate $r_t > 0$.

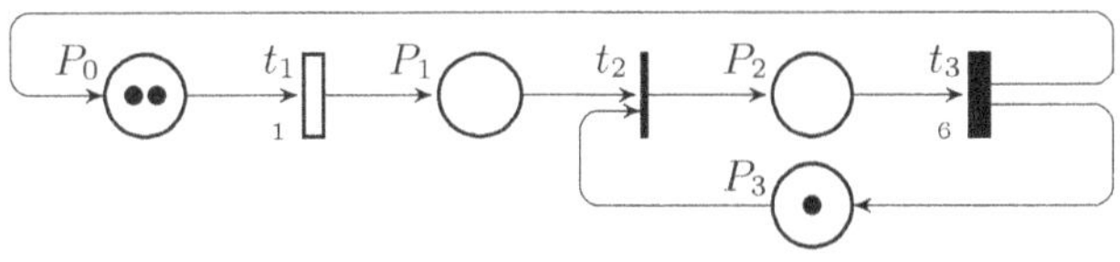

Figure 1: A DSPN example of an M/D/1/2 queue.

To illustrate the definitions, a model of an M/D/1/2 queue with memoryless arrival of jobs, one deterministic server, and capacity of the queue 2 is shown in Figure 1. Places are drawn as circles; immediate, deterministic, and exponential transitions are drawn as thin bars, thick black bars, and thick void bars, respectively (with their $d_t = 6$ or $r_t = 1$). Assume that the deterministic and the exponential transitions are enabled with time to fire 6 and 5, respectively. The exponential transition is fired first after 5 time units and the deterministic transition has at that moment time to fire 1.

In this paper, we focus on the long-run average behavior and discuss a crucial quantity, the *time-average limit* [2]

$$r(f) = \lim_{t\to\infty} \frac{1}{t} \int_0^t f(X(u))du,$$

of a function f that assigns a real number to each marking. Here, $X(u)$ denotes the marking of the net at time $u \geq 0$. Observe that in our example the utilization of the server can be expressed as a time-average limit by setting $f(m) = 1$ if the marking m has a token in P_2 and $f(m) = 0$, otherwise.

The time-average limits in DSPN have been widely studied. Yet, no one has addressed the existence of the limit $r(f)$. If the limit does not exist, its approximation, e.g. using simulation, makes no sense. In this paper, we show a simple example of a DSPN where the time-average limit $r(f)$ does not exist. We also provide a sufficient (but not necessary) syntactical restriction on DSPN guaranteeing existence of the limit $r(f)$.

*The work has been supported by the Czech Science Foundation, grant No. P202/12/P612.

[1] The firing policy we consider is *race with enabling memory.*

ICPE'13, April 21–24, 2013, Prague, Czech Republic.
ACM 978-1-4503-1636-1/13/04.

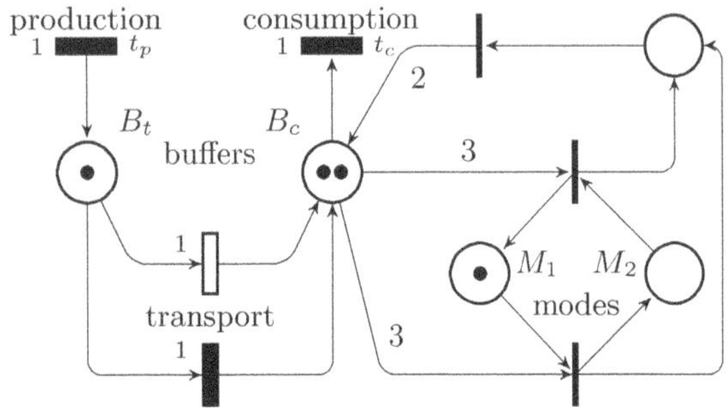

Figure 2: DSPN without the time-average limit.

2. INSTABILITY IN DSPN

In Figure 2, there is an example of a DSPN model with unstable behavior in the long-run, not having a time-average limit. It is a model of a simple producer-consumer system operated by one exponential and three deterministic transitions using buffers B_t and B_c. In addition, on the right, there is a controller with immediate transitions switching a token between modes M_1 and M_2.

Let us explain the behaviour of the model. The transition t_p produces a token exactly every time unit and places it to the transport buffer B_t. Due to the parallel competition of an exponential and a deterministic transition, the token is transported into the consumption buffer B_c in *at most* 1 time unit. The consumption of a token by the transition t_c takes again exactly 1 time unit. The consumption buffer B_c can hold at most two tokens. When a third token appears in the buffer, all three tokens are removed by switching the modes. The subsequent immediate transition returns two tokens back, re-enabling t_c.

The question to be analyzed is what is the average time spent in the mode M_1, i.e. the limit $r(f)$ where f equals 1 if there is a token in M_1 and 0, otherwise. This time-average limit $r(f)$ does not exist (for most runs). Let us explain the behavior on the following figure.

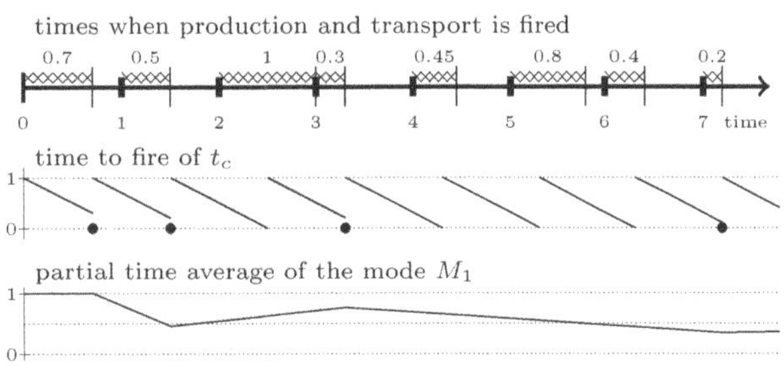

At the top, there is a time axis with thick and thin marks showing when the production and transport transitions are fired, respectively. The time spent transporting is written above (and also indicated by crosses). In the middle, the value of time to fire of the consumption transition t_c is plotted. Note that t_c is fired only when the graph reaches zero level. The bullets on the zero level mark re-enabling of t_c due to a switch of modes (caused by a third token being transported into B_c). Observe that the modes are switched whenever the current transport takes less time than it has ever taken. The lower the current minimal transport time is, the more time it takes to set a new minimum. Hence, the stays in a single mode get longer and longer. Due to this fact, the partial time-averages oscillate and never reach a limit (as indicated in the plot at the bottom).

Formal proof of the non-existence of the time-average limit follows from [1]. Intuitively, it stems from the absolute precision of the deterministic transitions causing that along each run of the model the beginning of the n-th consumption gets closer and closer to the end of the n-th production but never overtakes it. Since this behavior is rarely to be expected in reality, the validity of such a model is at least questionable. However, we can restrict DSPN in such a way that this form of instability does not appear.

3. CONDITIONS ON STABILITY

We say that a transition t *initiates* a transition u if there is a reachable marking m such that t is enabled in m and firing t in m (potentially followed by firing a sequence of immediate transitions) leads to a marking where u *is* enabled and either $u = t$ or u was not enabled in m (or in markings passed by the immediate transitions). For example, t_3 initiates t_3 in Figure 1, as from the marking with one token in both P_1 and P_2 we reach by firing t_3 and immediately t_2 the marking with one token in both P_0 and P_2 where t_3 is still enabled.

We say that a DSPN *initiates increasingly* if there is a function f assigning to each deterministic transition its level $l \in \mathbb{N}_0$ such that there is at most one deterministic transition with level 0 (that can initiate other transitions arbitrarily) and all other deterministic transitions with positive levels can initiate only deterministic transitions of strictly higher levels. Note that the DSPN of Figure 2 does not initiates increasingly as both t_p and t_c initiate themself.

THEOREM 3.1. *For a DSPN with finitely many reachable markings that initiates increasingly, the time-average limit $r(f)$ almost surely exist for any function f.*

The proof can be obtained as a straightforward modification of the proof of Theorem 4 in [1]. The markings and their enabled transitions correspond to states and their sheduled events in Generalized semi-Markov processes in [1], resp.

4. CONCLUSIONS

We show that a widely studied formalism of DSPN does not necessarily have a basic time-average limits which is caused by its unrealistic behavior. We provide syntactical restrictions on DSPN upon which this unrealistic behavior disappears and the time-average limits almost surely exist.

5. REFERENCES

[1] T. Brázdil, J. Krčál, J. Křetínský, and V. Řehák. Fixed-delay Events in Generalized Semi-Markov Processes Revisited. In *Proceedings of CONCUR'11*, volume 6901 of *LNCS*, pages 140–155. Springer, 2011.

[2] P.J. Haas. *Stochastic Petri Nets: Modelling, Stability, Simulation.* Springer, 2002.

[3] M. Ajmone Marsan and G. Chiola. On Petri nets with deterministic and exponentially distributed firing times. In *Advances in Petri Nets 1987*, volume 266 of *LNCS*, pages 132–145. Springer, 1987.

Model-Based Performance Testing in the Cloud Using the MBPeT Tool

Fredrik Abbors, Tanwir Ahmad, Dragos Truscan, Ivan Porres
Department of Information Technologies, Åbo Akademi University
Joukahaisenkatu 3-5 A, 20520, Turku, Finland
[Fredrik.Abbors, Tanwir.Ahmad, Dragos.Truscan, Ivan.Porres]@abo.fi

ABSTRACT

We present an approach for performance testing of software services. We use Probabilistic Timed Automata to model the workload of the system, by describing how different user types interact with the system. We use these models to generate load in real-time and we measure different performance indicators. An in-house developed tool, MBPeT, is used to support our approach. We exemplify with an auction web service case study and show how performance information about the system under test can be collected.

Categories and Subject Descriptors

D.4.8 [**Software**]: performance—*measurement, modeling, monitors*

Keywords

Model-Based Performance Testing, Workload Model, Load Generation, Probabilistic Timed Automata.

1. INTRODUCTION

The goal of performance testing is to validate the *system under test* (SUT) in terms of responsiveness, stability, and resource utilization when it is put under a certain synthetic workload [3]. The syntectic workload should mimic the real workload as generated when real users interact with the system, as closely as possible [5]. Traditionally, the synthetic workload is generated by simulating *virtual user* (VU) behavior with scripts or pre-recorded scenarios. However, real users do not behave like static scripts. Furthermore, these scrips or scenarios can be tedious to create and maintain. Therefore, we propose an approach, supported by the MBPeT tool [1], where load is generated from abstract graphical models describing VU behavior. As such, the models are on the one hand easier to create and update, and on the other hand easier to comprehend compared to scripts, due to their abstract and graphical nature, respectively.

2. PROBABILISTIC TIMED AUTOMATA

The MBPeT tool uses models described as *probabilistic timed automata* (PTA) [4] to generate synthetic workload. A PTA consist of locations and transitions. The transitions

ICPE'13, April 21–24, 2013, Prague, Czech Republic.
ACM 978-1-4503-1636-1/13/04.

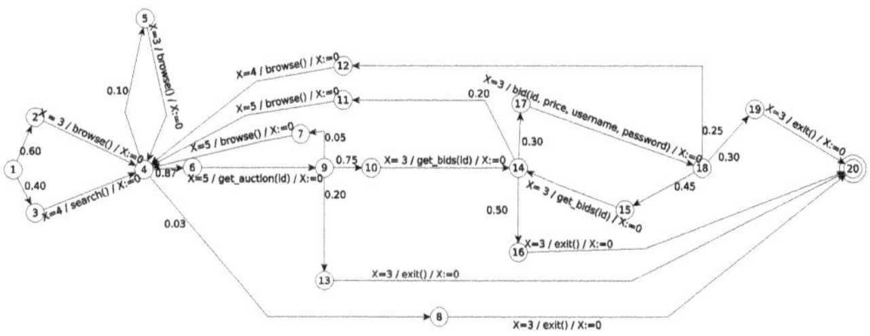

Figure 1: Example of a probabilistic timed automata

can be labeled with three different values: a probability value, an action, and a clock. The *probability* indicates the chance of that transition being taken. The *action* describes what action to take when the transition is fired, and the *clock* indicates how long to wait before firing the transition.

Figure 1 shows an example of a PTA describing a user profile for an online-auctioning system. Transitions have associated either a probability (e.g., 0.1) or a clock and a action to be executed. The former transitions can be used for making probabilistic choices between different paths in the graph. The later are used to send the actions corresponding to a given transition to the SUT. When a response is received, the clock is reset to zero and the next transition is fired. With the help of the probability values defined in the PTA some actions (or sequence of actions) are more likely to be chosen by MBPeT over another actions, whenever a choice is encountered in the PTA. Using PTA allows for a certain level of randomness in the generated workload which makes the later closer to the real workload.

3. MBPET TOOL

The MBPeT tool has a distributed architecture in which a master node controls and distributes the load generation on several slave nodes. Before the slaves start to generate load, the master node loads the PTA model, validates its consistency and initializes a test database with all the necessary test data, e.g., user names, password, form data, files, etc, that is needed to generate meaningful load.

The slave nodes are activated on-the-fly if more generation capacity is needed. If a slave node gets saturated, e.g., having the utilization local resources over a specified threshold, it notifies the master, while maintaining the current load capacity. The master then decides to continue load generation on another slave. The approach is especially beneficial if the slave nodes have heterogenous computation power, eliminating the need for a preliminary benchmarking process of the different nodes.

The PTA models are used by the slave nodes for generating load in real-time against the SUT, by creating traces from the PTA models. The number of VUs on each slave is decided by the master node based on the specified ramp function. A new process is started for each VU that is going to be run concurrently. Each process simulates an independent user based on the PTA. For each simulated user, a new transition is selected in the PTA and, after waiting for the specified time, the corresponding action is sent to the system. Whenever the number of concurrent VUs has to be decreased the corresponding number of processes on each slave are just stopped.

The actions generated from the models can not, as such, directly be sent to the SUT, because the actions are abstract. Therefore, each slave node sends the abstract actions through an adapter, which can be used to either translate on-the-fly every abstract action into a machine readable format (e.g., a HTTP request) and collect the response, or to write the actions to scripts that can be executed by other tools such as JMeter [7]. The adapter acts much like a Faban Driver would in the Faban framework [6]. Faban is a tool used for development of server benchmarks and for generating workloads.

Each slave node monitors different *Key Performance Indicators* (KPIs) of the SUT, such as response time and throughput. The values are reported to the master at the end of performance test. The master node then aggregates the values reported by the slaves, calculates different different statistical indicators, and creates an HTML test report. The test report shows information of how the response time of individual actions varied over time (Figure 2), and provides plots of error rates, average throughput and network utilization, etc. If one has access to the SUT, the MBPeT tool can collect monitoring information of how resources like, CPU, memory, disk, etc, were used during the test session and include them in the test report. In our experiments, we were able to monitor the SUT and collect information using for instance the *dstats* tool which was automatically processed and include in the test report. The MBPeT tool will also present statistics of the most executed traces in the model.

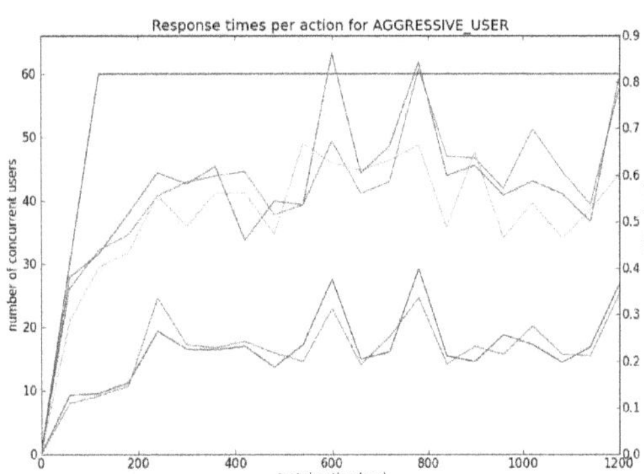

Figure 2: Response times for actions plotted over time

The MBPeT tool has been built to run in a private cloud or in a public ones, such as the Amazon EC2 cloud [2]. Since the master node does not require much computational power, it can run locally on a laptop, while the slave nodes can be set up to run in the cloud.

4. GOALS OF THE DEMONSTRATION

The goals of the demonstration is to show that using abstract models minimizes the effort of creating the workload profiles. Although the implementation the adapter requires an initial effort it can be afterwards reused without changes. Having available graphical models makes easier to visualize and change the workload profiles. With respect to load generation, using PTAs allows us to model the user profiles and generate syntectic load, via probabilistic-guided load generation, that mimics closer the real one workload. We also want to show how load is generated from the workload models and explain how the tool works. We plan to show the tool running the master node on a laptop and have the slave nodes running in the Amazon EC2 cloud. However, if it is not possible to connect properly to Amazon, we will demonstrate the tool running locally on two machines. During the demonstration we intend to show the tool in action on two different case studies, an auctioning web service and a web-based file storage service. We have done a preliminary evaluation of the MBPeT against the JMeter tool and we could conclude that for identical test configurations and load profiles, the results were similar in terms of throughput (number of actions generated per minute). We plan to briefly discuss our conclusions during the demo as well.

5. REFERENCES

[1] F. Abbors, T. Ahmad, D. Truscan, and I. Porres. MBPeT: A Model-Based Performance Testing Tool. *2012 Fourth International Conference on Advances in System Testing and Validation Lifecycle*, 2012.

[2] Amazon. Amazon elastic compute cloud (amazon ec2). http://aws.amazon.com/ec2/. Retrieved: October 2012.

[3] D. Ferrari. On the foundations of artificial workload design. In *Proceedings of the 1984 ACM SIGMETRICS conference on Measurement and modeling of computer systems*, SIGMETRICS '84, pages 8–14, New York, NY, USA, 1984. ACM.

[4] M. Jurdziński, M. Kwiatkowska, G. Norman, and A. Trivedi. Concavely-Priced Probabilistic Timed Automata. In M. Bravetti and G. Zavattaro, editors, *Proc. 20th International Conference on Concurrency Theory (CONCUR'09)*, volume 5710 of *LNCS*, pages 415–430. Springer, 2009.

[5] J. Shaw. Web Application Performance Testing – a Case Study of an On-line Learning Application. *BT Technology Journal*, 18(2):79–86, Apr. 2000.

[6] Sun. Faban harness and benchmark framework,. Online at http://java.net/projects/faban., 2013.

[7] The Apache Software Foundation. JMeter. Online at http://jmeter.apache.org/. Retrieved: October 2012.

SPECsip Infrastructure and Application Benchmarks

Yao-Min Chen
Azeem Jiva
Madhava Dass
Oracle
500 Oracle Pkwy,
Redwood City, CA 94065,
USA
yaomin.chen@oracle.com
azeem.jiva@oracle.com
madhav.dass@oracle.com

Victoria Roxas
Sam Warner
Intel
2200 Mission College
Blvd., Santa Clara,
CA 95054, USA
victoria.roxas@intel.com
samuel.r.warner@intel.com

Gary DeVal
IBM
4205 S. Miami Blvd.
Durham, NC 27709
USA
gjdeval@us.ibm.com

Wei Chen
Kenny Lee
Voxeo Labs
800 Menlo Ave Suite 210
Menlo Park, CA 94025
USA
wchen@voxeolabs.com
klee@voxeolabs.com

ABSTRACT
We describe two SIP benchmarks developed by SPEC. SPECsip_Infrastructure2011 is a SIP Proxy benchmark with realistic user behavior modeling. The second benchmark, not yet officially named, is for a JSR-289 conforming application server. We describe the design, usage and the underlying methodologies for both benchmarks.

Categories and Subject Descriptors
B.8.2 [**Performance Analysis and Design Aids**]; D.2.8 [**Metrics**] *Performance Measures;* D.4.8 [**Performance**] *Measurement, Modeling and Prediction*

Keywords
SPEC; SIP; SPECsip_Infrastructure2011; JSR289; RFC3261

1. INTRODUCTION
Marketing trends indicate that one of the key contributing factors to the expected and observed growth of VoIP and Internet multimedia industry is to unify the communication infrastructure and therefore lowering operating costs. The simplification and the reduction of costs are accelerating the acceptance and adoption in the worldwide enterprise today. The change is happening, thus validating the importance of having standard SIP benchmarks for both infrastructure and applications.

The SIP subcommittee under Open System Group (OSG) in SPEC was formed to research, plan, and initiate the creation of benchmarks in the converging area of multimedia communication over IP that use the IETF Session Initiation Protocol (SIP) [1]. The subcommittee has released one (infrastructure) benchmark and actively developing the second (application) benchmark.

SPECsip_Infrastructure2011 [2] was released in 2011 and a subsequently revised version is being worked on, towards a 2013 release. The benchmark provides for realistic set of workloads that permits fair and accurate analysis and evaluation of platform performance, at the Service Provider (ISP, NSP, and Carriers) and Enterprise level. The benchmark models a VoIP deployment supporting RFC3261 [1] and transport UDP. It is more a macro or full-system benchmark meant to capture user behavior and to be useful for capacity planning. Thus it uses "Simultaneous Number of Supported Subscribers" as its primary performance metric.

ICPE'13, April 21–24, 2013, Prague, Czech Republic.
ACM 978-1-4503-1636-1/13/04.

Currently also under active development is a benchmark that will consist of an application servlet to be run on JSR-289 [3] conforming application servers, a set of HTTP and SIP load generators to be scaled to distributed, multi-client testing rig, and a FABAN [4] based harness that controls over all benchmarking.

2. INFRASTRUCTURE BENCHMARK
SPECsip_Infrastructure2011 is both a specification and a released body of code that can be run and submitted for publication using SPEC's acceptance process. The benchmark focuses on a single node SIP server system under test, rather than complete network architecture such as IMS.

2.1 Design Goals
The main principle behind the design of the SPECsip_Infrastructure2011 benchmark is to emulate SIP user behavior in a realistic manner. Users are the main focus of how SIP traffic is generated. This means defining a set of assumptions for how users behave, applying these assumptions through the use of profiles, and implementing the benchmark to reproduce and emulate these profiles.

The main goal for the benchmark is to evaluate SIP server systems as a comparative benchmark in order to aid customers in analyzing core SIP operations within the infrastructure and to aid in capacity planning and provisioning of systems. Standard SIP scenarios and the traffic profile is included in the standard default workload but can be customized to fit one's own environment behavior in ISP, Telco, University, or Enterprise.

2.2 Components
The benchmark consists of several logical components that make up the benchmark. 1) *SUT* (SIP Server for an Application), which is the combination of hardware and software that provides the SIP application (in this case, VoIP) that the benchmark stresses. The SUT performs both SIP stateful proxying that forwards SIP requests and responses between users and registration that records user location provided by users via the REGISTER request. 2) *Clients* (Workload Generators and Consumers), which emulate the users sending and receiving SIP traffic. This is the load driver part of the benchmark. The load drivers use the SIPp tool [5] under the management and control of FABAN benchmark harness. 3) *Benchmark Harness*, which is responsible for coordinating the test: starting and stopping multiple client workload generators; collecting and aggregating the results from

the clients; determining if a run was successful and displaying the results. It is based on the FABAN framework.

2.3 Workloads and Metrics

The benchmark generates authenticated INVITE and REGISTER transactions and tracks transaction success rate and response-time histograms. The primary benchmark metric (score) is the number of supported subscribers. The secondary metrics include transaction success rate, call success rate, statistics on call mix, and INVITE and REGISTER latencies in terms of average and percentiles at different QoS levels (250ms, 500ms, and 600ms). The success of a run is dictated by whether the call mix, success rates, and QoS are met under a given subscriber population size. The benchmark has been validated on OpenSIPS [6] and Oracle Communications Converged Application Server (OCCAS) [7].

2.4 Benchmarking for the Cloud

The benchmark can be applied to test SIP servers deployed in the cloud. A SIP server, such as OCCAS, may consist of an engine tier and a data tier, where the engine tier performs transaction processing while the data tier maintains the session state. Each tier can be a cluster of managed application servers, where each server runs on its own virtual machine (VM). The VMs can be distributed on multiple blade servers for high availability and horizontal scaling. The workload drivers are used to drive SIP load to the engine tier, which communicates with data tier through inter-VM fabric. The authentication and authorization database can run on a database cluster that runs on VMs as well.

3. Application Benchmark

This benchmark targets SIP and HTTP converged applications as well as enterprise call scenarios. Besides Java EE based application servers such as Weblogic, Websphere and JBoss, the benchmark will be applicable to many JSR289-conforming application servers. For example, Voxeo provides cloud-based telephony application API which runs on JSR289 sip application servers. Developers can add sip services to web applications to bring real-time communication to the apps.

3.1 Design Goals

The benchmark is designed with the following goals. First, it should be a system-level benchmark that represents what industry does so we should avoid functional complexity if it does not add significantly to the workload. It is desired that non-SIP experts can relate to the use cases, which are from implementations and generalized, Second, the benchmark should be standard (JSR 289) compliant, which is for converged application that use HTTP and SIP session state. The benchmark should exercise multiple SIP sessions and HTTP transactions form a complete call flow. Third, the benchmark should use high level API and programming constructs like customers of an application server would. This will be fulfilled by using JPA and JSP. We made conscious decision about JPA such that database size is moderate so that the benchmark better illustrates application server performance, not limited by the database processing capacity. The database is used to store user login info, and operational info such as the number of voice mails, length and other attributes of the voice mails. The web page construction will be based on JSPs, which is fairly standard in enterprise applications.

3.2 Call Scenarios

The application benchmark will implement the call flows corresponding to the following scenarios: Scenario 1 (No Answer): There is no answer from called party so the call is forwarded to Voice Mail. Scenario 2 (Converged Applications): A user uses Web to browse the list of voice mails and play back a voice message using a SIP call. Scenario 3 (Three-way Conferencing): A call is established and then a third party is added to the existing call. Scenario 4 (Call Forwarding): A placed call receives no answer, or a busy tone, and gets forward to another number.

3.3 Benchmark Metrics

Besides the call and transaction level statistics in SPECsip_Infrastructure2011, the Application Benchmark will collect system capacity information such as the number of concurrent calls, as well as the call volume within a given period of time. Certainly, being a benchmark for a converged application including SIP and HTTP protocols, we will add transaction statistics such as throughput and latency statistics for the HTTP protocol.

4. CONCLUSIONS

SPECsip_Infrastructure2011 has been developed, and used in various situations such as sizing, benchmarking, and processor performance projection. The benchmark is used on traditional platforms as well as in cloud deployment. The actively developed Application benchmark targets enterprise applications and includes both HTTP and SIP loads, and a rich set of call scenarios. To reduce complexity and speed up development, we currently do not model Unified Messaging but may consider it in future enhancement. We also do not model RTP part of a VoIP call but may consider adding RTP modeling, based on JSR309 [8], in the future.

5. REFERENCES

[1] Rosenberg, J., Schulzrinne, H., Camarillo, G., Johnston, A., Peterson, J., Sparks, R., Handley, M., and E. Schooler, "SIP: Session Initiation Protocol", RFC 3261, June 2002.

[2] Standard Performance Evaluation Corporation, "SPECsip_Infrastructure2011", http://www.spec.org/sipinf2011.

[3] Java Community Process, "JSR-000289 SIP Servlet 1.1", http://jcp.org/aboutJava/communityprocess/final/jsr289.

[4] "FABAN Harness and Benchmark Framework", http://java.net/projects/faban.

[5] SIPp, http://sipp.sourceforge.net.

[6] "The OpenSIPS Project", http://www.opensips.org.

[7] "Oracle Communications Converged Application Server", http://www.oracle.com.

[8] Java Community Process, "JSR-000309 Media Server Control API", http://jcp.org/aboutJava/communityprocess/final/jsr309.

MockTell: Exploring Challenges of User Emulation in Interactive Voice Response Testing

Siddhartha Asthana
Indraprastha Institute of Information Technology, Delhi
New Delhi, India
siddharthaa@iiitd.ac.in

Pushpendra Singh
Indraprastha Institute of Information Technology, Delhi
New Delhi, India
psingh@iiitd.ac.in

Amarjeet Singh
Indraprastha Institute of Information Technology, Delhi
New Delhi, India
amarjeet@iiitd.ac.in

ABSTRACT

Increasing use of telephone devices has made the Interactive Voice Response (IVR), a technology for accessing information over phone, popular among the commercial organizations. IVR systems are used for critical applications like flight reservation, tele-banking, etc. which requires to have well tested IVR systems. Manual testing of an IVR application requires dialing number, listening and responding to voice prompts through key-press or speech.
Automating the tests for IVR applications requires mimicking the user behavior. We present MockTell, a generic tool for call emulation with ability to mimic user behavior. MockTell uses data generated from real world calls for call emulation that helps in optimizing and evaluating the performance of IVR applications. MockTell also allows simulation of calls to provide further testing of system.

Categories and Subject Descriptors

H.1.2 [**User/Machine Systems:**]: Human factors

General Terms

Design, Experimentation, Human Factors

Keywords

emulators, IVR, voice application, testing

1. INTRODUCTION

An Interactive Voice Response (IVR) system provides information and service through telephones and is used for automating the call handling in several commercial organizations. Many critical applications runs on IVR systems, so it is essential to have well tested IVR applications with all the necessary optimization done by the IVR developer.
IVR testing can be classified into infrastructure, system, and application level testing. At present, various industrial tools[1] are available to test the performance of IVR applications at infrastructure (voice gateway, routers) and system (Telephony servers, database servers) level, which helps to maintain the quality. However, application level tests like program flow test and error condition test require giving different user inputs to application under test. This requires testing tool to have the ability to mimic user behavior. Currently available testing tools lack the ability to mimic user behavior, many application level tests are usually done manually or through customized test scripts written by IVR developers. Moreover, IVR systems are evolving [1, 4, 2] and it will become increasingly difficult to test and optimize future IVR applications. MockTell addresses these challenges by mimicking user behavior in different contexts.

[1] `http://www.tmcnet.com/voip/0506/tech-roundup-voip-testing-tools.htm`

ICPE'13, April 21–24, 2013, Prague, Czech Republic.
ACM 978-1-4503-1636-1/13/04.

2. DESIGN AND FEATURES

MockTell has five functional components and two user models:
Logic Processor: This module co-ordinates with other modules and takes various decisions for emulating a call e.g. call initiation, responding to voice prompt etc. Two user models for emulation process are embedded into this and can be modified to emulate different user behavior for scenarios of different complexities.
Visualizer: This dedicated module handles graphical user interface of the MockTell. Tasks like taking user input for configuring different parameters of emulation process, showing current status of emulation etc. are done by this module.
SIP utility: This module provides API for initiating and releasing a call, generating a DTMF key-press and sending the audio file as speech utterance for communication with IVR under test.
Status Monitor: This module gathers the information about present state of IVR by communicating with IP-PBX software hosting the IVR application. This information helps MockTell to adjust the timing and value of events to be regenerated in testing dynamic IVR systems.
Data Handler: This module load and read the logs of IVR usage into a Java based object representing user model for emulation. Each object contains information about responses made by a user and their corresponding contextual information in a call. Logic Processor uses the information stored in these objects for emulation.
In MockTell, calls are emulated using two modes: emulation using previous call records and testing using random DTMF.

2.1 Emulation using previous call records

In this mode, MockTell reads the call logs, stored in XML files, which may be obtained from a real world deployment. MockTell supports two formats that capture user models:
Simple user emulation: This user emulation works for currently available IVR systems. MockTell replicates the call events like key-press and speech recording, as they happened in corresponding actual call. Real data stored for this emulation contains a time-stamp for each event relative to the start of the call. The events are generated based on the time-stamp captured in this model.
Intricate user emulation: With intricate user emulation model, MockTell can emulate calls for upcoming IVR systems where menu options sequence may change in order to give better services. Each menu option is a state which is assumed to be announced by IVR

application over the IPPBX console as they occur. In the current implementation, MockTell connects to command line interface of FreeSWITCH using fs_cli utility which comes with pre-installed FreeSWITCH binaries to capture the state information announced by IVR application. MockTell generates user responses to these states based on the data stored in the user model.
In both of the modes, MockTell can reorder the call sequence which changes the number of past calls and past system usage for a particular call. This helps in studying IVR system which adapts after every call based on the system usage. MockTell can also vary number of telephone lines to IVR application. With this feature, the number of telephone line connections can be increased to analyze the behavior of IVR application in handling multiple connections (e.g. accessing the same audio file for reading or writing the log).

2.2 Emulation using Random DTMF

In this mode, MockTell performs tests that are independent of user models and can be used to test IVR system's parameters. It generates events like key-press or speech utterance based on test parameters specified by the user. It supports 3 types of IVR testing.
Call Load Test: This test helps in measuring the number of simultaneous calls an IVR application can process and reports the integer value at which IVR application crashed. In our test, we found that FreeSWITCH was able to process 123 simultaneous calls for our sample IVR application. The 124th call failed because of too many connections open to the database (MySQL).
Sequence Test: In this test, MockTell generates random DTMF digits, each with a constant time delay between each generated digit and reports the sequence test case, if any, at which IVR application fails to respond. An IVR developer needs to specify the time delay and number of digits in the sequence (e.g. 4 digits). In our test, MockTell was able to check 100 sequence in 2,504 seconds of sequence length 5 and delay of 5 seconds. Initial 4 seconds out of 2,504 were taken by MockTell in setting up the call.
Sequence test with random delay: This test helps to detect erroneous DTMF input in more complex manner i.e. when the duration of played voice prompts varies for the announcement of menu options. This test generates random DTMF with different time gaps between each successive generated DTMF. The time gap is randomly decided in the range of t_1 and t_2 values specified by the tester. Ideally t_1 should correspond to the length of minimum voice prompt and t_2 be the length of maximum voice prompt in IVR.

3. PERFORMANCE EVALUATION

In this section, we evaluate the performance of MockTell through an experiment conducted using *call load Test* feature. For this experiment, we setup an IVR application written in JAVA and hosted on FreeSwitch. The FreeSwitch and MockTell were running on two different machines referred as Machine-I (HP Compaq dx7400: Ubuntu 10.04, Intel core 2 Duo, 3GB DDR2) and Machine-II (HP Probook 4520s: Ubuntu 12.04, Intel Core i5, 2GB DDR3) respectively, with IP connectivity (100Mbps) between them. We started *call load* test feature of the MockTell. It initiated a call every 2 seconds while keeping previously initiated calls alive till the end of experiment or terminated by FreeSwitch. A call in this experiment was always in one of the 3 states: active, dropped and timed-out.
Active State: refers to a call that is connected to FreeSwitch and is not being terminated.
Dropped State: refers to a call that was previously connected but terminated by FreeSwitch.
Timed-out: state refers to a call released by MockTell as it was not able to connect to FreeSwitch.
In total, we initiated 1024 real calls through MockTell. Initially, we did not observe dropped calls till 233rd call as the load was low on Machine-I. But after this, FreeSwitch started dropping calls at a higher rate which reduces the count of active calls and stabilizes around 145 calls that shows that the rate of dropping the calls becomes equal to rate of accepting the new calls. Our results for number of active calls (i.e. concurrent calls) for FreeSwitch running on hardware having configuration of Machine-I are conforming the results obtained by other developers as available on FreeSwitch website[2]. We measured the CPU and Memory load of Machine-I and found that it was saturated around 233rd call because rate of call dropping became nearly equal to rate of accepting new call. We also found that 3 calls were timed-out at 110th call because of network congestion between the two machines.
MockTell Load: We measured and categorized the CPU and memory usage of MockTell (running on Machine-II) in two categories: static load and running load. Static load refers to CPU usage in creating and holding the call objects. Running load refers to CPU load incurred due to handling of underlying SIP communication. We found that static load gradually increased from 8.1% to 9.1% (i.e. 1% increase) for emulating 1024 calls. Similarly running load varied from 17.9% to 18.5% for emulating 1024 real calls. Thus it shows initiating each call in MockTell has memory load of 20 KB (calculated as 1% of 2 GB system memory divided by 1024 calls).

4. CONCLUSION

We presented MockTell with two user models for testing IVR applications. Simple user model allows testing of currently available static IVR while Intricate user model enables testing of upcoming dynamic IVR applications. User model based testing provides characteristic to model different users easily [3].
User model based approach in MockTell helps in reducing manual investigation of IVR applications. The approach presented in this paper is independent of any voice or speech processing which eventually helps in avoiding any semantic processing as well. Experiences with trial run of MockTell for testing our basic IVR application suggests that apart from application level issues it can also help developers to find some system level issues like too many connections opened to MySQL.

5. ACKNOWLEDGMENTS

We acknowledge the support of Tata Consultancy Services (TCS) for this research.

6. REFERENCES

[1] Asthana, S., Singh, P., Kumaraguru, P., Singh, A., and Naik, V. Tring! tring! - an exploration and analysis of interactive voice response systems. *4th International Conference on Human Computer Interaction (IndiaHCI), Pune, India* (2012).

[2] Asthana, S., Singh, P., and Singh, A. A case for adaptive interface for interactive voice response system. *ACM DEV'13: Proceedings of the 3rd ACM Symposium on Computing for Development* (2013).

[3] Eckert, W., Levin, E., and Pieraccini, R. User modeling for spoken dialogue system evaluation. In *Proc. IEEE ASR Workshop* (1997), 80–87.

[4] Perugini, S., Anderson, T. J., and Moroney, W. F. A study of out-of-turn interaction in menu-based, IVR, voicemail systems. In *Proceedings of the SIGCHI conference on Human factors in computing systems*, CHI '07, ACM (New York, NY, USA, 2007), 961–970.

[2] `http://wiki.freeswitch.org/wiki/Real-world_results`

Introduction to Dynamic Program Analysis with DiSL

Lukáš Marek
Charles University
Prague, Czech Republic
lukas.marek@d3s.mff.cuni.cz

Yudi Zheng
University of Lugano
Lugano, Switzerland
first.last@usi.ch

Danilo Ansaloni
University of Lugano
Lugano, Switzerland
first.last@usi.ch

Lubomír Bulej
Charles University
Prague, Czech Republic
lubomir.bulej@d3s.mff.cuni.cz

Aibek Sarimbekov
University of Lugano
Lugano, Switzerland
first.last@usi.ch

Walter Binder
University of Lugano
Lugano, Switzerland
first.last@usi.ch

Zhengwei Qi
Shanghai Jiao Tong University
Shanghai, China
qizhwei@sjtu.edu.cn

ABSTRACT

DiSL is a new domain-specific language for bytecode instrumentation with complete bytecode coverage. It reconciles expressiveness and efficiency of low-level bytecode manipulation libraries with a convenient, high-level programming model inspired by aspect-oriented programming. This paper summarizes the language features of DiSL and gives a brief overview of several dynamic program analysis tools that were ported to DiSL. DiSL is available as open-source under the Apache 2.0 license.

Categories and Subject Descriptors

D.3.3 [**Programming Languages**]: Language Constructs and Features—*Frameworks*

Keywords

Dynamic program analysis, bytecode instrumentation, aspect-oriented programming, domain-specific languages, Java Virtual Machine

1. INTRODUCTION

Dynamic program analysis (DPA) techniques enable essential software engineering tools, such as profilers, debuggers, and testing tools. DPA tools often rely on code instrumentation, i.e., the insertion of analysis code at selected locations in the analyzed program. As nowadays many programming languages are compiled to bytecode for the Java Virtual Machine (JVM), bytecode instrumentation has become a common technique for building DPA tools for the JVM.

JVM bytecode instrumentation is supported by a variety of bytecode engineering libraries, such as ASM, BCEL, Javassist, or ShrikeBT, to mention some of them. Generally, these tools offer a rather low level of abstraction, which results in high development effort for building DPA tools; the sources of such tools tend to be verbose and difficult to maintain and to extend. On the other hand, bytecode manipulation libraries give the expert developer many opprtunities to optimize the inserted code, enabling efficient DPA tools.

Some researchers have proposed the use of aspect-oriented programming (AOP) for creating DPA tools. Thanks to its pointcut/advice mechanism, AOP offers a high abstraction level that allows to concisely specify certain kinds of instrumentations. Consequently, the use of AOP promises to significantly reduce the development effort for building DPA tools. Unfortunately, mainstream AOP languages such as AspectJ lack join points at the level of e.g. basic blocks of code or individual bytecodes that would be needed for certain DPA tools. In addition, aspect weavers often introduce high overhead due to complex code transformations involved. For example, in AspectJ, the access to dynamic reflective join point information may involve object allocations and iterations (e.g., to store the arguments of a method in an array) that are not visible in the instrumentation code.

DiSL [1, 3, 2] is a new domain-specific language for JVM bytecode instrumentation that combines the strengths of low-level bytecode manipulation libraries (i.e., expressiveness and efficiency) with the strength of AOP (i.e., high-level programming abstractions). DPA tools written in DiSL tend to be as concise as equivalent tools written in a mainstream AOP language (if this is possible). However, in contrast to prevailing AOP languages, DiSL features an open join point model where any region of bytecodes in a program can become a join point. Moreover, the DiSL weaver produces efficient code that does not incur any unexpected costs; the weaver introduces neither object allocations nor loops that were not written by the programmer.

Below we give a brief summary of DiSL's language features and mention some DPA tools that we successfully ported to DiSL.

2. LANGUAGE FEATURES

DiSL is a language hosted in Java and uses annotations to direct the insertion of the code. The language follows the Aspect-oriented paradigm, but is tailored specifically for bytecode instrumentation. To the developer, DiSL provides several key constructs that we now review in turn.

Marker is a DiSL construct responsible for capturing the join points and defining a bytecode region where the selected

ICPE'13, April 21–24, 2013, Prague, Czech Republic.
ACM 978-1-4503-1636-1/13/04.

instrumentation will be applied. For common tasks, DiSL provides a predefined set of markers that allow instrumenting method body, basic blocks of code, method invocations, or single bytecodes. However, a developer can take advantage of the open join point model of DiSL and implement markers that intercept arbitrary bytecode sequences.

To provide control over the selection of classes and methods for instrumentation, DiSL supports coarse-grained join-point filtering using a simplified scoping language, and fine-grained filtering using guards. The guards are written in Java, enabling the developer to express complex join-point filtering conditions evaluated at instrumentation time.

Another two constructs provide efficient data passing among the instrumented locations. The first is called Synthetic-local variable and allows passing data in the context of the instrumented method. The second, called Thread-local variable, provides thread local storage.

The mandatory feature of all instrumentation languages is access to static and dynamic context information. In DiSL, static context information is accessed through a library of predefined static context classes. The classes provides various information about the currently instrumented method or class. This includes method name, method visibility, method descriptor, class name, class signature or name of the super class. The developer is free to define custom static context classes, e.g., to provide access to the results of a custom static analysis. The static context data is inserted directly into the bytecode as a constants. This implies that static context information can only come in form of Java basic types or Strings.

DiSL also provides access to the dynamic context information, which includes the *this* object, method arguments and local variables, and operands on the stack.

DiSL is designed for application observation only. It effectively prevents the instrumentation from changing the application control flow. Besides prohibiting the insertion of the return or jump instructions, it also ensures that no uncaught exception escapes the instrumentation.[1]

Finally, DiSL enables full bytecode coverage and avoids disturbing the instrumented application. This is achieved by instrumeting the application classes in a separate JVM.[2]

3. RECASTED TOOLS

Many dynamic program analysis tools for the JVM have already been recasted in DiSL. We found that for tools that were originally implemented with a low-level bytecode manipulation library, the recasted tools were much more concise than the original versions. Regarding startup performance, the recasted tools were slightly slower, and concerning steady-state performance, they reached the same level of performance as the original versions. For tools that were originally implemented in a mainstream AOP language, the recasted tools were about the same size, but significantly outperformed the original versions.

Amongst others, we recasted the following tools with DiSL: **Cobertura**[3] is a tool for Java code coverage analysis. At runtime, Cobertura collects coverage information for every line of source code and for every branch. **EMMA**[4] is another coverage analysis tool for Java. At runtime, EMMA collects coverage information for every basic block of every method in every class. **HPROF** is a heap and CPU profiler for Java distributed with the HotSpot JVM. **JCarder**[5] is a tool for finding potential deadlocks in multi-threaded Java applications. **JP2**[6] is a calling-context profiler for Java that keeps the execution trace of the application in a calling-context tree. **JRat**[7] is a call graph profiler for Java. For each method, JRat collects the execution time of each invocation, grouped by the caller. **RacerAJ**[8] is a tool for finding potential data races in Java applications. **ReCrash**[9] is a tool for reproducing software failures. **TamiFlex**[10] is a tool that helps other (static analysis) tools deal with reflection and dynamically generated classes in Java.

4. CONCLUSION

DiSL is a domain-specific language for instrumentation-based dynamic program analysis of applications running in the JVM. It offers high-level language constructs tailored for dynamic analysis tasks. The design of DiSL has been inspired by AOP. However, in contrast to mainstream AOP languages, DiSL features an open join point model where any bytecode region can become a join point. DiSL builds on top of the popular bytecode manipulation library ASM. DiSL also offers interfaces for framework extensions based on ASM. DiSL is available open-source under the Apache 2.0 license on OW2: http://disl.ow2.org/

Acknowledgements: The research presented in this paper has been supported by the Swiss National Science Foundation (project CRSII2_136225), by the Scientific Exchange Programme NMS–CH (project code 11.109), by a Sino-Swiss Science and Technology Cooperation (SSSTC) Exchange Grant (project no. EG34–092011) and Institutional Partnership (project IP04–092010), by the European Commission (Seventh Framework Programme grant 287746), by the Czech Science Foundation (project GACR P202/10/J042), and International Cooperation Program of Shanghai (No.11530700500).

5. REFERENCES

[1] L. Marek, A. Villazón, Y. Zheng, D. Ansaloni, W. Binder, and Z. Qi. DiSL: A Domain-specific Language for Bytecode Instrumentation. In *Proc. 11th Intl. Conf. on Aspect-oriented Software Development (AOSD'12)*, pages 239–250. ACM, 2012.

[2] L. Marek, Y. Zheng, D. Ansaloni, A. Sarimbekov, W. Binder, P. Tůma, and Z. Qi. Java Bytecode Instrumentation Made Easy: The Disl Framework for Dynamic Program Analysis. In *Proc. 10th Asian Symposium on Programming Languages and Systems (APLAS 2012)*, pages 256–263. Springer, 2012.

[3] Y. Zheng, D. Ansaloni, L. Marek, A. Sewe, W. Binder, A. Villazón, P. Tuma, Z. Qi, and M. Mezini. Turbo DiSL: Partial Evaluation for High-Level Bytecode Instrumentation. In *Proc. 50th Intl. Conf. on Objects, Models, Components, Patterns (TOOLS'12)*, volume 7304 of *LNCS*, pages 353–368. Springer-Verlag, 2012.

[1] This feature can be deactivated if required.

[2] A native agent is used in the application JVM to exchange data with the instrumentation JVM.

[3] http://cobertura.sourceforge.net/

[4] http://emma.sourceforge.net

[5] http://www.jcarder.org/

[6] https://code.google.com/p/jp2/

[7] http://jrat.sourceforge.net/

[8] http://www.bodden.de/tools/raceraj/

[9] http://groups.csail.mit.edu/pag/reCrash/

[10] https://code.google.com/p/tamiflex/

FINCoS: Benchmark Tools for Event Processing Systems
— Invited Demo Paper —

Marcelo R. N. Mendes, Pedro Bizarro, and Paulo Marques
CISUC, University of Coimbra, Dep. Eng. Informática – Pólo II, Coimbra, Portugal
{mnunes, bizarro, pmarques}@dei.uc.pt

ABSTRACT
FINCoS is a set of benchmarking tools for load generation and performance measuring of event processing systems. It leverages the development of novel benchmarks by allowing researchers to create synthetic workloads, and enables users of the technology to evaluate candidate solutions using their own real datasets. An extensible set of adapters allows the framework to communicate with different CEP engines, and its architecture permits to distribute load generation across multiple nodes. In this paper we briefly review FINCoS, introducing its main characteristics and features, and discussing how it measures the performance of event processing platforms.

Categories and Subject Descriptors
C.4 [Performance of Systems]: Measurement techniques

General Terms
Experimentation, Measurement, Performance.

Keywords
Benchmarking, CEP, Event Processing, Framework, Tools.

1. INTRODUCTION
Recent years have witnessed the consolidation of *Complex Event Processing* (CEP) as an important research discipline and industrial trend. There is still, however, a lack of information about the performance of CEP engines as standard benchmarks are currently not available. The FINCoS framework aims at filling this gap by providing a flexible and neutral approach through which users, researchers and engineers can quickly run realistic performance tests on one or more event processing platforms without having to code themselves load generation, performance measurement and event conversion routines.

FINCoS development started in the end of 2007, as part of the BiCEP project [1]. At that time, there was very little information on how the CEP technology was being used in the real world and a great diversity of products, each with their own languages and implementation styles, which posed serious challenges for the development of novel benchmarks. Our goal then was to provide a flexible tool that allowed carrying out performance evaluation on CEP platforms independently on their structural differences or the workload employed. This would be beneficial for both users of the technology, which would be able to better evaluate the performance of their candidate platforms, and the academic community, which could more easily develop and experiment novel benchmarks for event processing systems.

The first version of the framework was released in 2008 [3], and since then it has been considerably extended and improved. Today, FINCoS can be used not only for stand-alone performance studies, but also as reusable load generation and measurement component in upcoming benchmarks kits. At the moment, the framework is in the final stages of the review process for being included in the SPEC Research Group tools repository [4].

2. OVERVIEW AND CORE FEATURES
The FINCoS framework is composed by five main components. *Drivers* simulate external event sources, submitting load to the system under test (SUT). On the opposite side, *Sinks* receive the results produced by the SUT, storing them in log files for subsequent answer validation and performance measurement. The communication with the CEP engine in both cases is made through an extensible set of *adapters*, which convert the events produced by the framework into a data format understood by the target system and vice-versa. A graphical application, denominated *Controller*, allows users to configure, execute, and monitor performance tests. The results of these performance runs can then be visualized both in real-time and after test completion, using the *Performance Monitor* component.

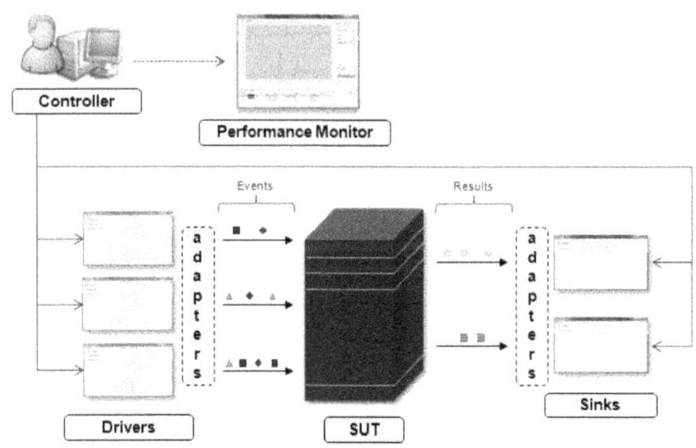

Figure 1: Overview of FINCoS components

FINCoS provides great flexibility and a wide range of options in the definition of experimental evaluations. For instance, the execution of *drivers* can be split into phases, each with its own workload characteristics (e.g., event submission rate, types and datasets). This is useful not only for breaking performance tests into well-defined parts (e.g., warm-up and measurement interval) but also for evaluating the ability of event processing platforms in adapting to changes in the load conditions. In addition, users can choose if events should be generated by the framework itself or read from files containing real-world event data. The former shall be useful for researchers developing novel benchmarks while the latter should help customers trying to mimic their environments. The workload can also be seamlessly scaled by simply adding more *drivers* and *sinks* to the configuration.

Besides enabling users to define arbitrarily complex and realistic workloads, the framework was also designed to be portable across different CEP products. FINCoS allows running performance tests with any CEP engine that is able to exchange events through a JMS middleware. In addition, the framework supports direct communication with event processing platforms through custom adapters (using products APIs)[1].

3. PERFORMANCE MEASUREMENT

After test completion, the performance of the system under test is measured using the *performance monitor* application and the log files produced by *sinks* (the framework also allows measuring performance while tests are running at the cost of a slight overhead). The tool presents performance stats in both tabular and graphical formats – the former displays a snapshot of throughput and latency for each query running at the SUT, while the latter shows the evolution of these metrics over time.

Response time is measured by computing the difference between the time the SUT emitted a given result and the timestamp of the incoming event that triggered it. For that, output tuples produced by the CEP engine must explicitly include the timestamp of the causer event (the timestamp of the result itself is automatically collected and appended by the framework).

Since there is a great variance in the way the several event processing platforms operate, FINCoS provides some flexibility for computing response time. Figure 2 illustrates the three possible definitions of response time supported by the framework.

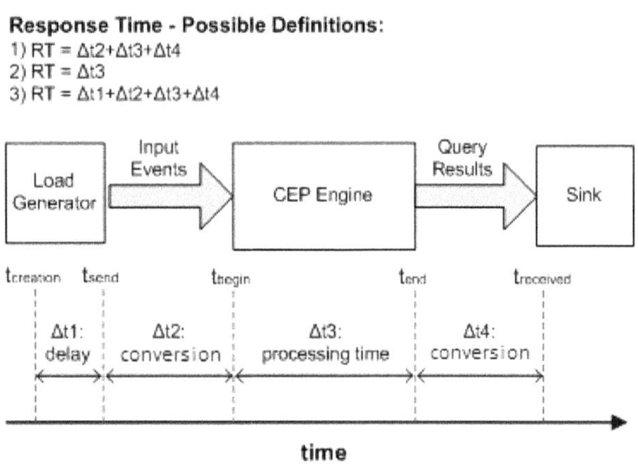

Figure 2: Latency measurement modes supported by FINCoS

---

[1] FINCoS is distributed with out-of-the-box support for the open-source engine Esper.

The first, which we denominate *end-to-end* latency, represents the time it takes for an output tuple to arrive at a *sink* after the corresponding event that triggered it is sent by a *driver*. Note that in this definition the time for converting the event from the internal representation of the framework to a format understood by the SUT (and vice-versa) is accounted as part of the response time. Alternatively, the second definition can be used if the user wants to measure only the *processing time* of events inside the CEP engine. In this case, the events are timestamped inside the *adapters*, immediately before and after sending and receiving events to the SUT. A third option is available and is intended for accounting for delays introduced when the dispatch of events blocks on their processing at CEP engine (the framework then uses the event scheduled time instead of a measured time).

In addition to the point where events are timestamped, FINCoS allows users to choose the resolution used to compute latency – either milliseconds or nanoseconds. Generally, the nanoseconds resolution should be preferred as it is more accurate and many CEP systems offer sub-millisecond processing latencies. Note, however, that response times can be measured in nanoseconds only if *drivers* and *sinks* run in the same machine. It also incurs in more overhead than a millisecond resolution, so users should balance the need for accurate response time measurement and high event submission rates.

4. SUMMARY

The FINCoS framework provides a starting point for users and researchers who desire to carry out performance evaluations on event processing platforms. Fully-customizable workloads can be configured and tests can be performed on virtually any event processing platform, via standard JMS middlewares or directly, through customized adapters. These two characteristics are also important for the development of novel benchmarks, as FINCoS can be reused as a portable component in multiple benchmark kits – this shall be the case for the benchmarks we are currently developing in the scope of the BiCEP project [1].

As future work, we plan to extend FINCoS to support automation of performance tests (the graphical nature of the framework greatly facilitates the definition and monitoring of experiments, but might become troublesome for large experiment sets). We also plan to extend performance measurement, collecting system-level metrics, such as memory consumption and CPU utilization, in addition to the already provided response time and throughput.

Acknowledgements. This research has been supported in part by the Portuguese Science and Technology Foundation (FCT), under grant N° 45121/2008.

5. REFERENCES

[1] BiCEP project web site: http://bicep.dei.uc.pt

[2] FINCoS web site: http://code.google.com/p/fincos/

[3] Mendes, M.R.N., Bizarro, P., Marques, P. A Framework for Performance Evaluation of Complex Event Processing Systems. In Proc. of the 2nd Conf. on Distributed Event-Based Systems (Rome, Italy, July 2008), 313-316. DOI= http://dx.doi.org/10.1145/1385989.1386030.

[4] SPEC Research Group tools repository: http://research.spec.org/tools.html

Author Index

www.ingramcontent.com/pod-product-compliance
Ingram Content Group UK Ltd.
Pitfield, Milton Keynes, MK11 3LW, UK
UKHW051412290926
15471UKWH00009B/1565

9 781450 324380